Green Synthesis of Silver Nanomaterials

Nanobiotechnology for Plant Protection
Green Synthesis of Silver Nanomaterials

Series Editor

Kamel A. Abd-Elsalam

Edited by

Kamel A. Abd-Elsalam
Research Professor, Agricultural Research Center, Plant Pathology Research Institute, Giza, Egypt

ELSEVIER

Elsevier
Radarweg 29, PO Box 211, 1000 AE Amsterdam, Netherlands
The Boulevard, Langford Lane, Kidlington, Oxford OX5 1GB, United Kingdom
50 Hampshire Street, 5th Floor, Cambridge, MA 02139, United States

Copyright © 2022 Elsevier Inc. All rights reserved.

No part of this publication may be reproduced or transmitted in any form or by any means, electronic or mechanical, including photocopying, recording, or any information storage and retrieval system, without permission in writing from the publisher. Details on how to seek permission, further information about the Publisher's permissions policies and our arrangements with organizations such as the Copyright Clearance Center and the Copyright Licensing Agency, can be found at our website: www.elsevier.com/permissions.

This book and the individual contributions contained in it are protected under copyright by the Publisher (other than as may be noted herein).

Notices
Knowledge and best practice in this field are constantly changing. As new research and experience broaden our understanding, changes in research methods, professional practices, or medical treatment may become necessary.

Practitioners and researchers must always rely on their own experience and knowledge in evaluating and using any information, methods, compounds, or experiments described herein. In using such information or methods they should be mindful of their own safety and the safety of others, including parties for whom they have a professional responsibility.

To the fullest extent of the law, neither the Publisher nor the authors, contributors, or editors, assume any liability for any injury and/or damage to persons or property as a matter of products liability, negligence or otherwise, or from any use or operation of any methods, products, instructions, or ideas contained in the material herein.

Library of Congress Cataloging-in-Publication Data
A catalog record for this book is available from the Library of Congress

British Library Cataloguing-in-Publication Data
A catalogue record for this book is available from the British Library

ISBN: 978-0-12-824508-8

For information on all Elsevier publications
visit our website at https://www.elsevier.com/books-and-journals

Publisher: Charlotte Cockle
Acquisitions Editor: Simon Holt
Editorial Project Manager: Rafael G. Trombaco
Production Project Manager: Selvaraj Raviraj
Cover Designer: Greg Harris

Typeset by STRAIVE, India

Contents

Contributors xix
Series Preface xxvii
Preface xxix

PART 1 Synthesis of various types of Silver NMs

CHAPTER 1 Nanosynthetic and ecofriendly approaches to produce green silver nanoparticles 3
Kamel A. Abd-Elsalam

    1 Introduction 3
    2 Green synthesis 5
        2.1 Synthesis of AgNPs from plants 5
        2.2 Microorganisms 6
        2.3 Biomolecules 8
    3 Advantages 9
    4 Challenges 10
    5 Future trends 13
    6 Conclusion 15
        References 15

CHAPTER 2 Chemical synthesis of silver nanoparticles 21
Belete Asefa Aragaw, Melisew Tadele Alula,
Stephen Majoni, and Cecil K. King'ondu

    1 Introduction 21
    2 Synthesis of silver nanoparticles 22
    3 Factors in synthesis of silver nanoparticles by chemical methods 24
    4 Formation mechanisms of silver nanoparticles 29
    5 Chemical reduction using reducing agents 30
        5.1 Reduction by sodium borohydride ($NaBH_4$) 31
        5.2 Reduction by hydrazine 32
        5.3 Reduction by trisodium citrate 34
        5.4 Polyol synthesis 37
        5.5 Tollens method 39
        5.6 Polysaccharide method 40
        5.7 Reduction using N,N-Dimethylformamide 41
        5.8 Photochemical synthesis 42
    6 Electrochemical reduction method 43

7 Microwave-assisted method 45
　　8 Conclusions and future perspectives 46
　　　References 46

CHAPTER 3　Chemical and green production of silver nanocomposites 55
Said Fatouh Hamed, Ayat F. Hashim, Heba H. Salama, and Kamel A. Abd-Elsalam

　1 Introduction 55
　2 Chemical production of silver nanocomposites 55
　　2.1 Chemical synthesis of silver nanoparticles 56
　　2.2 Chemical synthesis of silver-graphene nanocomposites 56
　　2.3 Chemical synthesis of chitosan/cellulose-silver nanocomposites 57
　　2.4 Chemical synthesis of clay-silver nanocomposites 57
　　2.5 Chemical synthesis of silver nanocomposites using different materials 58
　3 Green production of silver nanocomposites 59
　　3.1 Green synthesis of silver nanoparticles 59
　　3.2 Green synthesis of polymer-silver nanocomposites 60
　　3.3 Green synthesis of silver-graphene nanocomposites 62
　　3.4 Green synthesis of silver nanocomposites using different materials 62
　4 Characterization 63
　5 Mechanisms of AgNPs synthesis 64
　6 Conclusion 66
　　References 66

CHAPTER 4　Core-shell silver nanoparticles: Synthesis, characterization, and applications 75
Anbazhagan Sathiyaseelan, Kandasamy Saravanakumar, Murugesan Manikandan, Azeez Shajahan, Arokia Vijaya Anand Mariadoss, and Myeong-Hyeon Wang

　1 Introduction 75
　2 Synthesis of core-shell AgNPs 76
　3 Characterization of core-shell silver nanoparticles 79
　4 Application of silver nanoparticles 85
　　4.1 Biomedical application of core-shell silver nanoparticles 85
　5 Conclusion and future prospects 89

Acknowledgment 90
References 90

CHAPTER 5 Bimetallic silver nanoparticles: Green synthesis, characterization and bioefficacy........................... 99
Mukti Sharma, Ranjini Tyagi, Man Mohan Srivastava, and Shalini Srivastava

1 Introduction 99
2 Fabrication of silver nanoparticles 100
3 Bimetallic nanoparticles 103
4 Characterization of phytofabricated silver nanoparticles 104
5 Important bioefficacies of phytofabricated silver nanoparticles 106
   5.1 Antiinflammatory bioefficacy 106
   5.2 Wound healing bioefficacy 106
   5.3 Antimicrobial bioefficacy 107
   5.4 Antioxidant bioefficacy 108
   5.5 Anticancer bioefficacy 109
6 Conclusion and future perspectives 112
Acknowledgments 114
References 114

PART 2 Biogenic and plant-mediated synthesis

CHAPTER 6 Decoration of carbon nanomaterials with biogenic silver nanoparticles 127
Aswathi Shyam, S Smitha Chandran, R Divya Mohan, Sreedha Sambhudevan, and Bini George

1 Introduction 127
2 Graphene oxide-ag nanocomposites: synthesis and applications 133
   2.1 Green synthesis 133
   2.2 Ag and other carbon composites 139
3 Conclusion 141
References 142

CHAPTER 7 Bio-mediated synthesis of silver nanoparticles via microwave-assisted technique and their biological applications 149
Kondaiah Seku, K. Kishore Kumar, G. Narasimha, and G. Bhagavanth Reddy

1 Introduction 149

  2 Various methods for nanomaterials synthesis 150
   2.1 Physical methods 151
   2.2 Photochemical methods 152
   2.3 Chemical methods 152
   2.4 Biological synthesis of nanoparticles 153
   2.5 Microwave irradiation synthesis of silver
     nanoparticles 153
  3 Classification of NPs 154
  4 Antibacterial activity mechanism of silver nanoparticles 158
   4.1 General mechanisms established for antibacterial
     activity of nanoparticles 158
   4.2 Mechanisms proposed for antibacterial activity of
     silver nanoparticles 159
   4.3 Formation of Ag^+ ions from AgNPs 161
   4.4 Lipid and proteins interactions on the cell membrane 161
  5 Characterization of silver nanoparticles (AgNPs) 162
   5.1 UV-visible characterization 162
   5.2 FTIR analysis of AgNPs 164
   5.3 Powder XRD analysis of AgNPs 165
   5.4 Transmission electron microscopy analysis 166
   5.5 Dynamic light scattering studies 167
  6 Biological applications 167
   6.1 Antibacterial activity 167
   6.2 Antifungal activity 169
   6.3 Antiviral activity 170
   6.4 Antiprotozoal activity 171
   6.5 Anti-arthropods activity 171
   6.6 Anti-parasitic action 171
   6.7 Catalytic activity of AgNPs 172
   6.8 Biomedical applications of silver nanoparticles 174
  7 Future perspectives and recommendations 177
   References 178

CHAPTER 8 Biodegradable gum: A green source for silver nanoparticles 189
Tariq Khan, Husna Jalal, Kashmala Karam, and Mubarak Ali Khan

  1 Introduction 189
  2 Why plant-based synthesis of AgNPs? 190
  3 A brief account of different plant sources used for
   the synthesis of AgNPs 192

4 Natural biodegradable gums and their applications 192
   5 The role of biodegradable gum in bio-fabrication
     of AgNPs 193
         5.1 Gum Karaya 194
         5.2 Gum Ghatti 201
         5.3 Gum Acacia 201
         5.4 Gum Tragacanth (Astragalus Gummifer) 202
         5.5 Salmalia Malabarica 203
         5.6 Gum Kondagogu 204
         5.7 Cashew gum 205
         5.8 Gellan gum 206
         5.9 Oilbanum gum 206
         5.10 Locust bean gum 207
         5.11 Guar gum 208
         5.12 Neem gum 209
   6 Conclusion 210
     References 210

CHAPTER 9 Bio-mediated synthesis of silver nanoparticles via conventional and irradiation-assisted methods and their application for environmental remediation in agriculture 219

Lebea N. Nthunya, Leonardo Gutierrez, Sabelo D. Mhlanga, and Heidi L. Richards

   1 Introduction 219
   2 Green synthesis methods 221
         2.1 Green physical vapor deposition methods 221
         2.2 Green chemical and bio-mediated methods 222
         2.3 Microwave-assisted methods 226
         2.4 UV-assisted methods 226
   3 Applications in agro-ecosystems 230
   4 Conclusion 233
     References 233

CHAPTER 10 Biogenic silver nanoparticles: New trends and applications 241

Alexander Yu. Vasil'kov, Kamel A. Abd-Elsalam, and Andrei Yu. Olenin

   1 Introduction 241
   2 Methods for preparation of silver nanoparticles with
     use of bioorganic reductants 253

3 Preparation of silver nanoparticles using metal vapor synthesis 255
4 Applied aspects of green-synthesized silver nanoparticles 258
    4.1 The use of silver nanoparticles in chemical and biochemical analysis 258
    4.2 The use of silver nanoparticles obtained by green methods in catalysis 261
5 Conclusion 265
    Acknowledgments 266
    References 266

PART 3 Plant synthesis

CHAPTER 11 Plant extract and agricultural waste-mediated synthesis of silver nanoparticles and their biochemical activities 285

Pathikrit Saha and Beom Soo Kim

1 Introduction 285
2 Synthesis of AgNPs by plants and agricultural waste 287
    2.1 Synthesis by plants 287
    2.2 Synthesis by agricultural wastes 288
    2.3 Mechanism of plant and agricultural waste-mediated AgNP synthesis 289
3 Applications of AgNP 289
    3.1 Antibacterial activity 289
    3.2 Anticancer activity 296
    3.3 Antioxidant activity 299
    3.4 Antifungal activity 301
    3.5 Antiviral activity 302
4 Conclusions 304
    References 304

CHAPTER 12 Synthesis, properties, and uses of silver nanoparticles obtained from leaf extracts 317

Fiorella Tulli, Ana Belén Cisneros, Mauro Nicolás Gallucci, María Beatriz Espeche Turbay, Valentina Rey, and Claudio Darío Borsarelli

1 Introduction 317
2 Nanomaterials and the "nano-effect" 318
3 Green synthesis of silver nanoparticles 319

    4 Silver nanoparticles prepared with leaf extracts 327
    5 Properties and applications of phytosynthesized AgNPs 339
        5.1 Antioxidant/medicinal properties 339
        5.2 Antimicrobial activity... 341
        5.3 Antineoplastic properties... 343
        5.4 Molecular sensing and (photo)catalytic properties 344
    6 Conclusions 345
        Acknowledgments 346
        References... 346

CHAPTER 13 Synthesis of biogenic silver nanoparticles using medicinal plant extract: A new age in nanomedicine to combat multidrug-resistant pathogens 359

Sudip Some, Rittick Mondal, Paulami Dam, and Amit Kumar Mandal

    1 Introduction 359
    2 Medicinal plants: An overview.. 361
        2.1 Bioactive compounds or active phytochemicals in medicinal plants... 362
    3 Varied types of synthesis methods of nanoparticles 363
    4 Biosynthesis of silver nanoparticles using medicinal plants..... 365
        4.1 Protocol of synthesis.. 366
        4.2 Plausible mechanism of silver nanoparticle synthesis....... 368
    5 Characterization of biosynthesized silver nanoparticles 368
    6 Antimicrobial activity of biosynthesized silver nanoparticles .. 370
        6.1 Antibiotics: An overview 370
        6.2 Antibiotic resistance in bacteria... 371
        6.3 Efficacy of biogenic silver nanoparticles against MDR pathogens 373
    7 Conclusions 378
        References... 378

PART 4 Microbial synthesis

CHAPTER 14 Mycosynthesis of silver nanoparticles: Mechanism and applications 391

Jayshree Annamalai, Karuvelan Murugan, Jayashree Shanmugam, and Usharani Boopathy

    1 Introduction 391

- 2 Fungal species in silver nanoparticle synthesis 392
- 3 Mechanism and methods of synthesis 398
    - 3.1 Intracellular synthesis 399
    - 3.2 Extracellular synthesis 400
- 4 Advantages over other biosynthesis processes 402
- 5 Application and future prospects 402
- 6 Future prospects 405
- 7 Conclusion 406
    - References 406

CHAPTER 15 Synthesis of silver nanoparticles from mushroom: Safety and applications 413
Kanniah Paulkumar and Kasi Murugan

- 1 Introduction 413
- 2 Silver nanoparticles 415
- 3 Significance of mushrooms in nanoparticle production 416
- 4 Synthesis of silver nanoparticles using mushroom 417
- 5 Safety protocol 419
- 6 Versatile applications of silver nanoparticles 421
    - 6.1 Biological activity 421
    - 6.2 Environment 426
    - 6.3 Agriculture 428
    - 6.4 Food 429
- 7 Conclusion 430
    - Acknowledgments 430
    - References 430

CHAPTER 16 Microbially synthesized silver nanoparticles: Mechanism and advantages—A review 439
Antony V. Samrot, P.J. Jane Cypriyana, S. Saigeetha, A. Jenifer Selvarani, Sajna Keeyari Purayil, and Paulraj Ponnaiah

- 1 Introduction 439
- 2 Green biosynthesis of AgNPs 440
    - 2.1 Bacteria-mediated silver nanoparticle production 440
    - 2.2 Synthesis of AgNPs using fungi 442
    - 2.3 Synthesis of AgNP using algae 444
- 3 Mechanism 444
    - 3.1 Mechanism of bacterial-mediated synthesis 446
    - 3.2 Mechanism of fungal-mediated synthesis 447

3.3 Mechanism involved in algal-mediated synthesis 450
4 Bioactivity studies 452
5 Characterization 453
    5.1 Ultraviolet-visible spectroscopy 454
    5.2 Fourier transform infrared spectroscopy 454
    5.3 Scanning electron microscopy—Energy dispersive
        X-ray spectroscopy 455
    5.4 X-ray diffraction (XRD) 456
6 Applications of microbial-mediated silver nanoparticles 457
    6.1 Wound healing applications 458
    6.2 Biosensing applications 458
    6.3 Application in textiles 460
    6.4 Application in the food sector 460
    6.5 Agricultural applications 461
7 Conclusion 463
    References 463

CHAPTER 17 Synthesis of silver nanoparticles using *Actinobacteria* 479
Manik Prabhu Narsing Rao and Awalagaway Dhulappa

1 Introduction 479
2 Mechanism and synthesis of silver nanoparticles using
   actinobacteria 480
3 Synthesis of silver nanoparticles using
   actinobacterial strains 481
    3.1 Factors affecting the synthesis of silver
        nanoparticles 484
4 Application of silver nanoparticles synthesized using
   actinobacterial strains 485
5 Conclusions and future perspectives 487
   References 487

CHAPTER 18 Role of bacteria and actinobacteria in the biosynthesis of silver nanoparticles 493
Gonzalo Tortella, Olga Rubilar, María Cristina Diez,
Sergio Cuozzo, Joana Claudio Pieretti, and
Amedea Barozzi Seabra

1 Introduction 493
2 Synthesis of AgNPs by bacteria 496
3 Synthesis of AgNPs by actinobacteria 498

4 Antimicrobial properties of AgNPs 501
  5 Conclusion and perspectives 506
     Acknowledgment 506
     References 506

CHAPTER 19 Biogenic synthesis of silver nanoparticles using lichens 513
Debraj Dhar Purkayastha

  1 Introduction 513
  2 Lichen-mediated biogenic synthesis of silver nanoparticles 514
  3 Biomedical applications of silver nanoparticles 518
  4 Conclusions and future perspectives 519
     References 519

CHAPTER 20 Algae-mediated silver nanoparticles: Synthesis, properties, and biological activities 525
Emad A. Shalaby

  1 Introduction 525
  2 Silver nanoparticles synthesis/production from algae (algae-AgNPs) 526
  3 Practical methods for biosynthesis of AgNPs using algae 529
  4 Factors affecting AgNPs production by algae 530
     4.1 Methods/techniques 531
     4.2 Effect of extract concentration 531
     4.3 Effect of temperature 531
     4.4 Effect of pressure 532
     4.5 Algal extract composition 532
     4.6 Effect of contact time 532
     4.7 Effect of pH 532
  5 Properties of silver nanoparticles produced from algae 533
  6 Biological activities and applications of algae mediated nanoparticles 535
  7 Conclusion 539
     References 540

CHAPTER 21 Green synthesis of silver nanoparticles using actinomycetes 547
Zdenka Bedlovičová

  1 Introduction 547
  2 Silver nanoparticles green synthesis 548

3　Silver nanoparticles synthesized by actinomycetes...................551
　　4　Biological activity of AgNPs prepared by
　　　　actinomycetes 557
　　5　Conclusion and future perspectives 562
　　　　References.. 563

CHAPTER 22　Advantages of silver nanoparticles synthesized by microorganisms in antibacterial activity 571
Xixi Zhao, Xiaoguang Xu, Chongyang Ai, Lu Yan,
Chunmei Jiang, and Junling Shi

　　1　Introduction 571
　　2　Synthesis of AgNPs by microorganisms 572
　　　　2.1　Fungi-mediated synthesis of AgNPs 573
　　　　2.2　Bacteria-mediated synthesis of AgNPs............... 573
　　　　2.3　Other microorganisms 576
　　　　2.4　Mechanism of AgNPs synthesis by microorganisms 577
　　3　Antibacterial activity and mechanism of the AgNPs................579
　　　　3.1　Antibacterial activity of the AgNPs................... 579
　　　　3.2　Antibacterial mechanism of AgNPs................... 579
　　　　3.3　Advantages of AgNPs synthesized by microorganisms
　　　　　　in antibacterial activity 580
　　　　3.4　Future prospects................................. 581
　　4　Conclusion 581
　　　　References.. 582

PART 5　Mechanism, toxicity, fate, commercial production

CHAPTER 23　Methodologies for synthesizing silver nanoparticles 589
Asma Farheen

　　1　Introduction 589
　　2　Methodologies for synthesizing silver nanoparticles 589
　　　　2.1　Physical method to synthesize silver nanoparticles........... 591
　　　　2.2　Chemical methods to synthesize silver nanoparticles 591
　　　　2.3　Biological methods to synthesize silver nanoparticles 591
　　3　Applications...597
　　4　Conclusion and future perspectives 598
　　　　References.. 599

CHAPTER 24　Silver nanoparticles synthesis mechanisms......... 607
Kangkana Banerjee and V. Ravishankar Rai

　　1　Introduction 607

2 Mechanism of physical approach 609
  3 Chemical synthesis and its mechanism 611
  4 Mechanism of biosynthesis process 612
     4.1 Bacterial-mediated synthesis 614
     4.2 Fungal-mediated synthesis 614
     4.3 Other synthesis mechanisms 617
  5 Downstreaming and purification of biosynthesized
     nanoparticles 618
  6 Characterization methods 619
  7 Conclusion 620
     Acknowledgment 621
     References 621

CHAPTER 25 Toxicity of silver nanoparticles in the aquatic system 627

Muhammad Saleem Khan, Muhammad Shahroz Maqsud, Hasnain Akmal, and Ali Umar

  1 Introduction 627
  2 Causes of toxicity 628
  3 Routes of entry 628
  4 Toxicity to phytoplankton and microalgae 628
     4.1 Phytoplankton 628
     4.2 Microalgae 629
  5 Toxicity to fish 631
  6 Toxicity to other aquatic organisms 635
     6.1 Bivalves 635
     6.2 Annelids 636
     6.3 Shrimps 636
  7 LC_{50} value 636
  8 Fate and accumulation 637
  9 Mechanism of toxicity 638
 10 Conclusion and recommendations 640
     References 641

CHAPTER 26 Silver nanoparticles in natural ecosystems: Fate, transport, and toxicity 649

Parteek Prasher, Mousmee Sharma, Harish Mudila, Amit Verma, and Pankaj Bhatt

  1 Introduction 649

2 Sources of AgNPs release in the environment 650
3 Transport of AgNPs in the environment 651
4 Transformations of AgNPs 653
5 Sulfidation 655
6 Chlorination 657
7 Aggregation and dissolution 658
8 Photoinduced transformation 660
9 Conclusion 661
Acknowledgment 661
References 661

CHAPTER 27 Strategies for scaling up of green-synthesized nanomaterials: Challenges and future trends 669
Mohamed Amine Gacem and Kamel A. Abd-Elsalam

1 Introduction 669
2 Green biocatalytic synthesis of nanoscale materials 671
3 Selection of biological precursors for green biosynthesis and determination of their quantity 672
4 Selection of culture media 679
5 Development of information in biochemistry, molecular biology, and genetic engineering 680
6 Optimization of physicochemical factors 684
   6.1 Temperature effect 684
   6.2 Effect of pH 685
   6.3 Effect of reaction time 686
   6.4 Effect of $AgNO_3$ concentration 686
7 Challenges 687
8 Future perspective 688
9 Conclusion 689
References 690

CHAPTER 28 Enzymatic synthesis of silver nanoparticles: Mechanisms and applications 699
Anindita Behera, Sweta Priyadarshini Pradhan, Farah K. Ahmed, and Kamel A. Abd-Elsalam

1 Introduction 699
2 Methods of preparation of silver nanoparticles 700
3 Enzyme-assisted biosynthesis of silver nanoparticles 714
4 Characterization of silver nanoparticles 724
5 Applications of biogenic silver nanoparticles 727

      6 Challenges and limitations of enzyme assisted silver
         nanoparticles...734
      7 Conclusion and future perspective..735
         References... 736
         Further reading 756

Index 757

Contributors

Kamel A. Abd-Elsalam
Plant Pathology Research Institute, Agricultural Research Center (ARC), Giza, Egypt

Farah K. Ahmed
Biotechnology English Program, Faculty of Agriculture, Cairo University, Giza, Egypt

Chongyang Ai
Key Laboratory for Space Bioscience and Biotechnology, School of Life Sciences, Northwestern Polytechnical University, Xi'an, Shaanxi, China

Hasnain Akmal
Department of Zoology, Faculty of Life Sciences, University of Okara, Okara, Pakistan

Melisew Tadele Alula
Department of Chemical and Forensic Sciences, Faculty of Science, Botswana International University of Science and Technology, Palapye, Botswana

Jayshree Annamalai
Centre for Environmental Studies, Department of Civil Engineering, Anna University, CEG Campus, Chennai, India

Belete Asefa Aragaw
Department of Chemistry, College of Sciences, Bahir Dar University, Bahir Dar, Ethiopia

Kangkana Banerjee
Department of Studies in Microbiology, University of Mysore, Mysore, India

Zdenka Bedlovičová
Department of Chemistry, Biochemistry and Biophysics, University of Veterinary Medicine and Pharmacy, Košice, Slovakia

Anindita Behera
School of Pharmaceutical Sciences, Siksha 'O' Anusandhan Deemed to be University, Bhubaneswar, Odisha, India

G. Bhagavanth Reddy
Department of Chemistry, PG Centre Wanaparthy, Palamuru University, Mahbubnagar, India

Pankaj Bhatt
State Key Laboratory for Conservation and Utilization of Subtropical Agro-bioresources, Guangdong Laboratory for Lingnan Modern Agriculture, Integrative Microbiology Research Centre, South China Agricultural University, Guangzhou, China

Usharani Boopathy
Department of Biochemistry, Vels Institute of Science, Technology and Advanced Studies, Chennai, India

Claudio Darío Borsarelli
Instituto de Bionanotecnología del NOA (INBIONATEC), CONICET, Universidad Nacional de Santiago del Estero (UNSE); Facultad de Agronomía y Agroindustrias (FAyA)-UNSE, Santiago del Estero, Argentina

Ana Belén Cisneros
Instituto de Bionanotecnología del NOA (INBIONATEC), CONICET, Universidad Nacional de Santiago del Estero (UNSE), Santiago del Estero, Argentina

Sergio Cuozzo
Planta de Procesos industriales-Proimi-Conicet, San Miguel de Tucumán, Tucuman, Argentina

Paulami Dam
Chemical Biology Laboratory, Department of Sericulture, Raiganj University, Raiganj, West Bengal, India

Awalagaway Dhulappa
Department of Microbiology, Maharani Cluster University, Bangalore, India

María Cristina Diez
Centro de Excelencia en Investigación Biotecnológica Aplicada al Medio Ambiente, CIBAMA-BIOREN; Chemical Engineering Department, Universidad de La Frontera, Temuco, Chile

R. Divya Mohan
Department of Chemistry, Amrita School of Arts and Sciences, Kollam, Kerala, India

Asma Farheen
Department of Biotechnology, Khaja Bandanawaz University, Kalaburagi, India

Said Fatouh Hamed
Fats and Oils Department, Food Industry and Nutrition Research Division, National Research Centre, Cairo, Egypt

Mohamed Amine Gacem
Department of Biology, Faculty of Science, University of Amar-Tlidji, Laghouat, Algeria

Mauro Nicolás Gallucci
Instituto de Bionanotecnología del NOA (INBIONATEC), CONICET, Universidad Nacional de Santiago del Estero (UNSE), Santiago del Estero, Argentina

Bini George
Department of Chemistry, School of Physical Sciences, Central University of Kerala, Kasaragod, Kerala, India

Leonardo Gutierrez
Facultad del Mary Medio Ambiente, Universidad del Pacifico, Guayaquil, Ecuador; Particle and Interfacial Technology Group, Department of Green Chemistry and Technology, Ghent University, Ghent, Belgium

Ayat F. Hashim
Fats and Oils Department, Food Industry and Nutrition Research Division, National Research Centre, Cairo, Egypt

Husna Jalal
Department of Biotechnology, University of Malakand, Chakdara Dir Lower, Pakistan

P.J. Jane Cypriyana
Department of Biotechnology, School of Bio and Chemical Engineering, Sathyabama Institute of Science and Technology, Chennai, Tamil Nadu, India

A. Jenifer Selvarani
Department of Biotechnology, School of Bio and Chemical Engineering, Sathyabama Institute of Science and Technology, Chennai, Tamil Nadu, India

Chunmei Jiang
Key Laboratory for Space Bioscience and Biotechnology, School of Life Sciences, Northwestern Polytechnical University, Xi'an, Shaanxi, China

Kashmala Karam
Department of Biotechnology, University of Malakand, Chakdara Dir Lower, Pakistan

Mubarak Ali Khan
Department of Biotechnology, Faculty of Chemical and Life Sciences, Abdul Wali Khan University Mardan (AWKUM), Mardan, Pakistan

Muhammad Saleem Khan
Department of Zoology, Faculty of Life Sciences, University of Okara, Okara, Pakistan

Tariq Khan
Department of Biotechnology, University of Malakand, Chakdara Dir Lower, Pakistan

Beom Soo Kim
Department of Chemical Engineering, Chungbuk National University, Cheongju, Chungbuk, Republic of Korea

Cecil K. King'ondu
Department of Chemical and Forensic Sciences, Faculty of Science, Botswana International University of Science and Technology, Palapye, Botswana

K. Kishore Kumar
Department of Pharmaceutical Biology, Narayana Pharmacy College, JNTU-A, Nellore, India

Stephen Majoni
Department of Chemical and Forensic Sciences, Faculty of Science, Botswana International University of Science and Technology, Palapye, Botswana

Amit Kumar Mandal
Chemical Biology Laboratory, Department of Sericulture; Centre for Nanotechnology Sciences, Raiganj University, Raiganj, West Bengal, India

Murugesan Manikandan
Department of Biomedical Science, School of Biotechnology and Genetic Engineering, Bharathidasan University, Tiruchirappalli, Tamil Nadu, India

Muhammad Shahroz Maqsud
Department of Zoology, Faculty of Life Sciences, University of Okara, Okara, Pakistan

Arokia Vijaya Anand Mariadoss
Department of Bio-Health Convergence, Kangwon National University, Chuncheon, Republic of Korea

Sabelo D. Mhlanga
Research and Development Division, SabiNano (Pty) Ltd, Johannesburg, South Africa

Rittick Mondal
Chemical Biology Laboratory, Department of Sericulture, Raiganj University, Raiganj, West Bengal, India

Harish Mudila
Department of Chemistry, Lovely Professional University, Punjab, India

Karuvelan Murugan
Centre for Environmental Studies, Department of Civil Engineering, Anna University, CEG Campus, Chennai, India

Kasi Murugan
Department of Biotechnology, Manonmaniam Sundaranar University, Tirunelveli, Tamil Nadu, India

G. Narasimha
Department of Virology, Sri Venkateshwara University, Tirupati, India

Manik Prabhu Narsing Rao
State Key Laboratory of Biocontrol and Guangdong Provincial Key Laboratory of Plant Resources, School of Life Sciences, Sun Yat-Sen University, Guangzhou, PR China

Lebea N. Nthunya
Molecular Sciences Institute, School of Chemistry, University of Witwatersrand, Johannesburg, South Africa

Andrei Yu. Olenin
V.I. Vernadsky Institute of Geochemistry and Analytical Chemistry, Russian Academy of Sciences, Moscow, Russia

Kanniah Paulkumar
Department of Biotechnology, Manonmaniam Sundaranar University, Tirunelveli, Tamil Nadu, India

Joana Claudio Pieretti
Center for Natural and Human Sciences, Universidade Federal do ABC (UFABC), Santo André, SP, Brazil

Paulraj Ponnaiah
School of Bioscience, Faculty of Medicine, Bioscience and Nursing, MAHSA University, Jenjarom, Selangor, Malaysia

Sweta Priyadarshini Pradhan
School of Pharmaceutical Sciences, Siksha 'O' Anusandhan Deemed to be University, Bhubaneswar, Odisha, India

Parteek Prasher
Department of Chemistry, University of Petroleum & Energy Studies, Energy Acres, Dehradun, India

Sajna Keeyari Purayil
School of Bioscience, Faculty of Medicine, Bioscience and Nursing, MAHSA University, Jenjarom, Selangor, Malaysia

Debraj Dhar Purkayastha
Department of Chemistry, Cachar College, Silchar, Assam, India

V. Ravishankar Rai
Department of Studies in Microbiology, University of Mysore, Mysore, India

Valentina Rey
Instituto de Bionanotecnología del NOA (INBIONATEC), CONICET, Universidad Nacional de Santiago del Estero (UNSE); Facultad de Agronomía y Agroindustrias (FAyA)-UNSE, Santiago del Estero, Argentina

Heidi L. Richards
Molecular Sciences Institute, School of Chemistry, University of Witwatersrand, Johannesburg, South Africa

Olga Rubilar
Centro de Excelencia en Investigación Biotecnológica Aplicada al Medio Ambiente, CIBAMA-BIOREN; Chemical Engineering Department, Universidad de La Frontera, Temuco, Chile

Pathikrit Saha
Department of Chemical Engineering, Chungbuk National University, Cheongju, Chungbuk, Republic of Korea

S. Saigeetha
Department of Biotechnology, School of Bio and Chemical Engineering, Sathyabama Institute of Science and Technology, Chennai, Tamil Nadu, India

Heba H. Salama
Dairy Department, Food Industry and Nutrition Research Division, National Research Centre, Cairo, Egypt

Sreedha Sambhudevan
Department of Chemistry, Amrita School of Arts and Sciences, Kollam, Kerala, India

Antony V. Samrot
School of Bioscience, Faculty of Medicine, Bioscience and Nursing, MAHSA University, Jenjarom, Selangor, Malaysia

Kandasamy Saravanakumar
Department of Bio-Health Convergence, Kangwon National University, Chuncheon, Republic of Korea

Anbazhagan Sathiyaseelan
Department of Bio-Health Convergence, Kangwon National University, Chuncheon, Republic of Korea

Amedea Barozzi Seabra
Center for Natural and Human Sciences, Universidade Federal do ABC (UFABC), Santo André, SP, Brazil

Kondaiah Seku
Engineering Department, Civil Section, Applied Sciences–Chemistry, University of Technology and Applied Sciences, Shinas, Sultanate of Oman

Azeez Shajahan
Department of Biotechnology, PRIST (Deemed University), Puducherry Campus, Puducherry, Tamil Nadu, India

Emad A. Shalaby
Department of Biochemistry, Faculty of Agriculture, Cairo University, Giza, Egypt

Jayashree Shanmugam
Centre for Advanced Studies in Botany, School of Life Science, University of Madras, Guindy Campus, Chennai, India

Mousmee Sharma
Department of Chemistry, Uttaranchal University, Dehradun, India

Mukti Sharma
Department of Chemistry, Faculty of Science, Dayalbagh Educational Institute, Agra, India

Junling Shi
Key Laboratory for Space Bioscience and Biotechnology, School of Life Sciences, Northwestern Polytechnical University, Xi'an, Shaanxi, China

Aswathi Shyam
Department of Chemistry, Amrita School of Arts and Sciences, Kollam, Kerala, India

S. Smitha Chandran
Department of Chemistry, Amrita School of Arts and Sciences, Kollam, Kerala, India

Sudip Some
Department of Life Sciences, Chanchal Siddheswari Institution (H.S.), Chanchal, West Bengal, India

Man Mohan Srivastava
Department of Chemistry, Faculty of Science, Dayalbagh Educational Institute, Agra, India

Shalini Srivastava
Department of Chemistry, Faculty of Science, Dayalbagh Educational Institute, Agra, India

Gonzalo Tortella
Centro de Excelencia en Investigación Biotecnológica Aplicada al Medio Ambiente, CIBAMA-BIOREN, Universidad de La Frontera, Temuco, Chile

Fiorella Tulli
Instituto de Bionanotecnología del NOA (INBIONATEC), CONICET, Universidad Nacional de Santiago del Estero (UNSE), Santiago del Estero, Argentina

María Beatriz Espeche Turbay
Instituto de Bionanotecnología del NOA (INBIONATEC), CONICET, Universidad Nacional de Santiago del Estero (UNSE); Facultad de Agronomía y Agroindustrias (FAyA)-UNSE, Santiago del Estero, Argentina

Ranjini Tyagi
Department of Chemistry, Faculty of Science, Dayalbagh Educational Institute, Agra, India

Ali Umar
Department of Zoology, Faculty of Life Sciences, University of Okara, Okara, Pakistan

Alexander Yu. Vasil'kov
A.N. Nesmeyanov Institute of Organoelement Compounds, Russian Academy of Sciences, Moscow, Russia

Amit Verma
Department of Biochemistry, College of Basic Science and Humanities, SD Agricultural University, Gujarat, India

Myeong-Hyeon Wang
Department of Bio-Health Convergence, Kangwon National University, Chuncheon, Republic of Korea

Xiaoguang Xu
Key Laboratory for Space Bioscience and Biotechnology, School of Life Sciences, Northwestern Polytechnical University, Xi'an, Shaanxi, China

Lu Yan
Key Laboratory for Space Bioscience and Biotechnology, School of Life Sciences, Northwestern Polytechnical University, Xi'an, Shaanxi, China

Xixi Zhao
Key Laboratory for Space Bioscience and Biotechnology, School of Life Sciences, Northwestern Polytechnical University, Xi'an, Shaanxi, China

Series Preface

The field application of engineered nanomaterials (ENMs) has not yet been well investigated in terms of plant promotion and protection in the agro-environment, and many components have the most effective been taken into consideration theoretically or with prototypes, which make it hard to evaluate the utility of ENMs for plant promotion and protection. Nanotechnology is now invading the food enterprise and forming super potential. Nanotechnology applications in the food industry involve: encapsulation and delivery of materials in targeted sites; developing the flavor; introducing nano-antimicrobial agents into food; improvement of shelf life; sensing contamination; improved food preservatives; monitoring; tracing; and logo protection. The list of environmental problems that the world faces may be huge, but a few strategies for fixing them are small. Globally, scientists are developing nanomaterials that could use selected nanomaterials to capture poisonous pollutants from water and degrade solid waste into useful products. The market intake of nanomaterials is rapidly increasing, and the Freedonia Group predicts that nanostructures will grow to $100 billion by 2025. Nanotechnology research and development has been growing steeply across all scientific disciplines and industries. Against this background, the scientific series entitled "Nanobiotechnology for Plant Protection" was inspired by the desire of the Editor, Kamel A. Abd-Elsalam, to put together detailed, up-to-date, and applicable studies on the field of nanobiotechnology applications in agro-ecosystems, to foster awareness and extend our view of future perspectives.

The main appeal of this series is its specific focus on plant protection in agri-food and environment, which comprise one of the most topical nexus areas in the many challenges facing humanity today. The discovery and highlighting of new book inputs, based on nanobiotechnology, that can be used at lower application rates will be critical to eco-agriculture sustainability. The research carried out in the concerned fields is scattered and not available in a single place. This series will cover the applications in the agri-food and environment sectors, which is a new topic of research in the field of nanobiotechnology. This series will provide a comprehensive account of the literature on specific nanomaterials and their applications in agriculture, food, and environment. The audience will be able to gather information from a single book series. Students, teachers, and researchers from colleges, universities, and research institutes, as well as those working in industry will benefit from this book series. Four specific features make this series unique:

- The book series has a very specific editorial focus, and researchers can locate nanotechnology information precisely without looking into various additional full texts.
- More importantly, the series offers a crucial evaluation of the content material along with nanomaterials, technologies, applications, methods, equipment, and safety and regulatory aspects in agri-food and environmental sciences.

- The series offers readers a concise precision of the content material, it'll offer nano-scientists with clarity and deep information.
- Finally, it presents researchers with insights on new discoveries.

The current series gives the researchers a sense of what to do, both now and in future, and how to do it properly—by searching for others who have done it. This fifth book, entitled *Green Synthesis of Silver Nanomaterials*, has gathered the know-how, observations, and effective applications of zinc-based nanomaterials in the environmental and agri-food systems. The expected readership for this book series includes researchers in the fields of environmental science, food, and agriculture science. Some readers may also be chemists, green chemistry industry employees, material scientists, government regulatory agency workers, agro and food industry players, or academicians. Readers with an industrial background may also be interested in it. This series is useful to a wide audience of food, agriculture and environmental sciences research including undergraduate, graduate, and postgraduate students. In addition, agricultural producers could benefit from the applied knowledge that will be highlighted in the book, which otherwise would be buried in various journals. Both primary and secondary audiences are seeking up-to-date knowledge of nanotechnology applications in environmental science, agriculture, and food science. It is a trending area, and many new studies are published every week. Readers need some good summaries to help them learn the latest key findings, which could be reviews of articles and/or books. This series will help to put these pockets of knowledge together, and make the information more easily accessible globally.

Kamel A. Abd-Elsalam
Agricultural Research Center, Giza, Egypt

Preface

In the last decades, efforts have been made in developing new methods of green synthesis. Due to numerous applications of biogenic nanoparticles (NPs), the synthesis of silver nanoparticles using biological systems is considered an indispensable development. Green and sustainable AgNPs produced from algae, fungi, plants, and bacteria are fascinating in research because the process is quick, effective, simple, and nontoxic. The present book represents the fifth volume of a series titled "Nanobiotechnology for Plant Protection," the original book series approved by Elsevier. The present volume contains 27 chapters prepared by outstanding authors from Algeria, Argentina, Belgium, Botswana, Brazil, China, Chile, Ecuador, Egypt, Ethiopia, the Republic of Korea, Malaysia, India, Russia, Slovakia, the Sultanate of Oman, Pakistan, and South Africa. These chapters, written by professionals and experts, represent outstanding knowledge of green and eco-friendly production methods for synthesis silver-based nanostructures from different biological resources. The book title *Green Synthesis of Silver Nanomaterials* indicates that this book has collected in its 27 chapters the knowledge, discoveries, and fruitful findings of green synthesis silver nanomaterials via plants, agricultural waste, fungi, and microorganisms. Considering this, the recent volume is divided into five main parts. **Part 1** deals with the synthesis and characterization of chemical and green synthesis various types of silver nanomaterials is reflected in the five chapters. **Part 2** is focused on production of silver nanoparticles from agric-waste and plant-mediated and their characterization. **Part 3** reviews the ability of different fungal species such as filamentous fungi and mushrooms to produce silver nanoparticles, while **Part 4** comprises seven chapters providing detailed information related to microbial synthesis of silver NMs, etc.

One of the crucial difficulties in nanomaterials development is scaling up laboratory procedures to the commercial scale. Large-scale biosynthesis of nanoparticles has usually been a great obstacle, and in this book we will illustrate the most effective way of scaling up silver nanoparticle synthesis biologically. **Part 5** (containing four chapters) is focused on the biosynthesis mechanism, toxicity, fate, and commercial production of silver nanomaterials. Examples include greener pathways and mechanisms, toxicity of silver nanoparticles in aquatic life, silver nanoparticles in natural ecosystems, and strategies for scaling up of green synthesized nanomaterials.

The use of agro-waste, plant sources, and microbes not only minimizes the charge of nanomaterials production but also decreases the need to apply dangerous chemicals, encourages "green synthesis," and is a strong step toward promoting environmental sustainability. The current volume is multidisciplinary and will be of great benefit to students, teachers, and researchers, as well as agri-food and agroecosystem scientists working in biology, plant biotechnology, science materials, physics, technology, chemistry, microbiology, plant physiology, nanotechnology, and other interesting groups, such as the manufacturing sector. Our aim is to give not only experienced readers but also technical decision-makers, and people with limited knowledge of the field, a reliable, insightful, and innovative perspective in this regard. For the present book, the target audience comprises students from different fields in scientific and technological fields, and undergraduate and postgraduate students. The authors are very grateful to us for presenting this high-quality collection of manuscripts. We thank all the authors, who have provided useful suggestions and experiences in their chapters of this book. Without their participation and commitment, this book would not have been possible. We are also very grateful to Elsevier, since the project has benefited greatly from the publisher's experience, reliability, and tolerance.

We wish to express our sincere gratitude for the support they have provided and the efforts in publishing this book, notably Simon Holt, Rafael Trombaco, Editorial Project Manager for Intern, Nirmala Arumogam, and Narmathe Mohan. We thank all the reviewers who gave their valuable time to improve on each chapter. We also thank our family members for their continued care and encouragement.

Kamel A. Abd-Elsalam[a,b]

[a]*Agricultural Research Center, Giza, Egypt* [b]*Soils, Water and Environment Research Institute, Agricultural Research Center, Giza, Egypt*

PART 1

Synthesis of various types of Silver NMs

CHAPTER 1

Nanosynthetic and ecofriendly approaches to produce green silver nanoparticles

Kamel A. Abd-Elsalam
Plant Pathology Research Institute, Agricultural Research Center (ARC), Giza, Egypt

1 Introduction

The early uses of silver are in the Caldean hills and Pillsbury (Hill and Pillsbury, 1939) after 4000 BCE, and silver was used for biomedical and food storage purposes by Persian, Roman, and Egyptian citizens in ancient history (Alexander, 2009; Srikar et al., 2016). M.C. Lea published a summary of a citrate-stabilized silver colloid more than 120 years ago (Lea, 1889). The average particle diameter for this process is 7–9 nm (Frens and Overbeek, 1969). The total quantity of silver nanoparticles (AgNPs) generated each year is the subject of some debate because of the various applications, although several reports have been obtained. Today, approximately 320 tons of nanosilver per year are expected to be made and used globally (Gottschalk et al., 2010). The global production volume of the AgNPs at 210 tons per year and the global production volume (i.e., maximum, higher, or more promising estimate) at 530 tons per year for 2018 were viewed skeptically by Pulit-Prociak and Banach (2016) at a minimum, low, and pessimistic estimate (Temizel-Sekeryan and Hicks, 2020). The global AgNP market was worth around 1 billion dollars in 2015, and it is expected that by 2024 it will pass 3 billion dollars (Verma, 2018). The sector breakdown is estimated to be worth more than 1 billion dollars for medical applications (i.e., health and the life sciences), while textiles are predicted to reach 750 million dollars and applications for food and drinks will be more than 300 million dollars by 2024 (Temizel-Sekeryan and Hicks, 2020; Verma, 2018). The use of toxic chemicals or high energy demand, which is very complicated and includes inefficient purification, are chemical and physical ways of developing NPs. Therefore, it is essential to reduce, through the various chemicals used in physical and chemical processes, the possibility of environmental toxicity.

"Green synthesis" is the alternative approaches to developing NPs (Gour and Jain, 2019). There is an overwhelming need for the creation of environmentally sustainable and benign approaches to AgNP synthesis. The concept of "green chemicals" has grown since the mid-1990s, and researchers have searched for greener and more sustainable methods (Murphy, 2008; Moradi et al., 2021). The improved characteristic properties of silver bulk and their unique intrinsic characteristics including antimicrobial activity and electronics have been shown to attract great attention in recent decades (Sharma et al., 2009; Kaabipour and Hemmati, 2021). In the last decade, several authors have identified more than 500 different biological sources for AgNP synthetization (Patel, 2021). Many bio-sources have been exploited to manufacture AgNPs of various shapes and sizes, including plants or microbes such as fungi, yeasts, algae, or bacteria. Microorganisms can be used to synthesize metal NPs and several intra- or extracellular reports have been produced about this. Metal NPs are bacterial and fungal syntheses since organic solvents are unwanted, and minimal waste is generated under ambient temperature and pressure processes (Kato and Suzuki, 2020). Fungi appeared to have greater efficiency than algae, yeast, and bacteria in the case of microorganisms. Recently, the biogenic synthesis of silver NPs (AgNPs) using biomaterials such as plant extract and microbes as reducing agents, as well as their antimicrobial activity, has received a lot of attention (Rafique et al., 2017).

For the green synthesis of AgNPs, almost any section of plants (e.g., seeds, stems, flores, and leaves) (Remya et al., 2017). The secondary metabolites present in the plant are flavonoids, phenolics, polysaccharides, and terpenoids. These primarily account for the reduction of metal ions in bulk metal preparation NPs in redox response to obtain nanopolynes, which are nontoxic when M+ to M0 is transformed, and also act as a capping material for the obtained NPs (Aisida et al., 2019). The reduction of Ag^+ to Ag^0 is achieved by biological species or through bio-based components produced from a target plant or organism type during most green silver nanostructure synthesis processes (Sharma et al., 2009). The document stated indicated that growth parameters such as pH, temperature, and agent reduction: metal solutions have direct consequences for the synthesis of AgNPs and therefore influence their form and dimensions (Garg et al., 2020). It should be noted that the production of high-yield and single scattered colloidal parts is not effective in the biosynthesis process, in particular because of the mild reduction of the bio-sources (Moradi et al., 2020). The use of different capping agents regulates not only the size, agglomeration, and morphology of the substance, but also the flexibility of nanostructures over time (Restrepo and Villa, 2021). The synthesis process used for the AgNP may have a significant effect on the environmental impact of the future manufacture of these market-based nanomaterials (Temizel-Sekeryan and Hicks, 2020).

As mentioned in the Preface, there is no specific book available on "Green Synthesis of Silver Nanomaterials"; and hence this book provides a unique contribution to the field with almost no competition. Almost of chapters contents in **Part 1** Synthesis various types of silver nanomaterials, **Part 2** focused on different sources for synthesis biogenic silver NMs. **Part 3** reviews the ability of different plant species and parts to produce silver nanoparticles, while **Part 4** comprises seven chapters providing detailed information related to microbial synthesis of silver NMs. **Part 5** presents the synthesis mechanism, toxicity, fate, and commercial production of silver nanomaterials. There are no books available on green production of silver-based nanomaterials. This volume has shown the role of the various bio-sources for a simple, green, and sustainable AgNP synthesis. Ag the mechanism for the creation of nanostructures and how these variables can be adapted to maximize their production has been addressed. There have been present problems and future trends in 1D silver nanostructures green synthesis.

2 Green synthesis

Many methods have been used for the synthesis of silver NPs such as sol-gel processing, hydrothermal methods, chemical vapor deposition, thermal degradation, the microwave-assisted combustion method, etc. Traditional NP processing methods are expensive, harmful, and not environmentally sound. Thus, experts have introduced green approaches to NP synthesis to solve these problems (Ghaffari-Moghaddam et al., 2014). Anastas and Warner proposed the concepts of green chemistry, and established 12 principles describing green chemistry with eloquence (Anastas and Warner, 1998). Various classes of live organisms and biomolecules (prokaryotes and eukaryotes) are being used more and more as biotemplates for the construction of nanoscale materials (Rana et al., 2020). Bio-inspired NP synthesis technology has become a major branch of nanoscience and nanotechnology (Pathak et al., 2019).

Green chemistry should work to reduce pollution, use less oil, use recycled materials, and use risk-averse approaches. This synthesis can be carried out using fungi, algae, bacteria, and seeds, among other things. Because of the inclusion of phytochemicals in its extract, certain plant parts such as leaves, buds, roots, stems, and seeds have been used in the synthesis of different nanoparticles (Jadoun et al., 2020). The option of a solvent medium (preferably water), an environmentally friendly reducing agent, and a nontoxic material for the stabilization of nanoparticles are the three key concepts for preparing nanoparticles via green synthesis (Raveendran et al., 2003). In general, the natural reduction substance or various components are present as stabilizing or capping agents in the cell, thus minimizing the need to use these external agents (Sharma et al., 2009). Silver nitrate and a natural reducing agent form the basic requirement for green AgNP synthesis (Tarannum and Gautam, 2019). So far, several metal and oxide NPs using plant extract, microbial products, etc. have been synthesized (Ahmed and Ikram, 2015; Kato and Suzuki, 2020). Nanotechnology researchers nowadays consider green principles that allow minimal environmental and unintended health risks, and improved applications for next century (Ahmad et al., 2019). Fig. 1 depicts imperative methods for the synthesis of metal NPs as well as various approaches used for the synthesis of AgNPs.

2.1 Synthesis of AgNPs from plants

In comparison with the methods that use a microorganism, AgNPs are typically more synthesized on a plant-based basis as they can easily boost, are less biothreatening, and do not require cell growth steps (Lee and Jun, 2019). Several plant parts like

Green Methods	Chemical Methods	Physical Methods
Microorganisms	Chemical reductions	Pulsed Laser Ablations
Fungi	Sonochemical	Evaporation-Condensation
Bacteria	Microemulsion	Arc Discharge
Algae	Photochemical	Ball Milling
Actinobacteria	Electrochemicals	Vapor and Gas Phase
Plants	Pyrolysis	Photolithography
Agric-Waste	Microwave	Metal-Vapor Synthesis
Biomolecules	Solvothermal	
	Precipitation	
Non-Toxic	**Toxic**	**Toxic**

FIG. 1

Schematic representation of different methods employed for synthesis of AgNPs.

leaves, alkaloids, polysaccharides, tannins, terpenoids, phenols, and vitamins have been used successfully for the synthesis of AgNPs and include biomolecules, e.g., enzymes, alkaloids, polysaccharides, tannins, and phenols (Roy and Das, 2015; Sumitha et al., 2018). Specific parts of the plant are collected from various sources, thoroughly washed with common water, and then washed again with purified water so that waste and any other unnecessary content are excluded. The parts are then dried and ground, and produced in powder form or used for extracting as fresh. Green approaches to AgNPs with plant extracts include the production of stable goods in less time, reduced waste, a peaceful work climate, a friendly and environmentally friendly and low-cost approach green strategy (Kareem et al., 2020; Vanlalveni et al., 2021).

2.2 Microorganisms

In addition to plant extracts and macro algae, microorganisms such as bacteria, fungi, micro algae, and cyanobacteria may lead to necessary reductions in metal biosynthesis that provide environmentally sustainable, low-cost technology, as well as simplifying production for high production (Ahmed and Ikram, 2015; Kato and Suzuki, 2020). Fungi offer a good choice for large-scale biosynthesis of silver NMs. During the downstream treatment, they are simple to handle and secrete large quantities of enzymes required to reduce them. They also have filamentous metal resistance, high binding, and intracellular uptake (Rauwel et al., 2015). In addition, the bacterial synthesis and mechanism of AgNPs regarding their genetic aspect and function were discussed in cell walls, proteins, and bacterial enzymes. Extra- and intracellular AgNP bacterial synthesis was identified with biomass, cell-free extract, and derived intracellular extract (Singh et al., 2015). The bacterial organic materials are natural capping agents and stabilizing agents for AgNPs that prevent their aggregation and create long-term stability. This selection makes green synthesis based on bacteria more versatile, cheap, and an appropriate method for large-scale development (Samadi et al., 2009).

While the mechanism of extra- and intracellular synthesis is still not completely known, in AgNP biogenesis in bacteria, silver resistance machinery is of considerable importance. Two potential routes are available: (1) bacterial-emitted biomolecules for reducing ion silver into the external medium support AgNPs; and/or (2) cell-forming nanoparticles secreted externally (Mahdieh et al., 2012). Different fungal species benefit from the presence of large enzymes, proteins, and decrease components compared with other microorganisms, primarily because: (1) fungi are easily cultivable to obtain sufficient biomass for synthesis; (2) additional cellular sequestration of enzymes decreases postsynthesis; and (3) they have high wall binding ability. This makes fungi excellent candidates for the biosynthesis of AgNPs of various sizes among other microorganisms (Alghuthaymi et al., 2015). The microbial synthesis of nanoparticles is a time-consuming process that can take hours or days. To achieve the desired shape, size, and monodispersity of nanoparticles, factors such as microorganism form and growth stage, pH, temperature, growth medium, substrate concentration, and reaction time must be optimized.

Edible marine algae are popular due to their widespread availability and high concentration of bioactive compounds that can act as active stabilizing and reducing agents (Ulagesan et al., 2021). Algae-capped and stabilized NPs have recently gained widespread attention as a less hazardous, easy to handle, cost-effective, eco-friendly, nano-sized, safer to use, and greener method. In the conversion of metal salts to metal, metal oxide, or bimetallic NPs, the natural material from algae acts as a capping, reducing, and stabilizing agent (AlNadhari et al., 2021). Several algae species, including *Lyngbya majuscula*, *Pyropia yezoensis*, *Spirulina platensis*, *Padina pavonica*, and *Chlorella vulgaris*, have been used in silver nanoparticle syntheses (Govindaraju et al., 2008). Abdel-Raouf et al. (2017, 2019) reported that in bioreduction of marine macro algae metal nanoparts using maritime resources as *Padina pavonica* includes various natural products such as alkaloids, lipids and steroids, phenolic compounds and flavonoids, polysaccharides, and certain chemical (hydroxy, carboxylic, amino) functional cations as well as a number of chemical functional groups. Finally, the biosynthesis of AgNPs using algae extract is a simple, sustainable, and environmentally friendly process. Due to their special properties of rapid growth, high metal accumulation capacity, and plentiful organic material, various algae can be considered as candidates for biosynthesis of AgNPs under specific conditions.

Fungi have outstanding potential for certain compounds and can be used in various applications. About 6400 bioactive compounds (ascomycetes and incomplete fungi) and other fungal species are believed to be manufactured by microscopic filamentous fungi (Bérdy, 2005). Fungi are desirable agents for biogenic silver nanoparticles synthesis since they are metal-tolerant and simple to treat. They also secrete significant amounts of extracellular proteins that lead to nanoparticle stability (Du et al., 2015; Netala et al., 2016). The high amounts of proteins and enzymes produced by various fungal species can be used for quick and sustainable synthesis of nanoparticles compared to other microorganisms (Alghuthaymi et al., 2015). Fungi may be intracellular or extracellular in biogenic synthesis process of the nanoparticles. With intracellular synthesis, the metal precursor is inserted and integrated into the mycelial culture in biomass. The extraction of nanoparticles, using chemical treatment, centrifugation, and filtration after synthesis, is therefore needed in order to interrupt the biomass and release nanoparticles (Rajput et al., 2016; Molnár et al., 2018).

Extracellular synthesis involves the addition of a metal catalyst to an aqueous filtrate comprising only fungal biomolecules, culminating in the creation of free nanoparticles in the dispersion. This is the most commonly used approach since no procedures are needed to release the nanoparticles from the fungal cells (Silva et al., 2017; Gudikandula et al., 2017). There are, however, a range of drawbacks to solve for the efficient use of fungi for biogenic synthesis. These include the need to understand which fungi to use, their growth parameters, the need for sterile conditions, and the time needed to grow fungi and to complete a synthesis. There can be problems with scaling up, including the need for more research regarding mechanisms of capping layers' formation and their molecules (Guilger-Casagrande and Lima, 2019; Rai et al., 2021).

2.3 Biomolecules

The literature reported that the biomolecules generated from plants, fungi, bacteria, algals, and cyanobacteria, as polysaccharides, biopolymers, proteins, enzymes, and vitamins, were also successfully used to clog and cap nanoparctics synthesis agent to avoid aggregate, and to increase the stability and longevity of the biosynthesis. Another significant problem is the toxicity of silver nanoparticles. While biomolecular capsulation reduces toxicity in the green synthesis process, efficient capping and reduction agents for specific applications are essential (Roy et al., 2019). Fairly significant efforts were made to obtain regulated morphologically biosynthesized nanoparticles, with small dimensions, uniformity, and stability, which are favorable for high antibacterial and antimicrobial action (Mohanpuria et al., 2008; Rai et al., 2008; Ogar et al., 2015). Ecofriendly solvents, reagents, are used in the reduction of high energy consumption by nontoxic biomolecules such as nucleic acids and protein, enzymes, carbohydrates, and vegetable extracts (Samanta et al., 2019; Hou and O'Connor, 2020). Such natural biomolecules are relatively inexpensive, simple to process, and reusable, so their primary benefits such as their biodegradability and

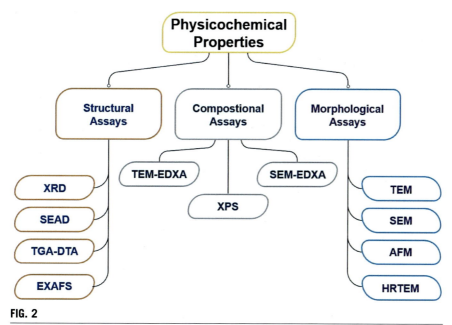

FIG. 2

Summary of the physicochemical methods that are used for silver nanoparticle characterization featured.

biocompatibility make them an attractive synthesis material for nanoparticles. Synthesis in controlled conditions using pure compounds can also provide a solution regarding the poly difference of green nanoparticles synthesis (Roy et al., 2019). Nanoparticle characterization is thus a critical step to comprehend fully the origins of nanoparticle activity and then translate its laboratory performance benefits into practical real-world applications. Scientists face a vital task today in determining the physicochemical properties of nanoparticles and investigating their structure-function relationships. Various measuring methods are used to characterize nanoparticles, which are summarized and outlined in Fig. 2.

3 Advantages

The benefits of green synthesis methodologies, such as low costs, low energy usage, capacity expansion, and simplicity, have been addressed, and these overwhelm their drawbacks in comparison with the methodological and physical syntheses. The biosynthesis approach for AgNPs synthesis has drawn considerable attention, since it is quick, easy to use, environmentally friendly, and cheaper than other methods. The current study demonstrated that a wide variety of biometric agents such as carbon hydrates, alkaloids, terpenoids, phenolic compounds, and enzymes have been

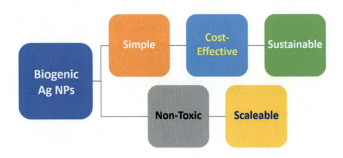

FIG. 3

Advantages of biogenic production of AgNPs produced from plants and microorganisms.

successfully employed in tabular form for the synthesis of AgNPs (Ijaz et al., 2020; Rajoriya et al., 2021). One of the main advantages of AgNPs green synthesis methods shows a wide range of shapes, sizes, and properties. As the reduction and stabilization of silver ions is governed by the combination of biomolecules already present in plant extracts, the main benefit of using the green pathway for the sum of silver nanoparticles is that no additional chemical is needed to stabilize the nanosubstance. These involve amino acids, enzymes, protein, polysaccharides, and various secondary metabolites such as tannins, phenol, saponins, terpenoids, alkaloids, etc. This includes a wide range of secondary metabolites. There are several advantages for biogenic productions of AgNPs, including: (1) green synthesis is simple and typically involves a one-pot reaction; (2) it is scalable; (3) the toxicity associated with hazardous chemicals is eliminated; (4) green biological entities may be used as reducing and capping agents; and (5) the process is cost-effective and requires little involvement or input (Fig. 3).

4 Challenges

There are few green and sustainable methods for 1D Ag nanostructures synthesis that are used for feasibility analysis in most of the syntheses. Batch synthesis also presents problems such as batch-to-batch reproductivity challenges. An important gap in green 1D silver nanometric synthesis is the small scale of processing, which has not been implemented outside the bank with low efficiency and predictability (Syafiuddin et al., 2017). This suggests a huge difference in the green synthesis of 1D Ag nanostructures and a need to increase the number of green reducing agents offered for 1D Ag nanostructure synthesis, especially plant-based ones. The other challenges posed by AgNP biosynthesis are linked to its long-term stability loss, which decreases the effectiveness of its properties. Changes to AgNPs' properties can increase their oxidation and aggregation rates, resulting in differences in dosage versus dosage provided, which may affect bioavailability (Izak-Nau et al., 2015).

4 Challenges

Due to the complex chemistry of silver in aqueous solutions and others, as well as the variety of reported findings (different media, various shapes, sizes, and coatings on the nanoparticles), the behavior of silver nanoparticles in solution is difficult to analyze (Reidy et al., 2013; Klitzke et al., 2015; Fernando and Zhou, 2019). Adsorptions, transformations, aggregations, and dissolution were observed for several different types of silver nanoparticles with various coatings. In this paper, we proposed a three-stage model to describe the development of new nanoparticles. Here we proposed a four-stage model, which explains the formation of new nanoparticles (Fig. 4).

The stability of AgNPs depends primarily on the electrical charge on the AgNP surface. The lack of these charges shows the loss of stability of AgNPs. A proper study of AgNPs' synthesis and storage conditions should be conducted to maintain the properties of AgNPs over the long term to resolve this issue (Velgosova et al., 2017). Furthermore, behind the synthesis reaction process there is an absence of fundamental knowledge and also poorly understood functionality and liability of the individual compounds. The size and morphology of the synthesized 1D Ag nanostructures cannot be monitored and adjusted with this basic information, and the distribution is also broad. By analyzing the importance of each functional group in the reduction, nuclearization, and development of silver nanostructures, the other novel vertebrates of silver and thus other 1D metal nanostructures in the natural environment can be predicted. However, fundamental information about the green synthesis process for 1D silver nanostructures, especially eco-friendly large-scale synthesis, a tool with a high potential to overcome the challenges of 1D silver nanostructures chemical synthesis, is lacking.

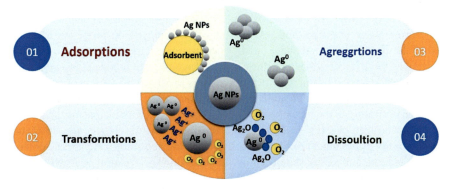

FIG. 4

Flowchart depicting the illustration of silver nanoparticle aggregations, dissolution, absorption, and transformations. Pathway proposed for the creation of new particles from parent nanoparticles.

Following the identification of the basic biochemical compound(s) or ingredient(s) in green reducing agents responsible for metal ion reduction and subsequent development of nanoparticles to a desired morphology, it is important to investigate the reaction process, which will be critical to regulating the reaction and thus controlling the size, shape, and morphology of the synthesized 1D silver particles. In contrast, the use of yeast and algae for AgNP biosynthesis did not achieve the same level of success as fungi and bacteria, and further research is required in this area (Moradi et al., 2021). To investigate the safe and effective application doses, a toxicity analysis of biosynthesized AgNPs on human cells is needed. AgNPs are potentially accumulating in the atmosphere and could be released by a variety of industrial goods.

Future tasks in the green manufacturing phase would be to determine how this approach can be extended uniformly to produce other forms such as truncated, decahedral, pyramidal shapes, cubical, and triangular. Furthermore, there is still a need to identify the best method for increasing efficiency in terms of product yield so that the biological approach can be used on a larger scale (Yaqoob et al., 2020). Although the fast green plant extract synthetic methods have shown great potential in the AgNPs, the method of synthesis and antimicroscopic inhibition are not yet fully understood in the mechanism by which plants' phytochemicals are involved. Moreover, it has remained largely unacceptable for the form of biosynthesized AgNPs to be regulated until now, while chemical methods for form controlled synthesis are already well-known. This problem may be due to the large number of different phytochemicals in the plant extract, making systematic interaction control with produced AgNPs difficult. Better insight into and interplay of each phytochemical would thus pave the way for shape-selective synthesis of biogenic NPs.

The production of marine algae green AgNPs is still in its early stages and cannot be extended to industrial applications and growth. This could occur because of a few limitations of NP algal biosynthesis—for example, lower kinetics (taking a matter of days or weeks), lower NP yield percentage, weak NP morphological characteristics, the marine algae selection, and less optimized NP preparation processes like pH influence, thermal characteristics, and concentration characteristics. As a result, addressing these issues will lead to the controlled synthesis of marine algae-mediated NPs, as well as a good forum for future researchers to carry out a variety of biological applications. It is difficult to scale up the development of biosynthesized AgNPs from the laboratory to the industrial level. Risks associated with increasing the output of biosynthesized AgNPs must be addressed carefully. To begin, the cost and dependability of biological resources such as reducing and stabilizing agents, potential application, and safety should all be considered in the creation of AgNPs. Second, when biosynthesized AgNPs are scaled up, their properties can change and decrease, particularly when dealing with large-scale productions. Certain parameters that affect stability and bioavailability control the stability of these NPs. As a result, further research is needed to determine the optimal experimental conditions for large-scale synthesis of AgNPs in order to maintain the stability and characteristics of biosynthesized AgNPs (Rosman et al., 2021). Finally, the following limitations or challenges were described: (1) stability in hostile environments; (2) lack of understanding in fundamental mechanism and modeling factors; (3) bioaccumulation/toxicity features; (4) extensive analysis requirements; (5) need for skilled operators; (6) system assembling and structures problems; and (7) recycle/reuse/regeneration. In the real world, it is preferable to develop the properties, behavior, and types of nanomaterials to meet the aforementioned criteria.

5 Future trends

Considering the numerous advantages of green synthesis of AgNPs using various biological tools such as microbes and plant extracts, as well as their excellent antimicrobial activities either bare or in conjugation with antibiotic drugs, there is no question that this research area has attracted a lot of attention in recent years. New research will focus on incorporating these in-situ characterization techniques into batch green synthesis of 1D silver nanostructures in order to evaluate the exact reaction mechanism and, as a result, not only control their size and morphology, but also increase the number of reducing and capping agents produced by nature for the synthesis of 1D silver nanostructures. Furthermore, future research will be focused on large-scale industry specific 1D silver nanostructure development that is green, sustainable, and continuous. The production and manufacturing of novel laboratory flow reactors has the potential to reduce waste, reduce building space and energy requirements, and produce more accurate predictive models. This anticipated future research path will allow a synthesis route in which 1D Ag nanostructure properties can be selected and tuned by simple changes in reaction parameters and millifluidic reactor design to manufacture these nanostructures on an industrial relative scale in an environmentally friendly and sustainable direction. Moreover, optimizing these green and sustainable strategies is important, in terms of not just scale-up capability, but also product quality and efficiency. Metals and their oxide materials/nanoparticles biosynthesis using marine algae, lichens, actinobacteria, and marine plants is a largely unexplored region. The synthesis mechanism of AgNPs using plant extracts is still unknown. More detailed studies revealing the precise molecular mechanism of AgNP formation by biological methods are needed to gain better control over the synthesis process (Fig. 5). A more thorough understanding of biochemical processes, surface chemistry, and the chemical nature of binding agents will lead to the discovery of new technologies that can be mass-produced on a large scale.

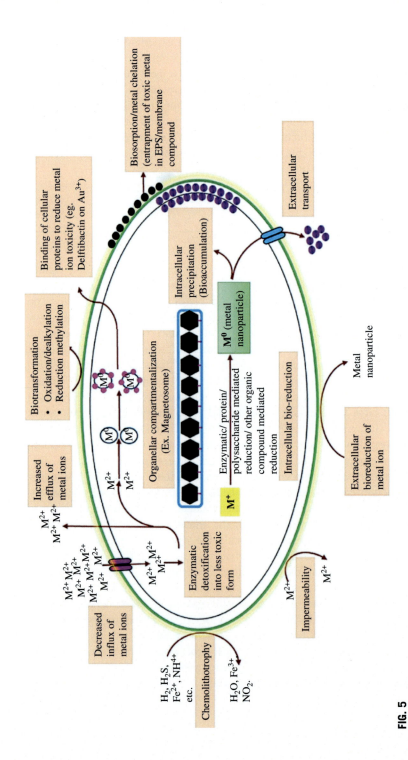

FIG. 5

Suggested mechanism of green synthesis of AgNPs using bioresources.

Reprinted from Rana, A., Yadav, K., Jagadevan, S., 2020. A comprehensive review on green synthesis of nature-inspired metal nanoparticles: Mechanism, application and toxicity. J. Clea. Prod., 122,880, with permission from Elsevier.

6 Conclusion

Standard AgNP syntheses techniques use a lot of resources and toxic chemicals (hydrazine or borohydride as reduction agents) and can result in the formation of hazardous by-products. Green methods for synthesis of AgNPs using bio-renewable materials appear promising because they require nontoxic silver salt reduction chemicals. Various microorganisms, plants, and biomolecules have enormous potential for producing nanomaterials that can be used in a variety of fields. AgNMs have piqued the attention of many researchers due to their peculiar properties and proven applicability in fields as diverse as medicine, catalysis, textile engineering, biotechnology, nanobiotechnology, bioengineering sciences, electronics, optics, and water treatment. Among all of the important techniques of biologically synthesized AgNPs in regular activities, further research is needed to untangle long-term chemical and physical properties as well as to resolve the toxicity of biosynthesized AgNPs.

In this volume, various biogenic methods for the synthesis of AgNPs using phytochemicals, a nontoxic, inexpensive, and environmentally friendly route, have been thoroughly reviewed. Green and sustainable synthesis methods were described, with the synthesis of Ag nanostructures using plant extracts deemed a promising route due to their ease of use, aqueous nature, and nontoxic properties. Plant extracts are expected to contain natural compounds such as polyphenols, flavonoids, alkaloids, and various functional groups such as hydroxy groups and carboxylic acids, many of which have good reducing and capping powers. Particle size and shape, as well as monodispersity, are critical parameters in the evaluation of NPs synthesis. As a result, effective control over the morphology and monodispersity of NPs must be investigated. The conditions of the reaction should be optimized. Well-characterized NPs can be obtained by synthesis processes that are faster or consistent with those of chemical and physical approaches by using screened species with high development capacity and regulating the reaction conditions. This book is about the green synthesis of silver nanoparticles using different living organisms (algae, actinobacteria, bacteria, fungi, plants, and agric-waste). The proposed pathways for AgNP synthesis as well as the mechanisms of action on target cells are illustrated.

References

Abdel-Raouf, N., Al-Enazi, N.M., Ibraheem, I.B., 2017. Green biosynthesis of gold nanoparticles using *Galaxaura elongata* and characterization of their antibacterial activity. Arab. J. Chem. 10, S3029–S3039.

Abdel-Raouf, N., Al-Enazi, N.M., Ibraheem, I.B.M., Alharbi, R.M., Alkhulaifi, M.M., 2019. Biosynthesis of silver nanoparticles by using of the marine brown alga *Padina pavonia* and their characterization. Saudi J. Biol. Sci. 26 (6), 1207–1215.

Ahmad, S., Munir, S., Zeb, N., Ullah, A., Khan, B., Ali, J., Bilal, M., Omer, M., Alamzeb, M., Salman, S.M., Ali, S., 2019. Green nanotechnology: a review on green synthesis of silver nanoparticles-an ecofriendly approach. Int. J. Nanomedicine 14, 5087.

Ahmed, S., Ikram, S., 2015. Silver nanoparticles: one pot green synthesis using *Terminalia arjuna* extract for biological application. J. Nanomed. Nanotechnol. 6 (4), 309.

Aisida, S.O., Ugwu, K., Akpa, P.A., Nwanya, A.C., Nwankwo, U., Botha, S.S., et al., 2019. Biosynthesis of silver nanoparticles using bitter leave (*Veronica amygdalina*) for antibacterial activities. Surf. Interfaces 17, 100359.

Alexander, J.W., 2009. History of the medical use of silver. Surg. Infect. (Larchmt.) 10 (3), 89–292.

Alghuthaymi, M.A., Almoammar, H., Rai, M., Said-Galiev, E., Abd-Elsalam, K.A., 2015. Myconanoparticles: synthesis and their role in phytopathogens management. Biotechnol. Biotechnol. Equip. 29 (2), 221–236.

AlNadhari, S., Al-Enazi, N.M., Alshehrei, F., Ameen, F., 2021. A review on biogenic synthesis of metal nanoparticles using marine algae and its applications. Environ. Res. 194, 110672.

Anastas, P.T., Warner, J.C., 1998. Green Chemistry: Theory and Practice. Oxford University Press, New York, NY, USA.

Bérdy, J., 2005. Bioactive microbial metabolites. J. Antibiot. 58, 1–26.

Du, L., Xu, Q., Huang, M., Xian, L., Feng, J.-X., 2015. Synthesis of small silver nanoparticles under light radiation by fungus *Penicillium oxalicum* and its application for the catalytic reduction of methylene blue. Mater. Chem. Phys. 160, 40–47.

Fernando, I., Zhou, Y., 2019. Impact of pH on the stability, dissolution and aggregation kinetics of silver nanoparticles. Chemosphere 216, 297–305.

Frens, G., Overbeek, J.T., 1969. Carey Leas colloidal silver. Kolloid Z. Z. Polym. 233 (1–2), 922.

Garg, D., Sarkar, A., Chand, P., Bansal, P., Gola, D., Sharma, S., Khantwal, S., Mehrotra, R., Chauhan, N., Bharti, R.K., 2020. Synthesis of silver nanoparticles utilizing various biological systems: mechanisms and applications-a review. Prog. Biomater. 9, 1–15.

Ghaffari-Moghaddam, M., Hadi-Dabanlou, R., Khajeh, M., Rakhshanipour, M., Shameli, K., 2014. Green synthesis of silver nanoparticles using plant extracts. Korean J. Chem. Eng. 31 (4), 548–557.

Gottschalk, F., Scholz, R.W., Nowack, B., 2010. Probabilistic material flow modeling for assessing the environmental exposure to compounds: methodology and an application to engineered nano-TiO_2 particles. Environ. Model. Softw. 25, 320–332.

Gour, A., Jain, N.K., 2019. Advances in green synthesis of nanoparticles. Artif. Cells Nanomed. Biotechnol. 47 (1), 844–851.

Govindaraju, K., Basha, S.K., Kumar, V.G., Singaravelu, G., 2008. Silver, gold and bimetallic nanoparticles production using single-cell protein (Spirulina platensis) Geitler. J. Mater. Sci. 43, 5115–5122.

Gudikandula, K., Vadapally, P., Charya, M.A.S., 2017. Biogenic synthesis of silver nanoparticles from white rot fungi: their characterization and antibacterial studies. Open Nano 2, 64–78.

Guilger-Casagrande, M., Lima, R.D., 2019. Synthesis of silver nanoparticles mediated by fungi: a review. Front. Bioeng. Biotechnol. 7, 287.

Hill, W.R., Pillsbury, D.M., 1939. Argyria: The Pharmacology of Silver. Williams & Wilkins.

Hou, D., O'Connor, D., 2020. Chapter 1-green and sustainable remediation: concepts, principles, and pertaining research. In: Hou, D. (Ed.), Sustainable Remediation of Contaminated Soil and Groundwater. Butterworth-Heinemann, Waltham, MA, USA, pp. 1–17. 978-0-12-817982-6.

Ijaz, M., Zafar, M., Iqbal, T., 2020. Green synthesis of silver nanoparticles by using various extracts: a review. Inorg. Nano-Metal Chem., 1–12. https://doi.org/10.1080/24701556.2020.1808680.

References

Izak-Nau, E., Huk, A., Reidy, B., Uggerud, H., Vadset, M., Eiden, S., Voetz, M., Himly, M., Duschl, A., Dusinska, M., Lynch, I., 2015. Impact of storage conditions and storage time on silver nanoparticles' physicochemical properties and implications for their biological effects. RSC Adv. 5 (102), 84172–84185.

Jadoun, S., Arif, R., Jangid, N.K., Meena, R.K., 2020. Green synthesis of nanoparticles using plant extracts: a review. Environ. Chem. Lett. 19, 1–20.

Kaabipour, S., Hemmati, S., 2021. A review on the green and sustainable synthesis of silver nanoparticles and one-dimensional silver nanostructures. Beilstein J. Nanotechnol. 12 (1), 102–136.

Kareem, M.A., Bello, I.T., Shittu, H.A., Awodele, M.K., Adedokun, O., Sanusi, Y.K., 2020. Green synthesis of silver nanoparticles (AgNPs) for optical and photocatalytic applications: a review. IOP Conf. Ser.: Mater. Sci. Eng. 805, 012020.

Kato, Y., Suzuki, M., 2020. Synthesis of metal nanoparticles by microorganisms. Crystals 10, 589.

Klitzke, S., Metreveli, G., Peters, A., Schaumann, G.E., Lang, F., 2015. The fate of silver nanoparticles in soil solution-sorption of solutes and aggregation. Sci. Total Environ. 535, 54–60.

Lea, M.C., 1889. On allotropic forms of silver. Am. J. Sci. 37, 476–491.

Lee, S.H., Jun, B.H., 2019. Silver nanoparticles: synthesis and application for nanomedicine. Int. J. Mol. Sci. 20, 865.

Mahdieh, M., Zolanvari, A., Azimee, A.S., 2012. Green biosynthesis of silver nanoparticles by Spirulina platensis. Sci. Iran. 19 (3), 926–929.

Mohanpuria, P., Rana, N.K., Yadav, S.K., 2008. Biosynthesis of nanoparticles: technological concepts and future applications. J. Nanopart. Res. 10, 507–517.

Molnár, Z., Bódai, V., Szakacs, G., Erdélyi, B., Fogarassy, Z., Sáfrán, G., Varga, T., Kónya, Z., Tóth-Szeles, E., Szűcs, R., Lagzi, I., 2018. Green synthesis of gold nanoparticles by thermophilic filamentous fungi. Sci. Rep. 8, 3943. https://doi.org/10.1038/s41598-018-22112-3.

Moradi, F., Sedaghat, S., Moradi, O., Arab Salmanabadi, S., 2020. Review on green nano-biosynthesis of silver nanoparticles and their biological activities: with an emphasis on medicinal plants. Inorg. Nano-Metal Chem. 51, 1–10.

Moradi, F., Sedaghat, S., Moradi, O., Arab Salmanabadi, S., 2021. Review on green nano-biosynthesis of silver nanoparticles and their biological activities: with an emphasis on medicinal plants. Inorg. Nano-Metal Chem. 51 (1), 133–142.

Murphy, C., 2008. Sustainability as an emerging design criterion in nanoparticle synthesis and applications. J. Mater. Chem. 18, 2173.

Netala, V.R., Bethu, M.S., Pushpalatha, B., Baki, V.B., Aishwarya, S., Rao, J.V., Tartte, V., 2016. Biogenesis of silver nanoparticles using endophytic fungus *Pestalotiopsis microspora* and evaluation of their antioxidant and anticancer activities. Int. J. Nanomedicine 11, 5683–5696.

Ogar, A., Tylko, G., Turnau, K., 2015. Antifungal properties of silver nanoparticles against indoor mould growth. Sci. Total Environ. 521, 305–314.

Patel, S., 2021. A review on synthesis of silver nanoparticles-a green expertise. Life Sci. Leaflets 132, 16–24.

Pathak, G., Rajkumari, K., Rokhum, L., 2019. Wealth from waste: *M. acuminata* peel waste-derived magnetic nanoparticles as a solid catalyst for the Henry reaction. Nanoscale Adv. 1, 1013–1020.

Pulit-Prociak, J., Banach, M., 2016. Silver nanoparticles—a material of the future…? Open Chem. 14, 76–91.

Rafique, M., Sadaf, I., Rafique, M.S., Tahir, M.B., 2017. A review on green synthesis of silver nanoparticles and their applications. Artif. Cells Nanomed. Biotechnol. 45, 1272–1291.

Rai, M., Bonde, S., Golinska, P., Trzcińska-Wencel, J., Gade, A., Abd-Elsalam, K., Shende, S., Gaikwad, S., Ingle, A., 2021. Fusarium as a novel fungus for the synthesis of nanoparticles: mechanism and applications. J. Fungi 7 (2), 139.

Rai, M., Yadav, A., Gade, A., 2008. CRC 675—current trends in phytosynthesis of metal nanoparticles. Crit. Rev. Biotechnol. 28 (4), 277–284.

Rajoriya, P., Barcelos, M.C., Ferreira, D.C., Misra, P., Molina, G., Pelissari, F.M., Shukla, P. K., Ramteke, P.W., 2021. Green silver nanoparticles: recent trends and technological developments. J. Polym. Environ., 1–27.

Rajput, S., Werezuk, R., Lange, R.M., Mcdermott, M.T., 2016. Fungal isolate optimized for biogenesis of silver nanoparticles with enhanced colloidal stability. Langmuir 32, 8688–8697.

Rana, A., Yadav, K., Jagadevan, S., 2020. A comprehensive review on green synthesis of nature-inspired metal nanoparticles: mechanism, application and toxicity. J. Clean. Prod. 272, 122880.

Rauwel, P., Küünal, S., Ferdov, S., Rauwel, E., 2015. A review on the green synthesis of silver nanoparticles and their morphologies studied via TEM. Adv. Mater. Sci. Eng. 2015. https://doi.org/10.1155/2015/682749, 682749.

Raveendran, P., Fu, J., Wallen, S.L., 2003. Completely "green" synthesis and stabilization of metal nanoparticles. J. Am. Chem. Soc. 125 (46), 13940–13941.

Reidy, B., Haase, A., Luch, A., Dawson, K.A., Lynch, I., 2013. Mechanisms of silver nanoparticle release, transformation and toxicity: a critical review of current knowledge and recommendations for future studies and applications. Materials 6 (6), 2295–2350.

Remya, V.R., Abitha, V.K., Rajput, P.S., Rane, A.V., Dutta, A., 2017. Silver nanoparticles green synthesis: a mini review. Chem. Int. 3 (2), 165–171.

Restrepo, C.V., Villa, C.C., 2021. Synthesis of silver nanoparticles, influence of capping agents, and dependence on size and shape: a review. Environ. Nanotechnol. Monit. Manag. 15, 100428.

Rosman, N.S.R., Harun, N.A., Idris, I., Wan Ismail, W.I., 2021. Nanobiotechnology: nature-inspired silver nanoparticles towards green synthesis. Energy Environ., 0958305X21989883.

Roy, A., Bulut, O., Some, S., Mandal, A.K., Yilmaz, M.D., 2019. Green synthesis of silver nanoparticles: biomolecule-nanoparticle organizations targeting antimicrobial activity. RSC Adv. 9 (5), 2673–2702.

Roy, S., Das, T.K., 2015. Plant mediated green synthesis of silver NPs—a review. Int. J. Plant Biol. Res. 3, 1044–1055.

Samadi, N., Golkaran, D., Eslamifar, A., Jamalifar, H., Fazeli, M.R., Mohseni, F.A., 2009. Intra/extracellular biosynthesis of silver nanoparticles by an autochthonous strain of proteus mirabilis isolated fromphotographic waste. J. Biomed. Nanotechnol. 5 (3), 247–253.

Samanta, S., Agarwal, S., Nair, K.K., Harris, R.A., Swart, H., 2019. Biomolecular assisted synthesis and mechanism of silver and gold nanoparticles. Mater. Res. Express 6 (8), 082009.

Sharma, V.K., Yngard, R.A., Lin, Y., 2009. AgNPs: green synthesis and their antimicrobial activities. Adv. Colloid Interface Sci. 145, 83–96.

Silva, L.P.C., Oliveira, J.P., Keijok, W.J., da Silva, A.R., Aguiar, A.R., Guimarães, M.C.C., Ferraz, C.M., Araújo, J.V., Tobias, F.L., Braga, F.R., 2017. Extracellular biosynthesis of silver nanoparticles using the cell-free filtrate of nematophagous fungus *Duddingtonia flagrans*. Int. J. Nanomedicine 12, 6373–6381.

Singh, R., Shedbalkar, U.U., Wadhwani, S.A., Chopade, B.A., 2015. Bacteriagenic silver nanoparticles: synthesis, mechanism, and applications. Appl. Microbiol. Biotechnol. 99 (11), 4579–4593.

Srikar, S.K., Giri, D.D., Pal, D.B., Mishra, P.K., Upadhyay, S.N., 2016. Green synthesis of silver nanoparticles: a review. Green Sustain. Chem. 6 (1), 34–56.

Sumitha, S., Vasanthi, S., Shalini, S., Chinni, S.V., Gopinath, S.C., Anbu, P., Bahari, M.B., Harish, R., Kathiresan, S., Ravichandran, V., 2018. Phyto-mediated photo catalysed green synthesis of silver nanoparticles using Durio zibethinus seed extract: antimicrobial and cytotoxic activity and photocatalytic applications. Molecules 23 (12), 3311.

Syafiuddin, A., Salim, M.R., Beng Hong Kueh, A., Hadibarata, T., Nur, H., 2017. A review of silver nanoparticles: research trends, global consumption, synthesis, properties, and future challenges. J. Chin. Chem. Soc. 64 (7), 732–756.

Tarannum, N., Gautam, Y.K., 2019. Facile green synthesis and applications of silver nanoparticles: a state-of-the-art review. RSC Adv. 9 (60), 34926–34948.

Temizel-Sekeryan, S., Hicks, A.L., 2020. Global environmental impacts of silver nanoparticle production methods supported by life cycle assessment. Resour. Conserv. Recycl. 156, 104676.

Ulagesan, S., Nam, T.J., Choi, Y.H., 2021. Biogenic preparation and characterization of Pyropia yezoensis silver nanoparticles (Py AgNPs) and their antibacterial activity against Pseudomonas aeruginosa. Bioprocess Biosyst. Eng. 44 (3), 443–452.

Vanlalveni, C., Lallianrawna, S., Biswas, A., Selvaraj, M., Changmai, B., Rokhum, S.L., 2021. Green synthesis of silver nanoparticles using plant extracts and their antimicrobial activities: a review of recent literature. RSC Adv. 11 (5), 2804–2837.

Velgosova, O., Čižmárová, E., Málek, J., Kavuličova, J., 2017. Effect of storage conditions on long-term stability of Ag nanoparticles formed via green synthesis. Int. J. Miner. Metall. Mater. 24 (10), 1177–1182.

Verma, V., 2018. Silver Nanoparticles Market Size by Application, Industry Analysis Report, Regional Outlook, Growth Potential, Price Trends, Competitive Market Share & Forecast, 2018–2024. No. GMI1118, Global Market Insights, Inc., USA.

Yaqoob, A.A., Umar, K., Ibrahim, M.N.M., 2020. Silver nanoparticles: various methods of synthesis, size affecting factors and their potential applications—a review. Appl. Nanosci., 1–10.

CHAPTER 2

Chemical synthesis of silver nanoparticles

Belete Asefa Aragaw[a], Melisew Tadele Alula[b], Stephen Majoni[b], and Cecil K. King'ondu[b]

[a]Department of Chemistry, College of Sciences, Bahir Dar University, Bahir Dar, Ethiopia
[b]Department of Chemical and Forensic Sciences, Faculty of Science, Botswana International University of Science and Technology, Palapye, Botswana

1 Introduction

Recently, particles in nanoscale ranges have received attention because of their unique properties compared to their bulk counterparts (Li et al., 2001). Properties including physical, chemical, electronic, electrical, mechanical, magnetic, thermal, and optical are characteristic of nanoparticles. The large surface area and large surface energy, as well as the reduced spatial confinement, are responsible for these unique physical properties. Their unique properties enable them to be applied in different areas including solar energy conversion, catalysis, medicine, water treatment, and sensor development (Dahl et al., 2007; Hutchison, 2008). Noble metals nanostructures have received attention in nanoscience and nanotechnology because of their wide range of applications in catalysis, sensors, optics, electronics, biotechnology, agriculture, and environment (Kearns et al., 2006; Smith et al., 2006). For these applications, size, shape, and crystallinity play a significant role. Therefore, synthesis of nanoparticles needs due attention to offer the required properties.

Interestingly, noble metals nanostructures exhibit remarkable optical properties in the visible region of electromagnetic radiation. This characteristic optical property arises from a phenomenon called surface plasmons. Surface plasmons are produced when the electromagnetic field in the visible range is coupled to the collective oscillations of conduction electrons of small particles (Tessier et al., 2000; Burda et al., 2005; Pastoriza-Santos and Liz-Marzán, 2008). When the size of particles are significantly smaller than the wavelength of the incident light, a resonance condition called localized surface plasmon resonance (LSPR) happens in a well-defined narrow spectral range in the UV/visible spectrum (Pastoriza-Santos and Liz-Marzán, 2008). The precise wavelength of the plasmon resonance depends on several parameters, among which particle size and shape, surface charge, and the nature of the environment are probably the most important (Mulvaney, 1996). The LSPRs are, however, characterized by strong field enhancement at the interface. Away from the surface, however, the strong electric field vector decays exponentially. These changes in particle size, shape, and interparticle properties give information about the analyte of contact based on the localized surface plasmon resonance (LSPR). Most of the above-mentioned applications of silver nanoparticles are due to their size- and shape-dependent unique chemical and physical properties.

The properties of AgNP and its performance in various applications are governed by the synthesis conditions. Hence, synthesis is considered as a determinant factor for controlling the properties of AgNPs and performance in the proposed purpose. The efficiency of the AgNPs in the proposed application is highly dependent on the property of the nanoparticles and for this purpose, a controlled synthesis is important to achieve the desired property. The size- and shape-controlled synthesis of AgNP is an attractive goal in developing highly active nanoparticle for a variety of purposes. In most of the employed methods, the particles' growth is controlled by a reducing agent and stabilizer. Hence the size and shape of the particles can be tuned by controlling these agents. For instance, the effects of sodium borohydride and ethylene glycol as reducing agents in the synthesis of silver nanoparticles have been reported (Zeng et al., 2010; Bastús et al., 2014). Utilization of sodium borohydride (Creighton) to reduce silver nitrate at room temperature resulted in particles of a quite narrow size distribution of around 10 nm (Bastús et al., 2014). Whereas, a broad range of sizes and geometries have been produced by using ethylene glycol as reducer and polyvinylpyrrolidone (PVP) as a stabilizer by controlling the nucleation and particle growth processes (Zeng et al., 2010). Chemical, physical, and biological methods are some of the synthesizing methods of silver nanoparticles. Among these, chemical synthesis has received massive attention because of its suitability in controlling particle size, morphology, and crystallinity. Importantly, the stabilizers in the chemical synthesis method play a significant role in preventing the aggregation of particles. In this chapter, some important chemical methods for the synthesis of silver nanoparticles are presented.

2 Synthesis of silver nanoparticles

Three approaches are generally used to synthesize nanoparticles: physical, biological, and chemical methods. The evaporation-condensation technique using a tube furnace at atmospheric pressure has been used as a physical method of synthesizing nanoparticles (Schmidt-Ott, 1988; Kruis et al., 2000). Rapidity, avoidance of hazardous chemicals, and utilization of radiation as reducing means are some of the advantages of physical methods. Low yield of nanoparticles, high energy consumption, solvent contamination, and poor size distribution can be considered as limitations of physical methods.

In chemical synthesis methods, size distribution of nanoparticles, morphology, and stability of nanoparticles are controlled by experimental conditions, which in turn affect the kinetics of interaction of metal ions and reducing molecules, and adsorption process of stabilizing agent with metal nanoparticles (Bastús et al., 2014).

Synthesis of silver nanoparticles of different sizes by, for example, monitoring the concentrations of precursor materials and stabilizing agent has been demonstrated by Liu et al. (2011). They produced monodispersed nanoparticles whose particle size increased from 40 to 70 nm then to 2 μm when 0.5, 2.5, and 10 mM silver salt concentrations were used, respectively. Increasing silver concentrations at constant PVP (stabilizing agent) concentrations reduced the ratio of PVP/Ag such that at 10 mM of silver nitrate, there was no enough PVP to stabilize the particles to the extent of the particles aggregating to the micrometer scale (Liu et al., 2011). In the same study, temperature was also found to be a key factor in controlling the size of particles, with large particles being obtained at higher temperatures. Temperature increases the rates of both the nucleation and growth process; hence, for a given amount of time, larger particles will be produced.

As much as varying reaction conditions can be used to control the size of the particles produced, techniques that allow control of the nucleation and growth process should be used to control the morphology in the synthesis of silver nanoparticles (Sun and Xia, 2002; Wiley et al., 2005; Wiley et al., 2006; Kilin et al., 2008; Dong et al., 2009; Tsuji et al., 2009; Xia et al., 2009; Jiang et al., 2010; Jiang et al., 2011; Kochkar et al., 2011). For example, Vo et al., used the seed-mediated technique to produce triangular silver nanoplates. The method was based on citrate stabilized seeds (average size of 6.22 nm) produced by the chemical reduction of Ag ions using sodium borohydride. The triangular nanoplates were produced by making use of a gelatin-chitosan solution in the presence of H_2O_2 and ascorbic acid. The formation of the triangular AgNPs was confirmed via UV-vis through the appearance of the out-of-plane quadrupole (340 nm), in-plane quadrupole (450 nm), and in-plane dipole (650 nm) surface plasmon resonance peaks, as shown in Fig. 1 (Vo et al., 2019). The triangular-shaped nanoplates were obtained by increasing the volume of $AgNO_3$ from 75 μL, where hexagonal particles where produced, to 200 μL. Gelatin and chitosan were essential in the production of triangular-shaped nanoplates, as without gelatin and chitosan, only quasi-spherical and spherical shapes were formed. pH also played a critical role, as nanoplates were only formed in the pH range 5–7; outside this range, only spherical particles were produced. Similarly, Rivero et al., (2013) varied the shapes of the obtained particles from triangular or hexagonal to spherical by varying the ratio of the reducing agent (dimethylaminoborane) to the silver nitrate precursor. Low ratios favored the production of triangular or hexagonal shaped particles while higher ratios resulted in the production of spherical particles. Therefore, interest has grown in controlling the size, morphology, stability, and different properties during designing of a synthesis method. Hence, in the following sections, we will present the principle and mechanism, the effect of synthesis parameters, and advantages and disadvantages of different chemical synthesis methods of AgNPs.

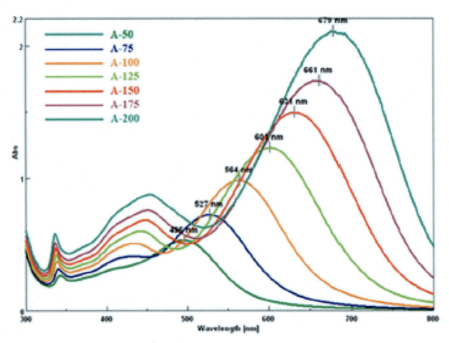

FIG. 1

UV-vis spectra of triangular AgNPs synthesized by different volumes (from 50 to 200 L) of AgNO₃ 0.01 M, 6.0 mL mixed solution of gelatin (0.2%, w/v)-chitosan (0.06%, w/v) (Vo et al., 2019).

3 Factors in synthesis of silver nanoparticles by chemical methods

The genesis of silver nanoparticle synthesis via the bottom-up approach entails the reduction of the metal ion precursors, with the subsequent formation of silver nuclei that grow into larger (nano/micro) particles, processes aptly termed crystallization and crystal growth. Close control of these key processes can be achieved to produce particles of the desired shapes and size with narrow particle size distribution (Abou El-Nour et al., 2010). Capping agents (stabilizers) that prevent particle agglomeration have been routinely used to control the crystal growth process, thereby controlling the particle size of synthesized nanomaterials. The reduction process can be achieved electrochemically through the sacrificial reduction of a silver anode (Schmidt-Ott, 1988; Jung et al., 2006; Harra et al., 2012; Tran et al., 2013) or reduction of aqueous silver ions using inert, mostly platinum, electrodes (Raffi et al., 2011; Nomoev and Bardakhanov, 2012). By far the most widely used method is the chemical reduction method that utilizes conventional chemicals as reducing agents. Control of reaction conditions during chemical reduction methods has been used to control the sizes and shapes of prepared nanomaterials. Large-scale synthesis of

3 Factors in synthesis of silver nanoparticles by chemical methods

AgNPs with different shapes and sizes can be obtained by changing different reaction parameters such as temperature, pH, reaction time, and varying ratio and concentration of reducing agents and precursor silver nitrate (Tripathi et al., 2009). Attaining the maximum yield with controlled size and shape of stable silver nanoparticles necessitates optimization of synthesis conditions. These include reaction temperature, pH of the reaction mixture, incubation time in preparation of the particles, and concentrations and ratio of silver salt and reducing agents. The type of particles depends on the requirements of the intended application of the silver nanoparticles.

The effect of synthesis conditions on size and shape of nanoparticles is illustrated in Fig. 2. Variation of the conditions are reflected in the size and shape of the particles and then in their optical properties. For example, synthesis of AgNPs at room temperature has the advantage of generating stable particles with uniform size distribution compared to autoclave or even microwave oven synthesis (Kora et al., 2012), whereas high temperature produces broad peaks, which indicates large-sized AgNPs (Zhang et al., 2011). Fig. 2 shows the effect of concentration of silver nitrate

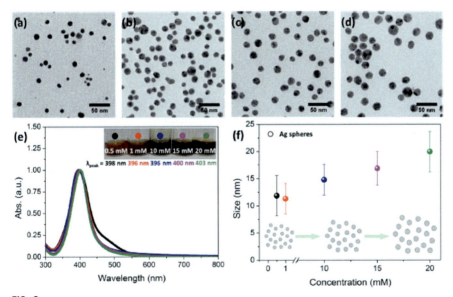

FIG. 2

Effect of the reaction concentration (at 10 min and 120 °C). (A–D) Representative TEM images of the different reaction concentrations analyzed (1, 10, 15, and 20 mM, respectively). (E) UV-vis measurements of the same samples. The inset contains digital images of the samples. (F) Mean size evolution with concentration (the error bars represent the standard deviation over the population mean). Inset: scheme of the AgNP evolution with the reaction concentration.

Reprinted from Vo, Q.K., et al., 2019. Controlled synthesis of triangular silver nanoplates by gelatin-chitosan mixture and the influence of their shape on antibacterial activity. Processes, 7(12), 873. Adapted with permission, ACS.

(0.5, 1, 10, 15, and 20 mM) synthesized using a microwave-assisted polyol method at a temperature of 120 °C using a 10 min reaction time. From this figure, it is clearly shown that the average size and the number of nanoparticles increase with concentration of silver nitrate. The TEM images are in agreement with this observation, i.e., sizes of silver nanoparticles increased with concentration. The variation in LSPR band maxima also supports this observation. As shown in Fig. 2e, the silver LSPR band maxima for silver nanoparticles prepared from 0.5, 1, 10, 15, and 20 mM $AgNO_3$ were red-shifted and located at 398, 396, 396, 400, and 403 nm, respectively. The average sizes of silver nanoparticles vs concentration of silver nitrate are given in Fig. 2f, which shows that the average sizes of silver nanoparticles increase with concentration.

Similar to the effect of concentration of the precursor silver nitrate solution, the effect of the reaction temperature was investigated. They took 1 mM $AgNO_3$ solution and the reaction time was fixed at 10 min. The reaction temperatures were then set at 60, 90, 120, 150, and 180 °C, and the synthesized nanoparticles were investigated. The TEM images, UV-vis spectra, and mean size evolution with temperature is given in Fig. 3. The absence of precipitation but a slight color change at 60 °C shows no

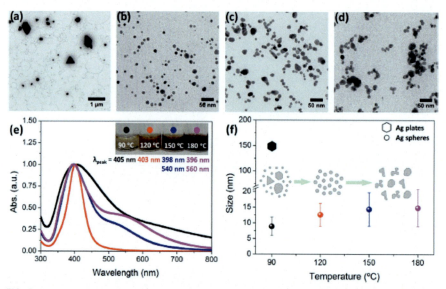

FIG. 3

Effect of the reaction temperature (at 10 min and 1 mM $AgNO_3$). (a–d) Representative TEM images of the different reaction temperatures analyzed (90, 120, 150, and 180 °C, respectively). (e) UV-vis measurements of the same samples. The inset contains digital images of the samples. (f) Mean size evolution with temperature (the error bars represent the standard deviation over the population mean). Inset: scheme of the AgNP evolution with the reaction temperature.

Reprinted from Torras, M., Roig, A., 2020. From silver plates to spherical nanoparticles: snapshots of microwave-assisted polyol synthesis. ACS Omega, 5(11), 5731–5738. Adapted with permission, ACS.

nucleation occurred at a lower temperature within 10 min. But the color changed to orange-yellow after the colloids were stored overnight. The representative TEM images for these temperatures are given in Fig. 3a–d. The TEM image for a temperature of 90 °C was found to show two types; polydispersed polyhedral shapes with an average size of around 150 nm and monodispersed spherical nanoparticles with an average size of 9±3 nm. Increasing the temperature further resulted in a loss of monodispersity and, significantly, the mean particle size increased. The loss of monodispersity is attributed to the fact that, at higher temperatures, particles start to fuse and aggregations of particles are observed, as shown in Fig. 3c and d. The UV-vis spectra are given in Fig. 3e. Broad LSPR band at 405 nm was observed at 90 °C while the band red-shifted with temperature and a narrower band with LSPR peak at 403 nm was obtained for 120 °C. For 150 and 180 °C, two broad peaks at 398 and 540 nm and at 396 and 560 nm, respectively, were found. Interestingly, these secondary and red-shifted bands are attributed to the bigger and nonisotropic nanoparticles formed, which are in agreement with the TEM images. The relationship between mean sizes and reaction temperatures is given in Fig. 3f (Torras and Roig, 2020).

The effect of pH in tuning the size of AgNPs is critical. Qin et al., demonstrated the effect of pH in tuning the size of spherical silver nanoparticles using ascorbic acid as the reducing agent. The reactivity of ascorbic acid in reducing silver ions is highly pH-dependent (Qin et al., 2010). This is via mediating the reduction rate and the number of nuclei formed. It was found that the average size of spherical nanoparticles decreases (73–31 nm) with the pH of the reaction mixture (6.0 to 10.5). Interestingly, intraparticle ripening was promoted and, as a result, more spherical-like silver nanoparticles were obtained as the pH of the reaction mixture was increased (Qin et al., 2010). The UV-vis absorption spectra and the corresponding TEM images obtained for AgNPs prepared from silver nitrate and ascorbic acid at different pH values clearly shows the effect of pH. The UV-vis spectra of the LSPR peaks at 480, 453, 442, 433, 422, and 412 nm were obtained at a pH of 6.0, 7.0, 8.0, 9.0, 10.0, and 10.5, respectively, as shown in Fig. 4. This shows that the peaks blue shifted and became narrower with a higher pH value. The TEM images (Fig. 5) show that all the particles prepared were quasi-spherical in shape. The shape of the product prepared under lower pH was less regular, especially for the one prepared at pH 6.0. The average sizes of the particles prepared at pH of 6.0, 7.0, 8.0, 9.0, 10.0, and 10.5 were 73 nm (±22%), 63 nm (±15%), 56 nm (±20%), 50 nm (±19%), 40 nm (±17%), and 31 nm (±19%), respectively. The average size decreased with elevated pH, which was consistent with the blue shift of the absorption peaks in the UV-vis spectra. Besides the concentration of silver seeds, the effect of pH in determining whether nanorods or nanowires are produced from silver seeds was demonstrated by Jana et al. (2001). They produced citrate-stabilized Ag seed particles using sodium borohydride as the reducing agent. The growth of the silver seeds into nanorods and nanowires was achieved utilizing ascorbic acid as the reducing agent in the presence of cetyltrimethylammonium bromide (CTAB) and NaOH. Nanorods were produced at pH values slightly above 11.8, which is the pK_a of the second proton of ascorbic acid, while nanowires were produced at pH values slightly lower.

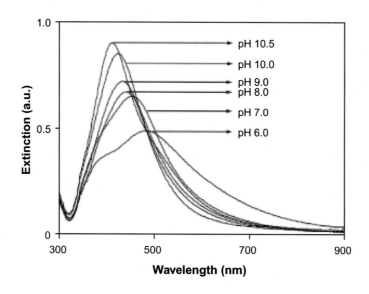

FIG. 4

UV-vis spectra of the silver nanoparticles synthesized at pH 6.0, 7.0, 8.0, 9.0, 10.0, and 10.5 at 30 °C.

Reprinted from Qin, Y., Ji, X., Jing, J., Liu, H., Wu, H., Yang, W., et al., 2010. Size control over spherical silver nanoparticles by ascorbic acid reduction. Colloids Surf, A Physicochem Eng Asp, 372(1), 172–176. Adapted with permission, Elsevier.

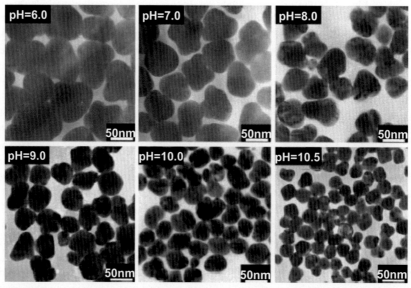

FIG. 5

TEM images of the silver nanoparticles prepared at pH 6.0, 7.0, 8.0, 9.0, 10.0, and 10.5 after 15 min of the reactions at 30 °C.

Reprinted from Qin, Y., Ji, X., Jing, J., Liu, H., Wu, H., Yang, W., et al., 2010. Size control over spherical silver nanoparticles by ascorbic acid reduction. Colloids Surf, A Physicochem Eng Asp, 372(1), 172–176. Adapted with permission, Elsevier.

The time effect on particle sizes is significant. For example, Pillai and Kamat (Pillai and Kamat, 2004) observed particles sizes of 3–5 nm after 10 min and 40–60 nm after 2 h reaction time in the citrate reduction method. For the chemical reduction method, time control can be achieved simply by the removal of particles from the reaction conditions and retaining a condition by which the reaction can be halted (Cheng et al., 2011; Alula et al., 2020).

4 Formation mechanisms of silver nanoparticles

To obtain particles with a certain property, precise control of size and shape during synthesis is critical. Therefore, understanding the particle growth mechanism is important in order to synthesize particles with the intended properties. In the chemical synthesis method, fine-tuning of different variables may help us to understand the formation of nanoparticles with well-defined morphologies. Nucleation and particle growth are two very important processes in synthesis of silver nanoparticles. The shape and sizes of nanoparticles are, therefore, controlled by these two processes. They are affected by parameters such as concentration of the precursor solutions, reaction temperature, type of reducing agent, and pH of the reaction mixture. Activation energies play a vital role in this regard. For example, high activation energy is required for nucleation to agglomerate the atoms together. Particle growth, on the contrary, requires low energy of activation for ordering the formation of particles (Jose Vega-Baudrit et al., 2019). The role of the stabilizing agent in controlling the size and shape of silver nanoparticles is immense. Importantly, stabilizing agents protect the nanoparticles from unexpected agglomeration such that size and shape can be controlled thermodynamically, kinetically, and stoichiometrically (Pastoriza-Santos and Liz-Marzán, 2008; Jose Vega-Baudrit et al., 2019). Similar to other particles formation, the key determinants of AgNPs formation are nucleation and particle growth. According to LaMer and Dinegar (1950), formation of colloids with narrow size distribution necessities a short nucleation burst followed by slowly controlled particle growth. Further nucleation after the onset of particle growth should be avoided. They proposed three phases for nucleation and particle growth. Initially, the point of supersaturation should be achieved by increasing the monomer concentration gradually. The second step is to increase the monomer concentration if the energy barrier to initiate homogeneous nucleation is high. In this case, the supersaturation finally reaches a critical value. Hence, homogeneous nucleation in the entire reaction system can occur by overcoming the energy barrier and results in a large number of nuclei simultaneously with no further nucleation. The third step is the growth of all the nuclei at the same time. Narrow size distribution is possible if the growth histories of the nuclei are identical (Khan et al., 2018).

5 Chemical reduction using reducing agents

Chemical reduction is the most commonly used method for preparation of AgNPs in liquid phase using a chemical reducing agent in aqueous or nonaqueous solvents. One of the key advantages of the chemical reduction method is the relative easiness of precise control over the size and shape of the nanoparticles, yielding a set of monodisperse nanoparticles, which is important in exhibiting consistent properties and reproducibility (García-Barrasa et al., 2011). Colloidal silver with particle diameters of several nanometers are the most commonly produced nanoparticles using reduction of silver ions. In the beginning, reduction of silver ions results in formation of silver atoms. The atoms are susceptible to agglomeration and finally result in oligomeric clusters. The formation of colloidal silver nanoparticles results from these clusters.

The size of silver nanoparticles depends on the strength of the reducing agent. A strong reducing agent like sodium borohydride results in small size but relatively monodispersed nanoparticles. A weaker reducing agent like citrate, on the other hand, reduces slowly and results in particle size with a large distribution. Therefore, controlled synthesis of AgNPs can be achieved in the presence of stabilizers so that unwanted agglomeration of the colloids can be prevented (Creighton et al., 1979; Lee and Meisel, 1982; Emory and Nie, 1997; Shirtcliffe et al., 1999).

Several reducing agents are available, such as sodium borohydride ($NaBH_4$), hydrazine (N_2H_4), ascorbic acid, trisodium citrate, polyols, aldehydes for Tollens test, N,N-dimethylformamide (DMF), sugars/polysaccharides, and others. The chemical reaction is an oxidation-reduction reaction where silver ions (Ag^+) get reduced to form silver atoms (Ag^0) and the reducing agent, which is the source of electrons, gets oxidized. The Ag atom monomers bind together to form oligomeric clusters and, subsequently, coalesces to form colloidal AgNPs. The colloidal AgNPs should be stabilized to stay dispersed in a solvent for a longer period to be used in a variety of applications. Hence, the use of stabilizing agents (also called surfactants, capping agents, and ligands) is necessary. These are molecules or ions that bind with the nanoparticle surface through chemisorption processes, electrostatic attractions, or hydrophobic interactions (Manojkumar et al., 2016). Surfactants such as citrate, polyvinylpyrrolidone (PVP), cetyltrimethylammonium bromide (CTAB), and polyvinyl alcohol (PVA) make interactions with particle surfaces that can not only protect particles from sedimentation and agglomeration but also control particle growth during synthesis (Kvítek et al., 2008; Liang et al., 2010; Wuithschick et al., 2013). These ligands create 3D steric or electrostatic repulsion between adjacent AgNPs so that precipitation of NPs from solution is prevented. The most common functional groups with strong surface interactions with AgNPs include —SH, —NH, —COOH, and —C=O. The strength of an interaction between the nanoparticle and ligand depends on the hard-soft acid-base nature of the Ag and ligand atom (Nath et al., 2006). Since the reducing agent alone plays an important role in determining the size, morphology, crystallinity, and physicochemical property of the AgNPs, and the large electropositive reduction potential of silver in water ($Ag^+ + 1e^- \rightarrow Ag^0$,

$E^0 = +0.799$ V) permits the use of several reducing agents such as sodium citrate ($E^0 = -0.180$ V), sodium borohydride ($E^0 = -0.481$ V), and hydrazine ($E^0 = -0.230$ V) (Pacioni et al., 2015), the synthesis of AgNPs based on different reducing agents can be discussed.

5.1 Reduction by sodium borohydride (NaBH$_4$)

The most common synthetic procedure to obtain colloidal AgNPs is the chemical reduction of silver precursor salts (AgNO$_3$, AgClO$_4$, AgSO$_4$, and Ag acetate) with sodium borohydride as a reducing agent in the presence or absence of stabilizers. Since all sodium salts and nitrates are soluble in water, AgNO$_3$ and sodium borohydride are two popular combinations during chemical reductions of silver ion in aqueous synthesis of AgNPs. The concentration and kind of the silver salt and surfactant used during the synthesis procedure determine the size, morphology, crystallinity, and physicochemical property of the AgNPs (Wiley et al., 2007). Particles with different shapes like polydispersed spherical (Angelescu et al., 2012) and nanoprism (25–400 nm) (Van Dong et al., 2012) AgNPs were obtained by the reduction of AgNO$_3$ with slight variations in the conditions with NaBH$_4$.

NaBH$_4$ is considered a strong reducing agent for the reduction of silver ions. Though a higher concentration of NaBH$_4$ is assumed to lead to a fast rate of production of silver nanoparticles, Zhang et al., reported the opposite effect. According to them, borohydride ions can slow down the formation of silver nanoparticles by stabilization, binding on the surface of the particles. Hence the initiation time required for nucleation increased with higher concentrations of NaBH$_4$ in a given reaction system (Zhang et al., 2011).

Wuithschick and coworkers investigated the size-controlled synthesis of colloidal AgNPs based on a mechanistic understanding and found that the chemical reduction of Ag$^+$ in an aqueous solution by NaBH$_4$ is a two-step reaction where the first step produces silver metal and boric acid (Wuithschick et al., 2013):

$$Ag^+ + BH_4^- + 3H_2O \rightarrow Ag^0 + B(OH)_3 + 3.5H_2$$

Using Raman spectroscopy, Edwards and coworkers showed that boric acid B(OH)$_3$ does not act as Brønsted acid in an aqueous solution (Edwards et al., 1955). Instead, it serves as Lewis acid, which leads to the formation of the tetrahydroxyborate anion, as shown in the following equation:

$$B(OH)_3 + H_2O \rightarrow B(OH)_4^- + H^+$$

Subsequently, the overall reaction can be expressed by:

$$Ag^+ + BH_4^- + 4H_2O \rightarrow Ag^0 + B(OH)_4^- + 3.5H_2 + H^+$$

Glavee et al. studied the reduction reaction of various metal ions with NaBH$_4$ and determined the stoichiometric factors in a balanced equation (Glavee et al., 1994). According to the above balanced equation, equimolar amounts of Ag$^+$ ion and BH$_4^-$ should be used. But to ensure the complete reduction of Ag$^+$, excess amounts of borohydride reducing agent is used, mostly sixfold excess, in the standard synthesis (Wuithschick et al., 2013). After the complete reduction of Ag$^+$ ion to Ag atom, the excess BH$_4^-$ hydrolyzes to form H$_2$ gas and B(OH)$_4^-$. The mechanism of reduction is that borotetrahydride (BH_4^-) ion can act as a source of nucleophilic hydride H^-, which can reduce a variety of metal ions (Glavee et al., 1994). The reducing species, which is the hydride ion, can reduce metal ions like Ag^+ within milliseconds (Polte et al., 2010). In the absence of metal ions, sodium borohydride can undergo a hydrolysis reaction with water in aqueous solutions as follows (Wuithschick et al., 2013):

$$BH_4^- + 4H_2O \rightarrow B(OH)_4^- + 4H_2$$

The Ag atom monomers formed in the reaction binds together to form oligomeric clusters and subsequently coalescences to form colloidal AgNPs. Some reports claim the involvement of BH_4^- in the stabilization of metastable states of Ag by creating a negative charge borohydride layer on surfaces so that electrostatic repulsion prevents nanoparticle agglomeration (Choi et al., 2016).

Wuithschick et al., (2013) made a detailed mechanistic investigation of the influence of the nanoparticle synthesis parameters during borohydride reduction. The reactant concentration, mixing conditions (addition of Ag^+ to BH_4^- or BH_4^- to Ag^+), stirring speed, the ratio of Ag salt to $NaBH_4$, and $NaBH_4$ solution age determines the size and size distribution of AgNPs. They systematically produced size-controlled AgNPs in the range of 4–8 nm in radius in aqueous solution and without addition of stabilizing agents. On the other hand, Ajitha et al., (2015) used variation in solution pH to control the size of silver nanoparticles produced via the reduction of silver ions by sodium borohydride. They observed preferential production of small, monodispersed, spherical nanoparticles at high pH values in ethanol using PVA as the stabilizing agent. Particles sizes of 14 nm were obtained at a pH of 12 while the largest particles with an average size of 31 nm were obtained at pH 6. Similarly results from face-centered central composite design, which has been used to optimize the synthesis process and control particle size and morphology, reveal an intricate interaction between the concentration of precursor silver ions and solution pH (Quintero-Quiroz et al., 2019). It should be noted that $NaBH_4$ is highly hygroscopic and a mass increase of up to 300% is demonstrated if not kept in a water-free environment (Beaird et al., 2010).

5.2 Reduction by hydrazine

Hydrazine (N_2H_4) is toxic and unstable unless made in solution form as hydrazine hydrate ($N_2H_4.xH_2O$). It is a known reducing agent with less reducing power than $NaBH_4$. The reduction of a metal ion with hydrazine hydrate produces harmless byproducts such as nitrogen gas and water. According to Audrieth and Ogg (1951), it can react with dissolved oxygen in the water and even undergo self-oxidation and reduction in both alkaline and acidic solution according to the following reactions:

$$N_2H_4 + O_2 \rightarrow N_2 + 2H_2O$$

$$3N_2H_4 \rightarrow N_2 + 4NH_3$$

The reaction with oxygen helps to remove the excess hydrazine in the form of nitrogen gas and water, according to the above reaction. In principle, hydrazine reduces any metal (first-row transition metal ions, second- and third-row transition metals ions, most posttransition elements, and a few nonmetals ions) with an E^0 more positive than -0.23 V at room temperature, given a sufficient excess of reducing agent and proper control of pH (Huang et al., 2010). The oxidation half-reaction of hydrazine is:

$$4OH^- + N_2H_4 \rightarrow N_2 + 4H_2O + 4e^- \quad E^0 = 1.17\,V$$

The $4e^-$ in the product side shows the oxidation of 1 mol of hydrazine reduces 4 monovalent metal ion to metal atom.

According to Lee (1996), the reduction of metal ions into an elemental metal atom with hydrazine can be represented as:

$$M^{m+} + N_2H_4 \rightarrow M^0 \downarrow + N_2 + H^+$$

In alkaline media, the reaction can be described as (Rich, 2007):

$$M^{m+} + N_2H_4 + OH^- \rightarrow M^0 \downarrow + NH_3 + H_2O$$

Nickel et al., (2000) studied the synthesis of a silver colloid using hydrazine as a reducing agent for surface-enhanced Raman spectroscopy measurement. A decrease in pH was observed after the reaction due to the release of H^+ and they proposed the following chemical reaction:

$$4Ag^+ + N_2H_4 \rightarrow 4Ag^0 \downarrow + N_2 + 4H^+$$

1 mol of hydrazine reduces $4Ag^+$ into $4Ag^0$ atoms. Even though the molar ratio of silver nitrate to hydrazine is 4:1, usually, excess hydrazine is used to confirm complete reduction. The product of the hydrazine reduction of metal ions is inert nitrogen gas, which can be completely removed from the reaction solution. This is considered as an advantage in using hydrazine as a reducing agent, whereas effects of the aqueous solution of hydrazine include corrosion to eyes, skin, and mucous membranes upon contact, and it is a probable human carcinogen. Furthermore, hydrazine fumes attack the nose and throat upon inhalation and can also irritate the eyes, causing temporary blindness (Audrieth and Ogg, 1951).

The solution pH is the most important factor and significant reduction can be achieved at higher pH of reaction mixture. The Ag^+ reduction with hydrazine in alkaline media is represented as (Chen and Lim, 2002):

$$2Ag^+ + 2N_2H_4 + 2OH^- \rightarrow 2Ag^0 \downarrow + 2NH_3 + 2H_2O$$

In addition to pH, other factors controlling the reduction reaction and property of the synthesized AgNPs include the concentration of reactant and surfactants, the solvents type, and the amount of dissolved oxygen.

Several works have reported on the preparation of AgNP via hydrazine reduction. For example, Tan et al., (2003) prepared hexagonal Ag nanocrystals by chemical reduction of silver ions in the presence of aniline using hydrazine monohydrate

($N_2H_4 \cdot H_2O$) as the reducing agent. Patil et al., (2012) reported a rapid synthesis of spherical morphology (10–60 nm diameter size) AgNPs at room temperature by using hydrazine hydrate as a reducing agent and polyvinyl alcohol as a stabilizing agent. Wang et al., (2012) also reported synthesis of porous submicron structures of AgNPs with high surface area via hydrazine-based reduction in glycerol-ethanol solution. The synthesis of uniform monodispersed crystalline silver nanoparticles in a water-ionic liquid system as solvent and stabilizer and hydrazine as a reducing agent was also reported (Patil et al., 2011).

5.3 Reduction by trisodium citrate

Following the Turkevich report (1951) of colloidal Au NP synthesis in aqueous solution at boiling temperature using trisodium citrate to reduce $AuCl_4^-$, the Turkevich method (Turkevich et al., 1951) was extended to the synthesis of AgNP in 1982 by Lee and Meisel (Lee and Meisel, 1982). Through these two important studies, citrate is known to be one of the common reductant and stabilizer in metal nanoparticle synthesis (especially Au, Ag, and Cu NPs). The weak interaction of sodium citrate molecules with metal surfaces enables the surfaces of the particles to be accessible (Bastús et al., 2014). In this method, citrate ions can act as a weak reducing agent and a strong complexing stabilizer. This affects the silver ions reduction kinetics that may affect nucleation and particle growth that would affect the morphologies. This can be considered as the drawback for citrate mediated synthesis of silver nanoparticles (Bastús et al., 2014). The formation of Ag_4^{2+}, which acts as a monomer for the formation of AgNPs, results in heterogeneous nucleation of final product of large and polydispersed particles. Controlling reaction conditions assists in controlling the complexation of silver ions and citrate, and hence, the nucleation and particle growth can be controlled. For example, Dong et al., investigated the role of pH in a reduction of silver ions by citrate (Dong et al., 2009).

The reduction mechanism of metal ions by citrate is still under investigation. One recent study demonstrated that oxidized citrate radical is a stronger reductant than citrate. According to this study, OH• radicals can abstract H from citrate, and the resulting citrate radicals can transfer one electron to silver ions that could initiate simultaneous nucleation and particle growth (Al Gharib et al., 2019). The formation of dicarboxy acetone (DCA) and CO_2 are claimed to be proof of the decarboxylation process (Al Gharib et al., 2019). This argument is very plausible as the thermodynamic reduction potential of Ag^+ to Ag atom is highly negative (<-1.2 V_{NHE}) and a strong reductant like Cit(—H)• is required to overcome the thermodynamic barrier compared to the weak citrate ion. According to this study, the DCA and CO_2 produced during the reaction are not the result of direct oxidation of citrate but the oxidation of the intermediate Cit(—H)• formed from citrate oxidation, confirming that the electron donor for Ag^+ reduction is Cit(—H)• and not citrate itself.

Most of the previous studies claimed that the reduction mechanism of Ag^+ ion by citrate starts by oxidation of citrate (considering citrate anion as a source of an electron to reduce Ag^+ ion) to dicarboxy acetone (DCA), CO_2, and H^+ as follows (Wuithschick et al., 2015):

$$Ag^+ + Cit^{3-} \rightarrow DCA^{2-} + CO_2 + H^+ + 2e^- \quad \text{(oxidation half-reaction)}$$

where the Cit^{3-} and DCA^{2-} represent the following structures:

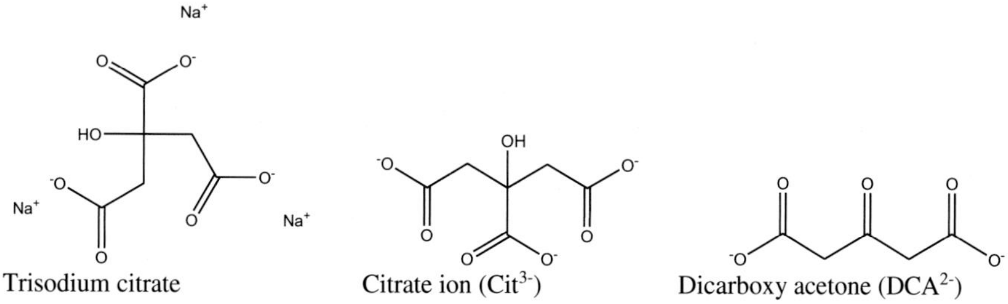

Trisodium citrate   Citrate ion (Cit^{3-})   Dicarboxy acetone (DCA^{2-})

Citrate can be considered as substituted tertiary alcohol, like tert-butanol, which, when oxidized, leads to the formation of dicarboxy ketone (e.g., DCA). The electron produced during the oxidation of citrate is used to drive the reduction of Ag^+ ion:

$$2Ag^+ + 2e^- \rightarrow 2Ag^0 \quad \text{reduction}$$

Subsequently, the overall redox reaction between Ag^+ ion and citrate could be represented as:

$$2Ag^+ + Na_3Cit \rightarrow 2Ag^0 + DCA^{2-} + CO_2 + H^+$$

Citrate ions are not only a reducing agent in the reaction but also control particle growth by complexing with positively charged Ag_2^+ dimers ($Ag^0 + Ag^+ \rightarrow Ag_2^+$) (Pillai and Kamat, 2004). Their steady-state and pulse-radiolysis experiments provide evidence for the multiple roles of citrate ions as a reductant, complexant, and stabilizer that collectively dictate the size and shape of silver nanocrystallites. They illustrated the formation of a complex between $(Ag_2)^+$ and citrate ion and AgNP formation as follows:

$$(Ag)_2^+ + Cit^- \leftrightarrow [(Ag)_2^+....Cit^-]$$

$$[(Ag)_2^+....Cit^-] \xrightarrow{\text{aggregation}} (Ag_4..Cit^-)\text{nanocluster}$$

$$(Ag_4..Cit^-)\text{nanocluster} \xrightarrow{\text{nucletion}} (Ag_n..Cit^-)\text{seeds}$$

$$(Ag_n..Cit^-) \xrightarrow{\text{growth}} AgNPs$$

In addition to reducing Ag^+ ion, the citrate ion acts as a capping agent for the AgNP to form stabilized AgNP. Since citrate is an anion, it acts as a negatively charged ligand covering the AgNPs' surface. The electrostatic repulsion of negatively charged citrate capped AgNPs prevents the particles from binding together to form an aggregate. This keeps the AgNPs dispersed in the solvent without aggregation for a long time.

Several factors affect the reduction of Ag^+ ion with citrates, such as the ratio of Ag salt to the citrate, pH, reaction temperature, and time of reaction. Even though the

Ag$^+$ ion to citrate reaction stoichiometry is 2:1, only a small fraction of Ag$^+$ gets reduced at equimolar concentration, showing a higher citrate concentration is required for complete reduction (five times or higher) (Pillai and Kamat, 2004). This is reasonable since the citrate ion is serving as both reductant and stabilizer. The reaction time is also influenced by the concentration of sodium citrate, where higher concentrations showed faster reduction and vice versa. The silver salt solution is set at reflux temperature for a better reduction reaction. A reaction time of about 50 min was found effective for achieving a complete reduction of Ag$^+$ ions at boiling temperature for a five-times citrate concentration for Ag$^+$ ion (Pillai and Kamat, 2004). The pH value of the solution highly determines the reduction activity of the citrate ion due to the H$^+$ involvement on the right side of its oxidation reaction, as given above. For example, Dong et al., employed trisodium citrate as the reducing agent and synthesized silver particles with average values of 30, 59, 69, and 96 nm at pH values of 11.1, 8.3, 6.1, and 5.7, respectively (Dong et al., 2009). They also demonstrated the effect of pH on the shapes of AgNPs. They used trisodium citrate as the reducing agent to produce differently shaped silver particles by varying the pH of the reaction solution. When the reaction pH was above 8.0, rod-like particles, in addition to spherical particles, were produced, while at pH values of 6.1 and lower, a mixture of spherical and triangle or polygon-shaped particles where observed. They were, however, able to produce only spherical shaped particles via a two-step process, whereby the nucleation and growth processes were separated. The nucleation process was conducted at pH 8.3 and then the pH reduced to 6.1 for the growth process. By utilizing a two-step instead of a one-step process, they managed to achieve control of the nucleation and growth processes, thereby controlling the shape of the particles produced. Formation of both spherical and rod-like silver nanoparticles at higher pH from AgNO$_3$ and sodium citrate was reported. On the other hand, tringles and polygons were formed at lower pH (Dong et al., 2009). Furthermore, an increase in size from ~30 to ~95 nm was observed when the pH was lowered from 11.1 to 5.7 (Dong et al., 2009). The stability of the dispersion of nanoparticles can be maintained by controlling the pH of the solution. Recently, it was reported that at pH > 7, citrate-stabilized small spherical silver nanoparticles remained stable. At lower pH, however, they aggregated. Similarly, pH-dependent agglomeration and dispersion of silver nanoparticles was reported (Ashraf et al., 2013). The effect of pH can be associated with the complexation and redox behaviors of the reducing molecules (Ghodake et al., 2010).

The redox potential of DCA^{2-}/Cit^{3-} could be calculated by the following equation:

$$E_{DCA^{2-}/Cit^{3-}} = E^o + \frac{RT}{nF} \ln \left\{ \frac{\alpha(DCA^{2-})\alpha(H^+)}{\alpha(Cit^{3-})} \right\}$$

$$E_{DCA^{2-}/Cit^{3-}} = E^o + \frac{RT}{nF} \ln \frac{\alpha(DCA^{2-})}{\alpha(Cit^{3-})} - \frac{RT \ln 10}{nF} pH$$

Based on the Le Chatelier principles, alkaline conditions shift the equilibrium to the product side and acidic media shifts the equilibrium to the reactant side. Therefore, the reduction ability of citrate could increase with increased pH value, based on the equation. One of the limitations of this method is the difficulty of determining the exact amount of citrate reductant due to its double role as reductant and stabilizer. In addition, oxidation products can be adsorbed on NP surfaces. The AgNP prepared by the citrate reduction method also produces relatively large-sized (50–100 nm) particles with polydispersed size and shape (Pillai and Kamat, 2004).

5.4 Polyol synthesis

Polyols are polyhydric alcohols such as ethanediol (ethylene glycol, EG), propanediol (propylene glycol), and ether glycols (ROR-OH). They are both mild reducing agents and nonaqueous solvents, which possesses hydrogen-bonding and a high value of relative permittivity. Due to their interesting properties, they have long been used as reducing agents for noble metal (Au, Ag) particle synthesis (Ducamp-Sanguesa et al., 1992; Silvert et al., 1996a; Silvert et al., 1996b; Silvert et al., 1997; Krutyakov et al., 2008). Polyol synthesis involves the reduction of a metal salt used as a precursor by a polyol, usually ethylene glycol (EG), at an elevated temperature ($\approx 160\,°C$ for EG), and to prevent agglomeration of the particles, a capping agent, commonly PVP, is used. The reducing power of polyols increases with temprature. The synthesis of AgNPs in polyols is usually carried out at a temperature of at least $100\,°C$ as the energy barrier required for the reduction of noble metal species is usually reached by increasing the reaction temperature (Fiévet et al., 2018). UV light can be also used in place of heat. For example, Alula and Yang synthesized silver nanoparticles using a UV light-assisted polyol process where ethylene glycol acted as a reducing agent and solvent (Tadele Alula and Yang, 2014). UV light exposure of the mixture containing $AgNO_3$ solution (dissolved in ethylene glycol) and ZnO/Fe_3O_4 composite resulted in the formation of silver nanostructures onto the composite. Despite the growing interest in using the polyol method for metal nanoparticle synthesis, the mechanism by which the metal ions are reduced is not well understood. The mechanism of polyol reduction (using eleyhlene glycol) with a focus on cobalt and nickel was studied. At $197\,°C$, diacetyl was detected and hence, acetaldehyde was considered as a possible reductant according to the following reaction (Blin et al., 1989):

$$HOCH_2CH_2OH \rightarrow CH_3CHO + H_2O$$
$$2M^+ + 2CH_3CHO \rightarrow CH_3COCOCH_3 + 2M^0 + 2H^+$$

Skrabalak et al., (2008) claimed that glycolaldehyde is the reducing agent for temperatures of 140–160 °C and not acetaldehyde (though at lower temperatures, EG may act as a significant reductant), considering that the mechanism given above fails to explain the following observations:

(i) reduction of many metal ions occured at 150 °C, though no diacetyl was detected; and
(ii) the reduction rate is highly dependent on the reaction atmosphere.

From these observations, they proposed that glycolaldehyde may be generated as a reductant of silver ions when EG is heated in air, according to the following reaction:

$2HOCH_2CH_2OH + O_2 \rightarrow 2HOCH_2CHO + 2H_2O$

$Ag^+ + HOCH_2CHO \rightarrow HOCCHO + Ag^0 + H^+$

Several parameters, including temperature, injection rate, and molar ratio of PVP to silver nitrate affect the redox reaction between Ag^+ and EG and the properties of AgNPs produced. Coskun et al., (2011) investigated an extensive parametric study for the polyol reduction of Ag^+ ion using PVP as a capping agent.

The reaction temperature is an important parameter for AgNPs synthesis by polyol method. The reaction temperature should be higher than 150 °C for the conversion of EG to glycolaldehyde with oxygen, which is the reducing agent for Ag^+ ion to Ag atom (Kim et al., 2004; Skrabalak et al., 2008). In addition, the reaction temperature used determines the diameter and length of Ag nanowire. Kim et al., (2006) prepared silver nanoparticles using the polyol method by either heating the precursor solution to the reaction temperature or injecting an aqueous silver salt solution into hot ethylene glycol, which was the reducing agent. They observed that they could control the particle size and particle size distribution by varying the precursor materials heating rate. A low heating rate resulted in large polydispersed particles, while they could synthesize smaller monodisperse particles at higher heating rates. The close control of the particle size distribution at high heating rates is identifiable in the SEM images (Kim et al., 2006).

In the precursor injection method, both the temperature of the ethylene glycol and the silver ion injection rate played an important role in determining the size of the particles. At an injection rate of $1\,mL\,s^{-1}$ and a reaction temperature of 150 °C, the average particle size was 19 nm as compared to the average size of 54 nm at 100 °C. Wojnicki et al., also showed that reaction conditions play an important role. Their results revealed that synthesis in microdroplet reactors resulted in smaller particle sizes as compared to batch synthesis (Wojnicki et al., 2019). Sun et al., (2002) investigated the role of temperature for face-controlled AgNPs synthesis. The PVP:$AgNO_3$ molar ratio also strongly determines the morphology of AgNPs in the polyol process, where lower ratio produced larger diameter Ag nanowires (Coskun et al., 2011).

Wiley et al., prepared nanoscale Ag particles via the polyol method utilizing ethylene glycol as the reducing agent and poly-(vinyl pyrrolidone) as the capping agent. They utilized oxygen in the presence of trace amounts of NaCl to direct the synthesis of a mixture of monodisperse cubes and tetrahedrons with truncated corners and edges (Wiley et al., 2004). The reaction produced nanowires or irregularly shaped materials in the absence of oxygen (under argon) or chloride ions, respectively. Oxygen in the presence of chloride ions was involved in the oxidative dissolution of initially formed twinned silver particles that favored the selective growth of single-crystal silver particles. Torras and Roig reported microwave-assisted polyol synthesis of AgNPs with a systematic and detailed study of several synthetic conditions including the reaction time, the reaction temperature, and the silver precursor concentration. The particle size, shape, and polydispersity of particles were found to

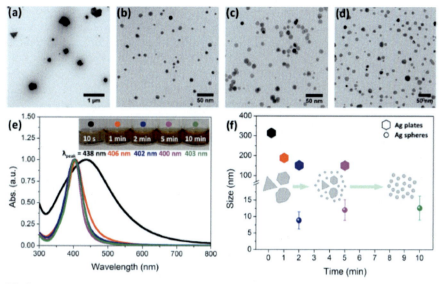

FIG. 6

Effect of the reaction time (at 120 °C and 1 mM AgNO3). (a–d) Representative TEM images of the different reaction times analyzed (10 s, 2 min, 5 min, and 10 min, respectively). (e) UV-vis measurements of the same samples. The inset contains digital images of the samples.
(f) Mean size evolution with time (the error bars represent the standard deviation over the population mean). Inset: scheme of the AgNP evolution with the reaction time.

Reprinted from Torras, M., Roig, A., 2020. From silver plates to spherical nanoparticles: snapshots of microwave-assisted polyol synthesis. ACS Omega, 5(11), 5731–5738. Adapted with permission, ACS.

be highly dependent on these conditions. At 120 °C reaction temperature, 1 mM $AgNO_3$ was used to prepare AgNPs. The effect of reaction time (10 s, 1, 2, 5, and 10 min) was observed as can be seen in Fig. 6. The variation in size and shapes at different reaction times was observed even at an extremely short reaction time of 10 s. The corresponding TEM images and UV/visible spectroscopy spectra are given in Fig. 6. A broad LSPR peak was obtained from 10 s reaction time. This was reflected in the polydispersed size distributions. Monodispersed size distributions were obtained from 2, 5, and 10 min reaction times and the maximum band at 438 nm, which was from 10 s red-shifted to 403 nm. This shows smaller size and isotropic shape particles formation. Fig. 6f shows the variation in shape and size of nanoparticles with reaction times (Torras and Roig, 2020).

5.5 Tollens method

This is a simple method for the reduction of silver ion to silver metal atom by using aldehyde as a reducing agent. It is named after the German chemist Bernhard Tollens, who first reported the method (Tollens, 1882). Interestingly, the one-step process in the Tollens synthesis method gives AgNPs with a controlled size

(Saito et al., 2003; Kvítek et al., 2005; He et al., 2006; Sarkar et al., 2007). Tollens' reagent contains diamine silver (I) complex, $[Ag(NH_3)_2]^+$, which oxidizes an aldehyde to carboxylic acid and in the process is reduced to elemental silver. Ammonia's strong affinity for silver ion provides $[Ag(NH_3)_2]^+$ as a stable complex ion. In Tollens' method, the nature of the reductant and ammonia concentration play a vital role in controlling the AgNPs' size (He et al., 2006). Tollens' procedure can be modified. For example, He et al. (2006) used saccharide in the presence of ammonia and produced silver nanoparticles films with particle sizes from 50 to 200 nm (He et al., 2006). The basic Tollens reaction involves the reduction of $Ag(NH_3)_2^+$ (aq), a Tollens reagent, by an aldehyde, Eq. (1).

$$Ag(NH_3)_2^+ (aq) + RCHO (aq) \rightarrow Ag (s) + RCOOH (aq) \quad (1)$$

Glucose is a common reducing agent used in Tollens reagent (Cheng et al., 2011; Alula et al., 2020). It is a common method used in the qualitative organic analysis that is used to distinguish aldehydes from other carbonyl functional groups. The electron in the aldehyde carbonyl double bond is available as a high lying HOMO to reduce the silver ion. Silver nitrate is first dissolved in water to form hydrated silver ion, $[Ag(H_2O)_x]^+$. When NH_3 is added to the aquo complex, a more stable silver-amine complex ion, $[Ag(NH_3)_2]^+$, is formed. The reduction potentials for the two metal ion complex is given as (Sarkar et al., 2007):

$$[Ag(H_2O)_x]^+ + e^- \leftrightarrow Ag + xH_2O \quad E^0 = 0.799\,V$$
$$[Ag(NH_3)_2]^+ + e^- \leftrightarrow Ag + 2NH_3 \quad E^0 = 0.373\,V$$

The above equations show that the ammonia complex is more difficult to reduce than the hydrated one. But the aldehyde reducing agent is easily oxidized in ammoniated solution based on the following reaction (Sarkar et al., 2007):

$$RCHO + H_2O \leftrightarrow RCOOH + 2H^+ + 2e^-$$

Since the oxidation produces H^+, the rate of aldehyde oxidation would be higher at high pH and in turn, the reduction rate of Ag^+ ion is higher.

The overall redox reaction in Tollens reaction is represented as:

$$2[Ag(NH_3)_2]^+ + RCHO + 3OH^- \rightarrow RCOO^- + 2Ag + 4NH_3 + 2H_2O$$

In addition to higher pH, higher temperature also increases the reduction of Ag^+ ion in the Tollens reduction method (Soukupová et al., 2008). Sarkar et al. prepared 15–260 nm-sized AgNPs using m-Hydroxy benzaldehyde as a reducing agent and sodium dodecyl sulphate as a surfactant in the temperature range of 80–86°C. The slow release of Ag^+ from $[Ag(NH_3)_2]^+$ helps to control the property and growth rate of the AgNPs (Sarkar et al., 2007).

5.6 Polysaccharide method

Polysaccharides can be used as capping agents, or in some cases, they may serve as both reducing and capping agents in an aqueous system that avoids the use of organic solvents that may be toxic. For instance, in a gently heated system, starch-capped silver nanoparticles were prepared using starch as a capping agent and β-D-glucose as a reducing agent (Huang and Yang, 2004). A negatively charged heparin was used as a reducing/stabilizing agent for AgNO$_3$ solution at 70°C for ~8h (de Barros et al., 2016). Heparin, in this case, acts as a nuclear controller and stabilizer.

5.7 Reduction using *N,N*-Dimethylformamide

N, *N*-Dimethylformamide (DMF) is one of the standard organic compounds used as a solvent for various processes, including the preparation of colloids. Properties including wide liquid temperature range, good chemical and thermal stability, high polarity, and wide solubility range for both organic and inorganic compounds give DMF a high synthetic value (Pastoriza-Santos and Liz-Marzán, 1999). Under suitable conditions and in the absence of an external agent, DMF reduces Ag$^+$ ions to zerovalent silver. Like other chemical reduction methods, coupling agents and temperature influence both the reaction rate and the morphology of the particles. Reduction of silver is possible simply by mixing aqueous solutions of AgNO$_3$ or AgClO$_4$ with DMF. A gradual appearance of yellow color with no stirring (after the initial shaking) signals the formation of silver nanoparticles (Wang et al., 2016). At room temperature, the following route has been proposed without evolution of H$_2$ or CO$_2$, which are commonly observed in the oxidation of DMF:

$$HCONMe_2 + 2Ag^+ + H_2O \rightarrow 2Ag^0 + Me_2NCOOH + 2H^+$$

Considering this equation, the increment in conductivity of the reaction mixture during the reaction indicates the silver ions were progressively exchanged for more mobile H$^+$ ions. This supports this proposed mechanism (Pastoriza-Santos and Liz-Marzán, 1999). Like other methods of synthesis of silver nanoparticles, colloidal stability can be achieved by using a capping agent together with DMF. Additionally, the capping agents can also play an important role in controlling the particle morphology like that of other chemical methods of synthesis of silver nanoparticles (Pastoriza-Santos and Liz-Marzán, 1999). A simple mixture of silver nitrate and DMF, however, may result in the deposition of particles on the wall of the reaction container, forming continuous metal films. For example, Wang et al. (Wang et al., 2016) and Pastoriza-Santos and Liz-Marza'n (Pastoriza-Santos and Liz-Marzán, 2008) reported the synthesis of silver nano triangles by reduction of AgNO$_3$ using DMF in the presence of the polymer poly(vinylpyrrolidone) (PVP). They argued that PVP was acting as a stabilizer and shape directing agent in the formation of triangular shaped silver nanoparticles. The dynamic adsorption and desorption of PVP on certain faces of the particles via kinetic growth control resulted in triangular shaped silver nanoparticles (Pastoriza-Santos and Liz-Marzán, 2008). The effect of concentration of PVP was reported by Tsuji et al. (Henglein and Tausch-Treml, 1981). They used *N*,*N*-dimethylformamide as the reducing agent and varied the concentrations of the capping agent, PVP, at constant AgNO$_3$ concentration to produce decahedra and hexagonal particles. A higher concentration of PVP (250mM), favored the production of hexagonal particles and their intermediates (41%) compared to decahedra and their intermediates (22%); whereas, at low concentration

PVP (63 mM), decahedra and their intermediates (32%) were favored as compared to hexagonal particles and their intermediates (27%).

5.8 Photochemical synthesis

In a typical chemical reaction process, the addition of the reducing agent to the reaction solution results in the reduction of metal salt. However, an external energy input (heat or light) may be required to produce the active species (reductant) that reduce the metal ion from the added chemicals. For example, in the citrate-assisted formation of Au and Ag, oxidized citrate radical [Cit(—H)$^•$], which is the active species to reduce the metal ion, is generated in situ at a higher temperature. Light or electromagnetic radiation is an energy source that induces a number of redox process in the medium to produce active species that later reduce metal ions. The energy of light is dependent on the wavelength of radiation. The three common light sources in the photochemical reduction of metal salts are gamma-radiation, UV light, and visible light, with decreasing order of energy. The photochemical metal ion reduction process could be either direct photoreduction of metal precursor (metal salt or complex) or an indirect reduction through photochemically produced species. For example, Alula and Yang synthesized AgNPs using UV light as a light source to irradiate a reaction mixture containing $AgNO_3$ and citrate. The particle size is highly dependent on the irradiation time (Alula and Yang, 2014; Tadele Alula and Yang, 2014). Radiolytic methods have been reported for the synthesis of AgNP, where hydrated electrons and hydroxyl radicals generated by radiolysis of water react with other molecules to produce a radical that reduces the metal salt (Henglein and Tausch-Treml, 1981). AgNPs have been synthesized with UV irradiation (lower energy light source compared to gamma radiation) induced excitation of Ag^+ from $AgClO_4$ causing H_2O oxidation, leading to reduction of Ag^+ to Ag atom, which grows to colloidal AgNPs according to the following equation (Yonezawa et al., 1987):

$$H_2O \xrightarrow{h\nu} H^+ + OH^• + e^-_{aq}$$
$$Ag^+ + e^-_{aq} \rightarrow Ag^0$$
$$nAg^0 \rightarrow (Ag^0)_n$$

Alternatively, the OH or H radical may attack the organic molecule (RH) present in the reaction medium, producing radical species (R), which in turn reduces the metal ion:

$$M^{n+} + nR^• \rightarrow M^0 + nR^• + nH^+$$

The metal atoms undergo nucleation and growth process to form the nanoparticle:

$$nM^0 \rightarrow (M^0)_n$$

If ligands are present, they could get involved in the photochemical reduction of metal ion, affecting the produced particle property. Huang et al. (1996) reported that both particle size and the UV-vis absorption peak are strongly dependent on the PVP concentration, showing the direct involvement of PVP in the UV photoreduction of silver nitrate. It is pointed out that the carbonyl group of PVP absorbs UV light and gets oxidized, leading to the reduction of Ag ion (Grzelczak and Liz-Marzán, 2014).

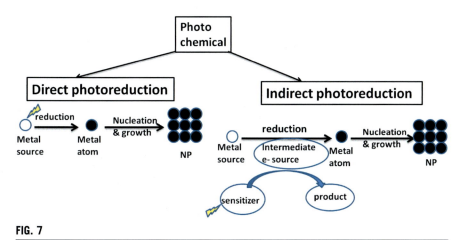

FIG. 7

Classification of photochemical reactions.

Stamplecoskie and Scaiano (2010) described a UV photochemical method for preparation of AgNPs with predictable and controllable size and morphology that are produced from a single source of photochemically grown AgNP seeds stabilized by sodium citrate (Fig. 7).

Since the localized surface plasmon resonance (LSPR) frequencies of noble metal nanoparticles can be tuned within the visible and NIR spectral ranges, visible light irradiation on a preprepared AgNP can be used to synthesize a different morphology and shape AgNP. For example, Jin et al. (2001) reported a photoinduced method for converting large quantities of presynthesized Ag nanospheres into triangular nanoprisms. Photomediated triangular Ag nanoprism growth involving reduction of silver cations by citrate on the silver particle surface and oxidative dissolution of small silver particles by O_2 was also reported (Xue et al., 2008; Millstone et al., 2009). The silver particles serve as photocatalysts and, under plasmon excitation, facilitate Ag^+ reduction by citrate:

$$Ag^0(\text{nanosphere}) + \frac{1}{2}O_2 + H_2O \rightarrow Ag^+ + 2OH^-$$

$$\text{Citrate} + Ag^+ + H_2O \xrightarrow{h\nu} Ag^0(\text{nanoprism})$$

6 Electrochemical reduction method

Electrochemical method is another approach to synthesize AgNPs. The advantage of electrochemical metal nanoparticle synthesis is that the rate of the reaction and, in turn, the particle geometry can be controlled by simply adjusting the electrolysis variables including current, potential, time, and charge. One of the electrochemical methods involves the use of a sacrificial anode, which is the silver precursor for obtaining AgNP at the cathode. This method does not require any chemical reducing agent, which produces byproduct contaminants and eliminates the use of metal precursor chemicals. The anode is made of a sacrificial Ag metal electrode that oxidizes to produce Ag^+ ion, acting as a source of silver in AgNPs. The electron and Ag^+ ion moves to the cathode and reduction of Ag^+ ion generates the Ag atom, which later agglomerates to AgNP on the cathode surface. The chemical reaction can be represented as (Henglein, 1989):

$$\text{Anode (oxidation)}: Ag_{metal} \rightarrow Ag^+ + 1e^- \quad E^0 = 0.799\,V\,vs.NHE$$
$$\text{Cathode (Reduction)}: Ag^+ + 1e^- \rightarrow Ag \quad E^0 = -1.8\,V\,vs.NHE$$
$$\text{Nucleation}: Ag + Ag^+ + 1e^- \rightarrow Ag_2$$
$$\vdots$$
$$\text{Growth}: Ag_n \rightarrow AgNP$$

The redox potential of the microelectrode (AgNP electrode) becomes more positive as the particles grow. For $n=1$ (free silver atom) the potential is $-1.8\,V$ and, finally, for $n \rightarrow \infty$, the potential of $0.799\,V$ of the conventional silver electrode is approached (Henglein, 1989).

The other electrochemical method uses soluble metal salt precursors (e.g., $AgNO_3$) instead of a sacrificial silver metal anode as a source of Ag atom in the AgNP synthesis. In this case, the anode could be a nonreactive electrode like Pt for oxidation half-reaction. The cathodic Ag^+ ion reduction could occur in the same way as described above. Reaction parameters such as the presence and nature of surfactant and electrolysis conditions affect the AgNP formation. The electroreduction of silver ions involves two competitive cathode surface processes: one is the formation of silver particles and the other is the deposition of a silver film on the cathode surface. Electrochemical reduction of Ag^+ ion usually results in deposition of silver film on cathode surfaces, which limits the yield of the particle synthesis due to complete surface coverage (Ma et al., 2004). The addition of PVP to enhance particle formation rate and use of Pt cathode for inhibition of film formation are the measures used to inhibit film formation, as the two atoms have a large difference in radius and lattice parameter (Rodríguez-Sánchez et al., 2000). Reetz and Helbig primary reported the electrochemical synthesis of metal clusters by anodically dissolving the metal sheet and reduction of the metal ion at the cathode producing tertaalkylammonium salt-stabilized nanoparticles (Reetz and Helbig, 1994). Yin et al., (2003) adopted this method for the synthesis of tertaalkylammonium-stabilized AgNP in acetonitrile.

Maria Starowicz and coworkers investigated silver polarization in an ethanol solution of $NaNO_3$ in ethanol to deposit metallic silver nanoparticles by both potentiostatic and galvanostatic method. The proposed mechanism assumes both the anodic dissolution of silver and its reduction to metallic state proceed during polarization. Yin et al., (2003) reported synthesis of spherical silver nanoparticles in aqueous phase using poly(N-vinylpyrrolidone) (PVP) as the stabilizer for the silver clusters, claiming PVP promotes the silver particle formation rate and significantly reduces the silver deposition rate. Kuntyi et al., (2019) synthesized silver sols that contained mainly small silver nanoparticles (up to 10 nm) by a method of electrolysis using silver electrodes under conditions of altered current polarity in sodium polyacrylate (NaPA) solutions. It is shown that the values of the current of the anode dissolution of silver increase with increasing of NaPA concentration and the dependence of i_{anode}—E is linear. In addition, an electrochemical method for the synthesis of AgNPs without surfactants has been reported (Rodríguez-Sánchez et al., 2000; Khaydarov et al., 2009). The method allows control over the rate of particle growth and hence particle size and shape, by controlling the reduction parameters such as current, potential, charge, and time more easily than other chemical synthesis parameters.

7 Microwave-assisted method

The above chemical synthesis methods could be carried out at higher temperatures to increase the kinetics of the reaction. The heat may be supplied through either heating plates/mantles, which is a rather slow and inefficient method for transferring energy to a reaction mixture, or microwave heating delivering homogeneous volume heating. Microwave is effective in heating matter and it provides uniform heating of the reaction solution. More importantly, the reaction time is short, resulting in low energy consumption and high product yield. This results in particles of narrow size distributions. Different materials have a specific capacity to absorb microwave radiation and this radiation can be changed into heat (Gerbec et al., 2005; Bilecka and Niederberger, 2010). Accordingly, a microwave-assisted reaction is based on the efficient heating of materials. The chemical reaction is not induced by absorption of high energy electromagnetic radiation (Bilecka and Niederberger, 2010). Irradiation of a reaction mixture with microwaves results in the alignment of the solvent dipoles in the electric field. Microwave-assisted synthesis of AgNP involves the chemical reduction of metal salts in a dipolar solvent such as water, ethanol, DMF etc., where the solvent converts the microwave energy to heat so that the reaction medium temperature is raised depending on the dielectric nature of the solvent. Microwave irradiation is a rapid heating method that increases reaction rates and product yield compared to conventional thermal heating in nanoparticle synthesis (Gerbec et al., 2005). Due to these advantages, it has been used in the synthesis of AgNPs (Dankovich, 2014). Hu et al., synthesized nearly monodispersed silver nanoparticles using L-lysine or L-arginine, and starch, as reducing and protecting agents, respectively. Pal and coworkers prepared AgNPs by microwave irradiation of silver nitrate ($AgNO_3$) solution in an ethanolic medium using polyvinylpyrrolidone (PVP) as a stabilizing agent. They proposed that ethanol was observed to act as a reducing agent in the presence of microwave. Masaharu et al., (Masaharu et al., 2008) also examined the significance of adsorption species in the formation of Ag nanostructures by a microwave-polyol route. The advantages of this method include shorter reaction times, lower energy consumption, better product yield, and lower chemical use. Nonpolar and low-dielectric solvents cannot be used in microwave-assisted synthesis as these materials are inefficient in converting microwave energy to heat.

8 Conclusions and future perspectives

Chemical synthesis of silver nanoparticles where different chemicals are used as reducing, stabilizing, and capping agents is relatively well established compared to biological and physical synthesis. This is because chemicals used in silver nanoparticles synthesis offer diverse reducing, stabilizing, and capping capabilities and that, unlike biological methods, chemical synthesis can be executed over wide temperature, concentration, and pH ranges.

Reaction parameters such as temperature, pH, time, type of reducing and capping agents and their concentrations, and ratio of silver source to the reducing/capping agent have been demonstrated to have a huge influence on nucleation extent and rate, particle growth, shape, size, dispersity, and stability. Manipulation of parameters, therefore, provides an avenue through which silver can readily be adapted to a targeted application, for instance, surface-enhanced Raman spectroscopy, via selective and accurate control of properties pertinent to this application.

Low reaction temperatures have been shown to afford relatively smaller but stable silver nanoparticles with uniform size distribution compared to higher temperatures. On the other hand, low pH favors the formation of smaller particles and sphere-like morphology while high pH translates to large particles and nanrod/nanowire morphology. In terms of reducing power, strong reducing agents like sodium borohydride furnish small-sized, relatively monodispersed nanoparticles compared to weak reducing agents like hydrazine, ascorbic acid, polyols, or aldehydes, among others.

Silver ions ratio to citrate is also a key consideration in silver nanoparticles synthesis since citrate ions serve as both reducing and capping agents. This necessitates the use of citrate ions concentration higher than the stoichiometric value to ensure that all silver ions are reduced to zerovalent silver nanoparticles.

Toxicity of reducing agents such as hydrazine, the multisteps nature of current synthesis methods, and the growing push for green synthetic strategies necessitates the search for solvents and reducing agents that are nontoxic, atom-efficient, and provide less energy-consuming procedures in silver nanoparticles synthesis.

References

Abou El-Nour, K.M.M., Eftaiha, A.A., Al-Warthan, A., Ammar, R.A.A., 2010. Synthesis and applications of silver nanoparticles. Arab. J. Chem. 3 (3), 135–140.

Ajitha, B., Ashok Kumar Reddy, Y., Sreedhara Reddy, P., 2015. Enhanced antimicrobial activity of silver nanoparticles with controlled particle size by pH variation. Powder Technol. 269, 110–117.

Al Gharib, S., Marignier, J.-L., El Omar, A.K., Naja, A., Le Caer, S., Mostafavi, M., Belloni, J., 2019. Key role of the oxidized citrate-free radical in the nucleation mechanism of the metal nanoparticle Turkevich synthesis. J. Phys. Chem. C 123 (36), 22624–22633.

Alula, M.T., Lemmens, P., Madiba, M., Present, B., 2020. Synthesis of free-standing silver nanoparticles coated filter paper for recyclable catalytic reduction of 4-nitrophenol and organic dyes. Cellul. 27 (4), 2279–2292.

Alula, M.T., Yang, J., 2014. Photochemical decoration of silver nanoparticles on magnetic microspheres as substrates for the detection of adenine by surface-enhanced Raman scattering. Anal. Chim. Acta 812, 114–120.

Angelescu, D.G., Vasilescu, M., Anastasescu, M., Baratoiu, R., Donescu, D., Teodorescu, V.S., 2012. Synthesis and association of Ag(0) nanoparticles in aqueous Pluronic F127 triblock copolymer solutions. Colloids Surf. A Physicochem. Eng. Asp. 394, 57–66.

Ashraf, S., Abbasi, A.Z., Pfeiffer, C., Hussain, S.Z., Khalid, Z.M., Gil, P.R., Hussain, I., 2013. Protein-mediated synthesis, pH-induced reversible agglomeration, toxicity and cellular interaction of silver nanoparticles. Colloids Surf. B Biointerfaces 102, 511–518.

Audrieth, L.F., Ogg, B.A., 1951. The Chemistry of Hydrazine. Wiley, New York.

Bastús, N.G., Merkoçi, F., Piella, J., Puntes, V., 2014. Synthesis of highly monodisperse citrate-stabilized silver nanoparticles of up to 200 nm: kinetic control and catalytic properties. Chem. Mater. 26 (9), 2836–2846.

Beaird, A.M., Davis, T.A., Matthews, M.A., 2010. Deliquescence in the hydrolysis of sodium borohydride by water vapor. Ind. Eng. Chem. Res. 49 (20), 9596–9599.

Bilecka, I., Niederberger, M., 2010. Microwave chemistry for inorganic nanomaterials synthesis. Nanoscale 2 (8), 1358–1374.

Blin, B., Fievet, F., Beaupere, D., Figlarz, M., 1989. Oxydation duplicative de l'éthylène glycol dans un nouveau procédé de préparation de poudres métalliques. Nouv. J. Chim. 13, 67–72.

Burda, C., Chen, X., Narayanan, R., El-Sayed, M.A., 2005. Chemistry and properties of nanocrystals of different shapes. Chem. Rev. 105 (4), 1025–1102.

Chen, J.P., Lim, L.L., 2002. Key factors in chemical reduction by hydrazine for recovery of precious metals. Chemosphere 49 (4), 363–370.

Cheng, M.-L., Tsai, B.-C., Yang, J., 2011. Silver nanoparticle-treated filter paper as a highly sensitive surface-enhanced Raman scattering (SERS) substrate for detection of tyrosine in aqueous solution. Anal. Chim. Acta 708 (1), 89–96.

Choi, S., Jeong, Y., Yu, J., 2016. Spontaneous hydrolysis of borohydride required before its catalytic activation by metal nanoparticles. Cat. Com. 84, 80–84.

Coskun, S., Aksoy, B., Unalan, H.E., 2011. Polyol synthesis of silver nanowires: an extensive parametric study. Cryst. Growth Des. 11 (11), 4963–4969.

Creighton, J.A., Blatchford, C.G., Albrecht, M.G., 1979. Plasma resonance enhancement of Raman scattering by pyridine adsorbed on silver or gold sol particles of size comparable to the excitation wavelength. J. Chem. Soc. Faraday Trans. 75 (0), 790–798.

Dahl, J.A., Maddux, B.L.S., Hutchison, J.E., 2007. Toward greener nanosynthesis. Chem. Rev. 107 (6), 2228–2269.

Dankovich, T.A., 2014. Microwave-assisted incorporation of silver nanoparticles in paper for point-of-use water purification. Environ. Sci. Nano 1 (4), 367–378.

de Barros, H.R., Piovan, L., Sassaki, G.L., de Araujo Sabry, D., Mattoso, N., Nunes, Á.M., Riegel-Vidotti, I.C., 2016. Surface interactions of gold nanorods and polysaccharides: from clusters to individual nanoparticles. Carbohydr. Polym. 152, 479–486.

Dong, X., Ji, X., Wu, H., Zhao, L., Li, J., Yang, W., 2009. Shape control of silver nanoparticles by stepwise citrate reduction. J. Phys. Chem. C 113 (16), 6573–6576.

Ducamp-Sanguesa, C., Herrera-Urbina, R., Figlarz, M., 1992. Synthesis and characterization of fine and monodisperse silver particles of uniform shape. J. Solid State Chem. 100 (2), 272–280.

Edwards, J.O., Morrison, G.C., Ross, V.F., Schultz, J.W., 1955. The structure of the aqueous borate ion. J. Am. Chem. Soc. 77 (2), 266–268.

Emory, S.R., Nie, S., 1997. Near-field surface-enhanced Raman spectroscopy on single silver nanoparticles. Anal. Chem. 69 (14), 2631–2635.
Fiévet, F., Ammar-Merah, S., Brayner, R., Chau, F., Giraud, M., Mammeri, F., Viau, G., 2018. The polyol process: a unique method for easy access to metal nanoparticles with tailored sizes, shapes and compositions. Chem. Soc. Rev. 47 (14), 5187–5233.
García-Barrasa, J., López-de-Luzuriaga, J., Monge, M., 2011. Silver nanoparticles: synthesis through chemical methods in solution and biomedical applications. Open Chem. 9 (1), 7.
Gerbec, J.A., Magana, D., Washington, A., Strouse, G.F., 2005. Microwave-enhanced reaction rates for nanoparticle synthesis. J. Am. Chem. Soc. 127 (45), 15791–15800.
Ghodake, G.S., Deshpande, N.G., Lee, Y.P., Jin, E.S., 2010. Pear fruit extract-assisted room-temperature biosynthesis of gold nanoplates. Colloids Surf. B Biointerfaces 75 (2), 584–589.
Glavee, G.N., Klabunde, K.J., Sorensen, C.M., Hadjipanayis, G.C., 1994. Borohydride reduction of Nickel and copper ions in aqueous and nonaqueous media. Controllable chemistry leading to nanoscale metal and metal boride particles. Langmuir 10 (12), 4726–4730.
Grzelczak, M., Liz-Marzán, L.M., 2014. The relevance of light in the formation of colloidal metal nanoparticles. Chem. Soc. Rev. 43 (7), 2089–2097.
Harra, J., Mäkitalo, J., Siikanen, R., Virkki, M., Genty, G., Kobayashi, T., Mäkelä, J.M., 2012. Size-controlled aerosol synthesis of silver nanoparticles for plasmonic materials. J. Nanopart. Res. 14 (6), 870.
He, Y., Wu, X., Lu, G., Shi, G., 2006. A facile route to silver nanosheets. Mater. Chem. Phys. 98 (1), 178–182.
Henglein, A., 1989. Small-particle research: physicochemical properties of extremely small colloidal metal and semiconductor particles. Chem. Rev. 89 (8), 1861–1873.
Henglein, A., Tausch-Treml, R., 1981. Optical absorption and catalytic activity of subcolloidal and colloidal silver in aqueous solution: a pulse radiolysis study. J. Colloid Interface Sci. 80 (1), 84–93.
Huang, H., Yang, X., 2004. Synthesis of polysaccharide-stabilized gold and silver nanoparticles: a green method. Carbohydr. Res. 339 (15), 2627–2631.
Huang, H.H., Ni, X.P., Loy, G.L., Chew, C.H., Tan, K.L., Loh, F.C., Xu, G.Q., 1996. Photochemical formation of silver nanoparticles in poly (N-vinylpyrrolidone). Langmuir 12 (4), 909–912.
Huang, P., Lin, J., Li, Z., Hu, H., Wang, K., Gao, G., Cui, D., 2010. A general strategy for metallic nanocrystals synthesis in organic medium. Chem. Commun. 46 (26), 4800–4802.
Hutchison, J.E., 2008. Greener nanoscience: a proactive approach to advancing applications and reducing implications of nanotechnology. ACS Nano 2 (3), 395–402.
Jana, N.R., Gearheart, L., Murphy, C.J., 2001. Wet chemical synthesis of silver nanorods and nanowires of controllable aspect ratio. Chem. Commun. (7), 617–618.
Jiang, X.C., Chen, C.Y., Chen, W.M., Yu, A.B., 2010. Role of citric acid in the formation of silver nanoplates through a synergistic reduction approach. Langmuir 26 (6), 4400–4408.
Jiang, X.C., Chen, W.M., Chen, C.Y., Xiong, S.X., Yu, A.B., 2011. Role of temperature in the growth of silver nanoparticles through a synergetic reduction approach. Nanoscale Res. Lett. 6 (1), 32.
Jin, R., Cao, Y., Mirkin, C.A., Kelly, K.L., Schatz, G.C., Zheng, J.G., 2001. Photoinduced conversion of silver nanospheres to nanoprisms. Science 294 (5548), 1901.
Jose Vega-Baudrit, S.M.G., Rojas, E.R., Martinez, V.V., 2019. Synthesis and characterization of silver nanoparticles and their application as an antibacterial agent. Int. J. Biosens. Bioelectron. 5, 166–173.

Jung, J.H., Cheol Oh, H., Soo Noh, H., Ji, J.H., Soo Kim, S., 2006. Metal nanoparticle generation using a small ceramic heater with a local heating area. J. Aerosol Sci. 37 (12), 1662–1670.

Kearns, G.J., Foster, E.W., Hutchison, J.E., 2006. Substrates for direct imaging of chemically functionalized SiO2 surfaces by transmission electron microscopy. Anal. Chem. 78 (1), 298–303.

Khan, M., Shaik, M.R., Adil, S.F., Khan, S.T., Al-Warthan, A., Siddiqui, M.R.H., Tremel, W., 2018. Plant extracts as green reductants for the synthesis of silver nanoparticles: lessons from chemical synthesis. Dalton Trans. 47 (35), 11988–12010.

Khaydarov, R.A., Khaydarov, R.R., Gapurova, O., Estrin, Y., Scheper, T., 2009. Electrochemical method for the synthesis of silver nanoparticles. J. Nanopart. Res. 11 (5), 1193–1200.

Kilin, D.S., Prezhdo, O.V., Xia, Y., 2008. Shape-controlled synthesis of silver nanoparticles: Ab initio study of preferential surface coordination with citric acid. Chem. Phys. Lett. 458 (1), 113–116.

Kim, D., Jeong, S., Moon, J., 2006. Synthesis of silver nanoparticles using the polyol process and the influence of precursor injection. Nanotechnology 17 (16), 4019–4024.

Kim, F., Connor, S., Song, H., Kuykendall, T., Yang, P., 2004. Platonic gold nanocrystals. Angew. Chem. Int. Ed. 43 (28), 3673–3677.

Kochkar, H., Aouine, M., Ghorbel, A., Berhault, G., 2011. Shape-controlled synthesis of silver and palladium nanoparticles using β-cyclodextrin. J. Phys. Chem. C 115 (23), 11364–11373.

Kora, A.J., Sashidhar, R.B., Arunachalam, J., 2012. Aqueous extract of gum olibanum (Boswellia serrata): a reductant and stabilizer for the biosynthesis of antibacterial silver nanoparticles. Process Biochem. 47 (10), 1516–1520.

Kruis, F.E., Fissan, H., Rellinghaus, B., 2000. Sintering and evaporation characteristics of gas-phase synthesis of size-selected PbS nanoparticles. Mater. Sci. Eng., B 69-70, 329–334.

Krutyakov, Y.A., Kudrinskiy, A.A., Olenin, A.Y., Lisichkin, G.V., 2008. Synthesis and properties of silver nanoparticles: advances and prospects. Russ. Chem. Rev. 77 (3), 233–257.

Kuntyi, O.I., Kytsya, A.R., Mertsalo, I.P., Mazur, A.S., Zozula, G.I., Bazylyak, L.I., Topchak, R.V., 2019. Electrochemical synthesis of silver nanoparticles by reversible current in solutions of sodium polyacrylate. Colloid Polym. Sci. 297 (5), 689–695.

Kvítek, L., Panáček, A., Soukupová, J., Kolář, M., Večeřová, R., Prucek, R., Zbořil, R., 2008. Effect of surfactants and polymers on stability and antibacterial activity of silver nanoparticles (NPs). J. Phys. Chem. C 112 (15), 5825–5834.

Kvítek, L., Prucek, R., Panáček, A., Novotný, R., Hrbáč, J., Zbořil, R., 2005. The influence of complexing agent concentration on particle size in the process of SERS active silver colloid synthesis. J. Mater. Chem. 15 (10), 1099–1105.

LaMer, V.K., Dinegar, R.H., 1950. Theory, production and mechanism of formation of monodispersed hydrosols. J. Am. Chem. Soc. 72 (11), 4847–4854.

Lee, J.D., 1996. Concise Inorganic Chemistry. CHAPMAN & HALL.

Lee, P.C., Meisel, D., 1982. Adsorption and surface-enhanced Raman of dyes on silver and gold sols. J. Phys. Chem. 86 (17), 3391–3395.

Li, L.-S., Hu, J., Yang, W., Alivisatos, A.P., 2001. Band gap variation of size- and shape-controlled colloidal CdSe quantum rods. Nano Lett. 1 (7), 349–351.

Liang, H., Wang, W., Huang, Y., Zhang, S., Wei, H., Xu, H., 2010. Controlled synthesis of uniform silver nanospheres. J. Phys. Chem. C 114 (16), 7427–7431.

Liu, T., Li, D., Yang, D., Jiang, M., 2011. Size controllable synthesis of ultrafine silver particles through a one-step reaction. Mater. Lett. 65, 628–631.

Ma, H., Yin, B., Wang, S., Jiao, Y., Pan, W., Huang, S., Meng, F., 2004. Synthesis of silver and gold nanoparticles by a novel electrochemical method. ChemPhysChem 5 (1), 68–75.

Manojkumar, K., Sivaramakrishna, A., Vijayakrishna, K., 2016. A short review on stable metal nanoparticles using ionic liquids, supported ionic liquids, and poly(ionic liquids). J. Nanopart. Res. 18 (4), 103.

Masaharu, T., Kisei, M., Peng, J., Ryoichi, M., Sachie, H., Xin-Ling, T., Nor, K.K.S., 2008. The role of adsorption species in the formation of Ag nanostructures by a microwave-polyol route. Bull. Chem. Soc. Jpn. 81 (3), 393–400.

Millstone, J.E., Hurst, S.J., Métraux, G.S., Cutler, J.I., Mirkin, C.A., 2009. Colloidal gold and silver triangular nanoprisms. Small 5 (6), 646–664.

Mulvaney, P., 1996. Surface plasmon spectroscopy of nanosized metal particles. Langmuir 12 (3), 788–800.

Nath, S., Ghosh, S.K., Kundu, S., Praharaj, S., Panigrahi, S., Pal, T., 2006. Is gold really softer than silver? HSAB principle revisited. J. Nanopart. Res. 8 (1), 111–116.

Nickel, U., Zu Castell, A., Pöppl, K., Schneider, S., 2000. A silver colloid produced by reduction with hydrazine as support for highly sensitive surface-enhanced Raman spectroscopy. Langmuir 16 (23), 9087–9091.

Nomoev, A.V., Bardakhanov, S.P., 2012. Synthesis and structure of Ag-Si nanoparticles obtained by the electron-beam evaporation/condensation method. Tech. Phys. Lett. 38 (4), 375–378.

Pacioni, N.L., Borsarelli, C.D., Rey, V., Veglia, A.V., 2015. Synthetic routes for the preparation of silver nanoparticles. In: Alarcon, E.I., Griffith, M., Udekwu, K.I. (Eds.), Silver Nanoparticle Applications: In the Fabrication and Design of Medical and Biosensing Devices. Springer International Publishing, Cham, pp. 13–46.

Pastoriza-Santos, I., Liz-Marzán, L.M., 1999. Formation and stabilization of silver nanoparticles through reduction by N,N-dimethylformamide. Langmuir 15 (4), 948–951.

Pastoriza-Santos, I., Liz-Marzán, L.M., 2002. Synthesis of silver nanoprisms in DMF. Nano Lett. 2 (8), 903–905.

Pastoriza-Santos, I., Liz-Marzán, L.M., 2008. Colloidal silver nanoplates. State of the art and future challenges. J. Mater. Chem. 18 (15), 1724–1737.

Patil, R.S., Kokate, M.R., Jambhale, C.L., Pawar, S.M., Han, S.H., Kolekar, S.S., 2012. One-pot synthesis of PVA-capped silver nanoparticles their characterization and biomedical application. Adv. Nat. Sci. Nanosci. Nanotechnol. 3 (1), 015013.

Patil, R.S., Kokate, M.R., Salvi, P.P., Kolekar, S.S., 2011. A novel one step synthesis of silver nanoparticles using room temperature ionic liquid and their biocidal activity. C. R. Chim. 14 (12), 1122–1127.

Pillai, Z.S., Kamat, P.V., 2004. What factors control the size and shape of silver nanoparticles in the citrate ion reduction method? J. Phys. Chem. B 108 (3), 945–951.

Polte, J., Erler, R., Thünemann, A.F., Sokolov, S., Ahner, T.T., Rademann, K., Kraehnert, R., 2010. Nucleation and growth of gold nanoparticles studied via in situ small angle X-ray scattering at millisecond time resolution. ACS Nano 4 (2), 1076–1082.

Qin, Y., Ji, X., Jing, J., Liu, H., Wu, H., Yang, W., 2010. Size control over spherical silver nanoparticles by ascorbic acid reduction. Colloids Surf. A Physicochem. Eng. Asp. 372 (1), 172–176.

Quintero-Quiroz, C., Acevedo, N., Zapata-Giraldo, J., Botero, L.E., Quintero, J., Zárate-Triviño, D., Pérez, V.Z., 2019. Optimization of silver nanoparticle synthesis by chemical reduction and evaluation of its antimicrobial and toxic activity. Biomater. Res. 23 (1), 27.

Raffi, M., Rumaiz, A.K., Hasan, M.M., Shah, S.I., 2011. Studies of the growth parameters for silver nanoparticle synthesis by inert gas condensation. J. Mater. Res. 22 (12), 3378–3384.

Reetz, M.T., Helbig, W., 1994. Size-selective synthesis of nanostructured transition metal clusters. J. Am. Chem. Soc. 116 (16), 7401–7402.

Rich, R., 2007. Inorganic Reactions in Water. Springer, Berlin Heidelberg.

Rivero, P.J., Goicoechea, J., Urrutia, A., Arregui, F.J., 2013. Effect of both protective and reducing agents in the synthesis of multicolor silver nanoparticles. Nanoscale Res. Lett. 8 (1), 101.

Rodríguez-Sánchez, L., Blanco, M.C., López-Quintela, M.A., 2000. Electrochemical synthesis of silver nanoparticles. J. Phys. Chem. B 104 (41), 9683–9688.

Saito, Y., Wang, J.J., Batchelder, D.N., Smith, D.A., 2003. Simple chemical method for forming silver surfaces with controlled grain sizes for surface plasmon experiments. Langmuir 19 (17), 6857–6861.

Sarkar, S., Jana, A.D., Samanta, S.K., Mostafa, G., 2007. Facile synthesis of silver nano particles with highly efficient anti-microbial property. Polyhedron 26 (15), 4419–4426.

Schmidt-Ott, A., 1988. New approaches to in situ characterization of ultrafine agglomerates. J. Aerosol Sci. 19 (5), 553–563.

Shirtcliffe, N., Nickel, U., Schneider, S., 1999. Reproducible preparation of silver sols with small particle size using borohydride reduction: for use as nuclei for preparation of larger particles. J. Colloid Interface Sci. 211 (1), 122–129.

Silvert, P.-Y., Herrera-Urbina, R., Duvauchelle, N., Vijayakrishnan, V., Elhsissen, K.T., 1996a. Preparation of colloidal silver dispersions by the polyol process. Part 1—synthesis and characterization. J. Mater. Chem. 6 (4), 573–577.

Silvert, P.-Y., Herrera-Urbina, R., Tekaia-Elhsissen, K., 1997. Preparation of colloidal silver dispersions by the polyol process. J. Mater. Chem. 7 (2), 293–299.

Silvert, P.Y., Vijayakrishnan, V., Vibert, P., Herrera-Urbina, R., Elhsissen, K.T., 1996b. Synthesis and characterization of nanoscale Ag-Pd alloy particles. Nanostruct. Mater. 7 (6), 611–618.

Skrabalak, S.E., Wiley, B.J., Kim, M., Formo, E.V., Xia, Y., 2008. On the polyol synthesis of silver nanostructures: glycolaldehyde as a reducing agent. Nano Lett. 8 (7), 2077–2081.

Smith, A.M., Duan, H., Rhyner, M.N., Ruan, G., Nie, S., 2006. A systematic examination of surface coatings on the optical and chemical properties of semiconductor quantum dots. Phys. Chem. Chem. Phys. 8 (33), 3895–3903.

Soukupová, J., Kvítek, L., Panáček, A., Nevěčná, T., Zbořil, R., 2008. Comprehensive study on surfactant role on silver nanoparticles (NPs) prepared via modified Tollens process. Mater. Chem. Phys. 111 (1), 77–81.

Stamplecoskie, K.G., Scaiano, J.C., 2010. Light emitting diode irradiation can control the morphology and optical properties of silver nanoparticles. J. Am. Chem. Soc. 132 (6), 1825–1827.

Sun, Y., Xia, Y., 2002. Shape-controlled synthesis of gold and silver nanoparticles. Science 298 (5601), 2176–2179.

Sun, Y., Yin, Y., Mayers, B.T., Herricks, T., Xia, Y., 2002. Uniform silver nanowires synthesis by reducing AgNO3 with ethylene glycol in the presence of seeds and poly (vinyl Pyrrolidone). Chem. Mater. 14 (11), 4736–4745.

Tadele Alula, M., Yang, J., 2014. Photochemical decoration of magnetic composites with silver nanostructures for determination of creatinine in urine by surface-enhanced Raman spectroscopy. Talanta 130, 55–62.

Tan, Y., Li, Y., Zhu, D., 2003. Preparation of silver nanocrystals in the presence of aniline. J. Colloid Interface Sci. 258 (2), 244–251.

Tessier, P.M., Velev, O.D., Kalambur, A.T., Rabolt, J.F., Lenhoff, A.M., Kaler, E.W., 2000. Assembly of gold nanostructured films templated by colloidal crystals and use in surface-enhanced Raman spectroscopy. J. Am. Chem. Soc. 122 (39), 9554–9555.

Tollens, B., 1882. Ueber ammon-alkalische Silberlösung als Reagens auf Aldehyd. Ber. Dtsch. Chem. Ges. 15 (2), 1635–1639.

Torras, M., Roig, A., 2020. From silver plates to spherical nanoparticles: snapshots of microwave-assisted polyol synthesis. ACS Omega 5 (11), 5731–5738.

Tran, Q.H., Nguyen, V.Q., Le, A.-T., 2013. Silver nanoparticles: synthesis, properties, toxicology, applications and perspectives. Adv. Nat. Sci. Nanosci. Nanotechnol. 4 (3), 033001.

Tripathi, A., Chandrasekaran, N., Raichur, A.M., Mukherjee, A., 2009. Antibacterial applications of silver nanoparticles synthesized by aqueous extract of *Azadirachta indica* (neem) leaves. J. Biomed. Nanotechnol. 5 (1), 93–98.

Tsuji, M., Maeda, Y., Hikino, S., Kumagae, H., Matsunaga, M., Tang, X.-L., Jiang, P., 2009. Shape evolution of octahedral and triangular platelike silver nanocrystals from cubic and right bipyramidal seeds in DMF. Cryst. Growth Des. 9 (11), 4700–4705.

Turkevich, J., Stevenson, P.C., Hillier, J., 1951. A study of the nucleation and growth processes in the synthesis of colloidal gold. Faraday Discuss. 11 (0), 55–75.

Van Dong, P., Ha, C.H., Binh, L.T., Kasbohm, J., 2012. Chemical synthesis and antibacterial activity of novel-shaped silver nanoparticles. Int. Nano Lett. 2 (1), 9.

Vo, Q.K., Phung, D.D., Vo Nguyen, Q.N., Hoang Thi, H., Nguyen Thi, N.H., Nguyen Thi, P.P., Van Tan, L., 2019. Controlled synthesis of triangular silver nanoplates by gelatin–chitosan mixture and the influence of their shape on antibacterial activity. Processes 7 (12), 873.

Wang, C., Liu, B., Dou, X., 2016. Silver nanotriangles-loaded filter paper for ultrasensitive SERS detection application benefited by interspacing of sharp edges. Sens. Actuators B 231, 357–364.

Wang, Y., Shi, Y.-F., Chen, Y.-B., Wu, L.-M., 2012. Hydrazine reduction of metal ions to porous submicro-structures of Ag, Pd, Cu, Ni, and Bi. J. Solid State Chem. 191, 19–26.

Wiley, B., Herricks, T., Sun, Y., Xia, Y., 2004. Polyol synthesis of silver nanoparticles: use of chloride and oxygen to promote the formation of single-crystal, truncated cubes and tetrahedrons. Nano Lett. 4 (9), 1733–1739.

Wiley, B., Sun, Y., Mayers, B., Xia, Y., 2005. Shape-controlled synthesis of metal nanostructures: the case of silver. Chem Eur J 11 (2), 454–463.

Wiley, B., Sun, Y., Xia, Y., 2007. Synthesis of silver nanostructures with controlled shapes and properties. Acc. Chem. Res. 40 (10), 1067–1076.

Wiley, B.J., Xiong, Y., Li, Z.-Y., Yin, Y., Xia, Y., 2006. Right bipyramids of silver: a new shape derived from single twinned seeds. Nano Lett. 6 (4), 765–768.

Wojnicki, M., Tokarski, T., Hessel, V., Fitzner, K., Luty-Błocho, M., 2019. Continuous, monodisperse silver nanoparticles synthesis using microdroplets as a reactor. J. Flow Chem. 9 (1), 1–7.

Wuithschick, M., Birnbaum, A., Witte, S., Sztucki, M., Vainio, U., Pinna, N., Polte, J., 2015. Turkevich in new robes: key questions answered for the most common gold nanoparticle synthesis. ACS Nano 9 (7), 7052–7071.

Wuithschick, M., Paul, B., Bienert, R., Sarfraz, A., Vainio, U., Sztucki, M., Polte, J., 2013. Size-controlled synthesis of colloidal silver nanoparticles based on mechanistic understanding. Chem. Mater. 25 (23), 4679–4689.

Xia, Y., Xiong, Y., Lim, B., Skrabalak, S.E., 2009. Shape-controlled synthesis of metal nanocrystals: simple chemistry meets complex physics? Angew. Chem. Int. Ed. 48 (1), 60–103.

Xue, C., Métraux, G.S., Millstone, J.E., Mirkin, C.A., 2008. Mechanistic study of photomediated triangular silver nanoprism growth. J. Am. Chem. Soc. 130 (26), 8337–8344.

Yin, B., Ma, H., Wang, S., Chen, S., 2003. Electrochemical synthesis of silver nanoparticles under protection of poly(N-vinylpyrrolidone). J. Phys. Chem. B 107 (34), 8898–8904.

Yonezawa, Y., Sato, T., Ohno, M., Hada, H., 1987. Photochemical formation of colloidal metals. J. Chem. Soc. Faraday Trans. 83 (5), 1559–1567.

Zeng, J., Zheng, Y., Rycenga, M., Tao, J., Li, Z.-Y., Zhang, Q., Xia, Y., 2010. Controlling the shapes of silver nanocrystals with different capping agents. J. Am. Chem. Soc. 132 (25), 8552–8553.

Zhang, Q., Li, N., Goebl, J., Lu, Z., Yin, Y., 2011. A systematic study of the synthesis of silver nanoplates: is citrate a "magic" reagent? J. Am. Chem. Soc. 133 (46), 18931–18939.

CHAPTER

Chemical and green production of silver nanocomposites

3

Said Fatouh Hamed[a], Ayat F. Hashim[a], Heba H. Salama[b], and Kamel A. Abd-Elsalam[c]

[a]*Fats and Oils Department, Food Industry and Nutrition Research Division, National Research Centre, Cairo, Egypt* [b]*Dairy Department, Food Industry and Nutrition Research Division, National Research Centre, Cairo, Egypt* [c]*Plant Pathology Research Institute, Agricultural Research Center (ARC), Giza, Egypt*

1 Introduction

Nanotechnology helps the production of polymer/silver nanocomposites with important physicochemical properties compared to their bulk materials (Ponnamma et al., 2018). Silver nanoparticles (AgNPs) have amazing popularity varying from their counterparts (gold, platinum, etc.) because they have unique properties chemically, physically, and biologically (Sharma et al., 2009). Polymers are considered good hosts for embedding nanoparticles and finish the growth of particles through controlling the nucleation process, and the silver nanoparticles improve the total performance (Li et al., 2012). In general, three various approaches have been applied to prepare polymer/silver nanocomposites: physical, chemical, and biological. This chapter briefly describes the chemical and biological methods for silver nanocomposites production. It also discusses characterization of polymer/silver nanocomposites using a variety of analytical tools and mechanism of synthesis of silver nanoparticles.

2 Chemical production of silver nanocomposites

In chemical methods, water or organic solvents are used to prepare the silver nanoparticles (Tao et al., 2006). This method generally utilizes three major components: metal precursors, capping/stabilizing agents, and reducing agents. Substantially, the reduction of the silver salt involves two steps: nucleation and subsequent growth. Commonly, silver nanomaterials can be obtained in two ways, categorized as "top-down" and "bottom-up" (Deepak et al., 2011). The top-down approach is the mechanical mashing of bulk metals with later stabilization using colloidal protecting agents (Mallick et al., 2004). The bottom-up approach includes chemical reduction, electrochemical methods, and sono-decomposition. Chemical methods include techniques such as laser ablation (Mafuné et al., 2000), laser irradiation (Abid et al., 2002), lithography (Hulteen et al., 1999), cryochemical synthesis (Sergeev et al., 1999), sono-decomposition (Talebi et al., 2010), thermal decomposition (Hosseinpour-Mashkani and Ramezani, 2014), electrochemical reduction (Zhu et al., 2001), and chemical reduction (Zhang et al., 2011).

Chemical reduction is the most popular approach for the preparation of silver NPs by reducing agents (organic or inorganic). Commonly, various reducing agents (Tollens reagent sodium borohydride ($NaBH_4$), ascorbate, elemental hydrogen, sodium citrate, polyol process, *N,N*-dimethyl formamide and poly (ethylene glycol)-block copolymers) are utilized to reduce silver ions (Ag^+) to metallic silver (Ag^0) in hydrous and anhydrous solutions, which is followed by agglomeration into oligomeric clusters. These clusters finally lead to the formation of metallic silver particles (Wiley et al., 2005). It is essential to use protective agents to stabilize dispersive nanoparticles during the synthesis of metal NPs, and to conserve the NPs that can be absorbed on or link into the surfaces of nanoparticle, avoiding their agglomeration (Oliveira et al., 2005). The use of surfactants containing functionalities (e.g., acids, alcohols) leads to stabilization of particle growth, protecting particles from agglomeration, sedimentation, and losing their surface properties. Polymeric compounds such as poly (vinyl alcohol), poly (vinylpyrrolidone), poly (ethylene glycol), poly (methacrylic acid), and polymethylmethacrylate have been reported as efficient protective agents to stabilize NPs.

2.1 Chemical synthesis of silver nanoparticles

Im et al. (2005) synthesized uniform silver nanocubes by reduction of silver nitrate using ethylene glycol at 140°C in the presence of poly(vinyl pyrrolidone) and HCl. Rashid et al. (2013) chemically prepared silver nanoparticles and silver nanocolloid solution by the reduction of silver salt. Mavani and Shah (2013) and Mehr et al. (2015) synthesized nano-Ag particles by the chemical reduction method of $AgNO_3$ using $NaBH_4$. Khan et al. (2017) prepared silver nanoparticles in the presence of cationic micelles of cetyltrimethylammonium bromide (CTAB) by a one-pot and simple chemical reduction method. Nishimoto et al. (2018) successfully prepared homogeneous Ag nanoparticles by chemical reduction using hydrogen peroxide (H_2O_2) as a reducing agent. Yari et al. (2021) synthesized efficient, eco-friendly, and low-cost adsorbent AgNPs from *Chenopodium botrys* extract.

2.2 Chemical synthesis of silver-graphene nanocomposites

Gurunathan et al. (2016) used $AgNO_3$ and pepsin to synthesis silver nanoparticle/graphene oxide nanocomposite. Bozkurt (2017) synthesized Ag-graphene nanocomposites by the one-step sonochemical method and used sodium citrate as a reducing agent. Chen et al. (2019) demonstrated a rapid method for the production of silver/reduced graphene oxide (Ag/rGO) nanocomposite by directly calcinating the mixture of silver nitrate and cellulose acetate precursors in a hydrogen (H_2) and argon (Ar) atmosphere. Beiranvand et al. (2019) demonstrated a simple and effective hydrothermal method in the preparation of a novel graphene oxide/hydroxyapatite/silver (rGO/HAP/Ag) ternary nanocomposite. Kumari et al. (2020) reported the fabrication of a graphene oxide-silver nanocomposite (GO-Ag) via the sonochemical method, which displays unique physiochemical properties. Potbhare et al. (2020) produced silver nanoparticles on the surface of graphene oxide nanosheets.

2.3 Chemical synthesis of chitosan/cellulose-silver nanocomposites

An et al. (2010) synthesized silver-N-carboxymethyl chitosan nanocomposites via Ag^+ reduction by $NaBH_4$ using water-soluble N-carboxymethyl chitosan (stabilizer). Govindan et al. (2012) synthesized chitosan-silver nanocomposite materials by a simple chemical method. Kaur et al. (2013) formed chitosan/silver nanocomposites by a chemical reduction method and embedding silver nanoparticles into a chitosan matrix. Adibelli et al. (2020) produced silver nanoparticles by a rapid and efficient method either in a solvent or in a polymer matrix. Tomke and Rathod (2020) developed a novel approach to synthesize an advanced magnetic nanocatalyst, Fe_3O_4@chitosan-AgNP nanocomposite.

Muthulakshmi et al. (2017) prepared silver nanocomposite films by immobilizing silver nanoparticles into cellulose composites via an in-situ approach and used bioflocculant as a reducing agent. Kokilavani et al. (2020) synthesized Ag/Cu-cellulose (NCs) by a chemical reduction method using $NaBH_4$ as a reducing agent. Yuan et al. (2021) used an impregnation-precipitation process to synthesize various photocatalysts with in-situ decorated Ag-based nanoparticles on the cellulosic paper.

2.4 Chemical synthesis of clay-silver nanocomposites

Praus et al. (2009) reported that silver cations were adsorbed on montmorillonite (Ag^+-MMT) by shaking in an $AgNO_3$ solution for 24 h. Thereafter, Ag^+-MMT was filtered out followed by drying. Silver cations were reduced by the addition of aqueous solutions of formaldehyde or sodium borohydride. Horue et al. (2020) demonstrated a nanocomposite based on bacterial cellulose (BC) including MMT modified with silver (BC-MMT-Ag). Jaime-Acuña et al. (2016) embedded a nanocomposite based on AgNPs on a zeolite matrix by a one-pot, template-, solvent-, and seed-free route. Abad-Álvaro et al. (2019) studied sepiolite-Ag and kaolinite-Ag nanocomposites as carriers for silver nanoparticles using two different clays (sepiolite and kaolinite).

2.5 Chemical synthesis of silver nanocomposites using different materials

Jiang et al. (2008) synthesized a silver/silica dioxide nanocomposite in a nanoreactor formed by an adsorption layer on a silica surface. Dulski et al. (2019) fabricated a nanocomposite with monodispersed silver (Ag^0 or Ag^+) embedded into a silica carrier. Sun et al. (2014) synthesized silver/zinc oxide (Ag/ZnO) nanocomposite photocatalysts with high photocatalytic performance via a facile sol-gel method. Allahyarzadeh et al. (2013) produced a titanium dioxide-silver nanoparticle (TiO_2/AgNPs) coated fabric with a solution having NaOH, CTAB, $AgNO_3$, and TiO_2 NPs using an in-situ chemical reduction synthesis method. Mahdieh et al. (2018) used a new in-situ synthesis method to prepare Ag-TiO_2 NPs-coated fabric by the direct photoreduction of the padded fabric. Saleh et al. (2013) described the radiation-induced synthesis of Ag/polyvinyl alcohol nanocomposites by a simple and easy route in the presence of polyvinyl alcohol as a stabilizer. Shimoga et al. (2019) loaded silver ions (2.5%–10.0%) into polyvinylchloride to prepare silver nanoparticle-polyvinylchloride (SNC-PVC) composites using a simple solution casting technique.

Hasan et al. (2019) used an in-situ free radical polymerization method to prepare a nanocomposite by silver nanoparticles using dextrin followed by grafting with poly (methyl methacrylate). Pozdnyakov et al. (2019) prepared a new functional insoluble silver-including nanocomposite based on 1-vinyl-1,2,4-triazole and acrylonitrile copolymer. Liu and Du (2020) described a facile mechanism of preparation of hybrid nanocomposite encompassing gold (Au) and Ag. Phuruangrat et al. (2020) synthesized heterostructure silver/bismuth molybdenum oxide (Ag/Bi_2MoO_6) nanocomposites using the hydrothermal method and followed this with precipitation-deposition of Ag nanoparticles in the solution having $NaBH_4$ as a reducing reagent at room temperature. Sabavath et al. (2020) synthesized silver-carbon (Ag-C) nanocomposites using a thermal plasma-assisted method by a single step. Khan et al. (2020) used a facile, economical, and ecofriendly method for the synthesis of silver/copper oxide (Ag/CuO) nanocomposites. Kokilavani et al. (2021) developed a simple visual technique for detecting mercury sensitively in aqueous samples using a Cu-sensitized Ag-dextran nanocomposite. Pham et al. (2021) prepared Ag/SiO_2 colloidal nanocomposites by the semicontinuous chemical reduction of silver ions on a silica surface. Abduraimova et al. (2021) successfully prepared a cetyltrimethylammonium loaded SiO_2-Ag mesoporous nanocomposite.

The main advantages of chemical methods are ease of production, high yield, and low cost. However, the drawbacks of these methods are that they produce materials not of expected purity, as their surfaces were found to contain a chemical residue. It is also very challenging to prepare AgNPs with well-defined size, demanding a further step for the prevention of particle aggregation (Malik et al., 2002). Moreover, chemical reducing agents are harmful to living organisms.

3 Green production of silver nanocomposites

To overcome the drawbacks of chemical methods, green techniques have appeared as applicable choices. Recently, biologically mediated preparation of nanoparticles has been identified as an easy, reliable, cost-effective, eco-friendly way resulting in high yield production. Different biological systems containing plant extracts, small biomolecules, bacteria, and fungi have been found to be substitutes for chemical methods (Gurunathan et al., 2014; Kalishwaralal et al., 2008). Plant extracts such as *Allophylus cobbe* (Gurunathan et al., 2014), *Artemisia princeps* (Gurunathan et al., 2015a,b), and *Typha angustifolia* (Gurunathan, 2015) are the most common reducing agents in the biological method (Mano et al., 2011). In addition, the silver nanoparticles are produced from Ag^+ ions by bioreduction with the help of plant metabolites (Nurani et al., 2015).

In addition, several biomolecules, such as biopolymers (Leung et al., 2010), starch (Kumar et al., 2014), fibrinolytic enzymes (Deepak et al., 2011), and amino acids (Shankar and Rhim, 2015), were utilized. The synthesis of silver nanoparticles happens through the reduction of silver ions with the help of microorganism extracts (Logeswari et al., 2015). The extracts of microorganisms can behave as both reducing agents and capping agents. In the green chemistry approach, several bacteria, containing *Pseudomonas stutzeri AG259* (Klaus et al., 1999), *Lactobacillus* strains (Nair and Pradeep, 2002), *Bacillus licheniformis* (Kalimuthu et al., 2008), *Escherichia coli* (Gurunathan et al., 2009), *Brevibacterium casei* (Kalishwaralal et al., 2010), fungi such as *Fusarium oxysporum* (Shankar et al., 2003), and *Ganoderma neo-japonicum Imazeki* (Gurunathan et al., 2013) were utilized.

Using plant extracts or bacterial protein as reducing agents can control the size, shape, and nanoparticles' monodispersity (Gurunathan et al., 2009). Biological methods lead to control of the morphology, size, and distribution of the nanoparticles produced compared to chemical methods, via optimization of several factors in the methods used. These factors include the amount of precursor, temperature, pH, and the amount of reducing and stabilizing factors (Khodashenas and Ghorbani, 2015). The other features of biological methods are the availability of a large order of biological resources, high consistency, a lower time requirement, stabilization, and the ready solubility of prepared nanoparticles in water (Thakkar et al., 2010). The biological synthesis of nanoparticles relies on three factors: (a) the solvent, (b) the nontoxic material, and (c) the reducing agent.

3.1 Green synthesis of silver nanoparticles

Green synthesis of silver NPs via latex of *Jatropha curcas* (Bar et al., 2009a), *Rosa rugosa* leaf extract (Dubey et al., 2010), *Citrus limon* (Prathna et al., 2011), sucrose and maltose (Filippo et al., 2010), banana peel extract (Bankar et al., 2010), *Hibiscus rosasinensis* (Philip, 2010a), *Cochlospermum gossypium* (Kora et al., 2010) and honey (Philip, 2010b) starch (Vigneshwaran et al., 2006), and seed extract of *Jatropha curcas* (Bar et al., 2009b) have been reported. Rodríguez-León et al. (2013) used extracts of

Rumex hymenosepalus (reducing agent) to prepare silver nanoparticles from silver nitrate solutions. Sampaio and Viana (2018) synthesized AgNPs by the reduction of silver salt using an extract of the artichoke (*Cynara scolymus* L.) flower.

3.2 Green synthesis of polymer-silver nanocomposites

Pandey et al. (2012) reported the biopolymer-silver nanoparticle nanocomposite synthesis from silver nitrate using polysaccharide as a template. Chen et al. (2014) improved a new technique to prepare green polymer composite spheres consisting of silver nanoparticles decorating the polymer colloids in a raspberry-like form. Susilowati et al. (2015) prepared colloidal silver-chitosan nanocomposites using a metal precursor (silver nitrate), a reducing agent (glucose), an accelerator reagent (sodium hydroxide), and a stabilizing agent (chitosan) via a chemical reduction process. Tork et al. (2014) described the preparation of hydrogel nanocomposite (silver nanoparticles/chitosan/polyvinyl alcohol/polyethene glycol) by in-situ green syntheses. Palem et al. (2018) created an eco-friendly way to synthesize crosslinked chitosan silver nanocomposites (CSHD-AgNCs) and biogenic rhubarb silver nanoparticles (RS-AgNPs). Abdallah et al. (2020) synthesized nanocomposite films of polyvinyl alcohol-silver (PVA-Ag) and chitosan-silver (CS-Ag) by augmentation with biosynthesized AgNPs, using *Enterobacter cloacae Ism* 26 (KP988024). Srikhao et al. (2021) incorporated green prepared AgNPs to bioactive packaging polyvinyl alcohol PVA/starch (PSt) films.

Ultra-small and stable silver nanoparticles on chitosan biopolymer (BP/AgP) were developed by in situ reduction of the diamine silver(I) complex $[Ag(NH_3)_2]^+$. The crystal type of small AgNP (3 nm) is a stable source of silver ions. TEM, SEM, and AFM verified the homogeneous distribution in the entire solid biofilm membrane. In addition, XPS and Auger analysis were used to determine the atomic structure, concentration, and chemical status of surface atoms (Fig. 1). Finally, using selected bacteria biofilms, the antibacterial efficacy of the BP/AgP nanocomposite was tested (Zienkiewicz-Strzałka and Deryło-Marczewska, 2020).

Sganzerla et al. (2020) produced poly (ethylene oxide) (PEO) nanocomposite films containing silver nanoparticles (AgNPs) using *Acca sellowiana* aqueous extract. Awad et al. (2015) reduced silver nitrate using orange peel extract to synthesize a silver/polystyrene nanocomposite. Motitswe and Fayemi (2019) used orange peel extracts as a reducing agent for the synthesis of silver nanoparticles. These nanoparticles were functionalized with polyacrylonitrile to form PAN/Ag nanofibers by electrospinning. Parmar et al. (2019) fabricated poly-D,L-lactide-*co*-glycolide (PLGA)-silver nanocomposites by a green chemistry method. Renu et al. (2020) elaborated the plant extract-mediated synthesized AgNPs-based (PLGA) nanocomposites preparation. Jayakumar et al. (2020) synthesized poly (3-hydroxybutyrate)-silver nanocomposite (PHB-AgNc) and biologically utilized a dairy industry by-product, cheese whey, which permeated as a substrate for *Bacillus megaterium*. Ortega et al. (2021) used lemon juice as a reducing and stabilizing agent to prepare nanocomposite starch-based films containing silver nanoparticles.

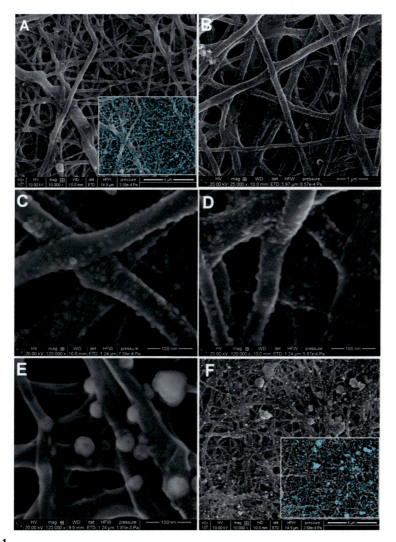

FIG. 1

SEM images of the investigated nanocomposites, (A) BP, (B–D) BP/AgP, and (E, F) BP/AgNP. *Insets* presented in panels (A) and (F) present the combination of SEM imaging with energy-dispersive X-ray spectroscopy (EDX). *Blue objects* represent the map of AgL signals.

Photo-plate reprinted from Zienkiewicz-Strzałka, M., Deryło-Marczewska, A., 2020. Small AgNP in the biopolymer nanocomposite system. Int. J. Mol. Sci. 21 (24), 9388, with permission from MDPI Publisher, Licensee MDPI, Basel, Switzerland. This article is an open-access article distributed under the terms and conditions of the Creative Commons Attribution (CC BY) license (http:/creativecommons.org/licenses/by/4.0).

Sarkandi et al. (2021) used green tea as a substrate to prepare bacterial cellulose/silver nanocomposites by innovative in-situ green methods. Valera-Zaragoza et al. (2021) obtained starch/AgNPs nanocomposites using a *Stevia rebaudiana* aqueous extract followed by their dispersion in starch. Thangaswamy et al. (2021) used an aqueous extract of *Chlorella acidophile* for green synthesis of mono and bimetallic alloy nanoparticles of gold and silver and their potential applications in sensors.

3.3 Green synthesis of silver-graphene nanocomposites

Salazar et al. (2019) reduced both Ag^+ cations and graphene oxide sheets using tea extract then synthesized silver nanoparticles-modified reduced graphene oxide nanocomposites (rGox/AgNPs) by a green synthesis method. Calderón-Ayala et al. (2020) synthesized a nanocomposite material of few-layer graphene (FLG) and silver nanoparticles by replacing the additional chemical reactant with the extract of the plant *Jatropha cordata* in the exfoliation and reduction process. Devi and Swain (2020) studied the preparation of reduced graphene oxide/Ag nanocomposites by a green synthesis method using *Tilia amurensis* plant extracts (TAPE). Chandu et al. (2020) reported a single-step in-situ synthesis of reduced graphene oxide sheets decorated with silver nanoparticles (CRG-Ag nanocomposite) using custard apple leaf extract as an effective reducing and stabilizing agent.

Nayak et al. (2020) reported an eco-friendly green synthesis of silver nanoparticles decorated on reduced graphene oxide (Ag-rGO) nanocomposites by using the aqueous fruit extract of *Phyllanthus acidus* in a basic medium. Mariadoss et al. (2020) used Lc-AgNPs to prepare *Lespedeza cuneata*-mediated silver nanoparticles (Lc-AgNPs) and graphene oxide-silver nanocomposites (GO-AgNComp). Karthik et al. (2020) developed a facile method for a reduced graphene oxide silver nanoparticle (rGO-AgNP) hybrid nanocomposite synthesis using an aqueous extract of *Brassica nigra*. Yu et al. (2021) prepared graphene oxide-silver nanocomposites embedded nanofiber core-spun yarns for durable antibacterial textiles. Dat et al. (2021) fabricated Ag/GO nanocomposites by coprecipitation with a green reducing agent and also used modern analytical techniques for characterization.

3.4 Green synthesis of silver nanocomposites using different materials

The most significant nanocomposites are those focused on metallic nanoparticles. Nanocomposites imply heterogeneous mixture at the nanoscale and multiphase structures with distinct properties for each component (Fig. 2). The need for materials that are more suited to developed technology has resulted in the research and production of new forms of nanocomposites. Composites based on biopolymers such as chitosan (CS) are becoming increasingly popular, with special attention paid to their biological properties (Yang et al., 2016). Pandey and Ramontja (2016) synthesized biopolymer-silver nanocomposites using guar gum as a capping and reducing agent. Roseline et al. (2019) synthesized green silver nanoparticles (AgNPs) using

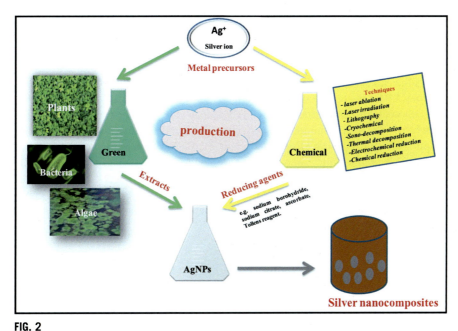

FIG. 2

Chemical and green production of silver nanocomposites.

the aqueous extracts of agar seaweeds (*Gracilaria corticata* and *G. edulis*) and carrageenan seaweeds (*Hypneamusci formis* and *Spyridia hypnoides*). Noohpisheh et al. (2020) utilized hydroalcoholic extract of fenugreek leaves to synthesize Ag-ZnO nanocomposites biologically.

Rao et al. (2018) synthesized nanocomposites of cotton fabrics within in-situ-generated AgNPs via *Pterocarpus santalinus* (red sanders) extract (reducing agent) in water. Omidi et al. (2018) used the aqueous tarragon extract (reducing agent) to synthesize silver-montmorillonite nanocomposites in the batch method. Allafchian et al. (2017) produced new green polymer support for silver by naturally coated magnetic nanocomposites using *Ocimum basilicum* mucilage.

Between all available methods, the basic method for the preparation of nanoparticles is realized by keeping in mind the parameters such as chemicals availability, natural extract and available instruments, the required size and morphology of nanoparticles, and most importantly the use of nanoparticles in a certain application.

4 Characterization

Characterization of polymer/silver nanocomposites is significant to assess the functional aspects of the prepared particles. It is performed using a variety of analytical tools including ultraviolet visible spectroscopy (UV-vis spectroscopy), which provides important information about the formation of metal nanoparticles in a solution, and Fourier transform infrared spectroscopy (FTIR), which is a method used to investigate the chemical composition. In addition, dynamic light scattering (DLS) measures the size distribution profiles of particles in the submicron range and studies the behavior of the nanoparticles in suspensions. X-ray diffraction (XRD) is an important technique to identify the crystalline phase, while X-ray photoelectron spectroscopy (XPS) is used to study the chemical modification of surfaces. The most commonly used techniques for the characterization of nanostructures are transmission electron microscopy (TEM), scanning electron microscopy (SEM), and atomic force microscopy (AFM) (Gurunathan et al., 2015a,b; Sapsford et al., 2011). Thermal gravimetric analysis (TGA) identifies the organic compounds on the NPs surfaces by determining the thermal stability of the compounds. An energy dispersive spectroscopy (EDS) profile displays a strong signal corresponding to silver fluorescence.

5 Mechanisms of AgNPs synthesis

A lot of research has looked at the mechanisms of action for silver nanoparticles. The analysis of nanocomposite samples before and after interaction with mercury was initiated to detect the stages formed in slow-scanning mode. The samples of postadsorption nano compounds contain AgCl (chlorargyrite) attributable to peaks at 32.23 degrees, 46.25 degrees for the K sample, and 46.21 degrees and 57.72 degrees for the M sample, while the initial AgNPs peaks have decreased significantly or almost disappeared, necessitating the conversion of Ag^0 to Ag^+. Tauanov et al. (2018, 2020) suggested that removal of mercury may be effective due to diluted oxidation followed by fusion reactions.

The green AgNPs can be synthesized via microorganisms (bacteria or fungi), biopolymers, and plants (plant extracts). Different components of bacteria, fungi, biopolymers, and plants are responsible for the reduction of $AgNO_3$ to AgNPs by various mechanisms. Fungi contain enzymes, NADH, NADPH, proteins, peptides, napthoquinones, anthraquinones, and nitrogenous biomacromolecules which can reduce silver ion using intracellular and extracellular mechanisms. In addition, biopolymers such as alginate, cellulose, chitosan, lignin, polypeptides and protein used electrostatic interaction between Ag^+ ion and polar groups attached to the polymer. Flavonoids, polyphenols, phenolic acids, terpenoids, alkaloids, alcohol, antioxidants, vitamins in plants used electrostatic interaction between the functional groups of a respective constituent of plant extract and Ag^+ ion for the reduction of silver nitrate to AgNPs.

Silver ions interact with groups of thyol proteins, inhibiting bacteria and inhibiting the reproduction of microorganisms. Ag^+ levels in $\mu mol\ L^{-1}$ have weakened the frequency of DNA due to the decoupling of electron transfer in the respiratory tract from oxidized fissures, inhibiting respiratory chain enzymes and/or interfering with membrane permeability (Durán et al., 2016).

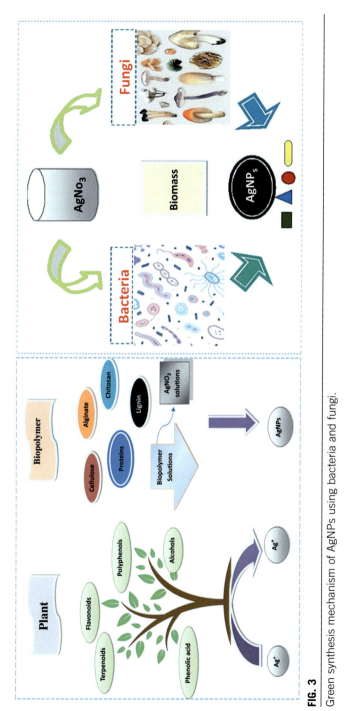

FIG. 3

Green synthesis mechanism of AgNPs using bacteria and fungi.

From the other side, silver ions interact with the thiol groups of many bioenzymes, disrupting them and generating reactive oxygen species (ROS) (Prabhu and Poulose, 2012). AgNPs can also act as a reservoir for monovalent silver types that are released in the presence of an oxidized agent (Le Ouay and Stellacci, 2015). It was found that the size of nanoparticles affects the release of silver ions which is related to the size of nanoparticles. The antibacterial activity of particles less than 10 nm is produced from the nanoparticles themselves and larger particles work according to the prevailing machine through silver ions (Durán et al., 2016). The above can be shortened to the mechanism of action of nanoparticles in Fig. 3. Silver nanoparticles have been drawn as antimicrobial agents in different fields, and although many references have described the mechanism of AgNPs, they are still unclear: (1) irreversible damage to the bacterial cell membrane through direct contact; (2) the generation of reactive oxygen species (ROS); and (3) interaction with DNA and proteins. Silver nanoparticles make structural changes that cause bacteria to be more permeable and affect breathing function (Morones et al., 2005). TEM also showed the presence of silver in the membranes of the treated bacteria as well as within them. In addition, the release of silver ions (Ag^+) has shown that a lower concentration of silver ions can give the same result in terms of toxicity (Russell and Hugo, 1994).

6 Conclusion

Silver nanocomposites can be prepared through chemical and green methods. In chemical method uses three main components: silver metal precursors, reducing agents, and stabilizing/capping agents. Various biological systems including plant extracts, small biomolecules, bacteria, and fungi provide alternative methods to chemical methods. In comparison to chemical methods, biological methods allow control of the shape, size, and distribution of the produced nanoparticles. Silver nanocomposites have become an interesting component due to their antimicrobial activities. This allows the development of active food packaging materials to prolong the shelf life of food products.

References

Abad-Álvaro, I., Trujillo, C., Bolea, E., Laborda, F., Fondevila, M., Latorre, M.A., Castillo, J.R., 2019. Silver nanoparticles-clays nanocomposites as feed additives: characterization of silver species released during in vitro digestions. Effects on silver retention in pigs. Microchem. J. 149, 104040.

Abdallah, O.M., EL-Baghdady, K.Z., Khalil, M.M., El Borhamy, M.I., Meligi, G.A., 2020. Antibacterial, antibiofilm and cytotoxic activities of biogenic polyvinyl alcohol-silver and chitosan-silver nanocomposites. J. Polym. Res. 27 (3), 1–9.

Abduraimova, A., Molkenova, A., Duisembekova, A., Mulikova, T., Kanayeva, D., Atabaev, T.S., 2021. Cetyltrimethylammonium bromide (CTAB)-loaded SiO$_2$–Ag mesoporous nanocomposite as an efficient antibacterial agent. Nanomaterials 11 (2), 477.

Abid, J.P., Wark, A.W., Brevet, P.F., Girault, H.H., 2002. Preparation of silver nanoparticles in solution from a silver salt by laser irradiation. Chem. Commun. 7, 792–793.

Adibelli, M., Ozcelik, E., Batibay, G.S., Arasoglu, T.O., Arsu, N., 2020. A facile and versatile route for preparation AgNp nanocomposite thin films via thiol-acrylate photopolymerization: determination of antibacterial activity. Prog. Org. Coat. 143, 105620.

Allafchian, A., Jalali, S.A.H., Hosseini, F., Massoud, M., 2017. Ocimum basilicum mucilage as a new green polymer support for silver in magnetic nanocomposites: production and characterization. J. Environ. Chem. Eng. 5 (6), 5912–5920.

Allahyarzadeh, V., Montazer, M., Nejad, N.H., NasrinSamadi, N., 2013. In situ synthesis of nanosilver on polyester using NaOH/Nano TiO2.PPL. Polym. Sci. 129, 892–900.

An, N.T., Dong, N.T., Hanh, P.T.B., Nhi, T.T.Y., Vu, D.A., et al., 2010. Silver-N-carboxymethyl chitosan nanocomposites: synthesis and its antibacterial activities. J. Bioterror. Biodef. 1, 1–4.

Awad, M.A., Mekhamer, W.K., Merghani, N.M., Hendi, A.A., Ortashi, K.M.O., FatimahAl-Abbas, F., Eisa, N.E., 2015. Green synthesis, characterization, and antibacterial activity of silver/polystyrene nanocomposite. J. Nanomater. 2015.

Bankar, A., Joshi, B., Kumar, A.R., Zinjarde, S., 2010. Banana peel extract mediated novel route for the synthesis of silver nanoparticles. Colloids Surf. A Physicochem. Eng. Asp. 368, 58–63.

Bar, H., Bhui, D.K., Sahoo, G.P., Sarkar, P., De, S.P., Misra, A., 2009a. Green synthesis of silver nanoparticles using latex of *Jatropha curcas*. Colloids Surf. A Physicochem. Eng. Asp. 339, 134–139.

Bar, H., Bhui, D.K., Sahoo, G.P., Sarkar, P., Pyne, S., Misra, A., 2009b. Green synthesis of silver nanoparticles using seed extract of *Jatropha curcas*. Colloids Surf. A Physicochem. Eng. Asp. 348, 212–216.

Beiranvand, M., Farhadi, S., Mohammadi, A., 2019. Graphene oxide/hydroxyapatite/silver (rGO/HAP/Ag) nanocomposite: synthesis, characterization, catalytic and antibacterial activity. Int. J. Nano Dimens. 10, 180–194.

Bozkurt, P.A., 2017. Sonochemical green synthesis of Ag/graphene nanocomposite. Ultrason. Sonochem. 35, 397–404.

Calderón-Ayala, G., Cortez-Valadez, M., Martínez-Núñez, C.E., Flores-Acosta, M., 2020. FLG/silver nanoparticles: nanocomposite by green synthesis. Diam. Relat. Mater. 101, 107618.

Chandu, B., Kurmarayuni, C.M., Kurapati, S., Bollikolla, H.B., 2020. Green and economical synthesis of graphene–silver nanocomposite exhibiting excellent photocatalytic efficiency. Carbon Lett. 30 (2), 225–233.

Chen, G., Lu, J., Lamb, C., Yub, Y., 2014. A novel green synthesis approach for polymer nanocomposites decorated with silver nanoparticles and their antibacterial activity. Analyst 139, 5793–5799.

Chen, L., Li, Z., Chen, M., 2019. Facile production of silver-reduced graphene oxide nanocomposite with highly effective antibacterial performance. J. Environ. Chem. Eng. 7, 103160.

Dat, N.M., Khang, P.T., Anh, T.N.M., Quan, T.H., Thinh, D.B., Thien, D.T., Nam, H.M., Phong, M.T., Hieu, N.H., 2021. Synthesis, characterization, and antibacterial activity investigation of silver nanoparticle-decorated graphene oxide. Mater. Lett. 285, 128993.

Deepak, V., Umamaheshwaran, P.S., Guhan, K., Nanthini, R.A., Krithiga, B., Jaithoon, N.M., Gurunathan, S., 2011. Synthesis of gold and silver nanoparticles using purified URAK. Colloids Surf. B 86, 353–358.

Devi, N.A., Swain, B.P., 2020. Green synthesis of rGO/Ag nanocomposite for clean energy storage application. In: Advances in Greener Energy Technologies. Springer, Singapore, pp. 337–355.

Dubey, S.P., Lahtinen, M., Sillanpää, M., 2010. Green synthesis and characterizations of silver and gold nanoparticles using leaf extract of *Rosa rugosa*. Colloids Surf. A Physicochem. Eng. Asp. 364 (1–3), 34–41.

Dulski, M., Peszke, J., Włodarczyk, J., Sułowicz, S., Piotrowska-Seget, Z., Dudek, K., Podwórny, J., Malarz, K., Mrozek-Wilczkiewicz, A., Zubko, M., Nowak, A., 2019. Physicochemical and structural features of heat treated silver-silica nanocomposite and their impact on biological properties. Mater. Sci. Eng. C 103, 109790.

Durán, N., Durán, M., de Jesus, M.B., Seabra, A.B., Fávaro, W.J., Nakazato, G., 2016. Silver nanoparticles: a new view on mechanistic aspects on antimicrobial activity. Nanomed. Nanotechnol. Biol. Med. 12, 789–799.

Filippo, E., Serra, A., Buccolieri, A., Manno, D., 2010. Green synthesis of silver nanoparticles with sucrose and maltose: morphological and structural characterization. J. Non-Cryst. Solids 356, 344–350.

Ortega, F., Arce, V.B., Garcia, M.A., 2021. Nanocomposite starch-based films containing silver nanoparticles synthesized with lemon juice as reducing and stabilizing agent. Carbohydr. Polym. 252.

Govindan, S., Nivethaa, E.A.K., Saravanan, R., Narayanan, V., Stephen, A., 2012. Synthesis and characterization of chitosan–silver nanocomposite. Appl. Nanosci. 2, 299–303.

Gurunathan, S., 2015. Biologically synthesized silver nanoparticles enhances antibiotic activity against Gram-negative bacteria. J. Ind. Eng. Chem. 29, 217–226.

Gurunathan, S., Han, J., Park, J.H., Kim, J.H., 2014. A green chemistry approach for synthesizing biocompatible gold nanoparticles. Nanoscale Res. Lett. 9, 248.

Gurunathan, S., Han, J.W., Kim, E.S., Park, J.H., Kim, J.H., 2015a. Reduction of graphene oxide by resveratrol: a novel and simple biological method for the synthesis of an effective anticancer nanotherapeutic molecule. Int. J. Nanomedicine 10, 2951–2969.

Gurunathan, S., Han, J.W., Dayem, A.A., Dayem, A.A., Eppakayala, V., Park, J.H., Cho, S.G., Lee, K.J., Kim, J.H., 2013. Green synthesis of anisotropic silver nanoparticles and its potential cytotoxicity in human breast cancer cells (MCF-7). J. Ind. Eng. Chem. 19, 1600–1605.

Gurunathan, S., Park, J.H., Y-J, C., Han, W.J., Kim, J.-H., 2016. Synthesis of graphene oxide-silver nanoparticle nanocomposites: an efficient novel antibacterial agent. Curr. Nanosci. 12, 762–773.

Gurunathan, S., Jeong, J.K., Han, J.W., Zhang, X.F., Park, J.H., Kim, J.H., 2015b. Multidimensional effects of biologically synthesized silver nanoparticles in *Helicobacter pylori*, *Helicobacter felis*, and human lung (L132) and lung carcinoma A549 cells. Nanoscale Res. Lett. 10, 1–17.

Gurunathan, S., Kalishwaralal, K., Vaidyanathan, R., Venkataraman, D., Pandian, S.R., Muniyandi, J., Hariharan, N., Eom, S.H., 2009. Biosynthesis, purification and characterization of silver nanoparticles using *Escherichia coli*. Colloids Surf. B Biointerfaces 74, 328–335.

Hasan, I., Qais, F.A., Husain, F.M., Khan, R.A., Alsalme, A., Alenazi, B., Usman, M., Jaafar, M.H., Ahmad, I., 2019. Eco-friendly green synthesis of dextrin based poly (methyl methacrylate) grafted silver nanocomposites and their antibacterial and antibiofilm efficacy against multi-drug resistance pathogens. J. Clean. Prod. 230, 1148–1155.

Horue, M., Cacicedo, M.L., Fernandez, M.A., Rodenak-Kladniew, B., Sánchez, R.M.T., Castro, G.R., 2020. Antimicrobial activities of bacterial cellulose-silver montmorillonite nanocomposites for wound healing. Mater. Sci. Eng. C 116, 111152.

Hosseinpour-Mashkani, S.M., Ramezani, M., 2014. Silver and silver oxide nanoparticles: synthesis and characterization by thermal decomposition. Mater. Lett. 130, 259–262.

Hulteen, J.C., Treichel, D.A., Smith, M.T., Duval, M.L., Jensen, T.R., van Duyne, R.P., 1999. Nanosphere lithography: size-tunable silver nanoparticle and surface cluster arrays. J. Phys. Chem. B 103, 3854–3863.

Im, S.H., Lee, Y.T., Wiley, B., Xia, Y., 2005. Large-scale synthesis of silver nanocubes: the role of HCl in promoting cube perfection and monodispersity. Angew. Chem. Int. Ed. 44, 2154–2157.

Jaime-Acuña, O.E., Meza-Villezcas, A., Vasquez-Peña, M., Raymond-Herrera, O., Villavicencio-Garcı́a, H., Petranovskii, V., Vazquez-Duhalt, R., Huerta-Saquero, A., 2016. Synthesis and complete antimicrobial characterization of CEOBACTER, an Ag-based nanocomposite. PLOS ONE. , e0166205.

Jayakumar, A., Prabhu, K., Shah, L., Radha, P., 2020. Biologically and environmentally benign approach for PHB-silver nanocomposite synthesis and its characterization. Polym. Test. 81, 106197.

Jiang, X., Chen, S., Mao, C., 2008. Synthesis of Ag/SiO_2 nanocomposite material by adsorption phase nanoreactor technique. Colloids Surf. A Physicochem. Eng. Asp. 320, 104–110.

Kalimuthu, K., Babu, R.S., Venkataraman, D., Bilal, M., Gurunathan, S., 2008. Biosynthesis of silver nanocrystals by *Bacillus licheniformis*. Colloid Surf. B 65, 150–153.

Kalishwaralal, K., Deepak, V., Pandian, S.R.K., Kottaisamy, M., BarathmaniKanth, S., Kartikeyan, B., Gurunathan, S., 2010. Biosynthesis of silver and gold nanoparticles using *Brevibacterium casei*. Colloids Surf. B Biointerfaces 77, 257–262.

Kalishwaralal, K., Deepak, V., Ramkumarpandian, S., Nellaiah, H., Sangiliyandi, G., 2008. Extracellular biosynthesis of silver nanoparticles by the culture supernatant of *Bacillus licheniformis*. Mater. Lett. 62, 4411–4413.

Karthik, C., Swathi, N., Pandi Prabha, S., Caroline, D.G., 2020. Green synthesized rGO-AgNP hybrid nanocomposite—an effective antibacterial adsorbent for photocatalytic removal of DB-14 dye from aqueous solution. J. Environ. Chem. Eng. 8.

Kaur, P., Choudhary, A., Thakur, R., 2013. Synthesis of chitosan-silver nanocomposites and their antibacterial activity. Int. J. Sci. Eng. Res. 4, 869–872.

Khan, A.U., Khan, A.U., Li, B., Mahnashi, M.H., Alyami, B.A., Alqahtani, Y.S., Tahir, K., Khan, S., Nazir, S., 2020. A facile fabrication of silver/copper oxide nanocomposite: an innovative entry in photocatalytic and biomedical materials. Photodiagn. Photodyn. Ther. 31, 101814. https://doi.org/10.1016/j.pdpdt.2020.101814.

Khan, Z., Hussain, J.I., Hashmi, A.A., AL-Thabaiti, S.A., 2017. Preparation and characterization of silver nanoparticles using aniline. Arab. J. Chem. 10, S1506–S1511.

Khodashenas, B., Ghorbani, H.R., 2015. Synthesis of silver nanoparticles with different shapes. Arab. J. Chem. 12, 1823–1838.

Klaus, T., Joerger, R., Olsson, E., Granqvist, C.G., 1999. Silver-based crystalline nanoparticles, microbially fabricated. Proc. Natl. Acad. Sci. U.S.A. 96, 13611–13614.

Kokilavani, S., Syed, A., Thomas, A.M., Elgorban, A.M., Bahkali, A.H., Marraiki, N., Raju, L.L., Das, A., Khan, S.S., 2021. Development of multifunctional Cu sensitized Ag-dextran nanocomposite for selective and sensitive detection of mercury from environmental sample and evaluation of its photocatalytic and anti-microbial applications. J. Mol. Liq. 321, 114742.

Kokilavani, S., Syed, A., Thomas, A.M., Elgorban, A.M., Raju, L.L., Das, A., Khan, S.S., 2020. Facile synthesis of Ag/Cu-cellulose nanocomposite for detection, photocatalysis and anti-microbial applications. Optik 220, 165218.

Kora, A.J., Sashidhar, R.B., Arunachalam, J., 2010. Gum kondagogu (Cochlospermum gossypium): a template for the green synthesis and stabilization of silver nanoparticles with antibacterial application. Carbohydr. Polym. 82, 670–679.

Kumar, B., Smita, K., Cumbal, L., Debut, A., Pathak, R.N., 2014. Sonochemical synthesis of silver nanoparticles using starch: a comparison. Bioinorg. Chem. Appl. 784268.

Kumari, S., Sharma, P., Yadav, S., Kumar, J., Vij, A., Rawat, P., Kumar, S., Sinha, C., Bhattacharya, J., Srivastava, C.M., Majumder, S., 2020. A novel synthesis of the graphene oxide-silver (GO-Ag) nanocomposite for unique physiochemical applications. ACS Omega 5, 5041–5047.

Le Ouay, B., Stellacci, F., 2015. Antibacterial activity of silver nanoparticles: a surface science insight. NanoToday 10 (3), 339–354.

Leung, T.C., Wong, C.K., Xie, Y., 2010. Green synthesis of silver nanoparticles using biopolymers, carboxymethylated-curdlan and fucoidan. Mater. Chem. Phys. 121, 402–405.

Li, H.J., Zhang, A.Q., Hu, Y., Sui, L., Qian, D.J., Chen, M., 2012. Large-scale synthesis and self-organization of silver nanoparticles with tween 80 as a reductant and stabilizer. Nanoscale Res. Lett. 7, 612.

Liu, F., Du, S.Y., 2020. Production of gold/silver doped carbon nanocomposites for effective photothermal therapy of colon cancer. Sci. Rep. 10 (1), 1–9.

Logeswari, P., Silambarasan, S., Abraham, J., 2015. Synthesis of silver nanoparticles using plants extract and analysis of their antimicrobial property. J. Saudi Chem. Soc. 19, 311–317.

Mafuné, F., Kohno, J.Y., Takeda, Y., Kondow, T., Sawabe, H., 2000. Formation and size control of silver nanoparticles by laser ablation in aqueous solution. J. Phys. Chem. B 104, 9111–9117.

Mahdieh, Z.M., Shekarriz, S., Taromi, F.A., Montazer, M., 2018. A new method for in situ synthesis of Ag–TiO_2 nanocomposite particles on polyester/cellulose fabric by photoreduction and self-cleaning properties. Cellulose 25, 2355–2366.

Malik, M.A., O'Brien, P., Revaprasadu, N., 2002. A simple route to the synthesis of core/shell nanoparticles of chalcogenides. Chem. Mater. 14, 2004–2010.

Mallick, K., Witcomb, M.J., Scurrell, M.S., 2004. Polymer stabilized silver nanoparticles: a photochemical synthesis route. J. Mater. Sci. 39, 4459–4463.

Mano, P.M., Selvi, B.K., Paul, J.A.J., 2011. Green synthesis of silver nanoparticles from the leaf extracts of *Euphorbia hirta* and *Nerium indicum*. Dig. J. Nanomater. Biostruct. 6, 869–877.

Mariadoss, A.V.A., Saravanakumar, K., Sathiyaseelan, A., Wang, M.H., 2020. Preparation, characterization and anti-cancer activity of graphene oxide-silver nanocomposite. J. Photochem. Photobiol. B Biol. 210, 111984.

Mavani, K., Shah, M., 2013. Synthesis of silver nanoparticles by using sodium borohydride as a reducing agent. Int. J. Eng. Res. Technol. 2, 1–5.

Mehr, F.P., Khanjani, M., Vatani, P., 2015. Synthesis of Nano-Ag particles using sodium borohydride. Orient. J. Chem. 31, 1831–1833.

Moeng, G., Motitswe, O., Fayemi, E., 2019. Characterization of green synthesized silver nanoparticles doped in polyacrylonitrile nanofibers. Am. J. Nanosci. Nanotechnol. Res. 7, 32–48.

Morones, J.L., Elichiguerra, A., Camacho, K., Holt, J.B., Kouri, J.T., Ramirez, M.Y., 2005. The bactericidal effect of silver nanoparticles. Nanotechnology 16 (10), 2346–2353.

Muthulakshmi, L., Rajini, N., Rajalu, A.V., Siengchin, S., Kathiresan, T., 2017. Synthesis and characterization of cellulose/silver nanocomposites from bioflocculant reducing agent. Int. J. Biol. Macromol. 103, 1113–1120.

Nair, B., Pradeep, T., 2002. Coalescence of nanoclusters and formation of submicron crystallites assisted by *Lactobacillus* strains. Cryst. Growth Des. 2 (4), 293–298.

Nayak, S.P., Ramamurthy, S.S., Kumar, J.K., 2020. Green synthesis of silver nanoparticles decorated reduced graphene oxide nanocomposite as an electrocatalytic platform for the simultaneous detection of dopamine and uric acid. Mater. Chem. Phys. 252, 123302.

Nishimoto, M., Abeb, S., Yonezawa, T., 2018. Preparation of Ag nanoparticles using hydrogen peroxide as a reducing agent. New J. Chem. 42, 14493–14501.

Noohpisheh, Z., Amiri, H., Farhadi, S., Mohammadi-gholami, A., 2020. Green synthesis of Ag-ZnO nanocomposites using *Trigonella foenum-graecum* leaf extract and their antibacterial, antifungal, antioxidant and photocatalytic properties. Spectrochim. Acta A Mol. Biomol. Spectrosc. 118595.

Nurani, S.J., Saha, C.K., Khan, MAR., Sunny, S.M.H., 2015. Silver nanoparticles synthesis, properties, applications and future perspectives: a short review. IOSR JEEE 10, 117–126.

Oliveira, M., Ugarte, D., Zanchet, D., Zarbin, A., 2005. Influence of synthetic parameters on the size, structure, and stability of dodecanethiol-stabilized silver nanoparticles. J. Colloid Interface Sci. 292, 429–435.

Omidi, S., Sedaghat, S., Tahvildari, K., Derakhshi, P., Motiee, F., 2018. Biosynthesis of silver nanocomposite with tarragon leaf extract and assessment of antibacterial activity. J. Nanostruct. Chem. 8, 171–178.

Palem, R.R., Ganesh, S.D., Kronekova, Z., Sláviková, M., Saha, N., Saha, P., 2018. Green synthesis of silver nanoparticles and biopolymer nanocomposites: a comparative study on physico-chemical, antimicrobial and anticancer activity. Bull. Mater. Sci. 41 (2), 1–11.

Pandey, S., Goswami, G.K., Nanda, K.K., 2012. Green synthesis of biopolymer–silver nanoparticle nanocomposite: an optical sensor for ammonia detection. Int. J. Biol. Macromol. 51, 583–589.

Pandey, S., Ramontja, J., 2016. Guar gum-grafted poly(acrylonitrile)-templated silica xerogel: nanoengineered material for lead ion removal. J. Anal. Sci. Technol. 7, 24 1-24.

Parmar, A., Kaur, G., Kapil, S., Sharma, V., Sachar, S., Sandhir, R., Sharma, S., 2019. Green chemistry mediated synthesis of PLGA-silver nanocomposites for antibacterial synergy: introspection of formulation parameters on structural and bactericidal aspects. React. Funct. Polym. 141, 68–81.

Pham, N.B.T., Le, V.K.T., Bui, T.T.T., Phan, N.G.L., Tran, Q.V., Nguyen, M.L., Dang, V.Q., Nguyen, T.T., Vo, T.N.H., Tran, C.K., 2021. Improved synthesis of Ag/SiO_2 colloidal nanocomposites and their antibacterial activity against *Ralstonia solanacearum* 15. J. Nanosci. Nanotechnol. 21 (3), 1598–1605.

Philip, D., 2010b. Honey mediated green synthesis of silver nanoparticles. Spectrochim. Acta A Mol. Biomol. Spectrosc. 75, 1078–1081.

Philip, D., 2010a. Green synthesis of gold and silver nanoparticles using *Hibiscus rosasinensis*. Physica E 42, 1417–1424.

Phuruangrat, A., Keereesaensuk, P.O., Karthik, K., Dumrongrojthanath, P., Ekthammathat, N., Thongtem, S., Thongtem, T., 2020. Synthesis of Ag/Bi_2MoO_6 nanocomposites using nabh4 as reducing agent for enhanced visible-light-driven photocatalysis of Rhodamine B. J. Inorg. Organomet. Polym. Mater. 30, 322–329.

Ponnamma, D., Erturk, A., Parangusan, H., Deshmukh, K., Ahamed, M.B., Al-Maadeed, M.A., 2018. Stretchable quaternary phasic PVDF-HFP nanocomposite films containing graphene-titania-SrTiO$_3$ for mechanical energy harvesting. Emergent Mater. 1 (1–2), 55–65.

Potbhare, A.K., Umekar, M.S., Chouke, P.B., Bagade, M.B., Aziz, S.T., Abdala, A.A., Chaudhary, R.G., 2020. Bioinspired graphene-based silver nanoparticles: fabrication, characterization and antibacterial activity. Mater. Today Proceed. 29, 720–725.

Pozdnyakov, A.S., Emel'yanov, A.I., Kuznetsova, N.P., Ermakova, T.G., Korzhova, S.A., Khutsishvili, S.S., Vakul'skaya, T.I., Prozorova, G.F., 2019. Synthesis and characterization of silver-containing nanocomposites based on 1-Vinyl-1,2,4-triazole and acrylonitrile copolymer. J. Nanomater. 1–7.

Prabhu, S., Poulose, E.K., 2012. Silver nanoparticles: mechanism of antimicrobial action, synthesis, medical applications, and toxicity effects. Int. Nano Lett. 2 (1), 32.

Prathna, T.C., Chandrasekaran, N., Raichur, A.M., Mukherjee, A., 2011. Biomimetic synthesis of silver nanoparticles by Citrus limon (lemon) aqueous extract and theoretical prediction of particle size. Colloids Surf B Biointerfaces 82, 152–159.

Praus, P., Turicováa, M., Klementováb, M., 2009. Preparation of silver-montmorillonite nanocomposites by reduction with formaldehyde and borohydride. J. Braz. Chem. Soc. 20, 1351–1357.

Tomke, P.D., Rathod, V.K., 2020. Facile fabrication of silver on magnetic nanocomposite (Fe$_3$O$_4$@chitosan–AgNP nanocomposite) for catalytic reduction of anthropogenic pollutant and agricultural pathogens. Int. J. Biol. Macromol. 149.

Rao, A.V., Ashok, B., Umamahesh, M., Chandrasekhar, V., Subbareddy, G.V., Rajulu, A.V., 2018. Preparation and properties of silver nanocomposite fabrics within situ-generated silver nanoparticles using red sanders powder extract as reducing agent. Int. J. Polym. Anal. Charact. 23, 493–501.

Rashid, M.U., Bhuiyan, M.K.H., Quayum, M.E., 2013. Synthesis of silver nanoparticles (Ag-NPs) and their uses for quantitative analysis of vitamin c tablets. Dhaka Univ. J. Pharm. Sci. 12, 29–33.

Renu, S., Shivashangari, K.S., Ravikumar, V., 2020. Incorporated plant extract fabricated silver/poly-D, l-lactide-co-glycolide nanocomposites for antimicrobial based wound healing. Spectrochim. Acta A Mol. Biomol. Spectrosc. 228, 117673.

Rodríguez-León, E., Iñiguez-Palomares, R., Navarro, R.E., Herrera-Urbina, R., Tánori, J., Iñiguez-Palomares, C., Maldonado, A., 2013. Synthesis of silver nanoparticles using reducing agents obtained from natural sources (*Rumex hymenosepalus* extracts). Nanoscale Res. Lett. 8, 318.

Roseline, T.A., Murugan, M., Sudhakar, M.P., Arunkumar, K., 2019. Nanopesticidal potential of silver nanocomposites synthesized from the aqueous extracts of red seaweeds. Environ. Technol. Innov. 13, 82–93.

Russell, A.D., Hugo, W.B., 1994. Antimicrobial activity and action of silver. Prog. Med. Chem. 31 (C), 351–370.

Sabavath, G., Rahman, M., Sarmah, T., Dihingia, P., Srivastava, D.N., Sharma, S., Pandey, L., Kakati, M., 2020. Single-step, DC thermal plasma assisted synthesis of Ag-C nanocomposites with less than ten nano-meter sizes for antibacterial applications. J. Phys. D. Appl. Phys. 53, 365201.

Salazar, P., Fernández, I., Rodríguez, M.C., Hernández-Creus, A., González-Mora, J.L., 2019. One-step green synthesis of silver nanoparticle-modified reduced graphene oxide nanocomposite for H$_2$O$_2$ sensing applications. J. Electroanal. Chem. 855, 113638.

Saleh, H.H., El-Hadedy, D.E., Meligi, G.A., Afify.T.A., 2013. Synthesis, characterization and antibacterial activity of Ag/PVA nanocomposite. J. Sci. Res. 5, 151–160.

Sampaio, S., Viana, J.C., 2018. Production of silver nanoparticles by green synthesis using artichoke (*Cynara scolymus* L.) aqueous extract and measurement of their electrical conductivity. Adv. Nat. Sci. Nanosci. Nanotechnol. 9, 045002 (9 pp).

Sapsford, K.E., Tyner, K.M., Dair, B.J., Deschamps, J.R., Medintz, I.L., 2011. Analyzing nanomaterial bioconjugates: a review of current and emerging purification and characterization techniques. Anal. Chem. 83, 4453–4488.

Sarkandi, A.F., Montazer, M., Harifi, T., Mahmoudi Rad, M., 2021. Innovative preparation of bacterial cellulose/silver nanocomposite hydrogels: in situ green synthesis, characterization, and antibacterial properties. J Appl. Polym. Sci. *138* (6), 49824.

Sergeev, B.M., Kasaikin, V.A., Litmanovich, E.A., Sergeev, G.B., Prusov, A.N., 1999. Cryochemical synthesis and properties of silver nanoparticle dispersions stabilised by poly (2-dimethylaminoethyl methacrylate). Mendeleev Commun. 4, 130–132.

Sganzerla, W.G., Longo, M., de Oliveira, J.L., da Rosa, C.G., de Lima Veeck, A.P., de Aquino, R.S., Masiero, A.V., Bertoldi, F.C., Barreto, P.L.M., Nunes, M.R., 2020. Nanocomposite poly (ethylene oxide) films functionalized with silver nanoparticles synthesized with *Acca sellowiana* extracts. Colloids Surf. A Physicochem. Eng. Asp. 125125.

Shankar, S., Rhim, J.W., 2015. Amino acid-mediated synthesis of silver nanoparticles and preparation of antimicrobial agar/silver nanoparticles composite films. Carbohydr. Polym. 130, 353–363.

Shankar, S.S., Ahmad, A., Sastry, M., 2003. Geranium leaf assisted biosynthesis of silver nanoparticles. Biotechnol. Prog. 19, 1627–1631.

Sharma, V.K., Yngard, R.A., Lin, Y., 2009. Silver nanoparticles: green synthesis and their antimicrobial activities. Adv. Colloid Surf. Interfaces 145, 83.

Shimoga, G., Shin, E.J., Sang-Youn Kim, S.Y., 2019. Silver nanoparticles incorporated PVC films: evaluation of structural, thermal, dielectric and catalytic properties. Polímeros. 29. , e2019032.

Srikhao, N., Kasemsiri, P., Ounkaew, A., Lorwanishpaisarn, N., Okhawilai, M., Pongsa, U., Hiziroglu, S., Chindaprasirt, P., 2021. Bioactive nanocomposite film based on cassava starch/polyvinyl alcohol containing green synthesized silver nanoparticles. J. Polym. Environ. 29 (2), 672–684.

Sun, L., Shao, R., Tang, L., Chen, Z., 2014. Synthesis of Ag/ZnO nanocomposite with excellent photocatalytic performance via a facile Sol-Gel method. Adv. Mater. Res. 875-877, 251–256.

Susilowati, E., Triyono, T., Santosa, S.J., IndrianaKartini, I., 2015. Synthesis of silver-chitosan nanocomposites colloidal by glucose as reducing agent. Indones. J. Chem. 15, 29–35.

Talebi, J., Halladj, R., Askari, S., 2010. Sonochemical synthesis of silver nanoparticles in Y-zeolite substrate. J. Mater. Sci. 45, 3318–3324.

Tao, A., Sinsermsuksakul, P., Yang, P., 2006. Polyhedral silver nanocrystals with distinct scattering signatures. Angew. Chem. Int. Ed. 45, 4597–4601.

Tauanov, Z., Lee, J., Inglezakis, V.J., 2020. Mercury reduction and chemisorption on the surface of synthetic zeolite-silver nanocomposites: equilibrium studies and mechanisms. J. Mol. Liq. 305, 112825.

Tauanov, Z., Tsakiridis, P.E., Mikhalovsky, S.V., Inglezakis, V.J., 2018. Synthetic coalfly ash-derived zeolites doped with silver nanoparticles for mercury (II) removal from water. J. Environ. Manag. 224, 164–171.

Thakkar, K.N., Mhatre, S.S., Parikh, R.Y., 2010. Biological synthesis of metallic nanoparticles. Nanomedicine 6, 257–262.

Thangaswamy, S.J.K., Mir, M.A., Muthu, A., 2021. Green synthesis of mono and bimetallic alloy nanoparticles of gold and silver using aqueous extract of *Chlorella acidophile* for potential applications in sensors. Prep. Biochem. Biotechnol. 1–10.

Tork, M.B., Nejad, N.H., Ghalehbagh, S., Bashari, A., Shakeri-Zadeh, A., Kamrava, S.K., 2014. In situ green synthesis of silver nanoparticles/chitosan/polyvinyl alcohol/polyethene glycol hydrogel nanocomposite for novel finishing of nasal tampons. J. Ind. Text. 45, 1399–1416.

Valera-Zaragoza, M., Huerta-Heredia, A.A., Peña-Rico, M.A., Juarez-Arellano, E.A., Navarro-Mtz, A.K., Ramírez-Vargas, E., Sánchez-Valdes, S., 2021. Morphological, structural and cytotoxic behavior of starch/silver nanocomposites with synthesized silver nanoparticles using Stevia rebaudiana extracts. Polym. Bull. 78 (3), 1683–1701.

Vigneshwaran, N., Nachane, R.P., Balasubramanya, R.H., Varadarajan, P.V., 2006. A novel one-pot 'green' synthesis of stable silver nanoparticles using soluble starch. Carbohydr. Res. 341, 2012–2018.

Wiley, B., Sun, Y., Mayers, B., Xi, Y., 2005. Shape-controlled synthesis of metal nanostructures: the case of silver. Chem. Eur. J. 11, 454–463.

Yang, C.-H., Wang, L.-S., Chen, S.-Y., Huang, M.-C., Li, Y.-H., Lin, Y.-C., Chen, P.-F., Shaw, J.-F., Huang, K.-S., 2016. Microfluidic assisted synthesis of silver nanoparticle–chitosan composite microparticles for antibacterial applications. Int. J. Pharm. 510, 493–500.

Yari, A., Yari, M., Sedaghat, S., Delbari, A.S., 2021. Facile green preparation of nano-scale silver particles using Chenopodium botrys water extract for the removal of dyes from aqueous solution. J. Nanostruct. Chem. 1–13.

Yu, W., Li, X., He, J., Chen, Y., Qi, L., Yuan, P., Ou, K., Liu, F., Zhou, Y., Qin, X., 2021. Graphene oxide-silver nanocomposites embedded nanofiber core-spun yarns for durable antibacterial textiles. J. Colloid Interface Sci. 584, 164–173.

Yuan, Z., Yang, H., Xu, P., Li, C., Jian, J., Zeng, J., Zeng, L., Sui, Y., Zhou, H., 2021. Facile in situ synthesis of silver nanocomposites based on cellulosic paper for photocatalytic applications. Environ. Sci. Pollut. Res. 28 (6), 6411–6421.

Zhang, Q., Li, N., Goebl, J., Lu, Z.D., Yin, Y.D., 2011. A systematic study of the synthesis of silver nanoplates: is citrate a "magic" reagent? J. Am. Chem. Soc. 133, 18931–18939.

Zhu, J.J., Liao, X.H., Zhao, X.N., Chen, H.Y., 2001. Preparation of silver nanorods by electrochemical methods. Mater. Lett. 49, 91–95.

Zienkiewicz-Strzałka, M., Deryło-Marczewska, A., 2020. Small AgNP in the biopolymer nanocomposite system. Int. J. Mol. Sci. 21 (24), 9388.

CHAPTER 4

Core-shell silver nanoparticles: Synthesis, characterization, and applications

Anbazhagan Sathiyaseelan[a], Kandasamy Saravanakumar[a], Murugesan Manikandan[b], Azeez Shajahan[c], Arokia Vijaya Anand Mariadoss[a], and Myeong-Hyeon Wang[a]

[a]*Department of Bio-Health Convergence, Kangwon National University, Chuncheon, Republic of Korea,* [b]*Department of Biomedical Science, School of Biotechnology and Genetic Engineering, Bharathidasan University, Tiruchirappalli, Tamil Nadu, India,* [c]*Department of Biotechnology, PRIST (Deemed University), Puducherry Campus, Puducherry, Tamil Nadu, India*

1 Introduction

The development of size-tunable smart nanomaterials with promising chemical properties are play a significant role in electronics, pharmaceutics, biomedicine, food, and agriculture industrial applications (Chatterjee et al., 2014; Patra et al., 2018; Yaqoob et al., 2020a,b). Metal nanoparticles (aluminum (Al), gold (Au), platinum (Pt), palladium (Pd), iron (Fe), silica (Si), silver (Ag), copper (Cu), cerium (Ce), manganese (Mn), nickel (Ni), titanium dioxide (TiO_2), and zinc (Zn)) form a strong metallic bonding and covalent bonding between the metal-metal and other organic materials, respectively (Berry and Lu, 2017; Yaqoob et al., 2020b). Among the metals, silver (Ag) is desired to manipulate size, shape, composition, structure, and surface properties subsequently to various forms, such as nanoparticles, nanoshells, nanotubes, nanogels, and nanofibers (Yaqoob et al., 2020a). Moreover, AgNPs are hetero-structured with multifunctional properties (catalysis, sensing, optoelectronics, medicine), that attractive behavior enable the metal to manipulate for improved applications (Cao et al., 2001; Mott et al., 2012).

Stabilized nanomaterials can be obtained by tweaking chemical, electrochemical, and photochemical reduction thereby influencing the physical, chemical, and morphological properties of nanomaterials (Jamkhande et al., 2019). In general, noble metallic nanoparticles are proven as a targeted delivery tool at a cellular level via the blood-brain barrier. However, numerous studies have indicated that AgNPs have high cytotoxic properties, but their cytotoxicity can be reduced through surface modifications using the secondary metabolites, proteins, enzymes, oligopeptides, and some other biomaterials (Gupta and Chhibber, 2019; Rai et al., 2015; Sethuram et al., 2019). The stability of these materials can be accompanied by covalent and noncovalent interactions. These modifications replace the toxicity of the materials that possess biological properties exhibiting synergistic activity, thereby frequently leading to several applications. Nanomaterials covered in single or multiple layers can be defined as core-shell nanoparticles (CSNPs). The CSNPs show principal advantages in the form of promising physicochemical properties such as stability, and preserve the chemical entity of the material (Fahmy et al., 2019; Jamkhande et al., 2019; Kalambate et al., 2019). The various physicochemical and characterization methods were employed to synthesis of CSNPs. Further, CSNPs are documented for various biomedical applications such as site-specific gene and drug delivery, bioimaging, and biosensors. Henceforth this chapter emphasizes Ag-based core-shell nanoparticles synthesis, characterization, and their biomedical applications.

2 Synthesis of core-shell AgNPs

The preparation of core-shell AgNPs can be achieved by either single, dual, and multiple steps using the methods of reduction, precipitation, deposition, or fabrication. In the case of metal oxide nanoparticles, the synthesis process is accomplished by solvothermal and sol-gel methods (Kalambate et al., 2019). AgNPs are functionalized in the core or shell material that is entirely depends on the applications of core-shell silver nanoparticles (CS-AgNPs; Fig. 1). Silver nitrate ($AgNO_3$) is used as a precursor for the synthesis of Ag core nanoparticles by the reduction method (Lee and Jun, 2019). Nanoparticles are synthesized through top-down or bottom-up approaches. Particularly, top-down approaches have been used widely for synthesis of nanoparticles and established by physical, chemical, and biological methods (Jamkhande et al., 2019). A physical method such as electrochemical, ultra-sonication, laser ablation, irradiation, or evaporation-condensation is used to synthesize AgNPs and

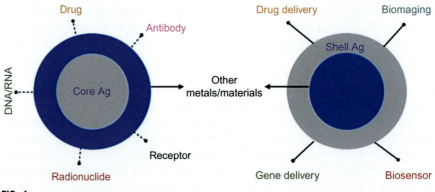

FIG. 1

Core/shell AgNPs and their biomedical application (created with BioRender.com).

CS-AgNPs (Syafiuddin et al., 2017). In the chemical methods, the inorganic and organic reducing agents such as sodium hydroxide (NaOH), sodium borohydride (NaBH$_4$), trisodium citrate (Na$_3$C$_6$H$_5$O$_7$), hydroxylamine hydrochloride, and sodium acetate are used for the preparation of AgNPs and CS-AgNPs. In the case of biological synthesis, the microbes (bacteria, algae, fungi) and plant extracts or molecules are employed as reduction or/and capping agents for the synthesis of AgNPs (Anbazhagan et al., 2017; Iravani et al., 2014; Patel et al., 2015; Sathiyaseelan et al., 2017; Syafiuddin et al., 2017; Taghizadeh et al., 2019).

Polymers are heavily utilized for the development of CSNPs, which are fabricated as the inner core or outer shell material (Huang et al., 2021; Sathiyaseelan et al., 2020; Soto-Quintero et al., 2019; Zhao et al., 2021). The polymeric substances are used to entrap the different metals, therapeutics, and theragnostic agents to enable improvements in medical applications. For example, antimicrobial peptide surface-functionalized CS-AgNPs have been synthesized in a single-step reduction of AgNO$_3$ using glutathione (GSH) and sodium borohydride (NaBH$_4$) for enhanced antibacterial activity (Gao et al., 2020). The Ag/cur (curcumin) and thermoresponsive monomer 2-(2-methoxyethoxy)ethyl methacrylate (MEO$_2$MA) core-shell hybrid NPs were synthesized by a one-pot two-step method with synergistic antibacterial, antiviral, and antioxidant activities (Soto-Quintero et al., 2019). Furthermore, Ag/SiO$_2$ CSNPs were synthesized by reduction of AgNO$_3$ using trisodium citrate (Na$_3$C$_6$H$_5$O$_7$) as the Ag core material and SiO$_2$ was synthesized by the reduction of tetraethyl orthosilicate using ethanol and ammonia along with Na$_3$C$_6$H$_5$O$_7$ as the shell material for the detection of anticancer drug valrubicin on plasmonic platforms (Synak et al., 2019). Ag/AIEgens (aggregation-induced emission luminogens) core-shell nanoparticles (AACSN) were synthesized by reduction of AgNO$_3$ and TPE-M$_2$OH using NaOH for multimodality cancer diagnosis (Fluorescence (FL), computed tomography (CT), radiation therapy (RT), and photothermal (PT), photoacoustic (PA), and synergistic therapy (He et al., 2020) (Fig. 2).

However, the usage and disposable vast number of dyes, therapeutics, and other organic and inorganic toxic materials were causing environmental toxicity. Nevertheless, the advancement of nanomaterials preparation also provided the solution for those challenges. For instance, Ag core capped inner and outer shell TiO$_2$ nanoparticles (TiO$_2$-Ag-TiO$_2$ CSNPs) are prepared by three steps to degrade tetracycline (Zhao et al., 2019). In the first step, cationic polystyrene spheres supporting TiO$_2$ are synthesized using tetrabutyl titanate (TBT) under calcination at 300°C. Then the second step, TiO$_2$-Ag spheres are prepared by AgNO$_3$, glycerol, and hallow titania (HT) under N$_2$ purging and photoirradiation. In the third step, allyltrimethoxysilane (ATS) used to fabricate TiO$_2$-Ag-TiO$_2$ under an aqueous medium and then calcinated at 450°C and reduced by NaBH$_4$ (Zhao et al., 2019). These TiO$_2$-Ag-TiO$_2$ CSNPs have reportedly degraded the antibiotic tetracycline through the photocatalysis process (Fig. 3).

In addition, bimetallic AgNPs and AuNCs (gold nanoclusters) synthesized by the two steps synthesis method for the improved catalytic application. In the first step, thiocholine (TCH) stabilized AgNPs were synthesized by the reduction of AgNO$_3$ by

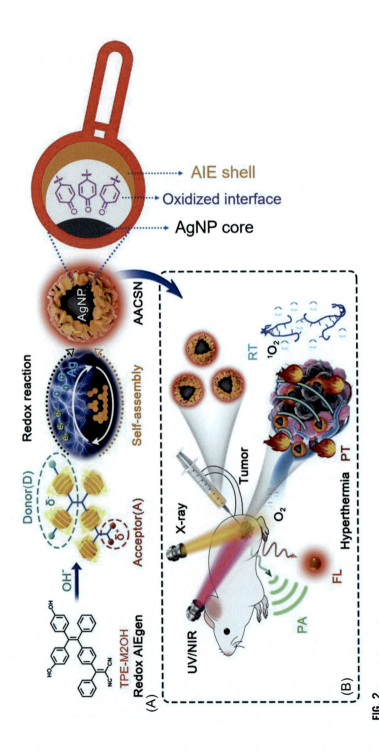

FIG. 2

Schematic illustrations of Ag/aggregation-induced emission luminogens core-shell nanoparticles (AACSN) synthesis for multifunctional cancer imaging and therapy. (A) AACSN synthesis via the redox reaction between Ag^+ and redox AIEgen and the self-assembly of AIEgens with D-A structure. The oxidized interface layer with quinone structures between the AgNP core and AIE shell was generated in situ through the redox reaction. This new functionality is assigned to the origin of PT and PA properties. (B) AACSNs were applied for five-modality tumor imaging (FL, PT, and PA) and synergistic therapy (PTT and RT).

Reprinted from He, X., Peng, C., Qiang, S., Xiong, L.H., Zhao, Z., Wang, Z., Kwok, R.T.K., Lam, J.W.Y., Ma, N., Tang, B.Z., 2020. Less is more: silver-AIE core@shell nanoparticles for multimodality cancer imaging and synergistic therapy. Biomaterials 238, 119834 with permission from MDPI Publisher, Licensee MDPI, Basel, Switzerland. This article is an open-access article distributed under the terms and conditions of the Creative Commons Attribution (CC BY) license (http://creativecommons.org/licenses/by/4.0).

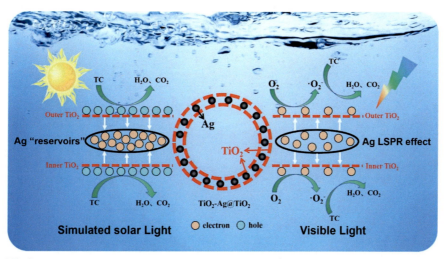

FIG. 3

Photocatalytic degradation mechanism of TiO_2-Ag-TiO_2 CSNPs.

Reprinted from Zhao, S., Chen, J., Liu, Y., Jiang, Y., Jiang, C., Yin, Z., Xiao, Y., Cao, S., 2019. Silver nanoparticles confined in shell-in-shell hollow TiO_2 manifesting efficiently photocatalytic activity and stability. Chem. Eng. J. 367, 249–259 with permission from Elsevier.

$NaBH_4$ and addition of acetylthiocholine (ATCh) with acetylcholinesterase (AchE). These AgNPs-AuNCs reduced the methyl orange (MO) and potassium ferricyanide $K_3[Fe(CN)_6]$ through the catalytic process (Vinotha Alex et al., 2020). Table 1 summarizes the various methods, techniques, and applications of CS-AgNPs.

3 Characterization of core-shell silver nanoparticles

Physicochemical characterization is an important study to validate the properties of nanomaterials (Lin et al., 2014). Furthermore, the characterization of CS-AgNPs offers knowledge of their environmental effects and their biological benefits (Syafiuddin et al., 2017). Moreover, evalualting the physicochemical properties (structure, size, optical, and crystal nature), and electrochemical properties of CS-AgNPs are the essential factors to determine the different applications. Nanoparticles including CSNPs are characterized and quantified by diverse of analytical techniques including UV-visible spectrophotometer, UV-visible NIR (near infrared) spectrophotometer, FTIR (Fourier transform infrared spectroscopy), XRD (X-ray diffractometer), SEM (scanning electron microscopy), TEM (transmission electron microscopy), Raman spectroscopy, AFM (atomic force microscopy), DLS (dynamic light scattering), zeta potential, TGA (thermogravimetric analysis), XPS (X-ray photoelectron spectroscopy), ICP-MS (inductively coupled plasma mass spectrometry),

Table 1 Some examples of the synthesis of core-shell AgNPs.

Synthesis of core-shell AgNPs

Core/shell	Core Method	Core Reagents	Core	Shell Method	Shell Reagents	Reference
Ag/P-13 (antimicrobial peptide)	Reduction by reduced glutathione (GSH) and sodium borohydride (NaBH4)	$AgNO_3$	Ag	Reduction by $NaBH_4$	P-13	Gao et al. (2020)
Ag/AIE (aggregation-induced emission)	Reduction by sodium hydroxide (NaOH)	$AgNO_3$	Ag	Reduction by NaOH	AIE(TPE-M$_2$OH)	He et al. (2020)
Ag/cur (curcumin)-G (P(MEO$_2$MA)	Oxidation and reduction by sodium citrate ($Na_3C_6H_5O_7$)	$AgNO_3$	Ag	Fabricated by cross-linking of MEO$_2$MA monomer using TEGDMA and SDS	Curcumin	Soto-Quintero et al. (2019)
ZVI (zero valent iron)/AgNPs	Reduction by cypress leaves extract	$AgNO_3$	Ag	Reduction by cypress leaves extract and Ag	$FeCl_3 \cdot 6H_2O$	Taghizadeh et al. (2019)
Ag/SiO$_2$	Reduction by ($Na_3C_6H_5O_7$)	$AgNO_3$	Ag	Reduction by $Na_3C_6H_5O_7$ with ethanol, water, and ammonia	Tetraethyl orthosilicate (TEOS)	Synak et al. (2019)
Ag/SiO$_2$/Ag	Fabricated and reduced by PVP and EG under 120°C	$AgNO_3$	Ag	Reduction by ethanol, water, and ammonia	TEOS	Manivannan et al. (2019)

Ag/SiO$_2$	Reduction by (Na$_3$C$_6$H$_5$O$_7$)	AgNO$_3$, aqueous ammonia, and PVP	Ag	Reduction by ethanol, water, and dimethylamine (DMA)	TEOS	Huang et al. (2021)
SiO$_2$/Ag	Reduction by ethanol, water, and ammonia	TEOS	SiO$_2$	Fabrication of PVP and reduction by heating at 75°C	[Ag(NH$_3$)$_2$]$^+$	Nur Kamilah et al. (2019)
AgNPs/MSNs (mesoporous silica nanoparticles)	Reduction by NaBH$_4$	AgNO$_3$ and CTAB		Fabrication and reduction by EG and TEOA at 60°C	TEOS	Xu et al. (2019)
Ag/ZrO$_2$	Reduction by Na$_3$C$_6$H$_5$O$_7$	AgNO$_3$	Ag	Reduction by (3-mercaptoprophyl) trimethoxysilane with L-arginine and cyclohexane	Zirconium (IV) propoxide	Zhou et al. (2020)
PEG-BSA-Ag/ICG	Reduction by NaBH$_4$	AgNO$_3$	Ag	Fabricated by PEGylation	BSA/ICG	Park et al. (2020)
FA-RBCm/Ag	—	Ag	Ag	Fabricated by PEGylation	FA (folic acid)-functionalized RBCm (red blood cell membrane)	Zhao et al. (2021)
Ag/PPy	Solid-state reaction addition with pyrrole	AgNO$_3$ and PVB	Ag	Solid-state reaction addition with pyrrole	Pyrrole (Py)	Ali et al. (2019)
Ag/TiO$_2$	Reduction by hydroxylamine hydrochloride	AgNO$_3$	Ag	Fabrication by heating at 150°C	Titanium (IV) butoxide (TBT)	Bartosewicz et al. (2020)

Continued

Table 1 Some examples of the synthesis of core-shell AgNPs—*con't*

Synthesis of core-shell AgNPs

Core/shell	Core Method	Core Reagents	Core	Shell Method	Shell Reagents	Reference
TiO_2-Ag-TiO_2	Reduction by irradiation and fabrication with HT (hallow titania)	$AgNO_3$	Ag	TBT reduction by calcination at 450°C and $NaBH_4$; inner shell TiO_2 fabricated with cationic polystyrene spheres	TBT	Zhao et al. (2019)
Ag/Ni	—	Ag	Ag	Reduction by NH_4 and H_3BO_3	$NiSO_4$	Kang et al. (2020)
Ag/Ni	Reduction by oleyl amine and octadecene at 230°C	Ag	$AgNO_3$	Reduction by oleyl amine and octadecene	$Ni(acac)_2$	Vykoukal et al. (2020)
Ag/Fe_3O_4	Reduction by sodium acetate	$AgNO_3$ and $Fe(NO_3)_3 \cdot 9H_2O$	Ag	Fabrication by heating at 200°C	—	Mazhani et al. (2020)
Ag/CuO	Reduction by carbon dots (CDs)	$AgNO_3$	Ag	Reduction CDs	$Cu(NO_3)_2$	Im et al. (2020)
AgNC (nanocubes)/Cu_2O	Fabrication and reduction by poly(vinyl) pyrrolidone, ethylene glycol (EG) at 150°C	$AgNO_3$, sodium sulfide (Na_2S), and EG	Ag	Reduction by NaOH and lactic acid	$Cu(NO_3)_2$	Gale-Mouldey et al. (2020)
Cu/Ag-rGO (reduced graphene oxide)	Reduction by $Na_3C_6H_5O_7$ and $NaBH_4$	$CuSO_4$ and $NABH_4$	Cu and Ag	rGO prepared by Hummers' method and fabricated with Cu/Ag	graphite powder	Li and Yao (2020)

Ag/MSNs	Reduction by $NaBH_4$	$AgNO_3$ and CTAB		Reduction by ethanol, water, and ammonia	TEOS and NH_4NO_3	Montalvo-Quirós et al. (2021)
Ag/Au/Pd	Reduction and fabrication by ascorbic acid, sodium alginate, and sodium carbonate	$AgNO_3$, $AuCl_3$, and $PdCl_2$	Ag	Reduction and fabrication by ascorbic acid, sodium alginate, and sodium carbonate	$AgNO_3$, $AuCl_3$, and $PdCl_2$	Ahmed and Emam (2020)
PLGA/Ag/Fe_3O_4	Synthesis of Fe by the reducing agent citric acid and Ag by $NaBH_4$	$FeSO_4 \cdot 7H_2O$, $Fe(NO_3)_3 \cdot 9H_2O$, and $AgNO_3$	Ag and Fe_3O_4	Fabricated by PVP, PEG, and PLGsA	Amphotericin	Sadat Akhavi and Moradi Dehaghi (2020)
Fe_3O_4/Ag	Reduction by NaOH	$FeCl_2$ and Fe_2Cl_3	Fe_3O_4	Reduction by honey	$AgNO_3$	Ramírez-Acosta et al. (2020)
Au/Ag	Reduction by $Na_3C_6H_5O_7$	$HAuCl_4$	Au	Polydopamine (PDA)	$AgNO_3$	Yilmaz and Yilmaz (2020)
Au/Ag	Reduction by $Na_3C_6H_5O_7$	$HAuCl_4$	Au	Reduction by NaOH and L-ascorbic acid	$AgNO_3$	Kovács et al. (2020)
AuNCs (nanoclusters)-AgNPs	Reduction by $NaBH_4$ and fabricated with thiocholine (TCh)	$AgNO_3$ and acetylcholinesterase (AChE)	Ag	Reduction by AChE and acetylthiocholine (ATCh)	Hydrogen tetrachloroaurate(III) trihydrate ($HAuCl_4 \cdot 3H_2O$)	Vinotha Alex et al. (2020)
Cu/Ag	Reduction by $NaBH_4$, PVP, and NaOH	$CuSO_4 \cdot 5H_2O$ and $Na_3C_6H_5O_7$	Cu	Reduction by NaOH and NH_4OH	$AgNO_3$	Huang et al. (2020)

and fluorescence spectrophotometer. Several instrumentation techniques are used to validate the properties of CSNPs because each instrument has its advantages and disadvantages. Hence, the required characterization techniques vary depending on the materials' properties and their applications. The detailed uses of the abovementioned instruments are given in Table 2.

Table 2 Physiochemical characterization of nanomaterials.

Characterization techniques	Investigation	Reference
UV-visible	Nanoparticles size, structural, optical properties, and surface chemistry	Gao et al. (2020), Ramírez-Acosta et al. (2020), Yilmaz and Yilmaz (2020)
UV-visible NIR	Optical properties of nanomaterials at NIR region	Das et al. (2020), Tomaszewska et al. (2013), Wu et al. (2019), Xia et al. (2020)
FTIR	Functional properties of nanoparticles	Park et al. (2020), Ramírez-Acosta et al. (2020), Sathiyaseelan et al. (2020); Shaheen et al. (2016)
XRD	Crystal structure, size	Lakshmanan et al. (2018), Mazhani et al. (2020), Sathiyaseelan et al. (2017)
SEM	Surface, size, shape, morphology, and particles distribution of the nanoparticles	Bartosewicz et al. (2020), Synak et al. (2019), Zhou et al. (2020)
TEM-EDX	External and internal structure, shape, particles distribution, and elemental composition	Gao et al. (2020), Park et al. (2020), Zhao et al. (2021), Zhou et al. (2020)
Raman spectroscopy	Nondestructive quantitative particle surface chemical analysis	Yilmaz and Yilmaz (2020), Zhou et al. (2020)
AFM	Morphology, size, surface structure, and mechanical properties	Yilmaz and Yilmaz (2020)
DLS	The hydrodynamic radius of the nanoparticles and particles distribution	Ramírez-Acosta et al. (2020)
Zeta potential	Nanoparticles charge correlated with drug interaction	Gao et al. (2020), Zhao et al. (2021)
TGA	Physical properties of the nanoparticles	Ramírez-Acosta et al. (2020), Sadat Akhavi and Moradi Dehaghi (2020)

Table 2 Physiochemical characterization of nanomaterials—cont'd

Characterization techniques	Investigation	Reference
XPS	Surface composition of the nanoparticles and their uniformity	Bartosewicz et al. (2020)
ICP-MS	Multielement analysis at lower limit detection	Aznar et al. (2017), Clark et al. (2019)
Fluorescence spectrophotometer	Fluorescence properties nanoparticles and quantification	Kang et al. (2020), Synak et al. (2019)

4 Application of silver nanoparticles

Due to nanotechnological advances, most of the materials including metals are used to prepare nanomaterials/nanoparticles by the top-down and bottom-up approaches. Metal-based nanoparticles have been substantially used in various applications, viz. environmental, textiles, catalysis, biomedical, health care, food, and agriculture (Chatterjee et al., 2014). However, this chapter focuses only on the biomedical application of CS-AgNPs.

4.1 Biomedical application of core-shell silver nanoparticles

CS-AgNPs are specially designed to augment biomedical applications such as antimicrobial (bacteria, fungi, and virus), wound healing, and anticancer through various mechanisms of action (Fig. 4). The following sections discuss details of each biomedical application of CS-AgNPs.

4.1.1 Antibacterial

AgNPs have attracted great potential in biomedical fields, for their high antimicrobial activity and cytotoxic effects (Sung et al., 2011; Singh et al., 2013). AgNPs showed significant antibacterial activity against human pathogenic bacteria, *Staphylococcus aureus*, *Bacillus subtilis*, *Streptococcus thermophilus*, *Escherichia coli*, and *Staphylococcus typhi* (El-Shanshoury et al., 2011; Marambio-Jones and Hoek, 2010). The major antimicrobial mechanisms of silver ions attained through the interaction of peptidoglycan cell wall, plasma membrane, and cytoplasmic DNA (Fig. 5), thereby inhibits the growth of microorganisms (Agnihotri et al., 2014; Dakal et al., 2016). Currently, AgNPs are used in burn ointment, antibacterial wear, and coatings for clinical units. Since ancient times, natural silver and mixes have been found to be effective against microorganisms including Gram-positive (*S. aureus*) and Gram-negative (*E. coli*), fungi, and viruses (Dakal et al., 2016; Qing et al., 2018). AgNPs could overcome bacterial resistance against antibiotics. In addition, Ag^+ interacts with DNA, and inversely disrupts the bacterial cell wall in contact with sulfur and

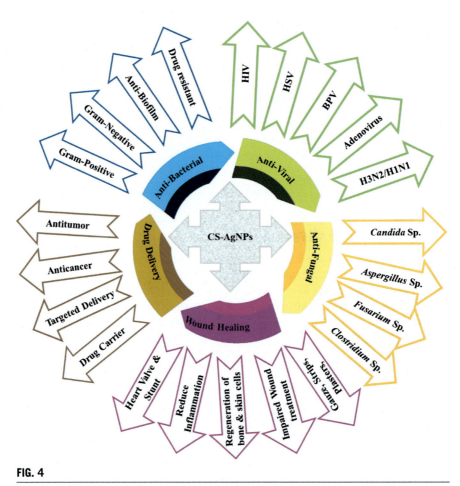

FIG. 4

Overview of the biomedical application of AgNPs and CS-AgNPs.

phosphorus, leading to inhibition of DNA replication. In addition, silver ion adhesion to the cell wall deactivates and produces reactive oxygen species (ROS) with the thiol organization of integral enzymes and directly destroys the mitochondrial membrane, leading to bacterial necrosis (El-Hussein and Hamblin, 2017; Flores-López et al., 2019; Jiao et al., 2014; Matsumura et al., 2003). The silver nanocrystalline chlorhexidine (Ag-CX) complex confirms a strong antibacterial effect against Gram-positive/negative and methicillin-resistant *S. aureus* (MRSA) strains (Khatoon et al., 2019; Percival et al., 2007). The efficacy of various antibiotics such as penicillin G, amoxicillin, erythromycin, clindamycin, and vancomycin against *E. coli* has been increased with the incorporation of AgNPs (Perveen et al., 2018; Kora and Rastogi, 2013; Shrivastava et al., 2007). Moreover, AgNPs exhibit potential exposure to antibiofilm activity formed by *P. aeruginosa* and *S. epidermidis*,

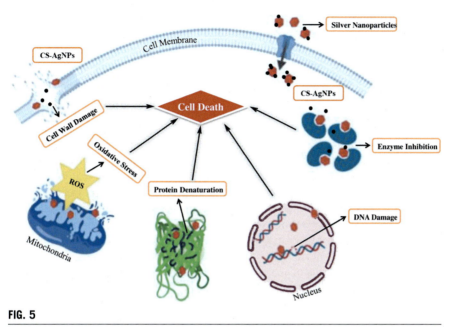

FIG. 5

Overview of cell death mechanisms of AgNPs and CS-AgNPs.

which is responsible for ocular-related infectious diseases including microbial keratitis (Salman et al., 2019).

4.1.2 Antiviral

Viral-mediated infections are normal and progressively more noticeable around the world; subsequently, it is vital to create antiviral agents. AgNPs have unique bacterial and viral interactions based on specific sizes and shapes. AgNPs have shown strong immunity against both HIV (human immunodeficiency virus) and hepatitis-B infection (HBV). AgNPs are perfectly linked to glycoprotein to HIV-1 and avoid a link between the virus and the host cells. In addition, following surgery, AgNPs may be used as anti-HIV-1 agents to prevent infections (Elechiguerra et al., 2005; Lara et al., 2010; Leu et al., 2008). Biosynthesized AgNPs inhibit adenovirus, herpes simplex virus (HSV) type 1 and 2 activations, respiratory syncytial virus (RSV), as well as human para-influenza viruses, type 3 (Chen et al., 2013; Galdiero et al., 2011; Leu et al., 2008). Studies found that tannic acid synthesizes AgNPs of various sizes, which can reduce HSV-2 infectiveness by direct interactions, blocking in-vivo and in-vitro adherence. Furthermore, an intranasal dose of AgNPs shows significantly improved survival, lower levels of lung viruses, and remarkable survival benefits following H3N2/H1N1 influenza infection. AgNPs are also integrated into ultrafiltration polysulfone membranes, and have significant antibacterial and antiviral properties in water treatment management (Tang et al., 2015; Zodrow et al., 2009).

4.1.3 Antifungal

Extreme fungal infections contribute significantly to an increased incidence of certain diseases and mortality in patients with immunodeficiency (Garraffo et al., 2017). One of the most common pathogens is the *Candida* species, which often causes nosocomial infections with a mortality rate of up to 40%. Several studies have shown that AgNPs are effective against *C. glabrata*, *C. albicans*, *C. krusei*, and 44 antifungal strains, including six fungal species. AgNPs were also successful against the laboratory cultivation of fungal organisms such as *Penicillium brevicompactum*, *Aspergillus fumigatus*, *Cladosporium cladosporoides*, *Cheetomium globosum*, and *Stachybotrys charterum* (Bocate et al., 2019; Khan et al., 2019; Ogar et al., 2015). AgNPs showed effective antifungal activity along with clinical isolate and "ATCC" strains of *Trichophyton mentagrophytes* and *Candida* species (Kim et al., 2011; Moazeni et al., 2012). Further studies revealed that AgNPs sepiolite fibers express significant antifungal action against *Issatchenkia orientalis* and an inert matrix containing AgNPs in the soda-lime glass which increases biocidal activity (Esteban-Tejeda et al., 2009). Furthermore, AgNPs have demonstrated strong antifungal activity with fluconazole against *Aspergillus niger*, *Fusarium semitectum*, *C. albicans*, and *Trichoderma* sp. (Gajbhiye et al., 2009; Majeed, 2017). The efficiency of AgNPs is measured in combination with "nystatin" (NYT) or "chlorhexidine" (CHX) against *C. albicans* and *C. glabrata* (Monteiro et al., 2013).

4.1.4 Wound healing

Prevention of microbial aggregation plays an important role in rapid wound healing treatment and in preventing secondary infections (Salas-Orozco et al., 2019). Recently, a topical metal dressing has been commonly used for the treatment of burns, wounds, dental composites, and chronic ulcers. Silver-based creams and ointments are available commercially for various medical applications and biomedical products for efficient wound repair. Silver-treated fabrics and surgical formulas showed improved wound healing properties both in vivo and in vitro (Li et al., 2012). Combining bioactive antimicrobial, antibacterial, and antiinflammatory properties together with AgNPs have great potential in chronic wound treatment, especially complication of diabetes mellitus, and reduce inflammation related to peritoneal catheters (Aflori, 2014; Alejandro Almonaci Hernández et al., 2017; Blanco et al., 2018). Researchers have found that the healing time of burns, Steven-Johnson syndrome, recurrent ulcers, and pemphigus could be shortened using AgNPs. Nowadays, AgNPs are included in tanning, gauze, strips, plasters, and many other creams and coatings (Balamurugan et al., 2017; Diniz et al., 2020). Chitosan-nanocrystalline silver dressing shows an excellent healing rate of up to 89%. Recent research has shown that hydrogels produced with β-chitin incorporated AgNPs that prevent the bacterial infection and improve the wound healing (Song et al., 2019). The path of research eventually leads to the creation of new medical materials with the incorporation of AgNPs (Ge et al., 2014; Marassi et al., 2018). The inclusion of AgNPs in poly-methyl methacrylate bone cement (PMMA) helped to establish knee-hip joints; additionally, AgNPs prevent microbial infection (Bistolfi et al., 2019;

Prokopovich et al., 2015). Furthermore, AgNPs incorporated into chitin scaffolds showed faster healing and improved aesthetic appearance. Another study revealed that poly(3-hydroxybutyrate-*co*-3-hydroxy valerate-PBHV/AgNPs hybrid nanofiber scaffold cultured osteoblasts (bone cells) and fibroblast (skin cells) were considerably enhanced the regeneration of bone and skin cells (Mehrabani et al., 2018; Xing et al., 2010). Interestingly, AgNPs can be used as coatings for devices like cardiac heart valves and stents (Andara et al., 2006).

4.1.5 Drug delivery system
Nanoparticles can overcome the struggle associated with the clinical availability of conventional small molecules or biomacromolecules. AgNPs have special high-performance vehicles for the delivery of treatment agents such as antisense oligonucleotides and other small molecules (Falzarano et al., 2014; Sun et al., 2017). In conjunction with the targeted treatment of tumors, biogenic synthesis of AgNPs may provide an alternative route for the successful therapeutic delivery of low-risk cancers (De Matteis et al., 2018; Raja et al., 2020). Drug imatinib-loaded green synthesized silver nanoparticles provided a stable delivering antitumor drug in breast cancer cells and facilitated controlled delivery in human myeloblastic leukemia cells and cervical cancer cells (Ghosn et al., 2019; Guo et al., 2014; Murugesan et al., 2019; Sadat Shandiz et al., 2017; Yuan et al., 2018). Furthermore, spherical-shaped AgNPs were synthesized from the *Aerva javanica* plant and mixed with the anticancer drug gefitinib, and showed higher anticancer activity in breast cancer cells (Khalid and Hanif, 2017). In addition, the incorporation of methotrexate into graphene oxide in silver nanoparticles (GO-AgNPs) targets folate receptors expressed by cancer cells and can be used synergistically for cancer treatment (Thapa et al., 2017). Similarly, adenosine $5'$-triphosphate ATP-capped silver nanoparticles (ATP-AgNPs) showed higher antitumor efficiency and induced cell death through the apoptosis process (Rajabnia and Meshkini, 2018). Also, AgNPs with internal photo-cleavable linkers and thiol terminated photo labeled DNAs benefits in targeted delivery including intracellular detection, nuclei protection, and effective photosensitive nanoparticles release (Steinbrück et al., 2008). In addition, AgNPs have been used as carriers for PEG-coated methotrexate in breast cancer patients (Muhammad et al., 2016). Silver nanoparticles are very useful for diagnosis and have potential to create a therapeutic drug delivery system for personalized treatment (Lee and Jun, 2019).

5 Conclusion and future prospects
The purpose of core-shell nanoparticles/nanomaterials development is to improve their functional and biological properties as well as their multifunctionality. The multifunctional biologically and chemically fabricated CS-AgNPs provide multidimensional properties which are used in various fields such as environmental, agriculture, dye industries, and biomedical. Due to their unique physicochemical and

biological properties, CS-AgNPs have been considered as smart materials that are substantially utilized in therapeutic and diagnostic approaches in biomedical applications. However, the overproduction of metal nanoparticles could create huge environmental and ecological challenges. Hence, this chapter indicated the requirement of the standard protocol for manufacture and evaluation of biodegradable and environmentally friendly CS-AgNPs that would be safe and ideal for future biomedical nanotechnology.

Acknowledgment

This work was supported by the National Research Foundation of Korea (2021R1I1A1A01057742; 2019R1A1055452).

References

Aflori, M., 2014. Surface characterization of peritoneal dialysis catheter containing silver nanoparticles. Rev. Roum. Chim. 59, 523–526.

Agnihotri, S., Mukherji, S., Mukherji, S., 2014. Size-controlled silver nanoparticles synthesized over the range 5-100 nm using the same protocol and their antibacterial efficacy. RSC Adv. 4, 3974–3983.

Ahmed, H.B., Emam, H.E., 2020. Seeded growth core-shell (Ag–Au–Pd) ternary nanostructure at room temperature for potential water treatment. Polym. Test. 89, 106720.

Alejandro Almonaci Hernández, C., Juarez-Moreno, K., Castañeda-Juarez, M.E., Almanza-Reyes, H., Pestryakov, A., Bogdanchikova, N., Oyuky Juarez-Moreno, K., 2017. Silver nanoparticles for the rapid healing of diabetic foot ulcers. Int. J. Med. Nano Res. 4, 2378–3664.

Ali, S.S.M., Eisa, W.H., Abouelsayed, A., 2019. Solvent-free and large-scale preparation of silver@polypyrrole core@shell nanocomposites; structural properties and terahertz spectroscopic studies. Compos. Part B Eng. 176, 107289.

Anbazhagan, S., Azeez, S., Morukattu, G., Rajan, R., Venkatesan, K., Thangavelu, K.P., 2017. Synthesis, characterization and biological applications of mycosynthesized silver nanoparticles. 3 Biotech 7, 1–9.

Andara, M., Agarwal, A., Scholvin, D., Gerhardt, R.A., Doraiswamy, A., Jin, C., Narayan, R.J., Shih, C.C., Shih, C.M., Lin, S.J., Su, Y.Y., 2006. Hemocompatibility of diamond like carbon-metal composite thin films. Diam. Relat. Mater. 15, 1941–1948.

Aznar, R., Barahona, F., Geiss, O., Ponti, J., José Luis, T., Barrero-Moreno, J., 2017. Quantification and size characterisation of silver nanoparticles in environmental aqueous samples and consumer products by single-particle-ICPMS. Talanta 175, 200–208.

Balamurugan, M., Saravanan, S., Soga, T., 2017. Coating of green-synthesized silver nanoparticles on cotton fabric. J. Coatings Technol. Res. 14, 735–745.

Bartosewicz, B., Liszewska, M., Budner, B., Michalska-Domańska, M., Kopczyński, K., Jankiewicz, B.J., 2020. Fabrication of Ag-modified hollow titania spheres via controlled silver diffusion in Ag-TiO2 core-shell nanostructures. Beilstein J. Nanotechnol. 11, 141–146.

Berry, J.F., Lu, C.C., 2017. Metal-metal bonds: from fundamentals to applications. Inorg. Chem. 56, 7577–7581.

References

Bistolfi, A., Ferracini, R., Albanese, C., Vernè, E., Miola, M., 2019. PMMA-based bone cements and the problem of joint arthroplasty infections: status and new perspectives. Materials (Basel) 12, 1–16.

Blanco, J., Tomás-Hernández, S., García, T., Mulero, M., Gómez, M., Domingo, J.L., Sánchez, D.J., 2018. Oral exposure to silver nanoparticles increases oxidative stress markers in the liver of male rats and deregulates the insulin signalling pathway and p53 and cleaved caspase 3 protein expression. Food Chem. Toxicol. 115, 398–404.

Bocate, K.P., Reis, G.F., de Souza, P.C., Oliveira Junior, A.G., Durán, N., Nakazato, G., Furlaneto, M.C., de Almeida, R.S., Panagio, L.A., 2019. Antifungal activity of silver nanoparticles and simvastatin against toxigenic species of Aspergillus. Int. J. Food Microbiol. 291, 79–86.

Cao, Y., Jin, R., Mirkin, C.A., 2001. DNA-modified core-shell Ag/Au nanoparticles. J. Am. Chem. Soc. 123, 7961–7962.

Chatterjee, K., Sarkar, S., Jagajjanani Rao, K., Paria, S., 2014. Core/shell nanoparticles in biomedical applications. Adv. Colloid Interf. Sci. 209, 8–39.

Chen, N., Zheng, Y., Yin, J., Li, X., Zheng, C., 2013. Inhibitory effects of silver nanoparticles against adenovirus type 3 in vitro. J. Virol. Methods 193, 470–477.

Clark, N.J., Clough, R., Boyle, D., Handy, R.D., 2019. Development of a suitable detection method for silver nanoparticles in fish tissue using single-particle ICP-MS. Environ. Sci. Nano 6, 3388–3400.

Dakal, T.C., Kumar, A., Majumdar, R.S., Yadav, V., 2016. Mechanistic basis of antimicrobial actions of silver nanoparticles. Front. Microbiol. 7, 1–17.

Das, N., Kumar, A., Kumar Roy, S., Kumar Satija, N., Raja Gopal, R., 2020. Bare plasmonic metal nanoparticles: synthesis, characterisation and in vitro toxicity assessment on a liver carcinoma cell line. IET Nanobiotechnol. 14, 851–857.

De Matteis, V., Cascione, M., Toma, C.C., Leporatti, S., 2018. Silver nanoparticles: synthetic routes, in vitro toxicity and theranostic applications for cancer disease. Nanomaterials (Basel) 8, 319.

Diniz, F.R., Maia, R.C.A.P., Rannier, L., Andrade, L.N., Chaud, M.V., da Silva, C.F., Corrêa, C.B., de Albuquerque Junior, R.L.C., da Costa, L.P., Shin, S.R., Hassan, S., Sanchez-Lopez, E., Souto, E.B., Severino, P., 2020. Silver nanoparticles-composing alginate/gelatine hydrogel improves wound healing in vivo. Nanomaterials (Basel) 10, 390.

Elechiguerra, J.L., Burt, J.L., Morones, J.R., Camacho-Bragado, A., Gao, X., Lara, H.H., Yacaman, M.J., 2005. Interaction of silver nanoparticles with HIV-1. J. Nanobiotechnol. 3, 1–10.

El-Hussein, A., Hamblin, M.R., 2017. ROS generation and DNA damage with photoinactivation mediated by silver nanoparticles in lung cancer cell line. IET Nanobiotechnology 11, 173–178.

El-Shanshoury, A.E.-R.R., ElSilk, S.E., Ebeid, M.E., 2011. Extracellular biosynthesis of silver nanoparticles using Escherichia coli ATCC 8739, Bacillus subtilis ATCC 6633, and Streptococcus thermophilus ESh1 and their antimicrobial activities. ISRN Nanotechnol. 2011, 1–7.

Esteban-Tejeda, L., Malpartida, F., Esteban-Cubillo, A., Pecharromán, C., Moya, J.S., 2009. The antibacterial and antifungal activity of a soda-lime glass containing silver nanoparticles. Nanotechnology 20, 085103.

Fahmy, H.M., Mosleh, A.M., Elghany, A.A., Shams-Eldin, E., Abu Serea, E.S., Ali, S.A., Shalan, A.E., 2019. Coated silver nanoparticles: synthesis, cytotoxicity, and optical properties. RSC Adv. 9, 20118–20136.

Falzarano, M.S., Passarelli, C., Ferlini, A., 2014. Nanoparticle delivery of antisense oligonucleotides and their application in the exon skipping strategy for duchenne muscular dystrophy. Nucleic Acid Ther. 24, 87–100.

Flores-López, L.Z., Espinoza-Gómez, H., Somanathan, R., 2019. Silver nanoparticles: electron transfer, reactive oxygen species, oxidative stress, beneficial and toxicological effects. Mini review. J. Appl. Toxicol. 39, 16–26.

Gajbhiye, M., Kesharwani, J., Ingle, A., Gade, A., Rai, M., 2009. Fungus-mediated synthesis of silver nanoparticles and their activity against pathogenic fungi in combination with fluconazole. Nanomedicine nanotechnology. Biol. Med. 5, 382–386.

Galdiero, S., Falanga, A., Vitiello, M., Cantisani, M., Marra, V., Galdiero, M., 2011. Silver nanoparticles as potential antiviral agents. Molecules 16, 8894–8918.

Gale-Mouldey, A., Jorgenson, E., Coyle, J.P., Prezgot, D., Ianoul, A., 2020. Hybridized plasmon resonances in core/half-shell silver/cuprous oxide nanoparticles. J. Mater. Chem. C 8, 1852–1863.

Gao, J., Na, H., Zhong, R., Yuan, M., Guo, J., Zhao, L., Wang, Y., Wang, L., Zhang, F., 2020. One step synthesis of antimicrobial peptide protected silver nanoparticles: the core-shell mutual enhancement of antibacterial activity. Colloids Surf B Biointerfaces 186, 110704.

Garraffo, A., Pilmis, B., Toubiana, J., Puel, A., Mahlaoui, N., Blanche, S., Lortholary, O., Lanternier, F., 2017. Invasive fungal infection in primary immunodeficiencies other than chronic granulomatous disease. Curr. Fungal Infect. Rep. 11, 25–34.

Ge, L., Li, Q., Wang, M., Ouyang, J., Li, X., Xing, M.M.Q., 2014. Nanosilver particles in medical applications: synthesis, performance, and toxicity. Int. J. Nanomedicine 9, 2399–2407.

Ghosn, Y., Kamareddine, M.H., Tawk, A., Elia, C., El Mahmoud, A., Terro, K., El Harake, N., El-Baba, B., Makdessi, J., Farhat, S., 2019. Inorganic nanoparticles as drug delivery systems and their potential role in the treatment of chronic myelogenous Leukaemia. Technol. Cancer Res. Treat. 18, 1–12.

Guo, D., Zhao, Y., Zhang, Y., Wang, Q., Huang, Z., Ding, Q., Guo, Z., Zhou, X., Zhu, L., Gu, N., 2014. The cellular uptake and cytotoxic effect of silver nanoparticles on chronic myeloid leukemia cells. J. Biomed. Nanotechnol. 10, 669–678.

Gupta, K., Chhibber, S., 2019. Biofunctionalization of silver nanoparticles with lactonase leads to altered antimicrobial and cytotoxic properties. Front. Mol. Biosci. 6, 63.

He, X., Peng, C., Qiang, S., Xiong, L.H., Zhao, Z., Wang, Z., Kwok, R.T.K., Lam, J.W.Y., Ma, N., Tang, B.Z., 2020. Less is more: silver-AIE core@shell nanoparticles for multimodality cancer imaging and synergistic therapy. Biomaterials 238, 119834.

Huang, Y., Wu, F., Zhou, Z., Zhou, L., Liu, H., 2020. Fabrication of fully covered Cu-Ag core-shell nanoparticles by compound method and anti-oxidation performance. Nanotechnology 31, 175601.

Huang, J., Han, X., Zhao, X., Meng, C., 2021. Facile preparation of core-shell Ag@SiO2 nanoparticles and their application in spectrally splitting PV/T systems. Energy 215, 119111.

Im, H., Noh, S., Shim, J.H., 2020. Spontaneous formation of core-shell silver-copper oxide by carbon dot-mediated reduction for enhanced oxygen electrocatalysis. Electrochim. Acta 329, 135172.

Iravani, S., Korbekandi, H., Mirmohammadi, S.V., Zolfaghari, B., 2014. Synthesis of silver nanoparticles: chemical, physical and biological methods. Res. Pharm. Sci. 9, 385–406.

Jamkhande, P.G., Ghule, N.W., Bamer, A.H., Kalaskar, M.G., 2019. Metal nanoparticles synthesis: an overview on methods of preparation, advantages and disadvantages, and applications. J. Drug Deliv. Sci. Technol. 53, 101174.

Jiao, Z.H., Li, M., Feng, Y.X., Shi, J.C., Zhang, J., Shao, B., 2014. Hormesis effects of silver nanoparticles at non-cytotoxic doses to human hepatoma cells. PLoS One 9, e102564.

Kalambate, P.K., Dhanjai, Huang, Z., Li, Y., Shen, Y., Xie, M., Huang, Y., Srivastava, A.K., 2019. Core@shell nanomaterials based sensing devices: a review. Trends Anal. Chem. 115, 147–161.

Kang, H., Choi, S.R., Kim, Y.H., Kim, J.S., Kim, S., An, B.S., Yang, C.W., Myoung, J.M., Lee, T.W., Kim, J.G., Cho, J.H., 2020. Electroplated silver-nickel core-shell nanowire network electrodes for highly efficient perovskite nanoparticle light-emitting diodes. ACS Appl. Mater. Interfaces 12, 39479–39486.

Khalid, S., Hanif, R., 2017. Green biosynthesis of silver nanoparticles conjugated to gefitinib as delivery vehicle. Int. J. Adv. Sci. Eng. Technol. 5, 59–63.

Khan, I., Bahuguna, A., Krishnan, M., Shukla, S., Lee, H., Min, S.H., Choi, D.K., Cho, Y., Bajpai, V.K., Huh, Y.S., Kang, S.C., 2019. The effect of biogenic manufactured silver nanoparticles on human endothelial cells and zebrafish model. Sci. Total Environ. 679, 365–377.

Khatoon, N., Alam, H., Khan, A., Raza, K., Sardar, M., 2019. Ampicillin silver nanoformulations against multidrug resistant bacteria. Sci. Rep. 9, 1–10.

Kim, H.R., Kim, M.J., Lee, S.Y., Oh, S.M., Chung, K.H., 2011. Genotoxic effects of silver nanoparticles stimulated by oxidative stress in human normal bronchial epithelial (BEAS-2B) cells. Mutat. Res. Genet. Toxicol. Environ. Mutagen. 726, 129–135.

Kovács, D., Igaz, N., Marton, A., Rónavári, A., Bélteky, P., Bodai, L., Spengler, G., Tiszlavicz, L., Rázga, Z., Hegyi, P., Vizler, C., Boros, I.M., Kónya, Z., Kiricsi, M., 2020. Core-shell nanoparticles suppress metastasis and modify the tumour-supportive activity of cancer-associated fibroblasts. J. Nanobiotechnol. 18, 18.

Kora, AJ., Rastogi, L., 2013. Enhancement of antibacterial activity of capped silver nanoparticles in combination with antibiotics, on model gram-negative and gram-positive bacteria. Bioinorg. Chem. Appl. 2013, 871097.

Lakshmanan, G., Sathiyaseelan, A., Kalaichelvan, P.T., Murugesan, K., 2018. Plant-mediated synthesis of silver nanoparticles using fruit extract of Cleome viscosa L.: assessment of their antibacterial and anticancer activity. Karbala Int. J. Mod. Sci. 4, 61–68.

Lara, H.H., Ayala-Nuñez, N.V., Ixtepan-Turrent, L., Rodriguez-Padilla, C., 2010. Mode of antiviral action of silver nanoparticles against HIV-1. J. Nanobiotechnol. 8, 1–10.

Lee, S.H., Jun, B.H., 2019. Silver nanoparticles: synthesis and application for nanomedicine. Int. J. Mol. Sci. 20.

Leu, L., Sun, R.W., Chen, R., Hui, C.K., Ho, C.M., Luk, J.M., Lau, G.K., CM, C., 2008. Silver nanoparticles inhibit hepatitis B virus replication. Antivir. Ther. 13, 253–262.

Li, C., Yao, Y., 2020. Synthesis of bimetallic core-shell silver-copper nanoparticles decorated on reduced graphene oxide with enhanced electrocatalytic performance. Chem. Phys. Lett. 761, 137726.

Li, J., Zuo, Y., Man, Y., Mo, A., Huang, C., Liu, M., Jansen, J.A., Li, Y., 2012. Fabrication and biocompatibility of an antimicrobial composite membrane with an asymmetric porous structure. J. Biomater. Sci. Polym. Ed. 23, 81–96.

Lin, P.C., Lin, S., Wang, P.C., Sridhar, R., 2014. Techniques for physicochemical characterization of nanomaterials. Biotechnol. Adv. 32, 711–726.

Majeed, S., 2017. Biosynthesis and characterization of nanosilver from *Alternata alternaria* and it antifungal and antibacterial activity in combination with fluconazole and gatifloxacin. Biomed. Pharmacol. J. 10, 1709–1714.

Manivannan, K., Cheng, C.C., Anbazhagan, R., Tsai, H.C., Chen, J.K., 2019. Fabrication of silver seeds and nanoparticle on core-shell ag@SiO2 nanohybrids for combined photothermal therapy and bioimaging. J. Colloid Interface Sci. 537, 604–614.

Marambio-Jones, C., Hoek, E.M.V., 2010. A review of the antibacterial effects of silver nanomaterials and potential implications for human health and the environment. J. Nanopart. Res. 12, 1531–1551.

Marassi, V., Di Cristo, L., Smith, S.G.J., Ortelli, S., Blosi, M., Costa, A.L., Reschiglian, P., Volkov, Y., Prina-Mello, A., 2018. Silver nanoparticles as a medical device in healthcare settings: a five-step approach for candidate screening of coating agents. R. Soc. Open Sci. 5, 171113.

Matsumura, Y., Yoshikata, K., Kunisaki, S.i., Tsuchido, T., 2003. Mode of bactericidal action of silver zeolite and its comparison with that of silver nitrate. Appl. Environ. Microbiol. 69, 4278–4281.

Mazhani, M., Alula, M.T., Murape, D., 2020. Development of a cysteine sensor based on the peroxidase-like activity of AgNPs@ Fe3O4 core-shell nanostructures. Anal. Chim. Acta 1107, 193–202.

Mehrabani, M.G., Karimian, R., Mehramouz, B., Rahimi, M., Kafil, H.S., 2018. Preparation of biocompatible and biodegradable silk fibroin/chitin/silver nanoparticles 3D scaffolds as a bandage for antimicrobial wound dressing. Int. J. Biol. Macromol. 114, 961–971.

Moazeni, M., Rashidi, N., Shahverdi, A.R., Noorbakhsh, F., Rezaie, S., 2012. Extracellular production of silver nanoparticles by using three common species of dermatophytes: Trichophyton rubrum, Trichophyton mentagrophytes and microsporum canis. Iran. Biomed. J. 16, 52–58.

Montalvo-Quirós, S., Gómez-Graña, S., Vallet-Regí, M., Prados-Rosales, R.C., González, B., Luque-Garcia, J.L., 2021. Mesoporous silica nanoparticles containing silver as novel antimycobacterial agents against mycobacterium tuberculosis. Colloids Surf B Biointerfaces 197, 111405.

Monteiro, D.R., Silva, S., Negri, M., Gorup, L.F., de Camargo, E.R., Oliveira, R., Barbosa, D.B., Henriques, M., 2013. Antifungal activity of silver nanoparticles in combination with nystatin and chlorhexidine digluconate against Candida albicans and Candida glabrata biofilms. Mycoses 56, 672–680.

Mott, D.M., Anh, D.T.N., Singh, P., Shankar, C., Maenosono, S., 2012. Electronic transfer as a route to increase the chemical stability in gold and silver core-shell nanoparticles. Adv. Colloid Interf. Sci. 185–186, 14–33.

Muhammad, Z., Raza, A., Ghafoor, S., Naeem, A., Naz, S.S., Riaz, S., Ahmed, W., Rana, N.F., 2016. PEG capped methotrexate silver nanoparticles for efficient anticancer activity and biocompatibility. Eur. J. Pharm. Sci. 91, 251–255.

Murugesan, K., Koroth, J., Srinivasan, P.P., Singh, A., Mukundan, S., Karki, S.S., Choudhary, B., Gupta, C.M., 2019. Effects of green synthesised silver nanoparticles (ST06-AgNPs) using curcumin derivative (ST06) on human cervical cancer cells (HeLa) in vitro and EAC tumour-bearing mice models. Int. J. Nanomedicine 14, 5257–5270.

Nur Kamilah, M., Khalik, W.M.A.W.M., Azmi, A.A., 2019. Synthesis and characterization of silica-silver core-shell nanoparticles. Malaysian J. Anal. Sci. 23, 290–299.

Ogar, A., Tylko, G., Turnau, K., 2015. Antifungal properties of silver nanoparticles against indoor mould growth. Sci. Total Environ. 521–522, 305–314.

Park, T., Lee, S., Amatya, R., Cheong, H., Moon, C., Kwak, H.D., Min, K.A., Shin, M.C., 2020. ICG-loaded pegylated BSA-silver nanoparticles for effective photothermal cancer therapy. Int. J. Nanomedicine 15, 5459–5471.

Patel, P., Agarwal, P., Kanawaria, S., Kachhwaha, S., Kothari, S.L., 2015. Plant-based synthesis of silver nanoparticles and their characterization. In: Nanotechnology and Plant Sciences: Nanoparticles and Their Impact on Plants. Springer international publishing, pp. 271–288.

Patra, J.K., Das, G., Fraceto, L.F., Campos, E.V.R., Rodriguez-Torres, M.D.P., Acosta-Torres, L.S., Diaz-Torres, L.A., Grillo, R., Swamy, M.K., Sharma, S., Habtemariam, S., Shin, H. S., 2018. Nano based drug delivery systems: recent developments and future prospects. J. Nanobiotechnol. 16, 1–33.

Percival, S.L., Bowler, P.G., Dolman, J., 2007. Antimicrobial activity of silver-containing dressings on wound microorganisms using an in vitro biofilm model. Int. Wound J. 4, 186–191.

Perveen, S., Safdar, N., Chaudhry, G.e.s., Yasmin, A., 2018. Antibacterial evaluation of silver nanoparticles synthesized from lychee peel: individual versus antibiotic conjugated effects. World J. Microbiol. Biotechnol. 34, 1–12.

Prokopovich, P., Köbrick, M., Brousseau, E., Perni, S., 2015. Potent antimicrobial activity of bone cement encapsulating silver nanoparticles capped with oleic acid. J. Biomed. Mater. Res. B Appl. Biomater. 103, 273–281.

Qing, Y., Cheng, L., Li, R., Liu, G., Zhang, Y., Tang, X., Wang, J., Liu, H., Qin, Y., 2018. Potential antibacterial mechanism of silver nanoparticles and the optimization of orthopedic implants by advanced modification technologies. Int. J. Nanomedicine 13, 3311–3327.

Rai, M., Ingle, A.P., Gupta, I., Brandelli, A., 2015. Bioactivity of noble metal nanoparticles decorated with biopolymers and their application in drug delivery. Int. J. Pharm. 496, 159–172.

Raja, G., Jang, Y.K., Suh, J.S., Kim, H.S., Ahn, S.H., Kim, T.J., 2020. Microcellular environmental regulation of silver nanoparticles in cancer therapy: a critical review. Cancers (Basel) 12, 1–33.

Rajabnia, T., Meshkini, A., 2018. Fabrication of adenosine 5′-triphosphate-capped silver nanoparticles: enhanced cytotoxicity efficacy and targeting effect against tumor cells. Process Biochem. 65, 186–196.

Ramírez-Acosta, C.M., Cifuentes, J., Cruz, J.C., Reyes, L.H., 2020. Patchy core/shell, magnetite/silver nanoparticles via green and facile synthesis: routes to assure biocompatibility. Nanomaterials (Basel) 10, 1857.

Sadat Akhavi, S., Moradi Dehaghi, S., 2020. Drug delivery of amphotericin B through core-shell composite based on PLGA/ag/Fe3O4: in vitro test. Appl. Biochem. Biotechnol. 191, 496–510.

Sadat Shandiz, S.A., Shafiee Ardestani, M., Shahbazzadeh, D., Assadi, A., Ahangari Cohan, R., Asgary, V., Salehi, S., 2017. Novel imatinib-loaded silver nanoparticles for enhanced apoptosis of human breast cancer MCF-7 cells. Artif. Cells. Nanomed. Biotechnol. 45, 1082–1091.

Salas-Orozco, M.F., Martinez, N.N., Martínez-Castañón, G.A., Méndez, F.T., Patiño-Marín, N., Ruiz, F., 2019. Detection of genes related to resistance to silver nanoparticles in bacteria from secondary endodontic infections. J. Nanomater. 2019, 8742975.

Salman, M., Rizwana, R., Khan, H., Munir, I., Hamayun, M., Iqbal, A., Rehman, A., Amin, K., Ahmed, G., Khan, M., Khan, A., Amin, F.U., 2019. Synergistic effect of silver nanoparticles and polymyxin B against biofilm produced by *Pseudomonas aeruginosa* isolates of pus samples in vitro. Artif. Cells. Nanomed. Biotechnol. 47, 2465–2472.

Sathiyaseelan, A., Shajahan, A., Kalaichelvan, P.T., Kaviyarasan, V., 2017. Fungal chitosan based nanocomposites sponges—an alternative medicine for wound dressing. Int. J. Biol. Macromol. 104, 1905–1915.

Sathiyaseelan, A., Saravanakumar, K., Mariadoss, A.V.A., Wang, M.H., 2020. Biocompatible fungal chitosan encapsulated phytogenic silver nanoparticles enhanced antidiabetic, antioxidant and antibacterial activity. Int. J. Biol. Macromol. 153, 63–71.

Sethuram, L., Thomas, J., Mukherjee, A., Chandrasekaran, N., 2019. Effects and formulation of silver nanoscaffolds on cytotoxicity dependent ion release kinetics towards enhanced excision wound healing patterns in Wistar albino rats. RSC Adv. 9, 35677–35694.

Shaheen, T.I., El-Naggar, M.E., Hussein, J.S., El-Bana, M., Emara, E., El-Khayat, Z., Fouda, M.M.G., Ebaid, H., Hebeish, A., 2016. Antidiabetic assessment; in vivo study of gold and core-shell silver-gold nanoparticles on streptozotocin-induced diabetic rats. Biomed. Pharmacother. 83, 865–875.

Shrivastava, S., Bera, T., Roy, A., Singh, G., Ramachandrarao, P., Dash, D., 2007. Characterization of enhanced antibacterial effects of novel silver nanoparticles. Nanotechnology 18, 225103.

Singh, M., Kumar, M., Kalaivani, R., Manikandan, S., Kumaraguru, A.K., 2013. Metallic silver nanoparticle: a therapeutic agent in combination with antifungal drug against human fungal pathogen. Bioprocess Biosyst. Eng. 36, 407–415.

Song, S., Zhao, Y., Yuan, X., Zhang, J., 2019. β-Chitin nanofiber hydrogel as a scaffold to in situ fabricate monodispersed ultra-small silver nanoparticles. Eng. Asp. 574, 36–43.

Soto-Quintero, A., Guarrotxena, N., García, O., Quijada-Garrido, I., 2019. Curcumin to promote the synthesis of silver NPs and their self-assembly with a thermoresponsive polymer in core-shell nanohybrids. Sci. Rep. 9, 1–14.

Steinbrück, A., Csaki, A., Ritter, K., Leich, M., Köhler, J.H., Fritzsche, W., 2008. Gold-silver and silver-silver nanoparticle constructs based on DNA hybridization of thiol- and aminofunctionalized oligonucleotides. J. Biophotonics 1, 104–113.

Sun, Y., Zhao, Y., Zhao, X., Lee, R.J., Teng, L., Zhou, C., 2017. Enhancing the therapeutic delivery of oligonucleotides by chemical modification and nanoparticle encapsulation. Molecules 22, 1724.

Sung, J.H., Ji, J.H., Song, K.S., Lee, J.H., Choi, K.H., Lee, S.H., Yu, I.J., 2011. Acute inhalation toxicity of silver nanoparticles. Toxicol. Ind. Health 27, 149–154.

Syafiuddin, A., Salmiati, S., Salim, M.R., Beng Hong Kueh, A., Hadibarata, T., Nur, H., 2017. A review of silver nanoparticles: research trends, global consumption, synthesis, properties, and future challenges. J. Chin. Chem. Soc. 64, 732–756.

Synak, A., Szczepańska, E., Grobelna, B., Gondek, J., Mońka, M., Gryczynski, I., Bojarski, P., 2019. Photophysical properties and detection of Valrubicin on plasmonic platforms. Dyes Pigments 163, 623–627.

Taghizadeh, S.M., Berenjian, A., Taghizadeh, S., Ghasemi, Y., Taherpour, A., Sarmah, A.K., Ebrahiminezhad, A., 2019. One-put green synthesis of multifunctional silver iron core-shell nanostructure with antimicrobial and catalytic properties. Ind. Crop. Prod. 130, 230–236.

Tang, L., Livi, K.J.T., Chen, K.L., 2015. Polysulfone membranes modified with bioinspired polydopamine and silver nanoparticles formed in situ to mitigate biofouling. Environ. Sci. Technol. Lett. 2, 59–65.

Thapa, R.K., Kim, J.H., Jeong, J.H., Shin, B.S., Choi, H.G., Yong, C.S., Kim, J.O., 2017. Silver nanoparticle-embedded graphene oxide-methotrexate for targeted cancer treatment. Colloids Surf B Biointerfaces 153, 95–103.

Tomaszewska, E., Soliwoda, K., Kadziola, K., Tkacz-Szczesna, B., Celichowski, G., Cichomski, M., Szmaja, W., Grobelny, J., 2013. Detection limits of DLS and UV-vis spectroscopy in characterization of polydisperse nanoparticles colloids. J. Nanomater. 2013, 313081.

Vinotha Alex, A., Chandrasekaran, N., Mukherjee, A., 2020. Novel enzymatic synthesis of core/shell AgNP/AuNC bimetallic nanostructure and its catalytic applications. J. Mol. Liq. 301, 112463.

Vykoukal, V., Zelenka, F., Bursik, J., Kana, T., Kroupa, A., Pinkas, J., 2020. Thermal properties of Ag@Ni core-shell nanoparticles. Calphad Comput. Coupling Phase Diagrams Thermochem. 69, 101741.

Wu, Y., Lin, Y., Xu, J., 2019. Synthesis of Ag-Ho, Ag-Sm, Ag-Zn, Ag-Cu, Ag-Cs, Ag-Zr, Ag-Er, Ag-Y and Ag-Co metal organic nanoparticles for UV-vis-NIR wide-range biotissue imaging. Photochem. Photobiol. Sci. 18, 1081–1091.

Xia, J., Liu, X., Gao, Y., Bai, L., 2020. Green synthesis of Ag/ZnO microplates by doping Ag ions on basic zinc carbonate for fast photocatalytic degradation of dyes. Environ. Technol. (United Kingdom) 41, 3584–3590.

Xing, Z.C., Chae, W.P., Baek, J.Y., Choi, M.J., Jung, Y., Kang, I.K., 2010. In vitro assessment of antibacterial activity and cytocompatibility of silver-containing phbv nanofibrous scaffolds for tissue engineering. Biomacromolecules 11, 1248–1253.

Xu, G.-Y., Zong, C.-H., Sun, Y.-A., Wang, X.-X., Zhang, N., Wang, F., Li, A.-X., Li, Q.-H., 2019. Preparation of core—shell silver nanoparticles@mesoporous silica Nanospheres with catalytic activities. J. Nanosci. Nanotechnol. 19, 5893–5899.

Yaqoob, S.B., Adnan, R., Rameez Khan, R.M., Rashid, M., 2020a. Gold, silver, and palladium nanoparticles: a chemical tool for biomedical applications. Front. Chem. 8, 376.

Yaqoob, A.A., Ahmad, H., Parveen, T., Ahmad, A., Oves, M., Ismail, I.M.I., Qari, H.A., Umar, K., Mohamad Ibrahim, M.N., 2020b. Recent advances in metal decorated nanomaterials and their various biological applications: a review. Front. Chem. 8, 341.

Yilmaz, A., Yilmaz, M., 2020. Bimetallic core-shell nanoparticles of gold and silver via bioinspired Polydopamine layer as surface-enhanced Raman spectroscopy (SERS) platform. Nanomaterials 10, 688.

Yuan, Y.G., Zhang, S., Hwang, J.Y., Kong, I.K., 2018. Silver nanoparticles potentiates cytotoxicity and apoptotic potential of camptothecin in human cervical cancer cells. Oxidative Med. Cell. Longev. 2018, 6121328.

Zhao, S., Chen, J., Liu, Y., Jiang, Y., Jiang, C., Yin, Z., Xiao, Y., Cao, S., 2019. Silver nanoparticles confined in shell-in-shell hollow TiO2 manifesting efficiently photocatalytic activity and stability. Chem. Eng. J. 367, 249–259.

Zhao, Q., Sun, X.Y., Wu, B., Shang, Y., Huang, X., Dong, H., Liu, H., Chen, W., Gui, R., Li, J., 2021. Construction of biomimetic silver nanoparticles in the treatment of lymphoma. Mater. Sci. Eng. C 119, 111648.

Zhou, Y., Liang, P., Zhang, D., Tang, L., Dong, Q., Jin, S., Ni, D., Yu, Z., Ye, J., 2020. A facile seed growth method to prepare stable ag@ZrO2 core-shell SERS substrate with high stability in extreme environments. Spectrochim. Acta A Mol. Biomol. Spectrosc. 228, 117676.

Zodrow, K., Brunet, L., Mahendra, S., Li, D., Zhang, A., Li, Q., Alvarez, P.J.J., 2009. Polysulfone ultrafiltration membranes impregnated with silver nanoparticles show improved biofouling resistance and virus removal. Water Res. 43, 715–723.

CHAPTER 5

Bimetallic silver nanoparticles: Green synthesis, characterization and bioefficacy

Mukti Sharma, Ranjini Tyagi, Man Mohan Srivastava, and Shalini Srivastava
Department of Chemistry, Faculty of Science, Dayalbagh Educational Institute, Agra, India

1 Introduction

The basic concept of nanotechnology is "The manipulation of matter at a molecular or atomic level to produce novel materials and devices with new extraordinary properties" (McNeil, 2005; Mamalis, 2007). The 21st century has seen the beginning of the application of nanotechnology. Nanoparticles have the advantages of their unique interaction with light owing to their surface plasmon resonance (SPR). The free available electrons of the metal nanoparticles, under the influence of the electromagnetic field of the light, encounter collective oscillation concerning the positive metallic lattice and start resonating at a particular frequency of the light (Kelly et al., 2003; Link and El-Sayed, 2003; Jain et al., 2006). The rewards of nanotechnology depend on the possibility to alter the structures of materials at tremendously reduced scales. Using the nanotech approach, the materials are made stronger or lighter, more durable, and more active. Nanotechnology is a revolutionizing technologies with applications in every sector of sustainable development, including transportation, environmental safety, and medicines (Ahmed and Ikram, 2016; Gong et al., 2021).

However, nanotechnology has not been found safe in all respects. Recently, inorganic nanoparticles synthesized by physicochemical methods have been found to be associated with toxicity issues and potential side effects, particularly when they are applied in a biological system (Thakkar et al., 2010; Das et al., 2011). Therefore, nanotechnology must be developed more sustainably. No new process, which might be worse than existing processes, can be acceptable to society just because it is based on nanotechnology. Any material and process must be assessed for safety, environmental effects, public health, and cost effectiveness. Physicochemical methods are popular for the fabrication of metallic nanoparticles. However, chemically fabricated metal nanoparticles have several constraints for their application in the biomedical area. The use of various toxic chemicals, nonpolar solvent, and synthetic additives in the fabrication of nanoparticles is a limitation on their use in medical related research areas (Ingale and Chaudhari, 2013; Huang et al., 2018). Thus, working with nanotechnology is quite risky. The consideration of the above facts would probably protect nanotechnology from negative public perceptions. Green nanotechnology is the new dimension and intersection of green chemistry with nanotechnology. The important principles of green chemistry given by American researchers Prof. Paul T. Anastas and Prof. John C. Warner have been widely utilized for mapping the greener design of nanotechnology.

The present chapter addresses the following subjects: the recent debate that nanotechnological development is not safe in all respects since it has been found to be associated with severe toxicity issues particularly when applied to biological systems; and modification in the trends of fabrication of nanoparticles, emphasizing phytomediated green synthesis of nanoparticles as a significant endeavor for the bulk production of nontoxic nanoparticles in a single step, easy, rapid, and cost effective manner with enhanced bioefficacies.

It highlights the journey of research with pros and cons, progressing from the use of crude plant extract to bioactive principle loaded metal nanoparticles in which synergistic redox potential of phytoactive constituents is used to alter ions to the nano state, possessing astonishing properties, suggesting such an ecofriendly approach would be for sustainable development in environmental safety, health, and biomedical applications.

2 Fabrication of silver nanoparticles

Among noble metals, gold and silver have gained considerable attention in biomedical research. Silver has specific inherent medicinal activities, including antiinflammatory, wound healing, antimicrobial, antioxidant, and anticancer, which make it an ideal candidate for the synthesis of nanoparticles (Chitturi et al., 2018; Dutta et al., 2020; Habibullah et al., 2021; Moradi et al., 2021). There are important characteristics and metabolic interactions of silver ions that govern their pharmacological efficacies. Silver is inert in its nonionized form, however, when dissolved in water, it generates silver ions. The charged silver ions are responsible for antimicrobial activity. Silver ions enter as granules in the cell wall, inhibit the cell division, damage the cellular contents, and finally prevent bacterial growth (Richards and Turner, 1984; Guggenbichler et al., 1999). Silver atoms strongly interact with thiol groups (-SH) and consequently cause enzymatic deactivation (Furr et al., 1994). Silver ion induced catalytic oxidation involves the hydrogen atom of the thiol group of enzymes and oxygen atom from cellular content forming disulfide bonds (R—S—S—R). During such structural changes in R—S—S—R bonds, the cellular enzymes get altered and have an impact on biological functions. Silver ions' interaction with the 30S ribosomal unit results in the disabling of the ribosomal complex and inhibits the translation of proteins (Yamanaka et al., 2005). Ag^+ interacts with base pairs (purine and pyrimidine), breaking the hydrogen bonds among the two antiparallel strands and finally denatures DNA. However, the exact mechanism is yet to be established. The transport of silver ions across the cell membrane takes place even in the absence of a silver transporter as its putative function is a copper transporter (Klueh et al., 2000). Silver is not toxic to the cardiovascular, nervous, immune, and reproductive systems. It is not carcinogenic in nature (ATSDR, 1990). The overexposure of silver, however, causes stomach irritation, fat degeneration of the liver and kidney, and low blood pressure and respiration rate (Venugopal and Luckey, 1978; Gulbranson et al., 2000).

There are two different fundamental principles of nanoparticle synthesis: top-down and bottom-up approaches, which are further subclassified (Iravani, 2011; Ahmed and Ikram, 2016). Green nanotechnology for the production of nanoparticles is a bottom-up approach and involves the use of biological entities like microorganisms and plant extracts. It lacks complex steps of chemical synthesis, and is easy to scale up, cost effective, and rapid. Microorganisms can reduce metal salts to metal nanoparticles (Geraldes et al., 2016; Das et al., 2017; Gahlawat and Choudhury, 2019; Crisan et al., 2021). The plant mediated nanoparticles are more strategic since they do not require complicated procedures like separation, or well conditioned culture and its maintenance. The plant mediated nanoparticles are economical and streamline the production of large quantities of nanoparticles straightforwardly and further alleviate the association of toxic substances (Kaabipour and Hemmati, 2021; Vanlalveni et al., 2021).

Plants are a storehouse of various phytochemicals known for a wide range of pharmacological efficacies. Phytochemicals are classified into primary and secondary metabolites. Secondary metabolites are formed under environmental stress and undergo chemical adaptation, therefore, are supposed to have some extraordinary bioefficacies (Weinberg, 1971). The crude herbal extracts are a cocktail of various phytochemicals and believed to work better, owing to synergism, which may not be additive but multiplicative. The chemical partnership of phytochemicals may reduce side effects due to possible nullifying effects while synergistic effects of medicinally important secondary metabolites enhance therapeutic effects. Working with the whole extract, pathogens may not gain resistance compared to a single compound recipe. On the other hand, the need for a bioactive constituent in disease management has also been justified. Why take a risk by swallowing the whole extract when recent analytical techniques are competent enough to isolate the active constituent and can be offered straight? The use of crude plant extracts poses several problems as they are easily digested, quickly removed from the body, and have short lived therapeutic efficacy, poor bioavailability, poor penetration, low water solubility, less stability, and first pass metabolic effects (Martinho et al., 2011; Patra et al., 2018). Therefore, research progressed from the use of crude plant extract to relevant bioactive constituents. However, it was realized that even a low dose of bioactive principle starts exhibiting severe toxicity. The easy development of resistance against a single chemical also raised another issue. Research attention was further made towards the improvement in bioefficacy without increasing the amount of dose with reduced possibility of the emergence of resistance.

Nowadays, the phytofabrication of silver metal nanoparticles has become an important tool for the enhancement of pharmacological efficacies and is widely popular in pharmacological research. The research was long drawnout from the direct use of bioactive constituent to bioactive loaded metal nanoparticles and explored for enhancement of various bioefficacies. The redox potential of the cocktail of the bioactive constituent (phenolics, proteins, polysaccharides, terpenoids, amino acids) was exploited to alter the oxidation states leading to the nano state of the metal as well as contribute in the rapid production of nontoxic nanoparticles (Moldovan et al., 2017; Lee et al., 2019; Javed et al., 2021; Saim et al., 2021). The polyphenolic biomolecules (OH) groups are capable of converting Ag^+ to Ag^0 (nano state). The mechanism of the phytofabrication of silver nanoparticles using polyphenolic biomolecules like flavonoids, saponins, and terpenoids has been put forward. The two OH groups of catechol moiety, present in the basic skeleton of flavonoid, possess 4.6–14.1 kcal mol^{-1} of bond dissociation energy, which is less than 89.0 kcal mol^{-1} of phenolic -OH group (Trouillas et al., 2006). This fact facilitates the displacement of $2H^+$ by $2Ag^+$ ions reducing silver into nano state (Ag^0). Thus, two Ag^+ may be reduced by one flavonoid molecule, resulting in the formation of one quinone moiety (Sharma et al., 2019). The oxidized quinone moiety, which is electron deficient, contributes additional free radical scavenging activity. In the terpenoid induced fabrication of silver nanoparticles, the cleavage of the OH group of the basic terpenoid structure involves the replacement of one H^+ by one Ag^+ ion. The Ag^+ ion is further reduced into the Ag^0 nano state (Fig. 1). Thus, one terpenoid moiety may reduce one Ag^+ along with the corresponding oxidized terpenoid moiety (Hongfanga et al., 2017).

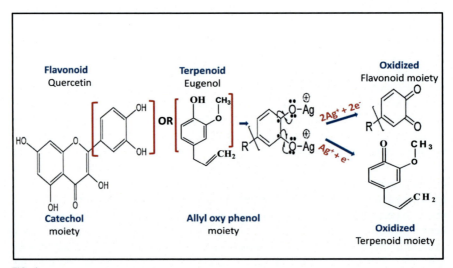

FIG. 1

Flavonoid and terpenoid induced fabrication of silver nanoparticles.

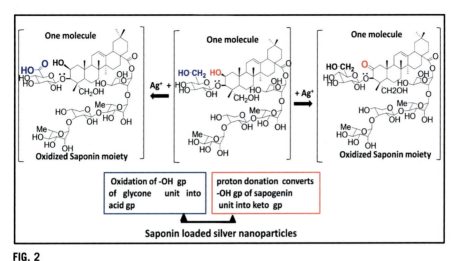

FIG. 2

Saponin induced fabrication of silver nanoparticles.

The phytofabrication of silver nanoparticles has also been carried out by saponins (plant secondary metabolite). Saponin exists as saponin glycoside, where the aglycone unit (sapogenin) and glycone unit (sugar) are both capable of reducing silver metal into its nano state (Fig. 2). The primary hydroxyl group of glycone and the secondary hydroxyl group of sapogenin moiety get oxidized into acid and keto groups, respectively, and convert the silver metal into silver nanoparticles (Palomo and Filice, 2016; Choi et al., 2018).

The pH has a significant role in the phytofabrication of silver nanoparticles by altering the charge on the molecule, affecting their properties and behavior. The change in the pH of the solution can alter peak absorption, wavelength, and intensity. The rate of formation of Ag nanoparticles is decreased with a lowering of the pH < 7 (acidic). With the increase in the pH from 9 to 13, the maximum shifts from 383 to 415 nm. However, at higher pH (13), the nanoparticles become unstable and start agglomerating. Thus, pH 11 has been found favorable for the fabrication of silver nanoparticles. Silver nanoparticles are prominently formed in alkaline conditions as at this pH phytochemicals are activated. The color change from pale yellow to brown color is characteristic of the formation of silver nanoparticles.

3 Bimetallic nanoparticles

Multimetallization is believed to enhance the catalytic activity, which may not be attained by single metal nanoparticles. The additive properties of two or more metals may enforce the synergistic effects of the two or more metals. Multimetallic nanostructures exhibit improved catalytic, optical-electronic, and pharmacological properties compared to monometallic counterparts (Sopoušek et al., 2014; Dutta et al., 2015; Nadeem et al., 2019). They display improved reactivity on account of the electronic and synergistic impacts of the various metals and their arrangement (Basavegowda and Baek, 2021). The enhancement in any bioefficacy in multimetallic nanoparticles has been ascribed to the amended compositional synergism, reinforced multimetallic surface, and elevated electron charge transfer. The structure of multimetallic nanoparticles is governed by the distribution patterns of the metals. The change in any individual metal causes an alteration in their geometrical structures and physicochemical properties, reorienting multimetallic nanoparticles into the form of an alloy with intermetallic, cluster, and core shell architecture (Kim et al., 2014; Ashishie et al., 2018). Phytofabrication of multimetallic nanoparticles has been carried out in two possible methods: co-reduction (simultaneous addition) and sequential reduction (stepwise addition). Simultaneous addition is used for the fabrication of bimetallic Au-Ag alloy in which both metals are reduced simultaneously. The second option is the stepwise addition of sequential reduction and used to form core shell structures of nanoparticles attached to another metal (Zhang et al., 2010; Ganaie et al., 2016). The simultaneous co-reduction method has been widely employed for the phytofabrication of multimetallic nanoparticles. The synthesis of the multimetallic (bimetallic) nanoparticles has been carried out by mixing silver nitrate and hydrogen tetrachloroaurate solutions with aqueous or aqueous alcoholic extracts in different ratios of plant extract as a function of pH (4–11) with sonication (Shankar et al., 2004; Naraginti et al., 2016; Sharma et al., 2019, 2020). The presence of two peaks at pH 4 and 6 have been reported. With an increase in pH, these peaks are merged into a single peak at pH (8 and 10). The prominent peak has been characterized at pH 10 (Abdel Hamid et al., 2013; Ganaie et al., 2016). The color change from pale yellow to pink has been considered as the characteristic of the fabrication of Au-Ag bimetallic nanoparticles. Mechanism of the fabrication of multimetallic nanoparticles has been given on grounds similar to that of monometallic nanoparticles. The phytofabrication of bimetallic nanoparticles has been tabulated exhibiting the enhancement in various pharmacological activities (Table 1).

4 Characterization of phytofabricated silver nanoparticles

The different astonishing properties of nanoparticles are characterized using advanced analytical techniques and include optical properties (UV-vis spectrophotometer), crystalline structure (XRD), surface morphology (FE-SEM), particles size (TEM), elemental identification and composition (EDX), surface roughness (AFM), and hydrodynamic size and zeta potential (Zetasizer). The internalization ability of nanoparticles is affected by various factors like shape, size, and surface charge. The capability of cellular uptake is influenced by the nanoparticle's shape. Rod shaped nanoparticles showed better competency of drug delivery and cellular uptake compared to sphere shape. These nanoparticles (<40nm) can pass through complex nuclear pores more easily than spheres of the same diameter (Hinde et al., 2017).

Table 1 Plant mediated bimetallic nanoparticles.

Plant	Part	Metals	Solvent	Size (nm) and shape	Bioefficacy	Reference
Syzygium aromaticum	Buds	Au-Ag	Deionized water	24.78 Spherical	Antioxidant	Sharma et al. (2020)
Salvadora persica	Whole	Ag-Ni	Deionized water	23.67 Cluster	Antioxidant	Riaz et al. (2020)
Solanum tuberosum	Rhizome	Ag-Au	Deionized water	9.7 Icosahedral	Antimicrobial, anticancer	Lomelí-Marroquín et al. (2019)
Stigmaphyllon ovatum	Leaves	Ag-Au	Deionized water	15 Triangular	Anticancer	Elemike et al. (2019)
Muntingia calabura	Leaves	Au-Ag	Deionized water	20–50 Spherical	Antioxidant	Hariharan et al. (2019)
Madhuca longifolia	Seed	Au-Ag	Aqueous alcohol	81.50 Spherical	Wound healing	Sharma et al. (2019)
Mirabilis jalapa	Leaves	ZnO-Ag	Deionized water	19–67 Spherical	Antibacterial, antioxidant	Nadeem et al. (2019)
Hamelia patens	Whole plant	Ag-Au	Deionized water	5–20 Irregular	Antioxidant	Chavez and Rosas (2019)
Phoenix dactylifera	Fruit	Ag-Fe	Deionized water	5–40 Irregular	Antioxidant, antimicrobial	Al-Asfar et al. (2018)
Borassus flabellifer	Fruit	Ag-Cu	Methanol	300 Spherical	Antioxidant, antibacterial	Chitturi et al. (2018)
Kigelia africana	Fruit	Cu-Ag	Deionized water	10 Spherical	Antimicrobial	Ashishie et al. (2018)
Gloriosa superba	Leaves	Ag-Au	Deionized water	10 Spherical	Antibacterial	Gopinath et al. (2016)
Antigonon leptopus	Whole plant	Au-Ag	Deionized water	10–60 Spherical	Antioxidant	Ganaie et al. (2016)
Chrysophyllum albidum	Leaves	Au-Ag	Deionized water	25–35 Spherical	Antioxidant	Sodeinde et al. (2016)
Punica granatum	Fruit juice	Au-Ag	Deionized water	12 Spherical	Antioxidant	Kumari et al. (2015)
Plumbago zeylanica	Root	Ag-Au	Deionized water	93.04 Polygonal	Antibacterial	Salunke et al. (2014)
Synthetic quercetin	–	Ag-Se	Deionized water	30–35 Irregular	Antimicrobial, anticancer	Mittal et al. (2014)
Potamogeton pectinatus	Whole	Ag-Au	Deionized water	6–16 Hexagonal	Antioxidant	Abdel Hamid et al. (2013)
Pueraria starch[a]	–	Au-Ag	Deionized water	32 Spherical	Catalytic activity	Xia et al. (2013)

[a]Bioactive principle.

Such nanoparticles have a longer retention time in the cells and are considered ideal for drug delivery (Kinnear et al., 2017; Cong et al., 2018). For cellular internalization, the size (<50 nm) of spherical nanoparticles has been considered highly suitable. The uptake of the charged nanoparticles is better than uncharged nanoparticles (He et al., 2010; Zhu et al., 2013; Kettler et al., 2014). The magnitude of negative charge on the fabricated nanoparticles might act as a repulsive barrier, avoiding aggregation of nanoparticles, and keeping them in nano state. Generally, most silver nanoparticles are found in different size ranges (2–50, 50–100, and 100–1000 nm) and are of spherical, rod, trigonal, rectangular, and hexagonal shapes.

5 Important bioefficacies of phytofabricated silver nanoparticles

The use of silver ions for their pharmacological properties diminished for some time owing to the discovery of newly designed chemo drugs. However, the emergence of resistance in pathogenic microorganisms against conventional chemo drugs became a point of immense concern (Desselberger, 2000). Hence, silver regained attention for various pharmacological applications.

5.1 Antiinflammatory bioefficacy

Plant based antiinflammatory drugs are currently gaining an appreciable position in pharmaceutical industries (Govindappa et al., 2018; Aafreen et al., 2019; Dawadi et al., 2021). Investigations on silver nanoparticles induced antiinflammatory bioefficacy have been tabulated (Table 2).

Saponin is one of the most important plant secondary metabolites, exhibiting strong antiinflammatory activity. Saponin suppresses the expression of NF-kβ regulated proteins (iNOS). It also inhibits the activation of NF-kβ, the release of cyclo-oxygenase II, and finally results in the suppression of the release of proinflammatory mediators. Silver nanoparticles enter the cell through endocytosis deeply and facilitate antiinflammatory bioefficacy (Fig. 3) in four possible ways: (a) suppression of the various proinflammatory cytokines; (b) decreasing vascular endothelial growth factor (VEGF) level; (c) blocking the translocation of NF-kβ into the nucleus; and (d) inhibition in the COX-2 gene expression (Ahn et al., 2005; Jang et al., 2013).

5.2 Wound healing bioefficacy

The wound healing bioefficacy has been successfully enhanced using plant mediated silver nanoparticles (Table 3).

Flavonoids are the most abundant plant bioactive constituent for wound healing and antimicrobial bioefficacy (Cyril et al., 2019; Odeniyi et al., 2020). Flavonoids as a wound healer promote the synthesis of collagen, fibroblast, and hydroxyproline content in the cell (Kumar and Pandey, 2013). They suppress lipid peroxidation, reactive oxygen species, and proinflammatory cytokines (Gopalakrishnan et al.,

Table 2 Antiinflammatory bioefficacy of plant mediated silver nanoparticles.

Plant	Part	Solvent	Phytochemicals/ *bioactive principle	Size (nm) and shape	Reference
Nigella sativa	Seeds	70% ethanol	Polyunsaturated, fatty acids	25.2 Spherical	Alkhalaf et al. (2020)
Zephyranthes rosea	Flowers	Deionized water	Flavonoids, lectins, alkaloids	10–30 Spherical	Maheshwaran et al. (2020)
Zingiber officinale	Oil	Deionized water	Sesquiterpenes	–	Aafreen et al. (2019)
Holoptelea integrifolia	Leaves	Deionized water	*Quercetin	32–38 Spherical	Kumar et al. (2019)
Calophyllum tomentosum	Leaves	Deionized water	Tannins, saponins, alkaloids	24 Spherical	Govindappa et al. (2018)
Madhuca longifolia	Seeds	Aqueous ethanol	*Saponin	20–48 Spherical	Sharma et al. (2018)
Phyllanthus acidus	Fruit	Deionized water	Amine, hydroxyl gp	10–46 Irregular	Manikandan et al. (2017)
Viburnum opulus	Fruit	Deionized water	*Polyphenols	25 Spherical	Moldovan et al. (2017)
Heritiera fomes	Leaves and bark	Deionized water	Alcoholic compounds	20–30	Thatoi et al. (2016)
Persea americana	Leaves	Deionized water	Alkaloids	28	Anitha and Sakthivel (2015)
Rosa indica	Petals	70% ethanol	Linolenic acid	24–61 Spherical	Manikandan et al. (2015)
Sambucus nigra	Fruit	Deionized water	Polyphenols	20–80 Spherical	David et al. (2014)
Terminalia catappa	Leaves	Deionized water	Flavonoids, polysaccharides	7–11 Spherical	El-Rafie and Hamed (2014)

2016). Inherently, silver has been found to suppress the emergence of resistance in pathogenic microorganisms. Silver compounds facilitate fibroblast formation enhancing the wound healing potential. Moreover, silver nanoparticles suppress the cytokines and IL-6 levels in the cell, which leads to the dampening of the inflammatory response and accelerate the rate of wound healing (Gunasekaran et al., 2011; Franková et al., 2016).

5.3 Antimicrobial bioefficacy

Silver and phytofabricated nanoparticles have been characterized for their antimicrobial nature with no toxicity to the host cells (Table 4).

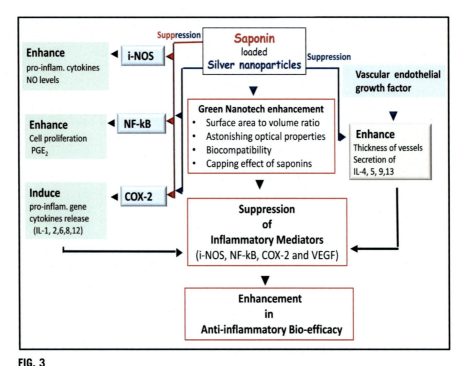

FIG. 3

Saponin loaded silver nanoparticles induced antiinflammatory bioefficacy.

5.4 Antioxidant bioefficacy

Different crude plant extracts and their bioactive constituents have been used for the phytofabrication of silver nanoparticles and assessed for antioxidant bioefficacy (Table 5).

Flavonoids have received enough attention in various epidemiological conditions to evaluate their beneficial effects. Flavonoids are a broad class of polyphenolic biomolecules with many hydroxyl groups. Depending upon the number and their positions, different pharmacological activities of flavonoids have been reported (Grigalius and Petrikaite, 2017). Free radicals scavenging bioefficacy of flavonoids is exhibited via two possible ways: hydrogen atom transfer and single electron transfer (Vrhovsek et al., 2004; Fraga et al., 2010). Silver nanoparticles loaded with flavonoids particularly facilitate antioxidant bioefficacy in three possible ways: (a) termination of free radical chain reaction via interaction between the electrons of the metallic conduction band and free radicals; (b) donation of the hydrogen atom by the flavonoid hydroxyl groups forming quinone radical; and (c) enhancement in free radical scavenging by quinone radicals (Fig. 4).

Table 3 Wound healing bioefficacy of plant mediated silver nanoparticles.

Plant	Part	Solvent	Phytochemicals/ *bioactive principle	Size (nm) and shape	Reference
Curcuma longa	Leaves	Deionized water	Flavonoids, terpenes	15–40 Spherical	Maghimaa and Alharbi (2020)
Cratoxylum formosum	Leaves	Ethanol	Phenolic compounds	9–35 Spherical	Ahn et al. (2019)
Madhuca longifolia	Seed	Aqueous alcohol	*Flavonoids	81.50 Spherical	Sharma et al. (2019)
Euphorbia milii	Leaves	Deionized water	Phenolic compounds	20–30 Spherical	Gong et al. (2018)
Cassia roxburghii	Leaves	Deionized water	Alkaloids	35 Spherical	Pannerselvam et al. (2017)
Coleus forskohlii	Roots	Deionized water	Terpenoids	5–15 Spherical	Naraginti et al. (2016)
Azadirachta indica	Leaves	Deionized water	Flavonoids	60–85 Spherical	Bhagavathy and Kancharla (2016)
Cucumis sativus	Leaves	Deionized water	Cucurbitacin	21–23 Spherical	Venkatachalam et al. (2015)
Lansium domesticum	Fruit peel	Deionized water	Triterpenoids	10–30 Spherical	Shankar et al. (2015)
Bryonia laciniosa	Leaves	Deionized water	Cucurbitacin	15	Dhapte et al. (2014)
Naringi crenulata	Leaves	Deionized water	Tannins, saponin	72–98 Spherical	Bhuvaneswari et al. (2014)

5.5 Anticancer bioefficacy

The strong relationship of antioxidant anticancer bioefficacy of flavonoids has been established. Flavonoids act as antioxidants in the homeostasis of ROS and prooxidants in cancer cells. They are able to arrest the cell proliferation cycle and initiate the apoptosis process (Jeon et al., 2019; Kopustinskiene et al., 2020). Plant mediated silver nanoparticles have also been assessed for anticancer bioefficacy (Table 6).

Plant mediated silver nanoparticles have modernized the treatment of cancerous tissues without interfering with the normal health of human cells (Raghunandan et al., 2011; Akhtar et al., 2013). They have unique applications in maintaining homeostatic of ROS level in cells and deep penetration in the cell membrane. They induce the generation of NO release, forming toxic peroxynitrite, which suppresses tumors via oxidative damage. It also regulates mitochondrial and death receptor mediated apoptosis via p53 and p21 pathways. The mitochondrial activity activates the enzymes (Bax and Bcl-2) and cytochrome c, which enhance the caspase activity and finally result in cellular apoptosis (Fig. 5).

Table 4 Antimicrobial bioefficacy of plant mediated silver nanoparticles.

Plant	Part	Solvent	Phytochemical/ *bioactive principle	Size (nm) and shape	Reference
Allium ampeloprasum	Arial parts	Deionized water	Phenolic compounds	8–50 Spherical	Jalilian et al. (2020)
Berberis asiatica	Root	Deionized water	Flavonoids, tannins, phenols	7–15 Spherical	Dangi et al. (2020)
Derris trifoliate	Seed	Deionized water	Alkaloids, flavonoid, saponin	11–21 Spherical	Cyril et al. (2019)
Mirabilis jalapa	Leaves	Deionized water	Phenolic, flavonoids	19–68 Spherical	Nadeem et al. (2019)
Calophyllum tomentosum	Leaves	Deionized water	Flavonoids, saponins, tannins, alkaloids	24 Spherical	Govindappa et al. (2018)
Diospyros montana	Bark	Methanol	Alkaloids, saponins, flavonoids	5–40 Spherical	Bharathi et al. (2018)
Erythrina suberosa	Leaves	Deionized water	Flavonoids, phenols	15–34 Circular	Mohanta et al. (2017)
Trigonella foenum-graecum	Seed	Deionized water	*Saponin	4–15 Spherical	Muniyan et al. (2017)
Bergenia ciliata	Rhizome	Methanol	Phenolics, flavonoids	35 Spherical	Phull et al. (2016)
Helicteres isora	Bark	Deionized water	Sapogenin	16–95 Oval	Bhakya et al. (2016)
Lantana camara	Leaves	Deionized water	Terpenoids	20	Ajitha et al. (2015)
Rosmarinus officinalis	Leaves	Deionized water	Polyphenols	10–33 Spherical	Ghaedi et al. (2015)
Chenopodium murale	Leaves	Deionized water	Flavonoids, phenolics	30–50 Spherical	Aziz Abdel et al. (2014)
Piper longum	Fruit	Deionized water	Polyphenol	48	Reddy et al. (2014)
Alternanthera sessilis	Leaves	Deionized water	Alkaloid, tannins	10–30 Spherical	Niraimathi et al. (2013)
Parthenium hysterophorus	Leaves	Deionized water	Flavonoids	5–25 Spherical	Kalaiselvi et al. (2013)
Dioscorea bulbifera	Tuber	Deionized water	Flavonoids	8–20 Spherical	Ghosh et al. (2012)
Desmodium triflorum	Whole plant	Deionized water	*Ascorbic acid	5–20	Ahmad et al. (2011)
Vitex negundo	Leaves	Methanol	Phenolic compounds	10–0 Spherical	Zargar et al. (2011)
Acalypha indica	Leaves	Deionized water	*Quercetin	10–30 Spherical	Krishnaraj et al. (2010)

Table 5 Antioxidant bioefficacy of plant mediated silver nanoparticles.

Plant	Part	Solvent	Phytochemicals/ *bioactive principle	Size (nm) and shape	Reference
Allium ampeloprasum	Arial parts	Deionized water	Phenolic compounds	8–50 Spherical	Jalilian et al. (2020)
Nigella sativa	Seeds	70% ethanol	Polyunsaturated, fatty acids	25.2 Spherical	Alkhalaf et al. (2020)
Allium cepa	Bulbs	Deionized water	Phenols, flavonoids, tannins	49–73 Spherical	Jini and Sharmila (2020)
Cratoxylum formosum	Leaves	Ethanol	Phenolic compounds	9–36 Spherical	Ahn et al. (2019)
Derris trifoliate	Seed	Deionized water	Alkaloids, flavonoids	11–21 Spherical	Cyril et al. (2019)
Passiflora edulis	Leaves	Deionized water	Terpenoids, flavonoids	3–7 Spherical	Thomas et al. (2019)
Calophyllum tomentosum	Leaves	Deionized water	Flavonoids, saponins, tannins	24 Spherical	Govindappa et al. (2018)
Andean blackberry	Fruit	Deionized water	Flavonoids, phenolic compounds	12–50 Spherical	Al-Asfar et al. (2018)
Agaricus bisporus	Whole	Deionized water	Flavoproteins	10–80 Spherical	Kumar et al. (2017)
Erythrina suberosa	Leaves	Deionized water	Flavonoids, phenols	15–34 Circular	Mohanta et al. (2017)
Bergenia ciliate	Rhizome	Methanol	Phenolic compounds, flavonoids	35 Spherical	Phull et al. (2016)
Helicteres isora	Bark	Deionized water	Sapogenin	16–95 Oval	Bhakya et al. (2016)
Citrus sinensis	Fruit juice	None	Phenolic compounds	4–10 Spherical	Khan et al. (2015)
Chenopodium murale	Leaves	Deionized water	Flavonoids, phenolic compounds	30–50 Spherical	Aziz Abdel et al. (2014)
Piper longum	Fruit	Deionized water	Polyphenol	48	Reddy et al. (2014)
Alternanthera sessilis	Leaves	Deionized water	Alkaloids, tannins	10–30 Spherical	Niraimathi et al. (2013)
Parthenium hysterophorus	Leaves	Deionized water	Flavonoids	5–25 Spherical	Kalaiselvi et al. (2013)
Iresine herbstii	Leaves	Ethanol	Phenolic compounds	44–64 Spherical	Dipankar and Murugan (2012)
Rhododendron dauricum	Flower	Deionized water	Phenolic compounds	25–40	Mittal et al. (2012)

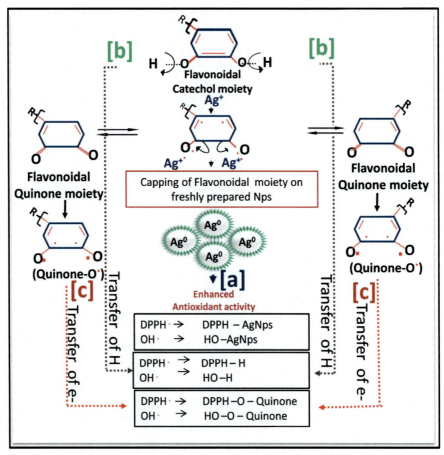

FIG. 4

Flavonoid loaded silver nanoparticles induced antioxidant bioefficacy.

6 Conclusion and future perspectives

Since ancient times, silver has been used for antimicrobial, wound dressing, and skin related ailments as it has inherent medicinal properties. Bioactive component loaded silver nanoparticles impart significant enhancement in pharmacological bioefficacies. The biocompatibility, astonishing optical properties, high surface area to volume ratio (nano sizing), and coating of medicinally important plant secondary metabolites on freshly generated nanoparticles has been attributed to the observed considerable enhancement. Such nanoparticles possess increased permeability in the cell membrane, and enhanced deep penetration and uptake in tissues. The increased drug accumulation in the cells largely enhances their bioactivity. Plant

Table 6 Anticancer bioefficacy of plant mediated silver nanoparticles.

Plant	Part	Solvent	Phytochemicals/ *bioactive principle	Size (nm) and shape	Reference
Allium ampeloprasum	Arial parts	Deionized water	Phenolic compounds	8–50 Spherical	Jalilian et al. (2020)
Cuminum cyminum	Seeds	Deionized water	Cumin aldehyde	100 Spherical	Dinparvar et al. (2020)
Tussilago farfara	Flower bud	Hexane, Methanol	*Sesquiterpenoids	10–15 Spherical	Lee et al. (2019)
Synthetic polyphenolics	–	Deionized water	*Gallic acid	10–25 Spherical	Thapliyal et al. (2019)
Nigella arvensis	Seed	Deionized water	*Flavonoids	2–15 Spherical	Chahardoli et al. (2017)
Helicteres isora	Bark	Deionized water	Sapogenin	16–95 Oval	Bhakya et al. (2016)
Bambusa arundinacea	Leaves	Deionized water	Flavonoids, phenolic acids	30–90 Spherical	Kalaiarasi et al. (2015)
Catharanthus roseus	Leaves	Deionized water	*Alkaloids	20–50 Spherical	Ghozali et al. (2015)
Helianthus annuus	Seed oil	Petroleum ether	*Linoleic acid	40–50 Hexagonal	Thakore et al. (2014)
Piper longum	Fruit	Deionized water	Polyphenol	48	Reddy et al. (2014)
Citrullus colocynthis	Leaves	Deionized water	Flavonoids	5–19 Spherical	Shawkey et al. (2013)
Azadirachta indica	Leaves	Deionized water	Flavonoids	10–100	Renugadevi and Aswini (2012)

mediated silver based nanomedicines have opened a new avenue for opposing multidrug resistance of microorganisms.

The futuristic research program on silver nanoparticles is in progress with the following specific objectives:

- Phytofabrication of silver nanoparticles with controlled physicochemical properties.
- Newer pharmacological responses to human cells by silver nanoparticles treatments.
- Enhanced beneficial payload to cells.
- Examination of possible resistance against silver nanoparticles.
- Considerations of cytotoxicity, genotoxicity, and genetic studies.

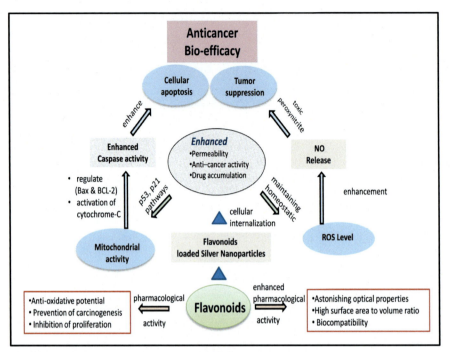

FIG. 5

Flavonoid loaded silver nanoparticles induced anticancer bioefficacy.

Acknowledgments

The authors are grateful to Prof. P.K. Kalra, Director, Dayalbagh Educational Institute, Agra, and Prof. S. Dass, Head of the Department, for extending all the necessary facilities and motivation. The financial support to Ms. Mukti Sharma given by the University Grants Commission (UGC), New Delhi (India), under UGC BSR/Fellowship No.ET/UGC/2136 is gratefully acknowledged.

References

Aafreen, M.M., Anitha, R., Preethi, R.C., Rajeshkumar, S., Lakshmi, T., 2019. Anti-inflammatory activity of silver nanoparticles prepared from ginger oil-an in vitro approach. Indian J. Public. Health Res. Dev. 10, 160–164.

Abdel Hamid, A.A., Al-Ghobashy, M.A., Fawzy, M., Mohamed, M.B., Abdel-Mottaleb, M.M., 2013. Phytosynthesis of Au, Ag, and Au–Ag bimetallic nanoparticles using aqueous extract of sago pondweed (*Potamogeton pectinatus* L.). ACS Sustain. Chem. Eng. 1, 1520–1529.

Agency for Toxic Substances and Disease Registry (ATSDR), 1990. Toxicological Profile for Silver. Atlanta, US.

Ahmad, N., Sharma, S., Singh, V.N., Shamsi, S.F., Fatma, A., Mehta, B.R., 2011. Biosynthesis of silver nanoparticles from *Desmodium triflorum*: a novel approach towards weed utilization. Biotechnol. Res. Int. 2011, 454090.

Ahmed, S., Ikram, S., 2016. Biosynthesis of gold nanoparticles: a green approach. J. Photochem. Photobiol. B 161, 141–153.

Ahn, K.S., Noh, E.J., Zhao, H.L., Jung, S.H., Kang, S.S., Kim, Y.S., 2005. Inhibition of inducible nitric oxide synthase and cyclooxygenase II by *Platycodon grandiflorum* saponins via suppression of nuclear factor-κB activation in RAW 264.7 cells. Life Sci. 76, 2315–2328.

Ahn, E.Y., Jin, H., Park, Y., 2019. Assessing the antioxidant, cytotoxic, apoptotic and wound healing properties of silver nanoparticles green-synthesized by plant extracts. Korean J. Couns. Psychother. 101, 204–216.

Ajitha, B., Reddy, Y.A.K., Reddy, P.S., 2015. Green synthesis and characterization of silver nanoparticles using *Lantana camara* leaf extract. Korean J. Couns. Psychother. 49, 373–381.

Akhtar, M.S., Panwar, J., Yun, Y.S., 2013. Biogenic synthesis of metallic nanoparticles by plant extracts. ACS Sustain. Chem. Eng. 1, 591–602.

Al-Asfar, A., Zaheer, Z., Aazam, E.S., 2018. Eco-friendly green synthesis of Ag@ Fe bimetallic nanoparticles: antioxidant, antimicrobial and photocatalytic degradation of bromothymol blue. J. Photochem. Photobiol. B 185, 143–152.

Alkhalaf, M.I., Hussein, R.H., Hamza, A., 2020. Green synthesis of silver nanoparticles by *Nigella sativa* extract alleviates diabetic neuropathy through anti-inflammatory and antioxidant effects. Saudi J. Biol. Sci. https://doi.org/10.1016/j.sjbs.2020.05.005.

Anitha, P., Sakthivel, P., 2015. Synthesis and characterization of silver nanoparticles using *Persea americana* (Avocado) and its anti-inflammatory effects on human blood cells. Int. J. Pharm. Sci. Rev. Res. 35, 173–177.

Ashishie, P.B., Anyama, C.A., Ayi, A.A., Oseghale, C.O., Adesuji, E.T., Labulo, A.H., 2018. Green synthesis of silver monometallic and copper-silver bimetallic nanoparticles using *Kigelia africana* fruit extract and evaluation of their antimicrobial activities. Int. J. Phys. Sci. 13, 24–32.

Aziz Abdel, M.S., Shaheen, M.S., El-Nekeety, A.A., Abdel-Wahhab, M.A., 2014. Antioxidant and antibacterial activity of silver nanoparticles biosynthesized using *Chenopodium murale* leaf extract. J. Saudi Chem. Soc. 18, 356–363.

Basavegowda, N., Baek, K.H., 2021. Multimetallic nanoparticles as alternative antimicrobial agents: challenges and perspectives. Molecules 26. https://doi.org/10.3390/molecules26040912.

Bhagavathy, S., Kancharla, S., 2016. Wound healing and angiogenesis of silver nanoparticle from *Azadirachta indica* in diabetes induced mice. Int. J. Herb. Med. 4, 24–29.

Bhakya, S., Muthukrishnan, S., Sukumaran, M., Grijalva, M., Cumbal, L., Benjamin, J.F., Kumar, T.S., Rao, M.V., 2016. Antimicrobial, antioxidant and anticancer activity of biogenic silver nanoparticles—an experimental report. RSC Adv. 6, 81436–81446.

Bharathi, D., Josebin, M.D., Vasantharaj, S., Bhuvaneshwari, V., 2018. Biosynthesis of silver nanoparticles using stem bark extracts of *Diospyros montana* and their antioxidant and antibacterial activities. J. Nanostructure Chem. 8, 83–92.

Bhuvaneswari, T., Thiyagarajan, M., Geetha, N., Venkatachalam, P., 2014. Bioactive compound loaded stable silver nanoparticle synthesis from microwave irradiated aqueous extracellular leaf extracts of *Naringi crenulata* and its wound healing activity in experimental rat model. Acta Trop. 135, 55–61.

Chahardoli, A., Karimi, N., Fattahi, A., 2017. Biosynthesis, characterization, antimicrobial and cytotoxic effects of silver nanoparticles using *Nigella arvensis* seed extract. Iran. J. Pharm. Res. 16, 1167–1175.

Chavez, K., Rosas, G., 2019. Green synthesis and characterization of Ag@ Au core-shell bimetallic nanoparticles using the extract of *Hamelia patens* plant. Microsc. Microanal. 25, 1102–1103.

Chitturi, K.L., Garimella, S., Marapaka, A.K., Kudle, K.R., Merugu, R., 2018. Single pot green synthesis, characterization, antitumor antibacterial, antioxidant activity of bimetallic silver and copper nanoparticles using fruit pulp of palmyra fruit. J. Bionanosci. 12, 284–289.

Choi, Y., Kang, S., Cha, S.H., Kim, H.S., Song, K., Lee, Y.J., Kim, K., Kim, Y.S., Cho, S., Park, Y., 2018. Platycodon saponins from Platycodi Radix (*Platycodon grandiflorum*) for the green synthesis of gold and silver nanoparticles. Nanoscale Res. Lett. 13, 1–10.

Cong, V.T., Gaus, K., Tilley, R.D., Gooding, J.J., 2018. Rod-shaped mesoporous silica nanoparticles for nanomedicine: recent progress and perspectives. Expert Opin. Drug Deliv. 15, 881–892.

Crisan, C.M., Mocan, T., Manolea, M., Lasca, L.I., Tăbăran, F.A., Mocan, L., 2021. Review on silver nanoparticles as a novel class of antibacterial solutions. Appl. Sci. 11, 1120. https://doi.org/10.3390/app11031120.

Cyril, N., George, J.B., Joseph, L., Raghavamenon, A.C., Salyas, V.P., 2019. Assessment of antioxidant, antibacterial and anti-proliferative (lung cancer cell line A549) activities of green synthesized silver nanoparticles from *Derris trifoliata*. Toxicol. Res. 8, 297–308.

Dangi, S., Gupta, A., Gupta, D.K., Singh, S., Parajuli, N., 2020. Green synthesis of silver nanoparticles using aqueous root extract of *Berberis asiatica* and evaluation of their antibacterial activity. Chem. Data Collect., 100411. https://doi.org/10.1016/j.cdc.2020.100411.

Das, R.K., Gogoi, N., Bora, U., 2011. Green synthesis of gold nanoparticles using *Nyctanthes arbortristis* flower extract. Bioprocess Biosyst. Eng. 34, 615–619.

Das, R.K., Pachapur, V.L., Lonappan, L., Naghdi, M., Pulicharla, R., Maiti, S., Cledon, M., Dalila, L.M., Sarma, S.J., Brar, S.K., 2017. Biological synthesis of metallic nanoparticles: plants, animals and microbial aspects. Nanotechnol. Environ. Eng. 2, 18.

David, L., Moldovan, B., Vulcu, A., Olenic, L., Perde-Schrepler, M., Fischer-Fodor, E., Florea, A., Crisan, M., Chiorean, I., Clichici, S., Filip, G.A., 2014. Green synthesis, characterization and anti-inflammatory activity of silver nanoparticles using European black elderberry fruits extract. Colloids Surf. B Biointerfaces 122, 767–777.

Dawadi, S., Katuwal, S., Gupta, A., Lamichhane, U., Thapa, R., Jaisi, S., Lamichhane, G., Bhattarai, D.P., Parajuli, N., 2021. Current research on silver nanoparticles: synthesis, characterization, and applications. J. Nanomater. https://doi.org/10.1155/2021/6687290.

Desselberger, U., 2000. Emerging and re-emerging infectious diseases. J. Inf. Secur. 40, 3–15.

Dhapte, V., Kadam, S., Moghe, A., Pokharkar, V., 2014. Probing the wound healing potential of biogenic silver nanoparticles. J. Wound Care 23, 431–441.

Dinparvar, S., Bagirova, M., Allahverdiyev, A.M., Abamor, E.S., Safarov, T., Aydogdu, M., Aktas, D., 2020. A nanotechnology-based new approach in the treatment of breast cancer: biosynthesized silver nanoparticles using *Cuminum cyminum* L. seed extract. J. Photochem. Photobiol. B Biol. 208, 111902.

Dipankar, C., Murugan, S., 2012. The green synthesis, characterization and evaluation of the biological activities of silver nanoparticles synthesized from *Iresine herbstii* leaf aqueous extracts. Colloids Surf. B Biointerfaces 98, 112–119.

Dutta, S., Ray, C., Sarkar, S., Roy, A., Sahoo, R., Pal, T., 2015. Facile synthesis of bimetallic Au-Pt, Pd-Pt, and Au-Pd nanostructures: enhanced catalytic performance of Pd-Pt analogue towards fuel cell application and electrochemical sensing. Electrochim. Acta 180, 1075–1084.

Dutta, T., Ghosh, N.N., Das, M., Adhikary, R., Mandal, V., Chattopadhyay, A.P., 2020. Green synthesis of antibacterial and antifungal silver nanoparticles using *Citrus limetta* peel extract. Experimental and Theoretical studies. J. Environ. Chem. Eng. 8, 104019.

Elemike, E.E., Onwudiwe, D.C., Nundkumar, N., Singh, M., Iyekowa, O., 2019. Green synthesis of Ag, Au and Ag-Au bimetallic nanoparticles using *Stigmaphyllon ovatum* leaf extract and their in vitro anticancer potential. Mater. Lett. 243, 148–152.

El-Rafie, H.M., Hamed, M.A., 2014. Antioxidant and anti-inflammatory activities of silver nanoparticles biosynthesized from aqueous leaves extracts of four *Terminalia* species. Adv. Nat. Sci. Nanosci. Nanotechnol. 5, 035008.

Fraga, C.G., Galleano, M., Verstraeten, S.V., Oteiza, P.I., 2010. Basic biochemical mechanisms behind the health benefits of polyphenols. Mol. Aspects Med. 31, 435–445.

Franková, J., Pivodová, V., Vágnerová, H., Juráňová, J., Ulrichová, J., 2016. Effects of silver nanoparticles on primary cell cultures of fibroblasts and keratinocytes in a wound-healing model. J. Appl. Biomater. Funct. Mater. 14, 137–142.

Furr, J.R., Russell, A.D., Turner, T.D., Andrews, A., 1994. Antibacterial activity of actisorb plus, actisorb and silver nitrate. J. Hosp. Infect. 27, 201–208.

Gahlawat, G., Choudhury, A.R., 2019. A review on the biosynthesis of metal and metal salt nanoparticles by microbes. RSC Adv. 9, 12944–12967.

Ganaie, S.U., Abbasi, T., Abbasi, S.A., 2016. Rapid and green synthesis of bimetallic Au–Ag nanoparticles using an otherwise worthless weed *Antigonon leptopus*. J. Exp. Nanosci. 11, 395–417.

Geraldes, A.N., Da Silva, A.A., Leal, J., Estrada-Villegas, G.M., Lincopan, N., Katti, K.V., Lugão, A.B., 2016. Green nanotechnology from plant extracts: synthesis and characterization of gold nanoparticles. Adv. Nanopart. 5, 176–185.

Ghaedi, M., Yousefinejad, M., Safarpoor, M., Khafri, H.Z., Purkait, M.K., 2015. *Rosmarinus officinalis* leaf extract mediated green synthesis of silver nanoparticles and investigation of its antimicrobial properties. J. Ind. Eng. Chem. 31, 167–172.

Ghosh, S., Patil, S., Ahire, M., Kitture, R., Kale, S., Pardesi, K., Cameotra, S.S., Bellare, J., Dhavale, D.D., Jabgunde, A., Chopade, B.A., 2012. Synthesis of silver nanoparticles using *Dioscorea bulbifera* tuber extract and evaluation of its synergistic potential in combination with antimicrobial agents. Int. J. Nanomedicine 7, 483–496.

Ghozali, S.Z., Vuanghao, L., Ahmad, N.H., 2015. Biosynthesis and characterization of silver nanoparticles using *Catharanthus roseus* leaf extract and its proliferative effects on cancer cell lines. J. Nanomed. Nanotechnol. 6, 2–6.

Gong, C.P., Li, S.C., Wang, R.Y., 2018. Development of biosynthesized silver nanoparticles-based formulation for treating wounds during nursing care in hospitals. J. Photochem. Photobiol. B 183, 137–141.

Gong, N., Sheppard, N.C., Billingsley, M.M., June, C.H., Mitchell, M.J., 2021. Nanomaterials for T-cell cancer immunotherapy. Nat. Nanotechnol. 16, 25–36.

Gopalakrishnan, A., Ram, M., Kumawat, S., Tandan, S.K., Kumar, D., 2016. Quercetin accelerated cutaneous wound healing in rats by increasing levels of VEGF and TGF-β1. Indian J. Exp. Biol. 54, 187–195.

Gopinath, K., Kumaraguru, S., Bhakyaraj, K., Mohan, S., Venkatesh, K.S., Esakkirajan, M., Kaleeswarran, P., Alharbi, N.S., Kadaikunnan, S., Govindarajan, M., Benelli, G., 2016. Green synthesis of silver, gold and silver/gold bimetallic nanoparticles using the *Gloriosa superba* leaf extract and their antibacterial and antibiofilm activities. Microb. Pathog. 101, 1–11.

Govindappa, M., Hemashekhar, B., Arthikala, M.K., Rai, V.R., Ramachandra, Y.L., 2018. Characterization, antibacterial, antioxidant, antidiabetic, anti-inflammatory and antityrosinase activity of green synthesized silver nanoparticles using *Calophyllum tomentosum* leaves extract. Results Phys. 9, 400–408.

Grigalius, I., Petrikaite, V., 2017. Relationship between antioxidant and anticancer activity of trihydroxyflavones. Molecules 22 (1–12), 2169.

Guggenbichler, J.P., Böswald, M., Lugauer, S., Krall, T., 1999. A new technology of microdispersed silver in polyurethane induces antimicrobial activity in central venous catheters. Infection 27, S16–S23.

Gulbranson, S.H., Hud, J.A., Hansen, R.C., 2000. Argyria following the use of dietary supplements containing colloidal silver protein. Cutis 66, 373–376.

Gunasekaran, T., Nigusse, T., Dhanaraju, M.D., 2011. Silver nanoparticles as real topical bullets for wound healing. J. Am. Coll. Clin. Wound Spec. 3, 82–96.

Habibullah, G., Viktorova, J., Ruml, T., 2021. Current strategies for noble metal nanoparticle synthesis. Nanoscale Res. Lett. 16. https://doi.org/10.1186/s11671-021-03480-8.

Hariharan, R.K., Pon Bharathi, A., Monisha, M., Vignesh Kumar, S., Devi, R., 2019. Green synthesis and characterization of bimetallic core-shell type nanoparticles. Personalized Medicine and Nutrition—An Insight at Coimbatore, India.

He, C., Hu, Y., Yin, L., Tang, C., Yin, C., 2010. Effects of particle size and surface charge on cellular uptake and biodistribution of polymeric nanoparticles. Biomaterials 31, 3657–3666.

Hinde, E., Thammasiraphop, K., Duong, H.T., Yeow, J., Karagoz, B., Boyer, C., Gooding, J.J., Gaus, K., 2017. Pair correlation microscopy reveals the role of nanoparticle shape in intracellular transport and site of drug release. Nat. Nanotechnol. 12, 81–89.

Hongfanga, G., Huia, Y., Chuangb, W., 2017. Controllable preparation and mechanism of nano-silver mediated by the microemulsion system of the clove oil. Results Phys. 7, 3130–3136.

Huang, Z., He, K., Song, Z., Zeng, G., Chen, A., Yuan, L., Li, H., Hu, L., Guo, Z., Chen, G., 2018. Antioxidative response of *Phanerochaete chrysosporium* against silver nanoparticle-induced toxicity and its potential mechanism. Chemosphere 211, 573–583.

Ingale, A.G., Chaudhari, A.N., 2013. Biogenic synthesis of nanoparticles and potential applications: an eco-friendly approach. J. Nanomed. Nanotechol. 4, 1–7.

Iravani, S., 2011. Green synthesis of metal nanoparticles using plants. Green Chem. 13, 2638–2650.

Jain, P.K., Lee, K.S., El-Sayed, I.H., El-Sayed, M.A., 2006. Calculated absorption and scattering properties of gold nanoparticles of different size, shape, and composition: applications in biological imaging and biomedicine. J. Phys. Chem. B 110, 7238–7248.

Jalilian, F., Chahardoli, A., Sadrjavadi, K., Fattahi, A., Shokoohinia, Y., 2020. Green synthesized silver nanoparticle from *Allium ampeloprasum* aqueous extract: characterization, antioxidant activities, antibacterial and cytotoxicity effects. Adv. Powder Technol. 31, 1323–1333.

Jang, K.J., Kim, H.K., Han, M.H., Oh, Y.N., Yoon, H.M., Chung, Y.H., Kim, G.Y., Hwang, H.J., Kim, B.W., Choi, Y.H., 2013. Anti-inflammatory effects of saponins derived from the roots of *Platycodon grandiflorus* in lipopolysaccharide-stimulated BV2 microglial cells. Int. J. Mol. Med. 31, 1357–1366.

Javed, B., Ikram, M., Farooq, F., Sultana, T., Raja, N.I., 2021. Biogenesis of silver nanoparticles to treat cancer, diabetes, and microbial infections: a mechanistic overview. Appl. Microbiol. Biotechnol. 105, 2261–2275.

Jeon, J.S., Kwon, S., Ban, K., Kwon Hong, Y., Ahn, C., Sung, J.S., Choi, I., 2019. Regulation of the intracellular ROS level is critical for the antiproliferative effect of quercetin in the hepatocellular carcinoma cell line HepG2. Nutr. Cancer 71, 861–869.

Jini, D., Sharmila, S., 2020. Green synthesis of silver nanoparticles from *Allium cepa* and its in vitro antidiabetic activity. Mater. Today: Proc. 22, 432–438.

Kaabipour, S., Hemmati, S., 2021. A review on the green and sustainable synthesis of silver nanoparticles and one-dimensional silver nanostructures. Beilstein J. Nanotechnol. 12, 102–136.

Kalaiarasi, K., Prasannaraj, G., Sahi, S.V., Venkatachalam, P., 2015. Phytofabrication of biomolecule-coated metallic silver nanoparticles using leaf extracts of in vitro-raised bamboo species and its anticancer activity against human PC3 cell lines. Turk. J. Biol. 39, 223–232.

Kalaiselvi, M., Subbaiya, R., Selvam, M., 2013. Synthesis and characterization of silver nanoparticles from leaf extract of *Parthenium hysterophorus* and its anti-bacterial and antioxidant activity. Int. J. Curr. Microbiol. App. Sci. 2, 220–227.

Kelly, K.L., Coronado, E., Zhao, L.L., Schatz, G.C., 2003. The optical properties of metal nanoparticles: the influence of size, shape, and dielectric environment. J. Phys. Chem. B 107, 668–677.

Kettler, K., Veltman, K., van de Meent, D., Van Wezel, A., Hendriks, A.J., 2014. Cellular uptake of nanoparticles as determined by particle properties, experimental conditions, and cell type. Environ. Toxicol. Chem. 33, 481–492.

Khan, A.U., Wei, Y., Khan, Z.U., Tahir, K., Khan, S.U., Ahmad, A., Khan, F.U., Cheng, L., Yuan, Q., 2015. Electrochemical and antioxidant properties of biogenic silver nanoparticles. Int. J. Electrochem. Sci. 10, 7905–7916.

Kim, N.R., Shin, K., Jung, I., Shim, M., Lee, H.M., 2014. Ag–Cu bimetallic nanoparticles with enhanced resistance to oxidation: a combined experimental and theoretical study. J. Phys. Chem. C 118, 26324–26331.

Kinnear, C., Moore, T.L., Rodriguez-Lorenzo, L., Rothen-Rutishauser, B., Petri-Fink, A., 2017. Form follows function: nanoparticle shape and its implications for nanomedicine. Chem. Rev. 117, 11476–11521.

Klueh, U., Wagner, V., Kelly, S., Johnson, A., Bryers, J.D., 2000. Efficacy of silver-coated fabric to prevent bacterial colonization and subsequent device-based biofilm formation. J. Biomed. Mater. Res. 53, 621–631.

Kopustinskiene, D.M., Jakstas, V., Savickas, A., Bernatoniene, J., 2020. Flavonoids as anticancer agents. Nutrients 12, 457.

Krishnaraj, C., Jagan, E.G., Rajasekar, S., Selvakumar, P., Kalaichelvan, P.T., Mohan, N.J., 2010. Synthesis of silver nanoparticles using *Acalypha indica* leaf extracts and its antibacterial activity against water borne pathogens. Colloids Surf. B Biointerfaces 76, 50–56.

Kumar, S., Pandey, A.K., 2013. Chemistry and biological activities of flavonoids: an overview. Sci. World J. https://doi.org/10.1007/978-3-0348-8482-2_8.

Kumar, B., Smita, K., Cumbal, L., Debut, A., 2017. Green synthesis of silver nanoparticles using Andean blackberry fruit extract. Saudi J. Biol. Sci. 24, 45–50.

Kumar, V., Singh, S., Srivastava, B., Bhadouria, R., Singh, R., 2019. Green synthesis of silver nanoparticles using leaf extract of *Holoptelea integrifolia* and preliminary investigation of its antioxidant, anti-inflammatory, antidiabetic and antibacterial activities. J. Environ. Chem. Eng. 7, 103094.

Kumari, M.M., Jacob, J., Philip, D., 2015. Green synthesis and applications of Au–Ag bimetallic nanoparticles. Spectrochim. Acta A Mol. Biomol. Spectrosc. 137, 185–192.

Lee, Y.J., Song, K., Cha, S.H., Cho, S., Kim, Y.S., Park, Y., 2019. Sesquiterpenoids from *Tussilago farfara* flower bud extract for the eco-friendly synthesis of silver and gold nanoparticles possessing antibacterial and anticancer activities. Nanomaterials 9, 819.

Link, S., El-Sayed, M.A., 2003. Optical properties and ultrafast dynamics of metallic nanocrystals. Annu. Rev. Phys. Chem. 54, 331–366.

Lomelí-Marroquín, D., Cruz, D.M., Nieto-Argüello, A., Crua, A.V., Chen, J., Torres-Castro, A., Webster, T.J., Cholula-Díaz, J.L., 2019. Starch-mediated synthesis of mono-and bimetallic silver/gold nanoparticles as antimicrobial and anticancer agents. Int. J. Nanomedicine 14, 2171.

Maghimaa, M., Alharbi, S.A., 2020. Green synthesis of silver nanoparticles from *Curcuma longa L.* and coating on the cotton fabrics for antimicrobial applications and wound healing activity. J. Photochem. Photobiol. B 204. https://doi.org/10.1016/j.jphotobiol.2020.111806.

Maheshwaran, G., Bharathi, A.N., Selvi, M.M., Kumar, M.K., Kumar, R.M., Sudhahar, S., 2020. Green synthesis of silver oxide nanoparticles using *Zephyranthes rosea* flower extract and evaluation of biological activities. J. Environ. Chem. Eng. 8, 104137.

Manikandan, R., Manikandan, B., Raman, T., Arunagirinathan, K., Prabhu, N.M., Basu, M.J., Perumal, M., Palanisamy, S., Munusamy, A., 2015. Biosynthesis of silver nanoparticles using ethanolic petals extract of *Rosa indica* and characterization of its antibacterial, anticancer and anti-inflammatory activities. Spectrochim. Acta A 138, 120–129.

Mamalis, A.G., 2007. Recent advances in nanotechnology. J. Mater. Process. Technol. 181, 52–58.

Manikandan, R., Beulaja, M., Thiagarajan, R., Palanisamy, S., Goutham, G., Koodalingam, A., Prabhu, N.M., Kannapiran, E., Basu, M.J., Arulvasu, C., Arumugam, M., 2017. Biosynthesis of silver nanoparticles using aqueous extract of *Phyllanthus acidus* L. fruits and characterization of its anti-inflammatory effect against H_2O_2 exposed rat peritoneal macrophages. Process Biochem. 55, 172–181.

Martinho, N., Damgé, C., Reis, C.P., 2011. Recent advances in drug delivery systems. J. Biomater. Nanobiotechnol. 2, 510.

McNeil, S.E., 2005. Nanotechnology for the biologist. J. Leukoc. Biol. 78, 585–594.

Mittal, A.K., Kaler, A., Banerjee, U.C., 2012. Free radical scavenging and antioxidant activity of silver nanoparticles synthesized from flower extract of *Rhododendron dauricum*. Nano Biomed. Eng. 4, 118–124.

Mittal, A.K., Kumar, S., Banerjee, U.C., 2014. Quercetin and gallic acid mediated synthesis of bimetallic (silver and selenium) nanoparticles and their antitumor and antimicrobial potential. J. Colloid Interface Sci. 431, 194–199.

Mohanta, Y.K., Panda, S.K., Jayabalan, R., Sharma, N., Bastia, A.K., Mohanta, T.K., 2017. Antimicrobial, antioxidant and cytotoxic activity of silver nanoparticles synthesized by leaf extract of *Erythrina suberosa* (Roxb.). Front. Mol. Biosci. 4, 14.

Moldovan, B., David, L., Vulcu, A., Olenic, L., Perde-Schrepler, M., Fischer-Fodor, E., Baldea, I., Clichici, S., Filip, G.A., 2017. In vitro and in vivo anti-inflammatory properties of green synthesized silver nanoparticles using *Viburnum opulus* L. fruits extract. Mater. Sci. Eng. C 79, 720–727.

Moradi, F., Sedaghat, S., Moradi, O., Arab Salmanabadi, S., 2021. Review on green nanobiosynthesis of silver nanoparticles and their biological activities: with an emphasis on medicinal plants. Inorg. Nano-Met. Chem. 51, 133–142.

Muniyan, A., Ravi, K., Mohan, U., Panchamoorthy, R., 2017. Characterization and in vitro antibacterial activity of saponin-conjugated silver nanoparticles against bacteria that cause burn wound infection. World J. Microbiol. Biotechnol. 33, 147.

Nadeem, A., Naz, S., Ali, J.S., Mannan, A., Zia, M., 2019. Synthesis, characterization and biological activities of monometallic and bimetallic nanoparticles using *Mirabilis jalapa* leaf extract. Biotechnol. Rep. 22, e00338.

Naraginti, S., Kumari, P.L., Das, R.K., Sivakumar, A., Patil, S.H., Andhalkar, V.V., 2016. Amelioration of excision wounds by topical application of green synthesized, formulated silver and gold nanoparticles in albino Wistar rats. Mater. Sci. Eng. C 62, 293–300.

Niraimathi, K.L., Sudha, V., Lavanya, R., Brindha, P., 2013. Biosynthesis of silver nanoparticles using *Alternanthera sessilis* (Linn.) extract and their antimicrobial, antioxidant activities. Colloids Surf. B Biointerfaces 102, 288–291.

Odeniyi, M.A., Okumah, V.C., Adebayo-Tayo, B.C., Odeniyi, O.A., 2020. Green synthesis and cream formulations of silver nanoparticles of *Nauclea latifolia* (African peach) fruit extracts and evaluation of antimicrobial and antioxidant activities. Sustain. Chem. Pharm. 15, 100197.

Palomo, J.M., Filice, M., 2016. Biosynthesis of metal nanoparticles: novel efficient heterogeneous nanocatalysts. Nanomaterials 6, 84.

Pannerselvam, B., Jothinathan, M.K., Rajenderan, M., Perumal, P., Thangavelu, K.P., Kim, H.J., Singh, V., Rangarajulu, S.K., 2017. An in vitro study on the burn wound healing activity of cotton fabrics incorporated with phytosynthesized silver nanoparticles in male Wistar albino rats. Eur. J. Pharm. Sci. 100, 187–196.

Patra, J.K., Das, G., Fraceto, L.F., Campos, E.V., del Pilar Rodriguez-Torres, M., Acosta-Torres, L.S., Diaz-Torres, L.A., Grillo, R., Swamy, M.K., Sharma, S., Habtemariam, S., 2018. Nano based drug delivery systems: recent developments and future prospects. J. Nanobiotechnology 16, 71.

Phull, A.R., Abbas, Q., Ali, A., Raza, H., Zia, M., Haq, I.U., 2016. Antioxidant, cytotoxic and antimicrobial activities of green synthesized silver nanoparticles from crude extract of *Bergenia ciliata*. Future J. Pharm. Sci. 2, 31–36.

Raghunandan, D., Ravishankar, B., Sharanbasava, G., Mahesh, D.B., Harsoor, V., Yalagatti, M.S., Bhagawanraju, M., Venkataraman, A., 2011. Anti-cancer studies of noble metal nanoparticles synthesized using different plant extracts. Cancer Nanotechnol. 2, 57–65.

Reddy, N.J., Vali, D.N., Rani, M., Rani, S.S., 2014. Evaluation of antioxidant, antibacterial and cytotoxic effects of green synthesized silver nanoparticles by *Piper longum* fruit. Mater. Sci. Eng. C 34, 115–122.

Renugadevi, K., Aswini, R.V., 2012. Microwave irradiation assisted synthesis of silver nanoparticle using *Azadirachta indica* leaf extract as a reducing agent and in vitro evaluation of its antibacterial and anticancer activity. Int. J. Nanomat. Bio. 2, 5–10.

Riaz, T., Mughal, P., Shahzadi, T., Shahid, S., Abbasi, M.A., 2020. Green synthesis of silver nickel bimetallic nanoparticles using plant extract of *Salvadora persica* and evaluation of their various biological activities. Mater. Res. Express 6, 1250k3.

Richards, S.R., Turner, R.J., 1984. A comparative study of techniques for the examination of biofilms by scanning electron microscopy. Water Res. 18, 767–773.

Saim, A.K., Kumah, F.N., Oppong, M.N., 2021. Extracellular and intracellular synthesis of gold and silver nanoparticles by living plants: a review. Nanotechnol. Environ. Eng. 6, 1. https://doi.org/10.1007/s41204-020-00095.

Salunke, G.R., Ghosh, S., Kumar, R.S., Khade, S., Vashisth, P., Kale, T., Chopade, S., Pruthi, V., Kundu, G., Bellare, J.R., Chopade, B.A., 2014. Rapid efficient synthesis and characterization of silver, gold, and bimetallic nanoparticles from the medicinal plant *Plumbago zeylanica* and their application in biofilm control. Int. J. Nanomedicine 9, 2635.

Shankar, S.S., Rai, A., Ahmad, A., Sastry, M., 2004. Rapid synthesis of Au, Ag, and bimetallic Au core–Ag shell nanoparticles using Neem (*Azadirachta indica*) leaf broth. J. Colloid Interface Sci. 275, 496–502.

Shankar, S., Jaiswal, L., Aparna, R.S., Prasad, R.G., Kumar, G.P., Manohara, C.M., 2015. Wound healing potential of green synthesized silver nanoparticles prepared from *Lansium domesticum* fruit peel extract. Mater. Express 5, 159–164.

Sharma, M., Yadav, S., Srivastava, M.M., Ganesh, N., Srivastava, S., 2018. Promising anti-inflammatory bio-efficacy of saponin loaded silver nanoparticles prepared from the plant *Madhuca longifolia*. Asian J. Nanosci. Mater. 1, 244–261.

Sharma, M., Yadav, S., Ganesh, N., Srivastava, M.M., Srivastava, S., 2019. Biofabrication and characterization of flavonoid-loaded Ag, Au, Au–Ag bimetallic nanoparticles using seed extract of the plant *Madhuca longifolia* for the enhancement in wound healing bio-efficacy. Prog. Biomater. 8, 51–63.

Sharma, C., Ansari, S., Ansari, M.S., Satsangee, S.P., Srivastava, M.M., 2020. Single-step green route synthesis of Au/Ag bimetallic nanoparticles using clove buds extract: enhancement in antioxidant bio-efficacy and catalytic activity. Mater. Sci. Eng. C 116, 111153.

Shawkey, A.M., Rabeh, M.A., Abdulall, A.K., Abdellatif, A.O., 2013. Green nanotechnology: anticancer activity of silver nanoparticles using *Citrullus colocynthis* aqueous extracts. Adv. Life Sci. Technol. 13, 60–70.

Sodeinde, K.O., Dare, E.O., Lasisi, A.A., Ndungu, P., Revaprasadu, N., 2016. Green synthesis of Ag, Au and Au–Ag bimetallic nanoparticles using *Chrysophyllum albidum* aqueous extract for catalytic application in electro-oxidation of methanol. J. Bionanosci. 10, 216–222.

Sopoušek, J., Pinkas, J., Brož, P., Buršík, J., Vykoukal, V., Škoda, D., Stýskalík, A., Zobač, O., Vřešťál, J., Hrdlička, A., Šimbera, J., 2014. Ag-Cu colloid synthesis: bimetallic nanoparticle characterisation and thermal treatment. J. Nanomater. 2014, 1–13.

Thakkar, K.N., Mhatre, S.S., Parikh, R.Y., 2010. Biological synthesis of metallic nanoparticles. Nanomedicine 6, 257–262.

Thakore, S., Rathore, P.S., Jadeja, R.N., Thounaojam, M., Devkar, R.V., 2014. Sunflower oil mediated biomimetic synthesis and cytotoxicity of monodisperse hexagonal silver nanoparticles. Mater. Sci. Eng. C 44, 209–215.

Thapliyal, A., Khar, R.K., Chandra, A., 2019. AgNPs loaded microemulsion using gallic acid inhibits MCF-7 breast cancer cell line and solid Ehrlich carcinoma. Int. J. Polym. Mater. 69, 292–316.

Thatoi, P., Kerry, R.G., Gouda, S., Das, G., Pramanik, K., Thatoi, H., Patra, J.K., 2016. Photo-mediated green synthesis of silver and zinc oxide nanoparticles using aqueous extracts of two mangrove plant species, *Heritiera fomes* and *Sonneratia apetala* and investigation of their biomedical applications. J. Photochem. Photobiol. B 163, 311–318.

Thomas, B., Vithiya, B., Prasad, T., Mohamed, S.B., Magdalane, C.M., Kaviyarasu, K., Maaza, M., 2019. Antioxidant and photocatalytic activity of aqueous leaf extract mediated green synthesis of silver nanoparticles using *Passiflora edulis* f. flavicarpa. J. Nanosci. Nanotechnol. 19, 2640–2648.

Trouillas, P., Marsal, P., Siri, D., Lazzaroni, R., Duroux, J.L., 2006. A DFT study of the reactivity of OH groups in quercetin and taxifolin antioxidants: the specificity of the 3-OH site. Food Chem. 97, 679–688.

Vanlalveni, C., Lallianrawna, S., Biswas, A., Selvaraj, M., Changmai, B., Rokhum, S.L., 2021. Green synthesis of silver nanoparticles using plant extracts and their antimicrobial activities: a review of recent literature. RSC Adv. 11, 2804–2837.

Venkatachalam, P., Sangeetha, P., Geetha, N., Sahi, S.V., 2015. Phytofabrication of bioactive molecules encapsulated metallic silver nanoparticles from *Cucumis sativus* L. and its enhanced wound healing potential in rat model. J. Nanomater. https://doi.org/10.1155/2015/753193.

Venugopal, B., Luckey, T.D., 1978. Metal Toxicity in Mammals. In Chemical Toxicology of Metals and Metalloids. Academic Press, New York, pp. 32–36.

Vrhovsek, U., Rigo, A., Tonon, D., Mattivi, F., 2004. Quantitation of polyphenols in different apple varieties. J. Agric. Food Chem. 52, 6532–6538.

Weinberg, E.D., 1971. Secondary metabolism: raison d'etre. Perspect. Biol. Med. 14, 565–577.

Xia, B., He, F., Li, L., 2013. Preparation of bimetallic nanoparticles using a facile green synthesis method and their application. Langmuir 29, 4901–4907.

Yamanaka, M., Hara, K., Kudo, J., 2005. Bactericidal actions of a silver ion solution on *Escherichia coli*, studied by energy-filtering transmission electron microscopy and proteomic analysis. Appl. Environ. Microbiol. 71, 7589–7593.

Zargar, M., Hamid, A.A., Bakar, F.A., Shamsudin, M.N., Shameli, K., Jahanshiri, F., Farahani, F., 2011. Green synthesis and antibacterial effect of silver nanoparticles using *Vitex negundo* L. Molecules 16, 6667–6676.

Zhang, Q., Xie, J., Yu, Y., Lee, J.Y., 2010. Monodispersity control in the synthesis of monometallic and bimetallic quasi-spherical gold and silver nanoparticles. Nanoscale 2, 1962–1975.

Zhu, M., Nie, G., Meng, H., Xia, T., Nel, A., Zhao, Y., 2013. Physicochemical properties determine nanomaterial cellular uptake, transport, and fate. Acc. Chem. Res. 46, 622–631.

PART 2

Biogenic and plant-mediated synthesis

CHAPTER 6

Decoration of carbon nanomaterials with biogenic silver nanoparticles

Aswathi Shyam[a], S. Smitha Chandran[a], R. Divya Mohan[a], Sreedha Sambhudevan[a], and Bini George[b]

[a]*Department of Chemistry, Amrita School of Arts and Sciences, Kollam, Kerala, India* [b]*Department of Chemistry, School of Physical Sciences, Central University of Kerala, Kasaragod, Kerala, India*

1 Introduction

For many centuries, both human life and technology have been using various forms of carbon. Since primitive times, different types of carbon like graphite, diamonds, charcoal, and carbon black have been extensively used due to their exclusive properties. In addition, carbon's functions are incredible as it can form bonds with itself and other light elements (Nasir et al., 2018). Carbon in the form of fuels, in the form of fossils to the costliest jewelry and the radioactive isotope, influenced and is continuing its influence in helping human beings to achieve better. Similarly, metals have also played a vital role in shaping a well-defined space for humans in the world. From weapons to the medical field, humans are forced to depend on metals. As science found more areas to root itself, new areas came into being which fascinate the whole community. One such area is nanoscience, but the world and the whole scientific community relies too much on this specific area with great enthusiasm.

Nanoscience is an interdisciplinary area which deals with extremely small particles with superior properties that can rewrite the fate of our planet. Noble metal nanoparticles possess incredible physical, chemical, mechanical, and biological properties due to their peculiar size, shape, and morphology, and are capable of performing many applications with more efficiency than their bulk counterparts. After dealing with individual metal and inorganic nanomaterials, the composite nanomaterials bearing the superior properties of the two constituent nanoparticles came into being. Silver-carbon nanocomposites are expected to revolutionize industries and research areas with their wide applications and the ability to be synthesized via green protocol methods that are both cost-effective and environmentally friendly. To understand clearly composites made of silver and carbon inorganic nanocomposites, it is essential to analyze these constituents individually. This chapter highlights graphene-based biogenic silver nanoparticles mainly along with other carbon-based silver nanocomposites and its applications.

Out of the noble metal nanoparticles, silver is the most exploited one due to its unique physicochemical features. Silver nanoparticles influence almost all areas of science and daily life, especially when they are green synthesized, and some of their vital applications are depicted in Fig. 1. *Morinda citrifolia* (Pai et al., 2015) and *Calotropis gigantea* (Ajith et al., 2019) are two of the reported plants among many that are applied for the synthesis of silver nanoparticles.

Silver, being very valuable and industrially important metal, found applications in the field of various industries like electronics, coin fabrication, and jewelry (Mijnendonckx and Leys, 2013). Along with the transition metal properties, silver itself has a wide range of antimicrobial properties. Green synthesis of silver nanoparticles is achievable through many simple and cost-effective methods. The roots, leaves, flowers, fruits, and stems of plants can act as reducing and capping agents in silver nanoparticle formation as they are rich in phytochemicals (Krishnaraj et al., 2010). Fig. 2 demonstrates the stages involved in the green synthesis of silver nanoparticles.

Materials with enormous differences in physical and chemical properties combine to form composites that possess properties different from any of their constituents (Fadiran et al., 2018). Nanoparticles as discussed earlier are the most potent substances influencing all areas of science, technology, and daily life. A new and astonishing area of materials can be developed by the supreme combination of composites and nanomaterials. Both of these materials are exceptional compared to their bulk counterparts and also exhibit much better properties. The synthesis and characterization of both organic and inorganic nanocomposites are identified as a rapidly growing area of science (Sadasivuni et al., 2019).

Nanocomposites are solid materials composed from different phases out of which any one of the phases has at least one of the dimensions in the range of nanometers. In these nanocomposites, any of the nanomaterials like nano clay, nanoparticles, or

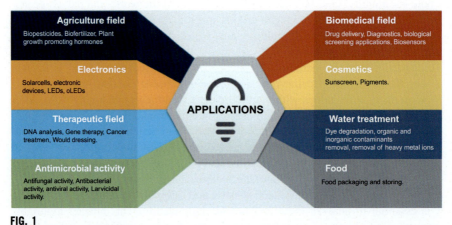

FIG. 1

Application of green AgNPs.

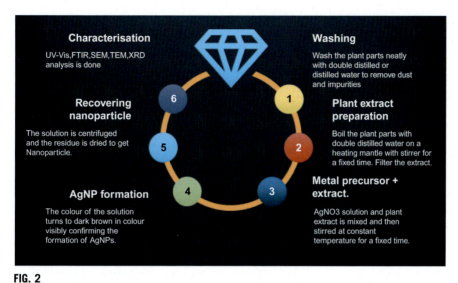

FIG. 2

Green synthesis of AgNPs.

nanofibers are incorporated into a standard matrix phase. It is composed of a matrix and filler material (Guo et al., 2018) and is depicted in Fig. 3.

According to the matrix material, nanocomposites can be classified into three different types:

1. Metal matrix nanocomposites;
2. Ceramic matrix nanocomposites; and
3. Polymer matrix nanocomposites.

Polymer matrix nanocomposites have attracted a great deal of attention due to their excellent properties and applications (Omanovi et al., 2019). Nano clay-reinforced

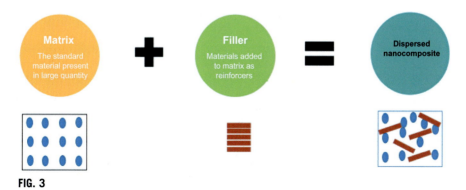

FIG. 3

Nanocomposite formation.

composite and expanded graphite-reinforced composite are important polymer matrix nanocomposites (Hussain et al., 2006). Later metal matrix nanocomposites attained more focus, and metal-organic frameworks and metal-inorganic composites came into being. A lot of biomedical and industrial applications made metal-inorganic nanocomposites a main topic of discussion. Metal-organic frameworks were the center of interest at one time as they can be employed for biomedical applications due to their higher compatibility and biodegradability (Li et al., 2014). However, their poor encapsulation capability and instability limited their applications. These limitations were possibly compensated by inorganic nanocomposites.

Graphite is the allotropic form of carbon and this graphitic layer forms the basic structure of all carbon nanoforms. When the graphite layer is in the form of a sphere, it is called zero-dimensional fullerene, and when it is rolled in correspondence to an axis, it forms one-dimensional carbon nanotubes. Graphene is the planar two-dimensional structure of the graphitic layer (Bhuyan et al., 2016). As graphene is a nano-dimensional material of carbon, it can be added to the category of nanomaterials and it paved the way for relevant investigations in the fields of nanoscience and technology (Rai et al., 2009). Graphene comprises a single sheet of graphite, i.e., a hexagonal sheet of carbon atoms is also bonded in a hexagonal lattice, forming three-dimensional hollow cylindrical structures that can be ellipsoids (C70 or C84), spheres (C60), or tubular (CNTs) (Szunerits and Boukherroub, 2016). However, only pristine graphene shows complete sp2 hybridization with no defects in a single layer of carbon atoms. Graphene may also be defined as layered materials that are held together by van der Waals force. The carbon atoms in graphene possess a two-dimensional arrangement in the honeycomb lattice. Graphene is unique because of its larger surface area, high thermal conductivity, zero bandgaps, excellent mobility of electrons at room temperature, and many other properties. Various synthetic methods are adopted for graphene preparation depending upon the desired morphology, size, purity, and quantity. Those with a high density of defects are called graphene oxides (GO) or reduced graphene oxides (rGO). Graphene oxides are obtained by the oxidation of graphene and consist of functional groups like epoxy, hydroxyl, carboxyl, and carbonyl groups. Moreover, reduced graphene oxide is a type of graphene oxide whose oxygen content is reduced by either chemical or thermal methods (Mmaduka Obodo et al., 2019). The formation of GO and rGO is depicted in Fig. 4.

The answer to the question of why graphene is preferable over other carbon-based materials is that it is free of any metallic impurities (Szunerits and Boukherroub, 2016). The proliferation of human and mammalian cells is restricted as only 50 ppm metallic impurities impart significant cytotoxicity within biological samples (Pumera and Miyahara, 2009). The low mammalian cell cytotoxicity of graphene-based materials fortified the investigations of their antibactericidal properties (Taylor et al., 2015; Nguyen and Lin, 2015; Zhou and Gao, 2014; Pinto et al., 2013). Thus, graphene-based research has become one of the main areas of nanoscience and technology due to graphene's two-dimensional honeycomb carbon architecture and also its distinguishing physicochemical behavior (Allen et al., 2010;

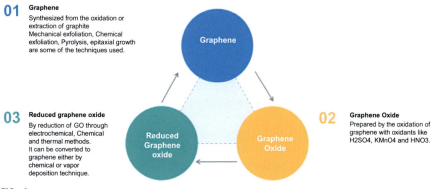

FIG. 4

Synthesis of graphene, GO, and rGO.

Choi et al., 2010). This has a wide range of applications including composites, energy storage/conversion devices, electronics, catalysis, sensors, medicine, and biology (Wei, 2012; Huang et al., 2012; Shao et al., 2010; Machado et al., 2012). Just like silver, GO can also be synthesized via a green pathway, and this method is more advantageous. Several works have reported the green synthesis of GO and these are depicted in Table 1.

Graphene can act as an important filling of composite materials due to its unique thermal, mechanical, and electrical properties. Very low addition of graphene material enhances the multifunctional properties and thus redefines the area of composites (Dhand et al., 2013). GN nanocomposites can be categorized into two types depending upon the architecture of the nanocomposites and the second component of the

Table 1 Plant-mediated synthesis of GO and its applications.

Plant	Applications	References
Aloe vera	Electrochemical and dye removal applications	Bhattacharya et al. (2017)
Rose	Fabrication of sensors and biosensors	Haghighi and Tabrizi (2013)
Eucalyptus	In high-performance supercapacitors	Manchala et al. (2019)
Lantana camara, Citrus limeta	Industrial applications, specifically in biologically sensitive areas	Medha et al. (2017)
Citrullus colocynthis	Anticancer activity against human prostate cancer	Zhu et al. (2017)
Chenopodium album	Antimicrobial and anticancer agent	Umar et al. (2020)
Salvadora persica	Technological and biological applications	Khan et al. (2015a)

nanocomposites. Based on the second component of the nanocomposites, GN nanocomposites can be classified into GN-inorganic nanocomposites and GN-polymer nanocomposites. GN-inorganic nanocomposites are of significant interest and include many types such as GN-metal nanocomposites, GN-nonmetal nanocomposites, GN-carbon nanocomposites, and GN-metal compound nanocomposites. GN-inorganic nanocomposites can be fabricated through in-situ and ex-situ techniques. In-situ techniques involve hydrothermal/solvothermal methods, template methods, direct decomposition of the metal precursor, solution deposition methods, etc. Ex situ techniques involve covalent interactions, noncovalent interactions, electrostatic interactions, etc. and the commonly used reducing agents include $NaBH_4$, ethylene glycol, and formaldehyde. The presence of randomly distributed oxygen groups on the GO sheets during reduction will obstruct the uniformity of the dispersed metal nanoparticles. In addition, the π-π stacking interactions result in low dispersibility during the GO reduction. These disadvantages can be resolved by employing a capping agent (Bai and Shen, 2012). Now that the need of applying a green reducing agent that can perform as a reducing, stabilizing, and capping agent became vigorous.

Greater potential is achieved when graphenes are integrated with noble metal nanoparticles like silver, gold, etc. This is because of the ability of GN to reduce the noble metal consumption and the potential electronic interaction with the noble metal. In these composites, GN nanosheets act as the continuous phase which is the substrate that holds the second metal component. However, silver, like graphite, has immense application possibilities in biology and medicine, and also has excellent electronic, catalytic, and antimicrobial properties. In addition, silver is much more cost-effective than gold and palladium. Integrating metal nanoparticles with graphene sheets has enormous potential particularly due to their electron distribution between metal and carbon (Subrahmanyam et al., 2010). This also results in distinctive physicochemical properties which are used in different applications like solar energy conversion, Raman scattering, conductive thin films, and antimicrobial products (Chang et al., 2014; Zhao et al., 2014; Tien et al., 2010; Nguyen et al., 2012).

It was recently reported that the antimicrobial property of GO sheets can be enhanced by controlling various parameters like time of exposure, incorporation of metallic NPs into GO sheets, concentration, and methodology of exposure. The AgNPs introduced into the sheets decreased the exposure time for complete inactivation of microbes to 30 min and a concentration of $100\,\mu g\,mL^{-1}$ was preferred (Montes-Duarte et al., 2021). A reductant- and surfactant-free facile method can be used for the preparation of Ag-GO nanocomposite with synergistic inhibitory capacity against gram-negative *E. coli* microbes. The SEM image of this synthesized Ag-GO nanocomposite showed the light-colored AgNPs decorated throughout the crumbled GO sheets (Ahmad et al., 2021).

However, the synthetic methods considered for the preparation of carbon-silver composites influence their properties and hence the applications. The unique properties of these graphene-silver nanocomposites come from the combined effect of the superior properties of both graphene and silver. As mentioned above, the morphologies of these composites are greatly influenced by the synthetic methods adopted (Abboud et al., 2016). Earlier chemical and physical approaches have been used for the preparation of graphene-Ag nanocomposites (Pandey et al., 2011; Zedan et al., 2012; Liu et al., 2012; Qiu et al., 2011; Liu et al., 2011; Tang et al., 2011). Two different methods to synthesize graphene/Ag nanosheets were used. In the first method, nanocomposites were obtained by the chemical reduction of presynthesized nanoparticles added to the surface of the prefunctionalized graphene sheets (Ren et al., 2011). The other method directly adds NPs on the graphene oxide sheets which were reduced separately using some inorganic precursors (Feng et al., 2013; Kumar et al., 2013). As this method was found to require a large number of copious chemical reagents and sophisticated instruments, scientists started to search for an alternative form of green synthesis. Green synthesis methods are cost-effective and the reducing/capping agents used are derived from bioresources such as microbes, plants, and other biomaterials (Narayanan and Sakthivel, 2010; Iravani, 2011; Vinod et al., 2011). Green synthesis is used not only for the metal nanoparticle synthesis but also for functionalizing the graphene sheets. Thus, a synthesis method that can be achieved in a single step came into being. This chapter focuses mainly on the green synthetic methods of Ag/GO nanocomposites. Along with this, green synthesis of silver and other carbon-based nanocomposites is also considered.

2 Graphene oxide-ag nanocomposites: synthesis and applications

2.1 Green synthesis

Decoration of graphene sheets with silver NPs was obtained by leaf extract-mediated synthesis which is an economically viable and environmentally friendly approach. In all plant-mediated synthesis, the extract should be prepared which is going to act as a reducing/stabilizing agent. The leaves should be washed carefully with distilled water to remove dust and impurities. Soya leaf extract was used in one study. The extract prepared was cooled and then added to the solution containing graphene, which was sonicated before $AgNO_3$ solutions of different concentrations were added. It was understood that soya leaf extract acts as the reducing agent and silver ions get converted to neutral Ag atoms (Abboud et al., 2016). When *Pulicaria glutinosa* leaf extract was used, the same procedure occurs (Al-marri et al., 2015). Graphene oxide sheets act as the center of nucleation. Plant extracts work as reducing agents along with their function as a ligand to bind AgNPs on graphene sheets. In these green protocol methods, aqueous solvents are used for preparation rather than organic solvents, which may be hazardous (Iravani, 2011) (Kumar and Yadav, 2009) (Rai et al., 2008). It should also be noted that the concentration of leaf extract, the pH, and the temperature at which the reaction is performed influence the size and morphology of the composite formed, which will then affect the properties and thus the application (Patra and Baek, 2014). *Colocosia esculenta* leaf extract was added to the ultrasonicated solution containing graphene oxide, followed by the addition of AgNO$_3$ (Barua et al., 2014).

These approaches completely follow the green chemistry principles, and better particle size control is also achieved by these methods. Not only do plant extracts serve as biological agents but also different types of microorganisms play a vital role in the bio-chelation and biomineralization of heavy metals (Gadd and Griffiths, 1977). Hence these can also be used for the preparation of metallic nanoparticles. This biosynthesis using microorganisms can be done either intra- or extracellularly (Ahmad et al., 2003). The *F. oxysprum* strain was used in one study for the development of bio-Ag nanoparticles. Graphene oxide was suspended in deionized water and the previously prepared bio-Ag was added to this solution (De Faria et al., 2014). The synthesis of bio-Ag-rGO composites is depicted in Fig. 5. The prepared graphene-Ag nanocomposites were characterized by UV-vis spectroscopy, TEM, SEM, and XRD. All these synthesized nanocomposites have varied applications that can influence human life significantly.

In another case of preparation of graphene oxide-silver nanoparticle nanohybrids, L-ascorbic acid was employed as a green reducing agent. GO and AgNO$_3$ underwent reduction simultaneously in the presence of L-ascorbic acid (Cobos et al., 2020). With the same L-ascorbic acid as the reducing agent, another paper reported the synthesis of Ag/GO nanocomposites possessing antimicrobial resistance against *P. aeruginosa* and *S. aureus*. A solution of 100 mL of AgNO$_3$/NH$_3$ was slowly added to 100 mL of GO; 100 mL of L-ascorbic acid was then added, with constant stirring. The resultant mixture was heated for 1 h at 50°C to obtain an Ag/GO nanocomposite. A TEM image of the nanocomposite showed uniformly scattered black dots of AgNPs with an average size of 17.65 ± 4.76 nm (Minh Dat et al., 2021).

Lespedeza cuneata leaf extract-mediated synthesis of GO-Ag nanocomposites was reported, and this exhibited excellent anticancer activity against brain tumors (LN229) and human cancer cells (A549). A powder of 50 mg of already synthesized

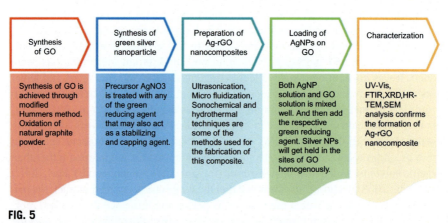

FIG. 5

Synthesis of bio-Ag-rGO composites.

GO was mixed with 50 mL of AgNO$_3$ solution and sonication occurred for about 30 min. To this solution, *L. cuneata* extract of 50 mL was added. After the overnight and constant stirring of the solution, the resultant nanocomposite was obtained by centrifugation. *Mangifera indica* flowers (Mariadoss et al., 2020) were also used as stabilizing and reducing agents for the synthesis of GO-Ag nanocomposites that can act as an efficient catalyst for the reduction of 4-nitrophenol and organic dyes (Kolya et al., 2019). Characterization techniques confirmed the formation of these composites by giving an absorption spectrum in the range 230–265 nm for L-ascorbic acid-mediated synthesis and demonstrating an AgNP peak at 404 nm (Cobos et al., 2020), while *L. cuneata*-mediated synthesis showed a peak of GO at 234 nm and an AgNP peak at 450 nm (Mariadoss et al., 2020), confirming the formation of AgNPs on GO sheets. The in-situ immobilization of AgNPs onto GO and rGO is achieved in the presence of green tea extract as a reducing agent. This nanocomposite is highly preferred for the degradation of methylene blue, which is a toxic pollutant in water resources. The presence of GO in GO-Ag nanocomposites is confirmed by the appearance of a peak at $2\theta = 10.7$ degree in XRD analysis. SEM analysis also displayed the morphological analysis that confirms the presence of spherical AgNPs on the layered structure of GO (Aboelfetoh et al., 2020).

Yet another method, using sodium citrate as an eco-friendly reducing and stabilizing agent, accomplishes the synthesis of Ag-GO nanocomposites in a greener way. The XRD pattern of this GNS/AgNPs gives four major peaks at $2\theta = 38.2, 44.4, 64.5$, and 77.5 degree, corresponding to the (111), (200), (220), and (311) planes of the Ag cubic crystal, confirming the reduction and formation of AgNPs. A corrugated morphology was found in SEM analysis which indicates the lamellar structure of the composite where the Ag crystallites act as spacers that separate the neighboring sheets from each other. TEM images gave the size of AgNPs on GNS as 20–25 nm and the IR spectra gave peaks at 1060 cm^{-1}, 1245 cm^{-1}, 1371 cm^{-1}, 1626 cm^{-1}, 1725 cm^{-1}, and 3414 cm^{-1}, which are assigned correspondingly to stretching vibrations of C—O, symmetrical ring deformation of an epoxy group, deformations of the O—H groups in C—OH groups, skeletal vibrations exhibited by unoxidized graphitic domains, and stretching vibrations of C=O groups in COOH (Acar Bozkurt, 2017). While the peaks at 1245 cm^{-1}, 1725 cm^{-1}, 1371 cm^{-1}, and 1060 cm^{-1} became weak, a new band was formed at 1570 cm^{-1} that was attributed to the graphene sheet's skeletal vibrations (Acar Bozkurt, 2017). The UV spectra analysis confirmed the red-shift of the characteristic peak of GO from 227 nm to 263 nm along with the disappearance of the peak at 302 nm, which indicated the restoring of extensive conjugation of the sp^2 carbon network (Amare et al., 2020). Moreover, the peak at 440 nm was attributed to the surface plasmon resonance of AgNPs.

Similarly, grape seed extract can serve as the reducing agent for the synthesis of the specified nanocomposites. The grape seed extract synthesized composite was characterized again with similar techniques. The morphology identified using TEM, structural analysis using XRD, and UV confirmed the peak formation, and FTIR sorted out the functional groups. Concentrations of 1, 2, and 3 mM of

AgNO$_3$ were mixed with 260 μL of GO solution. A final volume of 20 mL was made by adding 1 mL of grape seed extract to all three solutions and heating it for 20 min at 95°C. Possession of antimicrobial activity against bacteria like *Pseudomonas aeruginosa, Escherichia coli, Bacillus subtilis,* and *Staphylococcus aureus* and aquatic bacteria like *Vibrio alginolyticus, Vibrio anguillaru, Vibrio parahaemolyticus,* and *Aeromonas punctata* makes this GO/Ag nanocomposite an essential entity (Amare et al., 2020).

Designing and manufacturing of AgNPs-GO composites for the selective detection of As(III) in natural water are made possible by employing ascorbic acid as the reducing agent and β-cyclodextrin as the stabilizing agent. Direct reduction of AgNO$_3$ on graphene oxide results in composite formation. The stable of GO colloid generated after sonication was mixed with 5% AgNO$_3$ of 8.3 mL and 0.2 g β-cyclodextrin and stirred continuously. Then about 10 mL of 2.6% ascorbic acid was added, which formed a precipitate of the composite. This could then be used for the development of a glassy carbon electrode that can act as an electrochemical sensor (Dar et al., 2014). Following the modified Hummers method, the oxidation of natural graphite powder resulted in GO formation where Sapodilla fruit extract was applied as a reducing agent. A visible color change from brown to dark yellowish-brown was observed when 2 mL of Sapodilla fruit extract was added to the mixture of 20 mL silver ammonia solution and 1 mL of GO, indicating the composite formation (Dutta et al., 2020). For charge storage applications, metal nanoparticles-incorporated rGO was recently fabricated. A green reducing agent (NaOH + Al foil) was employed for the concurrent reduction of GO and Ag and Cu nitrates. The result obtained from the cyclic voltammetry measurements of the electrode synthesized using this material which was coated by Cu foil confirmed the energy storage applications of GRMs (Rawal et al., 2020).

Cost-effective and sustainable fabrication of Ag/RGO nanocomposites was achieved through a one-pot flavonoid-mediated method. The flavonoid naringenin was employed in this study as a reducing and stabilizing agent. UV-vis, FTIR, Raman, and HR-TEM were the techniques used to analyze the physicochemical properties of this particular nanohybrid. NAR@Ag/RGO nanocomposites also exhibited in-vitro bactericidal activities against *E. coli, V. cholerae, S. epidermidis, Salmonella typhi, S. aureus, Proteus mirabilis,* and *Rhodococcus rhodochrous* (Pugazhendhi, 2020).

A sonochemical green fabrication method was also experimented with for the fabrication of this composite of Ag and GO. Sodium citrate was the green reducing agent. A sonochemical method is much preferred as this is a simple process involving the physisorption of Ag ions on the GO sheets. The functional groups like epoxy, carbonyl, carboxylic, and hydroxyl groups acted as the site for Ag ions. Along with the electrostatic interaction between these functional groups and Ag ions, the sonification process and the action of sodium citrate resulted in the formation of Ag/GN composites. Usually, it is difficult to achieve a carbon-metal composite due to the low surface area of carbon and weak interactions. GO, with higher surface area and chemically functionalized epoxy and hydroxyl groups, made the Ag-carbon composite a reality (Acar Bozkurt, 2017). For the electrochemical detection of ascorbic acid, hydrothermal-assisted green synthesis of the Ag/Ni and GO composite is achievable by using *Punica granatum* juice. A concentration of 70 mM of silver nitrate and nickel nitrate was stirred with 20 mL of pomegranate juice for 1 h at 80°C. GO was then added and autoclaved for 20 h at a temperature of 220°C. The square wave voltammetry technique was used to detect AA. An Ni/Ag-GO nanocomposite has lower electrode resistance than that of Ag@rGO and rGO electrodes. The electron transfer process was facilitated more in Ni/Ag@rGO nanocomposites due to its larger electrode surface area and roughness factor (Das and Sharma, 2020).

Another hydrothermal synthesis method came forward with the manufacturing of an Ag-rGO-CNT nanocomposite. This composite can act as a tyrosine-based biosensor of dopamine. A biocatalytic layer of biosensor containing tyrosine which was mixed with GO metal, coated together on a carbon-fiber microelectrode, formed the biosensor. Ag-rGO-CNT also demonstrates good catalytic activity for metal NP reduction, due to the combined properties of rGO-CNT and AgNPs like high durability and stability and larger surface area. A four-probe method of measurement of conductivity for semiconductors was used to perform the electrical conductivity measurements of the nanocomposite under discussion. Metal in the framework was responsible for making it electrically conductive by allowing the GO to exist in its conductive state. Conductivity was found to be a function of temperature as it increased with an increase in temperature. The biosensor developed from this composite is rapidly used in biological systems and characterized in vitro (Khan et al., 2015b).

Thiourea dioxide is an excellent nontoxic reducing agent that is widely used in leather, textiles, and paper industries. Using this reducing agent, uniform loading of Ag ions on GO sheets at ambient temperature is obtained. The first principle calculation showed that the AgNP adsorption and enhanced conductivity were attributed to the low vacancy defects of ideal graphene (Dong et al., 2019). New approaches for the manufacture of composites came into being over time as more works concentrated on composite applications. A green ternary composite Ag/TiO$_2$/RGO is essential in removing water contaminants. Ultrasonic impregnation with a photoreduction method is utilized here for the synthesis of this particular composite. The photodegradation of tetracycline and photoreduction of Cr (VI) has been found to be superior and can be applied for water treatments (Hou et al., 2019). Micro fluidization is an emerging green nanotechnology that can produce highly monodispersed particles with the aid of high-pressure homogenization to disperse solid-liquid systems and liquid-liquid systems. It can also reduce the particle size remarkably and form more uniform particles than any other conventional techniques due to its high-velocity impact, cavitation, and extreme shear (Villalobos-Castillejos et al., 2018).

The celery seed extract is employed as a reducing agent and capping agent for the impregnation of biogenic silver nanoparticles on the surface of rGO through the micro fluidization technique. When AgNPs were synthesized using a heating mantle with a stirrer and the ultrasonication method, the particle sizes obtained were 36.8 and 28.4, respectively. However, the micro fluidization technique gave a much smaller nanoparticle size of 20.9 nm. It required 16 cycles of micro fluidization for dispersing the smaller AgNPs on the rGO surface. The nanocomposite synthesized in this way exhibited potential antimicrobial activity against gram-positive *Bacillus cereus* and *S. aureus* and gram-negative *E. coli* and *K. pneumonia* (Soleymani et al., 2020). The adsorption studies of Ag-RGO nanocomposites can be examined at different conditions of pH, time, dosage, and concentration of metal ions when they are synthesized using pomegranate peel extract as a reducing agent. This composite can efficiently act as the adsorbent for the removal of Zn (II) and Pb (II) ions that are toxic water contaminants. Silver nanoparticles of varied concentrations are embedded on the GO for evaluating the adsorption capacity at each concentration. Therefore, the three different Ag/RGO nanocomposites samples were synthesized such as 10 wt% Ag-RGO, 20 wt% Ag-RGO, and 30 wt% Ag-RGO. The results confirmed that the effective removal of Pb (II) and Zn (II) ions was observed for the 20 wt% Ag-RGO and 30 wt% Ag-RGO nanocomposites. Removal efficiencies of 100% for Pb (II) metal ions and 71% for Zn (II) metal ions were obtained from 20 wt% Ag-RGO nanocomposites when prepared with 50 mg L^{-1} of Ag metal ion concentration. It should be taken into consideration that the adsorption efficiency of Zn (II) ions decreased rapidly with the increase in loading due to the development of Ag film, which was attributed to the high density of silver nanoparticles on the RGO surface. These adsorption reactions followed the kinetics of pseudo-second-order reactions and also the data of equilibrium is found to fit well into the Freundlich isotherms. These Ag-RGO nanocomposites were prepared by the one-pot phytochemical method and exhibited greater adsorption efficiency of toxic heavy metal ions than the raw AgNPs and RGO sheets (El-Sabban et al., 2020).

Syzygium cumini fruit extract-mediated synthesis of an Ag-GO nanocomposite is another ecologically friendly method. The composite prepared in this way has high bactericidal activity against both gram-positive and gram-negative bacteria like *P. aeruginosa* and *E. coli* even though their cytotoxicity toward mammalian cells is very low. Their microbicidal property is higher than the drug cephalexin (Thomas et al., 2020). The ultrasonicated mixture of 25 mg GO solvated in 50 mL of DH$_2$O and 50 mL of AgNO$_3$ was mixed with 50 mL of walnut extract that can act as both a reducing and a stabilizing agent. The pH was maintained around 8 with the aid of NaOH. After these steps, Ag-GO I was produced. Ag-GO II was also produced by making slight changes in the steps suggested above. The extract was added to the GO mixture before the addition of the AgNO$_3$ solution. Both these mixtures were stirred at 30–35°C for 24 h. These composites exhibited excellent anticancer and antibacterial activity (Khorrami et al., 2019).

An antibacterial nonenzymatic glucose sensor Ag-grGO nanocomposite was fabricated using glucose/saccharides as a green reducing and crosslinking agent through a one-pot hydrothermal method. AgNPs were impregnated on the porous framework of grGO (3D reduced GO) when AgCl was formed by mixing AgNO$_3$ and NaCl which are sonicated and diluted while separated in 5 mL of DI water. In the presence of glucose, this solution of AgCl was added to 30 mL of GO slurry, which resulted in an Ag/grGO nanocomposite (Hoa et al., 2017).

Table 2 Green reducing agents for the synthesis of Ag/GO nanocomposites and their applications.

Reducing agent	Applications	References
Lemon juice	Antibacterial agent against E. coli	Alsharaeh et al. (2017)
Phyllanthus acidius	Detection of uric acid and dopamine	Nayak et al. (2020)
Potamogeton pectinatus	Antibacterial activity	Sedki et al. (2015)
Psidium guajava	Detection of MB dye through SERS	Chettri et al. (2017)
Tea extract	H_2O_2 sensing applications	Salazar et al. (2019)
Tilia amarensis	Anticancer agent	Gurunathan et al. (2015)
Pome peel extract	Organic dye degradation in water	Karthik et al. (2018)
No reducing agent or toxic organic surface agent	Antibacterial agent and antibiofilm application	Wu et al. (2020)

Again, in another case, L-arginine (an essential amino acid) was employed as a reducing and superior intercalating agent. When $AgNO_3$ was mixed with GO reduced with L-Arg in a 1:1 ratio and stirred for 4h at 50°C, the solution color changed from brown to grayish-black, confirming the formation of an rGO/Ag nanocomposite with good catalytic activity toward the reduction of 4-nitrophenol (Das et al., 2018). *Pinus densiflora* is another leaf extract that helps to achieve the preparation of an AgNPs-decorated GO nanocomposite. This can be used as a dual-responsive colorimetric sensor for the detection of Cu^{2+} ions and dopamine (Basiri et al., 2018). Citric acid was used as an ecofriendly reducing agent for fabricating AgNPs, which were in turn used to decorate the amine functionalized and nonfunctionalized GOs in paper and powder form (Rodríguez-Otamendi et al., 2021). Many more studies and experiments on the green synthesis of Ag/GO nanocomposites have been reported and a few of them are discussed in Table 2.

2.2 Ag and other carbon composites

An Ag/AC nanocomposite demonstrated better catalytic activity for 4-nitrophenol reduction. Neem leaf extract served as a reducing agent in this case for the production of AgNPs (Taha and Aissa, 2020). Yet another study of biogenic silver-activated charcoal nanocomposite that can be applied for the development of water filtration column was reported. High bactericidal activity imparted by the silver nanoparticles formed with *Malpighia emarginata* extracts as a reducing agent made this nanocomposite a more effective water disinfectant. Activated charcoal weighing 100g prepared from dried coconut shell washed with sulfuric acid was mixed with 100 mL of silver nanoparticle solution for 15 min at room temperature and left overnight. The solution was filtered the next day and dried to obtain the silver-activated charcoal nanocomposite. The PXRD spectrum of this composite exhibited a less sharp and broad peak, which indicated the spherical and smaller size of the composite. SEM images confirmed the homogenous distribution of AgNPs on the porous sites of activated charcoal. The antimicrobial activity was tested against *E. coli*, *B. subtilis*, *P. aeruginosa*, and *S. aureus*. The reusability of the water filter column prepared by this composite has also attracted interest (Kiruba and Dakshinamurthy, 2013).

The presence of Cd (II) and Pb (II) in discharged textile effluent was identified by using carbon paste electrodes modified with AgNPs prepared using *Ocimum sanctum* leaf extract. Making the amount of graphite powder and paraffin oil constant, i.e., 1 g and 0.429 g, and varying the amount of AgNPs added, four modified carbon paste electrodes were constructed. As Cd is recognized as a potential carcinogen and lead is capable of causing nerval disorders when consumed in food and drink, recognition of these heavy metal ions in water resources is necessary. Cyclic voltammogram studies are sufficient to understand the ability of CPE modified with biogenic AgNPs in the selective determination of these heavy metal ions. The anodic peak current for the CPE/AgNPs was found to be higher than that of the unmodified one for both of the metal ions. This can be attributed to the presence of green AgNPs with greater surface area and therefore greater electrocatalytic activity (Amare et al., 2020). An aldehyde reduction process was used for the fabrication of Ag/MWNT. On the external walls of CNT, AgNPs were dotted by this process carried out in a supercritical fluid of carbon dioxide. The synthesized composite may be used as a lubricant additive in engine oil. The Ag/MWCNT must go through a surface modification treatment with oleylamine. The study mainly focused on the tribological performance of the composite.

Morphology and structure analysis were carried out using XRD and SEM. The diffraction peaks of silver in S-Ag/MWCNT composites were broader, which indicated the smaller grain size of silver crystals. Strong diffraction signals of AgNPs hid the peaks of carbon and therefore in S-Ag/MWCNT there was no diffraction peak for carbon. TEM revealed the size of this composite to be in the range of 10–30 nm. Low wear scar diameter and friction coefficient were confirmed in S-Ag/MWCNT when compared to other 10w40 engine oil. This is because S-Ag/MWCNT is deposited in the valleys and grooves on the worn surface (Meng et al., 2018). A similar nanocomposite of AgNPs and MWCNTs that was capable of exhibiting enhanced catalytic and antimicrobial activity was synthesized obeying the green chemistry principles. Gelatin, due to its innate biodegradability and biocompatibility, was employed as a green reducing and stabilizing agent. Before entering into the synthesis, MWCNT was functionalized to possess a carboxylic acid group. Functionalization was done because the pristine MWCNTs chosen here are deficient in surface functional groups that prevent them from forming complexes. Characterization techniques confirmed the homogenous distribution of silver nanoparticles on the MWCNT. Antimicrobial activity against *E. coli* was higher for these Ag-MWCNT nanocomposites than pure silver nanoparticles and catalytic efficiency was studied by analyzing its capability to convert p-nitrophenol to p-aminophenol (Mohan et al., 2016).

To treat various chemical and biological contaminants, a green carbon composite was fabricated from the activated carbon obtained by the phosphoric acid treatment in the pods of *Delonix regia* biomass and decorated with green silver nanoparticles. *Tabernaemontana divaricata* leaf extract was applied as a reducing agent for the first time in the preparation of silver-carbon nanocomposites in this case. EDAX and HR-TEM analysis confirmed the deposition of 0.8 wt% silver nanoparticles on the DRP surface and spherical shape of AgNPs. Both the crystalline and amorphous structures due to the presence of silver NPs and DRP carbon were confirmed by SAED analysis (Louis et al., 2020). Carbonaceous material obtained from the sewage sludge through pyrolysis and biogenic silver nanoparticles was produced by the reduction of $AgNO_3$ with *Camellia sinensis* leaf extract. BET analysis showed a lower surface area of 18 m^2/g Ag-CM than the CM as the active sites of CM were filled with AgNPs. This is the reason for the enhanced degradation of methylene blue dye by this composite (Vilchis-Nestor et al., 2014).

3 Conclusion

The significance of silver nanoparticles is increasing day by day due to their applications in different fields ranging from electrical, microelectronics, medical, cosmetics, drug delivery, biomedical imaging to environmental applications. Focusing on the methods available for the synthesis of nanoparticles includes different routes like physical, chemical, and biological. In addition, the wide use of several toxic chemicals and laborious experimental conditions in these procedures has caused much concern regarding the toxicity of silver nanoparticles relating to the biota and environment as a whole. The biogenic method of nanoparticle preparation is less hazardous, simple, environmentally friendly, inexpensive, and using less equipment, environment friendly, and risk-free reagents and chemicals, and time for the preparation of these particles. The present work has focused on the decoration of carbon nanomaterials with biogenic silver nanoparticles. This review summarized the available synthetic procedures adopted for the synthesis of silver nanoparticles along with their prevalent applications in different fields. GO, with its immense potential to form a composite with a wide range of materials, gives new hope in composite studies and development. Synthesis of GO is also achievable through green methods that also stabilize it and help in the homogeneous and uniform dispersion of silver nanoparticles on their surface. GO, as mentioned earlier in this chapter, possesses outstanding properties even in small quantities when forming a nanocomposite.

Green synthesis of rGO/Ag nanocomposites has paved a new way for the scientific world to develop materials with superior properties of both graphite and silver. As both of these materials possess a large number of unique properties due to their structure, shape, and size, they can be applied to various applications that will help humans to lead a successful life in all manners. These nanohybrids were found to exhibit greater stability and enhanced activities. When the scientific community identified the disadvantages of the chemical and physical methods used to synthesize the nanocomposites, they started to apply the green chemistry principles to a synthetic strategy. Green synthesis is a one-pot method that is environmentally benign, eco-friendly, and cost-effective; in addition, the phytochemicals present in the biological entities will greatly enhance the ability of the nanohybrids. The review in this chapter assessed the possible mechanisms for incorporating silver nanoparticles into graphene oxide (GO) sheets that may generate highly efficient inorganic nanocomposites.

References

Abboud, Z., Vivekanandhan, S., Misra, M., Mohanty, A.K., 2016. Leaf extract mediated biogenic process for the decoration of graphene with silver nanoparticles. Mater. Lett. 178, 115–119.

Aboelfetoh, E.F., Gemeay, A.H., El-Sharkawy, R.G., 2020. Effective disposal of methylene blue using green immobilized silver nanoparticles on graphene oxide and reduced graphene oxide sheets through one-pot synthesis. Environ. Monit. Assess. 192.

Acar Bozkurt, P., 2017. Sonochemical green synthesis of Ag/graphene nanocomposite. Ultrason. Sonochem. 35, 397–404.

Ahmad, M.A., Aslam, S., Mustafa, F., Arshad, U., 2021. Synergistic antibacterial activity of surfactant free Ag–GO nanocomposites. Sci. Rep. 11, 1–9.

Ahmad, A., Mukherjee, P., Senapati, S., Mandal, D., Khan, M.I., Kumar, R., Sastry, M., 2003. Extracellular biosynthesis of silver nanoparticles using the fungus Fusarium oxysporum. Colloids Surf. B. Biointerfaces 28, 313–318.

Ajith, P., Murali, A.S., Sreehari, H., Vinod, B.S., Anil, A., Chandran, S., 2019. Green synthesis of silver nanoparticles using calotropis gigantea extract and its applications in antimicrobial and larvicidal activity. Mater. Today Proc. 18, 4987–4991.

Allen, M.J., Tung, V.C., Kaner, R.B., 2010. Honeycomb carbon: a review of graphene. Chem. Rev., 132–145.

Al-marri, A.H., Khan, M., Khan, M., Adil, S.F., 2015. Pulicaria glutinosa extract: a toolbox to synthesize highly reduced graphene oxide-silver nanocomposites. Int. J. Mol. Sci., 1131–1142.

Alsharaeh, E., Alazzam, S., Ahmed, F., Arshi, N., Al-Hindawi, M., Sing, G.K., 2017. Green synthesis of silver nanoparticles and their reduced graphene oxide nanocomposites as antibacterial agents: a bio-inspired approach. Acta Metall. Sin. (English Lett.) 30, 45–52.

Amare, M., Worku, A., Kassa, A., Hilluf, W., 2020. Green synthesized silver nanoparticle modified carbon paste electrode for SWAS voltammetric simultaneous determination of Cd(II) and Pb(II) in Bahir Dar textile discharged effluent. Heliyon 6, e04401.

Bai, S., Shen, X., 2012. Graphene-inorganic nanocomposites. RSC Adv. 2, 64–98.

Barua, S., Thakur, S., Aidew, L., Buragohain, A.K., Chattopadhyay, P., Karak, N., 2014. One step preparation of a biocompatible, antimicrobial reduced graphene oxide-silver nanohybrid as a topical antimicrobial agent. RSC Adv. 4, 9777–9783.

Basiri, S., Mehdinia, A., Jabbari, A., 2018. Green synthesis of reduced graphene oxide-Ag nanoparticles as a dual-responsive colorimetric platform for detection of dopamine and Cu^{2+}. Sensors Actuators B Chem. 262, 499–507.

Bhattacharya, G., Sas, S., Wadhwa, S., Mathur, A., McLaughlin, J., Roy, S.S., 2017. Aloe vera assisted facile green synthesis of reduced graphene oxide for electrochemical and dye removal applications. RSC Adv. 7, 26680–26688.

Bhuyan, M.S.A., Uddin, M.N., Islam, M.M., Bipasha, F.A., Hossain, S.S., 2016. Synthesis of graphene. Int. Nano Lett. 6, 65–83.

Chang, Q., Wang, Z., Wang, J., Yan, Y., Zhu, J., Shi, W., Chen, Q., Yu, Q., 2014. Graphene nanosheets inserted by silver nanoparticles as zero-dimensional nanospacers for dye sensitized solar cells. Nanoscale 10.

Chettri, P., Vendamani, V.S., Tripathi, A., Singh, M.K., Pathak, A.P., Tiwari, A., 2017. Green synthesis of silver nanoparticle-reduced graphene oxide using Psidium guajava and its application in SERS for the detection of methylene blue. Appl. Surf. Sci. 406, 312–318.

Choi, W., Lahiri, I., Seelaboyina, R., Kang, Y.S., Choi, W., Lahiri, I., Seelaboyina, R., 2010. Synthesis of graphene and its applications: a review synthesis of graphene and its applications: a review. Crit. Rev. Solid State Mater. Sci. 8436.

Cobos, M., De-La-pinta, I., Quindós, G., Fernández, M.D., Fernández, M.J., 2020. Graphene oxide–silver nanoparticle nanohybrids: synthesis, characterization, and antimicrobial properties. Nanomaterials 10.

Dar, R.A., Khare, N.G., Cole, D.P., Karna, S.P., Srivastava, A.K., 2014. Green synthesis of a silver nanoparticle-graphene oxide composite and its application for As(iii) detection. RSC Adv. 4, 14432–14440.

Das, T.K., Bhawal, P., Ganguly, S., Mondal, S., Das, N.C., 2018. A facile green synthesis of amino acid boosted Ag decorated reduced graphene oxide nanocomposites and its catalytic activity towards 4-nitrophenol reduction. Surf. Interfaces 13, 79–91.

Das, T.R., Sharma, P.K., 2020. Hydrothermal-assisted green synthesis of Ni/Ag@rGO nanocomposite using Punica granatum juice and electrochemical detection of ascorbic acid. Microchem. J. 156, 104850.

De Faria, A.F., De Moraes, A.C.M., Marcato, P.D., Martinez, D.S.T., Durán, N., Filho, A.G.S., Brandelli, A., Alves, O.L., 2014. Eco-friendly decoration of graphene oxide with biogenic silver nanoparticles: antibacterial and antibiofilm activity. J. Nanopart. Res. 16.

Dhand, V., Rhee, K.Y., Ju Kim, H., Ho Jung, D., 2013. A comprehensive review of graphene nanocomposites: research status and trends. J. Nanomater. 2013.

Dong, L.L., Ding, Y.C., Huo, W.T., Zhang, W., Lu, J.W., Jin, L.H., Zhao, Y.Q., Wu, G.H., Zhang, Y.S., 2019. A green and facile synthesis for rGO/Ag nanocomposites using one-step chemical co-reduction route at ambient temperature and combined first principles theoretical analyze. Ultrason. Sonochem. 53, 152–163.

Dutta, T., Barman, A., Majumdar, G., 2020. Green and sustainable manufacturing of metallic, ceramic and composite materials, encyclopedia of renewable and sustainable materials. Encycl. Renew. Sustain. Mater. https:/doi.org/10.1016/b978-0-12-803581-8.11023-9.

El-Sabban, H., Eid, M., Moustafa, Y., Abdel-Mottaleb, M., 2020. Pomegranate peel extract in situ assisted phytosynthesis of silver nanoparticles decorated reduced graphene oxide as superior sorbents for Zn (II) and Lead (II). Egypt. J. Aquat. Biol. Fish. 24, 525–539.

Fadiran, O.O., Girouard, N., Meredith, J.C., 2018. Pollen fillers for reinforcing and strengthening of epoxy composites. Emergent Mater. 1, 95–103.

Feng, H., Cheng, R., Zhao, X., Duan, X., Li, J., 2013. A low-temperature method to produce highly reduced graphene oxide. Nat. Commun. 4, 1537–1539.

Gadd, G.M., Griffiths, A.J., 1977. Microorganisms and heavy metal toxicity. Microb. Ecol. 4, 303–317.

Guo, Z., Chen, Y., Lu, N.L., Yan, X., Guo, Z., 2018. Introduction to nanocomposites. In: Multifunctional Nanocomposites for Energy and Environmental Applications, pp. 1–5.

Gurunathan, S., Han, J.W., Park, J.H., Kim, E., Choi, Y.J., Kwon, D.N., Kim, J.H., 2015. Reduced graphene oxide-silver nanoparticle nanocomposite: a potential anticancer nanotherapy. Int. J. Nanomedicine 10, 6257–6276.

Haghighi, B., Tabrizi, M.A., 2013. Green-synthesis of reduced graphene oxide nanosheets using rose water and a survey on their characteristics and applications. RSC Adv. 3, 13365–13371.

Hoa, L.T., Linh, N.T.Y., Chung, J.S., Hur, S.H., 2017. Green synthesis of silver nanoparticle-decorated porous reduced graphene oxide for antibacterial non-enzymatic glucose sensors. Ionics (Kiel) 23, 1525–1532.

Hou, Y., Pu, S., Shi, Q., Mandal, S., Ma, H., Xue, S., Cai, G., Bai, Y., 2019. Ultrasonic impregnation assisted in-situ photoreduction deposition synthesis of ag/TiO2/rGO ternary composites with synergistic enhanced photocatalytic activity. J. Taiwan Inst. Chem. Eng. 104, 139–150.

Huang, X., Qi, X., Zhang, H., 2012. Graphene-based composites. Chem. Soc. Rev., 41. https://doi.org/10.1039/c1cs15078b.

Hussain, F., Hojjati, M., Okamoto, M., Gorga, R.E., 2006. Review article: polymer-matrix nanocomposites, processing, manufacturing, and application: an overview. J. Compos. Mater. 40, 1511–1575.

Iravani, S., 2011. Green synthesis of metal nanoparticles using plants. Green Chem. 13, 2638–2650.

Karthik, G., Harith, A., Nazrin Thazleema, N., Vishal, S., Jayan Jitha, S., Saritha, A., 2018. One step approach towards the green synthesis of silver decorated graphene nanocomposites for the degradation of organic dyes in water. IOP Conf. Ser. Mater. Sci. Eng., 310.

Khan, M., Al-Marri, A.H., Khan, M., Shaik, M.R., Mohri, N., Adil, S.F., Kuniyil, M., Alkhathlan, H.Z., Al-Warthan, A., Tremel, W., Tahir, M.N., Siddiqui, M.R.H., 2015a. Green approach for the effective reduction of graphene oxide using Salvadora persica L. root (Miswak) extract. Nanoscale Res. Lett. 10, 1–9.

Khan, A., Aslam, A., Khan, P., Asiri, A.M., Abu-zied, B.M., 2015b. Green synthesis of thermally stable Ag-rGO-CNT nano composite with high sensing activity. Compos. Part B 86.

Khorrami, S., Abdollahi, Z., Eshaghi, G., Khosravi, A., Bidram, E., Zarrabi, A., 2019. An improved method for fabrication of ag-GO nanocomposite with controlled anti-Cancer and anti-bacterial behavior; a comparative study. Sci. Rep. 9, 1–10.

Kiruba, V.S.A., Dakshinamurthy, A., 2013. Green synthesis of biocidal silver-activated charcoal nanocomposite for disinfecting water. J. Exp. Nanosci., 37–41.

Kolya, H., Kuila, T., Kim, N.H., Lee, J.H., 2019. Bioinspired silver nanoparticles/reduced graphene oxide nanocomposites for catalytic reduction of 4-nitrophenol, organic dyes and act as energy storage electrode material. Compos. Part B Eng. 173, 106924.

Krishnaraj, C., Jagan, E.G., Rajasekar, S., Selvakumar, P., Kalaichelvan, P.T., Mohan, N., 2010. Synthesis of silver nanoparticles using Acalypha indica leaf extracts and its antibacterial activity against water borne pathogens. Colloids Surf. B Biointerfaces 76, 50–56.

Kumar, S.V., Huang, N.M., Lim, H.N., Zainy, M., Harrison, I., Chia, C.H., 2013. Preparation of highly water dispersible functional graphene/silver nanocomposite for the detection of melamine. Sensors Actuators B Chem. 181, 885–893.

Kumar, V., Yadav, S.K., 2009. Plant-mediated synthesis of silver and gold nanoparticles and their applications. J. Chem. Technol. Biotechnol. 84, 151–157.

Li, Z., Zheng, M., Guan, X., Xie, Z., Huang, Y., Jing, X., 2014. Unadulterated BODIPY-dimer nanoparticles with high stability and good biocompatibility for cellular imaging. Nanoscale 6, 5662–5665.

Liu, L., Liu, J., Wang, Y., Yan, X., Sun, D.D., 2011. Facile synthesis of monodispersed silver nanoparticles on graphene oxide sheets with enhanced antibacterial activity. New J. Chem. 35, 1418–1423.

Liu, C.H., Mao, B.H., Gao, J., Zhang, S., Gao, X., Liu, Z., Lee, S.T., Sun, X.H., Wang, S.D., 2012. Size-controllable self-assembly of metal nanoparticles on carbon nanostructures in room-temperature ionic liquids by simple sputtering deposition. Carbon N. Y. 50, 3008–3014.

Louis, M.R., Sorokhaibam, L.G., Chaudhary, S.K., Bundale, S., 2020. Silver-loaded biomass (Delonix regia) with anti-bacterial properties as porous carbon composite towards comprehensive water purification. Int. J. Environ. Sci. Technol. 17, 2415–2432.

Machado, B.F., Serp, P., Machado, B.F., 2012. Graphene-based materials for catalysis. Catal. Sci. Technol., 54–75. https://doi.org/10.1039/c1cy00361e.

Manchala, S., Tandava, V.S.R.K., Jampaiah, D., Bhargava, S.K., Shanker, V., 2019. Novel and highly efficient strategy for the green synthesis of soluble graphene by aqueous polyphenol extracts of eucalyptus bark and its applications in high-performance supercapacitors. ACS Sustain. Chem. Eng. 7, 11612–11620.

Mariadoss, A.V.A., Saravanakumar, K., Sathiyaseelan, A., Wang, M.H., 2020. Preparation, characterization and anti-cancer activity of graphene oxid—silver nanocomposite. J. Photochem. Photobiol. B Biol. 210, 111984.

Medha, G., Sharmila, C., Anil, G., 2017. Green synthesis and characterization of nanocrystalline graphene oxide. Int. Res. J. Sci. Eng., 29–34.

Meng, Y., Su, F., Chen, Y., 2018. Effective lubricant additive of nano-ag/MWCNTs nanocomposite produced by supercritical CO2 synthesis. Tribol. Int. 118, 180–188.

Mijnendonckx, K., Leys, N., 2013. Antimicrobial silver : uses, toxicity and potential for resistance. Biometals, 609–621.

Minh Dat, N., Tan Tai, L., Tan Khang, P., Ngoc Minh Anh, T., Minh Nguyet, D., Hoang Quan, T., Ba Thinh, D., Thi Thien, D., Minh Nam, H., Thanh Phong, M., Huu Hieu, N., 2021. Synthesis, characterization, and antibacterial activity investigation of silver nanoparticle-decorated graphene oxide. Mater. Lett. 285, 128993.

Mmaduka Obodo, R., Ahmad, I., Ifeanyichukwu Ezema, F., 2019. Introductory chapter: graphene and its applications. In: Graphene and Its Derivatives—Synthesis and Applications [Working Title], pp. 1–9.

Mohan, S., Oluwafemi, O.S., Songca, S.P., Rouxel, D., Miska, P., Lewu, F.B., Kalarikkal, N., Thomas, S., 2016. Completely green synthesis of silver nanoparticle decorated MWCNT and its antibacterial and catalytic properties. Pure Appl. Chem. 88, 71–81.

Montes-Duarte, G.G., Tostado-Blázquez, G., Castro, K.L.S., Araujo, J.R., Achete, C.A., Sánchez-Salas, J.L., Campos-Delgado, J., 2021. Key parameters to enhance the antibacterial effect of graphene oxide in solution. RSC Adv. 11, 6509–6516.

Narayanan, K.B., Sakthivel, N., 2010. Biological synthesis of metal nanoparticles by microbes. Adv. Colloid Interf. Sci. 156, 1–13.

Nasir, S., Hussein, M.Z., Zainal, Z., Yusof, N.A., 2018. Carbon-based nanomaterials/allotropes: a glimpse of their synthesis, properties and some applications. Materials (Basel) 11, 1–24.

Nayak, S.P., Ramamurthy, S.S., Kiran Kumar, J.K., 2020. Green synthesis of silver nanoparticles decorated reduced graphene oxide nanocomposite as an electrocatalytic platform for the simultaneous detection of dopamine and uric acid. Mater. Chem. Phys. 252, 123302.

Nguyen, V.H., Kim, B., Jo, Y., Shim, J., 2012. The journal of supercritical fluids preparation and antibacterial activity of silver nanoparticles-decorated graphene composites. J. Supercrit. Fluids 72, 28–35.

Nguyen, T.H.D., Lin, M., 2015. Toxicity of graphene oxide on intestinal bacteria and caco-2 cells. J. Food Prot. 78, 996–1002.

Omanovi, E., Badnjevi, A., Kazlagi, A., Hajlovac, M., 2019. Nanocomposites: a brief review. Health Technol. 10, 51–59.

Pai, A.R., Sasidharan, S., Kavitha, S., Shweta Raj, S., Priyanka, P., Vrinda, A., Vivin, T.S., Silpa, S., 2015. Green synthesis and characterizations of silver nanoparticles using fresh leaf extract of Morinda citrifolia and its anti-microbial activity studies. Int J Pharm Pharm Sci 7, 1–8.

Pandey, P.A., Bell, G.R., Rourke, J.P., Sanchez, A.M., Elkin, M.D., Hickey, B.J., Wilson, N.R., 2011. Physical vapor deposition of metal nanoparticles on chemically modified graphene: observations on metal-graphene interactions. Mater. Sci., 3202–3210.

Patra, J.K., Baek, K.H., 2014. Green Nanobiotechnology: factors affecting synthesis and characterization techniques. J. Nanomater. 2014.

Pinto, A.M., Gonçalves, I.C., Magalhães, F.D., 2013. Colloids Surf. B Biointerfaces. https://doi.org/10.1016/j.colsurfb.2013.05.022.

Pugazhendhi, A., 2020. Fabrication of naringenin functionalized-Ag/RGO nanocomposites for potential bactericidal effects. Integr. Med. Res. 9, 7013–7019.

Pumera, M., Miyahara, Y., 2009. What amount of metallic impurities in carbon nanotubes is small enough not to dominate their redox properties? Nanoscale, 260–265.

Qiu, J.D., Wang, G.C., Liang, R.P., Xia, X.H., Yu, H.W., 2011. Controllable deposition of platinum nanoparticles on graphene as an electrocatalyst for direct methanol fuel cells. J. Phys. Chem. C 115, 15639–15645.

Rai, M., Yadav, A., Gade, A., 2008. CRC 675 - current trends in phytosynthesis of metal nanoparticles. Crit. Rev. Biotechnol. 28, 277–284.

Rai, M., Yadav, A., Gade, A., 2009. Silver nanoparticles as a new generation of antimicrobials. Biotechnol. Adv. 27, 76–83.

Rawal, N., Solanki, S., Shah, D., 2020. ScienceDirect green synthesis of reduced graphene oxide with in situ decoration of metal nanoparticles for charge storage application. Mater. Today Proc. 21, 2066–2071.

Ren, W., Fang, Y., Wang, E., 2011. A binary functional substrate for enrichment and ultrasensitive SERS spectroscopic detection of folic acid using graphene oxide/Ag nanoparticle hybrids. ACS Nano 5, 6425–6433.

Rodríguez-Otamendi, D.I., Meza-Laguna, V., Acosta, D., Álvarez-Zauco, E., Huerta, L., Basiuk, V.A., Basiuk, E.V., 2021. Eco-friendly synthesis of graphene oxide–silver nanoparticles hybrids: the effect of amine derivatization. Diam. Relat. Mater. 111, 108208.

Sadasivuni, K.K., Rattan, S., Waseem, S., Brahme, S.K., Kondawar, S.B., Ghosh, S., Das, A.P., Chakraborty, P.K., Adhikari, J., Saha, P., Mazumdar, P., 2019. Silver Nanoparticles and its Polymer Nanocomposites—Synthesis, Optimization, Biomedical Usage, and Its Various Applications, Lecture Notes in Bioengineering., https://doi.org/10.1007/978-3-030-04741-2_11.

Salazar, P., Fernández, I., Rodríguez, M.C., Hernández-Creus, A., González-Mora, J.L., 2019. One-step green synthesis of silver nanoparticle-modified reduced graphene oxide nanocomposite for H2O2 sensing applications. J. Electroanal. Chem. 855, 113638.

Sedki, M., Mohamed, M.B., Fawzy, M., Abdelrehim, D.A., Abdel-Mottaleb, M.M.S.A., 2015. Phytosynthesis of silver-reduced graphene oxide (Ag-RGO) nanocomposite with an enhanced antibacterial effect using Potamogeton pectinatus extract. RSC Adv. 5, 17358–17365.

Shao, Y., Wang, J., Wu, H., Liu, J., Aksay, I.A., Lin, Y., 2010. Graphene based electrochemical sensors and biosensors: a review. Electroanalysis, 1027–1036.

Soleymani, A.R., Rafigh, S.M., Hekmati, M., 2020. Green synthesis of RGO/Ag: as evidence for the production of uniform mono-dispersed nanospheres using microfluidization. Appl. Surf. Sci. 518.

Subrahmanyam, K.S., Manna, A.K., Pati, S.K., Rao, C.N.R., 2010. A study of graphene decorated with metal nanoparticles. Chem. Phys. Lett. 497, 70–75.

Szunerits, S., Boukherroub, R., 2016. Antibacterial activity of graphene-based materials. J. Mater. Chem. B 4, 6892–6912.

Taha, A., Aissa, M.B., 2020. Green synthesis of an activated carbon-supported Ag and ZnO nanocomposite for photocatalytic degradation and its antibacterial activities. Molecules, 8–16.

Tang, X.Z., Cao, Z., Zhang, H.B., Liu, J., Yu, Z.Z., 2011. Growth of silver nanocrystals on graphene by simultaneous reduction of graphene oxide and silver ions with a rapid and efficient one-step approach. Chem. Commun. 47, 3084–3086.

Taylor, P., Nanda, S.S., Papaefthymiou, G.C., Yi, D.K., 2015. Functionalization of graphene oxide and its biomedical applications. Crit. Rev. Solid State Mater. Sci., 37–41.

Thomas, R., Unnikrishnan, J., Nair, A.V., Daniel, E.C., Balachandran, M., 2020. Antibacterial performance of GO–ag nanocomposite prepared via ecologically safe protocols. Appl. Nanosci. 1–13.

Tien, H., Huang, Y., Yang, S., Wang, J., Ma, C.M., 2010. The production of graphene nanosheets decorated with silver nanoparticles for use in transparent, conductive films. Carbon N. Y. 49, 1550–1560.

Umar, M.F., Ahmad, F., Saeed, H., Usmani, S.A., Owais, M., Rafatullah, M., 2020. Biomediated synthesis of reduced graphene oxide nanoparticles from chenopodium album: their antimicrobial and anticancer activities. Nano 10, 1–14.

Vilchis-Nestor, A.R., Trujillo-Reyes, J., Colín-Molina, J.A., Sánchez-Mendieta, V., Avalos-Borja, M., 2014. Biogenic silver nanoparticles on carbonaceous material from sewage sludge for degradation of methylene blue in aqueous solution. Int. J. Environ. Sci. Technol. 11, 977–986.

Villalobos-Castillejos, F., Granillo-Guerrero, V.G., Leyva-Daniel, D.E., Alamilla-Beltrán, L., Gutiérrez-López, G.F., Monroy-Villagrana, A., Jafari, S.M., 2018. Fabrication of nanoemulsions by microfluidization. Nanoemulsions Formul. Appl. Charact., 207–232.

Vinod, V.T.P., Saravanan, P., Sreedhar, B., Devi, D.K., Sashidhar, R.B., 2011. A facile synthesis and characterization of Ag, Au and Pt nanoparticles using a natural hydrocolloid gum kondagogu (*Cochlospermum gossypium*). Colloids Surf. B: Biointerfaces 83, 291–298.

Wei, Z., 2012. Nanoscale Tunable Reduction of Graphene Oxide for Graphene Electronics., p. 1373, https://doi.org/10.1126/science.1188119.

Wu, J., Zhu, B., Zhao, Y., Shi, M., He, X., Xu, H., Zhou, Q., 2020. One-step eco-friendly synthesis of ag-reduced graphene oxide nanocomposites for Antibiofilm application. J. Mater. Eng. Perform. 29, 2551–2559.

Zedan, A.F., Moussa, S., Terner, J., Atkinson, G., El-shall, M.S., 2012. Ultrasmall gold nanoparticles anchored to graphene and enhanced photothermal effects by laser irradiation of gold nanostructures in graphene oxide solutions. ACS Nano 7.

Zhao, N., Cheng, X., Zhou, Y., 2014. Synthesis of flexible free-standing silver nanoparticles-graphene films and their surface-enhanced raman scattering activity. J. Nanopart. Res. https:/doi.org/10.1007/s11051-014-2335-0.

Zhou, R., Gao, H., 2014. Cytotoxicity of Graphene : Recent., https:/doi.org/10.1002/wnan.1277.

Zhu, X., Xu, X., Liu, F., Jin, J., Liu, L., Zhi, Y., Chen, Z.W., Zhou, Z.S., Yu, J., 2017. Green synthesis of graphene nanosheets and their in vitro cytotoxicity against human prostate cancer (DU 145) cell lines. Nanomater. Nanotechnol. 7, 1–7.

CHAPTER 7

Bio-mediated synthesis of silver nanoparticles via microwave-assisted technique and their biological applications

Kondaiah Seku[a], K. Kishore Kumar[b], G. Narasimha[c], and G. Bhagavanth Reddy[d]

[a]*Engineering Department, Civil Section, Applied Sciences–Chemistry, University of Technology and Applied Sciences, Shinas, Sultanate of Oman,* [b]*Department of Pharmaceutical Biology, Narayana Pharmacy College, JNTU-A, Nellore, India,* [c]*Department of Virology, Sri Venkateshwara University, Tirupati, India,* [d]*Department of Chemistry, PG Centre Wanaparthy, Palamuru University, Mahbubnagar, India*

1 Introduction

Nanobiotechnology is a new science of nanoparticle synthesis using biotechnological applications. It is a multifaceted field and covers various facets of biology, physics, engineering, and chemistry. Richard Feynman (Feynman, 1959) first used the expression "nanotechnology," which was considered to mark the creation of contemporary nanotechnology (Kuhad et al., 2011). In recent decades, large varieties of nanoparticles (NPs) have been reported for several applications in chemical, electrical, medical, biotechnology, industries, pharmaceutical, and agricultural fields utilizing several chemical, physical, and biological techniques. Nano-inorganic antibacterial agents have become of interest for bacterial control, owing to outstanding safety, thermal stability, sustainability, and large surface area. Nanomaterials' applications, especially those of silver, zinc, copper, gold, and platinum, are emerging rapidly in major scientific domains, including catalysis, micro and nanoelectronics, optical devices, biomedicine, diagnostics, foods, agricultural, environmental, and chemical research (Rajat et al., 2021).

NPs are, in general, between 0.1 and 100nm in size and are synthesized utilizing either top-down or bottom-up methods (Schirmer, 1999). The bulk materials size is reduced to nano size when using top-down processing; while in bottom-up processing, molecules/atoms are brought together to form subatomic structures at the nanometer scale. The second approach is more commonly practiced for chemical synthesis and the biogenic synthesis of NPs. In contrast to bulk materials, nanomaterials have characteristic physical, chemical, optical, thermal, electronic, dielectric, magnetic, mechanical, and biological features. The characteristic features of nanomaterials, such as size, shape, size distribution, surface area, solubility, aggregation, etc., are considered for the synthesis of NPs. Several analytical techniques are employed to assess nanomaterials, including ultraviolet-visible spectroscopy (UV-vis), X-ray diffractometry, Fourier-transform infrared spectroscopy (FTIR), dynamic light scattering, X-ray photoelectron, scanning electron microscopy (SEM), atomic force microscopy (AFM), and transmission electron microscopy (TEM).

For the synthesis of NPs, chemical, physical methods are available, which are expensive and environmentally hazardous due to dangerous chemicals that produce a range of biological risks. The progression in the biomass-related experimental procedures noted for the biogenesis of nanoparticles becomes a branch of nanotechnology. In this chapter, several techniques used in the synthesis of NPs, and their advantages and disadvantages are described. Biogenic synthesis of AgNPs is been discussed, listing various works reported in the literature, and risks and regulations regarding applications in the agricultural field are also depicted. Various regulatory authorities around the world and their regulations, with specific examples, are discussed.

2 Various methods for nanomaterials synthesis

Various techniques and approaches practiced for nanoparticle synthesis are listed in Table 1. Unfortunately, many nanoparticle synthesis or production methods involve hazardous chemicals, low material conversions, high energy requirements, and are difficult and wasteful for further purification. Chemical and physical NP synthesis

Table 1 List of various techniques for the synthesis of nanoparticles (NPs).

Sl. No	Physical methods	Chemical methods	Biological methods
1	High energy ball mill	Sol-gel synthesis	Polymeric assisted synthesis
2	Inert gas condensation	Microemulsion method	Microorganisms assisted
3	Pulse vapor deposition	Hydrothermal synthesis	Fungus assisted
4	Laser pyrolysis	Polysol synthesis	Natural gums assisted
5	Flash spray pyrolysis	Chemical vapor synthesis	Plant extracts assisted
6	Electro-spraying	Plasma enhanced chemical vapor deposition	Bio-template assisted
7	Melt mixing	Photochemical method	
8	Microwave irradiation		

cannot be expanded easily to large-scale production because of many drawbacks, such as: (i) presence of toxic organic solvents, (ii) production of hazardous by-products and intermediary compounds, (iii) high energy utilization, (iv) particles may produce aggregates, and (v) they must use capping agents. In addition to the above, other issues occur, such as: (i) toxicity, (ii) decreased particle rate of synthesis, (iii) deformation in particle structure, and (iv) particle growth inhibition; furthermore, in chemical synthesis: (i) nanocomposites are formed, (ii) increased toxicity and reactivity of particles, (iii) harmful to human health and hazardous to the environment, and (iv) reaction conditions and reaction kinetics.

2.1 Physical methods

Metal nanoparticles are usually prepared through evaporation condensation, in which a tube furnace is utilized at atmospheric conditions. A vessel containing source material is placed and vaporized into the carrier gas in the center of the furnace. By using evaporation/condensation techniques, several nanoparticles/nanomaterials have been produced with Au, Ag, Pb, etc. (Gurav et al., 1994; Magnusson et al., 1999). Using a tube furnace for production of AgNPs requires immense energy to attain the material's desired temperature and thermal stability, and also occupies large areas in the facility; altogether, it is not economically feasible. Another physical process to produce AgNPs is laser ablation of silver bulk material in solution (solution may be water or ethanol). This laser beam produces a colloid for the formation of nanoparticles without adding a chemical reagent. The main factors affecting the formation of nanoparticles are beam power, duration of ablation, and pore spot (Mafuné et al., 2000; Dolgaev et al., 2002; Sylvestre et al., 2004). After the formation of narrow size-distributed particles, surface modification follows, and these could be used for several biosciences and biotechnological applications (Ansari and Al-Shaeri, 2019). Sonochemical synthesis is another method for forming AgNPs; long-chain polymers are used to bind with nanoparticles. In the presence of $NaBH_4$ as a reducing agent, spherical AgNPs are formed with 10 nm size (Wani et al., 2011). In this method, thermal stability and nanoparticle stability are well established without the formation of chemical pollutants. But in sonochemical methods reducing agents are required, much more time is taken to form small size nanoparticles, and buffering the chemical synthesis (pH 9–11) under sonication conditions leads to the formation of clusters of Ag nanoparticles being produced.

The thermal decomposition method can also produce AgNPs by decomposing silver-organic complexes at higher temperatures, leading to silver nanoparticles. These capping agents are compulsorily required to mix at the rigid condition in the presence of nitrogen gas. In this process, the energy required to heat for about 3 h with refluxing, in addition to the capping agents (Navaladian et al., 2007), leads to a higher cost of the process. Another physical method is an electrochemical technique used to produce AgNPs with high purity. Here two types of stabilizers are utilized: (i) electrostatic stabilizers (usually organic monomers), and (ii) steric stabilizers (polymeric compounds) (Yu et al., 1997; Yin et al., 2003). Some workers have used polyethylene glycol (PEG) stabilizer for AgNPs, which are easily permeable through diffusion into small molecules and are also very stable (Shkilnyy et al., 2009).

Microwave irradiation (MW) is another technique in which the frequency 300 MHz to 300 GHz is used to heat the sample with H_2O (reducing agent) to orient the electrical field. The dipolar atoms try to reorient in the electric field, losing energy by molecular friction as heat. Gao et al. (2006) synthesized AgNPs by *N,N*-dimethylformamide (DMF) as a reducing agent in the presence of PVP at 140°C, irradiated for 3 h, and achieved decahedron shaped NPs of 80 nm. Pal et al. (2009) prepared silver nanoparticles by microwave irradiation of $AgNO_3$ solution with ethanol as a reducing agent and polyvinylpyrrolidone (PVP) stabilizing agent, achieved spherical and monodispersible particles of 10 nm. Overall, the physical methods utilize tremendous energy, are laborious, the primary cost of the equipment is high, and furthermore, some methods may produce toxicants.

2.2 Photochemical methods

This technique is a physicochemical method in which light energy is the reduction process to produce AgNPs. Huang and Yang (2008) used inorganic clay as a stabilizing agent (suspension) to prevent the aggregation in the photoreduction of $AgNO_3$. Photo-irradiation caused disintegration into smaller sizes, and the same distribution of size was achieved. In this method, the equipment and experimental conditions are costly.

2.3 Chemical methods

In addition to the above, the chemical reduction of silver salt is a suitable method owing to the simple equipment required and its convenience. AgNPs can be synthesized by chemical reactions with higher yields even at low cost. For the chemical reaction process, metal precursors (silver salt), stabilizing agents (PVP, sodium oleate, PVA, etc.), and reducing agents (ethylene glycol, $NaBH_4$, glucose, etc.) are required, along with the parameters of pH and temperature (Li et al., 2012; Dang et al., 2012; Patil et al., 2012). The conversion of silver salt into a colloidal solution occurs in two stages: nucleation and consequent growth. In the formation of nuclei, the primary consideration is that all nuclei are formed at a similar time with monodispersed AgNPs with similar size distribution, then subsequent growth is likely to occur, provided all the above agents are present in the reduction process. By observing all the steps, including the agents, one can observe that there is a release of toxic chemicals and waste residue is also formed during the process (Song and Kim, 2009).

2.4 Biological synthesis of nanoparticles

Several microbes, such as fungi, bacteria, yeasts, viruses, and plant extracts, are used in the biological synthesis of nanoparticles. Biosynthesis of NPs is classified into various types according to the biological material used: (i) biosynthesis by microorganisms, (ii) biosynthesis by biomolecules as template, and (iii) biosynthesis by plant extracts. Biosynthesis by microorganisms is of two types, intracellular and extracellular, depending upon the location of the formation of NPs. Biosynthesis of NPs uses bacterium, fungi, and plant extracts (Sadowski et al., 2008; Saifuddin et al., 2009; Seku et al., 2018) is a simple, practicable alternative to physical and chemical approaches. Green synthesis provides (Seku et al., 2018) advancement over chemical and physical methods as it is cost-effective, environment-friendly, easily scaled up for large-scale synthesis, and further chemicals, energy, temperature, toxic chemicals, and high pressures are not needed. Using plants and plant materials for nanoparticle synthesis can be advantageous over other biological processes. It eliminates the elaborate process of maintaining cell cultures and can also be scaled for large-scale synthesis of nanoparticles under a nonaseptic environment. This chapter provides a few descriptions of the preparation methods, characterization, catalytic, and antibacterial properties of biogenic nanoparticles from our recent studies. Amazing characteristics and applications are noted for AgNPs in the literature, here, according to the activity of biogenic AgNPs, they are discussed as: (i) antibacterial, (ii) antifungal, (iii) antiviral, (iv) antiinflammatory, (v) antiplatelet, (vi) anti-angiogenesis, etc.

2.5 Microwave irradiation synthesis of silver nanoparticles

The use of microwave-assisted (MWA) techniques in metal nanoparticle synthesis is an auspicious approach. As one type of electromagnetic wave, microwaves range from 1 mm to 1 m wavelengths (frequency -0.3 to $300\,GHz$), which gives enough energy to break bonds and then produce metal nanoparticles. MWA is attracting more and more attention owing to its simple, short-time reaction, efficient, higher reaction rate and selectivity, uniform product formation, controllable reaction conditions and greater yields as compared to other heating approaches, and greater control over the shape and size of NPs (Seku et al., 2019). Hence, the MWA approach is rapidly turning into an attractive alternative in metal nanoparticle synthesis (Seku et al., 2018). There are many reports on the useful application of MWA to synthesis AgNPs, which demonstrate the viability and promising potential of this approach. At the beginning of microwave synthesis, chemical synthesis was used by microwave reactors. These days, domestic microwave ovens are typically applied in microwave synthesis. However, a scientific microwave is also a good choice because it improves reproducibility by employing microwave power control, temperature, a more significant security level, and advantageous programming and data management (Yin et al., 2004). It is a nonclassical source of energy that increases the quantity and quality of the nanoparticles.

Plant-based materials that support the synthesis of silver nanoparticles by the MWA method have been reported widely in recent years. The method of utilization of plant extracts to synthesize silver nanoparticles is simple (Saim et al., 2021). In this technique, $AgNO_3$ is mixed with a solution of plant materials at various temperatures and for different microwave irradiation times (Seku et al., 2018). The single-step biogenic reduction of Ag^+ to Ag^0 is fast and can be completed at atmospheric pressure and room temperature, meaning that scale-up is possible.

The development of a biological materials-mediated experimental procedure to produce AgNPs is now emerging as a substantial part of nanotechnology. In this green approach, extracts from plant-based materials and living organisms act as a reducing and stabilizing agent. Biogenic synthesis of AgNPs has been prepared through plant extracts, biopolymers, and microbial cell biomass or cell-free growth medium. Plant parts such as root, stem, bark, flower, and leaf have been utilized for the synthesis of AgNPs through the microwave-assisted method. Kahrilas et al. (2014) prepared AgNPs by the MWA technique using orange peel extract. Abboud et al. (2013) prepared AgNPs by an MWA method using aqueous onion (*Allium cepa*) extract. Kudle et al. (2013) synthesized AgNPs using flower extract of *Boswellia serrate* with the aid of microwave irradiation. Renugadevi et al. (2012) prepared AgNPs using leaf extract of *Baliospermum montanum* with microwave heating. The biogenic materials contain different functional groups, like –COOH, –CHO, –OH, –CONH₂, –NH₂, –COOR, polysaccharides, etc., which help in the reduction and capping of AgNPs (Moradi et al., 2021). Several other plant-based materials and microprobes that have also been utilized in the biosynthesis of AgNPs are listed in Table.2. Biosynthesized AgNPs showed greater stability for several months. The present chapter is focused on the biosynthesis of AgNPs through microwave-assisted synthesis (Fig. 1) and biomedical applications.

3 Classification of NPs

Based on their various dimensions and sizes, nanomaterials are classified as: (i) 0D (0 dimension), (ii) 1D (1 dimensional), (iii) 2D (2 dimensional), and (iv) 3D (3 dimensional) (Yu et al., 2015). The nanomaterials with 1–100 nm diameter and spherical shape are called 0D nanomaterials (quantum dots, fullerenes, nanoparticles, etc.). The nanomaterials with two dimensions of nano size (1–100 nm) are classified as 1D (nanofibers, nanowires, nanoribbons, nano-belts, and nanotubes). Whereas the 2D nanomaterials (graphene, nanosheets, nanolayers) have one dimension restricted to nano size (1–100 nm), and the other two dimensions larger than 100 nm. Lastly, the nanomaterials in all 3 dimensions are larger than 100 nm, but in nano size are called 3D nanomaterials (mesoporous carbon, MOF, Metal foams, and graphene aerogels) (Yu et al., 2015). European Commission defined nanomaterials as "natural, incidental/artificial matter comprising particles in aggregate or unbound or agglomerate state where the particles 50% or more in size distribution, number and external dimensions are in the size of 1–100 nm" (Potočnik, 2011).

Table 2 Silver nanoparticles are synthesized by using various biological resources.

Various biological resources	Shape	Size of AgNPs (nm)	References
Nervalia zeylanica	Spherical	34.2	Vijayan et al. (2018a)
Iota-carrageenan	Spherical	18.2	Jayaramudu et al. (2013)
Jatropha curcas L leaf	Spherical	20.4	Demissie and Lele (2013)
Zingiber officinale (ginger)	Spherical	10	Yang et al. (2017)
Carboxymethylated gum kondagogu	Spherical	9 ± 2	Seku et al. (2018)
Indigofera tinctoria leaf	Spherical	16.4	Vijayan et al. (2018b)
Salmalia malabarica	Spherical	7	Murali Krishna et al. (2016)
Marine micro algae	Spherical	–	Sathishkumar et al. (2019)
Cymbopogan Citratus leaf extract	Spherical	32	Masurkar et al. (2011)
Eichhornia crassipes plant shoot	Spherical	2.14	Oluwafemi et al. (2016)
Euphorbiaceae latex	Spherical	5–10	Rajkuberan et al. (2017)
Saccharum officinarum	Spherical	32	Chaudhari et al. (2012)
Eucalyptus globulus leaf	Spherical	1.9–4.2	Ali et al. (2015)
Streptomyces coelicolor klmp33	Irregular spherical	50	Manikprabhu and Lingappa (2014)
Areca Nut *(Areca catechu)*	Irregular in shape	20 ± 5	Bhat et al. (2013)
Starch	Spherical	12	Raghavendra et al. (2016)
Sodium alginate	Spherical	10	Shao et al. (2018)
Alkali hydrolyzed pectin	Spherical	2.9	Su et al. (2019)
Elephantopus scaber	Spherical	37.86	Kharat and Mendhulkar (2016)
Saraca Indica leaf extract	Spherical	5–50	Perugu et al. (2016)
Baliospermum montanum	–	5–60	Renugadevi et al. (2012)
Aerva lanata	Spherical	18.62	Kanniah et al. (2020)
Aqueous onion *(Allium cepa)*	Spherical	43.37	Saxena et al. (2010)
Azadirachta indica leaf	Spherical	40–130	Roy et al. (2017)

Continued

Table 2 Silver nanoparticles are synthesized by using various biological resources. *Continued*

Various biological resources	Shape	Size of AgNPs (nm)	References
Marine macro algae:			Bhuyar et al. (2020)
Ulva faciata		7	Rajesh et al. (2012)
Pterocladia		7	El Kassas and Attia (2014)
Jania rubins		12	
Colpmenia sinusa		20	Saber et al. (2017)
Aloe vera plant extract	Spherical	5–50	Logaranjan et al. (2016)
Pleurotus florida	Irregular spherical	5–50	Kaur et al. (2018)
Orange peel extract	Spherical	7.36–8.95	Awad et al. (2014)
Fraxinus excelsior leaf	Irregular spherical	25–40	Parveen et al. (2016)
L-cysteine	Spherical	10	Panhwar et al. (2018)
Carboxymethylated cashew gum	Spherical	92	Araruna et al. (2020)
Biophytum sensitivum	Spherical	19.6	Augustine et al. (2016)
Carboxymethyl cellulose	Spherical	15.6	Hebeish et al. (2010)
Guava (Psidium guajava) leaf extract		26±5	Bose and Chatterjee (2016)
Origanum majorana and Citrus sinensis leaf	Spherical and cubic shaped	40–70	Singh and Rawat (2016)
L-lysine	Spherical	26.3	Han et al. (2017)
Synedrella nodiflora	Spherical and triangular	14.6	Ogunsile et al. (2016)
Carboxymethyl cellulose-gelatin	Spherical	12	Pedroza-Toscano et al. (2017)
Rosa canina	Spherical	9.75	Gulbagca et al. (2019)
Averrhoa bilimbi leaf and plum fruit extract		20–50 and 47	Sagadevan et al. (2019)
Parkia speciosa Hassk pods	Spherical	20–50	Fatimah (2016)
Alpinia galanga		20±1	Joseph and Mathew (2014)
Mussaenda glabrata leaf extract		51.32	Francis et al. (2017)
Culture supernatant of bacteria		5–60	Saifuddin et al. (2009)

Table 2 Silver nanoparticles are synthesized by using various biological resources. *Continued*

Various biological resources	Shape	Size of AgNPs (nm)	References
Cauliflower waste		5–50	Kadam et al. (2020)
Juglans regia leaf		168	Nasar et al. (2019)
Carom seed (*Trachyspermum copticum*)		6–50	Raghunandan et al. (2011)
Carboxymethyl chitosan		2–20	Huang et al. (2016)
Carboxymethylcellulose		1–2	Hebeish et al. (2010)
Epilobium parviflorum green tea extract		19±2	Ertürk (2019)
Chitosan biopolymer		37	Wei et al. (2009)
Carbon dots		6–7	Jin et al. (2015)
Ludwigia octovalvis		28–50	Sarathi Kannan et al. (2020)
Cavendish banana peel extract		23–30	Kokila et al. (2015)

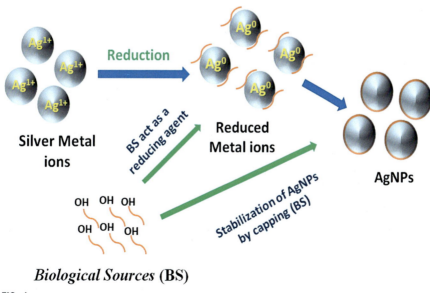

FIG. 1

The Synthetic mechanism for the formation of biogenic AgNPs.

4 Antibacterial activity mechanism of silver nanoparticles
4.1 General mechanisms established for antibacterial activity of nanoparticles

The mechanisms of metal oxide nanoparticles' antibacterial activity are inadequately explained. However, scientists have proposed the induction of oxidative stress, nonoxidative, and metal ion release mechanisms. Several studies have proposed that the electric charge on the bacterial membrane surface is rapidly neutralized by nanoparticles and changes its integrity and penetrability, eventually leading to bacterial death (Wang et al., 2017). Further, ROS (reactive oxygen species) are generated and inhibit the cell membrane's antioxidative defense mechanism, leading to mechanical damage. Existing research reports reveal the primary antibacterial effects of NPs to be: (i) cell membrane disruption; (ii) ROS generation; (iii) penetration into bacterial cell membrane; and (iv) interaction with cellular organelles, i.e., DNA, ribosomes, proteins, and plasmids. Biofilms (consortia of bacteria) prevention is attained by a nano size and greater surface area to mass ratio, and the particle shape of nanoparticles has a noteworthy effect on biofilm destruction (Lellouche et al., 2012).

The imperative antibacterial mechanism of nanoparticles is oxidative stress (through ROS), which has formidable redox potential utilizing molecules and active intermediates; several types of nanoparticles produce dissimilar varieties of reducing oxygen species (ROS). The reducing oxygen superoxide radicals are of four types, i.e., singlet oxygen (O_2), hydrogen peroxide (H_2O_2), hydroxyl radical (OH), and superoxide radical (O^{2-}), which exhibit various intensities of activities and dynamics. For a paradigm, magnesium oxide and calcium oxide nanoparticles can produce O_2, while zinc oxide nanoparticles can produce H_2O_2 and OH. Studies have explained that H_2O_2 and O^{2-} can produce low acute oxidative stress reactions that can be deactivated by endogenous antioxidants, catalase, and superoxide enzymes, whereas $-OH$ and O_2 can produce more significant acute oxidative stress reactions, leading to microbial death. Production of ROS by nanoparticles is intensified by defect sites, restructuring, and oxygen vacancies in the crystal (Malka et al., 2013). In general, ROS production and clearance in microorganisms is balanced; an inequitable state generates greater oxidative stress, which harms the cellular organelles (Li et al., 2012). Moreover, ROS is advantageous to escalating the gene representation intensity of oxidative proteins, an essential instrument in bacterial cell apoptosis (Wu et al., 2011; Srivastava et al., 2021). Further ROS can damage proteins and destroy the periplasmic enzymes and is vital to retaining normal physiology and morphology in microorganisms (Padmavathy and Vijayaraghavan, 2011).

Nanoparticles can release ROS by dissimilar mechanisms, and the photocatalytic hypothesis is most acceptable, nano-metal oxides releasing more electrons (e−) and holes (valency band, H) on and inside of the catalytic nanoparticles due to light irradiation energy superior to or equivalent to the bandgap leads to stimulation and transition to the conduction band (Panhwar et al., 2018). In the case of ZnO nanoparticles, hydroxyl radical (OH) is formed due to oxidation when interacted with H_2O or OH^{+-}

on the surface remains with H^+. After that, the superoxide radical (O_2^-) is formed due to electronic interaction utilizing O_2 on ZnO's surface. Here, the mechanism based on that the cell's vital active components responsible for sustaining the standard physiological and morphological functions in the microorganisms was destroyed (Yu et al., 2014), generating ROS, whereas in TiO_2, nano-particles produce pairs of electron-hole (Depan and Misra, 2014) after absorption of light, which immediately react with air and water on the surface and release extremely reactive ROS, which leads to damage of intracellular organic components of the bacteria.

Depan and Misra (2014) reported on Campylobacter jejuni cells for ROS mechanisms that, under UV and visible light, ZnO nanoparticles released extremely reactive ROS. Furthermore, hydroxyl and superoxide radicals (negatively charged) on the cell surface do not permeate into intracellular areas of bacteria, and H_2O_2 can pass right through the cell membrane. They reported that nanoparticles altered the shape of cells from spiral to spherical shape, causing cellular leakage. Seil and Webster (2012) applied ultrasonic treatment with ZnO nanoparticles and reported that H_2O split into H^+ and then reacted with dissolved oxygen in water, leading to the generation of H_2O_2, thereby inducing the formation of ROS. They explained that formed ROS, upon ultrasonic treatment, dissociated the colony-forming units and promoted the permeation of nanoparticles, then easily permeated the cell membrane, and the quick release of nanoparticles led to antimicrobial activity. Xu et al. (2013) demonstrated that NPs additionally have antibacterial action in obscurity. ZnO delivered a negligible OH, the primary antimicrobial species created by zinc oxide when invigorated by light. Oxygen vacancies situated on the exterior of ZnO assume a noteworthy function in delivering H_2O_2. In the case of heterogeneous catalysis, a nano-metal oxide relies upon the thickness of the dynamic site. Substantial deformities are regularly considered as the dynamic locales of heterogeneous catalysis. The crystal size effectively influenced nano ZnO antibacterial activity; the lattice constant, direction, and exterior concentration were reported.

4.2 Mechanisms proposed for antibacterial activity of silver nanoparticles

Silver in various forms has been used for wound management and in treating ulcers since the 18th century (Klasen, 2000). Identification of antibacterial activity of silver ions was achieved in the 19th century, and in the 1920s, the United States Food and Drug Administration (USFDA) accepted colloidal silver for ulcers and wound care management (Paladini and Pollini, 2019). Later, penicillin was discovered, and a few commercial products also came into the market. The term minimum inhibitory concentration (MIC) was established to determine the dosage regimen and compare the antibacterial activity of materials from various sources. Therefore, considering the antibacterial efficacy, susceptibility of microorganisms towards the exposure to systemic or topical antibacterial formulations, setting up of MIC and breakpoints came in the pharmacology. By setting up and adjusting the procedures, now all scientists, pharmacologists, and doctors are following new regulations to estimate and establish

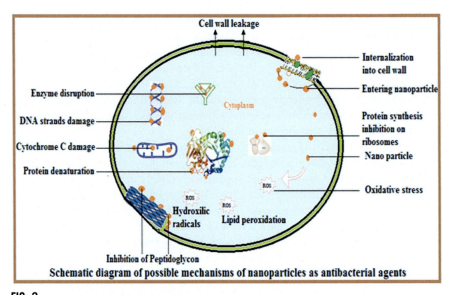

FIG. 2

Schematic representation of mechanisms proposed for antibacterial activity by nanoparticles (Mode of action of AgNPs on bacteria).

mechanisms of various antibacterial, antifungal agents, etc., estimation of microbial resistance incidents, and prediction of antimicrobial efficacy in the management of infections. In the case of AgNPs, the expected mechanism is due to the vast surface area, which easily attaches to the membrane of the microorganism and penetrates inside. After which, the entered AgNPs interfere in the respiratory chain in mitochondria, leading to cell death. In the above section, we explained all the literature's mechanisms along with diagrammatic representation in Fig. 2.

Here, for a clear understanding of the silver nanoparticles' process of action, is an in-depth explanation. On the microorganism's surface, if the cell wall is anchored to and penetrated, this leads to structural integrity damage, permeability results in leaking of the cellular components, which disorganizes the normal physiology, and cell death occurs. AgNPs are oxidized in aqueous environments in protons and oxygen, which dissolves the particle surfaces and releases Ag^+ ions. Release efficiency of Ag^+ ions mainly depends on the shape and size of AgNPs, their colloidal state, and type of capping agent. Significantly lower size AgNPs show a superior release pattern of Ag^+ onto gram-negative bacteria, and shape and temperature lead to more significant toxicity due to accelerated release of Ag^+ by effective dissolution (Stoehr et al., 2011). There is an interrelation between the size and surface area, as small size improves the surface area, and when the surface is large, dissolution and attachment is also superior. The release rate of ions is also greater, which leads to enhanced antimicrobial activity, which is required for management/treatment in clinical medicine (Abuayyash et al., 2018). In the existence of oxygen, sulfur, chlorine, and thiols synchronize the Ag^+ ions release in the microenvironment (Maurer and Meyer, 2016). Ag^+ ions readily react with thiols present in the bacterial proteins, and enzymes damage cellular respiration, leading to cell death.

4.3 Formation of Ag^+ ions from AgNPs

The antibacterial efficiency mainly depends on the release of Ag^+ ions from nanomaterials. The release of Ag^+ ions is highly dense where the considerable surface area of nanoparticles is available. Small size nanoparticles exhibit high surface area, and one can observe the proportional relationship with surface area and Ag^+ ions release. Ivask et al. (2014) investigated the toxicity of various sizes of AgNPs from 10 to 80 nm on bacteria, algae, yeast, crustacean, and mammalian cells in vitro. They explained that the lowest size, 10 nm AgNPs, show the highest surface area, that enhanced cell-particle interaction leads to superior intracellular bioavailability of Ag^+ ions compared to other sizes of AgNPs, and proved that 10 nm AgNPs are more toxic to *E. coli*. Zawadzka et al. (2014) also reported on AgNPs with a large surface area, which were characterized by enormous inhibition of *S. aureus* growth. Agnihotri et al. (2013) experimented and compared the Ag^+ ion release on immobilized AgNPs on amine-functionalized silica surface and colloidal AgNPs + AgCl surface on microorganisms, i.e., *E. coli* MTCC 443, and *B. subtilis* MTCC 441. They noted the release profile of Ag^+ ion release over time and concluded that immobilized AgNPs releases more Ag^+ ions than the colloidal form of AgNPs. This indicated that whatever the form, either salt or nanoparticles, the efficacy entirely depends on Ag+ions' release.

4.4 Lipid and proteins interactions on the cell membrane

The main function of cell membranes of bacteria and fungi is permitting nutrients into the cell for their growth; cellular components are suspended in the cytoplasm, facilitating normal physiology and protection from outer environmental challenges. Silver nanoparticles need to cross the cell membrane by interacting with the membrane, altering the membrane's integrity and fluidity, then entering the cell cytoplasm. The bacterial surface and silver nanoparticles or silver ions interact and interfere in the respiratory chain and obstruct the energy. Generally, the assembly and maintenance of the cell envelope or plasma membrane precisely depends on the energy. The bacteria are of two types depending on the cell wall: layers of lipopolysaccharides present are known as gram-negative; if layers of peptidoglycans are present, they are known as gram-positive. Collectively, the outer membranes are composed of proteins and lipids. Ag^+ ions can interact with the protein to form a complex, having electron donors containing phosphorous, nitrogen, sulfur, and oxygen atoms. An interaction of Ag^+ ions with proteins leads to the inactivation of membrane-bound proteins and enzymes. The lipid present in the bacterial membrane may also interact with Ag^+ ions and increase the trans/cis ratio of unsaturated fatty acids (Hachicho et al., 2014; Crisan et al., 2021). This isomerization of unsaturated membrane fatty acids and modification in membrane lipids leads to a change in fluidity and integrity. Therefore, penetrating the cell and leakage of cellular components leads to cell death (Rajesh et al., 2015). In this way, lipids and polysaccharides interact and damage the cell wall integrity and fluidity and release the Ag^+ ions in the presence of oxygen, phosphorous, and nitrogen.

5 Characterization of silver nanoparticles (AgNPs)

UV-visible Spectroscopy is a primary investigation tool for characterizing the biogenic synthesized AgNPs in the range of 200–700 nm against blank solution as a standard (Seku et al., 2018). In this work, FTIR analysis was performed to identify functional groups of carboxymethylated gum *Cochlospermum gossypium* (CMGK) involved in reducing and stabilizing synthesized NPs. CMGK and AgNPs samples were subjected to FTIR spectroscopy in the scan range of $4500–400\,cm^{-1}$ using a KBr pellet method. The crystallinity of CMGK-capped silver nanoparticles was studied by X-ray diffraction (Rigaku, Miniflex) method with CuKα ($\lambda = 1.5418\,Å$) radiation). Transmission electron microscopy (TEM) was used to investigate the morphology and size distribution of the CMGK stable AgNPs dispersion.

5.1 UV-visible characterization

The UV-visible spectrophotometer is a primary investigation technique for the characterization of metal nanoparticles. It is an easy and convenient method for the identification of metal nanoparticles. AgNPs exhibit a strong absorption band at 415–430 nm due to surface plasmon resonance in the visible region (Indana et al., 2016). The development of AgNPs is identified based on the change in color of the reaction mixture from a light-yellow color to dark yellow color. Optimization studies were carried out to significantly form silver nanoparticles by varying the concentration of natural gums and metal nanoparticles (Seku et al., 2018).

Fig. 3 shows the optimization studies of AgNPs. AgNPs were prepared using carboxymethylated gum *Cochlospermum gossypium* (CMGK). The effect of CMGK concentration on the development of AgNPs has been studied by changing the concentration of CMGK from 0.1% to 0.5%, keeping the concentration of $AgNO_3$ at a constant (Murali Krishna et al., 2016). Fig. 3A shows the corresponding changes in surface plasmon resonance peaks by increasing the concentration of CMGK. As the concentration of CMGK increased from 0.1% to 0.5%, the SPR peak intensity increased as the particle size of the AgNPs decreased. It reveals that with increase in the concentration of CMGK, formation of AgNPs also increased.

The formation of silver nanoparticles was increased by diminishing more Ag^+ ions. Furthermore, CMGK controls the formation of AgNPs and efficiently stabilizes the nanoparticles (Ibrahim, 2015). Another factor is the effect of $AgNO_3$ concentration on the formation of AgNPs. By keeping the concentration of CMGK (0.5%) as constant, we examined AgNO3 concentration's effect on the synthesis of AgNPs. Fig. 3B reveals that a more significant number of Ag^+ ions are available in solution

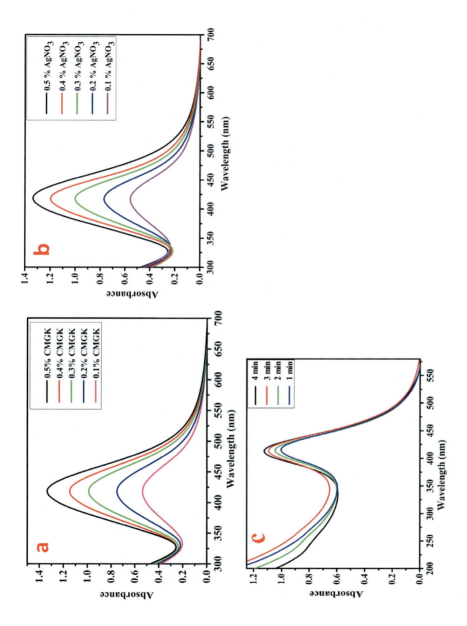

FIG. 3

(A) UV-visible spectra of CMGK solution at different concentrations at 0.5% $AgNO_3$; (B) UV-visible spectra of $AgNO_3$ solution at different concentrations at 0.5% CMGK; and (C) the UV-visible absorption spectra of AgNPs synthesized at varying the microwave irradiation time (Seku et al., 2018).

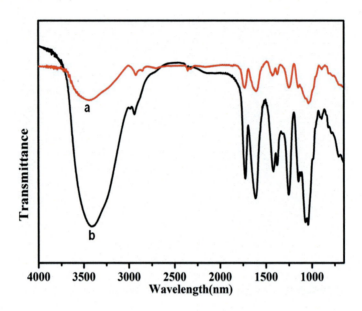

FIG. 4

FTIR spectra: (A) CMGK and (B) CMGK-capped AgNPs (Seku et al., 2018).

due to the increasing $AgNO_3$ concentration. It concluded that more Ag^+ ions were diminished and successfully capped by CMGK to generate more AgNPs.

The microwave irradiation time has an essential effect on the formation of AgNPs. The electronic spectra of synthesized AgNPs at different microwave irradiation times are presented in Fig. 4. From Fig. 4, it is observed that SPR peak intensity was increased by a gradual increase of the irradiation time up to 3 min. Further increase in the irradiation time to 4 min did not show any significant changes. Hence, 3 min of microwave irradiation time is taken as optimal for the biosynthesis of AgNPs by fruit extract (Pandian et al., 2015).

5.2 FTIR analysis of AgNPs

The formation of CMGK-stabilized silver nanoparticles was characterized by FTIR spectroscopy. Fig. 4 shows the CMGK and CMGK stabilized silver nanoparticles spectroscopic results. The significant peaks (curve (A) of Fig. 4) in the IR spectrum of CMGK peaks are observed at 3430, 2960, 1733, 1615, 1415, 1225, and 1035 cm^{-1}. The peaks at 3430 cm^{-1} and 2960 cm^{-1} are attributed to stretching vibrations of O—H and C—H groups. The other strong peaks observed at 1733, 1615, and 1416 cm^{-1} could be attributed to characteristic asymmetrical and symmetrical stretching vibrations of the COO (carboxylation) group connected with the carboxymethyl gum kondagogu. The peaks at 1225 and 1035 cm^{-1} are associated with the C—O—C stretching vibrations (Van Phu et al., 2014) of ether and alcohol groups of carboxymethyl gum kondagogu. Fig. 4 (curve b) shows the FTIR spectrum of carboxymethyl gum kondagogu capped AgNPs, which shows the characteristic frequencies at 3375, 1727, 1591, 1406, 1220, and 1026 cm^{-1}. A shift in the peaks of the FTIR spectrum of CMGK stabilized AgNPs was observed from 3430 to 3375 cm^{-1}, 1733 to 1727, 1615 to 1591 cm^{-1}, 1415 to 1410 cm^{-1}, and other peaks were found to remain unchanged. These shifts in the IR spectrum suggest the binding of AgNPs with hydroxyl and carboxylate groups of CMGK. Based on the peak shifts in the hydroxyl and carboxyl groups, it can be concluded that both hydroxyl and carbonyl groups of CMGK are involved in the synthesis and stabilization of AgNPs (Kora et al., 2010; Seku et al., 2018).

5.3 Powder XRD analysis of AgNPs

The formation of AgNPs was studied by the XRD technique. Fig. 5 shows the XRD pattern of the formed AgNPs. The diffraction peaks at 2θ theta values 38.32, 44.50, 64.75, and 77.05 were indexed as (111), (201), (222), and (312), correspondingly. According to the available literature on the XRD pattern, all peaks are indexed with face-centered cubic structure for AgNPs (JCPDS, file no. 04-0783). The peak (111) plane is the most intense compared to other peaks in the XRD spectrum, which indicates the preferred growth of AgNPs along (111) direction. The calculated lattice constant is $a = 0.4085$ nm, consistent with the standard value $a = 0.4086$ nm. Synthesized silver nanoparticles were evaluated by Scherer's formula (Rodríguez-León et al., 2013; Chung et al., 2016):

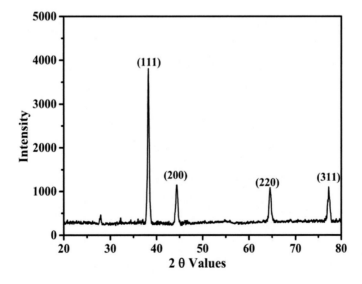

FIG. 5

XRD pattern of AgNPs (Seku et al., 2018).

$$D = 0.9\lambda/\beta\cos\theta$$

where D is the crystalline size, β is the full width at half maximum, λ is the wavelength of X-ray used (1.5406 A°), and θ is the Bragg's angle.

The crystallite size was calculated to be 8.6 nm, which is effectively matched with the TEM results.

5.4 Transmission electron microscopy analysis

Fig. 6A shows the TEM images of the synthesized AgNPs. Synthesized silver nanoparticles have spherical shapes with smooth surfaces and are well dispersed. The resultant AgNPs demonstrated that the CMGK could secure silver nanoparticles

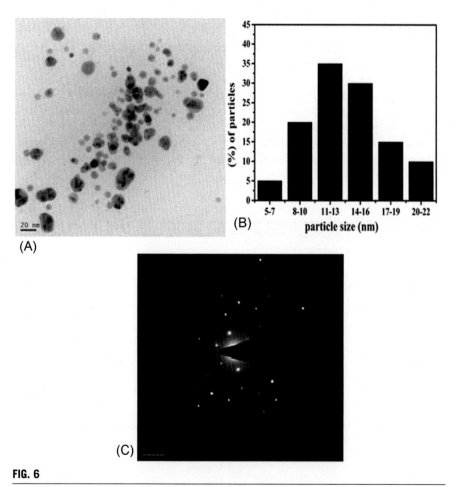

FIG. 6

(A) TEM image of synthesized AgNPs; (B) the histogram of AgNPs; and (C) SAED criterion of AgNPs (Seku et al., 2018).

from conglomeration successfully (Jain and Mehata, 2017). The histogram (Fig. 6B) was built considering 100 AgNPs; this demonstrates that the average size distribution is 9 ± 2 nm. The SAED pattern of AgNPs is shown in Fig. 6C. Crystal clear and identical lattice fringes are shown by AgNPs, which affirmed that the spherical particles are extremely crystalline in nature. Lattice space of 0.232 nm compares to planes of silver (111), which shows that the favored direction of AgNPs is along (111) planes. SEAD pattern and TEM confirmed that the microwave-assisted synthesized silver nanoparticles are single crystals (Seku et al., 2019).

5.5 Dynamic light scattering studies

The zeta potential and hydrodynamic particle size of AgNPs were quantified by DLS technology, and the zeta potential (ζ) was utilized to confirm the charge on AgNPs.

The confirmed Zeta potential indicates the presence of repulsive forces and can be utilized to assess the stability of nanoparticles (Reddy et al., 2015b). The determination of the hydrodynamic particle size of AgNPs using DLS is based on the laser diffraction method with multiple scattering techniques. In the suspension of small AgNPs, the large number of phytochemicals per particle is high, and subsequently, the hydrodynamic size acquired from DLS is more prominent than that aquired by TEM (Madhusudhan et al., 2019). Nanoparticles with severe positive or negative surface charges are stable. A flat-out zeta possible estimation of ± 30 mV is an overall sign that the colloidal solution is highly stable. In another report, the authors demonstrated that the *Solanum melongena L* stabilized (Das and Bhuyan, 2019) AgNPs have an average size of nanoparticles of about 29 nm and zeta potential value of -18.5 mV, and Seku et al. reported that the CMGK stabilized AgNPs had a zeta potential of -18.7 mV (Seku et al., 2018). These findings suggest that AgNPs stabilized with phytochemicals carried negatively charged groups and the AgNPs were stable due to the strong electrostatic repulsion between the nanoparticles.

6 Biological applications
6.1 Antibacterial activity

People are affected by chronic diseases due to bacterial infections, which lead to mortality. Antibiotics are preferred for the treatment of bacterial infections to control chronic illness. Antibiotics have been shown to be effective and provide powerful outcomes towards the control of chronic diseases. However, today, people are affected by multidrug resistance (MDR) bacterial strains because of over usage of antibiotics (Nikaido, 2009). Literature reports direct evidence for MDR that are resistant to the strongest antibiotics, which carry a super-resistance gene called NDM-1.

Most antibiotics attack bacteria in three targeted paths: cell wall synthesis, translational machinery, and DNA replication machinery. Regrettably, bacterial resistance can develop against each of these modes of action and modify or degrade antibiotics. The mode of action of nanoparticles on bacteria is different compared to that of antibiotics. Nanoparticles are used in direct attacks on the bacterial cell, and thus promote less bacterial resistance than do antibiotics. Present day researchers are focusing on developing novel and existing NP-based materials with good antibacterial properties (Hajipour et al., 2012; Franci et al., 2015). Hence, current-day NP-based antibacterial materials are more significant in controlling chronic illness (Kuppusamy et al., 2016).

Generally, silver in the form of $AgNO_3$ shows antimicrobial effects, whereas at the nano level antimicrobial activity increases due to its unique nature and increase in the surface area of silver nanoparticles (AgNPs). The effect of AgNPs on cellular metabolic activity and the damage to the membrane of cells results in the production of ROS and DNA dysfunction. The bacterial cell dysfunction is influenced by the size-dependent silver nanoparticles (Rai et al., 2012). Silver nanoparticles can attack bacterial cell walls and subsequently penetrate into the cell, increasing the accumulation of silver nanoparticles on the surface of the cell wall leading to the formation of fits, resulting in structural changes such as permeability and destruction of the cell. Electron spin resonance spectroscopy suggests that the interaction of silver nanoparticles with bacteria leads to production of silver radicals. These radicals damage the cell membrane, causing leakage of protein from the cell, and lead to the cell's death (Fig. 7).

Krishna et al. (2016) prepared AgNPs with *Salmalia malabarica* gum (SMG) as a stabilizing agent and studied the antibacterial activity by disc diffusion method using both *E. coli* and *S. aureus*. The zone of inhibition (ZOI) was measured for AgNPs with *Salmalia malabarica* gum (SMG) as a stabilizing agent and SMG alone, and the results showed 16 mm (1 mg/mL) and no inhibition activity, respectively. A ZOI of about 18 mm was observed for 5 mg/mL of SMG-capped AgNPs for both *E. coli* and *S. aureus*. The results demostrated that the ZOI decreases with the decrease in the concentration of AgNPs, which shows the antibacterial effectiveness of AgNPs. The AgNPs of size 1–60 nm effectively penetrated the cell and reacted with sulfur-containing proteins and phosphorus-containing DNA. AgNPs inactivate bacterial enzymes essential for their metabolic functions, produce hydrogen peroxide, and cause bacterial cell mortality. Seku et al. (2018) also reported similar results on carboxymethylated gum kondagogu (CMGK) as a capping agent and tested for antibacterial activity on gram-negative bacterial *P. aeruginosa* and *E. coli* and gram-positive bacterial *B. subtilis* and *B. cereus*. CMGK-AgNPs showed discrete antibacterial action against pathogenic microorganisms at 5 μg/mL concentration. AgNPs were truly lethal to *B. subtilis*, *B. cereus*, and *E. coli* with an inhibition zone of 28, 25, and 23 mm, respectively. The AgNPs contrasted positively with silver nitrate, CMGK, and standard antibiotic *streptomycin* at a concentration of 5 μg/mL. The findings revealed that AgNPs showed more activity than silver nitrate and standard antibiotic, but CMGK solution does not show a zone of inhibition.

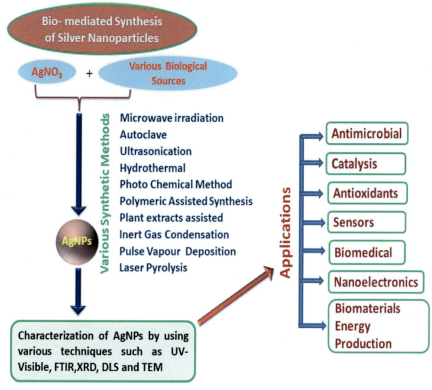

FIG. 7

Schematic representation of the synthesis of AgNPs using natural gum and their characterization and applications.

6.2 Antifungal activity

Antifungal activity of AgNPs against skin pathogens (ATCC strains of *Trichophyton mentagrophytes* and *Candida species*) was reported by Kim et al. (2008) who studied the comparison with *amphotericin-B* and *fluconazole*. In 2012 he supported the antifungal activity of nano-sized silver chitosan nanoformulation and silver/chitosan nanocomposite against seed-borne plant fungal pathogen-inhibiting mycelium growth. Xia et al. (2016) reported that silver nanoparticles have good antifungal activity against *T. asahii*. Based on our electron microscopy observations, silver nanoparticles may inhibit the growth of *T. asahii* by permeating the fungal cell and damaging the cell wall and cellular components.

Balashanmugam et al. (2016) suggested phytogenic synthesis of silver nanoparticles, optimization, and evaluation of in vitro antifungal action against human and plant pathogens. The AgNPs showed higher antifungal activity when compared with the conventional antifungal drug amphotericin B against all the tested human fungal pathogens such as *Aspergillus fumigatus*, *Aspergillus flavus*, *Aspergillus niger*, *Candida albicans*, *Penicillium* sp., and the plant pathogens such as *Fusarium oxysporum*, *Rhizoctonia solani*, and *Curvularia* sp. Mussin et al. (2019) reported silver nanoparticles' antifungal activity in combination with ketoconazole against *Malassezia furfur*. In this investigation, the authors tested the in vitro inhibitory effect of AgNPs against *M. furfur* medical separates. Herein analyze the interaction between the AgNPs—Malassezia and assess the synergism with ketoconazole (KTZ) and manufacture antimicrobial gel of AgNPs-KTZ. Potential improvement of topical therapy of superficial *Malassezia* is by AgNPs-KTZ gel and prevent the recurrence. Seku et al. (2021) reported that the ZJM fruit extract capped AgNPs showed good antifungal activity with ZOIs of 10, 11, 12, 14 mm, respectively, for *Aspergillus niger*, and 9, 10, 11, 13 mm respectively, for *Aspergillus fumigatus*. The antifungal activity increased with an increase in AgNPs concentration.

6.3 Antiviral activity

Today, people are affected by viral infections including influenza, hepatitis, chickenpox, common cold, herpes keratitis, infectious mononucleosis, human immunodeficiency virus, and viral encephalitis. In antiviral therapy, remarkable improvements have been made, but medicines cannot completely prevent all viral diseases. Therefore, it is necessary to develop potent antiviral agents against a broad range of viruses using biogenic metal nanoparticles (Dizaj et al., 2014).

Herpes simplex virus types 1 and 2 (HSV-1 and HSV-2, respectively) and human parainfluenza virus type 3 (HPIV-3) are important viruses because they cause severe infection in people. Antiviral action of mycosynthesized silver nanoparticles against human parainfluenza virus type 3 and herpes simplex virus was reported by Gaikwad et al. (2013), who suggest that the AgNPs had better antiviral activity (80%–90% inhibition) against HSV-1 and HPIV-3 viruses and were less cytotoxic to Vero cells. AgNPs prevented HIV infection via bound on a viral coat, inhibiting the HIV-1 infection cycle's initial stages through the blocking absorption and infectivity in a cell fusion assay method. Here the therapeutic index of AgNPs is 12 times greater compared with silver ions.

Lara et al. (2010) suggest that silver nanoparticles show an effective anti-HIV activity at an early stage of viral replication or as an inhibitor of viral entry. Furthermore, the postentry stages of the HIV-1 life cycle are inhibited by silver nanoparticles. Gaikwad et al. (2013) described mycosynthesized silver nanoparticles' antiviral action against human parainfluenza virus type 3 and the herpes simplex virus. They conclude that silver nanoparticles experience a size-dependent interaction with herpes simplex virus types 1 and 2 and with human parainfluenza virus type 3. The viral infection is reduced effectively by AgNPs, which block the interaction between the virus and cell. The viral infection control is dependent on the size and zeta potential of AgNPs. Reported literature confirms that the smaller silver nanoparticles were powerful enough to prevent the infectivity of viruses.

6.4 Antiprotozoal activity

Human beings, plants, animals, and some marine life suffer from different ailments due to protozoan infections. Protozoan infections cause African sleeping sickness, malaria, and amoebic ailments. Hassan et al. (2019) reported an antiprotozoal activity of silver nanoparticles against *Cryptosporidium parvum oocysts*, providing new insights into their feasibility as a water disinfectant. They studied the efficacy of AgNPs on viability and count of the *Cryptosporidium parvum* (CP) isolated from different tap water samples. Different dosages of AgNPs of 0.05, 0.1, and 1 ppm were exposed to oocysts for several contact times (30 min to 4 h). The dose-dependent manner shows a significant decrease in oocyst count and viability. The obtained results revealed that the shorter contact time was recommended for C. oocyst inactivation (either 30 min at 1 ppm, or 1 h at 0.1 ppm concentrations). AgNPs were successfully utilized in the treatment of wastewater as a disinfectant.

Pimentel-Acosta et al. (2019) reported a study to determine the in vitro anthelmintic effect of silver nanoparticles (AgNPs) against adults and eggs of monogenean fish parasites in freshwater using *Cichlidogyrus* spp. as a model organism. The study concludes that the adults and eggs of *Cichlid gyrus monogeneans* were ultimately affected by UTSA-AgNPs at a 36 µg/L concentration for 1 h.

6.5 Anti-arthropods activity

Biologically synthesized AgNPs/AuNPs can also be used as an eco-friendly approach to mosquito control (Adhikari et al., 2013). This innovative research is a new step to establish AgNPs as an insecticide.

6.6 Anti-parasitic action

Biogenic silver nanoparticles can be an active larvicidal agent against dengue vector *filariasis vector Culex quinquefasciatus*, *C. quinquefasciatus*, *Aedes aegypt*, and *Aedes aegypti*, *A. subpictu*, malarial vector *A. subpictus*, and other parasites (Suganya et al., 2014). The proper mechanism for the antiparasitic action of AgNPs has not yet been established. Denaturation of phosphorus-containing DNA and sulfur-containing proteins with AgNPs leads to the denaturation of organs and enzymes and is responsible for larvicidal action. Severe human diseases are spread by mosquito vectors, leading to millions of deaths every year. Vector control in developing counties is an essential problem with several aspects. Bhuvaneswari et al. (2016) reported the green synthesis of AgNPs using leaf extract and their larvicidal property against vector mosquitoes (*Anopheles stephensi* and *Aedes aegypti*). The green synthesized AgNPs shows an excellent larvicidal activity against *A. stephensi* ($LC_{50} = 78.4$; $LC_{90} = 144.7$ ppm) followed by *A. aegypti* ($LC_{50} = 84.2$; $LC_{90} = 117.3$ ppm). The AgNPs are promising materials to be used as an effective larvicidal agent.

6.7 Catalytic activity of AgNPs

Silver nanoparticles have been used as a catalyst in the electron exchange reaction between hexacyanoferrate (III) and sodium borohydride, which brings about the alignment of hexacyanoferrate (II) ions and dihydrogen borate ions. The redox reaction is as follows:

$$[BH_4]^- + 8[Fe(CN)_6]^{-3} 3H_2O \rightarrow H_2BO_3^- + 8[Fe(CN)_6]^{-4} + 8H^+$$

The advantage of the hexacyanoferrate ion for this redox research is that the oxidation states (+2 and +3) of the iron are very stable concerning decomposition and hydrolysis. The redox potential related to the reaction being evaluated, E_o (Fe^{2+}/Fe^{3+}), is +0.44 V versus a normal hydrogen electrode. The standard reduction potential of the borate particle is $E_o = -1.24$ V. Therefore, there is a substantial free energy change associated with the reaction (Reddy et al., 2015a, b; Kästner and Thünemann, 2016). However, even if this reaction is preexisting without the catalyst, it is considered a moderate reaction with zero-order kinetics.

The progression of the reaction reducing potassium hexacyanoferrate (III) to potassium hexacyanoferrate (II) was observed during changes in the UV-visible range. The typical absorption peak of hexacyanoferrate (III) was detected at 420 nm. Through the addition of nanoparticles, the magnitude of the absorption point is continuously reduced, which reveals the occurrence of catalytic reduction (Fig. 8). The redox reaction is a pseudo-first-order reaction to hexacyanoferrate (III) (Gangula et al., 2011). As the centralization of $NaBH_4$ was more significant than

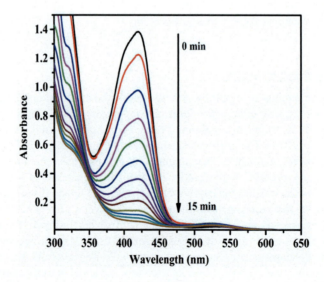

FIG. 8

Schematic representation of the synthesis of AgNPs using natural gum and their characterization and applications (Seku et al., 2018).

the hexacyanoferrate (III), the reducing rate can be thought to be independent of NaBH$_4$ quantity (Martínez et al., 2021). The kinetics of the reduction reaction of hexacyanoferrate (III) catalyzed by AgNPs have been investigated at different temperatures (30–70°C) and different amounts of catalysts. The reaction was carried out, setting all investigational conditions, such as starting concentration of hexacyanoferrate (III), borohydride concentration, and temperature, at constant.

The rate constants were calculated differently for the sum of the AgNPs (50–250 μL). The rated constant values (Table 3) are plotted against the different amounts of catalysts shown in Fig. 9A. Due to the improvement in the magnitude of the reaction surface sites, the reaction rate increases directly with the measurement expansion of AgNPs (Seku et al., 2018; Nemanashi and Meijboom, 2013). A linear relationship between the observed rate constant and the catalyst sum was found. The rate constant of the catalytic reaction was investigated at different temperatures (33°C, 42°C, 51°C, 62°C, and 73°C), and constant concentrations of

Table 3 Rate constants at different amounts of catalyst.

Sample	Amount of AgNPs (μL)	Calculated rate constant (min^{-1})
1	50	0.0721
2	100	0.123
3	150	0.164
4	200	0.210
5	250	0.254

Reprinted with permission from Seku, K., Gangapuram, B. R., Pejjai, B., Kadimpati, K. K., Golla, N. 2018. Microwave-assisted synthesis of silver nanoparticles and their application in catalytic, antibacterial and antioxidant activities. J. Nanostruct. Chem., 82, 179–188.

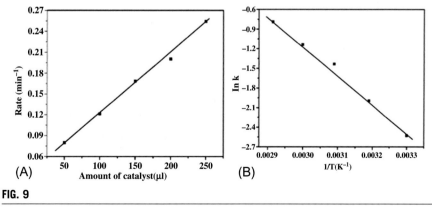

FIG. 9

(A) Plot of rate constant against different amounts of silver nanoparticles, and (B) plot of rate constant against $1/T$ (K^{-1}) for the electron transfer reaction between hexacyanoferrate (III) and sodium borohydride in the presence of AgNPs (Seku et al., 2018).

Table 4 Rate constants at different temperatures for AgNPs catalyzed by the reduction of hexacyanoferrate (III) by $NaBH_4$.

Temperature (K)	Rate constant (min^{-1})
303	0.0761
313	0.174
323	0.242
333	0.305
343	0.415

hexacyanoferrate (III), borohydride, and amount of catalyst. $1/T$ vs $\ln k$ Arrhenius plots are given in Fig. 9B. The active energy is found from the slope ($-E_a/R$). It has been observed that an increase in temperature expands the reaction rate (Table 4). Activation energy is calculated from a straight line ($-E_a/R$) and is determined as 7.2 kcal/mol.

6.8 Biomedical applications of silver nanoparticles

Nanotechnology has been playing a significant role in the medical field, revolutionizing medical treatments and therapies, and promising further biomedical applications. The advances of nanotechnology in the biomedical field include faster diagnosis, imaging, drug delivery, tissue regeneration, and the development of new medical products (Abbasi et al., 2016).

6.8.1 AgNPs in diagnosis and imaging

Currently, molecular diagnostics is of interest using nanomaterials in the assays of gases, DNA, metal ions, and protein markers for many diseases. Robust and highly sensitive and selective detection agents have been fueled by intense research that addresses conventional technologies' deficiencies. Chemists are playing a vital role in the design and fabrication of nanomaterials in the application of diagnosis. Overcoming the drawbacks of conventional diagnostic systems in selectivity and sensitivity is a significant advantage of nanomaterials (Ahmed et al., 2016).

Karthiga and Anthony (2013) reported selective colorimetric sensing of toxic metal cations by green synthesized silver nanoparticles over a wide pH range. AgNPs were prepared by using fresh neem and dried neem leaf extracts. Fresh neem (NF-AgNPs) silver nanoparticles selectively sensed the Hg^{2+} ions, whereas AgNPs selectively detected hg2+ and Pb2+ ions at micromolar concentrations. This study reveals that the AgNPs are a useful colorimetric sensor for the detection of health and environmentally hazardous metal ions such as Hg^{2+}, Pb^{2+}, and Zn^{2+} in an aqueous solution over a wide pH range. Eksin et al. (2019) described eco-friendly sensors developed by herbal-based silver nanoparticles for electrochemical detection of mercury (II) ion. The developed sensor showed an admirable selectivity for Hg^{2+} against other heavy metal ions such as Ca^{2+}, Cd^{2+}, Cr^{3+}, Cu^{2+}, Mg^{2+}, Ni^{2+}, Pb^{2+}, Zn^{2+}, Co^{2+}, and Mn^{2+}.

6.8.2 AgNPs in therapeutics (antibacterial coatings)

Colonization and adhesion of bacteria on the surfaces of medical devices causes infections and significant health issues. Silver-based antibacterial surface coatings can control bacterial infections. Therefore, we need to develop AgNPs as a surface coating to protect medical device surfaces from bacterial adherents. Recently, Marassi et al. (2018) reported silver nanoparticles as a medical device in healthcare settings: a five-step approach for candidate screening of coating agents. The efficacy of AgNPs was evaluated on luminescent strains of E. Coli and CFT073. Rodrigues et al. (2020) reported the biogenic synthesis and antimicrobial activity of silica-coated silver nanoparticles for esthetic dental applications. In this work, the synthesis of biogenic Ag@SiO_2 NPs with antimicrobial activity was evaluated against *S. mutans*. Ag@SiO_2 NPs are a promising agent against *S. mutans*. AgNPs@SiO2 NPs agents have prevented dental caries by inhibiting the formation of *S. mutans* biofilm. In dentistry, overcoming clinical problems has shown that the present research is a great step towards developing more interactive biomaterials.

Taheri et al. (2014) reported on substrate independent silver nanoparticles-based antibacterial coatings, developing AgNPs-based antibacterial surface coatings that can be applied to any type of material surface. The silver nanoparticles were surface engineered with a monolayer of 2-mercaptosuccinic acid, which reduces the rate of oxidation of nanoparticles, enhances the lifetime of the coatings, and enables the immobilization of the nanoparticles on the solid surface. These coatings show an excellent antibacterial efficacy against three clinically significant pathogenic bacteria: *Staphylococcus aureus*, *Staphylococcus epidermidis*, and *Pseudomonas aeruginosa*. Furthermore, findings with primary human fibroblast cells revealed that the coatings had no cytotoxicity in vitro.

6.8.3 Wound dressing

The integration of many different tissues and cell lineages in wound healing is regarded as a complex and multiple-step process. The use of silver nanoparticles is a recognized application in wound dressings. The first commercial wound dressing material is Acticoat, made up of two layers of polyamide ester membranes covered with AgNPs. To achieve better cosmetic results and promote healing silver has been used in the form of nanoparticles. Sibbald et al. (2001) reported that the use of AgNP dressings on various chronic nonhealing wounds was evaluated by conducting prospective studies. The study proves that the wound site is effectively protected from bacterial contamination with AgNP dressings. AgNPs have shown better results compared to silver compounds in promoting healing and achieved better cosmetic results. The AgNPs seem to hold the most significant promise in the aspects of wound healing.

Coutts and Gary Sibbald (2005) suggested a silver-containing Hydrofiber dressing on superficial wound beds and bacterial balance of chronic wounds, which increases bacterial load on chronic wounds causing drain healing without all the clinical signs of infection. Hence, silver dressings might be an alternative topical method to control the bacterial load on wounds. The present work aims to evaluate the clinical improvements of silver-containing Hydrofiber dressing in chronic wounds using different parameters such as the effect on wound size, maceration, resolution of the surface slough, and conversion to healthy granulation during a 4-week application.

6.8.4 Antioxidant activity

Oxidation is a chemical reaction that can produce free radicals, thereby leading to chain reactions that may damage organisms' cells. An antioxidant has been defined as any substance that significantly delays or prevents the substrate's oxidation when present at lower concentrations than those of an oxidizable substrate (Flores-López et al., 2019; Mohanta et al., 2017). The beneficial effects of antioxidants against radicals have received much attention in the oxidation process. A variety of antioxidants with different functions play their role in in vivo defense networks. Some antioxidants are proteins and enzymes, while others are natural or synthetic small molecules.

From a mechanical point of view, antioxidants can be classified into the following types: (i) preventive antioxidants that suppress the formation of reactive oxygen and nitrogen, e.g., by diminishing hydrogen peroxide and lipid hydroperoxide to water and lipid hydroxides, or by sequencing metal ions such as iron and copper; (ii) antioxidants that prevent radicals attacking biologically important molecules, e.g., scavenging antioxidants (many natural exogenous antioxidants use this mechanism); (iii) antioxidants that repair damages clearing the wastes and reconstituting the lost function, e.g., enzymes; and (iv) antioxidants that act as a cellular signaling messenger to regulate the level of antioxidant compounds and enzymes.

The term "antioxidant" refers to a chemical that inhibits the use of oxygen. The antioxidant agent inhibits the harmful effects of free radicals. Lee first used colloidal silver in medicine in 1889. Nsimba et al. (2008) reported the plant species' antioxidant activity in the Chenopodiaceae family, i.e., *Chenopodium quinoa* and *Chenopodium album*. They revealed the results that there was a slight increase in the antioxidant activity of plant extract capped AgNPs. Comparatively, alone, either plant extract or AgNPs showed less antioxidant activity than plant extract -AgNPs. Seku et al. (2018) described the antioxidant action of AgNPs. In this study, the antioxidant activity of AgNPs was assessed utilizing the DPPH measure utilizing ascorbic acid as reference. Inhibition action of AgNPs compared to ascorbic acid (standard) appears as an increment in DPPH radical inhibitory activity with the quantity of silver nanoparticles. Concentrations of 5, 10, 15, 20, 25, and 30 µL signify a scavenging rate 20.5%, 46.2%, 57.4%, 61.7%, 63.6%, and 64.9%, respectively. Compared to ascorbic acid, the AgNPs demonstrates superior action. The more satisfying rates of DPPH radicals show the greater antioxidation ability of the obtained AgNPs.

6.8.5 Food packaging

Nowadays, the food industry plays an essential role in providing a constant supply of safe, nutritious, and enjoyable food throughout the year. The main goal of nanotechnology and food research and development is food packaging and safety. Many food companies have already begun developing lighter and stronger packages with sensors that alert consumers to the presence of contaminants or pathogenic microorganisms. The use of antimicrobial food packaging materials is a new dimension of nanotechnology research in this rapidly expanding field. Nanotechnology offers a potential solution to discarding food before it reaches the end of its shelf life. Metal nanoparticles have been used as active packaging based on their powerful antimicrobial properties. Emerging metallic nanoparticles with biocidal properties include Cu, Zn, Au, Ti, and Ag (Toker et al., 2013). Among them, silver nanoparticles (AgNPs) have been shown to have highly effective bactericidal properties against a wide variety of pathogenic microorganisms, including bacteria, yeasts, fungi, and viruses (Carbone et al., 2016; Jena et al., 2012; Metak and Ajaal, 2013). As we have already seen, silver has a long history in the food and beverage industry. Silver has also been used in developing countries for the treatment of water containers and disinfection units. Nanoparticles doped plastic or polymer can prevent oxygen, carbon dioxide, and moisture entering into the diet, especially in particular meats.

Carbone et al. (2016) reported on silver nanoparticles in polymeric matrices for fresh food packaging. Silver nanoparticles with a mix of nondegradable and consumable polymers have been utilized as dynamic food packing, indicating antimicrobial, antifungal, hostile to yeast, and antiviral activities. Recent findings on the safety of proteins, organic products, and dairy items have shown that AgNPs doped nondegradable and palatable polymers and oils are the most effective against food microorganisms.

7 Future perspectives and recommendations

Biogenic synthesis of metal nanoparticles shows a wide range of applications in several sectors. Hence, researchers and scientists are paying attention to the development of various types of products containing nanomaterials for various applications. This chapter has addressed various synthetic methods, characterization, and biomedical applications of silver nanoparticles. The antioxidant activity of biogenic AgNPs has been proved; we can expect these nanoparticles to be used for antiinflammatory and anticancer therapies. Furthermore, AgNPs are nontoxic to human lungs after prolonged exposure. The use of silver nanoparticles is a recognized application in wound dressing. Multilayered AgNPs coatings on dressing can be preferably used for diabetic patients who suffer from diabetic foot problems and postoperative cases. To achieve better cosmetic results and promote healing, multilayered silver in the form of nanoparticles has been suggested. Hence, researchers must focus on this area of research to develop various formulations.

Nanobiotechnology is in the infancy stage in nanomedicine to develop nano-imaging, nano-drug delivery systems, and nano diagnosis. There is a need for research to develop a nanobiotechnological rapid diagnosis of viral infections and cancers in the early stages, and advances of nanobiotechnology in the biomedical field for the development of novel medical products and tissue generation. Recently AgNPs coatings have been applied as a thin film on the personal protective equipment (PPE) kits of health professionals. In this case, these silver nanoparticles thin film coatings are applied to the health infrastructure. In general, the process requires commercially high budgets. Researchers and scientists should focus on developing AgNPs thin film coatings technologies to large scale to reach the general public and health systems at low cost.

Due to the food-water nexus, to achieve sustainable development goals, food must be stored for a long time; AgNPs help preserve food in the form of packing materials or articles. Recent findings on the safety of proteins, organic products, and dairy items have shown that AgNPs doped nondegradable and palatable polymers and oils are the most effective against food microorganisms. There is a need for research for further developments to commercialize and achieve large-scale production at an affordable cost. Global food demand can be met by storing food in commercially developed packing equipment and achieving sustainable development goals (SDG 1). The funding agencies/governments must financially support researchers and scientists to solve societal problems. Researchers and scientists are also focusing on fulfilling risk assessment, developing standard operating procedures, life cycle assessment to follow the regulatory authorities' regulations and recommendations.

References

Abbasi, E., Milani, M., Fekri Aval, S., Kouhi, M., Akbarzadeh, A., Tayefi Nasrabadi, H., Samiei, M., 2016. Silver nanoparticles: synthesis methods, bio-applications, and properties. Criti. Rev. Microbiol. 42 (2), 173–180.

Abboud, Y., Eddahbi, A., El Bouari, A., Aitenneite, H., Brouzi, K., Mouslim, J., 2013. Microwave-assisted approach for rapid and green phytosynthesis of silver nanoparticles using aqueous onion Allium cepa extract and their antibacterial activity. J. Nanostruct. Chem. 3 (1), 84.

Abuayyash, A., Ziegler, N., Gessmann, J., Sengstock, C., Schildhauer, T.A., Ludwig, A., Köller, M., 2018. Antibacterial efficacy of sacrifical anode thin films combining silver with platinum group elements within a bacteria-containing human plasma clot. Adv. Eng. Mater. 20 (2), 1700493.

Adhikari, U., Ghosh, A., Chandra, G., 2013. Nano particles of herbal origin: a recent eco-friend trend in mosquito control. Asian Pac. J. Trop. Dis. 3 (2), 167.

Agnihotri, S., Mukherji, S., Mukherji, S., 2013. Immobilized silver nanoparticles enhance contact killing and show highest efficacy: elucidation of the mechanism of bactericidal action of silver. Nanoscale 5 (16), 7328–7340.

References

Ahmed, S., Ahmad, M., Swami, B.L., Ikram, S., 2016. A review on plants extract mediated synthesis of silver nanoparticles for antimicrobial applications: a green expertise. J. Adv. Res. 7 (1), 17–28.

Ali, K., Ahmed, B., Dwivedi, S., Saquib, Q., Al-Khedhairy, A.A., Musarrat, J., 2015. Microwave accelerated green synthesis of stable silver nanoparticles with Eucalyptus globulus leaf extract and their antibacterial and antibiofilm activity on clinical isolates. PLoS One. 10. (7), e0131178.

Ansari, S.A., Al-Shaeri, M., 2019. Biotechnological application of surface modified cerium oxide nanoparticles. Braz. J. Chem. Eng. 36 (1), 109–115.

Araruna, F.B., de Oliveira, T.M., Quelemes, P.V., de Araújo Nobre, A.R., Plácido, A., Vasconcelos, A.G., Mascarenhas, Y.P., 2020. Antibacterial application of natural and carboxymethylated cashew gum-based silver nanoparticles produced by microwave-assisted synthesis. Carbohydr. Polym. 241, 115260.

Augustine, R., Augustine, A., Kalarikkal, N., Thomas, S., 2016. Fabrication and characterization of biosilver nanoparticles loaded calcium pectinate nano-micro dual-porous antibacterial wound dressings. Prog Biomater 5 (3–4), 223–235.

Awad, M.A., Hendi, A.A., Ortashi, K.M., Elradi, D.F., Eisa, N.E., Al-Lahieb, L.A., Awad, A.A., 2014. Silver nanoparticles biogenic synthesized using an orange peel extract and their use as an anti-bacterial agent. Int. J. Phys. Sci. 9, 34–40.

Balashanmugam, P., Balakumaran, M.D., Murugan, R., Dhanapal, K., Kalaichelvan, P.T., 2016. Phytogenic synthesis of silver nanoparticles, optimization and evaluation of in vitro antifungal activity against human and plant pathogens. Microbiol. Res. 192, 52–64.

Bhat, R., Ganachari, S., Deshpande, R., Ravindra, G., Venkataraman, A., 2013. Rapid biosynthesis of silver nanoparticles using areca nut Areca catechu extract under microwave-assistance. J. Clust. Sci. 24 (1), 107–114.

Bhuvaneswari, R., Xavier, R.J., Arumugam, M., 2016. Larvicidal property of green synthesized silver nanoparticles against vector mosquitoes Anopheles stephensi and Aedes aegypti. J. King Saud Univ. Sci. 28 (4), 318–323.

Bhuyar, P., Rahim, M.H.A., Sundararaju, S., Ramaraj, R., Maniam, G.P., Govindan, N., 2020. Synthesis of silver nanoparticles using marine macroalgae Padina sp. and its antibacterial activity towards pathogenic bacteria. Beni-Suef Univ. J. Basic Appl. Sci. 9 (1), 1–15.

Bose, D., Chatterjee, S., 2016. Biogenic synthesis of silver nanoparticles using guava Psidium guajava leaf extract and its antibacterial activity against Pseudomonas aeruginosa. Appl. Nanosci. 6 (6), 895–901.

Carbone, M., Donia, D.T., Sabbatella, G., Antiochia, R., 2016. Silver nanoparticles in polymeric matrices for fresh food packaging. J. King Saudi Univ. 28, 273–279.

Chaudhari, P.R., Masurkar, S.A., Shidore, V.B., Kamble, S.P., 2012. Biosynthesis of silver nanoparticles using Saccharum officinarum and its antimicrobial activity. Micro Nano Lett. 7 (7), 646–650.

Chung, M., Park, I., Hyun, K.S., Thiruvengadam, M., Rajakumar, G., 2016. Plant-mediated synthesis of silver nanoparticles: their characteristic properties and therapeutic applications. Nanoscale Res. Lett. 11, 40.

Coutts, P., Gary Sibbald, R., 2005. The effect of a silver containing Hydrofiber® dressing on superficial wound bed and bacterial balance of chronic wounds. Int. Wound J. 2 (4), 348–356.

Crisan, C.M., Mocan, T., Manolea, M., Lasca, L.I., Tăbăran, F.-A., Mocan, L., 2021. Review on silver nanoparticles as a novel class of antibacterial solutions. Appl. Sci. 11, 1120.

Dang, T.M.D., Le, T.T.T., Fribourg-Blanc, E., Dang, M.C., 2012. Influence of surfactant on the preparation of silver nanoparticles by polyol method. Adv. Nat. Sci. Nanosci. Nanotechnol. 3. (3), 035004.

Das, R.K., Bhuyan, D., 2019. Microwave-mediated green synthesis of gold and silver nanoparticles from fruit peel aqueous extract of Solanum melongena L. and study of antimicrobial property of silver nanoparticles. Nanotechnol. Environ. Eng. 4 (1), 5.

Demissie, A.G., Lele, S.S., 2013. Phytosynthesis and characterization of silver nanoparticles using callus of jatropha curcas: a biotechnological approach. Int. J. Nanosci. 12 (02), 1350012.

Depan, D., Misra, R.D.K., 2014. On the determining role of network structure titania in silicone against bacterial colonization: mechanism and disruption of biofilm. Mater. Sci. Eng. C 34, 221–228.

Dizaj, S.M., Lotfipour, F., Barzegar-Jalali, M., Zarrintan, M.H., Adibkia, K., 2014. Antimicrobial activity of the metals and metal oxide nanoparticles. Mater. Sci. Eng. C 44, 278–284.

Dolgaev, S.I., Simakin, A.V., Voronov, V.V., Shafeev, G.A., Bozon-Verduraz, F., 2002. Nanoparticles produced by laser ablation of solids in liquid environment. Appl. Surf. Sci. 186 (1–4), 546–551.

Eksin, E., Erdem, A., Fafal, T., Kıvçak, B., 2019. Eco-friendly sensors developed by herbal based silver nanoparticles for electrochemical detection of mercury II ion. Electroanalysis 31 (6), 1075–1082.

El Kassas, H.Y., Attia, A.A., 2014. Bactericidal application and cytotoxic activity of biosynthesized silver nanoparticles with an extract of the red seaweed Pterocladiella capillacea on the HepG2 cell line. Asian Pac. J. Cancer Prev. 15, 1299–1306.

Ertürk, A.S., 2019. Biosynthesis of silver nanoparticles using Epilobium parviflorum green tea extract: analytical applications to colorimetric detection of Hg 2+ ions and reduction of hazardous organic dyes. J. Clust. Sci. 30 (5), 1363–1373.

Fatimah, I., 2016. Green synthesis of silver nanoparticles using extract of Parkia speciosa Hassk pods assisted by microwave irradiation. J. Adv. Res. 7 (6), 961–969.

Feynman, R.P., 1959. There is plenty of room at the bottom. Eng. Sci. 23.

Flores-López, L.Z., Espinoza-Gómez, H., Somanathan, R., 2019. Silver nanoparticles: electron transfer, reactive oxygen species, oxidative stress, beneficial and toxicological effects. Mini review. J. Appl. Toxicol. 39 (1), 16–26.

Franci, G., Falanga, A., Galdiero, S., Palomba, L., Rai, M., Morelli, G., Galdiero, M., 2015. Silver nanoparticles as potential antibacterial agents. Molecules 20 (5), 8856–8874.

Francis, S., Joseph, S., Koshy, E.P., Mathew, B., 2017. Green synthesis and characterization of gold and silver nanoparticles using Mussaenda glabrata leaf extract and their environmental applications to dye degradation. Environ. Sci. Pollut. Res. 24 (21), 17347–17357.

Gaikwad, S., Ingle, A., Gade, A., Rai, M., Falanga, A., Incoronato, N., Russo, L., Galdiero, S., Galdiero, M., 2013. Antiviral activity of mycosynthesized silver nanoparticles against herpes simplex virus and human parainfluenza virus type 3. Int. J. Nanomed. 8, 4303–4314. https://doi.org/10.2147/IJN.S50070.

Gangula, A., Podila, R., Karanam, L., Janardhana, C., Rao, A.M., 2011. Catalytic reduction of 4-nitrophenol using biogenic gold and silver nanoparticles derived from Breynia rhamnoides. Langmuir 27 (24), 15268–15274.

Gao, Y., Jiang, P., Song, L., Wang, J.X., Liu, L.F., Liu, D.F., Luo, S.D., 2006. Studies on silver nanodecahedrons synthesized by PVP-assisted N, N-dimethylformamide DMF reduction. J. Cryst. Growth 289 (1), 376–380.

Gulbagca, F., Ozdemir, S., Gulcan, M., Sen, F., 2019. Synthesis and characterization of Rosa canina-mediated biogenic silver nanoparticles for antioxidant, antibacterial, antifungal, and DNA cleavage activities. Heliyon. 5. (12), e02980.

Gurav, A.S., Kodas, T.T., Wang, L.M., Kauppinen, E.I., Joutsensaari, J., 1994. Generation of nanometer-size fullerene particles via vapor condensation. Chem. Phys. Lett. 218 (4), 304–308.

Hachicho, N., Hoffmann, P., Ahlert, K., Heipieper, H.J., 2014. Effect of silver nanoparticles and silver ions on growth and adaptive response mechanisms of Pseudomonas putida mt-2. FEMS Microbiol. Lett. 355 (1), 71–77.

Hajipour, M.J., Fromm, K.M., Ashkarran, A.A., de Aberasturi, D.J., de Larramendi, I.R., Rojo, T., Mahmoudi, M., 2012. Antibacterial properties of nanoparticles. Trends Biotechnol. 30 (10), 499–511.

Han, G.Z., Gao, K.L., Wu, S.R., Zhang, Y., 2017. A facile green synthesis of silver nanoparticles based on poly-L-lysine. J. Nanosci. Nanotechnol. 17 (2), 1534–1537.

Hassan, D., Farghali, M., Eldeek, H., Gaber, M., Elossily, N., Ismail, T., 2019. Antiprotozoal activity of silver nanoparticles against Cryptosporidium parvum oocysts: new insights on their feasibility as a water disinfectant. J. Microbiol. Methods 165, 105698.

Hebeish, A.A., El-Rafie, M.H., Abdel-Mohdy, F.A., Abdel-Halim, E.S., Emam, H.E., 2010. Carboxymethyl cellulose for green synthesis and stabilization of silver nanoparticles. Carbohydr. Polym. 82 (3), 933–941.

Huang, S., Wang, J., Zhang, Y., Yu, Z., Qi, C., 2016. Quaternized carboxymethyl chitosan-based silver nanoparticles hybrid: microwave-assisted synthesis, characterization, and antibacterial activity. Nanomaterials 6 (6), 118.

Huang, H., Yang, Y., 2008. Preparation of silver nanoparticles in inorganic clay suspensions. Compos. Sci. Technol. 68 (14), 2948–2953.

Ibrahim, H.M., 2015. Green synthesis and characterization of silver nanoparticles using banana peel extract and their antimicrobial activity against representative microorganisms. J. Radiat. Res. Appl. Sci. 8 (3), 265–275.

Indana, M.K., Gangapuram, B.R., Dadigala, R., Bandi, R., Guttena, V., 2016. A novel green synthesis and characterization of silver nanoparticles using gum tragacanth and evaluation of their potential catalytic reduction activities with methylene blue and Congo red dyes. J. Anal. Sci. Technol. 7 (1), 19.

Ivask, A., Kurvet, I., Kasemets, K., Blinova, I., Aruoja, V., Suppi, S., Visnapuu, M., 2014. Size-dependent toxicity of silver nanoparticles to bacteria, yeast, algae, crustaceans and mammalian cells in vitro. PLoS One. 9. (7), e102108.

Jain, S., Mehata, M.S., 2017. Medicinal plant leaf extract and pure flavonoid mediated green synthesis of silver nanoparticles and their enhanced antibacterial property. Sci. Rep. 7 (1), 1–13.

Jayaramudu, T., Raghavendra, G.M., Varaprasad, K., Sadiku, R., Ramam, K., Raju, K.M., 2013. Iota-Carrageenan-based biodegradable Ag0 nanocomposite hydrogels for the inactivation of bacteria. Carbohydr. Polym. 95 (1), 188–194.

Jena, P., Mohanty, S., Mallick, R., Jacob, B., Sonawane, A., 2012. Toxicity and antibacterial assessment of chitosancoated silver nanoparticles on human pathogens and macrophage cells. Int. J. Nanomedicine 7, 1805.

Jin, J.C., Xu, Z.Q., Dong, P., Lai, L., Lan, J.Y., Jiang, F.L., Liu, Y., 2015. One-step synthesis of silver nanoparticles using carbon dots as reducing and stabilizing agents and their antibacterial mechanisms. Carbon 94, 129–141.

Joseph, S., Mathew, B., 2014. Microwave assisted biosynthesis of silver nanoparticles using the rhizome extract of alpinia galanga and evaluation of their catalytic and antimicrobial activities. J. Nanopart. Res. 2014, 1–9.

Kadam, J., Dhawal, P., Barve, S., Kakodkar, S., 2020. Green synthesis of silver nanoparticles using cauliflower waste and their multifaceted applications in photocatalytic degradation of methylene blue dye and Hg 2+ biosensing. SN Appl. Sci. 2 (4), 1–16.

Kahrilas, G.A., Wally, L.M., Fredrick, S.J., Hiskey, M., Prieto, A.L., Owens, J.E., 2014. Microwave-assisted green synthesis of silver nanoparticles using orange peel extract. ACS Sustain. Chem. Eng. 2 (3), 367–376.

Kannan, D.S., Mahboob, S., Al-Ghanim, K.A., Venkatachalam, P., 2020. Antibacterial, antibiofilm and photocatalytic activities of biogenic silver nanoparticles from Ludwigia octovalvis. J. Clust. Sci. 32, 255–264.

Kanniah, P., Radhamani, J., Chelliah, P., Muthusamy, N., Joshua Jebasingh Sathiya Balasingh, E., Reeta Thangapandi, J., Shanmugam, R., 2020. Green synthesis of multifaceted silver nanoparticles using the flower extract of Aerva lanata and evaluation of its biological and environmental applications. ChemistrySelect 5 (7), 2322–2331.

Karthiga, D., Anthony, S.P., 2013. Selective colorimetric sensing of toxic metal cations by green synthesized silver nanoparticles over a wide pH range. RSC Adv. 3 (37), 16765–16774.

Kästner, C., Thünemann, A.F., 2016. Catalytic reduction of 4-nitrophenol using silver nanoparticles with adjustable activity. Langmuir 32 (29), 7383–7391.

Kaur, T., Kapoor, S., Kalia, A., 2018. Synthesis of silver nanoparticles from Pleurotus Florida, characterization and analysis of their antimicrobial activity. Int. J. Curr. Microbiol. App. Sci. 7 (7), 4085–4095.

Kharat, S.N., Mendhulkar, V.D., 2016. Synthesis, characterization and studies on antioxidant activity of silver nanoparticles using Elephantopus scaber leaf extract. Mater. Sci. Eng. C 62, 719–724.

Kim, K.J., Sung, W.S., Moon, S.K., Choi, J.S., Kim, J.G., Lee, D.G., 2008. Antifungal effect of silver nanoparticles on dermatophytes. J. Microbiol. Biotechnol. 18 (8), 1482–1484.

Klasen, H.J., 2000. A historical review of the use of silver in the treatment of burns. II. Renewed interest for silver. Burns 26 (2), 131–138.

Kokila, T., Ramesh, P.S., Geetha, D., 2015. Biosynthesis of silver nanoparticles from Cavendish banana peel extract and its antibacterial and free radical scavenging assay: a novel biological approach. Appl. Nanosci. 5 (8), 911–920.

Kora, A.J., Sashidhar, R.B., Arunachalam, J., 2010. Gum kondagogu Cochlospermum gossypium: a template for the green synthesis and stabilization of silver nanoparticles with antibacterial application. Carbohydr. Polym. 82 (3), 670–679.

Krishna, I.M., Reddy, G.B., Veerabhadram, G., Madhusudhan, A., 2016. Eco-friendly green synthesis of silver nanoparticles using Salmalia malabarica: synthesis, characterization, antimicrobial, and catalytic activity studies. Appl. Nanosci. 6 (5), 681–689.

Kudle, K.R., Donda, M.R., Merugu, R., Kudle, M.R., Rudra, M.P., 2013. Microwave assisted green synthesis of silver nanoparticles using Boswellia serrata flower extract and evaluation of their antimicrobial activity. Int. Res. J. Pharm. 4, 197–200 ISSN, 2230-8407.

Kuhad, R.C., Gupta, R., Singh, A., 2011. Microbial cellulases and their industrial applications. Enzyme. 2011.

Kuppusamy, P., Yusoff, M.M., Maniam, G.P., Govindan, N., 2016. Biosynthesis of metallic nanoparticles using plant derivatives and their new avenues in pharmacological applications—an updated report. Saudi Pharm. J. 24 (4), 473–484.

Lara, H.H., Ayala-Nuñez, N.V., Ixtepan-Turrent, L., Rodriguez-Padilla, C., 2010. Mode of antiviral action of silver nanoparticles against HIV-1. J. Nanobiotechnol. 8 (1), 1–10.

Lellouche, J., Friedman, A., Gedanken, A., Banin, E., 2012. Antibacterial and antibiofilm properties of yttrium fluoride nanoparticles. Int. J. Nanomedicine 7, 5611.

Li, Y., Zhang, W., Niu, J., Chen, Y., 2012. Mechanism of photogenerated reactive oxygen species and correlation with the antibacterial properties of engineered metal-oxide nanoparticles. ACS Nano 6 (6), 5164–5173.

Logaranjan, K., Raiza, A.J., Gopinath, S.C., Chen, Y., Pandian, K., 2016. Shape-and size-controlled synthesis of silver nanoparticles using Aloe vera plant extract and their antimicrobial activity. Nanoscale Res. Lett. 11 (1), 1–9.

Madhusudhan, A., Reddy, G.B., Krishana, I.M., 2019. Green synthesis of gold nanoparticles by using natural gums. In: Nanomaterials and Plant Potential. Springer, Cham, pp. 111–134.

Mafuné, F., Kohno, J.Y., Takeda, Y., Kondow, T., Sawabe, H., 2000. Formation and size control of silver nanoparticles by laser ablation in aqueous solution. J. Phys. Chem. Lett. 104 (39), 9111–9117.

Magnusson, M.H., Deppert, K., Malm, J.O., Bovin, J.O., Samuelson, L., 1999. Gold nanoparticles: production, reshaping, and thermal charging. J. Nanopart. Res. 1 (2), 243–251.

Malka, E., Perelshtein, I., Lipovsky, A., Shalom, Y., Naparstek, L., Perkas, N., … Gedanken, A., 2013. Eradication of multi-drug resistant bacteria by a novel Zn-doped CuO nanocomposite. Small 9 (23), 4069–4076.

Manikprabhu, D., Lingappa, K., 2014. Synthesis of silver nanoparticles using the Streptomyces coelicolor klmp33 pigment: an antimicrobial agent against extended-spectrum beta-lactamase ESBL producing Escherichia coli. Mater. Sci. Eng. C 45, 434–437.

Marassi, V., Di Cristo, L., Smith, S.G., Ortelli, S., Blosi, M., Costa, A.L., Prina-Mello, A., 2018. Silver nanoparticles as a medical device in healthcare settings: a five-step approach for candidate screening of coating agents. R. Soc. Open Sci. 5 (1), 171113.

Martínez, G., Merinero, M., Pérez-Aranda, M., Pérez-Soriano, E.M., Ortiz, T., Begines, B., Alcudia, A., 2021. Environmental impact of nanoparticles' application as an emerging technology: a review. Materials 14, 166.

Masurkar, S.A., Chaudhari, P.R., Shidore, V.B., Kamble, S.P., 2011. Rapid biosynthesis of silver nanoparticles using Cymbopogan citratus lemongrass and its antimicrobial activity. Nano-Micro Lett. 3 (3), 189–194.

Maurer, L.L., Meyer, J.N., 2016. A systematic review of evidence for silver nanoparticle-induced mitochondrial toxicity. Environ. Sci. Nano 3 (2), 311–322.

Metak, A.M., Ajaal, T.T., 2013. Investigation on polymer-based nano-silver as food packaging materials. Int. J. Biol. Food Vet. Agric. Eng. 7 (12), 772–778.

Mohanta, Y.K., Panda, S.K., Jayabalan, R., Sharma, N., Bastia, A.K., Mohanta, T.K., 2017. Antimicrobial, antioxidant and cytotoxic activity of silver nanoparticles synthesized by leaf extract of Erythrina suberosa Roxb. Front. Mol. Biosci. 4, 14.

Moradi, F., Sedaghat, S., Moradi, O., Salmanabadi, S.A., 2021. Review on green nanobiosynthesis of silver nanoparticles and their biological activities: with an emphasis on medicinal plants. Inorg. Nano-Met. Chem. 51 (1), 133–142.

Mussin, J.E., Roldán, M.V., Rojas, F., de los Ángeles Sosa, M., Pellegri, N., Giusiano, G., 2019. Antifungal activity of silver nanoparticles in combination with ketoconazole against Malassezia furfur. AMB Express 9 (1), 1–9.

Nasar, S., Murtaza, G., Mehmood, A., Bhatti, T.M., Raffi, M., 2019. Environmentally benign and economical phytofabrication of silver nanoparticles using juglans regia leaf extract for antibacterial study. J. Electron. Mater. 48 (6), 3562–3569.

Navaladian, S., Viswanathan, B., Viswanath, R.P., Varadarajan, T.K., 2007. Thermal decomposition as route for silver nanoparticles thermal decomposition as route for silver nanoparticles. Nanoscale Res. Lett. 2 (1), 44–48.

Nemanashi, M., Meijboom, R., 2013. Synthesis and characterization of Cu, Ag and Au dendrimer-encapsulated nanoparticles and their application in the reduction of 4-nitrophenol to 4-aminophenol. J. Colloid Interface Sci. 389 (1), 260–267.

Nikaido, H., 2009. Multidrug resistance in bacteria. Annu. Rev. Biochem. 78, 119–146.

Nsimba, R.Y., Kikuzaki, H., Konishi, Y., 2008. Antioxidant activity of various extracts and fractions of *Chenopodium quinoa* and *Amaranthus* spp. seeds. Food Chem. 106, 760–766.

Ogunsile, B.O., Labulo, A.H., Fajemilehin, A.M., 2016. Green synthesis of silver nanoparticles from leaf extracts of Parquetina nigrescens and Synedrella nodiflora and their antimicrobial activity. IFE J. Sci. 18 (1), 245–254.

Oluwafemi, O.S., Mochochoko, T., Leo, A.J., Mohan, S., Jumbam, D.N., Songca, S.P., 2016. Microwave irradiation synthesis of silver nanoparticles using cellulose from Eichhornia crassipes plant shoot. Mater. Lett. 185, 576–579.

Padmavathy, N., Vijayaraghavan, R., 2011. Interaction of ZnO nanoparticles with microbes—a physio and biochemical assay. J. Biomed. Nanotechnol. 7 (6), 813–822.

Pal, A., Shah, S., Devi, S., 2009. Microwave-assisted synthesis of silver nanoparticles using ethanol as a reducing agent. Mater. Chem. Phys. 114 (2-3), 530–532.

Paladini, F., Pollini, M., 2019. Antimicrobial silver nanoparticles for wound healing application: progress and future trends. Materials 12 (16), 2540.

Pandian, A.M.K., Karthikeyan, C., Rajasimman, M., Dinesh, M.G., 2015. Synthesis of silver nanoparticle and its application. Ecotoxicol. Environ. Saf. 121, 211–217.

Panhwar, S., Hassan, S.S., Mahar, R.B., Canlier, A., Arain, M., 2018. Synthesis of l-cysteine capped silver nanoparticles in acidic media at room temperature and detailed characterization. J. Inorg. Organomet. Polym. Mater. 28 (3), 863–870.

Parveen, M., Ahmad, F., Malla, A.M., Azaz, S., Alam, M., Basudan, O.A., … Silva, P.S.P., 2016. Acetylcholinesterase and cytotoxic activity of chemical constituents of Clutia lanceolata leaves and its molecular docking study. Nat. Prod. Bioprospect. 6 (6), 267–278.

Patil, R.S., Kokate, M.R., Jambhale, C.L., Pawar, S.M., Han, S.H., Kolekar, S.S., 2012. One-pot synthesis of PVA-capped silver nanoparticles their characterization and biomedical application. Adv. Nat. Sci. Nanosci. Nanotechnol. 3. (1), 015013.

Pedroza-Toscano, M.A., López-Cuenca, S., Rabelero-Velasco, M., Moreno-Medrano, E.D., Mendizabal-Ruiz, A.P., Salazar-Peña, R., 2017. Silver nanoparticles obtained by semicontinuous chemical reduction using carboxymethyl cellulose as a stabilizing agent and its antibacterial capacity. J. Nanomater. 2017, 1–7.

Perugu, S., Nagati, V., Bhanoori, M., 2016. Green synthesis of silver nanoparticles using leaf extract of medicinally potent plant Saraca indica: a novel study. Appl. Nanosci. 6 (5), 747–753.

Pimentel-Acosta, C.A., Morales-Serna, F.N., Chávez-Sánchez, M.C., Lara, H.H., Pestryakov, A., Bogdanchikova, N., Fajer-Ávila, E.J., 2019. Efficacy of silver nanoparticles against the adults and eggs of monogenean parasites of fish. Parasitol. Res. 118 (6), 1741–1749.

Potočnik, J., 2011. Commission recommendation of 18 October 2011 on the definition of nanomaterial 2011/696/EU. Off. J. Eur. Union 275, 38–40.

Raghavendra, G.M., Jung, J., Seo, J., 2016. Step-reduced synthesis of starch-silver nanoparticles. Int. J. Biol. Macromol. 86, 126–128.

Raghunandan, D., Borgaonkar, P.A., Bendegumble, B., Bedre, M.D., Bhagawanraju, M., Yalagatti, M.S., Abbaraju, V., 2011. Microwave-assisted rapid extracellular biosynthesis of silver nanoparticles using carom seed Trachyspermum copticum extract and in vitro studies. Am. J. Anal. Chem. 2 (04), 475.

Rai, M.K., Deshmukh, S.D., Ingle, A.P., Gade, A.K., 2012. Silver nanoparticles: the powerful nanoweapon against multidrug-resistant bacteria. J. Appl. Microbiol. 112 (5), 841–852.

Rajat, S., Rahul, M., Ankita, W., Simmy, G., Prince, C., Harish, K., Atul, T., Ravinder, K., Naveen, K., 2021. Colorimetric sensing approaches based on silver nanoparticles aggregation for determination of toxic metal ions in water sample: a review. Int. J. Environ. Anal. Chem. . https://doi.org/10.1080/03067319.2021.1873315.

Rajesh, S., Dharanishanthi, V., Kanna, A.V., 2015. Antibacterial mechanism of biogenic silver nanoparticles of Lactobacillus acidophilus. J. Exp. Nanosci. 10 (15), 1143–1152.

Rajesh, S., Raja, D.P., Rathi, J.M., Sahayaraj, K., 2012. Biosynthesis of silver nanoparticles using *Ulva fasciata* Delile ethyl acetate extract and its activity against Xanthomonas campestris pv. malvacearum. J. Biopestic. 5, 119.

Rajkuberan, C., Prabukumar, S., Sathishkumar, G., Wilson, A., Ravindran, K., Sivaramakrishnan, S., 2017. Facile synthesis of silver nanoparticles using Euphorbia antiquorum L. latex extract and evaluation of their biomedical perspectives as anticancer agents. J. Saudi Chem. Soc 21 (8), 911–919.

Reddy, G.B., Madhusudhan, A., Ramakrishna, D., Ayodhya, D., Venkatesham, M., Veerabhadram, G., 2015a. Green chemistry approach for the synthesis of gold nanoparticles with gum kondagogu: characterization, catalytic and antibacterial activity. J. Nanostruct. Chem. 5 (2), 185–193.

Reddy, G.B., Ramakrishna, D., Madhusudhan, A., Ayodhya, D., Venkatesham, M., Veerabhadram, G., 2015b. Catalytic reduction of p-nitrophenol and hexacyanoferrate III by borohydride using green synthesized gold nanoparticles. J. Chin. Chem. Soc. 62 (5), 420–428.

Renugadevi, K., Aswini, V., Raji, P., 2012. Microwave irradiation assisted synthesis of silver nanoparticle using leaf extract of Baliospermum montanum and evaluation of its antimicrobial, anticancer potential activity. Asian J. Pharm. Clin. Res. 5 (4), 283–287.

Rodrigues, M.C., Rolim, W.R., Viana, M.M., Souza, T.R., Gonçalves, F., Tanaka, C.J., … Seabra, A.B., 2020. Biogenic synthesis and antimicrobial activity of silica-coated silver nanoparticles for esthetic dental applications. J. Dent. 96. , 103327.

Rodríguez-León, E., Iñiguez-Palomares, R., Navarro, R.E., Herrera-Urbina, R., Tánori, J., Iñiguez-Palomares, C., Maldonado, A., 2013. Synthesis of silver nanoparticles using reducing agents obtained from natural sources Rumex hymenosepalus extracts. Nanoscale Res. Lett. 8 (1), 318.

Roy, P., Das, B., Mohanty, A., Mohapatra, S., 2017. Green synthesis of silver nanoparticles using Azadirachta indica leaf extract and its antimicrobial study. Appl. Nanosci. 7 (8), 843–850.

Saber, H., Alwaleed, E.A., Ebnalwaled, K.A., Sayed, A., Salem, W., 2017. Efficacy of silver nanoparticles mediated by Jania rubens and Sargassum dentifolium macroalgae; characterization and biomedical applications. Egypt. J. Chem. 4 (4), 249–255.

Sadowski, Z., Maliszewska, I.H., Grochowalska, B., Polowczyk, I., Kozlecki, T., 2008. Synthesis of silver nanoparticles using microorganisms. Mater. Sci. Pol. 26 (2), 419–424.

Sagadevan, S., Vennila, S., Singh, P., Lett, J.A., Johan, M.R., Muthiah, B., Lakshmipathy, M., 2019. Facile synthesis of silver nanoparticles using Averrhoa bilimbi L and plum extracts and investigation on the synergistic bioactivity using in vitro models. Green Processes Synth. 8 (1), 873–884.

Saifuddin, N., Wong, C.W., Yasumira, A.A., 2009. Rapid biosynthesis of silver nanoparticles using culture supernatant of bacteria with microwave irradiation. J. Chem. 6 (1), 61–70.

Saim, A.K., Kumah, F.N., Oppong, M.N., 2021. Extracellular and intracellular synthesis of gold and silver nanoparticles by living plants: a review. Nanotechnol. Environ. Eng. 6, 1.

Sathishkumar, R.S., Sundaramanickam, A., Srinath, R., Ramesh, T., Saranya, K., Meena, M., Surya, P., 2019. Green synthesis of silver nanoparticles by bloom forming marine microalgae Trichodesmium erythraeum and its applications in antioxidant, drug-resistant bacteria, and cytotoxicity activity. Saudi J. Chem. Sci. 23 (8), 1180–1191.

Saxena, A., Tripathi, R.M., Singh, R.P., 2010. Biological synthesis of silver nanoparticles by using onion Allium cepa extract and their antibacterial activity. J. Saudi Chem. Soc. 5 (2), 427–432.

Schirmer, W., 1999. Nanoparticles and nanostructured films, preparation, characterization and applications. Z. Phys. Chem. 213 (2), 226–227.

Seil, J.T., Webster, T.J., 2012. Antibacterial effect of zinc oxide nanoparticles combined with ultrasound. Nanotechnology 23 (49), 495101.

Seku, K., Gangapuram, B.R., Pejjai, B., Hussain, M., Hussaini, S.S., Golla, N., Kadimpati, K.K., 2019. Eco-friendly synthesis of gold nanoparticles using carboxymethylated gum Cochlospermum gossypium CMGK and their catalytic and antibacterial applications. Chem. Pap. 73 (7), 1695–1704.

Seku, K., Gangapuram, B.R., Pejjai, B., Kadimpati, K.K., Golla, N., 2018. Microwave-assisted synthesis of silver nanoparticles and their application in catalytic, antibacterial and antioxidant activities. J. Nanostruct. Chem. 8 (2), 179–188.

Seku, K., Hussaini, S.S., Pejjai, B., Al Balushi, M.M.S., Dasari, R., Golla, N., Reddy, G.B., 2021. A rapid microwave-assisted synthesis of silver nanoparticles using Ziziphus jujuba Mill fruit extract and their catalytic and antimicrobial properties. Chem. Pap. 75, 1341–1354.

Shao, Y., Wu, C., Wu, T., Yuan, C., Chen, S., Ding, T., … Hu, Y., 2018. Green synthesis of sodium alginate-silver nanoparticles and their antibacterial activity. Int. J. Biol. Macromol. 111, 1281–1292.

Shkilnyy, A., Soucé, M., Dubois, P., Warmont, F., Saboungi, M.L., Chourpa, I., 2009. Poly ethylene glycol-stabilized silver nanoparticles for bioanalytical applications of SERS spectroscopy. Analyst 134 (9), 1868–1872.

Sibbald, R.G., Browne, A.C., Coutts, P., Queen, D., 2001. Screening evaluation of an ionized nanocrystalline silver dressing in chronic wound care. Ostomy Wound Manage. 47 (10), 38–43.

Singh, D., Rawat, D., 2016. Microwave-assisted synthesis of silver nanoparticles from Origanum majorana and Citrus sinensis leaf and their antibacterial activity: a green chemistry approach. Bioresour. Bioprocess. 3 (1), 14.

Song, J.Y., Kim, B.S., 2009. Rapid biological synthesis of silver nanoparticles using plant leaf extracts. Bioprocess Biosyst. Eng. 32 (1), 79.

Srivastava, S., Zeba, U., Atanasov, A.G., Singh, V.K., Sharma, M., Bhargava, A., 2021. Biological nanofactories: using living forms for metal nanoparticle synthesis. Mini Rev. Med. Chem. 21, 245–265.

Stoehr, L.C., Gonzalez, E., Stampfl, A., Casals, E., Duschl, A., Puntes, V., Oostingh, G.J., 2011. Shape matters: effects of silver nanospheres and wires on human alveolar epithelial cells. Part Fibre Toxicol 8 (1), 36.

Su, D.L., Li, P.J., Ning, M., Li, G.Y., Shan, Y., 2019. Microwave assisted green synthesis of pectin based silver nanoparticles and their antibacterial and antifungal activities. Mater. Lett. 244, 35–38.

Suganya, G., Karthi, S., Shivakumar, M.S., 2014. Larvicidal potential of silver nanoparticles synthesized from Leucas aspera leaf extracts against dengue vector Aedes aegypti. Parasitol. Res. 1133, 875–880.

Sylvestre, J.P., Kabashin, A.V., Sacher, E., Meunier, M., Luong, J.H., 2004. Stabilization and size control of gold nanoparticles during laser ablation in aqueous cyclodextrins. J. Am. Chem. Soc. 126 (23), 7176–7177.

Taheri, S., Cavallaro, A., Christo, S.N., Smith, L.E., Majewski, P., Barton, M., … Vasilev, K., 2014. Substrate independent silver nanoparticle based antibacterial coatings. Biomaterials 35 (16), 4601–4609.

Toker, R.D., Kayaman-Apohan, N.İ.L.H.A.N., Kahraman, M.V., 2013. UV-curable nano-silver containing polyurethane based organic–inorganic hybrid coatings. Prog. Org. Coat. 76 (9), 1243–1250.

Van Phu, D., Duy, N.N., Lan, N.T.K., Du, B.D., Hien, N.Q., 2014. Study on antibacterial activity of silver nanoparticles synthesized by gamma irradiation method using different stabilizers. Nanoscale Res. Lett. 9 (1), 1–5.

Vijayan, R., Joseph, S., Mathew, B., 2018a. Green synthesis of silver nanoparticles using Nervalia zeylanica leaf extract and evaluation of their antioxidant, catalytic, and antimicrobial potentials. Part. Sci. Technol. 37, 809–819.

Vijayan, R., Joseph, S., Mathew, B., 2018b. Indigofera tinctoria leaf extract mediated green synthesis of silver and gold nanoparticles and assessment of their anticancer, antimicrobial, antioxidant and catalytic properties. Artif. Cells Nanomed. Biotechnol. 46 (4), 861–871.

Wang, L., Hu, C., Shao, L., 2017. The antimicrobial activity of nanoparticles: present situation and prospects for the future. Int. J. Nanomed. 12, 1227–1249. https://doi.org/10.2147/IJN.S121956.

Wani, I.A., Ganguly, A., Ahmed, J., Ahmad, T., 2011. Silver nanoparticles: ultrasonic wave assisted synthesis, optical characterization and surface area studies. Mater. Lett. 65 (3), 520–522.

Wei, D., Sun, W., Qian, W., Ye, Y., Ma, X., 2009. The synthesis of chitosan-based silver nanoparticles and their antibacterial activity. Carbohydr. Res. 344 (17), 2375–2382.

Wu, B., Zhuang, W.Q., Sahu, M., Biswas, P., Tang, Y.J., 2011. Cu-doped TiO_2 nanoparticles enhance survival of Shewanella oneidensis MR-1 under ultraviolet light UV exposure. Sci. Total Environ. 409 (21), 4635–4639.

Xia, Z.K., Ma, Q.H., Li, S.Y., Zhang, D.Q., Cong, L., Tian, Y.L., Yang, R.Y., 2016. The antifungal effect of silver nanoparticles on Trichosporon asahii. J. Microbiol. Immunol. Infect. 49 (2), 182–188.

Xu, X., Chen, D., Yi, Z., Jiang, M., Wang, L., Zhou, Z., Fan, X., Hui, D., 2013. Antimicrobial mechanism based on H_2O_2 generation at oxygen vacancies in ZnO crystals. Langmuir 29 (18), 5573.

Yang, N., Li, F., Jian, T., Liu, C., Sun, H., Wang, L., Xu, H., 2017. Biogenic synthesis of silver nanoparticles using ginger Zingiber officinale extract and their antibacterial properties against aquatic pathogens. Acta Oceanol. Sin. 36 (12), 95–100.

Yin, B.S., Ma, H.Y., Wang, S.Y., Chen, S.H., 2003. Electrochemical synthesis of gold nanocrystals and their 1D and 2D organization. J. Phys. Chem. B 107, 8898–8904.

Yin, H., Yamamoto, T., Wada, Y., Yanagida, S., 2004. Large-scale and size-controlled synthesis of silver nanoparticles under microwave irradiation. Mater. Chem. Phys. 83 (1), 66–70.

Yu, Y.Y., Chang, S.S., Lee, C.L., Wang, C.C., 1997. Gold nanorods: electrochemical synthesis and optical properties. J. Phys. Chem. B 101 (34), 6661–6664.

Yu, Z., Tetard, L., Zhai, L., Thomas, J., 2015. Supercapacitor electrode materials: nanostructures from 0 to 3 dimensions. Energy Environ. Sci. 8 (3), 702–730.

Yu, J., Zhang, W., Li, Y., Wang, G., Yang, L., Jin, J., Huang, M., 2014. Synthesis, characterization, antimicrobial activity, and mechanism of a novel hydroxyapatite whisker/nano zinc oxide biomaterial. Biomed. Mater. 10. (1), 015001.

Zawadzka, K., Kądzioła, K., Felczak, A., Wrońska, N., Piwoński, I., Kisielewska, A., Lisowska, K., 2014. Surface area or diameter—which factor really determines the antibacterial activity of silver nanoparticles grown on TiO 2 coatings. New J. Chem. 38 (7), 3275–3281.

CHAPTER 8

Biodegradable gum: A green source for silver nanoparticles

Tariq Khan[a], Husna Jalal[a], Kashmala Karam[a], and Mubarak Ali Khan[b]

[a]Department of Biotechnology, University of Malakand, Chakdara Dir Lower, Pakistan
[b]Department of Biotechnology, Faculty of Chemical and Life Sciences, Abdul Wali Khan University Mardan (AWKUM), Mardan, Pakistan

1 Introduction

Metal nanoparticles have gained much attention due to their size-dependent physicochemical properties compared to their counterparts (Ju-Nam and Lead, 2008). They have many applications in cosmetics, medicines, biomedical devices, and environmental remediation (Tran and Le, 2013). Among different metal nanoparticles, silver nanoparticles (AgNPs) have attracted attention in the past decade, because of their unique chemical, physical, and biological properties (Sharma et al., 2009). Higher thermal and electrical conductivity, the stability of chemicals, optics, and catalytic action are some of the properties of AgNPs (Krutyakov et al., 2008). In addition to these distinct properties, AgNPs possess enormous medicinal properties such as fungicidal and bactericidal activities (Ahamed et al., 2010; Abdelaziz et al., 2020).

Emerging infectious diseases that cause outbreaks and pandemics result in a significant burden on public health and global economics. In addition to improving the existing remedies against these infectious agents, alternatives are needed to prevent such infectious diseases and to eliminate the pathogens from the environment. Among metal nanoparticles, silver nanoparticles (AgNPs) have shown enormous potential to prevent fungal, viral, and bacterial infections due mainly to their well-studied and characterized disinfectant and antimicrobial properties (Tran and Le, 2013; Lok et al., 2006). Moreover, AgNPs are reported to have potential applications in health care and medicine such as reduction of infection in burn treatment, prevention of bacterial growth on dental materials, as antimicrobial agents on textile fabrics, and in water treatment (Panáček et al., 2006; Bo et al., 2009). AgNPs have also been used in different household activities in consumer products, water filters, clothing, cosmetics, antibacterial sprays, detergents, cutting boards, socks, cell phones, laptop keyboards, and children's toys due to their strong antimicrobial activity (Marambio-Jones and Hoek, 2010; Kaegi et al., 2010).

AgNPs are synthesized by the reduction of silver salts usually silver nitrate by UV-irradiation, gamma radiation, electrochemical reduction, microwave irradiation, and sodium borohydride (Starowicz et al., 2006; Yoksan and Chirachanchai, 2009; Darroudi et al., 2009). This means, to synthesize AgNPs, several routes such as physical, chemical, photochemical, and biological methods are applied (Tran and Le, 2013; Patel, 2021). Biological synthesis has been termed as a high-priority green synthesis method of AgNPs due to its environment-friendliness, ease of synthesis, and biocompatibility, which are explained further later in this chapter. Biological synthesis can be performed using extracts or live tissues of plants, or microbes and fungi, etc. wherein the biomolecules and secondary metabolites usually act as reducing agents for silver salts to synthesize AgNPs. Among the different biological agents, plants have been much explored due to their powerful secondary metabolism machinery. The biomolecules, the variety of secondary metabolites, the easy extraction process, abundance, and safety make plants ideal candidates for AgNPs synthesis. Therefore, many different plants have been employed for this purpose (Roy et al., 2019; Arib et al., 2021). A unique candidate among the plants is the gum producing plant. Biodegradable gum produced by plants provides an additional layer of safety regarding the use of compounds for synthesis methods. Many studies have reported the synthesis of AgNPs by reducing natural and biodegradable gums such as *gum kondagogu*, *gum acacia*, *gum karaya*, *gum cashew*, *gum ghatti*, *salmalia malabarica*, and *gum tragacanth*. These gums are nontoxic, renewable, natural polysaccharides, and are abundantly present in nature (Padil et al., 2016). This chapter aims to review the use of gums for the production of different types of AgNPs along with their importance and applications.

2 Why plant-based synthesis of AgNPs?

Apart from the high temperatures, pressures, electromagnetic radiations, and other stringent environmental conditions used for the synthesis of AgNPs through physical methods, the use of hazardous chemicals and their release into the environment is worrisome. For instance, to prevent agglomeration of AgNPs and thus to stabilize them, polymeric materials such as polyacrylonitrile (Zhang et al., 2010), polyethylene glycol (Shkilnyy et al., 2009), polyvinyl alcohol (Filippo et al., 2009), and polyvinyl pyrrolidone (Samadi et al., 2010) are usually used as stabilizers. This implies that these methods usually involve the use of hazardous and toxic chemicals, capping and reducing agents, are extremely expensive, and have a potential biological and environmental risk (Roldán et al., 2008). As already discussed, researchers have started to emphasize avoiding the use of toxic and hazardous chemicals and following the basic principles of green chemistry to produce nanomaterials (Anastas and Warner, 1998). Preparation methods used for the green synthesis of AgNPs should be facile, biocompatible, and environmentally friendly processes.

The green synthesis of AgNPs requires a silver metal ion solution and a biological reducing agent. There is usually no need to add stabilizing and capping agents

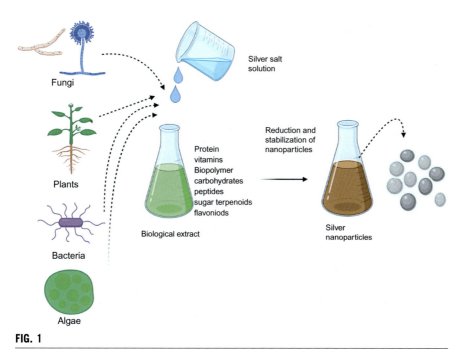

FIG. 1

A schematic representation of the different biosynthetic routes used for green synthesis of AgNPs.

because in most cases some constituents are present in the cells that act as capping and stabilizing agents (Srikar et al., 2016). In biological systems, reducing agents are widely distributed, and different biological sources are used for the synthesis of AgNPs. Plant extracts, biopolymers, microbial cell biomass, and part of plants like leaf, bark, root, and stems have been used for the green synthesis of AgNPs (Fig. 1). AgNPs were prepared by the mixing of silver nitrate solution and the biological reducing agent, which causes a yellow or brownish color in the reaction mixture. This is usually marked as an indication of the AgNPs synthesis (Sastry et al., 1997). The AgNPs synthesis by biological means is due to the presence of organic chemicals such as carbohydrates, fats, proteins, enzymes and coenzymes, phenols flavonoids, terpenoids, alkaloids, gum, etc. These organic chemicals are capable of reducing silver ions to AgNPs (Srikar et al., 2016; Aisida et al., 2019).

The synthesis of AgNPs is affected by several parameters such as reaction temperature, metal ion concentration, extract concentration, pH of the reaction mixture, duration of reaction, and agitation. The size, shape, and morphology of AgNPs are affected by the metal ion concentration, extract composition, and reaction period (Kora et al., 2010). The basic medium is most suitable for the AgNPs synthesis due to better stability, rapid growth rate, good yield, monodispersity, and enhanced reduction process (Srikar et al., 2016). AgNPs of small size and increased rate are synthesized at high temperatures (30–90°C) (Fayaz et al., 2009).

3 A brief account of different plant sources used for the synthesis of AgNPs

Many plants and their extracts have been used for the synthesis of AgNPs in plants and via plants. Among these plants, extracts from *Cavendish banana peel* (Kokila et al., 2015), *Dioscorea bulbifera tuber* (Ghosh et al., 2012). *Eclipta alba* (Jha et al., 2009), *Aloe vera* (Chandran et al., 2006), *Tinospora cordifolia* (Anuj and Ishnava, 2013), *Terminalia chebula* (Edison and Sethuraman, 2012), *Ocimum tenuiflorum* (Patil et al., 2012), *Emblica officinalis* (Ankamwar et al., 2005), *Piper nigrum* (Shukla et al., 2010), *Azadirachta indica* (Tripathi et al., 2009), and *Cinnamon zeylanicum* (Sathishkumar et al., 2009) have been utilized for the preparation of AgNPs. Many studies have attempted to pinpoint the compounds or classes of compounds involved in the fabrication of AgNPs. Studies have shown that different biomolecules and secondary metabolites play a role in the biosynthesis of these AgNPs. For instance, in a recent article, our group has attempted to show that terpenoids from plants could be regarded as important bioreducing agents during the synthesis of AgNPs. Similarly, other studies have highlighted the role of alkaloids, protein, chlorophyll, and metabolites present in the extract of plants act as reducing and capping agents for the AgNPs (Srikar et al., 2016; Kaabipour and Hemmati, 2021). Among the many different biomolecules, biopolymers such as dextran, alginates, chitosan, cellulose, and their derivatives present in tree gums are another class of natural biological sources for the stabilization/synthesis of AgNPs. The tree gums are rich sources of bioreducing agents for synthesis of AgNPs. An interesting property of these gums is their biodegradability, which among the many other properties adds another layer of safety and biocompatibility to the synthesis of AgNPs. We thus present an account of the important biodegradable gums and their significance in synthesis of AgNPs.

4 Natural biodegradable gums and their applications

Tree gums are exudates of different species of trees and are released as exudates either due to injury of the plant or in the case of adverse conditions such as drought or breakdown of the cell wall (accessory cellular formation gummosis) (Choudhary and Pawar, 2014). Gums owe their importance to being abundant, inexpensive, and nontoxic (Sashidhar et al., 2015). They are high molecular weight complicated structures mainly formed of heterogeneous polysaccharides, and are hydrophilic hydrocolloids in nature. Hydrocolloids are important polysaccharides and have a complex structure with glycosidic bonding. Having a large number of hydroxyl groups, water is held by hydrogen bonding within the molecular structure (Padil et al., 2016). The simple interaction of hydrocolloids with water makes them most suitable in foodstuffs. Gum hydrocolloids may be solubilized in water and are very effective water adsorbents. These gums are obtained from many plants, animals, and bacteria and are amenable to biological and chemical modifications. Due to this amenability, natural gums are biodegradable. The gum contains many hydroxyl groups, which are arranged to the backbone of the molecule. They form complex macrostructure due to the cross-linking of the hydrocolloid chain (Anderson and Wang, 1990; Anderson and Weiping, 1992; Khan et al., 2020b).

Major tree exudate gums include gum kondagogu, gum tragacanth, gum karaya, gum arabic, and gum ghatti. Research has been carried out on these tree gum polysaccharides to explore their different aspects such as molecular weight, distribution, availability, food, and nonfood applications, and chemical structure (Vinod, et al., 2008b; Wang et al., 2014; Vinod, 2008a). There is no structural formula for these gums because they have a complex chemical composition and vary depending on their age and the source. These gums have considerable food and nonfood applications (Verbeken et al., 2003).

Gums have found uses in a diversity of areas including emulsification, biosorption, bio-fabrication, bioremediation, biosensing, and medicinal applications. For instance, a serious biological problem in aquatic life is posed by heavy metals like Hg, Zn, Cu, Pb, Ni, and U. For the removal of these contaminants from the aquatic system, commonly used methods are filtration and adsorption. Different toxic heavy metals, industrial effluents, and radioactive materials are successfully removed by these gums because of their powerful biosorption capabilities (Saravanan et al., 2012; Vinod et al., 2010; Vinod et al., 2011b; Sahraei and Ghaemy, 2017). Toxic metals get adsorbed onto the gum structures through various ways like functional group interactions, high surface area, ion exchange, and modified surface properties (Fosso-Kankeu et al., 2016; Sashidhar et al., 2015).

For our purposes, it is important to note that biodegradable gums such as gum ghatti, gum acacia, gum kondagogu, and cashew gum are also reported to be used for bio-fabrication and production of stable metal nanoparticles (Padil et al., 2016). These gums have been used for the stabilization and synthesis of various metal and metal oxide nanoparticles like Ag, Fe, Se, CuO, and ZnO. For the preparation of nanoparticles, these natural gums have three main parameters: (1) they follow the basic principles of green chemistry—for example, the synthesis process uses water and ionic liquid-based green solvent; (2) due to the abundance of many functional groups in their structures, they act as reducing agents; and (3) and for the stabilization of nanoparticles, the gums are biodegradable and nontoxic materials (Padil et al., 2016). Gums are reported to be very potent bioreducing agents for the production of AgNPs (Kora et al., 2010).

5 The role of biodegradable gum in bio-fabrication of AgNPs

The various types of gums mentioned in the previous section play an important role in the synthesis and capping of AgNPs (Fig. 2). To give a detailed account of the studies done so far, each gum type with its role in the bio-fabrication of AgNPs is given below. Furthermore, a detailed account of many different gums along with the AgNPs bio-fabricated in their presence is given in Table 1.

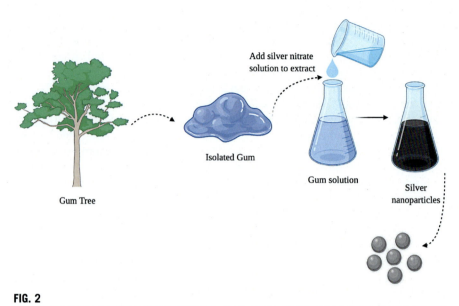

FIG. 2

Gum as a source of bioreducing agents for synthesis and stabilization of AgNPs.

5.1 Gum Karaya

The gum karaya is also known as Sterculia gum is derived from the *Sterculia urens* tree (Sterculiaceae family) (Venkatesham et al., 2014). The karaya gum tree is one of the most important industrial and commercial trees of the Indian subcontinent. The tree is medium-sized and deciduous with horizontally spreading branches, growing to a maximum height of about 15 m (49 ft) and producing smooth, fibrous and thick greenish-gray bark. Gum karaya, also known as gum kateera, consists of naturally occurring complex acidic polysaccharides, which are partially acetylated and obtained as magnesium and calcium salts (Padil et al., 2016). Gum karaya is a branched-structure molecule, partially acetylated polysaccharide, and high molecular weight was reported at about $\sim 16 \times 10^6$ Da (Padil et al., 2015b). The structure of this gum consists of β-D-galactose and β-D-glucuronic acid, and α-L-rhamnose and α-D-galacturonic acid (Silva et al., 2020). This gum is "generally regarded as safe" (GRAS) and functions as a Food and Drug Administration (FDA)-approved stabilizer, emulsifier, and thickener. Used for its bulk laxative, gastro-retentive, wound dressing, warts, analgesic, and antibiotic activities (Setia et al., 2010), gum karaya has attracted researchers for its bio-fabricating role during synthesis of AgNPs. Gum karaya has been employed to produce uniform-sized AgNPs through its reduction of silver nitrate solution. A study by Venkatesham et al. (2014) showed that reducing gum karaya could be harnessed for efficient biosynthesis of AgNPs. To avoid the use of any synthetic agent and to assist the reaction, autoclave (at a pressure of 15 psi and temperature of 120°C by varying the time) was employed during the process. The reaction was assisted by an autoclave method. Different

Table 1 Size and morphology/properties of AgNPs biosynthesized through different types of gums isolated from different plant sources.

Types of biodegradable gum	Source/plant name	Size	Morphology	Stability	References
Guar gum	Cyamopsis tetragonoloba	30–40 nm 100 ± 1 nm	Spherical in nature, highly crystalline in nature, and irregular shapes	Highly stable	Singh and Dhaliwal (2018)
Guar gum	Cyamopsis tetragonoloba	42 nm	Quasispherical and slightly aggregated	Highly stable for 60 days	Das et al. (2015)
Guar gum	Cyamopsis tetragonoloba	15–20 nm	Spherical in nature	Highly stable	Abdel-Halim et al. (2011)
Gum acacia	Acacia seyal, Senegal	81.45 ± 2.07 nm	Negative charge, spherical shape	Sufficiently stable	Rao et al. (2018b)
Gum acacia	Acacia Senegal, Acacia seyal	2.5 and 5.0 nm	Spherical	30-day stability	Singh and Ahmed (2012)
Gum arabic	Acacia Senegal	16.0 ± 2.0 nm	Face-centered cubic	Highly stable	Akele et al. (2015)
Gum arabic		7 nm	Predominantly spherical, face centered cubic, crystalline	Highly stable	Ansari et al. (2014)
Gum arabic		2–20 nm	Spherical shape, single-crystalline	Highly stable for 1 month	Padil et al. (2016)
Gum arabic		18 ± 7 nm	Spherical or near spherical in shape and nonagglomerated	Toxicity by surface coating up to 48 h	Newton et al. (2013)

Continued

Table 1 Size and morphology/properties of AgNPs biosynthesized through different types of gums isolated from different plant sources. *Continued*

Types of biodegradable gum	Source/plant name	Size	Morphology	Stability	References
Gum arabic	*Acacia Senegal*	5 nm	Face-centered cubic structures with crystalline	5-month stability	Mohan et al. (2007)
Gum karaya	*Sterculia urens*	12.5 ± 2.5 nm	Spherical particles, crystalline	Stable for 6 months	Padil and Černík (2015)
Gum karaya	*Sterculia urens*	7–10 nm	Spherical	Stable	Padil et al. (2015a)
Gum karaya	*Sterculia urens*	4 ± 2 nm	Face-centered cubic with crystalline structure	Highly stable	Venkatesham et al. (2014)
Gum kondagogu	*Cochlospermum gossypium*	5.5 ± 2.5 nm	Spherical, face-centered cubic	Stable for more than 6 months	Vinod et al. (2011a)
Gum kondagogu	*Cochlospermum gossypium*	3 nm	Spherical	Highly stable	(Kora et al., 2010)
Gum ghatti	*Anogeissus latifolia*	5.7 ± 0.2 nm	Spherical in shape, face-centered cubic	Highly stable	Kora and Sashidhar (2015)
Gum olibanum	*Boswellia serrata*	7.5 ± 3.8 nm	Spherical shaped, highly crystalline	Highly stable	Kora and Sashidhar (2015)
Gum olibanum	*Boswellia serrata*	7.5 ± 3.8 nm	Quasispherical shaped, highly crystalline nature nanoparticles	Highly stable	Kora et al. (2012a)

Gum cashew	*Anacardium occidentale*	4.38 ± 0.07 nm	Uniformly dispersed and spherical	Stable	Quelemes et al. (2013)
Gum cashew	*Anacardium occidentale*	4.5–6.5 nm	Spherical in shape, uniform dispersal	Highly stable	Araruna et al. (2013)
Gum tragacanth	*Astragalus gummifers*	5–10 nm	Dispersed and spherical in nature	Highly stable	Rao et al. (2018a)
Gum tragacanth	*Astragalus gummifers*	18 ± 2 nm	Spherical in nature and very well dispersed	Highly stable	Indana et al. (2016)
Gum tragacanth	*Astragalus gummifers*	13.0 nm	Narrow distributed	Highly stable	Sahraei and Ghaemy (2017)
Locust bean gum	*Ceratonia siliqua*	18–51 nm	Bell shape	Stable over 7 months	Tagad et al. (2013)
Neem gum	*Azadirachta indica*	<30 nm	Face-centered cubic crystalline	Highly stable	Velusamy et al. (2015)
Neem gum	*Azadirachta indica*	15–20 nm	Spherical	Highly stable	Mewada et al. (2013)
Gum karaya	*Sterculia urens*	2–4 nm	Spherical	Catalytic activity and antimicrobial activity against *Escherichia coli* and *Micrococcus luteus*	Venkatesham et al. (2014)
Cashew gum	*Anacardium occidentale*	96–145 nm	Spherical	Antimicrobial action against *Staphylococcus aureus* and *E. coli*	Araruna et al. (2020)
Guar gum	*Cyamopsis tetragonoloba*	100 ± 1 nm	Negative charge, highly crystalline in nature and irregular shapes	Highly stable	Singh and Dhaliwal (2020)
Gum ghatti	*Anogeissus latifolia*	5.7 nm	Spherical in shape	Antimicrobial action against *S. aureus*, *E. coli*, and *Pseudomonas aeruginosa*	Kora et al. (2012a)
Gum acacia	*Acacia Senegal*	5 nm			Mohan et al. (2007)

Continued

Table 1 Size and morphology/properties of AgNPs biosynthesized through different types of gums isolated from different plant sources. *Continued*

Types of biodegradable gum	Source/plant name	Size	Morphology	Stability	References
Gum tragacanth	*Astragalus gummifers*	13.1 ± 1.0 nm	Spherical	Antimicrobial action against *S. aureus*, *E. coli*, and *Pseudomonas aeruginosa*	Kora and Arunachalam (2012)
Salmalia malabarica	*Bombax ceiba*	7 nm	Spherical in shape	Catalytic activity and antimicrobial activity against *E. coli* and *S. aureus*	Krishna et al. (2016)
Gum kondagogu	*Cochlospermum gossypium*	5 nm	Spherical in shape	Catalytic, antibiofilm and antibactericidal activity	Kora et al. (2010)
Gellan gum		10–15 nm		Catalytic and antimicrobial activity against *Bacillus subtilis* and *E. coli*	
Ghatti gum	*Anogeissus latifolia*	10 nm	Spherical in shape, face-centered cubic	Antibacterial activity against gram-positive bacteria and gram-negative bacteria	Babaladimath and Badalamoole (2019)
Arabic gum	*Acacia Senegal*	10–100 nm	Spherical in shape, face centered cubic	Antibacterial activity against *S. aureus*, *E. coli*, and *Pseudomonas aeruginosa*	Eghbalifam et al. (2020)
Locust bean gum	*Ceratonia siliqua*	43.1 ± 1.1 nm	Spherical	Thermally stable	Matar and Andac (2020)
Gum tragacanth	*Astragalus gummifers*	<0.05 nm		Antibacterial activity against gram-positive bacteria and gram-negative bacteria	Bahrami et al. (2019)
Gum acacia	*Acacia Senegal*	20–80 nm	Spherical	Stable for 2 months	Sharma et al. (2020)

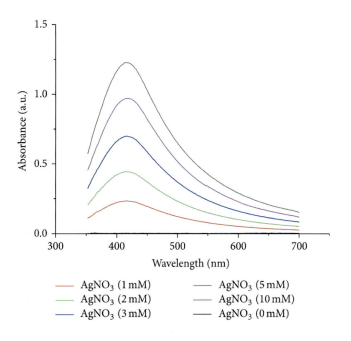

FIG. 3

Representative UV-vis spectra AgNPs and different concentrations of silver nitrate using gum karaya (Padil et al., 2015a).

Reprinted from Padil, V. V. T., Nguyen, N. H., Ševců, A. Černík, M., 2015a. Fabrication, characterization, and antibacterial properties of electrospun membrane composed of gum karaya, polyvinyl alcohol, and silver nanoparticles. J. Nanomater., 2015, 9. with permission from MDPI Publisher, Licensee MDPI, Basel, Switzerland. This article is an open-access article distributed under the terms and conditions of the Creative Commons Attribution (CC BY) license (http://creativecommons.org/licenses/by/4.0).

characterization techniques (FTIR, TEM, and XRD) have shown that gum karaya acts as a stabilizing as well as a reducing agent. The synthesized gum karaya-AgNPs were characterized by Fourier transform infrared spectroscopy (FTIR), transmission electron microscopy (TEM), UV-vis spectroscopy, and XRD techniques. UV-vis analysis confirmed the presence of AgNPs in the mixture (Fig. 3). Similarly, the peaks corresponding to hydroxyl, methyl, and carbonyl functional groups were detected via FTIR, which showed the presence of these groups in gum karaya. This ultimately suggests that these functional groups were involved in the formation and stabilization of AgNPs. Analysis of TEM images revealed that the AgNPs synthesized via gum karaya were spherical and the average size of the nanoparticles synthesized from gum karaya varied between 2 and 4 nm. Furthermore, it was also concluded that the AgNPs fabricated via gum karaya have significant catalytic and antibacterial activity on *Escherichia coli* and *Micrococcus luteus*.

Furthermore, the green synthesis of AgNPs was produced by mixing of silver nitrate solution with the gum karaya solution for a few minutes until the gum karaya was completely dissolved. Green synthesis of AgNPs was carried out at room temperature and after 12 h the silver nitrate solution was completely reduced by the gum karaya. For the green synthesis of AgNPs, gum karaya was used as a stabilizing as well as a reducing agent. The synthesized gum karaya-based AgNPs were characterized by UV visible spectroscopy techniques, TEM, X-ray analysis, and FTIR. The prepared AgNPs were characterized by a yellow color in the reaction which indicated the presence of AgNPs. This suggested that hydroxyl and carbonyl group functional group are present in gum karaya and involved and stability of AgNPs.

Similarly, for instance, Padil et al. (2015a) observed that AgNPs synthesized with gum karaya had a broad absorption peak centered around 3318–3350 cm^{-1}, which belongs to O—H stretching vibration in the hydrogen-bonded hydroxyl groups of the gum extracts (Fig. 4).

Furthermore, the study of Padil et al. that stabilized AgNPs by gum karaya show significantly antibacterial activity on gram-positive bacteria (*Staphylococcus aureus*) and gram-negative bacteria (*Escherichia coli*, *pseudomonas aeruginosa*) (Padil et al., 2015b).

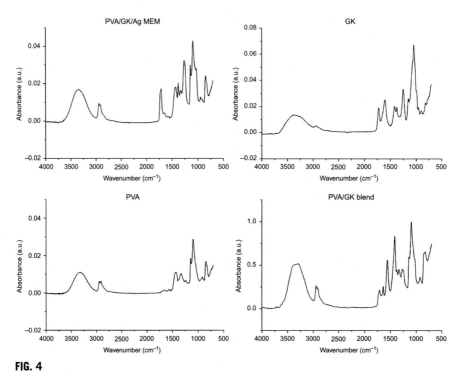

FIG. 4

FTIR spectrum of guar gum and dried AgNPs (Padil et al., 2015a).

Reprinted from Padil, V. V. T., Nguyen, N. H., Ševců, A. Černík, M., 2015a. Fabrication, characterization, and antibacterial properties of electrospun membrane composed of gum karaya, polyvinyl alcohol, and silver nanoparticles. J. Nanomater., 2015, 9. with permission from MDPI Publisher, Licensee MDPI, Basel, Switzerland. This article is an open-access article distributed under the terms and conditions of the Creative Commons Attribution (CC BY) license (http://creativecommons.org/licenses/by/4.0).

5.2 Gum Ghatti

Gum ghatti is a natural type of arabinogalactan gum derived from the native tree of India known as *Anogeissus latifolia* (*Combretaceae* family). Gum ghatti was given its name because of its transportation through mountain passes or ghats. This biopolymer is a protein-containing, high-arabinose, acidic polysaccharide that occurs in nature mixed with calcium, sodium, and magnesium salts. The structure of gum ghatti primarily consists of sugars like D-xylose, D-galactose, D-glucuronic acid, D-mannose, and L-arabinose. The molecular weight of this natural polymer was reported to be $\sim 8.94 \times 10^7$ g/mol. This gum is also regarded as safe under the function of food ingredients and an emulsifier by the FDA (Kora et al., 2012b). It is also used as a food product in Japan, China, South Korea, India, America, Australia, Russia, and many other countries (Kora and Rastogi, 2018) and for the treatment of diabetes, hypolipidemic, and diarrhea (Alam et al., 2017). Gum ghatti was marketed by the government after being collected among the tribes. About 1000–1500 million tons/year of gum ghatti is produced worldwide. The structural, physicochemical, solution, compositional, thermal, emulsifying, morphological, and rheological properties have been well studied and characterized.

Generally, the low cost, natural availability, and better stability are some of the attractive features of gum ghatti for the synthesis and stabilization of AgNPs (Kora et al., 2012a) synthesized AgNPs through reduction of silver nitrate salts with the help of gum ghatti. Just as reported previously, this study also employed an autoclave for the feasible green synthesis of AgNPs. Different characterization methods (FTIR, TEM, UVvis spectroscopy, and XRD) revealed that gum ghatti played an important role in the synthesis of AgNPs. FTIR studies suggested that hydroxyl and carboxylate functional groups present in gum ghatti were detected in AgNPs, pointing toward a prominent bioreducing role of gum ghatti. Similarly, TEM micrographs suggested that the average size of the nanoparticles synthesized from gum ghatti was 5.7 nm and that the nanoparticles were spherical. The AgNPs synthesized through gum ghatti were also potent antibacterial agents, as is evident from their effects on the growth of the gram-positive *Staphylococcus aureus* and gram-negative bacteria *Escherichia coli* and *Pseudomonas aeruginosa*.

5.3 Gum Acacia

Gum acacia is also known as gum arabic. This natural polymer is obtained from the acacia trees (Pal et al., 2019). Gum acacia is a highly branched complex and slightly acidic polysaccharide and is derived from mixed magnesium, calcium, and potassium salts. The molecular weight of gum acacia is about 2.5×10^5 Da. The structure of gum acacia is composed of D-galactopyranosyl, L-arabinofuranosyl, L-rhamnopyranosyl, and D-glucuronopyranosyl units (Padil et al., 2016). This gum is used for healing wounds, hydrogel, wound dressing, gastrointestinal tract infections, antiinflammatory activity, and drug delivery systems. Gum acacia is an eco-friendly, nontoxic, easily available, and biodegradable polysaccharide

(Aderibigbe et al., 2015). Moreover, this gum is used in cosmetic, pharmaceutical, and food industries because of its ability as an emulsifying, stabilizing, and thickening agent. This natural polymer not only has a very low cost and is abundantly available in nature but also has excellent emulsifying and surface-active properties, which would be beneficial for the synthesis of metal nanoparticles.

AgNPs were produced by the mixing of silver nitrate solution with the gum acacia solution to form a natural and renewable polymer. A magnetic stirrer was used to facilitate the reaction. The gum acacia acted as a stabilizing as well as a reducing agent for the preparation of AgNPs. The synthesized AgNPs were characterized by FTIR, X-ray diffraction, TEM, and ultraviolet-visible spectroscopy techniques. The prepared AgNPs were characterized by a yellow color appearing in the reaction mixture. It has been suggested that hydroxyl, methyl, and amines groups are present in gum acacia and are involved in the stability and formation of AgNPs. The size of the AgNPs synthesized by gum acacia is approximately 81.45 ± 2.07 nm and they are highly stable (Rao et al., 2018b).

5.4 Gum Tragacanth (Astragalus Gummifer)

Gum tragacanth occurs naturally and is acidic and complex polysaccharides. Gum tragacanth is obtained from the bark of the *Astragalus gummifer* (Fabaceae family), which is the native tree of western Asia. These natural gums are mostly produced in Turkey and Iron. This polymer is a natural type of arabinogalactan gum. The structural, physicochemical, compositional, solution, thermal, rheological, and emulsifying properties of these gums have been well-studied and characterized (Padil et al., 2016). The molecular weight of this natural polymer is 850 kDa (Zare et al., 2019). This natural polymer contains a mixture of tragacanthin (water-soluble) and bassorin (water-swellable) polysaccharide components (Rao et al., 2018a) About 30–40% of the gum is present in the water-soluble tragacanthin, which is a highly branched, neutral, and type II arabinogalactan. The structure of tragacanthin primarily consists of sugars like L-glucose, L-arabinose, D-xylose, D-mannose, D-galactose, and D-glucose (Padil et al., 2016).

Tragacanth gum is "generally regarded as safe" (GRAS). This gum, under the function of stabilizer, emulsifier, and thickener, has been approved by the US FDA. This drug has also been approved by the European Union under the function of food additive in the class of thickeners, stabilizers, emulsifiers, and gelling agents (Kora and Arunachalam, 2012). Many studies reported that tragacanth gums are nontoxic, nonmutagenic, nonteratogenic, and noncarcinogenic as food additives (Rao et al., 2018a). These gums have many potential applications as wound dressing, hydrogel, immobilizing agent, bone tissue engineering, rheumatoid arthritis, stomach pain, antimicrobial activity, and many other drug delivery systems. This gum is also used in cosmetics, textile, food, and pharmaceutical industries (Zare et al., 2019).

Gum tragacanth is used for the stabilization and synthesis of AgNPs due to some attractive features: (1) GRAS and nontoxic nature; (2) its availability is natural; (3) it has a long shelf life and higher resistance to microbes. For the production of AgNPs, a simple and green synthetic route is employed using a biodegradable, renewable, and natural plant polymer tragacanth gum. This gum is used as both a stabilizing and reducing agent in the synthesis of AgNPs.

AgNPs were synthesized by autoclaving the silver nitrate solutions containing various concentrations of gum extract for different durations of time. The synthesis was carried out in an aqueous medium by autoclaving to produce sterile AgNPs, free from microbes. After autoclaving, the silver nitrate-containing gum solution in the reaction mixture appeared to turn from a colorless solution to yellow. This is a clear indication of AgNPs by the gum. UV-visible spectroscopy, TEM, X-ray diffraction, FTIR, and Raman spectroscopy techniques were used for the characterization of AgNPs. The average potential size was about 13.1 ± 1.0 nm and spherical. Based on the results, it was concluded that the synthesized AgNPs had significant antibacterial action on both the gram-positive such as *Staphylococcus aureus* and gram-negative bacteria such as *Escherichia coli* and *Pseudomonas aeruginosa* (Kora and Arunachalam, 2012).

Furthermore, the synthesis of AgNPs was produced by silver nitrate solution reduction by gum tragacanth, an ecofriendly and natural available polymer. For green synthesis of AgNPs, gum tragacanth is used as a stabilizing as well as a reducing agent. An aqueous solution of gum tragacanth was mixed with silver nitrate solution by the ultrasonication method at 45°C for about 45 min. The synthesized AgNPs were characterized by FTIR, TEM, UV-vis spectroscopy, and XRD techniques. The prepared AgNPs were characterized by the noticeable yellow color that appeared in the reaction mixture from a colorless solution. The yellow color in the reaction mixture indicated the synthesis of AgNPs. It has been suggested that hydroxyl, methyl, carboxylate, and carbonyl functional groups are present in gum tragacanth and are involved in the stability and formation of AgNPs. The average size of the nanoparticles synthesized from gum tragacanth is 18 ± 2 nm, and the nanoparticles have a spherical shape. Moreover, it was concluded that the AgNPs stabilized by gum tragacanth have significant catalytic activity (Indana et al., 2016).

5.5 Salmalia Malabarica

Salmalia malabarica gum (SMG) occurs naturally and is a plant polysaccharide. SMG is obtained from a native tree of India, *Bombax ceiba*. *Salmalia malabarica* is a nontoxic, easily available, eco-friendly, cost-effective, renewable, and biodegradable polysaccharide (Krishna et al., 2016). This natural polymer was traditionally used in ayurvedic medicine as an antioxidant, antiinflammatory, diarrhea, analgesic, asthma, hepato-protective, hypotensive, anticancer, wound healing, antioxidant, and hypoglycemic activity (Jain and Verma, 2012; Saleem et al., 2003; Faizi et al., 2012). *Salmalia malabarica* is a complex polysaccharide, a negatively charged colloid, and has a higher molecular weight. Complete hydrolysis of the gum has revealed that it contains a mixture of various sugars such as D-galacturonic acid, D-galactose, D-galactose, D-galactose, L-arabinose, and L-arabinose (Krishna et al., 2016).

The AgNPs synthesis using SMG is a very cost-effective method and has many biomedical applications. This gum (silk cotton) is used as both a stabilizing and a reducing agent in the synthesis of AgNPs. The uniform AgNPs were prepared by the silver nitrate solution reduction using *Salmalia malabarica*. An external source of energy, for example, an autoclave, was used to facilitate the reaction. Autoclaving was used as a synthetic method to prepare sterile AgNPs which are completely free from viruses and bacteria. The *Salmalia malabarica* gum in the autoclave becomes expanded and more accessible at high pressures and temperatures for the silver ions to interact with functional groups that are present in the gum. For the green synthesis of AgNPs, *Salmalia malabarica* acts as a stabilizing as well as a reducing agent. The synthesized SMG-AgNPs were characterized by FTIR, TEM, UV-vis spectroscopy, and XRD techniques. The formation of AgNPs was characterized by the noticeable yellow color that appears in the reaction mixture from the colorless solution. The yellow color in the reaction mixture indicates the synthesized AgNPs. It has been suggested that hydroxyl and carboxyl functional groups are present in *Salmalia malabarica* and are involved in the stability and formation of AgNPs. The average size of the nanoparticles synthesized from *Salmalia malabarica* is 7 nm and they are spherical. Moreover, it was concluded that the stabilized AgNPs by *Salmalia malabarica* have significant catalytic activity and antibacterial activity on *E. coli* and *S. aureus* (Krishna et al., 2016).

5.6 Gum Kondagogu

Gum kondagogu is also known as rhamnogalacturonan gum (*Cochlospermum gossypium*) (Kora et al., 2010). The physicochemical, compositional, solution, morphological rheological, emulsifying, and metal biosorption properties have been well studied and characterized. The basic structural analysis of this hydrocolloid indicates that it consists of sugars like rhamnose, galactose, mannose, glucose glucuronic acid arabinose, and galacturonic acid (Padil et al., 2016).

AgNPs were produced by the silver nitrate solution reduction by gum kondagogu, an edible, natural, biodegradable, and renewable polymer. An autoclave is used as an external energy source to facilitate the reaction. Autoclaving was used as a synthetic method to prepare sterile AgNPs which are completely free from bacteria, viruses, and spores. For the green synthesis of AgNPs gum, kondagogu was used as a stabilizing as well as a reducing agent. The synthesized AgNPs were characterized by FTIR, TEM, UV-vis spectroscopy, and XRD techniques. The prepared AgNPs were characterized by a distinctive yellow color that appears in the reaction mixture from a colorless solution. The yellow color in the reaction mixture indicates the synthesis of AgNPs. The average size of the nanoparticles synthesized from gum kondagogu was about 5 nm and the nanoparticles were spherical. Furthermore, it was concluded that the AgNPs stabilized by gum kondagogu have significant catalytic and antibacterial activity. The AgNPs prepared by reducing gum kondagogu have bactericidal and antibiofilm activity. Cell surface changes, morphological activity, and cytoplasmic content leakage were also caused by the biogenic AgNPs. These nanoparticles showed promising antibactericidal activity for different biomedical and environmental applications such as packaging of food, textile, and purification of water, wound dressing, and topical gel (Rastogi et al., 2015).

The synthesis of AgNPs was produced by silver nitrate solution reduction by gum kondagogu have an ecofriendly and natural available polymer. For green synthesis of AgNPs, gum kondagogu was used as a stabilizing as well as a reducing agent. The solution of gum kondagogu mixed with a silver nitrate solution in a bubbling tube and blend was subjected to microwave radiation for 50–60 s at 750 Power. The synthesized AgNPs were characterized by FTIR, TEM, UV-vis spectroscopy, and XRD techniques. The prepared AgNPs were characterized by the distinctive yellow color that appears in the reaction mixture from a colorless solution. The yellow color in the reaction mixture indicates the synthesis of AgNPs. It has been suggested that hydroxyl, carboxyl, methyl, and carboxylate function groups are present in gum kondagogu and are involved in the stability and formation of AgNPs. The average size of the AgNPs synthesized from the gum kondagogu is about 9 ± 2 nm. Furthermore, it was concluded the stabilized AgNPs by gum kondagogu have significant catalytic and antibacterial activity on both gram-positive bacteria such as *Bacillus subtilis* and gram-negative bacteria such as *Escherichia coli* and *Pseudomonas aeruginosa* (Seku et al., 2018).

5.7 Cashew gum

Cashew gum is a natural product derived from the *Anacardium occidentale* L. tree, which is abundantly present in northwest Brazil. This gum has a low cost and is an easily available source of the plant. These gums are complex heteropolysaccharides; they are nontoxic and extracted from the trunk or the branches of the trees. Cashew gum is a resin yellowish in color and is a natural, renewable, nontoxic, and sustainable source of the plant. The basic structure of cashew gum is composed of D-galactose, D-glucose, arabinose, rhamnose, and glucuronic acid (Quelemes et al., 2013).

AgNPs were prepared by the silver nitrate solution reduction by cashew gum, an edible, natural, and renewable polymer. A magnetic stirrer was used to facilitate the reaction. A water bath was used as a synthetic method to prepare AgNPs at a controlled temperature for about 60 min. For the green synthesis of AgNPs, cashew gum as used as a stabilizing as well as a reducing agent. The synthesized cashew gum-based AgNPs were characterized by TEM, UV-vis spectroscopy, and dynamic light scattering techniques. The prepared AgNPs were characterized by a noticeable yellow color that appears in the reaction mixture from a colorless solution. The yellow color in the reaction mixture indicates the synthesis of AgNPs. The average size of the nanoparticles synthesized from cashew gum was about 4 nm and the nanoparticles were spherical. Furthermore, it was concluded that the stabilized AgNPs by cashew gum showed catalytic activity as well as a significant greater antibacterial activity on the gram-negative bacteria than the gram-positive bacteria (Quelemes et al., 2013).

The synthesis of AgNPs was prepared by the silver nitrate solution reduction by cashew gum, a natural and renewable polymer. A magnetic stirrer was used to facilitate the reaction. The solution of silver nitrate was mixed with the cashew gum solution using a microwave-assisted method with exposed radiation for 3 min. The used of microwave heating capable an improve the distribution and control size of the AgNPs. For the green synthesis of AgNPs, microwave heating has partially hydrolysis of polymer. For the green synthesis of AgNPs, cashew gum is used as a strong stabilizing as well as a weak reducing agent. The synthesized cashew gum-based AgNPs were characterized by TEM, UV-vis spectroscopy, FTIR, X-ray diffraction, and atomic absorption spectroscopy techniques. The prepared AgNPs were characterized by a distinctive yellow color that appears in the reaction mixture from a colorless solution. The yellow color in the reaction mixture indicates the synthesis of AgNPs. It has been suggested that hydroxyl, carbonyl, and carboxylate functional groups are present in cashew gum and are involved in the stability and formation of AgNPs. Furthermore, it was concluded that the stabilized AgNPs by gum cashew have significant antibacterial activity on both gram-positive bacteria such as *Staphylococcus aureus* and gram-negative bacteria such as *Escherichia coli* (Araruna et al., 2020).

5.8 Gellan gum

Gellan gum is a polysaccharide, in which each monomer is a tetra-saccharide. This tetra-saccharide contains two residues of D-glucose and one residue each of L-rhamnose and D-glucuronic acid. This gum has been approved under the function of food additive and thickening agents in many countries. AgNPs were produced by the silver nitrate solution reduction by gellan gum, an edible, natural, biodegradable, and renewable carbohydrate. For the green synthesis of AgNPs, gellan gum is used as a stabilizing as well as a reducing agent. An aqueous solution of gellan gum was mixed with the silver nitrate solution under the boiling condition and the pH was adjusted to 11–12. For the removal of the unreacted silver ions solution, the AgNPs were dialyzed for 24 h. The synthesized AgNPs were characterized by FTIR, TEM, UV-vis spectroscopy, and XRD techniques. The prepared AgNPs were characterized by the noticeable yellow color that appears in the reaction mixture from a colorless solution. The yellow color in the reaction mixture indicates the synthesis of AgNPs. The average size of the nanoparticles synthesized from gellan gum was about 10–15 nm and they are stable for 6 months. Furthermore, it was concluded that the stabilized AgNPs by gellan gum have significant catalytic and antibacterial activity against *Escherichia coli* (a gram-negative bacterium) and *Bacillus subtilis* (a gram-positive bacterium) (Dhar et al., 2012).

5.9 Oilbanum gum

Oilbanum gum occurs naturally gum oleoresin derived from the bark of Boswellia serrate (Burseraceae family), which is a native tree of India. Oilbanum gum is a nontoxic, cost-effective, eco-friendly, easily available, renewable, and biodegradable polysaccharide gum. This gum consists of water-soluble gum, volatile oil, lipophilic terpenes, and insoluble matter. The basic structure of oilbanum gum is composed of sugar such as arabinose, xylose, galactose, and D-glucuronic acid. The natural polymer was traditionally used in ayurvedic and Unani medicine for the treatment of reno-protective, hypolipidemic, hepatoprotective, and anticancer agents (Assefa et al., 2017) rheumatism, gastrointestinal problem, and relieve depression because of antiinflammatory and antiasthmatic properties are present in this gum (Patil et al., 2010).

This gum is also used in cosmetics, textile, paint, ceramic, food, and pharmaceutical industries. Low cost, nontoxic nature, and natural availability are some attractive features of oilbanum gum for the stabilization and synthesis of AgNPs. The AgNPs were produced by the silver nitrate solution reduction by oilbanum gum, a natural, edible, biodegradable, and renewable polymer. The AgNPs were synthesized by autoclaving the silver nitrate solution containing various gum extracts for different durations of time. The synthesis was carried out in an aqueous medium by autoclaving to produce sterile AgNPs which are completely free from microbes. For the green synthesis of AgNPs, oilbanum was used as a stabilizing as well as a reducing agent. The synthesized oilbanum gum-based AgNPs were characterized by UV-vis spectroscopy, X-ray diffraction, and TEM techniques. The prepared silver nanoparticle was characterized by a yellow color appearing in the reaction, which indicated the presence of AgNPs. It has been suggested that hydroxyl and carboxyl functional groups are present in oilbanum gum and are involved in the stability and formation of AgNPs. The size of the nanoparticles from oilbanum gum is 7.5 ± 3.8 nm and the nanoparticles have a spherical shape. Furthermore, it was concluded that the AgNPs synthesized by oilbanum gum showed significantly greater antibacterial activity on gram-positive bacteria (*Staphylococcus aureus*) and gram-negative bacteria (*Escherichia coli*, *Pseudomonas aeruginosa*) (Kora et al., 2012a).

5.10 Locust bean gum

Locust bean gum is extracted from the seed of the carob tree (*Ceratonia siliqua*) and mostly this tree is grown abundantly in the Mediterranean region. This natural gum is also produced in America, Asia, and South Africa. Locust bean gum is a nontoxic, easily available, renewable, and biodegradable polysaccharide. The molecular weight of locust bean gum is between 50 and 1000 kDa. The chemical structure of the locust bean gum is (1,4)-linked β-D-mannose backbone with (1–6) linked α-D-galactose (Dionísio and Grenha, 2012). Locust bean gum is used as a stabilizer, thickener, and a gelling agent in products such as dairy products, baked foods, ice cream, yoghurt, cheese, and beverages. Locust bean gum is "generally regarded as safe" (GRAS) and this gum is used for stabilizing, fat replacing properties, and thickening function, and has been approved by the FDA (Barak and Mudgil, 2014).

AgNPs were produced by mixing a silver nitrate solution with the locust bean gum solution for a few minutes until the locust bean gum was completely dissolved.

The green synthesis of AgNPs was carried out at room temperature and after 24 h the silver nitrate solution was completely reduced by the locust bean gum. The preparation of AgNPs was characterized by UV spectroscopy and atomic force microscopy (AFM) techniques. The locust bean gum acts as a stabilizing as well as a reducing agent for the preparation of AgNPs. When silver salt was added to the solution of locust bean gum, the reaction mixture changed from yellow to brown over time. This is a clear indication of the preparation of AgNPs. It has been suggested that hydroxyl functional groups are present in locust bean gum and are involved in the stability and formation of AgNPs. The size of the AgNPs synthesized by the locust bean gum ranged from 18 to 51 nm. The preparation of AgNPs by this method was stable for more than 7 months (Tagad et al., 2013).

5.11 Guar gum

Guar gum is obtained from the endosperm seed of *Cyamopsis tetragonolobus* (Leguminosae family) which is abundantly grown in Pakistan and India (Sharma et al., 2018). This gum is an eco-friendly, low-cost, biodegradable, biocompatible, hydrophilic, and easily available polysaccharide. The natural polymer of the guar gum is a water-soluble and swellable polysaccharide. Guar gum has a less stable, complex structure and is a slightly acid polysaccharide derived from mixed sodium chloride, calcium chloride, and iron chloride salts. The average molecular weight of guar gum is about 6.5×10^4. The structure of guar gum containing a straight chain of the mannose unit joined by β-D-(1,4) linked having a backbone with α-D-(1–6) linked galactopyranose subunits are attached side chain (Singh and Dhaliwal, 2018).

Guar gum is a thickening and stabilization agent most widely used in food, cosmetics, textile, printing, pharmaceuticals, and bakery industries. Guar gum is used for the treatment of diabetes, bowel problems, heart diseases, colon cancer (Khan et al., 2020a), hypolipidemic, hypoglycemic, antiproliferative, and antimicrobial activity (Gupta and Verma, 2014).

AgNPs were prepared by the silver nitrate solution reduction by guar gum, an edible natural biopolymer. A magnetic stirrer was used to facilitate the reaction. For the green synthesis of AgNPs, guar gum is used as both a stabilizing as well as a reducing agent. The synthesized guar gum-based AgNPs were characterized by UV-vis spectroscopy, FTIR, scanning electron microscopy (SEM), TEM, and X-ray diffraction techniques. The prepared AgNPs were characterized by a yellow color that appeared in a reaction mixture, changing it from a colorless solution. This is a clear indication of the synthesis of AgNPs. It has been suggested that hydroxyl, carbonyl, and amine functional groups are present in guar gum and are involved in the stability and formation of AgNPs (Das et al., 2015).

Moreover, it was concluded the synthesized AgNPs had significant bacterial action on both gram-positive *Staphylococcus aureus* and gram-negative *Escherichia coli* and *Pseudomonas aeruginosa* (Khan et al., 2020a). The green synthesis of AgNPs was produced by mixing silver nitrate solution with the guar gum solution for a few minutes until the guar gum was completely dissolved. The green synthesis of AgNPs was carried out at a different temperature and after being continuously stirred for 1 h, the silver nitrate was completely reduced by the guar gum. At the end of the reaction, the AgNPs were diluted five times and centrifuged at 10,000 rpm for 30 min. In the solution, the supernatant was carefully isolated and allowed to undergo sonication for a further 15 min. The synthesized AgNPs were characterized by FTIR, HRTEM, and UV-spectroscopy techniques. The prepared AgNPs were characterized by a distinctive yellow color that appeared in the reaction mixture, changing it from a colorless solution. The yellow color in the reaction mixture indicated the synthesis of AgNPs. It has been suggested that hydroxyl, carboxylic, carboxylate, and carbonyl functional group are present in guar gum and are involved in the stability and formation of AgNPs. The average size of AgNPs synthesized by guar gum was approximately 16–42 nm and they are stable for 60 days under refrigeration (Das et al., 2015).

5.12 Neem gum

Neem gum occurs naturally plant polysaccharide which is obtained from *Azadirachta indica* (Meliaceae family). Neem gum is an easily naturally available, non-toxic, water-dispersible, and biocompatible polysaccharide. The natural polymer of the neem gum is water-soluble and swellable polysaccharides (Mankotia et al., 2020).

The structural, physicochemical, thermal, compositional, morphological, emulsifying, and rheological properties have been well studied and characterized. The biopolymer is a protein-containing, high-arabinose, and acidic polysaccharide that occurs in nature in mixed calcium, sodium, and magnesium salts (Velusamy et al., 2015). The structure of neem gum consists of D-glucuronic acid, L-fructose, D-glucose, L-arabinose, xylose, mannose, and aspartic acid (Amrutham et al., 2020). Neem gum is mostly used in paper, cosmetic, pharmaceutical, and textile industries.

AgNPs were produced by the silver nitrate solution reduction by neem gum, a natural and renewable polymer. An autoclave was used as an external energy source to facilitate the reaction. The autoclave is a synthetic method to prepare sterile AgNPs which are completely free from bacteria, viruses, and spores. For the green synthesis of AgNPs, neem gum was used as a stabilizing as well as a reducing agent.

The synthesized neem gum AgNPs were characterized by UV-vis spectroscopy, FTIR, TEM, AFM, and X-ray diffraction techniques. The prepared AgNPs were characterized by a noticeable yellow color that appeared in the reaction mixture, changing it from a colorless solution. The yellow color in the reaction mixture indicated the synthesis of AgNPs. It has been suggested that carbonyl, hydroxyl, acetyl, and carboxylic functional groups are present in neem gum and are involved in the stability and formation of AgNPs. The average size of AgNPs is <30 nm and the nanoparticles are a spherical shape. It was concluded that the synthesis of AgNPs had significant antibacterial action on both gram-negative bacteria such as *Salmonella enteritidis* and gram-negative bacteria such as *Bacillus cereus* (Velusamy et al., 2015).

The AgNPs were produced by mixing silver nitrate solution with neem gum solution; they were then incubated in a dark condition overnight. The AgNPs were centrifuged at 10,000 rpm for 5 min and washed the pellet was twice with ion free water and lipolyzed. The synthesized AgNPs were characterized by FTIR, SEM, AFM, dynamic light scattering, and UV spectroscopy techniques. The formation of AgNPs was characterized by a distinctive yellow color that appeared in the reaction mixture. The yellow color in the reaction mixture indicated the synthesized AgNPs. It has been suggested that hydroxyl and carboxyl functional groups are present in neem gum and involved and the stability of AgNPs. Moreover, it was concluded that the AgNPs stabilized by neem gum have antimicrobial activity against the gram-positive bacteria *Streptococcus* and gram-negative bacteria *Escherichia coli* (Samrot et al., 2019).

6 Conclusion

In recent years, the green synthesis of AgNPs through biodegradable gums has gained global attraction. Though many approaches are available for the green synthesis of AgNPs, the AgNPs prepared through biodegradable gums involve environmentally friendly and facile processes, which do not require harsh and toxic chemicals. These biodegradable gums act as reducing and stabilizing agents during the synthesis process. These gums are abundant in nature and are nontoxic, renewable, and cost-effective. To gain more advantages from these gum-based AgNPs, more research should be carried out to exploit more potential applications besides antibacterial and catalytic activities.

References

Abdelaziz, D., Hefnawy, A., Al-Wakeel, E., El-Fallal, A., El-Sherbiny, I.M., 2020. New biodegradable nanoparticles-in-nanofibers based membranes for guided periodontal tissue and bone regeneration with enhanced antibacterial activity. J. Adv. Res. 28, 51–62.

Abdel-Halim, E., El-Rafie, M., Al-Deyab, S.S., 2011. Polyacrylamide/guar gum graft copolymer for preparation of silver nanoparticles. Carbohydr. Polym. 85, 692–697.

Aderibigbe, B., Varaprasad, K., Sadiku, E., Ray, S., Mbianda, X., Fotsing, M., Owonubi, S., Agwuncha, S.C., 2015. Kinetic release studies of nitrogen-containing bisphosphonate from gum acacia crosslinked hydrogels. Int. J. Biol. Macromol. 73, 115–123.

Ahamed, M., AlSalhi, M.S., Siddiqui, M., 2010. Silver nanoparticle applications and human health. Clin. Chim. Acta 411, 1841–1848.

Aisida, S.O., Ugwu, K., Akpa, P.A., Nwanya, A.C., Nwankwo, U., Botha, S.S., et al., 2019. Biosynthesis of silver nanoparticles using bitter leave (*Veronica amygdalina*) for antibacterial activities. Surf. Interfaces 17, 100359.

Akele, M.L., Assefa, A.G., Alle, M., 2015. Microwave-assisted green synthesis of silver nanoparticles by using gum acacia: synthesis, characterization and catalytic activity studies. Int. J. Green Energy 5, 21–27.

Alam, M.S., Garg, A., Pottoo, F.H., Saifullah, M.K., Tareq, A.I., Manzoor, O., Mohsin, M., Javed, M.N., 2017. Gum ghatti mediated, one pot green synthesis of optimized gold nanoparticles: investigation of process-variables impact using Box-Behnken based statistical design. Int. J. Biol. Macromol. 104, 758–767.

Amrutham, S., Maragoni, V., Guttena, V., 2020. One-step green synthesis of palladium nanoparticles using neem gum (Azadirachta Indica): characterization, reduction of rhodamine 6G dye and free radical scavenging activity. Appl. Nanosci. 10, 1–7.

Anastas, P.T., Warner, J.C., 1998. Principles of green chemistry. In: Green Chemistry: Theory and Practice, pp. 29–56.

Anderson, D., Wang, W., 1990. Composition of the gum from Combretum paniculatum and four other gums which are not permitted food additives. Phytochemistry 29, 1193–1195.

Anderson, D., Weiping, W., 1992. Gum arabic (Acacia Senegal) from Uganda: characteristic NMR spectra, amino acid compositions, and gum/soil cationic relationships. Int. Tree Crops J. 7, 167–179.

Ankamwar, B., Damle, C., Ahmad, A., Sastry, M., 2005. Biosynthesis of gold and silver nanoparticles using Emblica officinalis fruit extract, their phase transfer and transmetallation in an organic solution. J. Nanosci. Nanotechnol. 5, 1665–1671.

Ansari, M.A., Khan, H.M., Khan, A.A., Cameotra, S.S., Saquib, Q., Musarrat, J., 2014. Gum arabic capped-silver nanoparticles inhibit biofilm formation by multi-drug resistant strains of Pseudomonas aeruginosa. J. Basic Microbiol. 54, 688–699.

Anuj, S.A., Ishnava, K.B., 2013. Plant mediated synthesis of silver nanoparticles by using dried stem powder of Tinospora cordifolia, its antibacterial activity and comparison with antibiotics. Int. J. Pharma Bio Sci. 4, 849–863.

Araruna, F.B., de Oliveira, T.M., Quelemes, P.V., de Araújo Nobre, A.R., Plácido, A., Vasconcelos, A.G., de Paula, R.C.M., Mafud, A.C., de Almeida, M.P., Delerue-Matos, C., 2020. Antibacterial application of natural and carboxymethylated cashew gum-based silver nanoparticles produced by microwave-assisted synthesis. Carbohydr. Polym. 241, 115260.

Araruna, F.B., Quelemes, P.V., de Faria, B.E.F., Kuckelhaus, S.A.S., Marangoni, V.S., Zucolotto, V., da Silva, D.A., Júnior, J.R.S., Leite, J.R.S., Eiras, C., 2013. Green synthesis and characterization of silver nanoparticles reduced and stabilized by cashew tree gum. Adv. Sci. 5, 890–893.

Arib, C., Spadavecchia, J., de la Chapelle, M.L., 2021. Enzyme mediated synthesis of hybrid polyedric gold nanoparticles. Sci. Rep. 11 (1), 1–8.

Assefa, A.G., Mesfin, A.A., Akele, M.L., Alemu, A.K., Gangapuram, B.R., Guttena, V., Alle, M., 2017. Microwave-assisted green synthesis of gold nanoparticles using Olibanum gum (Boswellia serrate) and its catalytic reduction of 4-nitrophenol and hexacyanoferrate (III) by sodium borohydride. J. Clust. Sci. 28, 917–935.

Babaladimath, G., Badalamoole, V., 2019. Silver nanocomposite hydrogel of Gum Ghatti with potential antibacterial property. J. Macromol. Sci. A 56, 952–959.

Bahrami, A., Mokarram, R.R., Khiabani, M.S., Ghanbarzadeh, B., Salehi, R., 2019. Physico-mechanical and antimicrobial properties of tragacanth/hydroxypropyl methylcellulose/beeswax edible films reinforced with silver nanoparticles. Int. J. Biol. Macromol. 129, 1103–1112.

Barak, S., Mudgil, D., 2014. Locust bean gum: processing, properties and food applications—a review. Int. J. Biol. Macromol. 66, 74–80.

Bo, L., Yang, W., Chen, M., Gao, J., Xue, Q., 2009. A simple and 'green' synthesis of polymer-based silver colloids and their antibacterial properties. Chem. Biodivers. 6, 111–116.

Chandran, S.P., Chaudhary, M., Pasricha, R., Ahmad, A., Sastry, M., 2006. Synthesis of gold nanotriangles and silver nanoparticles using Aloevera plant extract. Biotechnol. Prog. 22, 577–583.

Choudhary, P.D., Pawar, H.A., 2014. Recently investigated natural gums and mucilages as pharmaceutical excipients: an overview. Int. J. Pharm. 2014.

Darroudi, M., Ahmad, M.B., Shameli, K., Abdullah, A.H., Ibrahim, N.A., 2009. Synthesis and characterization of UV-irradiated silver/montmorillonite nanocomposites. Solid State Sci. 11, 1621–1624.

Das, T., Yeasmin, S., Khatua, S., Acharya, K., Bandyopadhyay, A., 2015. Influence of a blend of guar gum and poly (vinyl alcohol) on long term stability, and antibacterial and antioxidant efficacies of silver nanoparticles. RSC Adv. 5, 54059–54069.

Dhar, S., Murawala, P., Shiras, A., Pokharkar, V., Prasad, B., 2012. Gellan gum capped silver nanoparticle dispersions and hydrogels: cytotoxicity and in vitro diffusion studies. Nanoscale 4, 563–567.

Dionísio, M., Grenha, A., 2012. Locust bean gum: exploring its potential for biopharmaceutical applications. J. Pharm. Bioallied. Sci. 4, 175.

Edison, T.J.I., Sethuraman, M., 2012. Instant green synthesis of silver nanoparticles using Terminalia chebula fruit extract and evaluation of their catalytic activity on reduction of methylene blue. Process Biochem. 47, 1351–1357.

Eghbalifam, N., Shojaosadati, S.A., Hashemi-Najafabadi, S., Khorasani, A.C., 2020. Synthesis and characterization of antimicrobial wound dressing material based on silver nanoparticles loaded gum Arabic nanofibers. Int. J. Biol. Macromol. 155, 119–130.

Faizi, S., Zikr-ur-Rehman, S., Naz, A., Versiani, M.A., Dar, A., Naqvi, S., 2012. Bioassay-guided studies on Bombax ceiba leaf extract: isolation of shamimoside, a new antioxidant xanthone C-glucoside. Chem. Nat. Compd. 48, 774–779.

Fayaz, A.M., Balaji, K., Kalaichelvan, P., Venkatesan, R., 2009. Fungal based synthesis of silver nanoparticles—an effect of temperature on the size of particles. Colloids Surf. B Biointerfaces 74, 123–126.

Filippo, E., Serra, A., Manno, D., 2009. Poly (vinyl alcohol) capped silver nanoparticles as localized surface plasmon resonance-based hydrogen peroxide sensor. Sensors Actuators B Chem. 138, 625–630.

Fosso-Kankeu, E., Mittal, H., Waanders, F., Ntwampe, I., Ray, S.S., 2016. Preparation and characterization of gum karaya hydrogel nanocomposite flocculant for metal ions removal from mine effluents. Int. J. Environ. Sci. Technol. 13, 711–724.

Ghosh, S., Patil, S., Ahire, M., Kitture, R., Kale, S., Pardesi, K., Cameotra, S.S., Bellare, J., Dhavale, D.D., Jabgunde, A., 2012. Synthesis of silver nanoparticles using Dioscorea bulbifera tuber extract and evaluation of its synergistic potential in combination with antimicrobial agents. Int. J. Nanomedicine 7, 483.

Gupta, A.P., Verma, D.K., 2014. Guar gum and their derivatives: a research profile. Int. J. Adv. Res. 2, 680–690.

Indana, M.K., Gangapuram, B.R., Dadigala, R., Bandi, R., Guttena, V., 2016. A novel green synthesis and characterization of silver nanoparticles using gum tragacanth and evaluation of their potential catalytic reduction activities with methylene blue and Congo red dyes. J. Anal. Sci. Technol. 7, 19.

Jain, V., Verma, S.K., 2012. Pharmacological investigations and toxicity studies. In: Pharmacology of Bombax ceiba Linn. Springer.

Jha, A.K., Prasad, K., Kumar, V., Prasad, K., 2009. Biosynthesis of silver nanoparticles using Eclipta leaf. Biotechnol. Prog. 25, 1476–1479.

Ju-Nam, Y., Lead, J.R., 2008. Manufactured nanoparticles: an overview of their chemistry, interactions and potential environmental implications. Sci. Total Environ. 400, 396–414.

Kaabipour, S., Hemmati, S., 2021. A review on the green and sustainable synthesis of silver nanoparticles and one-dimensional silver nanostructures. Beilstein J. Nanotechnol. 12 (1), 102–136.

Kaegi, R., Sinnet, B., Zuleeg, S., Hagendorfer, H., Mueller, E., Vonbank, R., Boller, M., Burkhardt, M., 2010. Release of silver nanoparticles from outdoor facades. Environ. Pollut. 158, 2900–2905.

Khan, M.A., Ali, A., Mohammad, S., Ali, H., Khan, T., Jan, A., Ahmad, P., 2020b. Iron nano modulated growth and biosynthesis of steviol glycosides in Stevia rebaudiana. Plant Cell Tiss. Org. Cult., 1–10.

Khan, N., Kumar, D., Kumar, P., 2020a. Silver nanoparticles embedded guar gum/gelatin nanocomposite: green synthesis, characterization and antibacterial activity. Colloids Interface Sci. Commun. 35, 100242.

Kokila, T., Ramesh, P., Geetha, D., 2015. Biosynthesis of silver nanoparticles from Cavendish banana peel extract and its antibacterial and free radical scavenging assay: a novel biological approach. Appl. Nanosci. 5, 911–920.

Kora, A.J., Arunachalam, J., 2012. Green fabrication of silver nanoparticles by gum Tragacanth (Astragalus gummifer): a dual functional reductant and stabilizer. J. Nanomater. 2012, 69.

Kora, A.J., Beedu, S.R., Jayaraman, A., 2012b. Size-controlled green synthesis of silver nanoparticles mediated by gum ghatti (Anogeissus latifolia) and its biological activity. Org. Med. Chem. Lett. 2, 1–10.

Kora, A.J., Rastogi, L., 2018. Green synthesis of palladium nanoparticles using gum ghatti (Anogeissus latifolia) and its application as an antioxidant and catalyst. Arab. J. Chem. 11, 1097–1106.

Kora, A.J., Sashidhar, R.B., 2015. Antibacterial activity of biogenic silver nanoparticles synthesized with gum ghatti and gum olibanum: a comparative study. J. Antibiot. 68, 88–97.

Kora, A.J., Sashidhar, R., Arunachalam, J., 2010. Gum kondagogu (Cochlospermum gossypium): a template for the green synthesis and stabilization of silver nanoparticles with antibacterial application. Carbohydr. Polym. 82, 670–679.

Kora, A.J., Sashidhar, R., Arunachalam, J., 2012a. Aqueous extract of gum olibanum (Boswellia serrata): a reductant and stabilizer for the biosynthesis of antibacterial silver nanoparticles. Process Biochem. 47, 1516–1520.

Krishna, I.M., Reddy, G.B., Veerabhadram, G., Madhusudhan, A., 2016. Eco-friendly green synthesis of silver nanoparticles using Salmalia malabarica: synthesis, characterization, antimicrobial, and catalytic activity studies. Appl. Nanosci. 6, 681–689.

Krutyakov, Y.A., Kudrinskiy, A.A., Olenin, A.Y., Lisichkin, G.V., 2008. Synthesis and properties of silver nanoparticles: advances and prospects. Russ. Chem. Rev. 77, 233.

Lok, C.-N., Ho, C.-M., Chen, R., He, Q.-Y., Yu, W.-Y., Sun, H., Tam, P.K.-H., Chiu, J.-F., Che, C.-M., 2006. Proteomic analysis of the mode of antibacterial action of silver nanoparticles. J. Proteome Res. 5, 916–924.

Mankotia, P., Choudhary, S., Sharma, K., Kumar, V., Bhatia, J.K., Parmar, A., Sharma, S., Sharma, V., 2020. Neem gum based pH responsive hydrogel matrix: a new pharmaceutical excipient for the sustained release of anticancer drug. Int. J. Biol. Macromol. 142, 742–755.

Marambio-Jones, C., Hoek, E.M., 2010. A review of the antibacterial effects of silver nanomaterials and potential implications for human health and the environment. J. Nanopart. Res. 12, 1531–1551.

Matar, G.H., Andac, M., 2020. Antibacterial efficiency of silver nanoparticles-loaded locust bean gum/polyvinyl alcohol hydrogels. Polym. Bull., 1–19.

Mewada, A., Thakur, M., Pandey, S., Oza, G., Shah, R., Sharon, M., 2013. A novel one pot synthesis of super stable silver nanoparticles using natural plant exudate from Azadirachta indica (Neem gum) and their inimical effect on pathogenic microorganisms. J. Bionanosci. 7, 296–299.

Mohan, Y.M., Raju, K.M., Sambasivudu, K., Singh, S., Sreedhar, B., 2007. Preparation of acacia-stabilized silver nanoparticles: a green approach. J. Appl. Polym. Sci. 106, 3375–3381.

Newton, K.M., Puppala, H.L., Kitchens, C.L., Colvin, V.L., Klaine, S.J., 2013. Silver nanoparticle toxicity to Daphnia magna is a function of dissolved silver concentration. Environ. Toxicol. Chem. 32, 2356–2364.

Padil, V.V.T., Černík, M., 2015. Poly (vinyl alcohol)/gum karaya electrospun plasma treated membrane for the removal of nanoparticles (Au, Ag, Pt, CuO and Fe3O4) from aqueous solutions. J. Hazard. Mater. 287, 102–110.

Padil, V.V.T., Nguyen, N.H., Ševců, A., Černík, M., 2015b. Fabrication, characterization, and antibacterial properties of electrospun membrane composed of gum karaya, polyvinyl alcohol, and silver nanoparticles. J. Nanomater. 2015, 9.

Padil, V.V.T., Senan, C., Černík, M., 2015a. Dodecenylsuccinic anhydride derivatives of gum karaya (Sterculia urens): preparation, characterization, and their antibacterial properties. J. Agric. Food Chem. 63, 3757–3765.

Padil, V.V.T., Wacławek, S., Černík, M., 2016. Green synthesis: nanoparticles and nanofibres based on tree gums for environmental applications. Ecol. Chem. Eng. S. 23, 533–557.

Pal, K., Roy, S., Parida, P.K., Dutta, A., Bardhan, S., Das, S., Jana, K., Karmakar, P., 2019. Folic acid conjugated curcumin loaded biopolymeric gum acacia microsphere for triple negative breast cancer therapy in invitro and invivo model. Mater. Sci. Eng. C 95, 204–216.

Panáček, A., Kvitek, L., Prucek, R., Kolář, M., Večeřová, R., Pizúrová, N., Sharma, V.K., Nevěčná, T.J., Zbořil, R., 2006. Silver colloid nanoparticles: synthesis, characterization, and their antibacterial activity. J. Phys. Chem. B 110, 16248–16253.

Patel, S., 2021. A review on synthesis of silver nanoparticles-a green expertise. Life sci. leafl. 132, 16–24.

Patil, R.S., Kokate, M.R., Kolekar, S.S., 2012. Bioinspired synthesis of highly stabilized silver nanoparticles using Ocimum tenuiflorum leaf extract and their antibacterial activity. Spectrochim. Acta A Mol. Biomol. Spectrosc. 91, 234–238.

Patil, J.S., Marapur, S.C., Kadam, D.V., Kamalapur, M.V., 2010. Pharmaceutical and medicinal applications of Olibanum gum and its constituents: a review. J. Pharm. Res. 3, 587–589.

Quelemes, P., Araruna, F., de Faria, B., Kuckelhaus, S., da Silva, D., Mendonça, R., Eiras, C., dos S Soares, M.J., Leite, J.R., 2013. Development and antibacterial activity of cashew gum-based silver nanoparticles. Int. J. Mol. Sci. 14, 4969–4981.

Rao, K., Aziz, S., Roome, T., Razzak, A., Sikandar, B., Jamali, K.S., Imran, M., Jabri, T., Shah, M.R., 2018b. Gum acacia stabilized silver nanoparticles based nano-cargo for enhanced anti-arthritic potentials of hesperidin in adjuvant induced arthritic rats. Artif. Cells Nanomed. Biotechnol. 46, 597–607.

Rao, K.M., Kumar, A., Rao, K.S.V.K., Haider, A., Han, S.S., 2018a. Biodegradable tragacanth gum based silver nanocomposite hydrogels and their antibacterial evaluation. J. Polym. Environ. 26, 778–788.

Rastogi, L., Kora, A.J., Sashidhar, R., 2015. Antibacterial effects of gum kondagogu reduced/stabilized silver nanoparticles in combination with various antibiotics: a mechanistic approach. Appl. Nanosci. 5, 535–543.

Roldán, M.V., Scaffardi, L.B., de Sanctis, O., Pellegri, N., 2008. Optical properties and extinction spectroscopy to characterize the synthesis of amine capped silver nanoparticles. Mater. Chem. Phys. 112, 984–990.

Roy, A., Bulut, O., Some, S., Mandal, A.K., Yilmaz, M.D., 2019. Green synthesis of silver nanoparticles: biomolecule-nanoparticle organizations targeting antimicrobial activity. RSC Adv. 9 (5), 2673–2702.

Sahraei, R., Ghaemy, M., 2017. Synthesis of modified gum tragacanth/graphene oxide composite hydrogel for heavy metal ions removal and preparation of silver nanocomposite for antibacterial activity. Carbohydr. Polym. 157, 823–833.

Saleem, R., Ahmad, S.I., Ahmed, M., Faizi, Z., Zikr-ur-Rehman, S., Ali, M., Faizi, S., 2003. Hypotensive activity and toxicology of constituents from Bombax ceiba stem bark. Biol. Pharm. Bull. 26, 41–46.

Samadi, N., Hosseini, S., Fazeli, A., Fazeli, M., 2010. Synthesis and antimicrobial effects of silver nanoparticles produced by chemical reduction method. DARU 18, 168.

Samrot, A.V., Angalene, J.L.A., Roshini, S., Raji, P., Stefi, S., Preethi, R., Selvarani, A.J., Madankumar, A., 2019. Bioactivity and heavy metal removal using plant gum mediated green synthesized silver nanoparticles. J. Clust. Sci. 30, 1599–1610.

Saravanan, P., Vinod, V., Sreedhar, B., Sashidhar, R., 2012. Gum kondagogu modified magnetic nano-adsorbent: an efficient protocol for removal of various toxic metal ions. Mater. Sci. Eng. C 32, 581–586.

Sashidhar, R., Selvi, S.K., Vinod, V., Kosuri, T., Raju, D., Karuna, R., 2015. Bioprospecting of gum kondagogu (Cochlospermum gossypium) for bioremediation of uranium (VI) from aqueous solution and synthetic nuclear power reactor effluents. J. Environ. Radioact. 148, 33–41.

Sastry, M., Mayya, K., Bandyopadhyay, K., 1997. pH dependent changes in the optical properties of carboxylic acid derivatized silver colloidal particles. Colloids Surf. A Physicochem. Eng. Asp. 127, 221–228.

Sathishkumar, M., Sneha, K., Won, S., Cho, C.-W., Kim, S., Yun, Y.-S., 2009. Cinnamon zeylanicum bark extract and powder mediated green synthesis of nano-crystalline silver particles and its bactericidal activity. Colloids Surf. B Biointerfaces 73, 332–338.

Seku, K., Gangapuram, B.R., Pejjai, B., Kadimpati, K.K., Golla, N., 2018. Microwave-assisted synthesis of silver nanoparticles and their application in catalytic, antibacterial and antioxidant activities. J. Nanostruct. Chem. 8, 179–188.

Setia, A., Goyal, S., Goyal, N., 2010. Applications of gum karaya in drug delivery systems: a review on recent research. Der Pharm. Lett. 2, 39–48.

Sharma, G., Sharma, S., Kumar, A., Ala'a, H., Naushad, M., Ghfar, A.A., Mola, G.T., Stadler, F.J., 2018. Guar gum and its composites as potential materials for diverse applications: a review. Carbohydr. Polym. 199, 534–545.

Sharma, S., Virk, K., Sharma, K., Bose, S.K., Kumar, V., Sharma, V., Focarete, M.L., Kalia, S., 2020. Preparation of gum acacia-poly (acrylamide-IPN-acrylic acid) based nanocomposite hydrogels via polymerization methods for antimicrobial applications. J. Mol. Struct. 1215, 128298.

Sharma, V.K., Yngard, R.A., Lin, Y., 2009. Silver nanoparticles: green synthesis and their antimicrobial activities. Adv. Colloid Interf. Sci. 145, 83–96.

Shkilnyy, A., Soucé, M., Dubois, P., Warmont, F., Saboungi, M.-L., Chourpa, I., 2009. Poly (ethylene glycol)-stabilized silver nanoparticles for bioanalytical applications of SERS spectroscopy. Analyst 134, 1868–1872.

Shukla, V.K., Singh, R.P., Pandey, A.C., 2010. Black pepper assisted biomimetic synthesis of silver nanoparticles. J. Alloys Compd. 507, L13–L16.

Silva, S.C.C.C., de Araujo Braz, E.M., de Amorim Carvalho, F.A., de Sousa Brito, C.A.R., Brito, L.M., Barreto, H.M., da Silva Filho, E.C., da Silva, D.A., 2020. Antibacterial and cytotoxic properties from esterified Sterculia gum. Int. J. Biol. Macromol. 164, 606–615.

Singh, V., Ahmed, S., 2012. Silver nanoparticle (AgNPs) doped gum acacia–gelatin–silica nanohybrid: an effective support for diastase immobilization. Int. J. Biol. Macromol. 50, 353–361.

Singh, J., Dhaliwal, A., 2018. Synthesis, characterization and swelling behavior of silver nanoparticles containing superabsorbent based on grafted copolymer of polyacrylic acid/Guar gum. Vacuum 157, 51–60.

Singh, J., Dhaliwal, A., 2020. Effective removal of methylene blue dye using silver nanoparticles containing grafted polymer of guar gum/acrylic acid as novel adsorbent. J. Polym. Environ. 29, 1–18.

Srikar, S.K., Giri, D.D., Pal, D.B., Mishra, P.K., Upadhyay, S.N., 2016. Green synthesis of silver nanoparticles: a review. Curr. Opin. Green Sustain. Chem. 6, 34.

Starowicz, M., Stypuła, B., Banaś, J., 2006. Electrochemical synthesis of silver nanoparticles. Electrochem. Commun. 8, 227–230.

Tagad, C.K., Dugasani, S.R., Aiyer, R., Park, S., Kulkarni, A., Sabharwal, S., 2013. Green synthesis of silver nanoparticles and their application for the development of optical fiber based hydrogen peroxide sensor. Sensors Actuators B Chem. 183, 144–149.

Tran, Q.H., Le, A.-T., 2013. Silver nanoparticles: synthesis, properties, toxicology, applications and perspectives. Adv. Nat. Sci. Nanosci. Nanotechnol. 4, 033001.

Tripathi, A., Chandrasekaran, N., Raichur, A., Mukherjee, A., 2009. Antibacterial applications of silver nanoparticles synthesized by aqueous extract of Azadirachta indica (Neem) leaves. J. Biomed. Nanotechnol. 5, 93–98.

Velusamy, P., Das, J., Pachaiappan, R., Vaseeharan, B., Pandian, K., 2015. Greener approach for synthesis of antibacterial silver nanoparticles using aqueous solution of neem gum (Azadirachta indica L.). Ind. Crops Prod. 66, 103–109.

Venkatesham, M., Ayodhya, D., Madhusudhan, A., Kumari, A.S., Veerabhadram, G., Mangatayaru, K.G., 2014. A novel green synthesis of silver nanoparticles using gum karaya: characterization, antimicrobial and catalytic activity studies. J. Clust. Sci. 25, 409–422.

Verbeken, D., Dierckx, S., Dewettinck, K., 2003. Exudate gums: occurrence, production, and applications. Appl. Microbiol. Biotechnol. 63, 10–21.

Vinod, V., Saravanan, P., Sreedhar, B., Devi, D.K., Sashidhar, R., 2011a. A facile synthesis and characterization of Ag, Au and Pt nanoparticles using a natural hydrocolloid gum kondagogu (Cochlospermum gossypium). Colloids Surf. B Biointerfaces 83, 291–298.

Vinod, V., Sashidhar, R., Sarma, V., Vijaya Saradhi, U., 2008a. Compositional analysis and rheological properties of gum kondagogu (Cochlospermum gossypium): a tree gum from India. J. Agric. Food Chem. 56, 2199–2207.

Vinod, V., Sashidhar, R., Sivaprasad, N., Sarma, V., Satyanarayana, N., Kumaresan, R., Rao, T.N., Raviprasad, P., 2011b. Bioremediation of mercury (II) from aqueous solution by gum karaya (Sterculia urens): a natural hydrocolloid. Desalination 272, 270–277.

Vinod, V., Sashidhar, R., Sukumar, A., 2010. Competitive adsorption of toxic heavy metal contaminants by gum kondagogu (Cochlospermum gossypium): a natural hydrocolloid. Colloids Surf. B Biointerfaces 75, 490–495.

Vinod, V., Sashidhar, R., Suresh, K., Rao, B.R., Saradhi, U.V., Rao, T.P., 2008b. Morphological, physico-chemical and structural characterization of gum kondagogu (Cochlospermum gossypium): a tree gum from India. Food Hydrocoll. 22, 899–915.

Wang, H., Williams, P.A., Senan, C., 2014. Synthesis, characterization and emulsification properties of dodecenyl succinic anhydride derivatives of gum Arabic. Food Hydrocoll. 37, 143–148.

Yoksan, R., Chirachanchai, S., 2009. Silver nanoparticles dispersing in chitosan solution: preparation by γ-ray irradiation and their antimicrobial activities. Mater. Chem. Phys. 115, 296–302.

Zare, E.N., Makvandi, P., Tay, F.R., 2019. Recent progress in the industrial and biomedical applications of tragacanth gum: a review. Carbohydr. Polym. 212, 450–467.

Zhang, C., Yang, Q., Zhan, N., Sun, L., Wang, H., Song, Y., Li, Y., 2010. Silver nanoparticles grown on the surface of PAN nanofiber: preparation, characterization and catalytic performance. Colloids Surf. A Physicochem. Eng. Asp. 362, 58–64.

CHAPTER 9

Bio-mediated synthesis of silver nanoparticles via conventional and irradiation-assisted methods and their application for environmental remediation in agriculture

Lebea N. Nthunya[a], Leonardo Gutierrez[b,c], Sabelo D. Mhlanga[d], and Heidi L. Richards[a]

[a]*Molecular Sciences Institute, School of Chemistry, University of Witwatersrand, Johannesburg, South Africa* [b]*Facultad del Mary Medio Ambiente, Universidad del Pacifico, Guayaquil, Ecuador* [c]*Particle and Interfacial Technology Group, Department of Green Chemistry and Technology, Ghent University, Ghent, Belgium* [d]*Research and Development Division, SabiNano (Pty) Ltd, Johannesburg, South Africa*

1 Introduction

Silver nanoparticles (AgNPs) exhibit remarkable physical and chemical properties, which are a product of their high-energy surface atoms (i.e., compared to their molecular counterparts) and the nanometer-scale mean free path of the electrons in their metallic state (Pal et al., 2007; Huang et al., 2016). The exploitation of AgNPs at the nanoscale has proven to be a fruitful field, where their properties (i.e., different from their bulk counterparts) are a result of surface and quantum size effects (Vanaja et al., 2014; Kobashigawa et al., 2019; Ameen et al., 2020). Due to their peculiar properties, the study of AgNPs is a fast-emerging field of science, encompassing a diverse range of applications, including medicine, agricultural farming, and molecular pharming (genetics) (Lamsal et al., 2011; Lara et al., 2011; Prabhu and Poulose, 2012; Moussa et al., 2013; Babu et al., 2014; Lee et al., 2013; Tang et al., 2017; Espino-pérez et al., 2018; Wang et al., 2018; Wu et al., 2018).

In the 20th century, AgNPs have been used in the therapy industry due to their efficient antimicrobial properties (Zhang et al., 2016). Furthermore, AgNPs have been applied as water and food disinfectants (Natsuki et al., 2015; Biswas and

Bandyopadhyaya, 2016). Based on their biocidal effects against microbes, the use of silver was approved by the Food and Drug Administration in the year 1920 (Mishra and Singh, 2015). While several pathogens (e.g., bacteria) have developed resistance against antibiotics, AgNPs have emerged as an alternative solution to address the challenges associated with these resistant pathogens (Armstrong et al., 1981; Ameen et al., 2020; Essawy et al., 2021). The effectiveness of AgNPs has been established by their potential to inhibit the growth of pathogenic microbes, including bacteria, viruses, fungi, and yeasts. Notably, AgNPs have been effectively used in domestic appliances, textiles, and surgical products and wound dressing (Budama et al., 2013; Wu et al., 2014; Zhang et al., 2016). Recently, AgNPs have received substantial consideration for the treatment of pests in the agricultural industry (Mishra and Singh, 2015). Their application in the agro-ecosystem was stimulated by the growing concern of pests developing resistance to existing pesticides.

AgNPs have demonstrated better performance in terms of growth inhibition against pathogens compared to their Ag^+ counterpart (Nthunya et al., 2019). Remarkably, Ag^+ ions and AgNPs have the same mechanism for cell inactivation. For instance, both AgNPs and Ag^+ ions interact with the thiol groups and nucleic acids of bacteria, bind to their enzymes, and damage their cell envelopes, ultimately resulting in cell inactivation (Pradeep and Anshup, 2009; Nangmenyi et al., 2011). During these processes, reactive species of oxygen are also generated, consequently, damaging the DNA of the bacteria. However, due to their high reactivity compared to that of AgNPs, Ag^+ ions react with bacterial proteins and form insoluble Ag(protein) complexes, which subsequently decrease their antibacterial activities (Nthunya et al., 2019). Therefore, to produce a strong antibacterial activity, Ag^+ ions would require a higher concentration compared to their AgNPs counterpart. The AgNPs' mechanism of action against pests is not yet well established. However, despite slight differences, the mechanism is similar to that of AgNPs against bacteria. The impact of AgNPs on the antioxidant and detoxifying enzymes of insects causes oxidative stress, leading to cell death (Lamsal et al., 2011; Zielinska et al., 2016; Benelli, 2018). Furthermore, AgNPs reduce the activity of acetylcholinesterase and inhibit CYP450 isoenzymes, which subsequently inactivate the oxidation of steroids, fatty acids, and xenobiotics, followed by the destruction to the defensive mechanism of the pests (Zielinska et al., 2016; Benelli, 2018). Finally, AgNPs bind to respective sulfur and phosphorus groups in proteins and nucleic acids, decrease the pests' cell membrane permeability, destroy their organelles and enzymes, and subsequently lead to cell death (Benelli, 2018).

Due to their remarkable biocidal properties, AgNPs are currently key for disease treatment in numerous fields, and the agricultural sector is no exception (Mishra and Singh, 2015; Kobashigawa et al., 2019). Their robust application in agro-ecosystems has enriched innovative nanotechnology to harness the challenges associated with various biotic and abiotic stresses. While crop production has been greatly affected by pests developing resistance against pesticides, causing notable losses in the agricultural industry, the use of AgNPs has given birth to promising sustainable crop production (Chu et al., 2012; Mavani and Shah, 2013). Despite their promising application in the agricultural sector, their transport, bioavailability, and toxicity require consistent monitoring. For instance, AgNPs not only have direct interaction with the soil and plants but also harm their biota (Pal et al., 2007; Xu and Zhang, 2018). Due to their nonselectivity towards microbes, AgNPs destroy denitrifying bacteria, thus destroying the nitrogen cycle (Mishra and Singh, 2015). However, an additional advantage inherent to this phenomenon is to ensure the maintenance of nitrate-rich soil (i.e., a fertile soil required for high agricultural production), where the conversion of nitrate to gaseous nitrogen is minimized. Contrariwise, the phytotoxicity of AgNPs towards plants has been recorded (Chu et al., 2012; Krishnaraj et al., 2012). However, these challenges are governed by the particulate sizes and concentrations of the AgNPs (Pal et al., 2007). in addition, the synthesis methods play a critical role in plant cytotoxicity. Therefore, carefully selected routes of synthesis need to be considered for such applications.

In this chapter, greener routes for the synthesis of AgNPs involving the use of conventional and irradiation-assisted methods are briefly discussed. Among other conventional methods, environmentally benign physical and chemical routes are examined. Likewise, microwave and UV photochemical irradiation-assisted procedures are discussed. Finally, bio-mediated synthesis routes will be briefly discussed, while the application of AgNPs in the agro-ecosystem is highlighted.

2 Green synthesis methods

2.1 Green physical vapor deposition methods

Numerous physical methods have been reported for the synthesis of different types of nanoparticles, including AgNPs (El-nour et al., 2010). In one method, a tube furnace at atmospheric pressure is used to synthesize the NPs through an evaporation-condensation method. A ceramic or quartz board containing silver is placed in a tube furnace and heated to produce silver vapor (Magnusson, 1999; Natsuki et al., 2015). Nitrogen is used as the carrier gas to transport the silver aerosols to the condensation nucleus counter. Notably, the aerosols are fractionated through differential mobility analyzers based on their charge level, mass, and shape (Magnusson, 1999; Einar et al., 2000). Nevertheless, to produce smaller nanoparticles, sintering is imperative. An evaporation-condensation method for the synthesis of NPs is shown in Fig. 1. This method produces uniform size distributed and high-purity nanoparticles (i.e., free of chemicals). However, the downsides of the technology are its low yield (Zhang et al., 2016), high energy requirements (Magnusson, 1999), and longer time required to achieve thermal stability (El-nour et al., 2010). Other physical methods used to synthesize AgNPs include laser ablation and arc-discharge. Their advantage compared to other conventional methods is the absence of chemicals, leading to high purity NPs (Tien et al., 2008; Natsuki et al., 2015).

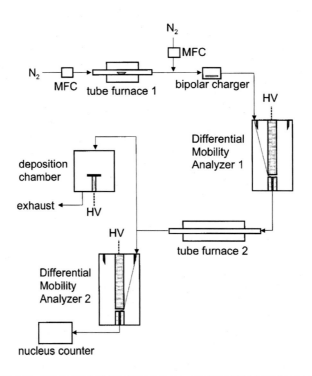

FIG. 1

Evaporation-condensation synthesis of NPs. The process consists of a tube furnace (1) for aerosol production, a bipolar charger, a differential mobility analyzer (DMA) for size fractionation, a tube furnace (2) for sintering, and a condensation nucleus counter.

Reprinted from Einar, F., Fissan, H., Rellinghaus, B., 2000. Sintering and evaporation characteristics of gas-phase synthesis of size-selected PbS nanoparticles. Mater. Sci. Eng. B 69–70, 329–334, with permssion from Springer.

2.2 Green chemical and bio-mediated methods

The synthesis via biological precursors has been extensively explored due to the environmental concerns imposed by the use of strong reducing agents. The biosynthesis of AgNPs is accomplished through algae, fungal, bacterial, and plant-mediated methods (Vigneshwaran et al., 2007; Logeswari et al., 2015; Ameen et al., 2020). Typically, the plant-based reducing agents, characterized by strong antioxidants, are extracted from fruits, leaves, seeds, peppers, and barks (Nayak et al., 2016; Elemike et al., 2017). Several plants, including garlic, green tea, turmeric, *Aloe vera*, *Cocos nucifera*, lemon, pine, magnolia, and ginkgo have also been investigated for the bio-mediated synthesis of AgNPs (Prabha et al., 2010; Kathiravan et al., 2014; Najimu Nisha et al., 2014; Paulkumar et al., 2014; Vanaja et al., 2014). In essence, the advantages of plants for the synthesis of the AgNPs primarily rely on their abundant availability; in addition, their extracts are characterized by a variety of active and reducing functional groups (Ahmed et al., 2016; Nayak et al., 2016; Rasheed

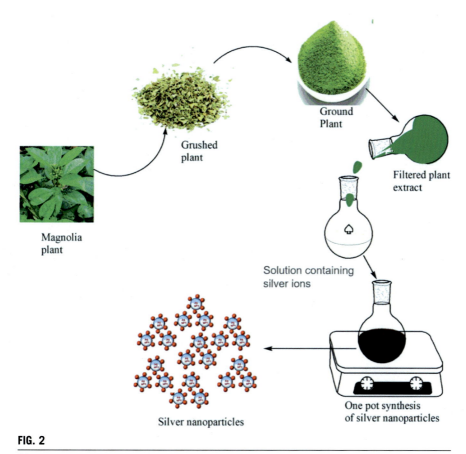

FIG. 2

Plant-mediated synthesis of AgNPs.

et al., 2017; Nthunya et al., 2019). Typically, the solution containing the reducing functional groups is extracted from dried plant powder. To reduce the Ag^+ to their AgNPs counterpart, the extract is mixed with the Ag^+ solution and stirred until a color change is observed. Typically, the color changes from light to dark brown. Fig. 2 presents the schematic illustration for the plant-mediated synthesis of AgNPs. The formation of the NPs is subsequently confirmed by a series of techniques including UV-Vis, XRD, TEM, and EDS (Gong et al., 2007; Ayaz et al., 2009; Nthunya et al., 2016, 2017, 2019; Fatima et al., 2020).

Several studies have reported the plant-mediated synthesis of AgNPs. For instance, Nayak et al. investigated the biosynthesis of AgNPs using bark extracts of *Ficus benghalensis* and *Azadirachta indica*. Interestingly, the synthesis was conducted without the use of any external reducing or capping agent. Partially agglomerated AgNPs with an average particle size of 40 and 50 nm were recorded for nanoparticles synthesized from *F. benghalensis* and *A. indica*, respectively

(Nayak et al., 2016). Similarly, Kathiravan et al. (2014) synthesized AgNPs from *Melia dubia* leaf extract. The NPs were characterized by irregular shapes, as seen by scanning electron microscopy (SEM) analysis. Some smaller particles were spherical, while other agglomerated particles showed undefined shapes. A wide range of nanoparticle size distribution was recorded, leading to an average particle size of 7.3 nm. Another study reported numerous synthesis conditions to address the agglomeration of AgNPs produced from the fruit extract of *Tanacetum vulgare* (Prabha et al., 2010). The pH of the solution was varied from acidic to basic conditions. Notably, smaller AgNPs were produced under basic conditions. These smaller particles are known to aggregate and form stable clusters in solution to reduce their surface free energy (Nthunya et al., 2019). Additionally, small-sized AgNPs were produced at lower metal ion concentrations. Therefore, partially stable and well dispersed AgNPs were produced under acidic conditions and higher ion concentrations, leading to the formation of triangular and spherical NPs with an average size of 16 nm (Nthunya et al., 2019). Therefore, the AgNPs synthesized from plant extract were energetically unstable and readily agglomerated.

While minimization of environmental concerns and the implementation of the sustainable process are desirable, the synthesis of stable and highly dispersed AgNPs is imperative. Besides the need for the application of the 12 principles of green chemistry and engineering, the novel properties of AgNPs does not only rely on their methods of synthesis and quantum size-effects, but also their morphological shapes and distribution (Ray, 2010). Therefore, an experimental approach under controlled conditions is of paramount importance. Some studies have investigated the use of polymeric materials acting as both reducing and capping agents (Hebeish and Emam, 2010). For instance, carboxymethyl cellulose (CMC) derivatives were used as a reducing and stabilizing agent at various conditions for the synthesis of AgNPs. Nonagglomerated AgNPs with narrow size distribution (average particle size ~15 nm) were obtained at a pH of 12 (Hebeish and Emam, 2010). Maciollek and Ritter evaluated one-pot synthesis of AgNPs using a copolymer from N-isopropylacrylamide (NIPAM) and mono-(1H-triazolylmethyl)-2-methylacryl-β-cyclodextrin acting as reductant and stabilizer (Maciollek and Ritter, 2014). The NPs were spherical, and their diameter was 12.5 ± 2.3 nm with a narrow size distribution (Maciollek and Ritter, 2014). However, these methods are affected by the low production rate of AgNPs.

The microorganism-based synthesis of AgNPs has emerged as an exciting green approach. Although a precise synthesis mechanism employing microorganisms for the synthesis of AgNPs has not been completely understood, the reduction process is believed to occur intracellularly or extracellularly (Nair and Pradeep, 2002; Natarajan et al., 2010; Kumar and Mamidyala, 2011; Das et al., 2014). During the intracellular synthesis, the Ag^+ ions migrate from the solution phase to the surface of the microorganism (e.g., bacteria) where the ions interact with the negative charge of the cell wall (Hulkoti and Taranath, 2014). The bacterial enzymes reduce the Ag^+ ions to their respective NPs. For instance, *Lactobacillus* strains from buttermilk were used to synthesize AgNPs (Nair and Pradeep, 2002). The growth of NPs occurred by the coalescence of clusters that accumulated within the bacteria. The formation of crystals through nucleation showed no effect on bacterial viability. The surface of the NPs was reduced by coalescence, thus effectively protecting the bacterial cell from damage (Nair and Pradeep, 2002). The harvested AgNPs were characterized by different shapes and sizes.

The microbial extracellular synthesis of AgNPs is essentially driven by nitrate reductase as the reducing agent (Gopinath and Velusamy, 2013; Hulkoti and Taranath, 2014). Briefly, the enzymatic reductase from fungi is released to the Ag^+ ions-containing solution, which subsequently reduces the latter to their respective NPs (Kumar and Mamidyala, 2011; Mishra and Singh, 2015). Similarly, metal-tolerant bacterial strains are capable of reducing Ag^+ ions to AgNPs. For instance, a strain of *Cupriavidus* isolated from soil successfully mediated the synthesis of AgNPs extracellularly (Ameen et al., 2020). The synthesis mechanism through extracellular enzymatic reduction was investigated by Fourier transform infrared (FTIR) analysis (Ameen et al., 2020). Interestingly, the spherical NPs were evenly distributed with particle sizes in the range of 10–50 nm. In another study, the supernatant extracted from *Pseudomonas aeruginosa* strain BS-161R was reported to biomediate the cost-effective green synthesis of AgNPs (Kumar and Mamidyala, 2011). The FT-IR results showed that the presence of a protein component (i.e., enzyme nitrate reductase and rhamnolipids) isolated from the culture supernatant was responsible for the reduction of Ag^+ ions into their respective NPs. The AgNPs synthesized via this method were monodispersed and spherical. The NPs were evenly distributed with an average particle size of 13 nm (Kumar and Mamidyala, 2011).

Although the microorganism-mediated synthesis of AgNPs offers a green route for the production of AgNPs, there are several challenges associated with this technology. For instance, while the supernatant is believed to increase the growth rate of the AgNPs, product stability is compromised. Similarly, this process requires the use of whole cells to avoid required coenzymes such as nicotinamide adenine dinucleotide hydride (NADH), nicotinamide adenine dinucleotide phosphate hydrogen (NADPH), and flavin adenine dinucleotide (FAD), which are costly (Kumar and Mamidyala, 2011; Prabhu and Poulose, 2012; Hulkoti and Taranath, 2014). Nonetheless, the extraction and purification of the intracellular or extracellular-produced metal NPs is imperative. Specifically, an additional treatment process or extraction with suitable detergent is required for the extraction of intracellularly produced AgNPs (Natarajan et al., 2010; Kumar and Mamidyala, 2011). The exploration of extraction processes such as heating and freeze-thawing has been observed to interfere with the microscopic properties of the NPs as well as the formation of aggregates and precipitates (Iravani, 2014). To address the low rate recoveries of AgNPs, Saifuddin et al. investigated a greener biosynthesis approach using combined culture supernatant of *Bacillus subtilis* and microwave irradiation in the presence of water (Saifuddin et al., 2009). Further details on irradiation-assisted bio-mediated synthesis of AgNPs are briefly discussed in the forthcoming sections.

2.3 Microwave-assisted methods

The fast synthesis of AgNPs has previously been conducted using strong and toxic reducing agents (e.g., sodium borohydride), leading to environmental concerns (Guzmán et al., 2009; Rai et al., 2009). The toxicity and environmental impacts imposed by these chemical methods led to the argument for greener methods. Briefly, green methods involve the use of plant extracts, which are mostly utilized in reducing and capping agents. However, the use of plant extracts leads to slow rates of NPs synthesis compared to strong reducing agents. Therefore, the microwave-assisted synthesis of AgNPs has proven to be a viable option. Notably, microwave irradiation induces rapid and uniform heat distribution within the reaction chamber, leading to homogeneous nucleation, growth, and quicker formation of monodispersed NPs (Pal et al., 2009; Wang et al., 2016; Nthunya et al., 2019). During the microwave-assisted method, the plant extract (e.g., citrate) and the silver nitrate solution are transferred to the microwave reaction vials and inserted into the microwave reaction chamber. The microwave is subsequently operated at high power and frequency (e.g., 800 W and 2450 MHz) until a solution color change is observed. During the synthesis process, the AgNPs form by a nucleation and growth process (Li and Kaner, 2006). When the plant extract is irradiated with the microwave rays, electrons are released. These electrons reduce Ag^+ ions to zero-valent Ag^0, leading to the formation of the Ag nuclei. The continuous reduction of the Ag^+ ions leads to the growth of AgNPs (Pal et al., 2009; Nthunya et al., 2019).

To study the effect of microwave irradiation on the rate of AgNPs, Nthunya et al. compared one-pot and microwave-assisted methods using apple extract (Nthunya et al., 2019; Reddy et al., 2021). The synthesis procedures are shown in Fig. 3. Remarkably, the microwave-assisted method produces highly dispersed AgNPs with smaller particle sizes compared to those produced using the thermally-assisted one-pot method (Nthunya et al., 2019). Interestingly, AgNPs were produced within 2.5 h during microwave-assisted reduction, while those produced in a thermally assisted one-pot reduction took 96 h (4 days) for complete reduction (Nthunya et al., 2019).

2.4 UV-assisted methods

The UV photochemical synthesis of AgNPs has been widely reported in the literature. Briefly, the UV synthesis of AgNPs is conducted in the absence of assisting chemical reducing agents. The reduction of Ag^+ ions to NPs is influenced by the UV irradiation of the $AgNO_3$ solution. Specifically, when water molecules are irradiated by UV light, solvated electrons are produced and subsequently reduce Ag^+ ions to AgNPs (Yu et al., 2019). This phenomenon has been confirmed in a previous study where vaporized water was irradiated with UV light to release electrons. These electrons are accepted by the positively-charged Ag^+ ions. During the electron acceptance process, the Ag^+ ions were reduced to Ag^0, thus forming AgNPs (Nthunya et al., 2016). The schematic illustration of this reduction method was reported by Nthunya et al., as shown in Fig. 4. Although this method is

2 Green synthesis methods   227

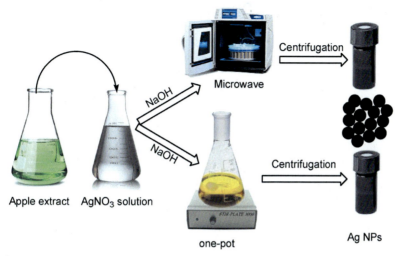

FIG. 3

Comparative synthesis of AgNPs using thermally-assisted one-pot and microwave-assisted reduction methods.

Reproduced from Nthunya, L.N., Derese, S., Gutrierrez, L., Verliefde, A.R., Mamba, B.B., Barnard, T.G., Mhlanga, S.D., 2019. Green synthesis of silver nanoparticles using one-pot and microwave-assisted methods and their subsequent embedment on PVDF nanofiber membranes for growth inhibition of mesophilic and thermophilic bacteria. New J. Chem. 43, 4168–4180, with permission from The Royal Society of Chemistry.

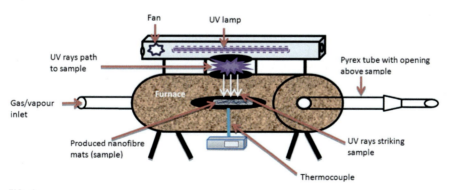

FIG. 4

UV/vapourised water synthesis of AgNPs.

Reproduced from Nthunya, L.N., Masheane, M.L., Malinga, S.P., Barnard, T.G., Nxumalo, E.N., Mamba, B.B., Mhlanga, S.D., 2016. UV-assisted reduction of in situ electrospun antibacterial chitosan-based nanofibres for removal of bacteria from water. RSC Adv. 6, 95,936–95,943, with permission from The Royal Society of Chemistry.

environmentally friendly, the rate of reduction is slow. For this reason, green chemicals-UV-assisted synthesis of AgNPs has been studied. In addition, capping reagents are simultaneously used for the reduction and dispersion of AgNPs. For instance, Omrania and Taghavinia (2012) reported the UV-assisted synthesis of AgNPs supported on cellulose fibers. During the UV irradiation on photo-absorptive cellulose, the active sites (i.e., oxygen bonds between the glucose rings) were broken down to form aldehydes. The produced aldehydes reduced the Ag^+ ions to their respective NPs while oxidized to carboxylic groups (Omrania and Taghavinia, 2012).

In addition to plant-extracted organic acids, photocatalysts such as ferritin are used to accelerate the synthesis of AgNPs (Nikandrov, 1997). Ferritin is characterized by a hollow interior and semiconductive properties. During UV irradiation, the electrons are excited from the valence to the conduction band of the semiconductor, where ferritin acts as a charge separator. The released electrons reduce the Ag^+ ions that bind on the exterior surface of ferritin (Watt et al., 1992). Remarkably, the plant-extracted organic acids are used to replenish the electron holes created by charge separation (Petrucci et al., 2019). Thus, the reduction process becomes continuous until Ag^+ ions are reduced to their NPs counterpart. This reduction process is faster compared to the sole use of extracted organic acids. The reaction process takes place within 1–5 min. Furthermore, this synthesis route avoids the use of toxic chemicals. An illustration of the photochemical synthesis of AgNPs using iron-containing ferritin and citric acid is presented in Fig. 5.

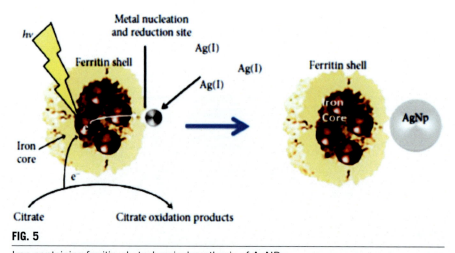

FIG. 5

Iron-containing ferritin photochemical synthesis of AgNPs.

Reprinted from Petrucci, O.D., Hilton, R.J., Farrer, J.K., Watt, R.K., 2019. A ferritin photochemical synthesis of monodispersed silver nanoparticles that possess antimicrobial properties. J. Nanomater (2019), 1–8, with permission from MDPI Publisher, Licensee MDPI, Basel, Switzerland. This article is an open-access article distributed under the terms and conditions of the Creative Commons Attribution (CC BY) license (http://creativecommons.org/licenses/by/4.0/).

The summary of different methods using bio-mediated synthesis of AgNPs is presented in Table 1. These methods include the use of algae, fungi, bacteria, plant extracts, and polysaccharides. Polysaccharides are known to act as both a reducing agent and a stabilizer. Based on microscopic analysis, the NPs synthesized by this method were predominantly spherical, with few indications of face-centered cubic

Table 1 Bio-mediated synthesis of AgNPs and their corresponding microscopic properties.

Bio-reagent (source of reducing agent) and assistive methods	Size (nm)	Shape	Max reduction time	λ_{max} (nm)	Reference
Algae: *Parachlorella kessleri*	5–20	Spherical	4 days	442	Velgosova et al. (2018)
Algae: *Portieria hornemannii*	70–75	Spherical	–	430	Fatima et al. (2020)
Green algae: *Caulerpa serrulata*	10±2	Spherical	8 days	412	Aboelfetoh et al. (2017)
Red marine algae: *Corallina elongata*	12–20	Spherical	20 days	420	Hamouda et al. (2019)
Bacterial exopolysaccharide	35	Spherical	30 days	425	Hamouda et al. (2019)
Supernatant: *Pseudomonas aeruginosa*	13	Spherical	–	430	Kumar and Mamidyala (2011)
Bacterial: *Bacillus sp.*	42–92	Spherical	3 days	450	Das et al. (2014)
Soil bacteria: *Cupriavidus sp.*	10–50	Spherical	–	500	Ameen et al. (2020)
Fungus: *Amylomyces rouxii*	5–27	Mixed	3 days	410	Musarrat et al. (2010)
Ligninolytic fungus: *Trametes trogii*	5–65	Clustered	5 days	430	Kobashigawa et al. (2019)
Fungus: *Alternaria alternata*	20–60	Spherical	2 days	430	Gajbhiye et al. (2009)
Fungus: *Aspergillus flavus*	8.92±1.61	Spherica	3 days	420	Vigneshwaran et al. (2007)
Melia dubia leaf extract	5–35	Flat plate-like	–	448	Kathiravan et al. (2014)
Lippia citriodora leaf extract	23.8	Platelets-like	3 h	417	Elemike et al. (2017)
Artemisia vulgaris leaf extract	25	Irregular-shaped	2 h	420	Rasheed et al. (2017)

Continued

Table 1 Bio-mediated synthesis of AgNPs and their corresponding microscopic properties. *Continued*

Bio-reagent (source of reducing agent) and assistive methods	Size (nm)	Shape	Max reduction time	λ_{max} (nm)	Reference
Maltose disaccharide sugar	3.7 ± 1.0	Spherical	24 h	436	Oluwafemi et al. (2013)
Heparin polysaccharide	10–28	Spherical	8 h	401	Huang and Yang (2004)
Carboxymethyl cellulose	10–25	Spherical	2 h	405	Hebeish and Emam (2010)
Glycyrrhiza polysaccharide	20–50	Spherical	48 h	423	Cai et al. (2019)
Cyclodextrin	30–100	Cubic	24 h	450	Maciollek and Ritter (2014)
Microwave-irradiated + apple extract	22.1 ± 1.0	Spherical	2 h	396	Nthunya et al. (2019)
UV-irradiated + vaporized water	38.8 ± 8.2	Spherical	3 h 30 min	399	Nthunya et al. (2017)
Sunlight-irradiated + peach gum polysaccharide	23.5 ± 7.8	Cubic	75 min	440.0	Yang et al. (2015)

shape. Additionally, the sizes of the NPs ranged between 5 and 50 nm, with few exceeding 60 nm. Based on the collected literature, the rate of AgNPs synthesis was slower in microorganism-bio-mediated synthesis methods. An improved rate of NPs production was observed in the use of plant extract-based methods. Interestingly, lower conversion rates were recorded on irradiation-assisted methods.

3 Applications in agro-ecosystems

The agricultural sector has been negatively affected by an increasing number of pests developing resistance against the existing chemical pesticides. Therefore, current research is driven to the exploitation and evaluation of AgNPs for their application in agro-farming. Although AgNPs present a possible breakthrough in addressing the challenges associated with persistent pests, the development of accurately-synthesized NPs of desired morphologies and very low impact on the biological properties of the biomolecular matrix of plants is imperative. In addition to pests developing resistance against pesticides, the excessive usage of pesticides affects soil biodiversity and kills birds and useful microbes. While AgNPs are also not selective to essential soil microbes, their controlled release is a key parameter for their successful use in agriculture. AgNPs have been widely investigated for their antimicrobial activities against a wide range of human pathogens (Son et al., 2006; Phong et al., 2009; Li et al., 2010; Lara et al., 2011; Abdelgawad et al., 2014; Elemike et al., 2017; Nthunya et al., 2019; Dawadi et al., 2021). However, their intensive application in the agricultural sector is not yet established. Nonetheless, due to their biocidal effects, AgNPs could be efficiently used in agriculture (e.g., plant disease management) (Mishra and Singh, 2015). The reviewed studies reported by Benelli et al. showed that AgNPs have been used for the treatment of agricultural pests and health-affecting ticks (Fig. 6). Interestingly, upon the treatment of aquatic environments using AgNPs, the growth of nontargeted mosquito larvae decreased, leading to the control of life-threatening mosquitos (Benelli, 2016; Benelli et al., 2017).

Several studies have advocated the antimicrobial activity of AgNPs to harness several plant pathogens (Lee et al., 2013; Mishra and Singh, 2015; Elemike et al., 2017; Das et al., 2018). Owing to their large surface areas relative to their volumes, AgNPs would increase their reactivity at the ecological level, causing adverse health effects to all sorts of insects. Afrasiabi et al. (2016) studied the effects of AgNPs on *Heliothis virescens* (tobacco budworm) and *Trichoplusia ni* (cabbage looper). The results showed no mortality of the investigated pests. However, delayed development, reduced weights, increased egg-laying timeframe, and decreased number of

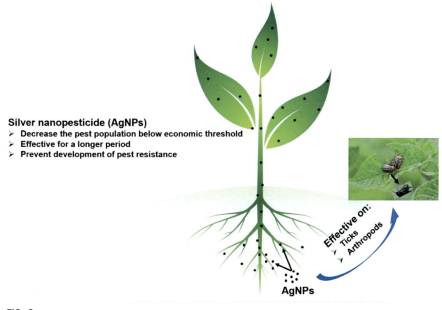

FIG. 6

The use of AgNPs for the treatment of pests and ticks.

laid eggs were reported. Park et al. (2007) studied the toxicity of Ag-based NPs against various phytopathogens fungi such as *Pseudomonas syringae*, *Pythium sp.*, *Xanthomonas compestris*, and *Colletotrichum* sp. Additional comprehensive and yet concise studies demonstrating the applicability of AgNPs for the management of plant pathogens are summarized in Table 2.

Table 2 Synthesis of AgNPs via greener approach for agricultural use.

Silver source and reduction method	Particle size and shape	Target plant pathogen	Reference
$AgNO_3$: PVP-capped AgNPs	25 ± 5 nm, spherical	Earthworm *Eisenia fetida*	Amorim et al. (2018)
$AgNO_3$: In vitro biosynthesis using marine bacteria *Shewanella algae bangaramma*	15 nm, spherical	Plant pest: *Lepidiota mansueta* larva	Babu et al. (2014)
$AgNO_3$: γ-irradiation	Spherical	Plant *Arabidopsis thaliana*	Chu et al. (2012)
$AgNO_3$: Leaf and stem extract of *Piper nigrum* using one-pot method	7–50 nm,	Plant bacteria: *Citrobacter freundii* and *Erwinia cacticida*	Paulkumar et al. (2014)
$AgNO_3$: Chitosan-capped AgNPs	100 nm,	Plant fungus: *Botrytis cinerea*	Moussa et al. (2013)
$AgNO_3$: Cow milk in one-pot method	30–90 nm, aggregated	Phytopathogens: *Colletotrichum coccodes*, *Monilinia sp.*, and *Pyricularia sp.*	Lee et al. (2013)
$AgNO_3$: In vitro biosynthesis using *Spirulina platensis*	50–77 nm, agglomerated	Phytogens: *P. solanocearum*, *P. syringae*, *X. malvacearum*, and *X. campestris*	Patra and Baek (2017)
$AgNO_3$: Biosynthesis using *Trichoderma viride*	5–40 nm, spherical and rod-like	Fruit preservation: *Escherichia coli* and *Staphylococcus aureus*	Ayaz et al. (2009)
$AgNO_3$: One-pot synthesis using sodium citrate	17 nm, spherical	Phytogenss: *Oryza sativa L.* and rhizosphere bacteria	Mirzajani et al. (2013)
$AgNO_3$: One-pot synthesis using *Thuja occidentalis* plant leaf extract	9.8 ± 0.15 nm,	Earthworm: *Eisenia fetida*	Das et al. (2018)
$AgNO_3$: Extracellular synthesis using *Bacillus* sp. GP-23	7–21 nm, spherical	Plant pathogenic fungus: *Fusarium oxysporum*	Gopinath and Velusamy (2013)
2 silver electrodes: High-voltage arc discharge	5–65 nm, spherical	Phytopathogen: *Fusarium culmorum*	Kasprowicz et al. (2010)

The research findings presented by Xu and Zhang demonstrated that the use of AgNPs in agriculture does not only combat pesticide-resistant pests but also induce adverse effects of the microbes beneficial to the soil. Upon 1-year soil exposure to 0.01 mg/kg AgNP, a significant decrease in microbial biomass, the abundance of total soil bacteria, and soil microbes responsible for nitrogen cycling were reported (Xu and Zhang, 2018). Therefore, the controlled release of AgNPs and the regulations governing their use in the environment must be exercised.

4 Conclusion

The synthesis of AgNPs using a variety of physical and chemical methods has been extensively reported. In previous decades, the synthesis of AgNPs was conducted using strong and toxic reducing agents. However, environmental and toxicological concerns demand the use of safer and cost-effective methods while maintaining the production of AgNPs with required morphologies and sizes. Therefore, this chapter provided an overview of bio-mediated and other greener syntheses of AgNPs. The reviewed methods include physical, chemical, in vitro and in vivo microorganism-based, plant extracts, and irradiation-assisted methods. The microorganism- and plant extracts-based methods showed a synthesis approach with partially agglomerated AgNPs. Upon electromagnetic irradiation of Ag^+ ions in the presence of green reducing agents, uniformly-distributed AgNPs were recorded. Likewise, the irradiation-assisted method decreased the overall time required for the synthesis of AgNPs, indicating that a higher output could be obtained in a shorter time.

The AgNPs produced via these greener techniques demonstrated the potential to combat pesticide-resistant agricultural pests, thus ensuring a considerably lower impact on agricultural production. However, the biocidal AgNPs were reported as nonselective and therefore affected the soil biota by killing denitrifying bacteria. Technically, the destruction of bacteria affects the nitrogen cycle, which includes the decomposition of organic matter into nutritional nitrogen required by the plants. Therefore, a controlled release of AgNPs to combat pests, as well as regulations and legislative bodies governing the environmental use of AgNPs, is imperative.

References

Abdelgawad, A.M., Hudson, S.M., Rojas, O.J., 2014. Antimicrobial wound dressing nanofiber mats from multicomponent (chitosan/silver-NPs/polyvinyl alcohol) systems. Carbohydr. Polym. 100, 166–178.

Aboelfetoh, E.F., El-shenody, R.A., Ghobara, M.M., 2017. Eco-friendly synthesis of silver nanoparticles using green algae (*Caulerpa serrulata*): reaction optimization, catalytic and antibacterial activities. Environ. Monit. Assess. 189, 1–15.

Afrasiabi, Z., Popham, H.J.R., Stanley, D., Suresh, D., Finley, K., Campbell, J., Kannan, R., Upendra, A., 2016. Dietary silver nanoparticles reduce fitness in a beneficial, but not pest, insect species. Arch. Insect Biochem. Physiol. 93, 190–201.

Ahmed, S., Ahmad, M., Swami, B.L., Ikram, S., 2016. A review on plants extract mediated synthesis of silver nanoparticles for antimicrobial applications: a green expertise. J. Adv. Res. 7, 17–28.

Ameen, F., Alyahya, S., Govarthanan, M., Aljahdali, N., Al-enazi, N., 2020. Soil bacteria *Cupriavidus* sp. mediates the extracellular synthesis of antibacterial silver nanoparticles. J. Mol. Struct. 1202, 127233.

Amorim, J.B., Jensen, J., Scott-Fordsm, J.J., Mariyadas, J., 2018. Earthworm avoidance of silver nanomaterials over time. Environ. Pollut. 239, 751–756.

Armstrong, J.L., Shigeno, D.S., Calomiris, J.J., Seidler, R.J., 1981. Antibiotic-resistant bacteria in drinking water. Appl. Environ. Microbiol. 42, 277–283.

Ayaz, A.M.O.F., Alaji, K.B., Irilal, M.G., Alaichelvan, P.T.K., Enkatesan, R.V., 2009. Myco-based synthesis of silver nanoparticles and their incorporation into sodium alginate films for vegetable and fruit preservation. J. Agric. Food Chem. 57, 6246–6252.

Babu, M.Y., Devi, V.J., Ramakritinan, C.M., Umarani, R., Taredahalli, N., 2014. Application of biosynthesized silver nanoparticles in agricultural and marine application of biosynthesized silver nanoparticles in agricultural and marine pest control. Curr. Nanosci. 10, 1–8.

Benelli, G., 2016. Green synthesized nanoparticles in the fight against mosquito-borne diseases and cancer — a brief review. Enzym. Microb. Technol. 95, 58–68.

Benelli, G., 2018. Mode of action of nanoparticles against insects. Environ. Sci. Pollut. Res. 25, 12329–12341.

Benelli, G., Pavela, R., Maggi, F., Petrelli, R., Nicoletti, M., 2017. Commentary : making green pesticides greener? The potential of plant products for nanosynthesis and pest control. J. Clust. Sci. 28, 3–10.

Biswas, P., Bandyopadhyaya, R., 2016. Water disinfection using silver nanoparticle impregnated activated carbon: *Escherichia coli* cell-killing in batch and continuous packed column operation over a long duration. Water Res. 100, 105–115.

Budama, L., Çakır, B.A., Topel, Ö., Hoda, N., 2013. A new strategy for producing antibacterial textile surfaces using silver nanoparticles. Chem. Eng. J. 228, 489–495.

Cai, Z., Dai, Q., Guo, Y., Wei, Y., Wu, M., Zhang, H., 2019. Glycyrrhiza polysaccharide-mediated synthesis of silver nanoparticles and their use for the preparation of nanocomposite curdlan antibacterial film. Int. J. Biol. Macromol. 141, 422–430.

Chu, H., Kim, H., Su, J., Kim, M., Yoon, B., Park, H., Young, C., 2012. A nanosized Ag-silica hybrid complex prepared by γ-irradiation activates the defense response in Arabidopsis. Radiat. Phys. Chem. 81, 180–184.

Das, P., Barua, S., Sarkar, S., Kumar, S., Mukherjee, S., Goswami, L., Das, S., Bhattacharya, S., Karak, N., 2018. Mechanism of toxicity and transformation of silver nanoparticles: inclusive assessment in earthworm-microbe-soil-plant system. Geoderma 314, 73–84.

Das, L.V., Thomas, R., Varghese, R.T., Soniya, E.V., Mathew, J., Radhakrishnan, E.K., 2014. Extracellular synthesis of silver nanoparticles by the Bacillus strain CS 11 isolated from industrialized area. Biotechnology 4, 121–126.

Dawadi, S., Katuwal, S., Gupta, A., Lamichhane, U., Thapa, R., Jaisi, S., Lamichhane, G., Bhattarai, D.P., Parajuli, N., 2021. Current research on silver nanoparticles: synthesis, characterization, and applications. J. Nanomater. 2021. https://doi.org/10.1155/2021/6687290, 6687290.

Einar, F., Fissan, H., Rellinghaus, B., 2000. Sintering and evaporation characteristics of gas-phase synthesis of size-selected PbS nanoparticles. Mater. Sci. Eng. B 69–70, 329–334.

Elemike, E.E., Onwudiwe, D.C., Ekennia, A.C., Ehiri, R.C., Nnaji, N.J., 2017. Photosynthesis of silver nanoparticles using aqueous leaf extracts of *Lippia citriodora* : antimicrobial, larvicidal and photocatalytic evaluations. Mater. Sci. Eng. C 75, 980–989.

El-nour, K.M.M.A., Al-warthan, A., Ammar, R.A.A., 2010. Synthesis and applications of silver nanoparticles. Arab. J. Chem. 3, 135–140.

Espino-pérez, E., Bras, J., Almeida, G., Plessis, C., Belgacem, N., Perré, P., Domenek, S., 2018. Designed cellulose nanocrystal surface properties for improving barrier properties in polylactide nanocomposites. Carbohydr. Polym. 183, 267–277.

Essawy, E., Abdelfattah, M.S., El-Matbouli, M., Saleh, M., 2021. Synergistic effect of biosynthesized silver nanoparticles and natural phenolic compounds against drug-resistant fish pathogens and their cytotoxicity: an in vitro study. Mar. Drugs 19 (1), 22.

Fatima, R., Priya, M., Indurthi, L., Radhakrishnan, V., Sudhakaran, R., 2020. Biosynthesis of silver nanoparticles using red algae *Portieria hornemannii* and its antibacterial activity against fish pathogens. Microb. Pathog. 138, 103780.

Gajbhiye, M., Kesharwani, J., Ingle, A., Gade, A., Rai, M., 2009. Fungus-mediated synthesis of silver nanoparticles and their activity against pathogenic fungi in combination with fluconazole. Nanomedicine 5, 382–386.

Gong, P., Li, H., He, X., Wang, K., Hu, J., Tan, W., Zhang, S., Yang, X., 2007. Preparation and antibacterial activity of Fe_3O_4@ag nanoparticles. Nanotechnology 18, 285604.

Gopinath, V., Velusamy, P., 2013. Extracellular biosynthesis of silver nanoparticles using Bacillus sp. GP-23 and evaluation of their antifungal activity towards *Fusarium oxysporum*. Spectrochim. Acta A Mol. Biomol. Spectrosc. 106, 170–174.

Guzmán, M., Dille, J., Godet, S., 2009. Synthesis of silver nanoparticles by chemical reduction method and their antibacterial activity. Int. J. Chem. 2, 104–111.

Hamouda, R.A., El-mongy, M.A., Eid, K.F., 2019. Comparative study between two red algae for biosynthesis silver nanoparticles capping by SDS : insights of characterization and antibacterial activity. Microb. Pathog. 129, 224–232.

Hebeish, A.A., Emam, H.E., 2010. Carboxymethyl cellulose for green synthesis and stabilization of silver nanoparticles. Carbohydr. Polym. 82, 933–941.

Huang, H., Yang, X., 2004. Synthesis of polysaccharide-stabilized gold and silver nanoparticles: a green method. Carbohydr. Res. 339, 2627–2631.

Huang, L., Zhao, S., Wang, Z., Wu, J., Wang, J., Wang, S., 2016. In situ immobilization of silver nanoparticles for improving permeability, antifouling and anti-bacterial properties of ultrafiltration membrane. J. Membr. Sci. 499, 269–281.

Hulkoti, N.I., Taranath, T.C., 2014. Biosynthesis of nanoparticles using microbes — a review. Colloids Surf. B Biointerfaces 121, 474–483.

Iravani, S., 2014. Bacteria in nanoparticle synthesis : current status. Int. Schol. Res. Not. 2014, 2–18.

Kasprowicz, M.J., Kozioł, M., Gorczyca, A., 2010. The effect of silver nanoparticles on phytopathogenic spores of *Fusarium culmorum*. Can. J. Microbiol. 56, 247–253.

Kathiravan, V., Ravi, S., Ashokkumar, S., 2014. Synthesis of silver nanoparticles from Melia dubia leaf extract and their in vitro anticancer activity. Spectrochim. Acta A Mol. Biomol. Spectrosc. 130, 116–121.

Kobashigawa, J.M., Robles, A.C., Martínez, L.M.R., Cristina, C., 2019. Influence of strong bases on the synthesis of silver nanoparticles (AgNPs) using the ligninolytic fungi *Trametes trogii*. Saudi J. Biol. Sci. 26, 1331–1337.

Krishnaraj, C., Jagan, E.G., Ramachandran, R., Abirami, S.M., Mohan, N., Kalaichelvan, P.T., 2012. Effect of biologically synthesized silver nanoparticles on *Bacopa monnieri* (Linn.) Wettst. Plant growth metabolism. Process Biochem. 47, 651–658.

Kumar, C.G., Mamidyala, S.K., 2011. Extracellular synthesis of silver nanoparticles using culture supernatant of Pseudomonas aeruginosa. Colloids Surf. B Biointerfaces 84, 462–466.

Lamsal, K., Kim, S.W., Jung, J.H., Kim, Y.S., Kim, K.S., Lee, Y.S., 2011. Application of silver nanoparticles for the control of colletotrichum species in vitro and pepper anthracnose disease in field. Mycobiology 39, 194–199.

Lara, H.H., Garza-Treviño, E.N., Ixtepan-Turrent, L., Singh, D.K., 2011. Silver nanoparticles are broad-spectrum bactericidal and virucidal compounds. J. Nanobiotechnol. 9, 1–8.

Lee, K., Park, S., Govarthanan, M., Hwang, P., Seo, Y., Cho, M., Lee, W., Lee, J., 2013. Synthesis of silver nanoparticles using cow milk and their antifungal activity against phytopathogens. Mater. Lett. 105, 128–131.

Li, D., Kaner, R.B., 2006. Shape and aggregation control of nanoparticles: not shaken, not stirred. J. Am. Chem. Soc. 128, 968–975.

Li, W.R., Xie, X.B., Shi, Q.S., Zeng, H.Y., Ou-Yang, Y.S., Chen, Y.B., 2010. Antibacterial activity and mechanism of silver nanoparticles on *Escherichia coli*. Appl. Microbiol. Biotechnol. 85, 1115–1122.

Logeswari, P., Silambarasan, S., Abraham, J., 2015. Synthesis of silver nanoparticles using plants extract and analysis of their antimicrobial property. J. Saudi Chem. Soc. 19, 311–317.

Maciollek, A., Ritter, H., 2014. One-pot synthesis of silver nanoparticles using a cyclodextrin containing polymer as reductant and stabilizer. Beilstein J. Nanotechnol. 5, 380–385.

Magnusson, H., 1999. Size-selected nanoparticles by aerosol technology. Nanostruct. Mater. 12, 45–48.

Mavani, K., Shah, M., 2013. Synthesis of silver nanoparticles by using sodium borohydride as a reducing agent. Int. J. Eng. Res. Technol. 2, 1–5.

Mirzajani, F., Askari, H., Hamzelou, S., Farzaneh, M., Ghassempour, A., 2013. Effect of silver nanoparticles on Oryza sativa L. and its rhizosphere bacteria. Ecotoxicol. Environ. Saf. 88, 48–54.

Mishra, S., Singh, H.B., 2015. Biosynthesized silver nanoparticles as a nanoweapon against phytopathogens: exploring their scope and potential in agriculture. Appl. Microbiol. Biotechnol. 99, 1097–1107.

Moussa, S.H., Tayel, A.A., Alsohim, A.S., Abdallah, R.R., 2013. Botryticidal activity of Nanosized silver-chitosan composite and its application for the control of gray Mold in strawberry. J. Food Sci. 78, 1589–1594.

Musarrat, J., Dwivedi, S., Raj, B., Al-khedhairy, A.A., Azam, A., 2010. Production of antimicrobial silver nanoparticles in water extracts of the fungus *Amylomyces rouxii* strain KSU-09. Bioresour. Technol. 101, 8772–8776.

Nair, B., Pradeep, T., 2002. Coalescence of nanoclusters and formation of submicron crystallites assisted by Lactobacillus strains. Cryst. Growth Des. 2, 293–298.

Najimu Nisha, S., Aysha, O.S., Syed Nasar Rahaman, J., Vinoth Kumar, P., Valli, S., Nirmala, P., Reena, A., 2014. Lemon peels mediated synthesis of silver nanoparticles and its antidermatophytic activity. Spectrochim. Acta A Mol. Biomol. Spectrosc. 124, 194–198.

Nangmenyi, G., Li, X., Mehrabi, S., Mintz, E., Economy, J., 2011. Silver-modified iron oxide nanoparticle impregnated fibreglass for disinfection of bacteria and viruses in water. Mater. Lett. 65, 1191–1193.

Natarajan, K., Selvaraj, S., Murty, V.R., 2010. Microbial production of silver nanoparticles. Dig. J. Nanomater. Biostruct. 5, 135–140.

Natsuki, J., Natsuki, T., Hashimoto, Y., 2015. A review of silver nanoparticles : synthesis methods, properties and applications. Int. J. Mater. Sci. Appl. 4, 325–332.

Nayak, D., Ashe, S., Rauta, P.R., Kumari, M., Nayak, B., 2016. Bark extract mediated green synthesis of silver nanoparticles : evaluation of antimicrobial activity and antiproliferative response against osteosarcoma. Mater. Sci. Eng. C 58, 44–52.

Nikandrov, V.V., 1997. Light-induced redox reactions involving mammalian ferfitin as photocatalyst. J. Photochem. Photobiol. B Biol. 41, 83–89.

Nthunya, L.N., Derese, S., Gutrierrez, L., Verliefde, A.R., Mamba, B.B., Barnard, T.G., Mhlanga, S.D., 2019. Green synthesis of silver nanoparticles using one-pot and microwave-assisted methods and their subsequent embedment on PVDF nanofibre membranes for growth inhibition of mesophilic and thermophilic bacteria. New J. Chem. 43, 4168–4180.

Nthunya, L.N., Masheane, M.L., Malinga, S.P., Barnard, T.G., Nxumalo, E.N., Mamba, B.B., Mhlanga, S.D., 2016. UV-assisted reduction of in situ electrospun antibacterial chitosan-based nanofibres for removal of bacteria from water. RSC Adv. 6, 95936–95943.

Nthunya, L.N., Masheane, M.L., Malinga, S.P., Nxumalo, E.N., Barnard, T.G., Kao, M., Tetana, Z.N., Mhlanga, S.D., 2017. Greener approach to prepare electrospun antibacterial β-cyclodextrin/cellulose acetate nanofibers for removal of bacteria from water. ACS Sustain. Chem. Eng. 5, 153–160.

Oluwafemi, O.S., Lucwaba, Y., Gura, A., Masabeya, M., Ncapayi, V., Olujimi, O.O., Songca, S.P., 2013. A facile completely 'green' size-tunable synthesis of maltose-reduced silver nanoparticles without the use of any accelerator. Colloids Surf. B: Biointerfaces 102, 718–723.

Omrania, A.A., Taghavinia, N., 2012. Photo-induced growth of silver nanoparticles using UV sensitivity of cellulose fibers. Appl. Surf. Sci. 258, 2373–2377.

Pal, A., Shah, S., Devi, S., 2009. Microwave-assisted synthesis of silver nanoparticles using ethanol as a reducing agent. Mater. Chem. Phys. 114, 530–532.

Pal, S., Tak, Y.K., Song, J.M., 2007. Does the antibacterial activity of silver nanoparticles depend on the shape of the nanoparticle ? A study of the gram-negative bacterium *Escherichia coli*. Appl. Environ. Microbiol. 73, 1712–1720.

Park, H.J., Kim, S.H., Kim, H.J., Choi, S.H., 2007. A new composition of nanosized silica-silver for control of various plant diseases. Plant Pathol. J. 22, 295–302.

Patra, J.K., Baek, K.-H., 2017. Antibacterial activity and synergistic antibacterial potential of biosynthesized silver nanoparticles against foodborne pathogenic bacteria along with its anticandidal and antioxidant effect. Front. Microbiol. 8, 1–14.

Paulkumar, K., Gnanajobitha, G., Vanaja, M., Rajeshkumar, S., Malarkodi, C., Pandian, K., Annadurai, G., 2014. *Piper nigrum* leaf and stem assisted green synthesis of silver nanoparticles and evaluation of its antibacterial activity against agricultural plant pathogens. Sci. World J. 2014, 1–9.

Petrucci, O.D., Hilton, R.J., Farrer, J.K., Watt, R.K., 2019. A ferritin photochemical synthesis of monodispersed silver nanoparticles that possess antimicrobial properties. J. Nanomater. 2019, 1–8.

Phong, N.T.P., Thanh, N.V.K., Phuong, P.H., 2009. Fabrication of antibacterial water filter by coating silver nanoparticles on flexible polyurethane foams. J. Phys. Conf. Ser. 187, 012079.

Prabha, S., Lahtinen, M., Sillanpää, M., 2010. Tansy fruit mediated greener synthesis of silver and gold nanoparticles. Process Biochem. 45, 1065–1071.

Prabhu, S., Poulose, E.K., 2012. Silver nanoparticles: mechanism of antimicrobial action, synthesis, medical applications, and toxicity effects. Int. Nano Lett. 2, 1–10.

Pradeep, T., Anshup, 2009. Noble metal nanoparticles for water purification: a critical review. Thin Solid Films 517, 6441–6478.

Rai, M., Yadav, A., Gade, A., 2009. Silver nanoparticles as a new generation of antimicrobials. Biotechnol. Adv. 27, 76–83.

Rasheed, T., Bilal, M., Iqbal, H.M.N., Li, C., 2017. Green biosynthesis of silver nanoparticles using leaves extract of *Artemisia vulgaris* and their potential biomedical applications. Colloids Surf. B: Biointerfaces 158, 408–415.

Ray, P.C., 2010. Size and shape-dependent second-order nonlinear optical properties of nanomaterials and their application in biological and chemical sensing. Chem. Rev. 110, 5332–5365.

Reddy, B., Dadigala, R., Bandi, R., Seku, K., Koteswararao, D., Shalan, A.E., 2021. Microwave-assisted preparation of a silver nanoparticles/N-doped carbon dots nanocomposite and its application for catalytic reduction of rhodamine B, methyl red and 4-nitrophenol dyes. RSC Adv. 11 (9), 5139–5148.

Saifuddin, N., Wong, C.W., Yasumira, A.A.N.U.R., 2009. Rapid biosynthesis of silver nanoparticles using culture supernatant of bacteria with microwave irradiation. E-J. Chem. 6, 61–70.

Son, W.K., Youk, J.H., Park, W.H., 2006. Antimicrobial cellulose acetate nanofibers containing silver nanoparticles. Carbohydr. Polym. 65, 430–434.

Tang, J., Sisler, J., Grishkewich, N., Tam, K.C., 2017. Journal of colloid and Interface science functionalization of cellulose nanocrystals for advanced applications. J. Colloid Interface Sci. 494, 397–409.

Tien, D.-C., Liao, C.-Y., Huang, J.-C., Tseng, K.-H., Lung, J.-K., Tsung, T.-T., Kao, W.-S., Tsai, T.-H., Cheng, T.-W., Yu, B.-S., Lin, H.-M., Stobinski, L., 2008. Novel technique for preparing a nano-silver water suspension by the arc-discharge method. Rev. Adv. Mater. Sci. 18, 750–756.

Vanaja, M., Paulkumar, K., Gnanajobitha, G., Rajeshkumar, S., Malarkodi, C., Annadurai, G., 2014. Herbal plant synthesis of antibacterial silver nanoparticles by *Solanum trilobatum* and its characterization. Int. J. Met. 2014, 1–8.

Velgosova, O., Mražíková, A., Ižmárová, E.Č., Málek, J., 2018. Green synthesis of ag nanoparticles : effect of algae life cycle on ag nanoparticle production and long-term stability. Trans. Nonferrous Metals Soc. China 28, 974–979.

Vigneshwaran, N., Ashtaputre, N.M., Varadarajan, P.V., Nachane, R.P., Paralikar, K.M., Balasubramanya, R.H., 2007. Biological synthesis of silver nanoparticles using the fungus *Aspergillus flavus*. Mater. Lett. 61, 1413–1418.

Wang, W., Liang, T., Zhang, B., Bai, H., Ma, P., Dong, W., 2018. Green functionalization of cellulose nanocrystals for application in reinforced poly (methyl methacrylate) nanocomposites. Carbohydr. Polym. 202, 591–599.

Wang, J., Sun, X., Yuan, Y., Chen, H., Wang, H., Hou, D., 2016. A novel microwave-assisted photo-catalytic membrane distillation process for treating the organic wastewater containing inorganic ions. J. Water Proc. Eng. 9, 1–8.

Watt, R.K., Frankeu, R.B., Watt, G.D., 1992. Redox reactions of Apo mammalian ferritin. Biochemistry 31, 9673–9679.

Wu, H., Nagarajan, S., Shu, J., Zhang, T., Zhou, L., Duan, Y., 2018. Green and facile surface modification of cellulose nanocrystal as the route to produce poly (lactic acid) nanocomposites with improved properties. Carbohydr. Polym. 197, 204–214.

Wu, J., Zheng, Y., Song, W., Luan, J., Wen, X., Wu, Z., Chen, X., Wang, Q., Guo, S., 2014. In situ synthesis of silver-nanoparticles/bacterial cellulose composites for slow-released antimicrobial wound dressing. Carbohydr. Polym. 102, 762–771.

Xu, J.Y., Zhang, H., 2018. Long-term effects of silver nanoparticles on the abundance and activity of soil microbiome. J. Environ. Sci. 69, 3–4.

Yang, N., Wei, X., Li, W., 2015. Sunlight irradiation-induced green synthesis of silver nanoparticles using peach gum polysaccharide and colorimetric sensing of H2O2. Mater. Lett. 154, 21–24.

Yu, Z., Wang, W., Dhital, R., Kong, F., Lin, M., Mustapha, A., 2019. Antimicrobial effect and toxicity of cellulose nanofibril/silver nanoparticle nanocomposites prepared by an ultraviolet irradiation method. Colloids Surf. B. Biointerfaces 180, 212–220.

Zhang, X.-F., Liu, Z.-G., Shen, W., Gurunathan, S., 2016. Silver nanoparticles: synthesis, characterization, properties, applications, and therapeutic approaches. Int. J. Mol. Sci. 17, 1534.

Zielinska, E., Tukaj, C., Radomski, M.W., Inkielewicz-Stepniak, I., 2016. Molecular mechanism of silver nanoparticles-induced human osteoblast cell death : protective effect of inducible nitric oxide synthase inhibitor. PLoS One 11 (10), e0164137.

CHAPTER

Biogenic silver nanoparticles: New trends and applications

10

Alexander Yu. Vasil'kov[a], Kamel A. Abd-Elsalam[b], and Andrei Yu. Olenin[c]

[a]*A.N. Nesmeyanov Institute of Organoelement Compounds, Russian Academy of Sciences, Moscow, Russia* [b]*Plant Pathology Research Institute, Agricultural Research Center (ARC), Giza, Egypt* [c]*V.I. Vernadsky Institute of Geochemistry and Analytical Chemistry, Russian Academy of Sciences, Moscow, Russia*

1 Introduction

Naturally occurring organic compounds are widely used to reduce silver compounds. This area of green chemistry is currently undergoing intensive development (Vaidyanathan et al., 2009; Sahayaraj and Rajesh, 2011; Ramya and Subapriya, 2012; Mittal et al., 2013; Rai et al., 2013, 2014; Makarov et al., 2014; Schröfel et al., 2014; Mashwani et al., 2015; Patel et al., 2015; Anjum et al., 2016; Banach and Pulit-Prociak, 2017; Barrocas et al., 2016; Chung et al., 2016; Singh et al., 2016; Corciova and Ivanescu, 2018; Khandel et al., 2018; Madkour, 2018; Sabela et al., 2018; Mikhailov and Mikhailova, 2019; Moradi et al., 2020). Two factors underlie this. First, silver has a high standard electrochemical potential in the reaction $Ag^+ + e^- \rightarrow Ag$. Many bioorganic compounds are effective Ag^+ ion reductants. The rate of the redox reaction is sufficient to carry out the reaction under normal conditions or with slight heating. Second, the functional groups of bioorganic compounds can interact with silver atoms and ions located on the surface of nanoparticles. A bioorganic reductant can also perform this function.

Bioorganic compounds contain many diverse reactive groups: alcohol and phenolic hydroxyls, carbonyl fragments of sugars, oligo- or polysaccharides, and polyphenols and their derivatives. Compounds involved in natural redox processes similar to the Krebs cycle can also produce metallic silver. Nitrogen- and phosphor-containing fragments of amino and nucleic acids, oligo- and polypeptides, nucleotides form strong complexes with silver ions both in solution and chemisorbed on the surface of metal nanoparticles (Xu et al., 2020).

Most authors working in this field of research use aqueous or organic water plant extracts. Some methods include a procedure for the isolation of bioorganic substances from extracts, while others use extracts without removing the solvent or combine extraction processes and a redox reaction. A review of the techniques is provided later in this chapter in Tables 1–3. Sources of bioorganic reductant

Table 1 Synthesis parameters and characteristics of Ag sols obtained by reduction of AgNO$_3$ with aqueous extracts of plant leaves.

Leaf extract	Temp. (°C)	Time	Additional action	λ_{max} SPR (nm)	Mean size of particles (nm)	Reference
Adiantum philippense L.	30	No data		420	20	Sant et al. (2013)
Aesculus hippocastanum (Horse Chestnut)	95	10 min		420–470	50	Küp et al. (2020)
Ageratum conyzoides L.	Room	24 h		443	~20	Chandraker et al. (2019)
Amorphophallus paeoniifolius	32	20 min		428	21–26 (XRD) 20 (TEM)	Gomathi et al. (2019)
Annona squamosa	Room	2 h		430	7–8	Ruddaraju et al. (2019)
Azadirachta indica	70	30–90 min		437	29	Haroon et al. (2019)
Butea monosperma	Room	2–4 h		450	~20	Patra et al. (2015)
Calotropis procera	20–90	No data		421	9	Borase et al. (2015)
Capparis zeylanica	Room	20 min		438	~5	Saranyaadevi et al. (2014)
Carica papaya	Room	15 min		430	15 (XRD)	Sathiyapriya et al. (2016)
Cassia alata	Room	No data		434	~30	Gaddam et al. (2014)
Cassia fistula (Linn.)	Room	12 h		421	80	Mohanta et al. (2016)

Coleus amboinicus	30	No data		437	18–35	Narayanan and Sakthivel (2011)
Combretum erythrophyllum	85	48 h		376	14	Jemilugba et al. (2019)
Croton bonplandianum	90	No data		425	32	Khanra et al. (2016)
Cuminum cyminum	4–60	0–4 h		420–450	20	Karamian and Kamalnejad (2019)
Diospyros lotus	Room	15 min		409	20	Hamedi and Shojaosadati (2019)
Diospyros Montana	Room	72 h		421	62	Siddiqi et al. (2019)
Eryngium campestre	25	30 min		390	25–30	Khodaie et al. (2019)
Extract of the A. scholaris	28	1 h		415	20–30	Ethiraj et al. (2016)
Geranium wallichianum	Room	24 h		405–425	1–2.5	Badoni et al. (2019)
Grevillea robusta	Room	5 h			~40 (TEM) 23 (DLS)	Brijnandan et al. (2019)
Indigofera barberi	50	4 h		440	3	Reddy et al. (2019)
Indoneesiella echioides	Room	No data		420	29	Kuppurangan et al. (2016)
Ipomoea pescaprae	Room	2 h		428	13	Arulmoorthy et al. (2015)
Iresine herbstii	Room	7 days		438	~50	Dipankar and Murugan (2012)

Continued

Table 1 Synthesis parameters and characteristics of Ag sols obtained by reduction of AgNO$_3$ with aqueous extracts of plant leaves—con't

Leaf extract	Temp. (°C)	Time	Additional action	λ_{max} SPR (nm)	Mean size of particles (nm)	Reference
Lemon	No data	1–30 min	Microwave irradiation	400–480	No data	Serdar et al. (2019)
Leucaena leucocephala L.	4	No data			~35	Ghotekar et al. (2018)
Limonia acidissima	Room	30–180 min		433	~30	Annavaram et al. (2015)
Mangifera indica	100	30 min		439	~30	Philip et al. (2011)
Mesua ferrea Linn.	No data	No data	US 24 kHz 460 W/cm^2	440	11.5	Konwarh et al. (2011a,b)
Murraya koenigii	No data	10 min		411–432	No data	Philip et al. (2011)
Nigella arvensis	No data	No data	Sunlight	416	No data	Chahardoli et al. (2018)
Ocimum sanctum (Tulsi)	30	8 min		413	14	Singhal et al. (2011)
Olax scandens	Room	1.5–7 h		410–430	~50	Mukherjee et al. (2014)
Origanum heracleoticum L.	60–90	90 min		450	30–35	Rajendran et al. (2015)
Origanum vulgare	60–90	10 min		440	63–85	Sankar et al. (2013)
Parkia speciosa	25	4 h		410	35	Ravichandran et al. (2019)
Parkinsonia florida	Room	24 h		444–450	10–57	López-Millán et al. (2019)

Passiflora caerulea	Room	5 h		379	24	Santhoshkumar et al. (2019)
Phlomis	Room	30 min		440	25	Allafchian et al. (2016)
Pistacia atlantica	80	No data		445–450	20	Golabiazar et al. (2019)
Psidium guajava (Guava)	No data	5 min	Microwave irradiation	490	26	Raghunandan et al. (2011)
Psidium Guajava L.	Room	0.5–3 h		438	No data	Wang et al. (2018)
Rosa chinensis L.	Room	24 h			~35	Bangale and Ghotekar (2019)
Salix nigra	30	48 h		430	54	Ahmad et al. (2018)
Salvadora persica	Room	24 h		426	3	Mamatha et al. (2017)
Saraca indica	60–80	10 min		412	~35 (DLS)	Tripathi et al. (2013)
Spent tea	Room	No data	Sunlight	385–435	30–100	Moosa et al. (2015)
Syzygium samarangense	37	24–48 h		425	20–25	Thampi and Jeyadoss (2015a)
Syzygium samarangense (Wax Apple)	60–80	1 h		418	~40	Thampi and Jeyadoss (2015b)
Tabebuia aurea	80	10 min		440–450	~25	Prathibha and Priya (2017)
Tecoma stans	27	4 h		No data	15	Arunkumar et al. (2013)

Continued

Table 1 Synthesis parameters and characteristics of Ag sols obtained by reduction of AgNO$_3$ with aqueous extracts of plant leaves—*con't*

Leaf extract	Temp. (°C)	Time	Additional action	λ_{max} SPR (nm)	Mean size of particles (nm)	Reference
Tinospora cordifolia	Room	10min–2h		420	55–80	Jayaseelan et al. (2011)
Tomato	50	24h		450	50–90	Santiago et al. (2019)
Tribulus terrestris	100	30min		440	25 (AFM)	Gopinath et al. (2015)
Trigonella foenum-graecum	Room			410	20–25	Khan et al. (2019a,b)
Tulsi	No data	No data	Sunlight Reflux Microwave	422–436	5–10	Singh et al. (2018)
Ziziphus spina-christi	60	15min	UV	440–455	20	Jasem et al. (2019)

Table 2 Synthesis parameters and characteristics of Ag sols obtained by reduction of AgNO$_3$ with aqueous extracts of fruits or seeds of plants.

Fruit or seed extract	Temp. (°C)	Time	Additional action	λ_{max} SPR (nm)	Mean size of particles (nm)	Reference
Coffee and tea	Room	No data			20–60	Nadagouda and Varma (2008)
Adansonia digitata L. fruits	60–80	No data		434	~20	Kumar et al. (2016a,b)
Allium cepa (onion)	Room	No data	Sunlight	418–427	~5	Chrislyn et al. (2016)
Alpinia nigra fruits	Room	2h		430	~35	Baruah et al. (2019)
Apple and grape	Room	12h		420	6	Parthibavarman et al. (2019)
Artocarpus heterophyllus (jackfruit)	Room	No data	Sunlight	418–427	~20	Chrislyn et al. (2016)
Avicennia marina seeds	121	5min		450	~35	Naidu et al. (2019)
Citrus sinensis peel	No data	45min		400–410	10	Konwarh et al. (2011a,b)
Citrus sinensis peel	25 60	45 10min		424 445	5	Kaviya et al. (2011)
Dark or white *Salvia hispanica* L. seeds	Room	15min		448	35 10	Hernández-Morales et al. (2019)
Dillenia indica fruits	Room	166h		421	7	Singh et al. (2013)
Dracocephalum moldavica seeds	Room	0–36min	Sunlight	450	~70 31	Pak et al. (2016)

Continued

Table 2 Synthesis parameters and characteristics of Ag sols obtained by reduction of AgNO$_3$ with aqueous extracts of fruits or seeds of plants—*con't*

Fruit or seed extract	Temp. (°C)	Time	Additional action	λ_{max} SPR (nm)	Mean size of particles (nm)	Reference
Essential oil of orange peels	70	48h		412	3	Veisi et al. (2019)
Ficus carica fruits	65–70	15min	Ultrasound	430–440	29–52	Kumar et al. (2016a,b)
Forsythia suspensa fruits	65	No data		430	47	Du et al. (2019)
Garlic	50–60	30min		408	10–15	Ahamed et al. (2011)
Grape seeds	20, 40, 60 or 80	No data		425	55	Yao et al. (2018)
Illicium verum (star anise) seeds	70	2h		410–480	12–15	Luna et al. (2015)
Jack fruit seeds	Room	5–90min		400	30	Chandhru et al. (2019)
Jalapeño Chili	30	12h		475	8–16	Luna-Sanchez et al. (2019)
Juglans regia (walnut fruits)	37–40	1–24h		417–424	31	Khorrami et al. (2018)
Linseed hydrogel	Room	15min–10h		397–413	10–35	Haseeb et al. (2017)
Malus domestica	80	20min		425	107	Anand et al. (2019)
Mimosa pudica seeds	25–100	1–4h		410–440	40	Iram et al. (2014)
Momordica charantia fruits	Room	30min–5h		440		Nahar et al. (2015a)

Momordica charantia fruits	Room	5 h		440	~25	Nahar et al. (2015b)
Nephrolepis exaltata (fern)	Room	No data	Sunlight	418–427	26–44	Chrislyn et al. (2016)
Persea americana seeds	Room	5 h		430	20	Girón-Vázquez et al. (2019)
Sapota fruit waste	Room	2 h		410–440	8–16	Vishwasrao et al. (2019)
Sida cordifolia seeds	45	No data		442	~20	Rajgovind et al. (2015)
Solanum melongena L.	No data	10–60 s at 1200 W	Domestic microwave oven	413	92	Das and Bhuyan (2019)
Spice blend	Room	15 min		410–450	~15	Otunola and Afolayan (2018)
Tamarindus indica fruits	45	2 h		450	~30	Gomathi et al. (2020)
Tansy fruits	100	10–15 min		452	16	Dubey et al. (2010)
Tectona grandis seeds	100	1 h		440	20	Rautela et al. (2019)

Table 3 Synthesis parameters and characteristics of Ag sols obtained by reduction of AgNO$_3$ with aqueous extracts of plant flowers.

Flower extract	Temp. (°C)	Time	λ_{max} SPR (nm)	Mean size of particles (nm)	Reference
Acmella oleracea	No data	No data	417	No data	Raj et al. (2016)
Cassia fistula	Room	15 min	422	27–32	Remya et al. (2015)
Datura inoxia	37	1 h	411	~40	Gajendran et al. (2019)
Lantana camara	Room	1 h	429–457	~90	Fatimah and Indriani (2018)
Lonicera hypoglauca	Room	25 min	437	~15	Jang et al. (2016)

include not only plants, but also algae, and microbiological media containing bacteria or fungi. The chemical composition of these does not differ fundamentally from plants, however, the specifics of working with them imposes some limitations (see Table 4).

Intensive absorption in the region of 400–450 nm is one of the main signs of the presence of nanosized metal silver particles in the studied object. The basis of this absorption is the effect of surface plasmon resonance (SPR) arising from the interaction of metal nanoparticles with an alternating electromagnetic field (Moores and Goettmann, 2006). The position of the absorption maximum, the line width, and its symmetry depend on factors such as the size and shape of the particles and the chemical composition of the surface layer. The silver dispersions obtained using bioorganic compounds usually contain fairly broad lines of SPR with a maximum having a shift in the long-wavelength region of the spectrum in comparison with analogues synthesized by sodium borohydride reducing.

Metal vapor synthesis (MVS) is an effective cryochemical method for producing extremely active metal nanoparticles and for modifying carriers of various natures (Ozin and Moskovits, 1976; Klabunde, 1980; Vitulli et al., 2008; Olenin et al., 2018). The large-scale application of MVS for obtaining atoms and clusters of metals and studying their reactivity began in the 1960s. Initially, the MVS method was used in organometallic chemistry to obtain metal complexes that are stable at low temperatures and in argon. However, with the advent of nanotechnology, the method received a new round of development. Now silver and other metal nanoparticles in various solvents have been obtained by the MVS method (Smetana et al., 2005; Vitulli et al., 2008; Bhaskar and Jagirdar, 2012; Bhattacharya and Jagirdar, 2018; Cárdenas-Triviño and Cruzat-Contreras, 2018). The size of the particles largely depends on the synthesis conditions—the nature and molar ratio of the

Table 4 Synthesis parameters and characteristics of Ag sols obtained by reduction of AgNO$_3$ in an aqueous medium containing microbiological objects.

Microbiological sample	Temp. (°C)	Time	λ_{max} SPR (nm)	Mean size of particles (nm)	Reference
Gram-positive bacteria					
Bacillus paralicheniformis extract	No data	24–72 h	443	~10	Allam et al. (2019)
Bacillus pumilus extracts	No data	24–72 h	449	~10	Allam et al. (2019)
Bacillus subtilis culture supernatant or wet biomass	Room	72–120 h	~400	No data	Omole et al. (2018)
Bacillus tequilensis	60	1 h	420	20–60	Gurunathan et al. (2015)
Brevibacterium casei	37	24 h	420	50	Kalishwaralal et al. (2010a,b)
Deinococcus radiodurans strain	32	24 h	400–450	17	Kulkarni et al. (2015)
Lactobacillus sp. from yoghurt	40	10–20 min	No data	20	Jha and Prasad (2010)
Streptomyces hygroscopicus	30	96 h	420–425	20	Sadhasivam et al. (2010)
Streptomyces spp. bact. culture	28	24 h	450	~15	Nejad et al. (2015)
Streptomyces violaceus MM72 extract	Room	1 h	420	30	Sivasankar et al. (2018)
Gram-negative bacteria					
Escherichia coli cell-free supernatant	Room	24 h	No data	13	Bigdeli et al. (2019)
Morganella psychrotolerans supernatant	4–25	20 h 24 h 5 days	350–530 650–950	70	Ramanathan et al. (2011)
Paracoccus sp. Arc7-R13 strain supernatant	37	48 h	416	2	Zhang et al. (2019a,b)
Pseudomonas aeruginosa strain BS-161R culture supernatant	Room	8 h	430	13	Kumar and Mamidyala (2011)

Continued

Table 4 Synthesis parameters and characteristics of Ag sols obtained by reduction of AgNO$_3$ in an aqueous medium containing microbiological objects—cont

Microbiological sample	Temp. (°C)	Time	λ_{max} SPR (nm)	Mean size of particles (nm)	Reference
Sphingomonas paucimobilis extract		24–72 h	423	~10	Allam et al. (2019)
Bacterial isolates	Room	7 days	350–460	~25	Zaki et al. (2011)
Silver-resistant bacteria	37	2 weeks	No data	4	Ahmad et al. (2019)
Cyanobacteria					
Nonliving strains of Spirulina platensis	Room	1–7 days	432	13	Verma et al. (2014)
Algae					
Calocybe indica	60	1 h	420	20–60	Gurunathan et al. (2015)
Chlorella vulgaris algal extract	Room	12 h	465 392 and 347 nm	20–60	Xie et al. (2007)
Chlorella vulgaris green alga	Room	3 h	490	~60	Mahajan et al. (2019)
Fungi					
Ganoderma neo-japonicum culture	Room	6 h	413	15	Philip (2009)
Edible mushroom extract	27	48 h	423	25	Fayaz et al., 2010
Fungus biomass	28	72 h	415	<50	Taha et al. (2019)
Fusarium oxysporum culture starch supernatant	Room	24 h	420	5–8	Gurunathan et al. (2013)
Penicillium cyclopium mycelium	Room	72 h	420	16–19	Wanarska and Maliszewska (2019)
Penicillium italicum fungal biomass	10–50		430–440	50	Ahluwalia et al. (2014)
Trichoderma harzianum extract	50	24 h	447	25	Soleimani et al. (2018)

organic reagent and metal used. The advantages of this method can be identified as: (i) the purity of the resulting metal nanoparticles; and (ii) the simplicity of modifying various types of carriers leading to new functional biomedical, tribological materials (Krasnov et al., 2009, 2010; Vasil'kov et al., 2018a,b, 2019; Rubina et al., 2019, 2020). The cryochemical method of generating silver nanoparticles is environmentally friendly, unlike most traditional methods, and can be easily integrated into various process cycles.

Transmission electron microscopy (TEM), scanning electron microscopy (SEM), dynamic laser light scattering (DLS), X-ray photoelectron spectroscopy (XPS), Fourier transform IR (FTIR) and UV-vis spectroscopy are used to study silver-containing colloids (Anukiruthika et al., 2020). The choice of research method is determined by the specifics of the final synthesis products.

The use of environmentally acceptable "green" technologies for producing silver nanoparticles can significantly reduce or eliminate negative environmental impacts. The production of silver nanoparticles using bioorganic reducing agents or the synthesis of metal vapor is an area of green chemistry that has been actively developed in the last decade.

2 Methods for preparation of silver nanoparticles with use of bioorganic reductants

Reduction of nitrate (Bozanic et al., 2010; Darroudi et al., 2010, 2011; Filippo et al., 2010, 2013; Vasileva et al., 2011; Madrakian et al., 2015), the ammonia hydroxyl complex (Chook et al., 2012) of silver glucose (Bozanic et al., 2010; Darroudi et al., 2010, 2011; Filippo et al., 2010, 2013; Chook et al., 2012; Jiang et al., 2021), maltose, sucrose (Filippo et al., 2010), or lactose (Contreras et al., 2011), in the presence of starch (Darroudi et al., 2010), gelatin (Darroudi et al., 2011), or chitosan (Bozanic et al., 2010), leads to the formation of silver colloids with an average particle size of 10–40nm. The reaction proceeds more intensively in an alkaline medium. In some cases, microwave (Bozanic et al., 2010; Chook et al., 2012; Filippo et al., 2013) or ultrasound (Vasileva et al., 2011) radiation is used to initiate or accelerate the process.

Aqueous extracts of leaves, flowers, or fruits of plants are most often used as bioorganic reducing agents for producing silver colloids. This is due to their availability for collection and ease of further processing. Efficient extraction of substances contained in natural objects is facilitated by increasing temperatures. Many authors raise the operating temperature to 80°C, and some work using boiling water. A review of techniques using aqueous extracts of plant leaves is given in Table 1. Leaves, fruits, seeds, flowers, and roots of plants can all be used to produce bioorganic compounds (see Tables 2 and 3).

Water extraction, including boiling, is not always an effective way to extract bioorganic compounds, as some of them have limited solubility in it. In these cases, water is replaced with a water-ethanol or organic solvents mixture (Khatami et al.,

2015, 2016; Dhand et al., 2016; Poggialini et al., 2018; Ahn et al., 2019; Choudhary et al., 2020), methanol (Khatami et al., 2016) or hexane (Mostafa et al., 2019).

The mechanism of the oxidation reaction for phenols contained in plant extracts does not differ from the classical one based on the principles of organic chemistry (Xie et al., 2007):

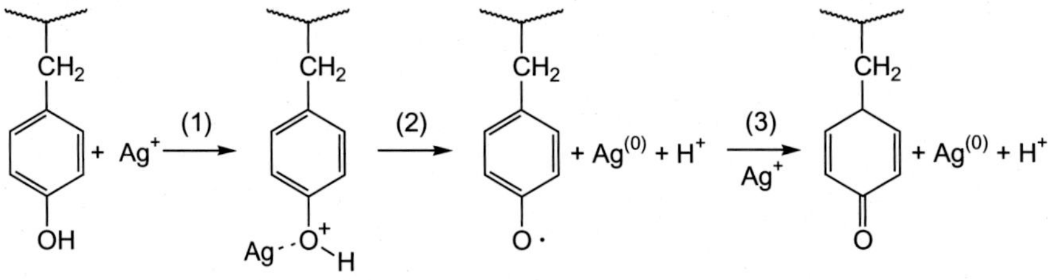

There are a number of consecutive transformations during the silver ion reaction with a phenolic group: from the formation of carbocations (1), through the radical particle (2), to the corresponding quinone (3). The formation of an alkaline medium leads, firstly, to the neutralization of protons and secondly, to the formation of phenolate ions, which contribute to the coordination of silver ions and have a higher value of the standard half-reaction potential of reduction.

The presence of chitosan in the reaction medium leads to the coordination of silver ions by its amino and hydroxyl groups. After exposure to UV-radiation, Ag nanoparticles are formed and stabilized on the polymer matrix surface (Shameli et al., 2010):

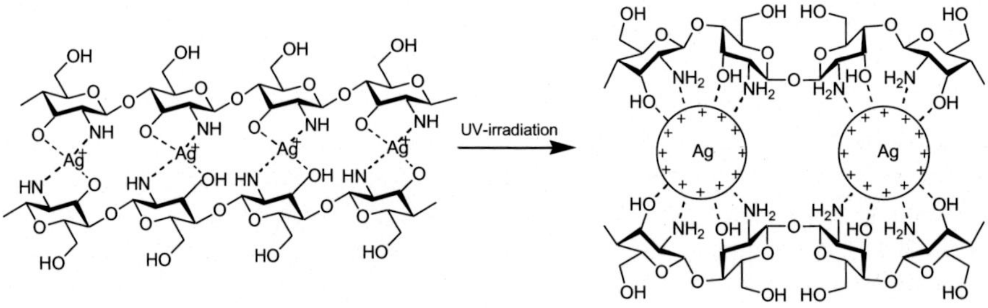

Not only plant extracts can be a source of bioorganic reducing agents for silver compounds, but also liquid media containing unicellular algae, bacteria, fungi, or their supernatants (Kato and Suzuki, 2020). The set of substances of natural origin in them is about the same as in plants. The chemical diversity of the cell walls of bacteria and fungi is significantly higher than that of plants. In plants, the main components of the cell wall are polysaccharides, mainly cellulose. The main components of the fungal cell wall are glucans and chitin. If glucans do not differ from cellulose in the set of functional groups, then chitin contains acetylamine fragments having a higher affinity for Ag^+ ions. This fact underlies the better regenerative ability of fungal cells compared to plants. Bacteria contain carboxylic acids and their derivatives, peptide fragments, in cell walls. The reduction rate of silver compounds on the membrane surface is higher for bacteria compared with fungi and plants. Equally important is the surface charge of the cell wall. This charge is determined by a set of functional groups and the pH of the environment. More effective interaction with Ag$^+$ ions occurs in the case of a negatively charged cell wall, such as in *Penicillium cyclopium* mycelium (Wanarska and Maliszewska, 2019). The data of IR spectroscopy obtained for the surface of *P. cyclopium* cells show a frequency shift of the amide group from 1638 to 1619 cm^{-1} and the disappearance of the absorption band at 1540 cm^{-1}. Anion in silver compounds such as nitrate may affect the functioning of cyanobacterial enzymatic systems (Lengke et al., 2007). Successive reduction of the NO_3^- ion to ammonium occurs. The utilization of ammonium fragments occurs through the formation of glutamine. Process characteristics are given in Table 4.

The redox reaction of silver ions with bioorganic reductants usually occurs either in the intercellular space or on the outer cell wall. Intracellular reduction of Ag$^+$ is possible, but is rarely realized for microbiological objects. The reason is the presence on the outer side of the cell membrane of a sufficient amount of bioorganic compounds capable of forming stable complexes with silver due to their functional groups. The localization of the formation of silver nanoparticles can be determined by the absorption of the SPR band in the sediment and supernatant.

3 Preparation of silver nanoparticles using metal vapor synthesis

The metal vapor synthesis (MVS) method is based on the interaction of extremely reactive atomic metals produced by evaporation under high vacuum conditions (10^{-4}–10^{-5} Torr) with organics during their co-condensation on cooled walls of a reaction vessel. From the synthesis perspective, the resistive or electron beam evaporator is preferable. Laboratory research is carried out in a 5–10 L glass reactor equipped with a resistive metal evaporator (Fig. 1).

Isopropanol was used as an organic dispersion medium in a typical experiment for silver organosol preparation. It was dried over 4Å zeolites and distilled under argon atmosphere before synthesis. Silver was evaporated by resistive heating from a tantalum foil vessel (Vasil'kov et al., 2018a,b; Rubina et al., 2020). The glass flask of the reactor was evacuated, immersed in a vessel with liquid nitrogen before synthesis started. Then the organic reagent and metal vapors were condensed together on the cooled walls of the reactor. After vapor co-condensation, the liquid nitrogen coolant was removed, argon was supplied to the reactor, and metal co-condensate and organics were heated to their melting point. The resulting organosol Ag/i-PrOH was placed in an evacuated flask and used to modify various materials. The procedure can be formally divided into three steps as shown in Fig. 2.

The currently known methods of obtaining metal nanoparticles in matrices of various natures mostly use the process of chemical reduction of metal salts. These methods have a number of significant limitations, which substantially complicate the use of the obtained materials: (i) the presence of a significant amount of

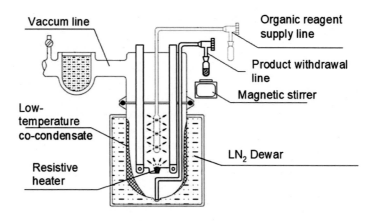

FIG. 1

The reactor for the metal vapor synthesis.

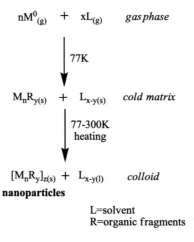

FIG. 2

Synthesis steps of silver nanoparticles using metal vapor.

surfactant impurities and residues of synthesis products, (ii) it is difficult to control the completeness of metal reduction and the formation of the internal structure of materials due to the presence of impurities in the matrix. Often, high temperatures are required for the recovery process. Such a factor can affect the partial destruction of used matrices. The advantages of the MVS method for producing metal nanoparticles include the absence of synthesis by-products during the formation of silver nanoparticles, which is especially important for the production of biomedical materials, and ease of modification of various types of carriers to give them new functional properties. This method allows the use of any metal or a combination of metals and can be used for any variation of organic reagents and metals. Table 5 summarizes data of obtaining silver nanoparticles by metal vapor synthesis.

Table 5 Cryochemical synthesis of silver nanoparticles using organic reagents of different nature.

Metal	Cryomatrix	Stabilizing medium	Average particle size (nm)	Reference
Ag	Ethyl methacrylate			Cárdenas and Acuna (1992)
Ag	n-Butyl methacrylates			Cárdenas and Salgado (1993)
Ag	Vinylacetate			Cárdenas and Muñoz (1993)
Ag	4-Methylstyrene			Cárdenas and Salgado (1994)
Ag	Styrene-co-4-methylstyrene			Cárdenas et al. (1995a)
Ag	Styrene-co-acrylonitrile			Cárdenas et al. (1995b)
Ag	Styrene-co-ethyl methacrylate			Cárdenas and Gonzales (1996)
Ag	Styrene-co-butyl methacrylate			Cárdenas and Gonzales (1996)
Ag	Styrene-co-ethyl methacrylate			Cárdenas and Salgado (1997)
Ag	Methacrylic acid			Cárdenas et al. (1999a)
Ag	Butilacrylate-co-butilmethacrylate			Cárdenas et al. (1999b)
Ag	Acrylic/methacrylic acid			Cárdenas et al. (2000)
Ag Ag/Pd	2-Propanol	Hyaluronic acid (HA)	50.41 (HA-Ag) 33.22 (HA-AgPd)	Cárdenas-Triviño et al. (2017)
Ag	2-Butanone	Dodecanethiol trioctylphosphine	6.6 ± 1.0 6.0 ± 2.0	Smetana et al. (2005)
Ag	2-Propanol	Microcrystalline and bacterial cellulose	10.2–17	Vasil'kov et al. (2019) Gromovykh et al. (2019)
Ag	2-Propanol	Chitosan aerogel	4	Rubina et al. (2020)
Pd/Ag	Acetone, 2-propanol, 2-methoxyethanol		3.9–45.4	Cardenas and Oliva (2003)
Ni/Ag	2-Propanol	Chitosan	49 ± 4	Contreras et al. (2011)
Au/Ag	2-Pentanone		6.9 ± 1.8	Bhattacharya and Jagirdar (2018)

The size and electronic state of Ag in organosols are mainly determined by such factors as the molar ratio metal; organic reagent, which can vary from 1:20 to 1:1000; the nature of the organic reagent, such as acetone, 2-propanol, 2-methoxyethanol, and others; temperature condition; and storage time of the obtained organosol (Cardenas and Oliva, 2003; Smetana et al., 2005; Bhattacharya and Jagirdar, 2018).

A process for the redispersion of silver nanoparticles obtained by the MVS method has been developed, which allows controlling the process of grinding silver aggregates into smaller particles of about 6 nm in size, which can then be used to form various nanostructures (Smetana et al., 2005). Analogous experiments were carried out to produce silver and gold nanoparticles in order to obtain bimetallic particles with a core-shell structure Ag @ Au. Particles with a size of 6.9 ± 1.8 nm were obtained, consisted of a gold core coated with a silver shell (Bhattacharya and Jagirdar, 2018).

Typically, the use of organosol Ag to produce metal-containing materials by modifying substrates of different nature leads to significant changes in the size of metal particles, their size distribution, as well as the chemical state of the surface of nanoparticles (Rubina et al., 2016; Vasil'kov et al., 2009).

4 Applied aspects of green-synthesized silver nanoparticles

Areas of practical use of green-synthesized silver nanoparticles are determined by two fundamental properties that combine in them. First, specific optical properties, such as surface plasmon resonance, near-surface enhancement of fluorescence, or Raman scattering, form the basis of applied work in the fields of chemical and biochemical analysis. Furthermore, highly dispersed silver promotes the occurrence of certain reactions, having a catalytic effect. Second, the biological activity of silver ions determines their use for biomedical and biological purposes. Consideration of this issue is beyond the scope of this review.

4.1 The use of silver nanoparticles in chemical and biochemical analysis

Chemical and biochemical analysis is now a successful and actively developing application of silver nanoparticles (Sahu et al., 2021). Two trends can be highlighted for silver nanoparticles. These are optical and electrochemical methods of analysis.

The phenomenon of surface plasmon resonance underlies the spectrophotometric direction (Vilela et al., 2012; Terenteva et al., 2017; Apyari et al., 2019; Olenin, 2019). The related principle of immunoassay is based on the same principle (Zhu et al., 2010; Liu et al., 2014). The effects of near-surface enhancement of fluorescence (Goldys and Xie, 2008; Demchenko, 2013) and Raman scattering (Eremina et al., 2018) are the basis for the development of these analysis methods. Silver nanoparticles are also used as chemiluminescent agents (Iranifam, 2016), and can be used for chemical analysis.

Sorption of chemically modified nanoparticles, including silver, increases the selectivity of electrochemical analysis methods (Welch and Compton, 2006; Vertelov et al., 2007). Metal nanoparticles combine two functions: electron transport and selective sorption of the analyte. This fact significantly increases the prospects for their use in this area.

Spectrophotometric methods of analysis are most often used for silver nanoparticles obtained by green chemistry methods. With changes in the composition of the surface layer, coagulation of particles can occur under the influence of an analyte. These processes affect the spectral properties of nanoparticles. A change in intensity, a shift in the absorption maximum, and the appearance of new bands are the analytical signals that underlie qualitative and quantitative chemical and biochemical analysis using silver nanoparticles. An overview of these types of techniques is given in Table 6.

Strong enough oxidizing agents, such as Hg^{2+} ions or hydrogen peroxide, cause the dissolution of silver nanoparticles (Farhadi et al., 2012; Chen et al., 2014; Ahmed et al., 2015; Kumar et al., 2016a,b; Lokhande et al., 2017; Chandraker et al., 2019; Choudhary et al., 2020):

$$Hg^{2+} + 2Ag \rightarrow Hg + 2Ag^+$$

$$Ag + 2H_2O_2 \rightarrow Ag^+ + O_2^{\bullet-} + 2H_2O$$

This process affects the intensity of the absorption band of surface plasmon resonance of silver nanoparticles. The simultaneous presence of dissolved oxygen in Cu^{2+} and $S_2O_3^{2-}$ ions in the presence of dissolved oxygen leads to the degradation of silver nanoparticles (Basiri et al., 2017):

$$Ag + Cu^{2+} + 5\,S_2O_3^{2-} \rightarrow Cu(S_2O_3)_3^{5-} + Ag(S_2O_3)_2^{3-}$$

$$Cu(S_2O_3)_3^{5-} + O_2 + H_2O \rightarrow Cu(S_2O_3)_2^{2-} + S_2O_3^{2-} + 2OH^-$$

Coagulation is the second process affecting the spectral properties of silver nanoparticles. The threshold effect of ionic strength on the stability of a colloid is known. Particle coagulation is observed when approaching a critical concentration of cations or anions. The nature of the charged particles affects the critical concentration of the destruction of the colloid. Such a process can be the basis for the determination of Cu^{2+} ions (Jeevika and Shankaran, 2014) or I^- (Niaz et al., 2018).

Functional groups of organic compounds can cause significant changes in the surface layer of silver nanoparticles. This affects the optical properties of colloids. A similar principle underlies spectrophotometric methods for determining organic analytes, such as glucose (Chen et al., 2014), cysteine (Borase et al., 2015; Shen et al., 2016; Raj and Sudarsanakumar, 2017), triethylamine (Filippo et al., 2013), aminobutyric acid (Jinnarak and Teerasong, 2016), dopamine (Raj et al., 2016), and berberine (Ling et al., 2008) using biologically synthesized silver nanoparticles.

Table 6 Parameters of spectrophotometric determination methods using silver nanoparticles.

Analyte	Analytical signal	Limit of detection (μM)	Linear range (μM)	Reference
Inorganic substances				
Hg^{2+}	ΔA_{481}	~100	No data	Lokhande et al. (2017)
Hg^{2+}	ΔA_{425}	~1	No data	Ahmed et al. (2015)
Hg^{2+}	ΔA_{408}	2.2	10–100	Farhadi et al. (2012)
Hg^{2+}	ΔA_{430}	0.013	0.1–1	Choudhary et al. (2020)
Cu^{2+}	ΔA_{425}	0.0011	0.005–0.1	Basiri et al. (2017)
Cu^{2+}	ΔA_{411}	0.1	0.01–100	Jeevika and Shankaran, 2014
H_2O_2	ΔA_{405}	0.033	0.1–100	Chen et al. (2014)
H_2O_2	ΔA_{425}	No data	0.1–100,000	Kumar et al. (2016a,b)
H_2O_2	ΔA_{443}	No data	No data	Chandraker et al. (2019)
I^-	ΔA_{405}	0.06	0.1–25	Niaz et al. (2018)
Organic substances				
Glucose	ΔA_{405}	0.017	0.5–400	Chen et al. (2014)
Cysteine	ΔA_{410}	0.04	0.1–100	Raj and Sudarsanakumar (2017)
Cysteine	ΔA_{410}	0.035	0.05–1	Shen et al. (2016)
Cysteine	ΔA_{421}	0.1	0.1–10	Borase et al. (2015)
Triethylamine	ΔA_{400}	0.15	0.5–1.7	Filippo et al. (2013)
Triethylamine	A_{580}/A_{400}	0.15	3300–43,000	Filippo et al. (2013)
3-Aminobutyric acid	ΔA_{390}	0.56	1000–5000	Jinnarak and Teerasong (2016)
Dopamine	ΔA_{445}	0.2	0.2–30	Raj et al. (2016)
Berberine	ΔA_{396}	0.013	0.05–0.5	Ling et al. (2008)

The formation of coagulates due to interparticle interaction caused by the functional groups of the analyte can also occur. Not the decrease in the intensity of the surface plasmon resonance line, but the ratio between the signals of individual nanoparticles and their agglomerates is more informative in the region of high analyte concentrations (Filippo et al., 2013).

4.2 The use of silver nanoparticles obtained by green methods in catalysis

The decrease in the activation energy of elementary acts of a chemical reaction underlies the catalysis phenomenon. Both $Ag^{(I)}$ compounds and silver nanoparticles have catalytic properties. Compounds of both $Ag^{(I)}$ individually and in the presence of metal-complex or organometallic co-catalysts contribute to many fine organic synthesis reactions (Weibel et al., 2008). Usually, these are coupling or condensation reactions. Most reactions accelerated by silver nanoparticles are related to redox (Dong et al., 2015; Lee et al., 2021).

The reasons for this lie in the features of silver nanoparticles. The first is a reversible redox reaction $Ag^{(0)}$-$Ag^{(I)}$. The standard potential of this reaction has a pronounced size effect (Ivanova and Zamborini, 2010). A significant decrease in this parameter is observed for particles of less than 20 nm. The second is the high electron mobility characteristic of metallic chemical bonds. Both an oxidizing agent and a reducing agent can be adsorbed simultaneously on one nanoparticle. The electron transfer between them occurs with the participation of a nanoparticle. The difference in the energies of the ground and excited levels for silver nanoparticles corresponds to the region of electromagnetic radiation in the visible and near UV regions. Around the same area are similar transitions of reactive oxygen species, many organic compounds. This contributes to the manifestation of the photocatalytic effect by silver nanoparticles. In addition, the geometry and energy of the unfilled electronic orbitals of silver atoms in nanoparticles contribute to a number of organic reactions. A review of the catalytic and photocatalytic reactions for biologically synthesized silver nanoparticles is presented in Tables 7 and 8.

The green synthesis of silver nanoparticles introduces specifics in their use as catalysts (Mahiuddin et al., 2020). They are not used practically in reactions of flow-through gas-phase heterogeneous catalysis, such as epoxidation. The supported silver catalysts obtained by traditional chemical methods are more effective here. The temperature of the catalytic reaction in most cases is room temperature or slightly higher. Very rarely, this parameter rises to values of 80°C and above. Most of the reactions presented in Table 8 are modeled and only show the potential use of biologically synthesized silver nanoparticles in redox reactions. Almost always, a catalytic reaction is carried out in an aqueous medium. This is probably due to two reasons. First, replacement of the reaction medium is not always possible due to the limited solubility of liquids and the formation of heterogeneous systems. Second, coagulation of aqueous sols of silver nanoparticles occurs when they are transferred to a nonaqueous organic medium.

The reduction of organic compounds with sodium borohydride in water is possible without the participation of a catalyst. The limiting stage of the process in this case is the solvation of electrons and their transport through the liquid medium. Metallic nanoparticles (including silver) are capable of both absorbing reaction components on the surface and quickly transfer electrons, as shown in Fig. 3.

Formal kinetic studies of the reduction of organic compounds with sodium borohydride show the first order of the reaction. Such values should be treated quite

Table 7 Catalytic properties of biosynthesized silver nanoparticles.

Substrate	Product	Rate constant ($s^{-1} \cdot 10^3$)	Mean size of particles (nm)	Reference
Aqueous borohydride reduction at room temperature				
4-nitrophenol	4-aminophenol	4.06	64	Gangula et al. (2011)
		34	5–10	Singh et al. (2018)
		3.97	12–26	Vilas et al. (2014)
		2.38	25	Ismail et al. (2019)
		No data	14–20	Li et al. (2017)
methylene blue	leucomethylene blue	1.97	12–26	Vilas et al. (2014)
		8–17	22–32	Vidhu and Philip (2014)
		16.9	10–15	Ahmed et al. (2015)
		2.0–7.4	20	Hamedi and Shojaosadati (2019)
eosin Y	reduced eosin Y	1.32–3.85	22–32	Vidhu and Philip (2014)
methyl orange	reduced methyl orange	3–11	22–32	Vidhu and Philip (2014)
		8.3	10–15	Ahmed et al. (2015)
methyl red	reduced methyl red	2.0	10–15	Ahmed et al. (2015)

	(structure: bis-azo naphthalene sulfonic acid amide)	(structure: amino-hydroxy-naphthalene sulfonic acid)	1.7	55	Haider et al. (2018)

Hydroxylation of phenylboronic acid at room temperature

	(phenylboronic acid)	(phenol)	No data	~20	Kandula et al. (2019)

A³ coupling

	(benzaldehyde, morpholine, phenylacetylene)	(propargylamine product)	No data	3	Veisi et al. (2019)

Table 8 Photocatalytic reactions of green-synthesized silver nanoparticles.

Substrate	Light source	Rate constant ($s^{-1} \cdot 10^3$)	Mean size of particles (nm)	Reference
	Halogen lamp with cutoff filters	5.8	12	Suganthi et al. (2019)
	Solar irradiation	0.16	35	Ravichandran et al. (2019)
	Halogen lamp with cutoff filters	6.6	12	Suganthi et al. (2019)
	Solar irradiation	0.27	6	Baruah et al. (2019)
	Halogen lamp	1.3–1.6	~20	Parthibavarman et al. (2019)
	Halogen lamp	0.17–3.6	~20	Parthibavarman et al. (2019)
	Solar irradiation	0.29	6	Baruah et al. (2019)
	Solar irradiation	0.23	6	Baruah et al. (2019)

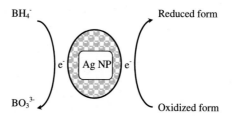

FIG. 3

Sodium borohydride as a reagent for the reduction of organic compounds without the participation of a catalyst.

critically. The reaction is heterogeneous and occurs on the surface of silver nanoparticles. We must normalize the reaction rate, not to the volume concentration of nanoparticles, but their surface. Most likely, the limiting stage is the rate of sorption of the initial substances. The formation of more dispersed particles often does not lead to an increase in the reaction rate, constant in some cases. The reason for this is the competitive sorption of bioorganic surface stabilizer molecules and redox reaction components.

A similar mechanism of the catalytic activity is also observed in oxidation reactions. There are also features of the nanoscale state of silver. First, the standard electrochemical potential of the oxidizing agent should not exceed the same parameter for silver. Degradation and complete dissolution of silver occur in case of excess. Secondly, the difference in the energies of the ground and excited states of silver nanoparticles is close to the active forms of oxygen and lies in the regions of near UV and visible light. Therefore, oxidative reactions in conjunction with lighting are described more often for silver nanoparticles. An overview of such photocatalytic reactions is given in Table 8.

The mechanism of photocatalytic oxidation of organic compounds is significantly different from that for reduction reactions. The main difference is that silver nanoparticles contribute to the excitation of organic molecules with the formation of an electron-hole pair (e^-/h^+) under the influence of a quantum of electromagnetic irradiation (Suganthi et al., 2019). This pair is capable of generating oxygen-containing free radical particles that directly carry out the redox reaction. Electrons of silver nanoparticles having the effect of surface plasmon resonance directly take part in the elemental catalytic act.

The rate constants of photocatalytic oxidation are approximately an order of magnitude lower than similar reactions of borohydride reduction of organic compounds. These values should also be taken critically for the same reasons, since the reaction is heterogeneous. The use of filters to cut off the long-wavelength emission of halogen lamps increases the efficiency of photochemical oxidation (Suganthi et al., 2019).

Biochemically synthesized silver nanoparticles are similar in catalytic properties to traditional ones. However, their synthesis is more environmentally friendly, does not require the use of highly toxic substances, and is not a source of hazardous waste.

5 Conclusion

A sharp increase in the number of publications in the field of synthesis of silver nanoparticles by methods of "green" chemistry has been observed in recent years. This is largely due to the need to develop new environmentally friendly and economical methods for the synthesis of metal nanoparticles and nanomaterials, which have prospects for use in a wide range of areas—medicine, chemical and biochemical analysis, catalysis, etc. The greatest success has been achieved in the area of biosynthesis of silver nanoparticles, which is largely based on the use of environmentally friendly and renewable products and is also quite simple to experiment with. The stability of biosynthesized silver nanoparticles under environmental conditions can be achieved by modifying their surface. This modification occurs during their formation. Materials with valuable properties for practical applications can be obtained due to the variety of reducing agents capable of chemisorption on the surface of nanoparticles. Biocompatibility is often provided by these reducing agents. However, an analysis of advances in the use of biosynthesis to produce silver nanoparticles shows that several process parameters, such as the completeness of metal precursor recovery, particle size and shape, as well as some other characteristics, cannot be fully reproducible. This stimulates research on the development of new principles and methodological approaches to biosynthesis, which would make it possible to gain knowledge not only about the mechanism of the process but also about the influence of the basic parameters of it—concentration, temperature, and time. Systematization of this information, which is not enough in most publications, will allow finding new approaches to increasing the yield of the target product, as well as obtaining nanomaterials with controlled parameters. It is hoped that these obstacles will be overcome and the "green" synthesis of metal nanoparticles will be a serious competitor to classical chemical methods.

Acknowledgments

This work was supported by RFBR (№20-53-18006). Access to electronic scientific resources was provided by INEOS RAS with the support of Ministry of Science and Higher Education of the Russian Federation.

References

Ahamed, M., Khan, M.A.M., Siddiqui, M.K.J., AlSalhi, M.S., Alrokayan, S.A., 2011. Green synthesis, characterization and evaluation of biocompatibility of silver nanoparticles. Phys. E 43, 1266–1271.

Ahluwalia, V., Kumar, J., Sisodia, R., Shakil, N.A., Walia, S., 2014. Green synthesis of silver nanoparticles by *Trichoderma harzianum* and their bio-efficacy evaluation against *Staphylococcus aureus* and *Klebsiella pneumonia*. Ind. Crop. Prod. 55, 202–206.

Ahmad, F., Ashraf, N., Zhou, R.-B., Chen, J.J., Liu, Y.-L., Zeng, X., Zhao, F.-Z., Yin, D.-C., 2019. Optimization for silver remediation from aqueous solution by novel bacterial isolates using response surface methodology: recovery and characterization of biogenic AgNPs. J. Hazard. Mater. 380, 120906.

Ahmad, W., Ahmad, M., Ali Khan, R., Mushtaq, N., 2018. Toxic effects of plant-mediated silver nanoparticles on human cancer cell lines and bacteria. Sci. Technol. Dev. 37, 62–68.

Ahmed, K.B.A., Senthilnathan, R., Megarajan, S., Anbazhagan, V., 2015. Sunlight mediated synthesis of silver nanoparticles using redox phytoprotein and their application in catalysis and colorimetric mercury sensing. J. Photochem. Photobiol. B 151, 39–45.

Ahn, E.-Y., Jin, H., Park, Y., 2019. Assessing the antioxidant, cytotoxic, apoptotic and wound healing properties of silver nanoparticles green-synthesized by plant extracts. Mater. Sci. Eng. C 101, 204–216.

Allafchian, A.R., Mirahmadi-Zare, S.Z., Jalali, S.A.H., Hashemi, S.S., Vahabi, M.R., 2016. Green synthesis of silver nanoparticles using *Phlomis* leaf extract and investigation of their antibacterial activity. J. Nanostruct. Chem. 6, 129–135.

Allam, N.G., Ismail, G.A., El-Gemizy, W.M., Salem, M.A., 2019. Biosynthesis of silver nanoparticles by cell-free extracts from some bacteria species for dye removal from wastewater. Biotechnol. Lett. 41, 379–389.

Anand, M.A.V., Vinayagam, R., Vijayakumar, S., Balupillai, A., Herbert, F.J., Kumar, S., Ghidan, A.Y., Al-Antary, T.M., David, E., 2019. Green synthesis, characterization and antibacterial activity of silver nanoparticles by *Malus domestica* and its cytotoxic effect on (MCF-7) cell line. Microb. Pathog. 135, 103609.

Anjum, S., Abbasi, B.H., Shinwari, Z.K., 2016. Plant-mediated green synthesis of silver nanoparticles for biomedical applications: challenges and opportunities. Pak. J. Bot. 48, 1731–1760.

Annavaram, V., Posa, V.R., Uppara, V.G., Jorepalli, S., Somala, A.R., 2015. Facile green synthesis of silver nanoparticles using *Limonia acidissima* leaf extract and its antibacterial activity. Bionanoscience 5, 97–103.

Anukiruthika, T., Priyanka, S., Moses, J.A., Anandharamakrishnan, C., 2020. Characterisation of green nanomaterials. In: Green Nanomaterials. Springer, Singapore, pp. 43–79.

Apyari, V.V., Dmitrienko, S.G., Gorbunova, M.V., Furletov, A.A., Zolotov, Y.A., 2019. Gold and silver nanoparticles in optical molecular absorption spectroscopy. J. Anal. Chem. 74, 21–32.

Arulmoorthy, M.P., Vasudevan, S., Vignesh, R., Rathiesh, A.C., Srinivasan, M., 2015. Green synthesis of silver nanoparticles using medicinal sand dune plant *Ipomoea pescaprae* leaf extract. IJSIT 4, 384–392.

Arunkumar, C., Nima, P., Astalakshmi, A., Ganesan, V., 2013. Green synthesis and characterization of silver nanoparticles using leaves of *Tecoma stans (L.)* kunth. Int. J. Nanotechnol. Appl. 3, 1–10.

Badoni, P.P., Kumar, G., Singh, M., Singh, N., Khajuria, A.K., Dutt, B., Dutt, S., 2019. *Geranium wallichianum* leaf extract mediated synthesis of silver nanoparticles: characterization and its antimicrobial activity. Asian J. Chem. 31, 1128–1132.

Banach, M., Pulit-Prociak, J., 2017. Proecological method for the preparation of metal nanoparticles. J. Clean. Prod. 141, 1030–1039.

Bangale, S., Ghotekar, S., 2019. Bio-fabrication of silver nanoparticles using *Rosa chinensis* L. extract for antibacterial activities. Int. J. Nano Dimens. 10, 217–224.

Barrocas, B., Nunes, C.D., Carvalho, M.L., Monteiro, O.C., 2016. Titanate nanotubes sensitized with silver nanoparticles: synthesis, characterization and in-situ pollutants photodegradation. Appl. Surf. Sci. 385, 18–27.

Baruah, D., Yadav, R.N.S., Yadav, A., Das, A.M., 2019. *Alpinia nigra* fruits mediated synthesis of silver nanoparticles and their antimicrobial and photocatalytic activities. J. Photochem. Photobiol. B 201, 111649.

Basiri, S., Mehdinia, A., Jabbari, A., 2017. Biologically green synthesized silver nanoparticles as a facile and rapid label-free colorimetric probe for determination of Cu^{2+} in water samples. Spectrochim. Acta A 171, 297–304.

Bhaskar, S.P., Jagirdar, B.R., 2012. Digestive ripening: a synthetic method *par excellence* for core-shell, alloy, and composite nanostructured materials. J. Chem. Sci. 124, 1175–1180.

Bhattacharya, C., Jagirdar, B.R., 2018. Monodisperse colloidal metal nanoparticles to core–shell structures and alloy nanosystems via digestive ripening in conjunction with solvated metal atom dispersion: a mechanistic study. J. Phys. Chem. C 122 (19), 10559–10574.

Bigdeli, R., Shahnazari, M., Panahnejad, E., Cohan, R.A., Dashbolaghi, A., Asgary, V., 2019. Cytotoxic and apoptotic properties of silver chloride nanoparticles synthesized using *Escherichia coli* cell-free supernatant on human breast cancer MCF 7 cell line. Artif. Cells Nanomed. Biotechnol. 47, 1603–1609.

Borase, H.P., Pati, C.D., Salunkhe, R.B., Suryawanshi, R.K., Kim, B.S., Bapat, V.A., Patil, S.V., 2015. Bio-functionalized silver nanoparticles: a novel colorimetric probe for cysteine detection. Appl. Biochem. Biotechnol. 175, 3479–3493.

Bozanic, D.K., Trandafilovic, L.V., Luyt, A.S., Djokovic, V., 2010. 'Green' synthesis and optical properties of silver-chitosan complexes and nanocomposites. React. Funct. Polym. 70, 869–873.

Brijnandan, D., Kumar, P., Meenu, S., A., 2019. Green synthesis of silver nanoparticles using *Grevillea robusta*. AIP Conf. Proc. 2142, 060008.

Cárdenas, T.G., Acuna, E.J., 1992. Synthesis and molecular weights of metal poly(ethyl methacrylates). IV. Polym. Bull. 29, 1–7.

Cárdenas, T.G., Acuna, E.J., Rodriguez, B.M., Carbacho, H.H., 1995a. Metal polymers: synthesis and molecular weights of metal poly(styrene-co-acrylonitrile). IX. Polym. Bull. 34 (1), 31–36.

Cárdenas, T.G., Gonzales, G.M., 1996. Metal polymers XIII. Synthesis and molecular weights of metal poly(styrene-co-butil methacrylate). Polym. Bull. 37, 175–182.

Cárdenas, T.G., Muñoz, C.D., 1993. Synthesis and molecular weights of metal poly(vinylacetate)s. Makromol. Chem. 194, 3377–3383.

Cárdenas, T.G., Muñoz, C.D., Carbacho, H.H., 2000. Thermal properties and TGA–FTIR studies of polyacrylic and polymethacrylic acid doped with metal clusters. Eur. Polym. J. 36, 1091–1099.

Cárdenas, T.G., Muñoz, C.D., Rodrigues, M., Morales, J., Soto, H., 1999a. Synthesis and properties of poly(methacrylic acid) doped with metal clusters. Eur. Polym. J. 35, 1017–1021.

Cardenas, G., Oliva, R., 2003. Synthesis and structure of colloids and films from Pd-Ag in organic solvents. Colloid Polym. Sci. 281, 27–35.

Cárdenas, T.G., Salgado, C.E., 1993. Metal polymers: synthesis and molecular weights of metal poly(n-butyl methacrylates). VI. Polym. Bull. 31, 23–28.

Cárdenas, T.G., Salgado, C.E., 1994. Metal polymers: synthesis and molecular weights of metal poly(4-methylstyrene)s with AIBN. VIII. Polym. Bull. 33, 629–634.

Cárdenas, T.G., Salgado, C.E., 1997. Metal polymers XV. Synthesis and molecular weights of metal poly(styrene-co-ethyl methacrylate). Polym. Bull. 38, 279–285.

Cárdenas, T.G., Salgado, C.E., Carbacho, H.H., 1999b. Metal polymers XII. Synthesis and molecular weights of metal poly(styrene-co-ethyl methacrylate). Polym. Bull. 36, 569–576.

Cárdenas, T.G., Salgado, C.E., Gonzalez, G.M., 1995b. Metal polymers: synthesis and characterization of metal poly (styrene-co-4-methylstyrene) copolymers with benzoyl peroxide. Xl. Polym. Bull. 35, 553–559.

Cárdenas-Triviño, G., Cruzat-Contreras, C., 2018. Study of aggregation of gold nanoparticles in chitosan. J. Clust. Sci. 29, 1081–1088.

Cárdenas-Triviño, G., Ruiz-Parra, M., Vergara-González, L., Ojeda-Oyarzún, J., Solorzano, G., 2017. Synthesis and bactericidal properties of hyaluronic acid doped with metal nanoparticles. J. Nanomater. 6, 1–9. ID 9573869.

Chahardoli, A., Karimi, N., Fattahi, A., 2018. *Nigella arvensis* leaf extract mediated green synthesis of silver nanoparticles: their characteristic properties and biological efficacy. Adv. Powder Technol. 29, 202–210.

Chandhru, M., Logesh, R., Rani, S.K., Ahmed, N., Vasimalai, N., 2019. One-pot green route synthesis of silver nanoparticles from jack fruit seeds and their antibacterial activities with *Escherichia coli* and *Salmonella* bacteria. Biocatal. Agric. Biotechnol. 20, 101241.

Chandraker, S.K., Lal, M., Shukla, R., 2019. DNA-binding, antioxidant, H_2O_2 sensing and photocatalytic properties of biogenic silver nanoparticles using *Ageratum conyzoides L.* leaf extract. RSC Adv. 9, 23408–23417.

Chen, S., Hai, X., Chen, X.-W., Wang, J.-H., 2014. *In situ* growth of silver nanoparticles on graphene quantum dots for ultrasensitive colorimetric detection of H_2O_2 and glucose. Anal. Chem. 86, 6689–6694.

Chook, S.W., Chia, C.H., Zakaria, S., Ayob, M.K., Chee, K.L., Huang, N.M., Neoh, H.M., Lim, H.N., Jamal, R., Rahman, R.M.F.R.A., 2012. Antibacterial performance of Ag nanoparticles and AgGO nanocomposites prepared via rapid microwave-assisted synthesis method. Nanoscale Res. Lett. 7, 541.

Choudhary, M.K., Garg, S., Kaur, A., Kataria, J., Sharma, S., 2020. Green biomimetic silver nanoparticles as invigorated colorimetric probe for Hg^{2+} ions: a cleaner approach towards recognition of heavy metal ions in aqueous media. Mater. Chem. Phys. 240, 122164.

Chrislyn, G., Prachi, S., Sangeeta, S., Priya, S., 2016. Green synthesis and characterization of silver nanoparticles. Int. J. Adv. Res. 4, 1563–1568.

Chung, I.-M., Park, I., Seung-Hyun, K., Thiruvengadam, M., Rajakumar, G., 2016. Plant-mediated synthesis of silver nanoparticles: their characteristic properties and therapeutic applications. Nanoscale Res. Lett. 11, 40.

Contreras, C.C., Peña, O., Meléndrez, M.F., Díaz-Visurraga, J., Cárdenas, G., 2011. Synthesis, characterization and properties of magnetic colloids supported on chitosan. Colloid Polym. Sci. 289, 21–31.

Corciova, A., Ivanescu, B., 2018. Biosynthesis, characterization and therapeutic applications of plant-mediated silver nanoparticles. J. Serb. Chem. Soc. 83, 515–538.

Darroudi, M., Ahmad, M.B., Abdullah, A.H., Ibrahim, N.A., 2011. Green synthesis and characterization of gelatin-based and sugar-reduced silver nanoparticles. Int. J. Nanomedicine 6, 569–574.

Darroudi, M., Ahmad, M.B., Abdullah, A.H., Ibrahim, N.A., Shameli, K., 2010. Effect of accelerator in green synthesis of silver nanoparticles. Int. J. Mol. Sci. 11, 3898–3905.

Das, R.K., Bhuyan, D., 2019. Microwave-mediated green synthesis of gold and silver nanoparticles from fruit peel aqueous extract of *Solanum melongena L.* and study of antimicrobial property of silver nanoparticles. Nanotechnol. Environ. Eng. 4, 5.

Demchenko, A.P., 2013. Nanoparticles and nanocomposites for fluorescence sensing and imaging. Methods Appl. Fluoresc. 1, 022001.

Dhand, V., Soumya, L., Bharadwaj, S., Chakra, S., Bhatt, D., Sreedhar, B., 2016. Green synthesis of silver nanoparticles using *Coffea arabica* seed extract and its antibacterial activity. Mater. Sci. Eng. C 58, 36–43.

Dipankar, C., Murugan, S., 2012. The green synthesis, characterization and evaluation of the biological activities of silver nanoparticles synthesized from *Iresine herbstii* leaf aqueous extracts. Colloids Surf. B 98, 112–119.

Dong, X.-Y., Gao, Z.-W., Yang, K.-F., Zhang, W.-Q., Xu, L.-W., 2015. Nanosilver as a new generation of silver catalysts in organic transformations for efficient synthesis of fine chemicals. Cat. Sci. Technol. 5, 2554–2574.

Du, J., Hu, Z., Yu, Z., Li, H., Pan, J., Zhao, D., Bai, Y., 2019. Antibacterial activity of a novel *Forsythia suspensa* fruit mediated green silver nanoparticles against food-borne pathogens and mechanisms investigation. Mater. Sci. Eng. C 102, 247–253.

Dubey, S.P., Lahtinen, M., Sillanpää, M., 2010. Tansy fruit mediated greener synthesis of silver and gold nanoparticles. Process Biochem. 45, 1065–1071.

Eremina, O.E., Semenova, A.A., Sergeeva, E.A., Brazhe, N.A., Maksimov, G.V., Shekhovtsova, T.N., Goodilin, E.A., Veselova, I.A., 2018. Surface-enhanced Raman spectroscopy in modern chemical analysis: advances and prospects. Russ. Chem. Rev. 87, 741–770.

Ethiraj, A.S., Jayanthi, S., Ramalingam, C., Banerjee, C., 2016. Control of size and antimicrobial activity of green synthesized silver nanoparticles. Mater. Lett. 185, 526–529.

Farhadi, K., Forough, M., Molaei, R., Hajizadeh, S., Rafipour, A., 2012. Highly selective Hg^{2+} colorimetric sensor using green synthesized and unmodified silver nanoparticles. Sensors Actuators B Chem. 161, 880–885.

Fatimah, I., Indriani, N., 2018. Silver nanoparticles synthesized using *Lantana camara* flower extract by reflux, microwave and ultrasound methods. Chem. J. Mold. 13, 95–102. General, Industrial and Ecological Chemistry.

Fayaz, A.M., Balaji, K., Girilal, M., Yadav, R., Kalaichelvan, P.T., Venketesan, R., 2010. Biogenic synthesis of silver nanoparticles and their synergistic effect with antibiotics: a study against gram-positive and gram-negative bacteria. Nanomedicine 6, 103–109.

Filippo, E., Manno, D., Buccolieri, A., Serra, A., 2013. Green synthesis of sucralose-capped silver nanoparticles for fast colorimetric triethylamine detection. Sensors Actuators B Chem. 178, 1–9.

Filippo, E., Serra, A., Buccolieri, A., Manno, D., 2010. Green synthesis of silver nanoparticles with sucrose and maltose: morphological and structural characterization. J. Non-Cryst. Solids 356, 344–350.

Gaddam, S.A., Kotakadi, V.S., Gopal, D.V.R.S., Rao, Y.S., Reddy, A.V., 2014. Efficient and robust biofabrication of silver nanoparticles by *Cassia alata* leaf extract and their antimicrobial activity. J. Nanostruct. Chem. 4, 82–90.

Gajendran, B., Durai, P., Varier, K.M., Liu, W., Li, Y., Rajendran, S., Nagarathnam, R., Chinnasamy, A., 2019. Green synthesis of silver nanoparticle from *Datura inoxia* flower extract and its cytotoxic activity. Bionanoscience 9, 564–572.

Gangula, A., Podila, R., Ramakrishna, M., Karanam, L., Janardhana, C., Rao, A.M., 2011. Catalytic reduction of 4-nitrophenol using biogenic gold and silver nanoparticles derived from Breynia rhamnoides. Langmuir 27, 15268–15274.

Ghotekar, S., Savale, A., Pansambal, S., 2018. Phytofabrication of fluorescent silver nanoparticles from *Leucaena leucocephala L.* leaves and their biological activities. J. Water Environ. Technol. 3, 95–105.

Girón-Vázquez, N.G., Gómez-Gutiérrez, C.M., Soto-Robles, C.A., Nava, O., Lugo-Medina, E., Castrejón-Sánchez, V.H., Vilchis-Nestor, A.R., Luque, P.A., 2019. Study of the effect of *Persea americana* seed in the green synthesis of silver nanoparticles and their antimicrobial properties. Results Phys. 13, 102142.

Golabiazar, R., Othman, K.I., Khalid, K.M., Maruf, D.H., Aulla, S.M., Yusif, P.A., 2019. Green synthesis, characterization, and investigation antibacterial activity of silver nanoparticles using *Pistacia atlantica* leaf extract. BioNanoScience 9, 323–333.

Goldys, E.M., Xie, F., 2008. Metallic nanomaterials for sensitivity enhancement of fluorescence detection. Sensors 8, 886–896.

Gomathi, A.C., Rajarathinam, S.R.X., Sadiq, A.M., Rajeshkumar, S., 2020. Anticancer activity of silver nanoparticles synthesized using aqueous fruit shell extract of *Tamarindus indica* on MCF-7 human breast cancer cell line. J. Drug Deliv. Sci. Technol. 55, 101376.

Gomathi, M., Prakasam, A., Rajkumar, P.V., 2019. Green synthesis, characterization and antibacterial activity of silver nanoparticles using *Amorphophallus paeoniifolius* leaf extract. J. Clust. Sci. 30, 995–1001.

Gopinath, V., Priyadarshini, S., Venkatkumar, G., Saravanan, M., Ali, D.M., 2015. *Tribulus terrestris* leaf mediated biosynthesis of stable antibacterial silver nanoparticles. Pharm. Nanotechnol. 3, 26–34.

Gromovykh, T.I., Vasil'kov, A.Y., Sadykova, V.S., Feldman, N.B., Demchenko, A.G., Lyundup, A.V., Butenko, I.E., Lutsenko, S.V., 2019. Creation of composites of bacterial cellulose and silver nanoparticles: evaluation of antimicrobial activity and cytotoxicity. Int. J. Nanotechnol. 16, 408–420.

Gurunathan, S., Park, J.H., Han, J.W., Kim, J.-H., 2015. Comparative assessment of the apoptotic potential of silver nanoparticles synthesized by *Bacillus tequilensis* and *Calocybe indica* in MDA-MB-231 human breast cancer cells: targeting p53 for anticancer therapy. Int. J. Nanomedicine 10, 4203–4223.

Gurunathan, S., Raman, J., Malek, S.N.A., John, P.A., Ikineswary, S., 2013. Green synthesis of silver nanoparticles using *Ganoderma neo-japonicum* Imazeki: a potential cytotoxic agent against breast cancer cells. Int. J. Nanomedicine 8, 4399–4413.

Haider, A., Haider, S., Kang, I.-K., Kumar, A., Kummara, M.R., Kamal, T., Han, S.S., 2018. A novel use of cellulose-based filter paper containing silver nanoparticles for its potential application as wound dressing agent. Int. J. Biol. Macromol. 108, 455–461.

Hamedi, S., Shojaosadati, S.A., 2019. Rapid and green synthesis of silver nanoparticles using *Diospyros lotus* extract: evaluation of their biological and catalytic activities. Polyhedron 171, 172–180.

Haroon, M., Zaidi, A., Ahmed, B., Rizvi, A., Khan, M.S., Musarrat, J., 2019. Effective inhibition of phytopathogenic microbes by eco-friendly leaf extract mediated silver nanoparticles (AgNPs). Indian J. Microbiol. 59, 273–287.

Haseeb, M.T., Hussain, M.A., Abbas, K., Youssif, B.G.M., Bashir, S., Yuk, S.H., Abbas Bukhari, S.N., 2017. Linseed hydrogel-mediated green synthesis of silver nanoparticles for antimicrobial and wound-dressing applications. Int. J. Nanomedicine 12, 2845–2855.

Hernández-Morales, L., Espinoza-Gómez, H., Flores-López, L.Z., Sotelo-Barrera, E.L., Núñez-Rivera, A., Cadena-Nava, R.D., Alonso-Núñez, G., Espinoza, K.A., 2019. Study of the green synthesis of silver nanoparticles using a natural extract of dark or white *Salvia hispanica* L. seeds and their antibacterial application. Appl. Surf. Sci. 489, 952–961.

Iram, F., Iqbal, M.S., Athar, M.M., Saeed, M.Z., Yasmeen, A., Ahmad, R., 2014. Glucoxylan-mediated green synthesis of gold and silver nanoparticles and their phyto-toxicity study. Carbohydr. Polym. 104, 29–33.

Iranifam, M., 2016. Chemiluminescence reactions enhanced by silver nanoparticles and silver alloy nanoparticles: applications in analytical chemistry. Trends Anal. Chem. 82, 126–142.

Ismail, M., Khan, M.I., Khan, M.A., Akhtar, K., Asiri, A.M., Khan, S.B., 2019. Plant-supported silver nanoparticles: efficient, economically viable and easily recoverable catalyst for the reduction of organic pollutants. Appl. Organomet. Chem. 33, e4971.

Ivanova, O.S., Zamborini, F.P., 2010. Size-dependent electrochemical oxidation of silver nanoparticles. J. Am. Chem. Soc. 132, 70–72.

Jang, S.J., Yang, I.J., Tettey, C.O., Kim, K.M., Shin, H.M., 2016. *In-vitro* anticancer activity of green synthesized silver nanoparticles on MCF-7 human breast cancer cells. Mater. Sci. Eng. C 68, 430–435.

Jasem, L.A., Hameed, A.A., Al-Heety, M.A., Mahmood, A.R., Karadağ, A., Akbaş, H., 2019. The mixture of silver nanosquare and silver nanohexagon: green synthesis, characterization and kinetic evolution. Mater. Res. Express 6, 0850f9.

Jayaseelan, C., Rahuman, A.A., Rajakumar, G., Kirthi, A.V., Santhoshkumar, T., Marimuthu, S., Bagavan, A., Kamaraj, C., Zahir, A.A., Elango, G., 2011. Synthesis of pediculocidal and larvicidal silver nanoparticles by leaf extract from heartleaf moonseed plant, *Tinospora cordifolia Miers*. Parasitol. Res. 109, 185–194.

Jeevika, A., Shankaran, D.R., 2014. Visual colorimetric sensing of copper ions based on reproducible gelatin functionalized silver nanoparticles and gelatin hydrogels. Colloids Surf. A Physicochem. Eng. Asp. *461*, 240–247.

Jemilugba, O.T., Sakho, E.H.M., Parani, S., Mavumengwana, V., Oluwafemi, O.S., 2019. Green synthesis of silver nanoparticles using *Combretum erythrophyllum* leaves and its antibacterial activities. Colloids Interface Sci. Commun. 31, 100191.

Jha, A.K., Prasad, K., 2010. Biosynthesis of metal and oxide nanoparticles using *Lactobacilli* from yoghurt and probiotic spore tablets. Biotechnol. J. 5, 285–291.

Jiang, C., Jiang, Z., Zhu, S., Amulraj, J., Deenadayalan, V.K., Jacob, J.A., Qian, J., 2021. Biosynthesis of silver nanoparticles and the identification of possible reductants for the assessment of in vitro cytotoxic and in vivo antitumor effects. J. Drug Deliv. Sci. Technol. 63, 102444.

Jinnarak, A., Teerasong, S., 2016. A novel colorimetric method for detection of gamma-aminobutyric acid based on silver nanoparticles. Sensors Actuators B Chem. 229, 315–320.

Lee, S.J., Begildayeva, T., Yeon, S., Naik, S.S., Ryu, H., Kim, T.H., Choi, M.Y., 2021. Eco-friendly synthesis of lignin mediated silver nanoparticles as a selective sensor and their catalytic removal of aromatic toxic nitro compounds. Environ. Pollut. 269, 116174.

Kalishwaralal, K., BarathManiKanth, S., Pandian, S.R.K., Deepak, V., Gurunathan, S., 2010a. Silver nano—a trove for retinal therapies. J. Control. Release 145, 76–90.

Kalishwaralal, K., Deepak, V., Pandian, S.R.K., Kottaisamy, M., BarathManiKanth, S., Kartikeyan, B., Gurunathan, S., 2010b. Biosynthesis of silver and gold nanoparticles using *Brevibacterium casei*. Colloids Surf. B 77, 257–262.

Kandula, V., Nagababu, U., Behera, M., Yennam, S., Chatterjee, A., 2019. A facile green synthesis of silver nanoparticles: an investigation on catalytic hydroxylation studies for efficient conversion of aryl boronic acids to phenol. J. Saudi Chem. Soc. 23, 711–717.

Karamian, R., Kamalnejad, J., 2019. Green synthesis of silver nanoparticles using *Cuminum cyminum* leaf extract and evaluation of their biological activities. J. Nanostruct. 9, 74–85.

Kaviya, S., Santhanalakshmi, J., Viswanathan, B., Muthumary, J., Srinivasan, K., 2011. Biosynthesis of silver nanoparticles using citrus sinensis peel extract and its antibacterial activity. Spectrochim. Acta A 79, 594–598.

Khan, A.U., Khan, M., Khan, M.M., 2019a. Antifungal and antibacterial assay by silver nanoparticles synthesized from aqueous leaf extract of *Trigonella foenum-graecum*. BioNanoScience 9, 597–602.

Khan, I., Bahuguna, A., Krishnan, M., Shukla, S., Lee, H., Min, S.-H., Choi, D.K., Cho, Y., Bajpai, V.K., Huh, Y.S., Kang, S.C., 2019b. The effect of biogenic manufactured silver nanoparticles on human endothelial cells and zebrafish model. Sci. Total Environ. 679, 365–377.

Khandel, P., Yadaw, R.K., Soni, D.K., Kanwar, L., Shahi, S.K., 2018. Biogenesis of metal nanoparticles and their pharmacological applications: present status and application prospects. J. Nanostruct. Chem. 8, 217–254.

Khanra, K., Panja, S., Choudhuri, I., Chakraborty, A., Bhattacharyya, N., 2016. Antimicrobial and cytotoxicity effect of silver nanoparticle synthesized by *Croton bonplandianum Baill.* leaves. Nanomed. J. 3, 15–22.

Khatami, M., Nejad, M.S., Pourseyedi, S., 2015. Biogenic synthesis of silver nanoparticles using mustard and its characterization. Int. J. Nanosci. Nanotechnol. 11, 281–288.

Khatami, M., Nejad, M.S., Salari, S., Almani, P.G.N., 2016. Plant-mediated green synthesis of silver nanoparticles using *Trifolium resupinatum* seed exudate and their antifungal efficacy on *Neofusicoccum parvum* and *Rhizoctonia solani*. IET Nanobiotechnol. 10, 237–243.

Khodaie, M., Ghasemi, N., Ramezani, M., 2019. Green synthesis of silver nanoparticles using (*Eryngium campestre*) leaf extract. Iran. Chem. Commun. 7, 502–511.

Khorrami, S., Zarrabi, A.N., Khaleghi, M., Danaei, M., Mozafari, M., 2018. Selective cytotoxicity of green synthesized silver nanoparticles against the MCF-7 tumor cell line and their enhanced antioxidant and antimicrobial properties. Int. J. Nanomedicine 13, 8013–8024.

Klabunde, K.J., 1980. Chemistry of Free Atoms and Particles. Academic-Press, United States.

Kato, Y., Suzuki, M., 2020. Synthesis of metal nanoparticles by microorganisms. Crystals 10, 589.

Konwarh, R., Gogoi, B., Philip, R., Laskar, M.A., Karak, N., 2011a. Biomimetic preparation of polymer-supported free radical scavenging, cytocompatible and antimicrobial "green" silver nanoparticles using aqueous extract of *Citrus sinensis* peel. Colloids Surf. B 84, 338–345.

Konwarh, R., Karak, N., Sawian, C.E., Baruah, S., Mandal, M., 2011b. Effect of sonication and aging on the templating attribute of starch for "green" silver nanoparticles and their interactions at bio-interface. Carbohydr. Polym. 83, 245–1252.

Krasnov, A.P., Aderikha, V.N., Afonicheva, O.V., Mit', V.A., Tikhonov, N.N., Vasil'kov, A.Y., Said-Galiev, E.E., Naumkin, A.V., Nikolaev, A.Y., 2010. Categorization system of nanofillers to polymer composites. J. Frict. Wear 31, 68–80.

Krasnov, A.P., Said-Galiev, E.E., Vasil'kov, A.Y., Nikolaev, A.J., Podshibikhin, V.L., Afonicheva, O.V., Mit', V.A., Khokhlov, A.R., Gavrjushenko, N.S., Bulgakov, V.G., Mironov, S.P., 2009. Method for Making Polymer Sliding Friction Components for Artificial Endoprostheses. Patent RF 2 354 668.

Kulkarni, R.R., Shaiwale, N.S., Deobagkar, D.N., Deobagkar, D.D., 2015. Synthesis and extracellular accumulation of silver nanoparticles by employing radiation-resistant *Deinococcus radiodurans*, their characterization, and determination of bioactivity. Int. J. Nanomedicine 10, 963–974.

Kumar, C.G., Mamidyala, S.K., 2011. Extracellular synthesis of silver nanoparticles using culture supernatant of *Pseudomonas aeruginosa*. Colloids Surf. B 84, 462–466.

Kumar, C.M.K., Yugandhar, P., Savithramma, N., 2016a. Biological synthesis of silver nanoparticles from *Adansonia digitata L.* fruit pulp extract, characterization, and its antimicrobial properties. J. Intercult. Ethnopharmacol. 5, 79–85.

Kumar, V., Gundampati, R.K., Singh, D.K., Bano, D., Jagannadham, M.V., Hasan, S.H., 2016b. Photoinduced green synthesis of silver nanoparticles with highly effective antibacterial and hydrogen peroxide sensing properties. J. Photochem. Photobiol. B 162, 374–385.

Küp, F.Ö., Coşkunçay, S., Duman, F., 2020. Biosynthesis of silver nanoparticles using leaf extract of *Aesculus hippocastanum* (horse chestnut): evaluation of their antibacterial, antioxidant and drug release system activities. Mater. Sci. Eng. C 107, 110207.

Kuppurangan, G., Karuppasamy, B., Nagarajan, K., Sekar, R.K., Viswaprakash, N., Ramasamy, T., 2016. Biogenic synthesis and spectroscopic characterization of silver nanoparticles using leaf extract of *Indoneesiella echioides*: *in vitro* assessment on antioxidant, antimicrobial and cytotoxicity potential. Appl. Nanosci. 6, 973–982.

Lengke, M.F., Fleet, M.E., Southam, G., 2007. Biosynthesis of silver nanoparticles by filamentous cyanobacteria from a silver(I) nitrate complex. Langmuir 23, 2694–2699.

Li, P., Li, S., Wang, Y., Zhang, Y., Han, G.-Z., 2017. Green synthesis of β-CD-functionalized monodispersed silver nanoparticles with ehanced catalytic activity. Colloids Surf. A Physicochem. Eng. Asp. 520, 26–31.

Ling, J., Sang, Y., Huang, C.Z., 2008. Visual colorimetric detection of berberine hydrochloride with silver nanoparticles. J. Pharm. Biomed. Anal. 47, 860–864.

Liu, R., Zhang, Y., Zhang, S., Qiu, W., Gao, Y., 2014. Silver enhancement of gold nanoparticles for biosensing: from qualitative to quantitative. Appl. Spectrosc. Rev. 49, 121–138.

Lokhande, A.C., Shinde, N.M., Shelke, A., Babar, P.T., Kim, J.H., 2017. Reliable and reproducible colorimetric detection of mercury ions (Hg^{2+}) using green synthesized optically active silver nanoparticles containing thin film on flexible plastic substrate. J. Solid State Electrochem. 21, 2747–2751.

López-Millán, A., Del Toro-Sánchez, C.L., Ramos-Enríquez, J.R., Carrillo-Torres, R.C., Zavala-Rivera, P., Esquivel, R., Álvarez-Ramos, E., Moreno-Corral, R., Guzmán-Zamudio, R., Lucero-Acuña, A., 2019. Biosynthesis of gold and silver nanoparticles using *Parkinsonia florida* leaf extract and antimicrobial activity of silver nanoparticles. Mater. Res. Express 6, 095025.

Luna, C., Chávez, V.H.G., Barriga-Castro, E.D., Núñez, N.O., Mendoza-Reséndez, R., 2015. Biosynthesis of silver fine particles and particles decorated with nanoparticles using the extract of Illicium verum (star anise) seeds. Spectrochim. Acta A 141, 43–50.

Luna-Sanchez, J.L., Jimenez-Perez, J.L., Carbajal-Valdez, R., Lopez-Gamboa, G., Perez-Gonzalez, M., Correa-Pacheco, Z.N., 2019. Green synthesis of silver nanoparticles using *Jalapeño Chili* extract and thermal lens study of acrylic resin nanocomposites. Thermochim. Acta 678, 178314.

Madkour, L.H., 2018. Ecofriendly green biosynthesized of metallic nanoparticles: bioreduction mechanism, characterization and pharmaceutical applications in biotechnology industry. Glob. Drugs Ther. 3, 1–11.

Madrakian, T., Alizadeh, S., Karamian, R., Asadbegy, M., Bahram, M., Soleimani, M.J., 2015. Green synthesis of silver nanoparticles using lactose sugar and evaluation of their antimicrobial activity. Der Pharma Chem. 7, 442–452.

Mahajan, A., Arya, A., Chundawat, T.S., 2019. Green synthesis of silver nanoparticles using green alga (*Chlorella vulgaris*) and its application for synthesis of quinolines derivatives. Synth. Commun. 49, 1926–1937.

Mahiuddin, M., Saha, P., Ochiai, B., 2020. Green synthesis and catalytic activity of silver nanoparticles based on *Piper chaba* stem extracts. Nano 10 (9), 1777.

Makarov, V.V., Love, A.J., Sinitsyna, O.V., Makarova, S.S., Yaminsky, I.V., Taliansky, M.E., Kalinina, N.O., 2014. "Green" nanotechnologies: synthesis of metal nanoparticles using plants. Acta Nat. 6, 35–44.

Mamatha, R., Khan, S., Salunkhe, P., Satpute, S., Kendurkar, S., Prabhune, A., Deval, A., Chaudhari, B.P., 2017. Rapid synthesis of highly monodispersed silver nanoparticles from the leaves of *Salvadora persica*. Mater. Lett. 205, 226–229.

Mashwani, Z.R., Khan, T., Ali, K.M., Nadhman, A., 2015. Synthesis in plants and plant extracts of silver nanoparticles with potent antimicrobial properties: current status and future prospects. Appl. Microbiol. Biotechnol. 99, 9923–9934.

Mikhailov, O.V., Mikhailova, E.O., 2019. Elemental silver nanoparticles: biosynthesis and bio applications. Materials 12, 3177.

Mittal, A.K., Chisti, Y., Banerjee, U.C., 2013. Synthesis of metallic nanoparticles using plant extracts. Biotechnol. Adv. 31, 346–356.

Mohanta, Y.K., Panda, S.K., Biswas, K., Tamang, A., Bandyopadhyay, J., De, D., Mohanta, D., Bastia, A.K., 2016. Biogenic synthesis of silver nanoparticles from *Cassia fistula* (Linn.): *in vitro* assessment of their antioxidant, antimicrobial and cytotoxic activities. IET Nanobiotechnol. 10, 438–444.

Moores, A., Goettmann, F., 2006. The plasmon band in noble metal nanoparticles: an introduction to theory and applications. New J. Chem. 30, 1121–1132.

Moosa, A.A., Ridha, A.M., Allawi, M.H., 2015. Green synthesis of silver nanoparticles using *Spent Tea* leaves extract with atomic force microscopy. Int. J. Curr. Eng. Technol. 5, 3233–3241.

Moradi, F., Sedaghat, S., Moradi, O., Arab Salmanabadi, S., 2020. Review on green nano-biosynthesis of silver nanoparticles and their biological activities: with an emphasis on medicinal plants. Inorg. Nano-Metal Chem., 1–10.

Mostafa, E., Fayed, M.A.A., Radwan, R.A., Bakr, R.O., 2019. *Centaurea pumilio* L. extract and nanoparticles: a candidate for healthy skin. Colloids Surf. B Biointerfaces 182, 110350.

Mukherjee, S., Chowdhury, D., Kotcherlakota, R., Patra, S., Vinothkumar, B., Bhadra, M.P., Sreedhar, B., Patra, C.R., 2014. Potential theranostics application of bio-synthesized silver nanoparticles (4-in-1 system). Theranostics 4, 316–335.

Nadagouda, M.N., Varma, R.S., 2008. Green synthesis of silver and palladium nanoparticles at room temperature using coffee and tea extract. Green Chem. 10, 859–862.

Nahar, M.K., Zakaria, Z., Hashim, U., Bari, M.F., 2015a. Green synthesis of silver nanoparticles using *Momordica charantia* fruit extracts. Adv. Mater. Res. 1109, 35–39.

Nahar, M.K., Zakaria, Z., Hashim, U., Bari, M.F., 2015b. *Momordica charantia* fruit mediated green synthesis of silver nanoparticles. Green Process. Synth. 4, 235–240.

Naidu, K.S.B., Murugan, N., Adam, J.K., Sershen, 2019. Biogenic synthesis of silver nanoparticles from *Avicennia marina* seed extract and its antibacterial potential. BioNanoScience 9, 266–273.

Narayanan, K.B., Sakthivel, N., 2011. Extracellular synthesis of silver nanoparticles using the leaf extract of *Coleus amboinicus Lour*. Mater. Res. Bull. 46, 1708–1713.

Nejad, M.S., Khatami, M., Bonjar, G.H.S., 2015. *Streptomyces somaliensis* mediated green synthesis of silver nanoparticles. Nanomed. J. 2, 217–222.

Niaz, A., Bibi, A., Huma, Zaman, M.I., Khan, M., Rahim, A., 2018. Highly selective and eco-friendly colorimetric method for the detection of iodide using green tea synthesized silver nanoparticles. J. Mol. Liq. 249, 1047–1051.

Olenin, A.Y., 2019. Chemically modified silver and gold nanoparticles in spectrometric analysis. J. Anal. Chem. 74, 355–375.

Olenin, A.Y., Leenson, I.A., Lisichkin, G.V., 2018. Cryochemical co-condensation of metal vapors and organic compounds. In: Kharisov, B.I. (Ed.), Direct Synthesis of Metal Complexes. Elsevier Inc., Netherlands.

Omole, R.K., Torimiro, N., Alayande, S.O., Ajenifuja, E., 2018. Silver nanoparticles synthesized from *Bacillus subtilis* for detection of deterioration in the post-harvest spoilage of fruit. Sustain. Chem. Pharm. 10, 33–40.

Otunola, G.A., Afolayan, A.J., 2018. *In vitro* antibacterial, antioxidant and toxicity profile of silver nanoparticles green synthesized and characterized from aqueous extract of a spice blend formulation. Biotechnol. Biotechnol. Equip. 32, 724–733.

Ozin, G.A., Moskovits, M., 1976. Cryochemistry. John Wiley, USA.

Pak, Z.H., Abbaspour, H., Karimi, N., Fattahi, A., 2016. Eco-friendly synthesis and antimicrobial activity of silver nanoparticles using *Dracocephalum moldavica* seed extract. Appl. Sci. 6, 69.

Parthibavarman, M., Bhuvaneshwari, S., Jayashree, M., Raja, R.B., 2019. Green synthesis of silver (Ag) nanoparticles using extract of apple and grape and with enhanced visible light photocatalytic activity. BioNanoScience 9, 423–432.

Patel, P., Agarwal, P., Kanawaria, S., Kachhwaha, S., Kothari, S.L., 2015. Plant-based synthesis of silver nanoparticles and their characterization. In: Siddiqui, M.H., et al. (Eds.), Nanotechnology and Plant Sciences. Springer International Publishing Switzerland, pp. 271–288.

Patra, S., Mukherjee, S., Barui, A.K., Ganguly, A., Sreedhar, B., Patra, C.R., 2015. Green synthesis, characterization of gold and silver nanoparticles and their potential application for cancer therapeutics. Mater. Sci. Eng. C 53, 298–309.

Philip, D., 2009. Biosynthesis of Au, Ag and Au–Ag nanoparticles using edible mushroom extract. Spectrochim. Acta A 73, 374–381.

Philip, D., Unni, C., Aromal, S.A., Vidhu, V.K., 2011. *Murraya koenigii* leaf-assisted rapid green synthesis of silver and gold nanoparticles. Spectrochim. Acta A 78, 899–904.

Poggialini, F., Campanella, B., Giannarelli, S., Grifoni, E., Legnaioli, S., Lorenzetti, G., Pagnotta, S., Safi, A., Palleschi, V., 2018. Green-synthetized silver nanoparticles for nanoparticle-enhanced laser-induced breakdown spectroscopy (NELIBS) using a mobile instrument. Spectrochim. Acta B 141, 53–58.

Prathibha, B.S., Priya, K.S., 2017. Green synthesis of silver nanoparticles using *Tabebuia aurea* leaf extract. J. Appl. Chem. 10, 23–29.

Raghunandan, D., Mahesh, B.D., Basavaraja, S., Balaji, S.D., Manjunath, S.Y., Venkataraman, A., 2011. Microwave-assisted rapid extracellular synthesis of stable biofunctionalized silver nanoparticles from guava (*Psidium guajava*) leaf extract. J. Nanopart. Res. 13, 2021–2028.

Rai, M., Ingle, A.P., Gupta, I.R., Birla, S.S., Yadav, A.P., Abd-Elsalam, K.A., 2013. Potential role of biological systems in formation of nanoparticles: mechanism of synthesis and biomedical applications. Curr. Nanosci. 9, 576–587.

Rai, M., Birla, M., Ingle, A.P., Gupta, I., Gade, A., Abd-Elsalam, K., Marcato, P.D., Duran, P.D., 2014. Nanosilver: an inorganic nanoparticle with myriad potential applications. Nanotechnol. Rev. 3, 281–309.

Raj, D.R., Sudarsanakumar, C., 2017. Surface plasmon resonance-based fiber optic sensor for the detection of cysteine using diosmin capped silver nanoparticles. Sensors Actuators A 253, 41–48.

Raj, D.R., Prasanth, S., Vineeshkumar, T.V., Sudarsanakumar, C., 2016. Surface plasmon resonance-based fiber optic dopamine sensor using green synthesized silver nanoparticles. Sensors Actuators B Chem. 224, 600–606.

Rajendran, R., Ganesan, N., Balu, S.K., Alagar, S., Thandavamoorthy, P., Thiruvengadam, D., 2015. Green synthesis, characterization, antimicrobial and cytotoxic effects of silver nanoparticles using *Origanum heracleoticum L.* leaf extract. Int J Pharm Pharm Sci 7, 288–293.

Rajgovind, Sharma, G., Jasuja, N.D., Gupta, D.K., Joshi, S.C., 2015. Probing antibacterial properties of *Sida cordifolia* sponsored silver nanoparticles. IJRRA 2, 72–76.

Ramanathan, R., O'Mullane, A.P., Parikh, R.Y., Smooker, P.M., Bhargava, S.K., Bansal, V., 2011. Bacterial kinetics-controlled shape-directed biosynthesis of silver nanoplates using *Morganella psychrotolerans*. Langmuir 27, 714–719.

Ramya, M., Subapriya, M.S., 2012. Green synthesis of silver nanoparticles. Int. J. Pharm. Med. Biol. Sci. 1, 54–61.

Rautela, A., Rani, J., Debnath (Das), M., 2019. Green synthesis of silver nanoparticles from *Tectona grandis* seeds extract: characterization and mechanism of antimicrobial action on different microorganisms. J. Anal. Sci. Technol. 10, 5.

Ravichandran, V., Vasanthi, S., Shalini, S., Ali Shah, S.A., Tripathy, M., Paliwal, N., 2019. Green synthesis, characterization, antibacterial, antioxidant and photocatalytic activity of *Parkia speciosa* leaves extract mediated silver nanoparticles. Results Phys. 15, 102565.

Reddy, G.S., Saritha, K.V., Reddy, Y.M., Reddy, N.V., 2019. Eco-friendly synthesis and evaluation of biological activity of silver nanoparticles from leaf extract of *Indigofera barberi Gamble*: an endemic plant of Seshachalam Biosphere Reserve. SN Appl. Sci. 1, 968.

Remya, R.R., Rajasree, S.R.R., Aranganathan, L., Suman, T.Y., 2015. An investigation on cytotoxic effect of bioactive AgNPs synthesized using *Cassia fistula* flower extract on breast cancer cell MCF-7. Biotechnol. Rep. 8, 110–115.

Rubina, M.S., Elmanovich, I.V., Shulenina, A.V., Peters, G.S., Svetogorov, R.D., Egorov, A.A., Naumkin, A.V., Vasil'kov, A.Y., 2020. Chitosan aerogel containing silver nanoparticles: from metal-chitosan powder to porous material. Polym. Test. 86, 106481.

Rubina, M.S., Kamitov, E.E., Zubavichus, Y.V., Peters, G.S., Naumkin, A.V., Suzer, S., Vasil'kov, A.Y., 2016. Collagen-chitosan scaffold modified with Au and Ag nanoparticles: synthesis and structure. Appl. Surf. Sci. 366, 365–371.

Rubina, M.A., Said-Galiev, E.E., Naumkin, A.V., Shulenina, A.V., Belyakova, O.A., Vasil'kov, A.Y., 2019. Preparation and characterization of biomedical collagen–chitosan scaffolds with entrapped ibuprofen and silver nanoparticles. Polym. Eng. Sci. 59, 2479–2487.

Ruddaraju, L.K., Pallela, P.N.V.K., Pammi, S.V.N., Padavala, V.S., Kolapalli, V.R.M., 2019. Synergetic antibacterial and anticarcinogenic effects of *Annona squamosa* leaf extract mediated silver nanoparticles. Mater. Sci. Semicond. Process. 100, 301–309.

Sabela, M.I., Makhanya, T., Kanchi, S., Shahbaaz, M., Idress, D., Bisetty, K., 2018. One-pot biosynthesis of silver nanoparticles using *Iboza riparia* and *Ilex mitis* for cytotoxicity on human embryonic kidney cells. J. Photochem. Photobiol. B Biol. 178, 560–567.

Sadhasivam, S., Shanmugam, P., Yun, K.S., 2010. Biosynthesis of silver nanoparticles by *Streptomyces hygroscopicus* and antimicrobial activity against medically important pathogenic microorganisms. Colloids Surf. B 81, 358–362.

Sahayaraj, K., Rajesh, S., 2011. Bionanoparticles: synthesis and antimicrobial applications. In: Méndez-Vilas, A. (Ed.), Science against Microbial Pathogens: Communicating Current Research and Technological Advances. Formatex Research Center, Spain, pp. 228–244.

Sahu, S., Sharma, S., Kant, T., Shrivas, K., Ghosh, K.K., 2021. Colorimetric determination of L-cysteine in milk samples with surface functionalized silver nanoparticles. Spectrochim. Acta A Mol. Biomol. Spectrosc. 246, 118961.

Sankar, R., Karthik, A., Prabu, A., Karthik, S., Shivashangari, K.S., Ravikumar, V., 2013. *Origanum vulgare* mediated biosynthesis of silver nanoparticles for its antibacterial and anticancer activity. Colloids Surf. B 108, 80–84.

Sant, D.G., Gujarathi, T.R., Harne, S.R., Ghosh, S., Kitture, R., Kale, S., Chopade, B.A., Pardesi, K.R., 2013. *Adiantum philippense L.* frond assisted rapid green synthesis of gold and silver nanoparticles. J. Nanopart. 2013, 182320.

Santhoshkumar, J., Sowmya, B., Kumar, S.V., Rajeshkumar, S., 2019. Toxicology evaluation and antidermatophytic activity of silver nanoparticles synthesized using leaf extract of *Passiflora caerulea*. S. Afr. J. Chem. Eng. 29, 17–23.

Santiago, T.R., Bonatto, C.C., Rossato, M., Lopes, C.A.P., Lopes, C.A., Mizubuti, E.S.G., Silva, L.P., 2019. Green synthesis of silver nanoparticles using tomato leaf extract and their entrapment in chitosan nanoparticles to control bacterial wilt. J. Sci. Food Agric. 99, 4248–4259.

Saranyaadevi, K., Subha, V., Ravindran, R.S.E., Renganathan, S., 2014. Green synthesis and characterization of silver nanoparticle using leaf extract of *Capparis zeylanica*. Asian J. Pharm. Clin. Res. 7, 44–48.

Sathiyapriya, R., Geetha, D., Ramesh, P.S., 2016. Nanofabrication and characterization of silver nanostructures from *Carica papaya* leaf extract and their antioxidant activity. Int. J. Adv. Sci. Eng. 3, 271–277.

Schröfel, A., Kratošová, G., Šafařík, I., Šafaříková, M., Raška, I., Shor, L.M., 2014. Applications of biosynthesized metallic nanoparticles—a review. Acta Biomater. 10, 4023–4042.

Serdar, G., Albay, C., Sökmen, M., 2019. Biosynthesis and characterization of silver nanoparticles from the lemon leaves extract. Cumhuriyet Sci. J. 40, 170–172.

Shameli, K., Ahmad, M.B., Yunus, W.M.Z.W., Rustaiyan, A., Ibrahim, N.A., Zargar, M., Abdollahi, Y., 2010. Green synthesis of silver/montmorillonite/chitosan bionanocomposites using the UV irradiation method and evaluation of antibacterial activity. Int. J. Nanomedicine 5, 875–887.

Shen, Z., Han, G., Liu, C., Wang, X., Sun, R., 2016. Green synthesis of silver nanoparticles with bagasse for colorimetric detection of cysteine in serum samples. J. Alloys Compd. 686, 82–89.

Siddiqi, K.S., Rashid, M., Tajuddin, Husen, A., Rehman, S., 2019. Biofabrication of silver nanoparticles from *Diospyros montana*, their characterization and activity against some clinical isolates. BioNanoScience 9, 302–312.

Singh, J., Kaur, G., Kaur, P., Bajaj, R., Rawat, M., 2016. A review on green synthesis and characterization of silver nanoparticles and their applications: a green nanoworld. World J. Pharm. Pharm. Sci. 5, 730–762.

Singh, J., Mehta, A., Rawat, M., Basu, S., 2018. Green synthesis of silver nanoparticles using sun dried tulsi leaves and its catalytic application for 4-nitrophenol reduction. J. Environ. Chem. Eng. 6, 1468–1474.

Singh, S., Saikia, J.P., Buragohain, A.K., 2013. A novel 'green' synthesis of colloidal silver nanoparticles (SNP) using *Dillenia indica* fruit extract. Colloids Surf. B 102, 83–85.

Singhal, G., Bhavesh, R., Kasariya, K., Sharma, A.R., Singh, R.P., 2011. Biosynthesis of silver nanoparticles using *Ocimum sanctum* (Tulsi) leaf extract and screening its antimicrobial activity. J. Nanopart. Res. 13, 2981–2988.

Sivasankar, P., Seedevi, P., Poongodi, S., Sivakumar, M., Murugan, T., Sivakumar, L., Sivakumar, K., Balasubramanian, T., 2018. Characterization, antimicrobial and antioxidant property of exopolysaccharide mediated silver nanoparticles synthesized by *Streptomyces violaceus* MM72. Carbohydr. Polym. 181, 752–759.

Smetana, A.B., Klabunde, K.J., Sorensen, C.M., 2005. Synthesis of spherical silver nanoparticles by digestive ripening, stabilization with various agents, and their 3-D and 2-D superlattice formation. J. Colloid Interface Sci. 284, 521–526.

Soleimani, F.F., Saleh, T., Shojaosadati, S.A., Poursalehi, R., 2018. Green synthesis of different shapes of silver nanostructures and evaluation of their antibacterial and cytotoxic activity. BioNanoScience 8, 72–80.

Suganthi, S., Vignesh, S., Mohanapriya, S., Sundar, J.K., Raj, V., 2019. Microwave-assisted synthesis of L-histidine capped silver nanoparticles for enhanced photocatalytic activity under visible light and effectual antibacterial performance. J. Mater. Sci. Mater. Electron. 30, 15168–15183.

Taha, Z.K., Hawar, S.N., Sulaiman, G.M., 2019. Extracellular biosynthesis of silver nanoparticles from *Penicillium italicum* and its antioxidant, antimicrobial and cytotoxicity activities. Biotechnol. Lett. 41, 899–914.

Terenteva, E.A., Apyari, V.V., Kochuk, E.V., Dmitrienko, S.G., Zolotov, Y.A., 2017. Use of silver nanoparticles in spectrophotometry. J. Anal. Chem. 72, 1138–1154.

Thampi, N., Jeyadoss, V.S., 2015a. Biogenic synthesis and characterization of silver nanoparticles using *Syzygium samarangense* (wax apple) leaves extract and their antibacterial activity. Int. J. PharmTech Res. 8, 426–433.

Thampi, N., Jeyadoss, V.S., 2015b. Bio-prospecting the in-vitro antioxidant and anti-cancer activities of silver nanoparticles synthesized from the leaves of *Syzygium Samarangense*. Int J Pharm Pharm Sci 7, 269–274.

Tripathi, R.M., Rana, D., Shrivastav, A., Singh, R.P., Shrivastav, B.R., 2013. Biogenic synthesis of silver nanoparticles using *Saraca indica* leaf extract and evaluation of their antibacterial activity. Nano-Biomed. Eng. 5, 50–56.

Vaidyanathan, R., Kalishwaralal, K., Gopalram, S., Gurunathan, S., 2009. Nanosilver—the burgeoning therapeutic molecule and its green synthesis. Biotechnol. Adv. 27, 924–937.

Vasileva, P., Donkova, B., Karadjova, I., Dushkin, C., 2011. Synthesis of starch-stabilized silver nanoparticles and their application as a surface plasmon resonance-based sensor of hydrogen peroxide. Colloids Surf. A Physicochem. Eng. Asp. 382, 203–210.

Vasil'kov, A.Y., Dovnar, R.I., Smotryn, S.M., Iaskevich, N.N., Naumkin, A.V., 2018a. Plasmon resonance of silver nanoparticles as a method of increasing their antibacterial action. Antibiotics 7 (80), 1–18.

Vasil'kov, A.Y., Dovnar, R.I., Smotryn, S.M., Iaskevich, N.N., Naumkin, A.V., 2018b. Healing of experimental aseptic skin wound under the influence of the wound dressing, containing silver nanoparticles. Am. J. Nanotechnol. Nanomed. 1 (2), 70–77.

Vasil'kov, A., Naumkin, A., Rubina, M., Gromovykh, T., Pigaleva, M., 2019. Hybrid materials based on metal-containing microcrystalline and bacterial cellulose: green synthesis and characterization. In: International Multidisciplinary Scientific GeoConference Surveying Geology and Mining Ecology Management, SGEM. 19(6.1), pp. 199–206.

Vasil'kov, A.Y., Nikitin, L.N., Naumkin, A.V., Volkov, I.O., Buzin, M.I., Abramchuk, S.S., Bubnov, Y.N., Tolstopyatov, E.M., Grakovich, P.N., Pleskachevskii, Y.M., 2009. Gold- and silver-containing fibroporous polytetrafluoroethylene obtained under laser irradiation, supercritical carbon dioxide treatment, and metal-vapor synthesis. Nanotechnol. Russ. 4, 834–840.

Veisi, H., Dadres, N., Mohammadi, P., Hemmati, S., 2019. Green synthesis of silver nanoparticles based on oil-water interface method with essential oil of orange peel and its application as nanocatalyst for A^3 coupling. Mater. Sci. Eng. C 105, 110031.

Verma, S., Kumari, B., Shrivastava, J.N., 2014. Green synthesis of silver nanoparticles using single cell protein of *Spirulina platensis*. Int J Pharm. Bio. Sci 5, B458–B464.

Vertelov, G.K., Olenin, A.Y., Lisichkin, G.V., 2007. Use of nanoparticles in the electrochemical analysis of biological samples. J. Anal. Chem. 62, 813–824.

Vidhu, V.K., Philip, D., 2014. Catalytic degradation of organic dyes using biosynthesized silver nanoparticles. Micron 56, 54–62.

Vilas, V., Philip, D., Mathew, J., 2014. Catalytically and biologically active silver nanoparticles synthesized using essential oil. Spectrochim. Acta A 132, 743–750.

Vilela, D., González, M.C., Escarpa, A., 2012. Sensing colorimetric approaches based on gold and silver nanoparticles aggregation: chemical creativity behind the assay. A review. Anal. Chim. Acta 751, 24–43.

Vishwasrao, C., Momin, B., Ananthanarayan, L., 2019. Green synthesis of silver nanoparticles using *Sapota* fruit waste and evaluation of their antimicrobial activity. Waste Biomass Valoriz. 10, 2353–2363.

Vitulli, G., Evangelisti, C., Caporusso, A., Pertici, P.N., Bertozzi, S., Salvadori, P., 2008. Metal vapor-derived nanostructured catalysts in fine chemistry. In: Corain, B., Array, B., Toshima, N. (Eds.), Metal Nanoclusters in Catalysis and Materials Science. Elsevier, Netherlands, pp. 437–451.

Wanarska, E., Maliszewska, I., 2019. The possible mechanism of the formation of silver nanoparticles by *Penicillium cyclopium*. Bioorg. Chem. 93, 102803.

Wang, L., Wu, Y., Xie, J., Wu, S., Wu, Z., 2018. Characterization, antioxidant and antimicrobial activities of green synthesized silver nanoparticles from *Psidium guajava* L. leaf aqueous extracts. Mater. Sci. Eng. C 86, 1–8.

Weibel, J.-M., Blanc, A., Pale, P., 2008. Ag-mediated reactions: coupling and heterocyclization reactions. Chem. Rev. 108, 3149–3173.

Welch, C.M., Compton, R.G., 2006. The use of nanoparticles in electroanalysis: a review. Anal. Bioanal. Chem. 384, 601–619.

Xie, J., Lee, J.Y., Wang, D.I.C., Ting, Y.P., 2007. Silver nanoplates: from biological to biomimetic synthesis. ACS Nano 1, 429–439.

Xu, L., Wang, Y.Y., Huang, J., Chen, C.Y., Wang, Z.X., Xie, H., 2020. Silver nanoparticles: synthesis, medical applications and biosafety. Theranostics 10 (20), 8996.

Yao, P., Zhang, J., Xing, T., Chen, G., Tao, R., Choo, K.-H., 2018. Green synthesis of silver nanoparticles using grape seed extract and their application for reductive catalysis of Direct Orange 26. J. Ind. Eng. Chem. 58, 74–79.

Zaki, S., El-Kady, M.F., Abd-El-Haleem, D., 2011. Biosynthesis and structural characterization of silver nanoparticles from bacterial isolates. Mater. Res. Bull. 46, 1571–1576.

Zhang, Z., Li, S., Gu, X., Li, J., Lin, X., 2019a. Biosynthesis, characterization and antibacterial activity of silver nanoparticles by the Arctic anti-oxidative bacterium *Paracoccus sp.* Arc 7-R13. Artif. Cells Nanomed. Biotechnol. 47, 1488–1495.

Zhang, Z., Xin, G., Zhou, G., Li, Q., Veeraraghavan, V.P., Mohan, S.K., Wang, D., Liu, F., 2019b. Green synthesis of silver nanoparticles from *Alpinia officinarum* mitigates cisplatin-induced nephrotoxicity via down-regulating apoptotic pathway in rats. Artif. Cells Nanomed. Biotechnol. 47, 3212–3221.

Zhu, S., Fu, Y., Hou, J., 2010. Topical review: metallic nanoparticles array for immunoassay. J. Comput. Theor. Nanosci. 7, 1855–1869.

PART 3

Plant synthesis

CHAPTER 11

Plant extract and agricultural waste-mediated synthesis of silver nanoparticles and their biochemical activities

Pathikrit Saha and Beom Soo Kim

Department of Chemical Engineering, Chungbuk National University, Cheongju, Chungbuk, Republic of Korea

1 Introduction

The small size, shape, and large surface-to-volume ratio are some of the key properties of nanoparticles that can interact with plants, animals, and microorganisms. These unique features are advantageous in several biotechnological applications such as biomedicine, environmental bioremediation, optics, catalyst, wastewater treatment, drug delivery, and bioimaging (Ahmad et al., 2003; Gardea-Torresday et al., 2003; Pugazhendhi et al., 2015; Rauwel et al., 2015; Ramesh et al., 2015; Siddiqi et al., 2018; Song et al., 2012). The most extensively studied metal nanoparticles are silver, platinum, gold, and palladium (Ramesh et al., 2015). Among these, silver nanoparticles (AgNPs) have been used largely due to their broad bactericidal and fungicidal activity, as well as their ability to coordinate with various ligands and macromolecules in microbial cells (Hamouda et al., 2019; Krishnaraj et al., 2010; Velsi et al., 2016; Vivek et al., 2011).

AgNPs have been formulated using chemical, physical, and biological methods (Gardea-Torresday et al., 2003; Li et al., 2007; Kosmala et al., 2015; Tien et al., 2008; Zhang et al., 2007). The toxicity of chemicals, high energy, and operating costs of chemical and physical methods are the main limitations that result in a need to explore other alternative green methods for AgNP synthesis (Rauwel et al., 2015). To increase energy efficiency, the synthesis process should be carried out at ambient temperature and pressure and at neutral pH (Rauwel et al., 2015). In recent decades, various kinds of green methods including plant extracts, fungi, bacteria, and agricultural wastes have been extensively studied for the synthesis of AgNPs (Ahmad et al., 2003; Gardea-Torresday et al., 2003; Klaus et al., 1999; Rauwel et al., 2015). One of the first studies reported the use of the prokaryotic bacterium *Pesudomonas stutzeri* AG259 as a silver-resistant strain cultured in high concentrations of silver nitrate. Silver-based crystals up to 200 nm in size were observed to accumulate in the periplasmic space (Klaus et al., 1999). The extracellular biosynthesis of AgNPs by fungi was first reported in *Fusarium oxysporum*, which secretes enzymes responsible for the reduction process (Ahmad et al., 2003). Since the obtained nanoparticles do not bind to biomass, extracellular biosynthesis is more advantageous than intracellular biosynthesis (Ahmad et al., 2003; Duran et al., 2005). Plant extract-mediated synthesis is faster and easier compared to using bacteria and fungi (Mittal et al., 2013). Plant extracts can be used for the reduction and stabilization of nanoparticles (Iravani, 2011; Mittal et al., 2013). Plant extracts also stabilize nanoparticles due to the presence of phytochemicals such as flavonoids, alkaloids, and phenols as components to prevent aggregation. Besides, the use of plant extracts for nanoparticle synthesis can be advantageous over other environmentally friendly biological processes as there is no cell culture maintenance (Bar et al., 2009). In addition, nanoparticle preparation using plant extracts is easily scalable compared to relatively expensive microbial methods (Dhillon et al., 2012; Iravani, 2011; Li et al., 2011). The main issue in this scenario is choosing plants that can use the extract (Rauwel et al., 2015). Most plant parts such as leaves, roots, latex, bark, stems, and seeds are used for nanoparticle synthesis (Siddiqi et al., 2018; Rauwel et al., 2015). Over the past 20 years, AgNPs have been synthesized using plant extracts and agricultural wastes and their biochemical activities have been investigated. Plant extract-based synthesis of AgNPs has several biochemical applications such as antibacterial, anticancer, antifungal, antioxidant, antiviral, and photocatalytic dye degradation (Anthony et al., 2014; Behravan et al., 2019; Jagtap and Bapat, 2013; Oliver et al., 2018; Philip et al., 2011; Siddiqi et al., 2018; Rajkumar et al., 2021). AgNPs showed potent antibacterial activity against a wide range of gram-positive and gram-negative bacteria (Jagtap and Bapat, 2013; Muthukrishnan et al., 2015) due to their interaction with the bacterial cell wall leading to inactivation (Fu et al., 2016; He et al., 2017; Jeeva et al., 2014; Salem et al., 2014). In addition, AgNPs showed effective antioxidant activity against various free radicals such as 2,2-diphenyl-1-picrylhydrazyl (DPPH), 2,2′-azino-bis-3-ethylbenzothiazoline-6-sulphonic acid (ABTS), etc. (Abdel-Aziz et al., 2014; Banerjee and Narendhirakanan, 2011; Dipankar and Murugan, 2012; Zangeneh, 2019). Similarly, a wide range of fungal spores has been reported to be killed by AgNPs by damaging membrane integrity (Krishnaraj et al., 2012; Medda et al., 2015; Velmurugan et al., 2015). In recent years, AgNPs have been utilized against several viral strains such as retroviridae, herpesviridae, poxviridae, orthomyxoviridae, hepadnaviridae, arenaviridae, and paramyxoviridae (Haggag et al., 2019; Sharma et al., 2019; Sujitha et al., 2015; Zeedan et al., 2020). In this chapter, we discuss in detail the plant and agricultural waste-based synthesis of AgNPs and their applications in various fields.

2 Synthesis of AgNPs by plants and agricultural waste

In general, metallic nanoparticles are formed via two approaches: bottom-up (material construction at the bottom: atom by atom, molecule by molecule, or cluster by cluster) and top-down (material in bulk form). The bottom-up approach is considered a superior choice as the synthesis of nanoparticles based on catalysts (e.g., reducing agents and enzymes) involves a homogeneous system controlled by the catalyst itself. In a top-down approach, the size reduction of nanoparticles is achieved through special ablation such as thermal decomposition, mechanical grinding, cutting, etching, and sputtering. The main drawback of the top-down approach is surface defects that can be observed in physical features and surface chemistry (Siddiqi et al., 2018). Chemical, physical, and biological methods are different techniques for AgNP formulation. Chemical reduction, electrochemistry, irradiation, and pyrolysis are some of the key techniques of chemical methods. Physical methods include arc discharge, physical vapor decomposition, and energy ball milling (Li et al., 2007; Zhang et al., 2007; Tien et al., 2008; Kosmala et al., 2015; Asanithi et al., 2012; Salunke et al., 2016). However, biological methods are used as an alternative approach due to the toxicity of chemicals and the high energy and cost requirements of chemical and physical methods. Additionally, the biological approach is cost-effective, environmentally friendly, and scalable (Siddiqi et al., 2018). Various biological resources such as bacteria, fungi, and plants have been utilized for the synthesis of metal nanoparticles (Rauwel et al., 2015). Green synthesis using plant extracts appears to be faster than using bacteria and fungi (Rauwel et al., 2015).

2.1 Synthesis by plants

Plant extracts are known as potent reducing and stabilizing agents for the synthesis of metal nanoparticles (Vidhu et al., 2011; Gul et al., 2021; Jeyaraj et al., 2013). One of the first approaches to utilizing plants as a source for AgNP synthesis was reported in alfalfa sprouts (Gardea-Torresday et al., 2003). Alfalfa roots can absorb Ag ions and transfer them to the shoots in a similar oxidation state. Plant parts such as leaves, roots, stems, shoot flowers, bark, seeds, and their metabolites have been successfully exploited for nanoparticle synthesis (Husen and Siddqui, 2014). The first approach using plant leaf extract was studied by Sastry's group (Sankar et al., 2003). Ag^+ was reduced to Ag^0 using geranium leaf extract (*Pelargonium graveolens*) to obtain a spherical average particle size of 27 nm. Likewise, five plant leaves such as pine (*Pinus desiflora*), persimmon (*Diopyros kaki*), ginko (*Ginko biloba*), magnolia (*Magnolia kobus*), and platanus (*Platanus orientalis*) were used, and the particle size could be controlled by changing the reaction temperature, leaf broth concentration, and $AgNO_3$ concentration (Song and Kim, 2009).

Plant extracts play an important role in the stabilization of AgNPs (Bar et al., 2009; Chen et al., 2008; Sankar et al., 2003). AgNPs synthesized by *Jatropha curcas* was reported to be stable after 1 month due to the presence of cyclic octapeptide (curcacycline A), cyclic nonapeptide, and enzyme (curcain) (Bar et al., 2009). On the other hand, *Abelmoschus esculentus* pulp extract was reported to hold AgNPs stable for 8 months and no significant changes were observed in the surface plasmon resonance (SPR) band at 403 nm or spectral shape (Mollick et al., 2019).

The reaction temperature plays a key role in the efficient conversion to AgNPs (Huang et al., 2011). The bioreduction rate was higher at 90°C and 60°C than at 30°C, which can be further demonstrated in terms of conversion. At 60°C and 90°C, conversions were 83.6% and 98.2%, respectively, compared to 27.8% at 30°C. The wavelengths of the maximum absorption peaks (433, 445, and 455 nm) indicated that AgNPs were obtained in a spherical shape with a size range of 50–100 nm. AgNPs synthesized using *Acacia leucophloea* extract at a reaction temperature of 90°C showed SPR spectra at 420 nm in the size range of 11–29 nm (Murugan et al., 2014). The bandwidth of the SPR decreased as the amount of plant extract increased and shifted to lower wavelengths (432 nm to 411 nm) (Philip et al., 2011). Likewise, pH acts as an important parameter for AgNP synthesis as observed by Chutrakulwong et al. (2020). As the pH increased to 8.5, the AgNP formation rate increased. This means that alkaline medium provided better conditions for particle growth compared to acidic medium due to the negatively charged OH^- ions. In addition, an alkaline medium enhanced the nucleation of silver from Ag^+ (Chutrakulwong et al., 2020). Moreover, the particle size was dependent on the amount of reducing agent to the precursor (Pugazhendhi et al., 2015). They detected two or three bands indicating the anisotropic shape of the particles.

2.2 Synthesis by agricultural wastes

Recently, agricultural wastes have been applied as reducing and stabilizing agents for AgNP synthesis (Baicco et al., 2016; Barros et al., 2018; Sinsinwar et al., 2018). Dried grass waste was used for the production of nanoparticles with a spherical-oblate profile and a particle size range of 4–34 nm. The formation of nanoparticles was confirmed by the appearance of a UV-Vis peak in the range of 430–450 nm (Khatami et al., 2018). Bilberry waste (BW) and red currant waste (RCW) were employed as reducing and stabilizing agents for AgNP synthesis (Zuorro et al., 2019). Alkaline pH favored reduction using both wastes and the hydrodynamic size of AgNPs obtained using BW was found to be smaller than that obtained using RCW. The effectiveness of BW over RCW was due to higher phenolic content and reducing capacity. Similarly, AgNPs using BW and spent coffee grounds were synthesized with a spherical size range of 10–20 nm (Baicco et al., 2016). Waste orange peel extract was also exploited for AgNP synthesis (Kahrilas et al., 2014; Barros et al., 2018; Palanisamy et al., 2021). Subsequently, hesperidin and nanocellulose were prepared from orange peel and orange bagasse, respectively, and used for nanoparticle synthesis. Nanocellulose acted as a stabilizer, while silver ions were reduced by hesperidin. The hydrodynamic size of the nanoparticles obtained using orange extract was 48 nm, while the hesperidin-based nanoparticle size was 25.4 nm. This study demonstrated that large amounts of proteins present in oranges such as germin-like protein with oxidase activity and glutathione S-transferase are responsible for the reduction process.

2.3 Mechanism of plant and agricultural waste-mediated AgNP synthesis

Plant extracts and agricultural wastes contain various types of metabolites and phytochemicals such as polyphenols, alkaloids, terpenoids, caffeic acid, p-coumaric acid, quinone, etc. Fourier transform infrared spectroscopy is an important tool for the characterization of the capping of nanoparticles by these biomolecules (Li et al., 2007; Bar et al., 2009; Mollick et al., 2019). These biomolecules can undergo prompt transformation while producing AgNPs (Jha et al., 2009). The authors suggested that three types of benzoquinones—cyperoquinone (type I), dietchequinone (type II), and remirin (type III)—can go through redial tautomerization under gentle warming and subsequent incubation, leading to the reduction of silver precursors. Furthermore, silver ions can be entrapped inside the core structure of the cyclic proteins and then reduced and stabilized by the amide group of the host peptide (Bar et al., 2009). Likewise, phenolic acids can reduce silver precursors due to their strong antioxidant properties (Vidhu et al., 2011). Phenolic acids are a large family of secondary metabolites containing a hydroxyl benzoic or cinnamic structure. These structures have a strong metal chelating ability that can be related to the high nucleophilic character of aromatic rings. Kumar et al. (2012) observed that the hydroxyl and ketone groups in plant extracts reduce silver precursors and form a layer on the synthesized nanoparticles, providing long-term stability. Water-soluble ascorbic acid has also been reported to participate in AgNP synthesis using cashew apple juice (*Anacardium occidentale* L) (Mukunthan and Balaji, 2012). Due to its strong reducing property, ascorbic acid decomposes to ascorbate radical and free electron. Then the ascorbate radical is reduced again by NADH or reduced glutathione. These ascorbate and glutathione are essential nonenzymatic antioxidants that maintain a reduced state and in turn convert Ag^+ to Ag^0. Furthermore, aldehyde group is responsible for the reduction and stabilization of AgNPs (Shameli et al., 2012). Fig. 1 illustrates a plausible schematic for AgNP biosynthesis.

3 Applications of AgNP
3.1 Antibacterial activity

The potent microbial biocidal activity of silver compounds along with their ability to serve as drugs without any microbial resistance makes them an alternative to chemical drugs (Anthony et al., 2014). AgNPs exhibited improved antibacterial activity compared to other silver compounds such as silver nitrate due to their small size, crystallinity, high surface-to-volume ratio, and morphology (Anthony et al., 2014;

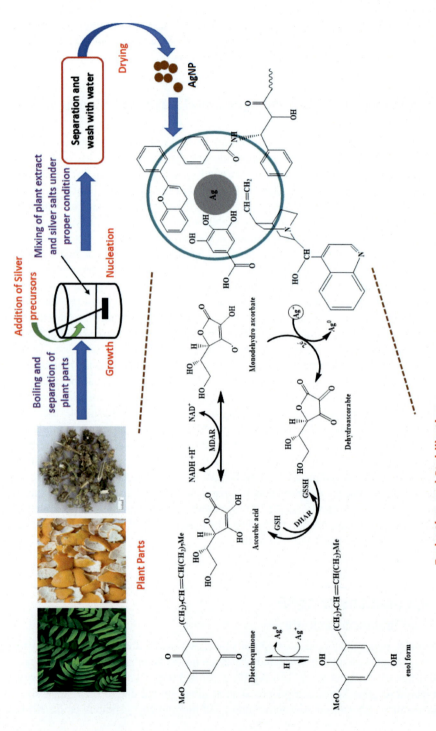

FIG. 1

Plausible schematics of AgNP biosynthesis.

Krishnaraj et al., 2010; Sun et al., 2014; Nagaich et al., 2016; Velsi et al., 2016). Due to these properties, AgNPs can increase membrane permeability and cell destruction (Krishnaraj et al., 2010; Nabikhan et al., 2010; Allafchian et al., 2016). Some studies observed the same minimum inhibitory concentration (MIC) and minimum bacterial concentration of AgNPs and silver nitrate against food-borne pathogens (Du et al., 2019).

Spherical AgNPs synthesized using *Ceropegia thwaitesii* as a reducing and stabilizing agent demonstrated antibacterial activity against a wide range of bacterial microorganism such as *E. coli*, *V. cholerae*, *Corynebacterium* sp., *S. aureus*, *P. aeruginosa*, *S. epidermidis*, *Mycobacterium* sp., *S. typhi*, *S. flexneri*, *K. pneumonia*, *B. subtilis*, *M. luteus*, and *P. mirabilis* (Muthukrishnan et al., 2015). Among them, *S. typhi* and *B. subtilis* showed maximum susceptibility, while *S. aureus* and *S. epidermis* showed minimal susceptibility. In another study, AgNPs produced using *Artocarpus heterophyllus* Lam seed extract showed antibacterial activity against *B. cereus*, *B. subtilis*, and *S. aureus*, while no inhibition zones were observed for *S. typhimurium* and *P. vulgaris* (Jagtap and Bapat, 2013). In the case of *P. aeruginosa*, the antibacterial activity was severely decreased after 7h of treatment, whereas the activities of *E. coli* and *S. aureus* remained the same from the beginning (He et al., 2017). Green synthesized AgNPs form electrostatic interactions with the structural components of bacteria, i.e., cell walls, plasma membranes, DNA, and proteins (Chaloupka et al., 2010; Jagtap and Bapat, 2013; Vanti et al., 2019). Table 1 summarizes the antibacterial activity of green synthesized AgNPs against a wide range of microorganisms.

Table 1 Antibacterial activity of plant extract or agricultural waste synthesized AgNPs.

No.	Plant extract or agricultural waste used	Size and shape	Bacteria tested	Inhibition zone/ minimum inhibitory concentration (MIC)	Reference
1	*Tamarix gallica*	12 nm, spherical	*E. coli*	9 mm	Lopez-Miranda et al. (2016)
2	*Sapindus mukorossi*	69±5 nm, spherical	*B. subtilis*, *S. aureus*, *E. coli*, and *P. aeruginosa*	4 µg/mL (*B. subtilis*) 3 µg/mL (*S. aureus*) 2 µg/mL (*E. coli*) 1 µg/mL (*P. aeruginosa*)	Dinda et al. (2017)
3	*Citrus sinensis*	10–35 nm, spherical	*E. coli*, *P. aeruginosa*, and *S. aureus*	14.3 mm (*E. coli*) 12.5 mm (*P. aeruginosa*) 8.5 mm (*S. aureus*)	Kaviya et al. (2011)
4	*Acacia leucophloea*	17–29 nm, spherical	*S. aureus*, *B. cereus*, *L. monocytogenes*, and *S. flexneri*	21 mm (*S. aureus*) 20 mm (*B. cereus*) 19 mm (*L. monocytogenes*) 17 mm (*S. flexneri*)	Murugan et al. (2014)

Continued

Table 1 Antibacterial activity of plant extract or agricultural waste synthesized AgNPs. Continued

No.	Plant extract or agricultural waste used	Size and shape	Bacteria tested	Inhibition zone/ minimum inhibitory concentration (MIC)	Reference
5	Salvia leriifolia	~27 nm, spherical	P. aeruginosa, E. coli, S. coagulase, C. frurdii, E. aerogenes, A. baumannii, S. marcescens, and K. pneumoniae	21.7 mm (P. aeruginosa) 14 mm (E. coli) 18.67 mm (S. coagulase) 19.33 mm (C. frurdii) 18.33 mm (E. aerogenes) 21.5 mm (A. baumannii) 12 mm (S. marcescens) 16.33 mm (K. pneumonia) 20 mm (S. pnemoniae)	Baghayeri et al. (2018)
6	Melissa officinalis	~12 nm	E. coli and S. aureus	12.5 mm (E. coli) 11.5 mm (S. aureus)	Ruiz-Baltazar et al. (2017)
7	Erythrina indica	20–118 nm	S. aureus, M. luteus, E. coli, B. subtilis, S. typhi, and S. paratyphi	16 mm (S. aureus, B. subtilis, and E. coli) 12 mm (M. luteus and S. paratyphi) 9.83 mm (S. typhi)	Sre et al. (2015)
8	Coffee bean	10–30 nm, spherical	E. coli and S. aureus	0.04 μmol/L (E. coli) 0.2 μmol/L (S. aureus)	Wang et al. (2017)
9	Egg white protein	2–20 nm, spherical	E. coli and S. typhimurium	4 μg/mL (E. coli) 6 μg/mL (S. typhimurium)	Thiyagarajan et al. (2018)
10	Ocimum tenuiflorum	15–25 nm, spherical	E. coli, C. bacterium, and B. substilus	20 mm (E. coli) 20 mm (C. bacterium) 26 mm (B. substilus)	Patil et al. (2012)
11	Ficus benghalensis	~16 nm, spherical	E. coli	25 mm	Saxena et al. (2012)
12	Parmotrema praesorediosum	19 nm, spherical	P. vulgaris, P. aeruginosa, S. marcescens, S. typhi, S. epidermidis, Methicillin-resistant S. aureus, B. subtilis, and S. faecalis	8 mm (P. vulgaris, P. aeruginosa, and S. marcescens) 7.5 mm (S. typhi) 6 mm (S. epidermidis, S. aureus, B. subtilis, and S. faecalis)	Mie et al. (2014)
13	Crataegus douglasii	29.28 nm, spherical	E. coli and S. aureus	9 mm (E. coli) 13 mm (S. aureus)	Ghafferi-Moggadam and Hadi-Dabanlou (2014)
14	Syzygium cumini	10–15 nm, spherical	E. coli, P. aeruginosa, B. subtilis, S. aureus, and M. smegmatis	24 mm (E. coli) 21 mm (P. aeruginosa and M. smegmatis) 26 mm (B. subtilis and S. aureus)	Gupta et al. (2014)

Table 1 Antibacterial activity of plant extract or agricultural waste synthesized AgNPs. *Continued*

No.	Plant extract or agricultural waste used	Size and shape	Bacteria tested	Inhibition zone/minimum inhibitory concentration (MIC)	Reference
15	*Cacumen Platycladi*	18.4 nm, spherical	*E. coli* and *S. aureus*	1.4 ppm (*E. coli*) 5.4 ppm (*S. aureus*)	Huang et al. (2011)
16	*Elaeagnus umbellate* fractionated in hexane (EUH), ethyl acetate (EUE), dichloromethane (EUD), butanol (EUB), and water (EUW)	20–100 nm, spherical with 40 nm (EUW)	*E. coli* and *S. aureus*	*S. aureus*: 70 mm (EUE and EUB), 75 mm (EUD), and 90 mm (EUW) *E. coli*: 70 mm (EUE), 74 mm (EUD), 71 mm (EUB), and 92 mm (EUE)	Ali et al. (2020)
17	*Lavandula stoechas*	20–50 nm, spherical	*S. aureus* and *P. aeruginosa*	125 µg/mL (*S. aureus*) 250 µg/mL (*P. aeruginosa*)	Mahmoudi et al. (2019)
18	*Phlomis*	25 nm, spherical	*S. typhi*, *E. coli*, *S. aureus*, and *B. cereus*	15 mm (*S. typhi*, *E. coli*) 14.7 mm (*S. aureus*) 12.1 mm (*B. cereus*)	Allafchian et al. (2016)
19	*Uncaria gambir*	6–41 nm, spherical	*E. coli* and *S. aureus*	34 mm (*E. coli*) 28 mm (*S. aureus*)	Labannia et al. (2019)
20	Waste grass	15 nm, spherical	*A. baumanni* and *P. aeruginosa*	1.56 ppm (*A. boumanni*) 0.78 ppm (*P. aeruginosa*)	Khatami et al. (2018)
21	Coconut shell	14.20–22.96 nm, spherical	*S. aureus*, *E. coli*, *S. typhi*, and *L. monocytogenes*	15 mm (*S. aureus*) 13 mm (*E. coli*, *S. typhi*) 10 mm (*L. monocytogenes*) MIC: 26 µg/mL (*S. aureus*) 53 µg/mL (*E. coli*) 106 µg/mL (*S. typhi*) 212 µg/mL (*L. monocytogenes*)	Sinsinwar et al. (2018)
22	Banana peel	23.7 nm, spherical	*E. coli*, *P. aeruginosa*, *B. subtilis*, and *S. aureus*	6.8 µg/mL (*E. coli*) 5.1 µg/mL (*P. aeruginosa*) 1.70 µg/mL (*B. subtilis*) 3.4 µg/mL (*S. aureus*)	Ibrahim (2015)
23	Teak waste leaves	26.36 nm, spherical	*E. coli* and *S. aureus*	~16 mm (*E. coli*) 18 mm (*S. aureus*) MIC: 25.6 µg/mL	Devadiga et al. (2015)
24	Cavendish banana	23–30 nm, spherical	*K. pneumoniae*, *E. coli*, *S. aureus*, and *B. subtilis*	22 mm (*K. pneumonia*) 20 mm (*E. coli*) 19 mm (*S. aureus*) 17 mm (*B. subtilis*)	Kokila et al. (2015)
25	*Kalopanax pictus*	10–30 nm, spherical	*E. coli*	30 µg/mL	Salunke et al. (2014)

Continued

Table 1 Antibacterial activity of plant extract or agricultural waste synthesized AgNPs. *Continued*

No.	Plant extract or agricultural waste used	Size and shape	Bacteria tested	Inhibition zone/ minimum inhibitory concentration (MIC)	Reference
26	Tamarind shell-husk extract	5–8 nm, spherical	*E. coli* and *S. aureus*	4.0 mm (*E. coli*) 4.6 mm (*S. aureus*)	Tade et al. (2020)

E. coli: Escherichia coli; *B. subtilis*: Bacillus subtilis; *S. aureus*: Staphylococcus aureus; *P. aeruginosa*: Pseudomonas aeruginosa; *L. monocytogenes*: Listeria monocytogenes; *S. flexneri*: Shigella flexneri; *S. coagulase*: Staphylococcus coagulase; *C. frudii*: Citrobacter freundii, *A. baumannii*: Acinetobacter baumannii; *E. aerogenes*: Enterobacter aerogenes; *S. marcescens*: Serratia marcescens; *K. pneumoniae*: Klebsiella pneumoniae; *M. luteus*: Micrococcus luteus; *S. typhi*: Salmonella typhi; *S. paratyphi*: Salmonella paratyphi; *S. typhimurium*: Salmonella typhimurium; *C. bacterium*: Corney bacterium; *S. epidermidis*: Staphylococcus epidermidis; *E. faecalis*: Enterococcus faecalis; *M. smegmatis*: Mycobacterium smegmatis; *B. cereus*: Bacillus cereus; *P. vulgaris*: Proteus vulgaris; *P. mirabilis*: Proteus mirabilis.

3.1.1 Mechanism of antibacterial activity

Silver ions (Ag^+) released from AgNPs can interact with proteins or components of the bacterial membrane, leading to inactivation of proteins and structural changes in the cell wall and nuclear membrane (Fu et al., 2016; He et al., 2017; Velsi et al., 2016). As a result, DNA molecules condense and lose their ability to replicate as silver ions penetrate. The thiol (–SH) groups of proteins are also inactivated when interacting with silver ions (Vanaja and Annadurai, 2013; Salem et al., 2014; Fu et al., 2016; Velsi et al., 2016). Likewise, the interaction of AgNPs with the bacterial cell wall can lead to structural damage by forming pits (Krishnaraj et al., 2010; Ulaeto et al., 2020). Moreover, reports indicated that oxidative stress is triggered by the generation of reactive oxygen species (ROS), leading to cell death (Jeeva et al., 2014; Koduru et al., 2018). Another mechanism has been proposed that R-S-S-R (disulfide) bonds are formed through reactions between silver, oxygen, and sulfhydryl groups on the bacterial cell wall, thereby blocking respiration and causing cell death (He et al., 2017). It has been suggested that AgNPs interrupt the bacterial growth signaling pathway by modulating the tyrosine phosphorylation of putative peptides critical for cell viability and division (Mubarakali et al., 2011). A plausible representation of the antibacterial mechanism is shown in Fig. 2.

Coating nonfunctional materials on nanoparticles can lead to significant differences in antibacterial activity (Rolim et al., 2019). AgNPs synthesized using green tea extract showed higher bactericidal activity than polyethylene glycol (PEG)-coated AgNPs. PEG created an additional layer that interfered with the interaction of AgNPs with the bacterial cell wall (Rolim et al., 2019).

Several studies have reported that AgNPs exhibited greater affinity for gram-negative bacteria compared to gram-positive bacteria due to a thin cell wall, the presence of high lipopolysaccharide, and a single or bilayer of the peptidoglycan in gram-negative bacteria (Kaviya et al., 2011; Mubarakali et al., 2011; Im et al.,

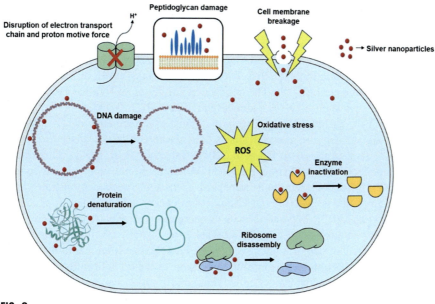

FIG. 2

Antibacterial mechanism of AgNPs.

Reproduced with permission from Roy A, Bulut O, Some S, Mandal AK, Yilmaz MD, Green synthesis of AgNPs: biomolecule-nanoparticle organizations targeting antimicrobial activity, RSC Adv., 9, 2673, 2019, with permission from The Royal Society of Chemistry.

2012; Prakash et al., 2013; Anthony et al., 2014; Huang et al., 2016; Ramesh et al., 2015; Behravan et al., 2019). For this reason, silver ions can easily bind to gram-negative bacteria. Gram-positive bacteria consist of multiple layers of peptidoglycans that can inhibit the interaction between AgNPs and bacterial cells (He et al., 2017). Conversely, AgNPs produced using gum olibanum showed a higher zone of inhibition for gram-positive bacteria (*S. aureus*) than gram-negative bacteria (*E. coli* and *P. aeruginosa*) (Kora et al., 2012). Since the negative charge of gram-negative bacterial cells is higher than that of gram-positive bacteria, it was suggested that the interaction between AgNPs and the bacterial cell wall of gram-positive bacteria is stronger. In addition, the cell wall of gram-negative bacteria is composed of lipids, proteins, and lipopolysaccharides that can be used as protective barriers against AgNPs (Kora et al., 2010). Hollow cellulose was applied as a reducing and synthesizing agent of AgNPs, and higher antibacterial activity was obtained in *S. aureus* than in *E. coli*. It was believed that gram-negative bacteria possess 10% peptidoglycans, which contain large amounts of protein, act as potential targets for silver ion-mediated cell wall disruption, making *E. coli* less susceptible to AgNPs (Fu et al., 2016).

3.2 Anticancer activity

It is believed that AgNPs have robust antiproliferative activity against different cancer cells. In addition, several studies have shown that green synthesized AgNPs have more inhibitory effects on cancerous cells than noncancerous cells (Arunachalam et al., 2015; Mukundan et al., 2015; Du et al., 2016; Jang et al., 2016; El-Naggar et al., 2017; Annu et al., 2018). The anticancer activity of AgNPs has been studied by several assays such as 3-(4, 5-dimethylthiazol-2-yl)-2, 5-diphenyl tetrazolium bromide (MTT), terminal deoxynucleotidyl transferase-mediated dUTP nick-end labelling, lactate dehydrogenase leakage, and western blotting analysis (Kajani et al., 2014; Jang et al., 2016; Pei et al., 2019; Satyavani et al., 2011). In addition, some components of plant extract alone exhibit antiproliferative properties against different cell lines. For example, triterpenoids from apple peels demonstrated anticancer activity against MCF-7 and Caco2 colon cancer cells (Sankar et al., 2013; Elangovan et al., 2015). Flavonoids such as quercetin were reported to display anticancer activity (Venkatesan et al., 2014). Carvacrol, thymol, sabinine, linolool, apigenin, etc. present in plants are believed to exhibit potent antitumor activity (Sankar et al., 2013). AgNPs obtained using sidr honey were tested for antiproliferative activity against human cervical cancer cells (Hela) and HepG2 cancer cell lines. It was reported that sidr honey containing AgNPs inhibited the growth of Hela and HepG2 cell lines, whereas sidr honey alone showed antiproliferative activity against HepG2 (Ghramh et al., 2020).

Three different leukemic cells—human epidemoid carcinoma (Hep2), human colon adenocarcinoma (COLO 205), and neuroblastoma (SH-SY5Y)—were studied against AgNPs synthesized using *Acorus calamus* (Nakkala et al., 2018). In all three, Hep2 cells were highly susceptible and had a 1.2-fold increase in malondialdehyde (MDA) levels. In addition, a decrease in the activity of natural defense enzymes such as superoxide dismutase, glutathione peroxidase (GPx), and catalase was found in Hep2 cells. The anticancer activity of AgNPs produced using dried longan (*Dimocarpus longan*) was performed against three cancer cells: prostate cancer (VcaP), pancreas cancer (BxPC-3), and lung cancer (H1299). MTT assay showed that AgNPs displayed the highest inhibitory effect on H1299 cells compared to BxPC-3 and VcaP cells. One of the plausible reasons is the characteristics of cancer cells, i.e., the higher metabolic rate and rapid cell division observed with the enhanced internalization of AgNPs leading to cell death (He et al., 2016). This can be corroborated with the other literature reports (Arunachalam et al., 2015; Du et al., 2016; Kummara et al., 2016; Annu et al., 2018; Cyril et al., 2019). Various studies related to the anticancer activity of AgNPs are highlighted in Table 2.

3.2.1 Proposed mechanism of anticancer activity

Apoptosis can be induced by two main signaling pathways: extrinsic and intrinsic, controlled through caspase 8 and caspase 9, respectively. The extrinsic pathway is mediated by the binding activity of extracellular ligands and membrane receptors, which further binds to the adapter proteins connected to the death domain of the

Table 2 Anticancer activity of plant extract or agricultural waste synthesized AgNPs.

No.	Plant extract or agricultural waste used	Tumor cells	AgNP concentration by MTT assay (IC$_{50}$)	Reference
1	Clerodendrum phlomidis	EAC and HT-29 cell	36.72 µg/mL (EAC) 32.69 µg/mL (HT 29)	Sriranjani et al. (2016)
2	Nepeta deflersiana	HeLa	100 µg/mL	Al-Sheddi et al. (2018)
3	Solanum trilobatum	MCF 7	50 µg/mL	Ramar et al. (2015)
4	Spinacia oleracea	C2C12	100 µg/mL	Ramachandran et al. (2017)
5	Bauhinia tomentosa	A549	50 µg/mL	Mukundan et al. (2015)
6	Beta vulgaris	MCF 7, A549, and Hep2	47.6 µg/mL (MCF 7) 48.2 µg/mL (A549) 47.1 µg/mL (Hep2)	Venugopal et al. (2017)
7	Anabaena doliolum	DL and Colo 205	20 µg	Singh et al. (2014)
8	Cynara scolymus	MCF 7	10 µg/mL	Erdogan et al. (2019)
9	Sucrose	HT 144 and H157	3.6 µM	Nazir et al. (2011)
10	Salacia chinensis	Hep G2, L-132, MIA-Pa-Ca-2, MDA-MB-231, KB, PC-3, and Hela	6.31 µg/mL (HepG2) 4.002 µg/mL (L-132) 5.228 µg/mL (MIA-Pa-Ca-2) 8.452 µg/mL (MDA-MB-231) 14.37 µg/mL (KB cells) 7.46 µg/mL (PC-3) 6.55 µg/mL (Hela)	Jadhav et al. (2018)
11	Rosa damascena	A549	80 µg/mL	Venkatesan et al. (2014)
12	Origanum vulgare	A549	100 µg/mL	Sankar et al. (2013)
13	Phoenix dactylifera, Ferula asafoetida, and Acacia nilotica	LoVo	69.73 µg/mL (Phoenix dactylifera) 46.15 µg/mL (Ferula asafoetida) 58.02 µg/mL (Acacia nilotica)	Mohammed et al. (2018)
14	Melia azedarach	Hela	300 µg/mL	Sukirtha et al. (2012)

Continued

Table 2 Anticancer activity of plant extract or agricultural waste synthesized AgNPs. *Continued*

No.	Plant extract or agricultural waste used	Tumor cells	AgNP concentration by MTT assay (IC$_{50}$)	Reference
15	Walnut fruits	MCF-7	60 µg/mL	Khorrami et al. (2018)
16	Melia dubia	MCF-7	31.2 µg/mL	Kathiravan et al. (2014)

EAC, *Ehrlich Ascites carcinoma*; HeLa, *human cervical cancer cells*; MCF 7, *human breast cancer cell line*; C2C12, *mouse myoblast cancer cells*; A549, *human lung cancer cell*; Hep2, *pharynx cell*; DL, *Dalton's lymphoma cell*; HT 144, *malignant melanoma of skin*; H157, *squamous cell lung carcinoma*; L-132, *lungs cell*; MIA-Pa-Ca-2, *pancreas cell*; MIDA-MB-231, *breast cell*; KB, *oral cell*; PC-3, *prostate cell.*

intracellular receptor. The death domain initiates the activation of caspase 8 which consequently triggers caspase 3, resulting in cell death. On the other hand, cytochrome C combines with procaspase 9, dATP, and apoptotic protease activating factor 1, triggering the activation of caspase 3. Upon activation, it cleaves substrates such as lamins, poly (ADP-ribose) polymerase, and DNA-associated proteins. This cleavage is a hallmark of apoptosis (Nakkala et al., 2018; Sriram et al., 2010). It has also been proposed that silver ions can induce death in tumor cells through ROS generation which leads to subsequent loss of mitochondrial membrane potential, resulting in apoptotic morphological changes (Pei et al., 2019). Furthermore, activation of intracellular caspase enzyme and oxidative stress increase internucleosomal DNA fragmentation and oxidative stress, leading to cell death. In addition, AgNP-treated cells showed decreased Bcl-2 expression levels while Bax levels increased, indicating that mitochondrial pathways resulted in death (Jeyaraj et al., 2013). Fig. 3 illustrates the possible mechanism for the anticancer action of AgNPs inside cells.

It was noticed that treatment of MCF-7 cells with AgNPs upregulated p53 mRNA (a potent indicator of tumor suppressor activity) and Bax mRNA, and downregulated Bcl-2 mRNA. It also significantly increased the relative expression of caspases 3 and 9, supporting the ability of AgNPs to induce apoptosis. In addition, AgNPs effectively inhibited tyrosine kinases of the Janus family, signal transducers and activators of transcription (STAT)-1, and STAT-3 mRNA expressions. Likewise, cytosolic cytochrome *c* levels increased significantly with exposure leading to an apoptotic effect on cells (Jang et al., 2016). It was reported that AgNPs elevate nitric oxide (NO) levels in leukemic cells due to several oxidative injuries and form peroxynitrate (ONOO$^-$), a potent tumoricidal agent (Mollick et al., 2019).

The viability of MCF-7, T47D cancer cells, and MCF10-A normal breast cells was studied for antitumor activity and decreased after 12 h incubation for MCF-7 and T47D cells, whereas MCF10-A cell viability was slightly affected. It was recognized that caspase 3/7 and caspase 9 activities were dose-dependently overexpressed in tumor cells while Bcl-2 expression decreased. On the other hand, in

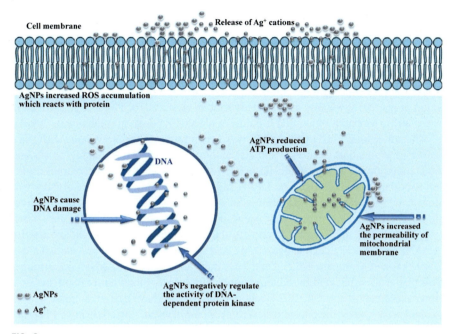

FIG. 3

Proposed mode of action of AgNPs on cancer cell.

Reprinted with permission from El-Naggar NE-A, Hussein MH, El-Sawah AA, Bio-fabrication of AgNPs by phycocyanin, characterization, in vitro anticancer activity against breast cancer cell line and in vivo cytotoxicity, Sci. Rep., 7, 10844, 2017, with permission from Springer Nature.

MCF10-A cells, subtle changes were observed in caspases and BCl-2 concentration (Ortega et al., 2015). Fluorescent DNA-specific binding dyes such as acridine orange and ethidium bromide were used to observe morphological changes after interacting with green synthesized AgNPs (Sreekanth et al., 2016; Kummara et al., 2016). Treated NCI-H460 cells displayed apoptotic features with condensed or fragmented chromatin. Ethidium bromide fluorescence was shown due to membrane damage along with squashed nuclei and apoptotic bodies. On the contrary, no significant apoptotic features were recognized in human skin fibroblast (HDFa) cells when treated with AgNPs (Kummara et al., 2016).

3.3 Antioxidant activity

ROS are harmful by-products of cellular respiration, causing oxidative damage to cells. ROS are very active due to the presence of unpaired valence shell electrons, free radicals besides molecules and ions (Kregal and Tornorth-Horsefield, 2015). The production of ROS includes superoxide anion, hydrogen peroxide, hydroxyl radical, and singlet oxygen (Kanpandian et al., 2014). Excessive ROS inside the body causes cancer, aging cataracts, and cardiovascular disease (He et al., 2017). Antioxidants have been reported to play a key role in terminating free radicals due to their scavenging properties (Kumar et al., 2016). Various assays such as DPPH, ABTS, ferric reducing antioxidant power (FRAP), nitric oxide scavenging activity (NOx), and hydroxyl radical scavenging assays have been commonly used (Dipankar and Murugan, 2012; Abdel-Aziz et al., 2014; Kumar et al., 2014; Rajan et al., 2015; Swamy et al., 2015; Kumar et al., 2016; Phull et al., 2016; He et al., 2017; Mohanta et al., 2017; Salari et al., 2019; Singh and Dhaliwal, 2018).

The antioxidant activity of AgNPs was first reported using *Syzygium cumini* as a source of production (Banerjee and Narendhirakanan, 2011). Reducing power and total antioxidant capacity were observed for capped AgNPs, seed extract alone, and ascorbic acid (control). The DPPH radical scavenging activity of AgNPs was almost the same as that of ascorbic acid and was higher than that of seed extract alone. The presence of a reducing agent triggered the conversion of the Fe^{3+}/ferricyanide complex used in FRAP assay to the ferrous form. The antioxidant activity of AgNPs was lower than that of ascorbic acid in DPPH free radical scavenging assay (Dipankar and Murugan, 2012). On the contrary, the activity of AgNPs was stronger than that of ascorbic acid in reducing power analysis and total antioxidant analysis. However, Abdel-Aziz et al. (2014) reported that AgNPs have a slight increase in antioxidant activity over the plant extract (*Chenopodium murale*) itself, using DPPH radical scavenging activity and beta-carotene leaching assay. AgNPs exhibited higher scavenging activity in DPPH and ABTS assays than plant extract alone, while OH radical scavenging assay showed higher antioxidant capacity for plant extract alone (Fafal et al., 2017). Furthermore, NOx assay was implemented to determine antioxidant property (Bhakya et al., 2016; Das et al., 2019). This activity is mainly due to the interaction between AgNPs and NO radical. The less stable and high electronegative NO radicals can accept electrons from AgNPs.

The DPPH antioxidant activity of AgNPs was observed in a dose-dependent manner and reached a maximum of 86% and IC_{50} value of 126.6 µg/mL (Kharat and Mendhulkar, 2016). The same trend was observed with AgNPs synthesized using apricot and black currant pomace waste products (Vasyliev et al., 2020). On the other hand, it was observed that the DPPH activity of the obtained AgNPs decreased with increasing the dose due to less solubility of AgNPs and insufficient DPPH content at higher concentration (Kumar et al., 2016).

Moldovan et al. (2016) studied both in vivo and in vitro methods to determine the antioxidant activity of AgNPs produced using *Sambucus nigra*. The in vitro assessment of antioxidant capacity was studied using the trolox equivalent antioxidant capacity. The antioxidant capacity of nanoparticles (177.09 µM trolox) was higher than that of plant extract alone (118.44 µM trolox). The in vivo activity of the obtained AgNPs was investigated for oxidative stress generated by carrageenan-induced inflammation in Wistar rats. Treatment of AgNPs increased GPx and catalase activity after 48 h (Moldovan et al., 2016). These studies demonstrate that AgNPs are potent antioxidants for nanomedicine and therapeutics.

3.4 Antifungal activity

Recently, the exploration of other biochemical activities such as antifungal and antiviral activity of green synthesized AgNPs has begun. However, unlike other activities, few studies have been reported on the antifungal and antiviral activity of AgNPs (Velmurugan et al., 2015).

AgNPs demonstrated strong antifungal activity against different kinds of fungal pathogens (Krishnaraj et al., 2012; Narayanan and Park, 2014; Medda et al., 2015; Hernández-Díaz et al., 2021). Vivek et al. (2011) first reported the antifungal activity of AgNPs synthesized using *Gelidiella acerosa* against four human pathogens: *Humicola insolens*, *Fusarium dimerun*, *Mucor indicus*, and *Trichoderma reesei*. Higher activity was recorded against *Mucor indicus* and *Trichoderma reesei*, while moderate activity was found in *Humicola insolens* and *Fusarium dimerun*. A wide range of phytopathogenic fungi such as *Alternaria alternata*, *Sclerotinia sclerotiorium*, *Macrophomina phaseolina*, *Rhizoctonia solani*, *Botrytis cinerea*, and *Curvularia lunata* have been studied for the antifungal activity of AgNPs (Krishnaraj et al., 2012). Of these, a maximum level of inhibition zone (2.3 mm) was observed for *Alternaria alternata* and *Macrophomina phaseolina*. The remaining plant pathogens exhibited moderate antifungal activity against AgNPs (1.8–2.0 mm). Various *Candida* species, such as *Candida guillermondii*, *Candida krusei*, *Candida albicans*, and *Candida glabarta*, were treated with biosynthesized AgNPs, of which *Candida krusei* showed the highest antifungal activity (Ahmeda et al., 2020). *Candida tropicalis* and *Candida albicans* was reported to be more sensitive than *Candida tropicalis* (Mallmann et al., 2015).

3.4.1 Mechanism of antifungal activity

The mechanism of antifungal properties is not clear. It has been hypothesized that AgNPs may kill fungal spores by damaging membrane integrity. Furthermore, AgNPs react with phosphorus and sulfur-containing compounds, and their interactions may disrupt DNA and proteins, resulting in membrane potential interference and cell death. Silver ions form complexes with bases contained in DNA and are strong inhibitors of DNase (Azizi et al., 2016; Devi and Bhimba, 2014). The inhibitory effect of silver ions resulted in inactivation of the expression of ribosomal subunit proteins and other cellular proteins/enzymes essential for ATP production (Al-Zubaidi et al., 2019). In addition, AgNPs had a detrimental effect on the germination of fungal hyphae and conidia. Like anticancer and antibacterial activities, this activity also depends in a concentration-dependent manner.

The mechanism behind the antifungal activity of AgNPs was further elucidated where staining assays validated necrotic and apoptotic effects on fungal cells, while R6G (substrate of fungal efflux pumps) influx/efflux assay and gas chromatographic (GC) analysis favor the cell rupture process (Prasher et al., 2018). The basis of the influx/efflux assay was a competition between AgNPs and R6G, which reduced the transportation of R6G out of fungal cells. The influx/

efflux of R6G was monitored by the absorbance/fluorescence spectra of the extracellular fluids. In the presence of AgNPs, much R6G instantly entered the cells and their influx did not change with time. On the other hand, R6G efflux in the presence of AgNPs was much faster at first but tends to decrease over time. AgNPs altered function and the decay began. Thus, this assay confirmed cell rupture and modulation in the efflux protein. For GC analysis, the effect of AgNPs was investigated for the enzyme 14-α-lanosterol (an important component for fungal membrane integrity). This enzyme converts lanosterol to ergosterol. Moreover, AgNPs successfully inhibited ergosterol biosynthesis and was believed to possess excellent antifungal properties (Mallmann et al., 2015).

3.5 Antiviral activity

Recently, many researchers have begun to focus their attention on the antiviral action of AgNPs (Haggag et al., 2019; Morris et al., 2019; Sharma et al., 2019; Zeedan et al., 2020). Chemically synthesized AgNPs have been extensively studied against a wide range of viruses such as retroviridae, herpesviridae, poxviridae, orthomyxoviridae, hepadnaviridae, arenaviridae, and paramyxoviridae over the past decade (Galdiero et al., 2011; Lara et al., 2010; Lu et al., 2008; Lv et al., 2014; Morris et al., 2019).

However, due to the toxicity of the chemical process, the focus is now on biosynthesized AgNPs and their antiviral activity. To date, few studies on the antiviral activity of biosynthesized AgNPs have been reported (Orlowski et al., 2014; Sujitha et al., 2015; Haggag et al., 2019; Sharma et al., 2019; Zeedan et al., 2020). One of the first reports of the antiviral activity of biosynthesized AgNPs is against dengue virus (DEN-2) (Sujitha et al., 2015). Low-dose AgNPs synthesized using *Moringa oleifera* seed extract reduced the dengue virus yield by 75% as determined by plaque assay. Three different plant extracts—*Andrographis paniculata*, *Phyllanthus niruri*, and *Tinospora cordifolia*—were exploited for AgNP synthesis, and antiviral activity was evaluated against chikungunya virus (CKV) (Sharma et al., 2019). The in vitro antiviral activity studied using the MTT assay showed that AgNPs from *Andrographis paniculata* exhibited maximum potential against CKV followed by *Tinospora cordifolia* and *Phyllanthus niruri*, signifying the role of plant extracts in antiviral activity. According to the authors, andrographolide is the main phytocomponent in *Andrographis paniculata* and is known as a potent antiviral agent (Gupta et al., 2017). On the other hand, *Tinospora cordifolia* possesses immunomodulatory and immunostimulatory properties due to the existence of different bioactive compounds consisting of alkaloids, diterpenoid lactones, glycosides, and steroids (Esteri et al., 2012). The extract of *P. niruri* was reported to inhibit woodchuck hepatitis virus, hepatitis B virus, HIV, and dengue virus (Lee et al., 2013). Similarly, aqueous and hexane extracts of *Lampranthus coccineus* and *Malephora lutea* were used for the green synthesis of AgNPs. AgNPs produced using the hexane extract of *L. coccineus* enhanced antiviral activity against herpes simplex virus, hepatitis

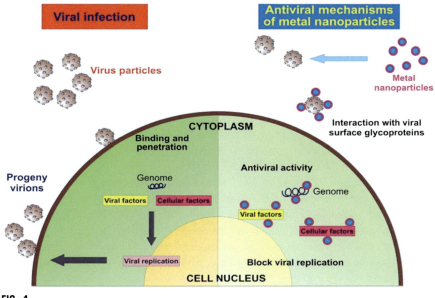

FIG. 4

Antiviral mechanism of AgNPs.

Reprinted with permission from Galdiero S, Falanga A, Vitiello M, Cantisani M, Marra V, Galdiero M, AgNPs as potential antiviral agents, Molecules, 16, 8894–8918, 2011, with permission from MDPI Publisher, Licensee MDPI, Basel, Switzerland. This article is an open-access article distributed under the terms and conditions of the Creative Commons Attribution (CC BY) license (http:/creativecommons.org/licenses/by/4.0).

A virus (HAV-10), and coxsackie B4 virus, while AgNPs produced using *M. lutea* exhibited significant inhibition of the viral infection against HAV-10 and coxsackie B4 virus.

It was hypothesized that the binding between AgNPs and the surface glycoproteins of the viral cell prevents the virus from penetrating the host cell. Moreover, interaction with the viral genome inhibits pathways essential for viral replication. A possible mechanism for antiviral activity is shown in Fig. 4.

In vitro and in vivo antiviral activity was also studied for bovine herpes virus-1 (BoHV-1) (Zeedan et al., 2020). In vitro studies showed that AgNPs slightly depleted Madin-Darby bovine kidney cells and suppressed BoHV-1 cells in a dose-dependent manner. Besides, AgNPs stuck to the sulfur in the gp120 glycoprotein knobs, inhibiting the normal activities of the virus. AgNPs were produced using clove extract and were examined for in vitro and in vivo antiviral activity against Newcastle virus disease (NDV). In vitro activity was assessed with NDV and chicken red blood cells by performing spot assay and microhemagglutination tests, while in vivo studies were assessed on embryonated chicken eggs aged 9–11 days. In spot assay, diluted AgNPs exhibited antiviral activity (Mehmood et al., 2020). Overall, biosynthesized AgNPs can be used as an antiviral drug in medicine.

4 Conclusions

Studies on the plant and agricultural waste-mediated synthesis of AgNPs and their applications in various fields were described in this chapter. Synthesis using plant and agricultural waste has been reported to be simpler, faster, and more cost-effective compared to other green synthesis techniques and chemical methods. Plant extracts also contain reducing and stabilizing agents. Thus, the obtained AgNPs show potent antibacterial activity due to their binding ability with the cell walls of gram-negative and gram-positive bacteria. In addition, the generation of ROS and cell wall pits contribute to the strong antibacterial activity of AgNPs. Cancer cell apoptosis can be induced by AgNPs, which indicates effective anticancer activity. A decrease in the Bcl-2 expression level was observed, whereas an increase in Bax levels was found in cancer cells after interacting with AgNPs. AgNPs have also been reported to demonstrate strong antioxidant and antifungal properties. Recently, the unexplored potential of AgNPs in terms of antiviral activity has been reported against a wide range of viruses such as retroviridae, herpesviridae, poxviridae, orthomyxoviridae, hepadnaviridae, arenaviridae, and paramyxoviridae. In the future, plant extract-based AgNPs could be utilized in the nanomedicine field due to the various biochemical properties that can be prepared on a large scale under optimized conditions.

References

Abdel-Aziz, M.S., Shaheen, M.S., El-Nekeety, A.A., Abdel-Wahhab, M.A., 2014. Antioxidant and antibacterial activity of silver nanoparticles biosynthesized using *Chenopodium murale* leaf extract. J. Saudi Chem. Soc. 18, 356–363.

Ahmad, A., Mukherjee, P., Senapati, S., Mandal, D., Khan, M.I., Kumar, R., Sastry, M., 2003. Extracellular biosynthesis of silver nanoparticles using the fungus *Fusarium oxysporum*. Colloids Surf. B. Biointerfaces 28, 313–318.

Ahmeda, A., Zangeneh, A., Kalbesi, R.J., Seydi, N., Zangeneh, M.M., Mansouri, S., Goorani, S., Moradi, R., 2020. Green synthesis of silver nanoparticles from aqueous extract of *Ziziphora clinopodioides* Lam and evaluation of their bio-activities under *in vitro* and *in vivo* conditions. Appl. Organometal. Chem. 34, e5358.

Ali, S., Praveen, S., Ali, M., Jiao, T., Sharma, A.S., Hassan, H., Devaraj, S., Li, H., Chen, Q., 2020. Bioinspired morphology-controlled silver nanoparticles for antimicrobial application. Mater. Sci. Eng. C 108, 110421.

Allafchian, A.R., Mirahmadi-Zare, S.Z., Jalali, S.A.H., Hashemi, S.S., Vahabi, M.R., 2016. Green synthesis of silver nanoparticles using *phlomis* leaf extract and investigation of their antibacterial activity. J. Nanostruct. Chem. 6, 129–135.

Al-Sheddi, E., Farshori, N.N., Al-Oqail, M.M., Al-Massarani, S.M., Saquib, Q., Wahab, R., Musarrat, J., Al-Khedhairy, A.A., Siddiqui, M.A., 2018. Anticancer potential of green synthesized silver nanoparticles using extract of *Nepeta deflersiana* against human cervical cancer cells (HeLA). Bioinorg. Chem. Appl. 2018, 9390784.

Al-Zubaidi, S., Al-Ayafi, A., Abdelkader, H., 2019. Biosynthesis, characterization and antifungal activity of silver nanoparticles by *Aspergillus niger* isolate. J. Nanotechnol. Res. 1, 23–36.

Annu, Ahemd, S., Kaur, G., Sharma, P., Singh, S., Ikram, S., 2018. Evaluation of antioxidant, antibacterial and anticancer (lung cancer cell line A549) activity of *Punica granatum* mediated silver nanoparticles. Toxicol. Res. 7, 923–930.

Anthony, K.J.B., Murugan, M., Gurunathan, S., 2014. Biosynthesis of silver nanoparticles from the culture supernatant of *Bacillus marisflavi* and their potential antibacterial activity. J. Ind. Eng. Chem. 20, 1505–1510.

Arunachalam, K.D., Arun, L.B., Annamalai, S.K., Arunachalam, A.M., 2015. Potential anticancer properties of bioactive compounds of *Gymnema sylvestre* and its biofunctionalized silver nanoparticles. Int. J. Nanomedicine 10, 31–41.

Asanithi, P., Chaiyakun, S., Limsuwan, P., 2012. Growth of silver nanoparticles by DC magnetron sputtering. J. Nanomater. 2012, 963609.

Azizi, Z., Pourseyedi, S., Khatam, M., Mohammadi, H., 2016. *Stachys lavandulifolia* and *Lathyrus* sp. mediated for green synthesis of silver nanoparticles and evaluation its antifungal activity against *Dothiorella sarmentorum*. J. Clust. Sci. 27, 1613–1628.

Baghayeri, M., Mahdavi, B., Abadi, Z.H.-M., Farhadi, S., 2018. Green synthesis of silver nanoparticles using water extract of *Salvia leriifolia*: antibacterial studies and applications as catalysts in the electrochemical detection of nitrite. Appl. Organomet. Chem. 32, e4057.

Baicco, D., Lavecchia, R., Natali, S., Zuorro, A., 2016. Production of metal nanoparticles by agro-industrial wastes: a green opportunity for nanotechnology. Chem. Eng. Trans. 47, 67–72.

Banerjee, J., Narendhirakanan, R.T., 2011. Biosynthesis of silver nanoparticles from *Syzygium cumini* (L.) seed extract and evaluation of their in vitro antioxidant activities. Dig. J. Nanomater. Biostruct. 6, 961–968.

Bar, H., Bhui, D.K., Sahoo, G.P., Sarkar, P., De, S.P., Misra, A., 2009. Green synthesis of silver nanoparticles using latex of *Jatropha curcas*. Colloids Surf. A Physicochem. Eng. Asp. 339, 134–139.

Barros, C.H.N., Cruz, G.C.F., Mayrink, W., Tasic, L., 2018. Bio-based synthesis of silver nanoparticles from orange waste: effects of distinct biomolecule coatings on size, morphology, and antimicrobial activity. Nanotechnol. Sci. Appl. 11, 1–14.

Behravan, M., Panahi, A.H., Naghizadeh, A., Ziaee, M., Mahdavi, R., Mirzapour, A., 2019. Facile green synthesis of silver nanoparticles using *Berberis vulgaris* leaf and root aqueous extract and its antibacterial activity. Int. J. Biol. Macromol. 124, 148–154.

Bhakya, S., Muthukrishnan, S., Sukumaran, M., 2016. Biogenic synthesis of silver nanoparticles and their antioxidant and antibacterial activity. Appl. Nanosci. 6, 755–766.

Chaloupka, K., Malam, Y., Seifalien, A.M., 2010. Nanosilver as a new generation of nanoproduct in biomedical applications. Trends Biotechnol. 28, 580–588.

Chen, J., Wang, J., Zhang, X., Jin, Y., 2008. Microwave-assisted green synthesis of silver nanoparticles by carboxymethyl cellulose sodium and silver nitrate. Mater. Chem. Phys. 108, 421–424.

Chutrakulwong, F., Thamaphat, K., Limsuwan, P., 2020. Photo-irradiation induced green synthesis of highly stable silver nanoparticles using durian rind biomass: effects of light intensity, exposure time and pH on silver nanoparticles formation. J. Phys. Commun. 4, 095015.

Cyril, N., George, J.B., Joseph, L., Raghavamenon, A.C., Sylas, V.P., 2019. Assessment of antioxidant, antibacterial and anti-proliferative (lung cancer line A549) activities of green synthesized silver nanoparticles from *Derris trifoliate*. Toxicol. Res. 8, 297–308.

Das, G., Patra, J.K., Debnath, T., Ansari, A., Shin, H.-S., 2019. Investigation of antioxidant, antibacterial, antidiabetic, and cytotoxicity potential of silver nanoparticles synthesized using the outer peel extract of *Ananas comosus* (L.). PLos One 14, e0220950.

Devadiga, A., Shetty, K.V., Saidutta, M.B., 2015. Timber industry waste-teak (*Tectona grandis* Linn.) leaf extract mediated synthesis of antibacterial silver nanoparticles. Int. Nano Lett. 5, 205–214.

Devi, J.S., Bhimba, B.V., 2014. Antibacterial and antifungal activity of silver nanoparticles synthesized using *Hypnea muciformis*. Biosci. Biotech. Res. Asia 11, 235–238.

Dhillon, G.S., Brar, S.K., Kaur, S., Verma, M., 2012. Green approach for nanoparticle biosynthesis by fungi: current trends and applications. Crit. Rev. Biotechnol. 32, 49–73.

Dinda, G., Halder, D., Mitra, A., Vázquez-Vázquez, C., López-Quintela, M.A., 2017. Study of the antibacterial and catalytic activity of silver colloids synthesized using the fruit of *Sapindus mukorossi*. New J. Chem. 41, 10703.

Dipankar, C., Murugan, S., 2012. The green synthesis, characterization and evaluation of the biological activities of silver nanoparticles synthesized from *Iresine herbstii* leaf aqueous extracts. Colloids Surf. B. Biointerfaces 98, 112–119.

Du, J., Hu, Z., Dong, W.-j., Wang, Y., Wu, S., Bai, Y., 2019. Biosynthesis of large-sized silver nanoparticles using *Angelica keiskei* extract and its antibacterial activity and mechanisms investigation. Microchem. J. 147, 333–338.

Du, J., Singh, H., Yi, T.-H., 2016. Antibacterial, anti-biofilm and anticancer potentials of green synthesized silver nanoparticles using benzoin gum (*Styrax benzoin*) extract. Bioprocess Biosyst. Eng. 39, 1923–1931.

Duran, N., Marcato, P.D., Alves, O.L., Souza, G.I.D., Esposito, E., 2005. Mechanistic aspects of biosynthesis of silver nanoparticles by several *Fusarium oxysporum* strains. J. Nanobiotechnol. 3, 8.

Elangovan, K., Elumalai, D., Anupriya, S., Shenbhagaraman, R., Kaleena, P.K., Murugesan, K., 2015. Phyto mediated biogenic synthesis of silver nanoparticles using leaf extract of *Andrographis echioides* and its bio-efficacy on anticancer and antibacterial activities. J. Photochem. Photobiol. B Biol. 151, 118–124.

El-Naggar, N.E.-A., Hussein, M.H., El-Sawah, A.A., 2017. Bio-fabrication of silver nanoparticles by phycocyanin, characterization, *in vitro* anticancer activity against breast cancer cell line and *in vivo* cytotoxicity. Sci. Rep. 7, 10844.

Erdogan, O., Abbak, M., Demirbolat, G.M., Birtekocak, F., Aksel, M., Pasa, S., Cevik, O., 2019. Green synthesis of silver nanoparticles via *Cynara scolymus* leaf extracts: the characterization, anticancer potential with photodynamic therapy in MCF7 cells. PLoS One 14, e0216496.

Esteri, M., Venkanna, L., Reddy, A.S., 2012. In vitro anti-HIV activity of crude extracts from *Tinospora cordifolia*. BMC Infect. Dis. 12, P10.

Fafal, T., Tastan, P., Tuzun, B.S., Ozyazici, M., Kivcak, B., 2017. Synthesis, characterization and studies on antioxidant activity of silver nanoparticles using *Asphodelus aestivus* Brot. aerial part extract. S. Afr. J. Bot. 112, 346–353.

Fu, L.-H., Deng, F., Ma, M.-G., Yang, J., 2016. Green synthesis of silver nanoparticles with enhanced antibacterial activity using holocellulose as a substrate and reducing agent. RSC Adv. 6, 28140–28148.

Galdiero, S., Falanga, A., Vitiello, M., Cantisani, M., Marra, V., Galdiero, M., 2011. AgNPs as potential antiviral agents. Molecules 16, 8894–8918.

Gardea-Torresday, J.L., Gomez, E., Peralta-Videa, J.R., 2003. Alfalfa sprouts: a natural source for the synthesis of silver nanoparticles. Langmuir 19, 1357–1361.

Ghafferi-Moggadam, M., Hadi-Dabanlou, R., 2014. Plant mediated green synthesis and antibacterial activity of silver nanoparticles using *Crataegus douglasii* fruit extract. J. Ind. Eng. Chem. 20, 739–744.

Ghramh, H.A., Ibrahim, E.H., Kilany, M., 2020. Study of anticancer, antimicrobial, immunomodulatory, and silver nanoparticles production by Sidr honey from three different sources. Food Sci. Nutr. 8, 445–455.

Gul, A.R., Shaheen, F., Rafique, R., Bal, J., Waseem, S., Park, T.J., 2021. Grass-mediated biogenic synthesis of silver nanoparticles and their drug delivery evaluation: a biocompatible anti-cancer therapy. Chem. Eng. J. 407, 127202.

Gupta, K., Barua, S., Hazarika, S.N., Manhar, A.K., Nath, D., Karak, N., Namsa, N.D., Mukhopadhyay, R., Kalia, V.C., Mandal, M., 2014. Green silver nanoparticles: enhanced antimicrobial and antibiofilm activity with effects on DNA replication and cell cytotoxicity. RSC Adv. 4, 52845–52855.

Gupta, S., Mishra, K.P., Ganju, L., 2017. Broad-spectrum antiviral properties of andrographolide. Arch. Virol. 162, 611–623.

Haggag, E.G., Elshamy, A.M., Rabeh, M.A., Gabr, N.M., Salem, M., Youssif, K.A., Samir, A., Muhsinah, A.B., Alsayari, A., Abdelmohsen, U.R., 2019. Antiviral potential of green synthesized silver nanoparticles of *Lampranthus coccineus* and *Malephora lutea*. Int. J. Nanomedicine 14, 6217–6229.

Hamouda, R.A., Hussein, M.H., Abo-elmagd, R.A., Bawazir, S.S., 2019. Synthesis and biological characterization of silver nanoparticles derived from the cyanobacterium *Oscillatoria limnetica*. Sci. Rep. 9, 13071.

He, Y., Du, Z., Ma, S., Liu, Y., Li, D., Huang, H., Jiang, S., Cheng, S., Wu, W., Zhang, K., Zheng, X., 2016. Effects of green-synthesized silver nanoparticles on lung cancer cells in vitro and grown as xenograft tumors in vivo. Int. J. Nanomedicine 11, 1879–1887.

He, Y., Wei, F., Ma, Z., Zhang, H., Yang, Q., Yao, B., Huang, Z., Li, J., Zeng, C., Zhang, Q., 2017. Green synthesis of silver nanoparticles using seed extract of *Alpinia katsumadai*, and their antioxidant, cytotoxicity, and antibacterial activities. RSC Adv. 7, 39842–39851.

Hernández-Díaz, J.A., Garza-García, J.J., Zamudio-Ojeda, A., León-Morales, J.M., López-Velázquez, J.C., García-Morales, S., 2021. Plant-mediated synthesis of nanoparticles and their antimicrobial activity against phytopathogens. J. Sci. Food Agric. 101 (4), 1270–1287.

Huang, X., Pang, Y., Liu, Y., Zhou, Y., Wang, Z., Hu, Q., 2016. Green synthesis of silver nanoparticles with high antimicrobial activity and low cytotoxicity using catechol-conjugated chitosan. RSC Adv. 6, 64357–64363.

Huang, J., Zhan, G., Zheng, B., Sun, D., Lu, F., Lin, Y., Chen, H., Zheng, Z., Zheng, Y., Li, Q., 2011. Biogenic silver nanoparticles by *Cacumen platycladi* extract: synthesis, formation mechanism, and antibacterial activity. Ind. Eng. Chem. Res. 50, 9095–9106.

Husen, A., Siddqui, K.S., 2014. Phytosynthesis of nanoparticles: concept, controversy and application. Nanoscale Res. Lett. 9, 229.

Ibrahim, H.M.M., 2015. Green synthesis and characterization of silver nanoparticles using banana peel extract and their antimicrobial activity against representative microorganisms. J. Radiat. Res. Appl. Sci. 8, 265–275.

Im, A.-R., Han, L., Kim, E.R., Kim, J., Kim, Y.S., Park, Y., 2012. Enhanced antibacterial activities of leonuri herba extracts containing silver nanoparticles. Phytother. Res. 26, 1249–1255.

Iravani, S., 2011. Green synthesis of metal nanoparticles using plants. Green Chem. 13, 2638.

Jadhav, K., Deore, S.L., Dhamecha, D., Rajeshwari, H.R., Jagwani, S., Jalalpure, S., Bohara, R., 2018. Phytosynthesis of silver nanoparticles: characterization, biocompatibility studies and anticancer activity. ACS Biomater Sci. Eng. 4, 892–899.

Jagtap, U.B., Bapat, V.A., 2013. Green synthesis of silver nanoparticles using *Artocarpus heterophyllus* lam. seed extract and its antibacterial activity. Ind. Crops Prod. 46, 132–137.

Jang, S.J., Yang, I.J., Tettey, C.O., Kim, K.M., Shin, H.M., 2016. In-vitro anticancer activity of green synthesized silver nanoparticles on MCF-7 human breast cancer cells. Mater. Sci. Eng. C 68, 430–435.

Jeeva, K., Thiyagarajan, M., Elangovan, V., Geetha, N., Venkatachalam, P., 2014. *Caesalpinia coriaria* leaf extracts mediated biosynthesis of metallic silver nanoparticles and their antibacterial activity against clinically isolated pathogens. Ind. Crop Prod. 52, 714–720.

Jeyaraj, M., Rajesh, M., Arun, R., MubarakAli, D., Sathishkumar, G., Sivanandan, G., Dev, G. K., Manickvasagam, M., Premkumar, K., Thajuddin, N., Ganapathi, A., 2013. An investigation on the cytotoxicity and caspase-mediated apoptotic effect of biologically synthesized silver nanoparticles using *Podophyllum hexandrum* on human cervical carcinoma cells. Colloids Surf. B. Biointerfaces 102, 708–717.

Jha, A.K., Prasad, K., Prasad, K., Kulkarni, A.R., 2009. Plant system: nature's nanofactory. Colloids Surf. B. Biointerfaces 73, 219–223.

Kahrilas, G.A., Wally, L.M., Fredrick, S.J., Hiskey, M., Prieto, A.L., Ownes, J.E., 2014. Microwave-assisted green synthesis of silver nanoparticles using orange peel extract. ACS Sustain. Chem. Eng. 2, 367–376.

Kajani, A.A., Bordbar, A.-K., Esfahani, S.H.Z., Khosropour, A.R., Razmjou, A., 2014. Green synthesis of anisotropic silver nanoparticles with potent anticancer activity using *Taxus baccata* extract. RSC Adv. 4, 61394–61403.

Kanpandian, N., Kannan, S., Ramesh, R., Subramanian, P., Thirumugan, R., 2014. Characterization, antioxidant and cytotoxicity evaluation of green synthesized silver nanoparticles using *Cleistanthus collinus* extract as surface modifier. Mater. Res. Bull. 49, 494–502.

Kathiravan, V., Ravi, S., Ashokkumar, S., 2014. Synthesis of silver nanoparticles from *Melia dubia* leaf extract and their in vitro anticancer activity. Spectrochim. Acta A Mol. Biomol. Spectrosc. 130, 116–121.

Kaviya, S., Santhanalakshmi, J., Viswanathan, B., Muthumary, J., Srinivasan, K., 2011. Biosynthesis of silver nanoparticles using citrus sinensis peel extract and its antibacterial activity. Spectrochim. Acta A Mol. Biomol. Spectrosc. 79, 594–598.

Kharat, S.N., Mendhulkar, V.D., 2016. Synthesis, characterization and studies on antioxidant activity of silver nanoparticles using *Elephantopus scaber* leaf extract. Mater. Sci. Eng. C 62, 719–724.

Khatami, M., Sharifi, I., Nobre, M.A.L., Zafarnia, N., Alfatoonian, M.R., 2018. Waste-grass-mediated green synthesis of silver nanoparticles and evaluation of their anticancer, antifungal and antibacterial activity. Green Chem. Lett. Rev. 11, 125–134.

Khorrami, S., Zarrabi, A., Khaleghi, M., Danaei, M., Mozafari, M.R., 2018. Selective cytotoxicity of green synthesized silver nanoparticles against the MCF-7 tumor cell line and their enhanced antioxidant and antimicrobial properties. Int. J. Nanomedicine 13, 8013–8024.

Klaus, T., Joerger, R., Olsson, E., Granqvist, C.-G., 1999. Silver-based crystalline nanoparticles, microbially fabricated. Proc. Natl. Acad. Sci. U. S. A. 96, 13611–13614.

Koduru, J.R., Kailasa, S.K., Bhamore, J.R., Kim, K.-H., Dutta, T., Vellingiri, K., 2018. Phytochemical-assisted synthetic approaches for silver nanoparticles antimicrobial applications: a review. Adv. Colloid Interface Sci. 256, 326–339.

Kokila, T., Ramesh, P.S., Geetha, D., 2015. Biosynthesis of silver nanoparticles from Cavendish banana peel extract and its antibacterial and free radical scavenging assay: a novel biological approach. Appl. Nanosci. 5, 911–920.

Kora, A.J., Sashidhar, R.B., Arunachalam, J., 2010. Gum kondagogu (*Cochlospermum gossypium*): a template for the green synthesis and stabilization of silver nanoparticles with antibacterial application. Carbohydr. Polym. 82, 670–679.

Kora, A.J., Sashidhar, R.B., Arunachalam, J., 2012. Aqueous extract of gum olibanum (*Boswellia serrata*): a reductant and stabilizer for the biosynthesis of antibacterial silver nanoparticles. Process Biochem. 47, 1516–1520.

Kosmala, A., Wright, R., Zhang, Q., Kirby, P., 2015. Synthesis of silver nano particles and fabrication of aqueous Ag inks for inkjet printing. Mater. Chem. Phys. 129, 1075–1080.

Kregal, U., Tornorth-Horsefield, S., 2015. Coping with oxidative stress. Science 347, 125–126.

Krishnaraj, C., Jagan, E.G., Rajasekar, S., 2010. Synthesis of silver nanoparticles using *Acalypha indica* leaf extracts and its antibacterial activity against water borne pathogens. Colloids Surf. B. Biointerfaces 76, 50–56.

Krishnaraj, C., Ramachandran, R., Mohan, K., Kalaichelvan, P.T., 2012. Optimization for rapid synthesis of silver nanoparticles and its effect on phytopathogenic fungi. Spectrochim. Acta A Mol. Biomol. Spectrosc. 93, 95–99.

Kumar, H.A.K., Mandal, B.K., Kumar, K.M., Maddinedi, S.B., Kumar, T.S., Madhiyazhagan, P., Ghosh, A.R., 2014. Antimicrobial and antioxidant activities of *Mimusops elengi* seed extract mediated isotropic silver nanoparticles. Spectrochim. Acta A Mol. Biomol. Spectrosc. 130, 13–18.

Kumar, R., Roopan, S.M., Prabhakarn, A., Khanna, V.G., Chakraborty, S., 2012. Agricultural waste *Annona squamosa* peel extract: biosynthesis of silver nanoparticles. Spectrochim. Acta A Mol. Biomol. Spectrosc. 90, 173–176.

Kumar, B., Smita, K., Seqqat, R., Benalcazar, K., Grijalva, M., Cumbal, L., 2016. *In vitro* evaluation of silver nanoparticles cytotoxicity on hepatic cancer (Hep-G2) cell line and their antioxidant activity: green approach for fabrication and application. J. Photochem. Photobiol. B Biol. 159, 8–13.

Kummara, S., Patil, M.B., Uriah, T., 2016. Synthesis, characterization, biocompatible and anticancer activity of green and chemically synthesized silver nanoparticles—a comparative study. Biomed. Pharmacother. 84, 10–21.

Labannia, A., Zulhadjria, Z., Handayani, D., Ohya, Y., Arief, S., 2019. The effect of monoethanolamine as stabilizing agent in *Uncaria gambir* Roxb mediated synthesis of silver nanoparticles and its antibacterial activity. J. Dispers. Sci. Technol. 41, 1480–1487.

Lara, H.H., Ayaa-Nunez, N.V., Ixtepan-Turrent, L., Rodriguez-Padilla, C., 2010. Mode of antiviral action of silver nanoparticles against HIV-1. J. Nanobiotechnol. 8, 1.

Lee, S.H., Tang, Y.Q., Rathkrishnan, A., Wang, S.M., Ong, K.C., Manikam, R., Payne, B.J., Jaganath, I.B., Sekaran, S.D., 2013. Effects of cocktail of four local Malaysian medicinal plants (*Phyllanthus* spp.) against dengue virus 2. BMC Complement. Altern. Med. 13, 192.

Li, S., Shen, Y., Xie, A., Yu, X., Qiu, L., Zhang, L., Zhang, Q., 2007. Green synthesis of silver nanoparticles using *Capsicum annuum* L. extract. Green Chem. 9, 852–858.

Li, W.-R., Xie, X.-B., Shi, Q.-S., Duan, S.-S., Ouyang, Y.-S., Chen, Y.-B., 2011. Antibacterial effect of silver nanoparticles on *Staphylococcus aureus*. BioMetals 24, 135–141.

Lopez-Miranda, J.L., Vazquez, M., Fletes, N., Esparza, R., Rosas, G., 2016. Biosynthesis of silver nanoparticles using a *Tamarix gallica* leaf extract and their antibacterial activity. Mater. Lett. 176, 285–289.

Lu, L., Sun, R.W.-Y., Chen, R., Hui, C.K., Ho, C.M., Luk, J.M., Lau, G.K., Che, C.M., 2008. Silver nanoparticles inhibit hepatitis B virus replication. Antivir. Ther. 13, 253–262.

Lv, X., Wang, P., Bai, R., Cong, Y., Suo, S., Ren, X., Chen, C., 2014. Inhibitory effect of silver nanomaterials on transmissible virus-induced host cell infections. Biomaterials 39, 4195–4203.

Mahmoudi, R., Aghaei, S., Salehpour, Z., Mousavizadeh, A., Khoramrooz, S.S., Sisakht, M.T., Christiansen, G., Baneshi, M., Karimi, B., Bardania, H., 2019. Antibacterial and antioxidant properties of phytosynthesized silver nanoparticles using *Lavandula stoechas* extract. Appl. Organomet. Chem. 34, e5394.

Mallmann, E.J.J., Cunha, A.F., Castro, B.M.N.F., Maciel, A.M., Menezes, E.A., Fechine, P.B.A., 2015. Antifungal activity of silver nanoparticles obtained by green synthesis. Rev. Inst. Med. Trop. Sao Paulo 57, 165–167.

Medda, S., Hajra, A., Dey, U., Bose, P., Mondal, N.K., 2015. Biosynthesis of silver nanoparticles from *Aloe vera* leaf extract and antifungal activity against *Rhizopus* sp. and *Aspergillus* sp. Appl. Nanosci. 5, 875–880.

Mehmood, Y., Farroq, U., Yousaf, H., Riaz, H., Mahmood, R.K., Nawaz, A., Abid, Z., Gondal, M., Malik, N.S., Barkat, K., Khalid, I., 2020. Antiviral activity of green silver nanoparticles produced using aqueous buds extract of *Syzygium aromaticum*. Pak. J. Pharm. Sci. 33, 839–845.

Mie, R., Samsudin, M.W., Din, L.B., Ahmad, A., Ibrahim, N., Adnan, S.N.A., 2014. Synthesis of silver nanoparticles with antibacterial activity using the lichen *Parmotrema praesorediosum*. Int. J. Nanomedicine 9, 121–127.

Mittal, A.K., Chisti, Y., Banerjee, U.C., 2013. Synthesis of metallic nanoparticles using plant extracts. Biotechnol. Adv. 31, 346–356.

Mohammed, A.E., Al-Qahtani, A., Al-Mutairi, A., Al-Shamiri, B., Aabed, K., 2018. Antibacterial and cytotoxic potential of biosynthesized silver nanoparticles by some plant extracts. Nanomaterials 8, 382.

Mohanta, Y.K., Panda, S.K., Jayabalan, R., Sharma, N., Bastia, A.K., Mohanta, T.K., 2017. Antimicrobial, antioxidant and cytotoxic activity of silver nanoparticles synthesized by leaf extract of *Erythrina suberosa* (Roxb.). Front. Mol. Biosci. 4, 14.

Moldovan, B., David, L., Achim, M., Clichici, S., Filip, G.A., 2016. A green approach to phytomediated synthesis of silver nanoparticles using *Sambucus nigra* L. fruits extract and their antioxidant activity. J. Mol. Liq. 221, 271–278.

Mollick, M.M.R., Rana, D., Dash, S.K., Chattapadhyay, S., Bhowmick, B., Maity, D., Mondal, D., Pattanayak, S., Roy, S., Chakraborty, M., Chattopadyay, D., 2019. Studies on green synthesized silver nanoparticles using *Abelmoschus esculentus* (L.) pulp extract having anticancer (in vitro) and antimicrobial applications. Arab. J. Chem. 12, 2572–2584.

Morris, D., Ansar, M., Speshock, J., Ivanciuc, T., Qu, Y., Casola, A., Garofalo, R.P., 2019. Antiviral and immunomodulatory activity of silver nanoparticles in experimental RSV infection. Viruses 11, 732.

Mubarakali, D., Thajuddin, N., Jeganathan, K., Gunasekaran, M., 2011. Plant extract mediated synthesis of silver and gold nanoparticles and its antibacterial activity against clinically isolated pathogens. Colloids Surf. B. Biointerfaces 85, 360–365.

Mukundan, D., Mohankumar, R., Vasanthakumari, R., 2015. Green synthesis of silver nanoparticles using leaves extract of *Bauhinia tomentosa* Llnn and its in-vitro anticancer potential. Mater. Today 2, 4309–4316.

Mukunthan, K.S., Balaji, S., 2012. Cashew apple juice (*Anacardium occidentale* L.) speeds up the synthesis of silver nanoparticles. Int. J. Green Nanotechnol. 4, 71–79.

Murugan, K., Senthilkumar, B., Senbagam, D., Al-Sohaibani, S., 2014. Biosynthesis of silver nanoparticles using *Acacia leucophloea* extract and their antibacterial activity. Int. J. Nanomedicine 9, 2431–2438.

Muthukrishnan, S., Bhakya, S., Kumar, T.S., Rao, M.V., 2015. Biosynthesis, characterization and antibacterial effect of plant-mediated silver nanoparticles using *Ceropegia thwaitesii*—an endemic species. Ind. Crops Prod. 63, 119–124.

Nabikhan, A., Kandasamy, K., Raj, A., Alikhunhi, N.M., 2010. Synthesis of antimicrobial silver nanoparticles by callus and leaf extracts from saltmarsh plant, *Sesuvium portulacastrum* L. Colloids Surf. B. Biointerfaces 79, 488–493.

Nagaich, U., Gulati, N., Chauhan, S., 2016. Antioxidant and antibacterial potential of silver nanoparticles: biogenic synthesis utilizing apple extract. J. Pharm. Sci. 2016, 7141523.

Nakkala, J.R., Mata, R., Raja, K., Chandra, V.K., Sadras, S.R., 2018. Green synthesized silver nanoparticles: catalytic dye degradation, in vitro anticancer activity and in vivo toxicity in rats. Mater. Sci. Eng. C 91, 372–381.

Narayanan, K.B., Park, H.H., 2014. Antifungal activity of silver nanoparticles synthesized using turnip leaf extract (*Brassica rapa* L.) against wood-rotting pathogens. Eur. J. Plant Pathol. 140, 185–192.

Nazir, S., Hussain, T., Iqbal, M., Mazar, K., Muzzam, A.G., Ismail, M., 2011. Novel and cost-effective green synthesis of silver nanoparticle and their in-vivo antitumor properties against human cancer cell lines. J. Biosci. Bioeng. 2, 425–430.

Oliver, S., Wagh, H., Liang, Y., Yang, S., Boyer, C., 2018. Enhancing the antimicrobial and antibiofilm effectiveness of silver nanoparticles prepared by green synthesis. J. Mater. Chem. B 6, 4124.

Orlowski, P., Tomaszewska, E., Gniadek, M., Baska, P., Nowakowska, J., Sokolowska, J., Nowak, Z., Donten, M., Celichowski, G., Grobelny, J., Krzyowska, M., 2014. Tannic acid modified silver nanoparticles show antiviral activity in herpes simplex virus type 2 infection. PLoS One 9, e104113.

Ortega, F.G., Fernandaz-Baldo, M.A., Fernandaz, J.G., Serrano, M.J., Sanz, M.I., Diaz-Mochon, J.J., Lorente, J.A., Raba, J., 2015. Study of antitumor activity in breast cell lines using silver nanoparticles produced by yeast. Int. J. Nanomedicine 10, 2021–2031.

Palanisamy, S., Subramaniam, B.S., Thangamuthu, S., Nallusamy, S., Rengasamy, P., 2021. Review on agro-based nanotechnology through plant-derived green nanoparticles: synthesis, application and challenges. J. Environ. Sci. Public Health 5 (1), 77–98.

Patil, R.S., Kokate, M.R., Kolekar, S.S., 2012. Bioinspired synthesis of highly stabilized silver nanoparticles using *Ocimum tenuiflorum* leaf extract and their antibacterial activity. Spectrochim. Acta A Mol. Biomol. Spectrosc. 91, 234–238.

Pei, J., Fu, B., Jiang, L., Sun, T., 2019. Biosynthesis, characterization, and anticancer effect of plant-mediated silver nanoparticles using *Coptis chinensis*. Int. J. Nanomedicine 14, 1969–1978.

Philip, D., Unni, C., Aromal, S.A., Vidhu, V.K., 2011. *Murraya Koenigii* leaf-assisted rapid green synthesis of silver and gold nanoparticles. Spectrochim. Acta A Mol. Biomol. Spectrosc. 78, 899–904.

Phull, A.-R., Abbas, Q., Ali, A., Raza, H., Kim, S.J., Zia, M., Haq, I.-u., 2016. Antioxidant, cytotoxic and antimicrobial activities of green synthesized silver nanoparticles from crude extract of *Bergenia ciliate*. Future J. Pharm. Sci. 2, 31–36.

Prakash, P., Gnanaprakasam, P., Emmanuel, R., Arokiyaraj, S., Saravan, M., 2013. Green synthesis of silver nanoparticles from leaf extract of *Mimusops elengi*, Linn for enhanced antibacterial activity against multi drug resistant clinical isolates. Colloids Surf. B. Biointerfaces 108, 255–259.

Prasher, P., Singh, M., Mudila, H., 2018. Green synthesis of silver nanoparticles and their antifungal properties. Bionanoscience 8, 254–263.

Pugazhendhi, S., Kirubha, E., Palanisamy, P.K., Gopalakrishnan, R., 2015. Synthesis and characterization of silver nanoparticles from *Alpinia calcarata* by green approach and its applications in bactericidal and nonlinear optics. Appl. Surf. Sci. 357, 1801–1808.

Rajan, A., Vilas, V., Philip, D., 2015. Catalytic and antioxidant properties of biogenic silver nanoparticles synthesized using *Areca catechu* nut. J. Mol. Liq. 207, 231–236.

Rajkumar, R., Ezhumalai, G., Gnanadesigan, M., 2021. A green approach for the synthesis of silver nanoparticles by *Chlorella vulgaris* and its application in photocatalytic dye degradation activity. Environ. Technol. Innov. 21, 101282.

Ramachandran, R., Krishnaraj, C., Sivakumar, A.S., Prasannakumar, P., Abhaykumar, V.K., Shim, K.S., Song, C.-G., Yun, S.-I., 2017. Anticancer activity of biologically synthesized silver and gold nanoparticles on mouse myoblast cancer cells and their toxicity against embryonic zebrafish. Mater. Sci. Eng. C 73, 674–683.

Ramar, M., Manikandan, B., Marimuthu, P.N., Raman, T., Mahalingam, A., Subramanian, P., Karthick, S., Munusamy, A., 2015. Synthesis of silver nanoparticles using *Solanum trilobatum* fruits extract and its antibacterial, cytotoxic activity against human breast cancer cell line MCF 7. Spectrochim. Acta A Mol. Biomol. Spectrosc. 140, 223–228.

Ramesh, P.S., Kokila, T., Geetha, D., 2015. Plant mediated green synthesis and antibacterial activity of silver nanoparticles using *Emblica officinalis* fruit extract. Spectrochim. Acta A Mol. Biomol. Spectrosc. 142, 339–343.

Rauwel, P., Kuunal, S., Ferdov, S., Rauwel, E., 2015. A review on the green synthesis of silver nanoparticles and their morphologies studied via TEM. Adv. Mater. Sci. Eng. 2015, 682749.

Rolim, W.R., Pelegrino, M.T., Lima, B.A., Ferraz, L.S., Costa, F.N., Bernardes, S., Rodigues, T., Brocchi, M., Seabra, A.B., 2019. Green tea extract mediated biogenic synthesis of silver nanoparticles: characterization, cytotoxicity evaluation and antibacterial activity. Appl. Surf. Sci. 463, 66–74.

Ruiz-Baltazar, A.J., Reyes-Lopez, S.Y., Larranaga, D., Estevez, M., Perez, R., 2017. Green synthesis of silver nanoparticles using a *Melissa officinalis* leaf extract with antibacterial properties. Results Phys. 7, 2639–2643.

Salari, S., Bahabadi, S.E., Samzadeh-Kemani, A., Yosefzai, F., 2019. In-vitro evaluation of antioxidant and antibacterial potential of green synthesized silver nanoparticles using *Prosopis farcta* fruit extract. Iran J. Pharm. Sci. 18, 430–445.

Salem, W.M., Haridy, M., Sayed, W.F., Hassan, N.H., 2014. Antibacterial activity of silver nanoparticles synthesized from latex and leaf extract of *Ficus sycomorus*. Ind. Crops Prod. 62, 228–234.

Salunke, B.K., Sawant, S.S., Lee, S.-l., Kim, B.S., 2016. Microorganisms as efficient biosystem for the synthesis of metal nanoparticles: current scenario and future possibilities. World J. Microbiol. Biotechnol. 32, 88.

Salunke, B.K., Shin, J., Sawant, S.S., Alkotaini, B., Lee, S., Kim, B.S., 2014. Rapid biological synthesis of silver nanoparticles using *Kalopanax pictus* plant extract and their antimicrobial activity. Korean J. Chem. Eng. 31, 2035–2040.

Sankar, S.S., Ahmed, A., Satry, M., 2003. Geranium leaf assisted biosynthesis of silver nanoparticles. Biotechnol. Prog. 19, 1627–1631.

Sankar, R., Karthik, A., Prabu, A., Karthik, S., Shivashangari, K.S., Ravikumar, V., 2013. *Origanum vulgare* mediated biosynthesis of silver nanoparticles for its antibacterial and anticancer activity. Colloids Surf. B. Biointerfaces 108, 80–84.

Satyavani, K., Gurudeeban, S., Ramanathan, T., Balasubramanian, T., 2011. Biomedical potential of silver nanoparticles synthesized from calli cells of *Citrullus colocynthis* (L.) Schrad. J. Nanobiotechnol. 9, 43.

Saxena, A., Tripathi, R.M., Zafar, F., Singh, P., 2012. Green synthesis of silver nanoparticles using aqueous solution of *Ficus benghalensis* leaf extract and characterization of their antibacterial activity. Mater. Lett. 67, 91–94.

Shameli, K., Ahmed, M.B., Zamanian, A., Sangpour, P., Shabanzadeh, P., Abdollahi, Y., Zargar, M., 2012. Green biosynthesis of silver nanoparticles using *Curcuma longa* tuber powder. Int. J. Nanomedicine 7, 5603–5610.

Sharma, V., Kaushik, S., Pandit, P., Dhull, D., Yadav, J.P., Kaushik, S., 2019. Green synthesis of silver nanoparticles from medicinal plants and evaluation of their antiviral potential against chikungunya virus. Appl. Microbiol. Biotechnol. 103, 881–891.

Siddiqi, K.S., Husen, A., Rao, R.A.K., 2018. A review on biosynthesis of silver nanoparticles and their biocidal properties. J. Nanobiotechnol. 16, 14.

Singh, G., Babele, P.K., Shahi, S.K., Sinha, R.P., Tyagi, M.B., Kumar, A., 2014. Green synthesis of silver nanoparticles using cell extracts of *Anabaena doliolum* and screening of its antibacterial and antitumor activity. J. Microbiol. Biotechnol. 24, 1354–1367.

Singh, J., Dhaliwal, A.S., 2018. Novel green synthesis and characterization of the antioxidant activity of silver nanoparticles prepared from *Nepeta leucophylla* root extract. Anal. Lett. 52, 213–230.

Sinsinwar, S., Sarkar, M.S., Suriya, K.R., Nitiyanand, P., Vadival, V., 2018. Use of agricultural waste (coconut shell) for the synthesis of silver nanoparticles and evaluation of their antibacterial activity against selected human pathogens. Microb. Pathog. 124, 30–37.

Song, J.Y., Kim, B.S., 2009. Rapid biological synthesis of silver nanoparticles using plant leaf extracts. Bioprocess Biosyst. Eng. 32, 79–84.

Song, J.Y., Kwon, E.-Y., Kim, B.S., 2012. Antibacterial latex foams coated with biologically synthesized silver nanoparticles using *Magnolia kobus* leaf extract. Korean J. Chem. Eng. 29, 1771–1775.

Sre, P.R.P., Reka, M., Poovazhagi, R., Kumar, M.A., Murugesan, M., 2015. Antibacterial and cytotoxic effect of biologically synthesized silver nanoparticles using aqueous root extract of *Erythrina indica* lam. Spectrochim. Acta A Mol. Biomol. Spectrosc. 135, 1137–1144.

Sreekanth, T.V.M., Pandurangan, M., Kim, D.H., Lee, Y.R., 2016. Green synthesis: in-vitro anticancer activity of silver nanoparticles on human cervical cancer cells. J. Clust. Sci. 27, 671–687.

Sriram, M.I., Kanth, S.B.M., Kalishwarlal, K., Gurunathan, S., 2010. Antitumor activity of silver nanoparticles in Dalton's lymphoma ascites tumor model. Int. J. Nanomedicine 5, 753–762.

Sriranjani, R., Srinithya, B., Vellingiri, V., Brindha, P., Anthony, S.P., Sivsubhramanian, A., Muthuraman, M.S., 2016. Silver nanoparticle synthesis using *Clerodendrum phlomidis* leaf extract and preliminary investigation of its antioxidant and anticancer activities. J. Mol. Liq. 220, 926–930.

Sujitha, V., Murugan, K., Paulpandi, M., Paneerselvam, C., Suresh, U., Roni, M., Nicoletti, M., Highuchi, A., Madhiyazhagan, P., Subramaniam, J., Dinesh, D., Vadivalagan, C., Chnadramohan, B., Alarfaj, A.A., Munusamy, M.A., Barnard, D.R., Benelli, G., 2015. Green-synthesized silver nanoparticles as a novel control tool against dengue virus (DEN-2) and its primary vector *Aedes aegypti*. Parasitol. Res. 114, 3315–3325.

Sukirtha, R., Priyanka, K.M., Antony, J.J., Kamalakkannan, S., Thagam, R., Gunasekaran, P., Krishnan, M., Achiraman, S., 2012. Cytotoxic effect of green synthesized silver nanoparticles using *Melia azedarach* against in vitro HeLa cell lines and lymphoma mice model. Process Biochem. 47, 273–279.

Sun, Q., Cai, X., Li, J., Zheng, M., Chen, Z., Yu, C.-P., 2014. Green synthesis of silver nanoparticles using tea leaf extract and evaluation of their stability and antibacterial activity. Colloids Surf. A Physicochem. Eng. Asp. 444, 226–231.

Swamy, M.K., Sudipta, K.M., Jayanta, K., Subramanya, S., 2015. The green synthesis, characterization, and evaluation of the biological activities of silver nanoparticles synthesized from *Leptadenia reticulata* leaf extract. Appl. Nanosci. 5, 73–81.

Tade, R.S., Nangare, S.N., Patil, P.O., 2020. Agro-industrial waste-mediated green synthesis of silver nanoparticles and evaluation of its antibacterial activity. Nano Biomed. Eng. 12, 57–66.

Thiyagarajan, K., Bharti, V.K., Tyagi, S., Tyagi, P.K., Ahuja, A., Kumar, K., Raj, T., Kumar, B., 2018. Synthesis of non-toxic, biocompatible, and colloidal stable silver nanoparticle using egg-white protein as capping and reducing agents for sustainable antibacterial application. RSC Adv. 8, 23213–23229.

Tien, D.-C., Tseng, K.-H., Liao, C.-Y., Huang, J.-C., Tsung, T.-T., 2008. Discovery of ionic silver in silver nanoparticle suspension fabricated by arc discharge method. J. Alloys Compd. 463, 408–411.

Ulaeto, S.B., Mathew, G.M., Pancrecious, J.K., Nair, J.B., Rajan, T.P.D., Maiti, K.K., Pai, B.C., 2020. Biogenic Ag nanoparticles from neem extract: their structural evaluation and antimicrobial effects against *Pseudomonas nitroreducens* and *Aspergillus unguis* (NII 08123). ACS Biomater Sci. Eng. 6, 235–245.

Vanaja, M., Annadurai, G., 2013. *Coleus aromaticus* leaf extract mediated synthesis of silver nanoparticles and its bactericidal activity. Appl. Nanosci. 3, 217–223.

Vanti, G.L., Nargund, V.B., Basavesha, K.N., 2019. Synthesis of *Gossypium hirsutum*-derived silver nanoparticles and their antibacterial efficacy against plant pathogens. Appl. Organomet. Chem. 33, e4630.

Vasyliev, G., Vorobyova, V., Skiba, M., Khrokalo, L., 2020. Green synthesis of silver nanoparticles using waste products (apricot and black currant pomace) aqueous extracts and their characterization. Adv. Mater. Sci. Eng. 2020, 4505787.

Velmurugan, P., Subpiramaniyam, S., Song, Y.-C., Jang, S.-H., Yi, P.-I., Suh, J.-M., Hong, S.-C., 2015. Synthesis and characterization comparison of peanut shell extract silver nanoparticles with commercial silver nanoparticles and their antifungal activity. J. Ind. Eng. Chem. 31, 51–54.

Velsi, H., Hemmati, S., Shrivani, H., Velsi, H., 2016. Green synthesis and characterization of monodispersed silver nanoparticles obtained using oak fruit bark extract and their antibacterial activity. Appl. Organomet. Chem. 30, 387–391.

Venkatesan, B., Subramanian, V., Tumala, A., Vellaichamy, E., 2014. Rapid synthesis of biocompatible silver nanoparticles using aqueous extract of *Rosa damascena* petals and evaluation of their anticancer activity. Asian Pac. J. Trop. Med. 7, S294–S300.

Venugopal, K., Ahmad, H., Manikandan, E., Arul, K.T., Kavitha, K., Moodley, M.K., Rajagopal, K., Balabhaskar, R., Bhaskar, M., 2017. The impact of anticancer activity upon *Beta vulgaris* extract mediated biosynthesized silver nanoparticles (ag-NPs) against human breast (MCF-7), lung (A549) and pharynx (Hep-2) cancer cell lines. J. Photochem. Photobiol. B Biol. 173, 99–107.

Vidhu, V.K., Aromal, S.A., Philip, D., 2011. Green synthesis of silver nanoparticles using *Macrotyloma uniflorum*. Spectrochim. Acta A Mol. Biomol. Spectrosc. 83, 392–397.

Vivek, M., Kumar, P.S., Steffi, S., SuDha, S., 2011. Biogenic silver nanoparticles by *Gelidiella acerosa* extract and their antifungal effects. Avicenna J. Med. Biotechnol. 3, 143–148.

Wang, M., Zhang, W., Zheng, X., Zhu, P., 2017. Antibacterial and catalytic activities of biosynthesized silver nanoparticles prepared by using an aqueous extract of green coffee bean as a reducing agent. RSC Adv. 7, 12144–12149.

Zangeneh, M.M., 2019. Green synthesis and chemical characterization of silver nanoparticles from aqueous extract of *Falcaria vulgaris* leaves and assessment of their cytotoxicity and antioxidant, antibacterial, antifungal and cutaneous wound healing properties. Appl. Organomet. Chem. 33, e4963.

Zeedan, G.S.G., El-Razik, K.A.A., Allam, A.M., Abdalhamed, A.M., Zeina, H.A.A., 2020. Evaluations of potential antiviral effects of green zinc oxide and silver nanoparticles against bovine Herpesvirus-1. Adv. Anim. Vet. Sci. 8, 433–443.

Zhang, W., Qiao, X., Chen, J., 2007. Synthesis of silver nanoparticles—effects of concerned parameters in water/oil microemulsion. Mater. Sci. Eng. B 142, 1–15.

Zuorro, A., Lannone, A., Natali, S., 2019. Green synthesis of silver nanoparticles using bilberry and red currant waste extracts. Processes 7, 193.

CHAPTER 12

Synthesis, properties, and uses of silver nanoparticles obtained from leaf extracts

Fiorella Tulli[a], Ana Belén Cisneros[a], Mauro Nicolás Gallucci[a], María Beatriz Espeche Turbay[a,b], Valentina Rey[a,b], and Claudio Darío Borsarelli[a,b]

[a]*Instituto de Bionanotecnología del NOA (INBIONATEC), CONICET, Universidad Nacional de Santiago del Estero (UNSE), Santiago del Estero, Argentina* [b]*Facultad de Agronomía y Agroindustrias (FAyA)-UNSE, Santiago del Estero, Argentina*

1 Introduction

The vegetable kingdom, with approximately 300,000 species of plants available on the planet, is an abundant source of phytomolecules with reducing and stabilizing properties suitable for the production of metal nanoparticles by eco-friendly and economical synthetic routes. This is the case for silver nanoparticles (AgNPs) synthesized using plant leaf extracts, which has become a developing research field due to optical, electrical, magnetic, catalytic, bio(medical), and antimicrobial applications of the resulting nanocomposites.

The goal of this chapter is to bring to the readers a general outlook and the current state of the art of the green synthesis of AgNPs using leaf extracts with different types of phytochemicals, considering experimental conditions to control the final size, shape, and homogeneity of the AgNPs. The current interest in the leaf extract-mediated synthesis of AgNPs is not only promoted by the optical, electrical, magnetic, and antimicrobial intrinsic properties of the metal nanoparticle, but also by the synergic benefits of the incorporation of suitable phytochemicals that allow the assembly of new green nanocomposites for utilization in electronic components, cryogenic superconducting materials, (bio)sensor materials, cosmetic products, and antimicrobial composites, among others (Abbasi et al., 2014; Burda et al., 2005; Chung et al., 2016; Jain et al., 2008; Mahmoudi et al., 2020; Ni and Wang, 2016; Siddiqi et al., 2018; Trotta and Mele, 2019; Wei et al., 2015).

In this chapter, we have revised and discussed recent literature describing the use of harvested plant leaves as a source of phytochemicals for green synthesis of silver nanoparticles (AgNPs) and describe synthetic methods, characteristics, properties, and applications of the green-prepared nanocomposites. Finally, it is noteworthy to mention that the use of leaf extracts as a naturally occurring media for the preparation of metal nanoparticles is a growing field with a large potential for multiple applications.

2 Nanomaterials and the "nano-effect"

Nanotechnology is the field that involves the manipulation and application of materials at the nanometer size scale, typically with at least one dimension, x, y, or z, ranging between 1 and 100 nm, as conventionally defined (Trotta and Mele, 2019). According to the International Organization for Standardization (ISO, https://www.iso.org/standard/44278.html), nanomaterials can be clasified as nanoobjects and nanostructured materials. The former are discrete pieces of a material or substance with one or more outer dimensions at the nanoscale. Therefore, nanoobjects can include nanoparticles (3D nanoscale, e.g., spheres, rods, cubes, prisms, etc.), nanofibers (2D nanoscale, e.g., fibrils, wires), and nanoplates (1D nanoscale, e.g., layers, films, networks) (Boverhof et al., 2015; Calderón-Jiménez et al., 2017; Krug and Wick, 2011); whereas nanostructured materials refers to 3D materials having some internal and/or surface structures in the nanoscale, such as nanocomposites, supramolecular aggregates, nanofoams, nanosponges, etc. (Calderón-Jiménez et al., 2017; Jeevanandam et al., 2018; Krug and Wick, 2011; Trotta and Mele, 2019). Nevertheless, the size limit for the definition of nanomaterials is not strictly restricted to this range since some molecules with <1 nm size (e.g., graphene and fullerene, among others), as well as materials of up to 500 nm, are also defined as nanomaterials due to the biological and/or physical chemistry "nano-effect" that they exert (Boverhof et al., 2015; Restrepo and Villa, 2021). The latter concept refers to the remarkable differences in material properties observed at the nano- and bulk scale, produced by the huge surface area/volume ratio value of the nanomaterials (Fig. 1A). Hence, the nano-effect increases the interactions of nanomaterials with surrounding molecules and also yields unique physicochemical responses to external environmental stimuli (e.g., light, temperature, pH, etc.). Thus, the control of atoms or molecules in the structure of the nanomaterials allows the tuning of their remarkable properties for specific applications in multiple technological fields (Adams and Barbante, 2013; Crisan et al., 2021; Dasgupta et al., 2015; Kargozar and Mozafari, 2018; Kaul et al., 2018; Mishra et al., 2017; Schaming and Remita, 2015; Silva et al., 2015).

The most surprising phenomenon produced at the nanoscale is the change of the optical properties compared with those of the same bulk material, due to the existence of the surface plasmon resonance (SPR) produced by the collective oscillation of surface electrons of the metal nanoparticle modulated by the incident electromagnetic radiation (Noginov et al., 2007) (Fig. 1B). As an example, the color difference perceived by the human eye between bulk metal solids and nano-colloidal solutions of gold and silver, respectively, is shown in Fig. 1C. The "new" yellowish and reddish coloration of each nanoparticle solution is the

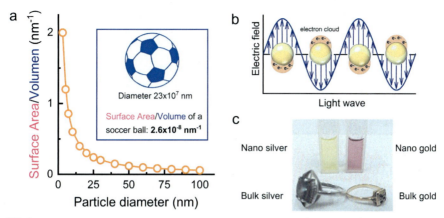

FIG. 1

The nano-effect: (A) Variation of the surface area/volume ratio with the nanoparticle diameter compared with the value of a "bulk" particle, such as a soccer ball. (B) Schematic representation of the surface resonance plasmon phenomenon produced by the interaction of the electric field of the incident electromagnetic radiation and the surface electron cloud of the metal nanoparticles. (C) Color comparison between solid bulk metals and colloidal solution of their nanoparticles, where physicochemical properties such as interaction with light are completely different for the bulk and nanomaterial of the same nature.

Credit: F. Tulli and C.D. Borsarelli.

consequence of the optical nano-effect, since the interaction of light with the AgNPs and AuNPs of about <20 nm diameter produces intense SPR absorptions bands at 405 nm and 520 nm, respectively. Typically, the SPR absorption band of metal nanoparticles is in the visible region, and its spectroscopic properties change with the atom's nature, the size and shape of the nanoparticles, and also by the surrounding stabilizing and solvent molecules (Burda et al., 2005; Islam et al., 2021; Restrepo and Villa, 2021). For instance, for nanoparticles with the same shape, the position of maximum absorption wavelength of the SPR band is red-shifted with the particle average size, whereas the SPR spectrum strongly changes for nanoparticles with similar size but with a different shape. Furthermore, the increment of the SPR band full width at half maximum (fwhm) indicates an increment in the size/shape dispersion of the particles. Consequently, all these spectroscopic attributes can be exploited in molecular sensing and analytical applications (Sabela et al., 2017).

3 Green synthesis of silver nanoparticles

Nanoparticle preparation can be conducted following top-down or bottom-up synthetic approaches (Iqbal et al., 2012) (Fig. 2). The first approach is equivalent to the demolition worker's task, breaking down a wall (bulk matter) into bricks

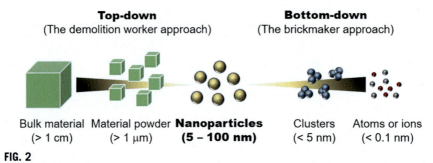

FIG. 2

Schematic representation of top-down and bottom-up methodologies for fabrication of nanoparticles, as compared with the demolition worker and brickmaker approaches, respectively.

Credit: F. Tulli and C.D. Borsarelli.

(nanoparticles), while bottom-up synthesis is similar to the task of the brickmaker, building the bricks (nanoparticles) from their components such as clay-bearing soil, sand, lime, and water (ions, atoms, and/or molecules).

Thus, in nanoscience and nanotechnology, top-down methods involve the use of mechanical, physical, and/or chemical destructive techniques applied to bulk materials to reduce them into particles at the nanoscale size. Typically, techniques of laser ablation, electron beam evaporation, ion beam lithography, arc discharge, and atomization are utilized to produce nanoobjects of defined size and shape (Iqbal et al., 2012). However, most of these instruments are expensive and such techniques are not usually available in most labs. Instead, the bottom-up approach is the most commonly chosen method for nanoparticles and nanoobjects preparation, mainly due to its versatility to control their size, shape, and functionalization at a relatively low cost and using the most common equipment and facilities available in labs (Boennemann and Richards, 2001; Iqbal et al., 2012; Pacioni et al., 2015).

Recently, the most commonly used bottom-up synthetic routes of silver nanoparticles (AgNPs) have been summarized and discussed, describing most chemical reduction, photophysical, photochemical, electrochemical, and ultrasonic utilized methods (Pacioni et al., 2015). However, although most of these methodologies are efficient and versatile for the preparation of AgNPs with different sizes and shapes, some of them could be economically expensive and environmentally unfriendly (Kaabipour and Hemmati, 2021).

Hence, considering the overwhelming growth of global anthropogenic contamination, synthetic approaches to nanoparticles based on green chemistry principles have massively increased since the beginning of this century (Dahl et al., 2007; Duan et al., 2015; Silva et al., 2015). Green chemistry is a new concept or approach to using chemistry in a fashion that maximizes the efficiency of the process to reduce costs and minimizes hazardous effects on human health and the environment. The green approach aims to improve the everyday lab practices of chemists in the manipulation and creation of new materials, and to eliminate or decreases the intrinsic chemical hazard using eco-friendly or naturally occurring reactants and low-energy consuming methodologies, and/or with limited steps/times of reaction. Furthermore, green synthetic methods can be applied to large-scale production (Warner et al., 2004).

Green chemistry concepts are currently applied to the bottom-up synthesis of silver nanoparticles (AgNPs) using as precursor silver salts, typically silver nitrate $AgNO_3$ (Abdelghany et al., 2018; Algebaly et al., 2020; Chaudhari et al., 2016; Mohammadlou et al., 2016; Mousavi et al., 2018; Rajeshkumar and Bharath, 2017; Roy and Das, 2015; Siddiqi et al., 2018). Fig. 3 compares the evolution of the number of scientific articles published in the database Scopus using the following searching phases: "synthesis of silver nanoparticles," "green synthesis of silver nanoparticles," and "green synthesis of silver nanoparticles with leaf extracts," respectively. Nowadays, the green synthesis of AgNPs represents almost 35% of the total publications for the synthesis of AgNPs, with approximately 2.5% increment per year during the last 12 years. It must be noted that this search also includes articles using all types of biogenic materials able to reduce silver ions (Ag^+) to silver atoms (Ag^0), including those produced by microorganisms (Siddiqi et al., 2018), and also those extracted from different plants parts (seeds, roots, leaves, flowers, etc.) (Algebaly et al., 2020; Bhakya et al., 2016; Rajeshkumar and Bharath, 2017). In the biogenic synthesis of AgNPs using bacteria, fungi, or yeast, the reduction of the Ag^+ occurs through the oxide-reductive machinery of the microorganism, both intra- and extracellular, or from reducing compounds released into the biological milieu as metabolism's products (Siddiqi et al., 2018). These mechanisms are discussed in other chapters of this book.

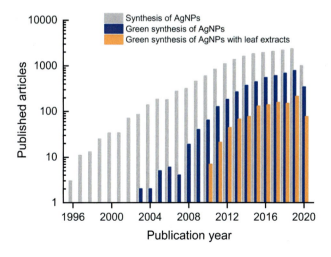

FIG. 3

Evolution of the number of published articles found in the database Scopus with the search keywords indicated in the figure.

Credit: F. Tulli and C.D. Borsarelli.

On the other hand, the use of plant extracts due to the very promising reductive and stabilizing functions of the phytomolecules contained in the plants has increased the research interest in this field, with the additional advantage in most cases of allowing a one-pot synthesis of AgNPs to obtain biocomposites with synergic beneficial properties provided by the phytochemicals adsorbed onto the silver metal core (Algebaly et al., 2020; Bhakya et al., 2016; Gour and Jain, 2019; Rajeshkumar and Bharath, 2017; Vanlalveni et al., 2021).

The bottom-up green synthesis of AgNPs using plant extracts initially involves the chemical redox reaction in aqueous media between silver ions (Ag^+) and reducing phytomolecules (Red) to yield silver atom seeds (Ag^0) and oxidized byproducts (Ox). Eventually, the Ag^0 seeds form intermediate clusters, which finally get larger to form stable colloidal AgNPs of different shapes and sizes through diverse nucleation and coalescence steps mechanisms depending on the reaction conditions such as pH, temperature, light, and stabilizing agents (Thanh et al., 2014) (Fig. 4).

As for any spontaneous reaction, the driving force of the Red+$Ag^+ \to Ag^0$+Ox process is determined by the standard reduction potential (E^0) values of the reaction partners to obtain an $\Delta E^0 > 0$, since $\Delta G^0 = -nF\Delta E^0$. For the silver ion ($Ag^+$), a relatively electropositive reduction potential of $E^0(Ag^+/Ag^0) = +0.799\,V$ vs NHE is needed in water, permitting the use of several natural reducing phytomolecules with lower standard reduction potential value. Hence, a large variety of primary and secondary phytomolecules present in leaves, principally organic acids, polyphenols, flavonoids, terpenoids, sterols, alkaloids, sugars, amino acids, saponins, etc., are involved as reducing and/or capping agents during green synthesis (Ahmed et al.,

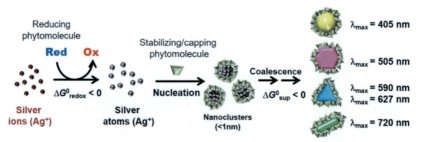

FIG. 4

Schematic illustration of the main steps for bottom-up chemical reduction of silver ions (Ag^+) by phytomolecules to form stable silver nanoparticles AgNPs with different sizes and/or shapes. In some cases, the same reducing phytomolecule acts as a capping/stabilizing agent. Note that the surface plasmon resonance (SPR) spectrum is dependent on the shape and size of the nanoparticle, and hence characteristic optical properties (color) of the colloidal solutions can be observed, as indicated by the absorption maximum wavelength (λ_{max}) for AgNPs with different geometrical shapes.

Credit: F. Tulli and C.D. Borsarelli.

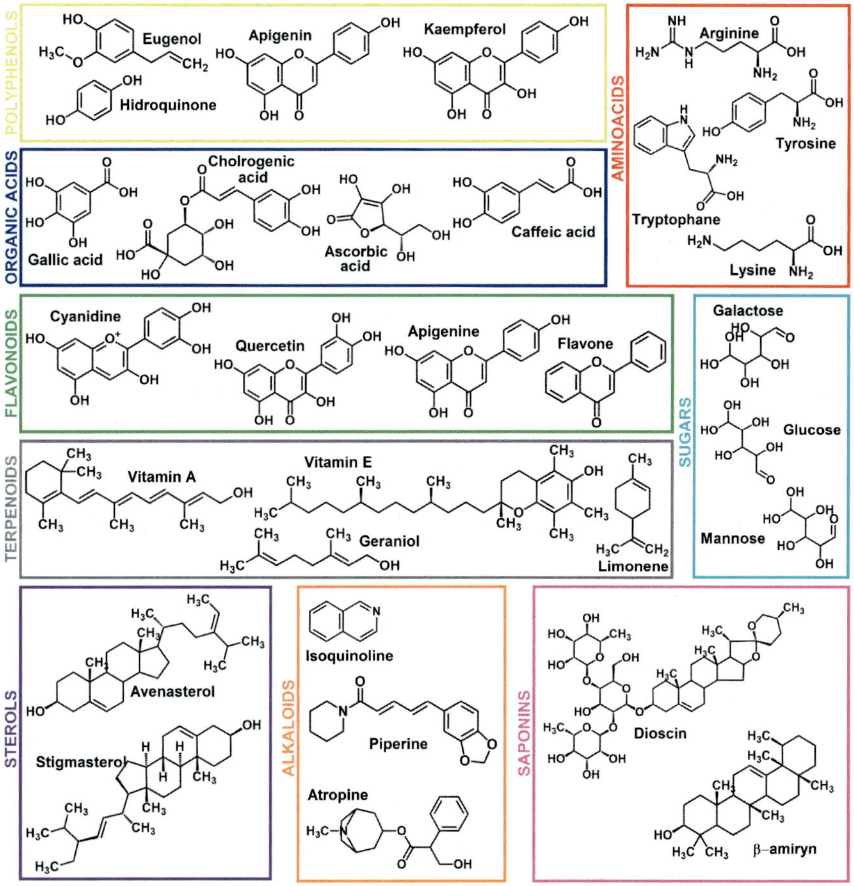

FIG. 5

Chemical structure of typical phytochemicals involved in the formation of AgNPs.

Credit: F. Tulli.

2016b; Anjum et al., 2016; Chung et al., 2016; Iravani, 2011; Makarov et al., 2014; Mohammadlou et al., 2016; Rajeshkumar and Bharath, 2017; Roy and Das, 2015) (Fig. 5).

Plant secondary metabolites are not required for essential cellular functions and are end-products of the primary metabolites (e.g., vitamins, carbohydrates, proteins, lipids), which are produced during the growth phase to develop and maintain vital cellular functions of the plant. The latter are present in all species, but the amount and type of secondary phytomolecules may vary according to the plant species and stress conditions for plant growth (Jha et al., 2009). Hence, the biological diversity of the secondary phytomolecules will give green chemists a versatile naturally occurring toolbox for obtaining AgNPs with different shapes, sizes, stability, and functionality (Gour and Jain, 2019; Hussain et al., 2016; Iravani, 2011; Makarov et al., 2014; Mohammadlou et al., 2016; Moradi et al., 2021; Roy and Das, 2015).

As depicted in Fig. 4, the general mechanism of (phyto)synthesis of AgNPs can be divided into three stages: (i) ion reduction, (ii) clustering, and (iii) further nanoparticle growth (Khan et al., 2019). The first stage is sensitive to the nature of reducing phytomolecules, pH, temperature, and concentration ratio Ag^+/phytomolecule. The rate of formation of the nanoparticles depends on the intrinsic reactivity of each phytomolecule with the metal ion and, therefore, is governed by the relative amount and redox potential of the phytochemical (Hussain et al., 2016; Iravani, 2011; Jha et al., 2009; Khan et al., 2019; Makarov et al., 2014).

A typical example is given by the keto-enol tautomeric equilibrium of secondary phytomolecules, which may release a reactive hydrogen atom, reducing metal ions to form nanoparticles (Jha et al., 2009; Makarov et al., 2014). Interestingly, depending on the plant type and stress conditions, different phytomolecules are involved in the keto-enol equilibrium (Fig. 6). For instance, in mesophyte plants, which are terrestrial plants requiring moderate water for their survival (e.g., cucurbits, privet, lilac, goldenrod, clover, and oxeye daisy), it has been claimed that the formation of AgNPs might have mainly resulted due to tautomerization of quinones, in addition to direct silver ion reduction by other secondary phytochemicals (e.g., terpenoids, organic acids, amines, flavones, etc.) (Jha et al., 2009).

In turn, xerophytes, which are plants adapted for survival in dry environments (e.g., such as cacti and succulents), in addition to a large pool of organic acids that reduce Ag^+, also contain emodin (6-methyl-1,3,8-trihydroxyanthraquinone), which can reduce silver ions by the tautomerization mechanism (Jain and Mehata, 2017; Jha et al., 2009).

Finally, hydrophytes or aquatic plants, like *Nymphaea alba* and *Victoria amazonica*, contain a potent antioxidant system against stressing reactive oxygen species (ROS) produced by environmental conditions (salinity, ammonia levels, pH), generating the ascorbate ion as ROS scavenger, which is enzymatically recovered by the dehydroascorbate reductase (DHAR) by the reduction of dehydroascorbate DHA. Furthermore, under alkaline conditions catechol can be transformed into protocatechualdehyde and finally oxidized to protocatechuic acid, releasing a hydrogen atom involved in the synthesis of AgNPs (Jain and Mehata, 2017; Jha et al., 2009) (Fig. 6).

Flavonoids comprise a large family of polyphenolic compounds (flavonols, isoflavonols, flavones, chalcones, flavanones, anthocyanins), which can actively chelate and reduce metal ions into nanoparticles either with their carbonyl groups or π-electrons (Ahmed et al., 2016b; Iravani, 2011; Jha et al., 2009; Makarov et al., 2014). Moreover, some hydroxyl substituents (—OH) existing in flavonoids are responsible for the reduction of Ag^+ and AgNPs formation (Jain and Mehata, 2017; Makarov et al., 2014). For instance, the —OH groups of the catechol moiety in the B-ring of the flavonol quercetin are involved in Ag^+ reduction by the releasing of two protons, i.e., two silver ions are reduced per molecule of quercetin (Jain and Mehata, 2017) (Fig. 6). Besides their reducing capability, flavonoids can also be involved as stabilizers in the early stages of AgNP formation (nucleation and

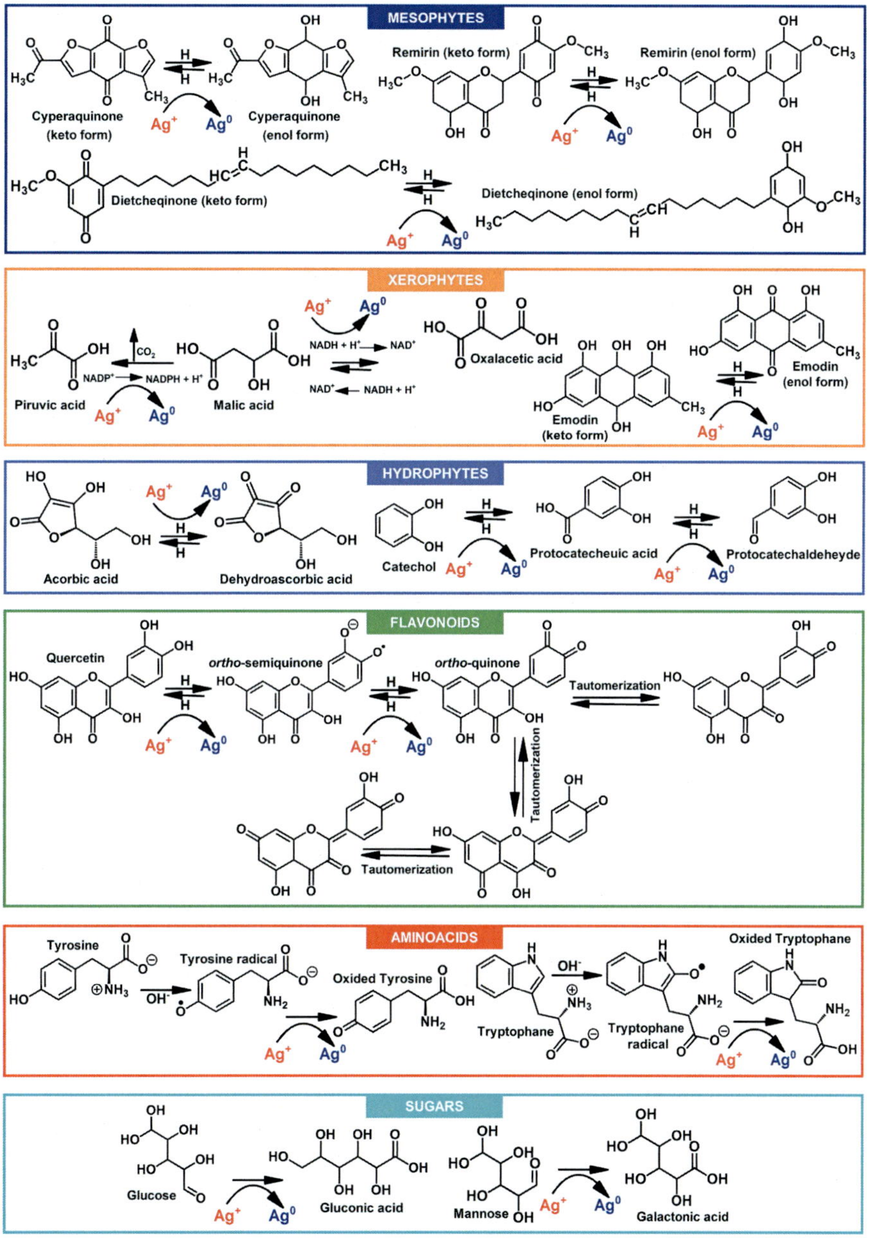

FIG. 6

Proposed mechanisms for the reduction of silver ion by secondary phytomolecules in different types of plants.

Credit: F. Tulli.

clustering) and further aggregation, because of their capability to be adsorbed onto the surface of the nascent AgNPs by the interaction of their —C=O and —OH groups (Makarov et al., 2014).

The reducing sugars present in the plants induce the formation of metal NPs (Fig. 6). Free aldehyde groups of open-ring monosaccharides such as glucose are strong reducing agents, and then, the reducing ability of disaccharides and polysaccharides depends strongly on the content of free aldehyde groups in their monosaccharide components (Makarov et al., 2014). Instead, for monosaccharides containing a keto-group, e.g., fructose, they become reducing agents after tautomeric transformations from ketone to the aldehyde (Korbekandi et al., 2013; Makarov et al., 2014; Soleimani et al., 2018).

Amino acids, peptides, and proteins are capable of reducing metal ions to produce nanoparticles (Daima et al., 2011; Khan et al., 2019; Kim et al., 2017; Tan et al., 2010; Zheng et al., 2013; Zhou et al., 2010). In a systematic study, it was discovered that the reduction capability of a peptide depends on the presence of certain reducing amino acid residues, whose activity may be regulated by neighboring residues with different metal-binding strengths (Tan et al., 2010). Another relevant finding was the effect of the peptide net charge on the nucleation and growth of the metal nanoparticles. They analyzed the reducing capability of 20 amino acids to generate gold nanoparticles, finding that tryptophan was the fastest reducing agent, while histidine was the strongest ion-complexing agent. However, in the peptide chain, the individual capability of a residue to bind and reduce metal ions may be modified, and only free side chains of the residues can interact and reduce the metal ions. Thus, it can be expected that the sequence of amino acids of the protein present in the plant extracts modulates the size and morphology as well as the overall yield of the produced nanoparticles (Tan et al., 2010).

In the second stage of the formation of AgNPs (Fig. 4), the resulting complex of silver atoms, ions, and phytomolecules suffer sudden nucleation to form small metal clusters (<1 nm). These metastable metal clusters spontaneously aggregate leading to the formation of polydispersed larger particles (2–3 nm). Further coalescence drives the formation of the final nanoparticles (5–50 nm), by the growth of larger particles at the expense of smaller ones (Ostwald ripening with single domain nanoparticle formation) and/or by coalescent growth where smaller particles come together to produce larger particles reducing the total number of particles within the solution, i.e., multiple domain nanoparticle formations (Thanh et al., 2014).

Finally, the termination phase, where the nanoparticle conformation is the most energetically favorable, is intensely influenced by the ability of the plant extract to stabilize the AgNPs. As the duration of the growth phase increases, the final stage of the synthesis is reached, where nanoparticles of different shapes such as nanotubes, nanoprisms, and other irregular shapes can be formed (Khan et al., 2019; Pacioni et al., 2015; Thanh et al., 2014). The more stable geometric shapes are those with the largest reduction of the surface Gibbs free energy change (ΔG^0_{surf}) during their formation. Spherical-shaped nanoparticles produce the largest $\Delta G^0_{surf} < 0$ and hence are the shape more frequently obtained (Burda et al., 2005; Soleimani et al., 2018). Nanotriangles require more energy, which makes them less stable, and if they are not energetically stabilized by a given component of the plant extract, then they will change to a more stable morphology to minimize ΔG^0_{surf} (Makarov et al., 2014). Therefore, the final shape and stabilization of the AgNPs will depend on the capability of the different types of plant extracts having different levels of ΔG^0 to ensure the growth, coalescence, and stabilization of the permanent size and shape of the nanoparticle. In this final stage, the presence of proteins and polysaccharides as capping and stabilizing macromolecules can provide additional stabilization of the AgNPs.

4 Silver nanoparticles prepared with leaf extracts

According to the article searching in the Scopus database shown in Fig. 3 it can be observed that among green synthesis approaches, the utilization of leaf extracts as raw materials is rather a new strategy that rapidly increased in the last decade to represent almost a quarter of the total amount of articles on the green synthesis of AgNPs. This recent interest probably is a consequence of the ease of the phytosynthetic method that does not require special and expensive lab facilities, in addition to the enormous diversity of primary and secondary phytomolecules available in the almost 300,000 species of plants existing on the planet (Christenhusz and Byng, 2016). As an example of exploitation of plant diversity for green synthesis of nanoparticles, Fig. 7 compares the number of plant genera existing in the Southern Cone of America (orange bars), which is the region comprised of Argentina, Chile, Paraguay, Uruguay, and southern states of Brazil, with the number of published articles in the Scopus database for synthesis AgNPs using leaf extracts from those botanical families (blue bars). The comparison indicates that some plant families are not yet fully explored in their capability to form AgNPs, suggesting that there is space for developing research in this field that can contribute to the rising value of regional botanical resources (Vanlalveni et al., 2021).

Regarding the ease of the method for green synthesis of AgNPs using leaf extracts, usually, the more frequent protocol encompasses the following steps (Abbasi et al., 2019; Ahmed et al., 2016a; Al-Shmgani et al., 2017; Ali et al., 2015; Amooaghaie et al., 2015; Anandalakshmi et al., 2016; Andersen, 2007; Arockia John Paul et al., 2015; Ashokkumar et al., 2015; Balan et al., 2016; Banerjee et al., 2014; Bhau et al., 2015; Dwivedi and Gopal, 2010; Elavazhagan and Arunachalam, 2011; Elemike et al., 2017; Gallucci et al., 2017; Garg, 2013; Gomathi et al., 2020; Gopinath et al., 2016; Gorbe et al., 2016; Gude et al., 2012; Jain and Mehata, 2017; Kathiravan et al., 2014; Kathireswari et al., 2014; Krishnaraj et al., 2010; Krithiga et al., 2015; Küp et al., 2020; Hajra et al., 2015; Nahar et al., 2020; Narayanan and Park, 2014; Pattanayak et al., 2013; Paulkumar et al., 2014; Raja et al., 2012, 2017; Rajakumar and Abdul Rahuman, 2011;

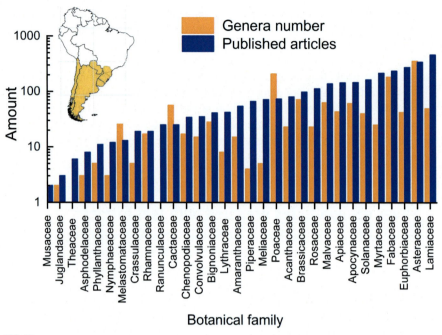

FIG. 7

Comparison of the number of published articles in Scopus using as keywords "silver nanoparticles + botanical family" *(blue bars)* with the number of genera per botanical family *(orange bars)* existing in the Southern Cone of America (Argentina, Chile, Paraguay, Uruguay, and south Brazil), as indicated in the inserted map.

Credit: A.B. Cisneros and C.D. Borsarelli.

Santhoshkumar et al., 2011; Saravanakumar et al., 2017; Siddiquee et al., 2020; Singh et al., 2012; Sun et al., 2014; Vanlalveni et al., 2021):

(i) collection of leaves from various sites of the plant;
(ii) washing of the collected leaves with a mild detergent solution, followed by two or three thorough rinses with distilled water;
(iii) the washed leaves are dried at room temperature and in darkness;
(iv) the chopped dried material is weighed and then dispersed with a certain volume of deionized water and heated to around 60–80°C for 10–15 min;
(v) the resulting solution is filtered through a nylon mesh cloth or with an appropriate paper filter and the filtrate kept in the fridge until the nanoparticle synthesis;
(vi) later, the filtrated leaf extract is mixed with the aqueous solution of a silver salt (e.g., $AgNO_3$ salt, because of its high solubility in water and the inert nature of the nitrate counter ion) to obtain a final concentration of silver ions around 1 mM and kept at room temperature or it is softly heated to activate the bioreduction reaction of Ag^+;

(vii) the final point of AgNPs formation is observed by the eye due to the color produced in the solution or by UV-vis absorption spectroscopy since the SPR band of AgNPs intensely absorbs visible light.

The stable AgNPs in the leaf extracts can be characterized by different available techniques, such as several spectroscopies (e.g., UV-vis-NIR absorption, photoluminescence, Fourier-transform infrared (FTIR), Raman, etc.) that allow interpreting molecular interactions between the metal core and the capping molecules, colloidal homogeneity, etc.; microscopies (e.g., atomic force microscopy (AFM), scanning or transmission electron microscopy (SEM/TEM), etc.) to give 2D/3D information with high topographic resolution (0.1–100 nm); dynamic (DLS) and static (SLS) light scattering techniques to determine the hydrodynamic size (>3 nm) and nanoparticle aggregation (50–1000 nm) in solution, respectively, and electrophoretic mobility (zeta-potential) to characterize superficial electrical charges and colloidal stability, among other modern techniques. For readers interested in a more detailed discussion about the scope of characterization techniques of metal nanoparticles, several more specific reviews can be consulted (Amooaghaie et al., 2015; Banerjee et al., 2014; Huo et al., 2019; Kaabipour and Hemmati, 2021; Okafor et al., 2013; Shnoudeh et al., 2019).

Despite the many advantages of the phytosynthetic approach of AgNPs from the point of view of the principles of green chemistry, a shortcoming to improve is the control of the monodispersity of the colloidal solutions obtained after the bioreduction. The intrinsic molecular complexity of the plant extracts is very helpful for reducing and stabilizing the nanoparticles, but it can also drive multiple pathways of nucleation and growth of particles resulting in polydisperse nano-colloids since the final uniformity and shape of the nanoparticles are controlled by these processes (Haider and Kang, 2015; Rey et al., 2018; Thanh et al., 2014).

As an example of the effect of the phytochemical diversity and effect on the final properties of the synthesized AgNPs, Fig. 8 compares the normalized UV-vis absorption spectra of AgNPs solutions prepared by chemical reduction of 1 mM $AgNO_3$ with sodium citrate (blue filled spectrum), with those obtained with leaf extracts of non- and autochthonous plant species growing in different regions of Argentina, as indicated in the figure text. The UV-vis absorption spectra of the AgNPs colloidal solution prepared by reduction with leaf extracts were different depending on the plant extract utilized, but all of them rendered red-shifted and broader SPR bands compared with that prepared with the citrate ion as the reducing agent. Hence, the color of the solutions turns from pale yellowish for the AgNPs prepared with citrate to a palette of brownish tones for those obtained with the leaf extracts, the latter as the result of the different degree of polydispersity of the AgNPs in the colloidal solutions produced by the different phytochemicals present in each leaf extract (Iravani, 2011).

Fig. 9 shows the Euclidean dendrogram obtained for the size distribution of the phytosynthesized AgNPs with the plant species reported in Table 1 (Zareh et al., 2018), which additionally summarizes reducing/stabilizing biomolecule(s), size

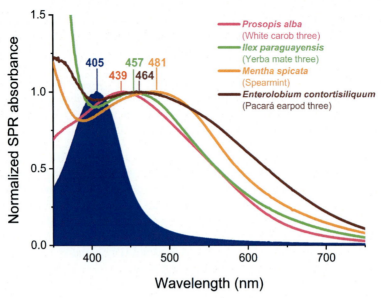

FIG. 8

Normalized absorption spectra of the surface plasmon resonance (SPR) band of AgNPs solutions prepared by chemical reduction of 1 mM AgNO₃ with sodium citrate (*blue* filled), and with leaf extracts from different three and herb species growing in Argentina. Numbers in the spectra plot indicated the absorption maximum λ_{max} of the SPR band.

Credit: N.M. Gallucci, F. Tulli, and C.D. Borsarelli.

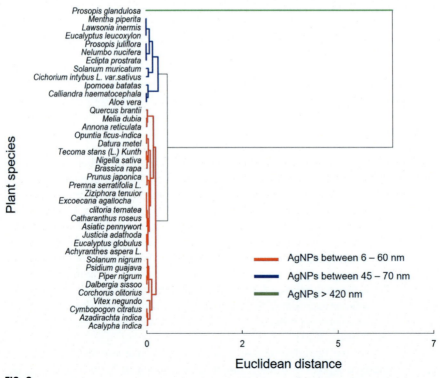

FIG. 9

Dendrogram for phytosynthesized silver nanoparticles AgNPs considering size particles and plant species; obtained by the hierarchal clustering method and the Euclidean distance from data presented in Table 1.

Credit: A.B. Cisneros.

Table 1 Plant families and species utilized for green synthesis of silver nanoparticles (AgNPs) with their leaf extracts as reducing and stabilizing agents, together with morphological properties and applications of the AgNPs solutions.

Botanical family	Species	Reducing/stabilizing agent(s)	Size (nm)	Shape/aspect	Applications	Reference
Acanthaceae	*Justicia adathoda*	Eugenol (allylbenzene), linalool (terpene alcohol), phenolic, and aromatic compounds	11–20	Spherical	Antibacterial	Rao Kudle et al. (2013)
	Barleria longiflora	Phenolic compounds and proteins	3	Spherical	Medical and photocatalytical	Cittrarasu et al. (2019)
Amaranthaceae	*Achyranthes aspera* L.	Ecdysterone, betaine, and pentatriacontanone	6–18	Spherical	N.R.	Gude et al. (2012)
Amaryllidaceae	*Allium rotundum*	N.R	21	Spherical	Antimicrobial	Hekmati et al. (2020)
Apiaceae	*Falcaria vulgaris* Bernh	N.R	21	Spherical	Antimicrobial	Hekmati et al. (2020)
Apocynaceae	*Catharanthus roseus*	Carboxylic acid and amide groups of proteins/enzymes	20	Spherical	Antioxidant, antimicrobial, and wound healing	Al-Shmgani et al. (2017)
	Nerium oleander	Alkaloids	380–420	N.R.	Antibacterial	Subbaiya et al. (2014)
	Gymnema sylvestre	Carboxylic acid and amide groups of proteins/enzymes	20–30	Spherical	Antibacterial	Gomathi et al. (2020)
Asphodelaceae	*Aloe vera*	Anthraquinones (emodin, chrysophanol, etc.) and carbonyl groups of proteins	70	Spherical, rectangular, cubic, and triangular	Antifungal	Hajra et al. (2015)

Continued

Table 1 Plant families and species utilized for green synthesis of silver nanoparticles (AgNPs) with their leaf extracts as reducing and stabilizing agents, together with morphological properties and applications of the AgNPs solutions. *Continued*

Botanical family	Species	Reducing/stabilizing agent(s)	Size (nm)	Shape/aspect	Applications	Reference
Asteraceae	*Cichorium intybus* L. var. *sativus*	Phenolic acids (chicoric acid and chlorogenic acid)	19–64	Spherical	Antibacterial activity	Gallucci et al. (2017)
	Parthenium hysterophorus L.	Peptides in amide	20–50	Spherical	Antibacterial and antifungal activity	Gallucci et al. (2017)
	Eclipta prostrata	Fatty acids, carbonyl groups, flavanones, and proteins.	35–60	Triangular, pentagonal, and hexagonal	Larvicidal activity and malaria vectors	Rajakumar and Abdul Rahuman (2011)
Bignoniaceae	*Tecoma stans* (L.) Kunth	Unsaturated carbonyl groups	5–30	Spherical	Biomedicine	Arunkumar et al. (2013)
Brassicaceae	*Brassica rapa*	Amines and aliphatic esters	6–24	Spherical	Antifungal against wood-rotting pathogens	Narayanan and Park (2014)
Cactaceae	*Opuntia ficus-indica*	Quercetin, 2,3-dihydroquercetin	12	Spherical	Antibacterial	Gade et al. (2010)
Chenopodiaceae	*Chenopodium album*	Aldehyde, alkaloids, apocarotenoids, flavonoids, and oxalic acid	10–30	Quasispherical	N.R.	Dwivedi and Gopal (2010)
Colchicaceae	*Gloriosa superba*	colchicine, ketones (gloriosine), and phytosterols (stigmasterol)	20	Triangular and spherical	Antibacterial	Gopinath et al. (2016)
Convolvulaceae	*Ipomoea batatas* (L.) Lam	Vitamins (A, B2, B6, C, and E), phenolic acids, flavonoids, anthraquinones, and aminoacyl sugars	40–67	N.R.	Medicinal and water treatment	Meva et al. (2016)
Ericaceae	*Rhododendron ponticum*	N.R	10–21	Amorphous	Antimicrobial, and cytotoxic activities	Nesrin et al. (2020)

Family	Species	Compounds	Size	Shape	Application	Reference
Euphorbiaceae	*Excoecaria agallocha L.*	Alkaloids, saponins, tannins, and flavonoids	23–42	Hexagonal and spherical	Antibacterial, antioxidant, and cytotoxic effects	Bhuvaneswari et al. (2017)
	Phyllanthus niruri	Proteins, polyphenols	30–60	Spherical aggregated	Mosquito control	Suresh et al. (2015)
	Acalypha indica	Quercetin (flavonoid)	20–30	Spherical	Antibacterial	Krishnaraj et al. (2010)
Fabaceae	*Delonix regia*	N.R	75	Spherical	Anticancer	Siddiquee et al. (2020)
	Prosopis glandulosa	Terpenoids and flavonoids	32–600	Spherical	Antimicrobial	Abdelmoteleb et al. (2017)
	Calliandra haematocephala	Gallic acid	14–91	Spherical	Antibacterial activity	Raja et al. (2017)
	Clitoria ternatea	Aromatic compounds, alkenes, and amines	20	Spherical	Antibacterial against common nosocomial pathogens	Krithiga et al. (2015)
	Dalbergia sissoo	Genistein, biochanin A, pratensein, caviunin (isoflavones), quercetin 3-O-β-D-glucopyranoside, and kampferol-3-O-rutinoside (flavonol glycosides)	5–55	Crystalline and spherical	Optical sensors and NIR absorbers	Singh et al. (2012)
	Saraca indica	Proteins, alcohols, and phenolic compounds	5–50	Spherical	Antibacterial	Garg (2013)
	Prosopis juliflora	Alkaloids, patulitrin (glucoflavonoid), prosogerin-D (flavonoid), procyanidins, and ellagic acid (polyphenol)	35–60	Triangles, pentagons, and hexagons	Antimicrobial	Raja et al. (2012)

Continued

Table 1 Plant families and species utilized for green synthesis of silver nanoparticles (AgNPs) with their leaf extracts as reducing and stabilizing agents, together with morphological properties and applications of the AgNPs solutions. *Continued*

Botanical family	Species	Reducing/stabilizing agent(s)	Size (nm)	Shape/aspect	Applications	Reference
Lamiaceae	*Lavandula stoechas*	Flavonoids and amino acids	20–50	Spherical	Biomedical application	Mahmoudi et al. (2020)
	Ocimum tenuiflorum	Triterpenes, flavonoids, and eugenol	15–25	Spherical	Antimicrobial	Patil et al. (2012)
	Premna serratifolia L.	Alkaloids and flavonoid	16–32	Cubical	Anticancer activity	Arockia John Paul et al. (2015)
	Ziziphora tenuior	Amino groups of proteins, phenols, ethers, esters, and carboxylic acids	8–40	Spherical	Antioxidant	Sadeghi and Gholamhoseinpoor (2015)
	Vitex negundo	Phenolic compounds and flavonoids	10–30	Spherical	Antibacterial	Zargar et al. (2011)
Lythraceae	*Lawsonia inermis*	Gallic, hennotannic, and tannic acids	50	Spherical	Photosensitizer	Kiruba Daniel et al. (2013)
Malvaceae	*Corchorus olitorius* Linn	Alkaloids, cardiac glycoside, flavonoids, phenols, saponins, tannins, terpenoids, and steroids	30–38	N.R.	Medicinal and water treatment	Meva et al. (2016)
Melastomataceae	*Memecylon edule*	Saponin	50–90	Square, triangular and irregular	N.R.	Elavazhagan and Arunachalam (2011)
Meliaceae	*Azadirachta indica*	Flavonoids and terpenoids	41–60	Spherical	Antimicrobial	Ahmed et al. (2016a)
			<200	Plates (triangles, pentagons, and hexagons) and spherical		Banerjee et al. (2014)
	Melia dubia	Alkaloids, carbohydrates, glycosides, phenolic compounds tannins, gums, and mucilages	5–35	Spherical	Human breast anticancer	Kathiravan et al. (2014)
	Swietenia mahagony	Polyols	20–100	Spherical	N.R.	Mondal et al. (2011)

Musaceae	*Musa balbisiana*	Flavonoids and terpenoids	<200	Spherical and plates (triangles, pentagons, and hexagons)	Antimicrobial	Banerjee et al. (2014)
Myrtaceae	*Eucalyptus globulus*	Flavonoids and terpenes (1,8-cineole, limonene, p-cymene, c-terpinene, α-pinene, α-terpineol, and α-phellandrene)	2–25	Spherical	Antibacterial and antibiofilm	Ali et al. (2015)
	Eucalyptus leucoxylon	Polyphenols	50	Spherical	Antioxidant activity	Rahimi-Nasrabadi et al. (2014)
	Psidium guajava	Flavonoids, terpenoids, and ascorbate	20–30	Spherical	N.R.	Raghunandan et al. (2011)
Nymphaeaceae	*Nelumbo nucifera*	Betulinic acid (a steroidal pentacyclic triterpenoid), liensinine, and isoliensinine (isoquinoline alkaloids)	25–80	Decahedral, truncated triangles, spherical and triangular	Larvicidal activity	Santhoshkumar et al. (2011)
Papaveraceae	*Fumaria parviflora*	Proteins, terpenoids, and flavonoids	35–40	Spherical	Anticancer	Sattari et al. (2020)
Pedaliaceae	*Pedalium murex*	Proteins and flavonoids	20–50	Spherical	Antimicrobial	Anandalakshmi et al. (2016)
Phyllanthaceae	*Securinega leucopyrus* (wild)	Carboxylic, hydroxylic, amine groups in proteins, and terpenoids.	11–20	Spherical and oval	Antibacterial	Rao Kudle et al. (2013)
Piperaceae	*Piper nigrum*	Piperine (alkaloid), carboxylic acids	7–50	Spherical and irregular shaped	Antibacterial	Paulkumar et al. (2014)
Poaceae	*Cymbopogon citratus*	Groups derived from heterocyclic compounds: alkaloids, flavonoids	15–65	Cuboidal and rectangular	Antibacterial	Geetha et al. (2014)

Continued

Table 1 Plant families and species utilized for green synthesis of silver nanoparticles (AgNPs) with their leaf extracts as reducing and stabilizing agents, together with morphological properties and applications of the AgNPs solutions. *Continued*

Botanical family	Species	Reducing/stabilizing agent(s)	Size (nm)	Shape/aspect	Applications	Reference
Ranunculaceae	*Nigella sativa*	Terpenoids, flavones, ketones, aldehydes, amides, and carboxylic acids	15	Spherical	Cytotoxicity	Amooaghaie et al. (2015)
Rhamnaceae	*Ziziphus joazeiro*	Flavonoids, terpenoids, soluble proteins, glycoside cells, and pyrenoid ring	5–50	N.R.	Antibacterial	Guimarães et al. (2020)
Rosaceae	*Prunus japonica*	Polysaccharides, flavonoids, triterpenoids, and polyphenols	24	Spherical, hexagonal, and irregular	Antibacterial and antioxidant	Saravanakumar et al. (2017)
Rubiaceae	*Oldenlandia umbellata*	Phenolic, flavonoids, and proteins	23	Spherical	Biomedical application	Subramanian et al. (2019)
Solanaceae	*Solanum muricatum*	Phenols and flavonoids	20–80	Rounded	Cytotoxicity on HeLa cells	Gorbe et al. (2016)
	Solanum nigrum	Aromatic compounds, alkenes, and amines	28	Spherical	Antibacterial for nosocomial pathogens	Krithiga et al. (2015)
	Datura metel	Alkaloids, flavonoids, phenols, tannins, saponins, and sterols	5–50	Spherical	Antimicrobial	Ojha et al. (2013)
Theaceae	*Tea*	Polyphenols, proteins, and amino acids	20–90	Spherical	Antibacterial	Sun et al. (2014)
Verbenaceae	*Lantana trifolia*	N.R	35–70	Amorphous	Antimicrobial	Madivoli et al. (2020)

N.R., *Not reported.*

and shape, and main application(s) of the colloidal solutions from bibliographic data. Three groups of AgNPs with associated average sizes can be observed: (i) *Prosopis glandulosa* showing the highest average size, about 420 nm, and well separated from the other two clusters with an average size ranging between (ii) 6–60 nm and (iii) 45–70 nm, respectively. The cluster with the smaller diameter of AgNPs includes the bulk of species but yields a larger polydispersity than the cluster with the intermediate size.

The above-summarized information confirms the concepts previously described for size, shape, and polydispersity properties of the colloidal solutions of AgNP obtained by phytosynthesis using leaf extracts. Therefore, a large effort is currently being made to tune the green synthesis steps to improve or control the morphological properties of the AgNPs. Some literature examples are briefly discussed as follows.

For instance, AgNPs prepared with *Mentha piperita* leaf extracts were homogenized from polydisperse nanoparticles between 10 and 50 nm to a final average diameter of 7($\pm$3) nm after 10 min of centrifugation at 1500 RPM (Parashar et al., 2009). A similar result was also reported for the AgNPs obtained using *Juglans regia* (walnut) leaf extract prepared in diverse ways (Korbekandi et al., 2013). They obtained quasispherical nanoparticles of 10 to 50 nm, but forming aggregates, by adding walnut leaf powder as reducing media into the silver ion solution. On the contrary, by using ethanolic leaf extracts, nonaggregated spherical AgNPs of approximately 5 nm diameter were observed.

In terms of synthesis conditions such as pH and temperature, it was reported that under acidic conditions the nanoparticle aggregation process is favored and the growth rate of AgNPs is inhibited (Guimarães et al., 2020; Krithiga et al., 2015), while in mild alkaline media, nucleation and growth are favored, causing an increase in absorbance and greater stability of AgNPs (Arshad et al., 2021; Bhutto et al., 2018; Parthiban et al., 2019; Sadeghi and Gholamhoseinpoor, 2015; Yan yu et al., 2019; Zuorro et al., 2019). In the case of AgNPs prepared with leaf extracts of *Psidium guajava* (guava) and *Eucalyptus globulus* by heating the solutions in a microwave oven, it was reported that at low pH and temperature less nucleation occurs, forming larger nanoparticles, but at basic pH the extension of nucleation and protection are accelerated by the thermal activation, obtaining a nanoparticle size similar to those observed by conventional heating (Ali et al., 2015; Raghunandan et al., 2011). Small spherical shape AgNPs with blue-shifted SPR absorption band and good antimicrobial activity were also obtained by fast microwave-heating using *Saraca indica* leaf extracts (Mock et al., 2002),

Concerning the concentration effect of metal ions, it was shown that the formation of larger-sized Au and Ag nanoparticles by reduction with *Chenopodium album* leaf extracts as the concentration of the precursor salt was increased (Dwivedi and Gopal, 2010). A similar behavior was also reported using *Cichorium intybus* L. extracts, obtaining spherical AgNPs with an average diameter of 19, 56, or 64 nm, as the concentration of $AgNO_3$ in a fixed volume of leaf extract was increased from 0.1 mM, 1 mM, to 10 mM, respectively, demonstrating that size-control depends on the ratio between the ion metal concentration-to-reducing agent volume (extract) ratio (Gallucci et al., 2017).

The effect of the concentration of leaf extract on the size distribution and aggregation of the AgNPs can be different depending on the nature of the leaf extract. For *Memecylon edule* (Elavazhagan and Arunachalam, 2011), *Corchorus olitorus Linn*, and *Ipomea batatas (L.) Lam* (Meva et al., 2016) leaf extracts, the increment of leaf extract increased the size of spherical-shaped AgNPs and also the tendency to aggregate. Nevertheless, the opposite effect was observed for the AgNPs prepared with the *Chenopodium album* (Dwivedi and Gopal, 2010), *Melia dubia* (Kathiravan et al., 2014), and *Azadirachta indica* (Ahmed et al., 2016a) leaf extracts, where the increases of extract concentration reduced the size of the formed AgNPs. Consequently, the concentration effect of the phytochemicals present in the leaf extract must be explored in each green synthesis of AgNPs.

The effect of the harvesting season on AgNPs characteristics prepared with leaves of *Pterodon emarginatus* (white sucupira) was also explored (Oliveira et al., 2019). Both types of aqueous extracts (summer and winter) reduced the silver ions, forming AgNPs with considerable colloidal stability. However, the seasons of leaf collection influenced the formation yield of the AgNPs solutions, and also with their hydrodynamic diameters and zeta potentials, but after solvent evaporation, both types of dry particles showed similar sizes (height and diameter) and crystallinity patterns.

Finally, it must be noted that more than one of the aforecited molecular mechanisms of reduction of silver ion by using leaf extracts can be operative simultaneously. For example, *Nerium oleander* belonging to the *Apocynaceae* family is one of the most poisonous ornamental plants commonly found in gardens, but its leaf extract was able to produce AgNPs at room temperature, without the need of additives, accelerators, or any template, and it was proposed that the hydroxyl groups of flavonoids present in the leaf extract together with other alkaloids may be responsible for the facile synthesis of silver nanorods (Subbaiya et al., 2014).

For *Psidium guajava*, a species of the *Myrtaceae* family, it was suggested that flavonoids, terpenoids, and ascorbates, which are abundant in the leaves of this species, are the reducing/stabilizing molecules for the synthesis of AgNPs (Raghunandan et al., 2011). The leaf extracts of *Prosopis juliflora* and *Prosopis glandulosa* of the *Fabaceae* family contain proteins and secondary metabolites such as terpenoids or flavonoids involved in the reduction step and capping of the AgNPs (Abdelmoteleb et al., 2017; Raja et al., 2012).

The stabilization of the AgNPs is strongly dependent on the characteristics of the capping molecules, also present in the leaf extracts. FTIR studies on *Eclipta prostrata* (Rajakumar and Abdul Rahuman, 2011), *Aloe vera* (Hajra et al., 2015), *Catharanthus roseus* (Al-Shmgani et al., 2017), *Securinega leucopyrus* (Rao Kudle et al., 2013), and *Ziziphora tenuior* (Sadeghi and Gholamhoseinpoor, 2015) have confirmed that amine ($-NH_2$) and carbonyl ($-C=O$) groups of protein residues have a greater capacity to bind metals, allowing the proteins to form a layer (corona) that covers the metallic core of the nanoparticles, avoiding agglomeration and improving stabilization (Rey et al., 2018). Additionally, for extracts of *Melia dubia* (*Meliaceae* family) and *Ocimum americanum* (*Lamiaceae* family), it is was also reported that the carbonyl group (—C=O) of reducing sugars assists the stabilization of AgNPs (Ashokkumar et al., 2015; Manikandan et al., 2021), but numerous reports claim that the —C=O group of amino acid residues and protein peptides, sugars, and gallic acid show strong binding to ion metals generating reduction and/or stabilization of the nanoparticles in the medium (Kathiravan et al., 2014; Makarov et al., 2014; Rajeshkumar and Bharath, 2017). Terpenoids are also prone to be adsorbed on the metal core surface by the interaction of the π electron system (Mashwani et al., 2016).

Altogether, the reported literature indicates that the phytomolecular complexity and concentration of the leaf extracts can produce several effects on the green synthesis of AgNPs, with consequences for formation yield, shape and size, colloidal stability, etc., basically by the role of one or more phytomolecules on the nucleation and growth kinetics of the AgNPs during the phytosynthesis. Hence, researchers must be aware of this intrinsic effect, and also those generated by external parameters, such as ion concentration, temperature, pH, organic co-solvents, light, etc.

5 Properties and applications of phytosynthesized AgNPs

5.1 Antioxidant/medicinal properties

Plants have been used from antiquity as a source of active components for the treatment of human diseases. Plant extracts are known to have multiple antioxidant metabolites (organic acids, polyphenols, flavonoids, vitamins, proteins, etc.) (Hu et al., 2003; Nimse and Pal, 2015; Robinson and Zhang, 2011), which can scavenge harmful free radicals due to their ability to donate electrons and/or hydrogen atoms, reducing oxidative stress injury (Nimse and Pal, 2015). According to the World Health Organization (WHO), between 65% and 80% of the population in developing countries currently use plants for traditional medicine treatments, with leaves as the most utilized plant organs either in a fragmented, chopped, or pulverized state (Robinson and Zhang, 2011).

Therefore, besides the intrinsic properties of AgNPs, it can be expected that phytosynthesized AgNPs also show antioxidant activity due to the presence of antioxidant phytomolecules attached to the metal core surface (Alahmad et al., 2021; Bhakya et al., 2016; Elemike et al., 2017; Küp et al., 2020; Moradi et al., 2021). For instance, phenolic and flavonoid compounds are potent chelators of redox-active metal ions and they can inactivate free radical chain reactions, acting as recognized antioxidants (Arockia John Paul et al., 2015; Nimse and Pal, 2015). The green synthesis of AgNPs using *Thymus kotschyanus* extracts resulted in nanoparticles that showed almost twice the antioxidant capability of butylated hydroxytoluene (BHT), used as a standard antioxidant (Hamelian et al., 2018).

Other example cases are the species of the genus *Eucalyptus*, which constitute an important source of many highly valued phytochemicals from the pharmacological and medicinal point of view as an analgesic, antifungal, antiallergic (asthma, bronchitis), antiinflammatory, antibacterial, antidiabetic, antioxidant, antiviral, antitumor, antihistamine, and hepatoprotective; and even their essential oils and terpenoids that have been used in aromatherapy (Patil and Nitave, 2014). The leaves of *Eucalyptus glóbulus*, *Eucalyptus hybrida*, and *Eucalyptus leucoxylon* were utilized for the formation of AgNPs since they present terpenes and terpenoids (e.g., 1,8-cineole, limonene, *p*-cymene, α-pinene, α-terpineol, α-phellandrene), proteins, alkaloids, flavonoids, and phenolic compounds that are the surface-active molecules that stabilize the AgNPs, in addition to their intrinsic therapeutic properties (Ali et al., 2015; Rahimi-Nasrabadi et al., 2014).

AgNPs also show antidiabetic activity, since they can decrease the levels of catalytic enzymes, such as α-glucosidase and α-amylase, involved in complex carbohydrates hydrolysis, favoring the control of sugar level in the blood (Alkaladi et al., 2014). Hence, due to the enhanced biocompatibility of green-synthesized AgNPs, its antidiabetic activity was evaluated (Balan et al., 2016; Govindappa et al., 2018; Rajaram et al., 2015; Saratale et al., 2018). Aqueous extracts of *Lonicera japonica* leaves showed inhibition of digestive carbohydrate enzymes, revealing the noncompetitive nature of the inhibition mechanism, with similar IC_{50} values for the inhibition of α-amylase and α-glucosidase (Balan et al., 2016). A similar result was observed by Saratale et al. (2018) for AgNPs prepared with *Punica granatum* leaf extract since similar inhibition IC_{50} value for α-amylase and α-glucosidase was also observed. Furthermore, they demonstrated good antioxidant activity.

In addition to the antibacterial, antioxidant, antiinflammatory and antityrosinase activities of AgNPs prepared using aqueous leaf extract of *Calophyllum tomentosum* being evaluated, the antidiabetic capability was evaluated by measuring the inhibition percentage of α-amylase ($\approx 20\%$), α-glucosidase ($\approx 50\%$), and dipeptidyl peptidase IV ($\approx 50\%$) (Govindappa et al., 2018). A study focused on AgNPs synthesis using aqueous extract of *Tephrosia tinctoria* evaluated the antidiabetic activity, and the results showed significant inhibition of the main digestive carbohydrate enzymes of diabetes, α-amylase, and α-glucosidase, increasing the rate of glucose absorption (Rajaram et al., 2015).

The wound healing capability of AgNPs prepared with *Catharanthus roseus* leaf extract was characterized, together with their antioxidant and antimicrobial activities. The green-synthesized AgNPs were used for the in vivo wound healing of mice, revealing good efficacy in wound cicatrization. (Al-Shmgani et al., 2017).

The AgNPs prepared with *Indigofera aspalathoides* medicinal plant extract were successfully used for wound healing in animal models with an electrospinning method for the incorporation of the AgNPs in the skin tissue (Arunachalam et al., 2013).

Finally, The biosynthesized AgNPs from *P. nigram* leaves extract were incorporated into the electrospun polycaprolactone (PCL) membrane and their wound dressing capability was reported (Augustine et al., 2016). The results showed that the material with 0.5 wt% AgNPs demonstrated higher tensile strength in comparison to the neat PCL membrane, with 427 and 250 MPa elongations at the break, respectively.

5.2 Antimicrobial activity

Since ancient times, silver in various forms (metallic, salt, and colloidal) has been utilized as an effective antibacterial agent, and in contrast to many natural and synthetic antibiotics has retained its effective antibacterial activity throughout the centuries (Ebrahiminezhad et al., 2016). This particularity gives silver-based materials, such as AgNPs, a relevant role in the current antibiotic crisis for the fighting of multiresistant bacteria, which are producing the largest mortality and morbidity around the world (De Oliveira et al., 2020; Peterson and Kaur, 2018). Indeed, the efficiency of AgNPs to kill efficiently a wide variety of gram-positive and gram-negative strains, in addition to pathogenic fungi, is well documented (Crisan et al., 2021; Franci et al., 2015; Tang and Zheng, 2018; Vanlalveni et al., 2021; Wang et al., 2017).

The mechanisms through which AgNPs exert the killing effect are variable, and cannot be associated with a single target at the level of the bacterial strain. However, the critical point necessary to ensure the biocide effect of the AgNPs is given their attachment to the onto the cell wall and membrane of the microorganism, and subsequent damage of intracellular biomolecules and structures together with the induced oxidative stress (i.e., excessive generation of reactive oxygen species, ROS) caused by both the AgNPs themselves and the released silver ions (Franci et al., 2015; Tang and Zheng, 2018; Wang et al., 2017). Hence, the nano-effect given by the high surface area/volume ratio of the nanoparticles is crucial to assure the arrival of the AgNPs at the cell spaces (Franci et al., 2015; Rajeshkumar and Bharath, 2017; Tang and Zheng, 2018; Wang et al., 2017). Therefore, the nonspecificity of kill-action of the AgNPs makes them an excellent alternative for antimicrobial therapies (Franci et al., 2015), due to the inability of microorganisms to generate changes in multiple action targets simultaneously prevents the development of biocidal multiresistance to AgNPs, contrary to what has occurred with many natural and synthetic antibiotics (Peterson and Kaur, 2018).

Nevertheless, the immoderate antibacterial application of AgNPs can also enhance bacterial resistance to typical antibiotics by promoting oxidative stress tolerance through the induction of intracellular ROS (Kaweeteerawat et al., 2017; Panáček et al., 2018; Wang et al., 2017). The microbial resistance mechanism exerted is the overexpression of the adhesive protein flagellin, which induces the aggregation of the nanoparticles and avoidance of the biocide action. The addition of polymers or surfactants for stabilization of the AgNPs did not abate the resistance mechanism, but interestingly, the flagellin production was strongly suppressed by the phytochemicals present in pomegranate rind extract (Panáček et al., 2018). Thus, a smart strategy can be the utilization of this type of phytochemical in the preparation of AgNPs.

It is believed that the presence of certain phytochemicals present in the leaf extracts as reducing and/or stabilizing agents of AgNPs can contribute to the enhancement of the antimicrobial activity against specific microorganisms, such as those obtained with *Carissa carandas* L. versus *Salmonella typhimurium*, *Enterobacter faecalis*, *Shigella flexneri*, *Citrobacter* spp., and *Gonococci* spp.

(Singh et al., 2021), *Gymnema sylvestre* against *Staphylococcus aureus* and *Escherichia coli*. (Gomathi et al., 2020), *Barleria longiflora* for *Enterococcus sp.*, *Streptococcus* sp., *Bacillus megaterium*, *Pseudomonas putida*, *Ps. Aeruginosa*, and *S. aureus* (Cittrarasu et al., 2019), and *Lavandula stoechas* against *S. aureus* and *P. aeruginosa* (Mahmoudi et al., 2020). In addition, proteins and flavonoids in the leaf extracts of *Pedalium murex* were important in the synthesis of AgNPs with a pronounced antimicrobial activity on bacteria such as *E. coli*, *K. pneumonia*, *M. flavus*, *P. aeruginosa*, *B. subtilis*, *B. pumilus*, and *S. aureus* (Anandalakshmi et al., 2016).

Shahryari et al. (2020) obtained nanocomposites formed with AgNPs from the aqueous extract of *Sumac* leaves, in combination with the naturally occurring polysaccharide chitosan, and it was evaluated against strains of *Pseudomonas psyringae pv. psyringae*, the causal agent of gangrene disease in fruit trees. Also, AgNPs synthesized with leaf extract of *Hypericum perforatum*, a grassland weed, performed antimicrobial activity on the pathogenic bacterium of the soil *Ralstonia solanacearum* (Tortella et al., 2019).

The shape and size of the AgNPs are also critical parameters to consider in optimizing the antibacterial activity of the resulting material (Tang and Zheng, 2018). As mentioned before, these characteristics can be controlled depending on several experimental factors, e.g., concentration and nature of the leaf extract, concentration of $AgNO_3$, temperature, pH, etc. (Alsalhi et al., 2016; Dahl et al., 2007; Das et al., 2017; Gour and Jain, 2019; Iravani, 2011; Manosalva et al., 2019; Rafique et al., 2017; Roy and Das, 2015; Soleimani et al., 2018). Small spherical AgNPs (<4 nm) obtained with *Solanum nigrum* leaf extract showed potent inhibition against *E. coli*. The FTIR spectra analysis confirmed polyphenols and antioxidants existing in the extract act as reducing and capping agents, respectively (Vijilvani et al., 2020). Furthermore, Deshmukh et al. (Singh et al., 2016) determined that nanoparticles with a hydrodynamic diameter of ≈20 nm exhibit a greater degree of interaction with the cell membrane, causing its rupture and death.

The green synthesis of the extracts of leaves and roots of *Ricinus communis* showed antibacterial activity against *Streptococcus pneumoniae*, *Klebsiella pneumoniae*, *E. coli*, and *S. aureus*, and antifungal in *Alternaria alternate* and *Aspergillus niger*, being higher in AgNP@roots than in AgNP@leaves, which may be associated with their smaller size, that is, 29 nm versus 38 nm, respectively. However, the lowest cytotoxic effect, measured as the hemolytic effect, was presented by AgNP@leaves, being of greater potential for use in medicine (Gul et al., 2021).

Several studies have shown that irregular-shaped AgNPs prepared with leaf extracts of plant species such as *Musa balbisiana* y *Azadirachta indica* (Banerjee et al., 2014), *Eclipta prostrata* (Rajakumar and Abdul Rahuman, 2011), *Nelumbo nucifera* (Santhoshkumar et al., 2011), *Ludwigia octovalvis* (Kannan et al., 2021), *Premna serratifolia* (Arockia John Paul et al., 2015), Aloe vera (Hajra et al., 2015), and *Prosopis juliflora* (Raja et al., 2012) interact more efficiently with cells than the spherical-shaped nanoparticles. Indeed, it has been shown that the minimum inhibitory concentration (MIC) on gram-positive and gram-negative strains of AgNPs synthesized by green methods to yield differently shaped particles, was lower for both cubic and rod-like than for spherical AgNPs (Soleimani et al., 2018). Moreover, changes in the stress conditions for the growth of *Tephrosia apollinea* modify the composition of the phytomolecules extracted from the leaves, resulting in two types of AgNPs (spherical and cubic) with the consequent modification of the antimicrobial effectiveness (Ali et al., 2019).

Studies carried out with the aqueous leaf extract of *Oldenlandia umbellata* showed that the AgNPs obtained not only have excellent antimicrobial activity on *S. mutans* and *E. coli* but also showed an interesting antioxidant activity and good compatibility with human pulmonary fibroblast WI-38, indicating that these AgNPs could be used as therapeutic materials for biomedical applications (Subramanian et al., 2019).

The green synthesis with leaves of *Leea coccinea* produced fluorescent AgNPs with an excellent antimicrobial against phytopathogenic bacteria of *Xanthomonas phaseoli* pv *phaseoli* (Travieso Novelles et al., 2021).

The biosynthesis of AgNPs using branch extract of *Pyrus betulifolia* Bunge, in which plant bioactive such as phenols, proteins, and sugars acted as reducing and plugging agents of AgNPs, produced nanoparticles that proved to be effective against bacteria and fungi as well as possessing excellent antioxidant activity (Li et al., 2021). The AgNPs prepared with extracts of leaves and roots of *Ricinus communis* not only demonstrated antibacterial and antifungal activity but also inhibited oxidase enzymes such as urease/xanthine (Gul et al., 2021).

Food contamination with pathogens is a current problem in the food industry, and for instance, a good antimicrobial activity against the foodborne pathogen *Listeria monocytogene* was obtained with AgNPs prepared with leaf extracts of the emerging medicinal herb *Angelica keiskei* (Du et al., 2019).

However, a good antibacterial activity of AgNPs prepared from leaf extracts does not always mean that a broad-spectrum antimicrobial action is obtained. For instance, AgNPs obtained from the aqueous extract of *Aesculus hippocastanum* leaves exhibited strong antibacterial activity against all tested bacterial species but does not affect fungal strains (Küp et al., 2020).

5.3 Antineoplastic properties

The intrinsic ability of AgNPs to generate ROS when they come in contact with microbial and cancer cells, producing the destruction of the mitochondrial respiratory chain and DNA damage, opened their application as antitumor agents (Hembram et al., 2018; Moradi et al., 2021). Furthermore, light absorption by the nanoparticles can induce the occurrence of photodynamic and photothermal effects that accumulate extra ROS and increases the temperature in the cell, respectively, conducive to cell-death mechanisms, such as apoptosis and necrosis (Jain et al., 2008; Soriano et al., 2017). In this sense, the anticancer and antimicrobial efficacies of biosynthesized metal nanoparticles have been demonstrated (Anjum et al., 2016; Crisan et al., 2021; Moradi et al., 2021; Ovais et al., 2016; Pedone et al., 2017; Rafique et al., 2017; Rajeshkumar and Bharath, 2017; Saif et al., 2016; Vanlalveni et al., 2021; Wei et al., 2015).

Another antineoplastic strategy is the use of green synthesized AgNPs as vehicles for anticancer drug delivery. For instance, the *Butea monosperma* leaf extract was utilized for the preparation of an AgNPs-based delivery system of doxorubicin, an anticancer drug. The nanocomposites showed significant inhibition of cancer cell proliferation (B16F10, MCF-7) compared to the single administration of doxorubicin, demonstrating its suitability as a drug delivery system. Furthermore, better biocompatibility toward ECV-304 and HUVEC cell lines using biosynthesized gold and silver nanoparticles compared with those obtained by the chemical reduction process was demonstrated (Patra et al., 2015).

The cytotoxic effect of the AgNPs synthesized with *Excoecaria agallocha* leaf extract was evaluated toward the MCF-7 cell lines for 24, 48, and 72 h. At the AgNPs concentration of 50 and 100 μg/mL, about 50% percent of cell death was reached at 48 and 24 h, respectively, while the total death of cells was reached at more than 72 h in both cases, suggesting a good cytotoxic activity in addition to antimicrobial effects (Bhuvaneswari et al., 2017).

The green synthesis of stable AgNPs using *Hypericum perforatum* L. (St. John's wort) herb extract showed high cytotoxicity by inhibiting the cell viability of Hela, Hep G2, and A549 cells in a short period of 2 h. The organic molecules that coat St. John's wort, which produce high surface charges, probably play an additional role in the toxicity of the AgNPs.

5.4 Molecular sensing and (photo)catalytic properties

AgNPs are extensively utilized as a part of nanocomposite devices for molecular sensing through electrochemical, fluorescence, and colorimetric responses, among others (Jain et al., 2008; Sabela et al., 2017). Typically, AgNPs synthesized from aqueous leaf extracts can be decorated with sensing or reporter molecules, spectroscopic properties of which are enhanced by the nanoparticle, as in the case of AgNPs prepared with *Azadirachta indica* extract and the fluorescent xanthenic dye Rhodamine 6G, producing a highly sensitive and selective colorimetric and spectrofluorometric nanosensor of copper (II) (Kirubaharan et al., 2012).

Another strategy for the selective monitoring of molecules is the color change produced in the AgNPs solution. For example, the UV-vis spectrophotometric detection of hydrogen peroxide (H_2O_2) with AgNPs prepared with *Calliandra haematocephala* leaf extract indicated the bleaching of the SPR band of the nanoparticle by H_2O_2, suggesting the formation of free radical intermediates that initiated the degradation of AgNPs (Raja et al., 2017).

The influence of extract preparation conditions and pH were analyzed for the green synthesis of AgNPs using *Hibiscus sabdariffa* (Gongura) leaf (HL) and stem (HS) plant extracts, and the prepared nanoparticles were successful for selective colorimetric sensing of toxic metal ions, e.g., Hg(II), Cd(II), and Pb(II), at ppm level in aqueous solution, although both selectivity and sensitivity dependent on the plant part utilized in the extract preparation (Vinod Kumar et al., 2014).

Nowadays, contamination of water with pollutant textile dyes has become a serious problem, since these compounds are difficult to remove using conventional methods. Therefore, green synthesized AgNPs are also received great attention due to their potential as a catalyst (Sintubin et al., 2012; Tahir et al., 2015). The catalytic reduction of several organic dyes, such as methyl red, methyl orange, methylene blue, and safranine-O by AgNPs obtained with *Zanthoxylum armatum* leaves extract was reported (Jyoti and Singh, 2016). The catalytic degradation was associated with the nano-effect exerted by the very high surface area of the AgNPs and a prompt migration rate of the electron and hole pairs to the nanoparticle surface allowing the complete redox degradation of the organic pollutants before 24 h at room temperature.

In addition, the photocatalytic degradation of the persistent methylene blue was evaluated using the phytosynthesis of AgNPs with *Barleria longiflora* L. leaf extracts (Cittrarasu et al., 2019) and *Salvadora persica* stem extracts (Tahir et al., 2015). In both cases, the photocatalytic activity of the AgNPs was enhanced to complete the dye bleaching within 45 and 80 min, respectively, as compared with the degradation under dark conditions (Jyoti and Singh, 2016).

Finally, the green synthesis of AgNPs using *Thymbra spicata* leaf extract provided an excellent photocatalytic nanomaterial for complete degradation at room temperature of the hazardous pollutant dyes 4-nitrophenol, methylene blue, and rhodamine B. Besides showing excellent photocatalytic activity with a shorter time required to complete the reaction, this biosynthesized nano-catalyst presented the additional advantage of its reuse after separation by centrifugation for up to eight successive cycles without photocatalytic activity loss (Veisi et al., 2018).

6 Conclusions

The great diversity of the vegetable kingdom, with approximately 300,000 species of plants available on the planet, is an incommensurable source of phytomolecules with powerful reducing and complexing capabilities to produce and stabilize metal nanoparticles by eco-friendly and economical synthetic routes, in particular silver ones, AgNPs, which are of particular interest for their intrinsic optical, electrical, and antimicrobial properties.

Typically, phytosynthesis with leaf extracts produces spherical shaped and relatively polydispersed AgNPs solutions by chemical reduction of silver ions, but many synthetic modifications and strategies have been developed to obtain AgNPs of multiple shapes and with more homogeneous and controlled size.

The use of harvested leaves and other aerial parts of plants as a source of phytochemicals for the green synthesis of AgNPs has the additional advantage that the plant resource is not destroyed, allowing decades of utilization and environmental care. Many countries rich in botanical resources, such as India and Brazil, are exploiting the local vegetal biodiversity to facilitate fabrication of AgNPs for diverse applications, most medical-related, such as antidiabetic, antioxidant, wound-healing, and anticancer, as well as antimicrobial, taking advantage of the unspecific kill mechanism induced by silver that does not generate a rapid antibiotic resistance by the microorganism (Banerjee et al., 2014). Finally, it must be noted that the utilization of leaf plant extracts for the green synthesis of nanoparticles is a relatively recent and developing research field with a large potential for multiple applications of AgNPs and many other metal nanoparticles.

Acknowledgments

The authors thank the following funding agencies of Argentina: Consejo Nacional de Investigaciones Científicas y Técnicas (CONICET, PUE-2018-0035), Universidad Nacional del Santiago del Estero (UNSE, CICyT-23A254), and Fondo para la Investigación Científica y Tecnológica (FONCyT, PICT-2019-02052).

References

Abbasi, B.A., Iqbal, J., Mahmood, T., Qyyum, A., Kanwal, S., 2019. Biofabrication of iron oxide nanoparticles by leaf extract of *Rhamnus virgata*: characterization and evaluation of cytotoxic, antimicrobial and antioxidant potentials. Appl. Organomet. Chem. 33, 1–15.

Abbasi, E., Milani, M., Fekri Aval, S., Kouhi, M., Akbarzadeh, A., Tayefi Nasrabadi, H., Nikasa, P., Joo, S.W., Hanifehpour, Y., Nejati-Koshki, K., Samiei, M., 2014. Silver nanoparticles: synthesis methods, bio-applications and properties. Crit. Rev. Microbiol. 42, 1–8.

Abdelghany, T.M., Al-Rajhi, A.M.H., Al Abboud, M.A., Alawlaqi, M.M., Ganash Magdah, A., Helmy, E.A.M., Mabrouk, A.S., 2018. Recent advances in green synthesis of silver nanoparticles and their applications: about future directions. A review. Bionanoscience 8, 5–16.

Abdelmoteleb, A., Valdez-Salas, B., Ceceña-Duran, C., Tzintzun-Camacho, O., Gutiérrez-Miceli, F., Grimaldo-Juarez, O., González-Mendoza, D., 2017. Silver nanoparticles from *Prosopis glandulosa* and their potential application as biocontrol of *Acinetobacter calcoaceticus* and *Bacillus cereus*. Chem. Speciat. Bioavailab. 29, 1–5.

Adams, F.C., Barbante, C., 2013. Nanoscience, nanotechnology and spectrometry. Spectrochim. Acta B At. Spectrosc. 86, 3–13.

Ahmed, S., Ahmad, M., Swami, B.L., Ikram, S., 2016b. A review on plants extract mediated synthesis of silver nanoparticles for antimicrobial applications: a green expertise. J. Adv. Res. 7, 17–28.

Ahmed, S., Saifullah, Ahmad, M., Swami, B.L., Ikram, S., 2016a. Green synthesis of silver nanoparticles using *Azadirachta indica* aqueous leaf extract. J. Radiat. Res. Appl. Sci. 9, 1–7.

Alahmad, A., Feldhoff, A., Bigall, N.C., Rusch, P., Scheper, T., Walter, J.-G., 2021. *Hypericum perforatum* L.-Mediated green synthesis of silver nanoparticles exhibiting antioxidant and anticancer activities. Nanomaterials 11, 487.

Algebaly, A.S., Mohammed, A.E., Abutaha, N., Elobeid, M.M., 2020. Biogenic synthesis of silver nanoparticles: antibacterial and cytotoxic potential. Saudi J. Biol. Sci. 27, 1340–1351.

References

Ali, K., Ahmed, B., Dwivedi, S., Saquib, Q., Al-Khedhairy, A.A., Musarrat, J., 2015. Microwave accelerated green synthesis of stable silver nanoparticles with *Eucalyptus globulus* leaf extract and their antibacterial and antibiofilm activity on clinical isolates. PLoS One 10, 1–20.

Ali, M., Mosa, K., El-Keblawy, A., Alawadhi, H., 2019. Exogenous production of silver nanoparticles by *Tephrosia apollinea* living plants under drought stress and their antimicrobial activities. Nanomaterials 9, 1716.

Alkaladi, A., Abdelazim, A.M., Afifi, M., 2014. Antidiabetic activity of zinc oxide and silver nanoparticles on streptozotocin-induced diabetic rats. Int. J. Mol. Sci. 15, 2015–2023.

Alsalhi, M.S., Devanesan, S., Alfuraydi, A.A., Vishnubalaji, R., Munusamy, M.A., Murugan, K., Nicoletti, M., Benelli, G., 2016. Green synthesis of silver nanoparticles using *Pimpinella anisum* seeds: antimicrobial activity and cytotoxicity on human neonatal skin stromal cells and colon cancer cells. Int. J. Nanomedicine 11, 4439–4449.

Al-Shmgani, H.S.A., Mohammed, W.H., Sulaiman, G.M., Saadoon, A.H., 2017. Biosynthesis of silver nanoparticles from *Catharanthus roseus* leaf extract and assessing their antioxidant, antimicrobial, and wound-healing activities. Artif. Cells Nanomed. Biotechnol. 45, 1234–1240.

Amooaghaie, R., Saeri, M.R., Azizi, M., 2015. Synthesis, characterization and biocompatibility of silver nanoparticles synthesized from *Nigella sativa* leaf extract in comparison with chemical silver nanoparticles. Ecotoxicol. Environ. Saf. 120, 400–408.

Anandalakshmi, K., Venugobal, J., Ramasamy, V., 2016. Characterization of silver nanoparticles by green synthesis method using *Pedalium murex* leaf extract and their antibacterial activity. Appl. Nanosci. 6, 399–408.

Andersen, F.A., 2007. Final report on the safety assessment of *Aloe andongensis* extract, *Aloe andongensis* leaf juice, *Aloe arborescens* leaf extract, *Aloe arborescens* leaf juice, *Aloe arborescens* leaf protoplasts, *Aloe barbadensis* flower extract, *Aloe barbadensis* leaf, *Aloe barbadensis* leaf extract, *Aloe barbadensis* leaf juice, *Aloe barbadensis* leaf polysacharides, *Aloe barbadensis* leaf water, *Aloe ferox* leaf extract, *Aloe ferox* leaf juice, and *Aloe ferox* leaf juice extract. Int. J. Toxicol. 26, 1–50.

Anjum, S., Abbasi, B.H., Shinwari, Z.K., 2016. Plant-mediated green synthesis of silver nanoparticles for biomedical applications: challenges and opportunities. Pak. J. Bot. 48, 1731–1760.

Arockia John Paul, J., Karunai Selvi, B., Karmegam, N., 2015. Biosynthesis of silver nanoparticles from *Premna serratifolia L.* leaf and its anticancer activity in CCl_4-induced hepato-cancerous Swiss albino mice. Appl. Nanosci. 5, 937–944.

Arshad, H., Sami, M.A., Sadaf, S., Hassan, U., 2021. Salvadora persica mediated synthesis of silver nanoparticles and their antimicrobial efficacy. Sci. Rep. 11, 5996.

Arunachalam, K.D., Annamalai, S.K., Arunachalam, A.M., Kennedy, S., 2013. Green synthesis of crystalline silver nanoparticles using *Indigofera aspalathoides*-medicinal plant extract for wound healing applications. Asian J. Chem. 25, 18–20.

Arunkumar, C., Nima, P., Astalakshmi, A., Ganesan, V., 2013. Green synthesis and characterization of silver nanoparticles using fructose. Int. J. Nanotechnol. Appl. 3, 1–10.

Ashokkumar, S., Ravi, S., Kathiravan, V., Velmurugan, S., 2015. Synthesis of silver nanoparticles using *A. indicum* leaf extract and their antibacterial activity. Spectrochim. Acta A Mol. Biomol. Spectrosc. 134, 34–39.

Augustine, R., Kalarikkal, N., Thomas, S., 2016. Electrospun PCL membranes incorporated with biosynthesized silver nanoparticles as antibacterial wound dressings. Appl. Nanosci. 6, 337–344.

Balan, K., Qing, W., Wang, Y., Liu, X., Palvannan, T., Wang, Y., Ma, F., Zhang, Y., 2016. Antidiabetic activity of silver nanoparticles from green synthesis using *Lonicera japonica* leaf extract. RSC Adv. 6, 40162–40168.

Banerjee, P., Satapathy, M., Mukhopahayay, A., Das, P., 2014. Leaf extract mediated green synthesis of silver nanoparticles from widely available Indian plants: synthesis, characterization, antimicrobial property and toxicity analysis. Bioresour. Bioprocess. 1, 1–10.

Bhakya, S., Muthukrishnan, S., Sukumaran, M., Muthukumar, M., 2016. Biogenic synthesis of silver nanoparticles and their antioxidant and antibacterial activity. Appl. Nanosci. 6, 755–766.

Bhau, B.S., Ghosh, S., Puri, S., Borah, B., Sarmah, D.K., Khan, R., 2015. Green synthesis of gold nanoparticles from the leaf extract of *Nepenthes khasiana* and antimicrobial assay. Adv. Mater. Lett. 6, 55–58.

Bhutto, A.A., Kalay, Ş., Sherazi, S.T.H., Culha, M., 2018. Quantitative structure–activity relationship between antioxidant capacity of phenolic compounds and the plasmonic properties of silver nanoparticles. Talanta 189, 174–181.

Bhuvaneswari, R., Xavier, R.J., Arumugam, M., 2017. Facile synthesis of multifunctional silver nanoparticles using mangrove plant *Excoecaria agallocha L.* for its antibacterial, antioxidant and cytotoxic effects. J. Parasit. Dis. 41, 180–187.

Boennemann, H., Richards, R.M., 2001. Nanoscopic metal particles—synthetic methods and potential applications. Eur. J. Inorg. Chem. 2001, 2455–2480.

Boverhof, D.R., Bramante, C.M., Butala, J.H., Clancy, S.F., Lafranconi, W.M., West, J., Gordon, S.C., 2015. Comparative assessment of nanomaterial definitions and safety evaluation considerations. Regul. Toxicol. Pharmacol. 73, 137–150.

Burda, C., Chen, X., Narayanan, R., El-Sayed, M.A., 2005. Chemistry and properties of nanocrystals of different shapes. Chem. Rev. 105, 1025–1102.

Calderón-Jiménez, B., Johnson, M.E., Montoro Bustos, A.R., Murphy, K.E., Winchester, M.R., Baudrit, J.R.V., 2017. Silver nanoparticles: technological advances, societal impacts, and metrological challenges. Front. Chem. 5, 1–26.

Chaudhari, S.P., Damahe, A., Kumbhar, P., 2016. Silver nanoparticles—a review with focus on green synthesis. Int. J. Pharma Res. Rev. 5, 14–28.

Christenhusz, M.J.M., Byng, J.W., 2016. The number of known plants species in the world and its annual increase. Phytotaxa 261, 201–217.

Chung, I.M., Park, I., Seung-Hyun, K., Thiruvengadam, M., Rajakumar, G., 2016. Plant-mediated synthesis of silver nanoparticles: their characteristic properties and therapeutic applications. Nanoscale Res. Lett. 11, 1–14.

Cittrarasu, V., Balasubramanian, B., Kaliannan, D., Park, S., Maluventhan, V., Kaul, T., Liu, W.C., Arumugam, M., 2019. Biological mediated ag nanoparticles from *Barleria longiflora* for antimicrobial activity and photocatalytic degradation using methylene blue. Artif. Cells Nanomed. Biotechnol. 47, 2424–2430.

Crisan, C.M., Mocan, T., Manolea, M., Lasca, L.L., Tăbăran, F.A., Mocan, L., 2021. Review on silver nanoparticles as a novel class of antibacterial solutions. Appl. Sci. 11, 1120.

Dahl, J.A., Maddux, B.L.S., Hutchison, J.E., 2007. Toward greener nanosynthesis. Chem. Rev. 107, 2228–2269.

Daima, H.K., Selvakannan, P., Homan, Z., Bhargava, S.K., Bansal, V., 2011. Tyrosine mediated gold, silver and their alloy nanoparticles synthesis: antibacterial activity toward gram positive and gram negative bacterial strains. In: 2011 International Conference on Nanoscience, Technology and Societal Implications. IEEE, pp. 1–6.

Das, B., Dash, S.K., Mandal, D., Ghosh, T., Chattopadhyay, S., Tripathy, S., Das, S., Dey, S. K., Das, D., Roy, S., 2017. Green synthesized silver nanoparticles destroy multidrug resistant bacteria via reactive oxygen species mediated membrane damage. Arab. J. Chem. 10, 862–876.

Dasgupta, N., Ranjan, S., Mundekkad, D., Ramalingam, C., Shanker, R., Kumar, A., 2015. Nanotechnology in agro-food: from field to plate. Food Res. Int. 69, 381–400.

De Oliveira, D.M.P., Forde, B.M., Kidd, T.J., Harris, P.N.A., Beatson, S.A., Paterson, D.L., Walker, J., 2020. Antimicrobial resistance in ESKAPE pathogens. Clin. Microbiol. Rev. 33, 1–49.

Du, J., Hu, Z., Dong, W.j., Wang, Y., Wu, S., Bai, Y., 2019. Biosynthesis of large-sized silver nanoparticles using *Angelica keiskei* extract and its antibacterial activity and mechanisms investigation. Microchem. J. 147, 333–338.

Duan, H., Wang, D., Li, Y., 2015. Green chemistry for nanoparticle synthesis. Chem. Soc. Rev. 44, 5778–5792.

Dwivedi, A.D., Gopal, K., 2010. Biosynthesis of silver and gold nanoparticles using *Chenopodium album* leaf extract. Colloids Surf. A Physicochem. Eng. Asp. 369, 27–33.

Ebrahiminezhad, A., Raee, M.J., Manafi, Z., Sotoodeh Jahromi, A., Ghasemi, Y., 2016. Ancient and novel forms of silver in medicine and biomedicine. J. Adv. Med. Sci. Appl. Technol. 2, 122–128.

Elavazhagan, T., Arunachalam, K.D., 2011. *Memecylon edule* leaf extract mediated green synthesis of silver and gold nanoparticles. Int. J. Nanomedicine 6, 1265–1278.

Elemike, E.E., Fayemi, O.E., Ekennia, A.C., Onwudiwe, D.C., Ebenso, E.E., 2017. Silver nanoparticles mediated by *Costus afer* leaf extract: synthesis, antibacterial, antioxidant and electrochemical properties. Molecules 22, 701.

Franci, G., Falanga, A., Galdiero, S., Palomba, L., Rai, M., Morelli, G., Galdiero, M., 2015. Silver nanoparticles as potential antibacterial agents. Molecules 20, 8856–8874.

Gade, A., Gaikwad, S., Tiwari, V., Yadav, A., Ingle, A., Rai, M., 2010. Biofabrication of silver nanoparticles by *Opuntia ficus-indica*: in vitro antibacterial activity and study of the mechanism involved in the synthesis. Curr. Nanosci. 6, 370–375.

Gallucci, M.N., Fraire, J.C., Ferreyra Maillard, A.P.V., Páez, P.L., Aiassa Martínez, I.M., Pannunzio Miner, E.V., Coronado, E.A., Dalmasso, P.R., 2017. Silver nanoparticles from leafy green extract of Belgian endive (*Cichorium intybus L. var. sativus*): biosynthesis, characterization, and antibacterial activity. Mater. Lett. 197, 98–101.

Garg, S., 2013. Microwave-assisted rapid green synthesis of silver nanoparticles using *Saraca indica* leaf extract and their antibacterial potential. Int. J. Pharm. Sci. Res. 4, 3615–3619.

Geetha, N., Geetha, T.S., Manonmani, P., Thiyagarajan, M., 2014. Green synthesis of silver nanoparticles using *Cymbopogan Citratus (Dc) Stapf*. extract and its antibacterial activity. Aust. J. Basic Appl. Sci. 8, 324–331.

Gomathi, M., Prakasam, A., Rajkumar, P.V., Rajeshkumar, S., Chandrasekaran, R., Anbarasan, P.M., 2020. Green synthesis of silver nanoparticles using *Gymnema sylvestre* leaf extract and evaluation of its antibacterial activity. South African J. Chem. Eng. 32, 1–4.

Gopinath, K., Kumaraguru, S., Bhakyaraj, K., Mohan, S., Venkatesh, K.S., Esakkirajan, M., Kaleeswarran, P., Alharbi, N.S., Kadaikunnan, S., Govindarajan, M., Benelli, G., Arumugam, A., 2016. Green synthesis of silver, gold and silver/gold bimetallic nanoparticles using the *Gloriosa superba* leaf extract and their antibacterial and antibiofilm activities. Microb. Pathog. 101, 1–11.

Gorbe, M., Bhat, R., Aznar, E., Sancenón, F., Marcos, M.D., Herraiz, F.J., Prohens, J., Venkataraman, A., Martínez-Máñez, R., 2016. Rapid biosynthesis of silver nanoparticles using pepino (*Solanum muricatum*) leaf extract and their cytotoxicity on HeLa cells. Materials (Basel) 9, 1–15.

Gour, A., Jain, N.K., 2019. Advances in green synthesis of nanoparticles. Artif. Cells Nanomed. Biotechnol. 47, 844–851.

Govindappa, M., Hemashekhar, B., Arthikala, M.K., Ravishankar Rai, V., Ramachandra, Y.L., 2018. Characterization, antibacterial, antioxidant, antidiabetic, anti-inflammatory and antityrosinase activity of green synthesized silver nanoparticles using *Calophyllum tomentosum* leaves extract. Results Phys. 9, 400–408.

Gude, V., Upadhyaya, K., Prasad, M.N.V., Rao, N.V.S., 2012. Green synthesis of gold and silver nanoparticles using *Achyranthes aspera L.* leaf extract. Adv. Sci. Eng. Med. 5, 223–228.

Guimarães, M.L., da Silva, F.A.G., da Costa, M.M., de Oliveira, H.P., 2020. Green synthesis of silver nanoparticles using *Ziziphus joazeiro* leaf extract for production of antibacterial agents. Appl. Nanosci. 10, 1073–1081.

Gul, A., Fozia, Shaheen, A., Ahmad, I., Khattak, B., Ahmad, M., Ullah, R., Bari, A., Ali, S.S., Alobaid, A., Asmari, M.M., Mahmood, H.M., 2021. Green synthesis, characterization, enzyme inhibition, antimicrobial potential, and cytotoxic activity of plant-mediated silver nanoparticle using *Ricinus communis* leaf and root extracts. Biomolecules 11, 1–15.

Haider, A., Kang, I.K., 2015. Preparation of silver nanoparticles and their industrial and biomedical applications: a comprehensive review. Adv. Mater. Sci. Eng. 2015, 165257.

Hamelian, M., Zangeneh, M.M., Amisama, A., Varmira, K., Veisi, H., 2018. Green synthesis of silver nanoparticles using *Thymus kotschyanus* extract and evaluation of their antioxidant, antibacterial and cytotoxic effects. Appl. Organomet. Chem. 32, 1–8.

Hekmati, M., Hasanirad, S., Khaledi, A., Esmaeili, D., 2020. Green synthesis of silver nanoparticles using extracts of *Allium rotundum l, Falcaria vulgaris Bernh*, and *Ferulago angulate Boiss*, and their antimicrobial effects in vitro. Gene Rep. 19, 100589.

Hembram, K.C., Kumar, R., Kandha, L., Parhi, P.K., Kundu, C.N., Bindhani, B.K., 2018. Therapeutic prospective of plant-induced silver nanoparticles: application as antimicrobial and anticancer agent. Artif. Cells Nanomed. Biotechnol. 46, S38–S51.

Hu, Y., Xu, J., Hu, Q., 2003. Evaluation of antioxidant potential of *Aloe vera* (*Aloe barbadensis Miller*) extracts. J. Agric. Food Chem. 51, 7788–7791.

Huo, C., Khoshnamvand, M., Liu, P., Yuan, C.G., Cao, W., 2019. Eco-friendly approach for biosynthesis of silver nanoparticles using *Citrus maxima* peel extract and their characterization, catalytic, antioxidant and antimicrobial characteristics. Mater. Res. Express 6, 015010.

Hussain, I., Singh, N.B., Singh, A., Singh, H., Singh, S.C., 2016. Green synthesis of nanoparticles and its potential application. Biotechnol. Lett. 38, 545–560.

Iqbal, P., Preece, J.A., Mendes, P.M., 2012. Nanotechnology: the "Top-Down" and "Bottom-Up" approaches. In: Supramolecular Chemistry. John Wiley & Sons, Ltd, Chichester, UK.

Iravani, S., 2011. Green synthesis of metal nanoparticles using plants. Green Chem. 13, 2638–2650.

Islam, M.A., Jacob, M.V., Antunes, E., 2021. A critical review on silver nanoparticles: from synthesis and applications to its mitigation through low-cost adsorption by biochar. J. Environ. Manag. 281, 111918.

Jain, P.K., Huang, X., El-Sayed, I.H., El-Sayed, M.A., 2008. Noble metals on the nanoscale: optical and photothermal properties and some applications in imaging, sensing, biology, and medicine. Acc. Chem. Res. 41, 1578–1586.

Jain, S., Mehata, M.S., 2017. Medicinal plant leaf extract and pure flavonoid mediated green synthesis of silver nanoparticles and their enhanced antibacterial property. Sci. Rep. 7, 1–13.

Jeevanandam, J., Barhoum, A., Chan, Y.S., Dufresne, A., Danquah, M.K., 2018. Review on nanoparticles and nanostructured materials: history, sources, toxicity and regulations. J. Nanotechnol. 9, 1050–1074.

Jha, A.K., Prasad, K., Prasad, K., Kulkarni, A.R., 2009. Plant system: nature's nanofactory. Colloids Surf. B: Biointerfaces 73, 219–223.

Jyoti, K., Singh, A., 2016. Green synthesis of nanostructured silver particles and their catalytic application in dye degradation. J. Genet. Eng. Biotechnol. 14, 311–317.

Kaabipour, S., Hemmati, S., 2021. A review on the green and sustainable synthesis of silver nanoparticles and one-dimensional silver nanostructures. J. Nanotechnol. 12, 102–136.

Kannan, D.S., Mahboob, S., Al-Ghanim, K.A., Venkatachalam, P., 2021. Antibacterial, antibiofilm and photocatalytic activities of biogenic silver nanoparticles from *Ludwigia octovalvis*. J. Clust. Sci. 32, 255–264.

Kargozar, S., Mozafari, M., 2018. Nanotechnology and nanomedicine: start small, think big. Mater. Today Proc. 5, 15492–15500.

Kathiravan, V., Ravi, S., Ashokkumar, S., 2014. Synthesis of silver nanoparticles from *Melia dubia* leaf extract and their in vitro anticancer activity. Spectrochim. Acta A Mol. Biomol. Spectrosc. 130, 116–121.

Kathireswari, P., Gomathi, S., Saminathan, K., 2014. Plant leaf mediated synthesis of silver nanoparticles using *Phyllanthus niruri* and its antimicrobial activity against multi drug resistant human pathogens. Int. J. Curr. Microbiol. App. Sci. 3, 960–968.

Kaul, S., Gulati, N., Verma, D., Mukherjee, S., Nagaich, U., 2018. Role of nanotechnology in cosmeceuticals: a review of recent advances. J. Pharm. 2018, 1–19.

Kaweeteerawat, C., Na Ubol, P., Sangmuang, S., Aueviriyavit, S., Maniratanachote, R., 2017. Mechanisms of antibiotic resistance in bacteria mediated by silver nanoparticles. J. Toxicol. Environ. Health A 80, 1276–1289.

Khan, M.R., Adam, V., Rizvi, T.F., Zhang, B., Ahamad, F., Jośko, I., Zhu, Y., Yang, M., Mao, C., 2019. Nanoparticle–plant interactions: two-way traffic. Small 15, 1–20.

Kim, D.Y., Shinde, S., Ghodake, G., 2017. Colorimetric detection of magnesium (II) ions using tryptophan functionalized gold nanoparticles. Sci. Rep. 7, 1–9.

Kiruba Daniel, S.C.G., Mahalakshmi, N., Sandhiya, J., Nehru, K., Sivakumar, M., 2013. Rapid synthesis of Ag nanoparticles using henna extract for the fabrication of photoabsorption enhanced dye sensitized solar cell (PE-DSSC). Adv. Mater. Res. 678, 349–360.

Kirubaharan, C.J., Kalpana, D., Lee, Y.S., Kim, A.R., Yoo, D.J., Nahm, K.S., Kumar, G.G., 2012. Biomediated silver nanoparticles for the highly selective copper(ii) ion sensor applications. Ind. Eng. Chem. Res. 51, 7441–7446.

Korbekandi, H., Asghari, G., Sadat Jalayer, S., Sadat Jalayer, M., Bandegani, M., 2013. Nano-silver particle production using *Juglans Regia L.(Walnut)* leaf extract. Jundishapur J. Nat. Pharm. Prod. 8, 20–26.

Krishnaraj, C., Jagan, E.G., Rajasekar, S., Selvakumar, P., Kalaichelvan, P.T., Mohan, N., 2010. Synthesis of silver nanoparticles using *Acalypha indica* leaf extracts and its antibacterial activity against water borne pathogens. Colloids Surf. B: Biointerfaces 76, 50–56.

Krithiga, N., Rajalakshmi, A., Jayachitra, A., 2015. Green synthesis of silver nanoparticles using leaf extracts of *Clitoria ternatea* and *Solanum nigrum* and study of its antibacterial effect against common nosocomial pathogens. J. Nanosci. 2015, 1–8.

Krug, H.F., Wick, P., 2011. Nanotoxicology: an interdisciplinary challenge. Angew. Chem. Int. Ed. 50, 1260–1278.

Küp, F.Ö., Çoşkunçay, S., Duman, F., 2020. Biosynthesis of silver nanoparticles using leaf extract of *Aesculus hippocastanum* (horse chestnut): evaluation of their antibacterial, antioxidant and drug release system activities. Mater. Sci. Eng. C 107, 110207.

Li, C., Chen, D., Xiao, H., 2021. Green synthesis of silver nanoparticles using Pyrus betulifolia Bunge and their antibacterial and antioxidant activity. Mater. Today Commun. 26, 102108.

Madivoli, E.S., Kareru, P.G., Gachanja, A.N., Mugo, S.M., Makhanu, D.S., Wanakai, S.I., Gavamukulya, Y., 2020. Facile synthesis of silver nanoparticles using *Lantana trifolia* aqueous extracts and their antibacterial activity. J. Inorg. Organomet. Polym. Mater. 30, 2842–2850.

Mahmoudi, R., Aghaei, S., Salehpour, Z., Mousavizadeh, A., Khoramrooz, S.S., Taheripour Sisakht, M., Christiansen, G., Baneshi, M., Karimi, B., Bardania, H., 2020. Antibacterial and antioxidant properties of phyto-synthesized silver nanoparticles using *Lavandula stoechas* extract. Appl. Organomet. Chem. 34, 1–9.

Makarov, V.V., Love, A.J., Sinitsyna, O.V., Makarova, S.S., Yaminsky, I.V., Taliansky, M.E., Kalinina, N.O., 2014. "Green" nanotechnologies: synthesis of metal nanoparticles using plants. Acta Nat. 6, 35–44.

Manikandan, D.B., Sridhar, A., Krishnasamy Sekar, R., Perumalsamy, B., Veeran, S., Arumugam, M., Karuppaiah, P., Ramasamy, T., 2021. Green fabrication, characterization of silver nanoparticles using aqueous leaf extract of *Ocimum americanum* (Hoary Basil) and investigation of its in vitro antibacterial, antioxidant, anticancer and photocatalytic reduction. J. Environ. Chem. Eng. 9, 104845.

Manosalva, N., Tortella, G., Cristina Diez, M., Schalchli, H., Seabra, A.B., Durán, N., Rubilar, O., 2019. Green synthesis of silver nanoparticles: effect of synthesis reaction parameters on antimicrobial activity. World J. Microbiol. Biotechnol. 35, 1–9.

Mashwani, Z., Khan, M.A., Khan, T., Nadhman, A., 2016. Applications of plant terpenoids in the synthesis of colloidal silver nanoparticles. Adv. Colloid Interf. Sci. 234, 132–141.

Medda, S., Hajra, A., Dey, U., Bose, P., Mondal, N.K., 2015. Biosynthesis of silver nanoparticles from *Aloe vera* leaf extract and antifungal activity against *Rhizopus sp.* and *Aspergillus sp.* Appl. Nanosci. 5, 875–880.

Meva, F.E., Segnou, M.L., Ebongue, C.O., Ntoumba, A.A., Steve, D.Y., Malolo, F.A.E., Ngah, L., Massai, H., Mpondo, E.M., 2016. Unexplored vegetal green synthesis of silver nanoparticles: a preliminary study with *Corchorus olitorus Linn* and *Ipomea batatas (L.) Lam.* Afr. J. Biotechnol. 15, 341–349.

Mishra, S., Keswani, C., Abhilash, P.C., Fraceto, L.F., Singh, H.B., 2017. Integrated approach of agri-nanotechnology: challenges and future trends. Front. Plant Sci. 8, 1–12.

Mock, J.J., Barbic, M., Smith, D.R., Schultz, D.A., Schultz, S., 2002. Shape effects in plasmon resonance of individual colloidal silver nanoparticles. J. Chem. Phys. 116, 6755–6759.

Mohammadlou, M., Maghsoudi, H., Jafarizadeh-Malmiri, H., 2016. A review on green silver nanoparticles based on plants: synthesis, potential applications and eco-friendly approach. Int. Food Res. J. 23, 446–463.

Mondal, S., Roy, N., Laskar, R.A., Sk, I., Basu, S., Mandal, D., Begum, N.A., 2011. Biogenic synthesis of Ag, Au and bimetallic Au/Ag alloy nanoparticles using aqueous extract of mahogany (*Swietenia mahogani JACQ*.) leaves. Colloids Surf. B: Biointerfaces 82, 497–504.

Moradi, F., Sedaghat, S., Moradi, O., Salmanabadi, S.A., 2021. Review on green nanobiosynthesis of silver nanoparticles and their biological activities: with an emphasis on medicinal plants. Inorg. Nano-Met. Chem. 51, 133–142.

Mousavi, S.M., Hashemi, S.A., Ghasemi, Y., Atapour, A., Amani, A.M., Savar Dashtaki, A., Babapoor, A., Arjmand, O., 2018. Green synthesis of silver nanoparticles toward bio and medical applications: review study. Artif. Cells Nanomed. Biotechnol. 46, S855–S872.

Nahar, K., Aziz, S., Bashar, M.S., Hague, M.A., Al-Reza, S.M., 2020. Synthesis and characterization of silver nanoparticles from *Cinnamomum tamala* leaf extract and its antibacterial potential. Int. J. Nano Dimens. 11, 88–98.

Narayanan, K.B., Park, H.H., 2014. Antifungal activity of silver nanoparticles synthesized using turnip leaf extract (*Brassica rapa L.*) against wood rotting pathogens. Eur. J. Plant Pathol. 140, 185–192.

Nesrin, K., Yusuf, C., Ahmet, K., Ali, S.B., Muhammad, N.A., Suna, S., Fatih, Ş., 2020. Biogenic silver nanoparticles synthesized from *Rhododendron ponticum* and their antibacterial, antibiofilm and cytotoxic activities. J. Pharm. Biomed. Anal. 179, 112993.

Ni, B., Wang, X., 2016. Chemistry and properties at a sub-nanometer scale. Chem. Sci. 7, 3978–3991.

Nimse, S.B., Pal, D., 2015. Free radicals, natural antioxidants, and their reaction mechanisms. RSC Adv. 5, 27986–28006.

Noginov, M.A., Zhu, G., Bahoura, M., Adegoke, J., Small, C., Ritzo, B.A., Drachev, V.P., Shalaev, V.M., 2007. The effect of gain and absorption on surface plasmons in metal nanoparticles. Appl. Phys. B Lasers Opt. 86, 455–460.

Ojha, A.K., Rout, J., Behera, S., Nayak, P.L., 2013. Green synthesis and characterization of zero valent iron nanoparticles from the leaf extract of *Syzygium aromaticum* (clove). Int. J. Pharm. Res. Allied Sci. 2, 31–35.

Okafor, F., Janen, A., Kukhtareva, T., Edwards, V., Curley, M., 2013. Green synthesis of silver nanoparticles, their characterization, application and antibacterial activity. Int. J. Environ. Res. Public Health 10, 5221–5238.

Oliveira, G.Z.S., Lopes, C.A.P., Sousa, M.H., Silva, L.P., 2019. Synthesis of silver nanoparticles using aqueous extracts of *Pterodon emarginatus* leaves collected in the summer and winter seasons. Int. Nano Lett. 9, 109–117.

Ovais, M., Khalil, A.T., Raza, A., Khan, M.A., Ahmad, I., Islam, N.U., Saravanan, M., Ubaid, M.F., Ali, M., Shinwari, Z.K., 2016. Green synthesis of silver nanoparticles via plant extracts: beginning a new era in cancer theranostics. Nanomedicine 12, 3157–3177.

Pacioni, N.L., Borsarelli, C.D., Rey, V., Veglia, A.V., 2015. Synthetic routes for the preparation of silver nanoparticles. In: Alarcon, E.I., Griffith, M., Udekwu, K.I. (Eds.), Silver Nanoparticle Applications in the Fabrication and Design of Medical and Biosensing Devices. Springer International Publishing, Cham, Switzerland, pp. 13–48.

Panáček, A., Kvítek, L., Smékalová, M., Večeřová, R., Kolář, M., Röderová, M., Dyčka, F., Šebela, M., Prucek, R., Tomanec, O., Zbořil, R., 2018. Bacterial resistance to silver nanoparticles and how to overcome it. Nat. Nanotechnol. 13, 65–71.

Parashar, U.K., Saxena, P.S., Srivastava, A., 2009. Bioinspired synthesis of silver nanoparticles. Dig. J. Nanomater. Biostruct. 4, 159–166.

Parthiban, E., Manivannan, N., Ramanibai, R., Mathivanan, N., 2019. Green synthesis of silver-nanoparticles from *Annona reticulata* leaves aqueous extract and its mosquito larvicidal and anti-microbial activity on human pathogens. Biotechnol. Rep. 21, e00297.

Patil, R.S., Kokate, M.R., Kolekar, S.S., 2012. Bioinspired synthesis of highly stabilized silver nanoparticles using *Ocimum tenuiflorum* leaf extract and their antibacterial activity. Spectrochim. Acta A Mol. Biomol. Spectrosc. 91, 234–238.

Patil, V.A., Nitave, S.A., 2014. A review on *Eucalyptus globulus*: a divine medicinal herb. World J. Pharm. Pharm. Sci. 3, 559–567.

Patra, S., Mukherjee, S., Barui, A.K., Ganguly, A., Sreedhar, B., Patra, C.R., 2015. Green synthesis, characterization of gold and silver nanoparticles and their potential application for cancer therapeutics. Mater. Sci. Eng. C 53, 298–309.

Pattanayak, M., Mohapatra, D., Nayak, P.L., 2013. Green synthesis and characterization of zero valent silver nanoparticles from the leaf extract of *Datura Metel*. Middle-East J. Sci. Res. 18, 623–626.

Paulkumar, K., Gnanajobitha, G., Vanaja, M., Rajeshkumar, S., Malarkodi, C., Pandian, K., Annadurai, G., 2014. *Piper nigrum* leaf and stem assisted green synthesis of silver nanoparticles and evaluation of its antibacterial activity against agricultural plant pathogens. Sci. World J. 2014, 829894.

Pedone, D., Moglianetti, M., De Luca, E., Bardi, G., Pompa, P.P., 2017. Platinum nanoparticles in nanobiomedicine. Chem. Soc. Rev. 46, 4951–4975.

Peterson, E., Kaur, P., 2018. Antibiotic resistance mechanisms in bacteria: relationships between resistance determinants of antibiotic producers, environmental bacteria, and clinical pathogens. Front. Microbiol. 9, 1–21.

Rafique, M.S., Sadaf, I., Rafique, M.S., Tahir, M.B., 2017. A review on green synthesis of silver nanoparticles and their applications. Artif. Cells Nanomed. Biotechnol. 45, 1272–1291.

Raghunandan, D., Mahesh, B.D., Basavaraja, S., Balaji, S.D., Manjunath, S.Y., Venkataraman, A., 2011. Microwave-assisted rapid extracellular synthesis of stable biofunctionalized silver nanoparticles from guava (*Psidium guajava*) leaf extract. J. Nanopart. Res. 13, 2021–2028.

Rahimi-Nasrabadi, M., Pourmortazavi, S.M., Shandiz, S.A.S., Ahmadi, F., Batooli, H., 2014. Green synthesis of silver nanoparticles using *Eucalyptus leucoxylon* leaves extract and evaluating the antioxidant activities of extract. Nat. Prod. Res. 22, 1964–1969.

Raja, S., Ramesh, V., Thivaharan, V., 2017. Green biosynthesis of silver nanoparticles using *Calliandra haematocephala* leaf extract, their antibacterial activity and hydrogen peroxide sensing capability. Arab. J. Chem. 10, 253–261.

Raja, K., Saravanakumar, A., Vijayakumar, R., 2012. Efficient synthesis of silver nanoparticles from *Prosopis juliflora* leaf extract and its antimicrobial activity using sewage. Spectrochim. Acta A Mol. Biomol. Spectrosc. 97, 490–494.

Rajakumar, G., Abdul Rahuman, A., 2011. Larvicidal activity of synthesized silver nanoparticles using *Eclipta prostrata* leaf extract against filariasis and malaria vectors. Acta Trop. 118, 196–203.

Rajaram, K., Aiswarya, D.C., Sureshkumar, P., 2015. Green synthesis of silver nanoparticle using *Tephrosia tinctoria* and its antidiabetic activity. Mater. Lett. 138, 251–254.

Rajeshkumar, S., Bharath, L.V., 2017. Mechanism of plant-mediated synthesis of silver nanoparticles—a review on biomolecules involved, characterisation and antibacterial activity. Chem. Biol. Interact. 273, 219–227.

Rao Kudle, K., Alwala, J., Donda, M.R., Miryala, A., Sreedhar, B., Pratap Rudra, M., 2013. Synthesis of silver nanoparticles using extracts of *Securinega leucopyrus* and evaluation of its antibacterial activity. Int. J. Curr. Sci. 7, E1–E8.

Restrepo, C.V., Villa, C.C., 2021. Synthesis of silver nanoparticles, influence of capping agents, and dependence on size and shape: a review. Environ. Nanotechnol. Monit. Manage. 15, 100428.

Rey, V., Gramajo Feijoo, M.E., Giménez, R.E., Tuttolomondo, M.E., Morán Vieyra, F.E., Sosa Morales, M.C., Borsarelli, C.D., 2018. Kinetics and growth mechanism of the photoinduced synthesis of silver nanoparticles stabilized with lysozyme. Colloids Surf. B: Biointerfaces 172, 10–16.

Robinson, M.M., Zhang, X., 2011. The World Medicines Situation 2011 Traditional Medicines: Global Situation, Issues, and Challenges, third ed. World Health Organization, pp. 1–14.

Roy, S., Das, T., 2015. Plant mediated green synthesis of silver nanoparticles-A review. Int. J. Plant Biol. Res. 3, 1044–1055.

Sabela, M., Balme, S., Bechelany, M., Janot, J.M., Bisetty, K., 2017. A review of gold and silver nanoparticle-based colorimetric sensing assays. Adv. Eng. Mater. 19, 1–24.

Sadeghi, B., Gholamhoseinpoor, F., 2015. A study on the stability and green synthesis of silver nanoparticles using *Ziziphora tenuior* (Zt) extract at room temperature. Spectrochim. Acta A Mol. Biomol. Spectrosc. 134, 310–315.

Saif, S., Tahir, A., Chen, Y., 2016. Green synthesis of iron nanoparticles and their environmental applications and implications. Nanomaterials 6, 1–26.

Santhoshkumar, T., Rahuman, A.A., Rajakumar, G., Marimuthu, S., Bagavan, A., Jayaseelan, C., Zahir, A.A., Elango, G., Kamaraj, C., 2011. Synthesis of silver nanoparticles using *Nelumbo nucifera* leaf extract and its larvicidal activity against malaria and filariasis vectors. Parasitol. Res. 108, 693–702.

Saratale, R.G., Shin, H.S., Kumar, G., Benelli, G., Kim, D.S., Saratale, G.D., 2018. Exploiting antidiabetic activity of silver nanoparticles synthesized using *Punica granatum* leaves and anticancer potential against human liver cancer cells (HepG2). Artif. Cells Nanomed. Biotechnol. 46, 211–222.

Saravanakumar, A., Peng, M.M., Ganesh, M., Jayaprakash, J., Mohankumar, M., Jang, H.T., 2017. Low-cost and eco-friendly green synthesis of silver nanoparticles using *Prunus japonica (Rosaceae)* leaf extract and their antibacterial, antioxidant properties. Artif. Cells Nanomed. Biotechnol. 45, 1165–1171.

Sattari, R., Khayati, G.R., Hoshyar, R., 2020. Biosynthesis and characterization of silver nanoparticles capped by biomolecules by *Fumaria parviflora* extract as green approach and evaluation of their cytotoxicity against human breast cancer MDA-MB-468 cell lines. Mater. Chem. Phys. 241, 122438.

Schaming, D., Remita, H., 2015. Nanotechnology: from the ancient time to nowadays. Found. Chem. 17, 187–205.

Shahryari, F., Rabiei, Z., Sadighian, S., 2020. Antibacterial activity of synthesized silver nanoparticles by sumac aqueous extract and silver-chitosan nanocomposite against *Pseudomonas syringae pv. syringae*. J. Plant Pathol. 102, 469–475.

Shnoudeh, A.J., Hamad, I., Abdo, R.W., Qadumii, L., Jaber, A.Y., Surchi, H.S., Alkelany, S.Z., 2019. Synthesis, characterization, and applications of metal nanoparticles. In: Biomaterials and Bionanotechnology. Elsevier, pp. 527–612.

Siddiqi, K.S., Husen, A., Rao, R.A.K., 2018. A review on biosynthesis of silver nanoparticles and their biocidal properties. J. Nanobiotechnol. 16, 14.

Siddiquee, M.A., Parray, M.u.d., Mehdi, S.H., Alzahrani, K.A., Alshehri, A.A., Malik, M.A., Patel, R., 2020. Green synthesis of silver nanoparticles from *Delonix regia* leaf extracts:

in-vitro cytotoxicity and interaction studies with bovine serum albumin. Mater. Chem. Phys. 242, 122493.

Silva, L.P., Reis, I.G., Bonatto, C.C., 2015. Green processes for nanotechnology. In: Basiuk, V.A., Basiuk, E.V. (Eds.), Green Processes for Nanotechnology: From Inorganic to Bioinspired Nanomaterials. Springer International Publishing, Cham, pp. 1–446.

Singh, C., Baboota, R.K., Naik, P.K., Singh, H., 2012. Biocompatible synthesis of silver and gold nanoparticles using leaf extract of *Dalbergia sissoo*. Adv. Mater. Lett. 3, 279–285.

Singh, R., Hano, C., Nath, G., Sharma, B., 2021. Green biosynthesis of silver nanoparticles using leaf extract of carissa carandas l. And their antioxidant and antimicrobial activity against human pathogenic bacteria. Biomolecules 11, 1–11.

Singh, P., Kim, Y.J., Zhang, D., Yang, D.C., 2016. Biological synthesis of nanoparticles from plants and microorganisms. Trends Biotechnol. 34, 588–599.

Sintubin, L., Verstraete, W., Boon, N., 2012. Biologically produced nanosilver: current state and future perspectives. Biotechnol. Bioeng. 109, 2422–2436.

Soleimani, F.F., Saleh, T., Shojaosadati, S.A., Poursalehi, R., 2018. Green synthesis of different shapes of silver nanostructures and evaluation of their antibacterial and cytotoxic activity. Bionanoscience 8, 72–80.

Soriano, J., Mora-Espí, I., Alea-Reyes, M.E., Pérez-García, L., Barrios, L., Ibáñez, E., Nogués, C., 2017. Cell death mechanisms in Tumoral and non-Tumoral human cell lines triggered by photodynamic treatments: apoptosis, necrosis and parthanatos. Sci. Rep. 7, 1–13.

Subbaiya, R., Shiyamala, M., Revathi, K., Pushpalatha, R., Selvam, M.M., 2014. Biological synthesis of silver nanoparticles from *Nerium oleander* and its antibacterial and antioxidant property. Int. J. Curr. Microbiol. App. Sci. 3, 83–87.

Subramanian, P., Ravichandran, A., Manoharan, V., Muthukaruppan, R., Somasundaram, S.S.N.S., Pandi, B., Krishnan, A., Marimuthu, P.N., Somasundaram, S.S.N.S., You, S.G., 2019. Synthesis of *Oldenlandia umbellata* stabilized silver nanoparticles and their antioxidant effect, antibacterial activity, and bio-compatibility using human lung fibroblast cell line WI-38. Process Biochem. 86, 196–204.

Sun, Q., Cai, X., Li, J., Zheng, M., Chen, Z., Yu, C.P., 2014. Green synthesis of silver nanoparticles using tea leaf extract and evaluation of their stability and antibacterial activity. Colloids Surf. A. Physiochem. Eng. Asp. 444, 226–231.

Suresh, U., Murugan, K., Benelli, G., Nicoletti, M., Barnard, D.R., Panneerselvam, C., Kumar, P.M., Subramaniam, J., Dinesh, D., Chandramohan, B., 2015. Tackling the growing threat of dengue: *Phyllanthus niruri*-mediated synthesis of silver nanoparticles and their mosquitocidal properties against the dengue vector *Aedes aegypti* (Diptera: Culicidae). Parasitol. Res. 114, 1551–1562.

Tahir, K., Nazir, S., Li, B., Khan, A.U., Khan, Z.U.H., Ahmad, A., Khan, F.U., 2015. An efficient photo catalytic activity of green synthesized silver nanoparticles using *Salvadora persica* stem extract. Sep. Purif. Technol. 150, 316–324.

Tan, Y.N., Lee, J.Y., Wang, D.I.C., 2010. Uncovering the design rules for peptide synthesis of metal nanoparticles. J. Am. Chem. Soc. 132, 5677–5686.

Tang, S., Zheng, J., 2018. Antibacterial activity of silver nanoparticles: structural effects. Adv. Healthc. Mater. 7, 1701503.

Thanh, N.T.K., Maclean, N., Mahiddine, S., 2014. Mechanisms of nucleation and growth of nanoparticles in solution. Chem. Rev. 114, 7610–7630.

Tortella, G., Navas, M., Parada, M., Durán, N., Seabra, A.B., Hoffmann, N., Rubilar, O., 2019. Synthesis of silver nanoparticles using extract of weeds and optimized by response surface methodology to the control of soil pathogenic bacteria *Ralstonia solanacearum*. J. Soil Sci. Plant Nutr. 19, 148–156.

Travieso Novelles, M.d.C., Ortega, A.R., Pita, B.A., López, M.C., Pérez, L.D., Medina, E.A., Pérez, O.P., 2021. Biosynthesis of fluorescent silver nanoparticles from *Leea coccinea* leaves and their antibacterial potentialities against *Xanthomonas phaseoli* pv *phaseoli*. Bioresour. Bioprocess. 8, 3.

Trotta, F., Mele, A., 2019. Nanomaterials: classification and properties. In: Nanosponges. Wiley-VCH Verlag GmbH & Co. KGaA, Weinheim, Germany, pp. 1–26.

Vanlalveni, C., Lallianrawna, S., Biswas, A., Selvaraj, M., Changmai, B., Rokhum, S.L., 2021. Green synthesis of silver nanoparticles using plant extracts and their antimicrobial activities: a review of recent literature. RSC Adv. 11, 2804–2837.

Veisi, H., Azizi, S., Mohammadi, P., 2018. Green synthesis of the silver nanoparticles mediated by *Thymbra spicata* extract and its application as a heterogeneous and recyclable nanocatalyst for catalytic reduction of a variety of dyes in water. J. Clean. Prod. 170, 1536–1543.

Vijilvani, C., Bindhu, M.R., Frincy, F.C., AlSalhi, M.S., Sabitha, S., Saravanakumar, K., Devanesan, S., Umadevi, M., Aljaafreh, M.J., Atif, M., 2020. Antimicrobial and catalytic activities of biosynthesized gold, silver and palladium nanoparticles from *Solanum nigurum* leaves. J. Photochem. Photobiol. B Biol. 202, 111713.

Vinod Kumar, V., Anbarasan, S., Christena, L.R., SaiSubramanian, N., Philip Anthony, S., 2014. Bio-functionalized silver nanoparticles for selective colorimetric sensing of toxic metal ions and antimicrobial studies. Spectrochim. Acta A Mol. Biomol. Spectrosc. 129, 35–42.

Wang, L., Hu, C., Shao, L., 2017. The antimicrobial activity of nanoparticles: present and prospects for the future. Int. J. Nanomedicine 12, 1227–1249.

Warner, J.C., Cannon, A.S., Dye, K.M., 2004. Green chemistry. Environ. Impact Assess. Rev. 24, 775–799.

Wei, L., Lu, J., Xu, H., Patel, A., Chen, Z.S., Chen, G., 2015. Silver nanoparticles: synthesis, properties, and therapeutic applications. Drug Discov. Today 20, 595–601.

Yan yu, R., Hui, Y., Tao, W., Chuang, W., 2019. Bio-synthesis of silver nanoparticles with antibacterial activity. Mater. Chem. Phys. 235, 121746.

Zareh, M.M., Nafady, N.A., Faried, A.M., Mohamed, M.H., 2018. Green synthesis of silver nanoparticles from capitula extract of some *Launaea (Asteraceae)* with notes on their taxonomic significance. Egypt. J. Bot. 58, 185–194.

Zargar, M., Hamid, A.A., Bakar, F.A., Shamsudin, M.N., Shameli, K., Jahanshiri, F., Farahani, F., 2011. Green synthesis and antibacterial effect of silver nanoparticles using *Vitex negundo L*. Molecules 16, 6667–6676.

Zheng, B., Kong, T., Jing, X., Odoom-Wubah, T., Li, X., Sun, D., Lu, F., Zheng, Y., Huang, J., Li, Q., 2013. Plant-mediated synthesis of platinum nanoparticles and its bioreductive mechanism. J. Colloid Interface Sci. 396, 138–145.

Zhou, Y., Lin, W., Huang, J., Wang, W., Gao, Y., Lin, L., Li, Q., Lin, L., Du, M., 2010. Biosynthesis of gold nanoparticles by foliar broths: roles of biocompounds and other attributes of the extracts. Nanoscale Res. Lett. 5, 1351–1359.

Zuorro, A., Iannone, A., Natali, S., Lavecchia, R., 2019. Green synthesis of silver nanoparticles using bilberry and red currant waste extracts. Processes 7, 193.

CHAPTER 13

Synthesis of biogenic silver nanoparticles using medicinal plant extract: A new age in nanomedicine to combat multidrug-resistant pathogens

Sudip Some[a], Rittick Mondal[b], Paulami Dam[b], and Amit Kumar Mandal[b,c]

[a]*Department of Life Sciences, Chanchal Siddheswari Institution (H.S.), Chanchal, West Bengal, India* [b]*Chemical Biology Laboratory, Department of Sericulture, Raiganj University, Raiganj, West Bengal, India* [c]*Centre for Nanotechnology Sciences, Raiganj University, Raiganj, West Bengal, India*

1 Introduction

The term "nano" is referred to as an SI prefix and comes from the Greek word "*nanos*," which means "dwarf" or very small. It represents one thousand millionth part of a meter (10^{-9} m). Nanoscience is the study of the chemical and physical properties of nanomaterials. The science and engineering of nanoscale objects are defined in nanotechnology. It is an interdisciplinary as well as a multidisciplinary area of applied science that modulates metals into nanoparticles (NPs) for a wide range of applications. Nowadays nanotechnology has a great impact on the biological, agricultural, medical, food processing, and environmental fields (Varadan et al., 2010; Zhang, 2017; Bayda et al., 2019). Nanobiotechnology has emerged as a new field of research and development; it is an amalgamation between nanotechnology and biotechnology for improving the efficacy of nanomaterials in the highly developed field of biotechnology (Beg et al., 2017) (Fig. 1). The efficacy of NPs is associated with the size, concentration, pH of the medium, and time of exposure against microorganisms (Teow et al., 2018; Siddiqi et al., 2018).

Silver is a lenient, pure, shiny, and versatile metal with the highest electrical and thermal conductivity due to its special photothermal, chemical, and optical properties (Cohen et al., 1973; Paffett et al., 1985; Austin et al., 2014). This metal or its

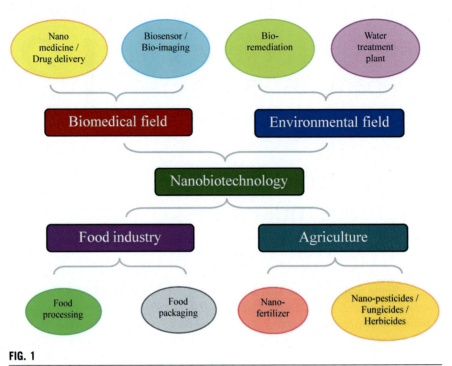

FIG. 1

Biotechnological applications of nanoparticles.

compound (silver nitrate) has been used for centuries to cure various illnesses or to prevent microbial infections (Politano et al., 2013). It was first introduced as the most significant antibacterial agent before the discovery of antibiotics (Alexander, 2009). Now it is commonly used in wound dressings as a wide-spectrum antimicrobial agent. *Rajat Bhasma* or *Chandi Bhasma* (silver powder) has been used as a targeted medication in the Indian Ayurveda (Lansdown, 2010; Singh, 2017). This metal has no harmful effect on higher organisms, but it is predominantly toxic to lower organisms (Medici et al., 2019). Now researchers have switched their intense attention from metals to NPs. Silver nanoparticles (AgNPs) are particles of metallic silver. They are known to be noble metallic nanoparticles (MNPs) having a range of 1–100 nm in size. In the past decade, AgNPs have been of great interest to scientists because of their unforeseen protection against a wide range of microorganisms (Prabhu and Poulose, 2012; Srikar et al., 2016).

Nanocrystalline silver has been used as a promising material in biotechnology for its excellent characteristics such as the large surface area to volume ratio, diverse morphology, significant biological properties which include catalytic activity, excellent permeability into the cells, and its remarkable biocidal activity (Alaqad and Saleh, 2016; Burduşel et al., 2018). In the agricultural sector, AgNPs are widely used for the management of plant diseases, enhancing nutrient uptake efficiency, improving plant growth, and sustaining the release of agrochemicals. The eco-friendly and cost-effective use of AgNPs for monitoring plant diseases is highly appreciated (Narware et al., 2019). Silver nanoparticles considerably reduce molds, coliform, and peroxide index or value in packaged foodstuffs. Therefore, nanosilver increases the shelf life of foods by preventing internal and external exploitation and microbial contaminations in the food packaging industry (Tavakoli et al., 2017).

AgNPs have also been used in drug delivery systems due to their excellent biocompatibility, durability, good absorption properties, low toxicity, and optical properties (Jahangirian et al., 2017; Lam et al., 2017; Ferdous and Nemmar, 2020). Plants and microbes such as bacteria, yeast, and fungi are widely used in AgNP biosynthesis. Various chemical functional groups (hydroxyl, carboxyl, amino, etc.) of bioactive compounds act as effective stabilizing agents as well as reducing agents in the biosynthesis process. (Gudikandula and CharyaMaringanti, 2016; Ndikau et al., 2017; Hamouda et al., 2019). Medicinal plant extract-mediated synthesized AgNPs are predominant, due to the presence of various phytochemicals in medicinal plants including terpenoids, polyphenols, alkaloids, phenolic acids, and vitamins (Makarov et al., 2014; Ovais et al., 2016; Rafique et al., 2016; Singh, 2019).

The size, shape, and distribution of AgNPs can be regulated by optimization of reaction conditions which include temperature, pH, and amount of plant extract, and it would be conceivable to improve manufacturing methods to produce significant amounts of stable and small-sized NPs (Sorbiun et al., 2018). Due to nontoxic issues, biogenic and biocompatible AgNPs have great advantages for using as an excellent therapeutic arsenal in different biomedical fields which include bio-imaging, disease diagnosis, and bio-sensing. Therefore, the strategy of effective and eco-accommodating methods for the synthesis of AgNPs is a vital step in the field of nano-biotechnology (Mousavi et al., 2018b; Lee and Jun, 2019). Biosynthesized AgNPs show better antibacterial activity against both gram-positive and gram-negative bacteria. The antibacterial activity of AgNP has been also recognized against multidrug-resistant (MDR) bacteria (Wang et al., 2016). Apart from that, it has been extensively used as antifungal, antiviral, antiinflammatory, antiangiogenic, and antitumorigenic agents (Lara et al., 2009; Srikar et al., 2016; Zhang et al., 2016).

In this context, various medicinal plants have been used in the biosynthesis of AgNPs due to their easy accessibility in nature and presence of a diverse group of bioactive compounds in plant material which can act as reducing and stabilizing agents. This article mainly discusses the medicinal plant extract mediated facile and eco-friendly synthesis and characterization of biomolecule capped AgNPs with their antimicrobial properties toward MDR pathogens.

2 Medicinal plants: An overview

Plants have an imperative role in food, medication, clothing, and shelter in human civilization. Modern awareness of medicinal plants has become an essential part of society. Indigenous civilization has been enhanced by the traditional knowledge of ancient medicinal plants. Rural and suburban people around the globe have a long tradition of using medicinal plants to cure various diseases. Leaves, barks, fruits, roots, and seeds of medicinal plants are commonly used as folk medicine by different ethnic communities (Some et al., 2017; Jima and Megersa, 2018). In developing countries, medicinal plants have been used as a crude form for the treatment of common diseases which include malaria, cholera, pneumonia, tuberculosis, and asthma. Therefore, further studies of medicinal plants are required, by isolating the active compounds that can be processed into new and potent drugs (Oteng Mintah et al., 2019). Ahvazi et al. (2012) reported that 16 plants belonging to 11 families have been used by the ethnic communities in the Alamut region of Iran for the curing of different ailments. Ahmed (2016) documented that 66 traditional medicinal plant species belonging to 63 genera and 34 families have been used to treat 99 different types of ailments and diseases in Sulaymaniyah province of Iraq. The earlier ethnobotanical study was noted that 30 plant species belong to 20 different families have been used for the treatment of respiratory, oral, and otorhinolaryngeal problems Ngadisari village of Indonesia (Jadid et al., 2020).

Jima and Megersa (2018) observed that plant roots have frequently been used as traditional medicine due to the presence of a maximum concentration of the bioactive compounds. Traditional healers at Koh-e-Safaid Range on the northern Pakistani-Afghan border have prescribed 92 plant species for the treatment of 53 ailments which include cancer, diabetes, hepatitis, gastrointestinal disorders, etc. (Hussain et al., 2018). *Andrographis paniculata* has been reported as highly used potential medicinal plants across the globe. The plant has been traditionally used for the treatment of gastrointestinal problems, respiratory infections, and cardiovascular disorders. It is also used as a blood purifier and an antioxidant. The bioactive compounds of the plant have been used to treat sexual dysfunctions and also serve as a contraceptive (Hossain et al., 2014). Anywar et al. (2020) reported the widespread use of a rich diversity of medicinal plants species which include *Erythrina abyssinica*, *Warburgia ugandensis*, *Zanthoxylum chalybeum*, *Acacia hockii*, *Mangifera indica*, *Aloe vera*, *Albizia coriaria*, *Azadirachta indica*, *Psorospermum febrifugum*, *Vernonia amygdalina*, and *Gymnosporia senegalensis* prescribed by traditional healers to manage opportunistic infections in HIV/AIDS patients in Uganda.

2.1 Bioactive compounds or active phytochemicals in medicinal plants

Due to the paramount availability of chemical variety, natural products derived from medicinal plants, either as standardized extracts or as pure compounds, offer unregulated prospects for new drug outcomes. Botanicals and herbal formulations for medicinal usage contain various types of bioactive compounds (Sasidharan et al., 2011). Plant metabolites have been classified into two types. Primary metabolites include protein, lipid, carbohydrate, chlorophyll molecule, and nucleic acids. Alkaloids, flavonoids, terpenoids, saponins, phenolic acids, and tannins are known as secondary metabolites (Segneanu et al., 2017; Takshak, 2018). Phytochemical analysis by techniques like liquid chromatography (LC), liquid chromatography-mass spectrometry (LC-MS), gas chromatography-mass spectrometry (GC-MS), and liquid chromatography-nuclear magnetic resonance (LC-NMR) provides a vast quantity of structural information about the known and unknown compounds from the crude plant extracts (Günther, 2013; Mustafa et al., 2017; Al Salhi et al., 2019; Swargiary et al., 2020).

Petropoulos et al. (2018) have quantified flavonoid and phenolic acid with the value of 34.4 and 28.7 mg/g in the leaf blade extract of *Cynara cardunculus*. The LC-MS/MS study explored the bioactive compounds in the leaf extract of the V1 variety of *Morus indica*. Among the phytochemicals, cyclomorusin, isoquercetin, gallic acid, sophoraisoflavanone A, kazinol B, mangiferin xanthonoid, and stigmasterol have been recorded as major bioactive compounds in aqueous leaf extract. Isoquercetin, sophoraisoflavanone A, and cyclomorusin are known as flavonoids. Mangiferin xanthonoid and gallic acid fall under the category of phenolic compounds. Kazinol B and stigmasterol are known as polyhydroxyflavan (benzopyran derivative) and steroid compounds (Some et al., 2019a). Mousavi et al. (2018a) have identified a group of polyphenols which include kaempferol, 3, 4-dimethoxycinnamic acid, diosmetin, rosmarinic acid, caffeic acid, luteolin, apigenin, and genistein in methanol extract of the leaves of *Ocimum tenuiflorum* exhibited excellent antidiabetic activity. The active phytochemicals have also antioxidant, antimicrobial, anticancer, antiviral, and antitumor activities to a greater or lesser extent (Kumar et al., 2019).

The alkaloids piperine, piperlongumine, guineensine, chabamide, and pellitorine of *Piper* sp. have excellent cytotoxicity against cancer cell lines. The bioactive compounds of the same plant such as pinoresinol and guineensine showed strong antibacterial activity against a different strain of *Vibrio* (Mgbeahuruike et al., 2017). An earlier study has shown that methyl gallate and fraxetin in the stem bark extract of *Jatropha podagrica* exhibited excellent antimicrobial activity against *Klebsiellapneumoniae*, *Staphylococcus aureus*, *Bacillus subtilis*, *Listeria monocytogenes*, *Escherichia coli*, and *Proteus mirabilis* with the minimum inhibitory concentration (MIC) values of 30, 5, 25, 20, 20, and 20 mg/mL, respectively (Minh et al., 2019).

3 Varied types of synthesis methods of nanoparticles

Two varied approaches are available for the synthesis of nanoparticles: top-down and bottom-up. These approaches are two potential practices of the fabrication of AgNPs. The top-down approach is characterized by the manufacturing of nanosilver from bulk content. The bottom-up approach is described by the fabrication of AgNPs from atomic or molecular condensation and assemblies of cluster by cluster. Among the two approaches, the bottom-up technique is more constructive than the top-down system for a higher chance of manufacturing AgNPs with fewer faults, more homogeneous chemical composition, and rapidity of synthesis (Abou El-Nour et al., 2010; Iqbal et al., 2012; Prasad, 2014).

Evaporation-condensation, gamma irradiation, electron irradiation, laser ablation, sputtering, and lithography are the most important routes in physical methods for the synthesis of AgNPs. In the evaporation-condensation process, AgNPs can be fabricated in a tube furnace at high atmospheric pressure. The large space of the tube furnace, the use of huge amounts of energy, the elevation of the ambient temperature around the source material, and a lot of time to maintain thermal stability are major disadvantages in physical methods. Chemical reduction, photochemical synthesis, solvothermal synthesis, hydrothermal, and sol-gel synthesis are recognized as common chemical methods (Fig. 2). Sodium citrate, ascorbate, sodium borohydride ($NaBH_4$), hydrazine, elemental hydrogen, Tollens reagent, N,N-dimethylformamide (DMF), 2-mercaptoethanol, and poly (ethylene glycol)-block copolymers are used as reducing agents in aqueous or nonaqueous mediums. However, chemically synthesized AgNPs have drawbacks due to the incorporation of some toxic chemicals such as sodium borohydride, thio-glycerol, 2-mercaptoethanol, etc. and it is hazardous to the environment (Arole and Munde, 2014; Iravani et al., 2014; Chouhan, 2018). Therefore, an ecologically and financially feasible method for synthesizing AgNPs is needed. Recently, different natural products have been used to synthesize biocompatible AgNPs, and these eco-friendly synthesis methods are known to be the green synthesis of MNPs. The enzymes, especially nitrate reductase in natural products,

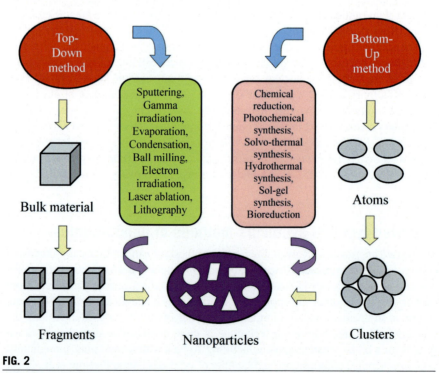

FIG. 2

Top-down and bottom-up approaches in the synthesis of silver nanoparticles.

reduce Ag^+ ions to form Ag^0 in the green synthesis route (Abbasi et al., 2016; Güzel and Erdal, 2018).

4 Biosynthesis of silver nanoparticles using medicinal plants

The synthesis of AgNPs using plant extract is a safe, eco-friendly, and cost-effective method that provides a contemporary and potential replacement of chemically synthesized NPs which minimizes the use and environmental security of hazardous and toxic chemicals (Hambardzumyan et al., 2020). Different parts of the medicinal plant which include leaf, root, stem bark, flower, fruit, and seed have been used in the biofabrication of AgNPs (Ahmad et al., 2019; Some et al., 2019b). The phytochemicals in plant extracts are reducing Ag^+ ions followed by attributing at the surface to produce stable AgNPs (Salari et al., 2019). The bioactive compounds also inhibit the oxidation of AgNPs through the formation of a protective layer around the NPs, thereby evading direct contact with O_2 (McShan et al., 2014). A step associated with medicinal plant-mediated biosynthesis of AgNPs is shown in Fig. 3.

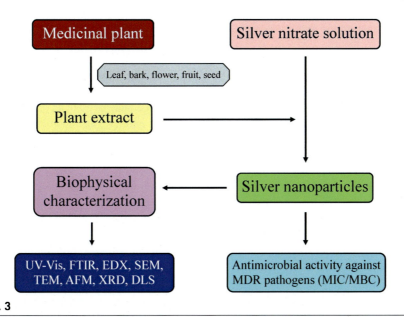

FIG. 3

Steps associated with medicinal plant-mediated biosynthesis of AgNPs.

4.1 Protocol of synthesis

Silver nanoparticles have been synthesized from their precursor compounds, silver nitrate ($AgNO_3$). The size distribution of AgNPs directly depends on the concentration of $AgNO_3$, pH, reaction time, and the quality and quantity of plant extract. An earlier study has shown that AgNPs have been synthesized using leaf and stem extract of *Piper nigrum*. This plant has been considered a common medicinal plant for the presence of a diverse group of bioactive compounds. The plant was collected from the medicinal plant garden, followed by washing with distilled water to eliminate foreign substances. Then finely dissected leaves were again cleaned with the detergent tween 20 followed by thoroughly washing in double-distilled water two or three times. Ten grams of leaf and stem were boiled with 100 mL of double-distilled water at 60–80°C for 10 min. The extract was allowed to cool at room temperature followed by filtering by nylon mesh. The extract was kept at 4°C for further use. For the biosynthesis of AgNPs, 10 mL of leaf and stem extract was added with 90 mL $AgNO_3$ (1 mM), separately. The reaction mixture was allowed to stay at room temperature. The color changes of solution from pale yellow to brown indicated the synthesis of biogenic AgNPs. The electron microscopic images exhibited the sizes with the values 4–50 nm for leaf extract-mediated AgNPs and 9–30 nm for stem extract-mediated AgNPs (Paulkumar et al., 2014).

In another experiment, Arunachalam et al. (2014) reported that the *Gymnema sylvestre* leaf extract mediated the green synthesis of AgNPs. Alkaloids, phenols, flavonoids, sterols, tannins, and triterpenes were recorded as predominant phytochemicals in leaf extract and played a pivotal role as reducing and stabilizing agents in the biofabrication of AgNPs. The leaves of *Gymnema sylvestre* were washed thoroughly then blot-dried with tissue paper and allowed to dry for 2 weeks at room temperature. The dried leaves were cut into small pieces and powdered in a mixer grinder, then passed through a 20-μm mesh sieve to obtain a uniform particle size range of leaf dust. Twenty grams of leaf dust was mixed with 100 mL of sterilized distilled water followed by boiling for 5 min. The mixer was placed in a shaker at 30°C under dark conditions. The extract was filtered and stored in an airtight container at room temperature. Five milliliters of plant extract was added to 15 mL of 1 mM aqueous $AgNO_3$ solution and placed on a magnetic stirrer at 150 rpm in dark conditions. The biosynthesis was confirmed by the reading of the UV-vis spectrophotometer at a specific time interval and completed within 24 h. Various parameters have been employed in *Pistia stratiotes* leaf extract-mediated biosynthesis of AgNPs which include the concentration of $AgNO_3$ solution (1–5 mM), pH value (4–10), reaction time (1–6 h), and light irradiations (red, green, and blue). This study noted that 5 mM, pH 10, and blue light irradiation are the optimum parameters for changing the color of the reaction mixture from yellow to brown within 15 min and this confirmed the data of UV-vis spectroscopy. TEM micrographs exhibited the size of AgNPs with the values of 14.79 ± 4.18 nm, 17.30 ± 4.36 nm, and 20.97 ± 7.97 nm at pH 4, 7, and 10, respectively (Traiwatcharanon et al., 2017). A list of medicinal plant extract-mediated biosyntheses of AgNPs are provided in Table 1.

Table 1 Biosynthesis of AgNPs using medicinal plants.

Name of plants	Used parts	Shape	Size (nm)	Reference
Allium sativum	Bulb	Spherical	3–6	Otunola et al. (2017)
Alpinia nigra	Fruit	Spherical	6	Baruah et al. (2019)
Bauhinia purpurea	Leaf	Spherical	–	Vijayan et al. (2019)
Callicarpa maingayi	Stem bark	–	12.40±3.27	Shameli et al. (2012a)
Capsicum frutescens	Fruit	Spherical	3–18	Otunola et al. (2017)
Costus afer	Leaf	Spherical	20	Elemike et al. (2017)
Curcuma longa	Tuber	–	6.30±2.64	Shameli et al. (2012b)
Decalepis hamiltonii	Root	Spherical	32.5	Rashmi and Sanjay (2017)
Emblica officinalis	Fruit	–	15	Ramesh et al. (2015)
Gymnema sylvestre	Leaf	Spherical	–	Arunachalam et al. (2014)
Mentha pulegium	Leaf	Anisotropic	5–50	Kelkawi et al. (2017)
Passiflora edulis	Leaf	Spherical	3–7	Thomas et al. (2019)
Piper longum	Leaf	Spherical	28.8	Yadav et al. (2019)
Pulicaria glutinosa	Aerial parts	Spherical	40–60	Khan et al. (2013)
Trifolium resupinatum	Seed	Spherical	17	Khatami et al. (2016)
Vitex negundo	Leaf	–	18.2±8.9	Zargar et al. (2011)
Zingiber officinale	Rhizome	Spherical	3–22	Otunola et al. (2017)
Salvia officinalis	Leaf	Spherical	41	Okaiyeto et al. (2021)
Carissa carandas	Leaf	–	35±2nm (25°C); 30±3nm (60°C)	Singh et al. (2021)
Ocimum americanum	Leaf	Face-centered cubic (fcc)	48.25	Manikandan et al. (2021)
Decaschistia crotonifolia	Leaf	Spherical	13.6	Palithya et al. (2021)

Continued

Table 1 Biosynthesis of AgNPs using medicinal plants—cont'd

Name of plants	Used parts	Shape	Size (nm)	Reference
Crataegus ambigua	Fruit and leaf	Spherical	30	Ojemaye et al. (2021)
Ixora brachypoda	Leaf	Spherical	27.76	Bhat et al. (2021)
Parthenium hysterophorus	Leaf	Spherical	10.3 ± 1.7	Sivakumar et al. (2021)
Syzygium guineense	Leaf	–	27.62	Desalegn et al. (2021)

4.2 Plausible mechanism of silver nanoparticle synthesis

Jain and Mehata (2017) described *Ocimum sanctum* leaf extract (flavonoids) and its derivative quercetin-mediated biosynthesis of AgNPs, respectively. In an aqueous solution, $AgNO_3$ dissociates into Ag^+ and NO_3^- ions. The hydroxyl and ketonic groups of flavonoid (quercetin) in leaf extract of *O. sanctum* performed the reduction of Ag^+ to form AgNPs. The enzymes in leaf extract of *O. sanctum* further attach with Ag^+ to form an enzyme-substrate complex with a charge transfer between quercetin and Ag^+, resulting in the formation of biomolecule encapsulated AgNPs. Some et al. (2020) described that the –OH and –COOH groups present in gallic acid in leaf extract of *Morus alba* S-1635 might be accountable for the bioreduction of Ag^+ to AgNPs. Gallic acid is oxidized to its corresponding quinine in the biosynthesis of AgNPs and absorbed on the surface of NPs by electrostatic interaction between carboxylic acid groups of gallic acid on the surface of the NPs.

5 Characterization of biosynthesized silver nanoparticles

Biosynthesized AgNPs have been characterized by spectroscopic, microscopic, diffractive, and light scattering methods (Mondal et al., 2021; Suriyakala et al., 2021; Rajkumar et al., 2021; Ratan et al., 2020; Muthukumar et al., 2020). The optical properties of AgNPs have been investigated by UV-vis spectrophotometer. Preliminary characterization of the synthesized AgNPs has been executed by a strong interaction between light and MNPs due to the conduction of electrons on the metal surface, which undergo a collective oscillation when they are exposed by light at specific wavelengths (300–600 nm) (Ashraf et al., 2016; Amirjani et al., 2020). Earlier investigations have shown the maximum absorbance bands at 434 nm (Devaraj et al., 2013) or 435 nm (Chahardoli et al., 2017) or 450 nm (Pirtarighat et al., 2019) correspond to the surface plasmon resonance (SPR) of AgNPs and the biomolecules functionalized AgNPs exhibited a blue to red shift in UV spectra (Some et al., 2020). Fourier transform infrared spectroscopy (FTIR) analysis has been performed to identify the presence of various functional groups in bioactive compounds responsible for the bioreduction of Ag^+. The spectrum of biogenic AgNPs displayed peaks at $1739\,cm^{-1}$, $1366\,cm^{-1}$, and $1217\,cm^{-1}$, indicating the characteristic functional groups [C=O, —CH(CH$_3$)$_2$] of the phytochemicals in the leaf extract of *Annona glabra* (Amarasinghe et al., 2020). Youssif et al. (2019) observed that the bands in the range of $1635.64\,cm^{-1}$, $1392.61\,cm^{-1}$, 1284.59–$1222.87\,cm^{-1}$, $1076.28\,cm^{-1}$, and 505.35–$428.20\,cm^{-1}$ indicate the stretching of C=C, tertiary amide, stretching of C—N, N—H bending, stretching of C—O, C—N, and alkyl halides bond stretching. The elemental composition, purity, and relative abundance of the biosynthesized AgNPs have been acquired from energy dispersive X-ray (EDX) spectroscopy. The other elemental spectra in EDX were attributed to stabilizing biomolecules bound to the surface of the AgNPs (Femi-Adepoju et al., 2019).

Scanning electron microscopy (SEM) has been used to understand the surface morphology of synthesized AgNPs but transmission electron microscopy (TEM) has been perfectly employed to recognize the shape, size, and distribution of MNPs (Rauwel et al., 2015; Rautela et al., 2019). The SEM images showed the biosynthesized AgNPs of tubular and cuboidal shape with an average particle size of 30–35 nm (Srirangam and Rao, 2017). Bhatnagar et al. (2019) observed the distribution of spherical, hexagonal, rod-shaped, and triangular-shaped AgNPs with the size of 4–41 nm using TEM. Atomic force microscopy (AFM) has been also applied to understand the topographic image or three-dimensional views of NPs. The study aims to determine the shape, size, and/or size distribution of MNPs (Rao et al., 2007). X-ray diffraction (XRD) analysis provides comprehensive information on the crystallographic structure and physical properties of synthesized AgNPs. The Bragg's diffraction peaks at 2θ angles of 32.5, 38.3, 44.4, 64.6, and 76.8 degree, and peaks assigned at the planes of (122), (111), (200), (220), and (311), confirm the face-centered, cubic, and crystalline nature of biosynthesized AgNPs (Jemal et al., 2017). The size of the crystal is calculated from the Debye-Scherrer equation, $D = (k\lambda)/(\beta \cos \theta)$, where D = size of crystallites (nm), $K = 0.9$ (shape factor), λ = wavelength of X-ray, β = full width at half maximum in radians of diffraction peak, and θ = Bragg diffraction angle (degree) (Meena et al., 2020).

Dynamic light scattering (DLS) study is used to determine the particle size by the random changes in the intensity of light scattered from a solution. This approach is often referred to as photon correlation spectroscopy (PCS) based on the laser diffraction method with multiple scattering processes used to investigate the average particle size of AgNPs (Devi et al., 2019). A study has shown that the hydrodynamic diameter of biosynthesized AgNPs in the range of 39–78.5 nm, which is larger than the size measured by TEM (24.2 ± 3.1 nm) due to attachment of ions or molecules at the surface of NPs (Erjaee et al., 2017). In terms of size distribution, the polydispersity index (PDI) is also an important quality indicator. A DLS study showed that the PDI values for the biogenic AgNPs ranged from 0.291 to 0.536 specify the polydisperse in nature of NPs (Hambardzumyan et al., 2020). The zeta potential (ξ) value ($>+30\,mV$ or $<-30\,mV$) is another significant parameter for understanding the surface charge and high degree of stability of NPs (Bhattacharjee, 2016).

6 Antimicrobial activity of biosynthesized silver nanoparticles

AgNPs have been used in a range of biomedical applications, from the production of unique antimicrobials as a coating on medical equipment, surgical devices, and nanomedicine to the use in drug targeted systems. The biosynthesized AgNPs exhibited excellent antimicrobial activity against a wide range of gram-negative and gram-positive bacteria which include *Escherichia coli*, *Pseudomonas aeruginosa*, *Salmonella typhimurium*, *Staphylococcus aureus*, methicillin-resistant *S. aureus*, and *Acinetobacter baumannii* (Peiris et al., 2017). A study has shown that biogenic AgNPs exhibited remarkable antibacterial properties against clinically isolated multidrug-resistant bacteria which include *Streptococcus pyogens*, *Pseudomonas aeruginosa*, *Escherichia coli*, *Bacillus subtilis*, and *Staphylococcus aureus* (Gopinath et al., 2012). Silver nanoparticles showed antimicrobial potential through various mode of action which include adhesion to microbial cells, penetration inside the cells, reactive oxygen species (ROS) and free radical generation, and modulation of microbial signal transduction pathways (Dakal et al., 2016). The generation of ROS by AgNPs has promising significance for cell toxicity through membrane disruption, DNA damage, protein denaturation, enzyme inhibition, and lipid peroxidation (Qing et al., 2018; Yu et al., 2020).

6.1 Antibiotics: An overview

Antibiotics or antimicrobial substances are produced by a group of microorganisms and are active against another group of microorganisms. They have been used to treat or prevent some types of bacterial infection but do not work against viral infections such as the common cold and flu. Antimicrobial substances work against their target through bacteriostatic and bactericidal effects. In the bacteriostatic mechanism, the antimicrobial substance prevents the growth of bacteria or it keeps them in the stationary phase of growth. However, it kills bacteria through the bactericidal method. Broad-spectrum antibiotics are active against both gram-positive and gram-negative bacteria. Antibiotics are powerful chemotherapeutic agents that work very well for certain types of infections (Banerjee and Banerjee, 2006). However, some antibiotics are now less effective due to an increase in antibiotic resistance. Antibiotics are classified based on their mode of action into five major types, which include inhibition of cell wall synthesis, inhibition of protein synthesis, damage of cell membrane, inhibition of nucleic acid synthesis, and inhibition of enzyme function. Natural antibiotics are produced by different microorganisms such as actinomycetes, bacteria, and fungi. Sometimes they are chemically modified to derive synthetic or semisynthetic form (Table 2) (Ullah and Ali, 2017).

Table 2 Different classes of antibiotics.

Producing organisms	Name of antibiotics	Mode of actions
Penicillium notatum	Penicillin	Inhibition of cell wall synthesis
Bacillus subtilis	Bacitracin	
Streptomyces orientalis	Vancomycin	
Acremonium strictum	Cephalosporin	
–	Ampicillin	
–	Amoxicillin	
Streptomyces griseus	Streptomycin	Inhibition of protein synthesis
Streptomyces aureofaciens Streptomyces rimosus	Tetracycline	
Streptomyces erythreus	Erythromycin	
Streptomyces fradiae	Neomycin	
Streptomyces lincolnensis	Lincomycin	
Streptomyces venezuelae	Chloramphenicol	
Bacillus polymyxa	Polymyxin B	Damage of cell membrane
Streptomyces nodosus	Amphotericin B	
Streptomyces mediterranei	Rifamycin	Inhibition of nucleic acid synthesis
–	Quinolone	
–	Co-trimoxazole (trimethoprim/sulfamethoxazole)	Inhibition of enzyme function

6.2 Antibiotic resistance in bacteria

The extensive use of antibiotics in human and veterinary medicine has emerged as a major selective force in the occurrence of drug resistance among microorganisms. Aslam et al. (2018) described how drug resistance in microorganisms developed by different mechanisms like drug efflux, enzymatic modification of antibiotics, and enzymatic degradation of antibiotics (Fig. 4). Antibiotic resistance in bacteria is mediated by extrachromosomal elements or R-plasmid (Tadesse et al., 2012). The emergence of antibiotic resistance among bacteria was first reported in 1950 from Japan. In this study, *Shigella* sp. was isolated from antibiotic-treated dysentery patient. The pathogen showed resistance to multiple antibiotics and able to transfer its R-plasmid to *Escherichia coli* (Schlegel, 1996). The horizontal gene transfer mechanism plays a pivotal role in increasing drug-resistant bacteria in the environment (Burmeister, 2015). Integrons are a type of mobile genetic element, consisting

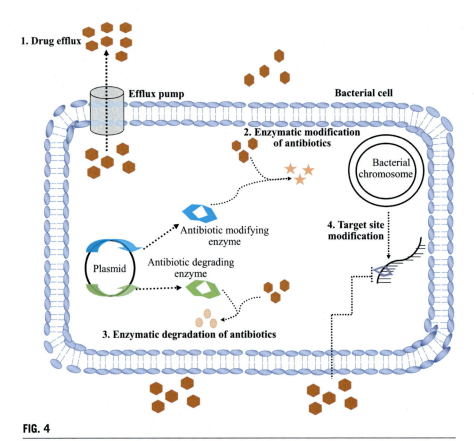

FIG. 4

Mechanisms of bacterial resistance to antibiotics.

of one or more gene cassettes located at a specific site of the genome, which is related to antibiotic resistance among bacteria (Hall and Collis, 1995). Hospital, clinical, and agricultural runoff may also increase the antibiotic resistance gene pool in the environment (Mishra et al., 2018). The high prevalence of antibiotic-resistant bacteria in drinking or surface water has created a serious issue in terms of public health (Mukherjee et al., 2005). An earlier study has shown that *Escherichia coli* in drinking water showed resistance against single or multiple antibiotics which include amoxicillin-clavulanic acid, ampicillin, azithromycin, cefotaxime, cefoxitin, chloramphenicol, ciprofloxacin, gentamicin, nalidixic acid, nitrofurantoin, sulfamethoxazole-trimethoprim, and tetracycline. The antibiotic resistance profile of isolated *Escherichia coli* exhibited the highest resistance against tetracycline (37.6%), followed by ampicillin (34.2%), sulfamethoxazole-trimethoprim (21.4%), and nalidixic acid (13%). Some isolates (19.7%) showed multidrug resistance against tested antibiotics (Larson et al., 2019). The clinical isolates of

Staphylococcus aureus showed the highest resistance against ampicillin and erythromycin (88% each) while resistances against oxacillin, fosfomycin, cefoxitin, and ciprofloxacin exhibited alarming results (Hanif and Hassan, 2019). The emergence of drug resistance in *Pseudomonas aeruginosa* was recorded against beta-lactams, aminoglycosides, and fluoroquinolones groups of antibiotics (Pachori et al., 2019). The opportunistic bacterial pathogen *Klebsiella pneumoniae* (84%) was recorded as MDR with high-level resistance to β-lactams, aminoglycosides, quinolones, tigecycline, and colistin (Ferreira et al., 2019).

Antimicrobial resistance is a serious problem in health care today across the globe. It hampers the effective prevention and treatment of an ever-increasing range of infections caused by microorganisms. Drug resistance has been gradually increased due to misuse or haphazard use of antibiotics. Single and multiple antibiotics lose their effectiveness against microorganisms. The rapid emergence of MAR bacteria might cause a severe epidemic of diseases. The public should not take antibiotics without consultation of physicians, and the dose of antibiotic drugs must be completed. Medicine shops must be introduced not to sell antibiotics without a prescription of a registered medical practitioner. Thus, it is enormously important to find a new class of antimicrobials to combat MDR pathogens.

6.3 Efficacy of biogenic silver nanoparticles against MDR pathogens

Silver nanoparticles have been widely used in an increasing number of medical and consumer products due to their excellent antibacterial properties. The efficacy of biogenic AgNPs against different bacterial pathogens is presented in Table 3. In an aqueous environment, AgNPs act primarily through a process known as oxidative dissolution, wherein Ag^+ ions are released through an oxidative mechanism and can have a severe impact on various organisms including microorganisms (Peretyazhko et al., 2014). A study noted that biogenic AgNPs (~11 nm) inhibited bacterial growth in a dose- and time-dependent manner. The minimum inhibitory concentration (MIC) values were recorded as 1 and 2 μg/mL and minimum bactericidal concentration (MBC) were documented as 2 and 4 μg/mL against MDR *Pseudomonas aeruginosa* and *Staphylococcus aureus*, respectively (Yuan et al., 2017). Sharma et al. (2018) measured the MIC value of *Coptidis rhizome*-mediated biogenic AgNPs (<30 nm) using a broth/tube dilution method and showed that the growth of *Escherichia coli* was completely inhibited until almost 12 h at a dose of 150 μg/mL. In another study, the antibacterial activity of biogenic AgNPs was explored against MDR strains of *Pseudomonas aeruginosa*. The zone of inhibition was recorded in the range of 10 ± 0.53 to 21 ± 0.11 mm using varied concentrations of AgNPs at 12.5–100 μg/mL. The zone of inhibition enlarged with a dose-dependent concentration of AgNPs. The MIC values of biosynthesized AgNPs were found in the range of 6.25–12.5 μg/mL compared to standard antibiotics (Singh et al., 2014).

Table 3 Efficacy of biogenic AgNPs against bacterial pathogens.

Biogenic source(s)	Size (nm)	Conc. of AgNP (µg or µl/disc or mL)	Tested Bacteria	MIC (µg/mL)	ZOI (mm)	Reference
Angelica pubescens	12.48	45	Escherichia coli ATCC 10798	–	14 ± 1.15	Markus et al. (2017)
			Staphylococcus aureus ATCC 6538	–	17.3 ± 0.44	
			Pseudomonas aeruginosa ATCC 27853	–	13 ± 0.5	
			Salmonella enterica ATCC 13076	–	12 ± 2	
Ananas comosus	–	50	Enterococcus faecium DB01	50	10.31 ± 0.68	Das et al. (2019)
			Listeria monocytogenes ATCC 19111	50	9.07 ± 0.11	
			Bacillus cereus KCTC 3624	100	8.91 ± 0.05	
			Staphylococcus aureus ATCC 13565	100	8.78 ± 0.08	
Azadirachta indica	183.2 ± 0.7	20	Pseudomonas aeruginosa ATCC 27584	–	19.9 ± 1.4	Algebaly et al. (2020)
			Escherichia coli ATCC 25922	–	22 ± 1.8	
			Staphylococcus aureus ATCC 29213	–	21.1 ± 1.2	

Calligonum comosum	90.8 ± 0.8	20	Pseudomonas aeruginosa ATCC 27584	—	18 ± 0.9
			Escherichia coli ATCC 25922	—	18.8 ± 1.2
			Staphylococcus aureus ATCC 29213	—	17.4 ± 0.6
Carya illinoinensis	12–30	—	Staphylococcus aureus ATCC 25923	128	—	Javan Bakht Dalir et al. (2020)
			Listeria monocytogenes ATCC 7644	64	
			Escherichia coli ATCC 25922	16	
			Pseudomonas aeruginosa ATCC 27853	32	
Coptis chinensis	6–45	100	Bacillus subtilis	—	18	Pei et al. (2019)
			Staphylococcus aureus	—	12
			Klebsiella pneumoniae	—	13
			Pseudomonas aeruginosa	—	16
Cucumis prophetarum	30–50	75	Staphylococcus aureus	—	18 ± 0.4	Hemlata et al. (2020)
			Salmonella typhi	—	20 ± 0.6
Ephedra procera	20.4	—	Escherichia coli ATCC 25922	11.12	—	Nasar et al. (2019)
			Bacillus subtilis ATCC 6633	11.33	—
Oedera genistifolia	10–60	—	Enterobacter cloacae ATCC 13047	500	—	Okaiyeto et al. (2019)
			Listeria ivanovic ATCC 19119	1000	—
			Streptococcus uberis ATCC 700407	500	—
			Staphylococcus aureus ATCC 29213	500	—

Continued

Table 3 Efficacy of biogenic AgNPs against bacterial pathogens—cont'd

Biogenic source(s)	Size (nm)	Conc. of AgNP (μg or μl/disc or mL)	Tested Bacteria	MIC (μg/mL)	ZOI (mm)	Reference
Origanum vulgare	1–25		Mycobacterium smergatis ATCC 19420	250	–	Hambardzumyan et al. (2020)
			Vibrio spp.	250	–	
			Staphylococcus aureus MDC 5233	13.75		
			Bacillus subtilis WT-A17	9.16		
			Escherichia coli VKPM-M17	18.35		
			Salmonella typhimurium MDC 1754	18.35		
			Escherichia coli DH5α-pUC18	18.35		
			Escherichia coli pARG-25	11		
Phlomis bracteosa	22.41	10	Escherichia coli ATCC 15224	–	13.2 ± 0.12	Anjum and Abbasi (2016)
			Staphylococcus aureus ATCC 6538	–	11.1 ± 0.10	
			Klebsiella pneumoniae ATCC 4619	–	10.3 ± 0.11	
Pisum sativum	10–25	100	Escherichia coli O157:H7 ATCC 2351	100	11.10 ± 0.11	Patra et al. (2019)
			Enterococcus faecium DB01	100	9.12 ± 0.3	
			Salmonella typhimurium KCTC 1925	100	8.88 ± 0.3	
			Salmonella enterica KCCM 11806	100	8.7 ± 0.21	

Terminalia mantaly	11–83	—	Streptococcus pneumoniae ATCC 49619	6.24	—	Majoumouo et al. (2019)
			Staphylococcus aureus NR-45003	6.24	—	
			Kbesiella pneumoniae ATCC 13883	6.24	—	
			Salmonella enterica enterica NR-4311	>12.50	—	
			Salmonella enterica NR-13555	6.24	—	
			Salmonella enterica NR-4294	6.24	—	
			Shigella flexineri NR-518	>12.50	—	
			Haemophilus influenzae ATCC 49247	3.12	—	
Ixora brachypoda	27.76	25, 50, and 100	B. subtilis	6	18, 20, 21	Bhat et al. (2021)
			P. aeruginosa	9	14, 14, 16	
			E. coli	8		
			S. aureus	7		
Parthenium hysterophorus	10.3 ± 1.7	80	E. coli	—	18	Sivakumar et al. (2021)
			P. aeruginosa		19	
			B. subtilis		13	
			S. aureus		17	
			E. faecalis		12	

7 Conclusions

Nanoparticles are a fundamental building block of nanotechnology and AgNPs are noble biosynthesized MNPs in nanobiotechnology. The medicinal plant extract-mediated synthesis method has been recognized as an alternative to conventional physical and chemical methods due to its simple, rapid, cost-effective, nontoxic, and environmentally friendly nature. Bioactive compounds or phytochemicals in different medicinal plant extracts act as both reducing and stabilizing agents in the synthesis of biogenic AgNPs. Biosynthesized AgNPs showed excellent antibacterial activity against common gram-positive and gram-negative bacteria compared to standard antimicrobial agents. Antimicrobial resistance in bacteria is a global problem in disease management. Drug resistance has gradually increased to commonly prescribed antibiotics. Therefore, biogenic AgNPs can widely use as nanomedicine in different biomedical sectors, particularly to combat MDR pathogens.

References

Abbasi, E., Milani, M., Fekri Aval, S., Kouhi, M., Akbarzadeh, A., Tayefi Nasrabadi, H., Nikasa, P., Joo, S.W., Hanifehpour, Y., Nejati-Koshki, K., Samiei, M., 2016. Silver nanoparticles: synthesis methods, bio-applications and properties. Crit. Rev. Microbiol. 42 (2), 173–180.

Abou El-Nour, K.M.M., Eftaiha, A., Al-Warthan, A., Ammar, R.A.A., 2010. Synthesis and applications of silver nanoparticles. Arab. J. Chem. 3 (3), 135–140.

Ahmad, S., Munir, S., Zeb, N., Ullah, A., Khan, B., Ali, J., Bilal, M., Omer, M., Alamzeb, M., Salman, S.M., Ali, S., 2019. Green nanotechnology: a review on green synthesis of silver nanoparticles—an ecofriendly approach. Int. J. Nanomedicine 14, 5087–5107.

Ahmed, H.M., 2016. Ethnopharmacobotanical study on the medicinal plants used by herbalists in Sulaymaniyah province, Kurdistan, Iraq. J. Ethnobiol. Ethnomed. 12, 8.

Ahvazi, M., Khalighi-Sigaroodi, F., Charkhchiyan, M.M., Mojab, F., Mozaffarian, V.A., Zakeri, H., 2012. Introduction of medicinal plants species with the most traditional usage in Alamut region. Iran. J. Pharm. Sci. 11 (1), 185–194.

Alaqad, K., Saleh, T.A., 2016. Gold and silver nanoparticles: synthesis methods, characterization routes and applications towards drugs. J. Environ. Anal. Toxicol. 6 (4), 1000384.

Alexander, J.W., 2009. History of the medical use of silver. Surg. Infect. (Larchmt.) 10 (3), 289–292.

Algebaly, A.S., Mohammed, A.E., Abutaha, N., Elobeid, M.M., 2020. Biogenic synthesis of silver nanoparticles: antibacterial and cytotoxic potential. Saudi J. Biol. Sci. 27 (5), 1340–1351.

Al Salhi, M.S., Elangovan, K., Ranjitsingh, A.J.A., Murali, P., Devanesan, S., 2019. Synthesis of silver nanoparticles using plant derived 4-N-methyl benzoic acid and evaluation of antimicrobial, antioxidant and antitumor activity. Saudi J. Biol. Sci. 26 (5), 970–978.

Amarasinghe, L.D., Wickramarachchi, P., Aberathna, A., Sithara, W.S., De Silva, C.R., 2020. Comparative study on larvicidal activity of green synthesized silver nanoparticles and *Annona glabra* (Annonaceae) aqueous extract to control *Aedes aegypti* and *Aedes albopictus* (Diptera: Culicidae). Heliyon 6 (6), e04322.

Amirjani, A., Firouzi, F., Haghshenas, D.F., 2020. Predicting the size of silver nanoparticles from their optical properties. Plasmonics 15, 1077–1082.

Anjum, S., Abbasi, B.H., 2016. Biomimetic synthesis of antimicrobial silver nanoparticles using in vitro-propagated plantlets of a medicinally important endangered species: *Phlomisbracteosa*. Int. J. Nanomedicine 11, 1663–1675.

Anywar, G., Kakudidi, E., Byamukama, R., Mukonzo, J., Schubert, A., Oryem-Origa, H., 2020. Indigenous traditional knowledge of medicinal plants used by herbalists in treating opportunistic infections among people living with HIV/AIDS in Uganda. J. Ethnopharmacol. 246, 112205.

Arole, V., Munde, S., 2014. Fabrication of nanomaterials by top-down and bottom-up approaches- An overview. J. Adv. Appl. Sci. Technol. 1 (2), 89–93.

Arunachalam, K.D., Arun, L.B., Annamalai, S.K., Arunachalam, A.M., 2014. Potential anticancer properties of bioactive compounds of *Gymnemasylvestre* and its biofunctionalized silver nanoparticles. Int. J. Nanomedicine 10, 31–41.

Ashraf, J.M., Ansari, M.A., Khan, H.M., Alzohairy, M.A., Choi, I., 2016. Green synthesis of silver nanoparticles and characterization of their inhibitory effects on AGEs formation using biophysical techniques. Sci. Rep. 6, 20414.

Aslam, B., Wang, W., Arshad, M.I., Khurshid, M., Muzammil, S., Rasool, M.H., Nisar, M.A., Alvi, R.F., Aslam, M.A., Qamar, M.U., Salamat, M., Baloch, Z., 2018. Antibiotic resistance: a rundown of a global crisis. Infect. Drug Resist. 11, 1645–1658.

Austin, L.A., Mackey, M.A., Dreaden, E.C., El-Sayed, M.A., 2014. The optical, photothermal, and facile surface chemical properties of gold and silver nanoparticles in biodiagnostics, therapy, and drug delivery. Arch. Toxicol. 88 (7), 1391–1417.

Banerjee, A.K., Banerjee, N., 2006. Fundamentals of Microbiology and Immunology. New Central Book Agency (P) Ltd, Kolkata, India.

Baruah, D., Yadav, R., Yadav, A., Das, A.M., 2019. *Alpinia nigra* fruits mediated synthesis of silver nanoparticles and their antimicrobial and photocatalytic activities. J. Photochem. Photobiol. B 201, 111649.

Bayda, S., Adeel, M., Tuccinardi, T., Cordani, M., Rizzolio, F., 2019. The history of nanoscience and nanotechnology: from chemical-physical applications to nanomedicine. Molecules 25 (1), 112.

Beg, M., Maji, A., Mandal, A.K., Das, S., Aktara, M.N., Jha, P.K., Hossain, M., 2017. Green synthesis of silver nanoparticles using *Pongamia pinnata* seed: characterization, antibacterial property, and spectroscopic investigation of interaction with human serum albumin. J. Mol. Recognit. 30 (1), 2565.

Bhat, M., Chakraborty, B., Kumar, R.S., Almansour, A.I., Arumugam, N., Kotresha, D., Pallavi, S.S., Dhanyakumara, S.B., Shashiraj, K.N., Nayaka, S., 2021. Biogenic synthesis, characterization and antimicrobial activity of *Ixora brachypoda* (DC) leaf extract mediated silver nanoparticles. J. King Saud Univ. Sci. 33 (2), 101296.

Bhatnagar, S., Kobori, T., Ganesh, D., Ogawa, K., Aoyagi, H., 2019. Biosynthesis of silver nanoparticles mediated by extracellular pigment from *Talaromycespurpurogenus* and their biomedical applications. J. Nanomater. 9 (7), 1042.

Bhattacharjee, S., 2016. DLS and zeta potential—what they are and what they are not? J. Control. Release 235, 337–351.

Burduşel, A.-C., Gherasim, O., Grumezescu, A.M., Mogoantă, L., Ficai, A., Andronescu, E., 2018. Biomedical applications of silver nanoparticles: an up-to-date overview. Nanomaterials 8 (9), 681.

Burmeister, A.R., 2015. Horizontal gene transfer. Evol. Med. Public Health 2015 (1), 193–194.

Chahardoli, A., Karimi, N., Fattahi, A., 2017. Biosynthesis, characterization, antimicrobial and cytotoxic effects of silver nanoparticles using *Nigella arvensis* seed extract. Iran. J. Pharm. Res. 16 (3), 1167–1175.

Chouhan, N., 2018. Silver nanoparticles: synthesis, characterization and applications. In: Silver Nanoparticles—Fabrication, Characterization and Applications. IntechOpen Limited, pp. 21–57.

Cohen, R.W., Cody, G.D., Coutts, M.D., Abeles, B., 1973. Optical properties of granular silver and gold films. Phys. Rev. B 8 (8), 3689–3701.

Dakal, T.C., Kumar, A., Majumdar, R.S., Yadav, V., 2016. Mechanistic basis of antimicrobial actions of silver nanoparticles. Front. Microbiol. 7, 1831.

Das, G., Patra, J.K., Debnath, T., Ansari, A., Shin, H.S., 2019. Investigation of antioxidant, antibacterial, antidiabetic, and cytotoxicity potential of silver nanoparticles synthesized using the outer peel extract of *Ananas comosus* (L.). PloS One 14 (8), e0220950.

Desalegn, T., Murthy, H.A., Limeneh, Y.A., 2021. Medicinal plant *Syzygium guineense* (Willd.) DC leaf extract mediated green synthesis of ag nanoparticles: investigation of their antibacterial activity. Ethiop. J. Sci. Sustain. Dev. 8 (1), 1–12.

Devaraj, P., Kumari, P., Aarti, C., Renganathan, A., 2013. Synthesis and characterization of silver nanoparticles using cannonball leaves and their cytotoxic activity against MCF-7 cell line. J. Nanotechnol. 2013, 598328.

Devi, M., Devi, S., Sharma, V., Rana, N., Bhatia, R.K., Bhatt, A.K., 2019. Green synthesis of silver nanoparticles using methanolic fruit extract of *Aegle marmelos* and their antimicrobial potential against human bacterial pathogens. J. Tradit. Complement. Med. 10 (2), 158–165.

Elemike, E.E., Fayemi, O.E., Ekennia, A.C., Onwudiwe, D.C., Ebenso, E.E., 2017. Silver nanoparticles mediated by *Costusafer* leaf extract: synthesis, antibacterial, antioxidant and electrochemical properties. Molecules 22 (5), 701.

Erjaee, H., Rajaian, H., Nazifi, S., 2017. Synthesis and characterization of novel silver nanoparticles using *Chamaemelum nobile* extract for antibacterial application. Adv. Nat. Sci. Nanosci. Nanotechnol. 8 (2).

Femi-Adepoju, A.G., Dada, A.O., Otun, K.O., Adepoju, A.O., Fatoba, O.P., 2019. Green synthesis of silver nanoparticles using terrestrial fern (*GleicheniaPectinata* (Willd.) C. Presl.): characterization and antimicrobial studies. Heliyon 5 (4), e01543.

Ferdous, Z., Nemmar, A., 2020. Health impact of silver nanoparticles: a review of the biodistribution and toxicity following various routes of exposure. Int. J. Mol. Sci. 21 (7), 2375.

Ferreira, R.L., da Silva, B.C.M., Rezende, G.S., Nakamura-Silva, R., Pitondo-Silva, A., Campanini, E.B., Brito, M.C.A., da Silva, E.M.L., Freire, C.C.d.M., da Cunha, A.F., Pranchevicius, M.-C.d.S., 2019. High prevalence of multidrug-resistant *Klebsiella pneumoniae* harboring several virulence and β-lactamase encoding genes in a Brazilian intensive care unit. Front. Microbiol. 9, 3198.

Gopinath, V., MubarakAli, D., Priyadarshini, S., Priyadharsshini, N.M., Thajuddin, N., Velusamy, P., 2012. Biosynthesis of silver nanoparticles from Tribulus terrestris and its antimicrobial activity: a novel biological approach. Colloids Surf. B Biointerfaces 96, 69–74.

Gudikandula, K., Charya Maringanti, S., 2016. Synthesis of silver nanoparticles by chemical and biological methods and their antimicrobial properties. J. Exp. Nanosci. 11 (9), 714–721.

Günther, H., 2013. NMR Spectroscopy: Basic Principles, Concepts and Applications in Chemistry. John Wiley and Sons.

Güzel, R., Erdal, G., 2018. Synthesis of silver nanoparticles. In: Silver Nanoparticles—Fabrication, Characterization and Applications. IntechOpen Limited.

Hall, R.M., Collis, C.M., 1995. Mobile gene cassettes and integrons: capture and spread of genes by site-specific recombination. Mol. Microbiol. 15, 593–600.

Hambardzumyan, S., Sahakyan, N., Petrosyan, M., Nasim, M.J., Jacob, C., Trchounian, A., 2020. *Origanum vulgare* L. extract-mediated synthesis of silver nanoparticles, their characterization and antibacterial activities. AMB Express 10 (1), 162.

Hamouda, R.A., Hussein, M.H., Abo-elmagd, R.A., Bawazir, S.S., 2019. Synthesis and biological characterization of silver nanoparticles derived from the cyanobacterium *Oscillatoria limnetica*. Sci. Rep. 9, 13071.

Hanif, E., Hassan, S.A., 2019. Evaluation of antibiotic resistance pattern in clinical isolates of *Staphylococcus aureus*. Pak. J. Pharm. Sci. 32 (4Supple), 1749–1753.

Hemlata, Meena, P.R., Singh, A.P., Tejavath, K.K., 2020. Biosynthesis of silver nanoparticles using *Cucumis prophetarum* aqueous leaf extract and their antibacterial and antiproliferative activity against cancer cell lines. ACS Omega 5 (10), 5520–5528.

Hossain, M.S., Urbi, Z., Sule, A., Hafizur Rahman, K.M., 2014. *Andrographis paniculata* (Burm. f.) Wall. ex Nees: a review of ethnobotany, phytochemistry, and pharmacology. Sci. World J. 2014, 274905.

Hussain, W., Badshah, L., Ullah, M., Ali, M., Ali, A., Hussain, F., 2018. Quantitative study of medicinal plants used by the communities residing in Koh-e-Safaid range, northern Pakistani-Afghan borders. J. Ethnobiol. Ethnomed. 14, 30.

Iqbal, P., Preece, J.A., Mendes, P.M., 2012. Nanotechnology: the "top-down" and "bottom-up" approaches. In: Supramolecular Chemistry. vol. 8. John Wiley & Sons Ltd, pp. 3589–3602.

Iravani, S., Korbekandi, H., Mirmohammadi, S.V., Zolfaghari, B., 2014. Synthesis of silver nanoparticles: chemical, physical and biological methods. Res. Pharm. Sci. 9 (6), 385–406.

Jadid, N., Kurniawan, E., Himayani, C.E.S., Andriyani, Prasetyowati, I., Purwani, K.I., Muslihatin, W., Hidayati, D., Tjahjaningrum, I.T.D., 2020. An ethnobotanical study of medicinal plants used by the Tengger tribe in Ngadisari village, Indonesia. PLoS One 15 (7), e0235886.

Jahangirian, H., Lemraski, E.G., Webster, T.J., Rafiee-Moghaddam, R., Abdollahi, Y., 2017. A review of drug delivery systems based on nanotechnology and green chemistry: green nanomedicine. Int. J. Nanomedicine 12, 2957–2978.

Jain, S., Mehata, M.S., 2017. Medicinal plant leaf extract and pure flavonoid mediated green synthesis of silver nanoparticles and their enhanced antibacterial property. Sci. Rep. 7 (1), 15867.

Javan Bakht Dalir, S., Djahaniani, H., Nabati, F., Hekmati, M., 2020. Characterization and the evaluation of antimicrobial activities of silver nanoparticles biosynthesized from *Carya illinoinensis* leaf extract. Heliyon 6 (3), e03624.

Jemal, K., Sandeep, B.V., Pola, S., 2017. Synthesis, characterization, and evaluation of the antibacterial activity of *Allophylus serratus* leaf and leaf derived callus extracts mediated silver nanoparticles. J. Nanomater. 2017, 4213275.

Jima, T.T., Megersa, M., 2018. Ethnobotanical study of medicinal plants used to treat human diseases in Berbere district, bale zone of Oromia regional state, south East Ethiopia. Evid. Based Complement. Alternat. Med. 2018, 8602945.

Kelkawi, A., Abbasi Kajani, A., Bordbar, A.K., 2017. Green synthesis of silver nanoparticles using *Mentha pulegium* and investigation of their antibacterial, antifungal and anticancer activity. IET Nanobiotechnol. 11 (4), 370–376.

Khan, M., Khan, M., Adil, S.F., Tahir, M.N., Tremel, W., Alkhathlan, H.Z., Al-Warthan, A., Siddiqui, M.R., 2013. Green synthesis of silver nanoparticles mediated by *Pulicariaglutinosa* extract. Int. J. Nanomedicine 8, 1507–1516.

Khatami, M., Nejad, M.S., Salari, S., Almani, P.G., 2016. Plant-mediated green synthesis of silver nanoparticles using *Trifolium resupinatum* seed exudate and their antifungal efficacy on *Neofusicoccum parvum* and *Rhizoctonia solani*. IET Nanobiotechnol. 10 (4), 237–243.

Kumar, A., Ahmad, F., Zaidi, S., 2019. Importance of bioactive compounds present in plant products and their extraction—a review. Agric. Rev. 40, 249–260.

Lam, P.L., Wong, W.Y., Bian, Z., Chui, C.H., Gambari, R., 2017. Recent advances in green nanoparticulate systems for drug delivery: efficient delivery and safety concern. Nanomedicine 12 (4), 357–385.

Lansdown, A.B.G., 2010. A pharmacological and toxicological profile of silver as an antimicrobial agent in medical devices. Adv. Pharmacol. Sci. 2010, 910686.

Lara, H.H., Ayala-Núñez, N.V., Ixtepan Turrent, L.d.C., Rodríguez Padilla, C., 2009. Bactericidal effect of silver nanoparticles against multidrug-resistant bacteria. World J. Microbiol. Biotechnol. 26 (4), 615–621.

Larson, A., Hartinger, S.M., Riveros, M., Salmon-Mulanovich, G., Hattendorf, J., Verastegui, H., Huaylinos, M.L., Mäusezahl, D., 2019. Antibiotic-resistant *Escherichia coli* in drinking water samples from rural Andean households in Cajamarca, Peru. Am. J. Trop. Med. Hyg. 100 (6), 1363–1368.

Lee, S.H., Jun, B.H., 2019. Silver nanoparticles: synthesis and application for nanomedicine. Int. J. Mol. Sci. 20 (4), 865.

Majoumouo, M.S., Sibuyi, N., Tincho, M.B., Mbekou, M., Boyom, F.F., Meyer, M., 2019. Enhanced anti-bacterial activity of biogenic silver nanoparticles synthesized from *Terminalia mantaly* extracts. Int. J. Nanomedicine 14, 9031–9046.

Makarov, V.V., Love, A.J., Sinitsyna, O.V., Makarova, S.S., Yaminsky, I.V., Taliansky, M.E., Kalinina, N.O., 2014. Green nanotechnologies: synthesis of metal nanoparticles using plants. Acta Nat. 6 (1), 35–44.

Manikandan, D.B., Sridhar, A., Sekar, R.K., Perumalsamy, B., Veeran, S., Arumugam, M., Karuppaiah, P., Ramasamy, T., 2021. Green fabrication, characterization of silver nanoparticles using aqueous leaf extract of *Ocimum americanum* (Hoary Basil) and investigation of its in vitro antibacterial, antioxidant, anticancer and photocatalytic reduction. J. Environ. Chem. Eng. 9 (1), 104845.

Markus, J., Wang, D., Kim, Y.J., Ahn, S., Mathiyalagan, R., Wang, C., Yang, D.C., 2017. Biosynthesis, characterization, and bioactivities evaluation of silver and gold nanoparticles mediated by the roots of Chinese herbal *Angelica pubescens* Maxim. Nanoscale Res. Lett. 12 (1), 46.

McShan, D., Ray, P.C., Yu, H., 2014. Molecular toxicity mechanism of nanosilver. J. Food Drug Anal. 22 (1), 116–127.

Medici, S., Peana, M., Nurchi, V.M., Zoroddu, M.A., 2019. The medical uses of silver: history, myths and scientific evidence. J. Med. Chem. 62 (13), 5923–5943.

Meena, R.K., Meena, R., Arya, D.K., Jadoun, S., Hada, R., Kumari, R., 2020. Synthesis of silver nanoparticles by *Phyllanthus emblica* plant extract and their antibacterial activity. Mat. Sci. Res. India 17, 170206.

Mgbeahuruike, E.E., Yrjönen, T., Vuorela, H.J., Holm, Y.M., 2017. Bioactive compounds from medicinal plants: focus on *Piper* species. S. Afr. J. Bot. 112, 54–69.

Minh, T.N., Xuan, T.D., Tran, H.-D., Van, T.M., Andriana, Y., Khanh, T.D., Quan, N.V., Ahmad, A., 2019. Isolation and purification of bioactive compounds from the stem bark of *Jatropha podagrica*. Molecules 24 (5), 889.

Mishra, M., Arukha, A.P., Patel, A.K., Behera, N., Mohanta, T.K., Yadav, D., 2018. Multidrug resistant coliform: water sanitary standards and health hazards. Front. Pharmacol. 9, 311.

Mondal, R., Yilmaz, M.D., Mandal, A.K., 2021. Green synthesis of carbon nanoparticles: characterization and their biocidal properties. Handbook of Greener Synthesis of Nanomaterials and Compounds, 1st. 2 Elsevier, Amsterdam, The Netherlands, pp. 277–306.

Mousavi, S.M., Hashemi, S.A., Ghasemi, Y., Atapour, A., Amani, A.M., Savar Dashtaki, A., Babapoor, A., Arjmand, O., 2018b. Green synthesis of silver nanoparticles toward bio and medical applications: review study. Artif. Cells Nanomed. Biotechnol. 46, S855–S872.

Mousavi, L., Salleh, R.M., Murugaiyah, V., 2018a. Phytochemical and bioactive compounds identification of *Ocimumtenuiflorum* leaves of methanol extract and its fraction with an anti-diabetic potential. Int. J. Food Prop. 21 (1), 2390–2399.

Mukherjee, S., Bhadra, B., Chakraborty, R., Gurung, A., Some, S., Chakraborty, R., 2005. Unregulated use of antibiotics in Siliguri city vis-à-vis occurrence of MAR bacteria in community waste water and river Mahananda, and their potential for resistance gene transfer. J. Environ. Biol. 26 (2), 229–238.

Mustafa, G., Arif, R., Atta, A., Sharif, S., Jamil, A., 2017. Bioactive compounds from medicinal plants and their importance in drug discovery in Pakistan. Matrix Sci. Pharma 1 (1), 17–26.

Muthukumar, H., Palanirajan, S.K., Shanmugam, M.K., Gummadi, S.N., 2020. Plant extract mediated synthesis enhanced the functional properties of silver ferrite nanoparticles over chemical mediated synthesis. Biotechnol. Rep. 26, e00469.

Narware, J., Yadav, R.N., Keswani, C., Singh, S.P., Singh, H.B., 2019. Silver nanoparticle-based biopesticides for phytopathogens: scope and potential in agriculture. In: Nano-Biopesticides Today and Future Perspectives. Elsevier Inc, pp. 303–314.

Nasar, M.Q., Khalil, A.T., Ali, M., Shah, M., Ayaz, M., Shinwari, Z.K., 2019. Phytochemical analysis, *Ephedra Procera* C. A. Mey. Mediated green synthesis of silver nanoparticles, their cytotoxic and antimicrobial potentials. Medicina 55 (7), 369.

Ndikau, M., Noah, N.M., Andala, D.M., Masika, E., 2017. Green synthesis and characterization of silver nanoparticles using *Citrullus lanatus* fruit rind extract. Int. J. Anal. Chem. 2017, 8108504.

Ojemaye, M.O., Okoh, S.O., Okoh, A.I., 2021. Silver nanoparticles (AgNPs) facilitated by plant parts of *Crataegus ambigua* Becker AK extracts and their antibacterial, antioxidant and antimalarial activities. Green Chem. Lett. Rev. 14 (1), 49–59.

Okaiyeto, K., Hoppe, H., Okoh, A.I., 2021. Plant-based synthesis of silver nanoparticles using aqueous leaf extract of *Salvia officinalis*: characterization and its antiplasmodial activity. J. Clust. Sci. 32, 101–109.

Okaiyeto, K., Ojemaye, M.O., Hoppe, H., Mabinya, L.V., Okoh, A.I., 2019. Phytofabrication of silver/silver chloride nanoparticles using aqueous leaf extract of *Oederagenistifolia*: characterization and antibacterial potential. Molecules 24 (23), 4382.

Oteng Mintah, S., Asafo-Agyei, T., Archer, M.-A., Atta-Adjei Junior, P., Boamah, D., Kumadoh, D., Appiah, A., Ocloo, A., Boakye, Y.D., Agyare, C., 2019. Medicinal plants for treatment of prevalent diseases. In: Pharmacognosy—Medicinal Plants. IntechOpen Limited, pp. 1–19.

Otunola, G.A., Afolayan, A.J., Ajayi, E.O., Odeyemi, S.W., 2017. Characterization, antibacterial and antioxidant properties of silver nanoparticles synthesized from aqueous extracts of *Allium sativum, Zingiber officinale*, and *Capsicum frutescens*. Pharmacogn. Mag. 13 (Suppl 2), S201–S208.

Ovais, M., Khalil, A.T., Raza, A., Khan, M.A., Ahmad, I., Islam, N.U., Saravanan, M., Ubaid, M.F., Ali, M., Shinwari, Z.K., 2016. Green synthesis of silver nanoparticles via plant extracts: beginning a new era in cancer theranostics. Nanomedicine 11 (23), 3157–3177.

Pachori, P., Gothalwal, R., Gandhi, P., 2019. Emergence of antibiotic resistance *Pseudomonas aeruginosa* in intensive care unit; a critical review. Genes Dis. 6 (2), 109–119.

Paffett, M.T., Campbell, C.T., Taylor, T.N., 1985. Surface chemical properties of silver/platinum (111): comparisons between electrochemistry and surface science. Langmuir 1 (6), 741–747.

Palithya, S., Gaddam, S.A., Kotakadi, V.S., Penchalaneni, J., Challagundla, V.N., 2021. Biosynthesis of silver nanoparticles using leaf extract of *Decaschistia crotonifolia* and its antibacterial, antioxidant, and catalytic applications. Green Chem. Lett. Rev. 14 (1), 137–152.

Patra, J.K., Das, G., Shin, H.S., 2019. Facile green biosynthesis of silver nanoparticles using *Pisum sativum* L. outer peel aqueous extract and its antidiabetic, cytotoxicity, antioxidant, and antibacterial activity. Int. J. Nanomedicine 14, 6679–6690.

Paulkumar, K., Gnanajobitha, G., Vanaja, M., Rajeshkumar, S., Malarkodi, C., Pandian, K., Annadurai, G., 2014. *Piper nigrum* leaf and stem assisted green synthesis of silver nanoparticles and evaluation of its antibacterial activity against agricultural plant pathogens. Sci. World J. 2014, 829894.

Pei, J., Fu, B., Jiang, L., Sun, T., 2019. Biosynthesis, characterization, and anticancer effect of plant-mediated silver nanoparticles using *Coptis chinensis*. Int. J. Nanomedicine 14, 1969–1978.

Peiris, M.K., Gunasekara, C.P., Jayaweera, P.M., Arachchi, N.D., Fernando, N., 2017. Biosynthesized silver nanoparticles: are they effective antimicrobials? Mem. Inst. Oswaldo Cruz 112 (8), 537–543.

Peretyazhko, T.S., Zhang, Q., Colvin, V.L., 2014. Size-controlled dissolution of silver nanoparticles at neutral and acidic pH conditions: kinetics and size changes. Environ. Sci. Technol. 48 (20), 11954–11961.

Petropoulos, S.A., Pereira, C., Tzortzakis, N., Barros, L., Ferreira, I.C.F.R., 2018. Nutritional value and bioactive compounds characterization of plant parts from *Cynara cardunculus* L. (Asteraceae) cultivated in central Greece. Front. Plant Sci. 9, 459.

Pirtarighat, S., Ghannadnia, M., Baghshahi, S., 2019. Biosynthesis of silver nanoparticles using *Ocimumbasilicum* cultured under controlled conditions for bactericidal application. Mater. Sci. Eng. C 98, 250–255.

Politano, A.D., Campbell, K.T., Rosenberger, L.H., Sawyer, R.G., 2013. Use of silver in the prevention and treatment of infections: silver review. Surg. Infect. (Larchmt.) 14 (1), 8–20.

Prabhu, S., Poulose, E.K., 2012. Silver nanoparticles: mechanism of antimicrobial action, synthesis, medical applications, and toxicity effects. Int. Nano Lett. 2 (1), 32.

Prasad, R., 2014. Synthesis of silver nanoparticles in photosynthetic plants. J. Nanopart. 2014, 963961.

Qing, Y., Cheng, L., Li, R., Liu, G., Zhang, Y., Tang, X., Wang, J., Liu, H., Qin, Y., 2018. Potential antibacterial mechanism of silver nanoparticles and the optimization of orthopedic implants by advanced modification technologies. Int. J. Nanomedicine 13, 3311–3327.

Rafique, M., Sadaf, I., Rafique, M.S., Tahir, M.B., 2016. A review on green synthesis of silver nanoparticles and their applications. Artif. Cells Nanomed. Biotechnol. 45 (7), 1272–1291.

Rajkumar, R., Ezhumalai, G., Gnanadesigan, M., 2021. A green approach for the synthesis of silver nanoparticles by *Chlorella vulgaris* and its application in photocatalytic dye degradation activity. Environ. Technol. Innov. 21, 101282.

Ramesh, P.S., Kokila, T., Geetha, D., 2015. Plant mediated green synthesis and antibacterial activity of silver nanoparticles using *Emblica officinalis* fruit extract. Spectrochim. Acta A Mol. Biomol. Spectrosc. 142, 339–343.

Rao, A., Schoenenberger, M., Gnecco, E., Glatzel, T., Meyer, E., Brändlin, D., Scandella, L., 2007. Characterization of nanoparticles using atomic force microscopy. J. Phys. Conf. Ser. 61, 971–976.

Rashmi, V., Sanjay, K.R., 2017. Green synthesis, characterisation and bioactivity of plant-mediated silver nanoparticles using *Decalepishamiltonii* root extract. IET Nanobiotechnol. 11 (3), 247–254.

Ratan, Z.A., Haidere, M.F., Nurunnabi, M., Shahriar, S.M., Ahammad, A., Shim, Y.Y., Reaney, M., Cho, J.Y., 2020. Green chemistry synthesis of silver nanoparticles and their potential anticancer effects. Cancers 12 (4), 855.

Rautela, A., Rani, J., Debnath (Das), M., 2019. Green synthesis of silver nanoparticles from *Tectona grandis* seeds extract: characterization and mechanism of antimicrobial action on different microorganisms. J. Anal. Sci. Technol. 10, 5.

Rauwel, P., Küünal, S., Ferdov, S., Rauwel, E., 2015. A review on the green synthesis of silver nanoparticles and their morphologies studied via TEM. Adv. Mater. Sci. Eng. 2015, 682749.

Salari, S., Esmaeilzadeh Bahabadi, S., Samzadeh-Kermani, A., Yosefzaei, F., 2019. *In-vitro* evaluation of antioxidant and antibacterial potential of green synthesized silver nanoparticles using *Prosopis farcta* fruit extract. Iran. J. Pharm. Res. 18 (1), 430–455.

Sasidharan, S., Chen, Y., Saravanan, D., Sundram, K.M., Yoga Latha, L., 2011. Extraction, isolation and characterization of bioactive compounds from plants' extracts. Afr. J. Tradit. Complement. Altern. Med. 8 (1), 1–10.

Schlegel, H.G., 1996. General Microbiology, seventh ed. Cambridge University Press.

Segneanu, A.E., Velciov, S.M., Olariu, S., Cziple, F., Damian, D., Grozescu, I., 2017. Bioactive molecules profile from natural compounds. In: Amino Acid—New Insights and Roles in Plant and Animal. IntechOpen Limited, pp. 209–228.

Shameli, K., Ahmad, M.B., Zamanian, A., Sangpour, P., Shabanzadeh, P., Abdollahi, Y., Zargar, M., 2012b. Green biosynthesis of silver nanoparticles using *Curcuma longa* tuber powder. Int. J. Nanomedicine 7, 5603–5610.

Shameli, K., Bin Ahmad, M., Jaffar Al-Mulla, E.A., Ibrahim, N.A., Shabanzadeh, P., Rustaiyan, A., Abdollahi, Y., Bagheri, S., Abdolmohammadi, S., Usman, M.S., Zidan, M., 2012a. Green biosynthesis of silver nanoparticles using *Callicarpa maingayi* stem bark extraction. Molecules 17 (7), 8506–8517.

Sharma, G., Nam, J.S., Sharma, A.R., Lee, S.S., 2018. Antimicrobial potential of silver nanoparticles synthesized using medicinal herb *Coptidis rhizome*. Molecules 23 (9), 2268.

Siddiqi, K.S., Husen, A., Rao, R., 2018. A review on biosynthesis of silver nanoparticles and their biocidal properties. J. Nanobiotechnol. 16 (1), 14.

Singh, J., 2017. Colloidal silver benefits, uses, dosage and side effects. In: Colloidal Silver. Retrieved September 19, 2018 from: https://www.ayurtimes.com/colloidal-silver.

Singh, I., 2019. Biosynthesis of silver nanoparticle from fungi, algae and bacteria. Eur. J. Biol. Res. 9 (1), 45–56.

Singh, R., Hano, C., Nath, G., Sharma, B., 2021. Green biosynthesis of silver nanoparticles using leaf extract of *Carissa carandas* L. and their antioxidant and antimicrobial activity against human pathogenic bacteria. Biomolecules 11 (2), 299.

Singh, K., Panghal, M., Kadyan, S., Chaudhary, U., Yadav, J.P., 2014. Green silver nanoparticles of *Phyllanthus amarus*: as an antibacterial agent against multi drug resistant clinical isolates of *Pseudomonas aeruginosa*. J. Nanobiotechnol. 12, 40.

Sivakumar, M., Surendar, S., Jayakumar, M., Seedevi, P., Sivasankar, P., Ravikumar, M., Anbazhagan, M., Murugan, T., Siddiqui, S.S., Loganathan, S., 2021. *Parthenium hysterophorus* mediated synthesis of silver nanoparticles and its evaluation of antibacterial and antineoplastic activity to combat liver cancer cells. J. Clust. Sci. 32 (1), 167–177.

Some, S., Bulut, O., Biswas, K., Kumar, A., Roy, A., Sen, I.K., Mandal, A., Franco, O.L., İnce, İ.A., Neog, K., Das, S., Pradhan, S., Dutta, S., Bhattacharjya, D., Saha, S., Das Mohapatra, P.K., Bhuimali, A., Unni, B.G., Kati, A., Mandal, A.K., Yilmaz, M.D., Ocsoy, I., 2019a. Effect of feed supplementation with biosynthesized silver nanoparticles using leaf extract of *Morus indica* L. V1 on *Bombyx mori* L. (Lepidoptera: Bombycidae). Sci. Rep. 9 (1), 14839.

Some, S., Ghosh, A., Mukherjee, J., Basu, D., 2017. Study of some traditionally important medicinal plants for primary healthcare by local people in the northern part of Malda district, West Bengal (India). Int. J. Curr. Res. Acad. Rev. 5 (7), 1–9.

Some, S., Sarkar, B., Biswas, K., Jana, T.K., Bhattacharjya, D., Dam, P., Mondal, R., Kumar, A., Deb, A.K., Sadat, A., Saha, S., Kati, A., Ocsoy, I., Franco, O.L., Mandal, A., Mandal, S., Mandal, A.K., İnce, İ.A., 2020. Bio-molecule functionalized rapid one-pot green synthesis of silver nanoparticles and their efficacy towards the multidrug resistant (MDR) gut bacteria of silkworms (*Bombyx mori*). RSC Adv. 10, 22742–22757.

Some, S., Sen, I.K., Mandal, A., Aslan, T., Ustun, Y., Yilmaz, E.S., Kati, A., Demirbas, A., Mandal, A.K., Ocsoy, I., 2019b. Biosynthesis of silver nanoparticles and their versatile antimicrobial properties. Mater. Res. Express 6, 012001.

Sorbiun, M., Shayegan Mehr, E., Ramazani, A., Mashhadi Malekzadeh, A., 2018. Biosynthesis of metallic nanoparticles using plant extracts and evaluation of their antibacterial properties. Nanochem. Res. 3 (1), 1–16.

Srikar, S., Giri, D., Pal, D., Mishra, P., Upadhyay, S., 2016. Green synthesis of silver nanoparticles: a review. Green Sustain. Chem. 6, 34–56.

Srirangam, G.M., Rao, K.P., 2017. Synthesis and characterization of silver nanoparticles from the leaf extract of *Malachra capitata* (L). Rasayan J. Chem. 10 (1), 46–53.

Suriyakala, G., Sathiyaraj, S., Gandhi, A.D., Vadakkan, K., Rao, U.M., Babujanarthanam, R., 2021. *Plumeria pudica* Jacq. flower extract-mediated silver nanoparticles: characterization and evaluation of biomedical applications. Inorg. Chem. Commun. 126, 108470.

Swargiary, G., Rawal, M., Singh, M., Mani, S., 2020. Molecular approaches to screen bioactive compounds from medicinal plants. In: Plant-Derived Bioactives. Springer, pp. 1–32.

Tadesse, D.A., Zhao, S., Tong, E., Ayers, S., Singh, A., Bartholomew, M.J., McDermott, P.F., 2012. Antimicrobial drug resistance in *Escherichia coli* from humans and food animals, United States, 1950-2002. Emerg. Infect. Dis. 18 (5), 741–749.

Takshak, S., 2018. Bioactive compounds in medicinal plants: a condensed review. SEJ Pharm. Nat. Med. 1 (1), 1–35.

Tavakoli, H., Rastegar, H., Taherian, M., Samadi, M., Rostami, H., 2017. The effect of nano-silver packaging in increasing the shelf life of nuts: an in vitro model. Ital. J. Food Saf. 6 (4), 6874.

Teow, S.Y., Wong, M.M., Yap, H.Y., Peh, S.C., Shameli, K., 2018. Bactericidal properties of plants-derived metal and metal oxide nanoparticles (NPs). Molecules 23 (6), 1366.

Thomas, B., Vithiya, B., Prasad, T., Mohamed, S.B., Magdalane, C.M., Kaviyarasu, K., Maaza, M., 2019. Antioxidant and photocatalytic activity of aqueous leaf extract mediated green synthesis of silver nanoparticles using *Passiflora edulis f. flavicarpa*. J. Nanosci. Nanotechnol. 19 (5), 2640–2648.

Traiwatcharanon, P., Timsorn, K., Wongchoosuk, C., 2017. Flexible room-temperature resistive humidity sensor based on silver nanoparticles. Mater. Res. Express 4, 085038.

Ullah, H., Ali, S., 2017. Classification of anti-bacterial agents and their functions. In: Antibacterial Agents. IntechOpen Limited.

Varadan, V.K., Pillai, A.S., Mukherji, D., Dwivedi, M., Chen, L., 2010. Nanoscience and Nanotechnology in Engineering. World Scientific, pp. 1–25.

Vijayan, R., Joseph, S., Mathew, B., 2019. Anticancer, antimicrobial, antioxidant, and catalytic activities of green-synthesized silver and gold nanoparticles using *Bauhinia purpurea* leaf extract. Bioprocess Biosyst. Eng. 42 (2), 305–319.

Wang, C., Kim, Y.J., Singh, P., Mathiyalagan, R., Jin, Y., Yang, D.C., 2016. Green synthesis of silver nanoparticles by *Bacillus methylotrophicus*, and their antimicrobial activity. Artif. Cells Nanomed. Biotechnol. 44 (4), 1127–1132.

Yadav, R., Saini, H., Kumar, D., Pasi, S., Agrawal, V., 2019. Bioengineering of *Piper longum* L. extract mediated silver nanoparticles and their potential biomedical applications. Mater. Sci. Eng. C 104, 109984.

Youssif, K.A., Haggag, E.G., Elshamy, A.M., Rabeh, M.A., Gabr, N.M., Seleem, A., Salem, M.A., Hussein, A.S., Krischke, M., Mueller, M.J., Abdelmohsen, U.R., 2019. Anti-Alzheimer potential, metabolomic profiling and molecular docking of green synthesized silver nanoparticles of *Lampranthuscoccineus* and *Malephora lutea* aqueous extracts. PLoS One 14 (11), e0223781.

Yu, Z., Li, Q., Wang, J., Yu, Y., Wang, Y., Zhou, Q., Li, P., 2020. Reactive oxygen species-related nanoparticle toxicity in the biomedical field. Nanoscale Res. Lett. 15 (1), 115.

Yuan, Y.G., Peng, Q.L., Gurunathan, S., 2017. Effects of silver nanoparticles on multiple drug-resistant strains of *Staphylococcus aureus* and *Pseudomonas aeruginosa* from mastitis-infected goats: an alternative approach for antimicrobial therapy. Int. J. Mol. Sci. 18 (3), 569.

Zargar, M., Hamid, A.A., Bakar, F.A., Shamsudin, M.N., Shameli, K., Jahanshiri, F., Farahani, F., 2011. Green synthesis and antibacterial effect of silver nanoparticles using *Vitex negundo* L. Molecules 16 (8), 6667–6676.

Zhang, F., 2017. Grand challenges for nanoscience and nanotechnology in energy and health. Front. Chem. 5, 80.

Zhang, X.F., Liu, Z.G., Shen, W., Gurunathan, S., 2016. Silver nanoparticles: synthesis, characterization, properties, applications, and therapeutic approaches. Int. J. Mol. Sci. 17 (9), 1534.

PART 4

Microbial synthesis

CHAPTER 14

Mycosynthesis of silver nanoparticles: Mechanism and applications

Jayshree Annamalai[a], Karuvelan Murugan[a], Jayashree Shanmugam[b], and Usharani Boopathy[c]

[a]Centre for Environmental Studies, Department of Civil Engineering, Anna University, CEG Campus, Chennai, India [b]Centre for Advanced Studies in Botany, School of Life Science, University of Madras, Guindy Campus, Chennai, India [c]Department of Biochemistry, Vels Institute of Science, Technology and Advanced Studies, Chennai, India

1 Introduction

In recent years, nanoparticles (NPs) have achieved a revolutionary change in nanoscience and nanotechnology in terms of "biosynthesis." Biosynthesis of NPs has been proven to be simple, cost-effective, size-controlled, and environmentally friendly approach involving biological systems and small biomolecules such as bacteria, fungi, algae, actinomycetes, plant extracts, amino acids, and vitamins (Zhao et al., 2018; Shahzad et al., 2019), while conventional chemical method of NP synthesis involve hazardous chemicals, high energy consumption, and increased manufacturing cost. Biosynthesis of NPs is a greener form of nanotechnology, providing greater surface area and higher catalytic activity to NPs than that of synthesized by chemical methods (Prasad, 2016). Apart from plant materials, microorganisms have been explored more in the past decade as biofactories for the synthesis of metal NPs upon reduction of metal ions. This is due to an effective and simple downstream NP recovery process where biomass extracts of microorganisms are used as reductants (Kubo et al., 2016; Romero et al., 2018; Paul and Roychoudhury, 2021).

Among most of the microorganisms used in biosynthesis process, fungi are considered to be more efficient and suitable for the synthesis of nanoscaled metal NPs. The reason for this might be the production of larger amount of proteins which in turn contributes to high productivity by acting as a reductant, stabilizing, and capping agent (Zhao et al., 2018). Other than this, fungal mycelia tends to withstand harsh environments, i.e., at diverse pH, temperatures, and substrate concentrations in a bioreactor or chamber, and is also easy to handle and harvest (Noor et al., 2020; Singh et al., 2014). Fungi are ubiquitous eukaryotic cells capable of surviving on a wide range of substrates as decomposers and are capable of producing numerous extracellular enzymes to hydrolyze complex macromolecules to simple ones (Bhargava et al., 2016). In addition, it is also reported that fungi isolated from metal-rich environment are a potential source for the reduction of metal salts, in turn initiating the biosynthesis of NPs (Jain et al., 2013).

Usually most of the indigenous microorganisms isolated from metal contaminated sites are capable of reducing different metal ions such as silver, gold, platinum, copper, zinc, titanium, and selenium to their NP forms; however, silver nanoparticles (AgNPs) have gained attention from many researchers over other NPs due to their wide application, cost-effectiveness, and unique properties (Iavicoli et al., 2012; Syed and Ahmad, 2012; Jain et al., 2013; Zhang et al., 2016; AbdelRahim et al., 2017; Song et al., 2017; Lv et al., 2018). AgNPs tend to have good conductivity, chemical stability, and catalytic and antibacterial activities, along with cytotoxic effect toward cancer cells (Lee et al., 2014; Borase et al., 2014). AgNPs find wide application in biomedical field (bandages, drug coating/delivery, and diagnosis), agriculture (pesticides), industries, cosmetics, textiles, water purification, and antibiotics (Ali et al., 2015; Khan et al., 2018; Danagoudar et al., 2021; Rai et al., 2021). Physicochemical properties and biological activities such as size, shape, surface morphology, particle composition, agglomeration, dissolution rate, particle reactivity, efficiency to release ions, biocompatibility, antibacterial potency, and anticancer and antioxidant activities determine the application suitability of AgNPs (Carlson et al., 2008).

Thus, in the present chapter, various fungal species are explored in the biosynthesis of AgNPs, along with the methods and mechanism involved, advantages of mycosynthesis over other biogenic sources, and application of AgNPs, and future prospects will be discussed.

2 Fungal species in silver nanoparticle synthesis

Fungi belong to the kingdom of multicellular eukaryotic organisms which are usually heterotrophs and have significant role in nutrient cycling, i.e., decomposition of waste and decaying matter in an ecosystem (Jasu et al., 2021). Extensive groups of fungi are used in the biosynthesis of AgNPs, such as *Aspergillus terreus*, *Aspergillus niger*, *Aspergillus clavatus*, *Aspergillus fumigatus*, *Aspergillus flavus*, *Agaricus bisporus*, *Alternaria alternata*, *Acremonium diospyri*, *Amylomyces rouxii*, *Chrysosporium keratinophilum*, *Cryptococcus humicola*, *Cladosporium cladosporioides*, *Chlamydomucor rouxii*, *Candida albicans*, *Candida glabrata*, *Epicoccum nigrum*, *Fusarium oxysporum*, *Fusarium semitectum*, *Fusarium solani*, *Ganoderma* sp., *Gloeophyllum abietinum*, *Neurospora intermedia*, *Pleurotus florida*, *Phomosis* sp., *Pleurotus sajor-caju*, *Penicillium verrucosum*, *Penicillium chrysogenum*, *Penicillium fellutanum*, *Phoma glomerata*, *Paecilomyces lilacinus*, *Phytophthora infestans*, *Pycnoporus sanguineus*, *Schizosaccharomyces pombe*, *Trichothecium roseum*, *Trichoderma harzianum*, *Trichoderma viride*, *Verticillium* sp., etc. (Khandel and Shahi, 2018; Khan et al., 2018). Among these fungal groups, recent research involving the mycogenic synthesis of AgNPs is discussed as follows.

In a study by Elamawi et al. (2018), *Trichoderma longibrachiatum* was used for an extracellular mycosynthesis of AgNPs where cell filtrate was used as a reducing and stabilizing agent. The NPs formed were monodispersed spherical in shape with mean diameter of 10 nm. In another study, AgNPs were also extracellularly biosynthesized using *Phanerochaete chrysosporium* MTCC-787. The synthesized NPs were spherical and oval shaped, with sizes ranging between 34 and 90 nm in diameter (Saravanan et al., 2018). Entomopathogenic fungi were also reported to be used in the synthesis of AgNPs and were efficiently used to control a rural malaria vector, *Anopheles culicifacies*. In a study by Amerasan et al. (2016), *Metarhizium anisopliae* filtrate was used to synthesize AgNPs upon reduction of $AgNO_3$ which were found to be rod shaped and 28–38 nm in size, while in a study by Qamandar and Shafeeq (2017), *Beauveria bassiana* (a parasite of insects) was used for an extracellular synthesis of AgNPs and in the study very effective secretion of extracellular enzymes served as a reducing agent. Synthesized NPs were found to be clustered and irregular in shape with sizes of around 89 nm. A saprophytic fungus, *Punctularia atropurpurascens* (H2126), was used to mediate extracellular biosynthesis of AgNPs, resulting in spherical NPs of ∼25 nm in diameter (Sanguinedo et al., 2018) (Fig. 1).

Studies on mycosynthesis of AgNPs reveals that most of the biogenic NPs are synthesized using fungal filtrate in presence of rich extracellular enzymes (Santos et al., 2021). As in the case of lignolytic fungi, *Trametes trogii* was used for the

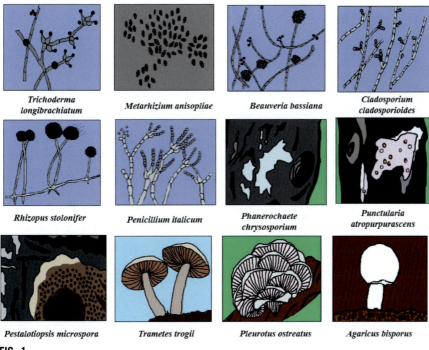

FIG. 1

Various groups of fungi in mycosynthesis of silver nanoparticles.

synthesis of AgNPs and the results revealed the formation of spherical, core shell, and elongated rod-shaped NPs. In a study by Hulikere and Joshi (2019), a marine endophytic fungus, *C. cladosporioides* of *Sargassum wightii*, was used and its extracellular biomolecules in fungal filtrate served as a reducing agent. The formed NPs were evenly spherical in shape with sizes ranging between 30 and 60 nm. In another study, AgNPs biosynthesized using an endophytic fungus, *Pestalotiopsis microspora*, mycelial filtrate were spherical and sizes ranged between 2 and 10 nm in diameter (Netala et al., 2016). Small spherical AgNPs of size 2.36 ± 0.3 nm were biosynthesized using mycelial filtrates of *Rhizopus stolonifer* by adjusting the temperature between 20°C and 60°C (AbdelRahim et al., 2017). In another study, *Penicillium italicum* was used to obtain extracellular AgNPs and these were assessed against multidrug-resistant pathogens; as a result, NPs of 33–47 nm in diameter were synthesized and also found to be effective against multidrug-resistant pathogens (Nayak et al., 2018). In the exploration of mycogenic synthesis of AgNPs, the edible mushroom *Pleurotus ostreatus* was also investigated and spherical NPs of <40 nm were synthesized (Al-Bahrani et al., 2017).

As described above, there are numerous fungal strains explored in the biosynthesis of AgNPs from the past decades to the time of writing. Reports on mycosynthesis of AgNPs suggest that most of synthesis process involve extracellular synthesis using mycelial filtrate of fungi. This might be due to rapid bioreduction of Ag^+ ions to Ag^0 in the presence of enzymes and bioactive molecules in mycelial extract and easy separation or recovery of AgNPs. In addition, enzymes and proteins in cell-free extracts act as capping and stabilizing agents for NPs (Netala et al., 2016). Table 1 illustrates the types of fungal species, optimal parameters, and nature of AgNPs mycosynthesized in recent years.

Table 1 Biosynthesis of silver nanoparticles using various types of fungi.

Fungi	Type of fungi	Mode of synthesis and optimal parameters for mycelial growth	Size and shape of nanoparticles	Reference
Bjerkandera sp. R1	White rot fungus	Mycelial filtrate was mixed with 1 mM $AgNO_3$ and reaction time was for about 144 h	Spherical shaped, 70–90 nm	Osorio-Echavarría et al. (2021)
Penicillium duclauxii	Seed-borne fungus isolated from corn grains	Fungal filtrate solution added to 1 mM $AgNO_3$ and incubated at 28°C, 200 rpm in dark	Mostly spherical, 3–32 nm in diameter	Almaary et al. (2020)

Table 1 Biosynthesis of silver nanoparticles using various types of fungi. *Continued*

Fungi	Type of fungi	Mode of synthesis and optimal parameters for mycelial growth	Size and shape of nanoparticles	Reference
Penicillium oxalicum	Soil fungus	Cell free fungal extract along with 1 mM AgNO$_3$ was incubated in dark for 72 h at 37°C along with agitation at 140 rpm	Spherical, 60–80 nm, few particles were agglomerated	Feroze et al. (2020)
Trametes trogii	White rot basidiomycete involved in wood decay, produces several ligninolytic enzymes	Extracellular, fungal mycelia grown at pH 8, fungal filtrate pH optimized to pH 13, 5 mM AgNO$_3$, incubated for 20 days at 28°C in dark, agitated at 100 rpm speed.	Clusters of NPs, spherical and elongated rods	Kobashigawa et al. (2019)
Cladosporium cladosporioides	Marine endophytic fungus of *Sargassum wightii* containing myriads of bioactive molecules	Fungal biomass grown in potato dextrose broth, fungal filtrate treated to AgNO$_3$ 1 mM, incubated for 48 h at room temperature	Even, spherical NPs, 30–60 nm	Hulikere and Joshi (2019)
Aspergillus niger	Soil-borne fungus isolated from rotten vegetables and fruits	Fungal filtrate added to 1 mM AgNO$_3$ at the ratio of 1:9, then incubated at 26°C for 48 h	Poly-dispersed nanoparticles with average size of 30 nm	Al-Zubaidi et al. (2019)

Continued

Table 1 Biosynthesis of silver nanoparticles using various types of fungi. *Continued*

Fungi	Type of fungi	Mode of synthesis and optimal parameters for mycelial growth	Size and shape of nanoparticles	Reference
Trichoderma longibrachiatum	Soil-borne fungus, adapted to industrial pollutants, secrete large amount of protein and metabolites	Extracellular, 10g of fungal biomass, 1mM of $AgNO_3$, temperature: 28°C, 72h of incubation without agitation	Spherical, ~10nm in diameter	Elamawi et al. (2018)
Punctularia atropurpurascens (H2126)	Saprophytic purplish fungus, grows on rotting wood. Often represented by aqueous exudate in the form of reddish droplets	Extracellular, fungal mycelia $0.1 g mL^{-1}$, 1mM $AgNO_3$, incubated for 72h at 28°C with agitation on an orbital shaker operating at 150rpm	Spherical, ~25nm	Sanguinedo et al. (2018)
Phanerochaete chrysosporium MTCC-787	White rot fungi, produces unique lignin degrading extracellular oxidative enzyme. This fungi is also involved in the degradation of toxic wastes, pesticides, and explosive contaminated materials	Extracellular, 10g of fungal biomass, 1mM of $AgNO_3$, temperature: 28°C, 72h of incubation in an orbital shaker operating at 200rpm	Spherical and oval, 34–90nm	Saravanan et al. (2018)
Penicillium italicum	Exhibits insect-microbe symbiosis, isolated from wasp nest soil	Mycelial filtrate incubated for 24h with 1mM $AgNO_3$	Clustered NPs, 33–46nm	Nayak et al. (2018)

Table 1 Biosynthesis of silver nanoparticles using various types of fungi. *Continued*

Fungi	Type of fungi	Mode of synthesis and optimal parameters for mycelial growth	Size and shape of nanoparticles	Reference
Rhizopus stolonifer	Saprophytic fungus isolated from naturally infected tomato fruits	Mycelial extract treated to 1 mM AgNO$_3$ and incubated at 20, 40 and 60°C for 2 days in orbital shaker agitated at 180 rpm	Small spherical monodispersed NPs, 25.89±3.8, 2.86±0.3, 48.43±5.2 nm at 20, 40, and 60°C	AbdelRahim et al. (2017)
Pleurotus ostreatus	Edible mushroom	Fresh basidiocarp aqueous extract was treated to 1 mM AgNO$_3$, incubated in dark for 40 h	Spherical, 10–40 nm NPs without aggregation	Al-Bahrani et al. (2017)
Beauveria bassiana	Soil fungus, causes muscadine disease in susceptible insects	Extracellular, fungal mycelia 10 g/L, 50 mM AgNO$_3$, incubated at 25°C for 120 h	Clustered, irregular shaped, 89 nm	Qamandar and Shafeeq (2017)
Agaricus bisporus	Edible mushroom	Fresh basidiocarp aqueous extract was treated to 1 mM AgNO$_3$, incubated in dark for 40 h	Spherical in shape with mean size of 27.45 nm	Awad et al. (2021)
Metarhizium anisopliae	Soil fungus capable of causing disease in insects; often termed as parasitoid	Extracellular, 10 g of fungal biomass, 1 mM of AgNO$_3$, temperature: 28°C, 72 h of incubation in an orbital shaker operating at 100 rpm	Rod, 28–38 nm	Amerasan et al. (2016)

Continued

Table 1 Biosynthesis of silver nanoparticles using various types of fungi. *Continued*

Fungi	Type of fungi	Mode of synthesis and optimal parameters for mycelial growth	Size and shape of nanoparticles	Reference
Pestalotiopsis microspora	Endophytic fungus isolated from the leaves of *Gymnema sylvestre*, bioactive compounds produced by fungi are alternative source for plant producing bioactive compounds	Macerated fungal biomass was treated to 1 mM $AgNO_3$ at the ratio of 1:9, incubated for 24 h at room temperature in dark	Monodispersed spherical NPs, 2–10 nm in diameter without aggregation	Netala et al. (2016)

3 Mechanism and methods of synthesis

Biosynthesis of AgNPs using fungi is evidenced to be a convenient, green, and ecofriendly method of synthesis. Since fungi biomasses are known to be easily developed and simpler to handle, mycosynthesis of NPs could be represented as economically viable compared to that of other microbial sources such as bacteria, algae, and actinomycetes. Studies suggest that treating fungal cells with metal salts reduces metal ions to metal NPs in response to their toxicity either by secreting reductase enzymes such as α-NADPH dependent reductases, nitrate-dependent reductases and others, or by biosorbing metal ions onto their cell surface (Alghuthaymia et al., 2015).

Biosynthesis of NPs using microbes could be illustrated as a defense mechanism exhibited by the cells to strive in metal-precursor amended environment, among which fungi are preferred for large-scale synthesis processes. The reason might be the diverse group of enzymes, proteins, peptides, and other biomolecules secreted by fungi, and these biomolecules readily acting as reducing agents. Napthoquinones, anthraquinones, and nitrate reductase tends to be specific to particular metal ions and exhibit their breakdown to the corresponding metal NPs (Kumar et al., 2007). Metal ions also tend to induce oxidative stress in fungal cells, which in turn secrete proteins as protective elements; however, these proteins determine the size and shape of NPs (Guilger-Casagrande and de Lima, 2019). Due to such dependency on proteins in controlling NP size and shape, in recent years there has been growing interest regarding such enzymatic processes involved in NP synthesis; research in synthesizing size-controlled NPs has also been carried out.

Mycosynthesis of AgNPs is of great potential and beneficial due to the elimination of toxic chemicals and its cost-effectiveness. The most common method in mycosynthesis of AgNPs involves treatment of either fungal mycelium or mycelial filtrate with $AgNO_3$ solution and incubation at optimal temperature and other optimal conditions; this could be termed as intracellular or extracellular synthesis based on accumulation of NPs inside or outside the fungal cell (Banerjee and Ravishankar Rai, 2018). These synthesis processes depend mainly on extracellular and intracellular enzymes, proteins, and other biomolecules that act as reducing and stabilizing agents. Intracellular synthesis involves binding of positively charged metal ions to negatively charged fungal cell walls, whereas extracellular synthesis involves responsible enzymes in the mycelium free medium, i.e., mycelial filtrate (Guilger-Casagrande and de Lima, 2019).

3.1 Intracellular synthesis

In the intracellular synthesis of NPs, initially the fungus is grown in a suitable medium at 25–28°C for 96h along with agitation (200rpm) in an orbital shaker. The grown mycelia is then separated from the culture broth by centrifuging at 5000rpm for 20min at 10°C and washed three times with distilled water. Ten grams of the harvested mycelial biomass is then resuspended in sterile distilled water, and in most of the studies 1mM of $AgNO_3$ is added to it. The reaction mixture is again incubated at 28°C for 72h along with agitation (200rpm) (Guilger-Casagrande and de Lima, 2019). This leads to the dispersion of Ag^+ ions all over the suspension, in turn being attracted toward the negatively charged functional groups along the cell wall. This results in nucleation and bioreduction of Ag^+ ions to Ag^0 ions, leading to NP synthesis (Fig. 2).

Several studies on intracellular synthesis of NPs have demonstrated the biosorption of Ag^+ ions onto the cell wall of fungus, resulting in diffusion of these metallic ions into the cytoplasm via active cellular pumps. This ATP-mediated transfer of Ag^+ ions is further followed by bioreduction of these metal ions to NPs. Formed NPs are also reported to be stabilized and capped in the presence of intracellular binding proteins such as open-amine groups and cysteine residues, which neutralizes the surface charges of the NPs (Asmathunisha and Kathiresan, 2013; Erasmus et al., 2014). It has also been suggested that metallic ions are heterogeneously distributed over the binding sites along the cell wall which determines the shape, size, and stability of the NPs (Kalabegishvili et al., 2015). The capping proteins prevent agglomeration of NPs and alter the properties of NPs, in turn significantly improving bioconjugation with other molecules. These capping proteins also provide stability to the NPs which is otherwise provided by the addition of toxic surfactants in the traditional method of NP synthesis (El-Deeb et al., 2013).

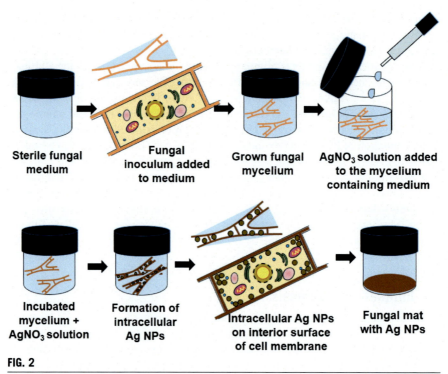

FIG. 2

Intracellular mycosynthesis of silver nanoparticles.

3.2 Extracellular synthesis

In the extracellular synthesis of AgNPs using mold type of fungi, initially the isolated or maintained fungal culture is grown aerobically in potato sucrose broth at 28°C along with agitation of 120 pm for 72 h. After the incubation period, fungal biomass is harvested by sieving it through a plastic sieve and filter paper. The resulting biomass is then washed three times using sterile distilled water to remove the residues of medium components. Biomass of about 20 g is mixed with 100 mL of double distilled water and again kept in agitation (120 rpm) for 72 h at 28°C. After the incubation period, mycelial filtrate is obtained by passing the mycelial suspension through Whatman filtrate paper no.1 and the filtrate is treated to 1 mM aqueous solution of $AgNO_3$ (Tyagi et al., 2019). The reaction mixture is incubated in the dark at 28°C for 48 h; the change from colorless to pale yellow followed by dark brown indicates the formation of AgNPs (Fig. 3).

In the case of basidiomycota fungus or mushrooms, basidiocarps are sliced, oven dried at 45°C for 48 h, and ground to obtain fine powder. Ten grams of ground powder is then boiled in distilled water for 30 min at 60°C at 1:10 ratio (w/v) (Raman et al., 2015). The boiled suspension is cooled, filtered, centrifuged at 10,000 rpm

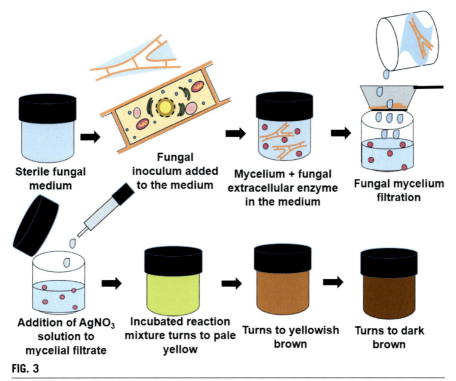

FIG. 3

Extracellular mycosynthesis of silver nanoparticles.

for 10 min, and finally filtered through Whatman filter paper. The filtrate thus obtained is freeze dried and termed a hot water extract; it is used for the mycosynthesis of AgNPs. Varying the concentration of the hot water extract ($1–5\,mg\,mL^{-1}$), 10 mM $AgNO_3$ solution is added and incubated at 25°C for 48 h under continuous stirring (Gurunathan et al., 2013; Owaid et al., 2015). A change in color from pale yellow to brownish yellow followed by dark brown indicates the formation of AgNPs.

In the extracellular synthesis of AgNPs, binding of metal ions onto the outer surface of the cells and reduction of these metal ions in the presence of enzymes and biomolecules occur. In comparison to intracellular synthesis, the extracellular synthesis method has been suggested as preferable; this might be due to simpler downstream processing (Siddiqi et al., 2018). Research studies also suggest that NPs synthesized by extracellular process are stabilized by the enzymes and proteins secreted by the fungal biomass. The most commonly detected enzyme is NADH-dependent reductase, which is found in its native form in aqueous mycelial extract as well as on the surface of NPs in bound form (Mohapatra et al., 2015).

4 Advantages over other biosynthesis processes

As silver nanoparticles are a beneficial and crucial element in the biomedical field, worldwide researchers aim to develop environmentally friendly and less toxic NPs retaining unique physicochemical and biological properties. In this search, biogenic syntheses of various metal NPs including AgNPs using biological and microbial entities such as plants, fungi, bacteria, algae, and actinomycetes have provided greater options for a green chemistry-based NP synthesis approach. Application of these plant materials and microbial cells as a whole provides a simple, cost-effective, reliable, ecofriendly, and large-scale production platform (Govindappa et al., 2016). However, fungal biomass proves to be convenient in nature for handling, culturing, and harvesting; in addition, mycosynthesis of NPs is an economically viable process compared to all other microbial sources. The mycelial mats provide a larger surface area during the intracellular synthesis process, which in turn improves the interactions of metal ions and cell wall-bound biomolecules (Espinosa-Ortiz et al., 2015). Fungal cells are also suggested to produce larger quantities of enzymes that are significantly responsible for the bioreduction of metal ions encountered during the NP synthesis process.

The most common fungal enzymes involved in bioreduction of metal ions include oxidoreductase enzymes: NADH-dependent reductase, nitrate reductase, hydrogenase, and others such as laccases, liginocellulase, tyrosinases, peroxidases, etc. (Balakumaran et al., 2015; Golinska et al., 2016; Vetchinkina et al., 2017). Apart from enzymes, proteins, peptides, quinones, hydroxyquinoline, and other biomolecules responsible for the stability of NPs, fungi also tend to show slower kinetics, offering better manipulation options in size- and shape-controlled synthesis of NPs along with long-term stability. However, in the intracellular mycosynthesis of NPs, most of the NPs attach to the biomass cell wall, which makes the separation or downstreaming process difficult (Zhao et al., 2018). In addition, NPs that are synthesized extracellularly have a wide range of particle sizes (1–200 nm), while those produced by the intracellular process have a shorter range of particle sizes (25–60 nm) (Korbekandi et al., 2013). Thus, the research reports state that extracellular mycosynthesis of AgNPs paves the way for large-scale production of NPs since they have higher bioaccumulation capacity for NPs due to elevated tolerance levels and allow simpler downstream processing (Abdel-Hafez et al., 2016).

5 Application and future prospects

Mycogenically synthesized nanoparticles in a wide range by many researchers globally have been assessed for their various applications in the medical and pest management fields. Applications and efficiency of AgNPs that have been synthesized in recent years using fungal biomass are discussed as follows.

In a study by Xue et al. (2016), *Arthroderma fulvum* isolated from soil was used for the biosynthesis of AgNPs and was assessed for fungicidal properties against fungal pathogens: *C. albicans*, *Candida parapsilosis*, *Candida krusei*, *Candida tropicalis*,

A. fumigatus, *A. flavus*, *A. terreus*, *F. solani*, *Fusarium moniliforme*, and *F. oxysporum*. Minimum inhibitory concentrations (MICs) of AgNPs against the fungal pathogens were in the range of 0.125–4.00 µg mL^{-1}, whereas MICs for flucanozole and itraconazole ranged between 0.250–16.00 and 0.030–0.250 µg mL^{-1}. The results strongly suggested that biosynthesized AgNPs were potent antifungal agents and were capable of treating fungal infections. Antibacterial activities of mycosynthesized AgNPs are also proven to be potential. Ottoni et al. (2017) isolated 20 different filamentous fungi from sugar cane plantation soil and assessed their antibacterial activity against pathogenic bacterial strains. The results revealed that AgNPs biosynthesized using three fungal strains—*Rhizopus arrhizus*, *Trichoderma gamsii*, and *A. niger*—were found to be active against *Escherichia coli*, *Staphylococcus aureus*, and *Pseudomonas aeruginosa*.

Antimicrobial activity exhibited by AgNPs tends to be the synergy of multiple bactericidal mechanisms contributing to a broad spectrum of activity against various bacterial cells. Certain mechanisms involved in encountering bacterial cells include: formation of insoluble compounds by inactivating sulfhydryl groups in the cell walls; binding to external proteins and creating pores, followed by disruption of membrane-bound enzymes and lipids causing cell lysis; interfering with the DNA repair mechanism and replication process; and inducing the formation of reactive oxygen species (hydrogen peroxide, superoxide anions, and hydroxyl radicals) leading to oxidative stress (Duncan, 2011; Duran et al., 2016) (Fig. 4). Apart from these mechanisms, the size of AgNPs also contributes significantly in terms of bacterial toxicity; the smaller the NPs, the greater the bactericidal efficiency (Rahimi et al., 2016; Ottoni et al., 2017). In a study by Ismail et al. (2021), AgNPs synthesized using *Penicillium expansum* of 9–18 nm in diameter were irregular, spherical, hexagonal, and monodispersed; these mycosynthesized NPs had significant antibacterial and anticancer properties against methicillin-resistant *S. aureus* RCMB 010010 and breast cancer (MCF-7) cell lines.

Antimicrobial property of AgNPs were also been extended in assessing the efficiency of NPs in dental root-canal disinfection and endodontic and periodontal management. In a study by Halkai et al. (2018), AgNPs were biosynthesized using *F. semitectum* isolated from healthy leaves of *Withania somnifera*. Antibacterial activity of mycosynthesized AgNPs against *Enterococcus faecalis* biofilm was assessed by determining the minimum inhibitory concentration. In the experiment, 30 root dentin blocks were prepared using human extracted single-rooted teeth and were inoculated with *E. faecalis* in trypticase soy broth for 2 weeks, resulting in biofilm formation. Later, the dentin blocks were rinsed in saline, and serial dilution was carried out and the blocks were treated with mycosynthesized AgNPs. The results revealed that NPs exhibited effective antibacterial activity against *E. faecalis*, suggesting their application in treating root-canal infection. These NPs also found applications against other endo-perio pathogens such as *Bacillus pumilus* and *Porphyromonas gingivalis* (Halkai et al., 2017).

In addition to antimicrobial activity, AgNPs are potential antioxidants and cytotoxic agents; this was well proven in vitro by performing antioxidant assays and 3-(4,5-dimethylthiazol-2-yl)-2,5-diphenyltetrazolium bromide (MTT) assays

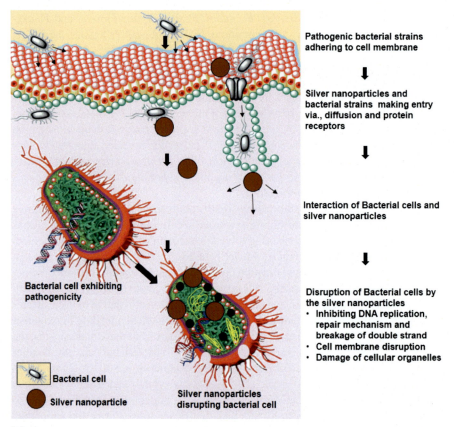

FIG. 4

Mechanism of toxicity exhibited by silver nanoparticles against pathogenic bacterial cells.

(Govindappa et al., 2016; Hulikere and Joshi, 2019). In a study by Taha et al. (2019), AgNPs were synthesized using *P. italicum* isolated from Iraqi lemon fruits. The sizes of the NPs ranged between 32 and 100 nm and the NPs were assessed for antimicrobial, antioxidant, and cytotoxic activity. Antimicrobial activity results revealed that AgNPs had both bactericidal and fungicidal properties against *S. aureus*, *E. coli*, *C. albicans*, and *C. tropicalis* in concentrations ranging from 40 to 100 $\mu g\,mL^{-1}$. Antioxidant activity was carried out by performing a 2,2-diphenyl-1-picrylhydrazyl (DPPH) assay, a resazurin dye scavenging assay, and a hydroxyl radical scavenging assay. The results revealed that 30 $\mu g\,mL^{-1}$ of AgNPs had 60% of DPPH radical scavenging compared to that of ascorbic acid, and showed 50% of inhibition activity upon resazurin radical scavenging and 53% upon H_2O_2 radical scavenging. Cytotoxic activity against breast cancer (MCF-7) cell lines was found to be maximal up to 87% at 80 $\mu g\,mL^{-1}$ of AgNPs, and 16% in the case of control at 5 $\mu g\,mL^{-1}$ concentration (Taha et al., 2019).

In a recent study on fermentable yeasts, *Pichia kudriavzevii* HA-NY2 and *Saccharomyces uvarum* HA-NY3 were used for extracellular biosynthesis of AgNPs and were also found to have biomedical applications. These biosynthesized AgNPs were well-dispersed round and cubic regular particles with sizes ranging between 12.4 and 20.6 nm. The nanoparticles exhibited significant antimicrobial properties against both gram-positive (*S. aureus* ATCC29213) and gram-negative (*P. aeruginosa* ATCC27953) bacterial strains and fungal strains (*C. tropicalis* ATCC750 and *F. oxysporum* NRC21). In addition, NPs also exhibited antiinflammatory properties by inhibiting paw-edema upon oral administration and anticancer activity against colon cancer (HCT-116) and prostate cancer (PC$_3$) cell lines, with IC$_{50}$ values ranging between 0.24–0.29 and 0.50–0.57 μg mL^{-1} (Ammar et al., 2021). These abilities of AgNPs make them suitable for application in drug formulation in the pharmaceutical industry; they would also serve in the management of various deleterious diseases (Govindappa et al., 2016; Netala et al., 2016; Rehman et al., 2021).

The role of mycosynthesized AgNPs in pest management has also been established to be noteworthy by several researchers during the past decade. In a study by Amerasan et al. (2016), AgNPs were synthesized extracellularly using an entomopathogenic fungus, *M. anisopliae*, and the NPs were evaluated to control a rural malaria vector, *A. culicifacies*. The results revealed that AgNPs were toxic and caused detrimental effects on larval and pupal development of *A. culicifacies* at lower dosage with EC$_{50}$ at 14.9 ppm; the larvicidal activity was proportional to the concentration of AgNPs, and LC$_{50}$ values ranged between 32.8 (I instar) and 60 ppm (pupa). In another study, mycosynthesized AgNPs using filamentous fungi, *P. verrucosum*, were also proven to exhibit toxicity against larvae and pupae of *Culex quinquefasciatus* with LC$_{50}$ of 4.91 ppm (I instar), 5.16 ppm (II instar), 5.95 ppm (III instar), 6.87 ppm (IV instar), and 7.83 ppm (pupa) (Kamalakannan et al., 2014). Other entomopathogenic fungi that were used in mycosynthesis of AgNPs and assessed for mosquitocidal effects against larval instars of *C. quinquefasciatus* and *Aedes aegypti* include *B. bassiana* and *Isaria fumosorosea* (Banu and Balasubramanian, 2014). The mosquitocidal properties of AgNPs might be due to the synergetic insecticidal action of AgNPs and capping molecules onto NPs that tend to cause structural deformation of genetic material, inhibit digestive enzyme secretion, and induce reactive oxygen species generation (Patil et al., 2012; Rajan et al., 2015). Thus, mycosynthesized AgNPs is a single-step approach that could serve the purpose of being larvicidal and pupicidal agent, in turn enhancing mosquito control programs and even other pest management programs in an ecofriendly manner.

6 Future prospects

As mycosynthesis of AgNPs is efficient and economically approachable, in future application of mycosynthesized NPs in various fields such as electronics, textile, food, cosmetics, agriculture, and optics will be possible, in addition to the biomedical and pest management fields. However, at present, applications of mycosynthesized

NPs are at laboratory scale and have only been demonstrated in in vitro conditions. There is a need to understand the mechanism of NP synthesis in and out of mycelial cells and other biological elements. A clear understanding on enzymes, proteins, and biomolecules in the biosynthesis of NPs will lead to optimization of mycosynthesis processes; in turn, NPs of desired sizes and shapes would be obtained. Compared to other microbial synthesis, understanding the mechanism of Ag and other metal NPs using fungi is complex, since in eukaryotic organisms, identification of genes involved in bioreduction of metal ions via specific enzymatic pathways is difficult. In the case of harvesting or downstream processing of Ag and other metal NPs synthesized using fungal mycelial filtrate, a purification step might be required. This is due to the presence of spores and myco-based biomolecule dispersion in filtrate, which may tend to cause infections, diseases, and inflammatory actions.

7 Conclusion

Mycosynthesis of AgNPs seems to be simple, cost-effective, and economically viable; however, the clear bioreduction mechanism, safety, and risk assessment of mycosynthesized NPs are not clearly understood. It is hypothesized that fungal mycelial cells are rich in proteins and biomolecules, and these act as stabilizing and capping agents; mycosynthesized NPs appear in clusters or agglomerate to larger-sized NPs. This hinders the application of mycosynthesized NPs due to size-dependent aspects. Apart from this, although mycosynthesis of AgNPs appears to be an ecofriendly method, and in comparison with other biosynthesis processes, it is preferred, disposal of all metal NPs in the environment and its life-threatening consequences remain the same. There is an urgent need for government bodies of all developing and developed nations to set strict regulations for environment and health protection agencies, and research institutes, in order to prevent over exposure and accumulation of NPs in biological tissues. Thus, ecofriendly mycosynthesis of silver and other NPs should be accomplished on the basis of product purity and safer disposal.

References

Abdel-Hafez, S.I., Nafady, N.A., Abdel-Rahim, I.R., Shaltout, A.M., Mohamed, M.A., 2016. Biogenesis and optimisation of silver nanoparticles by the endophytic fungus *Cladosporium sphaerospermum*. Int. J. Nano Chem. 2, 11–19.

AbdelRahim, K., Mahmoud, S.Y., Ali, A.M., Almaary, K.S., Mustafa, A.E.M.A., Husseiny, S.M., 2017. Extracellular biosynthesis of silver nanoparticles using *Rhizopus stolonifer*. Saudi J. Biol. Sci. 24, 208–216.

Al-Bahrani, R., Raman, J., Lakshmanan, H., Hassan, A.A., Sabaratnam, V., 2017. Green synthesis of silver nanoparticles using tree oyster mushroom *Pleurotus ostreatus* and its inhibitory activity against pathogenic bacteria. Mater. Lett. 186, 21–25.

Alghuthaymia, M.A., Almoammarb, H., Raic, M., Said-Galievd, E., Abd-Elsalam, K.A., 2015. Myconanoparticles: synthesis and their role in phytopathogens management. Biotechnol. Biotechnol. Equip. 29, 221.

Ali, S.M., Yousef, N.M.H., Nafady, N.A., 2015. Application of biosynthesized silver nanoparticles for the control of land snail *Eobania vermiculata* and some plant pathogenic fungi. J. Nanomater. 2015, 1–10.

Almaary, K.S., Sayed, S.R.M., Abd-Elkader, O.H., Dawoud, T.M., El-Orabi, N.F., Elgorban, A.M., 2020. Complete green synthesis of silver nanoparticles applying seed-borne *Penicillium duclauxii*. Saudi J. Biol. Sci. 27, 1333–1339.

Al-Zubaidi, S., Al-Ayafi, A., Abdelkader, H., 2019. Biosynthesis, characterization and antifungal activity of silver nanoparticles by *Aspergillus niger* isolate. J. Nanotechnol. Res. 1, 023–036.

Amerasan, D., Nataraj, T., Murugan, K., Panneerselvam, C., Madhiyazhagan, P., Nicoletti, M., Benelli, G., 2016. Myco-synthesis of silver nanoparticles using *Metarhizium anisopliae* against the rural malaria vector *Anopheles culicifacies* Giles (Diptera: Culicidae). J. Pestic. Sci. 89, 249–256.

Ammar, H.A., El Aty, A.A.A., El Awdan, S.A., 2021. Extracellular myco-synthesis of nano-silver using the fermentable yeasts *Pichia kudriavzevii* HA-NY2 and *Saccharomyces uvarum* HA-NY3, and their effective biomedical applications. Bioprocess Biosyst. Eng. https://doi.org/10.1007/s00449-020-02494-3.

Asmathunisha, N., Kathiresan, K., 2013. A review on biosynthesis of nanoparticles by marine organisms. Colloids Surf. B Biointerfaces 103, 283–287.

Awad, M., Yosri, M., Abdel-Aziz, M.M., Younis, A.M., Sidkey, N.M., 2021. Assessment of the antibacterial potential of biosynthesized silver nanoparticles combined with vancomycin against methicillin-resistant *Staphylococcus aureus*–induced infection in rats. Biol. Trace Elem. Res. https://doi.org/10.1007/s12011-020-02561-6.

Balakumaran, M., Ramachandran, R., Kalaichelvan, P., 2015. Exploitation of endophytic fungus, *Guignardia mangiferae* for extracellular synthesis of silver nanoparticles and their *in vitro* biological activities. Microbiol. Res. 178, 9–17.

Banerjee, K., Ravishankar Rai, V., 2018. A review on mycosynthesis, mechanism, and characterization of silver and gold nanoparticles. BioNanoScience 8, 17–31.

Banu, A.N., Balasubramanian, C., 2014. Optimization and synthesis of silver nanoparticles using *Isaria fumosorosea* against human vector mosquitoes. Parasitol. Res. 113, 3843–3851.

Bhargava, A., Jain, N., Khan, M.A., Pareek, V., Dilip, R.V., Panwar, J., 2016. Utilizing metal tolerance potential of soil fungus for efficient synthesis of gold nanoparticles with superior catalytic activity for degradation of rhodamine B. J. Environ. Manage. 183, 22–32.

Borase, H.P., Salunke, B.K., Salunkhe, R.B., Patil, C.D., Hallsworth, J.E., Kim, B.S., Patil, S.V., 2014. Plant extract: a promising biomatrix for ecofriendly, controlled synthesis of silver nanoparticles. Appl. Biochem. Biotechnol. 173, 1–29.

Carlson, C., Hussain, S.M., Schrand, A.M., Braydich-Stolle, L.K., Hess, K.L., Jones, R.L., Schlager, J.J., 2008. Unique cellular interaction of silver nanoparticles: size-dependent generation of reactive oxygen species. J. Phys. Chem. B 112, 13608–13619.

Danagoudar, A., Pratap, G.K., Shantaram, M., Ghosh, K., Kanade, S.R., Sannegowda, L.K., Joshi, C.G., 2021. Antioxidant, cytotoxic and anti-choline esterase activity of green silver nanoparticles synthesized using *Aspergillus austroafricanus* CGJ-B3 (Endophytic Fungus). J. Anal. Chem. Lett. 11 (1), 15–18.

Duncan, T.V., 2011. Applications of nanotechnology in food packaging and food safety: barrier materials, antimicrobials and sensors. J. Colloid Interface Sci. 363, 1–24.

Duran, N., Nakazato, G., Seabra, A.B., 2016. Antimicrobial activity of biogenic silver nanoparticles, and silver chloride nanoparticles: an overview and comments. Appl. Microbiol. Biotechnol. 100, 6555–6570.

Elamawi, R.M., Al-Harbi, R.E., Hendi, A.A., 2018. Biosynthesis and characterization of silver nanoparticles using *Trichoderma longibrachiatum* and their effect on phytopathogenic fungi. Egypt. J. Biol. Pest Co. 28, 1–18.

El-Deeb, B., Mostafa, N.Y., Altalhi, A., Gherbawy, Y., 2013. Extracellular biosynthesis of silver nanoparticles by bacteria *Alcaligenes faecalis* with highly efficient anti-microbial property. Int. J. Chem. Eng. 30, 1137–1144.

Erasmus, M., Cason, E.D., van Marwijk, J., Botes, E., Gericke, M., van Heerden, E., 2014. Gold nanoparticle synthesis using the thermophilic bacterium *Thermus scotoductus* SA-01 and the purification and characterization of its unusual gold reducing protein. Gold Bull. 47, 245–253.

Espinosa-Ortiz, E.J., Gonzalez-Gil, G., Saikaly, P.E., Van Hullebusch, E.D., Lens, P.N., 2015. Effects of selenium oxyanions on the white-rot fungus *Phanerochaete chrysosporium*. Appl. Microbiol. Biotechnol. 99, 2405–2418.

Feroze, N., Arshad, B., Younas, M., Afridi, M.I., Saqib, S., Ayaz, A., 2020. Fungal mediated synthesis of silver nanoparticles and evaluation of antibacterial activity. Microsc. Res. Tech. 83, 72–80.

Golinska, P., Rathod, D., Wypij, M., Gupta, I., Składanowski, M., Paralikar, P., et al., 2016. Mycoendophytes as efficient synthesizers of bionanoparticles: nanoantimicrobials, mechanism, and cytotoxicity. Crit. Rev. Biotechnol. 17, 1–14.

Govindappa, M., Farheen, H., Chandrappa, C.P., Rai, R.V., Raghavendra, V.B., 2016. Mycosynthesis of silver nanoparticles using extract of endophytic fungi, *Penicillium* species of *Glycosmis mauritiana*, and its antioxidant, antimicrobial, anti-inflammatory and tyrokinase inhibitory activity. Adv. Nat. Sci. Nanosci. Nanotechnol. 7, 035014.

Guilger-Casagrande, M., de Lima, R., 2019. Synthesis of silver nanoparticles mediated by fungi: a review. Front. Bioeng. Biotechnol. 7, 287.

Gurunathan, S., Raman, J., AbdMalek, S.N., John, P.A., Vikineswary, S., 2013. Green synthesis of silver nanoparticles using *Ganoderma neo-japonicum* Imazeki: a potential cytotoxic agent against breast cancer cells. Int. J. Nanomedicine 8, 4399–4413.

Halkai, K.R., Mudda, J.A., Shivanna, V., Rathod, V., Halkai, R.S., 2017. Evaluation of antibacterial efficacy of biosynthesized silver nanoparticles derived from fungi against endoperio pathogens *Porphyromonas gingivalis*, *Bacillus pumilus*, and *Enterococcus faecalis*. J. Conserv. Dent. 20, 398–404.

Halkai, K.R., Mudda, J.A., Shivanna, V., Rathod, V., Halkai, R., 2018. Antibacterial efficacy of biosynthesized silver nanoparticles against *Enterococcus faecalis* biofilm: an *in vitro* study. Contemp. Clin. Dent. 9, 237–241.

Hulikere, M.M., Joshi, C.G., 2019. Characterization, antioxidant and antimicrobial activity of silver nanoparticles synthesized using marine endophytic fungus-*Cladosporium cladosporioides*. Process Biochem. 82, 199–204.

Iavicoli, I., Leso, V., Bergamaschi, A., 2012. Toxicological effects of titanium dioxide nanoparticles: a review of in vivo studies. J. Nanomater. 2012, 1–36.

Ismail, R.S., El-Sharkawy, R.M., Amin, B.H., Swelim, M.A., 2021. Mycosynthesize of Ag-nanoparticles by *Penicillium expansum* and its antibacterial activity against bacterial pathogens. PCBMB 22 (9–10), 13–25.

Jain, N., Bhargava, A., Tarafdar, J.C., Singh, S.K., Panwar, J., 2013. A biomimetic approach towards synthesis of zinc oxide nanoparticles. Appl. Microbiol. Biotechnol. 97, 859–869.

Jasu, A., Lahiri, D., Nag, M., Ray, R.R., 2021. Fungi Bio-Prospects in Sustainable Agriculture, Environment and Nano-Technology. Academic Press, Elsevier.

Kalabegishvili, T.L., Murusidze, I.G., Kirkesali, E.I., Rcheulishvili, A.N., Ginturi, E.N., Gelagutashvili, E.S., Kuchava, N.E., Bagdavadez, N.V., Janjalia, M.V., Pataraya, D.T., Gurielidze, M.A., Frontasyeva, M.V., Zinicovacaia, I.I., Pavlov, S.S., Tsertsvadze, Gabunia, V.N., 2015. Possibilities of physical methods in development of microbial nanotechnology. Eur. Chem. Bull. 4, 43–49.

Kamalakannan, S., Gobinath, C., Ananth, S., 2014. Synthesis and characterization of fungus mediated silver nanoparticle for toxicity on filarial vector, *Culex quinquefasciatus*. Int. J. Pharm. Sci. Rev. Res. 24, 124–132.

Khan, A.U., Malik, N., Khan, M., Cho, M.H., Khan, M.M., 2018. Fungi-assisted silver nanoparticle synthesis and their applications. Bioprocess Biosyst. Eng. 41, 1–20.

Khandel, P., Shahi, S.K., 2018. Mycogenic nanoparticles and their bio-prospective applications: current status and future challenges. J. Nanostructure Chem. 8, 369–391.

Kobashigawa, J.M., Robles, C.A., Ricci, M.L.M., Carmran, C.C., 2019. Influence of strong bases on the synthesis of silver nanoparticles (AgNPs) using the ligninolytic fungi *Trametes trogii*. Saudi J. Biol. Sci. 26, 1331–1337.

Korbekandi, Z.A.H., Iravani, S., Abbasi, S., 2013. Optimization of biological synthesis of silver nanoparticles using *Fusarium oxysporum*. Iran J. Pharm. Res. 12, 289–298.

Kubo, A.M., Gorup, L.F., Amaral, L.S., Filho, E.R., Camargo, E.R., 2016. Kinetic control of microtubule morphology obtained by assembling gold nanoparticles on living fungal biotemplates. Bioconjug. Chem. 27, 2337–2345.

Kumar, S.A., Abyaneh, M.K., Gosavi, S.W., Kulkarni, S.K., Pasricha, R., Ahmad, A., Khan, M.I., 2007. Nitrate reductase mediated synthesis of silver nanoparticles from $AgNO_3$. Biotechnol. Lett. 29, 439.

Lee, S.H., Salunke, B.K., Kim, B.S., 2014. Sucrose density gradient centrifugation separation of gold and silver nanoparticles synthesized using *Magnolia kobus* plant leaf extracts. Biotechnol. Bioprocess Eng. 19, 378.

Lv, Q., Zhang, B., Xing, X., Zhao, Y., Cai, R., Wang, W., Gu, Q., 2018. Biosynthesis of copper nanoparticles using *Shewanella loihica* PV-4 with antibacterial activity: novel approach and mechanisms investigation. J. Hazard. Mater. 347, 141–149.

Mohapatra, B., Kuriakose, S., Mohapatra, S., 2015. Rapid green synthesis of silver nanoparticles and nanorods using *Piper nigrum* extract. J. Alloys Compd. 637, 119–126.

Nayak, B.K., Nanda, A., Prabhakar, V., 2018. Biogenic synthesis of silver nanoparticle from wasp nest soil fungus, *Penicillium italicum* and its analysis against multi drug resistance pathogens. Biocatal. Agric. Biotechnol. 16, 412–418.

Netala, V.R., Bethu, M.S., Pushpalatha, B., Baki, V.B., Aishwarya, S., Rao, J.V., Tartte, V., 2016. Biogenesis of silver nanoparticles using endophytic fungus *Pestalotiopsis microspora* and evaluation of their antioxidant and anticancer activities. Int. J. Nanomedicine 11, 5683–5696.

Noor, S., Shah, Z., Javed, A., Ali, A., Hussain, S.B., Zafar, S., Ali, H., Muhammad, S.A., 2020. A fungal based synthesis method for copper nanoparticles with the determination of anticancer, antidiabetic and antibacterial activities. J. Microbiol. Methods 174, 105966.

Osorio-Echavarría, J., Osorio-Echavarría, J., Ossa-Orozco, C.P., Gomez-Vanegasa, N.A., 2021. Synthesis of silver nanoparticles using white-rot fungus anamorphous *Bjerkandera* sp. R1: influence of silver nitrate concentration and fungus growth time. Sci. Rep. 11, 3842.

Ottoni, C.A., Simoes, M.F., Fernandes, S., dos Santos, J.M., da Silva, E.S., de Souza, R.F.B., Maiorano, A.E., 2017. Screening of filamentous fungi for antimicrobial silver nanoparticles synthesis. AMB Expr. 7, 31.

Owaid, M.N., Raman, J., Lakshmanan, H., Al-Saeedi, S.S.S., Sabaratnam, V., Abed, I.A., 2015. Mycosynthesis of silver nanoparticles by *Pleurotus cornucopiae* var. *citrinopileatus* and its inhibitory effects against *Candida* sp. Mater. Lett. 53, 186–190.

Patil, S.V., Borase, H.P., Patil, C.D., Salunke, B.K., 2012. Biosynthesis of silver nanoparticles using latex from few Euphorbian plants and their antimicrobial potential. Appl. Biochem. Biotechnol. 167, 776–790.

Paul, A., Roychoudhury, A., 2021. Go green to protect plants: repurposing the antimicrobial activity of biosynthesized silver nanoparticles to combat phytopathogens. Nanotechnol. Environ. Eng. 6, 10.

Prasad, R., 2016. Advances and applications through fungal nanobiotechnology. In: Fungal Biology. Springer International Publishing, Switzerland.

Qamandar, M.A., Shafeeq, M.A.A., 2017. Biosynthesis and properties of silver nanoparticles of fungus *Beauveria bassiana*. Int. J. ChemTech Res. 10, 1073–1083.

Rahimi, G., Alizadeh, F., Khodavandi, A., 2016. Mycosynthesis of silver nanoparticles from *Candida albicans* and its antibacterial activity against *Escherichia coli* and *Staphylococcus aureus*. Trop. J. Pharm. Res. 15, 371–375.

Rai, M., Bonde, S., Golinska, P., Trzinska-Wencel, J., Gade, A., Abd-Elsalam, K.A., Shend, S., Gaiwad, S., Ingle, A.P., 2021. Fusarium as a novel fungus for the synthesis of nanoparticles: mechanism and applications. J. Fungi 7 (2), 139.

Rajan, R., Chandran, K., Harper, S.L., Yun, S.I., Kalaichelvan, P.T., 2015. Plant extract synthesized nanoparticles: an ongoing source of novel biocompatible materials. Ind. Crop Prod. 70, 356–373.

Raman, J., Reddy, G.R., Lakshmanan, H., Selvaraj, V., Gajendran, B., Nanjian, R., Chinnasamy, A., Sabaratnam, V., 2015. Mycosynthesis and characterization of silver nanoparticles from *Pleurotus djamor* var. *roseus* and their *in vitro* cytotoxicity effect on PC3 cells. Process Biochem. 50, 140–147.

Rehman, S., Ansari, M.A., Al-Dossary, H.A., Fatima, Z., Hameed, S., Ahmad, W., Ali, A., 2021. Modeling and Control of Drug Delivery Systems. Academic Press, Elsevier.

Romero, C.M., Alvarez, A., Martínez, M.A., Chaves, S., 2018. Fungal nanotechnology: a new approach toward efficient biotechnology application. In: Fungal Nanobionics: Principles and Applications. Springer Nature Publication, pp. 117–143.

Sanguinedo, G., Fratila, R.M., Estevez, M.B., de la Fuente, J.M., Grazú, V., Albores, S., 2018. Extracellular biosynthesis of silver nanoparticles using fungi and their antibacterial activity. Nano Biomed. Eng. 10, 165–173.

Santos, T.S., dos Passos, E.M., de Jesus Seabra, M.G., Souto, E.B., Severino, P., da Costa Mendonca, M., 2021. Entamopathogenic fungi biomass production and extracellular biosynthesis of silver nanoparticles for bioinsecticide. Appl. Sci. 11 (6), 2465.

Saravanan, M., Arokiyaraj, S., Lakshmi, T., Pugazhendhi, A., 2018. Synthesis of silver nanoparticles from *Phenerochaete chrysosporium* (MTCC-787) and their antibacterial activity against human pathogenic bacteria. Microb. Pathog. 117, 68–72.

Shahzad, A., Saeed, H., Iqtedar, M., Hussain, S.A., Kaleem, A., Abdullah, R., Sharif, S., Naz, S., Saleem, F., Aihetasham, A., Chaudhary, A., 2019. Size-controlled production of silver nanoparticles by *Aspergillus fumigatus* BTCB10: likely antibacterial and cytotoxic effects. J. Nanomater. 2019, 1–14.

Siddiqi, K.S., Husen, A., Rao, R.A.K., 2018. A review on biosynthesis of silver nanoparticles and their biocidal properties. J. Nanobiotechnol. 16, 1–28.

Singh, D., Rathod, V., Ninganagouda, S., Kiremath, J., Singh, A.K., Methew, J., 2014. Optimization and characterization of silver nanoparticle by endophytic fungi Penicillium sp.

isolated from *Curcuma longa* (turmeric) and application studies against MDR *E. coli* and *S. aureus*. Bioinorg. Chem. Appl. 2014, 1–8.

Song, D., Li, X., Cheng, Y., Xiao, X., Lu, Z., Wang, Y., Wang, F., 2017. Aerobic biogenesis of selenium nanoparticles by *Enterobacter cloacae* Z0206 as a consequence of fumarate reductase mediated selenite reduction. Sci. Rep. 7, 19–21.

Syed, A., Ahmad, A., 2012. Extracellular biosynthesis of platinum nanoparticles using the fungus *Fusarium oxysporum*. Colloids Surf. B Biointerfaces 97, 27–31.

Taha, Z.K., Hawar, S.N., Sulaiman, G.M., 2019. Extracellular biosynthesis of silver nanoparticles from *Penicillium italicum* and its antioxidant, antimicrobial and cytotoxicity activities. Biotechnol. Lett. 41, 899–914.

Tyagi, S., Tyagi, P.K., Gola, D., Chauhan, N., Bharti, R.K., 2019. Extracellular synthesis of silver nanoparticles using entomopathogenic fungus: characterization and antibacterial potential. SN Appl. Sci. 1, 1545.

Vetchinkina, E.P., Loshchinina, E.A., Vodolazov, I.R., Kursky, V.F., Dykman, L.A., Nikitina, V.E., 2017. Biosynthesis of nanoparticles of metals and metalloids by basidiomycetes. Preparation of gold nanoparticles by using purified fungal phenol oxidases. Appl. Microbiol. Biotechnol. 101, 1047–1062.

Xue, B., He, D., Gao, S., Wang, D., Yokoyama, K., Wang, L., 2016. Biosynthesis of silver nanoparticles by the fungus *Arthroderma fulvum* and its antifungal activity against genera of *Candida*, *Aspergillus* and *Fusarium*. Int. J. Nanomedicine 11, 1899–1906.

Zhang, X.-F., Liu, Z.-G., Shen, W., Gurunathan, S., 2016. Silver nanoparticles: synthesis, characterization, properties, applications, and therapeutic approaches. Int. J. Mol. Sci. 17, 1–34.

Zhao, X., Zhou, L., Riaz Rajoka, M.S., Yan, L., Jiang, C., Shao, D., Zhu, J., Shi, J., Huang, Q., Yang, H., Jin, M., 2018. Fungal silver nanoparticles: synthesis, application and challenges. Crit. Rev. Biotechnol. 38, 817–835.

CHAPTER 15

Synthesis of silver nanoparticles from mushroom: Safety and applications

Kanniah Paulkumar and Kasi Murugan

Department of Biotechnology, Manonmaniam Sundaranar University, Tirunelveli, Tamil Nadu, India

1 Introduction

Recently, nanoparticles have attained increased advancement in various areas such as electronics, environment, medicine, pharmaceuticals, and agriculture (Schaming and Remita, 2015; Abou El-Nour et al., 2010; Mathur et al., 2018; Mishra and Singh, 2015) (Fig. 1). Nanoscale particles with a size of less than 100 nm exhibit improved applications compared to bulk materials due to superior physicochemical properties (Abou El-Nour et al., 2010; Gautam and Van Veggel, 2013). The main aim of nanoparticles utilization in divergent fields is to obtain high-quality products that are innovative, durable, and nontoxic. In electronics, researchers struggle with nanoscale particles to achieve a better energy storage material, and are focusing on sustainable energy using nanoparticles for solar and fuel cells applications (Zhang, 2017). In addition, the sensing and catalytic properties of nanoparticles are popularly used for wastewater treatment applications (Yaqoob et al., 2020). Biological activities, including antimicrobial, anticancer, and antiinflammatory, make nanoparticles suitable for smart medicine (Talapko et al., 2020; Augustine and Hasan, 2020; Mikhailova, 2020). Numerous research reports are available to exemplify the application of nanoparticles in agriculture by giving special attention to crop protection, crop improvement, and stress tolerance (Khan and Rizvi, 2017; Ghormade et al., 2011; Zhao et al., 2020). Moreover, the combination of nanoparticles with biomolecules including proteins, enzymes, and nucleic acids represents an area for exploration in biomedical nanotechnology (Gautam and Van Veggel, 2013; Jiao et al., 2018). Therefore, organic and inorganic nanoparticles are pivotal for many applications to strengthen the modern life style of humans. In inorganic nanoparticles, metal and metal oxide nanoparticles perform significant roles in various arenas such as environment, therapeutics, and human health care (Jiao et al., 2018; Augustine and Hasan, 2020).

FIG. 1

Applications of nanoparticles in various fields.

Nowadays, the production of metal and metal oxide nanoparticles by green protocols is a well-known goal of green nanotechnology. The term "green synthesis" refers to the hazardous chemicals and surplus compounds free, eco-friendly, and inexpensive process (Singh et al., 2018a). In the field of nanotechnology, the green synthesis process offers extensive applications in the production of biocompatible nanoparticles. Currently these biocompatible nanoparticles are being utilized for development of sustainable environment, energy, and healthcare medicine applications (Augustine and Hasan, 2020). The classical physical and chemicals routes have failed in the implementation of nanoparticle production at large scale. Generally, the morphologies such as size and shape of nanoparticles are important factors to design a nanoparticle from an application point of view. Hence, the modification in volume of chemical agents or environmental conditions (pH and temperature) is necessary to synthesize nanoparticles with desired size and shape. Apart from these problems, the production of surplus compounds along with nanoparticles may raise huge questions about eco-friendliness. In addition, the accumulation of nanoparticles with cellular biomoieties is also responsible for long-term side effects. Another possible feature is the recycling or reusability of nanoparticles may provide innovative technologies to reduce the cost globally. Therefore, an alternative protocol is needed to tackle the aforementioned issues, hence the green synthesis process using bio-resources gaining more interest in research and development of nanoscience and technology (Singh et al., 2018a; Augustine and Hasan, 2020).

2 Silver nanoparticles

Silver is a noble metal and there is evidences available that nanoscale silver has been used from the ancient period onwards. Mostly, silver materials are well known for their superior wide spectrum antimicrobial activity (Talapko et al., 2020). Compared to other inorganic nanoparticles, silver nanoparticles possess additional interest in various fields due to their unique optical, catalytic, and biological properties (Abou El-Nour et al., 2010). Ancient peoples including Romans, Egyptians, and Macedonians, and the Renaissance philosopher Paracelsus all used silver salts for the treatment of wounds, ulcers, and injuries (Alexander, 2009; Yu et al., 2013). In the past, silver has been used as a preservative/storage material for water, milk, vinegar, and wine to keep them free from contaminants (Alexander, 2009). Based on this historical evidence, in 1954, colloidal silver nanoparticles were registered as antibiotic material in the United States (Yu et al., 2013). Therefore, the medicinal property of silver is not a new discovery; however, exploration of the enhanced antimicrobial properties of various sized and shaped silver nanoparticles is a crucial development in modern medicine. Recently, the US Food and Drug Administration (FDA) approved silver for wound bandage applications (Paladini and Pollini, 2019). In drinking water, the accepted level of silver is 90 ppm announced by the World Health Organization (WHO) (Krishnaraj et al., 2010).

Currently, the world is experiencing exponential growth of silver nanoparticles research and production, leading to rapid advancement in industrial and commercial aspects. The electrochemical properties of silver nanoparticles are employed as nanosensors for rapid and sensitive detection of trace-amount contaminants (Prosposito et al., 2020). The sensing property of silver nanoparticles enables the production of nanomedical devices (Zhang et al., 2016). Nowadays, the world needs fast and early diagnosis of diseases that threaten mankind, through novel and innovative technologies. Due to the impressive physicochemical properties of silver nanoparticles, they fulfill the essential needs for smart detection, with positive results for human health (Zhang et al., 2016; Wong and Liu, 2010).

The antimicrobial property of silver nanoparticles is pivotal all areas such as food, agriculture, medicine, and the environment. As mentioned above, silver salts are popular for their biocidal activity and interesting research has been done suggesting that nano-sized silver ought to be involved in enhanced antimicrobial activity. Now, exciting reports are describing the role of the size and shape of silver nanoparticles in their activity against various microorganisms (Singh et al., 2018a). However, outside the application point of view, the production of silver nanoparticles creates huge questions about the involvement of toxic chemicals in the synthesis process. The classical synthesis methods have used chemicals, high temperature, and pressure to synthesize silver nanoparticles. However, these methods create a huge burden in terms of environment issues. Furthermore, the production of surplus compounds along with silver nanoparticles may outway the benefits and features of silver nanoparticles (Singh et al., 2018a). Therefore, demand has grown for an environmental care process to overcome the drawbacks found in classical methods. The silver nanoparticles synthesized by plant materials such as seed, flower, leaf, fruit, bark, stem, rind, and peel have gained considerable attention in green nanotechnology (Singh et al., 2018a; Augustine and Hasan, 2020). The numerous phytocompounds in plant materials are the key bio-resources actively participating in the production of silver nanoparticles. Apart from this, the preferential selection of plant materials is due to nontoxic, amenable for large scale, speedy synthesis, and eco-benevolent processes (Singh et al., 2018a). Generally, an oxidation and reduction reaction is involved between the precursor silver salts and electron-rich phytochemicals, including tannins, flavonoids, alkaloids, steroids, polyphenols, phenolic acids, terpenoids, etc. (Gudikandula and Charya Maringanti, 2016; Singh et al., 2018a). The biological method using bacteria and fungus for production of silver nanoparticles has obtained great attention in modern nanotechnology (Hebbalalu et al., 2013; Guilger-Casagrande and Lima, 2019). Like plants, the macro fungi edible mushrooms are also contributing to the synthesis of silver nanoparticles. The involvement of medicinally important edible mushroom extracts in nanoparticles production is generally known as "mycosynthesis." This process is inexpensive, and the silver nanoparticles synthesized by this approach are more biocompatible for medical and various industrial applications (Owaid, 2019).

3 Significance of mushrooms in nanoparticle production

In green nanotechnology, mycosynthesis—that is, synthesis of nanoparticles using mushrooms—has great advantages due to being a simple, risk-free, and environmentally safe process (Owaid, 2019). Historical evidence is available to show that mushrooms have been consumed by people around the world (Ren et al., 2014), and they are considered an excellent nutritional food stuff due to the presence of antioxidant-rich compounds and immune stimulatory molecules (Reis et al., 2012). The occurrence of bioactive secondary metabolites, including steroids, terpenoids, anthraquinones, and quinolones, may push the mushroom as a potential candidate for biotherapeutic applications (Valverde et al., 2015). The bioactive primary and secondary metabolites of mushroom extract are employed as green-reducers and green-stabilizers in the synthesis of silver nanoparticles (Owaid, 2019).

4 Synthesis of silver nanoparticles using mushroom

Currently, the utilization of silver nanoparticles in health care products, agriculture, pharmaceutics, and medicine is increasing due to their multifaceted properties. Therefore, the advance of synthesis protocol requires novel strategies for large scale-up with low cost, quick, and safe procedures (Chung et al., 2016). Generally, in green/biosynthesis, the silver metal ions are reduced to silver atoms, and then the silver atoms are grown by Ostwald ripening/coalescence, resulting in the formation of silver nanoparticles (Kanniah et al., 2020) (Fig. 2). Among green nanotechnologies, mushroom-mediated synthesis has gained much attention in production of silver nanoparticles due to the availability of bioactive phytomolecules including phenolic compounds, unsaturated fatty acids, carotenoids, ascorbic acids, and tocopherols. In addition, mushrooms are a rich reservoir of vitamins, fiber, and essential amino acids (Owaid, 2019) (Fig. 2).

The medicinally valuable mushroom *Ganoderma lucidum* was successfully employed as reducing and stabilizing agent for the production of silver nanoparticles. The synthesized silver nanoparticles were of spherical architecture with sizes of 15–22 nm. Traditionally, the *G. lucidum* mushroom was used as a medicinal product to improve the immune system and for its potent antiinflammatory activity. In addition, it offers excellent prevention agents against heart diseases, liver diseases, diabetes, skin diseases, human immune deficiency virus, and hypertension (Aygun et al., 2020). The edible mushroom *Pleurotus ostreatus* also includes medicinally important properties such as anticancer, antiinflammatory, and antimicrobial (Al-Bahrani et al., 2017). Therefore, Al-Bahrani and his co-workers (2017) have used this mushroom to synthesize silver nanoparticles. They obtained spherical shaped silver nanoparticles with size of about <40 nm. The biomolecules of *P. ostreatus* are possible agents for reduction and stabilization of synthesized silver nanoparticles (Al-Bahrani et al., 2017). Another mushroom, *Pleurotus florida*, is popularly best known for its potent antitumor and antioxidant properties (Bhat et al., 2011). The mushroom *P. florida* was used by Bhat et al. (2011) to synthesize silver nanoparticles under ambient conditions. Interestingly, they employed sunlight as a catalyst source in the reduction of silver metal ions into silver nanoparticles. The phenomenon that homolytic cleavage of OH bond from phenolic compounds results in the generation of H^+ free radicals and, in due course, the electron released from OH reacted with silver ions to from silver nanoparticles meant, as they expected, that compared to dark conditions, the silver nanoparticles were actively synthesized during sunlight exposure. The proteins present in the mushroom may act as capping agent to form a layer on the surface silver nanoparticles and also inhibit the particle agglomeration. Thus, the synthesized particles were polydispersed with a size of 20 nm (Bhat et al., 2011). Furthermore, Sen et al. (2013) reported the production of silver nanoparticles using the same mushroom *P. florida*. However, here they isolated the polysaccharide glucan from the *P. florida* mushroom and utilized it for silver nanoparticle synthesis. As a result, they efficaciously synthesized the particles that were mostly spherical in shape with an average size of about 2.4 nm.

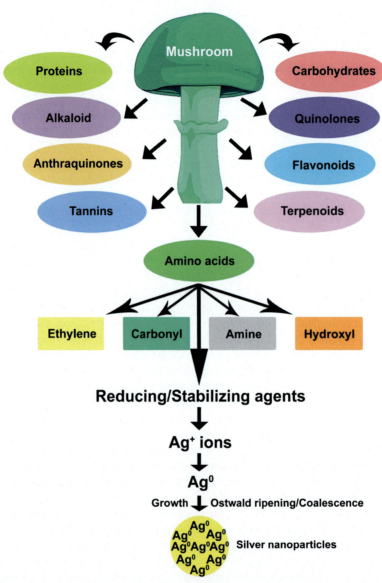

FIG. 2

General concept of synthesis of silver nanoparticles by mushroom extract.

The polysaccharide glucan plays a crucial role in both reduction and stabilization of synthesized nanoparticles (Sen et al., 2013). The *Pleurotus citrinopileatus* mushroom extract was reacted with silver nitrate to produce silver nanoparticles, reported by Maurya and his co-workers (2016). Ethylene is one of the possible functional groups present in biomolecules of *P. citrinopileatus* that may be responsible for the reduction of silver metal ions into silver nanoparticles. Another *Pleurotus* species, *Pleurotus giganteus*, was effectively applied as reducing agent in the fabrication of silver nanoparticles. The silver nanoparticles were in spherical shape with size ranging from 5 to 25 nm. The carbonyl, amine, and hydroxyl functional groups in biomolecules of mushroom extract may play a significant role in the silver metal ions reduction and silver nanoparticles stabilization process (Debnath et al., 2019). The two *Tricholoma* species *Tricholoma matsutake* and *Tricholoma crassum* were actively involved in the synthesis of silver nanoparticles (Anthony et al., 2014; Ray et al., 2011). The obtained silver nanoparticles from *T. matsutake* were predominantly spherical in shape with a size of about 10 nm (Anthony et al., 2014), whereas the *T. crassum* produced spherical and a few hexagonal shaped silver nanoparticles with size ranging from 5 to 50 nm (Ray et al., 2011).

The carbonyl and hydroxyl groups from protein and amino acid moieties possibly serve in reduction and stabilization of silver nanoparticles (Anthony et al., 2014; Ray et al., 2011). Sriramulu and Sumathi (2017) have used the mushrooms *G. lucidum* and *Agaricus bisporus* for effective production of silver nanoparticles. The *G. lucidum* and *A. bisporus* are wild and edible mushrooms, respectively, and are rich in biomolecules such as vitamins, amino acids, essential fatty acids, and proteins. The silver nanoparticles synthesized by *G. lucidum* expressed needle-like shape, whereas sponge shapes were observed in *A. bisporus* fabricated silver nanoparticles (Sriramulu and Sumathi, 2017). The *A. bisporus* mushroom was also utilized by Dhanasekaran et al. (2013) to examine their potency for the formation of silver nanoparticles. The role of capping agents, e.g., proteins molecules, in nanoparticle synthesis is to stabilize the fabricated particles and protect the particles from agglomeration (Dhanasekaran et al., 2013). Another edible mushroom, *Volvariella volvacea*, extract was effectively employed as reducing and capping agent to manufacture silver nanoparticles (Philip, 2009). Due to this nontoxic protocol, Philip (2009) obtained spherical shaped and approximately 15 nm sized particles. Furthermore, Philip (2009) postulated that carboxylate functional groups from amino acid moieties play a critical role as capping agents in the fabrication of nanoparticles synthesis process. Recently, Zhang et al. (2020) demonstrated the involvement of *Flammulina velutipes* mushroom extract in nanoparticle production. Initially, they have confirmed the formation of silver nanoparticles by arisen a peak at 450 nm in UV-Vis spectroscopy. Furthermore, transmission electron microscopic analysis showed the particles obtained from the reaction of *F. velutipes* extract and silver nitrate were spheres with monodispersity and size of around 22 nm (Zhang et al., 2020).

5 Safety protocol

In nano-structured and nano-sized materials synthesis, it is important to design ecosafer protocols and to pay more attention to biosafety and biocompatibility features of synthesized nanomaterials. The main aim of green nanotechnology is to produce nanomaterials that are low cost, nontoxic, and environmentally nonhazardous

(Singh et al., 2018a; Abbasi et al., 2016). Before green synthesis, physical and chemical methods were followed for nanoparticle production. However, these methods do not reliably provide the expected size, distribution, and shapes, are much expensive, and ultimately, the formation of hazardous by-products along with nanoparticles means the top-down approach is not suitable for biomedical and agriculture applications (Devatha and Thalla, 2018; Kumar et al., 2019). Green synthesis has built up novel strategies to construct nanoparticles by a bottom-up approach with confined parameters to achieve bio-secured and eco-benign nanoparticles (Singh et al., 2018a; Kumar et al., 2019). The bio/phytosynthesis process has gained great attention due to the avoidance of toxic chemicals in the synthesis protocol (Devatha and Thalla, 2018). In this process, the bio/phyto molecules are used in the reduction and stabilization process. The reducing and stabilizing agents are important for size- and shape-controlled synthesis of nanoparticles (Singh et al., 2018a; Abbasi et al., 2016; Kumar et al., 2019). The bio/green synthesis of nanomaterials concept was obtained from the metal reduction behavior of microorganisms and plant materials (Paulkumar et al., 2014). The nonhazardous proteins, enzymes, pigments, alkaloids, phenols, tannins, flavonoids, terpenoids, quinones, etc. have prominent input in the synthesis of nanomaterials (Silva et al., 2019; Noah, 2019; Kumar et al., 2019). Phytosynthesis is considered as more advantageous compared to the biosynthesis protocol. Maintenance of pure stock and culturing without contamination is a risky and time-consuming process. In addition, the synthesis of nanoparticles at higher volume may inhibit the bacterial culture and result in incomplete reduction of metal ions to metal nanoparticles (Ajitha et al., 2015; Devatha and Thalla, 2018). Therefore, researchers selectively used the plant-assisted process to achieve preferred shaped and sized nanoparticles. Mushrooms are macroscopic fungi that include essential proteins, vitamins, amino acids, flavonoids, steroids, alkaloids, and phenolic compounds. The utilization of these biomolecules in nanoparticle production is green, safer, rapid, and amenable for mass production. In general, the usage of fresh/powdered mushroom extract without toxic solvents in nanoparticles synthesis protocol results in less chance of contamination (Chhipa, 2019; Owaid, 2019).

Recently, there has been an increase in research worldwide to find novel strategies for the advancement of medical and environmental applications. Nanoparticles have received much attention and possibly satisfy the needs for the development of mankind. The emergence of silver nanoparticles has captured keen attention in all disciplines related to human healthcare, diagnosis, and therapeutic applications (Silva et al., 2019). Nowadays silver-based nanoproducts have revolutionized the pharmaceutical industries and surpass other synthetic products to occupy all the emerging areas, especially in medicine. Due to the potent microbial resistance features of silver nanoparticles, they could serve as excellent antimicrobial agents against various human and plant pathogens (Noah, 2019; Silva et al., 2019). In Asian countries, silver in combination with herbs is popularly known as "Rajata Bhasma," used in traditional Ayurveda medicine (Liu et al., 2019). The formulation of silver with herbal extract has given better treatment results for various diseases and disorders (Rai et al., 2019). Hence, strong evidence is available to show the usage of silver in ancient general medicine. However, proper scientific evaluation is needed to prove the medicinal value of Rajata Bhasma with modern tools and technology.

According to the World Health Organization (WHO) and United States Environmental Protection Agency (USEPA), the recommended level of silver is 90 ppb in cosmetics (Pedahzur et al., 1995). In drinking water, the WHO and USEPA advise the maximum occurrence of silver ions released from silver based filters such as silver/anion resin, silver/fiberglass, silver/sand, and silver/zeolite is $<100\,\mu g/L$ for human intake (Mpenyana-Monyatsi et al., 2012). The Israeli Ministry of Health has also recommended that the concentration of silver used as a disinfectant in drinking water is 80 ppb (Pedahzur et al., 1995). Silver nanoparticles react well against harmful bacteria, fungi, cancer, and tumors. However, controversial reports regarding the cytotoxicity of silver nanoparticles against cancer and normal cells have added confusion to understanding the exact phenomenon of anticancer activity (Khorrami et al., 2018; Syed et al., 2013; Funez et al., 2013). The exact mechanisms are also needed to explore the possible interaction of silver nanoparticles with the human system.

6 Versatile applications of silver nanoparticles

6.1 Biological activity

The intense biological properties of silver nanoparticle can push them toward use in nano-therapeutic applications (Fig. 3). Nowadays early diagnosis is required for pandemic diseases and is crucial to providing better treatment. Hence, smart technologies with nano-based strategies are needed for providing a sustainable eco-system for all living organisms. The potent inhibition behavior of silver nanoparticles against microorganisms has led to a revolution in next generation nanomedicine (Le Ouay and Stellacci, 2015; Loo et al., 2018). The following reports provide strong evidence that silver nanoparticles have excellent biological activities.

6.1.1 Antibacterial

Nowadays, the increased number of multidrug resistant bacteria results in human infectious diseases and is responsible for a huge number of deaths around the world. The exploitation of various antibiotics in bacteria may lead them to develop antibiotic resistance. The emergence of silver nanoparticles can provide considerable enhancement in the antibacterial system due to their unique adsorption effect. For example, Lara et al. (2010) have examined 100 nm-sized silver nanoparticles' activity against multidrug resistant Pseudomonas aeruginosa, erythromycin-resistant Streptococcus pyogenes, and ampicillin-resistant Escherichia coli O157:H7. They reported that the silver nanoparticles may break the cell wall protein synthesis and nucleic acid synthesis. Further, they suggested the broad spectrum antibacterial behavior of silver nanoparticles makes them a promising agent in medical applications to avoid drug resistant bacterial transmission. Ocimum gratissimum leaf extract synthesized silver nanoparticles exhibited potential effects on multidrug resistant

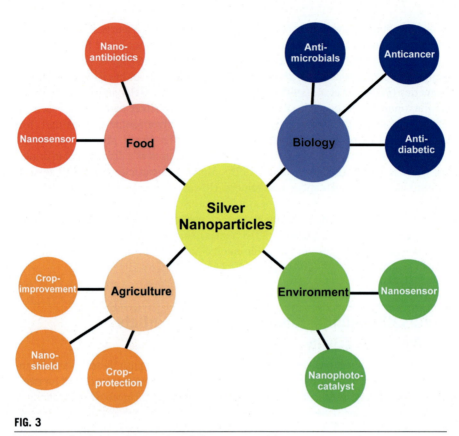

FIG. 3

Various applications and properties of silver nanoparticles.

E. coli and Staphylococcus aureus, with the minimum inhibitory concentration (MIC) of green synthesized silver nanoparticles of 4 μg/mL and 8 μg/mL for E. coli and S. aureus, respectively (Das et al., 2017). Similarly, Qais and his team (2019) reported that Murraya koenigii leaf extract synthesized silver nanoparticles effectively inhibit methicillin-resistant S. aureus (MRSA) and β-lactamase (BL) producing E. coli. The MIC value of MRSA and BL producing E. coli was 32 μg/mL and 32–64 μg/mL.

Like plant materials, mushroom extract also energetically acted to inhibit the growth of noxious bacterial pathogens. The silver nanoparticles produced with powdered extract of P. giganteus strongly control growth of bacteria including *Bacillus subtilis*, E. coli, S. aureus, and P. aeruginosa. The gram-negative bacteria P. aeruginosa and E. coli are more sensitive to silver nanoparticles and the MIC value of 10 μg/mL and 12 μg/mL, respectively. However, the MIC of gram-positive bacteria S. aureus and *B. subtilis* was found to be 15 μg/mL and 14 μg/mL, respectively

(Debnath et al., 2019). Compared to gram-negative bacteria, the gram-positive bacteria have a thick peptidoglycan layer, which may delay the antimicrobial activity (Debnath et al., 2019). Similarly, the silver nanoparticles synthesized with *T. matsutake* mushroom effectively inhibit the gram-negative E. coli compared to gram-positive B. subtilis. The viability of E. coli (∼ less than 10%) and B. subtilis (∼18%) cells progressively decreased while increasing concentration of silver nanoparticles from 0 to 10 μg/mL (Anthony et al., 2014). The *P. ostreatus* synthesized silver nanoparticles impressively act against S. aureus, P. aeruginosa, B. subtilis, E. coli and *Bacillus* cereus. The silver nanoparticles exhibited the MIC of all bacteria in the range of 13–27 μg/mL (Al-Bahrani et al., 2017). The biofilm forming aquatic pathogens Vibrio parahaemolyticus, *Vibrio splendidus*, *Aeromonas punctata*, *Vibrio alginolyticus*, *Vibrio harveyi*, and *Vibrio anguillarum* were also susceptible to silver nanoparticle fabricated by *F. velutipes* mushroom (Zhang et al., 2020).

6.1.2 Antifungal

Green synthesized silver nanoparticles also act as an excellent antagonist for fungi. Candida albicans is one of the infectious fungi that causes serious effects to humans. Silver nanoparticles fabricated with the edible mushroom *Pleurotus sajor-caju* actively impede the growth of C. albicans. The silver nanoparticles specifically downregulate the expression of isocitrate lyase, which is the main virulence factor in C. albicans (Musa et al., 2018). The mushroom *A. bisporus* synthesized silver nanoparticles reacted powerfully against the fungus *Aspergillums niger* (Narasimha et al., 2013). The extract of *Stachys lavandulifolia* sepals and *Lathyrus* sp. seeds synthesized silver nanoparticles exhibited potent antifungal activity against *Dothiorella sarmentorum*. More than 90% of mycelia growth was inhibited at the concentration of 40 μg/mL and 80 μg/mL of silver nanoparticles fabricated by *S. lavandulifolia* and *Lathyrus* sp. extracts, respectively (Azizi et al., 2016). Likewise, silver nanoparticles synthesized with seed extract of Sinapis arvensis exhibited potent antifungal activity against *Neofusicoccum parvum* (Khatami et al., 2015). Interestingly, Bahrami-Teimoori and his co-workers (2017) have examined the antifungal activity of Amaranthus retroflexus leaf extract synthesized silver nanoparticles on various fungi such as *Fusarium oxysporum*, *Macrophomina phaseolina*, and *Alternaria alternata*, in which, 50% MIC was found to be 328.05 ± 13.29, 159.80 ± 14.49, and 337.09 ± 19.72 for F. oxysporum, M. phaseolina, and A. alternata. The leaf extract of Melia azedarach successfully synthesized silver nanoparticles and was used as antifungal agent against eggplant fungus *Verticillium dahlia*. Under in vitro conditions, 60 ppm of silver nanoparticles was sufficient to inhibit the growth of V. dahlia. In contrast, 20 ppm of silver nanoparticles was enough to reduce the V. dahlia growth in in vivo studies (Jebril et al., 2020).

6.1.3 Anticancer

Silver nanoparticles have achieved considerable attention among researchers due to their increased applications in medicine, especially in cancer treatments (Yesilot and Aydin, 2019). Plenty of reports have stated that bio/green synthesized silver nanoparticles exhibited intense activity against cancer cells through their comprehensive apoptotic and antiproliferative properties (Vlasceanu et al., 2016). The silver nanoparticles produced with the mushroom *Inonotus obliquus* were successfully used as anticancer agents against human lung cancer (A549) and breast cancer (MCF-7) cell lines. About 51.61% and 59.11% of lung cancer and breast cancer cells, respectively, were inhibited after the exposure of silver nanoparticles (Nagajyothi et al., 2014). Similarly, Gurunathan et al. (2013) reported the green synthesized silver nanoparticles using *Ganoderma neo-japonicum* effectively showed their anticancer potential against breast cancer cells (MDA-MB-231). After 24h of silver nanoparticle exposure with MDA-MB-231 cells, the cell viability totally collapsed at the concentration from 6μg/mL to 10μg/mL. The cytotoxic activity was increased with increasing the concentration of silver nanoparticles from 1 to 10μg/mL. The silver nanoparticles induce the disruption of membrane structure, result in leakage, and also the apoptotic activity leads to the death of breast cancer cells (Gurunathan et al., 2013).

The *P. ostreatus*, which belongs to the edible oyster mushroom group, was used to synthesize silver nanoparticles and its anticancer property against MCF-7 cell lines explored. After 24h treatment of silver nanoparticles with MCF-7, in a dose-dependent manner, the cells' viability was inhibited from 5% to 78% at different concentrations from 10 to 640μg/mL (Yehia and Al-Sheikh, 2014). Another edible mushroom, *Pleurotus djamor* var. *roseus*, was used to produce 5–50nm sized silver nanoparticles and its cytotoxic effect against human prostate cancer cells (PC3) examined. The IC_{50} value of silver nanoparticles was achieved at the concentration of 10μg/mL. The shrunken nucleus and damaged DNA were observed through propidium iodide staining, which clearly revealed the silver nanoparticles have a higher toxic effect on PC3 cells (Raman et al., 2015). Apart from mushroom synthesized silver nanoparticles, plant materials-produced silver nanoparticles also exhibit potent cytotoxic activity against various cancer cells. The leaf extract of *Melia dubia* fabricated silver nanoparticles efficiently control the proliferation of human breast cancer cell lines (MCF-7). The IC_{50} of silver nanoparticles was obtained against MCF-7 cells lines at 31.2μg/mL. In contrast, the 50% cytotoxic concentration (CC_{50}) of silver nanoparticles attained against normal Vero cells was 500μg/mL. Indeed, silver nanoparticles showed significant cytotoxic effect on MCF-7 cancer cells at very low concentration. The normal Vero cells proliferation was inhibited only at higher concentration (Kathiravan et al., 2014). Another plant material, the flower extract of *Lonicera hypoglauca*, was used as reducing agent to synthesize silver nanoparticles and its cytotoxic behavior identified using MCF-7 cell lines. The results of the XTT assay postulated the silver nanoparticle mostly inhibited the cell proliferation on MCF-7 cell lines compared to normal macrophages cell lines RAW 264.7. The upregulated expression of p53 mRNA after the treatment of silver nanoparticles with MCF-7 cell lines indicated the promotion of apoptosis (Jang et al., 2016).

6.1.4 Antidiabetic

Worldwide, diabetes causes serious health complications and mortality, and the need to develop effective treatments is a major concern. The effective inhibitory activity of α-amylase and α-glucosidase enzymes by silver nanoparticles generates a promising action to control the hazardous *Diabetes mellitus* (Rajaram et al., 2015). These two enzymes play crucial roles in catalyzing the hydrolysis of polysaccharides and disaccharides. The blood glucose level may be regulated by controlling the activity of these two important enzymes (Rajaram et al., 2015; Balan et al., 2016). Rajaram et al. (2015) suggested the inhibition of carbohydrate-degrading enzymes by silver nanoparticles results in enhanced uptake of glucose. They controlled the hydrolyzing enzymes α-amylase and α-glucosidase using silver nanoparticles synthesized with the stem extract of *Tephrosia tinctoria*. While increasing the concentration of silver nanoparticles, the enzymes inhibition activity also increased, and higher inhibition of 83.52% and 94.76% was achieved for α-amylase and α-glucosidase, respectively, at 75 μg/mL of silver nanoparticles (Rajaram et al., 2015). Another interesting report by Balan and his co-workers (2016) exemplified the Lonicera japonica synthesized silver nanoparticles energetically collapse the polysaccharide and disaccharide digestive enzymes α-amylase and α-glucosidase with the IC50 value 54.56 μg/mL and 37.86 μg/mL, respectively. The silver nanoparticles synthesized from the leaf extract of *Calophyllum tomentosum* exhibited strong antidiabetic activity by inhibiting the activity of α-amylase and α-glucosidase (Govindappa et al., 2018).

Shah et al. (2020) reported the silver nanoparticles synthesized by seed extract of Silybum marianum inhibited α-amylase (38.74%) and α-glucosidase (32.62%) activity. Likewise, the silver nanoparticles obtained from the reaction of silver nitrate and Cleome viscosa successfully control the effect of α-amylase and α-glucosidase. After treatment with silver nanoparticles, the obtained α-amylase and α-glucosidase activity was 57.62% and 90.14%, respectively. The IC_{50} value of control (acarbose) was 14.06 μg/mL for α-amylase and 18.52 μg/mL for α-glucosidase. In contrast, the IC_{50} of silver nanoparticles was found to be 21.92 and 21.76 μg/mL for α-amylase and α-glucosidase, respectively (Yarrappagaari et al., 2020). The seagrass-assisted bio-fabrication of silver nanoparticles actively controlled α-glucosidase in a concentration-dependent manner from 10 to 100 μg/mL. The IC_{50} value of silver nanoparticles was 47 μg/mL and the control acarbose showed the IC_{50} value of 72 μg/mL. Another challenging study was carried out by Mahmoudi et al. (2021) to examine the antidiabetic activity of silver nanoparticles synthesized by green (Eryngium campestre *Boiss* extract) method (GM) and chemical (sodium borohydride and trisodium citrate) method (CM). Compared to GM synthesized silver nanoparticles, the silver nanoparticles synthesized by CM exhibited no difference in fasting blood sugar (FBS) level on the first day of treatment with silver nanoparticles. In control (without silver nanoparticle treated), there is no significant variance in

FBS observed at 7th and 15th day. In contrast, the CM synthesized silver nanoparticles showed considerable decrease in FBS level compared to GM synthesized silver nanoparticles. However, the silver nanoparticles synthesized by GM substantially decrease the level of liver enzymes such as alanine aminotransferase and aspartate aminotransferase in diabetic rats. Whereas the CM synthesized silver nanoparticles failed to decrease the liver enzymes level in diabetic rats. As a result, compared to CM synthesized silver nanoparticles, GM synthesized silver nanoparticles may protected the liver from damage and enhanced liver function in diabetic rats (Mahmoudi et al., 2021).

6.2 Environment

The massive growth of industrialization provides a modern lifestyle, but creates hazardous by-products, including organic dyes and heavy metals, which damage the ecosystem for all living things. Nowadays nano-enabled strategies have been employed in various industries for eradication of toxic heavy metals and dyes. Silver nanoparticles are effectively involved in the degradation/adsorption of environmental contaminants (heavy metals and dyes) due to their potent catalytic and sensing properties (Thivaharan et al., 2018) (Fig. 3). The forthcoming sections explain in detail about the sensing and photocatalytic application of green synthesized silver nanoparticles.

6.2.1 Nanophotocatalyst

Photocatalysis is a sustainable, eco-friendly process, which has captured significant attention for the eradication of toxic dyes. This process is triggered by a light source (UV-visible, sunlight, or mercury-xenon) and thereby forms electron-hole pairs on the surface of a photocatalyst, which can be appropriately used for materials oxidation/reduction processes (Lim et al., 2021; Lee et al., 2020). The narrow bandgap energy and high light absorption feature of silver nanoparticles in the visible light region shows significant application in catalytic processes (Mavaei et al., 2020). The silver nanoparticles fabricated with forest mushroom *G. lucidum* effectively degraded the direct blue (textile dye) within 150 min in UV light at room temperature (Sriramulu and Sumathi, 2017). Another report by Faisal and his team (2020) stated the edible mushroom *F. velutipes* synthesized silver nanoparticles were energetically involved in photocatalytic degradation of indigo carmine dye, at about 98.2% at 140 min. Thomas et al. (2019) reported the degradation of two dyes, methylene blue and methylene orange, using silver nanoparticles synthesized with the leaf extract of Passiflora edulis *f. flavicarpa*. The silver nanoparticle mixed dye solution was exposed to sunlight for 1–5 h and a gradual decrease in color observed. The 100% dye degradation was observed after 24 h exposure (Thomas et al., 2019). The UV treated solution of silver nanoparticle and methylene blue dye degraded the toxic dye by about 96% at 80 min. The dye degradation efficacy of 40%–96% was increased with increasing the concentration of silver nanoparticles from 2 to 8 mg. In contrast, above 8 mg (up to 12 mg) the observation of decreased degradation revealed that at low concentration of silver nanoparticles the particles are dispersed highly, with small sizes generating a number of absorbing active sites and a large surface area. The high concentration of silver nanoparticles resulted in large particle size and decreased surface are and active sites leads to low absorption of dyes (Tahir et al., 2015).

The Alpinia officinarum rhizome extract synthesized silver nanoparticles actively degrade methylene blue and malachite green by about 38% and 91%, respectively, at 2 h of UV light irradiation. In contrast, at the exposure of visible light, about 80% of methylene blue and 38% of malachite green degradation was achieved at 2 h irradiation (Li et al., 2020). Another report by Aravind et al. (2021) demonstrated about 78% of methylene blue degradation was obtained at 120 min by treating the dye solution with silver nanoparticles synthesized with flower extract of jasmine. Sunlight was used as irradiation source for the effective degradation of methylene blue. The *Alpinia nigra* fruit extract synthesized silver nanoparticles energetically participated in the degradation of three dyes: orange G, methyl orange, and rhodamine B. Under sunlight exposure, the silver nanoparticles degraded the orange G, methyl orange, and rhodamine B dyes by about 79.9%, 83.4%, and 85.9%, respectively, in 90 min (Baruah et al., 2019).

6.2.2 Nanosensor

Silver nanoparticles are one of the plasmonic nanoparticles with excellent optical properties, due to which, they became a potential candidate for sensing applications (Jouyban and Rahimpour, 2020). The 25 nm sized spherical shaped silver nanoparticles synthesized with leaf extract of Sonchus arvensis was successfully employed to sense selectively the Hg^{2+} and Fe^{3+} ions. Among monovalent, divalent, and trivalent metal ions (Pb^{2+}, Al^{3+}, Fe^{3+}, Li^+, Mn^{2+}, Hg^{2+}, Ni^{2+}, Co^{2+}, Cd^{2+}, Cr^{3+}, and Cu^{2+}), the green synthesized silver nanoparticles particularly decolorize the Hg^{2+} and Fe^{3+} ions with a detection limit of 10^{-3} M (Chandraker et al., 2019). Farhadi and his co-workers (2012) also demonstrated the selective sensing of Hg^{2+} using the soap-root plant extract synthesized silver nanoparticles. The Hg^{2+} in the silver nanoparticles solution can remove the biological moieties (stabilizing agents) of soap-root plant from the surface of silver nanoparticles and bind with its surface through oxidation-reduction reaction. The detection limit of Hg^{2+} was found to be 2.2×10^{-6} m/L. Similarly, Ravi et al. (2013) reported the selective detection of Hg^{2+} ions using silver nanoparticles produced with the Citrus limon fruit extract. At 3.2 and 8.5 pH, the silver nanoparticles specifically sense the Hg^{2+} ions. Another interesting report by Ihsan et al. (2015) documented the *Amomum subulatum* leaf extract synthesized silver nanoparticles proved its Zn^{2+} ions sensing ability by colorimetric method. The silver nanoparticles showed the maximum detection limit of Zn^{2+} ions was 3.5×10^{-6} M. An important report of Jabariyan and Zanjanchi (2019) demonstrated rapid and facile detection of Cd^{2+} ions using silver nanoparticles fabricated with juice extract of grapes. The minimum detection limit was 4.95 µmol/L. The silver nanoparticles modified with L-cysteine selectively sense the Ni^{2+} ions. The L-cysteine acts as probe for binding of Ni^{2+} ions on the surface of silver nanoparticles. The carboxyl and amine functional groups of L-cysteine facilitate the attachment of Ni^{2+} ions on the silver surface (Khwannimit et al., 2019).

The discharge of pollutant antibiotics from pharmaceuticals and household waste into the environment is another important risk to human health and the ecosystem. Worldwide, concern is increasing drastically about the presence of nonbiodegradable antibiotics in drinking water. Nowadays, nanotechnology-based strategies are being explored to eliminate these noxious antibiotics (Malakootian et al., 2019), in which, silver nanoparticles are being exploited in degradation and sensing of antibiotic contaminants due to their magnificent catalytic properties (Khan et al., 2016) (Fig. 3). Silver nanoparticles synthesized with the edible mushroom *P. ostreatus* effectively degraded the ampicillin antibiotic by about 96.5% under sunlight irradiation at 4h incubation. The required silver nanoparticles concentration was about 5 ppm to achieve 96.5% degradation at pH 6 (Jassal et al., 2020). The Epigallocatechin gallate (EGG) is one of the phenolic compounds mostly found in green tea, which was successfully used as reducer and stabilizer in silver nanoparticle production, and the EGG stabilized silver nanoparticles were employed as nanosensors for detection of industrially important kanamycin and ionic sulfide. Among various antibiotics (rifamycin, fluconazole, tetracycline, ampicillin, chloramphenicol, and ciprofloxacin), silver nanoparticles selectively sensed kanamycin and it was visually identified by red color formation from yellow. The silver nanoparticles detect the kanamycin from different molar concentration from 0.5 to 50 μM. In that, silver nanoparticles sense the kanamycin even at the volume of 1 μM. In addition, the minimum detection limit of sulfide ions of about 1.62 μM was achieved with silver nanoparticles (Singh et al., 2018b). Silver nanoparticles stabilized with natural stabilizer gallic acid were successfully used to detect aminoglycoside antibiotic with a detection rate of 39 pmol/L (Ghodake et al., 2020). Red cabbage extract synthesized silver nanoparticles sense the amikacin sulfate by forming composite with reduced graphene oxide using nickel foam electrode. The limit of detection of amikacin sulfate was about 38 nm in the range 0.015–15 μM (Sharma et al., 2020).

6.3 Agriculture

The beneficial application of silver nanoparticles in agricultural crops is still not fully understood in a detailed manner, in that reports are available with certain limitations. However, fewer field experiment trails are available to demonstrate the importance of silver nanoparticles on crop improvement. Due to potential antimicrobial activity of silver nanoparticles, they can act as a nanoshield to protect plants from hazardous phytopathogens (Mishra and Singh, 2015) (Fig. 3). Silver nanoparticles fabricated using mature pine cone extract were for antibacterial agents against plant pathogens including Bacillus thuringiensis, *Burkholderia glumae*, *Pseudomonas syringae*, and *Xanthomonas oryzae*. The gram-positive B. thuringiensis (zone on inhibition (ZOI) 10 mm) was more susceptible to silver nanoparticles compared to gram-negative *B. glumae* (ZOI 8.6 mm), *X. oryzae* (ZOI 4.2 mm), and *P. syringae* (ZOI 8.9 mm). This result proved the green synthesized silver nanoparticles have potent antibacterial activity against toxic crop pathogens (Velmurugan et al., 2013). Lee et al. (2013) synthesized silver nanoparticles using cow's milk as a source and utilized it against important plant fungal pathogens *Monilinia* sp., *Pyricularia* sp., and *Colletotrichum coccodes*. The silver nanoparticles, at about 2 mM concentration, were sufficient for the inhibition of fungal growth about 86.5%, 83.5%, and 87.1% for *Monilinia* sp., *Pyricularia* sp., and *C. coccodes*, respectively. Fouda et al. (2020) have done an interesting report on the effect of starch-stabilized silver nanoparticles in onion cultivation. They used the silver as nano-fertilizer with different concentration ranging from 5 to 100 ppm to enhance the onion yield. In three time intervals, 25, 40, and 55 days, they applied the silver nanoparticle by foliar spray technique. After 120 days observation, among different concentrations, the 20 ppm of silver nanoparticles showed highest values (mean) for morphology, harvest, and quality features. This study is the best evidence to substantiate the green synthesized silver nanoparticles' excellent agriculture application through increased crop production.

6.4 Food

Food safety is an important area in food and beverage industries, which includes production, storage, and preservation of food stuffs and products. The presence of noxious compounds along with food products creates great health concern and results in economic loss for food industries and risk to human health. Through food chain processes, contamination with polluted food products can cause severe damage to terrestrial and aquatic ecosystems. Hence, the food safety process has attained prodigious attention worldwide. In recent years, nanotechnology-based tools have been applied in food safety processes for advancement in innovation and as alternatives to traditional analytical approaches (Liu et al., 2018) (Fig. 3). Due to intrinsic catalytic and antimicrobial activities, the silver nanoparticles are prevalently used as sensing materials to detect toxic food contaminants (Selvaraj et al., 2019). The silver nanoparticles synthesized with D-glucose effectively sense the oxytetracycline antibiotic in honey samples. Among different concentrations of oxytetracycline (5, 10, 15, and 20 ppb) in spiked honey samples, the silver nanoparticles detect oxytetracycline, even at a low level of 5 ppb, with potassium carbonate by surface enhanced Raman spectroscopy (SERS) (Fa et al., 2019). Similarly, Pan et al. (2018) have developed a fast detection protocol using silver nanoparticles for identification of mycotoxin alternariol in pear fruit extract. The silver nanoparticles are coated with pyridine, due to its strong affinity with alternariol compound. Thereby, they have achieved the detection limit of 1.30 μg/L of alternariol by the pyridine-modified silver nanoparticles. Another food contaminant, 3-monochloropropane-1,2-diol, was successfully detected rapidly by cysteine-modified silver nanoparticles (C-Ag NPs). After addition of 3-monochloropropane-1,2-diol to the C-Ag NPs solution, the yellow color was changed to pink at 5 min with the pH 9.3 and temperature 100°C. The 3-monochloropropane-1,2-diol detection limit was 0.084 μg/mL achieved by C-Ag NPs (Martin et al., 2021). The silver nanoparticles, along with the combination of gold and mercaptooctane, were used for the detection of antifungal compounds thiram residues, and tricyclazole in the extract of pear fruits. The core gold and shell silver nanoparticles exhibited a rapid, increased sensitivity, and enhanced Raman scattering in the sensing of thiram residues and tricyclazole. The detection limit of thiram residues and tricyclazole was 0.003 and 0.005 ppm, respectively (Hussain et al., 2021). These reports suggest that modified and unmodified silver nanoparticles act as potential nanosensors to explore contaminants in food industries.

7 Conclusion

Novel nano-based strategies have been explored in various disciplines to develop innovative approaches and sustainable ecosystems for humankind. Nowadays, silver nanoparticles are the "magic word," providing multifaceted traits in all areas including medicine, environment, agriculture, food, and cosmetics. In this chapter, we discussed the synthesis of silver nanoparticles using mushroom extract, and also the safety protocols of green synthesized silver nanoparticles and its versatile applications in different fields. Mushroom extract is a source of phyto/biomolecules such as alkaloids, terpenoids, flavonoids, phenolic compounds, proteins, and amino acids. The green synthesis protocol has gained interest due to its toxic chemical-free, rapid, eco-friendly qualities. The multifunctional applications of silver nanoparticles should bring novel nanophotocatalysts, nano-antimicrobials, nanosensors, and nanoshields for the advancement of medicine, the environment, and agricultural applications.

Acknowledgments

The authors thank biorender and presentationgo for drawing in this chapter.

References

Abbasi, E., Milani, M., Fekri Aval, S., Kouhi, M., Akbarzadeh, A., Nasrabadi, H.T., Nikasa, P., Joo, S.W., Haniehpour, Y., Nejati-Koshki, K., Samiei, M., 2016. Silver nanoparticles: synthesis methods, bio-applications and properties. Crit. Rev. Microbiol. 42 (2), 173–180.

Abou El-Nour, K.M., Eftaiha, A.A., Al-Warthan, A., Ammar, R.A., 2010. Synthesis and applications of silver nanoparticles. Arab. J. Chem. 3 (3), 135–140.

Ajitha, B., Reddy, Y.A.K., Reddy, P.S., 2015. Green synthesis and characterization of silver nanoparticles using *Lantana camara* leaf extract. Mater. Sci. Eng. C 49, 373–381.

Al-Bahrani, R., Raman, J., Lakshmanan, H., Hassan, A.A., Sabaratnam, V., 2017. Green synthesis of silver nanoparticles using tree oyster mushroom *Pleurotus ostreatus* and its inhibitory activity against pathogenic bacteria. Mater. Lett. 186, 21–25.

Alexander, J.W., 2009. History of the medical use of silver. Surg. Infect. (Larchmt.) 10 (3), 289–292.

Anthony, K.J.P., Murugan, M., Jeyaraj, M., Rathinam, N.K., Sangiliyandi, G., 2014. Synthesis of silver nanoparticles using pine mushroom extract: a potential antimicrobial agent against *E. coli* and *B. subtilis*. J. Ind. Eng. Chem. 20 (4), 2325–2331.

Aravind, M., Ahmad, A., Ahmad, I., Amalanathan, M., Naseem, K., Mary, S.M.M., Parvathiraja, C., Hussain, S., Algarni, T.S., Pervaiz, M., Zuber, M., 2021. Critical green routing synthesis of silver NPs using jasmine flower extract for biological activities and photocatalytical degradation of methylene blue. J. Environ. Chem. Eng. 9 (1), 104877.

Augustine, R., Hasan, A., 2020. Emerging applications of biocompatible phytosynthesized metal/metal oxide nanoparticles in healthcare. J. Drug Delivery Sci. Technol. 56, 101516.

Aygun, A., Ozdemir, S., Gulcan, M., Cellat, K., Şen, F., 2020. Synthesis and characterization of Reishi mushroom-mediated green synthesis of silver nanoparticles for the biochemical applications. J. Pharm. Biomed. Anal. 178, 112970.

Azizi, Z., Pourseyedi, S., Khatami, M., Mohammadi, H., 2016. *Stachys lavandulifolia* and *Lathyrus* sp. mediated for green synthesis of silver nanoparticles and evaluation its antifungal activity against *Dothiorella sarmentorum*. J. Clust. Sci. 27 (5), 1613–1628.

Bahrami-Teimoori, B., Nikparast, Y., Hojatianfar, M., Akhlaghi, M., Ghorbani, R., Pourianfar, H.R., 2017. Characterisation and antifungal activity of silver nanoparticles biologically synthesised by *Amaranthus retroflexus* leaf extract. J. Exp. Nanosci. 12 (1), 129–139.

Balan, K., Qing, W., Wang, Y., Liu, X., Palvannan, T., Wang, Y., Ma, F., Zhang, Y., 2016. Antidiabetic activity of silver nanoparticles from green synthesis using *Lonicera japonica* leaf extract. RSC Adv. 6 (46), 40162–40168.

Baruah, D., Yadav, R.N.S., Yadav, A., Das, A.M., 2019. *Alpinia nigra* fruits mediated synthesis of silver nanoparticles and their antimicrobial and photocatalytic activities. J. Photochem. Photobiol. B Biol. 201, 111649.

Bhat, R., Deshpande, R., Ganachari, S.V., Huh, D.S., Venkataraman, A., 2011. Photoirradiated biosynthesis of silver nanoparticles using edible mushroom *Pleurotus florida* and their antibacterial activity studies. Bioinorg. Chem. Appl. 2011, 1–7.

Chandraker, S.K., Ghosh, M.K., Lal, M., Ghorai, T.K., Shukla, R., 2019. Colorimetric sensing of Fe^{3+} and Hg^{2+} and photocatalytic activity of green synthesized silver nanoparticles from the leaf extract of *Sonchus arvensis* L. New J. Chem. 43 (46), 18175–18183.

Chhipa, H., 2019. Mycosynthesis of nanoparticles for smart agricultural practice: a green and eco-friendly approach. In: Green Synthesis, Characterization and Applications of Nanoparticles. Elsevier, pp. 87–109.

Chung, I.M., Park, I., Seung-Hyun, K., Thiruvengadam, M., Rajakumar, G., 2016. Plant-mediated synthesis of silver nanoparticles: their characteristic properties and therapeutic applications. Nanoscale Res. Lett. 11 (1), 1–14.

Das, B., Dash, S.K., Mandal, D., Ghosh, T., Chattopadhyay, S., Tripathy, S., Das, S., Dey, S.K., Das, D., Roy, S., 2017. Green synthesized silver nanoparticles destroy multidrug resistant bacteria via reactive oxygen species mediated membrane damage. Arab. J. Chem. 10 (6), 862–876.

Debnath, G., Das, P., Saha, A.K., 2019. Green synthesis of silver nanoparticles using mushroom extract of *Pleurotus giganteus*: characterization, antimicrobial, and α-amylase inhibitory activity. Bionanoscience 9 (3), 611–619.

Devatha, C.P., Thalla, A.K., 2018. Green synthesis of nanomaterials. In: Synthesis of Inorganic Nanomaterials. Woodhead Publishing, pp. 169–184.

Dhanasekaran, D., Latha, S., Saha, S., Thajuddin, N., Panneerselvam, A., 2013. Extracellular biosynthesis, characterisation and in-vitro antibacterial potential of silver nanoparticles using *Agaricus bisporus*. J. Exp. Nanosci. 8 (4), 579–588.

Fa, A.G., Pignanelli, F., López-Corral, I., Faccio, R., Juan, A., Di Nezio, M.S., 2019. Detection of oxytetracycline in honey using SERS on silver nanoparticles. TrAC Trends Anal. Chem. 121, 115673.

Faisal, S., Khan, M.A., Jan, H., Shah, S.A., Shah, S., Rizwan, M., Ullah, W., Akbar, M.T., Redaina, 2020. Edible mushroom (*Flammulina velutipes*) as biosource for silver nanoparticles: from synthesis to diverse biomedical and environmental applications. Nanotechnology 32 (6), 065101.

Farhadi, K., Forough, M., Molaei, R., Hajizadeh, S., Rafipour, A., 2012. Highly selective Hg^{2+} colorimetric sensor using green synthesized and unmodified silver nanoparticles. Sensors Actuators B Chem. 161 (1), 880–885.

Fouda, M.M., Abdelsalam, N.R., El-Naggar, M.E., Zaitoun, A.F., Salim, B.M., Bin-Jumah, M., Allam, A., Abo-Marzoka, A.A., Kandil, E.E., 2020. Impact of high throughput green synthesized silver nanoparticles on agronomic traits of onion. Int. J. Biol. Macromol. 149, 1304–1317.

Funez, A.A., Isabel Haza, A., Mateo, D., Morales, P., 2013. In vitro evaluation of silver nanoparticles on human tumoral and normal cells. Toxicol. Mech. Methods 23 (3), 153–160.

Gautam, A., Van Veggel, F.C., 2013. Synthesis of nanoparticles, their biocompatibility, and toxicity behaviour for biomedical applications. J. Mater. Chem. B 1 (39), 5186–5200.

Ghodake, G., Shinde, S., Saratale, R.G., Kadam, A., Saratale, G.D., Syed, A., Marraiki, N., Elgorban, A.M., Kim, D.Y., 2020. Silver nanoparticle probe for colorimetric detection of aminoglycoside antibiotics: picomolar-level sensitivity toward streptomycin in water, serum, and milk samples. J. Sci. Food Agric. 100 (2), 874–884.

Ghormade, V., Deshpande, M.V., Paknikar, K.M., 2011. Perspectives for nano-biotechnology enabled protection and nutrition of plants. Biotechnol. Adv. 29 (6), 792–803.

Govindappa, M., Hemashekhar, B., Arthikala, M.K., Rai, V.R., Ramachandra, Y.L., 2018. Characterization, antibacterial, antioxidant, antidiabetic, anti-inflammatory and antityrosinase activity of green synthesized silver nanoparticles using *Calophyllum tomentosum* leaves extract. Results Phys. 9, 400–408.

Gudikandula, K., Charya Maringanti, S., 2016. Synthesis of silver nanoparticles by chemical and biological methods and their antimicrobial properties. J. Exp. Nanosci. 11 (9), 714–721.

Guilger-Casagrande, M., Lima, R.D., 2019. Synthesis of silver nanoparticles mediated by fungi: a review. Front. Bioeng. Biotechnol. 7, 287.

Gurunathan, S., Raman, J., Abd Malek, S.N., John, P.A., Vikineswary, S., 2013. Green synthesis of silver nanoparticles using *Ganoderma neo-japonicum* Imazeki: a potential cytotoxic agent against breast cancer cells. Int. J. Nanomedicine 8, 4399.

Hebbalalu, D., Lalley, J., Nadagouda, M.N., Varma, R.S., 2013. Greener techniques for the synthesis of silver nanoparticles using plant extracts, enzymes, bacteria, biodegradable polymers, and microwaves. ACS Sustain. Chem. Eng. 1 (7), 703–712.

Hussain, N., Pu, H., Sun, D.W., 2021. Core size optimized silver coated gold nanoparticles for rapid screening of tricyclazole and thiram residues in pear extracts using SERS. Food Chem. 350, 129025.

Ihsan, M., Niaz, A., Rahim, A., Zaman, M.I., Arain, M.B., Sharif, T., Najeeb, M., 2015. Biologically synthesized silver nanoparticle-based colorimetric sensor for the selective detection of Zn^{2+}. RSC Adv. 5 (111), 91158–91165.

Jabariyan, S., Zanjanchi, M.A., 2019. Colorimetric detection of cadmium ions using modified silver nanoparticles. Appl. Phys. A 125 (12), 1–10.

Jang, S.J., Yang, I.J., Tettey, C.O., Kim, K.M., Shin, H.M., 2016. In-vitro anticancer activity of green synthesized silver nanoparticles on MCF-7 human breast cancer cells. Mater. Sci. Eng. C 68, 430–435.

Jassal, P., Khajuria, R., Sharma, R., Debnath, P., Verma, S., Johnson, A., Kumar, S., 2020. Photocatalytic degradation of ampicillin using silver nanoparticles biosynthesised by *Pleurotus ostreatus*. BioTechnologia 101 (1), 5–14.

Jebril, S., Jenana, R.K.B., Dridi, C., 2020. Green synthesis of silver nanoparticles using *Melia azedarach* leaf extract and their antifungal activities: in vitro and in vivo. Mater. Chem. Phys. 248, 122898.

Jiao, M., Zhang, P., Meng, J., Li, Y., Liu, C., Luo, X., Gao, M., 2018. Recent advancements in biocompatible inorganic nanoparticles towards biomedical applications. Biomater. Sci. 6 (4), 726–745.

Jouyban, A., Rahimpour, E., 2020. Optical sensors based on silver nanoparticles for determination of pharmaceuticals: an overview of advances in the last decade. Talanta 217, 121071.

Kanniah, P., Radhamani, J., Chelliah, P., Muthusamy, N., Sathiya Balasingh Thangapandi, E. J.J., Thangapandi, J.S., Balakrishnan, S., Shanmugam, R., 2020. Green synthesis of multifaceted silver nanoparticles using the flower extract of *Aerva lanata* and evaluation of its biological and environmental applications. ChemistrySelect 5, 2322–2331.

Kathiravan, V., Ravi, S., Ashokkumar, S., 2014. Synthesis of silver nanoparticles from *Melia dubia* leaf extract and their in vitro anticancer activity. Spectrochim. Acta A Mol. Biomol. Spectrosc. 130, 116–121.

Khan, M.R., Rizvi, T.F., 2017. Application of nanofertilizer and nanopesticides for improvements in crop production and protection. In: Nanoscience and Plant–Soil Systems. Springer, Cham, pp. 405–427.

Khan, F.U., Chen, Y., Khan, N.U., Khan, Z.U.H., Khan, A.U., Ahmad, A., Tahir, K., Wang, L., Khan, M.R., Wan, P., 2016. Antioxidant and catalytic applications of silver nanoparticles using *Dimocarpus longan* seed extract as a reducing and stabilizing agent. J. Photochem. Photobiol. B Biol. 164, 344–351.

Khatami, M., Pourseyedi, S., Khatami, M., Hamidi, H., Zaeifi, M., Soltani, L., 2015. Synthesis of silver nanoparticles using seed exudates of *Sinapis arvensis* as a novel bioresource, and evaluation of their antifungal activity. Bioresour. Bioprocess. 2 (1), 1–7.

Khorrami, S., Zarrabi, A., Khaleghi, M., Danaei, M., Mozafari, M.R., 2018. Selective cytotoxicity of green synthesized silver nanoparticles against the MCF-7 tumor cell line and their enhanced antioxidant and antimicrobial properties. Int. J. Nanomedicine 13, 8013.

Khwannimit, D., Jaikrajang, N., Dokmaisrijan, S., Rattanakit, P., 2019. Biologically green synthesis of silver nanoparticles from *Citrullus lanatus* extract with L-cysteine addition and investigation of colorimetric sensing of nickel (II) potential. Mater. Today Proc. 17, 2028–2038.

Krishnaraj, C., Jagan, E.G., Rajasekar, S., Selvakumar, P., Kalaichelvan, P.T., Mohan, N.J.C. S.B.B., 2010. Synthesis of silver nanoparticles using *Acalypha indica* leaf extracts and its antibacterial activity against water borne pathogens. Colloids Surf. B Biointerfaces 76 (1), 50–56.

Kumar, I., Mondal, M., Sakthivel, N., 2019. Green synthesis of phytogenic nanoparticles. In: Green Synthesis, Characterization and Applications of Nanoparticles. Elsevier, pp. 37–73.

Lara, H.H., Ayala-Núnez, N.V., Turrent, L.D.C.I., Padilla, C.R., 2010. Bactericidal effect of silver nanoparticles against multidrug-resistant bacteria. World J. Microbiol. Biotechnol. 26 (4), 615–621.

Le Ouay, B., Stellacci, F., 2015. Antibacterial activity of silver nanoparticles: a surface science insight. Nano Today 10 (3), 339–354.

Lee, K.J., Park, S.H., Govarthanan, M., Hwang, P.H., Seo, Y.S., Cho, M., Lee, W.H., Lee, J.Y., Kamala-Kannan, S., Oh, B.T., 2013. Synthesis of silver nanoparticles using cow milk and their antifungal activity against phytopathogens. Mater. Lett. 105, 128–131.

Lee, S.Y., Kang, D., Jeong, S., Do, H.T., Kim, J.H., 2020. Photocatalytic degradation of rhodamine B dye by TiO_2 and gold nanoparticles supported on a floating porous polydimethylsiloxane sponge under ultraviolet and visible light irradiation. ACS Omega 5 (8), 4233–4241.

Li, J.F., Liu, Y.C., Chokkalingam, M., Rupa, E.J., Mathiyalagan, R., Hurh, J., Ahn, J.C., Park, J.K., Pu, J.Y., Yang, D.C., 2020. Phytosynthesis of silver nanoparticles using rhizome extract of *Alpinia officinarum* and their photocatalytic removal of dye under UV and visible light irradiation. Optik 208, 164521.

Lim, H., Yusuf, M., Song, S., Park, S., Park, K.H., 2021. Efficient photocatalytic degradation of dyes using photo-deposited Ag nanoparticles on ZnO structures: simple morphological control of ZnO. RSC Adv. 11 (15), 8709–8717.

Liu, J.M., Hu, Y., Yang, Y.K., Liu, H., Fang, G.Z., Lu, X., Wang, S., 2018. Emerging functional nanomaterials for the detection of food contaminants. Trends Food Sci. Technol. 71, 94–106.

Liu, J., Zhang, F., Ravikanth, V., Olajide, O.A., Li, C., Wei, L.X., 2019. Chemical compositions of metals in Bhasmas and Tibetan Zuotai are a major determinant of their therapeutic effects and toxicity. Evid. Based Complement. Alternat. Med. 2019, 1–13.

Loo, Y.Y., Rukayadi, Y., Nor-Khaizura, M.A.R., Kuan, C.H., Chieng, B.W., Nishibuchi, M., Radu, S., 2018. In vitro antimicrobial activity of green synthesized silver nanoparticles against selected gram-negative foodborne pathogens. Front. Microbiol. 9, 1555.

Mahmoudi, F., Mahmoudi, F., Gollo, K.H., Amini, M.M., 2021. Biosynthesis of novel silver nanoparticles using *Eryngium thyrsoideum Boiss* extract and comparison of their antidiabetic activity with chemical synthesized silver nanoparticles in diabetic rats. Biol. Trace Elem. Res. 199, 1967–1978.

Malakootian, M., Yaseri, M., Faraji, M., 2019. Removal of antibiotics from aqueous solutions by nanoparticles: a systematic review and meta-analysis. Environ. Sci. Pollut. Res. 26 (9), 8444–8458.

Martin, A.A., Fodjo, E.K., Marc, G.B.I., Albert, T., Kong, C., 2021. Simple and rapid detection of free 3-monochloropropane-1, 2-diol based on cysteine modified silver nanoparticles. Food Chem. 338, 127787.

Mathur, P., Jha, S., Ramteke, S., Jain, N.K., 2018. Pharmaceutical aspects of silver nanoparticles. Artif. Cells Nanomed. Biotechnol. 46 (Suppl. 1), 115–126.

Maurya, S., Bhardwaj, A.K., Gupta, K.K., Agarwal, S., Kushwaha, A., Chaturvedi, V., Pathak, R.K., Gopal, R., Uttam, K.N., Singht, A.K., Verma, V., Singh, M.P., 2016. Green synthesis of silver nanoparticles using *Pleurotus* and its bactericidal activity. Cell. Mol. Biol. 62, 131.

Mavaei, M., Chahardoli, A., Shokoohinia, Y., Khoshroo, A., Fattahi, A., 2020. One-step synthesized silver nanoparticles using isoimperatorin: evaluation of photocatalytic, and electrochemical activities. Sci. Rep. 10 (1), 1–12.

Mikhailova, E.O., 2020. Silver nanoparticles: mechanism of action and probable bio-application. J. Funct. Biomater. 11 (4), 84.

Mishra, S., Singh, H.B., 2015. Biosynthesized silver nanoparticles as a nanoweapon against phytopathogens: exploring their scope and potential in agriculture. Appl. Microbiol. Biotechnol. 99 (3), 1097–1107.

Mpenyana-Monyatsi, L., Mthombeni, N.H., Onyango, M.S., Momba, M.N., 2012. Cost-effective filter materials coated with silver nanoparticles for the removal of pathogenic bacteria in groundwater. Int. J. Environ. Res. Public Health 9 (1), 244–271.

Musa, S.F., Yeat, T.S., Kamal, L.Z.M., Tabana, Y.M., Ahmed, M.A., El Ouweini, A., Lim, V., Keong, L.C., Sandai, D., 2018. *Pleurotus sajor-caju* can be used to synthesize silver nanoparticles with antifungal activity against *Candida albicans*. J. Sci. Food Agric. 98 (3), 1197–1207.

Nagajyothi, P.C., Sreekanth, T.V.M., Lee, J.I., Lee, K.D., 2014. Mycosynthesis: antibacterial, antioxidant and antiproliferative activities of silver nanoparticles synthesized from *Inonotus obliquus* (Chaga mushroom) extract. J. Photochem. Photobiol. B Biol. 130, 299–304.

Narasimha, G., Papaiah, S., Praveen, B., Sridevi, A., Mallikarjuna, K., Deva Prasad Raju, B., 2013. Fungicidal activity of silver nanoparticles synthesized by *Agaricus bisporus* (white button mushrooms). J. Nanosci. Nanotechnol. 7 (3), 114–115.

Noah, N., 2019. Green synthesis: characterization and application of silver and gold nanoparticles. In: Green Synthesis, Characterization and Applications of Nanoparticles. Elsevier, pp. 111–135.

Owaid, M.N., 2019. Green synthesis of silver nanoparticles by *Pleurotus* (oyster mushroom) and their bioactivity. Environ. Nanotechnol. Monit. Manag. 12, 100256.

Paladini, F., Pollini, M., 2019. Antimicrobial silver nanoparticles for wound healing application: progress and future trends. Materials 12 (16), 2540.

Pan, T.T., Sun, D.W., Pu, H., Wei, Q., 2018. Simple approach for the rapid detection of alternariol in pear fruit by surface-enhanced Raman scattering with pyridine-modified silver nanoparticles. J. Agric. Food Chem. 66 (9), 2180–2187.

Paulkumar, K., Gnanajobitha, G., Vanaja, M., Rajeshkumar, S., Malarkodi, C., Pandian, K., Annadurai, G., 2014. *Piper nigrum* leaf and stem assisted green synthesis of silver nanoparticles and evaluation of its antibacterial activity against agricultural plant pathogens. Scientific World Journal 2014, 1–9.

Pedahzur, R., Lev, O., Fattal, B., Shuval, H.I., 1995. The interaction of silver ions and hydrogen peroxide in the inactivation of *E. coli*: a preliminary evaluation of a new long acting residual drinking water disinfectant. Water Sci. Technol. 31 (5–6), 123–129.

Philip, D., 2009. Biosynthesis of Au, Ag and Au–Ag nanoparticles using edible mushroom extract. Spectrochim. Acta A Mol. Biomol. Spectrosc. 73 (2), 374–381.

Prosposito, P., Burratti, L., Venditti, I., 2020. Silver nanoparticles as colorimetric sensors for water pollutants. Chem. Aust. 8 (2), 26.

Qais, F.A., Shafiq, A., Khan, H.M., Husain, F.M., Khan, R.A., Alenazi, B., Alsalme, A., Ahmad, I., 2019. Antibacterial effect of silver nanoparticles synthesized using *Murraya koenigii* (L.) against multidrug-resistant pathogens. Bioinorg. Chem. Appl. 2019, 1–11.

Rai, A.P., Tripathi, S., Tiwari, O.P., 2019. Evaluation of antidiabetic activity of Rajata bhasma. Int. J. Green Pharm. 13 (04), 359–363.

Rajaram, K., Aiswarya, D.C., Sureshkumar, P., 2015. Green synthesis of silver nanoparticle using *Tephrosia tinctoria* and its antidiabetic activity. Mater. Lett. 138, 251–254.

Raman, J., Reddy, G.R., Lakshmanan, H., Selvaraj, V., Gajendran, B., Nanjian, R., Chinnasamy, A., Sabaratnam, V., 2015. Mycosynthesis and characterization of silver nanoparticles from *Pleurotus djamor var. roseus* and their in vitro cytotoxicity effect on PC3 cells. Process Biochem. 50 (1), 140–147.

Ravi, S.S., Christena, L.R., SaiSubramanian, N., Anthony, S.P., 2013. Green synthesized silver nanoparticles for selective colorimetric sensing of Hg^{2+} in aqueous solution at wide pH range. Analyst 138 (15), 4370–4377.

Ray, S., Sarkar, S., Kundu, S., 2011. Extracellular biosynthesis of silver nanoparticles using the mycorrhizal mushroom *Tricholoma crassum* (Berk.) Sacc: its antimicrobial activity against pathogenic bacteria and fungus, including multidrug resistant plant and human bacteria. Dig. J. Nanomater. Biostruct. 6, 1289–1299.

Reis, F.S., Barros, L., Martins, A., Ferreira, I.C., 2012. Chemical composition and nutritional value of the most widely appreciated cultivated mushrooms: an inter-species comparative study. Food Chem. Toxicol. 50 (2), 191–197.

Ren, L., Hemar, Y., Perera, C.O., Lewis, G., Krissansen, G.W., Buchanan, P.K., 2014. Antibacterial and antioxidant activities of aqueous extracts of eight edible mushrooms. Bioact. Carbohydr. Diet. Fibre 3 (2), 41–51.

Schaming, D., Remita, H., 2015. Nanotechnology: from the ancient time to nowadays. Found. Chem. 17 (3), 187–205.

Selvaraj, V., Sagadevan, S., Muthukrishnan, L., Johan, M.R., Podder, J., 2019. Eco-friendly approach in synthesis of silver nanoparticles and evaluation of optical, surface morphological and antimicrobial properties. J. Nanostructure Chem. 9 (2), 153–162.

Sen, I.K., Mandal, A.K., Chakraborti, S., Dey, B., Chakraborty, R., Islam, S.S., 2013. Green synthesis of silver nanoparticles using glucan from mushroom and study of antibacterial activity. Int. J. Biol. Macromol. 62, 439–449.

Shah, M., Nawaz, S., Jan, H., Uddin, N., Ali, A., Anjum, S., Giglioli-Guivarch, N., Hano, C., Abbasi, B.H., 2020. Synthesis of bio-mediated silver nanoparticles from *Silybum marianum* and their biological and clinical activities. Mater. Sci. Eng. C 112, 110889.

Sharma, N., Selvam, S.P., Yun, K., 2020. Electrochemical detection of amikacin sulphate using reduced graphene oxide and silver nanoparticles nanocomposite. Appl. Surf. Sci. 512, 145742.

Silva, L.P., Pereira, T.M., Bonatto, C.C., 2019. Frontiers and perspectives in the green synthesis of silver nanoparticles. In: Green Synthesis, Characterization and Applications of Nanoparticles. Elsevier, pp. 137–164.

Singh, J., Dutta, T., Kim, K.H., Rawat, M., Samddar, P., Kumar, P., 2018a. 'Green' synthesis of metals and their oxide nanoparticles: applications for environmental remediation. J. Nanobiotechnol. 16 (1), 1–24.

Singh, R.K., Panigrahi, B., Mishra, S., Das, B., Jayabalan, R., Parhi, P.K., Mandal, D., 2018b. pH triggered green synthesized silver nanoparticles toward selective colorimetric detection of kanamycin and hazardous sulfide ions. J. Mol. Liq. 269, 269–277.

Sriramulu, M., Sumathi, S., 2017. Photocatalytic, antioxidant, antibacterial and antiinflammatory activity of silver nanoparticles synthesised using forest and edible mushroom. Adv. Nat. Sci. Nanosci. Nanotechnol. 8 (4), 045012.

Syed, A., Saraswati, S., Kundu, G.C., Ahmad, A., 2013. Biological synthesis of silver nanoparticles using the fungus *Humicola* sp. and evaluation of their cytoxicity using normal and cancer cell lines. Spectrochim. Acta A Mol. Biomol. Spectrosc. 114, 144–147.

Tahir, K., Nazir, S., Li, B., Khan, A.U., Khan, Z.U.H., Ahmad, A., Khan, F.U., 2015. An efficient photo catalytic activity of green synthesized silver nanoparticles using *Salvadora persica* stem extract. Sep. Purif. Technol. 150, 316–324.

Talapko, J., Matijevic, T., Juzbasic, M., Antolovic-Pozgain, A., Skrlec, I., 2020. Antibacterial activity of silver and its application in dentistry, cardiology and dermatology. Microorganisms 8 (9), 1400.

Thivaharan, V., Ramesh, V., Raja, S., 2018. Green synthesis of silver nanoparticles for biomedical and environmental applications. In: Green Metal Nanoparticles: Synthesis, Characterization and Their Applications. Wiley-Scrivener Publishing LLC, pp. 387–439.

Thomas, B., Vithiya, B., Prasad, T., Mohamed, S.B., Magdalane, C.M., Kaviyarasu, K., Maaza, M., 2019. Antioxidant and photocatalytic activity of aqueous leaf extract mediated green synthesis of silver nanoparticles using *Passiflora edulis f. flavicarpa*. J. Nanosci. Nanotechnol. 19 (5), 2640–2648.

Valverde, M.E., Hernández-Pérez, T., Paredes-Lopez, O., 2015. Edible mushrooms: improving human health and promoting quality life. Int. J. Microbiol. 2015, 1–14.

Velmurugan, P., Lee, S.M., Iydroose, M., Lee, K.J., Oh, B.T., 2013. Pine cone-mediated green synthesis of silver nanoparticles and their antibacterial activity against agricultural pathogens. Appl. Microbiol. Biotechnol. 97 (1), 361–368.

Vlasceanu, G.M., Marin, S., Tiplea, R.E., Bucur, I.R., Lemnaru, M., Marin, M.M., Grumezescu, A.M., Andronescu, E., 2016. Silver nanoparticles in cancer therapy. In: Nanobiomaterials in Cancer Therapy. William Andrew Publishing, pp. 29–56.

Wong, K.K., Liu, X., 2010. Silver nanoparticles—the real "silver bullet" in clinical medicine? Med. Chem. Commun. 1 (2), 125–131.

Yaqoob, A.A., Parveen, T., Umar, K., Mohamad Ibrahim, M.N., 2020. Role of nanomaterials in the treatment of wastewater: a review. Water 12 (2), 495.

Yarrappagaari, S., Gutha, R., Narayanaswamy, L., Thopireddy, L., Benne, L., Mohiyuddin, S.S., Vijayakumar, V., Saddala, R.R., 2020. Eco-friendly synthesis of silver nanoparticles from the whole plant of *Cleome viscosa* and evaluation of their characterization, antibacterial, antioxidant and antidiabetic properties. Saudi J. Biol. Sci. 27 (12), 3601–3614.

Yehia, R.S., Al-Sheikh, H., 2014. Biosynthesis and characterization of silver nanoparticles produced by *Pleurotus ostreatus* and their anticandidal and anticancer activities. World J. Microbiol. Biotechnol. 30 (11), 2797–2803.

Yesilot, S., Aydin, C., 2019. Silver nanoparticles; a new hope in cancer therapy? East. J. Med. 24 (1), 111–116.

Yu, S.J., Yin, Y.G., Liu, J.F., 2013. Silver nanoparticles in the environment. Environ. Sci. Process Impacts 15 (1), 78–92.

Zhang, F., 2017. Grand challenges for nanoscience and nanotechnology in energy and health. Front. Chem. 5, 80.

Zhang, X.F., Liu, Z.G., Shen, W., Gurunathan, S., 2016. Silver nanoparticles: synthesis, characterization, properties, applications, and therapeutic approaches. Int. J. Mol. Sci. 17 (9), 1534.

Zhang, L., Wei, Y., Wang, H., Wu, F., Zhao, Y., Liu, X., Wu, H., Wang, L., Su, H., 2020. Green synthesis of silver nanoparticles using mushroom *Flammulina velutipes* extract and their antibacterial activity against aquatic pathogens. Food Bioproc. Tech. 13 (11), 1908–1917.

Zhao, L., Lu, L., Wang, A., Zhang, H., Huang, M., Wu, H., Xing, B., Wang, Z., Ji, R., 2020. Nano-biotechnology in agriculture: use of nanomaterials to promote plant growth and stress tolerance. J. Agric. Food Chem. 68 (7), 1935–1947.

CHAPTER 16

Microbially synthesized silver nanoparticles: Mechanism and advantages—A review

Antony V. Samrot[a], P.J. Jane Cypriyana[b], S. Saigeetha[b], A. Jenifer Selvarani[b], Sajna Keeyari Purayil[a], and Paulraj Ponnaiah[a]

[a]School of Bioscience, Faculty of Medicine, Bioscience and Nursing, MAHSA University, Jenjarom, Selangor, Malaysia [b]Department of Biotechnology, School of Bio and Chemical Engineering, Sathyabama Institute of Science and Technology, Chennai, Tamil Nadu, India

1 Introduction

Nanotechnology is being recognized as a major interdisciplinary science evolving faster as it is contributing much in various fields including medicine, industry, etc. Nanomaterials commonly range between 1 and 100nm in size (Williams, 2008; Woldeamanuel et al., 2021) and these particles' properties vary highly from their corresponding bulk materials, enabling them to have many applications in various fields (Ahmed et al., 2016). Metal-based nanoparticles have earned attention due to their exceptional properties resulting from their size and shape. Metal-based nanoparticles include iron oxide nanoparticles, copper, silver, titanium, etc. Among these, silver nanoparticles (AgNPs) have received interest due to their unique properties including bioactivity, conductivity, catalytic activity, etc. (Ahmad et al., 2003; Klaus-Joerger et al., 2001). These silver nanoparticles are now used in many industries including medicine, health, textiles, etc. due to their better antibacterial activity and antiinflammatory activity (Abou El-Nour et al., 2010; Arshad et al., 2021; Samrot et al., 2018a, b). Furthermore, these nanosilver particles possess biosensing properties that depend on the nanoparticles' size, shape, and dielectric medium (Prabhu and Poulose, 2012). These properties and characteristics attracted keen interest, and the nanoparticles were initially produced using typical chemical methods, but chemically synthesized silver nanoparticles do possess toxicity (Samrot et al., 2018c). This forced researchers to find an alternate method of producing these nanoparticles and thus biological sources started to be used, including plant extracts which show various types of bioactivity including antibacterial activity, antioxidant activity, etc. (Samrot et al., 2018c,d, 2019a,c,d; Ahmed et al., 2016; Bennet et al., 2020; Raji et al., 2019). *Capsicum annum*-mediated green synthesized silver nanoparticles showed activity against *Pseudomonas aeruginosa* (Samrot et al., 2018d). These silver nanoparticles (AgNP) are used for coating dental/wound dressings (to aid antibacterial activity), cardiovascular implants, agriculture, deodorants, food preservations, coating agents in paints for hospitals, diagnostics, etc. (Rafique et al., 2017; Zivic et al., 2018). Biosynthesis of silver nanoparticles is also possible using various types of microorganisms including prokaryotes like bacteria and actinomycetes, and eukaryotes including yeast, algae, etc. (Rafique et al., 2017; Sadhasivam et al., 2010; Dhanker et al., 2021). In this chapter, biosynthesis of AgNPs using various microorganisms is detailed, and the mechanism of synthesis and the applications are also well explained.

2 Green biosynthesis of AgNPs

The term "green synthesis" stands for producing metal nanoparticles using biological sources including plants, microorganisms, biological products, etc. (Ahmad et al., 2019; Uddin et al., 2021). Green synthesis bridges nanotechnology and biotechnology and enables eco-friendly products to be made (Khalil et al., 2014; Mohan et al., 2014; Sharma et al., 2009; Awwad et al., 2013); moreover it is simple and cost-effective. The metal nanoparticles have varied activity (Philip, 2010; Aravind et al., 2021).

2.1 Bacteria-mediated silver nanoparticle production

AgNPs are said to be compatible with higher animals and have activity over bacteria, but some bacteria do resist the silver particles (Lara et al., 2010; Naqvi et al., 2013; Das et al., 2017; Shahbandeh et al., 2021). The change of color from yellow to brown confirms the formation of AgNPs, showing absorbance maxima at 420nm (Nanda and Saravanan, 2009; Thilagam et al., 2021). *Staphylococcus aureus* has been reported to produce AgNPs (Nanda and Saravanan, 2009). Kalishwaralal et al. (2010) produced AgNPs using *Brevibacterium casei* with the particle size between 10 and 50nm; during FTIR analysis, they found that the proteins were reducing the silver nitrate to AgNPs. A culture supernatant of two mesophilic and five psychrophilic bacteria was producing AgNPs of size 6–13nm (Shivaji et al., 2011). Mostly this bioreduction of silver nitrate to metallic AgNPs is due to sulfur-containing proteins, DNA, and reductase enzymes (Li et al., 2016; Samrot et al., 2019d; Shivaji et al., 2011; Srivastava et al., 2013). Fayaz et al. (2011) reported that *Geobacillus stearothermophilus*, a thermophilic bacteria is used for the production of AgNPs which had absorbane maxima at 423nm, and stated that the formation is due to some reducing enzymes and capping proteins present in the bacterium. The marine bacterium *Idiomarina* sp., PR_58-8, produced silver nanoparticles of 26nm and showed an absorption peak at 450nm (Seshadri et al., 2012); this *Idiomarina* sp. was reported to be resistant to heavy metals (Srivastava and Kowshik, 2016; Seshadri et al., 2012). Srivastava et al. (2013) reported haloarchaeal *Halococcus salifodinae BK3* grown in yeast extract broth and halophilic nitrate broth to produce spherical-shaped crystalline AgNPs with sizes of about 50 nm and 10 nm, respectively. Zonooz and Salouti (2011) biosynthesized uniform spherical-shaped AgNPs with an absorption peak at 430 nm extracellularly by using *Streptomyces* sp. ERI-3 where AgNPs were produced after 48 h of incubation and after 3 months of dark incubation the AgNPs self-assembled in the shape of flowers. *Bacillus* sp. GP-23 produced AgNPs with absorbance maxima of 420 nm and was found to possess antifungal activity against *Fusarium oxysporum* (Gopinath and Velusamy, 2013).

2.1.1 Cyanobacteria-mediated silver nanoparticle production

Cyanobacteria are photosynthetic prokaryotes (Whitton and Potts, 2012; Chen et al., 2021) where phycocyn and polysaccharides of cyanobacteria extracts are reported to be effective in forming AgNPs extracellularly (Rai et al., 2019; Patel et al., 2015). Extract of *Lyngbya majuscula* has been reported to reduce silver nitrate to nanoparticles with size between 20 and 50 nm, absorbance maxima 415 nm, and which showed good stability (Roychoudhury et al., 2016). Husain et al. (2015) reported 30 different cyanobacterial strains that could reduce silver nitrate solution to AgNP and proteinaceous substance as the reason for the formation. *Desertifilum* IPPAS B-1220 and *Microchaete* NCCU-342 are reported to produce AgNPs by reducing the silver nitrate with sizes of around 4.5–26 nm and 60–80 nm, respectively (Hamida et al., 2020; Husain et al., 2019). Sahoo et al. (2020) synthesized AgNP using *Chroococcus minutus*, where *C. minutus* was found to show a color change from yellow to brown on reducing the silver nitrate, which is indicative of AgNPs. Hamouda et al. (2019a) also produced using *Oscillatoria limnetica* and red algae having $AgNO_3$ as a substrate to produce AgNPs with size around 4–18 nm. Proteinaceous phycocyanin pigment extracted out of *Nostoc linckia* is reported to produce uniform-sized particles on reducing $AgNO_3$ (El-Naggar et al., 2017). Effective and controlled synthesis using cyanobacteria is still primitive, and more research is required to obtain a better protocol for synthesis of stable nanoparticles (Iravani et al., 2014).

2.1.2 Synthesis of AgNPs using endophytic bacteria

Endophytic bacteria colonize into parts of plants and have mutual relations with plants and are responsible for most secondary metabolites production (Brader et al., 2014; Lodewyckx et al., 2002; Ghasemnezhad et al., 2021). Recent days, endophytic bacteria are widely researched in the arena of nanotechnology (Rahman et al., 2019). *Bacillus cereus*, an endobacterium isolated from *Garcinia xanthochymus*, was reported to produce 11–16 nm sized AgNPs by reducing $AgNO_3$ but they were slightly aggregated (Sunkar and Nachiyar, 2012). *Pantoea ananatis* extracellularly synthesized AgNPs while being exposed to sunlight (Monowar et al., 2018), where it was reported that the stability was because of continuous exposure to sunlight. *Pseudomonas poae*, an endophytic bacterium of garlic, was able to reduce $AgNO_3$ to AgNPs (Ibrahim et al., 2020). An endophytic bacterium which was isolated from

Table 1 Synthesis of AgNPs using bacteria.

Bacteria	Precursor	Extracellular/intracellular	Size	Reference
Pseudomonas sp. THG-LS1.4	$AgNO_3$	Extracellular	50–150nm	Singh et al. (2018)
Klebsiella pneumoniae UVHC5, Escherichia coli ATCC 8739, and Pseudomonas jessinii UVKS19	$AgNO_3$	Extracellular	–	Müller et al. (2016)
Pseudomonas aeruginosa strain BS-161R	$AgNO_3$	Extracellular	13nm	Kumar and Mamidyala (2011)
Bacillus sp. strain (CS-11)	$AgNO_3$	Extracellular	42–90nm	Das et al. (2014)
Cupriavidus sp.	$AgNO_3$	Extracellular	10–50nm	Ameen et al. (2020)
Thermophilic Bacillus sp. AZ1	$AgNO_3$	Extracellular	7–13nm	Deljou and Goudarzi (2016)
Sporosarcina koreensis DC4 strain	$AgNO_3$	Extracellular	Average 102nm	Singh et al. (2016)

the drought-tolerant plant *Pennisetum setaceum* was utilized for producing AgNPs (Ahmed et al., 2019). Endophytic bacterial strain SYSU 333150 was able to synthesize AgNPs of size between 11 and 40nm by photo irradiation (Dong et al., 2017), which was an easy, fast, and efficient method for AgNPs synthesis. Endobacteria that are used to synthesize AgNP are listed in Table 1.

2.2 Synthesis of AgNPs using fungi

Fascinating opportunities are being provided by nature, and even fungi show competence in terms of green synthesis of AgNPs (Bhainsa and D'souza, 2006; Akther et al., 2019; Othman et al., 2021). Several factors are involved in synthesis of fungi, as they are known to secrete various enzymes and this might be the reason for reduction of $AgNO_3$ (Mandal et al., 2006). *Aspergillus terreus* was able to produce stable AgNPs of size around 1–20nm (Li et al., 2012). Hyphal intracellular synthesis of AgNPs was performed using cell-free extract by Arun et al. (2014). A cell-free filtrate of *Penicillium nalgiovense* has been reported to produce AgNPs where the formation of AgNPs was also reported because of cysteine-containing proteins. *Trichoderma harzianum* has been reported to biosynthesize AgNPs from $AgNO_3$ solution by Ahluwalia et al. (2014), where no detergents or capping agents were used. Biogenic polydispersed silver nanoparticles were synthesized using the fungus *Trichoderma viride* which was found to be 1–50 nm in size (Elgorban et al., 2016). Easy, convenient, and rapid synthesis of silver nanoparticles from *Phoma glomerata* was observed in the presence of sunlight (Gade et al., 2014). Vigneshwaran et al. (2007) observed the biosorption of silver nanoparticles on the cell wall of *Aspergillus flavus* when treated with silver nitrate solution; the nanoparticles were then separated by an ultrasonification process. Owaid et al. (2015) observed the biosynthesis of AgNPs by a macro fungus *Pleurotus* and the synthesized nanoparticles were 2–100 nm in size, crystalline, and highly stable in nature. Well-dispersed silver nanoparticles from an endophytic fungus *Andrographis paniculata* were synthesized, and TEM revealed the sizes covered a range of approximately 20–50 nm (Azmath et al., 2016). Extracellular mycosynthesis of silver nanoparticles from *Humicola* sp. when treated with silver nitrate solution was observed by Syed et al. (2013); the produced nanoparticles were highly dispersed and stable. Gudikandula et al. (2017) observed the synthesis of silver nanoparticles using two white rot fungal strains which had sizes of around 15 nm. Nayak et al. (2018) synthesized silver nanoparticles from a symbiont *Penicillium italicum* and observed the color change to yellowish brown when treated with silver nitrate solution. Silver nanoparticles can be synthesized both intracellularly and extracellularly. Green synthesis using fungi showed many potential applications (Zhao et al., 2018; Khandel and Shahi, 2018). Syntheses of nanoparticles by other fungal species are listed in Table 2.

Table 2 Synthesis of AgNPs using fungi.

Fungi	Precursor	Intracellular/extracellular	Size	Reference
Penicillium italicum	$AgNO_3$	Extracellular	32–100 nm	Taha et al. (2019)
Beauveria bassiana	$AgNO_3$	Extracellular	10–50 nm	Tyagi et al. (2019)
Duddingtonia flagrans	$AgNO_3$	Extracellular	11–38 nm	Silva et al. (2017)
Penicillium diversum	$AgNO_3$	Extracellular	5–45 nm	Ganachari et al. (2012)
Aspergillus foetidus	$AgNO_3$	Extracellular	20–40 nm	Roy et al. (2013)
Epicoccum nigrum	$AgNO_3$	Extracellular	1–22 nm	Qian et al. (2013)
Aspergillus tamarii	$AgNO_3$	Extracellular	25–50 nm	Kumar et al. (2012)
Neurospora intermedia	$AgNO_3$	Extracellular	30 nm (presence light); 24 nm (absence of light)	Hamedi et al. (2014)
Penicillium oxalicum	$AgNO_3$	–	60–80 nm	Feroze et al. (2020)

2.3 Synthesis of AgNP using algae

Algae is a fast-growing photosynthetic organism ranging micro to macro size, occurring in almost all types of environment and also a wide source of secondary metabolites (Khanna et al., 2019). AgNPs can be synthesized by both intracellular and extracellular mechanisms (Öztürk, 2019; Garg et al., 2020). Bhuyar et al. (2020) synthesized nanoparticles from marine algae *Padina* sp. and the reduction of $AgNO_3$ was observed by the color change to yellowish brown. Fatima et al. (2020) reported the synthesis of silver nanoparticles from the red algae *Portieria hornemannii* by treating the extract with silver nitrate solution and the synthesized AgNPs were found to be 30–50 nm in size. Haglan et al. (2020) synthesized *Enteromorpha intestinalis*-mediated silver nanoparticles by treating the aqueous extract of the culture with silver nitrate solution. Highly purified silver nanoparticles were synthesized from the aqueous extract of *Sargassum myriocystum* and were around 20–22 nm in size (Balaraman et al., 2020). Arya et al. (2019) demonstrated the synthesis of silver nanoparticles from green *Botryococcus braunii* which were around 40–90 nm. Mahajan et al. (2019) observed the synthesis of silver nanoparticles from green algae *Chlorella vulgaris* which were found to be highly stable. Silver nanoparticles were synthesized from *Corallina elongata* and *Gelidium amansii*, with SDS as a capping agent producing nanoparticles with high stability (Hamouda et al., 2019a,b). Sathishkumar et al. (2019) observed synthesis of highly stable silver nanoparticles from the aqueous extract of *Trichodesmium erythraeum* by treating it with silver nitrate solution. Stable and crystalline silver nanoparticles were synthesized from a *Caulerpa serrulata* aqueous extract treated with aqueous $AgNO_3$ (Aboelfetoh et al., 2017). Venkatesan et al. (2016) synthesized silver nanoparticles using *Ecklonia cava*, a marine algae, as a capping agent, and a color change was observed when treated with $AgNO_3$ solution. AgNPs were synthesized from polysaccharides of four different marine macro algae which acted as reducing agents and stabilizing agents for nanoparticle synthesis (El-Rafie et al., 2013). Silver nanoparticles were synthesized by Kathiraven et al. (2015), from *Caulerpa racemosa* and were found to be stable due to the protein present in the extract. Biosynthesis of AgNP from *Cystophora moniliformis* was observed by Prasad et al. (2013) which acted as a stabilizing agent in nanoparticle production. Fig. 1 indicates the synthesis of silver nanoparticles using various organisms. Synthesis of AgNP by several types of algae is given in Table 3.

3 Mechanism

The mechanism that leads to the synthesis of nanoparticles differs for every different method that is adopted (Lekamge et al., 2018; Samrot et al., 2018c; Iravani et al., 2014). Hence, an increase in the usage of organisms such as bacteria, fungi, and other microorganisms is observed in the synthesis of nanoparticles as they are said to be nontoxic and environmentally friendly that enable the formation of the nanoparticles (Banu and Balasubramanian, 2014). Hence, it is vital to comprehend the mechanism involved in the synthesis of nanoparticles using microorganisms.

3 Mechanism

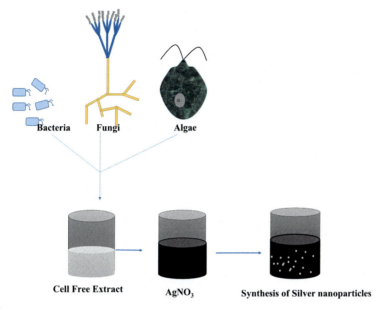

FIG. 1
Microbial-mediated synthesis of silver nanoparticles.

Table 3 Synthesis of AgNPs using algae.

Algae	Precursor	Extracellular/ intracellular	Size	Reference
Portieria hornemannii	AgNO$_3$	Extracellular	30–50 nm	Haglan et al. (2020)
Enteromorpha flexuosa	AgNO$_3$	Extracellular	1–100 nm	Yousefzadi et al. (2014)
Spyridia fusiformis	AgNO$_3$	Extracellular	5–50 nm	Murugesan et al. (2017)
Desmodesmus sp.	AgNO$_3$	Intracellular	15–30 nm	Dağlıoğlu and Öztürk (2019)
Gracilaria edulis	AgNO$_3$	Extracellular	55–90 nm	Priyadharshini et al. (2014)
Padina tetrastromatica	AgNO$_3$	–	Average 14 nm	Rajeshkumar et al. (2012)
Chlorococcum humicola	AgNO$_3$	Extracellular and intracellular	Average 100 nm	Jena et al. (2012)
Sargassum cinereum	AgNO$_3$	Extracellular	45–76 nm	Mohandass et al. (2013)

According to Prabhu and Poulose (2012), the biological synthesis of silver nanoparticles is believed to be a bottom-up approach that utilizes reduction/oxidation reactions to carry out the synthesis process. The microbial enzymes present in the microorganism possess reducing properties and are considered to be main elements that aid the formation of nanoparticles. However, the mechanism involved in the formation of nanoparticles differs with different types of microorganisms.

3.1 Mechanism of bacterial-mediated synthesis

The formation of nanoparticles by bacteria occurs due to their ability to reduce heavy metal ions. As a result of exposure to the adverse effects of metal particles, a resistance mechanism to toxic metal ion concentration occurs in bacterial species that also facilitates the organisms to thrive in the same conditions (Igiri et al., 2018). A few examples of such mechanisms of resistance are modification of solubility and toxicity via reduction or oxidation (Singh et al., 2017a,b), extracellular complex formation or precipitation of metals, efflux systems, and lack of specific metal transport systems (Aljerf and AlMasri, 2018). A wide range of bacterial species are involved in the biosynthesis of silver nanoparticles, including *Acinetobacter* sp., *Escherichia coli*, *Klebsiella pneumoniae*, *Lactobacillus* spp., *B. cereus*, *Corynebacterium* sp., and *Pseudomonas* sp. (Mohanpuria et al., 2008; Gowramma et al., 2015; Iravani, 2014). The first reported bacterial synthesis of silver nanoparticles was done using *Pseudomonas stutzeri* AG259 by Joerger et al. (2000) where silver nanoparticles with size less than 200 nm were produced. A diverse range of mechanisms are proposed to comprehend the synthesis of silver nanoparticles in a better manner.

3.1.1 Nitrate reductase enzyme

Most of the microorganisms such as bacteria or fungi are capable of secreting a significant amount of enzymes that aid in the production of metal nanoparticles through the process of enzymatic reduction of metal ions. In the case of silver nanoparticles, a general mechanism involved in bacterial-mediated synthesis of AgNPs occurs due to the presence of the enzyme nitrate reductase (Moteshafi et al., 2012; Kalimuthu et al., 2008). Nitrate reductase is defined as a multidomain enzyme that mainly comprises of three groups: molybdopterin, Fe-heme, and FAD (flavin adenine dinucleotide). They generally exist in 1:1:1 stoichiometry which facilitates the course of electron transfer from NAD(P)H to nitrate (Cannons et al., 1993). During the course of the reduction process, conversion of nitrate to nitrite occurs, which aids the transfer of electrons to the silver ions and thus induces the reduction of silver ions to silver nanoparticles (Ag^+ to Ag^0) (Mukherjee et al., 2018; Vaidyanathan et al., 2010). The nitrate reductase family mainly comprises of three kinds: cytoplasmic assimilatory nitrate reductase (Nas), membrane-bound respiratory nitrate reductase (Nar), and the periplasmic dissimilatory nitrate reductase (Nap) (Moreno-Vivián et al., 1999). The above-mentioned enzymes mediate transfer of nitrate (NO_3^-) to nitrite (NO_2^-), which is a vital part of nitrogen metabolism (Gonzalez et al., 2006). However, in an adverse metal stress environment, an oxidation reaction occurs on the NADH-dependent nitrate reductase, where in the presence of silver ions, the liberated electron is transferred to reduce Ag^+ to Ag^0, depriving NO_3^- of the electron (Kanehisa et al., 2017). A enzymatic strategy involved in the production of AgNPs was carried in silico by Mukherjee et al. (2018) for a better understanding of the contribution of NADPH-dependent nitrate reductase enzymes in the production of silver nanoparticles. The synthesis process was supported by the bacterial strain of *Bacillus clausii* where the color change to yellowish brown was taken as an indication of the formation of silver nanoparticles. The cause of the intense color change was explained through the phenomenon of plasmon resonance absorption (Kulkarni and Muddapur, 2014). The contribution of nitrate reductase enzyme from *B. clausii* in the formation of silver particles was confirmed by the revelation of the location of the ligand binding site in the enzyme, which was observed by docking analysis between the ligand (silver nitrate) and the receptor-nitrate reductase reduction (Mukherjee et al., 2018). Moreover, another study revealed the occurrence of a reduction process of Ag^+ ions in biosynthesis of AgNPs using an *E. coli* strain. This reaction was found to be favorable in anaerobic conditions. Accumulation of nanosized silver was observed in the periplasm of the *E. coli* (silver-resistance bacterial strain) that occurred upon prolonged exposure to the silver nitrate. This may be due to the resistance mechanism of the bacteria on exposure to the silver metal ions. Moreover, the role of nitrate reductase in the production of AgNPs was confirmed by reduced accumulation of silver nanoparticles upon deletion of mutant analysis of a subunit of periplasmic nitrate reductase (NapC) (Wing-ShanáLin et al., 2014). Similarly, Singh et al. (2018) confirmed the formation of AgNPs from *Pseudomonas* sp. (THG-LS1.4 strain) by the appearance of a deep brown color and various characterization studies. It was proposed that the accumulation of metal ions on the surface of the microbial cells was reduced to nanomaterials on coming in contact with the enzymes. Fig. 2 indicates the mechanism of silver nanoparticles synthesized from bacteria.

3.2 Mechanism of fungal-mediated synthesis

Fungi are eukaryotic organisms that are capable of digesting extracellular food by discharging particular enzymes to hydrolyze the larger extracellular molecules. These molecules are utilized as an energy source and hence they are also known

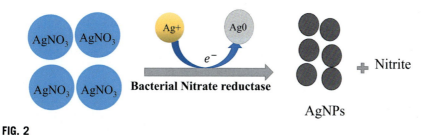

FIG. 2

Mechanism of bacterial-mediated synthesis.

as decomposer organisms. Recent data reveals that an estimated 5.1 million fungal species have been discovered, among which around 70,000 species have been known already (Blackwell, 2011). Due to the toleration and metal bioaccumulation capability of fungi, its role in the formation of metallic nanoparticles has been explored more in recent years (Sastry et al., 2003). Moreover, studies suggest that the fungi can secrete higher amounts of proteins and thus form stable nanoparticles, in comparison with bacteria (Mohanpuria et al., 2008; Netala et al., 2016). Moreover, fungi are said to provide good biomass production (Gade et al., 2008) and be more suitable for large-scale synthesis due to their resistance toward agitation and pressure (Velusamy et al., 2016). And due to their ability to tolerate heavy metal stress and to bioaccumulate the metal nanoparticles, they are widely recognized as both reducing and stabilizing agents (Azmath et al., 2016). The mechanism consists mainly of entrapment of silver ions on the fungal surface cells followed by the reduction of silver ions facilitated by the fungal-secreted enzymes (Mukherjee et al., 2001). For instance, in a study where the synthesis of monodispersed AgNPs of size range up to 25 ± 12 nm carried out by the fungus *Verticillium* sp., the mechanism of formation of silver nanoparticles was explored. On interaction of fungal biomass with the aqueous solution of $AgNO_3$, reduction of silver ions occurred due to the electrostatic interaction between the negative charge imposed by the enzymes and the positive charge of silver ions resulting in adsorption of Ag^+ ions on the surface of the fungal cells. As a result of contact between the enzyme and silver ions, Ag particles trapped onto the cell wall of fungi are reduced by the enzymes, leading to the production of nanoparticles. The formation of nanoparticles is confirmed by the color change of the solution to a darker shade and on observing the absorbance bands in the range of 400–450 nm by UV-visible spectroscopy (Samrot et al., 2018c, d; Singh et al., 2018). A surface plasmon phenomenon that alters the optical properties of the metal particles is said to be responsible for the color change (Elamawi et al., 2018). This phenomenon is also directly related to the color of the dispersion of the solution, which varies according to the size and absorbance of the nanoparticles (Lee and Jun, 2019). The same mechanism was also observed by Gudikandula et al. (2017) where two white rot fungal cultures, *Ganoderma enigmaticum* and *Trametes ljubarskyi*, were utilized as reducing agents that contributed to the formation of silver particles. The enzymes that mainly enable the reduction reaction of silver ions are naphthoquinones (Ovais et al., 2018), anthraquinones (Siddiqi and Husen, 2016), and NADPH-dependent nitrate reductase (Ahmad et al., 2003). The mechanism that comprises of nitrate reductase-dependent synthesis of nanoparticles is widely recognized and is similar to the mechanism responsible for bacterial-mediated synthesis mentioned earlier. Duran et al. (2005) had also proposed that the formation of silver nanoparticles by *F. oxysporum* was due to the action of the nitrate reductase enzyme and anthraquinones. In another study, the synthesis of AgNPs was carried using a fungus *A. alternata*-mediated synthesis of AgNPs showed the role of NADH-dependent nitrate reductase enzymes for reduction reaction of silver ions as they were secreted by the fungus extracellularly (Gajbhiye et al., 2009). However, in a recent study by Hietzschold et al. (2019), it was mentioned that the synthesis of nanoparticles can be carried out by the action of NADH alone, without imposing the necessary participation of reductase enzymes. Hence, the mechanism may differ with the involvement of different organisms responsible for the formation of nanoparticles.

There are two other types of mechanism involved in fungal-mediated synthesis of silver nanoparticles: intracellular and extracellular. In terms of intracellular, the metal ions supplement the mycelial culture which is further internalized in the biomass. The nanoparticles produced are further retrieved by disrupting the biomass through the process of centrifugation and filtration (Rajput et al., 2016; Molnár et al., 2018). In terms of extracellular synthesis, the aqueous solution that comprises the fungal biomass is added to by the metal ions. On contact with the biomolecules, formation of free nanoparticles occurs in the dispersion due to the reduction of silver ions facilitated by the extracellular enzymes present in the filtrate (Gudikandula et al., 2017; Silva et al., 2017). The resultant nanoparticles formed fail to bind to the biomass and no further process is required to retrieve the synthesized nanoparticles; thus, extracellular synthesis is considered to be most preferable (Balaji et al., 2009). However, a purification process is necessary to eliminate the fungal residues and impurities of the nanoparticle dispersed solution, and this can be achieved by membrane filtration, gel filtration, dialysis, and ultracentrifugation (Owaid et al., 2015; Yahyaei and Pourali, 2019). An extracellular mode of synthesis of silver nanoparticles was achieved using *Aspergillus niger* and particles of size range of 3–30 nm were obtained. They were also observed to be spherical in shape, showing antibacterial and antifungal activity (Jaidev and Narasimha, 2010). Likewise, biosynthesized AgNPs were attained extracellularly using *Fusarium graminearum* with an average size of 45 nm (Shatha et al., 2016). Moreover, Senapati et al. (2004) examined the formation of nanoparticles both intracellularly and extracellularly to comprehend the mechanism responsible for the process. *Verticillium* sp. was utilized to study the intracellular formation of nanoparticles where the silver ions trapped on the surface of the fungal biomass were exposed to a reduction process by the enzymes, leading to the formation of nanoparticles. The same mechanism was observed through TEM analysis, which revealed the formation of silver nanoparticles on the cytoplasmic membrane as well as within the cytoplasm. On the other hand, *F. oxysporum* was utilized to observe the extracellular mode of synthesis and the release of reducing agents into the solution by the species confirms the occurrence of a reduction process in the biomass, i.e., extracellularly. The same was further proved by preliminary gel electrophoresis of the aqueous solution, which showed the existence of four high molecular weight proteins released by the biomass, one of which was found to be an NADH-dependent reductase. Most of the literature studies reported the extracellular synthesis of silver nanoparticles as it is a more convenient method (Honary et al., 2013; Syed et al., 2013; Raheman et al., 2011; Fayaz et al., 2010). Fig. 3 shows the synthesis of fungal-mediated synthesis of silver nanoparticles.

3.3 Mechanism involved in algal-mediated synthesis

The contribution of algae to the formation of metallic nanoparticles has been increasingly explored in a more important manner as it provides a more rational method for large-scale production of nanoparticles (Lengke et al., 2007) and excludes the process of maintaining cell cultures. Algae are also said to have high metal uptake capacity that assists the production of metallic nanoparticles (Davis et al., 2003). Primarily, algal-mediated synthesis of nanoparticles was reported to be an intracellular process, but later the extracellular mode of synthesis was explored as well, due to a higher utilization of algae. Due to the utilization of hundreds of different species of micro and macro algae, it made us to understand the mechanism of synthesis of silver nanoparticles (Manivasagan and Kim, 2015). The ability of algal species to accumulate metal ions and to assist the optimization of metal ions into a more malleable form has made them a model organism to synthesize metallic nanoparticles (Fawcett et al., 2017). Furthermore, algal species consist of compounds like carbohydrates, minerals, proteins, fats, bioactive compounds, and pigments that can act as reducing and stabilizing agents, resulting in the production of nanoparticles (Michalak and Chojnacka, 2015). This arises when the metal ions interact with the algal extract. For instance, a rapid formation of silver nanoparticles of 34.03 nm in size was achieved within 5 min of interaction with the algal extract of *Pithophora oedogonia* (Sinha et al., 2015). Likewise, an aqueous extract of *C. vulgaris* contributed to the formation of silver nanoparticles 15–47 nm in size by acting as a reducing agent in the process (Annamalai and Nallamuthu, 2016). The presence of stable nanoparticles of different size and growth was confirmed through color change. Similar to that of bacteria and fungi, algal-mediated synthesis can also be carried out by both intracellular and extracellular methods.

In terms of an intracellular mode of synthesis, the formation of nanoparticles mostly takes place at the cell wall (Senapati et al., 2012), where reducing agents such as NADPH-dependent reductase produced during metabolic pathways such as respiration or nitrogen fixation react with the silver metal ions and get accumulated within the cell wall of algae. The bioactive components of algal species can also be responsible for the formation of silver nanoparticles (Khanna et al., 2019). Similarly, another study proposed that the functional groups such as carboxylic and aromatic groups present in *C. vulgaris* may have acted as reducing agents inducing the reduction of silver ions to silver nanoparticles (Satapathy et al., 2015). Aqueous extract of microalga *Scenedesmus* sp. (IMMTCC-25) was utilized for intracellular mode of formation of silver nanoparticles through the interaction of live cells of algal biomass with silver nitrate solution. Production of nanoparticles was evidenced by the color change of the solution to reddish-yellow color. In this study, the formation of more stable nanoparticles was possible and they were said to be directly dependent on shape, size, and conformation of protein molecules that contributed to the synthesis as reducing agents (Jena et al., 2012).

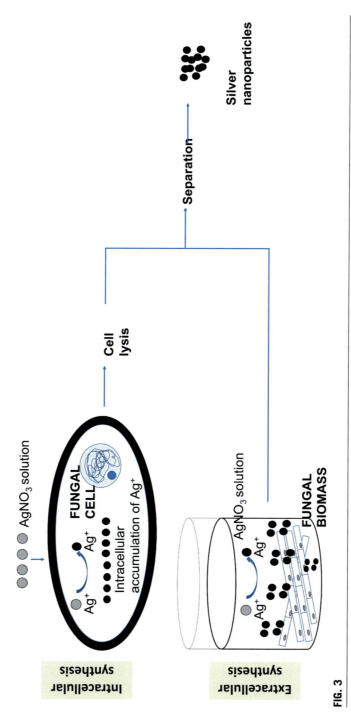

FIG. 3

Mechanism of fungal-mediated synthesis.

In terms of extracellular mode of synthesis, pretreatment of algal biomass is achieved through procedures such as purification, filtration, and centrifugation that induce the production of nanoparticles outside the cells (Vijayan et al., 2014). It was hypothesized that the combination of a metal precursor, number of cells, along with the active molecules in the surface of the cells is vital for the production of nanoparticles (Kalabegishvili et al., 2012). Similarly, an investigation of extracellular mode of production of silver nanoparticles was carried out by Jena et al. (2012) through the interaction of silver nitrate solution with the raw and the boiled extract of algal biomass of *Scenedesmus* sp. (IMMTCC-25). The same was also carried out intracellularly, as was mentioned earlier. It was observed that more stable nanoparticles were achieved through the interaction of raw and boiled extract, but it also led to the agglomeration of the particles due to the low concentration of reducing agents. Likewise, extracellular synthesis of silver nanoparticles was achieved by marine algal species *C. racemosa* which produced particles ranging from 5 to 25 nm (Kathiraven et al., 2015). Though a number of hypotheses have been proposed to comprehend the process responsible the formation of nanoparticles, the exact mechanism of synthesis of nanoparticles is still unknown.

4 Bioactivity studies

Silver nanoparticles synthesized using microorganisms are reported to exhibit various bioactivity studies against pathogenic organisms that have been extensively utilized for biomedical applications. There are various properties of nanosilver that contribute to a number of bioactivity studies such as antibacterial activity (Jeong et al., 2005; Raffi et al., 2008), antifungal activity (Alananbeh et al., 2017; Mussin et al., 2019), anticancer activity (Majeed et al., 2016; Abd-Elnaby et al., 2016), etc. One of the main reasons for nanosilver's extensive use in the medical field is the increasing threat of antibiotic resistance caused by the abuse of antibiotics in recent years. Moreover, silver nanoparticles are said to possess antiviral activity which can be used against currently lethal viruses (Mousavi et al., 2018; Mori et al., 2013).

Silver nanoparticles are reported to be capable of killing about 650 types of disease-causing organisms. For instance, stable silver nanoparticles synthesized by the filtrate of *T. viride* were evaluated for their increased antimicrobial activities with various antibiotics against gram-positive and gram-negative bacteria and an enhanced antibacterial effect was observed in the presence of AgNPs (Li et al., 2011). Ahluwalia et al. (2014) synthesized silver nanoparticles using *T. harzianum*, which were used to control the bacteria *S. aureus* and *K. pneumoniae* in vitro. The inhibition rates were concentration-dependent, with the gram-negative bacterium (*K. pneumoniae*) showing higher sensitivity. Similarly, the potential of AgNPs synthesized by the fungus *Guignardia mangiferae* was investigated against gram-negative bacteria, with effects including increased permeability, alteration of membrane transport, and release of nucleic acids (Balakumaran et al., 2015). It has been said that the decreased size of the nanoparticles leads to increased antibacterial efficacy due to their larger total surface area per volume and thus proves the advantage of nano-sized Ag particles (Jeong et al., 2005). Although the exact mechanism of antibacterial activity of silver nanoparticles is yet to be evaluated, several theories have been proposed for a better understanding of the mechanism. The most popular mechanism is explained by Li et al. (2008), where the occurrence of changes in cell wall due to the adhesion of nano-sized silver particles leads to the inhibition of bacterial growth and proliferation. The bacterial cell thus becomes susceptible to DNA damage and cell death due to the alteration of its normal functioning of bacterial DNA. The disruption of the bacterial cell wall is said to be caused by the interaction between Ag^+ ions and proteins containing sulfur in the cell, which leads to irreversible damage. Furthermore, the accumulation of AgNPs on bacterial strains such as trypanosomes and yeasts leading to the saturation of enzymes and protein in the cell can also be considered as evidence of antibacterial activity of AgNPs (Lansdown, 2004; Feng et al., 2000).

AgNPs are said to exhibit excellent antifungal activities against several fungal strains. For instance, Kim et al. (2008) observed good antifungal activity of silver nanoparticles against *Trichophyton mentagrophytes* and *Candida albicans* fungal strains and further stated that action of AgNPs on the cell wall fungi led to the abnormal functioning of the fungal cells, which may be the reason for vulnerability to the antifungal property. Similarly, several studies reveal the anticancer property of silver nanoparticles. A study reported that silver nanoparticles 5 nm in size exhibited higher cytotoxicity against p53-deficient tumor cells by inducing apoptosis-dependent programmed cell death in the absence of the tumor suppressor p53 (Melaiye et al., 2004). Likewise, AgNPs 5–35 nm in size primarily induced cell death through the mitochondrial structure and function targeting, and thus proved to be cytotoxic against tumor cells (Melaiye et al., 2005). Hence, the above evidence reveals the bioactivity of silver nanoparticles that can be of great use in medical applications.

5 Characterization

The characterization of silver nanoparticles aids in understanding of structural properties, chemical properties, morphology, and stability of nanoparticles. Silver nanoparticles can be characterized using various analyses such as ultraviolet-visible spectroscopy, Fourier transform infrared spectroscopy, and X-ray diffraction, along with microscopic techniques like scanning electron microscopy, transmission electron microscopy, atomic force microscopy, etc.

5.1 Ultraviolet-visible spectroscopy

UV-vis spectroscopy analysis can be used to monitor the absorbance properties of nanoparticles which can aid understanding of the formation and stability of the particles (Krasovskii and Karavanskii, 2008). Metallic nanoparticles exhibit a phenomenon known as surface plasmon resonance, which causes a strong absorption and scattering of visible and infrared light that explains the color of colloidal solutions of different metals. Each particle exhibits a different absorption spectrum depending upon the compounds present. In the case of silver nanoparticles, this phenomenon occurs at around 415–420 nm, which can be reflected in the band in the UV-vis spectrum (Baia and Simon, 2007; Samrot et al., 2018b). This can be used to verify the bioreduction of silver nitrate to silver nanoparticles (Samrot et al., 2018a, 2019b). A study also suggests that the characteristic absorbance peak of silver nanoparticles (420 nm) is due to the excitation of longitudinal plasmon vibrations that occur due to surface plasmon resonance (Basavaraja et al., 2008). For instance, Kumar and Ghosh (2016) performed biosynthesis of silver nanoparticles using *Bacillus* sp.-*Brevibacillus borstelensis*_MTCC10642 followed by characterization of silver NPs using UV-vis spectroscopy. The presence of nanoparticles was indicated using a surface plasmon resonance band that occurred at around 430 nm. Moreover, it was claimed that the biosynthesis of AgNPs was found to be more stable, which may be due to the presence of peaks in the spectrum. Likewise, silver nanoparticles were synthesized from three different growth culture media residues from *E. coli* using three different culture media residues: LB, LBN, and LBE. The absorption spectrum in UV-vis reveals different absorption properties depending upon the media used. For instance, LB media-mediated synthesis of AgNPs displayed a clear and prominent peak, in contrast with the absorption peaks produced using LBN and LBE media. It was explained that the composition of the media and the number of nutrients present in it may interfere with the metabolic changes in the bacteria, which leads to the formation of substances that affect the reduction process of the Ag^+ ions to Ag-NPs, preventing controlled reduction and homogeneous nanoparticle size. Thus, from this data it can be inferred that the other particles can also interfere with the UV-vis results that can be evidenced from the absorption peaks. The formation of nanoparticles can also be inferred from the types of peaks displayed in the spectrum because it was concluded from the abovementioned results that the LB culture-conditioned media was the most suitable to produce silver nanoparticles that displayed a narrow UV-vis spectrum absorption peak at 420 nm (Baltazar-Encarnación et al., 2019).

5.2 Fourier transform infrared spectroscopy

FTIR is one of the most powerful tools that aid in the determination of functional groups present in the compound and also in identifying the molecular bonds present between the chemical compounds. In the case of silver nanoparticles, the FTIR spectrum is vital in characterizing the protein binding between the nanoparticles and thus aiding in quantifying the secondary structure in metal nanoparticle-protein interaction (Gole et al., 2001). In a study, FTIR analysis of biosynthesized silver nanoparticles by the *Pseudomonas* strain isolated from *Euplotes focardii* revealed the possible interactions between the silver salts and proteins, which may have accounted for the reduction of Ag$^+$ ions with consequent stabilization of AgNPs. The amide linkages between the amino acids are revealed through the bands observed around 3280 cm^{-1} (Amide A) and 3070 cm^{-1} (Amide B) that are mainly assigned to the NH vibrations. It was also inferred from the absorption maximum at 1632 cm^{-1} that due to the C=O stretching of the peptide bond, Amide I was identified. Similarly, Amide II was identified due to the amide NH bending that peaked around 1538 cm^{-1}. Thus, the presence of capping proteins was confirmed in the nanoparticles, which may be responsible for the stabilization of AgNPs as they prevent aggregation (John et al., 2020). Moreover, the protein nanoparticles interaction occurs through the electrostatic attraction of negatively charged carboxylate groups, specifically in enzymes, proving the mechanism of enzymes in formation of nanoparticles as mentioned earlier (Gole et al., 2001). Thus, FTIR also revealed the stability of nanoparticles formed due to the presence of proteins.

Similarly, Huq (2020) performed FTIR for silver nanoparticles synthesized using *Pseudoduganella eburnea* MAHUQ-39 to identify various biological compounds that may have assisted in microbial synthesis and stabilization of AgNPs. The band found at 3439.16 cm^{-1} and 2922.29/2854.13 cm^{-1} was assigned to O—H (alcohol) and/or N—H (amine) stretching and C—H (alkane) stretching, respectively. The presence of the N—H (amine) bend was identified using peaks at 1636.01 cm^{-1}. The peak seen at 1636.01 cm^{-1} represents the N—H (amine) bend. The bands seen at 1457.49 cm^{-1} and 1057.74 cm^{-1} represent the C—H (alkane) bend and C—O (alcohol/ether) stretching, respectively. Thus, the presence of these bands in the FTIR spectra of silver nanoparticles dispersion indicates that the secondary structure of the proteins is not affected during the formation of silver nanoparticles or by its binding with the silver nanoparticles (Balaji et al., 2009).

5.3 Scanning electron microscopy—Energy dispersive X-ray spectroscopy

SEM analysis can be carried out to study the morphological characterization of silver nanoparticles. The size of the nanoparticles can also be observed in SEM analysis. This technique can be associated with energy diffraction analysis of X-rays (EDX), which allows the elemental analysis of isolated nanoparticles and particles to identify the compounds present in the particles (Vigneshwaran et al., 2007). SEM analysis uses a high electron beam that overlooks the backscattering of electrons by scanning over the surface of the particles, and thus provides characteristic information on the surface features of an object and its texture. Vigneshwaran et al. (2007) found that the size of the silver nanoparticles synthesized from white rot fungus, *Phanerochaete chrysosporium*, as 50–200 nm. They also identified that the reduction process occurred on the surface of the fungal mycelium without any clustering. EDX analysis also revealed the presence of Ag^0 apart from C, O, and P, which may be due to the fungal biomass background. Similarly, SEM analysis of silver nanoparticles synthesized by *E. focardii* revealed that the particles were spherical in shape and nonuniform in size, i.e., polydispersed in nature. The sizes of the nanoparticles were also found to be around 20–100 nm. The chemical composition of the nanoparticles was also studied using EDX, and an intense signal of Ag at 3 keV was observed. It was claimed that the metallic nanoparticles are reported to show a strong signal peak at 3 keV due to surface plasmon resonance. Other elements such as C, N, and O were also detected and it was explained that this was due to the emissions of capping proteins interaction with the nanoparticles (John et al., 2020). Therefore, along with size and shape, SEM analysis can also be carried out to observe the distribution of silver on the nanoparticle.

5.4 X-ray diffraction (XRD)

X-ray diffraction is a powerful method for characterization of nanoparticles as it reveals the crystalline nature of the particles, which is an important parameter to consider in terms of formation and stability of nanoparticles (Durán et al., 2007). The crystalline size of the particle can also be measured employing the Scherrer equation, using the broadening of the most intense peak of an XRD measurement for a specific sample (Mourdikoudis et al., 2018). The wavelength of X-rays is on the atomic scale, and thus intensities measured with XRD can provide quantitative, accurate information on the atomic arrangements at interfaces. Each different crystalline solid has a unique X-ray diffraction pattern which acts like a "fingerprint," and thus can be used to differentiate the compounds (Sharma et al., 2012). In the case of silver nanoparticles, usually, X-ray diffraction peaks observed at $2\theta = 38.00$ degrees, 44.16 degrees, 64.40 degrees, and 77.33 degrees, correspond to (111), (200), (220), and (311), Bragg's reflections of the face-centered cubic (fcc) structure of metallic silver, respectively (Durán et al., 2007). Sadhasivam et al. (2012) identified the presence of peaks centered at $\sim$38 degrees, $\sim$45 degrees, and $\sim$65 degrees which could be induced by the following crystalline planes of silver: 1 1 1, 2 0 0, and 2 2 0, respectively, for silver nanoparticles produced by *Streptomyces hygroscopicus*. The average grain size of the AgNPs formed in the bioreduction process was determined using Scherrer's formula. Similarly, the formation of metal silver nanoparticle was done using PVP as a protecting agent and silver nitrate as a precursor solution and XRD analysis revealed a face-centered cubic silver peaks. The crystalline size was also calculated using Scherrer's equation as 40 nm (Zielińska et al., 2009). Other peaks are also observed on the spectrum and it was claimed as crude nature of the extracts containing other metabolites and salts that may have assisted in the formation of the particles (Begam, 2016). Thus, these characterization techniques aid in understanding the nature and formation of the nanoparticles.

6 Applications of microbial-mediated silver nanoparticles

Biological synthesis of silver nanoparticles from microorganisms is a key field in nanotechnology that receives attention because of its eco-friendly nature and cost-effectiveness. This involves the concept of utilizing available biological microorganisms for synthesizing novel nanoparticles for environmental applications (Dar et al., 2013). The role of silver nanoparticles has contributed to applications in a wide range of products used by humans like pharmaceutical products, soaps, shampoos, deodorants, etc. (Banerjee et al., 2014; Singh et al., 2017a,b; Simbine et al., 2019; Garg et al., 2020). Due to its various antifungal (Bocate et al., 2019; Nguyen et al., 2020), antimicrobial (Cheon et al., 2019; Ahmad et al., 2019), antibacterial activities (Aisida et al., 2019; Ren et al., 2019; Rolim et al., 2019), and other potential applications in various fields are listed in Fig. 4.

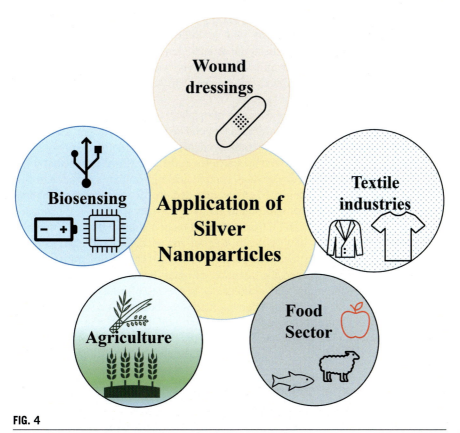

FIG. 4

Various applications of microbial-mediated silver nanoparticles.

6.1 Wound healing applications

The effective antimicrobial activity of silver nanoparticles enhanced its wound healing at a faster rate. Dressings impregnated with silver nanoparticles can be used for patients with chronic wounds as the AgNPs are considered to be noncytotoxic in nature (Kumar et al., 2018). Silver nanoparticles are known for their wound healing properties. In the presence of silver nanoparticles, wound healing takes place due to the local matrix metalloproteinase reduction and increase in the neutrophil apoptosis which is happening inside the wound. The important players in wound healing using silver nanoparticles are Acticoat 7, Acticoat Absorbent, Silvercel, Acticoat Moisture Control, Aquacel Ag, Contreet F, Actisorb, and Urgotol SSD (Prabhu and Poulose, 2012). Silver nanoparticles used in wound dressing are found to diminish the bacterial growth and clear the wound in 3.35 days (Rafique et al., 2017). Silver nanoparticles synthesized from a fungus named *T. viride* ranged size from 5 to 40 nm; they were found to have great and efficient antimicrobial properties. When silver nanoparticles were treated with various antibiotics like erythromycin, chloramphenicol, kanamycin, and ampicillin to check the antimicrobial properties of AgNPs, it was found that silver nanoparticles along with antibiotics increased the antibacterial effects of antibiotics against the bacterial strains. Among all the other antibiotics used, the highest enhancement of antibacterial activity was seen in ampicillin. Therefore, this study revealed that AgNPs can be used for wound dressing applications (Fayaz et al., 2010).

Silver nanoparticles synthesized using a nonpathogenic potato plant fungus *Phytophthora infestans* were used for observing wound healing properties on albino rats with standard silver sulfadiazine (generally available standard ointment for wounds) as a control. It was seen that silver nanoparticles at 0.125% (lowest) concentration healed the wound in around 8 days, which is rapid healing compared to standard silver sulfadiazine. The histological studies revealed the wound contraction and appearance of normal skin and hair follicle in rats after treating with silver nanoparticles (Thirumurugan et al., 2011). Cotton fabrics treated along with green synthesized silver nanoparticles play a major role in the medical field due to their efficient antimicrobial properties (Hebeish et al., 2014). The antimicrobial nature of silver nanoparticles attracted the attention of many researchers in designing a wound dressing for delivering active molecules to the site of infection. Intense literature studies in understanding the interaction between bacteria and silver nanoparticles in clinical fields are still needed (Paladini and Pollini, 2019). Fig. 5 illustrates the antimicrobial property and wound healing nature of silver nanoparticles.

6.2 Biosensing applications

Silver nanoparticles were used for designing biosensors with great flexibility and sensitivity (You et al., 2020). A sensitive electrochemical vanillin biosensor has been designed by Zheng et al. (2010) using Au—Ag alloy nanoparticles green synthesized from yeast with low detection limit of 40 nm; this biosensor was used to successfully

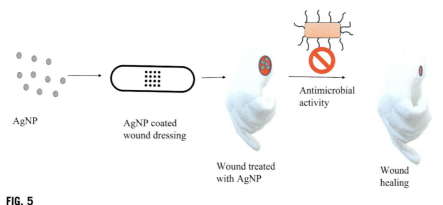

FIG. 5

Wound healing properties of microbial-mediated silver nanoparticles.

measure the vanillin in vanilla bean and vanilla and vanilla tea. Fayaz et al. (2010) synthesized silver nanoparticles from a microbe *T. viride* which ranged in size from 2 to 4 nm, and the nanoparticles were found to have biosensing applications. Due to the antimicrobial properties of AgNPs, they can be used as biosensors for environmental, biomedical, and nanomedical applications (Srikar et al., 2016; Singh et al., 2015). Biologically synthesized silver nanoparticles from various kinds of microorganisms are applied in designing nano-sized sensors for detecting biomolecules. Biocompatible nature along with the combination of silver nanoparticles provides a wide range of biosensing applications which takes research to the next level. However, more research is still needed based on biosensing applications combined with metal nanoparticles (Yun et al., 2019). Some biosensing applications of silver nanoparticles are presented in Fig. 6.

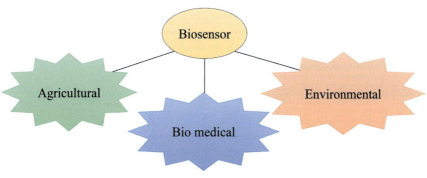

FIG. 6

Biosensing applications of microbial-mediated silver nanoparticles.

6.3 Application in textiles

Silver nanoparticles are considered to have great antimicrobial activities. Silver nanoparticles interact with the microbial enzymes, DNA, etc., and cause cell death. Therefore, usage of AgNPs is mostly recommended in textile industries (Perera et al., 2013) for the production of antiodor and self-cleaning clothes (Gade et al., 2010). Silver nanoparticles were green synthesized from a fungal species *A. terreus* and 25–100 ppm of silver nanoparticles were loaded into the cotton fiber, which was tested against some gram-positive bacterial populations like *B. cereus* and *S. aureus*, and some gram-negative bacterial populations involving *E. coli*, *P. aeruginosa*, and *Proteus vulgaris*. At the highest concentration (100 ppm), growth of both the gram-positive and gram-negative bacteria was inhibited. Silver embedded cotton fiber exhibited great antibacterial properties (Velhal et al., 2016). Silver nanoparticles synthesized from a fungus *Cunninghamella echinulata* and coated on the cotton fiber to check their antibacterial properties. The nanoparticles showed excellent and efficient antibacterial properties against both gram-positive and gram-negative bacteria like *Bacillus subtilis*, *S. aureus*, *E. coli*, and *K. pneumoniae*. Maximum inhibition of 30 mm was against *S. aureus*, followed by *B. subtilis* (27 mm) and *E. coli* (26 nm), and the lowest zone of inhibition was against *K. pneumonia* (23 mm) (Anbazhagan et al., 2017). Attia and Morsy (2016) reported that a 4.48 mm zone of inhibition was checked against *S. aureus* when silver nanoparticles of size 17–102 nm were synthesized and immobilized on fabrics. Organic cotton fabric was found to be resistant to UV radiations and possessed antibacterial activity when green synthesized silver nanoparticles were impregnated on textile fibers (Mahmud et al., 2020). AgNPs synthesized by Durán et al. (2007) from a fungus *F. oxysporum* were incorporated in a cotton fabric and tested for antimicrobial properties against *S. aureus*. The biosynthesized silver nanoparticles were found to have good antimicrobial properties as they inhibited the growth of *S. aureus* (Durán et al., 2007). Silver nanoparticles synthesized using dextran of *Leuconostoc mesenteroides* T3, impregnated onto the cotton fibers and antimicrobial activity was checked against *S. aureus*, *E. coli*, and a fungal species *C. albicans*, and were found to have a 99.9% inhibitory effect against all the microbes, proving silver nanoparticle to be efficient antimicrobial agents (Davidović et al., 2015). Antibacterial, antistatic, UV protection, and color fastness up to 20 washes were observed when the wool fabrics were treated with silver nanoparticles (Hassan, 2019). Fig. 7 demonstrates the role of silver nanoparticles in textile industries.

6.4 Application in the food sector

Silver nanoparticles have various applications in the food sector due to their antimicrobial properties (Mohd Yusof et al., 2020). Nanosilver can be used to increase the shelf life and quality maintenance for fruits and vegetables (Carbone et al., 2016; Ravindran et al., 2020. Containers containing AgNPs as coating can be used as a food packaging matrix and were useful in increasing its quality; the shelf life of the food

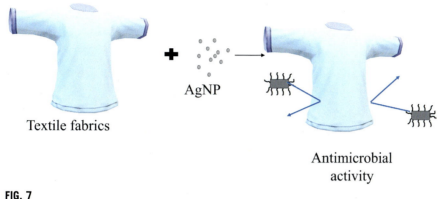

FIG. 7

Applications of microbial-mediated silver nanoparticles in textile industries.

also increased (Carbone et al., 2016). Silver nanoparticles synthesized and incorporated in an agar matrix exhibited great stability and antibacterial properties, and can be used for food packaging (Basumatary et al., 2018). Silver nanoparticles were biologically synthesized from *Serratia nematodiphila*. The green synthesized nanoparticles were found to have great antimicrobial properties against *P. aeruginosa*, *B. subtilis*, and *K. planticola*. The zone of inhibition increased as the AgNPs increased. Hence they can be used for food packaging because of their excellent antimicrobial effect (Malarkodi et al., 2013). Silver nanoparticles were synthesized from an exopolysaccharide of *Lactobacillus brevis* MSR104 isolated from Chinese koumiss. Biosynthesized silver nanoparticles were found to inhibit the growth of both gram-positive and gram-negative bacteria. On accessing the antioxidant property of AgNPs, a great efficient scavenging rate was shown against both nitric acid-free radicals and DPPH-free radicals, and excellent anticancer activity at highest concentration was shown by biosynthesized silver nanoparticles against HT-29 cell lines. Due to their antibacterial, antioxidant, and anticancer activity, AgNPs can be used as proper food packaging materials, protecting them from contaminants (Rajoka et al., 2020).

The shelf life of fruits and vegetables was increased when silver nanoparticles were incorporated in poly vinyl alcohol nanofiber, exhibiting its tremendous use in food packaging (Kowsalya et al., 2019). Fig. 8 demonstrates the quality packaging of fruits using microbial-mediated silver nanoparticles.

6.5 Agricultural applications

The usage of silver nanoparticles benefits the agricultural environment by penetrating and enriching the soil and land (Mishra and Singh, 2015; Manjunatha et al., 2016). For the detection of even trace amounts of pathogens, smart sensors have been designed (Chaudhry et al., 2018; Panpatte et al., 2016). Due to their stable and

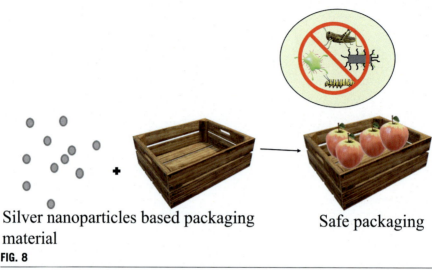

FIG. 8

Role of microbial-mediated silver nanoparticles in food packaging.

biodegradable nature, AgNPs are used as nanoformulations for producing small quantities of chemicals and act as nanofertilizers (Gupta et al., 2018; Gosavi et al., 2020; Muhammad et al., 2020). Antifungal activity against the pathogen causing spot blotch disease—*Bipolaris sorokiniana* was exhibited by the silver nanoparticles produced using *Serratia* sp. (Mishra et al., 2014). Green synthesis of silver nanoparticles from a green seaweed *Codium capitatum* found to have various applications in the agricultural sector (Kannan et al., 2013). Great antimicrobial properties were found in a combination of microorganism-mediated silver nanoparticles and microalgal biomass, a by-product obtained during microorganism-mediated synthesis. The combination of silver nanoparticles and microalgal biomass is used to eradicate plant pathogens and pests, and can be used in agricultural fields to improve pest control and crop production (Terra et al., 2019). When silver nanoparticles were used treat against two pathogenic fungi affecting plants, *Magnaporthe grisea* and *B. sorokiniana*, a strong inhibitory effect was revealed against the pathogens; they can also be used as pesticides in crop-producing fields (Jo et al., 2009). AgNPs were synthesized from *H. musciformis* of size 40–65 nm and tested for larvicidal and pupicidal activity; they were found to have great and effective activities against pests, and can be used in agricultural fields for pest control activities (Roni et al., 2015). Colloidal solutions of silver nanoparticles were found to have various fungicidal activity when tested against various plant pathogenic fungi and it was reported that they can be used as fungicide in agricultural fields (Kim et al., 2008). Silver nanoparticles were synthesized from *Streptomyces* sp. VSMGT1014 and were found to demonstrate inhibitory action against many phyto-fungi and to control several infections caused by the fungus to the crops. Hence it is proved that they can be used as fungicide to protect the crops (Shanmugaiah et al., 2015). Crops and plants can be made pest-resistant by treating with silver nanoparticles as pesticide (Fig. 9).

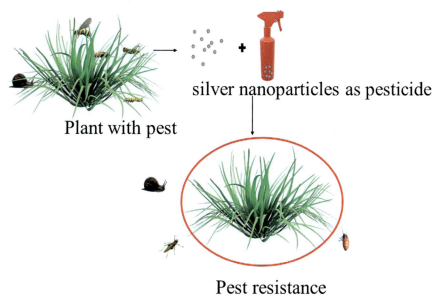

FIG. 9
Role of microbial-mediated silver nanoparticles for crop protection and usage as fungicide.

7 Conclusion

The synthesis of silver nanoparticles has been gaining importance in various fields due to their unique properties and their contribution to a number of applications. Therefore, an efficient and eco-friendly mode of synthesis has been imposed to overcome the overuse of chemicals in conventional methods of synthesis, and its implications in the environment and other areas. Production of silver nanoparticles using microorganisms is cost-effective and has been found to be more stable. Alternatively, microbial-mediated synthesis of silver nanoparticles is an efficient mode of synthesis and is proven to be more bio-acceptable and eco-friendly with multiple applications, as evidenced by the above literature studies.

References

Abd-Elnaby, H.M., Abo-Elala, G.M., Abdel-Raouf, U.M., Hamed, M.M., 2016. Antibacterial and anticancer activity of extracellular synthesized silver nanoparticles from marine *Streptomyces rochei* MHM13. Egypt. J. Aquat. Res. 42 (3), 301–312.

Aboelfetoh, E.F., El-Shenody, R.A., Ghobara, M.M., 2017. Eco-friendly synthesis of silver nanoparticles using green algae (*Caulerpa serrulata*): reaction optimization, catalytic and antibacterial activities. Environ. Monit. Assess. 189 (7), 349.

Abou El-Nour, K.M., Eftaiha, A.A., Al-Warthan, A., Ammar, R.A., 2010. Synthesis and applications of silver nanoparticles. Arab. J. Chem. 3 (3), 135–140.

Ahluwalia, V., Kumar, J., Sisodia, R., Shakil, N.A., Walia, S., 2014. Green synthesis of silver nanoparticles by *Trichoderma harzianum* and their bio-efficacy evaluation against *Staphylococcus aureus* and *Klebsiella pneumonia*. Ind. Crop Prod. 55, 202–206.

Ahmad, A., Mukherjee, P., Senapati, S., Mandal, D., Khan, M.I., Kumar, R., Sastry, M., 2003. Extracellular biosynthesis of silver nanoparticles using the fungus *Fusarium oxysporum*. Colloids Surf. B. Biointerfaces 28 (4), 313–318.

Ahmad, S., Munir, S., Zeb, N., Ullah, A., Khan, B., Ali, J., Ali, S., 2019. Green nanotechnology: a review on green synthesis of silver nanoparticles—an ecofriendly approach. Int. J. Nanomedicine 14, 5087.

Ahmed, S., Ahmad, M., Swami, B.L., Ikram, S., 2016. A review on plants extract mediated synthesis of silver nanoparticles for antimicrobial applications: a green expertise. J. Adv. Res. 7 (1), 17–28.

Ahmed, M.S., Soundhararajan, R., Akther, T., Kashif, M., Khan, J., Waseem, M., Srinivasan, H., 2019. Biogenic AgNPs synthesized via endophytic bacteria and its biological applications. Environ. Sci. Pollut. Res. 26 (26), 26939–26946.

Aisida, S.O., Ugwu, K., Akpa, P.A., Nwanya, A.C., Ejikeme, P.M., Botha, S., et al., 2019. Biogenic synthesis and antibacterial activity of controlled silver nanoparticles using an extract of *Gongronema latifolium*. Mater. Chem. Phys. 237, 121859.

Akther, T., Mathipi, V., Kumar, N.S., Davoodbasha, M., Srinivasan, H., 2019. Fungal-mediated synthesis of pharmaceutically active silver nanoparticles and anticancer property against A549 cells through apoptosis. Environ. Sci. Pollut. Res. 26 (13), 13649–13657.

Alananbeh, K.M., Al-Refaee, W.J., Al-Qodah, Z., 2017. Antifungal effect of silver nanoparticles on selected fungi isolated from raw and waste water. Indian J. Pharm. Sci. 79, 559–567.

Aljerf, L., AlMasri, N., 2018. A gateway to metal resistance: bacterial response to heavy metal toxicity in the biological environment. Ann. Adv. Chem. 2, 032–044. https://doi.org/10.29328/journal.aac.1001012.

Ameen, F., AlYahya, S., Govarthanan, M., ALjahdali, N., Al-Enazi, N., Alsamhary, K., et al., 2020. Soil bacteria *Cupriavidus sp.* mediates the extracellular synthesis of antibacterial silver nanoparticles. J. Mol. Struct. 1202, 127233.

Anbazhagan, S., Azeez, S., Morukattu, G., Rajan, R., Venkatesan, K., Thangavelu, K.P., 2017. Synthesis, characterization and biological applications of mycosynthesized silver nanoparticles. 3 Biotech 7 (5), 333.

Annamalai, J., Nallamuthu, T., 2016. Green synthesis of silver nanoparticles: characterization and determination of antibacterial potency. Appl. Nanosci. 6 (2), 259–265.

Aravind, M., Ahmad, A., Ahmad, I., Amalanathan, M., Naseem, K., Mary, S.M.M., et al., 2021. Critical green routing synthesis of silver NPs using jasmine flower extract for biological activities and photocatalytical degradation of methylene blue. J. Environ. Chem. Eng. 9 (1), 104877.

Arshad, H., Sami, M.A., Sadaf, S., Hassan, U., 2021. Salvadora persica mediated synthesis of silver nanoparticles and their antimicrobial efficacy. Sci. Rep. 11 (1), 1–11.

Arun, G., Eyini, M., Gunasekaran, P., 2014. Green synthesis of silver nanoparticles using the mushroom fungus Schizophyllum commune and its biomedical applications. Biotechnol. Bioprocess Eng. 19 (6), 1083–1090.

Arya, A., Mishra, V., Chundawat, T.S., 2019. Green synthesis of silver nanoparticles from green algae (*Botryococcus braunii*) and its catalytic behavior for the synthesis of benzimidazoles. Chem. Data Collect. 20, 100190.

Attia, N.F., Morsy, M.S., 2016. Facile synthesis of novel nanocomposite as antibacterial and flame retardant material for textile fabrics. Mater. Chem. Phys. 180, 364–372.

Awwad, A.M., Salem, N.M., Abdeen, A.O., 2013. Green synthesis of silver nanoparticles using carob leaf extract and its antibacterial activity. Int. J. Ind. Chem. 4 (1), 29.

Azmath, P., Baker, S., Rakshith, D., Satish, S., 2016. Mycosynthesis of silver nanoparticles bearing antibacterial activity. Saudi Pharm. J. 24 (2), 140–146.

Baia, L., Simon, S., 2007. UV-VIS and TEM assessment of morphological features of silver nanoparticles from phosphate glass matrices. In: Modern Research and Educational Topics in Microscopy. Badajoz, Spain: Formatex, pp. 576–583.

Balaji, D., Basavaraja, S., Deshpande, R., Mahesh, D.B., Prabhakar, B., Venkataraman, A., 2009. Extracellular biosynthesis of functionalized silver nanoparticles by strains of *Cladosporium cladosporioides* fungus. Colloids Surf. B Biointerfaces 68, 88–92.

Balakumaran, M.D., Ramachandran, R., Kalaicheilvan, P.T., 2015. Exploitation of endophytic fungus, *Guignardia mangiferae* for extracellular synthesis of silver nanoparticles and their in vitro biological activities. Microbiol. Res. 178, 9–17. https://doi.org/10.1016/j.micres.2015.05.009.

Balaraman, P., Balasubramanian, B., Kaliannan, D., Durai, M., Kamyab, H., Park, S., et al., 2020. Phyco-synthesis of silver nanoparticles mediated from marine algae *Sargassum myriocystum* and its potential biological and environmental applications. Waste Biomass. Valori. 11 (10), 5255–5271.

Baltazar-Encarnación, E., Escárcega-González, C.E., Vasto-Anzaldo, X.G., Cantú-Cárdenas, M.E., Morones-Ramírez, J.R., 2019. Silver nanoparticles synthesized through green methods using *Escherichia coli* top 10 (Ec-Ts) growth culture medium exhibit antimicrobial properties against nongrowing bacterial strains. J. Nanomater. 2019.

Banerjee, P., Satapathy, M., Mukhopahayay, A., Das, P., 2014. Leaf extract mediated green synthesis of silver nanoparticles from widely available Indian plants: synthesis, characterization, antimicrobial property and toxicity analysis. Bioresour. Bioprocess. 1 (1), 3.

Banu, A.N., Balasubramanian, C., 2014. Mycosynthesis of silver nanoparticles using *Beauveria bassiana* against dengue vector, *Aedes aegypti* (Diptera: Culicidae). Parasitol. Res. 113, 2869–2877. https://doi.org/10.1007/s00436-013-3656-0.

Basavaraja, S., Balaji, S.D., Lagashetty, A., Rajasab, A.H., Venkataraman, A., 2008. Extracellular biosynthesis of silver nanoparticles using the fungus *Fusarium semitectum*. Mater. Res. Bull. 43, 1164–1170.

Basumatary, K., Daimary, P., Das, S.K., Thapa, M., Singh, M., Mukherjee, A., Kumar, S., 2018. *Lagerstroemia speciosa* fruit-mediated synthesis of silver nanoparticles and its application as filler in agar based nanocomposite films for antimicrobial food packaging. Food Packag. Shelf Life 17, 99–106.

Begam, J.N., 2016. Biosynthesis and characterization of silver nanoparticles (AgNPs) using marine bacteria against certain human pathogens. Int. J. Adv. Sci. Res. 2 (07), 152–156.

Bennet, R.D., Raji, P., Divya, K.M., Sharma, K.V., Keerthana, D., Karishma, S., Antony, V.S., Thirumurugan, R., Purayil, S.K., Ponnaiah, P., Selvarani, J., 2020. Green synthesis and antibacterial activity studies of silver nanoparticles from the aqueous extracts of *Euphorbia hirta*. J. Pure Appl. Microbiol. 14, 301–306.

Bhainsa, K.C., D'souza, S.F., 2006. Extracellular biosynthesis of silver nanoparticles using the fungus *Aspergillus fumigatus*. Colloids Surf. B Biointerfaces 47 (2), 160–164.

Bhuyar, P., Rahim, M.H.A., Sundararaju, S., Ramaraj, R., Maniam, G.P., Govindan, N., 2020. Synthesis of silver nanoparticles using marine macroalgae *Padina sp.* and its antibacterial activity towards pathogenic bacteria. Beni Suef Univ. J. Basic Appl. Sci. 9 (1), 1–15.

Blackwell, M., 2011. The fungi: 1, 2, 3 ... 5.1 million species? Am. J. Bot. 98, 426–438.

Bocate, K.P., Reis, G.F., de Souza, P.C., Junior, A.G.O., Durán, N., Nakazato, G., et al., 2019. Antifungal activity of silver nanoparticles and simvastatin against toxigenic species of *Aspergillus*. Int. J. Food Microbiol. 291, 79–86.

Brader, G., Compant, S., Mitter, B., Trognitz, F., Sessitsch, A., 2014. Metabolic potential of endophytic bacteria. Curr. Opin. Biotechnol. 27, 30–37.

Cannons, A.C., Barber, M.J., Solomonson, L.P., 1993. Expression and characterization of the heme-binding domain of Chlorella nitrate reductase. J. Biol. Chem. 268, 3268–3271.

Carbone, M., Donia, D.T., Sabbatella, G., Antiochia, R., 2016. Silver nanoparticles in polymeric matrices for fresh food packaging. J. King Saud Univ. Sci. 28 (4), 273–279.

Chaudhry, N., Dwivedi, S., Chaudhry, V., Singh, A., Saquib, Q., Azam, A., Musarrat, J., 2018. Bio-inspired nanomaterials in agriculture and food: current status, foreseen applications and challenges. Microb. Pathog. 123, 196–200.

Chen, M.Y., Teng, W.K., Zhao, L., Hu, C.X., Zhou, Y.K., Han, B.P., et al., 2021. Comparative genomics reveals insights into cyanobacterial evolution and habitat adaptation. ISME J. 15 (1), 211–227.

Cheon, J.Y., Kim, S.J., Rhee, Y.H., Kwon, O.H., Park, W.H., 2019. Shape-dependent antimicrobial activities of silver nanoparticles. Int. J. Nanomed. 14, 2773.

Dağlıoğlu, Y., Öztürk, B.Y., 2019. A novel intracellular synthesis of silver nanoparticles using *Desmodesmus sp.* (Scenedesmaceae): different methods of pigment change. Rend. Lincei Sci. Fis. Nat. 30 (3), 611–621.

Dar, M.A., Ingle, A., Rai, M., 2013. Enhanced antimicrobial activity of silver nanoparticles synthesized by *Cryphonectria sp.* evaluated singly and in combination with antibiotics. Nanomed. Nanotechnol. Biol. Med. 9 (1), 105–110.

Das, V.L., Thomas, R., Varghese, R.T., Soniya, E.V., Mathew, J., Radhakrishnan, E.K., 2014. Extracellular synthesis of silver nanoparticles by the *Bacillus* strain CS 11 isolated from industrialized area. 3 Biotech 4 (2), 121–126.

Das, B., Dash, S.K., Mandal, D., Ghosh, T., Chattopadhyay, S., Tripathy, S., et al., 2017. Green synthesized silver nanoparticles destroy multidrug resistant bacteria via reactive oxygen species mediated membrane damage. Arab. J. Chem. 10 (6), 862–876.

Davidović, S., Miljković, M., Lazić, V., Jović, D., Jokić, B., Dimitrijević, S., Radetić, M., 2015. Impregnation of cotton fabric with silver nanoparticles synthesized by dextran isolated from bacterial species *Leuconostoc mesenteroides* T3. Carbohydr. Polym. 131, 331–336.

Davis, T.A., Volesky, B., Mucci, A., 2003. A review of the biochemistry of heavy metal biosorption by brown algae. Water Res. 37 (18), 4311–4330.

Deljou, A., Goudarzi, S., 2016. Green extracellular synthesis of the silver nanoparticles using thermophilic *Bacillus sp.* AZ1 and its antimicrobial activity against several human pathogenetic bacteria. Iran. J. Biotechnol. 14 (2), 25.

Dhanker, R., Hussain, T., Tyagi, P., Singh, K.J., Kamble, S.S., 2021. The emerging trend of bio-engineering approaches for microbial nanomaterial synthesis and its applications. Front. Microbiol. 12, 638003.

Dong, Z.Y., Narsing Rao, M.P., Xiao, M., Wang, H.F., Hozzein, W.N., Chen, W., Li, W.J., 2017. Antibacterial activity of silver nanoparticles against *Staphylococcus warneri* synthesized using endophytic bacteria by photo-irradiation. Front. Microbiol. 8, 1090.

Duran, N., Priscyla, D., Marcato, P.D., Alves, O., De Souza, G., Esposito, E., 2005. Mechanistic aspects of biosynthesis of silver nanoparticles by several *Fusarium oxysporum* strains. J. Nanobiotechnol. 3, 1–7.

Durán, N., Marcato, P.D., De Souza, G.I., Alves, O.L., Esposito, E., 2007. Antibacterial effect of silver nanoparticles produced by fungal process on textile fabrics and their effluent treatment. J. Biomed. Nanotechnol. 3 (2), 203–208.

Elamawi, R.M., Al-Harbi, R.E., Hendi, A.A., 2018. Biosynthesis and characterization of silver nanoparticles using *Trichoderma longibrachiatum* and their effect on phytopathogenic fungi. Egypt. J. Biol. Pest Control 28, 28. https://doi.org/10.1186/s41938-018-0028-1.

Elgorban, A.M., Al-Rahmah, A.N., Sayed, S.R., Hirad, A., Mostafa, A.A.F., Bahkali, A.H., 2016. Antimicrobial activity and green synthesis of silver nanoparticles using *Trichoderma viride*. Biotechnol. Biotechnol. Equip. 30 (2), 299–304.

El-Naggar, N.E.A., Hussein, M.H., El-Sawah, A.A., 2017. Bio-fabrication of silver nanoparticles by phycocyanin, characterization, in vitro anticancer activity against breast cancer cell line and in vivo cytotxicity. Sci. Rep. 7 (1), 1–20.

El-Rafie, H.M., El-Rafie, M., Zahran, M.K., 2013. Green synthesis of silver nanoparticles using polysaccharides extracted from marine macro algae. Carbohydr. Polym. 96 (2), 403–410.

Fatima, R., Priya, M., Indurthi, L., Radhakrishnan, V., Sudhakaran, R., 2020. Biosynthesis of silver nanoparticles using red algae *Portieria hornemannii* and its antibacterial activity against fish pathogens. Microb. Pathog. 138, 103780.

Fawcett, D., Verduin, J.J., Shah, M., Sharma, S.B., Poinern, G.E.J., 2017. A review of current research into the biogenic synthesis of metal and metal oxide nanoparticles via marine algae and seagrasses. J. Nanosci. Nanotechnol. 2017. https://doi.org/10.1155/2017/8013850, 8013850.

Fayaz, A.M., Balaji, K., Girilal, M., Yadav, R., Kalaichelvan, P.T., Venketesan, R., 2010. Biogenic synthesis of silver nanoparticles and their synergistic effect with antibiotics: a study against gram-positive and gram-negative bacteria. Nanomed. Nanotechnol. Biol. Med. 6 (1), 103–109.

Fayaz, A.M., Girilal, M., Rahman, M., Venkatesan, R., Kalaichelvan, P.T., 2011. Biosynthesis of silver and gold nanoparticles using thermophilic bacterium *Geobacillus stearothermophilus*. Process Biochem. 46 (10), 1958–1962.

Feng, J., Wu, G.Q., Chen, F.Z., Cui, T.N., Kim, J.O., Kim, A., 2000. A mechanistic study of the antibacterial effect of silver ions on *Escherichia coli* and *Staphylococcus aureus*. J. Biomed. Mater. Res. 52 (4), 662–668.

Feroze, N., Arshad, B., Younas, M., Afridi, M.I., Saqib, S., Ayaz, A., 2020. Fungal mediated synthesis of silver nanoparticles and evaluation of antibacterial activity. Microsc. Res. Tech. 83 (1), 72–80.

Gade, A.K., Bonde, P., Ingle, A.P., Marcato, P.D., Durán, N., Rai, M.K., 2008. Exploitation of *Aspergillus niger* for synthesis of silver nanoparticles. J. Biobaased Mater. Bioenergy 2, 243–247. https://doi.org/10.1166/jbmb.20 08.401.

Gade, A., Ingle, A., Whiteley, C., Rai, M., 2010. Mycogenic metal nanoparticles: progress and applications. Biotechnol. Lett. 32 (5), 593–600.

Gade, A., Gaikwad, S., Duran, N., Rai, M., 2014. Green synthesis of silver nanoparticles by *Phoma glomerata*. Micron 59, 52–59.

Gajbhiye, M., Kesharwani, J., Ingle, A., Gade, A., Rai, M., 2009. Fungus-mediated synthesis of silver nanoparticles and their activity against pathogenic fungi in combination with fluconazole. Nanomed. Nanotechnol. Biol. Med. 5 (4), 382–386.

Ganachari, S.V., Bhat, R., Deshpande, R., Venkataraman, A., 2012. Extracellular biosynthesis of silver nanoparticles using fungi *Penicillium diversum* and their antimicrobial activity studies. BioNanoScience 2 (4), 316–321.

Garg, D., Sarkar, A., Chand, P., Bansal, P., Gola, D., Sharma, S., et al., 2020. Synthesis of silver nanoparticles utilizing various biological systems: mechanisms and applications—a review. Prog. Biomater. 9, 81–95.

Ghasemnezhad, A., Frouzy, A., Ghorbanpour, M., Sohrabi, O., 2021. Microbial endophytes: new direction to natural sources. In: Endophytes: Mineral Nutrient Management. vol. 3. Springer, pp. 123–155.

Gole, A., Dash, C., Ramakrishnan, V., Sainkar, S.R., Mandale, A.B., Rao, M., Sastry, M., 2001. Pepsin—gold colloid conjugates: preparation, characterization, and enzymatic activity. Langmuir 17, 1674–1679.

Gonzalez, P.J., Correia, C., Moura, I., et al., 2006. Bacterial nitrate reductases: molecular and biological aspects of nitrate reduction. J. Inorg. Biochem. 100, 1015–1023. https://doi.org/10.1016/j.jinor gbio.2005.11.024.

Gopinath, V., Velusamy, P., 2013. Extracellular biosynthesis of silver nanoparticles using *Bacillus sp.* GP-23 and evaluation of their antifungal activity towards *Fusarium oxysporum*. Spectrochim. Acta A Mol. Biomol. Spectrosc. A106, 170–174.

Gosavi, V.C., Daspute, A.A., Patil, A., Gangurde, A., Wagh, S.G., Sherkhane, A., Anandrao Deshmukh, V., 2020. Synthesis of green nanobiofertilizer using silver nanoparticles of *Allium cepa* extract short title: green nanofertilizer from *Allium cepa*. Int. J. Chem. Stud. 8 (4), 1690–1694.

Gowramma, B., Keerthi, U., Rafi, M., Rao, D.M., 2015. Biogenic silver nanoparticles production and characterization from native stain of *Corynebacterium* species and its antimicrobial activity. Biotech 5, 195–201.

Gudikandula, K., Vadapally, P., Charya, M.S., 2017. Biogenic synthesis of silver nanoparticles from white rot fungi: their characterization and antibacterial studies. OpenNano 2, 64–78.

Gupta, N., Upadhyaya, C.P., Singh, A., Abd-Elsalam, K.A., Prasad, R., 2018. Applications of silver nanoparticles in plant protection. In: Nanobiotechnology Applications in Plant Protection. Springer, Cham, pp. 247–265.

Haglan, A.M., Abbas, H.S., Akköz, C., Karakurt, S., Aşikkutlu, B., Güneş, E., 2020. Characterization and antibacterial efficiency of silver nanoparticles biosynthesized by using green algae *Enteromorpha intestinalis*. Int. Nano Lett. 10 (3), 1–9.

Hamedi, S., Shojaosadati, S.A., Shokrollahzadeh, S., Hashemi-Najafabadi, S., 2014. Extracellular biosynthesis of silver nanoparticles using a novel and non-pathogenic fungus, *Neurospora intermedia*: controlled synthesis and antibacterial activity. World J. Microbiol. Biotechnol. 30 (2), 693–704.

Hamida, R.S., Abdelmeguid, N.E., Ali, M.A., Bin-Meferij, M.M., Khalil, M.I., 2020. Synthesis of silver nanoparticles using a novel cyanobacteria *Desertifilum sp.* extract: their antibacterial and cytotoxicity effects. Int. J. Nanomed. 15, 49.

Hamouda, R.A., Abd El-Mongy, M., Eid, K.F., 2019a. Comparative study between two red algae for biosynthesis silver nanoparticles capping by SDS: insights of characterization and antibacterial activity. Microb. Pathog. 129, 224–232.

Hamouda, R.A., Hussein, M.H., Abo-elmagd, R.A., Bawazir, S.S., 2019b. Synthesis and biological characterization of silver nanoparticles derived from the cyanobacterium *Oscillatoria limnetica*. Sci. Rep. 9 (1), 1–17.

Hassan, M.M., 2019. Enhanced colour, hydrophobicity, UV radiation absorption and antistatic properties of wool fabric multi-functionalised with silver nanoparticles. Colloids Surf. A Physicochem. Eng. Asp. 581, 123819.

Hebeish, A., El-Rafie, M.H., El-Sheikh, M.A., Seleem, A.A., El-Naggar, M.E., 2014. Antimicrobial wound dressing and anti-inflammatory efficacy of silver nanoparticles. Int. J. Biol. Macromol. 65, 509–515.

Hietzschold, S., Walter, A., Davis, C., Taylor, A.A., Sepunaru, L., 2019. Does nitrate reductase play a role in silver nanoparticle synthesis? Evidence for NADPH as the sole reducing agent. ACS Sustain. Chem. Eng. 7, 8070–8076. https://doi.org/10.1021/acssuschemeng.9b00506.

Honary, S., Barabadi, H., Gharaei-Fathabad, E., Naghibi, F., 2013. Green synthesis of silver nanoparticles induced by the fungus *Penicillium citrinum*. Trop. J. Pharm. Res. 12, 7–11.

Husain, S., Sardar, M., Fatma, T., 2015. Screening of cyanobacterial extracts for synthesis of silver nanoparticles. World J. Microbiol. Biotechnol. 31 (8), 1279–1283.

Huq, M., 2020. Green synthesis of silver nanoparticles using *Pseudoduganella eburnea* MAHUQ-39 and their antimicrobial mechanisms investigation against drug resistant human pathogens. Int. J. Mol. Sci. 21, 1510.

Husain, S., Afreen, S., Yasin, D., Afzal, B., Fatma, T., 2019. Cyanobacteria as a bioreactor for synthesis of silver nanoparticles-an effect of different reaction conditions on the size of nanoparticles and their dye decolorization ability. J. Microbiol. Methods 162, 77–82.

Ibrahim, E., Zhang, M., Zhang, Y., Hossain, A., Qiu, W., Chen, Y., et al., 2020. Green-synthesization of silver nanoparticles using endophytic bacteria isolated from garlic and its antifungal activity against wheat *Fusarium* head blight pathogen *Fusarium graminearum*. Nanomaterials 10 (2), 219.

Igiri, B.E., Okoduwa, S.I., Idoko, G.O., Akabuogu, E.P., Adeyi, A.O., Ejiogu, I.K., 2018. Toxicity and bioremediation of heavy metals contaminated ecosystem from tannery wastewater: a review. J. Toxicol. 2018. 16 pp.

Iravani, S., 2014. Bacteria in nanoparticle synthesis: current status and future prospects. Int. Sch. Res. Notices 2014. 18 pp.

Iravani, S., Korbekandi, H., Mirmohammadi, S.V., Zolfaghari, B., 2014. Synthesis of silver nanoparticles: chemical, physical and biological methods. Res. Pharm. Sci. 9, 385–406.

Jaidev, L.R., Narasimha, G., 2010. Fungal mediated biosynthesis of silver nanoparticles, characterization and antimicrobial activity. Colloids Surf. B Biointerfaces 81, 430–433.

Jena, J., Pradhan, N., Dash, B.P., Sukla, L.B., Panda, P.K., 2012. Biosynthesis and characterization of silver nanoparticles using microalgae *Chlorococcum humicola* and its antibacterial activity. Dig. J. Nanomater. Biostruct. 3 (1), 1–8.

Jeong, S.H., Yeo, S.Y., Yi, S.C., 2005. The effect of filler particle size on the antibacterial properties of compounded polymer/silver fibers. J. Mater. Sci. 40, 5407–5411.

Jo, Y.K., Kim, B.H., Jung, G., 2009. Antifungal activity of silver ions and nanoparticles on phytopathogenic fungi. Plant Dis. 93 (10), 1037–1043.

Joerger, R., Klaus, T., Granqvist, C.G., 2000. Biologically produced silver–carbon composite materials for optically functional thin-film coatings. Adv. Mater. 12 (6), 407–409.

John, M.S., Nagoth, J.A., Ramasamy, K.P., Mancini, A., Giuli, G., Natalello, A., Ballarini, P., Miceli, C., Pucciarelli, S., 2020. Synthesis of bioactive silver nanoparticles by a *Pseudomonas* strain associated with the antarctic psychrophilic protozoon *Euplotes focardii*. Mar. Drugs 18, 38.

Kalabegishvili, T.L., Kirkesali, E.I., Rcheulishvili, A.N., Ginturi, E.N., Murusidze, I.G., Pataraya, D.T., Gurielidze, M.A., Tsertsvadze, G.I., Gabunia, V.N., Lomidze, L.G., Gvarjaladze, D.N., 2012. Synthesis of gold nanoparticles by some strains of *Arthrobacter* genera. Proc. Inst. Mech. Eng. L J. Mat. Des. Appl. 7, 1–7.

Kalimuthu, K., Babu, R.S., Venkataraman, D., Bilal, M., Gurunathan, S., 2008. Biosynthesis of silver nanocrystals by *Bacillus licheniformis*. Colloids Surf. B Biointerfaces 65 (1), 150–153. https://doi.org/10.1007/978-3-030-16383-9_3.

Kalishwaralal, K., Deepak, V., Pandian, S.R.K., Kottaisamy, M., BarathManiKanth, S., Kartikeyan, B., Gurunathan, S., 2010. Biosynthesis of silver and gold nanoparticles using *Brevibacterium casei*. Colloids Surf. B Biointerfaces 77 (2), 257–262.

Kanehisa, M., Furumichi, M., Tanabe, M., et al., 2017. KEGG: new perspectives on genomes, pathways, diseases and drugs. Nucleic Acids Res. 45, D353–D361. https://doi.org/10.1093/nar/gkw1092.

Kannan, R.R.R., Stirk, W.A., Van Staden, J., 2013. Synthesis of silver nanoparticles using the seaweed *Codium capitatum* PC Silva (Chlorophyceae). S. Afr. J. Bot. 86, 1–4.

Kathiraven, T., Sundaramanickam, A., Shanmugam, N., Balasubramanian, T., 2015. Green synthesis of silver nanoparticles using marine algae *Caulerpa racemosa* and their antibacterial activity against some human pathogens. Appl. Nanosci. 5, 499–504.

Khalil, M.M., Ismail, E.H., El-Baghdady, K.Z., Mohamed, D., 2014. Green synthesis of silver nanoparticles using olive leaf extract and its antibacterial activity. Arab. J. Chem. 7 (6), 1131–1139.

Khandel, P., Shahi, S.K., 2018. Mycogenic nanoparticles and their bio-prospective applications: current status and future challenges. J. Nanostruct. Chem. 8 (4), 369–391.

Khanna, P., Kaur, A., Goyal, D., 2019. Algae-based metallic nanoparticles: synthesis, characterization and applications. J. Microbiol. Methods 163, 105656.

Kim, K.J., Sung, W.S., Moon, S.K., Choi, J.S., Kim, J.G., Lee, D.G., 2008. Antifungal effect of silver nanoparticles on dermatophytes. J. Microbiol. Biotechnol. 18 (8), 1482–1484.

Klaus-Joerger, T., Joerger, R., Olsson, E., Granqvist, C.G., 2001. Bacteria as workers in the living factory: metal-accumulating bacteria and their potential for materials science. Trends Biotechnol. 19 (1), 15–20.

Kowsalya, E., MosaChristas, K., Balashanmugam, P., Rani, J.C., 2019. Biocompatible silver nanoparticles/poly (vinyl alcohol) electrospun nanofibers for potential antimicrobial food packaging applications. Food Packag. Shelf Life 21, 100379.

Krasovskii, V.I., Karavanskii, V.A., 2008. Surface plasmon resonance of metal nanoparticles for interface characterization. Opt. Mem. Neural Netw. 17 (1), 8–14.

Kulkarni, N., Muddapur, U., 2014. Biosynthesis of metal nanoparticles: a review. J. Nanotechnol. 2014, 1–8.

Kumar, A., Ghosh, A., 2016. Biosynthesis and characterization of silver nanoparticles with bacterial isolate from gangetic-alluvial soil. Int. J. Biotechnol. Biochem. 12 (2), 95–102.

Kumar, C.G., Mamidyala, S.K., 2011. Extracellular synthesis of silver nanoparticles using culture supernatant of *Pseudomonas aeruginosa*. Colloids Surf. B Biointerfaces 84 (2), 462–466.

Kumar, R.R., Priyadharsani, K.P., Thamaraiselvi, K., 2012. Mycogenic synthesis of silver nanoparticles by the Japanese environmental isolate *Aspergillus tamarii*. J. Nanopart. Res. 14 (5), 1–7.

Kumar, S.S.D., Houreld, N.N., Kroukamp, E.M., Abrahamse, H., 2018. Cellular imaging and bactericidal mechanism of green-synthesized silver nanoparticles against human pathogenic bacteria. J. Photochem. Photobiol. B Biol. 178, 259–269.

Lansdown, A.B.G., 2004. A review of the use of silver in wound care: facts and fallacies. Br. J. Nurs. 13, P6–S19.

Lara, H.H., Ayala-Núñez, N.V., Turrent, L.D.C.I., Padilla, C.R., 2010. Bactericidal effect of silver nanoparticles against multidrug-resistant bacteria. World J. Microbiol. Biotechnol. 26 (4), 615–621.

Lee, S.H., Jun, B.H., 2019. Silver nanoparticles: synthesis and application for nanomedicine. Int. J. Mol. Sci. 20, E865. https://doi.org/10.3390/ijms20040865.

Lekamge, S., Miranda, A.F., Abraham, A., Li, V., Shukla, R., Bansal, V., Nugegoda, D., 2018. The toxicity of silver nanoparticles (AgNPs) to three freshwater invertebrates with different life strategies: *Hydra vulgaris*, *Daphnia carinata*, and *Paratya australiensis*. Front. Environ. Sci. 6, 152.

Lengke, M.F., Fleet, M.E., Southam, G., 2007. Biosynthesis of silver nanoparticles by filamentous cyanobacteria from a silver (I) nitrate complex. Langmuir 23, 2694–2699.

Li, Q., Mahendra, S., Lyon, D.Y., Brunet, L., Liga, M.V., Li, D., Alvarez, P.J., 2008. Antimicrobial nanomaterials for water disinfection and microbial control: potential applications and implications. Water Res. 42 (18), 4591–4602.

Li, X., Xu, H., Chen, Z.S., Chen, G., 2011. Biosynthesis of nanoparticles by microorganisms and their applications. J. Nanomater. 2011, 1–16.

Li, G., He, D., Qian, Y., Guan, B., Gao, S., Cui, Y., Wang, L., 2012. Fungus-mediated green synthesis of silver nanoparticles using *Aspergillus terreus*. Int. J. Mol. Sci. 13 (1), 466–476.

Li, J., Zhu, Z., Liu, F., Zhu, B., Ma, Y., Yan, J., et al., 2016. DNA-mediated morphological control of silver nanoparticles. Small 12 (39), 5449–5487.

Lodewyckx, C., Vangronsveld, J., Porteous, F., Moore, E.R., Taghavi, S., Mezgeay, M., der Lelie, D.V., 2002. Endophytic bacteria and their potential applications. Crit. Rev. Plant Sci. 21 (6), 583–606.

Mahajan, A., Arya, A., Chundawat, T.S., 2019. Green synthesis of silver nanoparticles using green alga (*Chlorella vulgaris*) and its application for synthesis of quinolines derivatives. Synth. Commun. 49 (15), 1926–1937.

Mahmud, S., Pervez, N., Taher, M.A., Mohiuddin, K., Liu, H.H., 2020. Multifunctional organic cotton fabric based on silver nanoparticles green synthesized from sodium alginate. Text. Res. J. 90 (11−12), 1224–1236.

Majeed, S., bin Abdullah, M.S., Nanda, A., Ansari, M.T., 2016. In vitro study of the antibacterial and anticancer activities of silver nanoparticles synthesized from *Penicillium brevicompactum* (MTCC-1999). J. Taibah Univ. Sci. 10 (4), 614–620.

Malarkodi, C., Rajeshkumar, S., Paulkumar, K., Vanaja, M., Jobitha, G.D.G., Annadurai, G., 2013. Bactericidal activity of bio mediated silver nanoparticles synthesized by *Serratia nematodiphila*. Drug Invent. Today 5 (2), 119–125.

Mandal, D., Bolander, M.E., Mukhopadhyay, D., Sarkar, G., Mukherjee, P., 2006. The use of microorganisms for the formation of metal nanoparticles and their application. Appl. Microbiol. Biotechnol. 69 (5), 485–492.

Manivasagan, P., Kim, S.K., 2015. Biosynthesis of nanoparticles using marine algae: a review. In: Kim, S.-.K., Chojnacka, K. (Eds.), Marine Algae Extracts., https://doi.org/10.1002/9783527679577.ch17.

Manjunatha, S.B., Biradar, D.P., Aladakatti, Y.R., 2016. Nanotechnology and its applications in agriculture: a review. J. Farm Sci. 29 (1), 1–13.

Melaiye, A., Simons, R.S., Milsted, A., Pingitore, F., Wesdemiotis, C., et al., 2004. Formation of water-soluble pincer silver(I)-carbene complexes: a novel antimicrobial agent. J. Med. Chem. 47 (4), 973–977.46.

Melaiye, A., Sun, Z., Hindi, K., Milsted, A., Ely, D., et al., 2005. Silver(I)-imidazole cyclophane gem-diol complexes encapsulated by electrospun tecophilic nanofibers: formation of nanosilver particles and antimicrobial activity. J. Am. Chem. Soc. 127 (7), 2285–2291.47.

Michalak, I., Chojnacka, K., 2015. Algae as production systems of bioactive compounds. Eng. Life Sci. 15 (2), 160–176.

Mishra, S., Singh, H.B., 2015. Biosynthesized silver nanoparticles as a nanoweapon against phytopathogens: exploring their scope and potential in agriculture. Appl. Microbiol. Biotechnol. 99 (3), 1097–1107.

Mishra, S., Singh, B.R., Singh, A., Keswani, C., Naqvi, A.H., Singh, H.B., 2014. Biofabricated silver nanoparticles act as a strong fungicide against *Bipolaris sorokiniana* causing spot blotch disease in wheat. PLoS One 9 (5), e97881.

Mohan, S., Oluwafemi, O.S., George, S.C., Jayachandran, V.P., Lewu, F.B., Songca, S.P., et al., 2014. Completely green synthesis of dextrose reduced silver nanoparticles, its antimicrobial and sensing properties. Carbohydr. Polym. 106, 469–474.

Mohandass, C., Vijayaraj, A.S., Rajasabapathy, R., Satheeshbabu, S., Rao, S.V., Shiva, C., DeMello, I., 2013. Biosynthesis of silver nanoparticles from marine seaweed *Sargassum cinereum* and their antibacterial activity. Indian J. Pharm. Sci. 75 (5), 606.

Mohanpuria, P., Rana, K.N., Yadav, S.K., 2008. Biosynthesis of nanoparticles: technological concepts and future applications. J. Nanopart. Res. 10, 507–517.

Mohd Yusof, H., Rahman, A., Mohamad, R., Zaidan, U.H., 2020. Microbial mediated synthesis of silver nanoparticles by *Lactobacillus plantarum* TA4 and its antibacterial and antioxidant activity. Appl. Sci. 10 (19), 6973.

Molnár, Z., Bódai, V., Szakacs, G., Erdélyi, B., Fogarassy, Z., Sáfrán, G., et al., 2018. Green synthesis of gold nanoparticles by thermophilic filamentous fungi. Sci. Rep. 8, 3943. https://doi.org/10.1038/s41598-018-22112-3.

Monowar, T., Rahman, M., Bhore, S.J., Raju, G., Sathasivam, K.V., 2018. Silver nanoparticles synthesized by using the endophytic bacterium *Pantoea ananatis* are promising antimicrobial agents against multidrug resistant bacteria. Molecules 23 (12), 3220.

Moreno-Vivián, C., Cabello, P., Martínez-Luque, M., Blasco, R., Castillo, F., 1999. Prokaryotic nitrate reduction: molecular properties and functional distinction among bacterial nitrate reductases. J. Bacteriol. 181 (21), 6573–6584.

Mori, Y., Ono, T., Miyahira, Y., et al., 2013. Antiviral activity of silver nanoparticle/chitosan composites against H1N1 influenza A virus. Nanoscale Res. Lett. 8, 93.

Moteshafi, H., Mousavi, S.M., Shojaosadati, S.A., 2012. The possible mechanisms involved in nanoparticles biosynthesis. J. Ind. Eng. Chem. 18, 2046–2050.

Mourdikoudis, S., Pallares, R.M., Thanh, N.T., 2018. Characterization techniques for nanoparticles: comparison and complementarity upon studying nanoparticle properties. Nanoscale 10 (27), 12871–12934.

Mousavi, S.M., Hashemi, S.A., Ghasemi, Y., Atapour, A., Amani, A.M., Savar Dashtaki, A., Babapoor, A., Arjmand, O., 2018. Green synthesis of silver nanoparticles toward bio and medical applications: review study. Artif. Cells Nanomed. Biotechnol. 46 (Suppl. 3), S855–S872.

Muhammad, Z., Inayat, N., Majeed, A., 2020. Application of nanoparticles in agriculture as fertilizers and pesticides: challenges and opportunities. In: New Frontiers in Stress Management for Durable Agriculture. Springer, Singapore, pp. 281–293.

Mukherjee, P., Ahmad, A., Mandal, D., Senapati, S., Sainkar, S.R., Khan, M.I., et al., 2001. Fungus-mediated synthesis of silver nanoparticles and their immobilization in the mycelial matrix: a novel biological approach to nanoparticle synthesis. Nano Lett. 1, 515–519.

Mukherjee, K., Gupta, R., Kumar, G., Kumari, S., Biswas, S., Padmanabhan, P., 2018. Synthesis of silver nanoparticles by *Bacillus clausii* and computational profiling of nitrate reductase enzyme involved in production. J. Genet. Eng. Biotechnol. 16 (2), 527–536. https://doi.org/10.1016/j.jgeb.2018.04.004.

Müller, A., Behsnilian, D., Walz, E., Gräf, V., Hogekamp, L., Greiner, R., 2016. Effect of culture medium on the extracellular synthesis of silver nanoparticles using *Klebsiella pneumoniae*, *Escherichia coli* and *Pseudomonas jessinii*. Biocatal. Agric. Biotechnol. 6, 107–115.

Murugesan, S., Bhuvaneswari, S., Sivamurugan, V., 2017. Green synthesis, characterization of silver nanoparticles of a marine red alga *Spyridia fusiformis* and their antibacterial activity. Int. J. Pharm. Pharm. Sci. 9 (5), 192–197.

Mussin, J.E., Roldán, M.V., Rojas, F., et al., 2019. Antifungal activity of silver nanoparticles in combination with ketoconazole against *Malassezia furfur*. AMB Express 9, 131. https://doi.org/10.1186/s13568-019-0857-7.

Nanda, A., Saravanan, M., 2009. Biosynthesis of silver nanoparticles from *Staphylococcus aureus* and its antimicrobial activity against MRSA and MRSE. Nanomed. Nanotechnol. Biol. Med. 5 (4), 452–456.

Naqvi, S.Z.H., Kiran, U., Ali, M.I., Jamal, A., Hameed, A., Ahmed, S., Ali, N., 2013. Combined efficacy of biologically synthesized silver nanoparticles and different antibiotics against multidrug-resistant bacteria. Int. J. Nanomed. 8, 3187.

Nayak, B.K., Nanda, A., Prabhakar, V., 2018. Biogenic synthesis of silver nanoparticle from wasp nest soil fungus, *Penicillium italicum* and its analysis against multi drug resistance pathogens. Biocatal. Agric. Biotechnol. 16, 412–418.

Netala, V.R., Bethu, M.S., Pushpalatah, B., Baki, V.B., Aishwarya, S., Rao, J.V., et al., 2016. Biogenesis of silver nanoparticles using endophytic fungus *Pestalotiopsis microspora* and evaluation of their antioxidant and anticancer activities. Int. J. Nanomed. 11, 5683–5696. https://doi.org/10.2147/IJN.S1128570.

Nguyen, D.H., Lee, J.S., Park, K.D., Ching, Y.C., Nguyen, X.T., Phan, V.H., Hoang Thi, T.T., 2020. Green silver nanoparticles formed by *Phyllanthus urinaria*, *Pouzolzia zeylanica*, and *Scoparia dulcis* leaf extracts and the antifungal activity. Nanomaterials 10 (3), 542.

Othman, A.M., Elsayed, M.A., Al-Balakocy, N.G., Hassan, M.M., Elshafei, A.M., 2021. Biosynthesized silver nanoparticles by *Aspergillus terreus* NRRL265 for imparting durable antimicrobial finishing to polyester cotton blended fabrics: statistical optimization, characterization, and antitumor activity evaluation. Biocatal. Agric. Biotechnol. 31, 101908.

Ovais, M., Khalil, A.T., Ayaz, M., Ahmad, I., Nethi, S.K., Mukherjee, S., 2018. Biosynthesis of metal nanoparticles via microbial enzymes: a mechanistic approach. Int. J. Mol. Sci. 19, 4100.

Owaid, M.N., Raman, J., Lakshmanan, H., Al-Saeedi, S.S.S., Sabaratnam, V., Abed, I.A., 2015. Mycosynthesis of silver nanoparticles by *Pleurotus cornucopiae* var. *citrinopileatus* and its inhibitory effects against Candida sp. Mater. Lett. 153, 186–190. https://doi.org/10.1016/j.matlet.2015.04.023.

Öztürk, B.Y., 2019. Intracellular and extracellular green synthesis of silver nanoparticles using *Desmodesmus sp.*: their antibacterial and antifungal effects. Caryol. Int. J. Cytol. Cytosyst. Cytogenet. 72 (1), 29–43.

Paladini, F., Pollini, M., 2019. Antimicrobial silver nanoparticles for wound healing application: progress and future trends. Materials 12 (16), 2540.

Panpatte, D.G., Jhala, Y.K., Shelat, H.N., Vyas, R.V., 2016. Nanoparticles: the next generation technology for sustainable agriculture. In: Microbial Inoculants in Sustainable Agricultural Productivity. Springer, New Delhi, pp. 289–300.

Patel, V., Berthold, D., Puranik, P., Gantar, M., 2015. Screening of cyanobacteria and microalgae for their ability to synthesize silver nanoparticles with antibacterial activity. Appl. Biotechnol. Rep. 5, 112–119.

Perera, S., Bhushan, B., Bandara, R., Rajapakse, G., Rajapakse, S., Bandara, C., 2013. Morphological, antimicrobial, durability, and physical properties of untreated and treated textiles using silver-nanoparticles. Colloids Surf. A Physicochem. Eng. Asp. 436, 975–989.

Philip, D., 2010. Green synthesis of gold and silver nanoparticles using *Hibiscus rosa sinensis*. Phys. E Low Dimens. Syst. Nanostruct. 42 (5), 1417–1424.

Prabhu, S., Poulose, E.K., 2012. Silver nanoparticles: mechanism of antimicrobial action, synthesis, medical applications, and toxicity effects. Int. Nano Lett. 2 (1), 32.

Prasad, T.N., Kambala, V.S.R., Naidu, R., 2013. Phyconanotechnology: synthesis of silver nanoparticles using brown marine algae *Cystophora moniliformis* and their characterisation. J. Appl. Phycol. 25 (1), 177–182.

Priyadharshini, R.I., Prasannaraj, G., Geetha, N., Venkatachalam, P., 2014. Microwave-mediated extracellular synthesis of metallic silver and zinc oxide nanoparticles using macro-algae (*Gracilaria edulis*) extracts and its anticancer activity against human PC3 cell lines. Appl. Biochem. Biotechnol. 174 (8), 2777–2790.

Qian, Y., Yu, H., He, D., Yang, H., Wang, W., Wan, X., Wang, L., 2013. Biosynthesis of silver nanoparticles by the endophytic fungus *Epicoccum nigrum* and their activity against pathogenic fungi. Bioprocess Biosyst. Eng. 36 (11), 1613–1619.

Raffi, M., Hussain, F., Bhatti, T.M., et al., 2008. Antibacterial characterization of silver nanoparticles against *E. coli* ATCC-15224. J. Mater. Sci. Technol. 24, 192–196.

Rafique, M., Sadaf, I., Rafique, M.S., Tahir, M.B., 2017. A review on green synthesis of silver nanoparticles and their applications. Artif. Cells Nanomed. Biotechnol. 45 (7), 1272–1291.

Raheman, F., Deshmukh, S., Ingle, A., Gade, A., Rai, M., 2011. Silver nanoparticles: novel antimicrobial agent synthesized from an endophytic fungus *Pestalotia sp.* isolated from leaves of *Syzygium cumini* (L). Nano Biomed. Eng. 3, 174–178.

Rahman, S., Rahman, L., Khalil, A.T., Ali, N., Zia, D., Ali, M., Shinwari, Z.K., 2019. Endophyte-mediated synthesis of silver nanoparticles and their biological applications. Appl. Microbiol. Biotechnol. 103 (6), 2551–2569.

Rai, S., Wenjing, W., Shrivastava, A.K., Singh, P.K., 2019. Cyanobacteria as a source of nanoparticles and their applications. In: Role of Plant Growth Promoting Microorganisms in Sustainable Agriculture and Nanotechnology. Woodhead Publishing, pp. 183–198.

Rajeshkumar, S., Kannan, C., Annadurai, G., 2012. Synthesis and characterization of antimicrobial silver nanoparticles using marine brown seaweed *Padina tetrastromatica*. Drug Invent. Today 4 (10), 511–513.

Raji, P., Samrot, A.V., Keerthana, D., Karishma, S., 2019. Antibacterial activity of alkaloids, flavonoids, saponins and tannins mediated green synthesised silver nanoparticles against *Pseudomonas aeruginosa* and *Bacillus subtilis*. J. Clust. Sci. 30 (4), 881–895. https://doi.org/10.1007/s10876-019-01547-2.

Rajoka, M.S.R., Mehwish, H.M., Zhang, H., Ashraf, M., Fang, H., Zeng, X., 2020. Antibacterial and antioxidant activity of exopolysaccharide mediated silver nanoparticle synthesized by *Lactobacillus brevis* isolated from Chinese koumiss. Colloids Surf. B Biointerfaces 186, 110734.

Rajput, S., Werezuk, R., Lange, R.M., Mcdermott, M.T., 2016. Fungal isolate optimized for biogenesis of silver nanoparticles with enhanced colloidal stability. Langmuir 32, 8688–8697. https://doi.org/10.1021/acs.langmuir.6b01813.

Ravindran, R.E., Subha, V., Ilangovan, R., 2020. Silver nanoparticles blended PEG/PVA nanocomposites synthesis and characterization for food packaging. Arab. J. Chem. 13, 6056–6060.

Ren, Y.Y., Yang, H., Wang, T., Wang, C., 2019. Bio-synthesis of silver nanoparticles with antibacterial activity. Mater. Chem. Phys. 235, 121746.

Rolim, W.R., Pelegrino, M.T., de Araújo Lima, B., Ferraz, L.S., Costa, F.N., Bernardes, J.S., 2019. Green tea extract mediated biogenic synthesis of silver nanoparticles: characterization, cytotoxicity evaluation and antibacterial activity. Appl. Surf. Sci. 463, 66–74.

Roni, M., Murugan, K., Panneerselvam, C., Subramaniam, J., Nicoletti, M., Madhiyazhagan, P., 2015. Characterization and biotoxicity of *Hypnea musciformis*-synthesized silver nanoparticles as potential eco-friendly control tool against *Aedes aegypti* and *Plutella xylostella*. Ecotoxicol. Environ. Saf. 121, 31–38.

Roy, S., Mukherjee, T., Chakraborty, S., Das, T.K., 2013. Biosynthesis, characterisation antifungal activity of silver nanoparticles synthesized by the fungus *Aspergillus foetidus* mtcc8876. Dig. J. Nanomater. Biostruct. 8 (1), 197–205.

Roychoudhury, P., Gopal, P.K., Paul, S., Pal, R., 2016. Cyanobacteria assisted biosynthesis of silver nanoparticles—a potential antileukemic agent. J. Appl. Phycol. 28 (6), 3387–3394.

Sadhasivam, S., Shanmugam, P., Yun, K., 2010. Biosynthesis of silver nanoparticles by *Streptomyces hygroscopicus* and antimicrobial activity against medically important pathogenic microorganisms. Colloids Surf. B Biointerfaces 81 (1), 358–362.

Sadhasivam, S., Shanmugam, P., Veerapandian, M., Subbiah, R., Yun, K., 2012. Biogenic synthesis of multidimensional gold nanoparticles assisted by *Streptomyces hygroscopicus* and its electrochemical and antibacterial properties. Biometals 25 (2), 351–360.

Sahoo, C.R., Maharana, S., Mandhata, C.P., Bishoyi, A.K., Paidesetty, S.K., Padhy, R.N., 2020. Biogenic silver nanoparticle synthesis with cyanobacterium *Chroococcus minutus* isolated from Baliharachandi sea-mouth, Odisha, and in vitro antibacterial activity. Saudi J. Biol. Sci. 27 (6), 1580–1586.

Samrot, A.V., Raji, P., Jenifer Selvarani, A., Nishanthini, P., 2018a. Antibacterial activity of some edible fruits and its green synthesized silver nanoparticles against uropathogen—*Pseudomonas aeruginosa* SU 18. Biocatal. Agric. Biotechnol. https://doi.org/10.1016/j.bcab.2018.08.014.

Samrot, A.V., Raji, P., Selvarani, A.J., Nishanthini, P., 2018b. Antibacterial activity of some edible fruits and its green synthesized silver nanoparticles against uropathogen–*Pseudomonas aeruginosa* SU 18. Biocatal. Agric. Biotechnol. 16, 253–270.

Samrot, A.V., Burman, U., et al., 2018c. A study on toxicity of chemically synthesised silver nanoparticle on *Eudrilus eugeniae*. Toxicol. Environ. Health Sci. 10, 162–167. https://doi.org/10.1007/s13530-018-0360-6.

Samrot, A.V., Shobana, N., Jenna, R., 2018d. Antibacterial and antioxidant activity of different staged ripened fruit of *Capsicum annuum* and its green synthesized silver nanoparticles. BioNanoScience 8 (2), 632–646.

Samrot, A.V., Angalene, J.L.A., Roshini, S.M., Raji, P., Stefi, S.M., Preethi, R., Selvarani, A.J., Madankumar, A., 2019a. Bioactivity and heavy metal removal using plant gum mediated green synthesized silver nanoparticles. J. Clust. Sci. https://doi.org/10.1007/s10876-019-01602-y.

Samrot, A.V., Saipriya, C., Angalene, J.L.A., Roshini, S.M., Cypriyana, P.J.J., Saigeetha, S., Raji, P., Kumar, S.S., 2019b. Evaluation of nanotoxicity of *Araucaria heterophylla* gum derived green synthesized silver nanoparticles on *Eudrilus eugeniae* and *Danio rerio*. J. Clust. Sci. https://doi.org/10.1007/s10876-019-01561-4b.

Samrot, A.V., Shobana, N., Kumar, S.S., Narendrakumar, G., 2019c. Production, optimization and characterisation of chitosanase of *Bacillus sp* and its applications in nanotechnology. J. Clust. Sci. 30 (3), 607–620.

Samrot, A.V., Silky, S., Ignatious, C.V., Raji, P., Saipriya, C., Selvarani, J.A., 2019d. Bioactivity studies of *Datura metel*, *Aegle marmelos*, *Annona reticulata* and *Saraca indica* and their green synthesized silver nanoparticle. J. Pure Appl. Microbiol. 13 (1). https://doi.org/10.22207/JPAM.13.1.

Sastry, M., Ahmad, A., Khan, M.I., Kumar, R., 2003. Biosynthesis of metal nanoparticles using fungi and actinomycete. Curr. Sci. 85, 162–170.

Satapathy, S., Shukla, S.P., Sandeep, K.P., Singh, A.R., Sharma, N., 2015. Evaluation of the performance of an algal bioreactor for silver nanoparticle production. J. Appl. Phycol. 27 (1), 285–291.

Sathishkumar, R.S., Sundaramanickam, A., Srinath, R., Ramesh, T., Saranya, K., Meena, M., Surya, P., 2019. Green synthesis of silver nanoparticles by bloom forming marine microalgae *Trichodesmium erythraeum* and its applications in antioxidant, drug-resistant bacteria, and cytotoxicity activity. J. Saudi Chem. Soc. 23 (8), 1180–1191.

Senapati, S., Mandal, D., Ahmad, A., Khan, M.I., Sastry, M., Kumar, R., 2004. Fungus mediated synthesis of silver nanoparticles: a novel biological approach. Indian J Phys 78, 101–105.

Senapati, S., Syed, A., Moeez, S., Kumar, A., Ahmad, A., 2012. Intracellular synthesis of gold nanoparticles using alga *Tetraselmis kochinensis*. Mater. Lett. 79, 116–118.

Seshadri, S., Prakash, A., Kowshik, M., 2012. Biosynthesis of silver nanoparticles by marine bacterium, *Idiomarina sp.* PR58-8. Bull. Mater. Sci. 35 (7), 1201–1205.

Shahbandeh, M., Eghdami, A., Moghaddam, M.M., Nadoushan, M.J., Salimi, A., Fasihi-Ramandi, M., et al., 2021. Conjugation of imipenem to silver nanoparticles for enhancement of its antibacterial activity against multidrug-resistant isolates of *Pseudomonas aeruginosa*. J. Biosci. 46 (1), 1–19.

Shanmugaiah, V., Harikrishnan, H., Al-Harbi, N.S., Shine, K., Khaled, J.M., Balasubramanian, N., Kumar, R.S., 2015. Facile synthesis of silver nanoparticles using Streptomyces sp. VSMGT1014 and their antimicrobial efficiency. Dig. J. Nanomater. Biostruct. 10 (1), 179–187.

Sharma, V.K., Yngard, R.A., Lin, Y., 2009. Silver nanoparticles: green synthesis and their antimicrobial activities. Adv. Colloid Interface Sci. 145 (1–2), 83–96.

Sharma, R., Bisen, D.P., Shukla, U., Sharma, B.G., 2012. X-ray diffraction: a powerful method of characterizing nanomaterials. Recent Res. Sci. Technol. 4 (8).

Shatha, A.S., Rana, H.A., Huda, Z.M., 2016. Study of biosynthesis silver nanoparticles by *Fusarium graminearum* and test their antimicrobial activity. Int. J. Innov. Appl. Stud. 15 (1), 43–50.

Shivaji, S., Madhu, S., Singh, S., 2011. Extracellular synthesis of antibacterial silver nanoparticles using psychrophilic bacteria. Process Biochem. 46 (9), 1800–1807.

Siddiqi, K.S., Husen, A., 2016. Fabrication of metal nanoparticles from fungi and metal salts: scope and application. Nanoscale Res. Lett. 11, 98. https://doi.org/10.1186/s11671-016-1311-2.

Silva, L.P.C., Oliveira, J.P., Keijok, W.J., da Silva, A.R., Aguiar, A.R., Guimarães, M.C.C., et al., 2017. Extracellular biosynthesis of silver nanoparticles using the cell-free filtrate of *nematophagous fungus Duddingtonia flagrans*. Int. J. Nanomedicine 12, 6373.

Simbine, E.O., Rodrigues, L.D.C., Lapa-Guimaraes, J., Kamimura, E.S., Corassin, C.H., Oliveira, C.A.F.D., 2019. Application of silver nanoparticles in food packages: a review. Food Sci Technol 39 (4), 793–802.

Singh, R., Shedbalkar, U.U., Wadhwani, S.A., Chopade, B.A., 2015. Bacteriagenic silver nanoparticles: synthesis, mechanism, and applications. Appl. Microbiol. Biotechnol. 99 (11), 4579–4593.

Singh, P., Singh, H., Kim, Y.J., Mathiyalagan, R., Wang, C., Yang, D.C., 2016. Extracellular synthesis of silver and gold nanoparticles by *Sporosarcina koreensis* DC4 and their biological applications. Enzyme Microb. Technol. 86, 75–83.

Singh, S.P., Bhargava, C.S., Dubey, V., Mishra, A., Singh, Y., 2017a. Silver nanoparticles: biomedical applications, toxicity, and safety issues. Int. J. Res Pharm Pharm Sci 4 (2), 01–10.

Singh, S., Singh, S.K., Chowdhury, I., Singh, R., 2017b. Understanding the mechanism of bacterial biofilms resistance to antimicrobial agents. Open Microbiol. J. 11, 53.

Singh, H., Du, J., Singh, P., Yi, T.H., 2018. Extracellular synthesis of silver nanoparticles by *Pseudomonas sp*. THG-LS1. 4 and their antimicrobial application. J. Pharm. Anal. 8 (4), 258–264.

Sinha, S.N., Paul, D., Halder, N., Dipta, S., Samir, K.P., 2015. Green synthesis of silver nanoparticles using fresh water green alga *Pithophora oedogonia* (Mont.) Wittrock and evaluation of their antibacterial activity. Appl. Nanosci. 5 (6), 703–709.

Srikar, S.K., Giri, D.D., Pal, D.B., Mishra, P.K., Upadhyay, S.N., 2016. Green synthesis of silver nanoparticles: a review. Curr. Opin. Green Sustain. 6 (1), 34–56.

Srivastava, P., Kowshik, M., 2016. Anti-neoplastic selenium nanoparticles from *Idiomarina sp*. PR58-8. Enzyme Microb. Technol. 95, 192–200.

Srivastava, P., Bragança, J., Ramanan, S.R., Kowshik, M., 2013. Synthesis of silver nanoparticles using haloarchaeal isolate *Halococcus salifodinae* BK 3. Extremophiles 17 (5), 821–831.

Sunkar, S., Nachiyar, C.V., 2012. Biogenesis of antibacterial silver nanoparticles using the endophytic bacterium *Bacillus cereus* isolated from *Garcinia xanthochymus*. Asian Pac. J. Trop. Biomed. 2 (12), 953–959.

Syed, A., Saraswati, S., Kundu, G.C., Ahmad, A., 2013. Biological synthesis of silver nanoparticles using the fungus *Humicola* sp. and evaluation of their cytoxicity using normal and cancer cell lines. Spectrochim. Acta A 114, 144–147.

Taha, Z.K., Hawar, S.N., Sulaiman, G.M., 2019. Extracellular biosynthesis of silver nanoparticles from *Penicillium italicum* and its antioxidant, antimicrobial and cytotoxicity activities. Biotechnol. Lett. 41 (8–9), 899–914.

Terra, A.L.M., Kosinski, R.D.C., Moreira, J.B., Costa, J.A.V., Morais, M.G.D., 2019. Microalgae biosynthesis of silver nanoparticles for application in the control of agricultural pathogens. J. Environ. Sci. Health B 54 (8), 709–716.

Thilagam, M., Esakkiammal, B., Mehalingam, P., 2021. Phytomediated synthesis and characterization of silver nanoparticles from the leaf extracts of *Begonia malabarica* Lam and its antimicrobial activity. Ann. Romanian Soc. Cell Biol. 25 (2), 3640–3649.

Thirumurugan, G., Veni, V.S., Ramachandran, S., Seshagiri Rao, J.V.L.N., Dhanaraju, M.D., 2011. Superior wound healing effect of topically delivered silver nanoparticle formulation using eco-friendly potato plant pathogenic fungus: synthesis and characterization. J. Biomed. Nanotechnol. 7 (5), 659–666.

Tyagi, S., Tyagi, P.K., Gola, D., Chauhan, N., Bharti, R.K., 2019. Extracellular synthesis of silver nanoparticles using entomopathogenic fungus: characterization and antibacterial potential. SN Appl. Sci. 1 (12), 1545.

Uddin, S., Safdar, L.B., Iqbal, J., Yaseen, T., Laila, S., Anwar, S., et al., 2021. Green synthesis of nickel oxide nanoparticles using leaf extract of *Berberis balochistanica*: characterization, and diverse biological applications. Microsc. Res. Tech. https://doi.org/10.1002/jemt.23756.

Vaidyanathan, R., Gopalram, S., Kalishwaralal, K., Deepak, V., Pandian, S.R., Gurunathan, S., 2010. Enhanced silver nanoparticle synthesis by optimization of nitrate reductase activity. Colloids Surf. B Biointerfaces 75, 335–341.

Velhal, S.G., Kulkarni, S.D., Latpate, R.V., 2016. Fungal mediated silver nanoparticle synthesis using robust experimental design and its application in cotton fabric. Int. Nano Lett. 6 (4), 257–264.

Velusamy, P., Kumar, G.V., Jeyanthi, V., Das, J., Pachaiappan, R., 2016. Bio-inspired green nanoparticles: synthesis, mechanism, and antibacterial application. Toxicol. Res. 32, 95–102. https://doi.org/10.5487/TR.2016.32.2.095.

Venkatesan, J., Kim, S.K., Shim, M.S., 2016. Antimicrobial, antioxidant, and anticancer activities of biosynthesized silver nanoparticles using marine algae *Ecklonia cava*. Nanomaterials 6 (12), 235.

Vigneshwaran, N., Ashtaputre, N.M., Varadarajan, P.V., Nachane, R.P., Paralikar, K.M., Balasubramanya, R.H., 2007. Biological synthesis of silver nanoparticles using the fungus *Aspergillus flavus*. Mater. Lett. 61 (6), 1413–1418.

Vijayan, S.R., Santhiyagu, P., Singamuthu, M., Kumari Ahila, N., Jayaraman, R., Ethiraj, K., 2014. Synthesis and characterization of silver and gold nanoparticles using aqueous extract of seaweed, *Turbinaria conoides*, and their antimicrofouling activity. Sci. World J. 2014.

Whitton, B.A., Potts, M., 2012. Introduction to the cyanobacteria. In: Ecology of Cyanobacteria II. Springer, Dordrecht, pp. 1–13.

Williams, D., 2008. The relationship between biomaterials and nanotechnology. Biomaterials 29 (12), 1737.

Wing-ShanáLin, I., Lok, C.N., Che, C.M., 2014. Biosynthesis of silver nanoparticles from silver (i) reduction by the periplasmic nitrate reductase c-type cytochrome subunit NapC in a silver-resistant *E. coli*. Chem. Sci. 5 (8), 3144–3150.

Woldeamanuel, K.M., Kurra, F.A., Roba, Y.T., 2021. A review on nanotechnology and its application in modern veterinary science. Int. J. Nanomater. Nanotechnol. Nanomed. 7 (1), 026–031.

Yahyaei, B., Pourali, P., 2019. One step conjugation of some chemotherapeutic drugs to the biologically produced gold nanoparticles and assessment of their anticancer effects. Sci. Rep. 9, 10242. https://doi.org/10.1038/s41598-019-46602-0.

You, Z., Qiu, Q., Chen, H., Feng, Y., Wang, X., Wang, Y., Ying, Y., 2020. Laser-induced noble metal nanoparticle-graphene composites enabled flexible biosensor for pathogen detection. Biosens. Bioelectron. 150, 111896.

Yousefzadi, M., Rahimi, Z., Ghafori, V., 2014. The green synthesis, characterization and antimicrobial activities of silver nanoparticles synthesized from green alga *Enteromorpha flexuosa* (wulfen) J. Agardh. Mater. Lett. 137, 1–4.

Yun, B.J., Kwon, J.E., Lee, K., Koh, W.G., 2019. Highly sensitive metal-enhanced fluorescence biosensor prepared on electrospun fibers decorated with silica-coated silver nanoparticles. Sens. Actuators B 284, 140–147.

Zhao, X., Zhou, L., Riaz Rajoka, M.S., Yan, L., Jiang, C., Shao, D., et al., 2018. Fungal silver nanoparticles: synthesis, application and challenges. Crit. Rev. Biotechnol. 38 (6), 817–835.

Zheng, D., Hu, C., Gan, T., Dang, X., Hu, S., 2010. Preparation and application of a novel vanillin sensor based on biosynthesis of Au–Ag alloy nanoparticles. Sensor Actuat. B Chem. 148, 247–252.

Zielińska, A., Skwarek, E., Zaleska, A., Gazda, M., Hupka, J., 2009. Preparation of silver nanoparticles with controlled particle size. Procedia Chem. 1, 1560–1566.

Zivic, F., Grujovic, N., Mitrovic, S., Ahad, I.U., Brabazon, D., 2018. Characteristics and applications of silver nanoparticles. In: Commercialization of Nanotechnologies—A Case Study Approach. Springer, Cham, pp. 227–273.

Zonooz, N.F., Salouti, M., 2011. Extracellular biosynthesis of silver nanoparticles using cell filtrate of *Streptomyces sp*. ERI-3. Sci. Iran. 18 (6), 1631–1635.

CHAPTER 17

Synthesis of silver nanoparticles using *Actinobacteria*

Manik Prabhu Narsing Rao[a] and Awalagaway Dhulappa[b]

[a]*State Key Laboratory of Biocontrol and Guangdong Provincial Key Laboratory of Plant Resources, School of Life Sciences, Sun Yat-Sen University, Guangzhou, PR China* [b]*Department of Microbiology, Maharani Cluster University, Bangalore, India*

1 Introduction

The first phylogenetic hierarchal clustering for the members of phylum *Actinobacteria* was provided by Stackebrandt et al. (1997). *Actinobacteria* represents one of the largest taxonomic units (Ventura et al., 2007) and recently, based on the 16S rRNA gene sequence phylogeny, the higher ranks in this phylum were updated to 6 classes, 46 orders, 79 families, and 425 genera (Salam et al., 2020). Following the discovery of streptomycin from actinobacteria (Schatz et al., 1944), the search for new drugs from them have made them a prolific producer of secondary metabolites. Approximately 22,000 secondary metabolites derived from microbes have been described, with actinobacteria accounting for half of them (Bérdy, 2012).

Nanotechnology is one of the most important technologies of the 21st century and represents an active area of research for particle sizes ranging from 1 to 100 nm (Manikprabhu and Lingappa, 2013a,b). Since ancient times, silver has been popular for its antimicrobial and wound healing properties (Manikprabhu and Li, 2015). Due to the wide applications of silver, it is one of the most widely studied types of nanomaterials (Manikprabhu and Lingappa, 2013a,b; Huang et al., 2020). It has been estimated that the world production of silver nanoparticles will be in the range of 360–450 tons per year by 2025, and it will grow to 800 tons per year (Syafiuddin et al., 2017; Huang et al., 2020).

Owing to the ability of actinobacteria to produce a wide range of bioactive molecules (Bérdy, 2012), researchers have started exploring them in the field of nanotechnology. In this chapter, we will discuss the synthesis and mechanism of silver nanoparticles using actinobacteria. In addition, an overview of the applications is given.

2 Mechanism and synthesis of silver nanoparticles using actinobacteria

The synthesis and mechanism for the formation of silver nanoparticles depends on which members of phylum *Actinobacteria* are used. Fig. 1 represents the overall steps for silver nanoparticles synthesis using actinobacteria.

In general, the actinobacterial strains will be grown in an appropriate liquid medium at optimal growth conditions. After the growth, it will be centrifuged and either the medium-free biomass (Shanmugasundaram et al., 2013) or the culture-free extract (Anasane et al., 2016; Fouda et al., 2019) will be treated with $AgNO_3$. The reaction mixture will be incubated until it turns brown. The color change (colorless to brown) of the solution is the preliminary confirmation for the silver nanoparticles synthesis (Manikprabhu and Lingappa, 2013a,b).

The possible mechanism for the synthesis of silver nanoparticles is either intracellular (Otari et al., 2015) or extracellular (Wypij et al., 2018a,b) mode. During intracellular synthesis, bacterial cells will be treated with $AgNO_3$ and incubated at appropriate growth conditions. Before the interaction, bacterial cells will be washed with sterilized

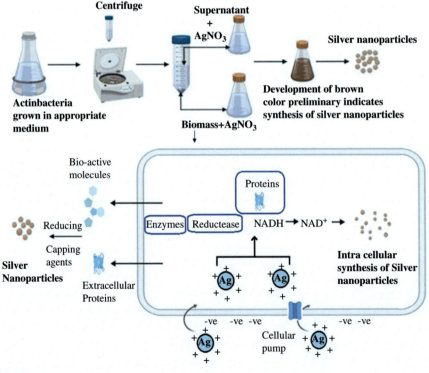

FIG. 1

Overall steps during actinobacteria-mediated synthesis of silver nanoparticles.

The figure was created with BioRender.com.

distilled water to avoid the involvement of medium components (Wypij et al., 2018a). During the interaction, Ag^+ ions will be trapped on the surface due to the electrostatic interaction between Ag^+ ions and the negatively charged cell wall, followed by reduction (by proteins, enzymes, etc.) and synthesis of silver nanoparticles underneath the cell wall (Lee and Jun, 2019). Intracellular synthesis of silver nanoparticles using actinobacteria was reported by Otari et al. (2015). *Rhodococcus* spp., when incubated in a medium containing 1 mM $AgNO_3$, produced spherical shaped silver nanoparticles of size less than 50 nm. The study suggested that biomolecules such as peptides, proteins, and carbohydrates may be responsible for the synthesis of silver nanoparticles. It was also proposed that proteins, along with other bioorganic molecules, may form a layer preventing the aggregation of silver nanoparticles.

Similarly, cell-free extract of actinobacteria (containing proteins, enzymes, and other bioactive molecules), when treated with $AgNO_3$, produced silver nanoparticles. The cell-free extract (containing proteins, enzymes, and other bioactive molecules) acted as a reducing and stabilizing agent (Dong et al., 2017; Manikprabhu and Lingappa, 2013a,b). Buszewski et al. (2018) reported the extracellular synthesis of silver nanoparticles using *Streptacidiphilus durhamensis* strain HGG16. When the culture-free extract and 3 mM $AgNO_3$ (1:1, v/v) were incubated at 26°C in a shaker for 7 days in the dark, silver nanoparticles were produced. The study suggested that the protein may be involved in the formation and stabilization of silver nanoparticles. Studies also suggested that when the actinobacterial supernatant and $AgNO_3$ mixture when exposed to sunlight, the rate of synthesis was enhanced. Endophytic actinobacteria, *Isoptericola* sp. culture-free extract (0.5 mL), and $AgNO_3$ (20 mL), when exposed to sunlight, silver nanoparticles were produced after 4 min. However, in the absence of sunlight, the synthesis was also observed, but after 3 h, suggesting sunlight acted as a catalyst (Dong et al., 2017).

Another possibility for synthesis is intracellular synthesis first, and transporting the synthesized nanoparticles outside the cell. In this case, synthesis of nanoparticles will be both extracellular and intracellular. A similar synthesis was reported by Prakasham et al. (2012), where the marine isolated *Streptomyces albidoflavus* strain produced silver nanoparticles both extra- and intracellularly.

3 Synthesis of silver nanoparticles using actinobacterial strains

Actinobacteria are ubiquitous and are the major producers of secondary metabolites (Zappelini et al., 2018; Nimaichand et al., 2015). The discovery of streptomycin (Schatz et al., 1944) from actinobacteria has initiated the antibiotic revolution and actinobacterial members, especially the genus *Streptomyces*, are acknowledged as a major producer of many bioactive compounds (Nimaichand et al., 2015; Law et al., 2019). *Streptomyces* has been documented as the major and dominant actinobacterial genus present in the various ecological niches (Zhang et al., 2006; Zappelini et al., 2018; Nimaichand et al., 2015). Synthesis of silver nanoparticles by various actinobacterial members has been reported (Table 1) but the majority of the reports

Table 1 List of actinobacterial strains synthesizing silver nanoparticles.

Sl. no.	Actinobacterial strain	Source/attribute	References
1	*Streptomyces* sp.	Soil	Saminathan (2015)
2	*Streptomyces noursei* H1-1	Soil	Alsharif et al. (2020)
3	*Streptomyces spongiicola* AS-3	Lawn soil	Devagi et al. (2020)
4	*Streptomyces* sp.	Acidic soil	Skladanowski et al. (2017)
5	*Streptacidiphilus* sp. strain CGG11n	Acidic soil	Railean-Plugaru et al. (2016)
6	*Streptomyces griseoplanus* SAI-25	Rhizospheric soil	Vijayabharathi et al. (2018)
7	*Streptomyces rochei* MHM13	Marine sediment	Abd-Elnaby et al. (2016)
8	*Nocardiopsis* sp. strain MBRC-1	Marine sediment	Manivasagan et al. (2013)
9	*Nocardiopsis dassonvillei* DS013	Marine sediment	Dhanaraj et al. (2020)
10	*Streptomyces intermedius*	Saline desert	Dayma et al. (2019)
11	*Streptomyces atrovirens*	Marine soil sample	Subbaiya et al. (2017)
12	*Streptomyces olivaceus* (MSU3)	Marine ecosystem	Sanjivkumar et al. (2019)
13	*Streptomyces coelicolor*	Pigment	Manikprabhu and Lingappa (2013a)
14	*Streptomyces* sp. NS-05	Pigment	Singh et al. (2021)
15	*Streptomyces* spp.	From marine sponge, *Crella cyathophora*	Hamed et al. (2020)
16	*Isoptericola* sp.	Endophyte	Dong et al. (2017)
17	*Pilimelia columellifera* subsp. *pallida*	Acidophilic	Golińska et al. (2016)
18	*Streptacidiphilus durhamensis* HGG16n	Acidophilic	Buszewski et al. (2018)
19	*Deinococcus radiodurans*	Radiation resistant	Kulkarni et al. (2015)
20	*Arthrobacter kerguelensis* and *Arthrobacter gangotriensis*	Psychrophilic	Shivaji et al. (2011)
21	*Sinomonas mesophila* MPKL26	Mesophilic	Manikprabhu et al. (2016)
22	*Streptomyces* sp. OSIP1 and *Streptomyces* sp. OSNP14	Cold-tolerant	Bakhtiari-Sardari et al. (2020)
23	*Dietzia maris* AURCCBT01	Pigment (canthaxanthin)	Venil et al. (2020)
24	*Streptomyces akiyoshiensis* GRG 6	Marine ecosystem	Rajivgandhi et al. (2020)
25	*Deinococcus wulumuqiensis* R12	Radiation resistant	Xiao et al. (2021)

suggest *Streptomyces* as a dominant group in the various ecological niches that could synthesis silver nanoparticles. For example, Ranjani et al. (2016) evaluated the diversity of actinobacteria synthesizing silver nanoparticles from nine different marine soil samples. A total of 1143 actinobacterial colonies were isolated, out of which 49 were morphologically distinct. Microscopic analysis revealed that the majority of them were *Streptomyces* (32.6%), followed by *Nocardiopsis* sp. (16.3%), *Kitasatospora* sp. (14.2%), *Thermoactinomyces* sp. (10.2%), *Actinomadura* sp. (8.1%), *Kibdelosporangium* sp. (6.1%), *Actinopolyspora* sp. (4.0%), *Saccharopolyspora* sp. (6.1%), and *Thermomonospora* sp. (2.0%). Out of 49 isolates, 25 can synthesize silver nanoparticles.

Different *Streptomyces* members isolated from various ecological niches are reported to synthesize silver nanoparticles. Fouda et al. (2019) reported the synthesis of silver nanoparticles using cell-free filtrates of three endophytic *Streptomyces* (*Streptomyces capillispiralis* Ca-1, *Streptomyces zaomyceticus* Oc-5, and *Streptomyces pseudogriseolus* Acv-11). The synthesized silver nanoparticles were spherical in shape with size ranging from 11.7 to 63.1 nm. The study suggested that the proteins present in the filtrate played a critical role in stabilizing the nanoparticles. Similarly, the ability of two haloalkaliphilic *Streptomyces calidiresistens* strains (IF11 and IF17) to synthesize silver nanoparticles has been evaluated (Wypij et al., 2018b). For synthesis, actinobacterial biomass was collected, washed, and resuspended for 2 days in 100 mL sterile deionized water. The cell pellet was discarded and the supernatant was mixed with silver nitrate solution (0.001 mol L^{-1} concentration). Spherical and polydisperse silver nanoparticles of 5–50 nm size were synthesized (Wypij et al., 2018b). An extremophilic *Streptomyces naganishii* MA7 biomass (5 g), when exposed to 1 mM AgNO$_3$ (50 mL), produced spherical shaped silver nanoparticles of size 5–50 nm. The study suggested that nanoparticles were stabilized through the binding of proteins to nanoparticles either through free residues or via electrostatic attraction of negatively charged carboxylate groups in the enzymes present in the cell wall of mycelia (Shanmugasundaram et al., 2013).

Apart from *Streptomyces*, other actinobacterial genera have also been reported to synthesize silver nanoparticles. Cell filtrate from acidophilic *Pilimelia* strains SF23 and C9, when treated with 1 mM silver nitrate, produced silver nanoparticles of size 4–60 nm. Proteins or peptides have been noticed on the particle surface, suggesting the probable stabilizing agent (Anasane et al., 2016). Similarly, when cell extract of alkaliphilic *Nocardiopsis valliformis* strain OT1 was treated with 1 mM silver nitrate solution, spherical and polydisperse silver nanoparticles of size 5–50 nm were synthesized. The study suggested that the proteins present in the cell extract might act as a capping agent (Rathod et al., 2016). Similarly, extracellular, spherical shape silver nanoparticles of size 20–40 nm were synthesized using *Thermoactinomyces* sp. mat extracts (Deepa et al., 2013).

Silver nanoparticles were also synthesized using actinobacterial pigments and bio-flocculant. Sowani et al., 2016 reported the synthesis of silver nanoparticles using cell-associated pigments of *Gordonia amicalis* HS-11. When the pigmented cells and culture-free extract were incubated with 1 mM AgNO$_3$ (pH 9.0, at 25°C), synthesis of silver nanoparticles was noticed via pigmented cells but not due to culture free extract, suggesting the synthesis might be due to cell-associated pigments. It was hypothesized that the carotenoids present in the cell-associated pigments caused the metal reduction. Similarly, silver nanoparticles were synthesized using melanin produced by *Nocardiopsis alba* MSA10. When 10 mL of melanin (20 μg/mL) treated with 40 mL of 1 mM AgNO$_3$ silver nanoparticles were synthesized (20–50 nm). It was reported that melanin played a dual role (reducing and capping agent) during the synthesis process. Manivasagan et al. (2015) reported the synthesis of silver nanoparticles using purified bioflocculant produced by *Streptomyces* sp. MBRC-91. The bioflocculant, when treated with silver nitrate (10^{-3} M) followed by incubation in the dark at 30°C for 96 h, produced spherical shape nanoparticles of size ranging from 10 to 60 nm (average particle size of 35 nm). It was suggested that O—H stretching might be responsible for reducing metal ions and, when the nanoparticles were synthesized, they were stabilized against the van der Waals forces of attraction.

3.1 Factors affecting the synthesis of silver nanoparticles

Reports on the effects of physicochemical parameters (Fig. 2) on the microbes-mediated synthesis of silver nanoparticles suggest they play a crucial role during the synthesis (Phanjom and Ahmed, 2017; Saeed et al., 2020).

Wypij et al. (2019) evaluated the physicochemical parameters factors affecting silver nanoparticles synthesis by four actinobacterial strains, *N. valliformis* strain OT1, *Streptomyces alkaliphilus* strain IF19, and *Streptomyces palmae* strains OF1 and OF2. The synthesis was noticed from 48 to 72 h, at pH 7 and/or 8, and at AgNO$_3$ concentration 0.001 mol L^{-1}; above and below these values the synthesis did not occur or was not efficient. For *S. alkaliphilus* IF19 and *S. palmae* OF1 and OF2 strains, small size particles were observed at 25°C; below and above this value, aggregation of nanoparticles was noticed. *N. valliformis* strain OT1 showed aggregation after 35°C, suggesting different actinobacterial members have different optimum parameters for synthesis. Similarly, the effect of pH on silver nanoparticles synthesis using acidophilic actinobacteria *Streptacidiphilus* sp. strain CGG11n was evaluated. It was reported that the size of the silver nanoparticles was pH-dependent. At pH 4.5–8.5, the size of silver nanoparticles was 100 ± 1.5 nm. At pH 2.0–4.0, the curve size distribution of nanoparticles decreased from 1500 to 200 nm, and above pH 10.0 an increase in particle size was noticed (Railean-Plugaru et al., 2016). Manikprabhu and Lingappa (2013a) reported a photo-irradiated synthesis of silver nanoparticles using *Streptomyces coelicolor* klmp33 pigment within 20 min. However, when microwave irradiation was used instead of photo-irradiation, silver nanoparticles were synthesized in just 90 s (Manikprabhu and Lingappa, 2013b), indicating the synthesis criteria are also important for the rapid formation of silver nanoparticles. The stability of the reducing agent plays an important role during silver nanoparticle synthesis. Kiran et al. (2014) used thermostable melanin from *N. alba* MSA10 to synthesize silver nanoparticles. As the reducing agent was

4 Silver nanoparticles synthesis using actinobacteria

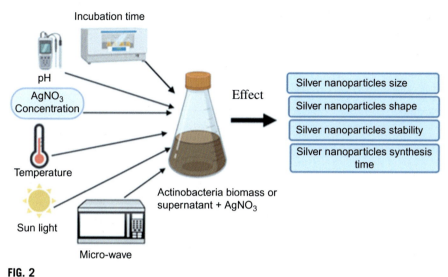

FIG. 2

Factors affecting silver nanoparticles synthesis.

The figure was created with BioRender.com.

thermostable, the synthesis was noticed even at high temperature, and with the increase in temperature up to 100°C stable and rapid synthesis of nanoparticles was observed.

4 Application of silver nanoparticles synthesized using actinobacterial strains

Silver's antimicrobial properties have been known from antiquity—it was used to preserve water and wine and in wound healing (Medici et al., 2019). Silver nanoparticles have distinct physical, chemical, and biological properties compared to their bulk parent materials (Manikprabhu and Lingappa, 2013a,b). Silver nanoparticles synthesized using actinobacterial strains have demonstrated diverse applications in various fields (Table 2). Silver nanoparticles synthesized using *S. albidoflavus* showed antimicrobial activity against both gram-negative and -positive bacterial strains (Prakasham et al., 2012). Samundeeswari et al. (2012) reported growth inhibition of food pathogens using *Streptomyces albogriseolus* synthesized silver nanoparticles. Silver nanoparticles synthesized using *Streptomyces spongiicola* showed broad-spectrum antibacterial activity against *Staphylococcus aureus*, *Bacillus cereus*, *Escherichia coli*, *Pseudomonas aeruginosa*, *Vibrio cholera*, *Shigella* sp., and *Salmonella typhi* (Devagi et al., 2020).

Reports also suggest that silver nanoparticles, when combined with antibiotics, show enhanced antimicrobial activity. Silver nanoparticles synthesized using *S.*

Table 2 Applications of silver nanoparticles synthesized using actinobacterial strains.

Sl. no.	Actinobacterial strain	Shape/size	Attributes	References
1	*Streptomyces griseorubens*	Spherical (5–20 nm)	Antioxidant activity	Baygar and Ugur (2017)
2	*Streptomyces naganishii* strain MA7	Spherical (5–50 nm)	Bactericidal, antibiofouling, cytotoxic, and antioxidant	Shanmugasundaram et al. (2013)
3	*Streptomyces* spp.	Spherical (8.6 ± 2 to 35 ± 2 nm)	Antibiofilm, antimicrobial, and cytotoxic	Hamed et al. (2020)
4	*Streptomyces* spp.	Spherical (11.3–63.1 nm)	Antimicrobial, antioxidant, and larvicidal activities	Fouda et al. (2019)
5	*Nocardiopsis dassonvillei* DS013	Circular and nonuniform (30–80 nm)	Bactericidal potential against clinical isolates	Dhanaraj et al. (2020)
6	*Streptomyces* sp. LK3	Spherical (5 nm)	Acaricidal activity	Karthik et al. (2014)
7	*Streptomyces* sp. M25	Spherical (10–35 nm)	Larvicidal activity	Shanmugasundaram and Balagurunathan (2015)
8	*Streptomyces coelicolor*	Irregular (28–50 nm)	Antibacterial against methicillin-resistant *Staphylococcus aureus*	Manikprabhu and Lingappa (2013a)
9	*Streptomyces coelicolor* klmp33	Irregular (28–50 nm)	Antibacterial against beta-lactamase (ESBL) producing *Escherichia coli*	Manikprabhu and Lingappa (2014)
10	*Isoptericola* sp.	Spherical (11–40 nm)	Antibacterial	Dong et al. (2017)
11	*Streptomyces* spp.	Spherical (8–15 nm)	Growth inhibitory effect against several pathogenic bacteria, biofilm inhibitor, and cytotoxic activity against carcinoma cells (CT26)	Bakhtiari-Sardari et al. (2020)
12	*Streptomyces* spp.	Spherical (16.4 ± 2.2 nm)	Antifungal activity against *Fusarium verticillioides*	Marathe et al. (2020)
13	*Streptomyces laurentii*	Spherical (7–15 nm)	Textile fabrics have antibacterial activity	Eid et al. (2020)
14	*Streptomyces noursei* H1-1	Spherical (6–30 nm)	Antibacterial, anticancer, and larvicidal activity	Alsharif et al. (2020)
15	*Rothia endophytica*	Cubic (47–72 nm)	Anticandidal activity	Elbahnasawy et al. (2021)

coelicolor and *Streptacidiphilus* sp. strain CGG11n not only showed good antimicrobial activity but their antimicrobial activity was enhanced when combined with antibiotics (Manikprabhu and Lingappa, 2013a; Railean-Plugaru et al., 2016).

Mosquitocidal and fungicidal activity has been reported using silver nanoparticles synthesized with *Streptomyces anulatus* (Soni and Prakash, 2015). *Streptomyces* sp. (09 PBT 005) synthesized silver nanoparticles showed 83.2% cytotoxicity activity against A549 adenocarcinoma lung cancer cell line (Saravana Kumar et al., 2015). *Streptomyces xinghaiensis* OF1 strain synthesized silver nanoparticles alone and in combination with antibiotics showed good cytotoxic effect against mouse fibroblasts and HeLa cell line (Wypij et al., 2018b). Silver nanoparticles synthesized using exopolysaccharide of *Streptomyces violaceus* MM72 showed antibacterial and antioxidant activities (Sivasankar et al., 2018).

5 Conclusions and future perspectives

Actinobacteria are a valuable source for bioactive molecules and in the past few years, they have been explored in the field of nanotechnology. Several studies have been carried out to synthesize silver nanoparticles using actinobacterial strains. Various synthesis mechanisms and process optimizations have been proposed. *Actinobacteria*-synthesized silver nanoparticles have been used in different fields and their results are promising. Although extensive studies have been carried out, only the genus *Streptomyces* was focused on regarding synthesis, and looking at the genera reported in the phylum *Actinobacteria*, only the tip of the iceberg of strains have been used to synthesize silver nanoparticles, and hence, further studies need to be carried out to explore other actinobacterial genera for synthesizing silver nanoparticles.

References

Abd-Elnaby, H.M., Abo-Elala, G.M., Abdel-Raouf, U.M., Hamed, M.M., 2016. Antibacterial and anticancer activity of extracellular synthesized silver nanoparticles from marine *Streptomyces rochei* MHM13. Egypt. J. Aquat. Res. 42, 301–312.

Alsharif, S.M., Salem, S.S., Abdel-Rahman, M.A., Fouda, A., Eid, A.M., El-Din Hassan, S., Awad, M.A., Mohamed, A.A., 2020. Multifunctional properties of spherical silver nanoparticles fabricated by different microbial taxa. Heliyon 6 (5), e03943.

Anasane, N., Golińska, P., Wypij, M., Rathod, D., Dahm, H., Rai, M., 2016. Acidophilic actinobacteria synthesised silver nanoparticles showed remarkable activity against fungi-causing superficial mycoses in humans. Mycoses 59, 157–166.

Bakhtiari-Sardari, A., Mashreghi, M., Eshghi, H., Behnam-Rasouli, F., Lashani, E., Shahnavaz, B., 2020. Comparative evaluation of silver nanoparticles biosynthesis by two cold-tolerant *Streptomyces* strains and their biological activities. Biotechnol. Lett. 42, 1985–1999.

Baygar, T., Ugur, A., 2017. Biosynthesis of silver nanoparticles by *Streptomyces griseorubens* isolated from soil and their antioxidant activity. IET Nanobiotechnol. 11, 286–291.

Bérdy, J., 2012. Thoughts and facts about antibiotics: where we are now and where we are heading. J. Antibiot. Res. 65, 385–395.

Buszewski, B., Railean-Plugaru, V., Pomastowski, P., Rafińska, K., Szultka-Mlynska, M., Golinska, P., Wypij, M., Laskowski, D., Dahm, H., 2018. Antimicrobial activity of bio-silver nanoparticles produced by a novel *Streptacidiphilus durhamensis* strain. J. Microbiol. Immunol. Infect. 51, 45–54.

Dayma, P.B., Mangrola, A.V., Suriyaraj, S.P., Dudhagara, P., Patel, R.K., 2019. Synthesis of bio-silver nanoparticles using desert isolated *Streptomyces intermedius* and its antimicrobial activity. J. Pharm. Chem. Biol. Sci. 7, 94–101.

Deepa, S., Kanimozhi, K., Panneerselvam, A., 2013. Antimicrobial activity of extracellularly synthesized silver nanoparticles from marine derived actinomycetes. Int. J. Curr. Microbiol. Appl. Sci. 2, 223–230.

Devagi, P., Suresh, T.C., Sandhiya, R.V., Sairandhry, M., Bharathi, S., Velmurugan, P., Radhakrishnan, M., Sathiamoorthi, T., Suresh, G., 2020. Actinobacterial-mediated fabrication of silver nanoparticles and their broad spectrum antibacterial activity against clinical pathogens. J. Nanosci. Nanotechnol. 20, 2902–2910.

Dhanaraj, S., Thirunavukkarasu, S., Allen John, H., Pandian, S., Salmen, S.H., Chinnathambi, A., Alharbi, S.A., 2020. Novel marine *Nocardiopsis dassonvillei*-DS013 mediated silver nanoparticles characterization and its bactericidal potential against clinical isolates. Saudi J. Biol. Sci. 27, 991–995.

Dong, Z.Y., Narsing Rao, M.P., Xiao, M., Wang, H.F., Hozzein, W.N., Chen, W., Li, W.J., 2017. Antibacterial activity of silver nanoparticles against *Staphylococcus warneri* synthesized using endophytic bacteria by photo-irradiation. Front. Microbiol. 8, 1090.

Eid, A.M., Fouda, A., Niedbała, G., Hassan, S.E., Salem, S.S., Abdo, A.M.F., Hetta, H., Shaheen, T.I., 2020. Endophytic *Streptomyces laurentii* mediated green synthesis of Ag-NPs with antibacterial and anticancer properties for developing functional textile fabric properties. Antibiotics 9 (10), E641.

Elbahnasawy, M.A., Shehabeldine, A.M., Khattab, A.M., Amin, B.H., Hashem, A.H., 2021. Green biosynthesis of silver nanoparticles using novel endophytic *Rothia endophytica*: characterization and anticandidal activity. J. Drug Delivery Sci. Technol. . https://doi.org/10.1016/j.jddst.2021.102401.

Fouda, A., Hassan, S.E., Abdo, A.M., El-Gamal, M.S., 2019. Antimicrobial, antioxidant and larvicidal activities of spherical silver nanoparticles synthesized by endophytic *Streptomyces* spp. Biol. Trace Elem. Res. . https://doi.org/10.1007/s12011-019-01883-4.

Golińska, P., Wypij, M., Rathod, D., Tikar, S., Dahm, H., Rai, M., 2016. Synthesis of silver nanoparticles from two acidophilic strains of *Pilimelia columellifera* subsp. *pallida* and their antibacterial activities. J. Basic Microbiol. 56, 541–556.

Hamed, A.A., Kabary, H., Khedr, M., Emam, A.N., 2020. Antibiofilm, antimicrobial and cytotoxic activity of extracellular green-synthesized silver nanoparticles by two marine-derived actinomycete. RSC Adv. 10, 10361–10367.

Huang, W., Yan, M., Duan, H., Bi, Y., Cheng, X., Yu, H., 2020. Synergistic antifungal activity of green synthesized silver nanoparticles and epoxiconazole against *Setosphaeria turcica*. J. Nanomater. 2020, 9535432.

Karthik, L., Kumar, G., Kirthi, A.V., Rahuman, A.A., Bhaskara Rao, K.V., 2014. *Streptomyces* sp. LK3 mediated synthesis of silver nanoparticles and its biomedical application. Bioprocess Biosyst. Eng. 37, 261–267.

Kiran, G.S., Dhasayan, A., Lipton, A.N., Selvin, J., Arasu, M.V., Al-Dhabi, N.A., 2014. Melanin-templated rapid synthesis of silver nanostructures. J. Nanobiotechnol. 12, 18.

Kulkarni, R.R., Shaiwale, N.S., Deobagkar, D.N., Deobagkar, D.D., 2015. Synthesis and extracellular accumulation of silver nanoparticles by employing radiation-resistant *Deinococcus radiodurans*, their characterization, and determination of bioactivity. Int. J. Nanomedicine 10, 963–974.

Law, J.W., Chan, K.G., He, Y.W., Khan, T.M., Ab Mutalib, N.S., Goh, B.H., Lee, L.H., 2019. Diversity of *Streptomyces* spp. from mangrove forest of Sarawak (Malaysia) and screening of their antioxidant and cytotoxic activities. Sci. Rep. 9, 15262.

Lee, S.H., Jun, B.H., 2019. Silver nanoparticles: synthesis and application for nanomedicine. Int. J. Mol. Sci. 20, 865.

Manikprabhu, D., Li, W.J., 2015. Microbe-mediated synthesis of silver nanoparticles a new drug of choice against pathogenic microorganisms. In: Dhanasekaran, D., Thajuddin, N., Panneerselvam, A. (Eds.), Antimicrobials: Synthetic and Natural Compounds. CRC Press, Boca Raton, FL, pp. 389–401.

Manikprabhu, D., Lingappa, K., 2013a. Antibacterial activity of silver nanoparticles against methicillin-resistant *Staphylococcus aureus* synthesized using model *Streptomyces* sp. pigment by photo-irradiation method. J. Pharm. Res. 6, 255–260.

Manikprabhu, D., Lingappa, K., 2013b. Microwave assisted rapid and green synthesis of silver nanoparticles using a pigment produced by *Streptomyces coelicolor* klmp33. Bioinorg. Chem. Appl. 2013, 341798.

Manikprabhu, D., Lingappa, K., 2014. Synthesis of silver nanoparticles using the *Streptomyces coelicolor* klmp33 pigment: an antimicrobial agent against extended-spectrum beta-lactamase (ESBL) producing *Escherichia coli*. Mater. Sci. Eng. C 45, 434–437.

Manikprabhu, D., Cheng, J., Chen, W., Sunkara, A.K., Mane, S.B., Kumar, R., Das, M., Hozzein, W.N., Duan, Y.Q., Li, W.J., 2016. Sunlight mediated synthesis of silver nanoparticles by a novel actinobacterium (*Sinomonas mesophila* MPKL 26) and its antimicrobial activity against multi drug resistant *Staphylococcus aureus*. J. Photochem. Photobiol. B 158, 202–205.

Manivasagan, P., Venkatesan, J., Senthilkumar, K., Sivakumar, K., Kim, S.K., 2013. Biosynthesis, antimicrobial and cytotoxic effect of silver nanoparticles using a novel *Nocardiopsis* sp. MBRC-1. Biomed. Res. Int. 2013, 287638.

Manivasagan, P., Kang, K.H., Kim, D.G., Kim, S.K., 2015. Production of polysaccharide-based bioflocculant for the synthesis of silver nanoparticles by *Streptomyces* sp. Int. J. Biol. Macromol. 77, 159–167.

Marathe, K., Naik, J., Maheshwari, V., 2020. Biogenic synthesis of silver nanoparticles using *Streptomyces* spp. and their antifungal activity against *Fusarium verticillioides*. J. Clust. Sci. . https://doi.org/10.1007/s10876-020-01894-5.

Medici, S., Peana, M., Nurchi, V.M., Zoroddu, M.A., 2019. Medical uses of silver: history, myths, and scientific evidence. J. Med. Chem. 62, 5923–5943.

Nimaichand, S., Devi, A.M., Tamreihao, K., Ningthoujam, D.S., Li, W.J., 2015. Actinobacterial diversity in limestone deposit sites in Hundung, Manipur (India) and their antimicrobial activities. Front. Microbiol. 6, 413.

Otari, S.V., Patil, R.M., Ghosh, S.J., Thorat, N.D., Pawar, S.H., 2015. Intracellular synthesis of silver nanoparticle by actinobacteria and its antimicrobial activity. Spectrochim. Acta A Mol. Biomol. Spectrosc. 136, 1175–1180.

Phanjom, P., Ahmed, G., 2017. Effect of different physicochemical conditions on the synthesis of silver nanoparticles using fungal cell filtrate of *Aspergillus oryzae* (MTCC No. 1846) and their antibacterial effect. Adv. Nat. Sci. Nanosci. Nanotechnol. 8, 045016.

Prakasham, R.S., Buddana, S.K., Yannam, S.K., Guntuku, G.S., 2012. Characterization of silver nanoparticles synthesized by using marine isolate *Streptomyces albidoflavus*. J. Microbiol. Biotechnol. 22, 614–621.

Railean-Plugaru, V., Pomastowski, P., Wypij, M., Szultka-Mlynska, M., Rafinska, K., Golinska, P., Dahm, H., Buszewski, B., 2016. Study of silver nanoparticles synthesized by acidophilic strain of Actinobacteria isolated from the of *Picea sitchensis* forest soil. J. Appl. Microbiol. 120, 1250–1263.

Rajivgandhi, G.N., Ramachandran, G., Li, J.L., Yin, L., Manoharan, N., Kannan, M.R., Velanganni, A.A.J., Alharbi, N.S., Kadaikunnan, S., Khaled, J.M., Li, W.J., 2020. Molecular identification and structural detection of anti-cancer compound from marine *Streptomyces akiyoshiensis* GRG 6 (KY457710) against MCF-7 breast cancer cells. J. King Saud Univ. Sci. 32, 3463–3469.

Ranjani, A., Gopinath, P., Rajesh, K., Dhanasekaran, D., Priyadharsini, P., 2016. Diversity of silver nanoparticle synthesizing actinobacteria isolated from marine soil, Tamil Nadu, India. Arab. J. Sci. Eng. 41, 25–32.

Rathod, D., Golinska, P., Wypij, M., Dahm, H., Rai, M., 2016. A new report of *Nocardiopsis valliformis* strain OT1 from alkaline Lonar crater of India and its use in synthesis of silver nanoparticles with special reference to evaluation of antibacterial activity and cytotoxicity. Med. Microbiol. Immunol. 205, 435–447.

Saeed, S., Iqbal, A., Ashraf, M.A., 2020. Bacterial-mediated synthesis of silver nanoparticles and their significant effect against pathogens. Environ. Sci. Pollut. Res. Int. . https://doi.org/10.1007/s11356-020-07610-0.

Salam, N., Jiao, J.Y., Zhang, X.T., Li, W.J., 2020. Update on the classification of higher ranks in the phylum *Actinobacteria*. Int. J. Syst. Evol. Microbiol. 70, 1331–1355.

Saminathan, K., 2015. Biosynthesis of silver nanoparticles using soil Actinomycetes Streptomyces sp. Int. J. Curr. Microbiol. Appl. Sci. 4, 1073–1083.

Samundeeswari, A., Dhas, S.P., Nirmala, J., John, S.P., Mukherjee, A., Chandrasekaran, N., 2012. Biosynthesis of silver nanoparticles using actinobacterium *Streptomyces albogriseolus* and its antibacterial activity. Biotechnol. Appl. Biochem. 59, 503–507.

Sanjivkumar, M., Vaishnavi, R., Neelakannan, M., Kannan, D., Silambarasan, T., Immanuel, G., 2019. Investigation on characterization and biomedical properties of silver nanoparticles synthesized by an actinobacterium *Streptomyces olivaceus* (MSU3). Biocatal. Agric. Biotechnol. 17, 151–159.

Saravana Kumar, P., Balachandran, C., Duraipandiyan, V., Ramasamy, D., Ignacimuthu, S., Al-Dhabi, N.A., 2015. Extracellular biosynthesis of silver nanoparticle using *Streptomyces* sp. 09 PBT 005 and its antibacterial and cytotoxic properties. Appl. Nanosci. 5, 169–180.

Schatz, A., Bugle, E., Waksman, S.A., 1944. Streptomycin, a substance exhibiting antibiotic activity against Gram-positive and Gram-negative bacteria. Proc. Soc. Exp. Biol. Med. 55, 66–69.

Shanmugasundaram, T., Balagurunathan, R., 2015. Mosquito larvicidal activity of silver nanoparticles synthesised using actinobacterium, *Streptomyces* sp. M25 against *Anopheles subpictus*, *Culex quinquefasciatus* and *Aedes aegypti*. J. Parasit. Dis. 39, 677–684.

Shanmugasundaram, T., Radhakrishnan, M., Gopikrishnan, V., Pazhanimurugan, R., Balagurunathan, R., 2013. A study of the bactericidal, anti-biofouling, cytotoxic and antioxidant properties of actinobacterially synthesised silver nanoparticles. Colloids Surf. B Biointerfaces 111, 680–687.

Shivaji, S., Madhu, S., Singh, S., 2011. Extracellular synthesis of antibacterial silver nanoparticles using psychrophilic bacteria. Process Biochem. 46, 1800–1807.

Singh, N., Naik, B., Kumar, V., Kumar, A., Kumar, V., Gupta, S., 2021. Actinobacterial pigment assisted synthesis of nanoparticles and its biological activity. J. Microbiol. Biotechnol. Food Sci. 10, 604–608.

Sivasankar, P., Seedevi, P., Poongodi, S., Sivakumar, M., Murugan, T., Sivakumar, L., Sivakumar, K., Balasubramanian, T., 2018. Characterization, antimicrobial and antioxidant property of exopolysaccharide mediated silver nanoparticles synthesized by *Streptomyces violaceus* MM72. Carbohydr. Polym. 181, 752–759.

Składanowski, M., Wypij, M., Laskowski, D., Golińska, P., Dahm, H., Rai, M., 2017. Silver and gold nanoparticles synthesized from *Streptomyces* sp. isolated from acid forest soil with special reference to its antibacterial activity against pathogens. J. Clust. Sci. 28, 59–79.

Soni, N., Prakash, S., 2015. Antimicrobial and mosquitocidal activity of microbial synthesized silver nanoparticles. Parasitol. Res. 114, 1023–1030.

Sowani, H., Mohite, P., Damale, S., Kulkarni, M., Zinjarde, S., 2016. Carotenoid stabilized gold and silver nanoparticles derived from the Actinomycete *Gordonia amicalis* HS-11 as effective free radical scavengers. Enzym. Microb. Technol. 95, 164–173.

Stackebrandt, E., Rainey, F.A., Ward-Rainey, N.L., 1997. Proposal for a new hierarchic classification system, *Actinobacteria* classis nov. Int. J. Syst. Bacteriol. 47, 479–491.

Subbaiya, R., Saravanan, M., Priya, A.R., Shankar, K.R., Selvam, M., Ovais, M., Balajee, R., Barabadi, H., 2017. Biomimetic synthesis of silver nanoparticles from *Streptomyces atrovirens* and their potential anticancer activity against human breast cancer cells. IET Nanobiotechnol. 11, 965–972.

Syafiuddin, A., Salmiati, M.R.S., Kueh, A.B.H., Hadibarata, T., Nur, H., 2017. A review of silver nanoparticles: research trends, global consumption, synthesis, properties, and future challenges. J. Chin. Chem. Soc. 64, 732–756.

Venil, C.K., Malathi, M., Velmurugan, P., Renuka Devi, P., 2020. Green synthesis of silver nanoparticles using canthaxanthin from *Dietzia maris* AURCCBT01 and their cytotoxic properties against human keratinocyte cell line. J. Appl. Microbiol. . https://doi.org/10.1111/jam.14889.

Ventura, M., Canchaya, C., Tauch, A., Chandra, G., Fitzgerald, G.F., Chater, K.F., van Sinderen, D., 2007. Genomics of Actinobacteria: tracing the evolutionary history of an ancient phylum. Microbiol. Mol. Biol. Rev. 71, 495–548.

Vijayabharathi, R., Sathya, A., Gopalakrishnan, S., 2018. Extracellular biosynthesis of silver nanoparticles using *Streptomyces griseoplanus* SAI-25 and its antifungal activity against *Macrophomina phaseolina*, the charcoal rot pathogen of sorghum. Biocatal. Agric. Biotechnol. 14, 166–171.

Wypij, M., Czarnecka, J., Świecimska, M., Dahm, H., Rai, M., Golinska, P., 2018a. Synthesis, characterization and evaluation of antimicrobial and cytotoxic activities of biogenic silver nanoparticles synthesized from *Streptomyces xinghaiensis* OF1 strain. World J. Microbiol. Biotchnol. 34, 23.

Wypij, M., Świecimska, M., Czarnecka, J., Dahm, H., Rai, M., Golinska, P., 2018b. Antimicrobial and cytotoxic activity of silver nanoparticles synthesized from two haloalkaliphilic actinobacterial strains alone and in combination with antibiotics. J. Appl. Microbiol. 124, 1411–1424.

Wypij, M., Świecimska, M., Dahm, H., Rai, M., Golinska, P., 2019. Controllable biosynthesis of silver nanoparticles using actinobacterial strains. Green Processes Synth. 8, 207–214.

Xiao, A., Wang, B., Zhu, L., Jiang, L., 2021. Production of extracellular silver nanoparticles by radiation-resistant *Deinococcus wulumuqiensis* R12 and its mechanism perspective. Process Biochem. 100, 217–223.

Zappelini, C., Alvarez-Lopez, V., Capelli, N., Guyeux, C., Chalot, M., 2018. *Streptomyces* dominate the soil under *Betula* trees that have naturally colonized a red gypsum landfill. Front. Microbiol. 9, 1772.

Zhang, H., Lee, Y.K., Zhang, W., Lee, H.K., 2006. Culturable actinobacteria from the marine sponge *Hymeniacidon perleve*: isolation and phylogenetic diversity by 16S rRNA gene-RFLP analysis. Antonie Van Leeuwenhoek 90, 159–169.

CHAPTER 18

Role of bacteria and actinobacteria in the biosynthesis of silver nanoparticles

Gonzalo Tortella[a], Olga Rubilar[a,b], María Cristina Diez[a,b], Sergio Cuozzo[c], Joana Claudio Pieretti[d], and Amedea Barozzi Seabra[d]

[a]Centro de Excelencia en Investigación Biotecnológica Aplicada al Medio Ambiente, CIBAMA-BIOREN, Universidad de La Frontera, Temuco, Chile, [b]Chemical Engineering Department, Universidad de La Frontera, Temuco, Chile, [c]Planta de Procesos industriales-Proimi-Conicet, San Miguel de Tucumán, Tucuman, Argentina, [d]Center for Natural and Human Sciences, Universidade Federal do ABC (UFABC), Santo André, SP, Brazil

1 Introduction

During recent years, nanotechnology has been one the most studied areas within science and technology worldwide, involving a group of small particles at nanoscale sizes ranged between 1 and 100 nm. Among all described nanomaterials (Klaessig et al., 2011), nanoparticles have gained considerable attention due to their unmatched physicochemical properties. For this same reason, the use of nanoparticles has increased in several applications, such as agriculture, medicine, energy, and electronics (Khan et al., 2017). Nanoparticles involve several types of carbon-based (fullerenes, graphene, carbon nanotubes, among others), organic-based (liposomes, dendrimers, micelles, among others), composite-based (combination of nanoparticles or with polymers), and metal-based materials (copper and copper oxides, titanium, gold, silver, among others). All the mentioned nanoparticles have been included in several applications. One of the most studied and amazing nanostructures is silver nanoparticles (AgNPs), due to their widespread uses in several applications, as shown in Fig. 1.

As a consequence of the widespread use of AgNPs worldwide, there are several commercial suppliers of AgNPs, and a rising global demand for these nanoparticles from different sectors of industry (Calderón-Jiménez et al., 2017; Rai et al., 2014a,b). In this regard, it is very interesting to know that industrial reports indicate the AgNPs market was valued at over USD 1.3 billion in 2017, and it was estimated that it will be >USD 3.0 billion for 2024 (Ahuja and Rawat, 2018). It is not difficult to imagine the reason for this huge number. One of the most explored properties of AgNPs is their

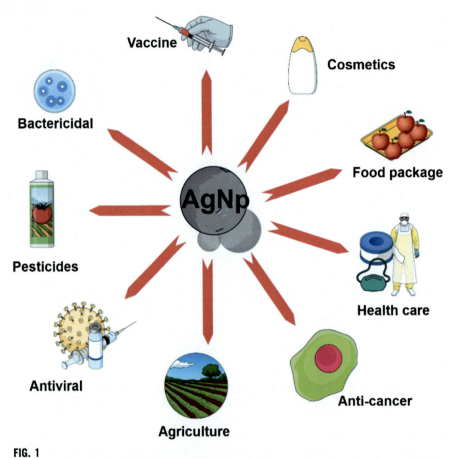

FIG. 1

Potential applications of silver nanoparticles (AgNPs).

inherent antimicrobial action (Deshmukh et al., 2018), which has led to their incorporation in many healthcare products, medical devices, textile products, and consumer products, such as washing machines. An interesting application of AgNPs is their use in food packaging, due to the effective nanoparticle protection against fungal and/or bacterial contamination, increasing fresh food shelf life (Carbone et al., 2016). Another important application of AgNPs is in medicine and health areas. For instance, the use of AgNPs in dentistry allows maintenance of better oral health (Rodrigues et al., 2020). Moreover, the combination of AgNPs with other materials can act as a powerful antimicrobial agent avoiding biofilm formation (Peng et al., 2012; Bapat et al., 2018). In addition, the use of AgNPs and biopolymer-based biomaterials for wound healing applications has shown interesting advances (Kumar et al., 2018). Interestingly, in a recent work, Zachar (2020) reported that AgNPs inhalation might be effective to combat viral diseases, such as the novel COVID-19, particularly in the early stages of treatment. Moreover, AgNPs associated with nitric oxide have been also reported as an efficient tool for COVID-19 treatment (Pieretti et al., 2020).

The synthesis of AgNPs has been reported by several physical, chemical, and biological methods. Physical methods include ultra-sonication (Wani et al., 2011; Anwar et al., 2021), laser ablation (Sportelli et al., 2018), irradiation (Sheikh et al., 2009), microwave (Faxian et al., 2017; Pletzer et al., 2021), and electrochemical (Rabinal et al., 2013). Chemical methods include chemical reduction (Wang et al., 2005), sol gel methods (Maharjan et al., 2019), inert gas condensation (Raffi et al., 2007), or photoinduced reduction (Shchukin et al., 2003). Physical methods are quick, but low yields are obtained, and high energy consumption is required (Zhang et al., 2016). On the contrary, chemical methods produce high yields, however toxic and hazardous substances might be required (Zhang et al., 2016). In this regard, given the great worldwide concern for the environment, biological methods, also known as green or biogenic methods, have been extensively studied (Ahmad et al., 2019). Biological synthesis of AgNPs involves the use of macro and micro algae (González-Ballesteros et al., 2019; Terra et al., 2019), fungi (Cuevas et al., 2015; Guilger-Casagrande and de Lima, 2019), plants (Chung et al., 2016), and bacteria or actinobacteria (Manivasagan et al., 2014; Ghashghaei and Emtiazi, 2015). The main advantages of biological methods over chemical and physical routes are: (i) low toxicity, since no toxic reagents are used; (ii) simplicity and low cost, since, in general, biological synthesis of nanoparticles is performed at room temperature and ambient atmosphere; and (iii) enhanced biocompatibility of the obtained nanoparticles, since the biological entities act as nanofactories leading to nanoparticle formation and acting as a coating agent on the nanoparticle surface (Das et al., 2017; Durán and Seabra, 2018).

Among all biological options to synthesize AgNPs, bacteria and actinobacteria are an amazing group of microorganisms that are present in many ecological niches worldwide, and can be found from aquatic to terrestrial ecosystems, including extreme habitats (Bauer et al., 2018; Qin et al., 2019). These extraordinary microorganisms have been recognized as an important source of secondary metabolites, which have been explored with a diversity of applications in agriculture, medicine, or industrial fields (Singh et al., 2017). Moreover, given their extraordinary versatility, both bacteria and actinobacteria have been used in the biosynthesis of nanoparticles including gold, silver, copper and copper oxides, titanium, and iron, among others (Siva Kumar et al., 2014; Prakash Patil et al., 2019; Ameen et al., 2020; Rasool and Hemalatha, 2017; Taran et al., 2018). Metal nanoparticles can be synthesized by bacteria or actinobacteria when these microorganisms take target ions from their surrounding environment and turn the metal ions into nanoparticles through enzymes or secondary metabolites generated by cell activity (Li et al., 2011; Marslin et al., 2018). The mechanism used for nanoparticle synthesis by bacteria and actinobacteria can be intracellular or extracellular, depending if the site is outside or inside the cell (Otari et al., 2015). Both mechanisms have reported the presence of extracellular or intracellular enzymes, such as NADH-dependent nitrate reductase

(Blasco et al., 2001), or other compounds, such as sugars, which act as a reducing agent (Li et al., 2011). Accordingly, this chapter highlights the main aspects related to the biosynthesis of AgNPs by bacteria and actinobacteria and their antimicrobial properties against microbial pathogens.

2 Synthesis of AgNPs by bacteria

Among the biological routes to synthesize AgNPs, bacterial-mediated synthesis has been growing, as it presents the following advantages: (i) safety, (ii) low-toxicity, (iii) low-cost, and (iv) easily scaling up for mass production (Javaid et al., 2017). Nowadays, numerous bacterial strains have been used in the synthesis of AgNPs and other noble-metal nanoparticles (Ali et al., 2019). The study of AgNPs synthesis using different strains has an important role in the biotechnological progress regarding biological synthesis, as each strain presents an individual property that might result in specific characteristics in the final AgNPs, such as size, or a distinct coating property, which will directly influence the nanoparticle applications (Javaid et al., 2017). To date, at least 70 different strains have been used to synthesize AgNPs and the number of papers published in this area continues to grow.

There are two different mechanisms for AgNPs synthesis using bacteria: intracellular or extracellular. Intracellular synthesis occurs inside the bacterial cell, in which the precursor (Ag^+) is reduced to Ag^0, and the obtained nanoparticle accumulates on the cell wall (Murugan et al., 2014). Usually, intracellular synthesis is performed in bacteria resistant to silver stress, as the silver ions are transported inside the bacteria cell. Among the strains reported to demonstrate efficient intracellular synthesis of AgNPs, cyanobacteria and few Pseudomonas strains stand out (Murugan et al., 2014; Roychoudhury et al., 2016; Srivastava and Constanti, 2012; Srivastava et al., 2013).

Even though some strains are able to perform intracellular synthesis, the majority of bacterial synthesis of AgNPs relies on extracellular pathways. Extracellular synthesis depends on secreted proteins or enzymes to reduce the silver ions, and this process is usually performed through the use of isolated reducing content, diminishing the biomass that would require further purification steps in the obtainment of AgNPs (Quinteros et al., 2019). A simple representation of the extracellular biosynthesis of AgNPs is shown in Fig. 2. Regarding extracellular synthesis, it is possible to highlight the use of *Bacillus* spp., however, a vast variety of strains has been used for extracellular production of AgNPs (Fouad et al., 2017).

The use of *Bacillus* covers a considerable portion of bacterial synthesis of AgNPs. For instance, AgNPs were synthesized by enzymatic extracts of *Bacillus amyloliquefaciens* and *Bacillus subtilis*, and nanoparticles with different characteristics were obtained, depending on the employed strain. The hydrodynamic size of AgNPs synthesized by *B. amyloliquefaciens* was smaller when compared to *B. subtilis*—74 and 136 nm, respectively (Ghiuță et al., 2018)—although the same crystalline pattern and morphology were confirmed. Similar synthesis was performed by Fouad et al.

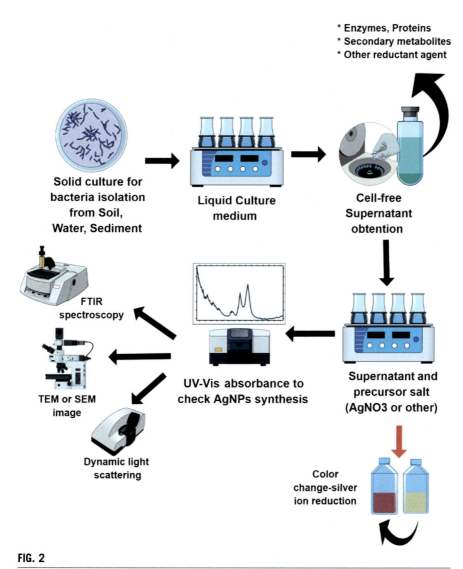

FIG. 2

Simple representation of steps for the extracellular biosynthesis of AgNPs by bacteria or actinobacteria.

employing the same strains, but at lower temperatures (Fouad et al., 2017). Similar results regarding the size and morphology of AgNPs were obtained, with a solid-state size from 16 to 80 nm and a spherical morphology, agreeing with the significative polydispersity also verified by the previous work. Results using *B. amyloliquefaciens* were also supported by Samuel and coworkers, although other *Bacillus* strains (*Bacillus brevis* and *Bacillus cereus*) have demonstrated better results regarding

AgNPs final properties, leading to spherical and less polydisperse AgNPs, presenting as 41–60 nm and 18–69 nm, respectively (Ahmed et al., 2020; Saravanan et al., 2018). Effective synthesis of AgNPs was also obtained with *Bacillus paramycoides*, which was isolated from mangrove sediment sample (Dharmaraj et al., 2021).

Another strain that has been widely used in the biosynthesis of AgNPs is Pseudomonas. The well-known *Pseudomonas aeruginosa* was used for extracellular synthesis of AgNPs (Quinteros et al., 2019). Interesting properties were obtained, leading to small spherical AgNPs (approximately 25 nm), with a corona layer composed of 20 identified proteins. A different profile was observed employing *Pseudomonas hibiscicola*, resulting in various shapes of AgNPs and a higher average diameter of 50 nm (Punjabi et al., 2017). Less common strains have also demonstrated interesting parameters of AgNPs (Divya et al., 2019; Viorica et al., 2017; Yumei et al., 2017).

Cytobacillus firmus, an endophytic bacterium isolated from the stem bark of *Terminalia arjuna*, has also been used to the synthesis of AgNPs (Sudarsan et al., 2021). The results demonstrated that small AgNPs were formed (average size between 30 and 45 nm), with Zeta potential of -5.5 mV, and antibacterial and antifungal activity against *Staphylococcus aureus*, *Escherichia coli*, and *Magnaporthe grisea*.

3 Synthesis of AgNPs by actinobacteria

Among prokaryotes, actinobacteria are a gram-positive aerobic bacterium, and like filamentous fungi, most actinobacteria produce mycelium, which is reproduced by sporulation (Barka et al., 2015). This interesting group of free-living microorganisms is widely distributed and can be isolated from soil, fresh water, or saline water, where this microorganism can live associated with marine sponges (Khan et al., 2012). Actinobacteria are a huge source of several bioactive compounds and metabolites, which have been drawing the attention for their applications in biotechnology, agriculture, and the food industry (Barka et al., 2015). Undoubtedly, given the great potential and versatility of actinobacteria, they have also been evaluated for use as nanofactories of AgNPs. The synthesis of AgNPs by actinobacteria can be carried out by extracellular or intracellular routes. However, extracellular synthesis (as shown in Fig. 2) is frequently reported.

One of the most studied actinobacteria genera for the AgNPs synthesis is the *Streptomyces* genus. In this regard, *Streptomyces hygroscopicus*, isolated from the Philippines, was evaluated as a source to biosynthesize AgNPs extracellularly by silver nitrate reduction (Sadhasivam et al., 2010). Transmission electron microscopy (TEM) and field emission scanning electron microscopy (FE-SEM) showed the formation of spherical AgNPs (70% purity) with average size of 20–30 nm. Similar results were obtained with cell-free supernatant of *Streptomyces albogriseolus*, isolated from mangrove sediment soil in India (Samundeeswari et al., 2012). The interaction of supernatant with 2 mM of $AgNO_3$ solution resulted in the biosynthesized of spherical and crystalline AgNPs of circa 16.25 nm. The authors attributed the extracellular synthesis to the presence of proteins and enzymes secreted by *S. albogriseolus*. Supernatant from *Streptomyces exfoliatus* ICN25, which was isolated from rhizome soil region of the mangrove *Rhizophora mucronata* on the west coast of India, was successfully used in the biosynthesis of AgNPs using silver nitrate as precursor salt (Iniyan et al., 2017). The obtained AgNPs presents spherical and rod shape with average size of 10–40 nm. In another work, Manivasagan et al. (2015) reported that *Streptomyces* sp. MBRC-91 isolated from marine sediment on the Busan coast of South Korea was cultivated to prepare polysaccharide-based bioflocculant, and used in the biosynthesis of AgNPs. The mixture of bioflocculant with aqueous $AgNO_3$ at 10^{-3} M resulted in the formation of coated spherical AgNPs with particle ranging in size from 10 to 60 nm. In a similar work, exopolysaccharide of the *Streptomyces violaceus* MM72 (isolated from the marine coastal sediment on the southeast coast of India) composed of carbohydrate (61.4%) and ash (16.1%) was demonstrated to be effective for the biosynthesis of AgNPs (Sivasankar et al., 2018). The use of 5 g of exopolysaccharide dissolved in 25 mL of distilled water and 3 mM of aqueous $AgNO_3$ resulted in the formation of AgNPs with spherical shape, and a size ranging from 10 to 60 nm (30 nm as average size).

In another work, the reduction of silver nitrate (1 mM) by cell-free supernatant from *Streptomyces olivaceus*, which was isolated in marine ecosystems off the southwest coast of India, resulted in the formation of spherical AgNPs, with a particle size of 12.3 nm, approximately (Sanjivkumar et al., 2018). Another marine actinobacteria that has been evaluated in the synthesis of AgNPs is *Streptomyces parvus* strain Al-Dhabi-91, isolated from Saudi Arabian marine regions (Al-Dhabi et al., 2019). The authors reported that the use of cell-free supernatant of *S. parvus* mixed with 10 mM of $AgNO_3$ led to AgNPs ranged between 2 and 5 nm. Two Streptomyces strains, denoted as *Streptomyces* sp. 192ANMG and *Streptomyces* sp. 17ANMG, which were isolated from marine sponge *Crella cyathophora*, were evaluated in the biosynthesis of AgNPs (Hamed et al., 2020). The incubation of culture of cell-free supernatant with 1 mM of $AgNO_3$ resulted in the formation of spherical monodisperse AgNPs of 8.66 nm in size and -24.6 mV of Z-potential with *Streptomyces* sp. 192ANMG, whereas *Streptomyces* sp. 17ANMG led to the formation of AgNPs with 35 nm in size, -25 mV of Z-potential, and polydisperse shapes (quasispherical and cubic shapes) (Hamed et al., 2020). The same authors reported that, according to the chemical composition of the isolated actinomycete functional groups ($-NH_2$ and $-COOH$), this might allow the trapping of silver ions on the surface of actinobacteria, and enzymes of the cell wall can act to form AgNPs. *Streptomyces xinghaiensis*, actinobacteria isolated from sediment in Lonar Crater of Maharashtra, India, was evaluated in the synthesis of AgNPs (Wypij et al., 2018). The cell-free supernatant of *S. xinghaiensis* combined with $AgNO_3$ (1 mM) produced spherical and polydisperse nanoparticles that were evaluated by TEM analysis, and with sizes ranged between 5 and 20 nm and a Z-potential of -15.7 mV. Moreover, a concentration of 2.7×10^7 particles/mL was determined. It was also isolated from sediment in the Suez Gulf, Red Sea, Egypt, when the cell-free supernatant of *Streptomyces rochei* MHM13 was mixed with 1 mM of AgNO$_3$, and the authors reported the formation of AgNPs with spherical shape, and sizes (20–80 nm) up to fourfold higher than that obtained with *S. xinghaiensis* (Abd-Elnaby et al., 2016). Interestingly, the authors reported that from 41 actinobacteria isolated from sediment, only 2, including *S. rochei*, showed the capacity to mediate the synthesis of AgNPs. This is particularly interesting given that species isolated from the same site apparently do not have the capacity to produce the same type or quantity of enzymes or secondary metabolites necessary for nanoparticle production.

In another work, carried out by Skaładanowski et al. (2016), cell-free supernatant of *Streptomyces* sp. NH21, an actinobacteria isolated from the humid layer of acidic pine forest soil in Poland, was mixed with 3 mM of AgNO$_3$ and incubated in a rotatory shaker for 72 h. The biomass was obtained from cell-free supernatant; after centrifugation it was washed and put again in distiller water, and after 48 h, cell-free supernatant was obtained and incubated with AgNO$_3$. The authors reported that the obtained AgNPs have spherical and oval shapes with size in the range of 44 nm and 8 nm from the first, and from the second syntheses, respectively (Skaładanowski et al., 2016). AgNPs have also been synthesized with direct contact with actinobacterial biomass. In this regard, Shanmugasundaram et al. (2013) incubated 5 g of wet biomass of *Streptomyces naganishii* (MA7) with 1 mM of AgNO$_3$ solution, and kept the final mixture in a rotatory shaker at 28°C. The AgNPs obtained by ultracentrifugation were visualized by TEM, which revealed the formation of nanoparticles with spherical shape, and sizes between 45 and 60 nm. Cell lysates of the actinobacteria *Streptomyces* sp. AOA21, which was isolated from Mersin soils (Turkey), was incubated with 1 mM solution of AgNO$_3$, obtaining spherical AgNPs with sizes between 35 and 60 nm (Adiguzel et al., 2018). The authors reported that pH 9.0, 1 mM of silver nitrate and twofold diluted cell lysate were the optimal parameters for the synthesis. Moreover, the same authors reported that protein, peptides, and amino acid, present in cell lysate, might be responsible by the reduction of silver ions, and the formation of AgNPs.

Other Streptomyces genera have also been evaluated for the synthesis of AgNPs. In this regard, Otari et al. (2012) evaluated the use of both cell-free supernatant and supernatant containing *Rhodococcus* NCIM 2891 combined with AgNO$_3$ solution at 3 mM. Interestingly, contrary to that reported with Streptomyces strains, incubation of silver nitrate with only cell-free supernatant did not show a typical resonance plasmon at 420 nm in UV-vis spectroscopy analysis, suggesting that nanoparticles were not produced. However, incubation of silver nitrate with supernatant containing biomass showed the formation of small spherical AgNPs, with an average size of 10 nm, as assayed by TEM analysis. Otari et al. (2015) reported that *Rhodococcus* NCIM 2891 mediated the intracellularly synthesis of AgNPs. The biomass of *Rhodococcus* NCIM 2891 combined with 1 mM of AgNO$_3$ at pH 7.0 allowed the formation of AgNPs inside the cell (cytoplasm), with sizes ranged between 5 at 50 nm, which were described as stable nanoparticles due to the presence of superficial proteins.

The genus *Nocardiopsis* has also been evaluated. In this sense, Dhanaraj et al. (2020) reported that *Nocardiopsis dassonvillei*-DS013, isolated from the coastal zone of India, was effective in the biosynthesis of AgNPs, through the mixture of fresh biomass with $AgNO_3$ solution. Circular shape and sizes ranged between 30 and 80 nm were formed, possibly by the presence of proteins in the culture supernatant, which were evaluated by Fourier-transform infrared (FT-IR) spectroscopy.

In another work, the *Nocardiopsis* sp. MBRC-1, actinobacteria isolated from marine sediment on the Busan coast, South Korea, showed that cell-free supernatant combined with $AgNO_3$ solution at 3 mM resulted in the formation of spherical AgNPs with an average particle size of 45 ± 0.15 nm (Manivasagan et al., 2013). Another evaluated genus has been *Streptacidiphilus*. Railean-Plugaru et al. (2016) reported that *Streptacidiphilus* sp. strain CGG1n, isolated from the mineral horizon of *Picea sitchensis* Carriere (Sitka spruce) forest, United Kingdom, demonstrated that free-cell supernatant combined with $AgNO_3$ solution (1:1 v/v) allowed the synthesis of spherical AgNP with sizes ranged between 4 and 45 nm. The same methodology was used to synthesize AgNPs by using *Streptacidiphilus durhamensis* HGG16n strain (Buszewski et al., 2018). Extracellular synthesis resulted in the formation of stable spherical nanoparticles with sizes ranged between 8 and 48 nm. Finally, although *Sinomonas* genus has been less evaluated in the biosynthesis of AgNPs, Manikprabhu et al. (2016) reported that using a photo-irradiation method, including cell-free supernatant of *Sinomonas mesophila* MPKL 26 and $AgNO_3$ solution at 3 mM, it was possible to obtain spherical nanoparticles ranged between 4 and 50 nm in size. According to the above information, it is clear that the Streptomyces genus has been widely evaluated for the synthesis of AgNPs. Other Actinobacteria genera have been less studied. However, the genera mentioned in this chapter provide a feasible way to biosynthesize small AgNPs, and principally by an extracellular method. Apparently, the isolation place does not have influence on AgNPs, given that strains isolated from marine environments, sediment, or soil, present the same capacity to synthesize AgNPs. Further works are important in this field.

4 Antimicrobial properties of AgNPs

The antimicrobial properties of silver have been recognized for many centuries, given that its applications for medical purposes in humans and other applications are known for at least 5000 years (Barillo and Marx, 2014). Antimicrobial capacity of silver or AgNPs has been reported through several mechanisms, such as silver ion release, cell membrane damage, DNA damage, or the generation of reactive oxygen species (ROS) (Durán et al., 2016). However, recently, it has been reported that silver ions induced DNA dehybridization, weakening the binding between H-NS proteins and DNA (Sadoon et al., 2020). In the synthesis of AgNPs, the formation of small size nanoparticles is desirable, enhancing the interaction of the nanoparticle with bacteria, virus, or fungus, due to the large surface area of AgNPs.

Moreover, it has been reported that the shape of AgNPs might have effects on their antimicrobial potential, due to the effective surface area available to interact with the microbial pathogens. Another important characteristic is to avoid the aggregation of AgNPs, given that it has been demonstrated that aggregation decreases the antimicrobial capacity of AgNPs (Panáček et al., 2017).

As mentioned above, both bacteria and actinobacteria have demonstrated great capacity to mediate the synthesis of AgNPs. Moreover, biosynthesized AgNPs have demonstrated a high antimicrobial capacity, and some relevant examples are shown in Table 1.

In an interesting recent work, AgNPs were biosynthesized by exopolysaccharides released by *Bacillus nitratireducens*, and their antimicrobial properties were evaluated against two drug-resistant fish pathogens; *Aeromonas hydrophila* and *Pseudomonas fluorescens* (Essawy et al., 2021). The results demonstrated that the treatment with AgNPs exhibited a dose-dependent antimicrobial activity against *A. hydrophila* and *P. fluorescens*. However, interestingly, when AgNPs were combined with a secondary metabolite (heliomycin), which was extracted from marine actinomycetes AB5, they displayed additive and synergistic effects against *A. hydrophila* and *P. fluorescens*. Similar results were reported by Sidhu and Nehra (2021), where AgNPs (31 nm) coated with bacteriocin Bac23 (plantaricin), which was isolated from *Lactobacillus plantarum*, showed a high antimicrobial capacity against *Shigella flexneri*, *S. aureus*, *P. aeruginosa*, *Listeria monocytogenes*, *B. cereus*, and *E. coli*, displaying a low MIC value (4 μg/mL) for two strains; *S. flexneri* and *P. aeruginosa*.

In another work, it is reported that small silver chloride nanoparticles (AgCl-NPs) (22.7 ± 10.2 nm), photosynthesized using *Stryphnodendron adstringens* (Martius) Covilleplant extracts, showed a remarkable antifungal activity against the pathogenic yeast *Cryptococcus neoformans*, and antibacterial activity against *P. aeruginosa*, displaying a MIC_{80} of 0.32 and 2.56 μg/mL respectively. Moreover, the authors reported synthesized AgNPs were nontoxic for mammalian Vero cells. Antimicrobial properties of AgNPs have also been used for applications in medical textiles, playing an important role for the prevention of infections in skin wounds, for example. In this sense, a multifunctional hydroxyapatite/silver nanoparticles/cotton gauze was designed and evaluated for its antimicrobial and biomedical applications (Said et al., 2021). AgNPs were biosynthesized in situ on cotton gauze fabrics by photosynthesis method using ginger oil as a green reductant, and taking advantage of their antiinflammatory properties. AgNPs deposition on cotton gauze fabrics was demonstrated to enhance antimicrobial properties against *Candida albicans*, as well as UV protection. Protection against pathogens associated with wound infection has also been reported. Biogenic AgNPs synthesized using the Indian folk medicinal plant, *Gnaphalium polycaulon* (Gp), leaves extract showed high antibacterial and antifungal activity against *S. aureus* and *C. albicans* (Shanmugapriya et al., 2020). Moreover, the authors reported that in vivo studies showed a fast rate of healing of a wound.

Table 1 Silver nanoparticles synthesized by bacterial and actinobacterial strains and evaluated as antimicrobials.

Strain	Size (nm)	Antimicrobial assays	Pathogens	Main results	Reference
Bacterial					
B. amyloliquefaciens	16–80	Agar diffusion method	Xanthomonas oryzae pv. oryzae	Approximately 15 mm inhibition at the highest evaluated concentration (20 μL)	Fouad et al. (2017)
B. subtilis	16–80	Agar diffusion method	Xanthomonas oryzae pv. oryzae	Approximately 25 mm inhibition at the highest evaluated concentration (20 μL)	Fouad et al. (2017)
B. cereus	18–39	Agar well-diffusion assay	Xanthomonas oryzae pv. oryzae (Xoo)	24.21 ± 1.01 mm of inhibition zone	Ahmed et al. (2020)
B. brevis	41–60	Nathan's agar well diffusion and Kirby-Bauer disk diffusion	Clinical isolates of S. aureus and S. typhi	Higher inhibition was observed for S. aureus, exhibiting 19 mm inhibition in comparison to 7.5 mm for S. typhi	Saravanan et al. (2018)
S. xinghaiensis	15–20	MIC and MBC determination followed by comparison with standard antibiotics	S. aureus, B. subtilis, E. coli, P. aeruginosa, K. pneumoniae, C. albicans, and M. furfur	Low MIC for P. aeruginosa (16 μg/mL) and highest MIC for K. pneumoniae (256 μg/mL)	Wypij et al. (2018)
Favites sp.	30–50	Well diffusion method, MIC and antibiofilm inhibition assay	Bacillus sp., E. coli, K. pneumonia, P. aeruginosa, S. aureus, and C. albicans	Antimicrobial activity against the isolates, and inhibition of bacterial and biofilm formation on urinary catheters	Divya et al. (2019)
Arthrobacter sp.	9–72	Agar diffusion method	P. aeruginosa, S. aureus, C. albicans, and Fusarium oxysporum	The highest antimicrobial activity was observed for P. aeruginosa, reaching almost 40 mm inhibition	Yumei et al. (2017)
Actinobacteria					
Sinomonas mesophila MPKL 26	4–50	Disk-diffusion method at 1.56 g/1000 mL	Multi drug-resistant (MDR) S. aureus	Zone of inhibition of 12 mm	Manikprabhu et al. (2016)

Continued

Table 1 Silver nanoparticles synthesized by bacterial and actinobacterial strains and evaluated as antimicrobials. *Continued*

Strain	Size (nm)	Antimicrobial assays	Pathogens	Main results	Reference
Streptacidiphilus durhamensis HGG16n	8–48	Sensitivity to the synthesized AgNPs and standard antibiotics, well and disk-diffusion methods, and MIC and MBC determination	*S. aureus*, *Bacillus subtilis*, *E. coli*, *P. aeruginosa*, *K. pneumoniae*, *P. mirabilis*, and *Salmonella infantis*	Highest antimicrobial activity against *P. aeruginosa*, *S. aureus*, and *P. mirabilis*. MIC <10 mg/mL and MBC <100 µg/mL	Buszewski et al. (2018)
Streptacidiphilus sp. CGG11n	4–45	Determination of MIC and MBC, and well-diffusion method	*S. aureus*, *S. infantis*, *B. subtilis*, *E. coli*, *P. aeruginosa*, *K. pneumoniae*, and *P. mirabilis*	MIC values of 6.25 µg/mL for all. MBC values of 50 µg/mL for *S. infantis*, *E. coli*, *P. aeruginosa*, and *B. subtilis* and 100 µg/mL for *P. mirabilis* and *S. aureus*	Railean-Plugaru et al. (2016)
Streptomyces griseoplanus SAI-25	19.5–20.9	Disk-diffusion method at 250, 500, and 1000 µg/mL	*Macrophomina phaseolina* fungus	Zone of inhibition of 13 mm at 1000 µg/mL	Vijayabharathi et al. (2018)
Streptomyces naganishii (MA7)	45–60	Disk-diffusion method at 1 mg/mL, and antibiofouling assay	For disk-diffusion: *S. aureus*, *B. cereus*, *E. coli*, *P. mirabilis*, and *P. aeruginosa*. For Antibiofouling: *Pseudomonas* sp. P1, *Aeromonas* sp. P26, *Bacillus* sp. P31, *Bacillus* sp. P46, *Alcaligenes* sp. P47, *Micrococcus* sp. P56, *Staphylococcus* sp. PP3, *Micrococcus* sp. PP5, *Aeromonas* sp. PP6, and *Alcaligenes* sp. PP8	Zone inhibition >8 mm. Decrease in biofilm establishment between 10% and 40% at 1 mg/mL, 40% and 70% at 5 mg/mL, and 50% and 90% at 10 mg/mL	Shanmugasundaram et al. (2013)

Organism	Size (nm)	Method	Test organisms	Results	Reference
Streptomyces sp. AOA21	35–60	MIC determination	*B. cereus, K. pneumoniae, E. coli,* and *S. aureus*	MIC of 8, 16, 16, and 32 µg/mL, respectively	Adiguzel et al. (2018)
Streptomyces intermedius	55	Resazurin dye as indicator of cell viability at 2.5, 5, 10, 20, 30, 40 µg/mL	*B. subtilis* and *E. coli*	Cell viability decreased with increased concentration of AgNPs, with 5%–10% of viability at 40 µg/mL IC50 value 10 µg/mL for *B. subtilis* and 20 µg/mL for *E. coli*.	Dayma et al. (2019)
Nocardiopsis dassonvillei-DS013	30–80	Well diffusion method	*E. coli, Enterococcus* sp., *Pseudomonas* sp., *Klebsiella* sp., *Proteus* sp., *Shigella* sp., *B. subtilis,* and *Streptococcus* sp.	Inhibition zone between 13 and 23 mm for all strains	Dhanaraj et al. (2020)
Rhodococcus NCIM 2891	5–50	Growth curve method and agar diffusion method at 30, 50, and 100 µg/mL	*S. aureus, K. pneumoniae, P. vulgaris, E. faecalis, P. aeruginosa,* and *E. coli*	At low tested AgNP concentrations, 10 µg/mL was the most effective for *K. pneumoniae*. At 50 µg/mL of AgNPs, the arrestment of the growth of the all organisms was observed	Otari et al. (2015)

5 Conclusion and perspectives

Biosynthesis of AgNPs has been carried out using several biological resources. Among them, many genera of bacteria and actinobacteria have been shown to play an important role in nanoparticle synthesis, as shown above. The reduction of silver ions can be carried out by these microorganisms in an extra or intracellular way, although most of the works reported the use of extracellular biosynthesis, possibly because nanoparticle recovery is less complex, and no postsynthesis process are necessary to separate the nanoparticles from the biomass. In general, it is reported that proteins, enzymes, or other metabolites could take part in the reduction of silver ions, but this has not yet been fully understood, and more studies are necessary. On the other hand, antimicrobial potentials of AgNPs biosynthesized by bacteria and actinobacteria have been widely demonstrated against a wide range of pathogenic microorganisms, even at lower concentrations than traditional antibiotics. In spite of these significant advances in the knowledge of the potential benefits of use bacterial and actinobacterial for AgNPs synthesis, more attention should be paid to the potential applications of AgNPs as well as their possible potential toxicity (Tortella et al., 2020), which can result in several unwanted side-effects to the environment, animals, or humans.

Acknowledgment

We thank ANID-REDES 180003, ANID-FAPESP (2018/08194-2), FAPESP (2018/08194-2, 2018/02832-7), CNPq (404815/2018-9, 313117/2019-5), and ANID/FONDAP/15130015.

References

Abd-Elnaby, H.M., Abo-Elala, G.M., Abdel-Raouf, U.M., Hamed, M.M., 2016. Antibacterial and anticancer activity of extracellular synthesized silver nanoparticles from marine *Streptomyces rochei* MHM13. Egypt. J. Aquat. Res. 42 (3), 301–312.

Adiguzel, A.O., Adiguzel, S.K., Mazmanci, B., Tunçer, M., Mazmanci, M.A., 2018. Silver nanoparticle biosynthesis from newly isolated streptomyces genus from soil. Mater. Res. Express 5 (4), 045402.

Ahmad, S., Munir, S., Zeb, N., Ullah, A., Khan, B., Ali, J., Bilal, M., Omer, M., Alamzeb, M., Salman, S.M., Ali, S., 2019. Green nanotechnology: a review on green synthesis of silver nanoparticles—an ecofriendly approach. Int. J. Nanomedicine 14, 5087–5107.

Ahmed, T., Shahid, M., Noman, M., Niazi, M.B.K., Mahmood, F., Manzoor, I., Zhang, Y., Li, B., Yang, Y., Yan, C., Chen, J., 2020. Silver nanoparticles synthesized by using *Bacillus cereus* SZT1 ameliorated the damage of bacterial leaf blight pathogen in rice. Pathogens 9, 160.

Ahuja, K., Rawat, A., 2018. Silver Nanoparticles Market Size by Application (Healthcare & Lifesciences, Textiles, Electronics & It, Food & Beverage), Industry Analysis Report, Regional Outlook, Growth Potential, Price Trends, Competitive Market Share & Forecast,

2018 – 2024. Global Market Insights (GMI). https://www.gminsights.com/industry-analysis/silver-nanoparticles-market. (Accessed 9 May 2020.

Al-Dhabi, N.A., Ghilan, A.-K.M., Esmail, G.A., Arasu, M.V., Duraipandiyan, V., Ponmurugan, K., 2019. Environmental friendly synthesis of silver nanomaterials from the promising *Streptomyces parvus* strain Al-Dhabi-91 recovered from the Saudi Arabian marine regions for antimicrobial and antioxidant properties. J. Photochem. Photobiol. B 197, 111529.

Ali, J., Ali, N., Wang, L., Waseem, H., Pan, G., 2019. Revisiting the mechanistic pathways for bacterial mediated synthesis of noble metal nanoparticles. J. Microbiol. Methods 159, 18–25.

Ameen, F., AlYahia, S., Govarthanan, M., ALjahdali, N., Al-Enazi, N., Alsamhari, K., Alshehri, W.A., Alwakeel, S.S., Alharbi, S.A., 2020. Soil bacteria Cupriavidus sp. mediates the extracellular synthesis of antibacterial silver nanoparticles. J. Mol. Struct. 15, 127233.

Anwar, M., Shukrullah, S., Haq, I.U., Saleem, M., AbdEl-Salam, N.M., Ibrahim, K.A., Mohamed, H.F., Khan, Y., 2021. Ultrasonic bioconversion of silver ions into nanoparticles with *Azadirachta indica* extract and coating over plasma-functionalized cotton fabric. ChemistrySelect 6, 1920–1928.

Bapat, R.A., Joshi, C.P., Bapat, P., Chaubal, T.V., Pandurangappa, R., Jnanendrappa, N., Gorain, B., Khurana, S., Kesharwani, P., 2018. The use of nanoparticles as biomaterials in dentistry. Drug Discov. Today 24 (1), 85–98.

Barillo, D.J., Marx, D.E., 2014. Silver in medicine: a brief history BC 335 to present. Burns 40, S3–S8.

Barka, E.A., Vatsa, P., Sanchez, L., Gaveau-Vaillant, N., Jacquard, C., Meier-Kolthoff, J.P., Klenk, H.P., Clément, C., Ouhdouch, Y., van Wezel, G.P., 2015. Taxonomy, physiology, and natural products of actinobacteria. Microbiol. Mol. Biol. Rev. 80 (1), 1–43.

Bauer, M.A., Kainz, K., Carmona-Gutierrez, D., Madeo, F., 2018. Microbial wars: competition in ecological niches and within the microbiome. Microb. Cell 5 (5), 215–219.

Blasco, R., Martínez-Luque, M., Madrid, M., Castillo, F., Moreno-Vivián, C., 2001. *Rhodococcus* sp. RB1 grows in the presence of high nitrate and nitrite concentrations and assimilates nitrate in moderately saline environments. Arch. Microbiol. 175 (6), 435–440.

Buszewski, B., Railean-Plugaru, V., Pomastowski, P., Rafińska, K., Szultka-Mlynska, M., Golinska, P., Dahm, H., 2018. Antimicrobial activity of biosilver nanoparticles produced by a novel *Streptacidiphilus durhamensis* strain. J. Microbiol. Immunol. Infect. 51 (1), 45–54.

Calderón-Jiménez, B., Johnson, M.E., Montoro Bustos, A.R., Murphy, K.E., Winchester, M.R., Vega Baudrit, J.R., 2017. Silver nanoparticles: technological advances, societal impacts, and metrological challenges. Front. Chem. 5, 6.

Carbone, M., Donia, D.T., Sabbatella, G., Antiochia, R., 2016. Silver nanoparticles in polymeric matrices for fresh food packaging. J. King Saud Univ. Sci. 28 (4), 273–279.

Chung, I., Park, I., Seung-Hyun, K., Thiruvengadam, M., Rajakumaret, G., 2016. Plant-mediated synthesis of silver nanoparticles: their characteristic properties and therapeutic applications. Nanoscale Res. Lett. 11, 40.

Cuevas, R., Durán, N., Diez, M.C., Tortella, G.R., Rubilar, O., 2015. Extracellular biosynthesis of copper and copper oxide nanoparticles by *Stereum hirsutum*, a native white rot fungus from Chilean forests. J. Nanomater. 2015, 789089.

Das, R.K., Pachapur, V.L., Lonappan, L., Naghdi, M., Pulicharla, R., Maiti, S., Cledon, M., Martinez, L., Sarma, S., Brar, S.K., 2017. Biological synthesis of metallic nanoparticles: plants, animals and microbial aspects. Nanotechnol. Environ. Eng. 2 (1), 18.

Dayma, P., Mangrola, A.V., Suriyaraj, S.P., Dudhagara, P., Rajesh, K., Patel, R.K., 2019. Synthesis of bio-silver nanoparticles using desert isolated *Streptomyces intermedius* and its antimicrobial activity. J. Pharm. Chem. Biol. Sci. 7 (2), 94–101.

Deshmukh, S.P., Patil, S.M., Mullani, S.B., Delekar, S.D., 2018. Silver nanoparticles as an effective disinfectant: a review. Mater. Sci. Eng. C 97, 954–965.

Dhanaraj, S., Thirunavukkarasu, S., Allen John, H., Pandian, S., Salmen, S.H., Chinnathambi, A., Ali Alharbi, S., 2020. Novel marine *Nocardiopsis dassonvillei*-DS013 mediated silver nanoparticles characterization and its bactericidal potential against clinical isolates. Saudi J. Biol. Sci. 27 (3), 991–995.

Dharmaraj, D., Krishnamoorthy, M., Rajendran, K., Karuppiah, K., Annamalai, J., Durairaj, K.R., Santhiyagu, P., Ethiraj, K., 2021. Antibacterial and cytotoxicity activities of biosynthesized silver oxide (Ag2O) nanoparticles using *Bacillus paramycoides*. J. Drug Delivery Sci. Technol. 61, 102111.

Divya, M., Kiran, G.S., Hassan, S., Selvin, J., 2019. Biogenic synthesis and effect of silver nanoparticles (AgNPs) to combat catheter-related urinary tract infections. Biocatal. Agric. Biotechnol. 18, 101037.

Durán, N., Seabra, A.B., 2018. Biogenic synthesized Ag/Au nanoparticles: production, characterization, and applications. Curr. Nanosci. 14, 82–94.

Durán, N., Durán, M., de Jesus, M.B., Seabra, A.B., Fávaro, W.J., Nakazato, G., 2016. Silver nanoparticles: a new view on mechanistic aspects on antimicrobial activity. Nanomedicine NBM 12 (3), 789–799.

Essawy, E., Abdelfattah, M.S., El-Matbouli, M., Saleh, M., 2021. Synergistic effect of biosynthesized silver nanoparticles and natural phenolic compounds against drug-resistant fish pathogens and their cytotoxicity: an in vitro study. Mar. Drugs 19, 22.

Faxian, L., Jie, L., Xueling, C., 2017. Microwave-assisted synthesis silver nanoparticles and their surface enhancement Raman scattering. Rare Met. Mater. Eng. 46 (9), 2395–2398.

Fouad, H., Hongjie, L., Yanmei, D., Baoting, Y., El-Shakh, A., Abbas, G., Jianchu, M., 2017. Synthesis and characterization of silver nanoparticles using *Bacillus amyloliquefaciens* and *Bacillus subtilis* to control filarial vector *Culex pipiens pallens* and its antimicrobial activity. Artif. Cells Nanomed. Biotechnol. 45, 1369–1378.

Ghashghaei, S., Emtiazi, G., 2015. The methods of nanoparticle synthesis using bacteria as biological nanofactories, their mechanisms and major applications. Curr. Bionanotechnol. 1 (1), 3–17.

Ghiuță, I., Cristea, D., Croitoru, C., Kost, J., Wenkert, R., Vyrides, I., Anayiotos, A., Munteanu, D., 2018. Characterization and antimicrobial activity of silver nanoparticles, biosynthesized using Bacillus species. Appl. Surf. Sci. 438, 66–73.

González-Ballesteros, N., Rodríguez-Argüelles, M.C., Prado-López, S., Lastra, M., Grimaldi, M., Cavazza, A., Bigi, F., 2019. Macroalgae to nanoparticles: study of *Ulva lactuca* L. role in biosynthesis of gold and silver nanoparticles and of their cytotoxicity on colon cancer cell lines. Mater. Sci. Eng. C 97, 498–509.

Guilger-Casagrande, M., de Lima, R., 2019. Synthesis of silver nanoparticles mediated by fungi: a review. Front. Bioeng. Biotechnol. 7, 287.

Hamed, A.A., Kabary, H., Khedr, M., Emam, A.N., 2020. Antibiofilm, antimicrobial and cytotoxic activity of extracellular green-synthesized silver nanoparticles by two marine-derived actinomycete. RSC Adv. 10 (17), 10361–10367.

Iniyan, A.M., Kannan, R.R., Joseph, F.-J.R.S., Mary, T.R.J., Rajasekar, M., Sumy, P.C., Vincent, S.G.P., 2017. In vivo safety evaluation of antibacterial silver chloride nanoparticles from *Streptomyces exfoliatus* ICN25 in zebrafish embryos. Microb. Pathog. 112, 76–82.

Javaid, A., Oloketuyi, S.F., Khan, M.M., Khan, F., 2017. Diversity of bacterial synthesis of silver nanoparticles. BioNanoScience 8, 43–69.

Khan, S.T., Takagi, M., Shin-ya, K., 2012. Actinobacteria associated with the marine sponges *Cinachyra* sp., *Petrosia* sp., and *Ulosa* sp. and their culturability. Microbes Environ. 27 (1), 99–104.

Khan, I., Saeed, K., Khan, I., 2017. Nanoparticles: properties, applications and toxicities. Arab. J. Chem. 12 (7), 908–931.

Klaessig, F., Marrapese, M., Abe, S., 2011. Current perspectives in nanotechnology terminology and nomenclature. In: Nanotechnology Standards. Springer, New York, pp. 21–52.

Kumar, S.S.D., Rajendran, N.K., Houreld, N.N., Abrahamse, H., 2018. Recent advances on silver nanoparticle and biopolymer-based biomaterials for wound healing applications. Int. J. Biol. Macromol. 115, 165–175.

Li, X., Xu, H., Chen, Z.-S., Chen, G., 2011. Biosynthesis of nanoparticles by microorganisms and their applications. J. Nanomater. 2011, 1–16.

Maharjan, S., Liao, K.-S., Wang, A.J., Zhu, Z., McElhenny, B.P., Bao, J., Curran, S.A., 2019. Synthesis of stabilized silver nanoparticles in organosiloxane matrix via sol-gel method and its optical nonlinearity study. Chem. Phys. 532, 110610.

Manikprabhu, D., Cheng, J., Chen, W., Sunkara, A.K., Mane, S.B., Kumar, R., Das, M., Hozzein, W.N., Duan, Y.-Q., Li, W.-J., 2016. Sunlight mediated synthesis of silver nanoparticles by a novel actinobacterium (*Sinomonas mesophila* MPKL 26) and its antimicrobial activity against multi drug resistant *Staphylococcus aureus*. J. Photochem. Photobiol. B 158, 202–205.

Manivasagan, P., Venkatesan, J., Senthilkumar, K., Sivakumar, K., Kim, S.-K., 2013. Biosynthesis, antimicrobial and cytotoxic effect of silver nanoparticles using a novel *Nocardiopsis* sp. MBRC-1. Biomed. Res. Int. 2013, 1–9.

Manivasagan, P., Venkatesan, J., Sivakumar, K., Kim, S.-K., 2014. Actinobacteria mediated synthesis of nanoparticles and their biological properties: a review. Crit. Rev. Microbiol. 42, 209–221.

Manivasagan, P., Kang, K.-H., Kim, D.G., Kim, S.-K., 2015. Production of polysaccharide-based bioflocculant for the synthesis of silver nanoparticles by *Streptomyces* sp. Int. J. Biol. Macromol. 77, 159–167.

Marslin, G., Siram, K., Maqbool, Q., Selvakesavan, R., Kruszka, D., Kachlicki, P., Franklin, G., 2018. Secondary metabolites in the green synthesis of metallic nanoparticles. Materials 11 (6), 940.

Murugan, K., Senthilkumar, B., Senbagam, D., Al-Sohaibani, S., 2014. Biosynthesis of silver nanoparticles using *Acacia leucophloea* extract and their antibacterial activity. Int. J. Nanomedicine 9, 2431–2438.

Otari, S.V., Patil, R.M., Nadaf, N.H., Ghosh, S.J., Pawar, S.H., 2012. Green biosynthesis of silver nanoparticles from an actinobacteria *Rhodococcus* sp. Mater. Lett. 72, 92–94.

Otari, S.V., Patil, R.M., Ghosh, S.J., Thorat, N.D., Pawar, S.H., 2015. Intracellular synthesis of silver nanoparticle by actinobacteria and its antimicrobial activity. Spectrochim. Acta A Mol. Biomol. Spectrosc. 136, 1175–1180.

Panáček, A., Kvítek, L., Smékalová, M., Večeřová, R., Kolář, M., Röderová, M., Zbořil, R., 2017. Bacterial resistance to silver nanoparticles and how to overcome it. Nat. Nanotechnol. 13 (1), 65–71.

Peng, J.J.-Y., Botelho, M.G., Matinlinna, J.P., 2012. Silver compounds used in dentistry for caries management: a review. J. Dent. 40 (7), 531–541.

Pieretti, J.C., Rubilar, O., Weller, R.B., Tortella, G.R., Seabra, A.B., 2020. Nitric oxide (NO) and nanoparticles—potential small tools for the war against COVID-19 and other human coronavirus infections. Virus Res. 291, 198202.

Pletzer, D., Asnis, J., Slavin, Y., Hancock, R., Bach, H., Saatchi, K., Häfeli, U., 2021. Rapid microwave-based method for the preparation of antimicrobial lignin-capped silver nanoparticles active against multidrug-resistant bacteria. Int. J. Pharm. 596, 120299.

Prakash Patil, M., Kang, M., Niyonizigiye, I., Singh, A., Kim, J.-O., Seo, Y.B., Kim, G.-D., 2019. Extracellular synthesis of gold nanoparticles using the marine bacterium *Paracoccus haeundaensis* BC74171T and evaluation of their antioxidant activity and antiproliferative effect on normal and cancer cell lines. Colloids Surf. B Biointerfaces 183, 110455.

Punjabi, K., Yedurkar, S., Doshi, S., Deshapnde, S., Vaidya, S., 2017. Biosynthesis of silver nanoparticles by *Pseudomonas* spp. isolated from effluent of an electroplating industry. IET Nanobiotechnol. 11, 584–590.

Qin, S., Li, W.-J., Klenk, H.-P., Hozzein, W.N., Ahmed, I., 2019. Editorial: actinobacteria in special and extreme habitats: diversity, function roles and environmental adaptations, second edition. Front. Microbiol. 10, 944.

Quinteros, M.A., Bonilla, J.O., Alborés, S.V., Villegas, L.P., Páez, P.L., 2019. Biogenic nanoparticles: synthesis, stability and biocompatibility mediated by proteins of *Pseudomonas aeruginosa*. Colloids Surf. B Biointerfaces 184, 110517.

Rabinal, M.K., Kalasad, M.N., Praveenkumar, K., Bharadi, V.R., Bhikshavartimath, A.M., 2013. Electrochemical synthesis and optical properties of organically capped silver nanoparticles. J. Alloys Compd. 562, 43–47.

Raffi, M., Rumaiz, A.K., Hasan, M.M., Shah, S.I., 2007. Studies of the growth parameters for silver nanoparticle synthesis by inert gas condensation. Mater. Res. 22 (12), 3378–3384.

Rai, M., Birla, S., Ingle, A.P., Gupta, I., 2014a. Nanosilver: an inorganic nanoparticle with myriad potential applications. Nanotechnol. Rev. 3, 281–309.

Rai, M., Kon, K., Ingle, A., Durán, N., Galdiero, S., Galdiero, M., 2014b. Broad-spectrum bioactivities of silver nanoparticles: the emerging trends and future prospects. Appl. Microbiol. Biotechnol. 98, 1951–1961.

Railean-Plugaru, V., Pomastowski, P., Wypij, M., Szultka-Mlynska, M., Rafinska, K., Golinska, P., Dahm, H., Buszewski, B., 2016. Study of silver nanoparticles synthesized by acidophilic strain of Actinobacteria isolated from the of *Picea sitchensis* forest soil. J. Appl. Microbiol. 120 (5), 1250–1263.

Rasool, U., Hemalatha, S., 2017. Marine endophytic actinomycetes assisted synthesis of copper nanoparticles (CuNPs): characterization and antibacterial efficacy against human pathogens. Mater. Lett. 194, 176–180.

Rodrigues, M.C., Viana, M.M., Souza, T.R., Tanaka, C.J., Bueno-Silva, B., Seabra, A.B., 2020. Biogenic synthesis and antimicrobial activity of silica-coated silver nanoparticles for esthetic dental applications. J. Dent. 96, 103327.

Roychoudhury, P., Gopal, P.K., Paul, S., Pal, R., 2016. Cyanobacteria assisted biosynthesis of silver nanoparticles—a potential antileukemic agent. J. Appl. Phycol. 28, 3387–3394.

Sadhasivam, S., Shanmugam, P., Yun, K., 2010. Biosynthesis of silver nanoparticles by *Streptomyces hygroscopicus* and antimicrobial activity against medically important pathogenic microorganisms. Colloids Surf. B Biointerfaces 81 (1), 358–362.

Sadoon, A.A., Khadka, P., Freeland, J., Gundampati, R.K., Manso, R., Ruiz, M., Krishnamurthi, V., Kumar, S., Chen, J., Wang, Y., 2020. Faster diffusive dynamics of histone-like nucleoid structuring proteins in live bacteria caused by silver ions. Appl. Environ. Microbiol. 86, e02479-19.

Said, M.M., Rehan, M., El-Sheikh, S.M., Zahran, M.K., Abdel-Aziz, M.S., Bechelany, M., Barhoum, A., 2021. Multifunctional hydroxyapatite/silver nanoparticles/cotton gauze for antimicrobial and biomedical applications. Nanomaterials 11 (2), 429.

Samundeeswari, A., Dhas, S.P., Nirmala, J., John, S.P., Mukherjee, A., Chandrasekaran, N., 2012. Biosynthesis of silver nanoparticles using actinobacterium *Streptomyces albogriseolus* and its antibacterial activity. Biotechnol. Appl. Biochem. 59 (6), 503–507.

Sanjivkumar, M., Vaishnavi, R., Neelakannan, M., Kannan, D., Silambarasan, T., Immanuel, G., 2018. Investigation on characterization and biomedical properties of silver nanoparticles synthesized by an actinobacterium *Streptomyces olivaceus* (MSU3). Biocatal. Agric. Biotechnol. 17, 151–159.

Saravanan, M., Barik, S.K., MubarakAli, D., Prakash, P., Pugazhendhi, A., 2018. Synthesis of silver nanoparticles from *Bacillus brevis* (NCIM 2533) and their antibacterial activity against pathogenic bacteria. Microb. Pathog. 116, 221–226.

Shanmugapriya, K., Palanisamy, S., Boomi, P., Subaskumar, R., Ravikumar, S., Thayumanavan, T., 2020. An eco-friendly *Gnaphalium polycaulon* mediated silver nanoparticles: synthesis, characterization, antimicrobial, wound healing and drug release studies. J. Drug Delivery Sci. Technol. 61, 102202.

Shanmugasundaram, T., Radhakrishnan, M., Gopikrishnan, V., Pazhanimurugan, R., Balagurunathan, R., 2013. A study of the bactericidal, anti-biofouling, cytotoxic and antioxidant properties of actinobacterially synthesised silver nanoparticles. Colloids Surf. B Biointerfaces 111, 680–687.

Shchukin, D.G., Radtchenko, I.L., Sukhorukov, G., 2003. Photoinduced reduction of silver inside microscale polyelectrolyte capsules. Chem. Phys. Chem. 4, 1101–1103.

Sheikh, N., Akhavan, A., Kassaee, M.Z., 2009. Synthesis of antibacterial silver nanoparticles by γ-irradiation. Phys. E Low Dimens. Syst. Nanostruct. 42 (2), 132–135.

Sidhu, P.K., Nehra, K., 2021. Purification and characterization of bacteriocin Bac23 extracted from *Lactobacillus plantarum* PKLP5 and its interaction with silver nanoparticles for enhanced antimicrobial spectrum against food-borne pathogens. LWT- Food Sci. Technol. 139, 110546.

Singh, R., Kumar, M., Mittal, A., Mehta, P.K., 2017. Microbial metabolites in nutrition, healthcare and agriculture. 3 Biotech 7 (1), 15.

Siva Kumar, K., Kumar, G., Prokhorov, E., Luna-Bárcenas, G., Buitron, G., Khanna, V.G., Sanchez, I.C., 2014. Exploitation of anaerobic enriched mixed bacteria (AEMB) for the silver and gold nanoparticles synthesis. Colloids Surf. A Physicochem. Eng. Asp. 462, 264–270.

Sivasankar, P., Seedevi, P., Poongodi, S., Sivakumar, M., Murugan, T., Sivakumar, L., Balasubramanian, T., 2018. Characterization, antimicrobial and antioxidant property of exopolysaccharide mediated silver nanoparticles synthesized by *Streptomyces violaceus* MM72. Carbohydr. Polym. 181, 752–759.

Skaładanowski, M., Wypij, M., Laskowski, D., Golińska, P., Dahm, H., Rai, M., 2016. Silver and gold nanoparticles synthesized from *Streptomyces* sp. isolated from acid forest soil with special reference to its antibacterial activity against pathogens. J. Clust. Sci. 28 (1), 59–79.

Sportelli, M.C., Clemente, M., Izzi, M., Volpe, A., Ancona, A., Picca, R.A., Palazzo, G., Cioffi, N., 2018. Exceptionally stable silver nanoparticles synthesized by laser ablation in alcoholic organic solvent. Colloids Surf. A Physicochem. Eng. Asp. 559, 148–158.

Srivastava, S.K., Constanti, M., 2012. Room temperature biogenic synthesis of multiple nanoparticles (Ag, Pd, Fe, Rh, Ni, Ru, Pt, Co, and Li) by *Pseudomonas aeruginosa* SM1. J. Nanopart. Res. 14, 1–10.

Srivastava, P., Braganca, J., Ramanan, S.R., Kowshik, M., 2013. Synthesis of silver nanoparticles using haloarchaeal isolate *Halococcus salifodinae* BK3. Extremophiles 17, 821–831.

Sudarsan, S., Kumar Shankar, M., Kumar Belagal Motatis, A., Shankar, S., Krishnappa, D., Mohan, C.D., Rangappa, K.S., Gupta, V.K., Siddaiah, C.N., 2021. Green synthesis of silver nanoparticles by *Cytobacillus firmus* isolated from the stem bark of *Terminalia arjuna* and their antimicrobial activity. Biomol. Ther. 11 (2), 259.

Taran, M., Rad, M., Alavi, M., 2018. Biosynthesis of TiO_2 and ZnO nanoparticles by *Halomonas elongata* IBRC-M 10214 in different conditions of medium. Bioimpacts 8 (2), 81–89.

Terra, A.L.M., Kosinski, R.D.C., Moreira, J.B., Costa, J.A.V., de Morais, M.G., 2019. Microalgae biosynthesis of silver nanoparticles for application in the control of agricultural pathogens. J. Environ. Sci. Health B 54, 1–8.

Tortella, G.R., Rubilar, O., Durán, N., Diez, M.C., Martínez, M., Parada, J., Seabra, A.B., 2020. Silver nanoparticles: toxicity in model organisms as an overview of its hazard for human health and the environment. J. Hazard. Mater. 390, 121974.

Vijayabharathi, R., Sathya, A., Gopalakrishnan, S., 2018. Extracellular biosynthesis of silver nanoparticles using *Streptomyces griseoplanus* SAI-25 and its antifungal activity against *Macrophomina phaseolina*, the charcoal rot pathogen of sorghum. Biocatal. Agric. Biotechnol. 14, 166–171.

Viorica, R., Pawel, P., Kinga, M., Michal, Z., Katarzyna, R., Boguslaw, B., 2017. *Lactococcus lactis* as a safe and inexpensive source of bioactive silver composites. Appl. Microbiol. Biotechnol. 101, 7141–7153.

Wang, H., Qiao, X., Chen, J., Ding, S., 2005. Preparation of silver nanoparticles by chemical reduction method. Colloids Surf. A Physicochem. Eng. Asp. 256 (2–3), 111–115.

Wani, I.A., Ganguly, A., Ahmed, J., Ahmad, T., 2011. Silver nanoparticles: ultrasonic wave assisted synthesis, optical characterization and surface area studies. Mater. Lett. 65 (3), 520–522.

Wypij, M., Czarnecka, J., Świecimska, M., Dahm, H., Rai, M., Golinska, P., 2018. Synthesis, characterization and evaluation of antimicrobial and cytotoxic activities of biogenic silver nanoparticles synthesized from *Streptomyces xinghaiensis* OF1 strain. World J. Microbiol. Biotechnol. 34, 23.

Yumei, L., Yamei, L., Qiang, L., Jie, B., 2017. Rapid biosynthesis of silver nanoparticles based on flocculation and reduction of an exopolysaccharide from Arthrobacter sp. b4: its antimicrobial activity and phytotoxicity. J. Nanomater. 2017, 9703614.

Zachar, O., 2020. Formulations for COVID-19 early stage treatment via silver nanoparticles inhalation delivery at home and hospital. ScienceOpen. https://doi.org/10.14293/S2199-1006.1.SOR-.PPHBJEO.v1. Preprints.

Zhang, X.F., Liu, Z.G., Shen, W., Gurunathan, S., 2016. Silver nanoparticles: synthesis, characterization, properties, applications, and therapeutic approaches. Int. J. Mol. Sci. 17 (9), 1534.

CHAPTER 19

Biogenic synthesis of silver nanoparticles using lichens

Debraj Dhar Purkayastha
Department of Chemistry, Cachar College, Silchar, Assam, India

1 Introduction

A crucial aspect of nanotechnology is to devise eco-friendly synthetic strategies for fabrication of materials. Physical and chemical methods of nanomaterials synthesis are quite costly and often generate toxic substances which are detrimental to the environment. Thus, biogenic synthesis of nanomaterials became quite appealing in comparison to the physical and chemical methods. Biogenic synthesis of nanomaterials is considered as a junction point between nanotechnology and biotechnology, which results in generation of new materials with a wide range of applications. At the time of writing, several researchers have put forward simple synthetic strategies for the metal nanoparticles using various biomaterials as sources of reductants and stabilizers (Chandran et al., 2006; Mohanpuria et al., 2008; Tripathy et al., 2010; Kumar et al., 2010; Sharma et al., 2014; Debnath et al., 2016; Paul et al., 2016). A comprehensive review on biogenic synthesis of nanoparticles by several marine organisms is also on record (Asmathunisha and Kathiresan, 2013). Among various metal nanoparticles, silver nanoparticles received significant attention because of their potent antibacterial, antifungal, and antiviral activities (Ouay and Stellacci, 2015; Panacek et al., 2009; Huy et al., 2017). Besides, silver nanoparticles have antiproliferative effects against various tumor cell lines (Vlasceanu et al., 2016). Extensive reviews on biogenic synthesis of silver nanoparticles have appeared recently (Javed et al., 2021; Moradi et al., 2021). At the time of writing, various biomaterials like bacteria, fungi, algae, lichens, different plant parts, etc. have been utilized effectively for the biogenic fabrication of silver nanoparticles (Ibrahim et al., 2019; Xue et al., 2016; Sajjadi et al., 2018; Kumar and Rajeshkumar, 2017; Siddiqi et al., 2018; Goga et al., 2021; Manikandan et al., 2021; Yari et al., 2021; Ciplak et al., 2018; Habibi et al., 2017; Paul et al., 2015). Although there has been wide utilization of various biomaterials for synthesizing silver nanoparticles in the last couple of years, reports on lichen-mediated synthesis are scanty.

The present chapter gives a detailed view on the lichen-mediated biogenic synthesis of silver nanoparticles. The potential applications of biosynthesized silver nanoparticles are also critically discussed.

2 Lichen-mediated biogenic synthesis of silver nanoparticles

Lichens are complex organisms that are formed from symbiotic association of a fungus and one or several algal or cyanobacterial species (Yuan et al., 2005). Lichens exist in different growth forms, such as: fruticose, which has tiny, leafless branches; foliose, which has a flat leaf-like structure; crustose, which has a flake-like structure and lies on the surface of its habitat; and leprose, which has a powder-like appearance. Lichens possess a number of secondary metabolites with antioxidant, antimicrobial, and anticancer actions (White et al., 2014). Use of lichens in the synthesis of silver nanoparticles has been very scarce. Biogenic synthesis of silver nanoparticles using various lichens and their applications are summarized in Table 1.

For instance, biosynthesis of silver nanoparticles was achieved using four different species of lichens: *Parmeliopsis ambigua*, *Punctelia subrudecta*, *Evernia mesomorpha*, and *Xanthoparmelia plitti* (Dasari et al., 2013). The antibacterial activities of the

Table 1 Biogenic synthesis of silver nanoparticles using various lichens and their applications.

Lichens used	Applications	References
Parmeliopsis ambigua, *Punctelia subrudecta*, *Evernia mesomorpha*, *Xanthoparmelia plitti*	Antibacterial, antioxidant	Dasari et al. (2013)
Parmotrema praesorediosum	Antibacterial	Mie et al. (2014)
Cetraria islandica (L) Ach	–	Yildiz et al. (2014)
Parmotrema clavuliferum	Antibacterial	Alqahtani et al. (2017)
Parmotrema tinctorum	Antibacterial	Khandel et al. (2018)
Usnea longissima	Antibacterial	Siddiqi et al. (2018)
Protoparmeliopsis muralis	Antibacterial, antibiofilm, antivirulence, antioxidant	Alavi et al. (2019)
Xanthoria elegans, *Cetraria islandica*, *Usnea antarctica*, *Leptogium puberulum*	Antibacterial	Balaz et al. (2020)
Pseudevernia furfuracea, *Lobaria pulmonaria*	Antibacterial, antioxidant	Goga et al. (2021)

as-synthesized nanoparticles have been assessed in vitro with the help of a disk diffusion technique against both gram-positive and gram-negative bacteria. The synthesized silver nanoparticles using the lichen *E. mesomorpha* showed the highest antibacterial activity against gram-negative bacteria. The antioxidant activities of the as-synthesized nanoparticles were also tested in vitro. The silver nanoparticles synthesized using *P. subrudecta* showed the highest antioxidant activity and were found to be very similar to those of standard antioxidant ascorbic acid. An aqueous extract of *Parmotrema praesorediosum* lichen was utilized for the synthesis of silver nanoparticles (Mie et al., 2014). The synthesized nanoparticles showed pronounced antibacterial activity against gram-negative bacteria. Biosynthesis of silver nanoparticles has been accomplished utilizing the ethanolic extract of lichen *Cetraria islandica* (L) Ach which acted both as a reductant and a stabilizer (Yildiz et al., 2014). Response surface methodology (RSM) has been utilized to study the role of reaction temperature, reaction time, and silver nitrate to lichen ratio during biosynthesis of silver nanoparticles. An aqueous extract of lichen *Parmotrema clavuliferum* served as a bio-mediator for the formation of antibacterial silver nanoparticles (Alqahtani et al., 2017). Biosynthesis and antibacterial properties of silver nanoparticles were reported using an aqueous extract of *Parmotrema tinctorum* lichen (Khandel et al., 2018). Biogenic fabrication of antibacterial silver nanoparticles has been accomplished using an aqueous ethanolic extract of lichen *Usnea longissima* (Siddiqi et al., 2018). The lichen *Protoparmeliopsis muralis* has been utilized for the synthesis of Ag, Cu, TiO_2, ZnO, and Fe_3O_4 nanoparticles (Alavi et al., 2019). Recently, a lichen-assisted biomechanochemical solid-state synthetic approach was developed for production of antibacterial silver nanoparticles (Balaz et al., 2020). The phytochemicals contained in the lichen extracts reduce silver ions and also stabilize synthesized silver nanoparticles. A schematic view of the lichen-assisted biomechanochemical solid-state synthesis of silver nanoparticles is provided in Fig. 1.

The synthesized silver nanoparticles were characterized by powder X-ray diffraction (XRD) and transmission electron microscopy (TEM). The reduction of silver nitrate into metallic silver nanoparticles was monitored using powder XRD. Fig. 2 shows the powder XRD patterns of the reaction mixtures of silver nitrate and the corresponding lichen species milled for different times. Based on the rate of decrease of the intensity of XRD peaks of silver nitrate (Fig. 2), the reducing ability of the utilized lichens were arranged in the order: *Xanthoria elegans* > *Cetraria islandica* ≫ *Usnea antarctica* > *Leptogium puberulum*.

Fig. 3 shows TEM images and the corresponding selected area electron diffraction (SAED) patterns of the silver nanoparticles synthesized in the presence of four different lichen species. Silver nanoparticles with bimodal particle size distribution were observed in all cases. This is the characteristic of silver nanoparticles synthesized mechanochemically in the presence of organic materials with reducing ability (Balaz et al., 2017). Initially, silver nitrate was reduced on the surface of the lichen biomatrix and formed small-sized silver nanoparticles (~5 nm) that were embedded within the biomatrix. During the ball milling, these small-sized nanoparticles merged into larger ones of size ~100 nm.

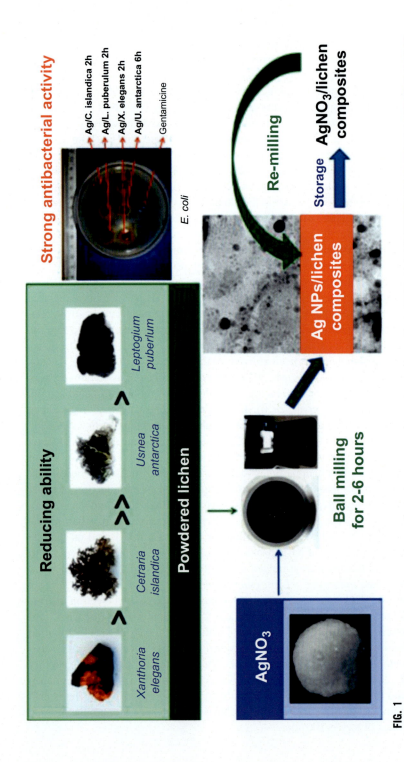

FIG. 1

Illustration of lichen-assisted biomechanochemical solid-state synthesis of silver nanoparticles.

Data from Balaz, M., Goga, M., Hegedus, M., Daneu, N., Kovacova, M., Tkacikova, L., Balazova, L., Backor, M., 2020. Bio-mechanochemical solid-state synthesis of silver nanoparticles with antibacterial activity using lichens. ACS Sustain. Chem. Eng. 8, 13945–13955 with permission from Copyright (2020) American Chemical Society.

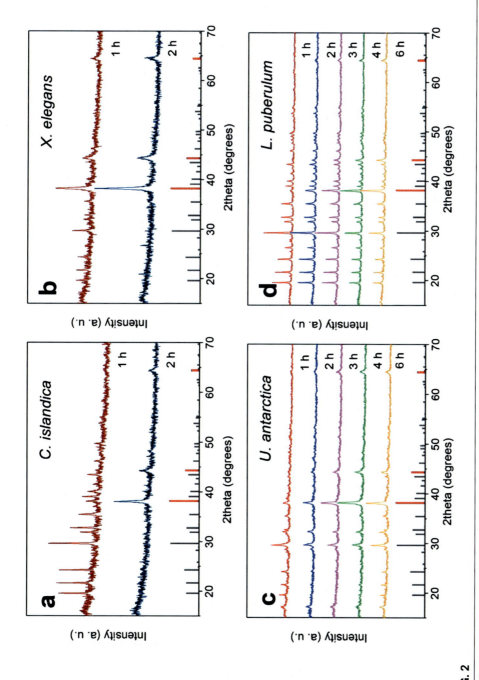

FIG. 2

XRD patterns of the reaction mixtures of silver nitrate and the corresponding lichen species milled for different times (calculated Bragg peak position for silver-*red*, and silver nitrate-*gray*).

Data from Balaz, M., Goga, M., Hegedus, M., Daneu, N., Kovacova, M., Tkacikova, L., Balazova, L., Backor, M., 2020. Bio-mechanochemical solid-state synthesis of silver nanoparticles with antibacterial activity using lichens. ACS Sustain. Chem. Eng. 8, 13945–13955 with permission from Copyright (2020) American Chemical Society.

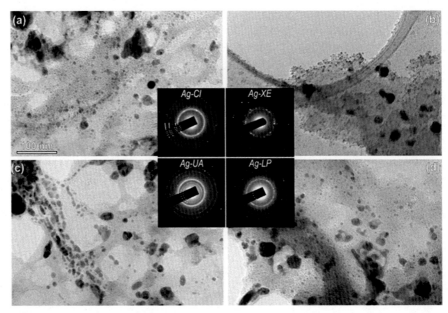

FIG. 3

TEM images of silver nanoparticles and the corresponding SAED patterns of the samples synthesized in the presence of (A) *Cetraria islandica*, (B) *Xanthoria elegans*, (C) *Usnea antarctica*, and (D) *Leptogium puberulum*.

Data from Balaz, M., Goga, M., Hegedus, M., Daneu, N., Kovacova, M., Tkacikova, L., Balazova, L., Backor, M., 2020. Bio-mechanochemical solid-state synthesis of silver nanoparticles with antibacterial activity using lichens. ACS Sustain. Chem. Eng. 8, 13945–13955 with permission from Copyright (2020) American Chemical Society.

3 Biomedical applications of silver nanoparticles

Over the years, antimicrobial and anticancer activities of silver nanoparticles have been extensively analyzed. Silver nanoparticles possess broad-spectrum antimicrobial actions against microorganisms which include bacteria, fungi, and viruses (Prabhu and Poulose, 2012; Jeong et al., 2014). Moreover, silver nanoparticles can effectively kill nematodes (Ahn et al., 2014) and worms (Li et al., 2015). Various factors like size, shape, dose, and stabilizer affects antimicrobial activities (Jeong et al., 2014; Burkowska-But et al., 2014; Oei et al., 2012). Silver nanoparticles may exhibit different antibacterial properties against both gram-positive and gram-negative bacteria (Abbaszadegan et al., 2015). Silver nanoparticles also have broad-spectrum anticancer activities. Various factors such as size, shape, dose, and exposure time affect the anticancer activities (Lin et al., 2014; Ishida, 2017; Gurunathan et al., 2015). The mechanism of antibacterial activity of silver nanoparticles is yet to be fully interpreted. However, it is anticipated that silver nanoparticles might attack the outer cell membrane of bacteria, which results in degradation and finally death of bacteria (Duran et al., 2016). Moreover, silver nanoparticles might liberate ionic silver (Ag^+), which is responsible for the production of reactive oxygen species (ROS) and singlet oxygen. These lead to oxidative stress and finally death of bacteria (Dakal et al., 2016). However, the mechanism of anticancer activity of silver nanoparticles is much more complex. At the time of writing, it has been agreed that silver nanoparticles might suppress cancer cells growth by destructing the cell membrane, causing ROS generation and damage of DNA (Wang et al., 2019; Lin et al., 2014; Pei et al., 2019; Zhang et al., 2017). Moreover, silver nanoparticles can either cause apoptosis of cancer cells by deactivating proteins and governing signaling tracts, or block cancer cell metastasis by suppressing growth within wounds (Fields, 2019; Yang et al., 2016). In addition to antimicrobial and anticancer applications, silver nanoparticles were also utilized in bone healing (Marsich et al., 2013), wound repair (Chowdhury et al., 2014), diabetes treatment (Sengottaiyan et al., 2016), dental materials (Bapat et al., 2018), vaccine adjuvant (Xu et al., 2013), biosensing (Anderson et al., 2017), etc.

4 Conclusions and future perspectives

The present study has showcased a detailed view on lichen-mediated biogenic synthesis of silver nanoparticles. The development of biogenic synthesis of silver nanoparticles over physical and chemical methods has been critically discussed. The phytochemicals contained in the lichen extracts served as reductants for silver ions as well as stabilizers for silver nanoparticles. Because of their biocompatible nature, as-synthesized silver nanoparticles find promising applications in a wide range of biomedical fields. Despite the environmental advantages of using biosynthesis strategy for silver nanoparticles over conventional methods, there are some issues yet to be resolved. These are consistency in nanoparticles size as well as shape, reproducibility of synthesis, and interpretation of exact mechanisms of nanoparticles synthesis by biomaterials. This is rather an unexplored area and needs further study to employ the biosynthesis strategy successfully for generation of silver nanoparticles.

References

Abbaszadegan, A., Ghahramani, Y., Gholami, A., Hemmateenejad, B., Dorostkar, S., Nabavizadeh, M., Sharghi, H., 2015. The effect of charge at the surface of silver nanoparticles on antimicrobial activity against gram-positive and gram-negative bacteria: a preliminary study. J. Nanomater. 2015, 720654.

Ahn, J.M., Eom, H.J., Yang, X., Meyer, J.N., Choi, J., 2014. Comparative toxicity of silver nanoparticles on oxidative stress and DNA damage in the nematode, *Caenorhabditis elegans*. Chemosphere 108, 343–352.

Alavi, M., Karimi, N., Valadbeigi, T., 2019. Antibacterial, antibiofilm, antiquorum sensing, antimotility, and antioxidant activities of green fabricated Ag, Cu, TiO_2, ZnO, and

Fe$_3$O$_4$ NPs via *Protoparmeliopsis muralis* lichen aqueous extract against multi-drug-resistant bacteria. ACS Biomater. Sci. Eng. 5, 4228–4243.

Alqahtani, M.A.M., Mohammed, A.E., Daoud, S.I., Alkhalifah, D.H.M., Albrahim, J.S., 2017. Lichens (*Parmotrema clavuliferum*) extracts: bio-mediator in silver nanoparticles formation and antibacterial potential. J. Bionanosci. 11, 410–415.

Anderson, K., Poulter, B., Dudgeon, J., Li, S.E., Ma, X., 2017. A highly sensitive nonenzymatic glucose biosensor based on the regulatory effect of glucose on electrochemical behaviors of colloidal silver nanoparticles on MoS$_2$. Sensors 17, 1807.

Asmathunisha, N., Kathiresan, K., 2013. A review on biosynthesis of nanoparticles by marine organisms. Colloids Surf. B 103, 283–287.

Balaz, M., Daneu, N., Balazova, L., Dutkova, E., Tkacikova, L., Briancin, J., Vargova, M., Balazova, M., Zorkovska, A., Balaz, P., 2017. Bio-mechanochemical synthesis of silver nanoparticles with antibacterial activity. Adv. Powder Technol. 28, 3307–3312.

Balaz, M., Goga, M., Hegedus, M., Daneu, N., Kovacova, M., Tkacikova, L., Balazova, L., Backor, M., 2020. Biomechanochemical solid-state synthesis of silver nanoparticles with antibacterial activity using lichens. ACS Sustain. Chem. Eng. 8, 13945–13955.

Bapat, R.A., Chaubal, T.V., Joshi, C.P., Bapat, P.R., Choudhury, H., Pandey, M., Gorain, B., Kesharwani, P., 2018. An overview of application of silver nanoparticles for biomaterials in dentistry. Mater. Sci. Eng. C 91, 881–898.

Burkowska-But, A., Sionkowski, G., Walczak, M., 2014. Influence of stabilizers on the antimicrobial properties of silver nanoparticles introduced into natural water. J. Environ. Sci. 26, 542–549.

Chandran, S.P., Chaudhary, M., Pasricha, R., Ahmad, A., Sastry, M., 2006. Synthesis of gold nanotriangles and silver nanoparticles using *Aloe vera* plant extract. Biotechnol. Prog. 22, 577–583.

Chowdhury, S., De, M., Guha, R., Batabyal, S., Samanta, I., Hazra, S.K., Ghosh, T.K., Konar, A., Hazra, S., 2014. Influence of silver nanoparticles on post-surgical wound healing following topical application. Eur. J. Nanomed. 6, 237–247.

Ciplak, Z., Gokalp, C., Getiren, B., Yildız, A., Yildız, N., 2018. Catalytic performance of Ag, Au and Ag-Au nanoparticles synthesized by lichen extract. Green Process. Synth. 7, 433–440.

Dakal, T.C., Kumar, A., Majumdar, R.S., Yadav, V., 2016. Mechanistic basis of antimicrobial actions of silver nanoparticles. Front. Microbiol. 7, 1831.

Dasari, S., Suresh, K.A., Rajesh, M., Samba Siva Reddy, C., Hemalatha, C.S., Wudayagiri, R., Valluru, L., 2013. Biosynthesis, characterization, antibacterial and antioxidant activity of silver nanoparticles produced by lichens. J. Bionanosci. 7, 237–244.

Debnath, R., Purkayastha, D.D., Hazra, S., Ghosh, N.N., Bhattacharjee, C.R., Rout, J., 2016. Biogenic synthesis of antioxidant, shape selective gold nanomaterials mediated by high altitude lichens. Mater. Lett. 169, 58–61.

Duran, N., Duran, M., de Jesus, M.B., Seabra, A.B., Favaro, W.J., Nakazato, G., 2016. Silver nanoparticles: a new view on mechanistic aspects on antimicrobial activity. Nanomedicine 12, 789–799.

Fields, G.B., 2019. Mechanisms of action of novel drugs targeting angiogenesis-promoting matrix metalloproteinases. Front. Immunol. 10, 1278.

Goga, M., Balaz, M., Daneu, N., Elecko, J., Tkacikova, L., Marcincinova, M., Backor, M., 2021. Biological activity of selected lichens and lichen-based Ag nanoparticles prepared by a green solid-state mechanochemical approach. Mater. Sci. Eng. C 119, 111640.

Gurunathan, S., Park, J.H., Han, J.W., Kim, J.H., 2015. Comparative assessment of the apoptotic potential of silver nanoparticles synthesized by *Bacillus tequilensis* and *Calocybe indica* in MDA-MB-231 human breast cancer cells: targeting p53 for anticancer therapy. Int. J. Nanomedicine 10, 4203–4223.

Habibi, B., Hadilou, H., Mollaei, S., Yazdinezhad, A., 2017. Green synthesis of silver nanoparticles using the aqueous extract of *Prangos ferulaceae* leaves. Int. J. Nano Dimens. 8, 132–141.

Huy, T.Q., Thanh, N.T.H., Thuy, N.T., Chung, P.V., Hung, P.N., Le, A.T., Hanh, N.T.H., 2017. Cytotoxicity and antiviral activity of electrochemical-synthesized silver nanoparticles against poliovirus. J. Virol. Methods 241, 52–57.

Ibrahim, E., Fouad, H., Zhang, M., Zhang, Y., Qiu, W., Yan, C., Li, B., Mo, J., Chen, J., 2019. Biosynthesis of silver nanoparticles using endophytic bacteria and their role in inhibition of rice pathogenic bacteria and plant growth promotion. RSC Adv. 9, 29293–29299.

Ishida, T., 2017. Anticancer activities of silver ions in cancer and tumor cells and DNA damages by Ag^+-DNA base-pairs reactions. MOJ Tumor Res. 1, 8–16.

Javed, B., Ikram, M., Farooq, F., Sultana, T., Mashwani, Z.R., Raja, N.I., 2021. Biogenesis of silver nanoparticles to treat cancer, diabetes, and microbial infections: a mechanistic overview. Appl. Microbiol. Biotechnol. 105, 2261–2275.

Jeong, Y., Lim, D.W., Choi, J., 2014. Assessment of size-dependent antimicrobial and cytotoxic properties of silver nanoparticles. Adv. Mater. Sci. Eng. 2014, 763807.

Khandel, P., Shahi, S.K., Kanwar, L., Yadaw, R.K., Soni, D.K., 2018. Biochemical profiling of microbes inhibiting silver nanoparticles using symbiotic organisms. Int. J. Nano Dimens. 9, 273–285.

Kumar, S.V., Rajeshkumar, S., 2017. Optimized production of silver nanoparticles using marine macroalgae *Sargassum myriocystum* for its antibacterial activity. J. Bionanosci. 11, 323–329.

Kumar, C.V., Yadav, S.C., Yadav, S.K., 2010. *Syzygium cumini* leaf and seed extract mediated biosynthesis of silver nanoparticles and their characterization. J. Chem. Technol. Biotechnol. 85, 1301–1309.

Li, L., Wu, H., Peijnenburg, W.J., van Gestel, C.A., 2015. Both released silver ions and particulate Ag contribute to the toxicity of AgNPs to earthworm *Eisenia fetida*. Nanotoxicology 9, 792–801.

Lin, J., Huang, Z., Wu, H., Zhou, W., Jin, P., Wei, P., Zhang, Y., Zheng, F., Zhang, J., Xu, J., Hu, Y., Wang, Y., Li, Y., Gu, N., Wen, L., 2014. Inhibition of autophagy enhances the anticancer activity of silver nanoparticles. Autophagy 10, 2006–2020.

Manikandan, D.B., Sridhar, A., Sekar, R.K., Perumalsamy, B., Veeran, S., Arumugam, M., Karuppaiah, P., Ramasamy, T., 2021. Green fabrication, characterization of silver nanoparticles using aqueous leaf extract of *Ocimum americanum* (Hoary Basil) and investigation of its *in vitro* antibacterial, antioxidant, anticancer and photocatalytic reduction. J. Environ. Chem. Eng. 9, 104845.

Marsich, E., Bellomo, F., Turco, G., Travan, A., Donati, I., Paoletti, S., 2013. Nano-composite scaffolds for bone tissue engineering containing silver nanoparticles: preparation, characterization and biological properties. J. Mater. Sci. Mater. Med. 24, 1799–1807.

Mie, R., Samsudin, M.W., Din, L.B., Ahmad, A., Ibrahim, N., Adnan, S.N.A., 2014. Synthesis of silver nanoparticles with antibacterial activity using the lichen *Parmotrema praesorediosum*. Int. J. Nanomedicine 9, 121–127.

Mohanpuria, P., Rana, K.N., Yadav, S.K., 2008. Biosynthesis of nanoparticles: technological concepts and future applications. J. Nanopart. Res. 10, 507–517.

Moradi, F., Sedaghat, S., Moradi, O., Salmanabadi, S.A., 2021. Review on green nano-biosynthesis of silver nanoparticles and their biological activities: with an emphasis on medicinal plants. Inorg. Nano-Metal Chem. 51, 133–142.

Oei, J.D., Zhao, W.W., Chu, L., DeSilva, M.N., Ghimire, A., Rawls, H.R., Whang, K., 2012. Antimicrobial acrylic materials with in situ generated silver nanoparticles. J Biomed Mater Res B Appl Biomater 100, 409–415.

Ouay, B.L., Stellacci, F., 2015. Antibacterial activity of silver nanoparticles: a surface science insight. Nano Today 10, 339–354.

Panacek, A., Kolar, M., Vecerova, R., Prucek, R., Soukupova, J., Krystof, V., Hamal, P., Zboril, R., Kvitek, L., 2009. Antifungal activity of silver nanoparticles against *Candida* spp. Biomaterials 30, 6333–6340.

Paul, B., Bhuyan, B., Purkayastha, D.D., Dhar, S.S., 2015. Green synthesis of silver nanoparticles using dried biomass of *Diplazium esculentum (retz.) sw.* and studies of their photocatalytic and anticoagulative activities. J. Mol. Liq. 212, 813–817.

Paul, B., Bhuyan, B., Purkayastha, D.D., Dhar, S.S., 2016. Photocatalytic and antibacterial activities of gold and silver nanoparticles synthesized using biomass of *Parkia roxburghii* leaf. J. Photochem. Photobiol. B 154, 1–7.

Pei, J., Fu, B., Jiang, L., Sun, T., 2019. Biosynthesis, characterization, and anticancer effect of plant-mediated silver nanoparticles using *Coptis chinensis*. Int. J. Nanomedicine 14, 1969–1978.

Prabhu, S., Poulose, E.K., 2012. Silver nanoparticles: mechanism of antimicrobial action, synthesis, medical applications, and toxicity effects. Int. Nano Lett. 2, 32.

Sajjadi, G., Amini, J., Arani, A.S., Nezammahalleh, H., 2018. Extracellular synthesis of silver nanoparticles using four fungal species isolated from lichens. IET Nanobiotechnol. 12, 64–70.

Sengottaiyan, A., Aravinthan, A., Sudhakar, C., Selvam, K., Srinivasan, P., Govarthanan, M., Manoharan, K., Selvankumar, T., 2016. Synthesis and characterization of *Solanum nigrum*-mediated silver nanoparticles and its protective effect on alloxan-induced diabetic rats. J. Nanostruct. Chem. 6, 41–48.

Sharma, B., Purkayastha, D.D., Hazra, S., Gogoi, L., Bhattacharjee, C.R., Ghosh, N.N., Rout, J., 2014. Biosynthesis of gold nanoparticles using a freshwater green alga, *Prasiola crispa*. Mater. Lett. 116, 94–97.

Siddiqi, K.S., Rashid, M., Rahman, A., Tajuddin, Husen, A., Rehman, S., 2018. Biogenic fabrication and characterization of silver nanoparticles using aqueous ethanolic extract of lichen (*Usnea longissima*) and their antimicrobial activity. Biomater. Res. 22, 23.

Tripathy, A., Raichur, A.M., Chandrasekaran, N., Prathna, T.C., Mukherjee, A., 2010. Process variables in biomimetic synthesis of silver nanoparticles by aqueous extract of *Azadirachta indica* (neem) leaves. J. Nanopart. Res. 12, 237–246.

Vlasceanu, G.M., Marin, S., Țiplea, R.E., Bucur, I.R., Lemnaru, M., Marin, M.M., Grumezescu, A.M., Andronescu, E., 2016. Silver nanoparticles in cancer therapy. In: Nanobiomaterials in Cancer Therapy. vol. 7. Elsevier, pp. 29–56.

Wang, Z.X., Chen, C.Y., Wang, Y., Li, F.X.Z., Huang, J., Luo, Z.W., Rao, S.S., Tan, Y.J., Liu, Y.W., Yin, H., Wang, Y.Y., He, Z.H., Xia, K., Wu, B., Hu, X.K., Luo, M.J., Liu, H.M., Chen, T.H., Hong, C.G., Cao, J., Liu, Z.Z., Long, Z., Gan, P.P., Situ, W.Y., Fan, R., Yuan, L.Q., Xie, H., 2019. Ångstrom scale silver particles as a promising agent for low toxicity broad spectrum potent anticancer therapy. Adv. Funct. Mater. 29, 1808556.

White, A.P., Oliveira, C.R., Oliveira, P.A., Serafini, R.M., Araujo, A.A., Gelain, P.D., Moreira, C.J., Almeida, R.J., Quintans, S.J., Quintans-Junior, J.L., Santos, R.M., 2014. Antioxidant activity and mechanisms of action of natural compounds isolated from lichens: a systematic review. Molecules 19, 14496.

Xu, Y., Tang, H., Liu, J.H., Wang, H., Liu, Y., 2013. Evaluation of the adjuvant effect of silver nanoparticles both *in vitro* and *in vivo*. Toxicol. Lett. 219, 42–48.

Xue, B., He, D., Gao, S., Wang, D., Yokoyama, K., Wang, L., 2016. Biosynthesis of silver NPs by the fungus *Arthroderma fulvum* and its antifungal activity against genera of *Candida*, *Aspergillus* and *Fusarium*. Int. J. Nanomedicine 11, 1899–1906.

Yang, T., Yao, Q., Cao, F., Liu, Q., Liu, B., Wang, X.H., 2016. Silver nanoparticles inhibit the function of hypoxia-inducible factor-1 and target genes: insight into the cytotoxicity and antiangiogenesis. Int. J. Nanomedicine 11, 6679–6692.

Yari, A., Yari, M., Sedaghat, S., Delbari, A.S., 2021. Facile green preparation of nano-scale silver particles using *Chenopodium botrys* water extract for the removal of dyes from aqueous solution. J. Nanostruct. Chem. https://doi.org/10.1007/s40097-020-00377-3.

Yildiz, N., Ates, C., Yilmaz, M., Demir, D., Yildiz, A., Calimli, A., 2014. Investigation of lichen based green synthesis of silver nanoparticles with response surface methodology. Green Process. Synth. 3, 259–270.

Yuan, X., Xiao, S., Taylor, T.N., 2005. Lichen-like symbiosis 600 million years ago. Science 308, 1017–1020.

Zhang, Y., Lu, H., Yu, D., Zhao, D., 2017. AgNPs and Ag/C225 exert anticancerous effects via cell cycle regulation and cytotoxicity enhancement. J. Nanomater. 2017, 7920368.

CHAPTER 20

Algae-mediated silver nanoparticles: Synthesis, properties, and biological activities

Emad A. Shalaby

Department of Biochemistry, Faculty of Agriculture, Cairo University, Giza, Egypt

1 Introduction

Nanoparticles or nanomaterials (NMs) or nanoproducts (NPs) are the products of nanotechnology, with a range of 1–100nm as dimensions. They have gained prominence in technological advancements due to their chemical, physical, and biological characteristics (Jeevanandam et al., 2018). Metal nanoparticles can be prepared in two ways: top-down (from original materials/cells) or bottom-up (from extracts or compounds) using various physical and chemical methods. Singh et al. (2018) reported that extracts and active compounds from different living organisms are extensively used as an alternative bottom-up method for preparation of metal NPs. Silver nanoparticles possess high conductivity and are very sensitive to metal surface absorption. Therefore, silver nanoparticles are commonly used in different application fields such as antimicrobial, food industries, cancer, wound recovery, electronics, and textiles (Fard et al., 2015). Green nanotechnology is a biotechnological tools to produce environmentally and nontoxic nanoparticles (Patra and Baek, 2014). AgNPs produced using these methods or tools have high stability and suitable dimensions. Various factors affect silver nanoparticles produced from biological cells, such as pH, temperatures, contact time, pressure, and concentration of silver ions.

Algae have a great ability to convert CO_2 and solar energy in the presence of chlorophyll to organic chemical compounds such as sugar, lipids, and carotenoids. Besides, algae are important biomass, cost-effective cell factories for the biosynthesis of natural products such as carbohydrates, protein, carotenoids, and lipids (Borowitzka, 2010, 2013). In addition, macro and microalgae or their extracts have various natural products of secondary metabolites (plant acids, phenolic, flavonoids, terpenoids, glycosides, and alkaloids) with different biological activities such as antioxidant, antitumor, antibacterial, and antiviral (Liu et al., 2019; Keeffe et al., 2019).

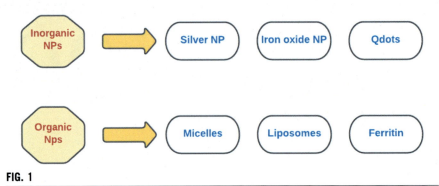

FIG. 1

Some inorganic and organic NPs.

In the field of nanotechnology, algae (micro and macro) in addition to cyanobacteria are being widely cultured and important living organisms in biological or green synthesis of nanomaterials, as reported by a significant number of articles in the last 10 years (Sharma et al., 2016). Various chemical compounds from algal cells act as reducing agents and were used for green synthesis of silver NPs such as polysaccharides (e.g., alginate, fucoidan and laminaran) from *Saccharina cichorioides* and *Fucus evanescens*, respectively. Yugay et al. (2020) found that the activity of tested polysaccharides could be arranged in the following order: alginate less than fucoidan, followed by laminaran. The obtained results demonstrate that various algal polysaccharides can be used for AgNPs synthesis with high antibacterial activity. Nowadays, different algal species (micro and macro) of capped and stabilized nanoparticles have received widespread attention due to their characteristics of low toxicity, easy harvesting, fast growing, cost-effective, and safer to use. The natural compounds from algae such as polysaccharides, phenolic, enzymes, and carboxylic compounds act as reducing, capping, and stabilizing agents (AlNadhari et al., 2021). Moreover, Thiruchelvi et al. (2021) found that silver nanoparticles synthesized from marine macroalgae were considered to exhibit antibacterial activity against positive and negative bacteria. Deoxy ribonucleic acids is considered as example of natural carbon-containing NPs. However, sea salt and magnetite considered as natural inorganic NPs, as shown in Fig. 1. This chapter comprehensively reviews work and activities done on algae-mediated biosynthesis of silver nanoparticles (AgNPs) and their methods of synthesis, followed by advances in characterization techniques with their application in different fields.

2 Silver nanoparticles synthesis/production from algae (algae-AgNPs)

Algal nanotechnology from the important approaches which can be used for the biosynthesis of metallic NPs (Jeffryes et al., 2015) with different advantages. This technology focuses on the production of nanoparticles using safe materials, rich with reducing agents and without any cytotoxicity on other living organisms, as shown in Fig. 2. These findings were in agreement with the results mentioned by Eroglu et al., 2013, Kharissova et al., 2013, and Mittal et al., 2013. Applications of biosynthesized silver nanotechnology in the different field have been reported by Pangi et al. (2003).

The biosynthesis of silver from algae can be categorized into four potential methods, as shown in Fig. 3.

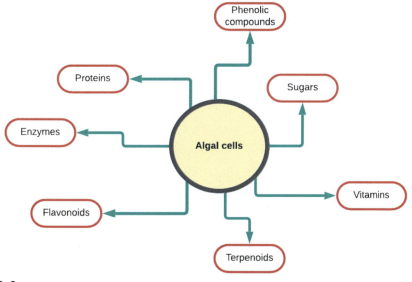

FIG. 2
Common reducing agents produced from different algal cells.

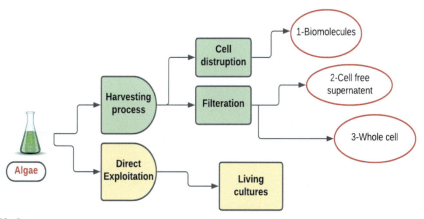

FIG. 3
Some suggested methods for silver nanoparticles production from microalgae.

- **Method No. 1:** This method focuses on using the extract or residues produced from algal cells.
- **Method No. 2:** This method focuses on the use of algal supernatants devoid of cells using centrifugation.
- **Method No. 3:** This method depends on the resuspension of whole microalgae cells without their culture media in distilled water, to promote nanoparticles' biosynthesis of different natures.
- **Method No. 4:** This method focuses on the use of living cells of microalgae under their natural culturing conditions.

As shown in Fig. 4A, silver cations added to the culture of microalgae, usually the silver salts ($AgNO_3$), are localized into the living cells and reduced inorganic silver

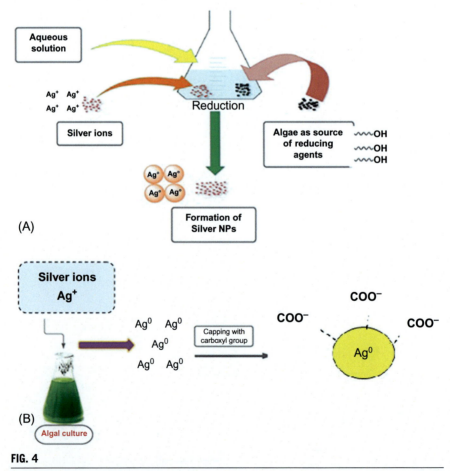

FIG. 4

(A) The role of reducing agents from algal species on the formation of silver. (B) The binding between some capping agents and nanoparticles.

(Ag^+) within the thylakoids via reducing enzymes especially dehydrogenase enzymes, leading to silver-NPs formation (Ag^0). These nanoparticles pass to the cell wall of algal cells and are capped by exopolysaccharides (EPS). The nanoparticles-EPS produced from the last step are released into the extracellular media and have a stable colloid formation as described by Dahoumane et al. (2012). In addition, Ramanathan et al. (2011) and Prasad et al. (2011) reported that eukaryotic systems (such as algae, plants, diatoms, fungi, etc.) have an excellent ability to reduce silver ions to silver nanoparticles (AgNPs), as shown in Fig. 4B. The nanomaterials synthesized by algal species are principally due to the reducing agents present in algal extracts such as phenolic, sugars, flavonoids, enzymes, etc. (Ponnuchamy and Jacob, 2016).

3 Practical methods for biosynthesis of AgNPs using algae

In 2009, the first methods for green synthesis of metal NPs were described by Raveendran et al. They used an aqueous solution of starch and exposed it to heating for solubilization, then add of silver nitrate and glucose as the green reducing factor (Raveendran et al., 2003). After this period, various scientists like Iravani produced high-quality review papers regarding the synthesis of silver nanoparticles using plant extracts as a green chemistry method (Iravani, 2011). Since then, biosynthesis of silver nanoparticles has been carried out by various research groups based on a variety of biological materials and their compositions. For example, Logeswari et al. (2013) created an eco-friendly production of silver nanoparticles from some plants such as *Solanum tricobatum* and *Citrus sinensis*, while Bagherzade et al. (2017) reported the antimicrobial action of silver nanoparticles produced through green biosynthesis using *Crocus sativus* L. extracts. Algal crude extracts in addition to different chemical compounds from algal cells can be used for preparation of silver nanoparticles such as phenolic compounds, carboxylic acid compounds, and polysaccharides. In brief, a known weight of algal extract was directly added to 1 mM aqueous solution of silver nitrate while stirring at 25°C (Fig. 5). The pH of the obtained mixture was adjusted to 10–11. After that, the reaction was kept for 30–69 min under heating. The reduction of silver ions to silver NPs was indicated by the color change of the mixture (Yugay et al., 2020).

Moreover, Kathiravan et al. (2015) mentioned that silver nanoparticles synthesis by add 20 mL of *Gelidiella acerosa* aqueous extract with 80 mL (1 mM $AgNO_3$) at 25°C and checking for 5 days.

However, Fatima et al. (2020) summarized the method of silver NPs from algae in the following steps: 45 mL of $AgNO_3$ solution (1 mM concentration) and around 5 mL of algal extract were added drop by drop with continuous stirring at 25°C. During this process, the color of the mixing solution converted to pale pink. With further incubation for 24–48 h, the color changed to dark brown, representing the formation of silver nanoparticles (AgNPs).

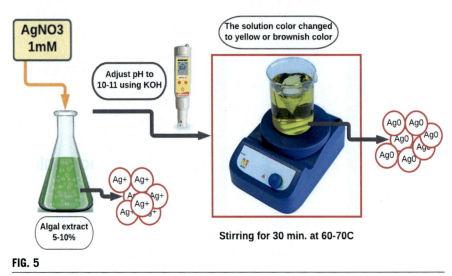

FIG. 5

Practical steps for the synthesis of silver nanoparticles by algae.

Another method described by Hamouda et al. (2019) was as follows: add 0.017 g of silver nitrate with 90 mL di-distilled water and then add algal aqueous extracts as drop by drop 10 mL with continued magnetic stirring at 60°C until the color becomes brownish.

4 Factors affecting AgNPs production by algae

Different conditions and parameters such as the techniques used for production, extract concentrations, pH, pressure, temperature, pore size, contact time, particle size, other environment conditions, and proximity greatly affect the quantity and quality of the biosynthesized silver nanoparticles and their properties and applications. The properties of the biosynthesized silver nanoparticles are the main factors regarding their potential use in different medicines or drug delivery and other pharmaceutical applications. This section illustrates major factors influencing the biosynthesis of silver nanoparticles by micro and macro algae. Current and new studies of algal nanotechnology will contribute to more complete skills and knowledge base regarding different parameters that affect green production of silver nanoparticles and the most advanced technology that can be used for characterization of the AgNPs for their activities and applications in biomedical and other fields industries. The following sections look at the factors affecting biosynthesis of silver nanoparticles from different micro and macro algae species and their extracts (Fig. 6).

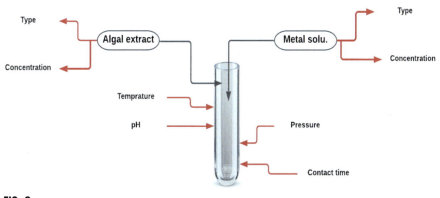

FIG. 6

Factors affecting biosynthesis of metal nanoparticles from algal species.

4.1 Methods/techniques

Different techniques can be used for biosynthesis of silver nanoparticles from algae, including (a) physical, (b) chemical, and (c) biological methods. Each procedure has specific advantages and disadvantages. However, biological methods (e.g., use of algal extracts or cultures) for synthesis of silver nanoparticles are nontoxic, ecofriendly, and more acceptable than other methods, as reported by Vadlapudi and Kaladhar (2014).

4.2 Effect of extract concentration

There is a significant correlation between the algal extract concentration and SPR band intensity of silver nanoparticles synthesized. The results revealed that the optimum concentration of algal concentration ranged between 10% and 20%.

4.3 Effect of temperature

Temperature is consider as common factors influence on silver nanoparticles formation from different algal species. There are three common methods for preparation of nanoparticles, as mentioned above (physical, chemical, and biological); the physical method requires a very high temperature (more than 350°C), whereas chemical methods need a temperature lower than 350°C. However, the preparation of silver nanoparticles by green technology needs temperatures below 100°C, as described by Rai et al. (2006). In addition, Jiang et al. (2011) reported that temperature can significantly affect the synthesis, growth, shape, and distribution of particle size. The authors found that (a) a low temperature decreases the formation of nanoparticles and increases the time needed to complete the reducing reaction; (b) with an increase in the temperature to 55°C, the synthesis rate increases, as well as the particle size; and (c) with an increase in the temperature to more than 60°C the reaction becomes more active in reducing silver ions (Pillai and Kamat, 2004). Moreover, Liu et al. (2020) mentioned that high temperatures are important for nucleation but low temperatures are required for growth in wet chemical conditions for silver nanoparticle synthesis. These phenomena due to the negative correlation between the size of nanoparticle and temperature used. However, larger silver nanoparticles can be produced at low temperatures. Therefore, the silver nanoparticles could be synthesized better and in larger sizes at low temperatures. The authors also found that with increasing temperature, the total reaction rate is increased. In general, the exposure to high temperature or low temperature (cooling) of the biosynthesis reaction will influence the surfactant adsorption/desorption, complexing stability, and biosynthesis rate in addition to the shape and size distributions of produced nanoparticles.

4.4 Effect of pressure

From the main factors affecting silver nanoparticles synthesis and their properties, the time needed or the rate of silver ions reduction using algal reducing agents is much faster at ambient pressure as mentioned by Tran et al. (2013). The shape and size of the synthesized silver nanoparticles are also affected by pressure applied during the biosynthesis.

4.5 Algal extract composition

The phytochemicals composition of algal extracts from the bioactive or its content from reducing agents in addition to the algal species, the type of solvent used for extraction and extraction methods and conditions are from important factors which affect silver nanoparticles biosynthesis and its properties (Park et al., 2011).

4.6 Effect of contact time

An increase in contact time leading to rapid biosynthesis of silver nanoparticles was reported by Darroudi et al. (2011), who also found that the type and properties of silver nanoparticle synthesized using algal extracts are greatly affected by the duration for which the reaction solution is incubated.

4.7 Effect of pH

pH values are consider as important factor which effect on biosynthesis of silver nanoparticles, using green technology methods. Various published articles emphasized that the pH of the metal solution affects the size and texture of the biosynthesized silver nanoparticles (Gardea-Torresdey et al., 1999; Armendariz et al., 2004; Patra and Baek, 2014). In addition, AgNPs' size can be modified by changing the pH of the metal solution. The same results were obtained by Soni and Prakash (2011), who reported that the pH influenced the shape and size of silver nanoparticle synthesized by algal species.

In this regard, Alqadi et al. (2014) mentioned that the silver nanoparticles synthesized at high pH (range 10–11) were distinguished by being more regular and smaller in size when compared to nanoparticles synthesized at low pH. The average radii of produced particles ranged 43.00 and 32.00 nm, and the shape of produced AgNPs is spherical. The same authors reported that increasing the pH led to the formation of spherical nanoparticles. When compared with the AgNPs' shape at low pH, the results indicate formation of rods and triangular shapes, which may be due to poor balance between growth processes and nucleation.

The following table summarizes the effect of pH in reaction media on the size of silver nanoparticles as described by Soni and Prakash (2011).

pH value	Wavelength (nm)	Average AgNP size (nm)
7.0	428.0 ± 3.0	74.0
8.0	418.0 ± 3.0	67.0
9.0	410.0 ± 3.0	54.0
10.0	404.0 ± 3.0	43.0
11.0	397.0 ± 3.0	32.0

5 Properties of silver nanoparticles produced from algae

The properties of algal silver nanoparticles (NPs) are usually determined using different spectroscopic, diffractographic, and microscopic techniques (Khanna et al., 2019).

The following points summarize the important indicators for biosynthesis of silver nanoparticles using algal cultures or extracts.

1. Biosynthesis of silver nanoparticles with algal extracts (as reducing agents sources) was ensured by the change of color from green, colorless or light yellow to dark yellowish or brown (Selvaraj et al., 2020).
2. These biosynthesized silver nanoparticles were characterized by ultraviolet-visible (UV-vis) spectrophotometry, which indicated the formation of silver nanoparticles at the absorbance range 390–450 nm (Fatima et al., 2020).
3. Fourier transform infrared (FTIR) spectroscopy evaluation of fine specimens was employed to investigate attainable active groups which interact between silver and algal extracts. The FTIR spectroscopy study was implemented to detect and identify the biomolecules responsible for reduction of silver ions into Ag^0 and observe any modification in groups' absorption peaks (Fatima et al., 2020).
4. The negative results or value of the zeta potential reported the efficiency of the reducing or capping chemical compounds of algal extracts in silver nanoparticles stabilization by given negative charges that keep all the particles away from each other, as found by Haider and Mehdi (2014). Table 1 illustrates the characteristics of silver nanomaterials prepared from different algal species.

Table 1 Physical properties of silver nanoparticles synthesized by some algal species.

Algae species	Type of NP	Shape of NP	Size of NP (nm)	References
Turbinaria conoides	Silver	Spherical	96	AlNadhari et al. (2021)
Gilidiella acerosa	Silver	Spherical	18–46	AlNadhari et al. (2021)
Padina pavonica	Silver	Spherical	10–72	AlNadhari et al. (2021)
Colpmenia sinusa	Silver	Spherical	20	AlNadhari et al. (2021)
Kappa phycus sp.	Silver	Spherical	52–104	AlNadhari et al. (2021)
Kappaphycus alverazii	Silver	FCC	73	AlNadhari et al. (2021)
Acanthophora specifera	Silver	Cubic	81	AlNadhari et al. (2021)
Porphyra vietnamensis	Silver	–	13–16	AlNadhari et al. (2021)
Chlamydomonas	Silver	Rectangular	1–15	AlNadhari et al. (2021)
Enteromorpha flexuosa	Silver	Circular	15	AlNadhari et al. (2021)
Pithophora oedogonia	Silver	Cube	25–44	AlNadhari et al. (2021)
Microcoleus sp.	Silver	Spherical	44–79	Sudha et al. (2013)
Arthrospira platensis	Silver	Spherical	17–25	Govindaraju et al. (2008)
Chlamydomonas reinhardtii	Silver	Rounded and rectangular	5–15	Barwal et al. (2011)
Caulerpa racemosa	Silver	Spherical and triangular	5–25	Kathiravan et al. (2015)
Ulva (Enteromorpha) compressa	Silver	Spherical	40–50	AlNadhari et al. (2021)
Ulva reticulata	Silver	Spherical	40–50	AlNadhari et al. (2021)
Ulva faciata	Silver	Spherical	7–20	El-Rafie et al. (2013)
Sargassum cinereum	Silver	Spherical	45–76	Mohandass et al. (2013)
Gracilaria dura	Silver	Spherical	6	Shukla et al. (2012)
Jania rubins	Silver	Spherical	12	AlNadhari et al. (2021)
Chlorococcum humicola	Silver	Spherical	16	AlNadhari et al. (2021)
Turbinaria ornata	Silver	Spherical	22	Khalil et al. (2014)
Sargassum polycystum	Silver	Spherical	7	Khanna et al. (2019)
Sargassum muticum	Silver	Spherical	43–79	Madhiyazhagan et al. (2015)
Sargassum wightiigrevilli	Silver	Spherical	8–27	Govindaraju et al. (2008)
Sargassum ilicifolium	Silver	Spherical	33–40	Li (2011)
Padina tetrastromatica	Silver	Spherical	4	Bhuyar et al. (2020)
Padina gymnospora	Silver	Spherical	25–40	Shiny et al. (2013)
Cystophora moniliformis	Silver	FCC	75	Prasad et al. (2013)
Scaberia agardhii	Silver	Polydispersed	40–50	Prasad and Elumalai (2013)

Table 1 Physical properties of silver nanoparticles synthesized by some algal species—*Con't*

Algae species	Type of NP	Shape of NP	Size of NP (nm)	References
Gracilaria edulis	Silver	Spherical	12–100	Pugazhendhi et al. (2018)
Gelidium amansii	Silver	Spherical		Kumar et al. (2013)
Pterocladia capillacae	Silver	Spherical	7	El-Rafie et al. (2013)
Amphiroa fragilissima	Silver	Crystalline	–	Sajidha Parveen and Lakshmi (2016)
Desmarestia menziesii	Silver	–	7	González-Ballesteros et al. (2018)
Oscillato riawillei	Silver	Spherical	10–25	Ali et al. (2011)
Spirulina platensis	Silver	Spherical	2–8	Doshi et al. (2007)
Plectonema boryanum	Silver	Octahedral	200	Lengke et al. (2007)
Aphanothece sp. and *Oscillitoria* sp.	Silver	Spherical	44–79	Sudha et al. (2013)
Microchaete	Silver	Polydispersed and spherical	80	Husain et al. (2019)
Cylindrospermum stagnale	Silver	Pentagonal	38–88	Husain et al. (2015)
Scencedesmus sp.	Silver	–	15–20	Vigneshwaran et al. (2007)
Pithophora oedogonia	Silver	Cubical and hexagonal	24–55	Sinha et al. (2015)
Chlorococcum humicola	Silver	Spherical	16	Jena et al. (2014)
Nannochloropsis oculata and *Chlorella vulgaris*	Silver	FCC	15	Mohseniazar et al. (2011)
Chlamydomonas reinhardtii	Silver	Rectangular and rounded	15	Barwal et al. (2011)
Chlorococcum humicola	Silver	Spherical	16	Jena et al. (2013)
Euglena gracilis	Silver	Polydispersed and spherical	15	Li et al. (2015)
Chlorella pyrenoidosa	Silver	FCC	5–20	Aziz et al. (2015)

6 Biological activities and applications of algae mediated nanoparticles

Micro and macro algae are the main biomaterials for the synthesis and large-scale production of different silver nanoparticles; this is due to the presence of various active ingredients with high reduction potential that includes phenolics, proteins, alkaloids, and terpenoids. They provide antioxidant, anticancer, antiinflammatory, antimicrobial, and antiviral properties (Table 2). Silver nanoparticles play a vital role in nanotechnology, biotechnology, and biomedicine technology. They are they preferred option because they are nonhazardous to human health at low concentrations and have a broad spectrum as biological activities with high antimicrobial properties. In addition, Meera et al. (2013) revealed that from all living organisms, micro and macro algae in various forms were used for the synthesis of AgNPs (live and dead

Table 2 Biological activities of some silver NPs produced from different algal species.

Algal species	Type	Capping agents	Biological activities	References
Ulva lactuca	Silver	Aromatic rings	Photocatalytic degradation	Kumar et al. (2013)
Ulva lactuca	Silver	Polyphenols	Malaria control	Murugan et al. (2015)
Gracillaria corticata	Silver	Polyphenols and tannins	Antifungal activity	Kumar et al. (2013)
Sargassum wightii	Silver	Carboxylic acid group	Antibacterial activity	Govindaraju et al. (2009)
Turbinaria conodies	Silver	Polyphenols, polysaccharides, primary amines	Antimicrobial	Rajeshkumar et al. (2013)
Sargassum swartzii	Silver	Alcohol, carboxylic and amide I group	Anticancer activity	AlNadhari et al. (2021)
Sargassum polycystum	Silver	Alkanes and alkyl alcohols	Gram-positive S. aureus	Thangaraju et al. (2012)
Padina tetrastromatica	Silver	Alkanes	Gram-positive Bacillus spp., B. subtilis	Rajeshkumar et al. (2012)
Sargassum longifolium	Silver	Terpenoids	Anticancer activity	Devi et al. (2013)
Turbinaria ornata	Silver	Organic compounds	Antibacterial activity against gram-positive strains	Krishnan et al. (2015)
Urospora sp.	Silver	Hydrogen bonded hydroxyl group, carbonyl and alcoholic groups	Antibacterial activity against gram-positive strains	Suriya et al. (2012)
Ulva lactuca	Silver	Release of protein molecules	Anticancer: Hep2, MCF7, and HT29 cancer cell lines	Devi and Bhimba (2012)
Gelidiella acerosa	Silver	Aromatic compound or alkanes or amines	Antifungal against Humicola insolens, Fusarium	Vivek et al. (2011)

Table 2 Biological activities of some silver NPs produced from different algal species—Con't

Algal species	Type	Capping agents	Biological activities	References
Chlorella vulgaris	Silver	Gold shape-directing protein of 28 kDa	Optical coatings and hyperthermia of cancer cells	Xie et al. (2007)
Plectonema boryanum UTEX 485	Silver	Utilizing nitrate by reducing nitrate to nitrite and ammonium, which is fixed as glutamine before death	Temperature-dependent size control of NPs	Lengke et al. (2007)
Microchaete NCCU-342	Silver	Cellular metabolites	Degradation of azo dye methyl red	Husain et al. (2019)
Euglena gracilis	Silver	Primary amines	Anticancer activity	Li et al. (2015)

dried biomasses). Silver nanoparticles produced from algal biomass have been widely used in antiviral, antibacterial, antioxidant, and antiinflammatory activities. Furthermore, Selvaraj et al. (2020) reported that *Sargassum wightii* AgNPs have high antibacterial activity against gram-positive and gram-negative bacteria: 25.66 ± 0.87; 24.33 ± 2.18 mm, respectively. In addition, Cavalli et al. (2021) found that Ag nanoparticles synthesized using water extracts of *Pterocladiella capillacea* presented antimicrobial activity against *S. aureus*. The same results were obtained by Chaudhary et al. (2020), who reported that the algal biosynthesized silver nanoparticles have different biological actions include antimicrobial, anticancer, biofilm prevention, bioremediation, antifouling, and biosensing activities.

From these species, *Chlorella* spp. (as micro algae) and *Sargassum* spp. (as macro algae) have been extensively used for the biosynthesis of silver nanoparticles with high antimicrobial and antitumor characteristics, and are easy to use as antibiotics. In addition, various published articles and data revealed that algal silver NPs possess high biological activities (Iravani et al., 2014; Khanna et al., 2019). They have been reported to have antibacterial (Sharma et al., 2015; Khanna et al., 2019), anticancerous (Govindaraju et al., 2015; Khanna et al., 2019), and antifungal activities (Azizi et al., 2014; Khanna et al., 2019).

The cytotoxicity and antimicrobial activity of silver nanoparticles (AgNPs) have received a lot of work by Hamouda et al. (2019) and Rahman et al. (2020). They found that capping AgNPs using red macro algae has been shown to increase antimicrobial activity against pathogenic bacteria when compared to noncapping AgNPs. The same results and findings were obtained by Thiruchelvi et al. (2021).

The antioxidant properties of silver nanoparticles prepared using *Microsorum pteropus* methanol extract were studied by Chick et al. (2020), who found that AgNPs have high antioxidant activity against DPPH radical assay with an IC_{50} value of 47.0 μg/mL and hydrogen peroxide assay with an IC_{50} value of 35.8 μg/mL, when compared with native extract (IC_{50} 174 and 11.5 μg/mL, respectively). In this field, AlNadhari et al. (2021) concluded that the strong antioxidant action of silver nanoparticles is greatly interlinked to the *Ecklonia cava* extract (rich with active ingredients) which act as capping agents on silver NPs. Free radical inhibition of *Ecklonia cava* extract and biosynthesized AgNPs was consistent, making use of the current antioxidant assay (DPPH). Moreover, the same authors mentioned that the anticancer activity of silver nanoparticles produced using *Ecklonia cava* was investigated by using tumor cervical cell lines. The silver nano-sized flecks cytotoxicity at various concentrations. The IC50 of the silver nano-sized flecks was recorded at 59 μg/mL.

In addition, Baker-Austin et al. (2006) reported that silver nanoparticles have many applications in diverse fields including biology and medicine, and revealed that silver nanoparticles can be used as a substitute for antibiotics. Moreover, Abdel-Raouf et al. (2013) mentioned that silver nanoparticles prepared by marine red alga *Gelidium amansii* are potential antimicrobial agents for prominent microfouling bacteria. However, AgNPs prepared using *Corallina elongata* and *Gelidium crinale* had a high cytotoxic activity for Ehrlich ascites carcinoma (EACC), as reported by Khalifa et al. (2016). Another study by González-Ballesteros et al. (2021) revealed that silver nanoparticles have significant biomedical potential for cancer immunostimulant treatment.

AgNPs synthesized by water extract of *Caulerpa racemosa* as green macro algae exhibited antimicrobial activity against various bacterial strains such as *K. pneumoniae*, *B. subtilis*, and *E. coli*, as mentioned by Kathiravan et al. (2015). The same authors found that there is a strong relationship between the inhibitory or lethal effect of silver nanoparticles toward bacteria and their concentrations. On this issue, Salvioni et al. (2017) indicated that there are strong correlation between the negative charge of AgNPs and the antimicrobial effect.

In another investigation, stable silver particles produced from marine macro algae *Caulerpa serrulata* displayed high and unique antimicrobial activity at low concentrations against *Staphylococcus aureus*, *Shigella* sp., *Pseudomonas aeruginosa*, and *Salmonella typhi*. In addition, silver nanoparticles using *Pithophora oedogonia* have shown possible antimicrobial activity against *Micrococcus luteus*, *B. subtilis*, and *Shigella flexneri*. The obtained results of this study show the remarkable antibacterial activity of silver nanoparticles against more resistant gram-negative bacteria (Aboelfetoh et al., 2017).

In the environmental application of silver nanoparticles, Kumar et al. (2015) reported that the AgNPs synthesized by green macro algae *Ulva lactuca* effectively degraded different kinds of artificial dyes such as methyl orange, chloroquine, and azo color methyl red. The same results were mentioned by Panja et al. (2020), who found that the synthesized silver nanoparticles degraded Congo red by 85% during 8 h at 200 μg/mL, and this led to possible application of the synthesized silver nanoparticles in water purification in the presence of sunlight. Recent data published by

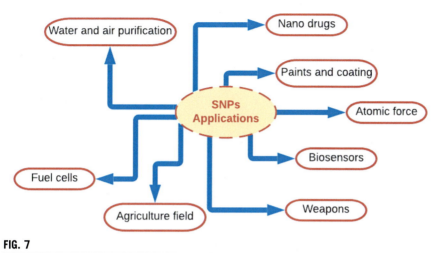

FIG. 7

Various applications of AgNPs in different fields.

AlNadhari et al. (2021) mentioned that due to the biocompatibility and excellent and physicochemical properties of silver nanoparticles synthesized using algae, these AgNPs have additionally been used for biomedical applications, which include antiviral, antioxidant, antiinflammatory, antibacterial, and anticancer activities (Fig. 7).

7 Conclusion

Various published reports employed several algal species for biosynthesis of metal nanoparticles. High yield, very low cultivation and harvesting cost, low time for production, and eco-friendliness with minimal toxic chemicals use (Fig. 8) make algal species an alternative method for the synthesis of silver NPs. Regarding the properties of produced silver nanoparticles from algae, different factors such as pH, temperature, pressure, time, and concentration decide the shape and size of the silver nanoparticles. Moreover, due to the biocompatibility and excellent and physicochemical properties of silver nanoparticles synthesized using algae, these AgNPs have additionally been used in biomedical applications, which include antiviral, antioxidant, antiinflammatory, antibacterial, and anticancer activities. Based on the results and data presented in this chapter, in the future, a remarkable boom may be witnessed in the biosynthesis of algae-based nanoparticles that will be likely to have potential in agriculture, pharmaceutics, and cosmetics.

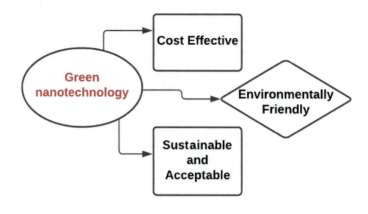

FIG. 8

Some advantages of green nanotechnology.

References

Abdel-Raouf, N., Al-Enazi, N.M., Ibraheem, N.M., 2013. Green biosynthesis of gold nanoparticles using *Galaxaura elongata* and characterization of their antibacterial activity. Arab. J. Chem. 10 (2), 23–29.

Aboelfetoh, E.F., El-Shenody, R.A., Ghobara, M.M., 2017. Eco-friendly synthesis of silver nanoparticles using green algae (*Caulerpa serrulata*): reaction optimization, catalytic and antibacterial activities. Environ. Monit. Assess. 189, 349.

Ali, A., Ali, M.A., Ali, M.U., Mohammad, S., 2011. Hospital outcomes of obstetrical-related acute renal failure in a tertiary care teaching hospital. Ren. Fail. 33, 285–290.

AlNadhari, S., Al-Enazi, N.M., Alshehrei, F., Ameen, F., 2021. A review on biogenic synthesis of metal nanoparticles using marine algae and its applications. Environ. Res. 194, 110672.

Alqadi, M.K., Abo Noqtah, O.A., ALzoubi, F.Y., ALzouby, J., ALJarrah, K., 2014. pH effect on the aggregation of silver nanoparticles synthesized by chemical reduction. Mater. Sci.-Pol. 32 (1), 107–111.

Armendariz, V., Herrera, I., Peralta-Videa, J.R., 2004. Size controlled gold nanoparticle formation by *Avena sativa* biomass: use of plants in nanobiotechnology. J. Nanopart. Res. 6 (4), 377–382.

Aziz, N., Faraz, M., Pandey, R., Shakir, M., Fatma, T., Varma, A., Barman, I., Prasad, R., 2015. Facile algae-derived route to biogenic silver nanoparticles: synthesis, antibacterial, and photocatalytic properties. Langmuir 31, 11605–11612.

Azizi, S., Ahmad, M.B., Namvar, F., Mohamad, R., 2014. Green biosynthesis and characterization of zinc oxide nanoparticles using brown marine macroalga *Sargassum muticum* aqueous extract. Mater. Lett. 116, 275–277.

Bagherzade, G., Tavakoli, M.M., Namaei, M.H., 2017. Green synthesis of silver nanoparticles using aqueous extract of saffron (*Crocus sativus* L.) wastages and its antibacterial activity against six bacteria. Asian Pac. J. Trop. Biomed. 7, 227–233.

Baker-Austin, C., Wright, M., Stepanauskas, R., McArthur, J.V., 2006. Co-selection of antibiotic and metal resistance. Trends Microbiol. 14, 176–182.

Barwal, I., Ranjan, P., Kateriya, S., Yadav, S.C., 2011. Cellular proteins of *Chlamydomonas reinhardtii* control the biosynthesis of silver nanoparticles oxido-reductive. J. Nanobiotechnol. 9, 1–12.

Bhuyar, P., Rahim, M.H.A., Sundararaju, S., Ramaraj, R., Maniam, G.P., Govindan, N., 2020. Synthesis of silver nanoparticles using marine macroalgae *Padina* sp. and its antibacterial activity towards pathogenic bacteria. Beni-Suef Univ. J. Basic Appl. Sci. 9, 3.

Borowitzka, M.A., 2010. In: Ratledge, C., Cohen, Z. (Eds.), Single Cell Oils. AOCS Publishing, Urbana, IL, USA, pp. 225–240.

Borowitzka, M.A., 2013. High-value products from microalgae—their development and commercialisation. J. Appl. Phycol. 25, 743–756.

Cavalli, P.A., Wanderlind, E.H., Hemmer, J.V., Gerlach, O.M., Emmerich, A.K., Bella-Cruz, A., Tamanaha, M., Almerindo, G.I., 2021. *Pterocladiella capillacea*-stabilized silver nanoparticles as a green approach toward antibacterial biomaterials. New J. Chem. 7.

Chaudhary, R., Nawaz, K., Komal Khan, A., Hano, C., Abbasi, B.H., Anjum, S., 2020. An overview of the algae-mediated biosynthesis of nanoparticles and their biomedical applications. Biomol. Ther. 10, 1498. https:/doi.org/10.3390/biom10111498.

Chick, C.N., Misawa-Suzuki, T., Suzuki, Y., Usuki, T., 2020. Preparation and antioxidant study of silver nanoparticles of *Microsorum pteropus* methanol extract. Bioorg. Med. Chem. Lett. 30, 127526.

Dahoumane, S.A., Djediat, C., Yepremian, C., Coute, A., Fievet, F., Coradin, T., Brayner, R., 2012. Recycling and adaptation of *Klebsormidium flaccidum* microalgae for the sustained production of gold nanoparticles. Biotechnol. Bioeng. 109, 284–288.

Darroudi, M., Ahmad, M.B., Zamiri, R., Zak, A.K., Abdullah, A.H., Ibrahim, N.A., 2011. Time-dependent effect in green synthesis of silver nanoparticles. Int. J. Nanomedicine 6 (1), 677–681.

Devi, J.S., Bhimba, B.V., 2012. Anticancer activity of silver nanoparticles synthesized by the seaweed *Ulva lactuca* invitro. Sci. Rep. 1, 242.

Devi, J.S., Bhimba, B.V., Peter, D.M., 2013. Production of biogenic silver nanoparticles using *Sargassum longifolium* and its applications. Indian J. Mar. Sci. 42, 125–130.

Doshi, H., Ray, A., Kothari, I., 2007. Bioremediation potential of live and dead Spirulina: spectroscopic, kinetics and SEM studies. Biotechnol. Bioeng. 96, 1051–1063.

El-Rafie, H., El-Rafie, M., Zahran, M., 2013. Green synthesis of silver nanoparticles using polysaccharides extracted from marine macro algae. Carbohydr. Polym. 96, 403–410.

Eroglu, E., Chen, X., Bradshaw, M., Agarwal, V., Zou, J., Stewart, S.G., Duan, X., Lamb, R.N., Smith, S.M., Raston, C.L., Iyer, K.S., 2013. Biogenic production of palladium nanocrystals using microalgae and their immobilization on chitosan nanofibers for catalytic applications. RSC Adv. 3, 1009–1012.

Fard, J.K., Jafari, S., Eghbal, M.A., 2015. A review of molecular mechanisms involved in toxicity of nanoparticles. Adv. Pharm. Bull. 5, 447.

Fatima, R., Priyaa, M., Indurthi, L., Radhakrishnanb, V., Sudhakaran, R., 2020. Biosynthesis of silver nanoparticles using red algae *Portieria hornemannii* and its antibacterial activity against fish pathogens. Microb. Pathog. 138, 103780.

Gardea-Torresdey, J.L., Tiemann, K.J., Gamez, G., Dokken, K., Pingitore, N.E., 1999. Recovery of gold (III) by alfalfa biomass and binding characterization using X-ray microfluorescence. Adv. Environ. Res. 3 (1), 83–93.

González-Ballesteros, N., González-Rodríguez, J., Rodríguez-Argüelles, M., Lastra, M., 2018. New application of two Antarctic macroalgae *Palmaria decipiens* and *Desmarestia menziesii* in the synthesis of gold and silver nanoparticles. Pol. Sci. 15, 49–54.

González-Ballesteros, N., Diego-González, L., Lastra-Valdor, M., Grimaldi, M., Cavazza, A., Bigi, F., Rodríguez-Argüelles, M.C., Simón-Vázquezb, R., 2021. *Saccorhiza polyschides* used to synthesize gold and silver nanoparticles with enhanced antiproliferative and Immunostimulant activity. Mater. Sci. Eng. C, 111960.

Govindaraju, K., Basha, S.K., Kumar, V.G., Singaravelu, G., 2008. Silver, gold and bimetallic nanoparticles production using single-cell protein (*Spirulina platensis*) Geitler. J. Mater. Sci. 43, 5115–5122.

Govindaraju, K., Kiruthiga, V., Kumar, V.G., Singaravelu, G., 2009. Extracellular synthesis of silver nanoparticles by a marine alga, *Sargassum wightii Grevilli* and their antibacterial effects. J. Nanosci. Nanotechnol. 9, 5497–5501.

Govindaraju, K., Krishnamoorthy, K., Alsagaby, S.A., Singaravelu, G., Premanathan, M., 2015. Green synthesis of silver nanoparticles for selective toxicity towards cancer cells. IET Nanobiotechnol. 9, 325–330.

Haider, M.J., Mehdi, M.S., 2014. Study of morphology and zeta potential analyzer for the silver nanoparticles. Int. J. Sci. Eng. Res. 5, 381–385.

Hamouda, R.A., El-Mongy, M.A., Eid, K.F., 2019. Comparative study between two red algae for biosynthesis silver nanoparticles capping by SDS: insights of characterization and antibacterial activity. Microb. Pathog. 129, 224–232.

Husain, S., Sardar, M., Fatma, T., 2015. Screening of cyanobacterial extracts for synthesis of silver nanoparticles. World J. Microbiol. Biotechnol. 31, 1279–1283.

Husain, S., Afreen, S., Yasin, D., Afzal, B., Fatma, T., 2019. Cyanobacteria as a bioreactor for synthesis of silver nanoparticles-an effect of different reaction conditions on the size of nanoparticles and their dye decolorization ability. J. Microbiol. Methods 162, 77–82.

Iravani, S., 2011. Green synthesis of metal nanoparticles using plants. Green Chem. 13, 2638–2650.

Iravani, S., Korbekandi, H., Mirmohammadi, S.V., Zolfaghari, B., 2014. Synthesis of silver nanoparticles: chemical, physical and biological methods. Res. Pharm. Sci. 9, 385–406.

Jeevanandam, J., Barhoum, A., Chan, Y.S., Dufresne, A., Danquah, M.K., 2018. Review on nanoparticles and nanostructured materials: history, sources, toxicity and regulations. Beilstein J. Nanotechnol. 9, 1050–1074.

Jeffryes, C., Agathos, S.N., Rorrer, G., 2015. Biogenic nanomaterials from photosynthetic microorganisms. Curr. Opin. Biotechnol. 33, 23–31.

Jena, J., Pradhan, N., Dash, B.P., Sukla, L.B., Panda, P.K., 2013. Biosynthesis and characterization of silver nanoparticles usingmicroalga *Chlorococcum humicola* and its antibacterial activity. Int. J. Nanomater. Biostruct. 3, 1–8.

Jena, J., Pradhan, N., Nayak, R.R., Dash, B.P., Sukla, L.B., Panda, P.K., Mishra, B.K., 2014. Microalga *Scenedesmus* sp.: a potential low-cost green machine for silver nanoparticle synthesis. J. Microbiol. Biotechnol. 24, 522–533.

Jiang, X.C., Chen, W.M., Chen, C.Y., Xiong, S.X., Yu, A.B., 2011. Role of temperature in the growth of silver nanoparticles through a synergetic reduction approach. Nanoscale Res. Lett. 6 (1), 32.

Kathiravan, T., Sundaramanickam, A., Shanmugam, N., Balasubramanian, T., 2015. Green synthesis of silver nanoparticles using marine algae *Caulerpa racemose* and their antibacterial activity against some human pathogens. Appl. Nanosci. 5 (4), 499–504.

Keeffe, O.E., Hughes, H., Mcloughlin, P., Sp, T., 2019. Antibacterial activity of seaweed extracts against plant pathogenic bacteria. J. Bacteriol. Mycol. 6 (3), 1105.

Khalifa, K.S., Hamouda, R.A., Hamza, H.A., 2016. In vitro antitumor activity of silver nanoparticles biosynthesized by marine algae. Dig. J. Nanomater. Biostruct. 11, 213–221.

Khalil, M.M., Ismail, E.H., El-Baghdady, K.Z., Mohamed, D., 2014. Green synthesis of silver nanoparticles using olive leaf extract and its antibacterial activity. Arab. J. Chem. 7, 1131–1139.

Khanna, P., Kaur, A., Goyal, D., 2019. Algae-based metallic nanoparticles: synthesis, characterization and applications. J. Microbiol. Methods 163, 105656.

Kharissova, O.V., Dias, H.V.R., Kharisov, B.I., Perez, B.O., Perez, V.M.J., 2013. The greener synthesis of nanoparticles. Trends Biotechnol. 31, 240–248.

Krishnan, M., Sivanandham, V., Hans-Uwe, D., Murugaiah, S.G., Seeni, P., Gopalan, S., Rathinam, A.J., 2015. Antifouling assessments on biogenic nanoparticles: a field study from polluted offshore platform. Mar. Pollut. Bull. 101, 816–825.

Kumar, P., Senthamilselvi, S., Govindaraju, M., 2013. Seaweed-mediated biosynthesis of silver nanoparticles using *Gracilaria corticate* for its antifungal activity against *Candida* spp. Appl. Nanosci. 3, 495–500.

Kumar, P.V., Shameem, U., Kollu, P., Kalyani, R.L., Pammi, S.V.N., 2015. Green synthesis of copper oxide nanoparticles using *Aloe vera* leaf extract and its antibacterial activity against fish bacterial pathogens. BioNanoScience 5, 135–139.

Lengke, M.F., Fleet, M.E., Southam, G., 2007. Biosynthesis of silver nanoparticles by filamentous cyanobacteria from a silver (I) nitrate complex. Langmuir 23, 2694–2699.

Li, X., 2011. Green Energy for Sustainability and Energy Security. Springer, New York, NY, USA, pp. 1–16.

Li, X., Schirmer, K., Bernard, L., Sigg, L., Pillai, S., Behra, R., 2015. Silver nanoparticle toxicity and association with the alga *Euglena gracilis*. Environ. Sci. Nano 2, 594–602.

Liu, Z., Gao, T., Yang, Y., Meng, F., Zhan, F., Jiang, Q., Sun, X., 2019. Anti-cancer activity of porphyran and carrageenan from red seaweeds. Molecules 24 (4286), 1–14.

Liu, H., Zhang, H., Wang, J., Wei, J., 2020. Effect of temperature on the size of biosynthesized silver nanoparticle: deep insight into microscopic kinetics analysis. Arab. J. Chem. 13 (1), 1011–1019.

Logeswari, P., Silambarasan, S., Abraham, J., 2013. Ecofriendly synthesis of silver nanoparticles from commercially available plant powders and their antibacterial properties. Sci. Iran. 20, 1049–1054.

Madhiyazhagan, P., Murugan, K., Kumar, A.N., Nataraj, T., Dinesh, D., Panneerselvam, C., Subramaniam, J., Kumar, P.M., Suresh, U., Roni, M., 2015. *Sargassum muticum*-synthesized silver nanoparticles: an effective control tool against mosquito vectors and bacterial pathogens. Parasitol. Res. 114, 4305–4317.

Meera, R.L., Baskar, P.V., Somasundaram, S.T., 2013. Evaluating antioxidant property of brown alga *Copromania sinuosa* (Derb. Et sol). Afr. J. Food Sci. 2 (11), 126–130.

Mittal, A.K., Chisti, Y., Banerjee, U.C., 2013. Synthesis of metallic nanoparticles using plant extracts. Biotechnol. Adv. 31, 346–356.

Mohandass, C., Vijayaraj, A.S., Rajasabapathy, R., Satheeshbabu, S., Rao, S.V., Shiva, C., De-Mello, L., 2013. Biosynthesis of silver nanoparticles from marine seaweed *Sargassum cinereum* and their antibacterial activity. Indian J. Pharm. Sci. 75, 606–610.

Mohseniazar, M., Barin, M., Zarredar, H., Alizadeh, S., Shanehbandi, D., 2011. Potential of microalgae and Lactobacilli in biosynthesis of silver nanoparticles. Bioimpacts 1, 149.

Murugan, K., Samidoss, C.M., Panneerselvam, C., Higuchi, A., Roni, M., Suresh, U., Chandramohan, B., Subramaniam, J., Madhiyazhagan, P., Dinesh, D., Rajaganesh, R., Alarfar, A.A., Nioletti, M., Kumar, S., Wei, H., Canale, A., Mehlhorn, H., Benelli, G., 2015. Sea weed synthesized silver nanoparticles: an eco-friendly tool in the fight against *Plasmodium falciparum* and its vector *Anopheles stephensi*? Parasitol. Res. 114, 4087–4097.

Pangi, Z., Beletsi, A., Evangelatos, K., 2003. PEG-ylated nanoparticles for biological and pharmaceutical application. Adv. Drug Deliv. Rev. 24, 403–419.

Panja, S., Choudhuri, I., Khanra, K., Pati, B., Bhattacharyya, N., 2020. Biological and photocatalytic activity of silver nanoparticle synthesized from *Ehretia laevis* Roxb. leaves extract. Nano Biomed. Eng. 12 (1), 104–113.

Park, Y., Hong, Y.N., Weyers, A., Kim, Y.S., Linhardt, R.J., 2011. Polysaccharides and phytochemicals: a natural reservoir for the green synthesis of gold and silver nanoparticles. IET Nanobiotechnol. 5 (3), 69–78.

Patra, J.K., Baek, K., 2014. Green nanobiotechnology: factors affecting synthesis and characterization techniques. J. Nanomater., 417305. 12 pages.

Pillai, Z.S., Kamat, P.V., 2004. What factors control the size and shape of silver nanoparticles in the citrate ion reduction method? J. Phys. Chem. B 108 (3), 945–951.

Ponnuchamy, K., Jacob, J.A., 2016. Metal nanoparticles from marine seaweeds—a review. Nanotechnol. Rev. 5 (6), 589–600.

Prasad, T., Elumalai, E., 2013. Marine algae mediated synthesis of silver nanopaticles using *Scaberia agardhü* Greville. J. Biol. Sci. 13, 566.

Prasad, N.V.K.V., Subba Rao Kambala, V., Naidu, R.A., 2011. Critical review on biogenic silver nanoparticles and their antimicrobial activity. Curr. Nanosci. 7, 531–544.

Prasad, T.N., Kambala, V.S.R., Naidu, R., 2013. Phyconanotechnology: synthesis of silver nanoparticles using brown marine algae *Cystophora moniliformis* and their characterisation. J. Appl. Phycol. 25, 177–182.

Pugazhendhi, A., Prabakar, D., Jacob, J.M., Karuppusamy, I., Saratale, R.G., 2018. Synthesis and characterization of silver nanoparticles using *Gelidium amansii* and its antimicrobial property against various pathogenic bacteria. Microb. Pathog. 114, 41–45.

Rahman, A., Kumar, S., Nawaz, T., 2020. Biosynthesis of nanomaterials using algae. In: Microalgae Cultivation for Biofuels Production. Academic Press, pp. 265–279.

Rai, A., Singh, A., Ahmad, A., Sastry, M., 2006. Role of halide ions and temperature on the-morphology of biologically synthesized gold nanotriangles. Langmuir 22 (2), 736–741.

Rajeshkumar, S., Kannan, C., Annadurai, G., 2012. Synthesis and characterization of antimicrobial silver nanoparticles using marine brown seaweed *Padina tetrastromatica*. Drug Invent. Today 4, 511–513.

Rajeshkumar, S., Malarkodi, C., Gnanajobitha, G., Paulkumar, K., Vanaja, M., Kannan, C., Annadurai, G., 2013. Seaweed-mediated synthesis of gold nanoparticles using *Turbinaria conoides* and its characterization. J. Nanostruct. Chem. 3, 44.

Ramanathan, R., O'Mullane, A.P., Parikh, R.Y., Smooker, P.M., Bhargava, S.K., Bansal, V., 2011. Bacterial kinetics-controlled shapedirected biosynthesis of silver nanoplates using *Morganella psychrotolerans*. Langmuir 27, 714–719.

Raveendran, P., Fu, J., Wallen, S.L., 2003. Completely "green" synthesis and stabilization of metal nanoparticles. J. Am. Chem. Soc. 125, 13940–13941.

Sajidha Parveen, K., Lakshmi, D., 2016. Biosynthesis of silver nanoparticles using red algae, *Amphiroa fragilissima* and its antibacterial potential against grampositive and gramnegative bacteria. Int. J. Curr. Sci. 19, 93–100.

Salvioni, L., Galbiati, E., Collico, V., Alessio, G., Avvakumova, S., Corsi, F., Tortora, P., Prosperi, D., Colombo, M., 2017. Negatively charged silver nanoparticles with potent antibacterial activity and reduced toxicity for pharmaceutical preparations. Int. J. Nanomedicine 12, 2517–2530.

Selvaraj, P., Neethu, E., Rathika, P., Jayaseeli, J.P.R., Jermy, B.R., AbdulAzeez, S., Borgio, J.F., Dhas, T.S., 2020. Antibacterial potentials of methanolic extract and silver nanoparticles from marine algae. Biocatal. Agric. Biotechnol. 28, 101719.

Sharma, D., Kanchi, S., Bisetty, K., 2015. Biogenice synthesis of nanoparticles: a review. Arab. J. Chem. https://doi.org/10.1016/j.arabjc.2015.11.002.

Sharma, A., Sharma, S., Sharma, K., Chetri, S.P.K., Vashishtha, A., Singh, P., Kumar, R., Rathi, B., Agrawal, V., 2016. Algae as crucial organisms in advancing nanotechnology: a systematic review. J. Appl. Phycol. 28, 1759–1774.

Shiny, P., Mukherjee, A., Chandrasekaran, N., 2013. Marine algae mediated synthesis of the silver nanoparticles and its antibacterial efficiency. Int. J. Pharm. Pharm. Sci. 5, 239–241.

Shukla, M.K., Singh, R.P., Reddy, C.R.K., Jha, B., 2012. Synthesis and characterization of agar-based silver nanoparticles and nanocomposite film with antibacterial applications. Bioresour. Technol. 107, 295–300.

Singh, J., Dutta, T., Kim, K.H., Rawat, M., Samddar, P., Kumar, P., 2018. "Green" synthesis of metals and their oxide nanoparticles: applications for environmental remediation. J. Nanobiotechnol. 16, 84.

Sinha, S.N., Paul, D., Halder, N., Sengupta, D., Patra, S.K., 2015. Green synthesis of silver nanoparticles using fresh water green alga *Pithophora oedogonia* (Mont.) Wittrock and evaluation of their antibacterial activity. Appl. Nanosci. 5, 703–709.

Soni, N., Prakash, S., 2011. Factors affecting the geometry of silver nanoparticles synthesis in *Chrysosporium tropicum* and *Fusarium oxusporum*. Am. J. Nanotechnol. 2 (1), 112–121.

Sudha, S.S., Rajamanickam, K., Rengaramanujam, J., 2013. Microalgae mediated synthesis of silver nanoparticles and their antibacterial activity against pathogenic bacteria. Indian J. Exp. Biol. 51, 393–399.

Suriya, J., Bharathi Raja, S., Sekar, V., Rajasekaran, R., 2012. Biosynthesis of silver nanoparticles and its antibacterial activity using seaweed *Urospora* sp. Afr. J. Biotechnol. 11, 12192–12198.

Thangaraju, N., Venkatalakshmi, R.P., Chinnasamy, A., Kannaiyan, P., 2012. Synthesis of silver nanoparticles and the antibacterial and anticancer activities of the crude extract of *Sargassum polycystum C. Agardh*. Nano Biomed. Eng. 4, 89–94.

Thiruchelvi, R., Jayashree, P., Mirunaalini, K., 2021. Synthesis of silver nanoparticle using marine red seaweed *Gelidiella acerosa*—a complete study on its biological activity and its characterisation. Mater. Today Proc. 37, 1693–1698.

Tran, Q.H., Nguyen, V.Q., Le, A.T., 2013. Silver nanoparticles: synthesis, properties, toxicology, applications and perspectives. Adv. Nat. Sci.: Nanosci. Nanotechnol. 4, 033001.

Vadlapudi, V., Kaladhar, D.S.V.G.K., 2014. Review: green synthesis of silver and gold nanoparticles. Middle-East J. Sci. Res. 19 (6), 834–842.

Vigneshwaran, N., Ashtaputre, N., Varadarajan, P., Nachane, R., Paralikar, K., Balasubramanya, R., 2007. Biological synthesis of silver nanoparticles using the fungus *Aspergillus flavus*. Mater. Lett. 61, 1413–1418.

Vivek, M., Kumar, P.S., Steffi, S., Sudha, S., 2011. Biogenic silver nanoparticles by *Gelidiella acerosa* extract and their antifungal effects. Avicenna J. Med. Biotechnol. 3, 143–148.

Xie, J., Lee, J.Y., Wang, D., Ting, Y.P., 2007. Identification of active biomolecules in the high-yield synthesis of single-crystalline gold nanoplates in algal solutions. Small 3, 672–682.

Yugay, Y.A., Usoltseva, R.V., Silant'ev, V.E., Egorova, A.E., Karabtsov, A.A., Kumeiko, V.V., Ermakova, S.P., Bulgakov, V.P., Shkryl, Y.N., 2020. Synthesis of bioactive silver nanoparticles using alginate, fucoidan and laminaran from brown algae as a reducing and stabilizing agent. Carbohydr. Polym. 245, 116547.

CHAPTER 21

Green synthesis of silver nanoparticles using actinomycetes

Zdenka Bedlovičová

Department of Chemistry, Biochemistry and Biophysics, University of Veterinary Medicine and Pharmacy, Košice, Slovakia

1 Introduction

Silver nanoparticles (AgNPs) represent a promising development in the fight against bacterial infections. Due to the small size of particles, different shapes, and high surface area, silver nanoparticles offer unique biological, chemical, and physical properties, providing various applications in spectroscopy, electronics, catalysis, sensors, and the pharmaceutical field (Abbasi et al., 2014).

Silver and its compounds is well known for its ability to inhibit various bacterial strains and microorganisms, so it is commonly used in the medical industry, e.g., in soaps, pastes, plastics, catheters, medical cosmetics with the addition of silver for protection of open wounds and localized inflammation after burns, or dressings containing silver (García-Barrasa et al., 2011; Kumar et al., 2018; Paladini and Pollini, 2019; Sim et al., 2018).

Synthesis of silver nanoparticles (AgNPs) is usually provided by chemical, physical, and biological methods. The biological methods of Ag nanoparticles synthesis are becoming more popular due to them being environmentally friendly without producing toxic chemicals that are harmful for living organisms. The "green" process of Ag nanoparticles preparation means using biological methods for reduction of Ag^+ by various natural sources of reducing agents including not only microorganisms, plants/plant extracts, fungi, or yeasts, but also biomolecules (amino acids, polysaccharides, vitamins) (Fernández et al., 2016; Gudikandula et al., 2017; Natesan et al., 2020; Öztürk et al., 2020; Rajeshkumar and Bharath, 2017; Samuel et al., 2020; Siddiqi et al., 2018).

As was mentioned before, microorganisms are also being used as an eco-friendly method of Ag nanoparticles production (Plaza et al., 2014). Some microbes are able to accumulate metals in ionic form from the environment, followed by changing these ionic forms into elemental metals controlled by the biochemical processes and the genetic nature of microorganisms used for the biosynthesis. Another advantage of using bacteria for nanoparticles synthesis is easy manipulation, making it a good choice for preparation of shape- and size-specific metallic nanoparticles (Abdeen et al., 2014; Zhang et al., 2011). This chapter deals with an overview of current knowledge, centered on the biological synthesis of silver nanoparticles using actinomycetes and their biological activity.

2 Silver nanoparticles green synthesis

Synthesis of silver nanoparticles is usually achieved by chemical, physical, and biological means. The chemical approach represents an easy way to produce Ag nanoparticles in solution using reducing agents of organic or inorganic origin, including H_2, sodium borohydride, and N,N-dimethylformamide. These mentioned examples of reducing agents lead to reduction of silver ion (Ag^+) to elemental silver (Ag^0) followed by agglomeration (Evanoff et al., 2004). When using chemical method of silver ions reduction, it is necessary to use stabilizing agents to avoid agglomeration (polymeric compounds, e.g., polyvinylpyrrolidone [PVP], polyvinyl alcohol [PVA], or polyethylene glycol [PEG]) (Alqadi et al., 2014; Chen and Zhang, 2012; Oliveira et al., 2005; Sun and Xia, 2002). On the other hand, physical approaches use energy consuming methods, such as thermal decomposition (Lee and Kang, 2004), spray pyrolysis (Pluym et al., 1993), ultrasound (Manjamadha and Muthukumar, 2016), or mechanochemical methods (Hernández et al., 2020; Szczęśniak et al., 2020; Tsuzuki and McCormick, 2004).

In order to prevent the formation of toxic, hazardous chemicals, or high energy exhaustion, increased attention has been paid to green methods of synthesis in recent years. Green synthesis is an environmentally friendly method presenting a different way of thinking in chemistry intended to eliminate toxic waste, reduce energy consumption, and to use ecological solvents (water, ethanol, ethyl acetate, etc.) and chemicals (de Marco et al., 2018). These principles are valuable for AgNPs preparation by biological methods. Using a green approach in synthesis, the silver ions reduction is realized by naturally occurring sources such as plants, plant extracts, bacteria, fungi, and yeasts (Baláž et al., 2017; Sastry et al., 2003; Sowani et al., 2016a, b). Biological methods, with a couple of examples, are briefly summarized in Table 1.

In general, synthesis of silver nanoparticles via biological systems is based on reduction of silver ions into elemental silver (Fig. 1), continued by growth and stabilization of fabricated nanoparticles (Marchiol et al., 2014).

Preparation of nanoparticles using biological methods offers an ecological and nontoxic method for synthesis of AgNPs with various sizes, shapes, compositions, and properties (Mohanpuria et al., 2008). The advantage of AgNPs biosynthesis is the ability of biological system to act as reducing as capping agent. The reduction of silver ions is normally carried out in an environmentally friendly aqueous medium. According to source of reducing agent, Ag^+ ions are reduced in different ways. For example, the process of biomolecules' (saccharides, amino acids, biosurfactants, etc.) isolation from naturally occurred material requires more steps of

Table 1 Biological methods of Ag nanoparticles synthesis.

Source of reducing agents	Example	References
Algae	*Chlorella pyrenoidosa*	Aziz et al. (2015)
	Chaetoceros calcitrans	Merin et al. (2010)
	Chlorella salina	
	Isochrysis galbana	
	Tetraselmis gracilis	
	Ulva lactuca L.	González-Ballesteros et al. (2019)
Biomolecules	Exopolysaccharide from *Streptomyces violaceus*	Sivasankar et al. (2018)
	Pectin	Pallavicini et al. (2017)
	Chitosan	Shahid-ul-Islam et al. (2019)
	Biosurfactants	Plaza et al. (2014)
Essential oil	*Coleus aromaticus*	Vilas et al. (2016)
	Orange peel	Veisi et al. (2019)
	Clove	Hongfang et al. (2017)
Microorganisms	*Trichoderma atroviride*	Saravanakumar and Wang (2018)
	Lactobacillus strains	Nair and Pradeep (2002)
	Actinomycetes	Abdeen et al. (2014)
Mushroom	*Volvariella volvacea*	Philip (2009)
	Pleurotus ostreatus	Al-Bahrani et al. (2017)
	Ganoderma neo-japonicum	Gurunathan et al. (2013)
Plant	Alfalfa	Gardea-Torresdey et al. (2003)
	Brassica juncea	Marchiol et al. (2014)
	Festuca rubra	
	Medicago sativa	
Plant extract	*Ougeinia oojeinensis*	Kumar et al. (2019)
	Bergenia ciliata	Phull et al. (2016)
	Prosopis farcta	Salari et al. (2019)
	Pongamia pinnata	Priya et al. (2016)
	Tropaeolum majus	Valsalam et al. (2019)
Yeasts	*Cornitermes cumulans*	Eugenio et al. (2016)
	Saccharomyces cerevisiae	Niknejad et al. (2015)
	Cryptococcus laurentii	Fernández et al. (2016)
	Rhodotorula glutinis	

FIG. 1

Schematic representation of silver nanoparticles biosynthesis.

extraction and purification, and is time consuming. If the simple biomolecules, also serving as stabilizing agent, such as α-D-glucose (Raveendran et al., 2003), dextrin (Konował et al., 2017), cellulose (Liu et al., 2011), glutathione (Murariu et al., 2014), biosurfactants (Plaza et al., 2014), or oleic acid (Le et al., 2010) are used, the process of green synthesis is simplified and more economically advantageous.

Another way to prepare AgNPs by biological approach is using an extract of various organisms (plants, fungi, yeasts, microorganisms, algae). The most used and popular method is Ag nanoparticles synthesis by plant extracts containing compounds able to reduce silver ion, but also able to serve as capping agent, such as aldehydes, terpenoids, flavonoids, or carboxylic acids (Prabhu and Poulose, 2012). There is a need to notice that different parts of plants can be used for extract preparation—root, leaves, seed, fruit, peel, flower, bark, and stem. The extracts are typically completed by filtration of milled parts of the plant suspended and macerated in solvents of polar nature (water, ethanol, and methanol). Extraction is carried out under different conditions (time, pH, concentration, temperature) (Ahmed et al., 2016). The use of microorganism extract is considered a simple method of Ag nanoparticles synthesis. Extracts obtained from microorganisms (bacteria, actinomycetes, fungi, yeasts) may act as reducing and capping agent. The process of Ag^+ ions reduction is performed by molecules occurring in the extract, including proteins, enzymes, polysaccharides, and amino acids. Microbes are usually extracted by washing of the biomass followed by dissolving the cells in buffer solution or water (Binupriya et al., 2010), another method is using a medium in which the microorganism is grown (Nanda and Saravanan, 2009).

Another valuable technique is using living organisms for nanoparticles preparation. It is known that both plants and microorganisms are able to absorb and accumulate inorganic metal ions from the environment in which they occur. These properties make biological systems effective "bio-factories" able to reduce environmental pollution. The very first article dealing with AgNPs synthesis using living plant alfalfa (*Medicago salvia*) was published by Gardea-Torresdey. The root of alfalfa absorbs Ag^0 from the medium and then bring it through all the plant tissues to produce silver nanoparticles (Gardea-Torresdey et al., 2003). Marchiol et al. (2014) also observed successful in vivo synthesis of AgNPs by the plants *Brassica juncea*, *Medicago sativa*, and *Festuca rubra*. The authors demonstrated the presence of various AgNPs in all the plant tissues (Marchiol et al., 2014).

Microorganisms including bacteria, fungi, actinomycetes, and yeast are widely used for biogenic synthesis of nanoparticles due to their ability to use biochemical processes to transform metal ions into metal nanoparticles. These processes may be provided intracellularly or extracellularly by both unicellular and multicellular microorganisms. The specific mechanisms of nanoparticle growth have been studied, but not all biological entities are able to synthesize nanoparticles due to their enzymatic activities and essential metabolic processes (Zhang et al., 2011). This fact may lead to careful selection of a biological system with the potential to accumulate metal ions producing nanoparticles. Optimization of the microbes culturing conditions (pH, temperature, nutrients, size of inoculum, buffer, time, and light) is very important and the presence of the substrate from the beginning of the growth would increase the enzymatic activity (Durán et al., 2005; Shah et al., 2015; Siavash, 2011; Singh et al., 2011; Sowani et al., 2016b). In addition, the control of the size and monodispersity of NPs is important. Each species of microbe forms a different size and shape of nanoparticle. Actinomycetes and yeast produce better controlled monodisperse nanoparticles, yeast has also shown better control of size (Ahmad et al., 2003; Sastry et al., 2003). The studies of silver ion concentration suggest that high concentration inhibits the process of formation of AgNPs (Kathiresan et al., 2009). Selection of an optimal microorganism and concentration of silver nitrate is key for successful biosynthesis of AgNPs, due to excess Ag^+ ions being toxic to microbes and the rate of the reaction being decreased.

3 Silver nanoparticles synthesized by actinomycetes

Actinomycetes are aerobic, gram-positive bacteria possessing properties of bacteria and fungi, the name coming from the original Greek words "aktis" (ray) and "mykes" (fungus). In general, actinomycetes are considered to be bacteria, due to the chemical composition of cell walls being similar to gram-positive bacteria, although they have well developed morphology (hyphae) and cultural characteristics, meaning they are a separate group of bacteria (Bhatti et al., 2017; Das et al., 2008). Actinomycetes had been originally assumed as largely of dry alkaline soil origin, but they have also been isolated from aquatic ecosystems including sediments from deep seas (Colquhoun et al., 1998).

These bacteria are known for their importance as producers of biologically active substances, such as antibiotics, vitamins, and enzymes; they are an important source of diverse antimicrobial metabolites, anticancer, and insecticidal compounds (Gupta et al., 2019).

In addition to the properties already mentioned, actinomycetes are good candidates for producing small, polydisperse, and stable metal nanoparticles by both intracellular and extracellular methods. Intracellular biosynthesis of silver nanoparticles takes place on the mycelia surface as a consequence of electrostatic binding of Ag^+ ions to the carboxylate anions of enzymes present on the mycelia cell wall, resulting in neutral Ag^0 and then forming silver nuclei followed by accumulation and fabrication of silver nanoparticles (Abdeen et al., 2014; Sunitha et al., 2013). Extracellular AgNPs preparation depends on the presence of the nitrogen cycle providing enzymes involved in electron transfer during the reduction of metal ions (Abdeen et al., 2014; Karthik et al., 2014) (Fig. 2). The stabilization of biosynthesized nanoparticles is provided by proteins secreted by the microorganism (Abdeen et al., 2014; Shah et al., 2015).

The AgNPs prepared by actinomycetes are briefly presented in Table 2, including precursors of Ag^+ ions, morphology, and size of prepared silver nanoparticles. The presented data are obtained from the last decade, between the years 2010 and 2020.

Extracellular biosynthesis of silver nanoparticles using *Streptomyces laurentii* was carried out by using cell-free extract of biomass and 1 mM solution of silver nitrate and incubation for 24 h. The first evidence of successful synthesis was a color change and presence of surface plasmon resonance (SPR) peak at the wavelength of 420 nm. The size (7–15 nm) and morphology (spherical) of prepared nanoparticles were measured by TEM analysis (Eid et al., 2020). Tropical soil actinomycete *Gordonia amicalis* HS-11 mediated synthesis of nanoparticles was successfully proceeded using colorless 1 mM silver nitrate solution as precursor, which is usually used (Sowani et al., 2016b). The cell-free supernatant was used in an extracellular manner. Prepared nanoparticles were polydispersed. Small spherical Ag nanoparticles (10–20 nm) were prepared using *Streptomyces* and *Rhodococcus* species using 1 mM silver nitrate solution added to cell filtrate (Abdeen et al., 2014). *Streptomyces griseoplanus* SAI-25 was also used for extracellular biosynthesis of silver nanoparticles using 1 mM $AgNO_3$ as a source of Ag^+ ions. Color change of the reaction mixture indicated the formation of nanoparticles, as well as standard characterization methods being used (Vijayabharathi et al., 2018).

The extracellular method for AgNPs preparation was also used by Alsharif et al. (2020). The authors prepared silver nanoparticles from different microbial taxa including actinomycetes strain *Streptomyces noursei* H1-1. The filtrate of strain isolate biomass was incubated with 1 mM silver nitrate solution for 24 h in the dark,

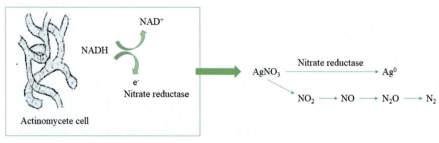

FIG. 2

Probable mechanism of AgNPs extracellular biosynthesis.

Modified from Karthik, L., Kumar, G., Kirthi, A.V., Rahuman, A.A., Rao, K.V.B., 2014. Streptomyces sp. LK3 mediated synthesis of silver nanoparticles and its biomedical application. Bioprocess Biosyst. Eng. 37, 261–267.

Table 2 Silver nanoparticles prepared by actinomycetes.

Actinomycete spp.	Precursor of silver ions	Size of nanoparticles	Characterization technique[a]	Morphology	Extracellular/ intracellular	References
Streptomyces spp.	1 mM AgNO$_3$	42.5 nm	UV-vis SEM FT-IR	Spherical	E	Singh et al. (2021)
Streptomyces spp.	1 mM AgNO$_3$	8.66–35 nm	UV-vis DLS ZP TEM	Spherical, polydispersed, respectively	E	Hamed et al. (2020)
Streptomyces spp.	1 mM AgNO$_3$	16.4 ± 2.2 nm	UV-vis SEM TEM XRD ZP FT-IR	Spherical	E	Marathe et al. (2020)
Nocardiopsis dassonvillei-DS013	AgNO$_3$ solution of unknown concentration	30–80 nm	UV-vis SEM XRD FT-IR	Circular	E	Dhanaraj et al. (2020)
Streptomyces laurentii	1 mM AgNO$_3$	7–15 nm	UV-vis TEM FT-IR XRD	Spherical	E	Eid et al. (2020)
Streptomyces sp. RAB 10	1 mM AgNO$_3$	5–40 nm	UV-vis FT-IR TEM DLS XRD EDX	Spherical	E	Praveena et al. (2020)

Continued

Table 2 Silver nanoparticles prepared by actinomycetes—cont'd

Actinomycete spp.	Precursor of silver ions	Size of nanoparticles	Characterization technique[a]	Morphology	Extracellular/ intracellular	References
Streptomyces sp.	1 mM AgNO$_3$	51.2 nm	UV-vis FT-IR HR-TEM EDX	Spherical	E	Natesan et al. (2020)
Kitasatospora albolonga	AgNO$_3$ solution of unknown concentration	10–50 nm	UV-vis DLS XRD SEM HR-TEM	Spherical	E	D'Lima et al. (2020)
Streptomyces noursei H1-1	1 mM AgNO$_3$	6–40 nm	UV-vis XRD TEM FT-IR EDX DLS	Spherical	E	Alsharif et al. (2020)
Streptomyces spp. Al-Dhabi-89	10 mM AgNO$_3$	9.7–17.25 nm	UV-vis XRD SEM TEM FT-IR EDX	Crystalline	E	Al-Dhabi et al. (2019)
Nocardiopsis alba	0.1 M AgNO$_3$	20–60 nm	UV-vis FT-IR TEM XRD	Polydispersed Spherical	E	Avilala and Golla (2019)
Nocardiopsis sp. GRG1	1 mM AgNO$_3$	20–50 nm	UV-vis FT-IR XRD TEM	Spherical	E	Rajivgandhi et al. (2019)

Streptomyces sp.	1 mM AgNO$_3$	75–85 nm	UV-vis SEM EDX	Polydispersed	E	Sheik et al. (2019)
Streptomyces intermedius	2 mM AgNO$_3$	55 nm	UV-vis FT-IR XRD SEM HR-TEM DLS ZP	Spherical	E	Dayma et al. (2019)
Streptomyces spp. Al-Dhabi-89	1; 2, 3, 4, and 5 mM AgNO$_3$	11–21 nm	UV-vis XRD SEM TEM FT-IR EDX	Cubic	E	Al-Dhabi et al. (2018)
Streptacidiphilus durhamensis HGG16n	3 mM AgNO$_3$	8–48 nm	FT-IR DLS ZP TEM UV-vis EA ICP-MS	Spherical	E	Buszewski et al. (2018)
Streptomyces griseoplanus SAI-25	1 mM AgNO$_3$	19.5–20.5 nm	UV-vis FT-IR XRD HR-TEM DLS	Spherical	E	Vijayabharathi et al. (2018)
Streptomyces rochei MHM13	1 mM AgNO$_3$	22–85 nm	UV-vis EDX FT-IR SEM	Spherical	E	Abd-Elnaby et al. (2016)
Streptomyces graminofaciens Streptomyces catenulae	1% AgNO$_3$	10–20 nm	UV-vis FT-IR TEM	Spherical	E	Kamel et al. (2016)

Continued

Table 2 Silver nanoparticles prepared by actinomycetes—cont'd

Actinomycete spp.	Precursor of silver ions	Size of nanoparticles	Characterization technique[a]	Morphology	Extracellular/ intracellular	References
Gordonia amicalis HS-11	1 mM AgNO$_3$	5–25 nm	UV-vis XRD TEM FT-IR	Polydispersed	E	Sowani et al. (2016b)
Rhodococcus spp. NCIM 2981	1 mM AgNO$_3$	5–50 nm	UV-vis XRD TEM FT-IR EDS	Spherical	I	Otari et al. (2015)
Streptomyces spp. *Rhodococcus* spp.	1 mM AgNO$_3$	10–20 nm	UV-vis FT-IR SEM	Spherical	E	Abdeen et al. (2014)
Streptomyces sp. VITPK1	1 mM AgNO$_3$	20–45 nm	UV-vis XRD FT-IR EDX AFM	Spherical	E	Sanjenbam et al. (2014)
Streptomyces sp. VITSTK7	1 mM AgNO$_3$	20–60 nm	UV-vis XRD AFR TEM	Spherical	E	Thenmozhi et al. (2013)
Streptomyces spp. ERI-3	1 mM AgNO$_3$	10–100 nm	UV-vis XRD TEM SEM	Spherical	E	Zonooz and Salouti (2011)
Streptomyces glaucus 71MD	1 mM AgNO$_3$	4–25 nm	UV-vis XRD TEM SEM EDX	Spherical	E	Tsibakhashvili et al. (2011)

[a]*AFM*, atomic force microscopy; *DLS*, dynamic light scattering; *EA*, elemental analysis; *EDX*, energy-dispersive X-ray spectroscopy; *FT-IR*, Fourier transform infrared spectroscopy; *HR-TEM*, high resolution transmission electron microscopy; *ICP-MS*, inductively coupled plasma mass spectroscopy; *SEM*, scanning electron microscopy; *TEM*, transmission electron microscopy; *UV-vis*, ultraviolet and visible spectroscopy; *XRD*, X-ray diffraction; *ZP*, zeta potential.

resulting in color change from colorless to brown. Spherical nanoparticles with size of 6–40 nm were obtained and characterized. New strains of actinomycetes *Streptomyces* sp. Al-Dhabi-89 and Al-Dhabi-91 obtained from Saudi Arabia were successfully used for extracellular AgNPs biosynthesis (Al-Dhabi et al., 2018, 2019). The reducing ability of cell surface extract led to formation of cubic nanoparticles of size 11–21 nm for Al-Dhabi-89 strain, and for Al-Dhabi-91 strain, the authors obtained nanocrystals. *Streptomyces intermedius* was used for reduction of Ag^+ ions to prepare silver nanoparticles by the biogenic method. Filtrate of actinomycete solution was added to 2 mM $AgNO_3$, and the resulting AgNPs were spherical with average size of 55 nm (Dayma et al., 2019). Isolate DW102 of *Streptomyces* sp. was also used for extracellular biological synthesis of silver nanoparticles by Sheik et al. Using 1 mM silver nitrate as precursor they obtained polydispersed nanostructures of 75–85 nm (Sheik et al., 2019).

Intracellular biosynthesis of Ag nanoparticles was performed using actinomycete *Rhodococcus* NCIM 2981 strain. After cultivation in growing medium, the culture was used as the inoculum for the synthesis using 1 mM silver nitrate. Color change from white to dark brown was observed as the indication of forming Ag nanoparticles. Using TEM (transmission electron microscopy) the authors confirmed spherical AgNPs of 5–50 nm (Otari et al., 2015).

Generally, it can be stated that extracellular biosynthesis of Ag nanoparticles is more popular due to shorter reaction time and better manipulation of the samples. The precursor usually used is 1 mM silver nitrate ($AgNO_3$) in various ratios of culture supernatant and Ag^+ ions. The reaction process is mostly monitored visually by change of color (from white/yellowish to brown) and spectrophotometrically using UV-vis (ultraviolet and visible) spectroscopy based on the presence of SPR (surface plasmon resonance) band. For detailed analysis of the shape, size, and other characteristics of synthesized nanoparticles, various techniques are used, most commonly TEM (transmission electron microscopy), FT-IR (Fourier transform infrared) spectroscopy, and XRD (X-ray diffraction). The methods used for characterization of AgNPs prepared by actinomycetes are listed in Table 2.

4 Biological activity of AgNPs prepared by actinomycetes

Silver nanoparticles are widely studied due to a variety of biological activities, including antimicrobial, anticancer, antifungal, or antioxidant (Abdeen et al., 2014; Khorrami et al., 2018; Otari et al., 2015; Sowani et al., 2016b; Vijayabharathi et al., 2018). Silver is well known for its antimicrobial activity and serves as a good choice for nanoparticles preparation (Jiang et al., 2004). Actinomycetes are capable of producing secondary metabolites with antimicrobial activity (streptomycin, gentamicin) as nanoparticles (Franco-Correa et al., 2010). Silver nanoparticles synthesized using actinomycetes and their biological properties are summarized in Table 3.

Table 3 Biological activity of silver nanoparticles prepared by actinomycetes.

Actinomycete spp. used for AgNPs synthesis	Biological activity	Method of evaluation	References
Gordonia amicalis HS-11	Antioxidant	Nitric oxide Hydroxyl radical	Sowani et al. (2016b)
Streptomyces sp. NS-5	Antibacterial	Agar well diffusion method	Singh et al. (2021)
Streptomyces sp. RAB 10	Antioxidant Antibacterial	DPPH[a] assay Agar well diffusion method	(Praveena et al., 2020)
Streptomyces sp. Rhodococcus sp.	Antibacterial	Disk diffusion method	Abdeen et al. (2014)
Nocardiopsis dassonvillei-DS013	Antibacterial	Agar well diffusion method	(Dhanaraj et al., 2020)
Streptomyces sp.	Antibacterial	Agar well diffusion method	Sheik et al. (2019)
Streptomyces intermedius	Antibacterial	Resazurin assay	Dayma et al. (2019)
Nocardiopsis sp. GRG1	Antibacterial Antibiofilm	Agar well diffusion method Microtiter plates	Rajivgandhi et al. (2019)
Kitasatospora albolonga	Antibacterial	Disk diffusion method	(D'Lima et al., 2020)
Nocardiopsis alba	Antibacterial Antiviral	Agar well diffusion method Embryonated chicken eggs	Avilala and Golla (2019)
Streptomyces griseoplanus SAI-25	Antifungal	Agar well diffusion method	Vijayabharathi et al. (2018)
Streptomyces sp.	Antifungal	In vitro poisoned food technique	Natesan et al. (2020)
Rhodococcus spp. NCIM 2891	Antibacterial	Agar well diffusion method	Otari et al. (2015)
Streptomyces laurentii	Antibacterial Cytotoxicity	Agar well diffusion method MTT assay	Eid et al. (2020)
Streptomyces noursei H1-1	Antibacterial Insecticidal	Agar well diffusion method	Alsharif et al. (2020)
Streptomyces spp. Al-Dhabi-89	Antibacterial	Disk diffusion method	Al-Dhabi et al. (2018)
Streptacidiphilus durhamensis HGG16n	Antibacterial	Agar well diffusion method	Buszewski et al. (2018)

Table 3 Biological activity of silver nanoparticles prepared by actinomycetes—cont'd

Actinomycete spp. used for AgNPs synthesis	Biological activity	Method of evaluation	References
Streptomyces rochei MHM13	Antibacterial	Agar well diffusion method	Abd-Elnaby et al. (2016)
	Anticancer	MTT assay	
	Antifouling		
Streptomyces graminofaciens Streptomyces catenulae	Antibacterial	Agar well diffusion method	Kamel et al. (2016)
Streptomyces sp.	Antifungal	Agar well diffusion method	(Marathe et al., 2020)
Streptomyces sp. VITSTK7	Antifungal	Poison plate technique	Thenmozhi et al. (2013)
Streptomyces antimycoticus	Antibacterial	Agar well diffusion method	Salem et al. (2020)
Streptomyces sp. VITPK1	Anticandidal	Kirby-Bauer	Sanjenbam et al. (2014)
Streptomyces sp. MHM38	Antifungal Antibacterial	Agar well diffusion method	Bukhari et al. (2021)

[a]2,2-Diphenyl-1-picrylhydrazyl radical.

As we can see, the most studied biological effect of silver nanoparticles prepared by actinomycetes is antibacterial activity. For antimicrobial activity evaluation, different methods are used. Common techniques of antimicrobial activity determination include diffusion methods (agar disk diffusion method, E-test, and agar well diffusion method) and dilution techniques (agar dilution method, time-kill test). The antibacterial properties of AgNPs are usually studied by dilution methods for obtaining the MIC values, expressed in $mg\,mL^{-1}$, representing the lowest concentration of silver nanoparticles that avoids increase of the microorganism population. On the other hand, the MBC method is based on the evaluation of the lowest studied antimicrobial compound concentration inhibiting 99.9% of microorganisms (Balouiri et al., 2016).

Antibacterial activity of silver nanoparticles prepared by *Rhodococcus* spp. NCIM 2891 has been studied by agar well diffusion method. Different concentrations of AgNPs (10, 30, and $50\,\mu g\,mL^{-1}$) were used against gram-positive bacteria *Staphylococcus aureus* and gram-negative bacteria *Klebsiella pneumoniae*, *Proteus vulgaris*, *Enterococcus faecalis*, *Pseudomonas aeruginosa*, and *Escherichia coli*. The bactericidal and bacteriostatic effect was studied. The authors demonstrated that the lowest concentrations ($10\,\mu g\,mL^{-1}$) of Ag nanoparticles were the most effective for *Klebsiella pneumoniae*, where the microorganism showed growth after 48 h of incubation. The concentration of $30\,\mu g\,mL^{-1}$ of silver nanoparticles after incubation of 50 h showed some growth of *E. coli*, *S. aureus*, and *P. aeruginosa*. The AgNPs concentration of $50\,\mu g\,mL^{-1}$ effectively avoided the growth of all the studied microorganisms (Otari et al., 2015).

Antibacterial activity of biogenic synthesizes Ag nanoparticles using *Streptomyces* sp. and *Rhodococcus* sp. was investigated by disk diffusion method. Human pathogens, both gram-negative (*E. coli*, *P. aeruginosa*, *K. pneumoniae*, and *P. vulgaris*) and gram-positive (*S. aureus*) were used. The highest antibacterial effect was determined against *P. aeruginosa* followed by *S. aureus* and *K. pneumoniae*. The lowest antibacterial activity was observed against *P. vulgaris* and *E. coli* (Abdeen et al., 2014).

Resazirin assay for antibacterial activity evaluation was used by Dayma and his colleagues. This assay is based on reduction of resazurin dye to indicate the cell viability to monitor living cells. Different concentrations of silver nanoparticles prepared using *Streptomyces intermedius* were tested against *E. coli* and *B. subtilis* pathogens. The authors showed that AgNPs at all concentrations inhibit the growth of both studied microorganisms. The viability of the cells decreased with increasing concentration level. The values of IC_{50} were 10 ppm for *B. subtilis* and 20 ppm for *E. coli* (Dayma et al., 2019).

The MIC of silver nanoparticles prepared using marine *Streptomyces* spp. Al-Dhabi-89 were studied by disk diffusion method. Against gram-positive *S. epidermis* and *B. subtilis* bacteria the authors observed MIC values of $250\,\mu g\,mL^{-1}$ and $125\,\mu g\,mL^{-1}$, respectively. *S. aureus* showed a $62.5\,\mu g\,mL^{-1}$ value of MIC. Gram-negative pathogens showed the following values of minimal inhibitory concentrations: *P. aeruginosa* ($15.6\,\mu g\,mL^{-1}$); *E. coli* ($31.25\,\mu g\,mL^{-1}$); *K. pneumoniae* ($62.5\,\mu g\,mL^{-1}$) (Al-Dhabi et al., 2018).

Streptacidiphilus durhamensis HGG16n synthesized silver nanoparticles were tested against *P. aeruginosa*, *S. aureus*, *P. mirabilis*, *E. coli*, *K. pneumoniae*, and *B. subtilis*. In addition, the synergistic effect of Ag nanoparticles with commercial antibiotics (streptomycin, gentamicin, kanamycin, ampicillin, tetracycline, and neomycin) was studied. The best MIC was observed for *S. aureus* at a concentration of $6.25\,mg\,mL^{-1}$. Synergistic effect was observed against *K. pneumoniae*, *S. aureus*, and *P. aeruginosa* in the presence of kanamycin, ampicillin, and tetracycline (Buszewski et al., 2018).

Extracellularly biosynthesized AgNPs using *Streptomyces laurentii* were effective against gram-positive (*B. subtilis*, *S. aureus*) and gram-negative (*E. coli*, *P. aeruginosa*, *S. typhimurium*) bacteria in concentration-dependent manner. Nanoparticles of all studied concentrations were effective (6.25, 12.5, 25, 50, 75, and 100 ppm). The best inhibition of pathogens was obtained using silver nanoparticles at the highest concentration (100 ppm) against all studied microorganisms (Eid et al., 2020). The authors also provided cytotoxicity tests of prepared AgNPs using MTT assay. They found out that bio-fabricated silver nanoparticles were selectively cytotoxic against cancer cell lines (CACO-2) at low concentrations (IC_{50} value was 4.66 ± 0.21 ppm) (Eid et al., 2020).

Antibacterial, antifouling, and anticancer activity of AgNPs prepared by *Streptomyces rochei* MHM13 was studied. The authors declared very good antimicrobial effect against diverse pathogens (*Vibrio fluvialis*, *P. aeruginosa*, *Salmonella typhimurium*, *Vibrio damsela*, *E. coli*, *B. subtilis*, *S. aureus*, *B. cereus*), where zones of inhibition were 16–19 mm and 13–18 mm, respectively. Anticancer activity was determined using MTT assay. Eight cell lines were tested (HepG2—hepatocellular, MCF-7—breast, HCT-116—colon, PC-3—prostate, A-549—lung, CACO—intestinal, HEP-2—larynx, and HELA—cervical carcinoma cells). According to measured data, the cytotoxic effect of AgNPs is dose-dependent. Silver nanoparticles exhibited anticancer activity after a 24 h exposure and the most sensitive cell lines were Hep-G2, HCT-116, A-549, MCF-7, and PC-3. The CACO, HEP-2, and HELA were the most resistant. Antifouling activity was also tested. Ag nanoparticles inhibited formation of biofilm, so they can be applicable as an antibiofilm agent (Abd-Elnaby et al., 2016).

Kamel et al. (2016) studied antimicrobial activity of two *Streptomyces* strains. They observed that all the tested bacteria were inhibited by silver nanoparticles. Inhibition zones were 11–19 mm.

Antibiofilm study of silver nanoparticles prepared using *Nocardiopsis* sp. GRG1 was introduced by Rajivgandhi et al. (2019). The tests were conducted after the detection phase of biofilm formation (MR-CoNS strain). The experiments showed that silver nanoparticles are able to inhibit formation of the biofilm at the concentration value of $55\,\mu g\,mL^{-1}$ (Rajivgandhi et al., 2019).

Antibacterial and insecticidal effects were tested with *Streptomyces noursei* H1-1 synthesized AgNPs. AgNPs demonstrated potential antibacterial activity, the most sensitive was strain *Pseudomonas aeruginosa*. Insecticidal assay were investigated on the third larval instar of *Aedes aegypti*. Obtained results showed that no significant difference between concentration of applied silver nanoparticles and larval mortality has been achieved (74.7%–100% for concentrations of 20–100 ppm) (Alsharif et al., 2020).

Antiviral activities of silver nanoparticles were also studied. The study was carried out in 9-day-old embryonated chicken eggs by inoculation with the mixture of New Castile virus NDV and silver nanoparticles prepared by *Nocardiaopisis alba* using hemagglutination test (HA). The best results were observed for 0.3 and 0.5 mL of nanosuspension amounts when 2/3 of NDV were inhibited (Avilala and Golla, 2019).

Antioxidant properties of different samples of biological and/or chemical origin, or compounds obtained from biological materials including AgNPs have been widely studied. Free radicals are generated in living biological systems as a result of oxidative stress and lead to cell death (Apak et al., 2016). *Gordonia amicalis* HS-11 biosynthesized silver nanoparticles showed antioxidant activity against hydroxyl radicals (88.5%) and also inhibited nitric oxide radicals (61.5%) (Sowani et al., 2016b). In vitro antioxidant properties were also studied. The silver NPs prepared using *Streptomyces parvus* strain Al-Dhabi-91 underwent the DPPH and nitric oxide inhibition assay. As a standard, the ascorbic acid was used. Antioxidant activity increased in concentration dependent manner, but all the tested AgNPs showed lower scavenging activity than standard (Al-Dhabi et al., 2019).

Finally, the last, but not least, activity of silver nanoparticles produced by actinomycetes, is antifungal and anticandidal activities. *M. phaseolina* was inhibited using biosynthesized silver nanoparticles by *Streptomyces* strain. The inhibition zone was 13 mm at a concentration of 1 mg mL^{-1} (Vijayabharathi et al., 2018). Antifungal activity against *A. niger*, *A. fumigatus*, and *A. flavus* was determined by silver nanoparticles prepared by *Streptomyces* sp. VITSTK7. The authors concluded that silver nanoparticles showed significant antifungal activity against all mentioned fungi at a concentration of 50 mg mL^{-1}: *A. fumigatus* (75.25 ± 1.61), *A. niger* (67.22 ± 1.32), and *A. flavus* (62.30 ± 2.47) (Thenmozhi et al., 2013). The fungi *Poria hypolateritia* and *Phomopsis theae* affecting tea (*Camellia sinensis*) health were used to study the antifungal activity of AgNPs prepared by various *Streptomyces* sp. strains. The results showed good activity against mentioned fungi: 69.9% of *P. hypolateritia* and 67.47% of *P. theae* was inhibited after 15 days of incubation with 5 ppm of Ag nanoparticles (Natesan et al., 2020).

Silver nanoparticles obtained from *Streptomyces* sp. VITPK1 showed prominent effects against *Candida* strains. The inhibition zones in millimeters of AgNPs (50 mg mL^{-1}) were: *C. tropicalis* (18 ± 0.23), *C. krusei* (16 ± 1.07), and *C. albicans* (20 ± 0.067). The results were compared with cell-free supernatant and the zones of inhibition of AgNPs were higher than those of cell-free supernatant of isolate (Sanjenbam et al., 2014).

5 Conclusion and future perspectives

A brief overview of silver nanoparticles using microorganisms—actinomycetes—has been achieved. The mechanisms of producing silver nanoparticles have been intensively studied, but for successful biogenic synthesis, the assumption of presence of negatively charged functional groups (such as carboxylate ions) on the cell surface is necessary. The electrostatic interactions lead to the attraction of silver ion on the cell wall. The silver ions then undergo reduction via enzymes produced by microbes into their elemental form. This is the definitive role in nanoparticles biosynthesis. There have been an increasing amount of articles dealing with the development of microbe-producing NPs and their biological purposes during recent years. However, a lot of study is necessary to optimize the regulation of synthetic approaches for better controlling of size, morphology, and stability of prepared nanoparticles. It is a fact that preparation of silver NPs by microbes is a very lengthy process (from a few hours to several days) in comparison to physical or chemical methods. But, on the other hand, the better results of size and morphology control are noticed in microbes-mediated biosynthesis. This is a valuable advantage for preparation of AgNPs by microorganisms. Variations of parameters and conditions, such as pH, synthetic conditions, substrate concentration, growth medium, and temperature may provide a good control of size and monodispersity of particles. The biological methods of nanoparticles production are still at the development stage. The main problem is to improve the stability of prepared nanoparticles, so there is a need of further study.

Both methods, intracellular and extracellular, have been studied. The extracellular approach to nanoparticles synthesis is more used due to shorter synthesis time and easier preparation, compared to intracellular approach that often requires additional operations, such as further chemical reactions or ultrasound application to obtain NPs from cells. When the mechanisms of biosynthesis are well-understood, there will be great progress on the applications.

References

Abbasi, E., Milani, M., Fekri Aval, S., Kouhi, M., Akbarzadeh, A., Tayefi Nasrabadi, H., Nikasa, P., Joo, S.W., Hanifehpour, Y., Nejati-Koshki, K., Samiei, M., 2014. Silver nanoparticles: synthesis methods, bio-applications and properties. Crit. Rev. Microbiol. 42, 1–8.

Abdeen, S., Geo, S., Sukanya, S., Praseetha, P.K., Dhanya, R.P., 2014. Biosynthesis of silver nanoparticles from *Actinomycetes* for therapeutic applications. Int. J. Nano Dimens. 5, 155–162.

Abd-Elnaby, H.M., Abo-Elala, G.M., Abdel-Raouf, U.M., Hamed, M.M., 2016. Antibacterial and anticancer activity of extracellular synthesized silver nanoparticles from marine *Streptomyces rochei* MHM13. Egypt. J. Aquat. Res. 42, 301–312.

Ahmad, A., Senapati, S., Khan, M.I., Kumar, R., Ramani, R., Srinivas, V., Sastry, M., 2003. Intracellular synthesis of gold nanoparticles by a novel alkalotolerant actinomycete, *Rhodococcus* species. Nanotechnology 14, 824–828.

Ahmed, S., Ahmad, M., Swami, B.L., Ikram, S., 2016. A review on plants extract mediated synthesis of silver nanoparticles for antimicrobial applications: a green expertise. J. Adv. Res. 7, 17–28.

Al-Bahrani, R., Raman, J., Lakshmanan, H., Hassan, A.A., et al., 2017. Green synthesis of silver nanoparticles using tree oyster mushroom *Pleurotus ostreatus* and its inhibitory activity against pathogenic bacteria. Mater. Lett. 186, 21–25. https://doi.org/10.1016/j.matlet.2016.09.069.

Al-Dhabi, N.A., Ghilan, A.M., Arasu, M.V., 2018. Green biosynthesis of silver nanoparticles produced from marine *Streptomyces* sp. Al-Dhabi-89 and their potential applications against wound infection and drug resistant clinical pathogens. J. Photochem. Photobiol. B Biol. 189, 176–184.

Al-Dhabi, N.A., Ghilan, A.-K.M., Esmail, G.A., Arasu, M.V., Duraipandiyan, V., Ponmurugan, K., 2019. Environmental friendly synthesis of silver nanomaterials from the promising *Streptomyces parvus* strain Al-Dhabi-91 recovered from the Saudi Arabian marine regions for antimicrobial and antioxidant properties. J. Photochem. Photobiol. B Biol. 197, 111529.

Alqadi, M.K., Noqtah, O.A.A., Alzoubi, F.Y., Alzouby, J., Aljarrah, K., 2014. pH effect on the aggregation of silver nanoparticles synthesized by chemical reduction. Mater. Sci.-Pol. 32, 107–111.

Alsharif, S.M., Salem, S.S., Abdel-rahman, M.A., Fouda, A., Mohamed, A., Hassan, S.E., Awad, M.A., Mohamed, A.A., 2020. Multifunctional properties of spherical silver nanoparticles fabricated by different microbial taxa. Heliyon 6, e03943.

Apak, R., Özyürek, M., Güçlü, K., Çapanolu, E., 2016. Antioxidant activity/capacity measurement. 1. Classification, physicochemical principles, mechanisms, and electron transfer (ET)-based assays. J. Agric. Food Chem. 64, 997–1027.

Avilala, J., Golla, N., 2019. Antibacterial and antiviral properties of silver nanoparticles synthesized by marine Actinomycetes. Int. J. Pharm. Sci. Res. 10, 1223–1228.

Aziz, N., Faraz, M., Pandey, R., Shakir, M., et al., 2015. Facile algae-derived route to biogenic silver nanoparticles: synthesis, antibacterial, and photocatalytic properties. Langmuir 31 (42), 11605–11612. https://doi.org/10.1021/acs.langmuir.5b03081.

Baláž, M., Balážová, Ľ., Daneu, N., Dutková, E., Balážová, M., Bujňáková, Z., Shpotyuk, Y., 2017. Plant-mediated synthesis of silver nanoparticles and their stabilization by wet stirred media milling. Nanoscale Res. Lett. 12, 83.

Balouiri, M., Sadiki, M., Ibnsouda, S.K., 2016. Methods for in vitro evaluating antimicrobial activity: a review. J. Pharm. Anal. 6, 71–79.

Bhatti, A.A., Haq, S., Bhat, R.A., 2017. Actinomycetes benefaction role in soil and plant health. Microb. Pathog. 111, 458–467.

Binupriya, A.R., Sathishkumar, M., Yun, S.I., 2010. Myco-crystallization of silver ions to nanosized particles by live and dead cell filtrates of *Aspergillus oryzae* var. *viridis* and its bactericidal activity toward *Staphylococcus aureus* KCCM 12256. Ind. Eng. Chem. Res. 49, 852–858.

Bukhari, S.I., Hamed, M.M., Al-Agamy, M.H., Gazwi, H.S., Radwan, H.H., Youssif, A.M., 2021. Biosynthesis of copper oxide nanoparticles using *Streptomyces* MHM38 and its biological applications. J. Nanomater. 2021.

Buszewski, B., Railean-Plugaru, V., Pomastowski, P., Rafińska, K., Szultka-Mlynska, M., Golinska, P., Wypij, M., Laskowski, D., Dahm, H., 2018. Antimicrobial activity of biosilver nanoparticles produced by a novel *Streptacidiphilus durhamensis* strain. J. Microbiol. Immunol. Infect., 5145–5154.

Chen, S.-F., Zhang, H., 2012. Aggregation kinetics of nanosilver in different water conditions. Adv. Nat. Sci. Nanosci. Nanotechnol. 3, 035006.

Colquhoun, J.A., Mexson, J., Goodfellow, M., Ward, A.C., Horikoshi, K., Bull, A.T., 1998. Novel rhodococci and other mycolate actinomycetes from the deep sea. Antoine van Leeuwenhoek 74, 27–40.

Das, S., Lyla, P.S., Khan, S.A., 2008. Distribution and generic composition of culturable marine actinomycetes from the sediments of Indian continental slope of Bay of Bengal. Chinese J. Oceanol. Limnol. 26, 166–177.

Dayma, P.B., Mangrola, A.V., Suriyaraj, S.P., Dudhagara, P., Patel, R.K., 2019. Synthesis of bio-silver nanoparticles using desert *Streptomyces intermedius* and its antimicrobial activity isolated. J. Pharm. Chem. Biol. Sci. 7, 94–101.

de Marco, B.A., Rechelo, B.S., Tótoli, E.G., Kogawa, A.C., Salgado, H.R.N., 2018. Evolution of green chemistry and its multidimensional impacts: a review. Saudi Pharm. J. 27, 1–8.

Dhanaraj, S., Thirunavukkarasu, S., John, H.A., Pandian, S., et al., 2020. Novel marine *Nocardiopsis dassonvillei*-DS013 mediated silver nanoparticles characterization and its bactericidal potential against clinical isolates. Saudi J. Biol. Sci. 27 (3), 991–995. https://doi.org/10.1016/j.sjbs.2020.01.003.

Durán, N., Marcato, P.D., Alves, O.L., De, G.I.H., Esposito, E., 2005. Mechanistic aspects of biosynthesis of silver nanoparticles by several *Fusarium oxysporum* strains. J. Nanobiotechnol. 3, 1–7.

D'Lima, L., Phadke, M., Ashok, V.D., 2020. Biogenic silver and silver oxide hybrid nanoparticles: a potential antimicrobial against multi drug-resistant *Pseudomonas aeruginosa*. New J. Chem. 44, 4935–4941. https://doi.org/10.1039/C9NJ04216D.

Eid, A.M., Fouda, A., Niedbała, G., Hassan, S.E.-D., Salem, S.S., Abdo, A.M., Hetta, H.F., Shaheen, T.I., 2020. Endophytic mediated *Streptomyces laurentii* green synthesis of Ag-NPs with antibacterial and anticancer properties for developing functional. Antibiotics 9, 1–18.

Eugenio, M., Müller, N., Frasés, S., Almeida-Paes, R., et al., 2016. Yeast-derived biosynthesis of silver/silver chloride nanoparticles and their antiproliferative activity against bacteria. RSC Adv. 6 (12), 9893–9904. https://doi.org/10.1039/C5RA22727E.

Evanoff, D., et al., 2004. Size-controlled synthesis of nanoparticles. 2. Measurement of extinction, scattering, and absorption cross sections. J. Phys. Chem. B 108, 13957–13962.

Fernández, J.G., Fernández-baldo, M.A., Berni, E., Camí, G., Durán, N., Raba, J., Sanz, M.I., 2016. Production of silver nanoparticles using yeasts and evaluation of their antifungal activity against phytopathogenic fungi. Process Biochem. 51, 1306–1313.

Franco-Correa, M., Quintana, A., Duque, C., Suarez, C., Rodríguez, M.X., Barea, J., 2010. Evaluation of actinomycete strains for key traits related with plant growth promotion and mycorrhiza helping activities. Appl. Soil Ecol. 45, 209–217.

García-Barrasa, J., López-de-Luzuriaga, J.M., Monge, M., 2011. Silver nanoparticles: synthesis through chemical methods in solution and biomedical applications. Cent. Eur. J. Chem. 9, 7–19.

Gardea-Torresdey, J.L., Gomez, E., Peralta-Videa, J.R., Parsons, J.G., Troiani, H., Jose-Yacaman, M., 2003. Alfalfa sprouts: a natural source for the synthesis of silver nanoparticles. Langmuir 19, 1357–1361.

González-Ballesteros, N., Rodríguez-Argüelles, M., Prado-López, S., et al., 2019. Macro algae to nanoparticles: study of *Ulva lactuca* L. role in biosynthesis of gold and silver nanoparticles and of their cytotoxicity on colon cancer cell lines. Mater. Sci. Eng. C 97, 498–509. https://doi.org/10.1016/j.msec.2018.12.066.

Gudikandula, K., Vadapally, P., Charya, M.A.S., 2017. Biogenic synthesis of silver nanoparticles from white rot fungi: their characterization and antibacterial studies. OpenNano 2, 64–78.

Gupta, A., Singh, D., Singh, S.K., Singh, V.K., Singh, A.V., Kumar, A., 2019. Role of actinomycetes in bioactive and nanoparticle synthesis. In: Role of Plant Growth Promoting Microorganisms in Sustainable Agriculture and Nanotechnology. Elsevier Inc. (Chapter 10).

Gurunathan, S., Raman, J., Malek, S., John, P., Vikineswary, S., 2013. Green synthesis of silver nanoparticles using *Ganoderma neojaponicum* Imazeki: a potential cytotoxic agent against breast cancer cells. Int. J. Nanomed. 8, 4399–4413. https://doi.org/10.2147/IJN.S51881.

Hamed, A.A., Kabary, H., Khedr, M., Emam, A.N., 2020. Antibiofilm, antimicrobial and cytotoxic activity of extracellular green-synthesized silver nanoparticles by two marine-derived actinomycete. RSC Adv. 10, 10361–10367.

Hernández, J.G., Halasz, I., Crawford, D.E., Krupička, M., Baláž, M., André, V., Vella-Zarb, L., Niidu, A., García, F., Maini, L., Colacino, E., 2020. European research in focus: mechanochemistry for sustainable industry (COST action MechSustInd). Eur. J. Org. Chem., 8–9.

Hongfang, G., Hui, Y., Chuang, W., 2017. Controllable preparation and mechanism of nanosilver mediated by the microemulsion system of the clove oil. Results Phys. 7, 3130–3136. https://doi.org/10.1016/j.rinp.2017.08.032.

Jiang, H., Manolache, S., Wong, A.C.L., Denes, F.S., 2004. Plasma-enhanced deposition of silver nanoparticles onto polymer and metal surfaces for the generation of antimicrobial characteristics. J. Appl. Polym. Sci. 93, 1411–1422.

Kamel, Z., Saleh, M., El Namoury, N., 2016. Biosynthesis, characterization, and antimicrobial activity of silver nanoparticles from actinomycetes. Res. J. Pharm. Biol. Chem. Sci. 7, 119–127.

Karthik, L., Kumar, G., Kirthi, A.V., Rahuman, A.A., Rao, K.V.B., 2014. *Streptomyces* sp. LK3 mediated synthesis of silver nanoparticles and its biomedical application. Bioprocess Biosyst. Eng. 37, 261–267.

Kathiresan, K., Manivannan, S., Nabeel, M.A., Dhivya, B., 2009. Biointerfaces studies on silver nanoparticles synthesized by a marine fungus, *Penicillium fellutanum* isolated from coastal mangrove sediment. Colloids Surf. B. Biointerfaces 71, 133–137.

Khorrami, S., Zarrabi, A., Khaleghi, M., Danaei, M., Mozfari, M., 2018. Selective cytotoxicity of green synthesized silver nanoparticles against the MCF-7 tumor cell line and their enhanced antioxidant and antimicrobial properties. Int. J. Nanomedicine 13, 8013–8024.

Konował, E., Sybis, M., Modrzejewska-Sikorska, A., Milczarek, G., 2017. Synthesis of dextrin-stabilized colloidal silver nanoparticles and their application as modifiers of cement mortar. Int. J. Biol. Macromol. 104, 165–172.

Kumar, D., Arona, S., Abdullah, Danish, M., 2019. Plant based synthesis of silver nanoparticles from *Ougeinia oojeinensis* leaves extract and their membrane stabilizing, antioxidant and antimicrobial activities. Mater. Proc. 17 (1), 313–320. https://doi.org/10.1016/j.matpr.2019.06.435.

Kumar, S.S.D., Rajendran, N.K., Houreld, N.N., Abrahamse, H., 2018. Recent advances on silver nanoparticle and biopolymer-based biomaterials for wound healing applications. Int. J. Biol. Macromol. 115, 165–175.

Le, A.T., Tam, L.T., Tam, P.D., Huy, P.T., Huy, T.Q., Van Hieu, N., Kudrinskiy, A.A., Krutyakov, Y.A., 2010. Synthesis of oleic acid-stabilized silver nanoparticles and analysis of their antibacterial activity. Mater. Sci. Eng. C 30, 910–916.

Lee, D.K., Kang, Y.S., 2004. Synthesis of silver nanocrystallites by a new thermal decomposition method and their characterization. ETRI J. 26, 252–256.

Liu, H., Wang, D., Song, Z., Shang, S., 2011. Preparation of silver nanoparticles on cellulose nanocrystals and the application in electrochemical detection of DNA hybridization. Cellulose 18, 67–74.

Manjamadha, V.P., Muthukumar, K., 2016. Ultrasound assisted green synthesis of silver nanoparticles using weed plant. Bioprocess Biosyst. Eng. 39, 401–411.

Marathe, K., Naik, J., Maheshwari, V., 2020. Biogenic synthesis of silver nanoparticles using *Streptomyces* spp. and their antifungal activity against *Fusarium verticillioides*. J. Clust. Sci. https://doi.org/10.1007/s10876-020-01894-5.

Marchiol, L., Mattiello, A., Pošćić, F., Giordano, C., Musetti, R., 2014. In vivo synthesis of nanomaterials in plants: location of silver nanoparticles and plant metabolism. Nanoscale Res. Lett. 9, 101.

Merin, D., Prakash, S., Bhimba, B., 2010. Antibacterial screening of silver nanoparticles synthesized by marine micro algae. Asian Pac. J. Trop. Med., 797–799.

Mohanpuria, P., Rana, N.K., Yadav, S.K., 2008. Biosynthesis of nanoparticles: technological concepts and future applications. J. Nanopart. Res. 10, 507–517.

Murariu, M., Stoica, I., Gradinaru, R., Drochioiu, G., Mangalagiu, I., 2014. Glutathione-based silver nanoparticles with dual biomedical activity. Rev. Roum. Chim. 59, 867–874.

Nair, B., Pradeep, T., 2002. Coalescence of nanoclusters and formation of submicron crystallites assisted by *Lactobacillus* strains. Cryst. Growth Des. 2 (4), 293–298. https://doi.org/10.1021/cg0255164.

Nanda, A., Saravanan, M., 2009. Biosynthesis of silver nanoparticles from *Staphylococcus aureus* and its antimicrobial activity against MRSA and MRSE. Nanomed. Nanotechnol. Biol. Med. 5, 452–456.

Natesan, K., Ponmurugan, P., Gnanamangai, B., Suganya, M., Kavitha, S., 2020. Biosynthesis of silver nanoparticles from *Streptomyces* spp., characterization and evaluating of its efficacy against *Phomopsis theae* and *Poria hypolateria* in tea. Nano Biomed. Eng. 12, 272–280.

Niknejad, F., Nabili, M., Daie Ghazvini, R., Moazeni, M., 2015. Green synthesis of silver nanoparticles: advantages of the yeast *Saccharomyces cerevisiae* model. Curr. Med. Mycol. 1 (3), 17–24. https://doi.org/10.18869/acadpub.cmm.1.3.17.

Oliveira, M.M., Ugarte, D., Zanchet, D., Zarbin, A.J.G., 2005. Influence of synthetic parameters on the size, structure, and stability of dodecanethiol-stabilized silver nanoparticles. J. Colloid Interface Sci. 292, 429–435.

Otari, S.V., Patil, R.M., Ghosh, S.J., Thorat, N.D., Pawar, S.H., 2015. Intracellular synthesis of silver nanoparticle by actinobacteria and its antimicrobial activity. Spectrochim. Acta A Mol. Biomol. Spectrosc. 136, 1175–1180.

Öztürk, F., Ço, S., Duman, F., 2020. Biosynthesis of silver nanoparticles using leaf extract of *Aesculus hippocastanum* (horse chestnut): evaluation of their antibacterial, antioxidant and drug release system activities. Mater. Sci. Eng. C 107, 1–11.

Paladini, F., Pollini, M., 2019. Antimicrobial silver nanoparticles for wound healing application: progress and future trends. Materials (Basel) 12, 1–16.

Pallavicini, P., Arciola, C.R., Bertoglio, F., Curtosi, S., et al., 2017. Silver nanoparticles synthesized and coated with pectin: an ideal compromise for anti-bacterial and anti-biofilm action combined with wound-healing properties. J. Colloid Interface Sci. 498, 271–281. https://doi.org/10.1016/j.jcis.2017.03.062.

Philip, D., 2009. Biosynthesis of Au, Ag and Au–Ag nanoparticles using edible mushroom extract. Spectrochim. Acta A Mol. Biomol. Spectrosc. 73 (2), 374–381. https://doi.org/10.1016/j.saa.2009.02.037.

Phull, A.-R., Abbas, Q., Ali, A., Raza, H., et al., 2016. Antioxidant, cytotoxic and antimicrobial activities of green synthesized silver nanoparticles from crude extract of *Bergenia ciliata*. Future J. Pharm. Sci. 2 (1), 31–36. https://doi.org/10.1016/j.fjps.2016.03.001.

Plaza, G.A., Chojniak, J., Banat, I.M., 2014. Biosurfactant mediated biosynthesis of selected metallic nanoparticles. Int. J. Mol. Sci. 15, 13720–13737.

Pluym, T.C., Powell, Q.H., Gurav, A.S., Ward, T.L., Kodas, T.T., Wang, L.M., Glicksman, H.D., 1993. Solid silver particle production by spray pyrolysis. J. Aerosol Sci. 24, 383–392.

Prabhu, S., Poulose, E.K., 2012. Silver nanoparticles: mechanism of antimicrobial action, synthesis, medical applications, and toxicity effects. Int. Nano Lett. 2, 32.

Praveena, G., Yagnam, S., Banoth, L., Trivedi, R., Prakasham, R.S., 2020. Bacterial biosynthesis of nanosilver: a green catalyst for the synthesis of (amino pyrazolo)-(phenyl)methyl naphth-2-ol derivatives and their antimicrobial potential. New J. Chem. 44, 13046–13061. https://doi.org/10.1039/D0NJ01924K.

Priya, R.S., Geetha, D., Ramesh, P.S., 2016. Antioxidant activity of chemically synthesized AgNPs and biosynthesized *Pongamia pinnata* leaf extract mediated AgNPs—a comparative study. Ecotoxicol. Environ. Saf. 134 (2), 308–318. https://doi.org/10.1016/j.ecoenv.2015.07.037.

Rajeshkumar, S., Bharath, L.V., 2017. Mechanism of plant-mediated synthesis of silver nanoparticles—a review on biomolecules involved, characterisation and antibacterial activity. Chem. Biol. Interact. 273, 219–227. https://doi.org/10.1016/j.cbi.2017.06.019.

Rajivgandhi, G., Maruthupandy, M., Muneeswaran, T., Anand, M., Quero, F., Manoharan, N., Li, W., 2019. Biosynthesized silver nanoparticles for inhibition of antibacterial resistance and biofilm formation of methicillin-resistant coagulase negative *Staphylococci*. Bioorg. Chem. 89, 103008.

Raveendran, P., Fu, J., Wallen, S.L., 2003. Completely "green" synthesis and stabilization of metal nanoparticles. J. Am. Chem. Soc. 125, 13940–13941.

Salari, S., Bahabadi, S.E., Samzahed-Kermani, A., Yosefzaei, F., 2019. In-vitro evaluation of antioxidant and antibacterial potential of green synthesized silver nanoparticles using *Prosopis farcta* fruit extract. Iran. J. Pharm. Res. 18 (1), 430–455.

Salem, S.S., El-Belely, E.F., Niedbała, G., Alnoman, M.M., Hassan, S.E.D., Eid, A.M., Shaheen, T.I., Elkelish, A., Fouda, A., 2020. Bactericidal and in-vitro cytotoxic efficacy of silver nanoparticles (Ag-NPs) fabricated by endophytic actinomycetes and their use as coating for the textile fabrics. Nanomaterials 10 (10), 2082.

Samuel, M.S., Jose, S., Selvarajan, E., Mathimani, T., Pugazhendhi, A., 2020. Biosynthesized silver nanoparticles using *Bacillus amyloliquefaciens*; application for cytotoxicity effect on A549 cell line and photocatalytic degradation of p-nitrophenol. J. Photochem. Photobiol. B Biol. 202, 111642.

Sanjenbam, P., Gopal, J.V., Kannabiran, K., 2014. Anticandidal activity of silver nanoparticles synthesized using *Streptomyces* sp. VITPK1. J. Mycol. Med. 24, 211–219.

Saravanakumar, K., Wang, M.-H., 2018. Trichoderma based synthesis of anti-pathogenic silver nanoparticles and their characterization, antioxidant and cytotoxicity properties. Microb. Pathog. 114, 269–273. https://doi.org/10.1016/j.micpath.2017.12.005.

Sastry, M., Ahmad, A., Khan, M.I., Kumar, R., 2003. Biosynthesis of metal nanoparticles using fungi and actinomycete. Curr. Sci. 85, 162–170.

Shah, M., Fawcett, D., Sharma, S., Tripathy, S.K., 2015. Green synthesis of metallic nanoparticles via biological entities. Materials (Basel) 8 (11), 7278–7308. https://doi.org/10.3390/ma8115377.

Shahid-ul-Islam, Butola, B.S., Verma, D., 2019. Facile synthesis of chitosan-silver nanoparticles onto linen for antibacterial activity and free-radical scavenging textiles. Int. J. Biol. Macromol. 133, 1134–1141. https://doi.org/10.1016/j.ijbiomac.2019.04.186.

Sheik, G.B., Ismail, A., Abdel, A., Alzeyadi, Z.A., Ibrahim, M., 2019. Extracellular synthesis, characterization and antibacterial activity of silver nanoparticles by a potent isolate *Streptomyces* sp. DW102. Asian J. Biol. Life Sci. 8, 89–96.

Siavash, I., 2011. Green synthesis of metal nanoparticles using plants. Green Chem. 13, 2638–2650.

Siddiqi, K.S., Husen, A., Rao, R.A.K., 2018. A review on biosynthesis of silver nanoparticles and their biocidal properties. J. Nanobiotechnol. 16, 1–28.

Sim, W., Barnard, R.T., Blaskovich, M.A.T., Ziora, Z.M., 2018. Antimicrobial silver in medicinal and consumer applications: a patent review of the past decade. Antibiotics, 1–15.

Singh, R., Gautam, N., Mishra, A., Gupta, R., 2011. Heavy metals and living systems: an overview. Indian J. Pharmacol. 43, 246–253.

Singh, N., Naik, B., Kumar, V., Kumar, A., Kumar, V., Gupta, S., 2021. Actinobacterial pigment assisted synthesis of nanoparticles and its biological activity. J. Microbiol. Biotechnol. Food Sci. 10, 604–608.

Sivasankar, P., Seedevi, P., Poongodi, S., Sivakumar, M., et al., 2018. Characterization, antimicrobial and antioxidant property of exopolysaccharide mediated silver nanoparticles synthesized by *Streptomyces violaceus* MM72. Carbohydr. Polym. 181, 752–759. https://doi.org/10.1016/j.carbpol.2017.11.082.

Sowani, H., Mohite, P., Damale, S., Kulkarni, M., Zinjarde, S., 2016a. Enzyme and microbial technology carotenoid stabilized gold and silver nanoparticles derived from the Actinomycete *Gordonia amicalis* HS-11 as effective free radical scavengers. Enzyme Microb. Technol. 95, 164–173.

Sowani, H., Mohite, P., Munot, H., Shouche, Y., Bapat, T., 2016b. Green synthesis of gold and silver nanoparticles by an actinomycete *Gordonia amicalis* HS-11: mechanistic aspects and biological application. Process Biochem. 51, 374–383.

Sun, Y.G., Xia, Y.N., 2002. Shape-controlled synthesis of gold and silver nanoparticles. Science 298, 2176–2179.

Sunitha, A., Isaac, R.S.R., Sweetly, G., Sornalekshmi, S., Arsula, R., Praseetha, P.K., 2013. Evaluation of antimicrobial activity of biosynthesized iron and silver nanoparticles using the fungi *Fusarium* and *Actinomycete* sp. on human pathogens. Nano Biomed. Eng. 5, 39–45.

Szczęśniak, B., Borysiuk, S., Choma, J., Jaroniec, M., 2020. Mechanochemical synthesis of highly porous materials. Mater. Horiz. 7, 1457–1473.

Thenmozhi, M., Kannabiran, K., Kumar, R., Gopiesh Khanna, V., 2013. Antifungal activity of *Streptomyces* sp. VITSTK7 and its synthesized Ag_2O/Ag nanoparticles against medically important *Aspergillus* pathogens. J. Mycol. Med. 23, 97–103.

Tsibakhashvili, N.Y., Kirkesali, E.I., Pataraya, D.T., Gurielidze, M.A., et al., 2011. Microbial synthesis of silver nanoparticles by *Streptomyces glaucus* and *Spirulina platensis*. Adv. Sci. Lett. 4, 1–10. https:/doi.org/10.1166/asl.2011.1915.

Tsuzuki, T., McCormick, P.G., 2004. Mechanochemical synthesis of nanoparticles. J. Mater. Sci. 39, 5143–5146.

Valsalam, S., Agastian, P., Arasu, M.V., et al., 2019. Rapid biosynthesis and characterization of silver nanoparticles from the leaf extract of *Tropaeolum majus* L. and its enhanced in-vitro antibacterial, antifungal, antioxidant and anticancer properties. J. Photochem. Photobiol. B Biol. 191, 65–74. https:/doi.org/10.1016/j.jphotobiol.2018.12.010.

Veisi, H., Dadres, N., Mohammadi, P., Hemmati, S., 2019. Green synthesis of silver nanoparticles based on oil-water interface method with essential oil of orange peel and its application as nanocatalyst for A^3 coupling. Mater. Sci. Eng. C 105, 110031. https:/doi.org/10.1016/j.msec.2019.110031.

Vijayabharathi, R., Sathya, A., Gopalakrishnan, S., 2018. Biocatalysis and agricultural biotechnology extracellular biosynthesis of silver nanoparticles using *Streptomyces griseoplanus* SAI-25 and its antifungal activity against *Macrophomina phaseolina*, the charcoal rot pathogen of sorghum. Biocatal. Agric. Biotechnol. 14, 166–171.

Vilas, V., Philip, D., Mathew, J., 2016. Biosynthesis of Au and Au/Ag alloy nanoparticles using *Coleus aromaticus* essential oil and evaluation of their catalytic, antibacterial and antiradical activities. J. Mol. Liq. 221, 179–189. https:/doi.org/10.1016/j.molliq.2016.05.066.

Zhang, X., Yan, S., Tyagi, R.D., Surampalli, R.Y., 2011. Chemosphere synthesis of nanoparticles by microorganisms and their application in enhancing microbiological reaction rates. Chemosphere 82, 489–494.

Zonooz, F.N., Salouti, M., 2011. Extracellular biosynthesis of silver nanoparticles using cell filtrate of *Streptomyces* sp. ERI-3. Sci. Iran. 18 (6), 1631–1635. https:/doi.org/10.1016/j.scient.2011.11.029.

CHAPTER 22

Advantages of silver nanoparticles synthesized by microorganisms in antibacterial activity

Xixi Zhao, Xiaoguang Xu, Chongyang Ai, Lu Yan, Chunmei Jiang, and Junling Shi

Key Laboratory for Space Bioscience and Biotechnology, School of Life Sciences, Northwestern Polytechnical University, Xi'an, Shaanxi, China

1 Introduction

Since ancient times, silver has been applied in wound healing due to its excellent antibacterial activity (Chernousova and Epple, 2013). The rapidly growing popularity and evolution of nanotechnology since its discovery a few decades ago has led scientists to develop silver nanoparticles with distinctive physical, chemical, and biological properties that are suitable for a wide variety of applications, including medicine, healthcare, electronics, and agriculture (Borase et al., 2014; Lee et al., 2014). Silver nanoparticles have been proven to exhibit excellent electrical conductivity, chemical stability, catalytic and antimicrobial properties, as well as cytotoxic effects on cancer cells.

AgNPs can be synthesized using chemical reduction, physical synthesis, and biological synthesis (Zhao et al., 2018). Chemical synthesis is the most commonly used method and can be achieved on a larger scale and at lower cost compared to the other methods, but it involves the use of toxic chemicals that could be detrimental to the environment and animals (Borase et al., 2014). Physical synthesis, which includes methods such as laser ablation, lithography, and evaporation-condensation, gives rise to AgNPs with high purity and high thermal stability. However, this method has several drawbacks, such as requiring the use of large machinery and manufacturing equipment, consuming excessive energy, and requiring a considerable amount of time (Zhao et al., 2020). The biological synthesis of silver nanoparticles has only been explored in recent years. Many biological organisms, including plants, fungi, algae, and bacteria, have been employed in the synthesis of nanoparticles (Barapatre et al., 2016; Devanesan et al., 2017; Nakkala et al., 2017; Soto et al., 2019; Gómez-Garzón et al., 2021). This has led to the development of "green nanotechnology," which combines biological principles with physical and chemical procedures to generate eco-friendly, nano-sized particles with specific functions.

2 Synthesis of AgNPs by microorganisms

Biosynthetic methods employing microorganisms, such as fungi, bacteria (Singh et al., 2016), and algae (Terra et al., 2019), have received increasing attention as being simpler and more eco-friendly compared to the customary chemical and physical methods (Hulkoti and Taranath, 2014). Microorganisms can synthesize AgNPs either intracellularly or extracellularly, according to the location of AgNPs synthesis (Fig. 1). Various microorganisms are cultured and centrifuged to separate the biomass and supernatant. Extracellular synthesis is achieved by adding silver salts, such as $AgNO_3$, to the supernatant and incubating for a suitable time. The color changes to yellow, indicating the formation of the AgNPs. Finally, the AgNPs are isolated by centrifugation and washed thoroughly. For the intracellular synthesis, the biomass

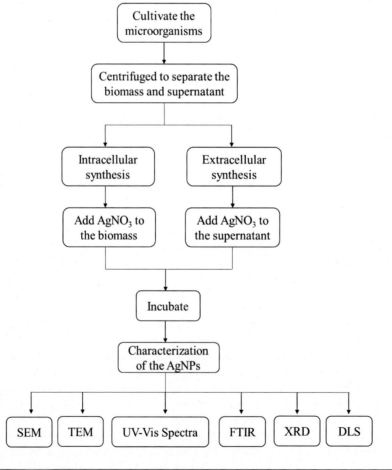

FIG. 1

Synthetic route of AgNPs by microorganisms.

is washed with water and supplemented with the silver salt. The AgNPs are obtained after breaking down the cell well of the biomass to release the nanoparticles from the microorganisms after synthesis. The obtained AgNPs are characterized by UV-vis spectroscopy, scanning electron microscopy, transmission electron microscopy, Fourier transform infrared spectroscopy, X-ray diffraction, and energy dispersive X-ray analysis (Zhao et al., 2018). Compared to the intracellular synthesis of silver nanoparticles, extracellular biosynthesis is cheaper, faster, and simpler (Das et al., 2014). Therefore, extracellular methods have attracted more attention from researchers.

2.1 Fungi-mediated synthesis of AgNPs

Mycelia and fungal cell-free filtrate have been used in the intracellular and extracellular synthesis of AgNPs. AgNPs synthesized both intracellularly and extracellularly by *Fusarium oxysporum* were 5–15 nm in size (Ahmad et al., 2003). Extracellular synthesis of AgNPs by *Phoma glomerata*, *Aspergillus terreus*, *Penicillium notatum*, *Phanerochaete chrysosporium*, *Trichoderma asperellum*, and *Aspergillus clavatus* were spherical in shape with sizes ranging from 10 to 60 nm (Vigneshwaran et al., 2006; Mukherjee et al., 2001; Basavaraja et al., 2008; Mukherjee et al., 2008; Birla et al., 2009; Gange, 2010; Li et al., 2012; Ravi et al., 2010). Their size, shape, and structure are dependent on the reaction conditions (Ottoni et al., 2017). Mohanta et al. demonstrated an economical and eco-friendly method for the synthesis of silver nanoparticles (AgNPs) using the wild mushroom *Ganoderma sessiliforme* (Mohanta et al., 2018). These AgNPs were 45 nm in size and exhibited high toxicity against common foodborne bacteria (Mohanta et al., 2018).

Compared to AgNPs derived from plants and other microorganisms, fungi-mediated synthesis of AgNPs is notably more efficient, higher yielding, and more stable (Singh et al., 2014). Moreover, fungal mycelia can withstand the harsh environments in bioreactors or chambers and are easy to handle and fabricate in downstream processing (Table 1).

2.2 Bacteria-mediated synthesis of AgNPs

The potential to biosynthesize AgNPs using bacteria has been well-documented. Bacteria are capable of synthesizing AgNPs either extracellularly or intracellularly by means of bioreduction. In recent research, both gram-positive and gram-negative bacteria have been used for the extracellular and intracellular biosynthesis of AgNPs (Singh et al., 2016; Siddiqi et al., 2018). For instance, *Escherichia coli* ATCC 8739, *Bacillus subtilis* ATCC 6633, and *Streptococcus thermophilus* ESh1 have been employed for the extracellular biosynthesis of AgNPs in a fast and effective way (El-Shanshoury et al., 2011). Despite silver ions being toxic to bacteria, the organisms overcome the toxicity through bioreduction mechanisms. Fiaz et al. synthesized AgNPs using three silver-resistant strains of bacteria, *Enterobacter cloacae*, *Cupriavidus necator*, and *Bacillus megaterium*, which were isolated from a silver mining

Table 1 Silver nanoparticles synthesized by different fungi.

Fungus species	Size (nm)	Time (h)	Temperature (°C)	References
Curvularia lunata	10–50	24	Room temperature	Parthasarathy Ramalingmam and Thangaraj (2015)
Schizophyllum commune	42.12	Immediately	30	Yen San and Mashitah (2012)
Lentinus sajor-caju	89.76	24	30	Yen San and Mashitah (2012)
Pycnoporus sanguineus	120.6	24	30	Yen San and Mashitah (2012)
Schizophyllum commune	50–60	Immediately	30	Yen San and Mashitah (2012)
Lentinus sajor-caju	50–60	24	30	Yen San and Mashitah (2012)
Pycnoporus sanguineus	50–60	24	30	Yen San and Mashitah (2012)
Fusarium oxysporum	5–15	24	–	Ahmad et al. (2003)
C. cladosporioides	10–100	78	27	Balaji et al. (2009)
Fusarium semitectum	10–60	72	27	Basavaraja et al. (2008)
Phoma glomerata	60–80	24	Room temperature	Birla et al. (2009)
Phoma sp.3.2883	71.06 ± 3.46	50	28	Chen et al. (2003)
Fusarium oxysporum	1.6	–	28	Durán et al. (2007)
Aspergillus niger	20	24	25	Gade et al. (2008)
Aspergillus clavatus	10–25	70	Room temperature	Gange (2010)
Penicillium	18	24	Room temperature	Govindappa et al. (2016)
Trichoderma harzianum	20–30	–	–	Guilger et al. (2017)
Aspergillus terreus	1–20	24	Room temperature	Li et al. (2012)
Penicillium aculeatum Su1	4–55	72	37	Ma et al. (2017)
Verticillium AAT-TS-4	25 ± 12	72	28	Mukherjee et al. (2001)
Trichoderma asperellum	13–18	48	25	Mukherjee et al. (2008)
Aspergillus flavus	8.92 ± 1.61	72	37	Vigneshwaran et al. (2007)
Trichoderma viride	5–40	48	27	Fayaz et al. (2010)
	50–200	72	37	Vigneshwaran et al. (2006)

Table 1 Silver nanoparticles synthesized by different fungi. *Continued*

Fungus species	Size (nm)	Time (h)	Temperature (°C)	References
Phaenerochaete chrysosporium				
Fusarium solani	5–35	24	25 ± 2	Ingle et al. (2009)
Aspergillus fumigatus	5–25	72	25	Bhainsa and D'Souza (2006)
Emericella nidulans EV4	10–20	–	45	Subha Rajam et al. (2017)
Pleurotus platypus	20–50	–	Room temperature	Prabhu (2017)
Anamorphous Bjerkandera	10–30	144	Room temperature	Osorio-Echavarria et al. (2021)
Penicillium oxalicum	60–80	72	37	Feroze et al. (2020)

site (Ahmad et al., 2019). The bacteria *Pseudomonas stutzeri* exhibited resistance to silver toxicity by reducing silver ions to silver nanoparticles (Hulkoti and Taranath, 2014).

It has been reported that the shape, size, and monodispersity of AgNPs synthesized by bacteria could be controlled by varying the microorganism or the synthetic parameters, such as temperature, pH, and silver-ion concentration (Table 2) (Monowar et al., 2018).

2.3 Other microorganisms

Similar to fungi and bacteria, yeasts are also widely used in the synthesis of AgNPs (Singh et al., 2016). *S. cerevisiae* has been used in traditional or industrial fermentation and also has the ability to synthesize AgNPs that exhibit high antifungal activity against *Candida albicans* (Niknejad et al., 2015). Among the biological systems explored, algae attract significant attention because they have the ability to bioremediate toxic metals by subsequently converting them to more amenable forms. Algae have also been shown to produce nanoparticles not only of silver but also

Table 2 Silver nanoparticles synthesized by different bacteria.

Bacterium species	Size (nm)	Time (h)	Temperature (°C)	References
P. aeruginosa	40	24	37	Quinteros et al. (2018)
Bacillus endophyticus	5	72	28	Gan et al. (2018)
Bacillus CS 11	42–92	24	Room temperature	Das et al. (2014)
Bacillus megaterium	80–98	48	Room temperature	Saravanan et al. (2011)
Pantoea ananatis	8–91	24	32	Monowar et al. (2018)
Bacillus cereus	26	15 min	121	Amir Rahimirad et al. (2019)
E. coli				
Staphylococcus aureus				
Salmonella enterica subsp. *enterica*				
Lactobacillus paracasei	18 ± 2.4	48	26	Viorica et al. (2021)
Lysinibacillus sphaericus	14–21	48	30	El-Bendary et al. (2021)
Streptomyces sp.	8–15	72	20	Bakhtiari-Sardari et al. (2020)

Table 3 Silver nanoparticles synthesized by algae and yeast.

Alga species	Size (nm)	Time	Temperature (°C)	References
Padina pavonia	49.58–86.37	3 h	Room temperature	Abdel-Raouf et al. (2019)
Ulva lactuca	8–19	10 min	55	Abdellatif et al. (2016)
Turbinaria turbinata	8–19	10 min	55	Abdellatif et al. (2016)
Sargassum wightii	15–20	24 h	28	Shanmugam et al. (2013)
Botryococcus Chlamidomonas sp. Ev-29 Chlorella Coelastrum Scenedesmus	13–31	72 h	25 ± 1	Patel et al. (2015)
Saccharomyces cerevisiae	5–20	72 h	Room temperature	Niknejad et al. (2015)

of other metal ions, such as gold, cadmium, and platinum (Patel et al., 2015). AgNPs synthesized by *Streptomyces* sp. were spherical in shape with an average size of 5 nm, and the reductase enzyme from *Streptomyces* sp. was found to play a critical role in the synthesis of the nanoparticles (Table 3) (Karthik et al., 2014).

2.4 Mechanism of AgNPs synthesis by microorganisms

The mechanism of AgNPs synthesis by different microorganisms is not well understood yet. It has been reported that cellular components such as the cell wall, cell membrane, various enzymes and proteins, and other organic molecules play important roles in the synthesis of AgNPs (Fig. 2) (Zhao et al., 2018). The positively charged silver ions electrostatically interact with the negatively charged groups on the extracellular surface of bacteria, which contains enzymes that can facilitate the reduction of silver ions (Salunke et al., 2016). It was revealed that NADH-dependent reductase uses electrons from NADH to reduce silver ions to AgNPs (Hulkoti and Taranath, 2014). Using *F. oxysporum*, phytochelatin, and α-NADPH-dependent nitrate reductase were also found to facilitate reduction and stabilization of AgNPs (Kumar et al., 2007). However, another study showed that naphthoquinones and anthraquinones in *F. oxysporum* function as electron shuttles to reduce Ag^+ to Ag^0 (Duran et al., 2005). Proteins and peptides from white rot fungus and *Trichoderma asperellum* have been reported to be responsible for the stabilization of AgNPs due to electrostatic attractions (Vigneshwaran et al., 2006; Mukherjee et al., 2008).

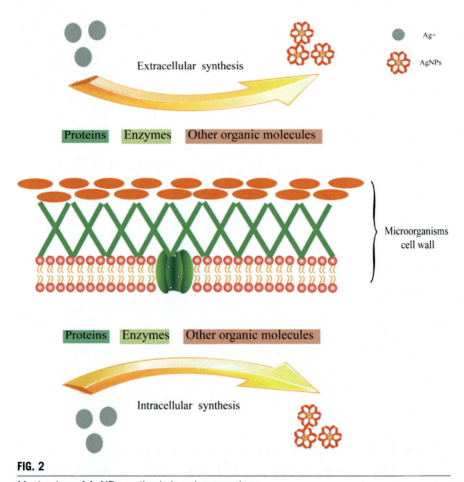

FIG. 2

Mechanism of AgNPs synthesis by microorganisms.

Research also suggests that enzymes in the bacteria play a critical role in the synthesis of AgNPs (Singh et al., 2016). Supplementing an aqueous $AgNO_3$ solution with bacteria resulted in the reduction of Ag^+ to Ag^0, which eventually formed AgNPs under the cell wall surface, indicating that enzymes that catalyze the reduction of Ag^+ to Ag^0 may be present in the cell membrane (Saravanan et al., 2018). Moreover, it has been suggested that sugars and NADH-dependent enzymes secreted by *Lactobacillus kimchicus* on the cell surface might be responsible for the reduction of gold ions, while proteins (or peptides) and amino acid residues inside the cells act as stabilizers of the NPs (Markus et al., 2016).

3 Antibacterial activity and mechanism of the AgNPs
3.1 Antibacterial activity of the AgNPs

With the increasing use of antibiotics, bacteria are developing resistance mechanisms that target the mechanisms of action of the antibiotics. It has been reported that the AgNPs have bactericidal properties against more than 650 pathogenic bacteria (Dakal et al., 2016; Terra et al., 2019). Several studies have demonstrated that the bactericidal properties of the AgNPs are strongly influenced by the nanoparticles' shape, size, concentration, and colloidal state (Dakal et al., 2016). AgNPs synthesized by *Bacillus siamensis* at different concentrations resulted in varying reduction in the growth of the bacteria, indicating that the inhibitory effect on bacterial growth was highly dependent on the concentrations of AgNPs (Ezzeldin Ibrahim et al., 2019). It has been suggested that the antibacterial activity of AgNPs is size-dependent, such that smaller nanoparticles have better antimicrobial activity than larger AgNPs (Agnihotri et al., 2014). Since smaller nanoparticles have a larger surface area, they are more likely to agglomerate together near the cell wall and inhibit cell division (Siddiqi et al., 2018). The size of the AgNPs also depends on the different bacteria used to synthesize them (Das et al., 2014). It was also suggested that AgNPs show greater antimicrobial action against gram-negative bacteria due to the electrostatic interactions between the negatively charged polysaccharides and the positively charged AgNPs, which allow them to attach to the surface of cell membrane and prevent permeation and respiration of the bacterial cells (Siddiqi et al., 2018).

3.2 Antibacterial mechanism of AgNPs

The properties of the nanoparticles (size, shape, structure, etc.), release of the silver ions from the nanoparticles inside the cell, oxidative stress from generation of reactive oxygen species, and damage to proteins and DNA are all important roles that the nanoparticles play in the antibacterial activity of AgNPs (Fig. 3). Multiple mechanisms underlying the antibacterial activity of AgNPs have been determined. Due to the larger surface area, silver nanoparticles can act by adhesion to the surface of the cell membrane, affecting permeability and respiratory functions (Siddiqi et al., 2018). It has been reported that the effectiveness of the bactericidal activities of AgNPs is due to the prolonged release of silver ions (Terra et al., 2019). The Ag^+ that is released from the AgNPs penetrates the cell membrane of the bacteria, reacting with the thiol groups of the proteins and inactivating the protein. This causes a reduction in membrane permeability and ultimately, cell death (Tripathi et al., 2017). Creating oxidative stress is another mechanism by which AgNPs exhibit antibacterial activity. It has been reported that AgNPs caused oxidative stress in *Staphylococcus aureus*, *Escherichia coli*, and *Pseudomonas aeruginosa* through the increased generation of reactive oxygen species (ROS), and this increase correlated with a better antimicrobial activity (Quinteros et al., 2016). This increase in the levels of ROS

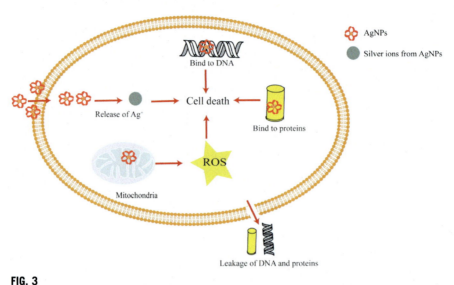

FIG. 3

Antibacterial mechanisms of the AgNPs.

was associated with the oxidation of different types of biologically relevant macromolecules, such as proteins and lipids (Quinteros et al., 2018). The AgNPs can also bind to proteins and DNA and damage them by inhibiting replication (Zhao et al., 2018).

3.3 Advantages of AgNPs synthesized by microorganisms in antibacterial activity

The antibacterial activity of AgNPs is dependent on the particle size and shape of the AgNPs. Although AgNPs synthesized by many different methods result in varying shapes and sizes, there are still limitations to the properties that the AgNPs feature. For example, adding an excess of strong reducing agent allows the synthesis of AgNPs with controlled particle size and shape. However, the additional reagents needed to achieve this create additional chemical waste and are harmful to the environment (Zhang et al., 2016). It has been reported that AgNPs synthesized by microorganisms have a controlled particle size and shape, which is accomplished through optimization of the synthetic variables, such as the type of microorganism used, the concentration of the precursors, the temperature, and the pH (Zhang et al., 2016). The AgNPs synthesized by microorganisms exhibited a higher antimicrobial activity compared with AgNPs synthesized by other methods (Sintubin et al., 2011). In addition, it was shown that AgNPs biosynthesized by *Penicillium aculeatum* display higher antimicrobial activity to *E. coli*, *P. aeruginosa*, *S. aureus*, *B. subtilis*, and *C. albicans* than $AgNO_3$ alone (Ma et al., 2017). Sintubin et al. demonstrated that the minimum inhibitory concentration of AgNPs produced by *Lactobacillus fermentum* to prevent growth of gram-positive and gram-negative bacteria was much lower than that of the chemically produced AgNPs, indicating that the biologically synthesized AgNPs exhibit a higher antimicrobial activity (Sintubin et al., 2011). This was found to be due to the aggregation of the chemically synthesized AgNPs (Sintubin et al., 2011). It has also been reported that the AgNPs synthesized by the algae *P. kessleri* have better activity toward biofilm inhibition and are less toxic in aquatic environments compared to AgNPs synthesized by chemical methods (Anna et al., 2017).

3.4 Future prospects

In recent years, there have been significant developments in biosynthetic methods of AgNPs by microorganisms, which have drawn increasing attention from researchers in this interdisciplinary field and beyond. However, despite the growing amount of research that is being published surrounding the biosynthesis of AgNPs, significantly more work is needed to improve the biosynthetic methods involved in preparation of the nanoparticles. First, compared to the other physical and chemical methods, the biosynthesis of AgNPs using microorganisms requires more time, which inevitably decreases the synthetic efficiency. Second, particle size and stability are very important properties for the application of AgNPs as an antibacterial agent. Therefore, the ability to control the size and improve the stability of the nanoparticles must be factored in when developing the synthetic methods of AgNPs using microorganisms. Last but not least, because the biosynthesis of AgNPs or other nanoparticles is relatively new and underexplored, synthesis of the nanoparticles using microorganisms is still on a laboratory scale. Therefore, more attention needs to be placed on the practical application of microorganisms in AgNPs production. Achieving a more thorough understanding of the biosynthetic mechanisms of AgNPs production by microorganisms may be a feasible way to accomplish this. If the mechanisms used by microorganisms to produce AgNPs can be thoroughly understood, the synthetic efficiencies and scale of AgNPs by microorganisms will be remarkably improved.

4 Conclusion

Synthesis of silver nanoparticles using microorganisms such as bacteria, fungi, and yeast has been discussed in this chapter. The synthetic mechanisms of the microorganism-produced AgNPs involve the use of enzymes, cell wall polymers, electron shuttle quinones, proteins, and other biological organic molecules. The AgNPs display bactericidal properties against many pathogenic organisms, with the antibacterial activity of the AgNPs being attributed to the intracellular release of silver ions, oxidative stress, and damage to proteins and DNA. The biological method of synthesizing nanoparticles by microorganisms also results in more controlled size and shape of the nanoparticles and does not involve the use of toxic and harmful substances, which are advantages over chemical and physical methods.

More importantly, compared to the AgNPs synthesized by other methods, the microorganism-produced AgNPs demonstrate a higher antimicrobial activity. All in all, the biological methods to synthesize AgNPs by microorganisms are still in the development stage, and significant efforts need to be made to improve the biosynthetic process in order to more practically synthesize AgNPs more efficiently and on a larger scale.

References

Abdellatif, K.F., Abdelfattah, R.H., El-Ansary, M.S.M., 2016. Green nanoparticles engineering on root-knot nematode infecting eggplants and their effect on plant DNA modification. Iran. J. Biotechnol. 14, 250–259.

Abdel-Raouf, N., Al-Enazi, N.M., Ibraheem, I.B.M., Alharbi, R.M., Alkhulaifi, M.M., 2019. Biosynthesis of silver nanoparticles by using of the marine brown *alga Padina pavonia* and their characterization. Saudi J. Biol. Sci. 26, 1207–1215.

Agnihotri, S., Mukherji, S., Mukherji, S., 2014. Size-controlled silver nanoparticles synthesized over the range 5–100 nm using the same protocol and their antibacterial efficacy. RSC Adv. 4, 3974–3983.

Ahmad, A., Mukherjee, P., Senapati, S., Mandal, D., Khan, M.I., Kumar, R., Sastry, M., 2003. Extracellular biosynthesis of silver nanoparticles using the fungus *Fusarium oxysporum*. Colloids Surf. B 28, 313–318.

Ahmad, F., Ashraf, N., Zhou, R.B., Chen, J.J., Liu, Y.L., Zeng, X., Zhao, F.Z., Yin, D.C., 2019. Optimization for silver remediation from aqueous solution by novel bacterial isolates using response surface methodology: recovery and characterization of biogenic AgNPs. J. Hazard. Mater. 380, 120906.

Amir Rahimirad, A.J., Mirzaei, H., Anarjan, N., Jafarizadeh-Malmiri, H., 2019. Biosynthetic potential assessment of four food pathogenic bacteria in hydrothermally silver nanoparticles fabrication. Green Process. Synth. 8, 629–634.

Anna, M., Oksana, V., Jana, K., 2017. Effect of chemically and biologically synthesized Ag nanoparticles on the algae growth inhibition. AIP Conf. Proc. 1919, 020008.

Bakhtiari-Sardari, A., Mashreghi, M., Eshghi, H., Behnam-Rasouli, F., Lashani, E., Shahnavaz, B., 2020. Comparative evaluation of silver nanoparticles biosynthesis by two coldtolerant *Streptomyces* strains and their biological activities. Biotechnol. Lett. 42, 1985–1999.

Balaji, D.S., Basavaraja, S., Deshpande, R., Mahesh, D.B., Prabhakar, B.K., Venkataraman, A., 2009. Extracellular biosynthesis of functionalized silver nanoparticles by strains of *Cladosporium cladosporioides* fungus. Colloids Surf. B 68, 88–92.

Barapatre, A., Aadil, K.R., Jha, H., 2016. Synergistic antibacterial and antibiofilm activity of silver nanoparticles biosynthesized by lignin-degrading fungus. Bioresour. Bioprocess. 3, 1–13.

Basavaraja, S., Balaji, S.D., Lagashetty, A., Rajasab, A.H., Venkataraman, A., 2008. Extracellular biosynthesis of silver nanoparticles using the fungus *Fusarium semitectum*. Mater. Res. Bull. 43, 1164–1170.

Bhainsa, K.C., D'souza, S.F., 2006. Extracellular biosynthesis of silver nanoparticles using the fungus *Aspergillus fumigatus*. Colloids Surf. B 47, 160–164.

Birla, S.S., Tiwari, V.V., Gade, A.K., Ingle, A.P., Yadav, A.P., Rai, M.K., 2009. Fabrication of silver nanoparticles by *Phoma glomerata* and its combined effect against *Escherichia coli, Pseudomonas aeruginosa* and *Staphylococcus aureus*. Lett. Appl. Microbiol. 48, 173–179.

Borase, H.P., Salunke, B.K., Salunkhe, R.B., Patil, C.D., Hallsworth, J.E., Kim, B.S., Patil, S.V., 2014. Plant extract: a promising biomatrix for ecofriendly, controlled synthesis of silver nanoparticles. Appl. Biochem. Biotechnol. 173, 1–29.

Chen, J.C., Lin, Z.H., Ma, X.X., 2003. Evidence of the production of silver nanoparticles via pretreatment of *Phoma* sp.3.2883 with silver nitrate. Lett. Appl. Microbiol. 37, 105–108.

Chernousova, S., Epple, M., 2013. Silver as antibacterial agent: ion, nanoparticle, and metal. Angew. Chem. Int. Ed. 52, 1636–1653.

Dakal, T.C., Kumar, A., Majumdar, R.S., Yadav, V., 2016. Mechanistic basis of antimicrobial actions of silver nanoparticles. Front. Microbiol. 7, 1831.

Das, V.L., Thomas, R., Varghese, R.T., Soniya, E.V., Mathew, J., Radhakrishnan, E.K., 2014. Extracellular synthesis of silver nanoparticles by the *Bacillus strain* CS11 isolated from industrialized area. 3 Biotech 4, 121–126.

Devanesan, S., Alsalhi, M.S., Vishnubalaji, R., Alfuraydi, A.A., Alajez, N.M., Alfayez, M., Murugan, K., Sayed, S.R.M., Nicoletti, M., Benelli, G., 2017. Rapid biological synthesis of silver nanoparticles using plant seed extracts and their cytotoxicity on colorectal cancer cell lines. J. Clust. Sci. 28, 595–605.

Duran, N., Marcato, P.D., Alves, O.L., Souza, G.I., Esposito, E., 2005. Mechanistic aspects of biosynthesis of silver nanoparticles by several *Fusarium oxysporum* strains. J. Nanobiotechnol. 3, 1–8.

Durán, N., Marcato, P.D., De Souza, G.I.H., Alves, O.L., Esposito, E., 2007. Antibacterial effect of silver nanoparticles produced by fungal process on textile fabrics and their effluent treatment. J. Biomed. Nanotechnol. 3, 203–208.

El-Bendary, M.A., Abdelraof, M., Moharam, M.E., Elmahdy, E.M., Allam, M.A., 2021. Potential of silver nanoparticles synthesized using low active mosquitocidal *Lysinibacillus sphaericus* as novel antimicrobial agents. Prep. Biochem. Biotechnol., 1–10.

El-Shanshoury, A.E.-R.R., Elsilk, S.E., Ebeid, M.E., 2011. Extracellular biosynthesis of silver nanoparticles using *Escherichia coli ATCC 8739, Bacillus subtilis ATCC 6633*, and *Streptococcus thermophilus ESh1* and their antimicrobial activities. ISRN Nanotechnol. 2011, 1–7.

Ezzeldin Ibrahim, H.F., Zhang, M., Zhang, Y., Qiu, W., Yan, C., Li, B., Mo, J., Chen, J., 2019. Biosynthesis of silver nanoparticles using endophytic bacteria and their role in inhibition of rice pathogenic bacteria and plant growth promotion. RSC Adv. 9, 29293–29299.

Fayaz, A.M., Balaji, K., Girilal, M., Yadav, R., Kalaichelvan, P.T., Venketesan, R., 2010. Biogenic synthesis of silver nanoparticles and their synergistic effect with antibiotics: a study against gram-positive and gram-negative bacteria. Nanomedicine 6, 103–109.

Feroze, N., Arshad, B., Younas, M., Afridi, M.I., Saqib, S., Ayaz, A., 2020. Fungal mediated synthesis of silver nanoparticles and evaluation of antibacterial activity. Microsc. Res. Tech. 83, 72–80.

Gade, A.K., Bonde, P., Ingle, A.P., Marcato, P.D., Duran, N., Rai, M.K., 2008. Exploitation of *Aspergillus niger* for synthesis of silver nanoparticles. J. Biobased Mater. Bioenergy 2, 243–247.

Gan, L., Zhang, S., Zhang, Y., He, S., Tian, Y., 2018. Biosynthesis, characterization and antimicrobial activity of silver nanoparticles by a *halotolerant Bacillus endophyticus* SCU-L. Prep. Biochem. Biotechnol. 48, 582–588.

Gange, V.V.R.K.A., 2010. Biosynthesis of antimicrobial silver nanoparticles by the endophytic fungus *Aspergillus clavatus*. Nanomedicine 5, 33–40.

Gómez-Garzón, M., Gutiérrez-Castañeda, L.D., Gil, C., Escobar, C.H., Rozo, A.P., González, M.E., Sierra, E.V., 2021. Inhibition of the filamentation of *Candida albicans* by *Borojoa patinoi* silver nanoparticles. SN Appl. Sci. 3.

Govindappa, M., Farheen, H., Chandrappa, C.P., Channabasava, Rai, R.V., Raghavendra, V.B., 2016. Mycosynthesis of silver nanoparticles using extract of endophytic fungi, *Penicillium* species of *Glycosmis mauritiana*, and its antioxidant, antimicrobial, anti-inflammatory and tyrokinase inhibitory activity. Adv. Nat. Sci. Nanosci. 7, 035014.

Guilger, M., Pasquoto-Stigliani, T., Bilesky-Jose, N., Grillo, R., Abhilash, P.C., Fraceto, L.F., Lima, R., 2017. Biogenic silver nanoparticles based on *Trichoderma harzianum*: synthesis, characterization, toxicity evaluation and biological activity. Sci. Rep. 7, 44421.

Hulkoti, N.I., Taranath, T.C., 2014. Biosynthesis of nanoparticles using microbes-a review. Colloids Surf. B 121, 474–483.

Ingle, A., Rai, M., Gade, A., Bawaskar, M., 2009. *Fusarium solani*: a novel biological agent for the extracellular synthesis of silver nanoparticles. J. Nanopart. Res. 11, 2079–2085.

Karthik, L., Kumar, G., Kirthi, A.V., Rahuman, A.A., Bhaskara Rao, K.V., 2014. *Streptomyces* sp. LK3 mediated synthesis of silver nanoparticles and its biomedical application. Bioprocess Biosyst. Eng. 37, 261–267.

Kumar, S.A., Abyaneh, M.K., Gosavi, S.W., Kulkarni, S.K., Ahmad, A., Khan, M.I., 2007. Sulfite reductase-mediated synthesis of gold nanoparticles capped with phytochelatin. Biotechnol. Appl. Biochem. 47, 191–195.

Lee, S.H., Salunke, B.K., Kim, B.S., 2014. Sucrose density gradient centrifugation separation of gold and silver nanoparticles synthesized using *Magnolia kobus* plant leaf extracts. Biotechnol. Bioprocess. Eng. 19, 378.

Li, G.Q., He, D., Qian, Y.Q., Guan, B.Y., Gao, S., Cui, Y., Yokoyama, K., Wang, L., 2012. Fungus-mediated green synthesis of silver nanoparticles using *Aspergillus terreus*. Int. J. Mol. Sci. 13, 466–476.

Ma, L., Su, W., Liu, J.-X., Zeng, X.-X., Huang, Z., Li, W., Liu, Z.-C., Tang, J.-X., 2017. Optimization for extracellular biosynthesis of silver nanoparticles by *Penicillium aculeatum* Su1 and their antimicrobial activity and cytotoxic effect compared with silver ions. Sci. Eng. Compos. Mater. 1, 963–971.

Markus, J., Mathiyalagan, R., Kim, Y.J., Abbai, R., Singh, P., Ahn, S., Perez, Z.E.J., Hurh, J., Yang, D.C., 2016. Intracellular synthesis of gold nanoparticles with antioxidant activity by probiotic *Lactobacillus kimchicus* DCY51(T) isolated from *Korean kimchi*. Enzym. Microb. Technol. 95, 85–93.

Mohanta, Y.K., Nayak, D., Biswas, K., Singdevsachan, S.K., Abd Allah, E.F., Hashem, A., Alqarawi, A.A., Yadav, D., Mohanta, T.K., 2018. Silver nanoparticles synthesized using wild mushroom show potential antimicrobial activities against food borne pathogens. Molecules 23, 655–673.

Monowar, T., Rahman, M.S., Bhore, S.J., Raju, G., Sathasivam, K.V., 2018. Silver nanoparticles synthesized by using the endophytic bacterium *Pantoea ananatis* are promising antimicrobial agents against multidrug resistant bacteria. Molecules 23, 1–17.

Mukherjee, P., Ahmad, A., Mandal, D., Senapati, S., Sainkar, S.R., Khan, M.I., Parishcha, R., Ajaykumar, P.V., Alam, M., Kumar, R., Sastry, M., 2001. Fungus-mediated synthesis of silver nanoparticles and their immobilization in the mycelial matrix: a novel biological approach to nanoparticle synthesis. Nano Lett. 1, 515–519.

Mukherjee, P., Roy, M., Mandal, B.P., Dey, G.K., Mukherjee, P.K., Ghatak, J., Tyagi, A.K., Kale, S.P., 2008. Green synthesis of highly stabilized nanocrystalline silver particles by a non-pathogenic and agriculturally important fungus *T-asperellum*. Nanotechnology 19, 075103–075110.

Nakkala, J.R., Mata, R., Sadras, S.R., 2017. Green synthesized nano silver: synthesis, physicochemical profiling, antibacterial, anticancer activities and biological in vivo toxicity. J. Colloid Interface Sci. 499, 33–45.

Niknejad, F., Nabili, M., Daie Ghazvini, R., Moazeni, M., 2015. Green synthesis of silver nanoparticles: another honor for the yeast model *Saccharomyces cerevisiae*. Curr. Med. Mycol. 1, 17–24.

Osorio-Echavarria, J., Osorio-Echavarria, J., Ossa-Orozco, C.P., Gomez-Vanegas, N.A., 2021. Synthesis of silver nanoparticles using white-rot fungus *Anamorphous Bjerkandera* sp. R1: influence of silver nitrate concentration and fungus growth time. Sci. Rep. 11, 3842.

Ottoni, C.A., Simoes, M.F., Fernandes, S., Dos Santos, J.G., Da Silva, E.S., De Souza, R.F., Maiorano, A.E., 2017. Screening of filamentous fungi for antimicrobial silver nanoparticles synthesis. AMB Express 7, 31–41.

Parthasarathy Ramalingmam, S.M., Thangaraj, P., 2015. Biosynthesis of silver anoparticles using an endophytic fungus *Curvularia lunata* and its antimicrobial potential. J. Nanosci. Nanoeng. 4, 241–247.

Patel, V., Berthold, D., Puranik, P., Gantar, M., 2015. Screening of cyanobacteria and microalgae for their ability to synthesize silver nanoparticles with antibacterial activity. Biotechnol. Rep. (Amst) 5, 112–119.

Prabhu, K.K.P., 2017. Biosynthesis of silver nanoparticles using *Lactobacillus acidophilus* and white rot fungus-a comparative study. Int. J. Adv. Res. 3, 299–306.

Quinteros, M.A., Cano Aristizabal, V., Dalmasso, P.R., Paraje, M.G., Paez, P.L., 2016. Oxidative stress generation of silver nanoparticles in three bacterial genera and its relationship with the antimicrobial activity. Toxicol. In Vitro 36, 216–223.

Quinteros, M.A., Viviana, C.A., Onnainty, R., Mary, V.S., Theumer, M.G., Granero, G.E., Paraje, M.G., Paez, P.L., 2018. Biosynthesized silver nanoparticles: decoding their mechanism of action in *Staphylococcus aureus* and *Escherichia coli*. Int. J. Biochem. Cell Biol. 104, 87–93.

Ravi, G.P., Chirag, K.P., Vimal, I.P., Sen, D.J., Panigrahi, B., Patel, C.N., 2010. Synthesis and biological evaluation of antitubercular activity of some synthesised pyrazole derivatives. J. Chem. Pharm. Res. 2, 112–117.

Salunke, B.K., Sawant, S.S., Lee, S.I., Kim, B.S., 2016. Microorganisms as efficient biosystem for the synthesis of metal nanoparticles: current scenario and future possibilities. World J. Microbiol. Biotechnol. 32, 88.

Saravanan, M., Vemu, A.K., Barik, S.K., 2011. Rapid biosynthesis of silver nanoparticles from *Bacillus megaterium* (NCIM 2326) and their antibacterial activity on multi drug resistant clinical pathogens. Colloids Surf. B Biointerfaces 88, 325–331.

Saravanan, M., Arokiyaraj, S., Lakshmi, T., Pugazhendhi, A., 2018. Synthesis of silver nanoparticles from *Phenerochaete chrysosporium* (MTCC-787) and their antibacterial activity against human pathogenic bacteria. Microb. Pathog. 117, 68–72.

Shanmugam, N., Rajkamal, P., Cholan, S., Kannadasan, N., Sathishkumar, K., Viruthagiri, G., Sundaramanickam, A., 2013. Biosynthesis of silver nanoparticles from the marine seaweed *Sargassum wightii* and their antibacterial activity against some human pathogens. Appl. Nanosci. 4, 881–888.

Siddiqi, K.S., Husen, A., Rao, R. a. K., 2018. A review on biosynthesis of silver nanoparticles and their biocidal properties. J. Nanobiotechnol. 16, 14–42.

Singh, D., Rathod, V., Ninganagouda, S., Hiremath, J., Singh, A.K., Mathew, J., 2014. Optimization and characterization of silver nanoparticle by endophytic fungi *Penicillium* sp. isolated from *Curcuma longa* (turmeric) and application studies against MDR *E. coli* and *S. aureus*. Bioinorg. Chem. Appl. 2014, 408021.

Singh, P., Kim, Y.J., Zhang, D., Yang, D.C., 2016. Biological synthesis of nanoparticles from plants and microorganisms. Trends Biotechnol. 34, 588–599.

Sintubin, L., De Gusseme, B., Van Der Meeren, P., Pycke, B.F., Verstraete, W., Boon, N., 2011. The antibacterial activity of biogenic silver and its mode of action. Appl. Microbiol. Biotechnol. 91, 153–162.

Soto, K.M., Quezada-Cervantes, C.T., Hernández-Iturriaga, M., Luna-Bárcenas, G., Vazquez-Duhalt, R., Mendoza, S., 2019. Fruit peels waste for the green synthesis of silver nanoparticles with antimicrobial activity against foodborne pathogens. LWT 103, 293–300.

Subha Rajam, K., Rani, M.E., Gunaseeli, R., Hussain Munavar, M., 2017. Extracellular synthesis of silver nanoparticles by the fungus *Emericella nidulans* EV4 and its application. Indian J. Exp. Biol. 55, 262–265.

Terra, A.L.M., Kosinski, R.D.C., Moreira, J.B., Costa, J.A.V., Morais, M.G., 2019. Microalgae biosynthesis of silver nanoparticles for application in the control of agricultural pathogens. J. Environ. Sci. Health B 54, 709–716.

Tripathi, D.K., Tripathi, A., Shweta, Singh, S., Singh, Y., Vishwakarma, K., Yadav, G., Sharma, S., Singh, V.K., Mishra, R.K., Upadhyay, R.G., Dubey, N.K., Lee, Y., Chauhan, D.K., 2017. Uptake, accumulation and toxicity of silver nanoparticle in autotrophic plants, and heterotrophic microbes: a concentric review. Front. Microbiol. 8, 1–16.

Vigneshwaran, N., Kathe, A.A., Varadarajan, P.V., Nachane, R.P., Balasubramanya, R.H., 2006. Biomimetics of silver nanoparticles by white rot fungus, *Phaenerochaete chrysosporium*. Colloids Surf. B 53, 55–59.

Vigneshwaran, N., Ashtaputre, N.M., Varadarajan, P.V., Nachane, R.P., Paralikar, K.M., Balasubramanya, R.H., 2007. Biological synthesis of silver nanoparticles using the fungus *Aspergillus flavus*. Mater. Lett. 61, 1413–1418.

Viorica, R.P., Pawel, P., Boguslaw, B., 2021. Use of *Lactobacillus paracasei* isolated from whey for silver nanocomposite synthesis: antiradical and antimicrobial properties against selected pathogens. J. Dairy Sci.

Yen San, C., Mashitah, M.D., 2012. Instantaneous biosynthesis of silver nanoparticles by selected macro fungi. Aust. J. Basic Appl. Sci. 6, 86–88.

Zhang, X.F., Liu, Z.G., Shen, W., Gurunathan, S., 2016. Silver nanoparticles: synthesis, characterization, properties, applications, and therapeutic approaches. Int. J. Mol. Sci. 17, 1534–1568.

Zhao, X., Zhou, L., Riaz Rajoka, M.S., Yan, L., Jiang, C., Shao, D., Zhu, J., Shi, J., Huang, Q., Yang, H., Jin, M., 2018. Fungal silver nanoparticles: synthesis, application and challenges. Crit. Rev. Biotechnol. 38, 817–835.

Zhao, X., Zhao, H., Yan, L., Li, N., Shi, J., Jiang, C., 2020. Recent developments in detection using noble metal nanoparticles. Crit. Rev. Anal. Chem. 50, 97–110.

PART 5

Mechanism, toxicity, fate, commercial production

CHAPTER 23

Methodologies for synthesizing silver nanoparticles

Asma Farheen

Department of Biotechnology, Khaja Bandanawaz University, Kalaburagi, India

1 Introduction

The term *nano* is adapted from the Greek word meaning "dwarf" (Manikprabhu and Li, 2015). Nanoparticles are particles with a size of up to 100 nm (Manikprabhu and Lingappa, 2014); they act as a bridge between bulk materials and atomic or molecular structures (Manikprabhu and Li, 2015). Due to the tunable physical, chemical, and biological properties of nanoparticles, they have gained applications in different areas (Jeevanandam et al., 2018). Among various nanoparticles, silver nanoparticles in particular have gained worldwide attention due to their assessment as antimicrobial agents (Burduşel et al., 2018). Silver nanoparticles have been used unknowingly for thousands of years; for example, in ancient India, people practiced the use of Rajat Bhasma (Rajat is silver and Bhasma means fine powder) for treating diseases (Chaturvedi and Jha, 2011). Silver nanoparticles are one of the most commercially used nanomaterials used in different fields such as antimicrobial agents, textiles, water purifiers, medical devices, sensors, home cosmetics, electronics devices, and household appliances (Singh et al., 2015a,b; Manikprabhu and Li, 2015). There are two approaches to synthesize silver nanoparticles: top-down and bottom-up, and in general they can be categorized into physical, chemical, and bio-assisted methods (Dhand et al., 2015). In this chapter, we will discuss all those approaches, and emphasize their advantages and disadvantages. The applications of silver nanoparticles in different areas will also be discussed.

2 Methodologies for synthesizing silver nanoparticles

Fabrication of nanoparticles typically employs two general strategies (Fig. 1): bottom-up and top-down (Yu et al., 2013). The top-down approach is considered the oldest method to synthesize nanoparticles wherein the bulk material was reduced to nanoparticles (Yusuf, 2019). The most frequently used methods based on the top-down approach are attrition or milling, micropatterning, and pyrolysis

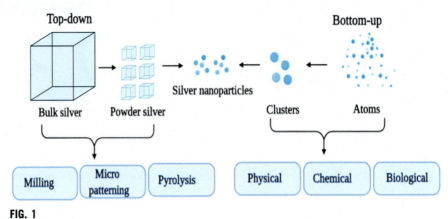

FIG. 1

Strategies to synthesize silver nanoparticles.

(Pareek et al., 2017). Attrition or milling includes grinding of macro- and microscale materials in a ball mill to generate particles of nano-size range (Pareek et al., 2017). Micropatterning is defined as the art of miniaturization of patterns and is the most used technique in electronics for fabricating nanoparticles (Chen et al., 2009). Recently, many techniques have been reported apart from the traditional photolithography approach (Pareek et al., 2017). For example, Ammosova et al. (2017), with the micro working robot technique, have demonstrated a simple yet versatile method for the preparation of silver nanoparticle micropatterns on polymer substrates. Silver nanoparticle micropatterns on plastic substrates have been performed using silver nanoparticle ink, containing silver cations and polyethylene glycol as a reducing agent. Similarly, Choi et al. (2009) reported micropatterning of poly(vinyl pyrrolidone)/silver nanoparticle thin films by ion irradiation.

Ultrasonic spray pyrolysis is a convenient and innovative tool to synthesize silver nanoparticles. In this process, a metal-containing solution is cold-atomized, which forms an aerosol and results in the production of particles with a narrow size distribution (Yusuf, 2019). Silver nanoparticles synthesis of size less than 20 nm was reported by Pingali et al. (2005), by pyrolysis of ultrasonically atomized spray of highly dilute aqueous silver nitrate solution at temperatures above 650°C and below the melting point of silver. Although these methods were efficient, they have many drawbacks, such as particle agglomeration. An enormous amount of energy will be needed for the synthesis and the process depends on various parameters that influence the particle size and shape (Pareek et al., 2017).

In contrast to top-down approaches, the bottom-up approach includes the construction of nanostructures atom by atom or particle by particle (Gour and Jain, 2019). In the past few decades, this method became very popular because it produces stable, homogenous nanostructures with perfect crystallographic and surface structures (Pareek et al., 2017). The bottom-up approach includes physical, chemical, and biological reduction (Iravani et al., 2014).

2.1 Physical method to synthesize silver nanoparticles

Physical methods for the synthesis of silver nanoparticles include evaporation, condensation, laser ablation, electrical irradiation, gamma irradiation, and microwave irradiation (Sreeram et al., 2008; Yaqoob et al., 2020). Condensation and evaporation are among the commonly used physical methods for synthesizing nanoparticles. Although these methods have been commonly used, they have their own merits and demerits. Some of the common drawbacks of these methods are higher energy requirement and time required (Haider and Kang, 2015). Some other physical methods have been developed to overcome these drawbacks. Laser-mediated synthesis is an upcoming method to synthesize silver nanoparticles. Compared to other methods, laser-medicated synthesis does not include a chemical reagent in solution and pure colloids can be synthesized; it will therefore be more useful in many applications (Tsuji et al., 2002). Similarly, microwave-mediated synthesis of silver nanoparticles has also been introduced; the advantage of this method is rapid synthesis (Sreeram et al., 2008).

2.2 Chemical methods to synthesize silver nanoparticles

The most common approach for synthesizing silver nanoparticles by chemical reduction is using organic and inorganic reducing agents. Different reducing agents such as sodium citrate, ascorbate, sodium borohydride, tannic acid, etc. (Ranoszek-Soliwoda et al., 2017; Chekin and Ghasemi, 2014; Agnihotri et al., 2014) have been used to synthesize silver nanoparticles. The major drawback of this method is the agglomeration of silver nanoparticles, but this can be avoided by the addition of stabilizing agents such as poly (vinyl alcohol), gelatin, poly (vinylpyrrolidone), etc. (Iravani et al., 2014; Chekin and Ghasemi, 2014).

2.3 Biological methods to synthesize silver nanoparticles

The biological synthesis of silver nanoparticles uses multiple living forms, such as plants (Zhang et al., 2019; Satsangi and Preet, 2021; Okaiyeto et al., 2021) and microbes (Dhanaraj et al., 2020; Ibrahim et al., 2021). In the past, biological methods (Fig. 2) have been tremendously used to synthesize silver nanoparticles because chemical methods are tedious, expensive, and may be harmful (Vaid et al., 2020).

The use of plants to synthesize silver nanoparticles has drawn attention, because of its rapid, eco-friendly, nonpathogenic, economical protocol and provision of a single-step technique for the biosynthetic processes (Ahmed et al., 2016). Various plants and their parts have been reported to synthesize silver nanoparticles (Table 1). A general protocol (Fig. 2) for the synthesis includes the collection of the plant part of interest and washing (to remove associated debris). The plant part is either boiled directly or dried, powdered (using a domestic blender), and boiled. The resultant extract is appropriately treated with $AgNO_3$ and incubated until it turns to brown. The color change (colorless to brown) of the solution is the preliminary

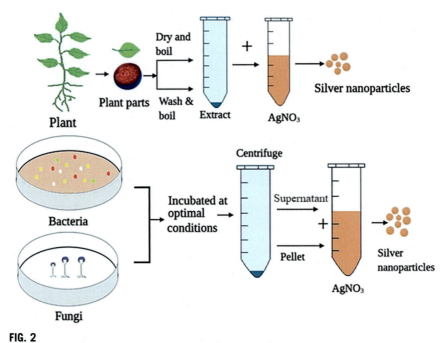

FIG. 2
Biological methods to synthesize silver nanoparticles.

Table 1 Synthesis of silver nanoparticles using various plant parts.

Sl. no.	Plant	Plant part used for the synthesis	Size (shape)	References
1	Withania coagulans	Leaf extract	14 nm (spherical)	Tripathi et al. (2019)
2	Arisaema flavum	Tuber extract	12–20 nm (spherical)	Rahman et al. (2019)
3	Cydonia oblonga	Seed extract	38 nm (face-centered cubic)	Zia et al. (2016)
4	Terminalia chebula	Fruit extract	30 nm (distorted spherical)	Edison et al. (2016)
5	Madhuca longifolia	Flower extract	30–50 nm (spherical and oval)	Patil et al. (2018)
6	Picea abies	Bark extract	44 nm (spherical or rarely polygonal)	Tanase et al. (2019)
7	Berberis vulgaris	Root extract	30–70 nm (spherical)	Behravan et al. (2019)

Table 1 Synthesis of silver nanoparticles using various plant parts—cont'd

Sl. no.	Plant	Plant part used for the synthesis	Size (shape)	References
8	Orange waste	Peel extract	48.1±20.5nm (not uniform)	Barros et al. (2018)
9	*Citrus sinensis*	Peel extract	10–35nm (spherical)	Kaviya et al. (2011)
10	*Thevetia peruviana*	Latex	10–30nm (spherical)	Rupiasih et al. (2013)
11	*Ruta graveolens*	Ethanol extract from leaves	40–45nm (spherical)	Ghramh et al. (2020)
12	*Achillea millefolium*	Aqueous, ethanol, and methanol plant extract	20.77, 18.53, and 14.27nm (spherical, rectangular, and cubical)	Yousaf et al. (2020)
13	*Brillantaisia patula*, *Crossopteryx febrifuga*, and *Senna siamea*	Aqueous leaf extracts	45nm (with polydispersity of 32.1%), 115nm (with polydispersity of 10.8%), and 47nm (with polydispersity of 46.0%)	Kambale et al. (2020)
14	Sea buckthorn	Berry extract	27.3±0.2nm (spherical)	Wei et al. (2020)
15	*Lysimachia foenum-graecum*	Dried leaves	10–20nm (quasispherical)	Chartarrayawadee et al. (2020)
16	*Cestrum nocturnum*	Aqueous leaf extract	20nm (mostly spherical)	Keshari et al. (2020)
17	*Cinnamomum camphora*	Callus culture	5.47–9.48nm (spherical)	Aref and Salem (2020)
18	*Phyllanthus emblica*	Fruit extract	19–45nm (hexagonal)	Renuka et al. (2020)
19	*Carica papaya*	Leaf	10–70nm (spherical shape)	Jain et al. (2020)
20	*Annona muricata*	Ethanolic extracts of fruits	51nm (spherical)	Gavamukulya et al. (2020)
21	*Salvadora persica*	Root extract	37.5nm (rod)	Arshad et al. (2021)
22	*Poa annua*	Leaves extract	36.66±7.85nm (oblate spheroid)	Gul et al. (2021)

Continued

Table 1 Synthesis of silver nanoparticles using various plant parts—cont'd

Sl. no.	Plant	Plant part used for the synthesis	Size (shape)	References
23	Aaronsohnia factorovskyi	Plant extract	104–140 nm (spherical)	Al-Otibi et al. (2020)
24	Areca catechu	Hydrothermal extract of Areca catechu	20–30 nm (round)	Choi et al. (2021)
25	Carica papaya	Leaf extract	10–20 nm (spherical)	Singh et al. (2021)

confirmation for the silver nanoparticles (Tripathi et al., 2019). Plants and their parts contain carbohydrates, fats, proteins, nucleic acids, pigments, and several types of secondary metabolites used as reducing and stabilizing agents to synthesize nanoparticles (Siddiqi et al., 2018).

Manik et al. (2020) reported the synthesis of silver nanoparticles using leaf extract of *Artocarpus heterophyllus* and *Azadirachta indica*. Before the synthesis, leaves were washed with water and dried in the sun. The leaves extracts obtained have been treated with 0.05 M $AgNO_3$ solution. The solution was heated continuously at 50°C for 15 min and a color change of the solution to brown was noticed, and further characterization of the solution suggested spherical shaped silver nanoparticles having size of 20–40 nm. Similarly, Odeniyi et al. (2020) reported irregular-shaped silver nanoparticles synthesis using fresh fruits of *Nauclea latifolia*. The fruits were washed, rinsed, cut into pieces, and blended. The sample was extracted using methanol and water and concentrated using a rotary evaporator at 40°C. A solution of 1 mM silver nitrate ($AgNO_3$) was added to 40 mL extract. A gradual color change of the reaction mixture was noticed after 72 h of incubation. The study suggested that alcohol, aromatic alcohol, amine, amide, carbonyl, carboxylic, and alkyl halide groups might be responsible for the bioreduction of silver ions. Similarly, ricinoleic acid extracted from the castor bean plant has been reported to synthesize silver nanoparticles. Ricinoleic acid was reported to act as a reducing and stabilizing agent and to synthesize silver nanoparticles at room temperature (Viana et al., 2020).

Although plants have been reported to synthesize silver nanoparticles, the usage of this source has many drawbacks; for example, nonavailability throughout the year and large-scale synthesis may lead to loss of valuable species (Manikprabhu and Lingappa, 2014). Apart from plants, microorganisms have been considered as an important source to synthesize silver nanoparticles in a cost-effective and eco-friendly way (Manikprabhu and Lingappa, 2013a). They have an enormous advantage over plant-mediated synthesis, including easy and rapid growth in low-cost medium, easy processing, and being independent of weather conditions (Manikprabhu and Lingappa, 2014).

To synthesize silver nanoparticles, microorganisms require the ability to withstand silver ions. Although silver nanoparticles producing microorganisms show resistance to silver ions, at high concentrations they become inactive and this is why silver is called a "moiety with two functions." One is inducing the microorganism to synthesize nanoparticles at lower concentrations; another is the induction of cell death at higher concentrations (Manikprabhu and Li, 2015). Microorganisms synthesize silver nanoparticles using various enzymes (Elegbede et al., 2018; Dhanaraj et al., 2020), proteins (Gudikandula et al., 2017), and pigment (Manikprabhu and Lingappa, 2013a). To date, several microorganisms, including bacteria (Beeler et al., 2020), fungi (Guilger-Casagrande and Lima, 2019), and yeast (Fernández et al., 2016), have been reported to synthesize silver nanoparticles. The mode of synthesis includes either intracellular or extracellular (Beeler et al., 2020; Otari et al., 2015). Bacteria have been preferred to synthesize silver nanoparticles because of the ease of manipulating them. The first evidence of bacteria-synthesizing nanoparticles was *Pseudomonas stutzeri* (Haefeli et al., 1984). Many bacterial strains have been reported to synthesize silver nanoparticles of different sizes and shapes (Table 2). A general protocol (Fig. 2) for the synthesis includes the growth of bacterial strains in an appropriate liquid medium maintained at optimal growth conditions. After the growth, the medium will be centrifuged and either the medium-free biomass or the culture-free extract will be treated with the appropriate amount of $AgNO_3$. The reaction mixture will be incubated until it turns brown, indicating the preliminary synthesis of the silver nanoparticles (Manikprabhu and Lingappa, 2013a,b). A wide variety of mechanisms have been proposed for the synthesis of silver nanoparticles. The most widely accepted mechanism includes reduction using nitrate reductase enzymes (Prabhu and Poulose, 2012).

Mukherjee et al. (2018) reported silver nanoparticles synthesis using nitrate reductase enzyme produced by *Bacillus clausii*. Apart from enzymes, proteins were also reported to synthesize silver nanoparticles. Shanmugasundaram et al. (2013) reported the synthesis of silver nanoparticles using proteins produced by *Streptomyces naganishii*. Manikprabhu and Lingappa (2013a) reported synthesis of silver nanoparticles using pigments while Manivasagan et al. (2015) reported synthesis using a bio-flocculant produced by actinobacterial strains. Some studies also suggested that organic functional groups of microbial cell walls could be responsible for the synthesis of silver nanoparticles (Baker et al., 2013); for example, *Lactobacillus* sp. A09 has been reported to synthesize silver nanoparticles by the interaction of the silver ions with the groups on the microbial cell wall (Fu et al., 2000).

Fungi have also been considered as an excellent source to synthesize silver nanoparticles. They offer high tolerance to metals, secrete large quantities of extracellular proteins that provide stability to the nanoparticles, and provide an excellent amount of biomass that can be easily separated (Guilger-Casagrande and Lima, 2019). Fungi have been reported to synthesize silver nanoparticles both intracellularly (Mukherjee et al., 2001) and extracellularly (Ahmad et al., 2003). The mechanism of synthesis is similar to bacterial synthesis. Several fungal strains have been reported to synthesize silver nanoparticles of different sizes and shapes (Table 3).

Table 2 Bacteria-mediated synthesis of silver nanoparticles.

Sl. no.	Bacterial strains	Shape	Size	References
1	*Nocardiopsis dassonvillei*	Circular and nonuniform	30–80 nm	Dhanaraj et al. (2020)
2	*Bacillus thuringiensis*	Spherical	42 nm	Khaleghi et al. (2019)
3	*Bacillus brevis*	Spherical	41–68 nm	Saravanan et al. (2018)
4	*Streptomyces coelicolor*	Irregular	28–50 nm	Manikprabhu and Lingappa (2013a)
5	*Bhargavaea indica*	Anobar, pentagon, spherical, icosahedron, hexagonal, truncated triangle, and triangular	30–100 nm	Singh et al. (2015a)
6	*Isoptericola* sp.	Spherical	11–40 nm	Dong et al. (2017)
7	*Aneurinibacillus migulanus*	Spherical, oval, hexagonal, cubic and triangular	20–25 nm	Syed et al. (2019)
8	*Streptomyces griseorubens*	Spherical	5–20 nm	Baygar and Ugur (2017)
9	*Sinomonas mesophila*	Spherical	4–50 nm	Manikprabhu et al. (2016)
10	*Bacillus siamensis*	Spherical	25–50 nm	Ibrahim et al. (2019)
11	*Cupriavidus* sp.	Spherical	10–50 nm	Ameen et al. (2020)
12	*Bacillus cereus* A1-5 and *Streptomyces noursei* H1-1	Spherical	6–50 nm, 6–30 nm	Alsharif et al. (2020)
13	*Bacillus safensis*	Spherical	22.77–45.98 nm	Ahmed et al. (2020)
14	*Streptomyces spongiicola*	Spherical	22 nm	Devagi et al. (2020)
15	*Pseudomonas poae*	Spherical	19.8–44.9 nm	Ibrahim et al. (2020)

Table 3 Synthesis of silver nanoparticles using fungi.

Sl. no.	Fungal strains	Size (shape)	References
1	White rot fungi	15–25 nm (spherical to round)	Gudikandula et al. (2017)
2	*Trichoderma longibrachiatum*	1–25 nm (spherical)	Elamawi et al. (2018)
3	*Rhizopus arrhizus* IPT1011, *Rhizopus arrhizus* IPT1013, *Trichoderma gamsii* IPT853, and *Aspergillus niger* IPT856	30–100 nm (round)	Ottoni et al. (2017)
4	*Ganoderma lucidum*	15–22 nm (spherical)	Aygün et al. (2020)
5	*Guignardia mangiferae*	5–30 nm (spherical)	Balakumaran et al. (2015)
6	*Rhizopus stolonifera*	9.47 nm (spherical)	AbdelRahim et al. (2017)
7	*Fusarium oxysporum*	1–50 nm (spherical)	Srivastava et al. (2019)
8	*Aspergillus flavus*	50 nm (spherical and hexagonal)	Fatima et al. (2016)
9	*Raphanus sativus*	4–30 nm (spherical)	Singh et al. (2017)
10	*Penicillium atramentosum*	5–25 nm (spherical)	Sarsar et al. (2015)
11	*Rhizopus stolonifer* A6-2	6–40 nm (spherical)	Alsharif et al. (2020)
12	*Setosphaeria rostrata*	2–20 nm (spherical)	Akther et al. (2020)
13	*Penicillium oxalicum*	60–80 nm (spherical)	Feroze et al. (2020)
14	*Aspergillus brunneoviolaceus*	0.72–15.21 nm (spherical)	Mistry et al. (2021)
15	*Cladosporium perangustum*	30–40 nm (spherical)	Govindappa et al. (2020)
16	*Penicillium verrucosum*	10–12 nm (polydispersed)	Yassin et al. (2021)

3 Applications

The effectiveness of silver nanoparticles as promising antimicrobial agents has attracted tremendous interest in their use in biomedicinal applications. Silver nanoparticles show good antimicrobial activity against both gram-positive and gram-negative bacteria. The silver nanoparticles synthesized using *Streptomyces coelicolor* pigment has been reported to show good antibacterial activity against gram-positive bacteria (Manikprabhu and Lingappa, 2013a). The silver nanoparticles synthesized using pu-erh tea leaves extract has been reported to demonstrate good antibacterial activity against gram-negative bacteria (Loo et al., 2018). Silver nanoparticles also showed antibacterial against multidrug-resistant pathogens. Silver nanoparticles synthesized using *Murraya koenigii* leaves showed good antibacterial activity against extended-spectrum β-lactamase and methicillin-resistant bacteria (Qais et al., 2019). Abalkhil et al. (2017) reported that the antibacterial activity of silver nanoparticles (synthesized using *Aloe vera*, *Portulaca oleracea*, and *Cynodon dactylon* extract) depends on the kind of test pathogens used. The study suggested that gram-positive bacteria were more susceptible than gram-negative bacteria. Reports also suggest that the antimicrobial activity of silver nanoparticles depends on the size, shape, and concentration (Pal et al., 2007; Dong et al., 2019). Some studies propose that silver nanoparticles, when combined with antibiotics, showed increased activity. Adil et al. (2019) reported that silver nanoparticles, when combined with an antibiotic, showed better antibacterial activity compared to silver nanoparticles alone.

Silver nanoparticles have also been reported to have antifungal activity. Lara et al. (2015) reported the antifungal activity of silver nanoparticles against *Candida albicans*. Silver nanoparticles can also be used for wound healing. Silver nanoparticles hydrogel synthesized using *Arnebia nobilis* root extract showed wound-healing potential due to silver nanoparticles' antimicrobial potential (Garg et al., 2014). Apart from antibacterial activity, Nesrin et al. (2020) reported antibiofilm and cytotoxic activities of silver nanoparticles synthesized using *Rhododendron ponticum*. Silver nanoparticles synthesized using *Tropaeolum majus* showed antibacterial, antifungal, antioxidant, and anticancer activity (Valsalam et al., 2019). Duong et al. (2016) suggested using silver nanoparticles to control *Cyanobacteria* bloom. *Musa paradisiaca*-synthesized silver nanoparticles showed mosquitocidal, antimalarial, and antidiabetic potential (Anbazhagan et al., 2017). Silver nanoparticles have also been used to control environmental pollutants. Silver nanoparticles synthesized using exopolysaccharide produced by *Leuconostoc lactis* showed the degradation ability of azo-dyes (Saravanan et al., 2017).

4 Conclusion and future perspectives

Silver nanoparticles are a boon to humankind and are being used in different fields. Various methods have been reported to synthesize silver nanoparticles, and biological methods have been considered as a safe, low-cost, and eco-friendly approach. Biological synthesis using microbes has several advantages over plant-mediated synthesis. Although a large number of microorganisms have been reported to synthesize silver nanoparticles, there is still plenty of scope for research. Hence studies have been conducted to synthesize silver nanoparticles and identify their applications in different areas.

References

Abalkhil, T.A., Alharbi, S.A., Salmen, S.H., Wainwright, M., 2017. Bactericidal activity of biosynthesized silver nanoparticles against human pathogenic bacteria. Biotechnol. Biotechnol. Equip. 31, 411–417.

AbdelRahim, K., Mahmoud, S.Y., Ali, A.M., Almaary, K.S., Mustafa, A.E., Husseiny, S.M., 2017. Extracellular biosynthesis of silver nanoparticles using *Rhizopus stolonifer*. Saudi J. Biol. Sci. 24, 208–216.

Adil, M., Khan, T., Aasim, M., Khan, A.A., Ashraf, M., 2019. Evaluation of the antibacterial potential of silver nanoparticles synthesized through the interaction of antibiotic and aqueous callus extract of *Fagonia indica*. AMB Express 9, 75.

Agnihotri, S., Mukherji, S., Mukherji, S., 2014. Size-controlled silver nanoparticles synthesized over the range 5–100 nm using the same protocol and their antibacterial efficacy. RSC Adv. 4, 3974–3983.

Ahmad, A., Mukherjee, P., Senapati, S., Mandal, D., Khan, M.I., Kumar, R., Sastry, M., 2003. Extracellular biosynthesis of silver nanoparticles using the fungus *Fusarium oxysporum*. Colloids Surf. B. Biointerfaces 28, 313–318.

Ahmed, S., Ahmad, M., Swami, B.L., Ikram, S., 2016. A review on plants extract mediated synthesis of silver nanoparticles for antimicrobial applications: a green expertise. J. Adv. Res. 7, 17–28.

Ahmed, T., Shahid, M., Noman, M., Bilal Khan Niazi, M., Zubair, M., Almatroudi, A., Khurshid, M., Tariq, F., Mumtaz, R., Li, B., 2020. Bioprospecting a native silver-resistant *Bacillus safensis* strain for green synthesis and subsequent antibacterial and anticancer activities of silver nanoparticles. J. Adv. Res. 17, 475–483.

Akther, T., Khan, M.S., Hemalatha, S., 2020. Biosynthesis of silver nanoparticles via fungal cell filtrate and their anti-quorum sensing against *Pseudomonas aeruginosa*. J. Environ. Chem. Eng. 8, 104365. https://doi.org/10.1016/j.jece.2020.104365.

Al-Otibi, F., Al-Ahaidib, R.A., Alharbi, R.I., Al-Otaibi, R.M., Albasher, G., 2020. Antimicrobial potential of biosynthesized silver nanoparticles by *Aaronsohnia factorovskyi* extract. Molecules 26 (1), 130.

Alsharif, S.M., Salem, S.S., Abdel-Rahman, M.A., Fouda, A., Eid, A.M., El-Din Hassan, S., Awad, M.A., Mohamed, A.A., 2020. Multifunctional properties of spherical silver nanoparticles fabricated by different microbial taxa. Heliyon 6 (5), e03943.

Ameen, F., AlYahya, S., Govarthanan, M., ALjahdali, N., Al-Enazi, N., Alsamhary, K., Alshehri, W.A., Alwakeel, S.S., Alharbi, S.A., 2020. Soil bacteria *Cupriavidus* sp. mediates the extracellular synthesis of antibacterial silver nanoparticles. J. Mol. Struct. 1202, 127233. https://doi.org/10.1016/j.molstruc.2019.127233.

Ammosova, L., Jiang, Y., Suvanto, M., Pakkanen, T.A., 2017. Precise micropatterning of silver nanoparticles on plastic substrates. Appl. Surf. Sci. 401, 353–361.

Anbazhagan, P., Murugan, K., Jaganathan, A., Sujitha, V., Samidoss, C.M., Jayashanthani, S., Amuthavalli, P., Higuchi, A., Kumar, S., Wei, H., Nicoletti, M., Canale, A., Benelli, G., 2017. Mosquitocidal, antimalarial and antidiabetic potential of *Musa paradisiaca*-synthesized silver nanoparticles: in vivo and in vitro approaches. J. Clust. Sci. 28, 91–107.

Aref, M.S., Salem, S.S., 2020. Bio-callus synthesis of silver nanoparticles, characterization, and antibacterial activities via *Cinnamomum camphora* callus culture. Biocatal. Agric. Biotechnol. https://doi.org/10.1016/j.bcab.2020.101689.

Arshad, H., Sami, M.A., Sadaf, S., Hassan, U., 2021. *Salvadora persica* mediated synthesis of silver nanoparticles and their antimicrobial efficacy. Sci. Rep. 11, 5996.

Aygün, A., Özdemir, S., Gülcan, M., Cellat, K., Şen, F., 2020. Synthesis and characterization of Reishi mushroom-mediated green synthesis of silver nanoparticles for the biochemical applications. J. Pharm. Biomed. Anal. 178, 112970.

Baker, S., Harini, B.P., Rakshith, D., Satish, S., 2013. Marine microbes: invisible nanofactories. J. Pharm. Res. 6, 383–388.

Balakumaran, M.D., Ramachandran, R., Kalaichelvan, P.T., 2015. Exploitation of endophytic fungus, *Guignardia mangiferae* for extracellular synthesis of silver nanoparticles and their in vitro biological activities. Microbiol. Res. 178, 9–17.

Barros, C.H.N., Cruz, G.C.F., Mayrink, W., Tasic, L., 2018. Bio-based synthesis of silver nanoparticles from orange waste: effects of distinct biomolecule coatings on size, morphology, and antimicrobial activity. Nanotechnol. Sci. Appl. 11, 1–14.

Baygar, T., Ugur, A., 2017. Biosynthesis of silver nanoparticles by *Streptomyces griseorubens* isolated from soil and their antioxidant activity. IET Nanobiotechnol. 11, 286–291.

Beeler, E., Choy, N., Franks, J., Mulcahy, F., Singh, O.V., 2020. Extracellular synthesis and characterization of silver nanoparticles from alkaliphilic *Pseudomonas* sp. J. Nanosci. Nanotechnol. 20, 1567–1577.

Behravan, M., Hossein Panahi, A., Naghizadeh, A., Ziaee, M., Mahdavi, R., Mirzapour, A., 2019. Facile green synthesis of silver nanoparticles using *Berberis vulgaris* leaf and root aqueous extract and its antibacterial activity. Int. J. Biol. Macromol. 124, 148–154.

Burduşel, A.C., Gherasim, O., Grumezescu, A.M., Mogoantă, L., Ficai, A., Andronescu, E., 2018. Biomedical applications of silver nanoparticles: an up-to-date overview. Nanomaterials 8 (9), 681.

Chartarrayawadee, W., Charoensin, P., Saenma, J., Rin, T., Khamai, P., Nasomjai, P., On Too, C., 2020. Green synthesis and stabilization of silver nanoparticles using *Lysimachia foenum-graecum* Hance extract and their antibacterial activity. Green Proces. Synth. 9, 107–118.

Chaturvedi, R., Jha, C.B., 2011. Standard manufacturing procedure of Rajata Bhasma. AYU 32, 566–571.

Chekin, F., Ghasemi, S., 2014. Silver nanoparticles prepared in presence of ascorbic acid and gelatin, and their electrocatalytic application. Bull. Mater. Sci. 37, 1433–1437.

Chen, J., Mela, P., Möller, M., Lensen, M.C., 2009. Microcontact deprinting: a technique to pattern gold nanoparticles. ACS Nano 3, 1451–1456.

Choi, J.H., An, M.Y., Lee, B.M., Kim, D.K., Jung, C.H., Hwang, I.T., Lee, J.S., Nho, Y.C., Shin, K., Huh, K.M., Hong, S.K., 2009. Micropatterning of poly(vinyl pyrrolidone)/silver nanoparticle thin films by ion irradiation. J. Nanosci. Nanotechnol. 9 (12), 7090–7093.

Choi, J.S., Jung, H.C., Baek, Y.J., Kim, B.Y., Lee, M.W., Kim, H.D., Kim, S.W., 2021. Antibacterial activity of green-synthesized silver nanoparticles using *Areca catechu* extract against antibiotic-resistant bacteria. Nanomaterials 11 (1), 205.

Devagi, P., Suresh, T.C., Sandhiya, R.V., Sairandhry, M., Bharathi, S., Velmurugan, P., Radhakrishnan, M., Sathiamoorthi, T., Suresh, G., 2020. Actinobacterial-mediated fabrication of silver nanoparticles and their broad spectrum antibacterial activity against clinical pathogens. J. Nanosci. Nanotechnol. 20, 2902–2910.

Dhanaraj, S., Thirunavukkarasu, S., Allen John, H., Pandian, S., Salmen, S.H., Chinnathambi, A., Alharbi, S.A., 2020. Novel marine *Nocardiopsis dassonvillei*-DS013 mediated silver nanoparticles characterization and its bactericidal potential against clinical isolates. Saudi J. Biol. Sci. 27, 991–995.

Dhand, C., Dwivedi, N., Loh, X.J., Ying, A.N.J., Verma, N.K., Beuerman, R.W., Lakshminarayanan, R., Ramakrishna, S., 2015. Methods and strategies for the synthesis of diverse nanoparticles and their applications: a comprehensive overview. RSC Adv. 5, 105003–105037.

Dong, Z.Y., Narsing Rao, M.P., Xiao, M., Wang, H.F., Hozzein, W.N., Chen, W., Li, W.J., 2017. Antibacterial activity of silver nanoparticles against *Staphylococcus warneri* synthesized using endophytic bacteria by photo-irradiation. Front. Microbiol. 8, 1090.

Dong, Y., Zhu, H., Shen, Y., Zhang, W., Zhang, L., 2019. Antibacterial activity of silver nanoparticles of different particle size against *Vibrio natriegens*. PLoS One 14 (9), e0222322.

Duong, T.T., Le, T.S., Tran, T.T.H., Nguyen, T.K., Ho, C.T., Dao, T.H., Le, T.P.Q., Nguyen, H.C., Dang, D.K., Le, T.T.H., 2016. Inhibition effect of engineered silver nanoparticles to bloom forming cyanobacteria. Adv. Nat. Sci. Nanosci. Nanotechnol. 7, 035018.

Edison, T.N.J.I., Apchudan, R., Lee, Y.R., 2016. Optical sensor for dissolved ammonia through the green synthesis of silver nanoparticles by fruit extract of *Terminalia chebula*. J. Clust. Sci. 27, 683–690.

Elamawi, R.M., Al-Harbi, R.E., Hendi, A.A., 2018. Biosynthesis and characterization of silver nanoparticles using *Trichoderma longibrachiatum* and their effect on phytopathogenic fungi. Egypt. J. Biol. Pest. Control. https://doi.org/10.1186/s41938-018-0028-1.

Elegbede, J.A., Lateef, A., Azeez, M.A., Asafa, T.B., Yekeen, T.A., Oladipo, I.C., Adebayo, E.A., Beukes, L.S., Gueguim-Kana, E.B., 2018. Fungal xylanases-mediated synthesis of silver nanoparticles for catalytic and biomedical applications. IET Nanobiotechnol. 12, 857–863.

Fatima, F., Verma, S.R., Pathak, N., Bajpai, P., 2016. Extracellular mycosynthesis of silver nanoparticles and their microbicidal activity. J. Glob. Antimicrob. Resist. 7, 88–92.

Fernández, J.G., Fernández-Baldo, M.A., Berni, E., Camí, G., Durán, N., Raba, J., Sanz, M.I., 2016. Production of silver nanoparticles using yeasts and evaluation of their antifungal activity against phytopathogenic fungi. Process Biochem. 51, 1306–1313.

Feroze, N., Arshad, B., Younas, M., Afridi, M.I., Saqib, S., Ayaz, A., 2020. Fungal mediated synthesis of silver nanoparticles and evaluation of antibacterial activity. Microsc. Res. Tech. 83, 72–80.

Fu, J.K., Liu, Y.Y., Gu, P.Y., Tang, D.L., Lin, Z.Y., Yao, B.X., Weng, S.Z., 2000. Spectroscopic characterization on the biosorption and bioreduction of Ag(I) by *Lactobacillus* sp. A09. Acta Phys. Chim. Sin. 16, 770–782.

Garg, S., Chandra, A., Mazumder, A., Mazumder, R., 2014. Green synthesis of silver nanoparticles using *Arnebia nobilis* root extract and wound healing potential of its hydrogel. Asian J. Pharm. 8, 95–101.

Gavamukulya, Y., Maina, E.N., Meroka, A.M., Madivoli, E.S., El-Shemy, H.A., Wamunyokoli, F., Magoma, G., 2020. Green synthesis and characterization of highly stable silver nanoparticles from ethanolic extracts of fruits of *Annona muricata*. J. Inorg. Organomet. Polym. 30, 1231–1242.

Ghramh, H.A., Ibrahim, E.H., Kilnay, M., Ahmad, Z., Alhag, S.K., Khan, K.A., Taha, R., Asiri, F.M., 2020. Silver nanoparticle production by *Ruta graveolens* and testing its safety, bioactivity, immune modulation, anticancer, and insecticidal potentials. Bioinorg. Chem. Appl. 2020. https://doi.org/10.1155/2020/5626382, 5626382.

Gour, A., Jain, N.K., 2019. Advances in green synthesis of nanoparticles. Artif. Cells Nanomed. Biotechnol. 47, 844–851.

Govindappa, M., Lavanya, M., Aishwarya, P., Pai, K., Lunked, P., Hemashekhar, B., Arpitha, B.M., Ramachandra, Y.L., Raghavendra, V.B., 2020. Synthesis and characterization of endophytic fungi, *Cladosporium perangustum* mediated silver nanoparticles and their antioxidant, anticancer and nano-toxicological study. BioNanoScience. https://doi.org/10.1007/s12668-020-00719-z.

Gudikandula, K., Vadapally, P., Charya, M.A.S., 2017. Biogenic synthesis of silver nanoparticles from white rot fungi: their characterization and antibacterial studies. OpenNano 2, 64–78.

Guilger-Casagrande, M., Lima, R., 2019. Synthesis of silver nanoparticles mediated by fungi: a review. Front. Bioeng. Biotechnol. 7, 287. https://doi.org/10.3389/fbioe.2019.00287.

Gul, A.R., Shaheen, F., Rafique, R., Bal, J., Waseem, S., Park, T.J., 2021. Grass-mediated biogenic synthesis of silver nanoparticles and their drug delivery evaluation: a biocompatible anti-cancer therapy. Chem. Eng. Technol. https://doi.org/10.1016/j.cej.2020.127202.

Haefeli, C., Franklin, C., Hardy, K., 1984. Plasmid-determined silver resistance in *Pseudomonas stutzeri* isolated from silver mine. J. Bacteriol. 158, 389–392.

Haider, A., Kang, I.K., 2015. Preparation of silver nanoparticles and their industrial and biomedical applications: a comprehensive review. Adv. Mater. Sci. Eng. https://doi.org/10.1155/2015/165257.

Ibrahim, E., Fouad, H., Zhang, M., Zhang, Y., Qiu, W., Yan, C., Li, B., Mo, J., Chen, J., 2019. Biosynthesis of silver nanoparticles using endophytic bacteria and their role in inhibition of rice pathogenic bacteria and plant growth promotion. RSC Adv. 9, 29293–29299.

Ibrahim, E., Zhang, M., Zhang, Y., Hossain, A., Qiu, W., Chen, Y., Wang, Y., Wu, W., Sun, G., Li, B., 2020. Green-synthesization of silver nanoparticles using endophytic bacteria isolated from garlic and its antifungal activity against wheat *Fusarium* head blight pathogen *Fusarium graminearum*. Nanomaterials 10, 219.

Ibrahim, S., Ahmad, Z., Manzoor, M.Z., Mujahid, M., Faheem, Z., Adnan, A., 2021. Optimization for biogenic microbial synthesis of silver nanoparticles through response surface methodology, characterization, their antimicrobial, antioxidant, and catalytic potential. Sci. Rep. 11 (1), 770.

Iravani, S., Korbekandi, H., Mirmohammadi, S.V., Zolfaghari, B., 2014. Synthesis of silver nanoparticles: chemical, physical and biological methods. Res. Pharm. Sci. 9, 385–406.

Jain, A., Ahmad, F., Gola, D., Malik, A., Chauhan, N., Dey, P., Tyagi, P.K., 2020. Multi dye degradation and antibacterial potential of papaya leaf derived silver nanoparticles. Environ. Nanotechnol. Monit. Manag. 14, 100337.

Jeevanandam, J., Barhoum, A., Chan, Y.S., Dufresne, A., Danquah, M.K., 2018. Review on nanoparticles and nanostructured materials: history, sources, toxicity and regulations. Beilstein J. Nanotechnol. 9, 1050–1074.

Kambale, E.K., Nkanga, C.I., Mutonkole, B.I., Bapolisi, A.M., Tassa, D.O., Liesse, J.I., Krause, R.W.M., Memvanga, P.B., 2020. Green synthesis of antimicrobial silver nanoparticles using aqueous leaf extracts from three Congolese plant species (*Brillantaisia patula*, *Crossopteryx febrifuga* and *Senna siamea*). Heliyon 6 (8), e04493.

Kaviya, S., Santhanalakshmi, J., Viswanathan, B., Muthumary, J., Srinivasan, K., 2011. Biosynthesis of silver nanoparticles using *Citrus sinensis* peel extract and its antibacterial activity. Spectrochim. Acta A Mol. Biomol. Spectrosc. 79, 594–598.

Keshari, A.K., Srivastava, R., Singh, P., Yadav, V.B., Nath, G., 2020. Antioxidant and antibacterial activity of silver nanoparticles synthesized by *Cestrum nocturnum*. J. Ayurveda Integr. Med. 11, 37–44.

Khaleghi, M., Khorrami, S., Ravan, H., 2019. Identification of *Bacillus thuringiensis* bacterial strain isolated from the mine soil as a robust agent in the biosynthesis of silver nanoparticles with strong antibacterial and anti-biofilm activities. Biocatal. Agric. Biotechnol. 18, 101047. https://doi.org/10.1016/j.bcab.2019.101047.

Lara, H.H., Romero-Urbina, D.G., Pierce, C., Lopez-Ribot, J.L., Arellano-Jimenez, M.J., Jose-Yacaman, M., 2015. Effect of silver nanoparticles on *Candida albicans* biofilms: an ultrastructural study. J. Nanobiotechnol. 13, 91.

Loo, Y.Y., Rukayadi, Y., Nor-Khaizura, M.A.R., Kuan, C.H., Chieng, B.W., Nishibuchi, M., Radu, S., 2018. In vitro antimicrobial activity of green synthesized silver nanoparticles against selected gram-negative foodborne pathogens. Front. Microbiol. 9, 1555.

Manik, U.P., Nande, A., Raut, S., Dhoble, S.J., 2020. Green synthesis of silver nanoparticles using plant leaf extraction of *Artocarpus heterophyllus* and *Azadirachta indica*. Results Mater. 6, 100086.

Manikprabhu, D., Li, W.J., 2015. Microbe-mediated synthesis of silver nanoparticles a new drug of choice against pathogenic microorganisms. In: Dhanasekaran, D., Thajuddin, N., Panneerselvam, A. (Eds.), Antimicrobials: Synthetic and Natural Compounds. CRC Press, Boca Raton, FL, pp. 389–401.

Manikprabhu, D., Lingappa, K., 2013a. Antibacterial activity of silver nanoparticles against methicillin-resistant *Staphylococcus aureus* synthesized using model *Streptomyces* sp. pigment by photo-irradiation method. J. Pharm. Res. 6, 255–260.

Manikprabhu, D., Lingappa, K., 2013b. Microwave assisted rapid and green synthesis of silver nanoparticles using a pigment produced by *Streptomyces coelicolor* klmp33. Bioinorg. Chem. Appl. 2013, 341798.

Manikprabhu, D., Lingappa, K., 2014. Synthesis of silver nanoparticles using the *Streptomyces coelicolor* klmp33 pigment: an antimicrobial agent against extended-spectrum beta-lactamase (ESBL) producing *Escherichia coli*. Mater. Sci. Eng. C 45, 434–437.

Manikprabhu, D., Cheng, J., Chen, W., Sunkara, A.K., Mane, S.B., Kumar, R., das, M., Hozzein, W.N., Duan, Y.Q., Li, W.J., 2016. Sunlight mediated synthesis of silver nanoparticles by a novel actinobacterium (*Sinomonas mesophila* MPKL 26) and its antimicrobial activity against multi drug resistant *Staphylococcus aureus*. J. Photochem. Photobiol. B 158, 202–205.

Manivasagan, P., Kang, K.H., Kim, D.G., Kim, S.K., 2015. Production of polysaccharide-based bioflocculant for the synthesis of silver nanoparticles by *Streptomyces* sp. Int. J. Biol. Macromol. 77, 159–167.

Mistry, H., Thakor, R., Patil, C., Trivedi, J., Bariya, H., 2021. Biogenically proficient synthesis and characterization of silver nanoparticles employing marine procured fungi *Aspergillus brunneoviolaceus* along with their antibacterial and antioxidative potency. Biotechnol. Lett. 43, 307–316.

Mukherjee, P., Ahmad, A., Mandal, D., Senapati, S., Sainkar, S.R., Khan, M.I., Parishcha, R., Kumar, A.P.V., Alam, M., Kumar, R., Sastry, M., 2001. Fungus-mediated synthesis of silver nanoparticles and their immobilization in the mycelial matrix: a novel biological approach to nanoparticles synthesis. Nano Lett. 1, 515–519.

Mukherjee, K., Gupta, R., Kumar, G., Kumari, S., Biswas, S., Padmanabhan, P., 2018. Synthesis of silver nanoparticles by *Bacillus clausii* and computational profiling of nitrate reductase enzyme involved in production. J. Genet. Eng. Biotechnol. 16, 527–536.

Nesrin, K., Yusuf, C., Ahmet, K., Ali, S.B., Muhammad, N.A., Suna, S., Fatih, S., 2020. Biogenic silver nanoparticles synthesized from *Rhododendron ponticum* and their antibacterial, antibiofilm and cytotoxic activities. J. Pharm. Biomed. Anal. 179, 112993.

Odeniyi, M.A., Okumah, V.C., Adebayo-Tayo, B.C., Odeniyi, O.A., 2020. Green synthesis and cream formulations of silver nanoparticles of *Nauclea latifolia* (African peach) fruit extracts and evaluation of antimicrobial and antioxidant activities. Sustain. Chem. Pharm. 15, 100197.

Okaiyeto, K., Hoppe, H., Okoh, A.I., 2021. Plant-based synthesis of silver nanoparticles using aqueous leaf extract of *Salvia officinalis*: characterization and its antiplasmodial activity. J. Clust. Sci. 32, 101–109.

Otari, S.V., Patil, R.M., Ghosh, S.J., Thorat, N.D., Pawar, S.H., 2015. Intracellular synthesis of silver nanoparticle by actinobacteria and its antimicrobial activity. Spectrochim. Acta A Mol. Biomol. Spectrosc. 5, 1175–1180.

Ottoni, C.A., Simões, M.F., Fernandes, S., Dos Santos, J.G., da Silva, E.S., de Souza, R.F.B., Maiorano, A.E., 2017. Screening of filamentous fungi for antimicrobial silver nanoparticles synthesis. AMB Express 7 (1), 31.

Pal, S., Tak, Y.K., Song, J.M., 2007. Does the antibacterial activity of silver nanoparticles depend on the shape of the nanoparticle? A study of the Gram-negative bacterium *Escherichia coli*. Appl. Environ. Microbiol. 73, 1712–1720.

Pareek, V., Bhargava, A., Gupta, R., Jain, N., Panwar, J., 2017. Synthesis and applications of noble metal nanoparticles: a review. Adv. Sci. Eng. Med. 9, 527–544.

Patil, M.P., Singh, R.D., Koli, P.B., Patil, K.T., Jagdale, B.S., Tipare, A.R., Kim, G.D., 2018. Antibacterial potential of silver nanoparticles synthesized using *Madhuca longifolia* flower extract as a green resource. Microb. Pathog. 121, 184–189.

Pingali, K.C., Rockstraw, D.A., Deng, S., 2005. Silver nanoparticles from ultrasonic spray pyrolysis of aqueous silver nitrate. Aerosol Sci. Technol. 39, 1010–1014.

Prabhu, S., Poulose, E.K., 2012. Silver nanoparticles: mechanism of antimicrobial action, synthesis, medical applications, and toxicity effects. Int. Nano Lett. 32, 1–10.

Qais, F.A., Shafiq, A., Khan, H.M., Husain, F.M., Khan, R.A., Alenazi, B., Alsalme, A., Ahmad, I., 2019. Antibacterial effect of silver nanoparticles synthesized using *Murraya koenigii* (L.) against multidrug-resistant pathogens. Bioinorg. Chem. Appl. 2019, 4649506.

Rahman, A.U., Khan, A.U., Yuan, Q., Wei, Y., Ahmad, A., Ullah, S., Khan, Z.U.H., Shams, S., Tariq, M., Ahmad, W., 2019. Tuber extract of *Arisaema flavum* eco-benignly and effectively synthesize silver nanoparticles: photocatalytic and antibacterial response against multidrug resistant engineered *E. coli* QH4. J. Photochem. Photobiol. B 193, 31–38.

Ranoszek-Soliwoda, K., Tomaszewska, E., Socha, E., Krzyczmonik, P., Ignaczak, A., Orlowski, P., Krzyzowska, M., Celichowski, G., Grobelny, J., 2017. The role of tannic acid and sodium citrate in the synthesis of silver nanoparticles. J. Nanopart. Res. 19 (8), 273.

Renuka, R., Devi, K.R., Sivakami, M., Thilagavathi, T., Uthrakumar, R., Kaviyarasu, K., 2020. Biosynthesis of silver nanoparticles using *Phyllanthus emblica* fruit extract for antimicrobial application. Biocatal. Agric. Biotechnol. 24, 101567. https://doi.org/10.1016/j.bcab.2020.101567.

Rupiasih, N.N., Aher, A., Gosavi, S., Vidyasagar, P.B., 2013. Green synthesis of silver nanoparticles using latex extract of *Thevetia peruviana*: a novel approach towards poisonous plant utilization. J. Phys. Conf. Ser. 423, 1–8.

Saravanan, C., Rajesh, R., Kaviarasan, T., Muthukumar, K., Kavitake, D., Shetty, P.H., 2017. Synthesis of silver nanoparticles using bacterial exopolysaccharide and its application for degradation of azo-dyes. Biotechnol. Rep. (Amst.) 15, 33–40.

Saravanan, M., Barik, S.K., MubarakAli, D., Prakash, P., Pugazhendhi, A., 2018. Synthesis of silver nanoparticles from *Bacillus brevis* (NCIM 2533) and their antibacterial activity against pathogenic bacteria. Microb. Pathog. 116, 221–226.

Sarsar, V., Selwal, M.K., Selwal, K.K., 2015. Biofabrication, characterization and antibacterial efficacy of extracellular silver nanoparticles using novel fungal strain of *Penicillium atramentosum* KM. J. Saudi Chem. Soc. 19, 682–688.

Satsangi, N., Preet, S., 2021. Phyto-synthesis and characterization of potent larvicidal silver nanoparticles using *Nyctanthes arbortristis* leaf extract. J. Plant Biochem. Biotechnol. https://doi.org/10.1007/s13562-020-00639-9.

Shanmugasundaram, T., Radhakrishnan, M., Gopikrishnan, V., Pazhanimurugan, R., Balagurunathan, R., 2013. A study of the bactericidal, anti-biofouling, cytotoxic and antioxidant properties of actinobacterially synthesised silver nanoparticles. Colloids Surf. B Biointerfaces 111, 680–687.

Siddiqi, K.S., Husen, A., Rao, R.A.K., 2018. A review on biosynthesis of silver nanoparticles and their biocidal properties. J. Nanobiotechnol. 16 (1), 14.

Singh, P., Kim, Y.J., Singh, H., Mathiyalagan, R., Wang, C., Yang, D.C., 2015a. Biosynthesis of anisotropic silver nanoparticles by *Bhargavaea indica* and their synergistic effect with antibiotics against pathogenic microorganisms. J. Nanomater. https://doi.org/10.1155/2015/234741.

Singh, R., Shedbalkar, U.U., Wadhwani, S.A., Chopade, B.A., 2015b. Bacteriagenic silver nanoparticles: synthesis, mechanism, and applications. Appl. Microbiol. Biotechnol. 99, 4579–4593.

Singh, T., Jyoti, K., Patnaik, A., Singh, A., Chauhan, R., Chandel, S.S., 2017. Biosynthesis, characterization and antibacterial activity of silver nanoparticles using an endophytic fungal supernatant of *Raphanus sativus*. J. Genet. Eng. Biotechnol. 15, 31–39.

Singh, S.P., Mishra, A., Shyanti, R.K., Singh, R.P., Acharya, A., 2021. Silver nanoparticles synthesized using *Carica papaya* leaf extract (AgNPs-PLE) causes cell cycle arrest and apoptosis in human prostate (DU145) cancer cells. Biol. Trace Elem. Res. 199 (4), 1316–1331.

Sreeram, K.J., Nidhin, M., Nair, B.U., 2008. Microwave assisted template synthesis of silver nanoparticles. Bull. Mater. Sci. 31, 937–942.

Srivastava, S., Bhargava, A., Pathak, N., Srivastava, P., 2019. Production, characterization and antibacterial activity of silver nanoparticles produced by *Fusarium oxysporum* and monitoring of protein-ligand interaction through *in-silico* approaches. Microb. Pathog. 129, 136–145.

Syed, B., Prasad, M.N.N., Satish, S., 2019. Synthesis and characterization of silver nanobactericides produced by *Aneurinibacillus migulanus* 141, a novel endophyte inhabiting *Mimosa pudica* L. Arab. J. Chem. 12, 3743–3752.

Tanase, C., Berta, L., Coman, N.A., Roşca, I., Man, A., Toma, F., Mocan, A., Nicolescu, A., Jakab-Farkas, L., Biró, D., Mare, A., 2019. Antibacterial and antioxidant potential of silver nanoparticles biosynthesized using the spruce bark extract. Nanomaterials 9 (11), 1541.

Tripathi, D., Modi, A., Narayan, G., Rai, S.P., 2019. Green and cost effective synthesis of silver nanoparticles from endangered medicinal plant *Withania coagulans* and their potential biomedical properties. Mater. Sci. Eng. C 100, 152–164.

Tsuji, T., Iryo, K., Watanabe, N., Tsuji, M., 2002. Preparation of silver nanoparticles by laser ablation in solution: influence of laser wavelength on particle size. Appl. Surf. Sci. 202, 80–85.

Vaid, P., Raizada, P., Saini, A.K., Saini, R.V., 2020. Biogenic silver, gold and copper nanoparticles—a sustainable green chemistry approach for cancer therapy. Sustain. Chem. Pharm. 16, 100247.

Valsalam, S., Agastian, P., Arasu, M.V., Al-Dhabi, N.A., Ghilan, A.M., Kaviyarasu, K., Ravindran, B., Chang, S.W., Arokiyaraj, S., 2019. Rapid biosynthesis and characterization of silver nanoparticles from the leaf extract of *Tropaeolum majus* L. and its enhanced in-vitro antibacterial, antifungal, antioxidant and anticancer properties. J. Photochem. Photobiol. B 191, 65–74.

Viana, A.D., Nobrega, E.T.D., Moraes, E.P., Wanderley Neto, A.O., Menezes, F.G., Gasparotto, L.H.S., 2020. Castor oil derivatives in the environmentally friendly one-pot synthesis of silver nanoparticles: application in cysteine sensing. Mater. Res. 124, 110755.

Wei, S., Wang, Y., Tang, Z., Hu, J., Su, R., Lin, J., Zhou, T., Guo, H., Wang, N., Xu, R., 2020. A size-controlled green synthesis of silver nanoparticles by using the berry extract of *Sea Buckthorn* and their biological activities. New J. Chem. 44, 9304–9312.

Yaqoob, A.A., Umar, K., Ibrahim, M.N.M., 2020. Silver nanoparticles: various methods of synthesis, size affecting factors and their potential applications—a review. Appl. Nanosci. 10, 1369–1378.

Yassin, M.A., Elgorban, A.M., El-Samawaty, A.E.R.M.A., Almunqedhi, B.M.A., 2021. Biosynthesis of silver nanoparticles using *Penicillium verrucosum* and analysis of their antifungal activity. Saudi J. Biol. Sci. https://doi.org/10.1016/j.sjbs.2021.01.063.

Yousaf, H., Mehmood, A., Ahmad, K.S., Raffi, M., 2020. Green synthesis of silver nanoparticles and their applications as an alternative antibacterial and antioxidant agents. Mater. Sci. Eng. C Mater. Biol. Appl. 112, 110901. https://doi.org/10.1016/j.msec.2020.110901.

Yu, H.D., Regulacio, M.D., Ye, E., Han, M.Y., 2013. Chemical routes to top-down nanofabrication. Chem. Soc. Rev. 42, 6006–6018.

Yusuf, M., 2019. Silver nanoparticles: synthesis and applications. In: Martínez, L., Kharissova, O., Kharisov, B. (Eds.), Handbook of Ecomaterials. Springer, Cham, pp. 2343–2356.

Zhang, K., Liu, X., Samuel Ravi, S., Ramachandran, A., Aziz Ibrahim, I.A., Nassir, A.M., Yao, J., 2019. Synthesis of silver nanoparticles (AgNPs) from leaf extract of *Salvia miltiorrhiza* and its anticancer potential in human prostate cancer LNCaP cell lines. Artif. Cells Nanomed. Biotechnol. 47, 2846–2854.

Zia, F., Ghafoor, N., Iqbal, M., Mehboob, S., 2016. Green synthesis and characterization of silver nanoparticles using Cydonia oblong seed extract. Appl. Nanosci. 6, 1023.

CHAPTER 24

Silver nanoparticles synthesis mechanisms

Kangkana Banerjee and V. Ravishankar Rai
Department of Studies in Microbiology, University of Mysore, Mysore, India

1 Introduction

Silver nanoparticles are well-known for their traditional bactericidal properties against a wide range of microorganisms. However, their applications are also seen in optical and electronic devices, medical equipment, bio-imaging, environmental remediation, and so on. Hence, their synthesis from different perspectives has become extremely important. Considering their end application, silver nanoparticles of a varied range of sizes and morphologies are synthesized by either physical or chemical methods. Another well-known approach of synthesis is a green synthesis or the biogenic method involving bacteria, fungi, actinomycetes, or plant extracts. Green synthesis systems use eco-friendly agents for the reduction of material components like silver salts. Several studies have been conducted globally on the biosynthesis of silver nanoparticles (Siddiqi et al., 2018; Banerjee and Rai, 2017). Silver has natural antimicrobial properties that have been significantly used in skin care, wound healing, and water purification, apart from other minor applications in human health and electronics. Interaction of silver with the microbial cell membrane can destroy it and nullify infections in the human body (Sastry et al., 2003; Korbekandi et al., 2012; Sunkar and Nachiyar, 2012). However, there are microbes that are becoming silver-resistant. This has led to study of nanoparticles, which involves silver ions in healing ailments. Silver toxicity has been a major drawback in using silver nanoparticles for several therapies, but Castro-Longoria et al. (2011) explained the lower toxic threshold in the case of silver ions. Therefore, study of nanosilver synthesis has gained importance in the past few decades.

Various physical methods used in nanosilver synthesis are vapor condensation method involving rapid vaporization and condensation process, arc discharge method that applies direct current arc voltage, laser ablation method, and physical deposition method. Beside physical synthesis procedures, silver nanomaterials are also chiefly produced using chemical methods that include components like a precursor, reducing agent, and stabilizing agent. Photochemical and electrochemical methods are also often applied for nanosilver synthesis. The most studied and diversified synthesis method is biological. Silver nanomaterials synthesized in eco-friendly ways such as by plants or microbes are very relevant. Microbes and plants are easily available and are considered very low-cost materials. Plant or microbial extracts are extensively used in this process of green synthesis to reduce the metal salt of silver into particles generally measuring >100 nm. These natural extracts contain biomolecules like polysaccharides, vitamins, amino acids, proteins, and enzymes that help in the reduction process of metal ions to metal nanoparticles (Güzel and Erdal, 2018). Few bacterial and fungal extracts have the potential of reducing as well as stabilizing these particles. All the abovementioned methods are known to have two main approaches, "top-down" and "bottom-up," whereas the other two processes are intracellular and extracellular.

Biological processes of nanosilver synthesis have shown the formation of predominantly spherical-shaped particles; however, we can also control the size of the nanomaterial by simply increasing the incubation temperature for a specific period of time (Khodashenas and Ghorbani, 2019). A diverse group of endophytic fungi and fungal pathogens has been selected for synthesizing AgNPs. Scientists carried out diverse formulations to demonstrate the production of silver by fungi such as *Penicillium citrinum* (Bharde et al., 2006), *Rhizoctonia* sp. (Mukherjee et al., 2008), *Fusarium semitectum* (KSU-4) (Usha et al., 2010), *Aspergillus flavus* (Vahabi et al., 2011), and *Fusarium oxysporum* (Alghuthaymi et al., 2015). Furthermore, fungi like *Trichoderma viride* (Kumar et al., 2007), *Coriolus versicolor* (Thakkar et al., 2010), and *Trichoderma asperellum* (Nair et al., 2013) were also investigated for their ability to synthesize silver nanoparticles.

The mechanism of silver nanobiosynthesis involves reduction by electron transfer from NADH with the help of NADH-dependent reductase enzymes, which act as electron carriers (Nair et al., 2013). Proteins like α-NADPH-dependent nitrate reductase, phytochelatin, or hydrogenase enzymes are largely responsible for AgNP synthesis in most instances. Silver nitrate is widely used for bioreduction methods of AgNPs as silver salts, and the NPs are produced because of the aforementioned reducing agents, and also the stabilizing agents created by fungi. According to researchers, the negatively charged cell wall containing carboxylate groups interacts with the positively charged silver metals and thus produces nanosilver particles (Agnihotria et al., 2009). Fungal species like *Phanerochaete chrysosporium* showed secretion of reducing sugars that have also been identified to assist in the synthesis of AgNPs (Sen et al., 2011).

Nanoparticle characterization is mainly based on the size, surface charge, and morphology using such advanced microscopic techniques and light scattering systems like UV spectroscopy, X-ray diffraction (XRD), dynamic light scattering (DLS), scanning electron microscopy (SEM), transmission electron microscopy (TEM), atomic force microscopy (AFM), etc. The different properties of nanoparticles after synthesis such as particle diameter, size distribution, charge, etc. affect the stability and their in vivo distribution. In addition, physical stability and dispersibility in various solvents as well as their in vivo performance are influenced by the charge on the nanoparticles (Webster, 2006). Researchers have concluded that

UV-visible spectroscopy is very useful for analyzing different nanoparticles such as gold and silver nanoparticles. Strong surface plasmon peaks are observed for various metal nanoparticles in the visible range depending on the size of the particles. Two main standard methods for imaging and measurements of nanostructures include scanning and transmission electron microscopy. Furthermore, to understand the crystalline nature of these particles, X-ray diffraction is carried out.

This chapter focuses on the plausible mechanism of nanosilver synthesis routes as depicted by different scientific groups. The broadly suggested mechanism explains biosynthesis is the simplest process, involving less energy and time when compared to other methods. However, mechanisms of formation in bioreduction methods are more complicated when compared to physical or chemical methods (Fig. 1). In addition, the knowledge of this systemic process brings numerous opportunities for the utilization of biodegradable materials.

2 Mechanism of physical approach

The physical method of synthesis of nanoparticles involves an evaporation-condensation process. Light-assisted methods are also applied in the physical approach. In the photo-physical method, the size of the particles is controlled by absorption of the radiation, thus leading to fragmentation, whereas the evaporation-condensation method utilizes temperature to reduce particle size.

i. *Physical vapor condensation method*

Here, silver salts are vaporized by heat sources and rapidly condensed to achieve formation of silver nanoparticles. Jung et al. (2006) described this method using a small ceramic heater that produced monodispersed spherical stable silver nanoparticles.

ii. *Laser ablation method*

Tsuji et al. (2000) demonstrated a procedure of synthesis of silver nanoparticles using a laser beam ablated at a radiation wavelength of 1064 nm. Silver salt in water was reduced by Cl^- ions, and formed long-term stable AgNP particles. With the increase in absorption signal, an increase of silver nanoparticles was observed, detected by a plasmon absorbance band. Other researchers also explained processes where silicon was used instead of NaCl, in an $AgNO_3$ solution, where $\sim$11 nm diameter of AgNP was formed due to irradiation. Different laser wavelengths were also explored in other studies in order to control nanoparticle size (Tsuji et al., 2003).

iii. *Arc discharge method*

Apart from the laser ablation method and condensation process, the DC arc-discharge system is a technique employed by Tien et al. (2008). A pulse voltage was applied to silver wire immersed in an aqueous medium at specified intervals. This led to evaporation and condensation of the surface of the silver wire in water and thus formation of silver nanoparticles.

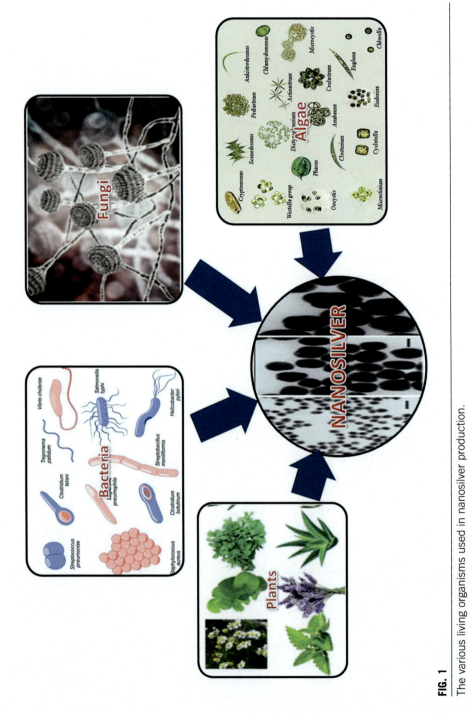

FIG. 1
The various living organisms used in nanosilver production.

3 Chemical synthesis and its mechanism

i. *Chemical reduction*

This synthesis process involves three major components: metal precursors, stabilizing/capping agents, and reducing agents. Several steps like reduction, nucleation, coarsening, and agglomeration take place, leading to the coprecipitation phenomenon. Tiny crystalline particles are formed and further capping agents are added to ensure the stability of these particles. The main components that are used as reducing agents are $NaBH_4$, polyols, sodium citrate (TSC), tri-sodium citrate (TSC), and N,N-dimethylformamide (DMF). The main stabilizing agents in silver nanoparticles preparation are sodium dodecyl sulfate, polymethylmethacrylate, polymethacrylic acid, and polyvinyl pyrrolidone (PVP). Researchers reported silver nanoparticle synthesis by various mechanisms of action. Gallic acid-mediated reduction of silver salts was carried out by Martínez-Castañón et al. (2008). This group attempted synthesis of different sizes, applying UV light to ionize the phenol particles. Here, gallic acid acted as a stabilizing agent as well. Similarly, another group of researchers explained the production mechanism of AgNPs synthesized by a polyol process catalyzed by supercritical carbon dioxide ($SCCO_2$) present in silver nitrate salt and stabilized by polyvinyl pyrrolidone (PVP) (Liang et al., 2009). Reduction, agglomeration, and stabilization are basic mechanisms involved in chemical synthesis of different shapes, sizes, and morphologies of silver nanoparticles.

ii. *Electrochemical methods*

Electro deposition of silver particles is achieved by this method. A vital system of mechanism in this case includes use of a cathode and anode that contributes in the reduction process of silver salts. A controlled temperature is an important factor in this process. PVP-like agents are used in the stabilization of the particles formed. Synthesis of silver nanoparticles using varied stabilizing agents by the electrochemical process are reported by different researchers (Rodriguez-Sanchez et al., 2000; Reetz and Helbig, 1994).

iii. *Photochemical method*

In this case, no reducing agents are required; however, the reduction takes place on irradiation of silver salts. Photoradiation or photolysis of salts leads to formation of nanoparticles using a laser or lamp followed by surfactant stabilization. Zaarour et al. (2014) demonstrated photochemical synthesis of polydispersed and small silver nanoparticles by microwaves in 45 min.

iv. *Others*

Nanocage-assisted synthesis and solid-assisted synthesis are types of chemical synthesis methods of silver nanoparticle formation. Nanocage-assisted production is an electrochemical technique that was used to form uniquely small-sized (1–2 nm), stable silver particles with the help of polymeric molecular cages. A certain manner of arrangements of these micropores serves as a sound barrier against agglomeration of nanoparticles, whereas graphene oxide is used as solid support for formation of stable silver particles by the microwave-assisted method. The platform provides an anchoring site for reduction of silver ions. A silica matrix was also used for the same purpose by other researchers (Dawadi et al., 2021).

4 Mechanism of biosynthesis process

Green chemistry is the science that relates to plant- or microbe-mediated synthesis of nanoparticles. Naturally biodegradable compounds found in bacteria and plant act as components that have potential in the bioremediation of toxic metals. As mentioned previously, two major approaches are "top-down," which deals with breaking down of bulk materials into tiny nanostructures, and "bottom-up," which involves building nanosized particles from single atoms or molecules in three main steps: nucleation, seeding, and growth (Khodashenas and Ghorbani, 2019; Lee and Jun, 2019). Green synthesis occurs through the former process, i.e., the "top-down" route. This includes the reduction of silver salts like silver nitrate into silver ions (Ag^+) and eventually to silver nuclei (Ag^0). This procedure results in the production of stable capped silver nanoparticles. The three perspectives that are important in green synthesis are an eco-friendly reducing agent, the solvent medium used, and a nontoxic capping agent for the stabilization process. Commonly employed microorganisms in nanosilver synthesis are *Escherichia coli*, *Pseudomonas* sp., *Staphylococcus* sp., *Bacillus* sp., *Fusarium* sp., *Aspergillus* sp., and *Bacillus cereus* (Sheoran and Kaur, 2018; Ibrahim et al., 2021a,b) (Table 1).

The mechanism behind the green synthesis of nanoparticles has not been thoroughly investigated. The technique of synthesis varies from microbe to microbe but the main principle is the reduction reaction. Mainly cellular peptides and polysaccharides are involved in both intracellular and extracellular routes which lead to enzymatic oxidation, absorption, reduction, and chelation. Extracellular formations of nanoparticles take place due to intermembranous transport and consequent nucleation (Saravanan and Nanda, 2010). It is also reported that the production of AgNPs occurs by protonated anionic functional groups present on the cell wall in lactic acid bacteria. In the case of the fungus *F. oxysporum*, silver ions are reduced with the help of an NADPH-dependent nitrate reductase enzyme by transferring an electron to it using NADP as a cofactor. AgNPs are also reduced by quinine derivatives of naphthoquinones and anthraquinones (Mukherjee et al., 2001). On the other hand, *Stenotrophomonas maltophilia*, a gram-negative bacterium, reduces gold ions Au^{3+} to Au^0 (Gade et al., 2008). Zinc sulfide (ZnS) nanoparticles synthesized by the bacteria *Rhodobacter sphaeroides* intracellularly secrete enzymes like sulfate permease, sulfite reductase, and sulfurylase (Namasivayam and Avimanyu, 2011); cadmium sulfide (CdS) nanoparticles are produced using a protein named cysteine desulfhydrase extracellularly secreted by *Rhodopseudomonas palustris* (Du et al., 2015). Extracellular synthesis of AgNPs is explained using *Verticillium* sp. and *F. oxysporum* (Banerjee and Rai, 2015a, 2016).

Table 1 List of fungal species synthesizing silver nanoparticles.

Species	Size (nm)	Morphology	Process	Reference
Aspergillus fumigatus BTCB10	~94	Cube shaped	Extracellular	Shahzad et al. (2019)
Pseudoduganella eburnea MAHUQ-39	8–24	Spherical	Extracellular	Huq (2020)
Tectona grandis	10–30	Oval, spherical	–	Rautela et al. (2019)
Botryococcus braunii	5	Spherical	Extracellular	Gallón et al. (2019)
Chlorella pyrenoidosa	15	Spherical	Extracellular	
Bacillus sp. KFU36	5–15	Spherical	–	Almalki and Khalifa (2020)
Cupriavidus sp.	10–50	Spherical, crystalline	Extracellular	Ameen et al. (2020)
Sphingobium sp. MAH-11	7–22	Spherical	–	Akter and Huq (2020)
Lactobacillus plantarum TA4	4–14	Spherical	–	Mohd Yusof et al. (2020)
Lactobacillus sp. strain LCM5	3–35	Spherical	Extracellular	Matei et al. (2020)
Bacillus sp. LBF-01	19	Spherical	–	

The silver ions in this case are found trapped on the fungal cell wall and thus reduced to nuclei, facilitated by the reductase enzyme present on the cell surfaces, and accumulation of these silver ions on the generated nuclei then occurs. In yeast, however, an oxidoreductase mechanism is observed. *Saccharomyces cerevisiae* produces cadmium telluride quantum dots (CdTe QDs) extracellularly with the help of protein ligands and phytochelatin synthase, which is used in the reduction of Cd^+ (Goswami et al., 2013; Banerjee and Rai, 2015b). Melanin is employed by the yeast *Yarrowia lipolytica* during the extracellular formation of silver as well as gold nanoparticles (Shugaba et al., 2012). To conclude, certain naturally produced enzymes and chemicals like nitrate reductase enzyme, sulfate permease, quinine derivatives of anthraquinones and naphthoquinones, cysteine desulfhydrase, sulfite reductase, phytochelatin synthase, sulfurylase, and myco-based melanin can be extracted from various microorganism autolysis or with lysis-promoting agents to produce a varied range of bio-nanoparticles in a simpler technique (Lee and Jun, 2019; Kim et al., 2018).

Basically, trapping, bioreduction, and capping of the nuclei produced are seen in cases utilizing the intracellular method. This method also involves ion transportation and electrostatic interaction between microbial cells and metal ions, which further leads to the formation of nanoparticles; alternatively the extracellular method includes the secretion of enzymes, bioreduction, and capping of particles. The most common enzyme isolated to date is nitrate reductase, which may be responsible for the majority of AgNP production methods (Moharrer et al., 2012). Nonetheless, Bharathidasan and Panneerselvam (2012) clarified that the extracellular method is always preferable for the synthesis of nanoparticles because downstreaming and purification processes are easier when compared to intracellular processes. Intracellular processes are very time-consuming and biosynthesis is likely to be costly. The extracellular method of synthesis can be economical and suitable compared to the intracellular method, which remains the chief method for biosynthesis (Fig. 2).

4.1 Bacterial-mediated synthesis

Bacterial tendency of reacting to stresses because of toxic heavy metals leads to reduction processes of metals. A few common resistance mechanisms include metal efflux systems, alteration of solubility, efflux pumps, inactivation of silver, changes in redox states of silver ions, impermeability to silver, extracellular precipitation of silver, absence of silver transport systems, and volatilization by enzymatic reactions (Iravani, 2014). Intracellular as well as extracellular approaches have been seen in bacterial synthesis mechanisms for silver nanoparticles. Due to electrostatic interactions in intracellular synthesis, the bacterial cell wall attracts the silver metal ions. The enzymes present on the cell wall subsequently reduce the metal ions into silver nanoparticles. Accumulation of the produced metallic nanoparticles takes place in the cell wall and the cytoplasmic or periplasmic components of the cell. This has been revealed during intracellular processes under a microscope. In the case of the extracellular mechanism of synthesis, production of nanoparticles happens outside the cells, where a reductase enzyme from the bacterial cell is secreted to reduce the metal ions. It was also observed by Deplanche et al. (2010) that mutant strains of the same bacterial species undergo altered mechanisms due to the presence or absence of certain genes in them. Similarly, different growth phases of bacteria determine the mechanism of silver reduction. Furthermore, it was evident that only enzymes present in bacteria do not contribute to biomineralization of silver; certain specific genes and proteins play a critical role in the method of biosynthesis (Schüler, 1999). In other cases, photoreduction of silver ions is significantly detected as the reaction fails to occur in dark conditions even in the presence of the required enzymes (Mokhtari et al., 2009; Garg et al., 2020).

4.2 Fungal-mediated synthesis

Fungi and yeast have been extensively used in silver nanoparticle synthesis by various researchers, so the mechanism has been studied in detail. The reports on fungal-mediated synthesis claim that nanoparticle formation takes place mainly extracellularly, which makes the downstream processing of silver nanoparticles that are formed incredibly simple. The silver nanoparticles are found to be formed on the surface of fungal mycelia more often than in the solution. This explains the

4 Mechanism of biosynthesis process

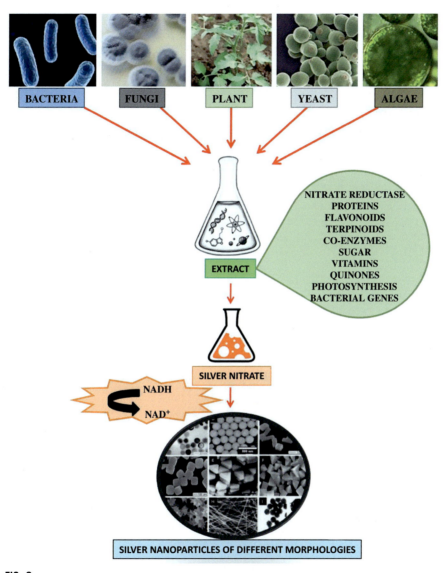

FIG. 2

Flowchart showing the mechanism of formation of nanoparticles.

electrostatic interaction of the silver ions toward the mycelial cells, followed by adsorption of the ions and reduction of these ions due to the carboxylate groups in enzymes (Table 2). The reports also confirm that fungi secrete higher amounts of enzymes or proteins involved in silver nanoparticle synthesis when compared to bacteria and thus increase the productivity of this approach. Furthermore, fungi

Table 2 List of yeast species synthesizing silver nanoparticles.

Species	Size (nm)	Morphology	Process	Reference
Kluyveromyces marxianus	3–12	Spherical	Extracellular	Ashour (2014)
Candida utilis	6–20	Spherical	Extracellular	Ashour (2014)
Candida utilis	20–80	Spherical	Extracellular	Waghmare et al. (2015)
Trichoderma longibrachiatum	≥ 20	Spherical	Extracellular	Elamawi et al. (2018)
Trichoderma harzianum	~180	–	Extracellular	Guilger-Casagrande et al. (2019)
Yeast extract	13.8	Spherical	–	Shu et al. (2020)
Saccharomyces cerevisiae	16.7	Oval	Extracellular	Olobayotan and Akin-Osanaiye (2019)

can be used for production of larger amount of silver particles. The extracellular synthesis process efficiently substitutes the chemical methods involving use of oxides and nitrides as well. NADH-dependent nitrate reductase and NADH (nicotinamide adenine dinucleotide) are the most significant biomolecules used in the fungal-mediated synthesis mechanism (Fig. 3). Moreover, in nearly every fungal species, nitrate reductase enzyme and anthraquinones mainly lead to silver nanoparticles synthesis (Mariana and de Renata, 2019).

i. *Plant-mediated synthesis*

The proteins present in plant extract material play a major role in the silver nanoparticle synthesis mechanism. It is hypothesized that silver ions are trapped on the protein because of electrostatic interactions. Secondary changes occur in silver ions and thus silver nuclei are formed. A further reduction process leads to growth of the silver nuclei, resulting in production of silver nanoparticles. In certain plants, like xerophytes, it has been observed that phytochemicals undergo a tautomerization process that leads to nanoparticle formation. Similarly, in other plants, reduction takes place with the help of different phytochemicals like flavones, ketones, amides, organic acids, carboxylic acids, etc. Stabilization and capping of the silver nanoparticles also occur due to these phytochemicals from the same plant extract, which cuts down the requirement of supplementary stabilizing or capping agents. In other cases, stem extracts of plant material containing aldehyde groups cause reduction of silver ions. This mechanism is known to be chiefly extracellular and thus easy to purify (Shanmugam and Bharath, 2017). Recently, a one-step, easy, and simple synthesis method for silver nanoparticle production using an extract of *Salvia officinalis* was demonstrated by a research group (Okaiyeto et al., 2021). In addition, Ibrahim et al. (2021a,b) exhibited silver nanoparticle synthesis by a *Moringa oleifera* extract which

4 Mechanism of biosynthesis process

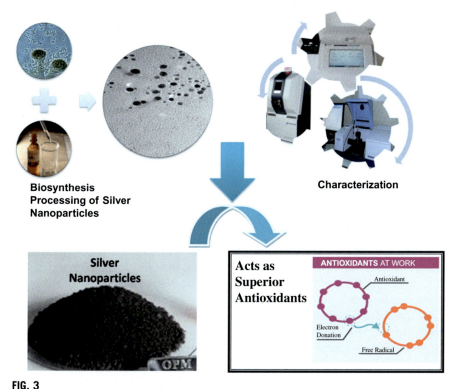

FIG. 3

Schematic diagram of fungal-mediated biosynthesis of nanoparticles, their characterization, and their applications.

is further used in cellulose-based fabrics. Reducing agents employed for the process were polyvinyl alcohol and glucose.

4.3 Other synthesis mechanisms

Researchers have proposed several mechanisms for algae, actinomycetes, and other bioreduction that takes place in the environment. It is suggested that a two-step mechanism is involved where the metal ions adhere to the surface of algal cells due to electrostatic attraction and algal reductase enzymes end up reducing the ions into silver nanoparticles. This process is very similar to that of bacterial and fungal mechanisms. In some cases, studies proved a light-mediated method as well, by either a direct or an indirect Photo System II (Note: PSII or water-plastoquinone oxidoreductase is the first protein complex in the light-dependent reactions of oxygenic photosynthesis). Hamouda et al. (2019) reported a reduction mechanism involving photosynthesis process. This study explained a crude extract of *Oscillatoria limnetica* initiated photosynthesis due to its phycobiliprotein content and thylakoids rich in chlorophyll. The reducing effect of flavonoids, proteins, or enzymes helps in absorption of light energy as a result of different conformational changes. Photosynthesis in the algae leads to illumination of molecules/electrons in chromospheres, which reduces silver nitrate to silver nanoparticles due to electrons that leap between energy levels. This explains how the presence of metabolites like saponins, terpenoids, flavonoids, and quinines plays a similar role in algal-mediated silver nanoparticles synthesis.

5 Downstreaming and purification of biosynthesized nanoparticles

In today's nano-biotechnology research, fungi are the most preferred organisms for synthesizing nanoparticles due to their simple and streamlined process. The synthesized nanoparticles by microbes require a plethora of equipment such as a sonicator, a centrifuge, and harsh chemical agents such as methanol. Meanwhile, fungal filtrate can be smoothly segregated from the mycelial mats by a simple filtration technique. This saves time and uses less intricate instruments, making it an overall convenient procedure (Ingle et al., 2009; Navazi et al., 2010; Iravani, 2014). In the fungal synthesis method, downstreaming involves the separation of biomass from the filtrate, which can then be used in the formulation of metal salts and microbial extracts. After the synthesis of the particles in the colloidal formulation, they are freeze-dried and purified for characterization. Lyophilizers are used for freeze-drying and the filtrate obtained is further purified by washing with Millipore water multiple times, or heating at high temperatures (within the melting point of the concerned metal), or treating with agents (phosphoric acid, hydrogen peroxide) to remove organic materials.

Although the metal nanoparticles purified from fungal precursors had no proven protocol, some researchers have developed different techniques for this purpose. Previous studies have verified ways to purify nanoparticles to filter impurities that might remain in the media, or the biomass used, or due to aggregation of organic molecules. One way involved the resultant colloidal solution of AgNPs being spun at 10,000 rpm for 10 min, two or more times to purify the substance fully. It was then washed to eliminate impurities by centrifuging at 14,000 rpm for 30 min or passed through microfilters to acquire similarly sized particles, and then separated them by a discontinuous sucrose density gradient with the help of the ultracentrifugation technique (Kumar et al., 2008; Li et al., 2012). To filter the monodispersed solution of stable gold particles, Al-Kazazz et al. (2013) exploited the dialysis membrane contrary to other researchers who had employed an organic solvent. Since the dialysis process does not require harmful chemicals, it is the preferred method. Again, the supernatant containing the colloidal particles can be separated from the concentrated pellets by centrifuging samples at high speeds (2000, 5000, 10,000, and 15,000 rpm for 20 min) (Basavaraja et al., 2008). The large aggregates are removed first from the supernatant and gradually the pellets are washed and then rewashed to obtain better-quality and pure particles. Consequently, these procedures help collect pure nanoparticles, which can then be further exploited for biomedical functions.

6 Characterization methods

The formation of nanosilver is followed by the separation of these materials from the resultant formulation. This is a major step to purify the crudely synthesized particles and get rid of the media components, solvents, or any unbound side groups. The centrifugation technique can be used to obtain the powdered materials. These metal particles can be further taken up for a series of characterization methods starting with UV visible spectrophotometry, followed by X-ray diffraction, Fourier transform infrared analysis, electron microscopy, energy dispersive spectroscopy, and dynamic light scattering (Srikar et al., 2016; Huq, 2020).

i. Spectrophotometry initially detects the presence of nanoparticles based on the SPR (surface plasmon resonance) peak, which is in the range of 400–450 nm for silver. It was also noted that the width of the SPR band was proportional to the size of nanoparticles. Also, the higher the silver concentration, the sharper or wider the peak. The study of the SPR behavior of silver nanoparticles can be depicted by the optical spectra of nanoprisms. Elaborate photography of silver nanoprism has been described by Lee and Jun (2019).

ii. Air-dried samples are also analyzed by the X-ray diffraction method and the peaks corresponding to lattice planes of nanosilver confirm its formation. This method helps analyze the face-centered cubic (FCC) crystalline structure of the silver nanoparticles.

iii. Fourier transform infrared (FTIR) analysis helps in observation of the functional groups involved in the stabilization of the silver particles following development. Infrared (IR) tools help to detect the common side groups like carboxylic acid, polysaccharides, aldehydes, and alkaloids beside amide linkages between amino acid residues in proteins. This study gives a brief idea of biomolecules involved in the capping of the silver nanoparticles (Siddiqi et al., 2018; Singh and Jain, 2014).

iv. Energy dispersive spectroscopy (EDS) is a method of determining the purity and composition of nanoparticles that reveals the homogenous distribution of silver atoms. It shows an absorption peak around 3 keV for nanosilver.

v. Dynamic light scattering (DLS), also known as photon correlation spectroscopy, verifies the dispersity aspects of the particles based on intensity, volume, and number. It also demonstrates the size of nanoparticles and charges on them. This measures the Brownian movement of particles in solution state, which is related to the size of particles. The technique detects aggregation in colloidal samples as well (Singh and Jain, 2014). However, the exact size of the particles can be measured by other instruments like SEM (scanning electron microscopy), TEM (transmission electron microscopy), AEM (atomic electron microscopy), etc.

vi. Scanning or transmission electron microscopy (SEM/TEM) gives a clear picture of nanomaterials including their size, shape, and morphology. Silver has been observed in various forms and shapes such as spherical, triangular, rod, flakes, plates, polyhedral, isotropic, etc.

There are several other characterization tools used for the analysis of purity, size morphology, and shape of silver nanoparticles. A few such well-known advanced analytical techniques are atomic force microscopy (AFM), surface-enhanced Raman scattering (SERS), selected area electron diffraction (SAED for crystallinity), X-ray photon spectroscopy (XPS), nuclear magnetic resonance (NMR), energy-dispersive X-ray spectroscopy (EDX), which is in relation with SEM and TEM, small-angle X-ray scattering (SAXS), scanning tunneling microscopy (STM), and electron energy loss spectroscopy (EELS) (Chouhan, 2018; Güzel and Erdal, 2018).

7 Conclusion

Nanoparticles have several functions, ranging from electronics and catalysis to the prevention of infectious diseases and medical diagnosis, and are thus one of the most important materials. Silver nanoparticles have been known as excellent antimicrobial and antiinflammatory agents, and thus have been used to improve wound healing.

However, concern has been raised regarding the toxicity of chemical agents used in nanoparticle synthesis. Thus, it is essential to develop a green approach for their production without using hazardous substances to human health and the environment. The ability to synthesize nanoparticles as prospective bio-agents using fungal species can be a gifted process for the sustainable and green production of nanomaterials.

The partially understood mechanism of silver nanoparticle formation involves certain drawbacks and challenges. Researchers have not been able to establish ways to control specific morphologies and monodispersity of the particles in the solution phase. Mechanistic aspects have to be comprehended thoroughly before they become a successful alternative for industrial silver nanoparticle production. There are several points that must be addressed when an organism is being selected for production of nanoparticles, including:

(1) correct selection of biocatalyst state;
(2) selection of best organism depending on growth rate, biochemical pathways, and enzyme activities;
(3) optimal cell growth, reaction conditions, and enzyme activity considering cleaner mediums and accurate parameters like pH, temperature, concentrations, mixing speed, light, etc.;
(4) careful choosing of purification processes so as to increase feasibility and easy analysis, and reduce cost of production;
(5) stability of nanoparticles based on enzymes and proteins that might be secreted by an organism; and
(6) the scaling-up process from laboratory to industrial level that can be verified by considering genetic properties and harvesting methods of organisms used.

Acknowledgment

The authors wish to acknowledge the Department of Studies in Microbiology, University of Mysore for supporting this publication.

References

Agnihotria, M., Joshia, S., Ravi, A.K., Zinjardea, S., Kulkarni, S., 2009. Biosynthesis of gold nanoparticles by the tropical marine yeast *Yarrowia lipolytica* NCIM 3589. Mater. Lett. 63, 1231.

Akter, S., Huq, M.A., 2020. Biologically rapid synthesis of silver nanoparticles by *Sphingobium* sp. MAH-11^T and their antibacterial activity and mechanisms investigation against drug-resistant pathogenic microbes. Artif. Cells Nanomed. Biotechnol. 48, 672–682.

Alghuthaymi, M.A., Almoammar, H., Rai, M., Said-Galiev, E., Abd-Elsalam, K.A., 2015. Myconanoparticles: synthesis and their role in phytopathogens management. Biotechnol. Biotechnol. Equip. 29, 221.

Al-Kazazz, F.F.M., Al-Imarah, K.A.F., Al-Hasnawi, S.I.A., Agelmashotjafar, L., Abdul-Majeed, B.A., 2013. A simple method for synthesis, purification and concentration stabilized gold nanoparticles. Int. J. Eng. Res. Appl. 3, 21.

Almalki, M.A., Khalifa, A.Y.Z., 2020. Silver nanoparticles synthesis from *Bacillus sp* KFU36 and its anticancer effect in breast cancer MCF-7 cells via induction of apoptotic mechanism. J. Photochem. Photobiol. B Biol. 204, 111786.

Ameen, F., AlYahya, S., Govarthanan, M., ALjahdali, N., Al-Enazi, N., Alsamhary, K., Alshehri, W.A., Alwakeel, S.S., Alharbi, S.A., 2020. Soil bacteria *Cupriavidus sp.* mediates the extracellular synthesis of antibacterial silver nanoparticles. J. Mol. Struct. 1202, 127233.

Ashour, S.M., 2014. Silver nanoparticles as antimicrobial agent from *Kluyveromyces marxianus* and *Candida utilis*. Int. J. Curr. Microbiol. App. Sci. 3, 384.

Banerjee, K., Rai, V.R., 2015a. Biofilm inhibitory activity of mycosynthesized silver nanoparticles against plaque forming bacteria *Pseudomonas aeruginosa*. J. Nanopharm. Drug Deliv. 3, 63–69.

Banerjee, K., Rai, V.R., 2015b. Preliminary screening of mycochemicals in *Aspergillus fischeri* for synthesizing silver nanoparticles and their antioxidant activity. Mater. Focus 4 (3), 252–258.

Banerjee, K., Rai, V.R., 2016. Study on green synthesis of gold nanoparticles and their potential applications as catalysts. J. Clust. Sci. 27, 1307–1315.

Banerjee, K., Rai, V.R., 2017. A review on mycosynthesis, mechanism, and characterization of silver and gold nanoparticles. BioNanoScience 8, 24.

Basavaraja, S., Balaji, S.D., Lagashetty, A., Rajasab, A.H., Venkataraman, A., 2008. Extracellular biosynthesis of silver nanoparticles using the fungus *Fusarium semitectum*. Mater. Res. Bull. 43, 1164.

Bharathidasan, R., Panneerselvam, A., 2012. Biosynthesis and characterization of silver nanoparticles using endophytic fungi *Aspergillus concius*, *Penicillium janthinellum* and *Phomosis* sp. Int. J. Pharm. Sci. Res. 3, 3163.

Bharde, A., Rautaray, D., Bansal, V., Ahmad, A., Sarkar, I., 2006. Extracellular biosynthesis of magnetite using fungi. Small 2, 135.

Castro-Longoria, E., Vilchis-Nestor, A.R., Avalos-Borja, M., 2011. Biosynthesis of silver, gold and bimetallic nanoparticles using the filamentous fungus *Neurospora crassa*. Colloids Surf. B Biointerfaces 83, 42.

Chouhan, N., 2018. Silver nanoparticles: synthesis, characterization and applications. In: Maaz, K. (Ed.), Silver Nanoparticles—Fabrication, Characterization and Applications. IntechOpen, https://doi.org/10.5772/intechopen.75611.

Dawadi, S., Katuwal, S., Gupta, A., Lamichhane, U., Thapa, R., Jaisi, S., Lamichhane, G., Bhattarai, D., Parajuli, N., 2021. Current research on silver nanoparticles: synthesis, characterization, and applications. J. Nanomater. 2021, 23.

Deplanche, K., Caldelari, I., Mikheenko, I.P., Sargent, F., Macaskie, L.E., 2010. Involvement of hydrogenases in the formation of highly catalytic Pd(0) nanoparticles by bioreduction of Pd(II) using *Escherichia coli* mutant strains. Microbiology 156 (9), 2630–2640.

Du, L., Xu, Q., Huang, M., Xian, L., Feng, J.X., 2015. Synthesis of small silver nanoparticles under light radiation by fungus *Penicillium oxalicum* and its application for the catalytic reduction of methylene blue. Mater. Chem. Phys. 160, 40.

Elamawi, R.M., Al-Harbi, R.E., Hendi, A.A., 2018. Biosynthesis and characterization of silver nanoparticles using *Trichoderma longibrachiatum* and their effect on phytopathogenic fungi. Egypt J. Biol. Pest Control 28, 28. https://doi.org/10.1186/s41938-018-0028-1.

Gade, A.K., Bonde, P., Ingle, A.P., Marcato, P.D., Durán, N., Rai, M.K., 2008. Exploitation of *Aspergillus niger* for synthesis of silver nanoparticles. J. Biobaased Mater. Bioenergy 2, 243.

Gallón, S.M.N., Alpaslan, E., Wang, M., Larese-Casanova, P., Londoño, M.E., Atehortúa, L., Pavón, J.J., Webster, T.J., 2019. Characterization and study of the antibacterial mechanisms of silver nanoparticles prepared with microalgal exopolysaccharides. Mater. Sci. Eng. C 99, 685–695.

Garg, D., Sarkar, A., Chand, P., et al., 2020. Synthesis of silver nanoparticles utilizing various biological systems: mechanisms and applications—a review. Prog. Biomater. https://doi.org/10.1007/s40204-020-00135-2.

Goswami, A.M., Sarkar, T.S., Ghosh, S., 2013. An ecofriendly synthesis of silver nano-bioconjugates by *Penicillium citrinum* (MTCC9999) and its antimicrobial effect. AMB Express 3, 16.

Guilger-Casagrande, M., Germano-Costa, T., Pasquoto-Stigliani, T., et al., 2019. Biosynthesis of silver nanoparticles employing *Trichoderma harzianum* with enzymatic stimulation for the control of *Sclerotinia sclerotiorum*. Sci. Rep. 9, 14351. https://doi.org/10.1038/s41598-019-50871-0.

Güzel, R., Erdal, G., 2018. Synthesis of silver nanoparticles. In: Maaz, K. (Ed.), Silver Nanoparticles—Fabrication, Characterization and Applications. IntechOpen, https://doi.org/10.5772/intechopen.75363.

Hamouda, R.A., Hussein, M.H., Abo-Elmagd, R.A., Bawazir, S.S., 2019. Synthesis and biological characterization of silver nanoparticles derived from the cyanobacterium *Oscillatoria limnetica*. Sci. Rep. 9 (1), 13071.

Huq, M.A., 2020. Green synthesis of silver nanoparticles using *Pseudoduganella eburnea* MAHUQ-39 and their antimicrobial mechanisms investigation against drug resistant human pathogens. Int. J. Mol. Sci. 21, 1510.

Ibrahim, S., Ahmad, Z., Manzoor, M.Z., Mujahid, M., Faheem, Z., Adnan, A., 2021a. Optimization for biogenic microbial synthesis of silver nanoparticles through response surface methodology, characterization, their antimicrobial, antioxidant, and catalytic potential. Sci. Rep. 11, 770.

Ibrahim, H.M., Zaghloul, S., Hashem, M., El-Shafei, A., 2021b. A green approach to improve the antibacterial properties of cellulose based fabrics using *Moringa oleifera* extract in presence of silver nanoparticles. Cellul. 28, 549–564.

Ingle, A., Rai, M., Gade, A., Bawaskar, M., 2009. *Fusarium solani*: a novel biological agent for the extracellular synthesis of silver nanoparticles. J. Nanopart. Res. 11, 2079.

Iravani, S., 2014. Bacteria in nanoparticle synthesis: current status and future prospects. Int. Sch. Res. Notices 2014, 1.

Jung, J.H., Cheol Oh, H., Soo Noh, H., Ji, J.H., Soo Kim, S., 2006. Metal nanoparticle generation using a small ceramic heater with a local heating area. Aerosol Sci. 37, 1662.

Khodashenas, B., Ghorbani, H.R., 2019. Synthesis of silver nanoparticles with different shapes. Arab. J. Chem. 12, 1823–1838.

Kim, T.Y., Kim, M.G., Lee, J., Hur, H., 2018. Biosynthesis of nanomaterials by *Shewanella* species for application in lithium-ion batteries. Front. Microbiol. 9, 2817.

Korbekandi, H., Iravani, S., Abbasi, S., 2012. Optimization of biological synthesis of silver nanoparticles using *Lactobacillus casei* subsp *casei*. J. Chem. Technol. Biotechnol. 87, 932.

Kumar, S.A., Abyaneh, M.K., Gosavi, S.W., Kulkarni, S.K., Pasricha, R., Ahmad, A., Khan, M.I., 2007. Nitrate reductase-mediated synthesis of silver nanoparticles from AgNO3. Biotechnol. Lett. 29, 439.

Kumar, S.K., Peter, Y.A., Nadeau, J.L., 2008. Facile biosynthesis, separation and conjugation of gold nanoparticles to doxorubicin. Nanotechnology 19, 4951.

Lee, S.H., Jun, B.H., 2019. Silver nanoparticles: synthesis and application for nanomedicine. Int. J. Mol. Sci. 20, 865.

Li, G., He, D., Qian, Y., Guan, B., Gao, S., Cui, Y., Yokoyama, K., Wang, L., 2012. Fungus-mediated green synthesis of silver nanoparticles using *Aspergillus terreus*. Int. J. Mol. Sci. 13, 466.

Liang, H., Li, Z., Wang, W., Wu, Y., Xu, H., 2009. Highly surface-roughened "flower-like" silver nanoparticles for extremely sensitive substrates of surface-enhanced Raman scattering. Adv. Mater. 21 (45), 4614–4618.

Mariana, G.-C., de Renata, L., 2019. Synthesis of silver nanoparticles mediated by fungi: a review. Front. Bioeng. Biotechnol. 7, 287.

Martínez-Castañón, G.A., Niño-Martínez, N., Martínez-Gutierrez, F., Martínez-Mendoza, J.R., Ruiz, F., 2008. Synthesis and antibacterial activity of silver nanoparticles with different sizes. J. Nanopart. Res. 10, 1343.

Matei, A., Matei, S., Matei, G., Cogălniceanu, G., Cornea, C.P., 2020. Biosynthesis of silver nanoparticles mediated by culture filtrate of lactic acid bacteria, characterization and antifungal activity. EuroBiotech J. 4 (2), 97–103.

Moharrer, S., Mohammadi, B., Gharamohammadi, R.A., Yargoli, M., 2012. Biological synthesis of silver nanoparticles by *Aspergillus flavus*, isolated from soil of Ahar copper mine. Indian J. Sci. Technol. 5, 2443.

Mohd Yusof, H., Abdul Rahman, N., Mohamad, R., Zaidan, U.H., 2020. Microbial mediated synthesis of silver nanoparticles by *Lactobacillus plantarum* TA4 and its antibacterial and antioxidant activity. Appl. Sci. 10, 6973.

Mokhtari, N., Daneshpajouh, S., Seyedbagheri, S., et al., 2009. Biological synthesis of very small silver nanoparticles by culture supernatant of *Klebsiella pneumonia*: the effects of visible-light irradiation and the liquid mixing process. Mater. Res. Bull. 44 (6), 1415–1421.

Mukherjee, P., Ahmad, A., Mandal, D., Senapati, S., Sainkar, S.R., 2001. Bioreduction of AuCl(4)(−) ions by the fungus, *Verticillium* sp. and surface trapping of the gold nanoparticles formed. Angew. Chem. Int. Ed. 40, 3585.

Mukherjee, P., Roy, M., Mondal, B.P., Dey, G.K., Mukherjee, P.K., Ghatak, J., Tyagi, A.K., Kale, S.P., 2008. Green synthesis of highly stabilized nanocrystalline silver particles by a non-pathogenic and agriculturally important fungus *T. asperellum*. Nanotechnology 19, 075–103.

Nair, V., Sambre, D., Joshi, S., Bankar, A., Kumar, A.R., Zinjarde, S., 2013. Yeast-derived melanin mediated synthesis of gold nanoparticles. J. Bionanosci. 7, 159.

Namasivayam, S.K.R., Avimanyu, 2011. Silver nanoparticles synthesis from *Lecanicillium lecanii* and evolutionary treatment on cotton fabrics by measuring their improved antibacterial activity with antibiotics against *Staphylococcus aureus* (ATCC 29213) and *E. coli* (ATCC 25922) strains. Int. J. Pharm. Pharm. Sci. 3, 190.

Navazi, Z.R., Pazouki, M., Halek, F.S., 2010. Investigation of culture conditions for biosynthesis of silver nanoparticles using *Aspergillus fumigates*. Iran. J. Biotechnol. 8, 56.

Okaiyeto, K., Hoppe, H., Okoh, A.I., 2021. Plant-based synthesis of silver nanoparticles using aqueous leaf extract of *Salvia officinalis*: characterization and its antiplasmodial activity. J. Clust. Sci. 32, 101–109.

Olobayotan, I., Akin-Osanaiye, B.C., 2019. Biosynthesis of silver nanoparticles using baker's yeast, *Saccharomyces cerevisiae* and its antibacterial activities. Access Microbiol. 1. https://doi.org/10.1099/acmi.ac2019.po0316.

Rautela, A., Rani, J., Das, M.D., 2019. Green synthesis of silver nanoparticles from *Tectona grandis* seeds extract: characterization and mechanism of antimicrobial action on different microorganisms. J. Anal. Sci. Technol. 10, 1–10.

Reetz, M.T., Helbig, W., 1994. Size-selective synthesis of nanostructured transition metal clusters. J. Am. Chem. Soc. 116 (16), 7401–7402.

Rodriguez-Sanchez, L., Blanco, M.C., Lopez-Quintela, M.A., 2000. Electrochemical synthesis of silver nanoparticles. J. Phys. Chem. B 104 (41), 9683–9688.

Saravanan, M., Nanda, A., 2010. Extracellular synthesis of silver bionanoparticles from *Aspergillus clavatus* and its antimicrobial activity against MRSA and MRSE. Colloids Surf. B Biointerfaces 77, 214.

Sastry, M., Ahmad, A., Khan, M.I., Kumar, R., 2003. Biosynthesis of metal nanoparticles using fungi and actinomycetes. Curr. Sci. 85, 162.

Schüler, D., 1999. Formation of magnetosomes in magnetotactic bacteria. J. Mol. Microbiol. Biotechnol. 1, 79–86.

Sen, K., Sinha, P., Lahiri, S., 2011. Time dependent formation of gold nanoparticles in yeast cells: a comparative study. Biochem. Eng. J. 55, 1.

Shahzad, A., Saeed, H., Iqtedar, M., Hussain, S.Z., Kaleem, A., Abdullah, R., Sharif, S., Naz, S., Saleem, F., Aihetasham, A., Chaudhary, A., 2019. Size-controlled production of silver nanoparticles by *Aspergillus fumigatus* BTCB10: likely antibacterial and cytotoxic effects. J. Nanomater. 5168698, 14.

Shanmugam, R., Bharath, L.V., 2017. Mechanism of plant-mediated synthesis of silver nanoparticles—a review on biomolecules involved, characterization and antibacterial activity. Chem. Biol. Interact. 273, 10.

Sheoran, N., Kaur, P., 2018. Biosynthesis of nanoparticles using eco-friendly factories and their role in plant pathogenicity: a review. Biotechnol. Res. Innov. 2. https://doi.org/10.1016/j.biori.2018.09.003.

Shu, M., He, F., Li, Z., et al., 2020. Biosynthesis and antibacterial activity of silver nanoparticles using yeast extract as reducing and capping agents. Nanoscale Res. Lett. 15, 14. https://doi.org/10.1186/s11671-019-3244-z.

Shugaba, A., Buba, F., Kolo, B.G., Nok, A.J., Ameh, D.A., Lori, J.A., 2012. Uptake and reduction of hexavalent chromium by *Aspergillus niger* and *Aspergillus parasiticus*. J. Pet. Environ. Biotechnol. 3, 1.

Siddiqi, K.S., Husen, A., Rao, R.A.K., 2018. A review on biosynthesis of silver nanoparticles and their biocidal properties. J. Nanobiotechnol. 16, 14.

Singh, P., Jain, S.K., 2014. Biosynthesis of nanomaterials: growth and properties. Rev. Adv. Sci. Eng. 3, 1–8.

Srikar, S.K., Giri, D.D., Pal, D.B., Mishra, P.K., Upadhyay, S.N., 2016. Green synthesis of silver nanoparticles: a review. Green Sustain. Chem. 6, 34–56.

Sunkar, S., Nachiyar, C.V., 2012. Biogenesis of antibacterial silver nanoparticles using the endophytic bacterium *Bacillus cereus* isolated from *Garcinia xanthochymus*. Asian Pac. J. Trop. Biomed. 12, 953.

Thakkar, K.N., Mhatre, S.S., Parikh, R.Y., 2010. Biological synthesis of metallic nanoparticles. Nanomed. Nanotechnol. Biol. Med. 6, 257.

Tien, D.C., Liao, C.Y., Huang, J.C., Tseng, K.H., Lung, J.K., Tsung, T.T., Kao, W.S., Tsai, T.H., Cheng, T.W., Yu, B.S., Lin, H.M., Stobinski, L., 2008. Novel technique for preparing a nano-silver water suspension by the arc-discharge method. Rev. Adv. Mater. Sci. 18, 750.

Tsuji, T., Iryo, K., Ohta, H., Nishimura, Y., 2000. Preparation of metal colloids by a laser ablation technique in solution: influence of laser wavelength on the efficiencies of colloid formation. Japan J. Appl. Phys. 39, 981.

Tsuji, T., Kakita, T., Tsuji, M., 2003. Laser induced morphology change of silver colloids: formation of nano-size wires. Appl. Surf. Sci. 206, 314.

Usha, R., Prabu, E., Palaniswamy, M., Venil, C.K., Rajendran, R., 2010. Synthesis of metal oxide nano particles by *Streptomyces* sp for development of antimicrobial textiles. Glob. J. Biotechnol. Biochem. 5, 153.

Vahabi, K., Mansoori, G.A., Karimi, S., 2011. Biosynthesis of silver nanoparticles by fungus *Trichoderma reesei*. Insciences J. 11, 65–79.

Waghmare, S.R., Mulla, M.N., Marathe, S.R., Sonawane, K.D., 2015. Ecofriendly production of silver nanoparticles using *Candida utilis* and its mechanistic action against pathogenic microorganisms. 3 Biotech 5, 33.

Webster, T.J., 2006. Nanomedicine: what's in a definition? Int. J. Nanomedicine 1 (2), 115–116.

Zaarour, M., El-Roz, M., Dong, B., Retoux, R., Aad, R., Cardin, J., Dufour, C., Gourbilleau, F., Gilson, J.P., Mintova, S., 2014. Photochemical preparation of silver nanoparticles supported on zeoliten crystals. Langmuir 30 (21), 6250–6256.

CHAPTER 25

Toxicity of silver nanoparticles in the aquatic system

Muhammad Saleem Khan, Muhammad Shahroz Maqsud, Hasnain Akmal, and Ali Umar

Department of Zoology, Faculty of Life Sciences, University of Okara, Okara, Pakistan

1 Introduction

Both domestic and industrial uses of silver nanoparticles (AgNPs) produce wastes that discharge into aquatic habitats (Khan et al., 2016). These particles contaminate the water or air through different sources such as weathering rocks, the cement industry, and fossil fuel. Rain also transfers them to the ground land water reservoirs for human use (Wijnhoven et al., 2009; Khan et al., 2018).

Silver exists in several oxidation states including Ag, Ag^+, Ag^{+2}, and Ag^{+3}. Among them, Ag and Ag^+ are the most commonly exiting ionization forms Moreover, its metallic form is usually insoluble and salts can be soluble. It is present as colloidal particles (Fabrega et al., 2011).

In the presence of available data, it can be hypothesized "AgNPs can be toxic to all living forms" (Khan et al., 2020). Its nanoforms are more toxic because these are readily available in the aquatic environment and can enter ecosystems, altering their fitness and population densities (Luoma et al., 2008). Their fate or behavior can vary due to many aspects, including particle size, surface area interaction, and composition (Raza et al., 2017). Water and lipid solubility, vapor pressure, and coagulation can also influence their properties (Shakeel et al., 2018). Numerous studies on the subject, i.e., toxicological aspects had been done; however, these studies showed huge variations and harmful effects due to poor particle characterization. (Gliga et al., 2014). Some studies showed production of reactive oxygen species (ROS), because of AgNPs, apoptosis (Khan et al., 2015a), reduction in mitochondrial performance (Park et al., 2007; Bressan et al., 2013), lipid peroxidase (LPO) formation, and alterations in stress markers (Khan et al., 2016, 2018). Furthermore, Larese et al. (2009) investigated and proved AgNPs to be capable of passing through the stratum corneum and also through the blood-brain barrier (Arora et al., 2009; Sung et al., 2008). This chapter gives an introduction to AgNPs along with their toxic effects on aquatic life. Various routes through which AgNPs may enter the body of organisms and the source of these toxic particles are also discussed. This chapter also explains the mechanism and nature of toxicity to various forms of aquatic life.

2 Causes of toxicity

Thomson Reuters database (Web of Science) showed that research has been published on toxicological aspects since 1990. Since then, more than 3000 research articles have been published each year. The previously published research suggests that many characteristics of particles are responsible for toxicity. These characteristics include the size of particles, the shape of particles, surface area, and chemical composition. Most studies suggest the properties of particles change when size decreases to the nanoscale (Khan et al., 2017). One other proposed argument is that smaller size particles can enter an organism more easily than larger ones. To prove this, Ivask et al. (2014) used four different AgNPs with various sizes range from 10 to 80 nm. Toxicity assay revealed that 10 nm size particles were more toxic. Secondly, Ivask and colleagues coated the silver particles with different organic coating agents, which also influenced the fate, toxicity, and stability of particles. Further studies also revealed that some organic coated AgNPs also damage the cell membrane, interfere in the replication process, ATP generation, mutate the expression of genes, and produce ROS in metabolic pathways.

3 Routes of entry

AgNPs use different routes to gain entry to an organism's body (Fig. 1). These routes may be oral absorption, inhalation, damaged skin, diffusion, or endocytosis (Asharani et al., 2008). Colloidal AgNPs and other forms of silver can enter through the ingestion of contaminated food as preservatives, contaminated water, and children's toys, dust, or fumes inhalation in industrial and jewelry makings. It can also take the route through damaged skin when present in burn creams or liquids, cosmetics, and textiles (Marambio-Jones and Hoek, 2010). The female genitalia can also be a route of entry through use of silver-containing hygiene products (Chen and Schluesener, 2008; Schrand et al., 2008). Acupuncture needles, dental amalgams, contact of jewelry with the body (Catsakis and Sulica, 1978; Drake and Hazelwood, 2005), or through the placenta can provide routes of AgNPs' entry (Melnik et al., 2013). Once they enter the body AgNPs accumulate in organs and can cause hepatotoxicity or renal toxicity (Sung et al., 2008).

4 Toxicity to phytoplankton and microalgae
4.1 Phytoplankton

Phytoplankton is the primary producer in the aquatic environment. Their productivity and photosynthetic efficiency reduce when AgNPs level is increased up to $500\,mg\,L^{-1}$ (Baptista et al., 2015). Toxicity of $5\,mg\,L^{-1}$ of AgNPs and silver ions

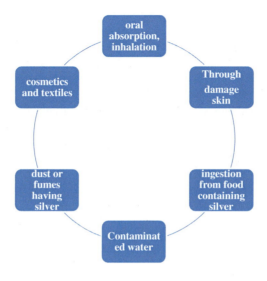

FIG. 1

Routes of entry into the body.

were compared through short-term exposure to aquatic microorganisms by Boenigk and coworkers in 2014. Silver ion from $AgNO_3$ showed more alternations in transcriptomic responses than AgNPs with the same concentration. Furthermore, AgNP exposure showed less influence on phytoplankton than silver ionic treatment. Das et al. (2014) also inspected phytoplankton to analyze the effects and levels of influence of AgNPs. They exposed phytoplankton to AgNPs of four different concentrations (each of 10 nm). They also exposed algal flora to five concentrations. Bacterioplankton was exposed to AgNPs and phosphorus for 3 days. Results showed that values of EC_{50} for algal blooms 2.21 µg L^{-1}. Results showed that naturally occurring bacterial communities were less sensitive to toxicity caused by AgNPs than algae (Das et al., 2014). The authors also observed a decrease in cellular levels of carbon and nitrogen. Growth rates were also inhibited; however, at the saturated level of phosphorus, specific growth rates were found to be higher.

4.2 Microalgae

Algae are ecologically the most important organisms of the aquatic environment and produce 40% of total biomass around the globe. Their ability to respond to pollutants quickly makes them bioindicators of water pollution. Traces of toxic substances alter the composition and densities of algal species. Due to this nature, algal species are frequently used in ecotoxicological investigations to study and explore the toxicity of various substances.

De Aragao et al. (2019) Investigated the toxicity of nanoparticles on bacterial strains, protozoan communities, and algal communities. Among these organisms, the algal community showed higher sensitivity than others only because of higher growth inhibition. Gonzalez et al. (2006) studied various algal species to analyze photosynthetic as well as metabolic activities of *Chlorophyceae*, *Diatomophyceae*, and *Cyanophyceae* to assess how much biofilm was sensitive because of the dominance of the above-mentioned species to silver nanoparticles and silver ions. They used a high concentration level of AgNPs ($100\,mg\,L^{-1}$) and exposed biofilm to it. Concentrations vary for different algal classes, being much higher for green algae. Results showed significant effects of AgNPs on algal classes. Alternations in the structure of biofilm were observed. The productivity of biomass also reduced and the concentration of chlorophyll decreased. The authors mentioned a negative charge on AgNPs as the basic cause of these toxic effects. Due to these negative charges they can move through water channels in the extracellular matrix of biofilm, whereas Ag ions can only interact with the surface of biofilm and do not show toxic effects. *Ulva lactuca*, which belongs to the marine macro-algal community, was exposed to a higher concentration level of AgNPs and Ag ion. As fishes live in aquatic habitats, so they are more vulnerable to silver toxicity. The most important route for silver residues for entrance into the body of fish is the gills (Hawkins et al., 2015).

Among fish species, the zebrafish (*Danio rerio*) is the most employed model for toxicological assay (Wu et al., 2010). The toxicological parameters ($AgNO_3$) was studied for 2 days. At $15\,mg\,L^{-1}$ concentration of AgNPs, quenching of chlorophyll was decreased. Similar results were obtained from $AgNO_3$ treatment but the concentration level was much lower.

Yang et al. (2021) harvested extracellular polymeric substances from *Chlorella vulgaris* and explored their impacts on sulfidation and dissolution of silver nanoparticles (AgNPs) by using kinetic estimation and spectral analyses. Both phenomena—sulfidation and dissolution of AgNPs—were found to be facilitated when various concentration levels of algal EPS were applied. The absorption of algal EPS on the surface of AgNPs was aided by the presence of excessive aromatic and nitrogen-containing groups.

Govindan et al. (2021) described silver nanoparticles synthesis through the application of the aqueous seaweed extract *Chondrococcus hornemannii* by the reduction of silver ions in a solution of silver nitrate. The synthetic nanoparticles were distinguished using UV, FTIR, XRD, and Zeta. *C. hornemannii* marine algae extract, having antibacterial activity from biosynthesized nanoparticles, was investigated against bacteria that cause pathogenicity such as *Pseudomonas aeruginosa*, *Streptococcus aureus*, and *Escherichia coli*. Marinho et al. (2021) investigated the effects of 96 h of exposure to various ANP concentrations on *D. rerio* (zebrafish). They looked at how AgNP affected different parts of the zebrafish (brain, gills, and muscles). Secondary lamellar fusion, rotation, marginal channel, and epithelial lifting were observed in zebrafish, which showed morphological changes.

5 Toxicity to fish

The residue of silver present in the aquatic medium enters inside fish's cause-effect include increasing heart rate, interruption in hatching, and high mortality (Asharani et al., 2008). The embryo of this species has about 250 mg L^{-1} LC$_{50}$ value (Choi et al., 2010). Another group of researchers, Bar-Ilan et al. (2009), reported other ranges of LC$_{50}$ values for small to large particles: 3–100 nm. Some recorded values include 93.31 μM for 3 nm and 137.26 μM for 100 nm AgNPs particle size. These findings suggest larger particles have less toxicity compared to smaller particles. Another study also supported that Ag$^+$ had the same toxicity level as in AgNPs. Both ionic and nano-size Ag showed almost the same LC$_{50}$ values in zebrafish (Kim et al., 2007). Zebrafish model studies also suggested different storage sites for AgNPs accumulations such as gills, intestinal blood, liver, and the brain (Handy et al., 2008). The liver showed maximum accumulation of between 0.29 and 2.4 ng/mg when fish were treated with 30–120 mg L^{-1} of nano-sized silver (Choi et al., 2010). Accumulation of silver nanoparticles caused severe changes in the cellular structure of the cell. The alterations might be in haptic cell cords, apoptotic cell formation, chromatin pyknosis (Gonzalez et al., 2006), circulatory system, and morphology (Asharani et al., 2008). The exposure of about 2–4 mg L^{-1} for 14 days caused decline in activity of gills sodium and potassium-ATPase and acetylcholinestrase activity in erythrocytes (Katuli et al., 2014). In animals treated with AgNPs, many changes occurred, such as oxidative stress on hepatic cells, double strand of DNA breaks, and lesions in cells.

Silver carp, *Hypophthalmichthys molitrix*, has also been studied for the toxic effects of AgNPs. Hedayati et al. (2012) compared the toxicity of AgNPs with metallic silver in silver carp and recorded that the AgNPs were more toxic at all stages in the life cycle. The LC$_{50}$ value was 0.34 ppm for Nanocid and 66.4 ppm in Nanosil (Jahanbakhshi et al., 2012). Continuous exposure to particles increased the concentration and ultimately led to the mortality of organisms—100% mortality was recorded at 1 ppm for 96 h of exposure. Many researchers described different toxicity related to the size of particles, the time duration of exposure, physiological condition, and age (Rathore and Khangarot, 2002). Shaluei et al. (2013) reported various observations at different periods. At 24 h, value is 0.810 mg L^{-1} LC$_{50}$, 0.64, 0.383, and 0.202 mg L^{-1} for 48, 72, and 96 h, respectively. AgNPs also decreased the concentration of RBC, hemoglobin, and hematocrit levels.

In common carp (*Cyprinus carpio*), the results indicated that the toxicities of AgNPs are high as compared to Ag ions. Firstly AgNPs affect the metabolic enzymes in specific organs, i.e., liver, gills, kidney, and brain (Reddy et al., 2013). Like other fish species, the liver is a major organ and its function altered as compared to other tissues as AgNPs concentration increased. Some other studies also suggested higher accumulation in gills, gastrointestinal tract, skeletal muscle, brain, and blood. Furthermore, AgNPs decrease the activities of metabolic enzymes like CAT, GST, and SOD. Many chemicals are used in toxicological studies; silver salts (AgNO$_3$), Nanocid, and

Nanosil are among the of AgNPs used to test toxicity. Hedayati and coworkers found that average recorded values of LC_{50} at 96 h exposure are 0.49 ± 0.90 ppm for Nanocid, 73.8 ± 0.38 for Nanosil, and 0.33 ± 0.3 ppm for $AgNO_3$.

Very little work has been done on the toxic effect of AgNPs in *Catla catla*. Reddy et al. (2013) described a significantly increased level of lipid peroxidation, found in the gills, at $100\,\mu g\,L^{-1}$ of AgNPs for 48 h duration. However, after 96 h, lipid peroxidation level in gills had declined at the same concentration because the gill endogenous antioxidant system has started to produce free radicals. Taju et al. (2014) also reported an increased level in lipid peroxidation and a decrease in antioxidant enzymatic level because of AgNPs exposure.

Rohu, *Labeo rohita*, is an extensively studied model for silver toxicity in fish classes. For example, Rajkumar et al. (2015) described that AgNPs show dose-dependent toxicity. It was found that an amount of $100\,mg\,kg^{-1}$ has a mortality rate of 50%, while a mortality rate of 100% was caused by $500\,mg\,kg^{-1}$. Stressful conditions are created by AgNPs, which cause alterations in various serum contents like total protein level, RBC, and WBC. In treated tissues of AgNPs acid phosphate (ACP) and alkane, phosphate level increased. Orally administered AgNPs also caused a reduction of GST, SOD, and CAT activities (Rajkumar et al., 2015).

AgNPs cause heamotoxicity, histotoxicity, i.e., damage to hepatic parenchyma (Khan et al., 2018, 2020), and significant damage to genetic makeup (Khan et al., 2017). Khan and coworkers investigated the influence of amine-coated silver nanoparticles on rohu fish. The size of the NPs was 15.78 nm. They found that NPs had an effect that depends on the influence of hematological parameters. Higher concentration levels of aminated AgNPs caused more alternations. Hepatocytes reduced in size and apoptotic and necrotic bodies were also seen due to the effect of these silver nanoparticles. Gills were also affected through proliferation of bronchial chloride tissue cells and aneurism formation (Khan et al., 2018). AgNPs cause severe genetic damage in rohu fish. These particles induce alternation in the nuclear material of red blood cells.

High concentrations of AgNPs caused great damage to the structure of living organisms. Oxidative stress in *L. rohita* by AgNPs is also reported (Khan et al., 2018). Later, in 2020, how *L. rohita* liver is influenced by AgNPs was investigated in a study conducted by Dr. Khan and his team. They used garlic oil as the source of silver NPs. They had found that AgNPs are responsible for inflammation due to the formation of yellow pigment in the liver and necrosis of hepatic tissues. Damage in hepatic parenchyma led to venous congestion. Gills were found to be damaged, i.e., fusion of secondary lamella and necrosis, in rohu fish treated with AgNPs (Khan et al., 2020).

Decreased tolerance in response to hypoxia as well as oxygen diffusion by the epithelium of gills is caused by exposure of crucian carp (*Perca fluviatilis*) to nano-silver particles (Bilberg et al., 2011). Interferences in the mechanism of odor detection and olfactory responses are suppressed by an amount of $45\,mg\,L^{-1}$ of nanoparticles of silver. It happens because the surface of AgNPs releases free ions of Ag that do not let the odor attach to olfactory receptors (Klaprat et al., 1992).

In studies on rainbow trout (*Oncorhynchus mykiss*), the authors showed the size consequences of AgNPs. Scown et al. (2010) treated this animal model with 10, 35, and 600–1600 nm through a water medium for a short period of about 10 days. A very low level of uptake was found. The concentration of 10 nm NPs was also found to high compared to other concentrations. Moreover, its concentration was also found to be high in gills as compared to the liver and kidney. The weight of the liver was also decreased significantly in treated fish ($P < .05$). Exposure of AgNPs causes a size reduction of local congestion in hepatic parenchyma (Monfared and Soltani, 2013). Silver nanoparticles covered by citrate and polyvinylpyrrolidone gets accumulated in the gills of rainbow trout, resulting in cytotoxicity (Farkas et al., 2010).

When alevins were exposed for 96 h to different forms AgNPs, it was found that the concentrations of $0.25\,mg\,L^{-1}$ and $28.25\,mg\,L^{-1}$ were the LC_{50} for colloidal form and suspended powder form, respectively, while in juveniles, a concentration of $2.16\,mg\,L^{-1}$ was found to be LC_{50} for the colloidal form of AgNPs. The powdered form of AgNPs caused no mortality in alevins (Kalbassi et al., 2013). The LC_{50} value of silver nanoparticles' colloidal form was measured by Johari et al. (2013). Various concentrations were measured by them, and they calculated the concentration in eleuthero embryo to be $0.25\,mg\,L^{-1}$, in larvae $0.71\,mg\,L^{-1}$, and juveniles $2.16\,mg^{-1}$. A compound should be termed as very toxic if the value of its LC_{50} is $<1\,mg\,L^{-1}$ for 96 h, according to European Union legislation (EC, 2008) and according to European Union Directive (EC, 2008) number, 67/548/EEC dated 27 June 1967, there must be a prolonged toxic effect of this compound on aquatic organisms as well. Johari et al. (2013) found that the colloidal form of AgNPs in eleuthero embryos, larvae, and juveniles of rainbow trout is highly toxic.

The serum level of total protein decreases with an increase in the amount of AgNPs (Monfared and Soltani, 2013). AgNPs exposure causes various changes in blood plasma such as the increased concentration of cholinesterase and cortisol while a reduced concentration of potassium and chloride ions is seen in blood plasma (Johari et al., 2013). Glucose concentration was also increased (Webb and Wood, 1998). All of these changes depend on the dose of silver nanoparticles. Lipid peroxidation is increased when exposed to AgNPs, as a result, the amount of Ag^+ causes increased DNA damage (Massarsky et al., 2014) along with inhibition of sodium and potassium ions and ATPase activity (Schultz et al., 2012).

An investigation conducted by Kim et al. (2009) showed that more silver ions are released by aged AgNPs as compared to fresh ones, therefore, older AgNPs are more toxic. A Japanese rice fish, medaka (*Oryzias latipes*), was used and a toxicity assay was performed. An LC_{50} value of $1.44\,mg\,L^{-1}$ was found for aged AgNPs as compared to $3.53\,mg\,L^{-1}$ for fresh AgNPs. No mortality was seen with $0.5\,mg\,L^{-1}$ concentration of AgNPs, while a concentration of $2.0\,mg\,L^{-1}$ AgNPs caused 100% mortality in a toxicity test that was carried out for 48 h by Wu et al. (2010).

Reduced optic tectum width and pigmentation along with development retardation were observed by exposure to a concentration of $400\,\mu g\,L^{-1}$ of AgNPs (Wu et al., 2010). Further developmental errors included edema, abnormalities of the spinal cord, and eye defects, as well as deformation of the heart by exposure to different concentrations of AgNPs (Wu et al., 2010). A concentration of 0.8 mg L^{-1} induces heartbeat retardation, low body length, and drop in young emerging rate (Choi et al., 2010). It was found that the silver nanoparticles can enter the chorion of the embryo first, then penetrate the membrane of the skin, and eventually, they are distributed to other tissues (Lee et al., 2012). The gills of *O. latipes* are the principal sites of uptake (Kwok et al., 2012).

The toxicity caused by AgNPs and their nanoproducts includes hemorrhages to parts of the head, yolk sac edema, and gills variations in the embryo of fathead minnows (*Pimephales promelas*) (Laban et al., 2010). Prominent changes in gills include disruption in the circulation of blood (Hawkins et al., 2015). The total goblet cells of mucous membranes are increased by a concentration of $1.3\,\mu g\,L^{-1}$ $AgNO_3$. Both $AgNO_3$ and AgNPs decrease the immune reactivity of Na^+/K^+-ATPase in gills and are found to be decreased by the effect of AgNPs and $AgNO_3$ (Hawkins et al., 2015). $10.6\,mg\,L^{-1}$ and $9.4\,mg\,L^{-1}$ were LC_{50} calculated values for AgNPs and NanoAmor, respectively. Similarly, LC_{50} calculated value for sonicated particles was $1.36\,mg\,L^{-1}$ as compared to $1.25\,mg\,L^{-1}$ for stirred particles. AgNPs were found to be less toxic as compared to $AgNO_3$ in the case of *P. promelas* (Laban et al., 2010).

Kakakhel et al. (2021) studied toxicity, mortality, bioaccumulation, and histological changes in *C. carpio* when it was exposed to silver nanoparticles, which were blood-mediated. The toxic effect of silver nanoparticles made from animal blood serum was investigated in common carp at various concentration levels (0.03, 0.06, and $0.09\,mg\,L^{-1}$). Blood-induced silver nanoparticles had little effect on fish behavior at the highest concentration ($0.09\,mg\,L^{-1}$). In various organs of fish, bioaccumulation of blood-mediated silver nanoparticles has been observed. The bioaccumulation of AgNPs was highest in the liver. Silver nanoparticles have an effect on aquatic life, according to this study. To fully comprehend the effects of AgNPs on aquatic life, more systematic research is required.

The effects of AgNPs and TiO_2 on Ag bioaccumulation, oxidative stress, and gill histopathology in common carp were studied by using common carp as an aquatic animal model. In 96-h exposure, the acute toxicity tests showed decreased LC_{50}, indicating that TiO_2NPs increased AgNP toxicity. The chronic toxicity test was administered after a 10-day recovery period. Ag- and TiO_2-NPs cause more severe histological damage than individual AgNPs do in most cases (Haghighat et al., 2021).

Abdel-Khalek et al. (2021) evaluated the effectiveness of orange peel (OP) and banana peel (BP) in removing AgNPs from water and reducing their toxicity in Nile tilapia. For 24h, 48h, and 96h, fish were divided into four groups: control (dechlorinated tap water), AgNPs ($4\,mg\,L^{-1}$)+OP ($40\,mg\,L^{-1}$) group, AgNPs ($4\,mg\,L^{-1}$) exposed group, and AgNPs ($4\,mg\,L^{-1}$)+BP ($40\,mg\,L^{-1}$) group. After exposure to AgNPs, globulin, plasma glucose, liver enzymes (ALT, ALP, and AST), total proteins, creatinine, and uric acid levels gradually increased, while albumin and total lipid levels decreased significantly. Moreover, bioavailability of AgNPs and its toxicological impact could be reduced by the adsorbent abilities of both peels.

Rhamdia quelen was exposed to a single or combined mixture of toxic metals (Mn, Pb, Hg, or AgNPs) after the hatching stages of native catfish to study the effects of toxic metals. In fertilized eggs, hatching and survival rates, malformation frequencies, and neuromast structural damages were evaluated for 24, 48, 72, and 96 h. The results showed alterations in hatching rates after single and combined metal exposure, but mixtures showed more severe effects compared to single exposure (Nagamatsu et al., 2021).

Natural organic matter has a significant impact on nanoparticles because it directly influences the fate as well as transport of these nanoparticles. Ale and his coworkers, therefore, conducted a study to check how humic acids influence the toxicity of AgNPs. 100 μg L^{-1} concentrations of AgNPs were used in ex vivo exposure of gills of *Piaractus mesopotamicus*. Ag bioaccumulation was analyzed in the tissues of fish along with lipid peroxidation and antioxidant activities of enzymes. Results of the study showed that Ag^+ released from AgNPs decreased about 28% when humic acids were present in the medium. Higher accumulation of silver in gills was found when gills were exposed to AgNPs alone than in any other form. The findings suggested that HAs could be mitigated when exposed to both types of Ag, providing important information on the fate and behavior of this emerging pollutant (Ale et al., 2021).

6 Toxicity to other aquatic organisms

Kang and Park (2021) exposed four different aquatic species, including *Pseudokirchneriella subcapitata*, *D. rerio*, *Daphnia magna*, and *Hydra vulgaris*, to Ag_2SNPs—a transformed from of AgNPs. In algae, crustaceans, and fish, acute toxicity of Ag_2SNPs was rarely observed, but it was significantly restored in cnidarians. Despite the lower dissolution rates of Ag_2SNPs, higher levels of Ag^+ were found in *H. vulgaris*. To understand the mechanism of Ag_2SNP toxicity in the cnidarians the transcriptional profiles of *H. vulgaris* were exposed to $AgNO_3$, AgNPs, and Ag_2SNPs.

Guo et al. (2021) studied the interactions of Au^{3+} and Ag^+ ions in mixed solutions (Ag = 0.2, 0.5, and 0.8) in the presence of humic acids under artificial sunlight, as well as the mechanism of how bimetallic Ag-Au NPs are formed. Thermally dynamic dark processes produced the mixture of prepared NPs, which was compositionally and morphologically isolated.

6.1 Bivalves

Bivalves are marine and freshwater mollusks. Two species of mollusks accumulate a significant amount of AgNPs and protein contents in their digestive organs. Inhibition of process phagocytosis in hemolymph in mussels is a result of AgNPs exposure. Sodium potassium ATPase activity is found to impair gill cells because gill tissues are the most sensitive target site of AgNPs in mollusks (Adriano and Pelletier, 2018). ROS formation, DNA damage, oxidative stress, disrupted actin filaments, activation of antioxidant catalase enzyme, and some developmental impairments are caused in bivalves due to AgNPs exposures (Ringwood et al., 2010). Feeding habits are a major factor that directly influences AgNPs toxicity. *Mytilus galloprovincialis* is a suspension feeder mollusk, which was treated with AgNPs for 15 days at the temperature of ±17°C and showed significant toxic effects of AgNPs, i.e., lipid peroxidation, ROS formation, DNA damage, oxidative stress, and downregulation of Hsp 70 (Bebianno et al., 2015); whereas, *Macoma balthica* is a deposit feeder and was exposed to AgNPs at 15°C for 35 days and did not show toxic effects like *M. galloprovincialis* (McGreer, 1979).

6.2 Annelids

Annelids are parasitic invertebrates found in wet places. Ringworms, earthworms, and leeches belong to annelids. AgNPs have significant toxicity to annelids. In a short time exposure to AgNPs coated with humic acid, Polychaetes show toxic effects, i.e., lipid peroxidation, ROS formation, DNA damage, and oxidative stress, compared to soluble silver exposure (García-Alonso et al., 2014). *Nereis virens* demonstrates silver speciation in its tissues in the form of silver metal (AgCl, AgS) (Wang et al., 2014). When coelomocytes of *Nereis* were exposed to AgNPs coated with polyvinylpyrrolidone for 28 days at a temperature of 15°C, it experienced DNA damage and lysosomal membrane permeability. The toxicity of AgNPs to this annelid was reported to be concentration-dependent. Low concentrations cause less toxicity while higher amounts of AgNPs cause more toxicity (Cong et al., 2014).

6.3 Shrimps

The use of AgNPs is increasing day by day, which is a key factor in increased risk of silver contamination in natural environments, especially in the aquatic environment, leading to ecosystem disruption. Kachenton et al. (2018) investigated the toxic effects of AgNPs in brine shrimps. They exposed 10 adults of *Artemia salina*, a species of brine shrimp, to solutions of AgNPs of various concentrations for 24 h. They found the AgNPs concentration of $3521.13\,mg\,L^{-1}$ is the lethal concentration (LC_{50}) for *A. salina*. Intestinal lumen presented obstruction of AgNPs. Blabbing, increased thickening of mucus, necrosis, hyperplasia, and detachment from muscle lining was demonstrated by epithelial cells of the intestine. Other negative impacts of AgNPs, i.e., oxidative stress, DNA damage, and some developmental abnormalities, were also reported in shrimps.

7 LC_{50} value

Investigators have extensively studied AgNPs for acute toxicity in different fish groups. The survey of literature revealed that AgNPs had different acute toxicity in each group due to the type, size, and conditions of the test animal. Furthermore,

Table 1 The 96 h LC_{50} values of nano-silver in different fish groups.

No.	Form of NPS	Fish	LC_{50}	Reference
1	Nanocid[a]	Striped catfish	37.32 µg L^{-1}	Ghazanfari et al. (2020)
2	AgNO$_3$	Major carp	0.035 mg L^{-1}	Shobana et al. (2021)
3	SiO$_2$	Rainbow trout	8.9 mg L^{-1}	Shabrangharehdasht et al. (2020)
4	Nanocid	Persian sturgeon	0.89 mg L^{-1}	Banan et al. (2020)
5	AgNPs (50 nm)	Major carp	0.09 mg L^{-1}	Kakakhel et al. (2021)
6	AgNPs (1.25 ppm)	Guppy fish	0.528 ppm	Mohsenpour et al. (2020)
7	AgNPs	Zebrafish	125 µg L^{-1}	Kumari et al. (2020)

[a]This is a nanosilver antibiotic product with colloidal particles.

the test medium might also influence the toxicity of these particles (Khan et al., 2015b). The LC_{50} results from the studies of different authors are presented in Table 1.

In the body, it is either beneficial or harmful. In moderate concentration, superoxide anion along with nitric oxide and other ROS plays a critical part in signaling. In higher concentration, ROS causes oxidative stress from which arises conditions of pathology like diseases related to the cardiovascular system, cancer conditions (Sosa et al., 2013), diabetes mellitus (Moussa, 2008), inflammation, various neurodegenerative disorders, and onset of the aging process (Oyinloye et al., 2015).

The body has several enzymatic and nonenzymatic systems. These systems include CAT, along with another enzyme superoxide dismutase, GSH-Px and GST, glutathione peroxidase, vitamin E components, and glutathione (Memısogullari et al., 2003; Shafaq, 2012). Several synthetic and elements such as cerium also mimic the natural antioxidant activity, scavenging hazardous radicals and ameliorating stress in the body (Khan et al., 2015a).

8 Fate and accumulation

Less knowledge is present in literature when we talk about the circulation and fate of silver nanoparticles in an aqueous environment like freshwater bodies. The ionic form of silver plays a significant role in the effects of AgNPs and the stability of NPs. Both of these are significantly related to toxicity caused by AgNPs. After release in the water body AgNPs' transformation depends upon conditions and circumstances of the aquatic environment. Absorption of both organic compounds and inorganic substances and alterations in the oxidation number has a greater role in the determination of AgNPs' fate. Changes in both chemical and physical properties of AgNPs influence their circulation and toxicity in the aquatic environment. Metallic silver in an aqueous medium is oxidized at a greater level and interacts with other substances (Xiu et al., 2011). Numerous experimental studies reported that AgNPs undergo oxidative dissolution. Not only Ag nanoparticles exist in superficial waters, but their other forms (AgCl, AgS, other ionic forms, etc.) also exist (Ribeiro et al., 2014). When AgNPs undergo sulfidation, and insoluble layer of AgS is formed, which covers the surface of AgNPs; because of this, a lower number of Ag ions are released in water and toxicity caused by AgNPs is decreased in the aquatic environment. Additionally, chloride, if present in the waterbody, also causes a reduction in toxicity of AgNPs due to the formation of an organic material coating on AgCl. Layer of organic material makes AgCl stable, which leads to a reduction in toxicity. Natural freshwater bodies contain chelating agents. Among all chelating agents, ethylenediaminetetraacetic acid (EDTA) significantly influences the physical as well as chemical properties of AgNPs. EDTA has a great influence on size, surface, shape, charge bioavailability, and even toxic effects of AgNPs. The ability of AgNPs to aggregate depends upon the concentration of nanoparticles and pH levels. The work of Kittler and coworkers reported a modification of proteins binding with ionic silver releasing from AgNPs in an aqueous medium and how this phenomenon decreases AgNPs toxicity (Keller et al., 2010).

9 Mechanism of toxicity

Both vertebrate and nonvertebrate cell lines are affected by the toxic effects of AgNPs in the aquatic environment. Various processes are involved in toxicity induction, for example, increased lipid peroxidation, oxidative stress, and reduction of oxidative stress markers (Khan et al., 2015a). Reduction in mitochondrial function, alternation in membrane integrity, damage to skin, lungs, liver, and apoptosis of olfactory bulbs are also involved in toxicity induction (Schrand et al., 2008). Among all organs, the liver and gills are the most liable cites to suffer toxic effects. Recent studies indicate that AgNPs cause toxic effects in the liver of common carp and gills of zebrafish (Lee et al., 2012).

Toxicity caused by AgNPs depends upon their transformation in biological and environmental media (Fig. 2). Oxidation takes place on the surface of AgNPs and silver ions release and interact with biomolecules. It is a challenging (McShan et al., 2014) task to distinguish which part of toxicity is caused by AgNPs and which part is caused by Ag ions (Yang et al., 2012). Asharani and her coworkers investigated the antiproliferating activity of AgNPs (Asharani et al., 2008). They described the mechanism of toxicity caused by AgNPs. These nanoparticles can react with membrane proteins—this interaction activates signaling pathways causing inhibition of cell proliferation. Diffusion and endocytosis are other ways through which AgNPs enter the cell and cause dysfunction of mitochondria. These nanoparticles are also responsible for the generation of reactive oxygen species and damage to proteins

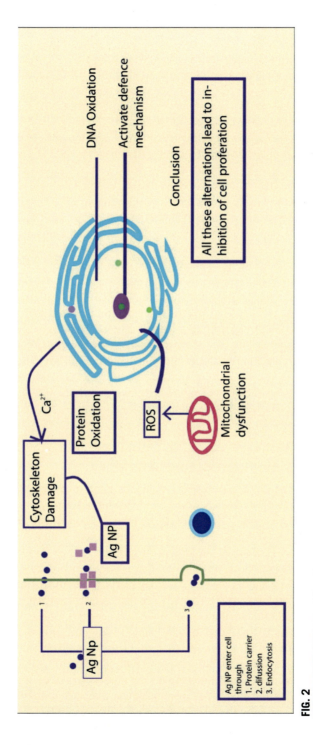

FIG. 2

Mechanism of toxicity.

and nucleic acid within the cell. Cell proliferation inhibition is also caused by the toxic effects of AgNPs.

Oxidative stress, which is one of the key mechanisms of AgNPs toxicity, occurs when cellular antioxidation increases due to the enhanced ability of the defense system due to the formation of ROS within the cell. Oxidative damage is caused by glutathione depletion and protein-bound sulfhydryl, leading to malfunctioning of various antioxidant enzymes (Awasthi et al., 2013). Furthermore, the interaction of AgNPs and silver ions with sulfur-containing macro biomolecules is an important toxicity mechanism because both ionic and nanoforms of silver have a great affinity for sulfur.

Mitochondria are a more sensitive target site for AgNPs within any cell. To know the mechanism of toxicity caused by AgNPs leading to mitochondrial dysfunction, Bressan and his coworkers investigated human dermal fibroblast. They found an accumulation of AgNPs outside the mitochondria, which directly influenced mitochondrial activity by damaging it. Mitochondrial damage caused dysfunction of the respiratory chain leading to the formation of ROS and oxidative stress. All these phenomena led to the interruption of ATP synthesis and DNA damage (Bressan et al., 2013). Hsin and his colleagues investigated NIH3T3 fibroblasts to study the mechanism of Ag NPs toxicity. They found that cytochrome c is released, due to the exposure of AgNPs, in cytosol leading to translocation of Bax to mitochondria. In this phenomenon, apoptosis is caused by nanoparticles when AgNPs act through ROS and C Jun N terminal kinase through the mitochondrial pathway. Interaction between AgNPs and DNA causes interruptions in the cell cycle. Due to the action of nanoparticles, the cell cycle arrested in the G1 phase and S phase blocked completely, which leads to apoptosis (Park et al., 2007).

10 Conclusion and recommendations

AgNPs attract the attention of researchers due to their toxicity to aquatic organisms and all other types of organisms, which can cause serious damage to different tissues in aquatic flora and fauna. There are various toxicity endpoints in the different studied models. As AgNPs have extensive uses, they produce huge amounts of waste, which is ultimately disposed of in aquatic habitats. These particles contaminate the water or air through different sources such as weathering rocks, the cement industry, and fossil fuel. AgNPs can be toxic to living forms but their nanoforms are more toxic because these are radially available in the aquatic environment and can enter ecosystems, altering their fitness and population densities. These particles magnify in the food chain and produce toxicity, which is significantly dangerous or even lethal to all kinds of life but is more toxic and lethal to aquatic life. Both vertebrate and invertebrate cell lines are affected by the toxic effects of AgNPs in the aquatic environment. In conclusion, nanoparticles are highly dangerous and even lethal for life, therefore, it is essential to follow precautions when using AgNPs. It is therefore recommended to prohibit the release of waste of silver nanoparticles into the aquatic environment. Applicable laws about the use and release of waste of silver nanoparticles of AgNPs should be formulated and applied by environmental protection authorities. AgNPs waste can also be recycled for various uses. It is further recommended that the use of AgNPs in human products should be below the threshold level.

References

Abdel-Khalek, A.A., Hamed, A., Hasheesh, W.S., 2021. Does the adsorbent capacity of orange and banana peels toward silver nanoparticles improve the biochemical status of *Oreochromis niloticus*? Environ. Sci. Pollut. Res. 28, 1–16.

Adriano, M., Pelletier, É., et al., 2018. Cytotoxicity and physiological effects of silver nanoparticles on marine invertebrates. Cell. Mol. Toxicol. Nanoparticles, 285–309. https://doi.org/10.1007/978-3-319-72041-8_17.

Ale, A., Galdoporpora, J.M., Mora, M.C., De La Torre, F.R., Desimone, M.F., Cazenave, J., 2021. Mitigation of silver nanoparticle toxicity by humic acids in gills of *Piaractus mesopotamicus* fish. Environ. Sci. Pollut. Res. Int. 28, 1–11.

Arora, S., Jain, J., Rajwade, J.M., Paknikar, K.M., 2009. Interactions of silver nanoparticles with primary mouse fibroblasts and liver cells. Toxicol. Appl. Pharmacol. 236, 310–318.

Asharani, P.V., Wu, Y.L., Gong, Z., Valiyaveettil, S., 2008. Toxicity of silver nanoparticles in zebrafish models. Nanotechnology 19, 255102.

Awasthi, K.K., Awasthi, A., Kumar, N., Roy, P., Awasthi, K., John, P., 2013. Silver nanoparticle induced cytotoxicity, oxidative stress, and DNA damage in CHO cells. J. Nanopart. Res. 15, 1898.

Banan, A., Kalbassi, M.R., Bahmani, M., Sotoudeh, E., Johari, S.A., Ali, J.M., Kolok, A.S., 2020. Salinity modulates biochemical and histopathological changes caused by silver nanoparticles in juvenile *Persian sturgeon* (*Acipenser persicus*). Environ. Sci. Pollut. Res. Int. 27, 10658–10671.

Baptista, M.S., Miller, R.J., Halewood, E.R., Hanna, S.K., Almeida, C.M., Vasconcelos, V.M., Keller, A.A., Lenihan, H.S., 2015. Impacts of silver nanoparticles on a natural estuarine plankton community. Environ. Sci. Technol. 49, 12968–12974.

Bar-Ilan, O., Albrecht, R.M., Fako, V.E., Furgeson, D.Y., 2009. Toxicity assessments of multisized gold and silver nanoparticles in zebrafish embryos. Small 5, 1897–1910.

Bebianno, M., Gonzalez-Rey, M., Gomes, T., Mattos, J., Flores-Nunes, F., Bainy, A., 2015. Is gene transcription in mussel gills altered after exposure to Ag nanoparticles? Environ. Sci. Pollut. Res. 22, 17425–17433.

Bilberg, K., Doving, K.B., Beedholm, K., Baatrup, E., 2011. Silver nanoparticles disrupt olfaction in Crucian carp (*Carassius carassius*) and Eurasian perch (*Perca fluviatilis*). Aquat. Toxicol. 104, 145–152.

Bressan, E., Ferroni, L., Gardin, C., Rigo, C., Stocchero, M., Vindigni, V., Cairns, W., Zavan, B., 2013. Silver nanoparticles and mitochondrial interaction. Int. J. Dent. 2013, 312747.

Catsakis, L.H., Sulica, V.I., 1978. Allergy to silver amalgams. Oral Surg. Oral Med. Oral Pathol. 46, 371–375.

Chen, X., Schluesener, H., 2008. Nanosilver: a nanoproduct in medical application. Toxicol. Lett. 176, 1–12.

Choi, J.E., Kim, S., Ahn, J.H., Youn, P., Kang, J.S., Park, K., Yi, J., Ryu, D.Y., 2010. Induction of oxidative stress and apoptosis by silver nanoparticles in the liver of adult zebrafish. Aquat. Toxicol. 100, 151–159.

Cong, Y., Banta, G.T., Selck, H., Berhanu, D., Valsami-Jones, E., Forbes, V.E., 2014. Toxicity and bioaccumulation of sediment-associated silver nanoparticles in the estuarine polychaete, *Nereis (Hediste) diversicolor*. Aquat. Toxicol. 156, 106–115.

Das, P., Metcalfe, C.D., Xenopoulos, M.A., 2014. Interactive effects of silver nanoparticles and phosphorus on phytoplankton growth in natural waters. Environ. Sci. Technol. 48, 4573–4580.

De Aragao, A.P., De Oliveira, T.M., Quelemes, P.V., Perfeito, M.L.G., Araujo, M.C., Santiago, J.D.A.S., Cardoso, V.S., Quaresma, P., De Almeida, J.R.D.S., Da Silva, D.A., 2019. Green synthesis of silver nanoparticles using the seaweed *Gracilaria birdiae* and their antibacterial activity. Arab. J. Chem. 12, 4182–4188.

Drake, P.L., Hazelwood, K.J., 2005. Exposure-related health effects of silver and silver compounds: a review. Ann. Occup. Hyg. 49, 575–585.

EC, 2008. Regulation (EC) no. 1272/2008 of the European Parliament and of the Council of 16 December 2008 on classification, labelling and packaging of substances and mixtures, amending and repealing Directives 67/548/EEC and 1999/45/EC, and amending Regulation (EC) no. 1907/2006. OJEU 353, 1195–1197.

Fabrega, J., Luoma, S.N., Tyler, C.R., Galloway, T.S., Lead, J.R., 2011. Silver nanoparticles: behaviour and effects in the aquatic environment. Environ. Int. 37, 517–531.

Farkas, J., Christian, P., Urrea, J.A., Roos, N., Hassellov, M., Tollefsen, K.E., Thomas, K.V., 2010. Effects of silver and gold nanoparticles on rainbow trout (*Oncorhynchus mykiss*) hepatocytes. Aquat. Toxicol. 96, 44–52.

García-Alonso, J., Misra, S.K., Valsami-Jones, E., Croteau, M.-N., Luoma, S.N., Rainbow, P.S., Rodriguez-Sanchez, N., et al., 2014. Toxicity and accumulation of silver nanoparticles during development of the marine polychaete *Platynereis dumerilii*. Sci. Total Environ. 476–477, 688–695. https://doi.org/10.1016/j.scitotenv.2014.01.039.

Ghazanfari, S., Rahimi, R., Zamani-Ahmadmahmoodi, R., Momeninejad, A., Abed-Elmdoust, A., 2020. Impact of silver nanoparticles on hepatic enzymes and thyroid hormones in striped catfish, *Pangasianodon hypophthalmus* (Pisces: Pangasiidae). Casp. J. Environ. Sci. 18, 265–275.

Gliga, A.R., Skoglund, S., Wallinder, I.O., Fadeel, B., Karlsson, H.L., 2014. Size-dependent cytotoxicity of silver nanoparticles in human lung cells: the role of cellular uptake, agglomeration and Ag release. Part. Fibre Toxicol. 11, 11.

Gonzalez, P., Baudrimont, M., Boudou, A., Bourdineaud, J.P., 2006. Comparative effects of direct cadmium contamination on gene expression in gills, liver, skeletal muscles and brain of the zebrafish (*Danio rerio*). Biometals 19, 225–235.

Govindan, P., Murugan, M., Pitchaikani, S., Venkatachalam, P., Gopalakrishnan, A.V., Kandasamy, S., Shakila, H., 2021. Synthesis and characterization of bioactive silver nanoparticles from red marine macroalgae *Chondrococcus hornemannii*. Mater. Today. https://doi.org/10.1016/j.matpr.2021.02.497. In press.

Guo, B., Alivio, T.E., Fleer, N.A., Feng, M., Li, Y., Banerjee, S., Sharma, V.K., 2021. Elucidating the role of dissolved organic matter and sunlight in mediating the formation of Ag–Au bimetallic alloy nanoparticles in the aquatic environment. Environ. Sci. Technol. 55, 1710–1720.

Haghighat, F., Kim, Y., Sourinejad, I., Yu, I.J., Johari, S.A., 2021. Titanium dioxide nanoparticles affect the toxicity of silver nanoparticles in common carp (*Cyprinus carpio*). Chemosphere 262, 127805.

Handy, R.D., Owen, R., Valsami-Jones, E., 2008. The ecotoxicology of nanoparticles and nanomaterials: current status, knowledge gaps, challenges, and future needs. Ecotoxicology 17, 315–325.

Hawkins, A.D., Thornton, C., Kennedy, A.J., Bu, K., Cizdziel, J., Jones, B.W., Steevens, J.A., Willett, K.L., 2015. Gill histopathologies following exposure to nanosilver or silver nitrate. J. Toxicol. Environ. Health A 78, 301–315.

Hedayati, A., Kolangi, H., Jahanbakhshi, A., Shaluei, F., 2012. Evaluation of silver nanoparticles ecotoxicity in silver carp (*Hypophthalmichthys molitrix*) and goldfish (*Carassius auratus*). Bulg. J. Vet. Med. 15, 172.

Ivask, A., Kurvet, I., Kasemets, K., Blinova, I., Aruoja, V., Suppi, S., Vija, H., Kakinen, A., Titma, T., Heinlaan, M., Visnapuu, M., Koller, D., Kisand, V., Kahru, A., 2014. Size-dependent toxicity of silver nanoparticles to bacteria, yeast, algae, crustaceans and mammalian cells in vitro. PLoS One 9, e102108.

Jahanbakhshi, A., Shaluei, F., Hedayati, A., 2012. Detection of silver nanoparticles (nanosil) LC in silver carp®. World J. Zool. 7, 126–130.

Johari, S., Kalbassi, M., Soltani, M., Yu, I., 2013. Toxicity comparison of colloidal silver nanoparticles in various life stages of rainbow trout (*Oncorhynchus mykiss*). Iran. J. Fish. Sci. 12, 76–95.

Kachenton, S., Whangpurikul, V., Kangwanrangsan, N., Tansatit, T., Jiraungkoorskul, W., 2018. Siver nanoparticles toxicity in brine shrimp and its histopathological analysis. Int. J. Nanosci. 17, 1850007.

Kakakhel, M.A., Wu, F., Sajjad, W., Zhang, Q., Khan, I., Ullah, K., Wang, W., 2021. Long-term exposure to high-concentration silver nanoparticles induced toxicity, fatality, bioaccumulation, and histological alteration in fish (*Cyprinus carpio*). Environ. Sci. Eur. 33, 1–11.

Kalbassi, M.R., Johari, S.A., Soltani, M., Yu, I., 2013. Particle size and agglomeration affect the toxicity levels of silver nanoparticle types in aquatic environment. ECOPERSIA 1, 273–290.

Kang, J.S., Park, J.W., 2021. Silver ion release accelerated in the gastrovascular cavity of *Hydra vulgaris* increases the toxicity of silver sulfide nanoparticles (Ag_2S-NPs). Environ. Toxicol. Chem. 40, 1662–1672. https://doi.org/10.1002/etc.5017.

Katuli, K.K., Massarsky, A., Hadadi, A., Pourmehran, Z., 2014. Silver nanoparticles inhibit the gill Na(+)/K(+)-ATPase and erythrocyte AChE activities and induce the stress response in adult zebrafish (*Danio rerio*). Ecotoxicol. Environ. Saf. 106, 173–180.

Keller, A.A., Wang, H., Zhou, D., Lenihan, H.S., Cherr, G., Cardinale, B.J., Miller, R., Ji, Z., 2010. Stability and aggregation of metal oxide nanoparticles in natural aqueous matrices. Environ. Sci. Technol. 44, 1962–1967.

Khan, M.S., Jabeen, F., Asghar, M.S., Qureshi, N., Shakeel, M., Noureen, A., Shabbir, S., 2015a. Role of nao-ceria in the amelioration of oxidative stress: current and future applications in medicine. Int. J. Biosci. 6, 89–109.

Khan, M.S., Jabeen, F., Qureshi, N.A., Asghar, M.S., Shakeel, M., Noureen, A., 2015b. Toxicity of silver nanoparticles in fish: a critical review. J. Biol. Environ. Sci. 6, 211–227.

Khan, M.S., Qureshi, A.N., Jabeen, F., et al., 2017. Assessment of toxicity in fresh water fish *Labeo rohita* treated with silver nanoparticles. Appl. Nanosci. 7 (5), 167–179.

Khan, M.S., Qureshi, N.A., Jabeen, F., 2018. Ameliorative role of nano-ceria against amine coated Ag-NP induced toxicity in *Labeo rohita*. Appl. Nanosci. 8, 323–337.

Khan, M.S., Qureshi, N.A., Jabeen, F., Wajid, M., Sabri, S., Shakir, M., 2020. The role of garlic oil in the amelioration of oxidative stress and tissue damage in rohu *Labeo rohita* treated with silver nanoparticles. Fish. Sci. 86, 255–269.

Khan, M.U., Qurashi, N.A., Khan, M.S., Jabeen, F., Umar, A., Yaqoob, J., Wajid, M., 2016. Generation of reactive oxygen species and their impact on the health related parameters: a critical review. Int. J. Biosci. 9, 303–323.

Kim, J., Lee, J., Kwon, S., Jeong, S., 2009. Preparation of biodegradable polymer/silver nanoparticles composite and its antibacterial efficacy. J. Nanosci. Nanotechnol. 9, 1098–1102.

Kim, J.S., Kuk, E., Yu, K.N., Kim, J.H., Park, S.J., Lee, H.J., Kim, S.H., Park, Y.K., Park, Y.H., Hwang, C.Y., Kim, Y.K., Lee, Y.S., Jeong, D.H., Cho, M.H., 2007. Antimicrobial effects of silver nanoparticles. Nanomedicine (Lond.) 3, 95–101.

Klaprat, D.A., Evans, R.E., Hara, T.J., 1992. Environmental contaminants and chemoreception in fishes. In: Fish Chemoreception. Springer.

Kumari, S., Kumari, P., Panda, P.K., Patel, P., Jha, E., Mallick, M.A., Verma, S.K., 2020. Biocompatible biogenic silver nanoparticles interact with caspases on an atomic level to elicit apoptosis. Nanomedicine 15, 2119–2132.

Kwok, K.W., Auffan, M., Badireddy, A.R., Nelson, C.M., Wiesner, M.R., Chilkoti, A., Liu, J., Marinakos, S.M., Hinton, D.E., 2012. Uptake of silver nanoparticles and toxicity to early life stages of Japanese medaka (*Oryzias latipes*): effect of coating materials. Aquat. Toxicol. 120–121, 59–66.

Laban, G., Nies, L.F., Turco, R.F., Bickham, J.W., Sepulveda, M.S., 2010. The effects of silver nanoparticles on fathead minnow (*Pimephales promelas*) embryos. Ecotoxicology 19, 185–195.

Larese, F.F., D'agostin, F., Crosera, M., Adami, G., Renzi, N., Bovenzi, M., Maina, G., 2009. Human skin penetration of silver nanoparticles through intact and damaged skin. Toxicology 255, 33–37.

Lee, H.J., Reimann, J., Huang, Y., Ädelroth, P., 2012. Functional proton transfer pathways in the heme–copper oxidase superfamily. Biochim. Biophys. Acta 1817, 537–544.

Luoma, S.N., Rainbow, P.S., Luoma, S., 2008. Metal Contamination in Aquatic Environments: Science and Lateral Management. Cambridge University Press.

Marambio-Jones, C., Hoek, E.M., 2010. A review of the antibacterial effects of silver nanomaterials and potential implications for human health and the environment. J. Nanopart. Res. 12, 1531–1551.

Marinho, C.S., Matias, M.V.F., Toledo, E.K.M., Smaniotto, S., Ximenes-Da-Silva, A., Tonholo, J., Santos, E.L., Machado, S.S., Zanta, C., 2021. Toxicity of silver nanoparticles on different tissues in adult *Danio rerio*. Fish Physiol. Biochem. 47, 1–11.

Massarsky, A., Abraham, R., Nguyen, K.C., Rippstein, P., Tayabali, A.F., Trudeau, V.L., Moon, T.W., 2014. Nanosilver cytotoxicity in rainbow trout (*Oncorhynchus mykiss*) erythrocytes and hepatocytes. Comp. Biochem. Physiol. C Toxicol. Pharmacol. 159, 10–21.

McGreer, E.R., 1979. Sublethal effects of heavy metal contaminated sediments on the bivalve *Macoma balthica* (L.). Mar. Pollut. Bull. 10 (9), 259–262. https:/doi.org/10.1016/0025-326X(79)90482-X.

McShan, D., Ray, P.C., Yu, H., 2014. Molecular toxicity mechanism of nanosilver. J. Food Drug Anal. 22, 116–127.

Melnik, E., Buzulukov, Y.P., Demin, V., Demin, V., Gmoshinski, I., Tyshko, N., Tutelyan, V., 2013. Transfer of silver nanoparticles through the placenta and breast milk during in vivo experiments on rats. Acta Nat. 5, 107.

Memısogullari, R., Taysi, S., Bakan, E., Capoglu, I., 2003. Antioxidant status and lipid peroxidation in type II diabetes mellitus. Cell Biochem. Funct. 21, 291–296.

Mohsenpour, R., Mousavi-Sabet, H., Hedayati, A., Rezaie, A., 2020. Determination of lethal concentration (LC50 96h) of silver nanoparticles in guppy fish (*Poecilia reticulate*). Iran. J. Mar. Sci. Technol. 24, 59–67.

Monfared, A.L., Soltani, S., 2013. Effects of silver nanoparticles administration on the liver of rainbow trout (*Oncorhynchus mykiss*): histological and biochemical studies. Eur. J. Exp. Biol. 3, 285–289.

Moussa, S., 2008. Oxidative stress in diabetes mellitus. Rom. J. Biophys. 18, 225–236.

Nagamatsu, P.C., Garcia, J.R.E., Esquivel, L., Da Costa Souza, A.T., De Brito, I.A., De Oliveira Ribeiro, C.A., 2021. Post hatching stages of tropical catfish *Rhamdia quelen* (Quoy Gaimard, 1824) are affected by combined toxic metals exposure with risk to population. Chemosphere 277, 130199.

Oyinloye, B.E., Adenowo, A.F., Kappo, A.P., 2015. Reactive oxygen species, apoptosis, antimicrobial peptides and human inflammatory diseases. Pharmaceuticals (Basel) 8, 151–175.

Park, B., Martin, P., Harris, C., Guest, R., Whittingham, A., Jenkinson, P., Handley, J., 2007. Initial in vitro screening approach to investigate the potential health and environmental hazards of Enviroxtrade mark—a nanoparticulate cerium oxide diesel fuel additive. Part. Fibre Toxicol. 4, 12.

Rajkumar, K., Kanipandian, N., Thirumurugan, R., 2015. Toxicity assessment on haemotology, biochemical and histopathological alterations of silver nanoparticles-exposed freshwater fish *Labeo rohita*. Appl. Nanosci. 6, 19–29. https://doi.org/10.1007/s13204-015-0417-7.

Rathore, R.S., Khangarot, B., 2002. Effects of temperature on the sensitivity of sludge worm *Tubifex tubifex* Müller to selected heavy metals. Ecotoxicol. Environ. Saf. 53, 27–36.

Raza, A., Javed, S., Qureshi, M.Z., Khan, M.S., 2017. Synthesis and study of catalytic application of l-methionine protected gold nanoparticles. Appl. Nanosci. 7, 429–437.

Reddy, T.K., Reddy, S.J., Prasad, T., 2013. Effect of silver nanoparticles on energy metabolism in selected tissues of *Aeromonas hydrophila* infected Indian major carp, *Catla catla*. IOSR J. Pharm. 3, 49–55.

Ribeiro, F., Gallego-Urrea, J.A., Jurkschat, K., Crossley, A., Hassellov, M., Taylor, C., Soares, A.M., Loureiro, S., 2014. Silver nanoparticles and silver nitrate induce high toxicity to *Pseudokirchneriella subcapitata*, *Daphnia magna* and *Danio rerio*. Sci. Total Environ. 466–467, 232–241.

Ringwood, A.H., Mccarthy, M., Bates, T.C., Carroll, D.L., 2010. The effects of silver nanoparticles on oyster embryos. Mar. Environ. Res. 69 (Suppl), S49–S51.

Schrand, A.M., Braydich-Stolle, L.K., Schlager, J.J., Dai, L., Hussain, S.M., 2008. Can silver nanoparticles be useful as potential biological labels? Nanotechnology 19, 235104.

Schultz, A.G., Ong, K.J., Maccormack, T., Ma, G., Veinot, J.G., Goss, G.G., 2012. Silver nanoparticles inhibit sodium uptake in juvenile rainbow trout (*Oncorhynchus mykiss*). Environ. Sci. Technol. 46, 10295–10301.

Scown, T.M., Santos, E.M., Johnston, B.D., Gaiser, B., Baalousha, M., Mitov, S., Lead, J.R., Stone, V., Fernandes, T.F., Jepson, M., Van Aerle, R., Tyler, C.R., 2010. Effects of aqueous exposure to silver nanoparticles of different sizes in rainbow trout. Toxicol. Sci. 115, 521–534.

Shabrangharehdasht, M., Mirvaghefi, A., Farahmand, H., 2020. Effects of nanosilver on hematologic, histologic and molecular parameters of rainbow trout (*Oncorhynchus mykiss*). Aquat. Toxicol. 225, 105549.

Shafaq, N., 2012. An overview of oxidative stress and antioxidant defensive system. Open Access Sci. Rep. 01, 1–9.

Shakeel, M., Jabeen, F., Iqbal, R., Chaudhry, A.S., Zafar, S., Ali, M., Khan, M.S., Khalid, A., Shabbir, S., Asghar, M.S., 2018. Assessment of titanium dioxide nanoparticles (TiO2-NPs) induced hepatotoxicity and ameliorative effects of *Cinnamomum cassia* in Sprague-Dawley rats. Biol. Trace Elem. Res. 182, 57–69.

Shaluei, F., Hedayati, A., Jahanbakhshi, A., Kolangi, H., Fotovat, M., 2013. Effect of subacute exposure to silver nanoparticle on some hematological and plasma biochemical indices in silver carp (*Hypophthalmichthys molitrix*). Hum. Exp. Toxicol. 32, 1270–1277. https://doi.org/10.1177/0960327113485258.

Shobana, C., Rangasamy, B., Hemalatha, D., Ramesh, M., 2021. Bioaccumulation of silver and its effects on biochemical parameters and histological alterations in an Indian major carp *Labeo rohita*. Environ. Chem. Ecotoxicol. 3, 51–58.

Sosa, V., Moline, T., Somoza, R., Paciucci, R., Kondoh, H., Me, L.L., 2013. Oxidative stress and cancer: an overview. Ageing Res. Rev. 12, 376–390.

Sung, J.H., Ji, J.H., Yoon, J.U., Kim, D.S., Song, M.Y., Jeong, J., Han, B.S., Han, J.H., Chung, Y.H., Kim, J., Kim, T.S., Chang, H.K., Lee, E.J., Lee, J.H., Yu, I.J., 2008. Lung function changes in Sprague-Dawley rats after prolonged inhalation exposure to silver nanoparticles. Inhal. Toxicol. 20, 567–574.

Taju, G., Abdul Majeed, S., Nambi, K.S., Sahul Hameed, A.S., 2014. In vitro assay for the toxicity of silver nanoparticles using heart and gill cell lines of *Catla catla* and gill cell line of *Labeo rohita*. Comp. Biochem. Physiol. C Toxicol. Pharmacol. 161, 41–52.

Wang, H., Ho, K.T., Scheckel, K.G., Wu, F., Cantwell, M.G., Katz, D.R., Horowitz, D.B., Boothman, W.S., Burgess, R.M., 2014. Toxicity, bioaccumulation, and biotransformation of silver nanoparticles in marine organisms. Environ. Sci. Technol. 48, 13711–13717.

Webb, N.A., Wood, C.M., 1998. Physiological analysis of the stress response associated with acute silver nitrate exposure in freshwater rainbow trout (*Oncorhynchus mykiss*). Environ. Toxicol. Chem. 17, 579–588.

Wijnhoven, S.W.P., Peijnenburg, W.J.G.M., Herberts, C.A., Hagens, W.I., Oomen, A.G., Heugens, E.H.W., Roszek, B., Bisschops, J., Gosens, I., Van De Meent, D., Dekkers, S., De Jong, W.H., Van Zijverden, M., Sips, A.J.A.M., Geertsma, R.E., 2009. Nano-silver—a review of available data and knowledge gaps in human and environmental risk assessment. Nanotoxicology 3, 109–138.

Wu, Y., Zhou, Q., Li, H., Liu, W., Wang, T., Jiang, G., 2010. Effects of silver nanoparticles on the development and histopathology biomarkers of Japanese medaka (*Oryzias latipes*) using the partial-life test. Aquat. Toxicol. 100, 160–167.

Xiu, Z.M., Ma, J., Alvarez, P.J., 2011. Differential effect of common ligands and molecular oxygen on antimicrobial activity of silver nanoparticles versus silver ions. Environ. Sci. Technol. 45, 9003–9008.

Yang, X., Gondikas, A.P., Marinakos, S.M., Auffan, M., Liu, J., Hsu-Kim, H., Meyer, J.N., 2012. Mechanism of silver nanoparticle toxicity is dependent on dissolved silver and surface coating in *Caenorhabditis elegans*. Environ. Sci. Technol. 46, 1119–1127.

Yang, Y., Zheng, S., Li, R., Chen, X., Wang, K., Sun, B., Zhang, Y., Zhu, L., 2021. New insights into the facilitated dissolution and sulfidation of silver nanoparticles under simulated sunlight irradiation in aquatic environments by extracellular polymeric substances. Environ. Sci. Nano 8, 748–757. https://doi.org/10.1039/D0EN01142H.

CHAPTER 26

Silver nanoparticles in natural ecosystems: Fate, transport, and toxicity

Parteek Prasher[a], Mousmee Sharma[b], Harish Mudila[c], Amit Verma[d], and Pankaj Bhatt[e]

[a]*Department of Chemistry, University of Petroleum & Energy Studies, Energy Acres, Dehradun, India* [b]*Department of Chemistry, Uttaranchal University, Dehradun, India* [c]*Department of Chemistry, Lovely Professional University, Punjab, India* [d]*Department of Biochemistry, College of Basic Science and Humanities, SD Agricultural University, Gujarat, India* [e]*State Key Laboratory for Conservation and Utilization of Subtropical Agro-bioresources, Guangdong Laboratory for Lingnan Modern Agriculture, Integrative Microbiology Research Centre, South China Agricultural University, Guangzhou, China*

1 Introduction

The ubiquitous applications of AgNPs in materials, renewable energy, medicine, disinfectants, healthcare, fertilizers, pesticides, microbicides, and cosmetics increase the likelihood of environmental exposure (El-Nour et al., 2010; Prasher et al., 2018a,b, 2020). The plasmonic, electrical, magnetic, optical, and biocidal properties of AgNPs lay the basis of these applications and overexploitation. In addition to this, the natural sources of AgNPs exposure in the ecosystem encompass the natural conversion of silver salts to nanoparticles by humic acid in the presence of irradiation (Yusuf, 2019; Prasher et al., 2021). The presence of natural organic matter at the point of exposure serves as the natural capping agent for naturally synthesized AgNPs (Prasher et al., 2019; Burdusel et al., 2018; Koser et al., 2017). Importantly, natural organic matter promotes the stabilization of AgNPs in natural ecosystems, which further augments the persistence of nanoparticles in the environment (Bhatt et al., 2020; Verma et al., 2021).

Natural ecosystems act as a sink for AgNPs, which undergo various physicochemical transformations before meeting their fate (Dong et al., 2021; Zhang et al., 2018a,b). These include aggregation, dissolution, oxidation, sulfidation, and chlorination that decide the transport, uptake, and accumulation of AgNPs in natural ecosystems, where deleterious effects appear pertaining to silver-stress in plants, microbiota, and microflora (Yu et al., 2012). The over-accumulation of AgNPs in ecosystems lowers the soil productivity by killing useful microbes and interfering with the uptake of micronutrients (Prasher et al., 2018a,b; Courtois et al., 2021a, b; Herrera et al., 2021; Souza et al., 2021). The successive bioaccumulation of AgNPs in autotrophs, their biomagnification in further trophic levels, and integration to the food chains adversely effects the various components in an ecosystem (Huang et al., 2019a,b; Ferdous et al., 2021). The silver stressed soils become deficient in the essential metal ions needed for plants to thrive. A number of physical, chemical, and biological factors determine the fate of AgNPs during their transformation process, which include the size, shape, surface charge, and pH of parent media (Kwak et al., 2016). In addition, the fate of AgNPs depends on the nature of surface functionalities, ionic strength of parent media, concentration of natural organic matter in the parent media, and the presence of competing cations in the parent media (Tripathi et al., 2017). The present chapter deals with the transport, toxicity, and fate of AgNPs in natural ecosystems.

2 Sources of AgNPs release in the environment

Humic acids are a prevalent source for the reduction of naturally occurring silver ions to AgNPs in the environmentally relevant conditions (Yan and Chen, 2019; Jianchang et al., 2008). The humic acids enriched with diverse functional groups including hydroxyls, quinines, phenolics, and carbonyls facilitate the reduction process of AgNPs (Milne et al., 2017). The presence of phenolic group on humic acids generates superoxide ions that mediate the photoreduction of silver ions to AgNPs in a pH-dependent manner (Kulikova et al., 2020; Peng et al., 2021). As such, the presence of dissolved oxygen facilitates the humic acid-mediated photoreduction of silver ions to AgNPs (Ding et al., 2019). In addition, the dissolved organic matter present in natural waters readily reduce silver ions to AgNPs in the presence of sunlight. Nevertheless, the AgNPs so formed possess lower stability in natural waters due to the presence of inorganic cations that cause the nanoparticles to coalesce (Bundschuh et al., 2018).

AgNPs are involved in various processes in manufacturing units and production facilities such as packaging and mixing, mechanical grinding, and solution drying, which cause their eventual release in the environment. The oligodynamic effect of AgNPs offers applications such as disinfectant, and antiodor sprays that promote their direct emission to the atmosphere (Prasher et al., 2018a,b). Similarly, the widespread utilization of surface disinfectants, paints, varnishes, and polishes containing AgNPs increases the probability of their release in the atmosphere during spraying (Quadros and Marr, 2010). Owing to their ultrafine size, AgNPs undergo a rapid diffusion in the atmosphere due to a high diffusion coefficient, as framed by Fick's law of diffusion (Kostic et al., 2016). However, the distance traveled by the AgNPs after diffusion to the atmosphere depends on their stability parameters. AgNP-impregnated textile garments and bandage materials pose a high probability of AgNPs release to the aquatic ecosystem during laundering and washing (Rearick et al., 2018). Since AgNPs possess biocidal potency, their application as pesticides and microbicidal agents in agricultural sector causes release directly to the agroecosystems, followed by a rapid uptake by autotrophs, and dumping to water bodies during irrigation operations (Furtado et al., 2015).

Owing to a higher surface area, AgNPs act as a template for attaching pollutants and organic contaminants, which further aggravates their hazardous effects (Pallavi et al., 2016). Similarly, during their transportation, the aerosol suspension of AgNPs can coalesce to generate large aggregates that tend to settle on various surfaces under gravitational force, followed by being washed down to the terrestrial and aquatic ecosystems by precipitation or washing of these surfaces (Yoo-iam et al., 2014). The direct disposing of AgNP-containing waste products, followed by their incineration on land causes the suspension of nanoparticles in terrestrial ecosystem. Further utilization of this AgNP-containing sewage sludge as a fertilizer in agriculture causes the infiltration of AgNPs into agroecosystems. During their transport through the ecosystems, AgNPs notably adsorb on their surface various organic contaminants, which further augments their toxic effect.

The AgNPs functionalized with charged molecules and capping agents demonstrate considerable electrostatic interactions with the soil types that results in an altered mobility of nanoparticles and hence causes their accumulation, thereby resulting in silver stress (Klitzke et al., 2015). As such, the positively charged AgNPs display higher retention in the oppositely charged soil type, and vice versa, demonstrating a higher mobility in the soil type having a similar charge (Li et al., 2017). Similarly, the natural organic materials present in aquatic ecosystems readily adsorb on the surface of AgNPs, thereby stabilizing the latter and improving their motility in aqueous environment (Lekamge et al., 2018). Nevertheless, the nature of the natural organic matter, their molecular weight, concentration in the aquatic body, and the nature of functional groups determine the stability of AgNPs (Huang et al., 2019a,b). The presence of divalent metal ions customarily present in the water bodies also plays an important role in the induction of AgNPs aggregation. These factors contribute toward increasing bioavailability of AgNPs to the aquatic environment; however, the deposition of AgNPs in sediments results in comparatively lower bioavailability to the aquatic ecosystem. Fig. 1 illustrates the factors governing the fate of AgNPs and their transformations in natural ecosystems.

3 Transport of AgNPs in the environment

The environmental release of AgNPs by different natural and anthropogenic sources results in several transformations, such as aggregation owing to a high ionic strength, or remaining of individual AgNPs in suspension for their long-distance delivery. After coming in contact with dioxygen and oxidants, the onset of oxidation and dissolution of AgNPs takes place to transform them to silver ions (Steinmetz et al., 2020). The sulfidation and chlorination of AgNPs occurs more frequently, which alters their characteristic physicochemical properties. Factors such as ionic strength

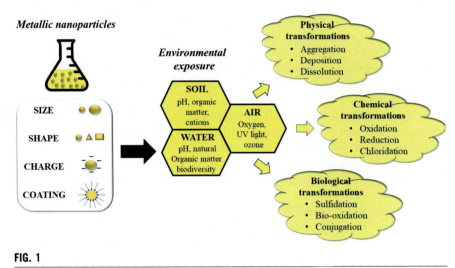

FIG. 1

Fate of AgNPs in the ecosystem, and the factors governing the same.

of parent solution, pH, composition of electrolyte, nature and composition of organic matter, and the surface energy of AgNPs determine the behavior and fate of AgNPs in natural ecosystems (Hou et al., 2017). The large surface area, owing to smaller, finer size, and high surface energies cause the coalescence of AgNPs to form large clusters. The capping agent therefore plays a critical role in controlling the size of AgNPs by electrostatic repulsion, and steric repulsions caused by the presence of functional head groups on these capping agents. In addition, these interactions play an important role in deciding the stability of colloidal solution of AgNPs (Ellis et al., 2018).

A high ionic strength acts as an important factor in deciding the AgNPs stability. The presence of divalent cations increase the AgNPs coagulation by weakening of electrostatic repulsive forces between the AgNPs by eroding the electrical double layer present on the surface of AgNPs that causes aggregation of nanoparticles in the colloidal solution (Sagee et al., 2012; Gutierrez et al., 2020). Similarly, pH plays a paramount role in maintaining the surface potential of AgNPs and in determining their coagulation properties. The AgNPs demonstrate various aggregation states over pH ranges. As the pH approaches the point of zero charge, the aggregation size of AgNPs increases. Eventually, the nature and composition of organic matter plays a key role in determining the fate of AgNPs, especially in aquatic ecosystems (Chinnapongse et al., 2011). The functional groups present on the organic materials promote the surface binding of AgNPs, thereby causing colloidal stabilization of the nanoparticle suspension. The carbonate ions, sulfate ions, and sulfide ions present in fresh water ecosystems associate with the positively charged silver ions to form the corresponding salts. However, in the marine water ecosystem, the silver ions undergo chlorination as the main transformation process, owing to the presence of NaCl in high concentration. The chlorination of silver ions in marine water results in the formation of $AgCl_2^-$, $AgCl_3^{2-}$, and $AgCl_4^{3-}$ species that further improve the mobility of AgNPs in marine water ecosystems.

The AgNPs serve as a plant strengthening agent, which requires their application to soil systems. Mainly, the porous media such as quartz sand, ferrihydrite-coated sand, kaolin-coated sand, ceramics, unmodified silica glass beads, porous sandstones, and hematite-modified silica glass beads promote the transportation of AgNPs in ground water aquifers, sand filters, and soils. The nature of surface coating of AgNPs determines their mobility or deposition in a porous medium. As such, the branched polyethylenimine-coated AgNPs displayed a higher mobility in quartz sand, kaolin-coated sand, and ferrihydrite-coated sand, thereby indicating the significance of electrostatic coating in determining the mobility and stability of AgNPs. Generally, the citrate-coated and PVP-coated AgNPs showed maximum deposition in kaolin-coated sand, followed by ferrihydrite-coated and, and eventually the quartz sand demonstrated a minimum deposition of AgNPs. This difference arises due to the difference in the zeta potential of the porous media due to which the extent of interactions between the surface of porous media and AgNPs differ.

In consistence to the zeta potential, the pH also plays a considerable role in deciding the retention of AgNPs to the porous media at a fixed ionic strength. As such, the PVP-coated AgNPs displayed improved retention on the glass beads from pH 4 to 6.5, while further increase in pH lowered the retention of nanoparticles. Furthermore, the increase in the ionic strength enhanced the retention of AgNPs, mainly due to the induction of aggregation of AgNPs that enhanced the physical filtration of the nanoparticle clusters through the porous medium. Similarly, the ionic strength lowers the repulsive forces between the AgNPs and the surface of porous media, hence causing the retention of AgNPs. The natural organic material, such as humic acid, constituting high molecular weight polymers, reduced the retention of PVP-coated AgNPs on the surface of glass beads, mainly due to the increased electrostatic stabilization caused by the humic acid (Liu et al., 2020). Cysteine achieved a similar effect at lower ionic strengths, mainly due to the increase in the interactions with the electric double layer of PVP-coated AgNPs. The dissolved oxygen presented a marked effect on the overall mobility of AgNPs. The reduction in the dissolved oxygen at pH 4 from 8.9 to 0.2 mg/L increased the mobility of citrate capped AgNP to 15%, mainly due to the highly negative zeta potential of citrate capped AgNPs at low levels of dissolved oxygen.

4 Transformations of AgNPs

The AgNPs exposed to the ecosystem undergo physical, chemical, and biological transformations to meet their ultimate fate (Lowry et al., 2012; Pang et al., 2020). The physical transformations involve aggregation, deposition, and dissolution; chemical transformations include oxidation, reduction, chloridation; and the biological transformations include bio-oxidation, sulfidation, and conjugation (Potter et al.,

2019; Zhang et al., 2018a,b; Bao et al., 2021). These transformations occur mainly due to the interactions of AgNPs with the components of the environment, mainly soil, air, water, and microbiota; and the redox potential of the parent medium where the AgNPs accumulate (Mohan et al., 2019). The transformations of AgNPs mainly result in dissolution, aggregation, and adsorption. The dissolution of AgNPs includes the oxidation of metallic Ag to silver ions in the presence of oxidants (Choi et al., 2008). Dissolution enhances the toxicity of AgNPs as the silver ions present deleterious side effects by interacting with biological systems where they stimulate oxidative stress (Shen et al., 2015). The aggregation of AgNPs occurs when the nanoparticles lose their surface coatings and capping agents that result in the collapse of the colloidal suspension (Zhang et al., 2019). The loss of colloidal AgNPs further results in the aggregation of AgNPs. Adsorption occurs due to the appending of various natural capping agents, such as organic matter, or pollutants in the atmosphere (Nason et al., 2012). Adsorption causes mitigation of AgNP-mediated toxicity by preventing the formation of silver ions; however, it presents a key mechanism for the transference of adsorbent to far-off places, which may prove hazardous, depending on the nature of the adsorbent (Ferdous and Nemmar, 2020). Fig. 2 demonstrates the pathways for the environmental transformation of AgNPs.

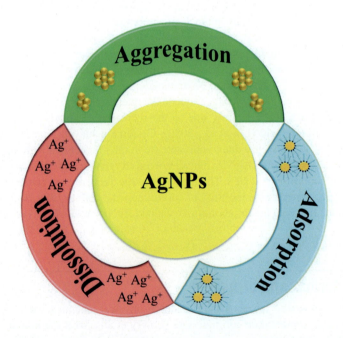

FIG. 2

Transformations of AgNPs in the environment.

5 Sulfidation

The surface sulfidation of AgNPs occurs in the presence of SO_2, COS, H_2S (Franey et al., 1985), and other sulfur-containing solutions (Johnston et al., 2019). The sulfidation results in the formation of an Ag_2S layer on the nanoparticle surface, in addition to forming the Ag-Ag_2S core-shell nanostructures (Shahjamali et al., 2016; Liu et al., 2011). Importantly, the Ag_2S displays a prominent presence in the sewage sludge of wastewater treatment plants, mainly due to the sulfidation of disposed silver ions in the presence of sulfur-rich materials (Kraas et al., 2017). The sulfidation of AgNPs mainly takes place via Eqs. (1)–(4).

Sulfidation (without oxygen):

$$Ag + H_2S \rightarrow Ag_2S + H_2 \tag{1}$$

Sulfidation (with metal sulfides):

$$2Ag^+ + MS \rightarrow Ag_2S + M^{2+} \quad (M = Zn, Cu) \tag{2}$$

Oxysulfidation (direct):

$$\begin{aligned} 4Ag + O_2 + 2H_2S &\rightarrow 2Ag_2S + 2H_2O \\ 4Ag + O_2 + 2HS^- &\rightarrow 2Ag_2S + 2OH^- \end{aligned} \tag{3}$$

Oxysulfidation (indirect/oxidative dissolution):

$$\begin{aligned} 4Ag + O_2 + 4H^+ &\rightarrow 4Ag^+ + 2H_2O \\ 4Ag^+ + S^{2-} &\rightarrow 2Ag_2S \end{aligned} \tag{4}$$

Mainly, the oxysulfidation of AgNPs occurs in the presence of a higher concentration of sulfides, which directly generate Ag_2S by particle-fluid reaction, whereas the presence of a lower sulfide concentration results in the preliminary oxidation of AgNPs to silver ions, which eventually react with sulfide ions to generate Ag_2S (Liu et al., 2011; Kent et al., 2014).

The degree of sulfidation processes plays a major role in controlling the concentrations of silver ions in the environment and regulating their hazardous effects (Lee et al., 2016; Yang et al., 2021). The direct oxysulfidation of AgNPs occurs at higher concentrations of sulfides, whereas the indirect oxysulfidation termed as precipitation or oxidative dissolution occurs at comparatively lower concentrations of sulfide (Liu et al., 2018; Hao et al., 2021). The metal sulfides serve as an excellent source of sulfide ions for carrying out the indirect oxysulfidation process, which mainly occur in oxygenated surroundings, owing to the higher stability of metal sulfides than sulfide ions (Thalmann et al., 2014). Notably, the direct sulfidation of AgNPs occurs in the absence of oxygen, provided a high concentration of hydrogen sulfide is present. However, this sulfidation process depends on the pH of the parent medium, and occurs optimally at pH <9.6. Importantly, the lowering of pH from basic to neutral type ameliorates the extent of sulfidation. Similarly, the presence of dissolved oxygen plays a significant role in the sulfidation process at basic pH, while the absence of oxygen significantly inhibits the sulfidation process (Vazquez et al., 1989). Conversely, the direct sulfidation of AgNPs present in wastewater treatment plants occurs optimally in anaerobic conditions, mainly because the consumption of sulfide ions in aerobic background decelerates the rate of sulfidation (Gottschalk et al., 2009; Blaser et al., 2008).

The process of sulfidation is size-dependent, where the diameter of AgNPs and their aggregation state play a critical role in the rate of sulfidation. Generally, the rate of sulfidation lowers with increasing AgNP size, and with larger nanoparticle aggregates. Similarly, the increase in the ratio of S/Ag enhances the rate of sulfidation of AgNPs, where the AgNPs strongly aggregate, thereby generating chain-like structures due to the formation of Ag_2S bridges between the AgNPs. The sulfidation processes play an important role in determining the transport and ultimate fate of the AgNPs. The conversion of AgNPs or silver ions to sulfides significantly lowers their toxicity, thereby improving their bioavailability (Bianchini et al., 2002; Bianchini and Wood, 2008; Cao et al., 2021). The sulfidation of AgNPs lowers their zeta potential compared to the nonsulfidated AgNPs, which improves their aggregation and sedimentation from the parent solution while decreasing their dissolution by altering the colloidal dynamics of AgNPs (Levard et al., 2012). Sulfidation of AgNPs hence represents a prominent transformational process that plays a central role in regulating the toxicity profile of environmentally dumped AgNPs (Levard et al., 2013; Courtois et al., 2021a,b). Fig. 3 illustrates the various pathways for AgNPs sulfidation.

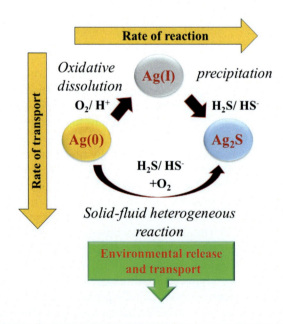

FIG. 3

Various pathways for the sulfidation of AgNPs.

6 Chlorination

The chlorination of AgNPs presents a beneficial effect in mitigating silver toxicity, owing to the lower solubility of the final products (Andryushechkin et al., 2007). The chlorination of AgNP occurs directly in the presence of chlorine gas, whereas the environmental chlorination of AgNPs occurs in the presence of hypochlorous or detergent that oxidizes AgNPs to silver ions that react with the chloride ions to yield AgCl. The chlorination of AgNPs mainly occurs via Eqs. (5) and (6) (Jones, 1988; Graedel, 1992).

Direct chlorination:

$$4Ag + Cl_2 \rightarrow 2AgCl \tag{5}$$

Environment-related chlorination:

$$\begin{aligned} 4Ag + Cl_2 + HOCl &\rightarrow AgCl + OH^- + Ag^+ \\ Ag^+ + Cl^- &\rightarrow AgCl \end{aligned} \tag{6}$$

AgCl forms a main constituent of the corrosion layer formed on the surface of AgNPs, thereby indicating chlorination to be a common phenomenon for the oxidative dissolution of AgNPs (Levard et al., 2011). The presence of chloride and HCl in the particulate matter promotes the corrosion of silver ions to AgCl in the presence of gaseous hydrogen peroxide (Garg et al., 2016). The surface treatment of AgNPs with chlorine gas in vacuum at room temperature causes nucleation to AgCl, which occurs more rapidly owing to the larger surface area of nanoparticles, compared to the bulk metallic silver. Mainly, the chlorination of AgNPs occurs due to laundering activities, where the AgNPs impregnated fabrics undergo oxidative dissolution of nanoparticle to AgCl in the presence of hypochlorite and detergent (Rutala and Weber, 1997).

Further transformations of AgCl obtained from the oxidative dissolution of AgNPs include their conversion to sulfides, and reduction to elemental silver in the presence of UV irradiation. Similarly, the presence of chloride ions in aqueous bodies promotes the photochemical transformation of AgNPs to AgCl, which further converts to sulfide due to the presence of competitive S^{2-} ions in the parent water body or the sewage sludge. The oxidative dissolution of AgNPs results in the formation of toxic silver ions, however their subsequent chlorination to AgCl mitigates the toxicity (Choi et al., 2008). The presence of excess amounts of chloride ions causes the formation of dissolved species such as $AgCl_2^-$, $AgCl_3^{2-}$, and $AgCl_4^{3-}$ in marine water ecosystems that further minimize the silver ions-related toxicity (Gupta et al., 1998; Silver, 2003). However, a direct transformation of AgNPs to AgCl does not take place even in high concentration of NaCl in the parent media. Generally, it takes place via a two-step mechanism where the silver ions formed by the oxidation of AgNPs in the first step later undergo chlorination to give AgCl. Fig. 4 highlights the pathways for the chlorination of AgNPs.

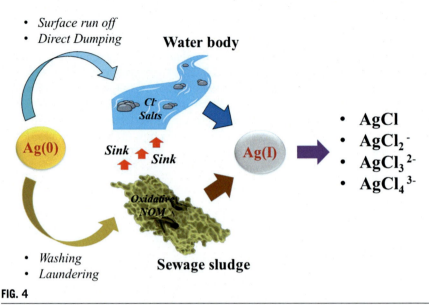

FIG. 4

Various pathways for the chlorination of AgNPs.

7 Aggregation and dissolution

Dissolution and aggregation of AgNPs occurs due to the destabilization of their colloidal suspension, mainly due to the loss of surface functionalities or zeta potential that maintains a colloidal dispersibility of AgNPs (Li et al., 2012). Several physicochemical factors play an important role in deciding the oxidative dissolution and aggregation of AgNPs (Zhou et al., 2016; Zhao et al., 2021). The pH alters the surface charge of AgNPs due to a differential behavior in acidic and basic media, thereby causing the onset of oxidative dissolution (Fernando and Zhou, 2019a,b). Mainly, the rise in pH causes an increase in the particle size that discourages aggregation and improves the stability of the AgNPs. The dissolution of AgNPs depends on the percentage of dissolved oxygen, and the protons present in the parent solution (Molleman and Hiemstra, 2017). The destabilization of AgNPs occurs at acidic and neutral pH, while the phenomenon reverses at basic pH in the presence of hydroxyl ions that considerably stabilize the colloidal suspensions of AgNPs. The dissolved oxygen promotes the aggregation of AgNPs, however, with a much lower intensity compared to the effect of pH (Zhang et al., 2011).

In the aquatic ecosystem, the colloidal stability of AgNPs decides their persistence and transport. The interactions of the natural organic matter with the surface of AgNPs depend on the chemical composition of the organic material containing the nitrogen and sulfur-rich species, which enhance the interactions with nanoparticles (Delay et al., 2011). Similarly, the affinity of AgNPs toward the capping agent decides the dissolution or stability of AgNPs in aquatic systems. The capping agents loosely bound to the AgNPs stabilize more effectively with the natural organic matter. Mainly, the natural organic matter rich in N or S species causes dissolution of AgNPs, thereby affecting their stability and toxicity (Afshinnia et al., 2018). The exposure of AgNPs to natural organic matter discourages the onset of aggregation and supports the stabilization of colloidal suspension of AgNPs. The nature of surface coating plays a vital role in the determination of the rate of dissolution of AgNPs (Badawy et al., 2010). The exposure of AgNPs to environmental factors results in the replacement of their original surface coatings with natural macromolecules (Cumberland and Lead, 2013). These surface coatings place a restriction on the reactants to reach the nanoparticle surface. Reportedly, the present of polyethylene glycol as surface coating on AgNPs offers steric barriers for the mass transferring of the reactants to AgNPs' surface, thereby resulting in the diminishing of the rate of dissolution (Sharma et al., 2014). The presence of bovine serum albumin coating on AgNPs surface however, causes a preliminary enhancement in the rate of dissolution but overall offers protection against the dissolution phenomenon over an extended period (Ernest et al., 2013). The biomolecules containing poly-thiol functionalities considerably improve the binding with silver ions obtained by the oxidative dissolution of AgNPs in the environment, as compared to the monothiol molecules (Marchioni et al., 2018). Molecules such as glutathione and phytochelatins containing more than one —SH group provide higher stability to the silver ions obtained from the AgNPs.

The particle size and dispersibility of AgNPs play a key role in their dissolution, toxicity, and mobility (Fernando and Zhou, 2019a,b). The clusters of AgNPs become vulnerable to sedimentation, which lowers their mobility compared to the original nanoparticles (Thio et al., 2011). Interestingly, the presence of sunlight also plays a role in the induction of aggregation of AgNPs, eventually leading to reduction in cytotoxicity. Similarly, the percentage of dissolved oxygen plays an important role in regulating the stability of AgNPs in aqueous environments, by enhancing the rate of aggregation of AgNPs by 3–8 times, compared to oxygen-free environments (Zou et al., 2017). Some studies reveal that the capping agents cause suspension stabilization that influences the surface behavior and aggregated particle size of AgNPs. The association between the capping agent and the aggregation behavior of AgNPs provides information regarding the mobility of nanoparticles in natural environments.

In addition, the pH, electrolyte composition, and ionic strength play a major role in influencing the fate and transport of AgNPs (Li et al., 2010). Acidic environments with high ionic strength containing mainly divalent cations which cause the aggregation and settling of AgNPs (Baalousha et al., 2013). These observations validate the instability of coated and electrostatically stabilized AgNPs in landfills, groundwater, surface water, wastewaters, and soils. The AgNPs that stabilize electrostatically demonstrate improved stability of the suspension in the environments with a high ionic strength. Nevertheless, the positively charged polymer-stabilized AgNPs that possess the ability to interact with oppositely charged surfaces exhibit a decreased transport and mobility (Badawy et al., 2010). Mainly, these AgNPs display restricted mobility in the ecosystem containing negatively charged clay and metal oxides. Polymers such as PVP that bear no charge and hence display no alterations in response to pH, ionic strength, and nature of the electrolyte, pose a significant impact on the transport and mobility of AgNPs in the environment (Mitzel and Tufenkji, 2014).

8 Photoinduced transformation

When released to the aquatic ecosystem, the photosensitive AgNPs undergo transformations in response to solar irradiation or light (Li et al., 2014) that also influences their aggregation, dissolution, and the generation of reactive oxygen species. As compared to dark conditions, simulated light, solar radiation, or UV lamps cause an enhancement in the rate of the release of silver ions from AgNPs (Nie et al., 2020). Mainly, solar light or simulated radiation drives the dipole-dipole interactions between the AgNPs, thereby causing their aggregation, followed by sedimentation (Hou et al., 2013). The incident irradiation induces localized surface plasmon resonance of AgNPs, hence inducing a strong adsorption of the photons of incident radiation, thereby promoting the generation of free radicals that influence the transport properties of AgNPs in the environment (He et al., 2011; Jiang et al., 2017). In aquatic environments, the suspended AgNPs interact with persistent organic pollutants or recalcitrant organic pollutants, mainly due to the high surface area of the former (Bundschuh et al., 2016). These agents modify the electrostatic interactions occurring between the AgNPs that result in altered stability, toxicity, and the photochemical activity (Loza et al., 2014). Mainly, the UV-365 radiation causes the dissolution of AgNPs to silver ions with simultaneous generation of reactive oxygen species that prove detrimental to the aquatic microbiota (Poda et al., 2013).

Similarly, 300nm UV radiation causes a loss of the surface plasmon resonance in citrate capped AgNPs due to their dissolution and aggregation (Gorham et al., 2012). The presence of UV radiation produces a synergistic effect with chloride ions in the parent solution for instigating the dissolution of AgNPs (Mittelman et al., 2015). Apart from their oxidative dissolution, the irradiation plays an important role in mediating the reduction of ionic silver to nano-metallic form in the aquatic ecosystems in the presence of dissolved organic matter that acts as the capping and reducing agent (Zhang et al., 2016). Moreover, the incident irradiation plays a crucial role in regulating the morphological transformation of AgNPs present in the aqueous environments (Stamplecoskie and Scaiano, 2010). In terrestrial ecosystems, in addition to the irradiation, fulvic acid and humic acid play an imperative role in determining the physicochemical transformation of AgNPs, including the morphological changes (Gunolus et al., 2015; Sritong et al., 2018; Fernando and Zhou, 2019a,b; Ale et al., 2021). In this process, the nature of parent solvent and the presence of other metal ions in the soil or water also contribute in regulating the transformational changes in the AgNPs morphology, and physical and chemical characteristics (Gupta et al., 2017; Li et al., 2012).

9 Conclusion

AgNPs, owing to their vast applications, are prone to cause environmental distress when released incessantly. After coming in contact with the various components of an ecosystem, the AgNPs undergo various transformations depending on the physical and chemical properties of the environment the nanoparticles come into immediate contact with. The presence of natural organic materials and metal ion species mediate the transformational behavior of AgNPs by acting as capping agents or reducing agents. Mainly, the components acting as oxidants prove quite harmful due to the generation of oxidative species. These transformations prove useful sometimes by mitigating the toxicity of the AgNPs due to their conversion to harmless forms. However, the oxidative, and dissolution processes enhance the toxicity profile of AgNPs by supporting the release of hazardous silver ions, which prove detrimental to the microbiota, microflora, and essential mineral in soils. The drainage of AgNPs in agroecosystems proves detrimental to crops by causing the silver stress. Hence, an effective AgNPs utilization program requires a comprehensive appraisal of their absorption in the environment for managing the associated harmful effects on the ecosystem.

Acknowledgment

The authors thank Guru Nanak Dev University for providing the research infrastructure. SEED support from the University of Petroleum and Energy Studies is duly acknowledged.

References

Afshinnia, K., Marrone, B., Baalousha, M., 2018. Potential impact of natural organic ligands on the colloidal stability of silver nanoparticles. Sci. Total Environ. 625, 1518–1526.

Ale, A., Galdoporpora, J.M., Mora, M.C., Torre, F.R., Desimone, M.F., Cazenave, J., 2021. Mitigation of silver nanoparticle toxicity by humic acids in gills of *Piaractus mesopotamicus* fish. Environ. Sci. Pollut. Res. Int. https://doi.org/10.1007/s11356-021-12590-w.

Andryushechkin, B.V., Eltsov, K.N., Shevlyuga, V.M., 2007. Local structure of thin AgCl films on silver surface. Phys. Wave Phenom. 15, 116–125.

Baalousha, M., Nur, Y., Romer, I., Tejamaya, M., Lead, J.R., 2013. Effect of monovalent and divalent cations, anions and fulvic acid on aggregation of citrate-coated silver nanoparticles. Sci. Total Environ. 454, 119–131.

Badawy, A.M.E., Luxton, T.P., Silva, R.G., Scheckel, K.G., Suidan, M.T., Tolaymat, T.M., 2010. Impact of environmental conditions (pH, ionic strength, and electrolyte type) on the surface charge and aggregation of silver nanoparticles suspensions. Environ. Sci. Technol. 44, 1260–1266.

Bao, S., Xu, J., Tang, W., Fang, T., 2021. Effect and mechanism of silver nanoparticles on nitrogen transformation in water-sediment system of a hypereutrophic lake. Sci. Total Environ. 761, 144182.

Bhatt, P., Verma, A., Verma, S., Anwar, S., Prasher, P., Mudila, H., Chen, S., 2020. Understanding photomicrobiome: a potential reservoir for better crop management. Sustainability 12, 5446.

Bianchini, A., Wood, C.M., 2008. Does sulfide or water hardness protect against chronic silver toxicity in *Daphnia magna*? A critical assessment of the acute-to-chronic toxicity ratio for silver. Ecotoxicol. Environ. Saf. 71, 32–40.

Bianchini, A., Bowles, K.C., Brauner, C.J., Gorsuch, J.W., Kramer, J.R., Wood, C.M., 2002. Evaluation of the effect of reactive sulfide on the acute toxicity of silver (I) to *Daphnia magna*. Part 2: toxicity results. Environ. Toxicol. Chem. 21, 1294–1300.

Blaser, S.A., Scheringer, M., MacLeod, M., Hungerbuhler, K., 2008. Estimation of cumulative aquatic exposure and risk due to silver: contribution of nano-functionalized plastics and textiles. Sci. Total Environ. 390, 396–409.

Bundschuh, M., Seitz, F., Rosenfeldt, R.R., Schulz, R., 2016. Effects of nanoparticles in fresh waters: risks, mechanisms and interactions. Freshw. Biol. 61, 2185–2196.

Bundschuh, M., Filser, J., Luderwald, S., McKee, M.S., Metreveli, G., Schaumann, G.E., Schulz, R., Wagner, S., 2018. Nanoparticles in the environment: where do we come from, where do we go to? Environ. Sci. Eur. 30, 6.

Burdusel, A.-C., Gherasim, O., Grumezescu, A.M., Mogoanta, L., Ficai, A., Andronescu, E., 2018. Biomedical applications of silver nanoparticles: an up-to-date overview. Nanomaterials 8, 681.

Cao, C., Huang, J., Yan, C., Zhang, X., Ma, Y., 2021. Impacts of Ag and Ag2S nanoparticles on the nitrogen removal within vertical flow constructed wetlands treating secondary effluent. Sci. Total Environ. 777, 45171.

Chinnapongse, S.L., MacCuspie, R.I., Hackley, V.A., 2011. Persistence of singly dispersed silver nanoparticles in natural freshwaters, synthetic seawater, and simulated estuarine waters. Sci. Total Environ. 409, 2443–2450.

Choi, O., Deng, K.K., Kim, N.J., Ross, L., Surampalli, R.Y., Hu, Z.Q., 2008. The inhibitory effects of silver nanoparticles, silver ions, and silver chloride colloids on microbial growth. Water Res. 42, 3066–3074.

Courtois, P., Rorat, A., Lemiere, S., Levard, C., Chaurand, P., Grobelak, A., Lors, C., Vandenbulcke, F., 2021a. Accumulation, speciation and localization of silver nanoparticles in the earthworm *Eisenia fetida*. Environ. Sci. Pollut. Res. 28, 3756–3765.

Courtois, P., Vaufleury, A., Grosser, A., Lors, C., Vandenbulcke, F., 2021b. Transfer of sulfidized silver from silver nanoparticles, in sewage sludge, to plants and primary consumers in agricultural soil environment. Sci. Total Environ. 777, 145900.

Cumberland, S.A., Lead, J.R., 2013. Synthesis of NOM-capped silver nanoparticles: size, morphology, stability, and NOM binding characteristics. ACS Sustain. Chem. Eng. 1, 817–825.

Delay, M., Dolt, T., Woellhaf, A., Sembritzki, R., Frimmel, F.H., 2011. Interactions and stability of silver nanoparticles in the aqueous phase: influence of natural organic matter (NOM) and ionic strength. J. Chromatogr. A 1218, 4206–4212.

Ding, Y., Bai, X., Ye, Z., Gong, D., Cao, J., Hua, Z., 2019. Humic acid regulation of the environmental behavior and phytotoxicity of silver nanoparticles to *Lemna minor*. Environ. Sci. Nano 6, 3712–3722.

Dong, F., Wu, C., Miao, A.-J., Pan, K., 2021. Reduction of silver ions to form silver nanoparticles by redox-active organic molecules: coupled impact of the redox state and environmental factors. Environ. Sci. Nano 8, 269–281.

Ellis, L.-J.A., Baalousha, M., Jones, E.V., Lead, J.R., 2018. Seasonal variability of natural water chemistry affects the fate and behaviour of silver nanoparticles. Chemosphere 191, 616–625.

El-Nour, K.M.M.A., Eftaiha, A., Warthan, A.A., Ammar, R.A.A., 2010. Synthesis and applications of silver nanoparticles. Arab. J. Chem. 3, 135–140.

Ernest, V., Chandrasekaran, N., Mukherjee, A., 2013. Colloidal silver nanoparticles and bovine serum albumin. In: Kretsinger, R.H., Uversky, V.N., Permyakov, E.A. (Eds.), Encyclopedia of Metalloproteins. Springer, New York, NY, https://doi.org/10.1007/978-1-4614-1533-6_526.

Ferdous, Z., Nemmar, A., 2020. Health impact of silver nanoparticles: a review of the biodistribution and toxicity following various routes of exposure. Int. J. Mol. Sci. 21, 2375.

Ferdous, Z., Al-Salam, S., Yuvaraju, P., Ali, B.H., Nemmar, A., 2021. Remote effects and biodistribution of pulmonary instilled silver nanoparticles in mice. NanoImpact 22, 100310.

Fernando, I., Zhou, Y., 2019a. Impact of pH on the stability, dissolution and aggregation kinetics of silver nanoparticles. Chemosphere 216, 297–305.

Fernando, I., Zhou, Y., 2019b. Concentration dependent effect of humic acid on the transformations of silver nanoparticles. J. Mol. Liq. 284, 291–299.

Franey, J.P., Kammlott, G.W., Graedel, T.E., 1985. The corrosion of silver by atmospheric sulfurous gases. Corros. Sci. 25, 133–143.

Furtado, L.M., Norman, B.C., Xenopoulos, M.A., Frost, P.C., Metcalfe, C.D., Hintelmann, H., 2015. Environmental fate of silver nanoparticles in boreal lake ecosystems. Environ. Sci. Technol. 49, 8441–8450.

Garg, S., Rong, H., Miller, C.J., Waite, T.D., 2016. Oxidative dissolution of silver nanoparticles by chlorine: implications to silver nanoparticle fate and toxicity. Environ. Sci. Technol. 50, 3890–3896.

Gorham, J.M., MacCuspie, R.I., Klein, K.L., Fairbrother, D.H., Holbrook, R.D., 2012. UV-induced photochemical transformations of citrate-capped silver nanoparticle suspensions. J. Nanopart. Res. 14, 1139.

Gottschalk, F., Sonderer, T., Scholz, R.W., Nowack, B., 2009. Modeled environmental concentrations of engineered nanomaterials (TiO2, ZnO, Ag, CNT, fullerenes) for different regions. Environ. Sci. Technol. 43, 9216–9222.

Graedel, T.E., 1992. Corrosion mechanisms for silver exposed to the atmosphere. J. Electrochem. Soc. 139, 1963–1970.

Gunolus, I.L., Mousavi, M.P.S., Hussein, K., Buhlmann, P., Haynes, C.L., 2015. Effects of humic and fulvic acids on silver nanoparticle stability, dissolution, and toxicity. Environ. Sci. Technol. 49, 8078–8086.

Gupta, A., Maynes, M., Silver, S., 1998. Effects of halides on plasmid mediated silver resistance in *Escherichia coli*. Appl. Environ. Microbiol. 64, 5042–5045.

Gupta, S., Kumar, V., Joshi, K.B., 2017. Solvent mediated photo-induced morphological transformation of AgNPs-peptide hybrids in water-EtOH binary solvent mixture. J. Mol. Liq. 236, 266–277.

Gutierrez, L., Schmid, A., Zaouri, N., Garces, D., Croue, J.-P., 2020. Colloidal stability of capped silver nanoparticles in natural organic matter-containing electrolyte solutions. NanoImpact 19, 100242.

Hao, Z., Li, F., Liu, R., Zhou, X., Mu, Y., Sharma, V.K., Liu, J., Jiang, G., 2021. Reduction of ionic silver by sulfur dioxide as a source of silver nanoparticles in the environment. Environ. Sci. Technol. https://doi.org/10.1021/acs.est.0c08790.

He, D., Jones, A.M., Garg, S., Pham, A.N., Waite, T.D., 2011. Silver nanoparticle–reactive oxygen species interactions: application of a charging–discharging model. J. Phys. Chem. C 115, 5461–5468.

Herrera, A.L., Borja, M.A., Nava, J.R.G., Tellez, L.I.T., Alarcon, A., Soberano, A.P., Martinez, V.C., Mancera, H.A.Z., 2021. Interaction of silver nanoparticles with the aquatic fern *Azolla filiculoides*: root structure, particle distribution, and silver accumulation. J. Nanopart. Res. 23, 13.

Hou, W.-C., Stuart, B., Howes, R., Zepp, R.G., 2013. Sunlight-driven reduction of silver ions by natural organic matter: formation and transformation of silver nanoparticles. Environ. Sci. Technol. 47, 7713–7721.

Hou, J., Zhang, M., Wang, P., Wang, C., Miao, L., Xu, Y., You, G., Lv, B., Yang, Y., Liu, Z., 2017. Transport, retention, and long-term release behavior of polymer-coated silver nanoparticles in saturated quartz sand: the impact of natural organic matters and electrolyte. Environ. Pollut. 229, 49–59.

Huang, Y.-N., Qian, T.-T., Dang, F., Yin, Y.-G., Li, M., Zhou, D.M., 2019a. Significant contribution of metastable particulate organic matter to natural formation of silver nanoparticles in soils. Nat. Commun. 10, 3775.

Huang, J., Xiao, J., Chen, M., Cao, C., Yan, C., Ma, Y., Huang, M., Wang, M., 2019b. Fate of silver nanoparticles in constructed wetlands and its influence on performance and microbiome in the ecosystems after a 450-day exposure. Bioresour. Technol. 281, 107–117.

Jianchang, L., Luoping, Z., Yuzhen, Z., Huasheng, H., Hongbing, D., 2008. Validation of an agricultural non-point source (AGNPS) pollution model for a catchment in the Jiulong River watershed. Chin. J. Environ. Sci. 20, 599–606.

Jiang, H.S., Yin, L.Y., Ren, N.N., Zhao, S.T., Li, Z., Zhi, Y., Shao, H., Li, W., 2017. Silver nanoparticles induced reactive oxygen species via photosynthetic energy transport imbalance in an aquatic plant. Nanotoxicology 11, 157–167.

Johnston, K.A., Stabryla, L.M., Gilbertson, L.M., Millstone, J.E., 2019. Emerging investigator series: connecting concepts of coinage metal stability across length scales. Environ. Sci. Nano 6, 2674–2696.

Jones, R.G., 1988. Halogen adsorption on solid surfaces. Prog. Surf. Sci. 27, 25–160.

Kent, R.D., Oser, J.G., Vikesland, P.J., 2014. Controlled evaluation of silver nanoparticle sulfidation in a full-scale wastewater treatment plant. Environ. Sci. Technol. 48, 8564–8572.

Klitzke, S., Metreveli, G., Peters, A., Schaumann, G.E., Lang, F., 2015. The fate of silver nanoparticles in soil solution—sorption of solutes and aggregation. Sci. Total Environ. 535, 54–60.

Koser, J., Engelke, M., Hoppe, M., Nogwski, A., Filser, J., Thoming, J., 2017. Predictability of silver nanoparticle speciation and toxicity in ecotoxicological media. Environ. Sci. Nano 4, 1470–1483.

Kostic, D., Vidovic, S., Obradovic, B., 2016. Silver release from nanocomposite Ag/alginate hydrogels in the presence of chloride ions: experimental results and mathematical modeling. J. Nanopart. Res. 18, 76.

Kraas, M., Schlich, K., Knopf, B., Wege, F., Kagi, R., Terytze, K., Hund-Rinke, K., 2017. Long-term effects of sulfidized silver nanoparticles in sewage sludge on soil microflora. Environ. Toxicol. Chem. 36, 3305–3313.

Kulikova, N.A., Volkov, D.S., Valikov, A.B., Abroskin, D.P., Krepak, A.I., Perminova, I.V., 2020. Silver nanoparticles stabilized by humic substances adversely affect wheat plants and soil. J. Nanopart. Res. 22, 100.

Kwak, J., Cui, R., Nam, S.-H., Kim, S.W., Chae, Y., An, Y.-J., 2016. Multispecies toxicity test for silver nanoparticles to derive hazardous concentration based on species sensitivity distribution for the protection of aquatic ecosystems. Nanotoxicology 10, 521–530.

Lee, S.-W., Park, S.-Y., Kim, Y., Im, H., Choi, J., 2016. Effect of sulfidation and dissolved organic matters on toxicity of silver nanoparticles in sediment dwelling organism, *Chironomus riparius*. Sci. Total Environ. 553, 565–573.

Lekamge, S., Miranda, A.F., Abraham, A., Li, V., Shukla, R., Bansal, V., Nugegoda, D., 2018. The toxicity of silver nanoparticles (AgNPs) to three freshwater invertebrates with different life strategies: *Hydra vulgaris*, *Daphnia carinata*, and *Paratya australiensis*. Front. Environ. Sci. 6, 152.

Levard, C., Michel, F.M., Wang, Y., Choi, Y., Eng, P., Brown, G.E., 2011. Probing Ag nanoparticle surface oxidation in contact with (in)organics: an X-ray scattering and fluorescence yield approach. J. Synchrotron Radiat. 18, 871–878.

Levard, C., Hotze, E.M., Lowry, G.V., Brown, G.E., 2012. Environmental transformations of silver nanoparticles: impact on stability and toxicity. Environ. Sci. Technol. 46, 6900–6914.

Levard, C., Hotze, E.M., Colman, B.P., Dale, A.L., Truong, L., Yang, X.Y., Bone, A.J., Brown, G.E., Tanguay, R.L., Giulio, R.T., Bernhardt, E.S., Meyer, J.N., Wiesner, M.R., Lowry, G.V., 2013. Sulfidation of silver nanoparticles: natural antidote to their toxicity. Environ. Sci. Technol. 47, 13440–13448.

Li, X., Lenhart, J.J., Walker, H.W., 2010. Dissolution-accompanied aggregation kinetics of silver nanoparticles. Langmuir 26, 16690–16698.

Li, X., Lenhart, J.J., Walker, H.W., 2012. Aggregation kinetics and dissolution of coated silver nanoparticles. Langmuir 28, 1095–1104.

Li, Y., Niu, J., Shang, E., Crittenden, J., 2014. Photochemical transformation and photoinduced toxicity reduction of silver nanoparticles in the presence of perfluorocarboxylic acids under UV irradiation. Environ. Sci. Technol. 48, 4946–4953.

Li, M., Wang, P., Dang, F., Zhou, D.-M., 2017. The transformation and fate of silver nanoparticles in paddy soil: effects of soil organic matter and redox conditions. Environ. Sci. Nano 4, 919–928.

Liu, J., Pennell, K.G., Hurt, R.H., 2011. Kinetics and mechanism of nanosilver oxysulfidation. Environ. Sci. Technol. 45, 7345–7353.

Liu, J., Zhang, F., Allen, A.J., Peck, A.C.J., Pettibone, J.M., 2018. Comparing sulfidation kinetics of silver nanoparticles in simulated media using direct and indirect measurement methods. Nanoscale 10, 22270–22279.

Liu, Y., Li, C., Luo, S., Wang, X., Zhang, Q., Wu, H., 2020. Inter-transformation between silver nanoparticles and Ag^+ induced by humic acid under light or dark conditions. Ecotoxicology. https://doi.org/10.1007/s10646-020-02284-3.

Lowry, G.V., Gregory, K.B., Apte, S.C., Lead, J.R., 2012. Transformation of nanomaterials in the environment. Environ. Sci. Technol. 46, 6893–6899.

Loza, K., Diendorf, J., Sengstock, C., Gonzalez, L.R., Calmet, J.M.G., Regi, M.V., Koller, M., Epple, M., 2014. The dissolution and biological effects of silver nanoparticles in biological media. J. Mater. Chem. B 2, 1634–1643.

Marchioni, M., Gallon, T., Worms, I., Jouneau, P.-H., Lebrun, C., Veronesi, G., Boutry, D.T., Mintz, E., Delangle, P., Deniaud, A., Soret, I.M., 2018. Insights into polythiol-assisted AgNP dissolution induced by bio-relevant molecules. Environ. Sci. Nano 5, 1911–1920.

Milne, C.J., Lapworth, D.J., Gooddy, D.C., Elgy, C.N., Jones, E.-V., 2017. Role of humic acid in the stability of Ag nanoparticles in suboxic conditions. Environ. Sci. Technol. 51, 6063–6070.

Mittelman, A.M., Fortner, J.D., Pennell, K.D., 2015. Effects of ultraviolet light on silver nanoparticle mobility and dissolution. Environ. Sci. Nano 2, 683–691.

Mitzel, M.R., Tufenkji, N., 2014. Transport of industrial PVP-stabilized silver nanoparticles in saturated quartz sand coated with *Pseudomonas aeruginosa* PAO1 biofilm of variable age. Environ. Sci. Technol. 48, 2715–2723.

Mohan, S., Princz, J., Ormeci, B., DeRosa, M.C., 2019. Morphological transformation of silver nanoparticles from commercial products: modeling from product incorporation, weathering through use scenarios, and leaching into wastewater. Nanomaterials 9, 1258.

Molleman, B., Hiemstra, T., 2017. Time, pH, and size dependency of silver nanoparticle dissolution: the road to equilibrium. Environ. Sci. Nano 4, 1314–1327.

Nason, J.A., McDowell, S.A., Callahan, T.W., 2012. Effects of natural organic matter type and concentration on the aggregation of citrate-stabilized gold nanoparticles. J. Environ. Monit. 14, 1885–1892.

Nie, Y., Wang, J., Dai, H., Wang, J., Wang, M., Cheng, L., Yang, Z., Chen, S., Zhao, G., Wu, L., Xu, A., 2020. UV-induced over time transformation of AgNPs in commercial wound dressings and adverse biological effects on *Caenorhabditis elegans*. NanoImpact 17, 100193.

Pallavi, Mehta, C.M., Srivastava, R., Arora, S., Sharma, A.K., 2016. Impact assessment of silver nanoparticles on plant growth and soil bacterial diversity. 3 Biotech 6, 254.

Pang, C., Zhang, P., Mu, Y., Ren, J., Zhao, B., 2020. Transformation and cytotoxicity of surface-modified silver nanoparticles undergoing long-term aging. Nanomaterials 10, 2255.

Peng, H., Guo, H., Gao, P., Zhou, Y., Pan, B., Xing, B., 2021. Reduction of silver ions to silver nanoparticles by biomass and biochar: mechanisms and critical factors. Sci. Total Environ. 779, 146326.

Poda, A.R., Kennedy, A.J., Cuddy, M.F., Bednar, A.J., 2013. Investigations of UV photolysis of PVP-capped silver nanoparticles in the presence and absence of dissolved organic carbon. J. Nanopart. Res. 15, 1673.

Potter, P.M., Navratilova, J., Rogers, K.R., Al-Abed, S.R., 2019. Transformation of silver nanoparticle consumer products during simulated usage and disposal. Environ. Sci. Nano 6, 592–598.

Prasher, P., Singh, M., Mudila, H., 2018a. Oligodynamic effects of silver nanoparticles: a review. BioNanoScience 8, 951–962.

Prasher, P., Singh, M., Mudila, H., 2018b. Silver nanoparticles as antimicrobial therapeutics: current perspectives and future challenges. 3 Biotech 8, 411.

Prasher, P., Sharma, M., Mudila, H., Khati, B., 2019. Uptake, accumulation, and toxicity of metal nanoparticles in autotrophs. In: Panpatte, D., Jhala, Y. (Eds.), Nanotechnology for Agriculture. Springer, Singapore.

Prasher, P., Sharma, M., et al., 2020. Emerging trends in clinical implications of bioconjugated silver nanoparticles in drug delivery. Colloid Interf. Sci. Commun. 35, 100244.

Prasher, P., Mudila, H., Sharma, M., 2021. Biosorption and bioaccumulation of pollutants for environmental remediation. In: Panpatte, D.G., Jhala, Y.K. (Eds.), Microbial Rejuvenation of Polluted Environment. Microorganisms for Sustainability, vol 26. Springer, Singapore, https://doi.org/10.1007/978-981-15-7455-9_15.

Quadros, M.E., Marr, L.C., 2010. Environmental and human health risks of aerosolized silver nanoparticles. J. Air Waste Manag. Assoc. 60, 770–781.

Rearick, D.C., Telgmann, L., Hintelmann, H., Frost, P.C., Xenopoulos, M.A., 2018. Spatial and temporal trends in the fate of silver nanoparticles in a whole-lake addition study. PLoS One 13, e0201412.

Rutala, W.A., Weber, D.J., 1997. Uses of inorganic hypochlorite (bleach) in health-care facilities. Clin. Microbiol. Rev. 10, 597–610.
Sagee, O., Dror, I., Berkowitz, B., 2012. Transport of silver nanoparticles (AgNPs) in soil. Chemosphere 88, 670–675.
Shahjamali, M.M., Zhou, Y., Zaraee, N., Xue, C., Wu, J., Large, N., McGuirk, M., Boey, F., Dravid, V., Cui, Z., Schatz, G.C., Mirkin, C.A., 2016. Ag–Ag2S hybrid nanoprisms: structural versus plasmonic evolution. ACS Nano 10, 5362–5373.
Sharma, V.K., Siskova, K.M., Zboril, R., Torresdey, J.L.G., 2014. Organic-coated silver nanoparticles in biological and environmental conditions: fate, stability and toxicity. Adv. Colloid Interface Sci. 204, 15–34.
Shen, M.-H., Zhou, X.-X., Yang, X.-Y., Chao, J.-B., Liu, R., Liu, J.F., 2015. Exposure medium: key in identifying free Ag+ as the exclusive species of silver nanoparticles with acute toxicity to *Daphnia magna*. Sci. Rep. 5, 9674.
Silver, S., 2003. Bacterial silver resistance: molecular biology and uses and misuses of silver compounds. FEMS Microbiol. Rev. 27, 341–353.
Souza, L.R.R., Correa, T.Z., Bruni, A.T., Veiga, M.A.M.S., 2021. The effects of solubility of silver nanoparticles, accumulation, and toxicity to the aquatic plant *Lemna minor*. Environ. Sci. Pollut. Res. 28, 16720–16733.
Sritong, N., Chumsook, S., Siri, S., 2018. Light emitting diode irradiation induced shape conversion of DNA-capped silver nanoparticles and their antioxidant and antibacterial activities. Artif. Cell Nanomed. Biotechnol. 46, 955–963.
Stamplecoskie, K.G., Scaiano, J.C., 2010. Light emitting diode irradiation can control the morphology and optical properties of silver nanoparticles. J. Am. Chem. Soc. 132, 1825–1827.
Steinmetz, L., Geers, C., Balog, S., Bonmarin, M., Lorenzo, L.R., Blanco, P.T., Ratishauser, B.R., Fink, A.P., 2020. A comparative study of silver nanoparticle dissolution under physiological conditions. Nanoscale Adv. 2, 5760–5768.
Thalmann, B., Voegelin, A., Sinnet, B., Morgenroth, E., Kaegi, R., 2014. Sulfidation kinetics of silver nanoparticles reacted with metal sulfides. Environ. Sci. Technol. 48, 4885–4892.
Thio, B.J.R., Montes, M.O., Mahmoud, M.A., Lee, D.-W., Zhou, D., Keller, A.A., 2011. Mobility of capped silver nanoparticles under environmentally relevant conditions. Environ. Sci. Technol. 46, 6985–6991.
Tripathi, D.K., Tripathi, A., Shweta, Singh, S., Singh, Y., Vishwakarma, K., Yadav, G., Sharma, S., Singh, V.K., Mishra, R.K., Upadhyay, R.G., Dubey, N.K., Lee, Y., Chauhan, D.K., 2017. Uptake, accumulation and toxicity of silver nanoparticle in autotrophic plants, and heterotrophic microbes: a concentric review. Front Microbiol. 8, 7.
Vazquez, F., Zhang, J.Z., Millero, F.J., 1989. Effect of metals on the rate of the oxidation of H2S in seawater. Geophys. Res. Lett. 16, 1363–1366.
Verma, S., Bhatt, P., Verma, A., Mudila, H., Prasher, P., Rene, E.R., 2021. Microbial technologies for heavy metal remediation: effect of process conditions and current practices. Clean Technol. Environ. Policy. https://doi.org/10.1007/s10098-021-02029-8.
Yan, A., Chen, Z., 2019. Impacts of silver nanoparticles on plants: a focus on the phytotoxicity and underlying mechanism. Int. J. Mol. Sci. 20, 1003.
Yang, Y., Zheng, S., Li, R., Chen, X., Wang, K., Sun, B., Zhang, Y., Zhu, L., 2021. New insights into the facilitated dissolution and sulfidation of silver nanoparticles under simulated sunlight irradiation in aquatic environments by extracellular polymeric substances. Environ. Sci. Nano. https://doi.org/10.1039/D0EN01142H.

Yoo-iam, M., Chaichana, R., Satapanajaru, T., 2014. Toxicity, bioaccumulation and biomagnification of silver nanoparticles in green algae (Chlorella sp.), water flea (*Moina macrocopa*), blood worm (*Chironomus spp.*) and silver barb (*Barbonymus gonionotus*). Chem. Speciat. Bioavailab. 26, 257–265.

Yu, S.-J., Yin, Y.-G., Liu, J.-F., 2012. Silver nanoparticles in the environment. Environ. Sci. Process. Impacts 15, 78–92.

Yusuf, M., 2019. Silver nanoparticles: synthesis and applications. In: Martínez, L., Kharissova, O., Kharisov, B. (Eds.), Handbook of Ecomaterials. Springer, Cham, https://doi.org/10.1007/978-3-319-68255-6_16.

Zhang, W., Yao, Y., Li, K., Huang, Y., Chen, Y., 2011. Influence of dissolved oxygen on aggregation kinetics of citrate-coated silver nanoparticles. Environ. Pollut. 159, 3757–3762.

Zhang, X., Yang, C.-W., Yu, H.-Q., Sheng, G.-P., 2016. Light-induced reduction of silver ions to silver nanoparticles in aquatic environments by microbial extracellular polymeric substances (EPS). Water Res. 106, 242–248.

Zhang, Z., Shen, W., Xue, J., Liu, Y., Yan, P., Liu, J., Tang, J., 2018a. Recent advances in synthetic methods and applications of silver nanostructures. Nanoscale Res. Lett. 13, 54.

Zhang, W., Xiao, B., Fang, T., 2018b. Chemical transformation of silver nanoparticles in aquatic environments: mechanism, morphology and toxicity. Chemosphere 191, 324–334.

Zhang, F., Allen, A.J., Peck, A.C.J., Liu, J., Pettibone, J.M., 2019. Transformation of engineered nanomaterials through the prism of silver sulfidation. Nanoscale Adv. 1, 241–253.

Zhao, J., Li, Y., Wang, X., Xia, X., Shang, E., Ali, J., 2021. Ionic-strength-dependent effect of suspended sediment on the aggregation, dissolution and settling of silver nanoparticles. Environ. Pollut. 279, 116926.

Zhou, W., Liu, Y.-L., Stallworth, A.M., Ye, C., Lenhart, J.J., 2016. Effects of pH, electrolyte, humic acid, and light exposure on the long-term fate of silver nanoparticles. Environ. Sci. Technol. 50, 12214–12224.

Zou, X., Li, P., Lou, J., Fu, X., Zhang, H., 2017. Stability of single dispersed silver nanoparticles in natural and synthetic freshwaters: effects of dissolved oxygen. Environ. Pollut. 230, 674–682.

CHAPTER 27

Strategies for scaling up of green-synthesized nanomaterials: Challenges and future trends

Mohamed Amine Gacem[a] and Kamel A. Abd-Elsalam[b]

[a]*Department of Biology, Faculty of Science, University of Amar-Tlidji, Laghouat, Algeria* [b]*Plant Pathology Research Institute, Agricultural Research Center (ARC), Giza, Egypt*

1 Introduction

Nanomaterials (NMs) manufacturing is one of the most exciting sectors in the manufacturing section. "Nanomanufacturing" is becoming a thriving industry (Busnaina et al., 2013). The gross global of various nanomaterial intake of all forms amounts to several million tons per year, worth $3.4 billion on the world market, and is estimated to hit >$10 billion by 2020, considering continuous development in the sector (Intelligence, 2016). According to some observations, the market is expected to grow more and more and reach 55 billion dollars in 2022 (Domagalski et al., 2021). Green AgNP synthesis methods offer impressive advantages in terms of capacity and speed of synthesis reactions. However, the majority of published research focuses on the biosynthesis of AgNPs on a small scale. From the results of these studies, it appears that there are good opportunities to develop green synthesis on a large scale (El-Moslamy et al., 2017; Baranitharan et al., 2021). This problem has not yet been resolved due to the significant lack of scientific reports and research allowing the establishment of industrial production.

The majority of green processes do not require heat to run the reactions, which minimizes both production costs and carbon dioxide emissions. On the basis of a study carried out on silica, the synthetic pathway using the green method may prove to be economically comparable to current techniques. In addition, the green process can easily be adapted to existing unit operations in the production of silica, without requiring additional investments. In addition, the green synthesis of NPs offers other additional advantages: it makes it possible to generate products with much more diversified and easily controllable characteristics (Drummond et al., 2014). However, bringing AgNPs biosynthesis to industrial scale requires in-depth knowledge of production costs, availability of bioresources, necessary downstream treatments, and scalability predictions. As the benefits of small-scale green synthesis of AgNPs are encouraging, they can prove to be costly, inefficient, and nonscalable on a large scale, and therefore fail to meet market demands due to unforeseen problems (Patwardhan et al., 2018). It is essential to keep in mind the beneficial aspects in order to properly put the necessary steps and solve the problems at the industrial stage. According to the databases consulted, few fundamental studies have started to appear recently and are limited to a single nanomaterial.

Important problems remain to be solved, even at the laboratory scale, in particular the purification and separation of the AgNPs produced. These treatments are completely ignored on a small scale whereas on an industrial scale they are compulsory. They can appear expensive due to the excessive use of solvents for washing and energy for centrifugation (Ndikau et al., 2017), whereas it is essential to carry out purification in order to remove impurities and generate high-quality functional products.

The use of plants at the laboratory scale for the biosynthesis of AgNPs requires the organic extraction of biomolecules via solvents (Baranitharan et al., 2021), and this compulsory extraction process for the reduction reactions of silver ions can prove to be expensive in an industrial scale. Extending the biosynthesis of AgNPs using suspended fungal or bacterial biomasses or their cell extracts will require bioreactors where microbial biomass or their extracts are conveniently immobilized to minimize losses and contamination as well as the time required for biosynthesis. Retrieving the formed NPs should also be easy and less expensive.

Currently, no report is available on microbial biomass reactors for AgNPs biosynthesis. However, bioreactors (fermenter) have already described the use of microbial biomass as a catalyst for well-defined bioprocesses (Li et al., 2015; Svobodová and Novotný, 2018). Otherwise, the achievement of biomass productivity is strongly linked to the ideal fermentation conditions (dissolved oxygen, gas flow, agitation, temperature, pH, etc.) and fermentation modes (batch, substrate fed-batch, whole nutrient fed-batch, etc.). The improvement of these conditions is achievable by the application of some strategies such as the experimental design method of Taguchi demonstrated in the biosynthesis of nano-MgO carried out on a semiindustrial scale (7-L fermenter), in which the synthesis yield increased more than twice compared to basal conditions (El-Moslamy, 2018). New developed reactors were applied to the production of silver nanoparticles and liposomes. The advantage of the work was the development of a cost-effective and easy-to-perform fabrication method for continuous flow reactors production (Cristaldi et al., 2018; Długosz and Banach, 2019). If biogenic AgNPs are available on an industrial scale, they will have great applications in different applications in numerous consumer goods due to their broad spectrum of antimicrobial and antiviral properties. The application of silver nanoparticles (AgNPs) colonized all fields due to the beneficial effects they bring (Fig. 1). Without a question, the present distance between scientific research on nanomaterials and their use in real life will be bridged in the coming decade (Charitidis et al., 2014). Therefore, the objective of this chapter is to propose and discuss basic strategies allowing the implementation of large-scale production of AgNPs by green chemistry pathways. In addition, we outline the core challenges and commercialization of future trends in silver-based nanomaterials bioproduction.

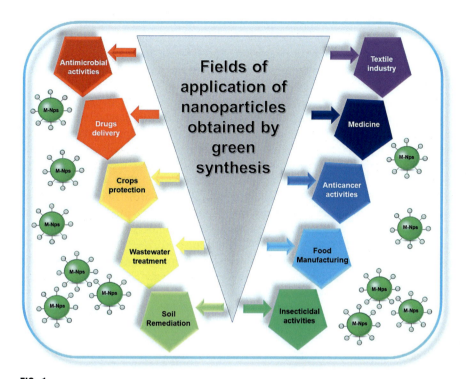

FIG. 1

Fields of application of silver nanoparticles produced by green methods.

2 Green biocatalytic synthesis of nanoscale materials

Among these pathways, microorganisms have provided a new platform for greener and more profitable biological synthesis of nanoparticles. NPs of microbial origin contain unique and dynamic characteristics that make them applicable in multiple fields of agriculture, medicine, and biotechnology (Ovais et al., 2018a). The bioresources of green synthesis are not limited to bacteria and their enzymes which act as NPs reducers and stabilizers; medicinal plants with their very complex photochemical constituents also have a huge tendency to produce NPs (Ovais et al., 2018b). The fungal cell system also contains a favorable and efficient environment for the synthesis of NPs. Fungal cells have distinctive characters compared to bacterial cells due to their high tolerance to metals, their absorption capacity and the binding capacity of their walls, their cultures, and the easy manipulation of cell biomasses (Yadav et al., 2015). The fungal mycelial mass can also withstand the pressure, agitation, and other conditions of bioreactors more than plant material and bacterial cells (Velusamy et al., 2016). The ability of microalgae to synthesize AgNPs is also very promising, as demonstrated in *Spirulina platensis* (Mahdieh et al., 2012).

When the microorganism is incubated with silver ions, a defense mechanism is put in place by the microbial cell; this mechanism prevents the toxicity of metal ions and transforms them into nanoparticles. The defense mechanism is currently exploited as a method of synthesizing nanoparticles and offers immense advantages over traditional synthetic pathways (Suresh et al., 2010). The transformation of ions into NPs is carried out by intracellular and extracellular pathways. In the case of positively charged ions, the electrostatic forces of the negatively charged bacterial wall have the ability to attract them; at this stage the bacterial enzymes reduce them to their respective nanoparticles. In the opposite case, the bacterial cell secretes reductase enzymes to reduce metal ions into corresponding NPs (Ovais et al., 2018a). The ability of cyanobacteria and algae to adsorb or absorb metal ions is linked to intracellular polyphosphates and extracellular polysaccharides from living algae; these compounds participate in metal sequestrations. Alternatively, the binding of metal ions is possible through the formation of metallothionein or metal-binding proteins that bind to metal ions as metal thiolate (Govindaraju et al., 2008). The mechanism of reduction of metal ions into NPs begins after the capture of silver ions at the surface of the cell, by electrostatic interaction between the ions and the carboxylate groups, and thereafter the enzymes catalyze the reduction of the ions. This process first generates nuclei which accumulate and eventually form NPs. Microalgae also release enzymes into the culture medium, the reductases of which are responsible for reducing silver ions (Terra et al., 2019). The carbonyl groups resulting from residues of amino acids, peptides, or proteins bind strongly to the metal; they act as a capping agent for AgNPs to subsequently prevent their agglomeration by stabilizing them in the medium (AbdelRahim et al., 2017). The ability of microorganisms to absorb toxic metals efficiently makes them attractive for more advanced applications in order to transform them into high-value products. This bio-inspiration approach entails first understanding the chemical rules underlying biological routes and then constructing synthetic molecules with the desired motifs. Such synthetic molecules help to preserve the mild, green essence of biological synthesis while also enabling lab-scale regulation of key properties of nanomaterials such as particle size, crystallinity, and porosity (Walsh and Knecht, 2017). Table 1 presents some biological precursors involved in the synthesis of AgNPs as well as their biological activities.

The control of nanoparticles and nanostructures biosynthesis and the adjustment of the parameters in large-scale biocatalysis is a crucial point, as it allows the development and the success of the production and the success of NPs activities during their applications. The following sections of this chapter present some basic strategies for scaling up the production of AgNPs by biological methods.

3 Selection of biological precursors for green biosynthesis and determination of their quantity

The choice of plant extracts and microbial biomasses as bioresources for the synthesis of AgNPs is a profitable, easy, and environmentally friendly decision. However, the concentration of bioreducers is one of the important parameters that affect the

Table 1 Some biological precursors involved in the green biosynthesis of AgNPs.

Bacteria

Biological species	NPs	Size and shape of the NPs	Biological activity of the formed NPs	References
Citrobacter spp.	AgNPs	5–15 nm, spherical	– The NPs exhibited anticandidal and antibacterial affects against extended spectrum β-lactamase producing multidrug resistant gram-negative bacteria	Mondal et al. (2020a)
Shewanella sp.	AgNPs	38 nm, spherical	– AgNPs exhibited excellent antibacterial activity against gram-negative bacteria	Mondal et al. (2020b)
Streptomyces capillispiralis Streptomyces zaomyceticus Streptomyces pseudogriseolus	AgNPs	3.77–63.14 nm, spherical 11.32–36.72 nm, spherical 11.70–44.73 nm, spherical	– AgNPs exhibited antifungal activity against plant pathogenic fungi represented by Alternaria alternata, Fusarium oxysporum, Pythium ultimum, and Aspergillus niger	Fouda et al. (2020)
Pythium ultimum	AgNPs	7–15 nm, spherical	– AgNPs exhibited broad-spectrum antibacterial activity – AgNPs had a highly cytotoxic effect against cancerous caco-2 cell line – When used on cotton AgNPs demonstrated a broad-spectrum activity of nano-finished fabrics against pathogenic bacteria, even after 5 and 10 washing cycles	Eid et al. (2020)
Bacillus cereus Streptomyces noursei	AgNPs	6–50 nm, spherical 6–30 nm, spherical	– AgNPs exhibited broad-spectrum antibacterial activity – AgNPs exhibited a cytotoxic effect against cancer cell line (Caco-2)	Alsharif et al. (2020)
Nocardiopsis dassonvillei	AgNPs	30–80 nm	– AgNPs exhibited broad-spectrum antibacterial activity	Dhanaraj et al. (2020)
Streptomyces antimycoticus	AgNPs	3–40 nm, spherical	– AgNPs exhibited excellent antibacterial activity – AgNPs displayed high efficacy against the Caco-2 cancerous cell line	Salem et al. (2020)

Continued

Table 1 Some biological precursors involved in the green biosynthesis of AgNPs. *Continued*

Bacteria

Biological species	NPs	Size and shape of the NPs	Biological activity of the formed NPs	References
Enterobacter cloacae	AgNPs	7–25 nm, spherical	– AgNPs inhibited both gram-positive and gram-negative bacteria	El-Baghdady et al. (2018)
Streptomyces xinghaiensis	AgNPs	5–20 nm, spherical	– AgNPs demonstrated higher antimicrobial activity with MIC in the range of 16–256 µg mL^{-1}	Wypij et al. (2018a)
Streptomyces calidiresistens	AgNPs	5–50 nm, spherical	– AgNPs demonstrated higher antimicrobial activity with MIC in the range of 8–256 µg mL^{-1}	Wypij et al. (2018b)
Microbacterium marinilacus Stenotrophomonas maltophilia	Ag-O-NP	30–145 nm	– AgNPs inhibited biofilm formation	Mukherjee (2017)
Pilimelia columellifera	AgNPs	12.7 nm, spherical	– AgNPs demonstrated good antifungal activity against pathogenic fungi, namely *Candida albicans*, *Malassezia furfur*, and *Trichophyton erinacei*	Wypij et al. (2017)
Lactococcus lactis	AgNPs	5–50 nm, spherical	– AgNPs demonstrated good antibacterial activity with MIC90 in the range of 3.125–12.5 µg mL^{-1}	Viorica et al. (2017)
Nocardiopsis valliformis	AgNPs	5–50 nm, spherical	– AgNPs demonstrated higher antibacterial activity with MIC in the range of 30–80 µg mL^{-1} – An IC$_{50}$ value of 100 µg mL^{-1} was found against cancer HeLa cell line	Rathod et al. (2016)

Fungi

Biological species	NPs	Size and shape of the NPs	Biological activity of the formed NPs	References
Rhizopus stolonifer	AgNPs	6–40 nm, spherical	– AgNPs exhibited broad-spectrum antibacterial activity – AgNPs exhibited a cytotoxic effect against cancer cell line (Caco-2)	Alsharif et al. (2020)
Penicillium italicum	AgNPs	<50 nm, spherical	– The NPs exhibited potential antioxidants activity against 2, 2-diphenyl-1-picrylhydrazyl, hydroxyl radical, and resazurin – The NPs showed potent antimicrobial activity against various pathogens	Taha et al. (2019)
Pestalotiopsis microspora	AgNPs	2–10 nm, spherical	– The NPs exhibited potential antioxidants activity against 2,2'-diphenyl-1-picrylhydrazyl and H$_{2}$O$_{2}$ radicals – The NPs exhibited significant cytotoxic effects against B16F10, SKOV3, A549, and PC3 cell lines	Netala et al. (2016)

Fusarium oxysporum	AgNPs	3.4–64.9 nm, spherical	– The NPs showed a high antifungal activity against *Candida* and *Cryptococcus*	Ishida et al. (2014)
Aspergillus terreus	AgNPs	1–20 nm, spherical	– The NPs showed potent antimicrobial activity against various pathogens	Li et al. (2012)
Algae				
Chlorella vulgaris	AgNPs	9–11 nm	– The NPs inhibited the growth of *Staphylococcus aureus* and alpha hemolysin expression	Soleimani and Habibi-Pirkoohi (2017)
Chlorella vulgaris	Ag-Cl-NPs	9.8 ± 5.7 nm, spherical	– The NPs inhibited the growth of *Staphylococcus aureus* and *Klebsiella pneumoniae*	da Silva Ferreira et al. (2017)
Chlamydomonas reinhardtii	AgNPs	10 nm	– The NPs exhibited inhibitory effects on *Listeria monocytogenes* growth and antagonist activity on the expression of listeriolysin O	Shahriari Ahmadi et al. (2016)
Plant				
Momordica charantia Psidium guajava	AgNPs	5–29 nm, spherical 5–53 nm	– The synthetized NPs exhibited an antifungal effect against *Aspergillus niger*, *Aspergillus flavus*, and *F. oxysporum* as dose-dependence	Nguyen et al. (2020)
Prunus cerasifera	AgNPs	57–144 nm, spherical	– The NPs were effective toward *Salmonella typhi*	Jaffri and Ahmad (2020)
Mentha piperita	AgNPs	15 nm	– The NPs have a remarkable antiproliferative activity toward A549, MDA-MB-231, HepG2, and MCF-7 cell lines	Sharbaf Moghadas et al. (2020)
Boswellia dalzielii	AgNPs	12–101 nm	– The NPs inhibited Kasumi-1 cell proliferation with IC_{50}s of 49.5 and 13.25 µg mL^{-1} at 48 and 72 h	Adebayo et al. (2020)
Cucumis prophetarum	AgNPs	30–50 nm, with polymorphic shapes	– The NPs demonstrated good antioxidant activity – The NPs exhibited excellent antibacterial effect against *Bacillus subtilis*, *Staphylococcus aureus*, *Klebsiella*, and *Escherichia coli*, antimycotic effect against *Aspergillus niger*, *Trichoderma harzianum*, and *Aspergillus flavus*, and biofilm inhibition	Hemlata et al. (2020)
Clerodendrum inerme	AgNPs			Khan et al. (2020)

Continued

Table 1 Some biological precursors involved in the green biosynthesis of AgNPs. *Continued*

Bacteria

Biological species	NPs	Size and shape of the NPs	Biological activity of the formed NPs	References
Momordica charantia	AgNPs	10–40 nm	– The NPs showed complete death of *Pheretima posthuman*, and 85% mortality in larvae of *Aedes albopictus* and *Aedes aegypti*	Shelar et al. (2019)
Aqueous garlic, green tea, and turmeric extracts	AgNPs	8 nm, spherical	– The NPs exhibited cytotoxicity activity on four cancer cell lines: human breast adenocarcinoma (MCF-7), cervical (HeLa), epithelioma (Hep-2), and lung (A549) cell line	Arumai Selvan et al. (2018)
Psidium guajava	AgNPs	20–35 nm	– The NPs showed significant antioxidant and antifungal effect against *Saccharomyces cerevisiae*, *Aspergillus niger*, and *Rhizopus. oryzae*	Wang et al. (2018)
Allium sativum *Zingiber officinale* *Capsicum frutescens*	AgNPs	3–6 nm 3–22 nm, spherical 3–18 nm, spherical	– AgNPs exhibited antimicrobial and antioxidant activity	Otunola et al. (2017)

characteristics of AgNPs by controlling their size and shape; it is a parameter among the operational parameters of green synthesis that needs to be resolved (Dada et al., 2018). The extraction method also plays an important role in the synthesis of NPs (Geetha et al., 2013). This suggests that the characterization of extracts should be discussed from a qualitative and quantitative point of view before the start of biosynthesis. NPs synthesis reactions may be faster under certain conditions, which are highly dependent on the phytochemicals of the plant (Sadeghi et al., 2015). The antioxidant activity of the extracts plays a key role in the kinetics of biosynthesis, by affecting the size and stability of NPs. The presence of antioxidants in the extract accelerates the formation of phyto-NPs. It is also noted that the presence of antioxidants leads to a reduction in the size of the NPs (Stozhko et al., 2019). When using *Eucalyptus oleosa* as a reducing agent, an increase in the amount of plant from 1% to 5% (w/w) decreases the size of AgNPs; however, an increase of more than 10% results in an increase in the size of AgNPs (Pourmortazavi et al., 2015).

When studying phytochemicals present in mango leaves extracts by GC-MS chromatography, the results identified the presence of the phenolic compound pyrogallol and oleic acid. These compounds are the reducing and stabilizing agents of AgNPs (Martínez-Bernett et al., 2016), in pepper leaves extract; the presence of protein acts as capping agents around AgNPs (Mallikarjuna et al., 2014). The richness of *Rhynchotechum ellipticum* leaves by phytochemicals, namely polyphenols, alkaloids, terpenoids, flavonoids, carbohydrates, and steroids, leads to the formation of spherical AgNPs with an average size between 0.51 and 0.73 μm (Hazarika et al., 2014). Using bark extract of *Saraca asoca* as a reducing agent for AgNPs synthesis at an ambient temperature of 45°C allows spherical NPs to be obtained with a size between 2 and 10 nm. FTIR analysis confirmed that the carboxyl, hydroxyl, and amine functional groups of the phytochemicals present in the extract are involved in the reduction, capping, and stabilization of AgNPs (Banerjee and Nath, 2015).

Biosynthetic factors, in particular the ambient temperature of the reaction, and the concentration and volume of solutions containing silver ions and a plant extract (Dada et al., 2018), must be well stabilized on a small scale before moving to a large scale. Temperature has an important effect on the NPs size (Liu et al., 2020). A study carried out on leaf aromatic plant extracts from basil, mint, marjoram, and peppermint resulted in the synthesis of quasispherical AgNPs in a range of sizes between 2 and 80 nm. Regardless of the species used, the increase in the heating rate resulted in a statistically significant reduction in the size of the NPs formed (Hernández-Pinero et al., 2016). These results confirm that the extraction method, the composition of the plant extract, and its concentration play key roles in the final characteristics of AgNPs. Thorough studies are needed before including plant material in large-scale processes in order to obtain high-quality AgNPs.

The use of microbial biomasses as a reducing agent also requires very strict precautions. The amount of biomass used to catalyze the reaction of reducing silver ions to AgNPs needs to be measured, because some studies have reported that AgNP biosynthesis is higher when the biomass concentration is lower, while others have noted that the rate of synthesis is higher in the presence of higher biomass concentrations

(Guilger-Casagrande and de Lima, 2019). The extracellular synthesis of AgNPs by *Guignardia mangiferae*, endophytic fungi isolated from medicinal plant leaves, demonstrated that the lowest biomass concentration (10 g/100 mL) promotes better synthesis of AgNPs compared to other concentrations (20 g/100 mL and 30 g/100 mL). AgNPs were well dispersed and extremely stable with sizes ranging between 5 and 30 nm under optimized reaction conditions (Balakumaran et al., 2015). The use of biomass from *Sclerotinia sclerotiorum* MTCC 8785 as a reducing agent has shown that the biosynthesis of NPs increases with increasing biomass. The increase in synthesis is due to the proportional relationship between the amount of biomass and the release of the agent responsible for the synthesis of AgNPs (Saxena et al., 2016). Another study confirmed that an increase in the biomass concentration (5–25 g of *Penicillium oxalicum* GRS-1) leads to a high production of NPs characterized by a small and uniform size (Rose et al., 2019). The increased volume of cyanobacterium extract (*Oscillatoria limnetica*) induces a high production of Ag-O-NPs. This result is attributed to the high reducing power of *O. limnetica* extract which promotes silver ions reduction (Hamouda et al., 2019). There are several pathways for the green synthesis of NPs (Fig. 2).

The published results indicate a large difference between the biomasses used; this confirms that the differences in the amount of microbial biomass used depend on the species. Otherwise, a good success of the synthesis process requires the establishment of a balance between the quantity of selected microbial biomass and the concentration of the metal precursor.

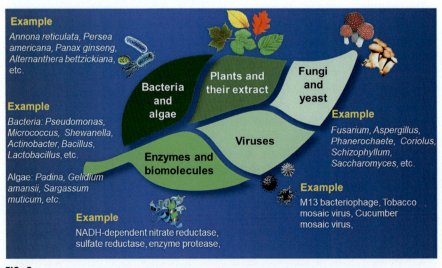

FIG. 2

Cells and biological molecules used in the green synthesis of silver nanoparticles.

4 Selection of culture media

Bacterial fermentation is an extremely important biological process in industry. In nanotechnology, microbial cells appear as a real fermentation machine with many nanometric advantages: they produce high-quality NPs with low structural defects, high reproducibility, and flexibility when recovering synthesized nanostructures (Moon et al., 2010). However, the shift from the synthesis of nanostructures in the laboratory to large-scale production requires the selection of the culture medium. The selection of culture media favorable to the growth of microorganisms and to the synthesis of NPs must be done in a studied manner in order to limit costs and to have high profitability in NPs. The components constituting the culture medium used for the growth of anaerobic or facultative anaerobic microorganisms must be fermentable in order to ensure fermentation (Kim and Roh, 2015). In addition, bacterial culture media play an essential role in the synthesis of silver nanoparticles in terms of size and shape. In their study, Luo and his team demonstrated that the choice of medium affects the yield of NPs synthesis, which suggests that the catalytic activity of reducing silver ions is strongly related to the composition of the culture medium. The culture medium may contain components which contain a deprotonated form of hydroxyl and carboxylic groups, and by oxidation of these groups to carbonyl form, complexation and reduction of Ag ions to AgNPs are carried out (Luo et al., 2018).

The move to industrial scale requires less expensive and available culture media. The selection of wastes is a better choice to produce NPs and to treat wastes by bioprocessing synthesis (Pat-Espadas and Cervante, 2018). Wastewater contaminating the receiving natural environment poses a major problem for the environment. The decontamination of polluted water containing metal contaminants involves two crucial objectives: the first is to make the water less polluted, while the second objective focuses on the recovery of metals in NPs form. A report explored this type of water for these purposes, and demonstrated that it is possible to recover the palladium contaminating the solution in NPs form after its reduction by gram-negative bacteria of the genus *Cupriavidu* (Gauthier et al., 2010).

Using biodegradable food waste for the production of NP is a good way, as food waste contains a wide range of organic compounds. Some organic compounds such as polyphenols, flavonoids, carotenoids, and vitamins act as reducing agents (Ndikau et al., 2017). The functional groups that characterize these compounds are key factors in the process of metal ion reduction. Synthesis of gold nanoparticles from mango peel extract was effective; the NPs formed have a size between 6.03 ± 2.77 and 18.01 ± 3.67 nm. It should be noted that at a concentration of $160 \mu g\, mL^{-1}$, no biological toxicity was noted in normal kidney cells of the African green monkey (CV-1), or in normal human fetal lung fibroblast cells (WI-38) (Yang et al., 2014). Grape seeds are produced by several food industries, especially those of fermentation. The extract obtained from the papins was effective as a reagent to reduce and stabilize silver ions and achieve their transformation into NPs. AgNPs between 25 and 35 nm in size with high stability were formed during 10 min at 95°C. The antimicrobial activity of the NPs was significant against four pathogenic bacteria, namely: *Vibrio alginolyticus*, *V. anguillarum*, *V. parahaemolyticus*, and *Aeromonas punctata* (Xu et al., 2015). The aqueous extracts of orange peels are also used for the synthesis of AgNPs under microwave irradiation. The NPs are nanospheres with an average diameter of 7.36 ± 8.06 nm. The components of orange extract served as a capping agent (Kahrilas et al., 2014). Synthesis of AgNPs is also possible by eggshells. The egg contains two biomembranes: the limiting membrane and the outer shell membrane. Under solar radiation, a research team was able to synthesize mainly spherical Ag nanobiocomposites with average sizes of 10.4 ± 5.6 and 23.6 ± 7.8 nm on the limiting membrane and the membrane of the outer shell, respectively. The NPs are endowed with important antibacterial activity (Huang et al., 2016). Table 2 presents details of some food waste used in the synthesis of AgNPs.

Great challenges remain to be overcome for the waste and industrial effluents selected to recover metals in the form of NPs using biological processes. Waste and effluent can contain very high concentrations of toxic metals, and the latter can create unfavorable conditions by negatively affecting the biosynthesis reactions of NPs (Igiri et al., 2018). The pH is another constraint that must be checked before any use of waste or effluent. The use of acidophilic microbes can overcome this problem. The interactions of NPs with medium components or their agglomerations are other problems; in these cases, the recovery and separation methods must be well studied (Pat-Espadas and Cervante, 2018).

5 Development of information in biochemistry, molecular biology, and genetic engineering

Knowledge and disclosure of the genetic and molecular aspects regulating the metabolic pathways involved in nanosynthesis is a key step for the success of large-scale NPs production. Identification of the enzymes and molecules involved in the synthesis of NPs can lead to better control of the reactions, structures, and sizes of NPs synthesis. Otherwise, the genomic sequences of microorganisms published on the databases can be used for genetic manipulations aimed at improving the biosynthesis of NPs (Grasso et al., 2019). Using molecular biology, genetic engineering, and bioinformatics tools, the development of new genetically modified strains could be an effective solution allowing a reduction of the costs of biosynthesis, rapidity of the chemical reactions, and an improvement of the yield and the quality of the NPs (Jacek et al., 2019a).

As demonstrated in *Komagataeibacter oboediens* MSKU3, the restoration of the ability of bacteria to produce BNC (bacterial nanocellulose) is carried out by adaptive mutation by repetitive static culture; the result shows one revertant produce BNC with a finer structure and narrower range of fibrils width (Taweecheep et al., 2019). In *Komagataeibacter hansenii* ATCC 53582, the genetic modification of two genes encoding proteins with homology to the MotA and MotB proteins, which participate in motility and energy transfer, leads to an improvement in the mechanical properties of nanocellulose fibers (Jacek et al., 2019b). In another study, the genetically

Table 2 Some food waste used in the synthesis of AgNPs.

NPs	Food waste	Morphology of synthesized NPs	Size of the NPs	Biological activity of the produced NPs	References
AgNPs	*Persea americana* peel (PAP)	AgNPs were small dots-like and nearly round-shaped entities.		– AgNPs displayed higher anticancer activities and moderate antioxidant effects – AHS-AgNPs exhibited a higher antidiabetic effect	Das et al. (2020)
	Beta vulgaris peel (BVP)				
	Raw *Arachis hypogaea* shell (AHS)				
	Tomato skins	Rounded NPs (reproduced with permission)	21–93 nm	– AgNPs displayed excellent catalytic properties in the reduction of methylene	Carbone et al. (2020)

Continued

Table 2 Some food waste used in the synthesis of AgNPs. *Continued*

NPs	Food waste	Morphology of synthesized NPs	Size of the NPs	Biological activity of the produced NPs	References
AgNPs	Tomato leaf extract		35 nm	– AgNPs inhibited the growth of *Ralstonia solanacearum*	Santiago et al. (2019)
AgNPs	Banana peel extract			– AgNPs displayed antimicrobial activities	Bankar et al. (2010)
AgNPs	Outer peel extract of *Ananas comosus*	Spherical NPs (under an open access Creative Commons CC BY 4.0 license)		– AgNPs are effective with high antidiabetic potential at a very low concentration and exhibited higher cytotoxic activity against the HepG2 cancer cells – AgNPs exhibited antioxidant activity and moderate antibacterial activity against foodborne pathogenic bacteria	Das et al. (2019)
AgNPs	*Pisum sativum* L. outer peel aqueous extract		10–25 nm	– AgNPs exhibited promising antidiabetic activity as determined by inhibition of α-glucosidase – AgNPs exhibited cytotoxicity effect against HepG2 cells – AgNPs exhibited moderate antioxidant and antibacterial activity against human pathogenic bacteria	Patra et al. (2019)

AgNPs	Orange (*Citrus sinensis*) peel	(under an open access Creative Commons license)	47–63 nm	– AgNPs exhibited excellent catalytic properties in the reduction of methylene blue under solar irradiation	Skiba and Vorobyova (2019)
AgNPs	Grape seed extract	(reproduced with permission)	54.8 nm	– Reductive degradation of Direct Orange 26 was noted	Yao et al. (2018)
AgNPs	Apple extract	Spherical NPs (under an open access Creative Commons license)	30.25 ± 5.26 nm	– The NPs showed antibacterial activities against gram-negative and gram-positive bacteria	Ali et al. (2016)
AgNPs	Satsuma mandarin (*Citrus unshiu*) peel extract	Spherical NPs (reproduced with permission)	5–20 nm		Basavegowda and Lee (2013)

modified *Thalassiosira pseudonana* diatom capable of displaying an IgG-binding domain of protein G on the biosilica surface, allowing the binding of antibodies targeting the cancer cell, demonstrated that neuroblastoma and B lymphoma cells are selectively targeted and killed by nanoporous biosilica (Delalat et al., 2015).

The study carried out on *Saccharomyces cerevisiae* clearly demonstrated that intracellular concentrations of ATP during the process of biosynthesis are necessary; however, a deficit in ATP synthesis leads to a decrease in the synthesis of quantum dots (QDs). The ATP affected the accumulation of the seleno-precursor, and helped the uptake of Cd and the formation of QDs. Genetic modification can improve metabolism and ATP levels in yeast cells; it positively affects QD synthesis (Zhang et al., 2017). Another interesting study has shown that transformation of the yeast *Saccharomyces cerevisiae* expressing hydrogenase on their cell membranes by constructed plasmids has great success in precipitating platinum and producing NPs (Ito et al., 2016). Transformation of *Escherichia coli* DH5α cells by insertion of the vector pUC19 carrying the metallothionein gene from *Candida albicans* revealed the tolerance of transformed bacterial cells to metal ions and an increase in the yield of AgNP synthesis. NPs formed by recombinant cells appeared to be more homogeneous in shape and size than those produced by cells lacking the pUC19 vector (Yuan et al., 2019). The use of genetic engineering to enhance the reduction potential of silver ions in plant cells is also proven. The transformation of the *Nicotiana tabacum* callus by transfer of the silicatein gene LoSilA1 from the marine sponge *Latrunculia oparinae* demonstrated that the transgenic plant has a more than threefold capacity to reduce silver ions and to biosynthesize AgNPs. The ions are reduced to NPs and stabilized by capping agents, which are probably proteins present in the plant extract (Shkryl et al., 2018). These results suggest that molecular biology and genetic engineering may play key roles in intensifying the production of AgNPs.

6 Optimization of physicochemical factors

The physicochemical parameters (temperature, pH, light, and agitation) applied during the biosynthesis process must be optimized in order to obtain good monodispersity, high stability, and biocompatibility. Optimization of the physicochemical parameters of AgNP biosynthesis must be established for each biological entity used in the process (Guilger-Casagrande and de Lima, 2019). This step must be well established in order to avoid an alteration in the reduction of metal ions and their structure. Table 3 demonstrates the main physicochemical parameters controlling the synthesis of AgNPs.

6.1 Temperature effect

The temperature of NP biosynthesis significantly affects the speed of the process, size, and stability of NPs. The increase in temperature during the synthesis of AgNPs by *Fusarium oxysporum* filtrate demonstrated a progressive increase in the rate of

Table 3 The main parameters controlling the characteristics of AgNPs during biosynthesis.

Parameter	Effects noted on NPs	References
Reaction time	Size of NPs	Balavandy et al. (2014) and Abdullah et al. (2015)
Amount of extract	Size and shape	Pourmortazavi et al. (2015)
Temperature of the reaction	Size and shapes	Hernández-Pinero et al. (2016) and Hassan Basri et al. (2020)
Effect of $AgNO_3$ concentration	Size and dispersion	Phanjom and Ahmed (2017)
pH	Morphology, shape, and size	Nayak et al. (2011)

synthesis; the increase is due to a higher secretion of proteins by the fungal biomass. However, at temperatures between 80°C and 100°C, a start of NP aggregation is noted due to denaturation of proteins capping at high temperature (Birla et al., 2013). Another study showed that the increase in temperature during the synthesis of AgNPs by *Aspergillus oryzae* filtrate generates stable AgNPs of different sizes. However, the size of AgNPs decreases with increasing temperature (Phanjom and Ahmed, 2017). When synthesizing AgNPs by *Aspergillus fumigatus* BTCB10, a size of 322.8 nm is obtained at 25°C, while at 55°C, the size reaches 1073.45 nm. The results suggest that at higher temperatures, an aggregation of nanoparticles is recorded (Shahzad et al., 2019). Another study noted a completely different result, as Cobos and his team obtained a small nanaohybrid composed of graphene oxide decorated with silver nanoparticles at lower concentrations of silver precursor and at lower temperatures (Cobos et al., 2020). According to these results, uncontrolled temperatures lead to the generation of large or small unstable NPs, and this is strongly due to the denaturation or to the low activity of the enzymes involved in the biosynthesis.

6.2 Effect of pH

The pH of the medium plays a key role in the synthesis of NPs. The morphology and size of NPs are linked to the proton gradient, as demonstrated during the synthesis of NPs by *Penicillium purpurogenum*; polydispersed NPs are observed at low pH, while at high pH, highly monodispersed spherical NPs are obtained (Nayak et al., 2011). The proton gradient affects the structure and activity of reductase enzymes. Another study carried out involving lactic acid bacteria biomasses clearly demonstrated that the amount of AgNPs and the biomass are affected by variations in pH. The more the pH increases, the more the quantity of NPs formed increases. In this study, the maximum recovery of NPs was recorded at a pH of 11.5 (Sintubin et al., 2009). The study of

Azmath et al. also demonstrated that the generation of NP was faster by Colletotrichum at 50°C and at alkaline pH (Azmath et al., 2016). This result was also observed during the synthesis of AgNPs by endophytic bacteria; under optimized conditions (70°C and an alkaline pH), the synthesis of NPs was rapid (Syed et al., 2016).

6.3 Effect of reaction time

According to the literature, the optimum reaction time is completed when the color of the solution is stable. Increasing reaction time significantly affects NPs by increasing their sizes (Balavandy et al., 2014; Abdullah et al., 2015).

6.4 Effect of AgNO$_3$ concentration

NPs sizes are strongly influenced by variations in the concentration of AgNO$_3$. A change in NPs sizes is noted in the study by Phanjom and Ahmed using *Aspergillus oryzae* as a reducer. In the concentration range between 1 and 8 mM of AgNO$_3$, AgNPs of size between 17.06 and 7.22 nm are generated. However, at a concentration between 9 and 10 mM, AgNPs of size between 45.93 and 62.12 nm are generated. This suggests that the reduction of ions depends on the capacity of extracellular proteins secreted by fungal biomass. The availability of functional groups for the reaction leads to the formation of small particles; however, increasing the concentration of AgNO$_3$ leads to the formation of larger NPs due to the unavailability of functional groups responsible for the reduction (Phanjom and Ahmed, 2017). In another study, the application of different concentrations of AgNO$_3$ (10^{-1}, 10^{-2}, and 10^{-3} M), small AgNPs (2.86 ± 0.3 nm) are detected at a concentration of 10^{-2} M of AgNO$_3$; however, large AgNPs (54.67 ± 4.1 and 14.23 ± 1.3 nm) are obtained at concentrations of 10^{-1} and 10^{-3} M of AgNO$_3$, respectively (AbdelRahim et al., 2017). Another process carried out using bacteria recommends a concentration of 1 mM during the biosynthesis of AgNPs (Gurunathan et al., 2009). During the synthesis of AgNPs by *Acinetobacter calcoaceticus*, the highest concentration applied (5 mM AgNO$_3$) showed the least bioreduction of silver ions. Saturation of the active sites of enzymes or biomolecules responsible for ion reduction decreases biosynthesis. a concentration of 0.7 mM AgNO$_3$ was ideal for the formation of AgNPs by *Acinetobacter calcoaceticus* (Singh et al., 2013). When using *Eucalyptus oleosa* as a reducing agent, increasing the concentration of silver ions from 1 to 5 mmol L^{-1} leads to the generation of larger NPs, while a further increase in silver ions (5–10 mmol L^{-1}) leads to a reverse trend (Pourmortazavi et al., 2015). Adding an excessive concentration of metal ions generates NPs of large size and irregular structure. These results imply that there is a limit of AgNO$_3$ concentration, which must be respected in order to generate AgNPs with satisfactory characteristics (AbdelRahim et al., 2017). An optimal combination of all parameters is recommended in order to generate high-quality AgNPs.

7 Challenges

Over the last two to three decades, our ability to nano-engineer and produce nanostructures and nanosystems with a high degree of control has skyrocketed. Due to the expected economic and social effects of nanotechnology devices, as well as the fact that many fields of use are still being explored, it seems fair to conclude that industrial use of nanomaterials will continue to grow in the near future. Today, the majority of nanosilver processing processes are batch-based in the laboratory. Bad reproductivity and long phase times are the drawbacks of this approach. The reactants are continuously fed into the reactor (Długosz and Banach, 2019) as an alternative to a batch operation. Often, the optimal characteristics of nanoparticles may be altered and lost during the laboratory scale procedure (Colombo et al., 2001). Thus, nanoparticle processing strategies must be carefully chosen based not just on the scalability and cost of production but also on the characteristics of nanoparticles required for particular purposes (Tsuzuki, 2009). These stringent criteria on nanomaterial characteristics face major challenges for their mass processing. In addition, many of the most advanced nanomaterials are currently produced at the lab level by complicated multistage batch processes that cannot be produced at large scales (Sebastian et al., 2014). Phytoextract-mediated synthesis of NPs generates more scale, form, and structure variants. In addition, there is a lack of holistic awareness for the development of green plant NPs. Maximum salt to ion conversion can also be achieved. The biggest challenge is to transform salt into ion. Also, the tedious purification steps and inadequate knowledge of pathways are some drawbacks of the synthesis of microbial nanoparticles. In addition, it is essential in the solution process to monitor form, size, and monodispersion. The processing of output levels for industrial applications is an important challenge. It involves the collection, according to growth rates, enzyme activity, and biochemical pathways, the selection of the biocatalyst condition (bacterial enzymes) of the whole bacterium, raw or distilled enzymes that can enhance cell growth and enzyme activity. Optimization is needed for the synthesis of higher biomass, optimized conditions for reaction to remove more effectively unwanted residual and metabolite nutrients, better extraction and purification processes of the nanoparticles, and better stabilization of the processed NPs (without aggregation, heating processes, and osmotic shock) (Iravani, 2014). Most of the proposed methods do not relate either to large-scale processing or to the preservation of the manufactured NM with the desired characteristics. The fact that these new production technologies need to be monitored in situ to provide high-quality goods is another important issue facing scientists. One potential solution is to track product quality electronically, thus reducing the likelihood of product failure. The entire process is continuously monitored. Good collaboration must therefore be established in order to resolve these challenges, both for the scientific community and for the end users. Significant difficulties in large-scale biogenic nanoparticles for production and development are shown in Fig. 3. In large-scale nanomaterials development, there are promising alternative strategies to avoid such challenges. Biotechnology and biologic-related processes for the manufacture of nanomaterials

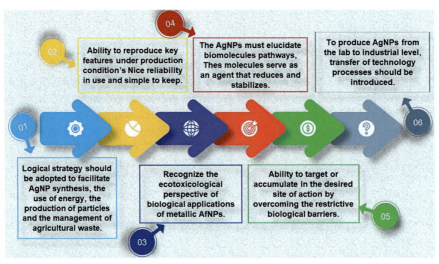

FIG. 3

The biggest challenge of scaling up procedures for synthesis biogenic nanoparticles for industrial applications in different sectors.

are used in the industrial nanobiotechnology field. In addition, most of these techniques remain under development and problems must be addressed.

8 Future perspective

Manufactured nanostructures are mostly manufactured with state-of-the-art traditional technology, while the massive installations required for large-scale manufacturing have a negative environmental effect (Malanowski et al., 2008). Though still in research laboratories, these recently established methods of producing bionanomaterials can provide high process output and low product costs for nanoparticles and nanostructure. Rough estimates indicate that more than 100 kg a day of NPs will be synthesized using advanced processing processes 10 times more than today's quantities. Clearly, these techniques will greatly minimize the existing barriers to nanoproduction and contribute to the advancement of radically new applications. Furthermore, present and future research has also been proposed as motivation for designing sustainable and scalable methods for bionanomaterials to imitate natural designs and concepts of the eco-design field (Hutchison, 2008). For example, specific fungi, microalgae, and lichens produce nanostructured biosilica, which is achieved by the use of the functionalized biomolecules (Patwardhan et al., 2018; Rai et al., 2021). However, bio-based nanomaterials syntheses have significant limitations for industrial uptake, which are mainly associated with prohibitive costs and a limited supply of biomolecules (Hernandez, 2015; Jacquemart et al., 2016). For this reason, great focus has been devoted to understand molecular behavior of biomolecules in the search for alternative synthetic analogues that can be available in large quantities at reasonable costs in order to enable large-scale manufacturing of silver nanomaterials.

9 Conclusion

The use of high-quality and stable nanomaterials is growing, making them the next wave of enticing tools with exciting applications in the industry. The aim of this chapter was to identify how to supply high-purity green AgNPs in commercial quantities while still being able to scale up the process to keep pace with demand as the industrial market grows. In addition, we have concentrated closely on the complexities and prospects of nanoparticle synthesis and characterization. Many methods for producing nanoparticles of desired properties have been developed. Emergent approaches like membrane extrusion, supercritical fluid technology, and microfluidizer technology have the capacity to expand, and few of these technologies' devices are on the market. However, the use of these approaches to produce target- and large-scale surface-functional nanoparticles is still questionable. The current synthesis of silver-based nanomaterials is carried out by chemical techniques, which are characterized by high cost, waste of chemicals, long manufacturing time, and significant environmental pollution. The only way to alleviate these problems is to move toward bio-based synthesis. Biosynthesis techniques of silver nanomaterials may mitigate those problems; however, because of the lack of biological agents available, these methods are not feasible for industrial use. Bio-inspired approaches incorporate the advantages of new bio-based methods with existing chemicals to avoid their shortcomings, which is an excellent way to combine existing manufacturing processes with the refined essence of these techniques. Factors influencing the process efficiency include: the type of precursor for nanosilver used, reductions and stabilizers, physicochemical parameters including solution temperature and pH, and more.

In order to ensure the transfer of the biosynthesis of AgNPs on a large scale, additional work must be set up in order to model the processes. Modeling will increase cascading biosynthesis until the successful completion of large-scale production. Despite the immense benefits noted for green synthesis of AgNPs, the combination of biosynthetic and chemical pathways makes scaling up a difficult step due to the lack of recommended strategies. In order to educate the public about existing needs and potential problems and to design bio-AgNPs directly transferred in the industrial field, the progress of nano-manufacturing depends on the close collaboration between academia and industry. Regulatory directives compel producers to guarantee the quality and stability of the final biogenic NP formulations in order to produce inorganic bionanoparticles on a wide scale. Otherwise, additional work will highlight the specifications, properties of bionanomaterials, and the costs necessary for production. Fig. 4 summarizes the strategies proposed for the large-scale production of AgNPs.

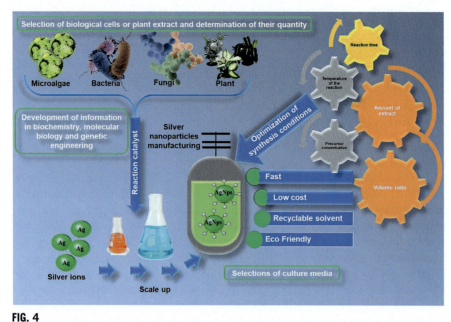

FIG. 4
Proposed strategies for scaling up of green-synthesized nanomaterials.

In summary, it is crucial that the formulation process is robust to ensure high replicability and simplified in order to facilitate the development of scale-up. Multicomponent systems are often needed to produce nanoparticles across multiple process stages. While smaller processes can achieve reproducibility with well-decorated parts, reproducibility and durability of the structures remain constant challenges for the scale-up and production process once beyond the early prototype. The production plan must establish appropriate limits for key nanoparticles and identify process conditions that are essential to achieving these essential characteristics and functions. These essential conditions must be detected on a small scale by comprehensive experiments in order to obtain a full understanding of how physical and biochemical conditions will affect the product. In order to be able to discover inconsistencies in the sample, the physicochemical and biological tests must be sufficiently responsive. Multistage processes are also highly insightful for the monitoring of the critical parameters using a fast and efficient system. In order to ensure manufacturing success, building a knowledge base with properly targeted in-process tests may be crucial.

References

AbdelRahim, K., Mahmoud, S.Y., Ali, A.M., Almaary, K.S., Mustafa, A.E., Husseiny, S.M., 2017. Extracellular biosynthesis of silver nanoparticles using *Rhizopus stolonifer*. Saudi J. Biol. Sci. 24, 208–216.

Abdullah, N.I., Ahmad, M.B., Shameli, K., 2015. Biosynthesis of silver nanoparticles using *Artocarpus elasticus* stem bark extract. Chem. Cent. J. 9, 61.

Adebayo, I.A., Usman, A.I., Shittu, F.B., Ismail, N.Z., Arsad, H., Muftaudeen, T.K., Samian, M.R., 2020. *Boswellia dalzielii*-mediated silver nanoparticles inhibited acute myeloid leukemia AML Kasumi-1 cells by inducing cell cycle arrest. Bioinorg. Chem. Appl. 2020, 8898360.

Ali, Z.A., Yahya, R., Sekaran, S.D., Puteh, R., 2016. Green synthesis of silver nanoparticles using apple extract and its antibacterial properties. Adv. Mater. Sci. Eng. 2016, 4102196.

Alsharif, S.M., Salem, S.S., Abdel-Rahman, M.A., Fouda, A., Eid, A.M., El-Din Hassan, S., Awad, M.A., Mohamed, A.A., 2020. Multifunctional properties of spherical silver nanoparticles fabricated by different microbial taxa. Heliyon 6, e03943.

Arumai Selvan, D., Mahendiran, D., Senthil Kumar, R., Kalilur Rahiman, A., 2018. Garlic, green tea and turmeric extracts-mediated green synthesis of silver nanoparticles: phytochemical, antioxidant and in vitro cytotoxicity studies. J. Photochem. Photobiol. B Biol. 180, 243–252.

Azmath, P., Baker, S., Rakshith, D., Satish, S., 2016. Mycosynthesis of silver nanoparticles bearing antibacterial activity. Saudi Pharm. J. 24, 140–146.

Balakumaran, M.D., Ramachandran, R., Kalaichelvan, P.T., 2015. Exploitation of endophytic fungus, *Guignardia mangiferae* for extracellular synthesis of silver nanoparticles and their in vitro biological activities. Microbiol. Res. 178, 9–17.

Balavandy, S.K., Shameli, K., Biak, D.R., Abidin, Z.Z., 2014. Stirring time effect of silver nanoparticles prepared in glutathione mediated by green method. Chem. Cent. J. 8, 11.

Banerjee, P., Nath, D., 2015. A phytochemical approach to synthesize silver nanoparticles for non-toxic biomedical application and study on their antibacterial efficacy. Nanosci. Technol. 2, 1–14.

Bankar, A., Joshi, B., Kumar, A.R., Zinjarde, S., 2010. Banana peel extract mediated novel route for the synthesis of silver nanoparticles. Colloids Surf. A 368, 58–63.

Baranitharan, M., Alarifi, S., Alkahtani, S., Ali, D., Elumalai, K., Pandiyan, J., Krishnappa, K., Rajeswary, M., Govindarajan, M., 2021. Phytochemical analysis and fabrication of silver nanoparticles using *Acacia catechu*: an efficacious and ecofriendly control tool against selected polyphagous insect pests. Saudi J. Biol. Sci. 28, 148–156.

Basavegowda, N., Lee, Y.R., 2013. Synthesis of silver nanoparticles using Satsuma mandarin *Citrus unshiu* peel extract: a novel approach towards waste utilization. Mater. Lett. 109, 31–33.

Birla, S.S., Gaikwad, S.C., Gade, A.K., Rai, M.K., 2013. Rapid synthesis of silver nanoparticles from *Fusarium oxysporum* by optimizing physicocultural conditions. Sci. World J. 2013, 796018.

Busnaina, A.A., Mead, J., Isaacs, J., Somu, S., 2013. Nanomanufacturing and sustainability: opportunities and challenges. J. Nanoparticle Res. 15, 1984.

Carbone, K., De Angelis, A., Mazzuca, C., Santangelo, E., Macchioni, V., Cacciotti, I., Petrella, G., Cicero, D.O., Micheli, L., 2020. Microwave-assisted synthesis of catalytic silver nanoparticles by hyperpigmented tomato skins: a green approach. LWT 133, 110088.

Charitidis, C.A., Georgiou, P., Koklioti, M.A., Trompeta, A.F., Markakis, V., 2014. Manufacturing nanomaterials: from research to industry. Manuf. Rev. 1, 11.

Cobos, M., De-La-Pinta, I., Quindós, G., Fernández, M.J., Fernández, M.D., 2020. Graphene oxide-silver nanoparticle nanohybrids: synthesis, characterization, and antimicrobial properties. Nanomaterials 10, 376.

Colombo, A.P., Briançon, S., Lieto, J., Fessi, H., 2001. Project, design, and use of a pilot plant for nanocapsule production. Drug Dev. Ind. Pharm. 27, 1063–1072.

Cristaldi, D.A., Yanar, F., Mosayyebi, A., García-Manrique, P., Stulz, E., Carugo, D., Zhang, X., 2018. Easy-to-perform and cost-effective fabrication of continuous-flow reactors and their application for nanomaterials synthesis. N. Biotechnol. 47, 1–7.

da Silva Ferreira, V., Conz Ferreira, M.E., Lima, L.M., Frasés, S., de Souza, W., Sant'Anna, C., 2017. Green production of microalgae-based silver chloride nanoparticles with antimicrobial activity against pathogenic bacteria. Enzyme Microb. Technol. 97, 114–121.

Dada, A.O., Inyinbor, A.A., Idu, E.I., Bello, O.M., Oluyori, A.P., Adelani-Akande, T.A., Okunola, A.A., Dada, O., 2018. Effect of operational parameters, characterization and antibacterial studies of green synthesis of silver nanoparticles using *Tithonia diversifolia*. PeerJ 6, e5865.

Das, G., Patra, J.K., Debnath, T., Ansari, A., Shin, H.S., 2019. Investigation of antioxidant, antibacterial, antidiabetic, and cytotoxicity potential of silver nanoparticles synthesized using the outer peel extract of *Ananas comosus* L. PLoS One 14, e0220950.

Das, G., Shin, H.S., Patra, J.K., 2020. Comparative assessment of antioxidant, anti-diabetic and cytotoxic effects of three peel/shell food waste extract-mediated silver nanoparticles. Int. J. Nanomedicine 15, 9075–9088.

Delalat, B., Sheppard, V.C., Rasi Ghaemi, S., Rao, S., Prestidge, C.A., McPhee, G., Rogers, M.L., Donoghue, J.F., Pillay, V., Johns, T.G., Kröger, N., Voelcker, N.H., 2015. Targeted drug delivery using genetically engineered diatom biosilica. Nat. Commun. 6, 8791.

Dhanaraj, S., Thirunavukkarasu, S., Allen John, H., Pandian, S., Salmen, S.H., Chinnathambi, A., Alharbi, S.A., 2020. Novel marine *Nocardiopsis dassonvillei*-DS013 mediated silver nanoparticles characterization and its bactericidal potential against clinical isolates. Saudi J. Biol. Sci. 27, 991–995.

Długosz, O., Banach, M., 2019. Continuous production of silver nanoparticles and process control. J. Clust. Sci. 30, 541–552.

Domagalski, J.T., Xifre-Perez, E., Marsal, L.F., 2021. Recent advances in nanoporous anodic alumina: principles, engineering, and applications. Nanomaterials 11, 430.

Drummond, C., McCann, R., Patwardhan, S.V., 2014. A feasibility study of the biologically inspired green manufacturing of precipitated silica. Chem. Eng. J. 244, 483–492.

Eid, A.M., Fouda, A., Niedbała, G., Hassan, S.E., Salem, S.S., Abdo, A.M., Hetta, H.F., Shaheen, T.I., 2020. Endophytic *Streptomyces laurentii* mediated green synthesis of Ag-NPs with antibacterial and anticancer properties for developing functional textile fabric properties. Antibiotics 9, 641.

El-Baghdady, K.Z., El-Shatoury, E.H., Abdullah, O.M., Khalil, M., 2018. Biogenic production of silver nanoparticles by *Enterobacter cloacae* Ism26. Turk. J. Biol. 42, 319–321.

El-Moslamy, S.H., 2018. Bioprocessing strategies for cost-effective large-scale biogenic synthesis of nano-MgO from endophytic *Streptomyces coelicolor* strain E72 as an anti-multidrug-resistant pathogens agent. Sci. Rep. 8, 3820.

El-Moslamy, S.H., Elkady, M.F., Rezk, A.H., Abdel-Fattah, Y.R., 2017. Applying Taguchi design and large-scale strategy for mycosynthesis of nano-silver from endophytic *Trichoderma harzianum* SYA.F4 and its application against phytopathogens. Sci. Rep. 7, 45297.

Fouda, A., Hassan, S.E., Abdo, A.M., El-Gamal, M.S., 2020. Antimicrobial, antioxidant and larvicidal activities of spherical silver nanoparticles synthesized by endophytic *Streptomyces spp.* Biol. Trace Elem. Res. 195, 707–724.

Gauthier, D., Søbjerg, L.S., Jensen, K.M., Lindhardt, A.T., Bunge, M., Finster, K., Meyer, R.L., Skrydstrup, T., 2010. Environmentally benign recovery and reactivation of palladium from industrial waste by using gram-negative bacteria. ChemSusChem 3, 1036–1039.

Geetha, A.R., George, E., Srinivasan, A., Shaik, J., 2013. Optimization of green synthesis of silver nanoparticles from leaf extracts of *Pimenta dioica* allspice. Sci. World J. 2013, 362890.

Govindaraju, K., Basha, S.K., Kumar, V.G., Singaravelu, G., 2008. Gold and bimetallic nanoparticles production using single-cell protein *Spirulina platensis* Geitler. J. Mater. Sci. 43, 5115–5122.

Grasso, G., Zane, D., Dragone, R., 2019. Microbial nanotechnology: challenges and prospects for green biocatalytic synthesis of nanoscale materials for sensoristic and biomedical applications. Nanomaterials 10, 11.

Guilger-Casagrande, M., de Lima, R., 2019. Synthesis of silver nanoparticles mediated by fungi: a review. Front. Bioeng. Biotechnol. 7, 287.

Gurunathan, S., Kalishwaralal, K., Vaidyanathan, R., Venkataraman, D., Pandian, S.R., Muniyandi, J., Hariharan, N., Eom, S.H., 2009. Biosynthesis, purification and characterization of silver nanoparticles using *Escherichia coli*. Colloids Surf. B Biointerfaces 74, 328–335.

Hamouda, R.A., Hussein, M.H., Abo-Elmagd, R.A., Bawazir, S.S., 2019. Synthesis and biological characterization of silver nanoparticles derived from the cyanobacterium *Oscillatoria limnetica*. Sci. Rep. 9, 13071.

Hassan Basri, H., Talib, R.A., Sukor, R., Othman, S.H., Ariffin, H., 2020. Effect of synthesis temperature on the size of ZnO nanoparticles derived from pineapple peel extract and antibacterial activity of ZnO-starch nanocomposite films. Nanomaterials 10, 1061.

Hazarika, D., Phukan, A., Saikia, E., Chetia, B., 2014. Phytochemical screening and synthesis of silver nanoparticles using leaf extract of *Rhynchotechum ellipticum*. Int J Pharm Pharm Sci 6, 672–674.

Hemlata, Meena, P.R., Singh, A.P., Tejavath, K.K., 2020. Biosynthesis of silver nanoparticles using *Cucumis prophetarum* aqueous leaf extract and their antibacterial and antiproliferative activity against cancer cell lines. ACS Omega 5, 5520–5528.

Hernandez, R., 2015. Continuous manufacturing: a changing processing paradigm. Biopharm. Int. 28, 20–27.

Hernández-Pinero, J.L., Terrón-Rebolledo, M., Foroughbakhch, R., Moreno-Limón, S., Melendrez, M.F., Solís-Pomar, F., Pérez-Tijerina, E., 2016. Effect of heating rate and plant species on the size and uniformity of silver nanoparticles synthesized using aromatic plant extracts. Appl. Nanosci. 6, 1183–1190.

Huang, M., Du, L., Feng, J.-X., 2016. Photochemical synthesis of silver nanoparticles/eggshell membrane composite, its characterization and antibacterial activity. Sci. Adv. Mater. 8, 1641–1647.

Hutchison, J.E., 2008. Greener nanoscience: a proactive approach to advancing applications and reducing implications of nanotechnology. ACS Nano 2, 395–402.

Igiri, B.E., Okoduwa, S., Idoko, G.O., Akabuogu, E.P., Adeyi, A.O., Ejiogu, I.K., 2018. Toxicity and bioremediation of heavy metals contaminated ecosystem from tannery wastewater: a review. J. Toxicol. 2018, 2568038.

Intelligence, M., 2016. Global Nanomaterials Market-Segmented by Product Type, End-User Industry, and Geography-Trends and Forecasts (2015–2020).

Iravani, S., 2014. Bacteria in nanoparticle synthesis: current status and future prospects. Int. Sch. Res. Notices. 44, 235–239.

Ishida, K., Cipriano, T.F., Rocha, G.M., Weissmüller, G., Gomes, F., Miranda, K., Rozental, S., 2014. Silver nanoparticle production by the fungus *Fusarium oxysporum*: nanoparticle characterisation and analysis of antifungal activity against pathogenic yeasts. Mem. Inst. Oswaldo Cruz 109, 220–228.

Ito, R., Kuroda, K., Hashimoto, H., Ueda, M., 2016. Recovery of platinum 0 through the reduction of platinum ions by hydrogenase-displaying yeast. AMB Express 6, 88.

Jacek, P., Dourado, F., Gama, M., Bielecki, S., 2019a. Molecular aspects of bacterial nanocellulose biosynthesis. J. Microbial. Biotechnol. 12, 633–649.

Jacek, P., Kubiak, K., Ryngajłło, M., Rytczak, P., Paluch, P., Bielecki, S., 2019b. Modification of bacterial nanocellulose properties through mutation of motility related genes in *Komagataeibacter hansenii* ATCC 53582. N. Biotechnol. 52, 60–68.

Jacquemart, R., Vandersluis, M., Zhao, M., Sukhija, K., Sidhu, N., Stout, J., 2016. A single-use strategy to enable manufacturing of affordable biologics. Comput. Struct. Biotechnol. J. 14, 309–318.

Jaffri, S.B., Ahmad, K.S., 2020. Biomimetic detoxifier *Prunus cerasifera* Ehrh. silver nanoparticles: innate green bullets for morbific pathogens and persistent pollutants. Environ. Sci. Pollut. Res. Int. 27, 9669–9685.

Kahrilas, G.A., Wally, L.M., Fredrick, S.J., Hiskey, M., Prieto, A.L., Owens, J.E., 2014. Microwave-assisted green synthesis of silver nanoparticles using orange peel extract. ACS Sustain. Chem. Eng. 2, 367–376.

Khan, S.A., Shahid, S., Lee, C.S., 2020. Green synthesis of gold and silver nanoparticles using leaf extract of *Clerodendrum inerme*; characterization, antimicrobial, and antioxidant activities. Biomol. Ther. 10, 835.

Kim, Y., Roh, Y., 2015. Microbial synthesis and characterization of superparamagnetic Zn-substituted magnetite nanoparticles. J. Nanosci. Nanotechnol. 15, 6129–6132.

Li, G., He, D., Qian, Y., Guan, B., Gao, S., Cui, Y., Yokoyama, K., Wang, L., 2012. Fungus-mediated green synthesis of silver nanoparticles using *Aspergillus terreus*. Int. J. Mol. Sci. 13, 466–476.

Li, J., Jaitzig, J., Lu, P., Süssmuth, R.D., Neubauer, P., 2015. Scale-up bioprocess development for production of the antibiotic valinomycin in *Escherichia coli* based on consistent fed-batch cultivations. Microb. Cell Fact. 14, 83.

Liu, H., Zhang, H., Wang, J., Wei, J., 2020. Effect of temperature on the size of biosynthesized silver nanoparticle: deep insight into microscopic kinetics analysis. Arab. J. Chem. 13, 1011–1019.

Luo, K., Jung, S., Park, K.H., Kim, Y.R., 2018. Microbial biosynthesis of silver nanoparticles in different culture media. J. Agric. Food Chem. 66, 957–962.

Mahdieh, M., Zolanvari, A., Azimee, A.S., Mahdieh, M., 2012. Green biosynthesis of silver nanoparticles by *Spirulina platensis*. Sci. Iran. 19, 926–929.

Luther, W., Werner, M., 2008. In: Malanowski, N., Heimer, T. (Eds.), Growth Market Nanotechnology: An Analysis of Technology and Innovation. John Wiley Sons.

Mallikarjuna, K., Sushma, N.J., Narasimha, G., Manoj, L., Raju, B.D.P., 2014. Phytochemical fabrication and characterization of silver nanoparticles by using pepper leaf broth. Arab. J. Chem. 7, 1099–1103.

Martínez-Bernett, D., Silva-Granados, A., Correa-Torres, S., Herrera, A., 2016. Chromatographic analysis of phytochemicals components present in *Mangifera indica* leaves for the synthesis of silver nanoparticles by $AgNO_3$ reduction. J. Phys. Conf. Ser. 687, 012033.

Mondal, A.H., Yadav, D., Ali, A., Khan, N., Jin, J.O., Haq, Q., 2020a. Anti-bacterial and anti-candidal activity of silver nanoparticles biosynthesized using *Citrobacter spp.* MS5 culture supernatant. Biomolecules 10, 944.

Mondal, A.H., Yadav, D., Mitra, S., Mukhopadhyay, K., 2020b. Biosynthesis of silver nanoparticles using culture supernatant of *Shewanella sp.* ARY1 and their antibacterial activity. Int. J. Nanomedicine 15, 8295–8310.

Moon, J.W., Rawn, C.J., Rondinone, A.J., Love, L.J., Roh, Y., Everett, S.M., Lauf, R.J., Phelps, T.J., 2010. Large-scale production of magnetic nanoparticles using bacterial fermentation. J. Ind. Microbiol. Biotechnol. 37, 1023–1031.

Mukherjee, P., 2017. *Stenotrophomonas* and *Microbacterium*: mediated biogenesis of copper, silver and iron nanoparticles-proteomic insights and antibacterial properties versus biofilm formation. J. Clust. Sci. 28, 331–358.

Nayak, R.R., Pradhan, N., Behera, D., Pradhan, K.M., Mishra, S., Sukla, L.B., Mishra, B.K., 2011. Green synthesis of silver nanoparticle by *Penicillium purpurogenum* NPMF: the process and optimization. J. Nanopart. Res. 13, 3129–3137.

Ndikau, M., Noah, N.M., Andala, D.M., Masika, E., 2017. Green synthesis and characterization of silver nanoparticles using *Citrullus lanatus* fruit rind extract. Int. J. Anal. Chem. 2017, 8108504.

Netala, V.R., Bethu, M.S., Pushpalatha, B., Baki, V.B., Aishwarya, S., Rao, J.V., Tartte, V., 2016. Biogenesis of silver nanoparticles using endophytic *fungus Pestalotiopsis microspora* and evaluation of their antioxidant and anticancer activities. Int. J. Nanomedicine 11, 5683–5696.

Nguyen, D.H., Vo, T., Nguyen, N.T., Ching, Y.C., Hoang Thi, T.T., 2020. Comparison of biogenic silver nanoparticles formed by *Momordica charantia* and *Psidium guajava* leaf extract and antifungal evaluation. PLoS One 15, e0239360.

Otunola, G.A., Afolayan, A.J., Ajayi, E.O., Odeyemi, S.W., 2017. Characterization, antibacterial and antioxidant properties of silver nanoparticles synthesized from aqueous extracts of *Allium sativum, Zingiber officinale*, and *Capsicum frutescens*. Pharmacogn. Mag. 13, S201–S208.

Ovais, M., Khalil, A.T., Ayaz, M., Ahmad, I., Nethi, S.K., Mukherjee, S., 2018a. Biosynthesis of metal nanoparticles via microbial enzymes: a mechanistic approach. Int. J. Mol. Sci. 19, 4100.

Ovais, M., Khalil, A.T., Islam, N.U., Ahmad, I., Ayaz, M., Saravanan, M., Shinwari, Z.K., Mukherjee, S., 2018b. Role of plant phytochemicals and microbial enzymes in biosynthesis of metallic nanoparticles. Appl. Microbiol. Biotechnol. 102, 6799–6814.

Pat-Espadas, A.M., Cervante, F.J., 2018. Microbial recovery of metallic nanoparticles from industrial wastes and their environmental applications. J. Chem. Technol. Biotechnol. 93, 3091–3112.

Patra, J.K., Das, G., Shin, H.S., 2019. Facile green biosynthesis of silver nanoparticles using *Pisum sativum* L. outer peel aqueous extract and its antidiabetic, cytotoxicity, antioxidant, and antibacterial activity. Int. J. Nanomedicine 14, 6679–6690.

Patwardhan, S.V., Manning, J.R.H., Chiacchia, M., 2018. Bioinspired synthesis as a potential green method for the preparation of nanomaterials: opportunities and challenges. Curr. Opin. Green Sustain. Chem. 12, 110–116.

Phanjom, P., Ahmed, G., 2017. Effect of different physicochemical conditions on the synthesis of silver nanoparticles using fungal cell filtrate of *Aspergillus oryzae* MTCC no. 1846 and their antibacterial effects. Adv. Nat. Sci. Nanosci. Nanotechnol. 8, 1–13.

Pourmortazavi, S.M., Taghdiri, M., Makari, V., Rahimi-Nasrabadi, M., 2015. Procedure optimization for green synthesis of silver nanoparticles by aqueous extract of *Eucalyptus oleosa*. Spectrochim. Acta A Mol. Biomol. Spectrosc. 136, 1249–1254.

Rai, M., Bonde, S., Golinska, P., Trzcinska-Wencel, J., Gade, A., Abd-Elsalam, K., Shende, S., Gaikwad, S., Ingle, A.P., 2021. *Fusarium* as a novel fungus for the synthesis of nanoparticles: mechanism and applications. J. Fungi 7, 139.

Rathod, D., Golinska, P., Wypij, M., Dahm, H., Rai, M., 2016. A new report *of Nocardiopsis valliformis* strain OT1 from alkaline Lonar crater of India and its use in synthesis of silver nanoparticles with special reference to evaluation of antibacterial activity and cytotoxicity. Med. Microbiol. Immunol. 205, 435–447.

Rose, G.K., Soni, R., Rishi, P., Soni, S.K., 2019. Optimization of the biological synthesis of silver nanoparticles using *Penicillium oxalicum* GRS-1 and their antimicrobial effects against common food-borne pathogens. Green Processes Synth. 8, 144–156.

Sadeghi, B., Rostami, A., Momeni, S.S., 2015. Facile green synthesis of silver nanoparticles using seed aqueous extract of *Pistacia atlantica* and its antibacterial activity. Spectrochim. Acta A Mol. Biomol. Spectrosc. 134, 326–332.

Salem, S.S., El-Belely, E.F., Niedbała, G., Alnoman, M.M., Hassan, S.E., Eid, A.M., Shaheen, T.I., Elkelish, A., Fouda, A., 2020. Bactericidal and in-vitro cytotoxic efficacy of silver nanoparticles Ag-NPs fabricated by endophytic actinomycetes and their use as coating for the textile fabrics. Nanomaterials 10, 2082.

Santiago, T.R., Bonatto, C.C., Rossato, M., Lopes, C., Lopes, C.A., Mizubuti, E.S.G., Silva, L.P., 2019. Green synthesis of silver nanoparticles using tomato leaf extract and their entrapment in chitosan nanoparticles to control bacterial wilt. J. Sci. Food Agric. 99, 4248–4259.

Saxena, J., Sharma, P.K., Sharma, M.M., Singh, A., 2016. Process optimization for green synthesis of silver nanoparticles by *Sclerotinia sclerotiorum* MTCC 8785 and evaluation of its antibacterial properties. Springer Plus 5, 861.

Sebastian, V., Arruebo, M., Santamaria, J., 2014. Reaction engineering strategies for the production of inorganic nanomaterials. Small 10, 835–853.

Shahriari Ahmadi, F., Tanhaeian, A., Habibi Pirkohi, M., 2016. Biosynthesis of silver nanoparticles using *Chlamydomonas reinhardtii* and its inhibitory effect on growth and virulence of *Listeria monocytogenes*. Iran. J. Biotechnol. 14, 163–168.

Shahzad, A., Saeed, H., Iqtedar, M., Hussain, S.Z., Kaleem, A., Abdullah, R., 2019. Size-controlled production of silver nanoparticles by *Aspergillus fumigatus* BTCB10: likely antibacterial and cytotoxic effects. J. Nanomater. 2019, 5168698.

Sharbaf Moghadas, M.R., Motamedi, E., Nasiri, J., Naghavi, M.R., Sabokdast, M., 2020. Proficient dye removal from water using biogenic silver nanoparticles prepared through solid-state synthetic route. Heliyon 6, e04730.

Shelar, A., Sangshetti, J., Chakraborti, S., Singh, A.V., Patil, R., Gosavi, S., 2019. Helminthicidal and larvicidal potentials of biogenic silver nanoparticles synthesized from medicinal plant *Momordica charantia*. Med. Chem. 15, 781–789.

Shkryl, Y.N., Veremeichik, G.N., Kamenev, D.G., Gorpenchenko, T.Y., Yugay, Y.A., Mashtalyar, D.V., Nepomnyaschiy, A.V., Avramenko, T.V., Karabtsov, A.A., Ivanov, V.V., Bulgakov, V.P., Gnedenkov, S.V., Kulchin, Y.N., Zhuravlev, Y.N., 2018. Green synthesis of silver nanoparticles using transgenic *Nicotiana tabacum* callus culture expressing silicatein gene from marine sponge *Latrunculia oparinae*. Artif. Cells. Nanomed. Biotechnol. 46, 1646–1658.

Singh, R., Wagh, P., Wadhwani, S., Gaidhani, S., Kumbhar, A., Bellare, J., Chopade, B.A., 2013. Synthesis, optimization, and characterization of silver nanoparticles from *Acinetobacter calcoaceticus* and their enhanced antibacterial activity when combined with antibiotics. Int. J. Nanomedicine 8, 4277–4290.

Sintubin, L., De Windt, W., Dick, J., Mast, J., van der Ha, D., Verstraete, W., Boon, N., 2009. Lactic acid bacteria as reducing and capping agent for the fast and efficient production of silver nanoparticles. Appl. Microbiol. Biotechnol. 84, 741–749.

Skiba, M.S., Vorobyova, V.I., 2019. Synthesis of silver nanoparticles using orange peel extract prepared by plasmochemical extraction method and degradation of methylene blue under solar irradiation. Adv. Mater. Sci. Eng. 2019, 8.

Soleimani, M., Habibi-Pirkoohi, M., 2017. Biosynthesis of silver nanoparticles using *Chlorella vulgaris* and evaluation of the antibacterial efficacy against *Staphylococcus aureus*. Avicenna J. Med. Biotechnol. 9, 120–125.

Stozhko, N.Y., Bukharinova, M.A., Khamzina, E.I., Tarasov, A.V., Vidrevich, M.B., Brainina, K.Z., 2019. The effect of the antioxidant activity of plant extracts on the properties of gold nanoparticles. Nanomaterials 9, 1655.

Suresh, A.K., Pelletier, D.A., Wang, W., Moon, J.W., Gu, B., Mortensen, N.P., Allison, D.P., Joy, D.C., Phelps, T.J., Doktycz, M.J., 2010. Silver nanocrystallites: biofabrication using *Shewanella oneidensis*, and an evaluation of their comparative toxicity on gram-negative and gram-positive bacteria. Environ. Sci. Technol. 44, 5210–5215.

Svobodová, K., Novotný, Č., 2018. Bioreactors based on immobilized fungi: bioremediation under non-sterile conditions. Appl. Microbiol. Biotechnol. 102, 39–46.

Syed, B., Yashavantha Rao, H.C., Nagendra-Prasad, M.N., Prasad, A., Harini, B.P., Azmath, P., Rakshith, D., Satish, S., 2016. Biomimetic synthesis of silver nanoparticles using endosymbiotic bacterium inhabiting *Euphorbia hirta* L. and their bactericidal potential. Scientifica 2016, 9020239.

Taha, Z.K., Hawar, S.N., Sulaiman, G.M., 2019. Extracellular biosynthesis of silver nanoparticles from *Penicillium italicum* and its antioxidant, antimicrobial and cytotoxicity activities. Biotechnol. Lett. 41, 899–914.

Taweecheep, P., Naloka, K., Matsutani, M., Yakushi, T., Matsushita, K., Theeragool, G., 2019. Superfine bacterial nanocellulose produced by reverse mutations in the bcsC gene during adaptive breeding of *Komagataeibacter oboediens*. Carbohydr. Polym. 226, 115243.

Terra, A., Kosinski, R., Moreira, J.B., Costa, J., Morais, M.G., 2019. Microalgae biosynthesis of silver nanoparticles for application in the control of agricultural pathogens. J. Environ. Sci. Health B 54, 709–716.

Tsuzuki, T., 2009. Commercial scale production of inorganic nanoparticles. Int. J. Nanotechnol. 6, 567–578.

Velusamy, P., Kumar, G.V., Jeyanthi, V., Das, J., Pachaiappan, R., 2016. Bio-inspired green nanoparticles: synthesis, mechanism, and antibacterial application. Toxicol. Res. 32, 95–102.

Viorica, R.P., Pawel, P., Kinga, M., Michal, Z., Katarzyna, R., Boguslaw, B., 2017. *Lactococcus lactis* as a safe and inexpensive source of bioactive silver composites. Appl. Microbiol. Biotechnol. 101, 7141–7153.

Walsh, T.R., Knecht, M.R., 2017. Biointerface structural effects on the properties and applications of bioinspired peptide-based nanomaterials. Chem. Rev. 117, 12641–12704.

Wang, L., Wu, Y., Xie, J., Wu, S., Wu, Z., 2018. Characterization, antioxidant and antimicrobial activities of green synthesized silver nanoparticles from *Psidium guajava* L. leaf aqueous extracts. Korean J. Couns. Psychother. 86, 1–8.

Wypij, M., Czarnecka, J., Dahm, H., Rai, M., Golinska, P., 2017. Silver nanoparticles from *Pilimelia columellifera subsp*. pallida SL19 strain demonstrated antifungal activity against fungi causing superficial mycoses. J. Basic Microbiol. 57, 793–800.

Wypij, M., Czarnecka, J., Świecimska, M., Dahm, H., Rai, M., Golinska, P., 2018a. Synthesis, characterization and evaluation of antimicrobial and cytotoxic activities of biogenic silver nanoparticles synthesized from *Streptomyces xinghaiensis* OF1 strain. World J. Microbiol. Biotechnol. 34, 23.

Wypij, M., Świecimska, M., Czarnecka, J., Dahm, H., Rai, M., Golinska, P., 2018b. Antimicrobial and cytotoxic activity of silver nanoparticles synthesized from two haloalkaliphilic actinobacterial strains alone and in combination with antibiotics. J. Appl. Microbiol. 124, 1411–1424.

Xu, H., Wang, L., Su, H., Gu, L., Han, T., Meng, F., Liu, C., 2015. Making good use of food wastes: green synthesis of highly stabilized silver nanoparticles from grape seed extract and their antimicrobial activity. Food Biophys. 10, 12–18.

Yadav, A., Kon, K., Kratosova, G., Duran, N., Ingle, A.P., Rai, M., 2015. Fungi as an efficient mycosystem for the synthesis of metal nanoparticles: progress and key aspects of research. Biotechnol. Lett. 37, 2099–2120.

Yang, N., WeiHong, L., Hao, L., 2014. Biosynthesis of au nanoparticles using agricultural waste mango peel extract and its in vitro cytotoxic effect on two normal cells. Mater. Lett. 134, 67–70.

Yao, P., Zhang, J., Xing, T., Chen, G., Tao, R., Choo, K.-H., 2018. Green synthesis of silver nanoparticles using grape seed extract and their application for reductive catalysis of direct Orange 26. J. Ind. Eng. Chem. 58, 74–79.

Yuan, Q., Bomma, M., Xiao, Z., 2019. Enhanced silver nanoparticle synthesis by *Escherichia coli* transformed with *Candida Albicans* metallothionein gene. Materials 12, 4180.

Zhang, R., Shao, M., Han, X., Wang, C., Li, Y., Hu, B., Pang, D., Xie, Z., 2017. ATP synthesis in the energy metabolism pathway: a new perspective for manipulating CdSe quantum dots biosynthesized in *Saccharomyces cerevisiae*. Int. J. Nanomedicine 12, 3865–3879.

CHAPTER 28

Enzymatic synthesis of silver nanoparticles: Mechanisms and applications

Anindita Behera[a], Sweta Priyadarshini Pradhan[a], Farah K. Ahmed[b], and Kamel A. Abd-Elsalam[c]

[a]*School of Pharmaceutical Sciences, Siksha 'O' Anusandhan Deemed to be University, Bhubaneswar, Odisha, India* [b]*Biotechnology English Program, Faculty of Agriculture, Cairo University, Giza, Egypt* [c]*Plant Pathology Research Institute, Agricultural Research Center (ARC), Giza, Egypt*

1 Introduction

In the current scenario, the need for nanotechnology has been urged due to its universal application in the field of electronics, biosensors, diagnosis, treatment, and many more. The term "nano" signifies a structure with a dimension of 10^{-9}th of a meter and was first introduced by Norio Taniguchi of the Tokyo University of Science in 1974 following the concept of Nobel laureate physicist Richard Feynman that "There's plenty of room at the bottom." The concept described the possibilities of synthesis by direct manipulation of atoms (Hochella, 2002; Sharma et al., 2019). Nanoparticles are superior in their properties as compared to the bulk material from which they are made (Behera et al., 2020). Recently, nanomaterials have been applied for their significant unique characteristic properties, architecture, and functions in comparison to conventional formulations (Padhi et al., 2015). The uniqueness in the properties of nanomaterials is due to small particle size, high surface area, and stability in colloidal state with improved solubility, pharmacokinetic and pharmacodynamic profiles, and reduced side effects with enhanced therapeutic efficacy (Padhi et al., 2018). Metal nanoparticles (NPs) are considered to be most useful as compared to other nanostructures due to increased strength, flexibility, electromagnetic property, and surface to volume ratio. Metal NPs can be categorized as magnetic or noble metal NPs depending on the metal involved in their synthesis. Magnetic NPs are generally made up of iron, cobalt, or nickel, whereas noble metal NPs are made up of silver, gold, rhenium, ruthenium, palladium, or platinum (Behera et al., 2020).

Silver nanoparticles are the oldest known metal NPs, having excellent antimicrobial properties, and hence are applied in diverse fields of nanofabrication, health care, agriculture, and waste management (Zhang et al., 2016; Krutyakov et al., 2008). Among the noble metal NPs, silver NPs (AgNPs) are superior as they are nonpoisonous, with high electrical and thermal stability. Surface plasmon propagation is negligible for AgNPs during the change in the optical frequency (Gong et al., 2009). AgNPs are stable in both ambient condition and in different chemical environments. The process for preparation of AgNPs is cheaper and easily available as compared to gold and platinum NPs.

In synthesis of metal nanoparticles, it is a great challenge to get stable NPs. Physical and chemical methods used for synthesis of metal nanoparticles result in short-lived NPs, and use of toxic reducing and stabilizing agents produces toxic intermediates (Tyagi et al., 2020). Consequently, green synthesis is preferred due to its simplicity, easy and fast synthesis, easy scale up with no or very little toxicity, and low cost (Jain et al., 2020). Green synthesis of silver nanoparticles of controlled size involves optimization of various factors like selection of a suitable biological system for reduction and stabilization, along with pH, temperature of reaction medium, ratio of biomass to metal ion concentration, reaction time, rate of agitation, etc. (Garg et al., 2020). Green synthesis of silver NPs by plant extracts or with microorganisms is favored due to the presence of different phytoconstituents and enzyme systems, which act as both reducing and stabilizing agents (Roy et al., 2019). Although we find extensive literature for biological methods for synthesis of different NPs, this chapter deals with enzyme-based synthesis of AgNPs and its application in different fields.

2 Methods of preparation of silver nanoparticles

"Top-down" and "bottom-up" methods are two approaches adopted for preparation of nanoparticles. In top-down methods, the bulk material is subjected to fragmentation by application of external force by different physical methods like pulsed laser ablation, evaporation-condensation method, pulsed wire discharge method, etc. (Behera et al., 2020; Rafique et al., 2016). In bottom-up methods, chemical or biological methods are involved at atomic or molecular level, in which the size and shape of the nanoparticles can be controlled by controlling the reaction conditions, whereas in top-down methods, the control on size and shape can't be achieved resulting in heterogeneous nanoparticles (Behera et al., 2020); however, the speed of nanoparticle formation in bottom-up methods is slower as compared to top-down methods.

Overall, the methods used in preparation of AgNPs can be classified as physical, chemical, photochemical, and biological methods. Physical methods utilize some external energies like thermal energy, alternating current, or arc discharge, etc. Among all the physical methods, evaporation-condensation is the most used method for synthesis of AgNPs, however, it has limitations, such as use of large tube furnace, energy requirement to establish the required temperature, and long time

(Tran et al., 2013). To get the specific size and shape of AgNPs, different physical methods have been developed, as listed in Table 1.

Chemical methods of synthesis of AgNPs are performed either in colloid solution in aqueous or organic solvent medium using reducing and capping agents. In presence of reducing agents, the metal precursor of silver gets reduced from Ag^+ to Ag^0 oxidation state, whereas stabilizing or capping agents keep the AgNPs in dispersed state. Optimization and choice of reducing and capping agents can produce AgNPs of required dimension and dispersion (Pillai and Kamat, 2003). The shape, size, and dispersion affect the optical properties of AgNPs, and ultimately, its application. Although the chemical methods of synthesis of AgNPs are advantageous due to low cost, ease of synthesis, and high yield, they compromise the purity of the NPs due to capping of NPs' surface with chemical stabilizers. Sometimes the chemical reducing and capping agents may show toxicity to living organisms (Zhang et al., 2016). During the synthesis of AgNPs, the nucleation of AgNPs seeds is followed by growth. The rate of synthesis is controlled by temperature of the reaction medium, pH, concentration and type of metal precursor, reductants, and a stabilizer (Dakal et al., 2016). A wide range of reducing agents are chosen for synthesis of AgNPs, including glucose, hydrazine, hydrazine hydrate, ascorbic acid, citrate (Turkevich

Table 1 Physical methods used for preparation of silver nanoparticles.

Methods used for preparation of silver nanoparticles	Size of silver nanoparticles (nm)	References
Thermal decomposition method	9.5; <10; 5–15; 3–4.5; 7–16	Lee and Kang (2004), Navaladian et al. (2006), Goudarzi et al. (2016), Jeevanandam et al. (2010), and Togashi et al. (2020)
Heating and evaporation of pure silver on small ceramic heater	6.2±1.23–21.5±1.88	Jung et al. (2007)
Electrical discharge arc method	10; 33, 220; 23.45±0.16–26.83±0.44, 8–12	Tien et al. (2008), Zhang et al. (2017), El-Khatib et al. (2020), and Jabłońska et al. (2021)
Sputtering of metals directly	3.5±2.4; 2.7–7.4	Siegel et al. (2012), Ishida et al. (2017), Staszek et al. (2017)
Sputtering without catalyst at controlled temperature 200–300°C	Nanorods of diameter 40–55 nm	Hu et al. (2012)
Laser ablation technique	2–5; 8, 20±3	Pyatenko et al. (2004), Sadrolhesseini et al. (2013), Mostafa and Menazea (2020), and Sportelli et al. (2018)
Pulsed photoacoustic method	10	Valverde-Alva et al. (2015)

method) (Turkevich and Kim, 1970; Turkevich, 1985), and sodium borohydride (Brust-Schiffrin synthesis or BSS method) (Goulet and Lennox, 2010). Sodium borohydride, being a strong reducing agent, forms smaller NPs at a faster rate, whereas citrates produce NPs slowly, hence the size and shape of NPs can be controlled (Bai et al., 2007).

Stabilizing or capping agents prevent the agglomeration of AgNPs by adsorbing on the surface of NPs. Commonly used capping agents include chitosan, oleyl amine gluconic acid, cellulose, or polymers like polyvinylpyrrolidone (PVP), polyethylene glycols (PEG), and polymethacrylic acid (PMAA) (Tylkowski et al., 2017). Stabilizing agents also prevent oxidation of AgNPs by either electrostatic or steric repulsion (Lee and Jun, 2019). The size and shape of the NPs also depend on type of surfactant or stabilizer that regulates the crystal lattice pattern. Various shaped AgNPs synthesized are sphere shaped, cubic, nanorods, nanowires, nanobars, pyramidal, flower shaped, etc. (Lee and Jun, 2019; Khodashenas and Ghorbani, 2019). Polyol method is one of the chemical methods where ethylene glycol is used as both solvent and reducing agent and polyvinylpyrrolidine (PVP) is commonly used as stabilizing agent (Chouhan and Meena, 2015). Precursor injection technique is another chemical method, which produces small and uniform sized AgNPs. The factors affecting the size and shape of AgNPs produced in this method are rate of injection and the temperature at which the process is done. In this method, precursor solution is injected into a hot solution, which facilitates the nucleation of NPs seeds in a short time. This results in formation of small and narrow size-distribution of AgNPs (Chouhan and Meena, 2015). The different reports for synthesis of different shaped AgNPs are illustrated in Table 2.

The photochemical method is a bottom-up method for synthesis of AgNPs, in which the bulk metal ion (Ag^+) is converted to zero oxidation state (Ag^0) by exposure to energy of photons from sunlight or laser radiation. In this method, the power of irradiation affects the size of AgNPs (Kshirsagar et al., 2011). A photoreducing agent or a photoactive species transfers an electron to Ag^+ and reduces it to Ag^0 at a specific wavelength (Zaarour et al., 2014). The AgNPs synthesized by this method are pure with high three-dimensional resolution and can be synthesized in various mediums like micelles, emulsions, glasses, cells, etc. (Sakamoto et al., 2009).

Apart from physical, chemical, and photochemical methods, AgNPs are preferred to be synthesized by biological methods. The biological methods have certain advantages over the abovementioned methods, such as lower cost, ease of synthesis, and no use of toxic reducing and capping agents. Two approaches are adopted for biosynthesis of AgNPs, the first is the use of microorganisms like algae, fungi, yeast, or bacteria, and the second is by use of extracts of different parts of plants, as represented in Fig. 1 (Behera et al., 2020).

Biosynthesis of AgNPs by microorganisms like fungi, algae, yeast, and bacteria involves selection of the correct strain of microorganism and use of a suitable medium for the growth and production of metal nanoparticles inside or outside the cell. Accordingly, the biosynthetic methods are termed as "intracellular" and "extracellular" methods of synthesis. The intracellular synthesis involves the

Table 2 Preparation of different shaped silver nanoparticles by chemical reduction method.

Shape of Ag nanoparticles	Metal precursor	Reducing agent used	Surfactant/ stabilizer used	Size (nm)	References
Sphere shaped	Silver nitrate	Sodium borohydride	PVP	9	Li et al. (2012b)
		Trisodium citrate	Sodium borohydride	5–50	Quintero-Quiroz et al. (2019)
		Trisodium citrate and tannic acid		30	Ranoszek-Soliwoda et al. (2017)
		Trisodium citrate/sodium borohydride/ ascorbic acid	Sodium borohydride		Rashid et al. (2013)
		PEG	PVP	54	Liang et al. (2010)
		Ascorbic acid	Citrate	Decrease in size 73 to 31	Qin et al. (2010)
		Glucose	Gelatin	Less than 20 (average size 5–24)	Aguilar-Méndez et al. (2010)
		Ascorbic acid	Dodecylbenzene sulfonic acid sodium and PVP		Moghimi-Rad et al. (2011)
		Sodium thiosulfate	Cetyltrimethyl ammonium bromide (CTAB) and polyvinyl alcohol (PVA)	Without stabilizer – nanorods With CTAB – sphere shaped With PVA – hexagonal	ALThabaiti et al. (2013)

Continued

Table 2 Preparation of different shaped silver nanoparticles by chemical reduction method—cont'd

Shape of Ag nanoparticles	Metal precursor	Reducing agent used	Surfactant/ stabilizer used	Size (nm)	References
Cubical	Silver nitrate	Ethylene glycol	PVP	30–50	Sun et al. (2020) Young et al. (2006) Tao et al. (2006)
		Ethylene glycol with pentanediol	PVP	High molar ratio produced nanocubes and lower molar ratio produced nanowires	Skrabalak et al. (2007)
	Silver trifluoro acetate (CF_3COOAg)	Ethylene glycol	PVP with traces of sulfide	Edge length of 30–70	Zhang et al. (2010)
Nanorods	Silver nitrate ($AgNO_3$)	Potassium tartaric acid	PVP		Gu et al. (2006)
		Acetaldehyde		Diameter of nanorods depends on PAM template and length depends on reaction temperature	Xu et al. (2010)
	$AgNO_3$	Ethylene glycol	PVP	Diameter 30–40 nm and length 50 µm	Jana et al. (2001)
		Trisodium citrate	Sodium dodecylsulfonate (SDSN)	Diameter 150–200 nm and length 50–100 µm	Hu et al. (2004)
Nanowires		Ethylene glycol	PVP	Pentagonal nanowires	Jiang et al. (2004) Wiley et al. (2006), Hsieh et al. (2012), and Chen et al. (2011)
				The concentration of $AgNO_3$ was adjusted to	Zhang et al. (2008) and Nghia et al. (2012)

Shape	Precursor	Reducing agent	Capping agent	Notes	Reference
Triangular/pyramidal AgNPs	Silver nitrate (AgNO$_3$)	PEG	PVP	maintain the diameter of the nanowires 100–300	Cong et al. (2012)
		Ethylene glycol	Different molecular weight PVP	Branched nanowires with Y and K shaped	Zhu et al. (2011)
		Sodium borohydride (NaBH$_4$) and sodium citrate		Increase in molecular weight of PVP increased the yield of nanowires	Dong et al. (2010)
		NaBH$_4$, sodium citrate, Ascorbic acid, PVA		Triangular AgNPs	Kelly et al. (2012)
		Hydrazine hydrate	PVP		Li et al. (2012a,b)
Nanoprism	AgNO$_3$	Ethylene glycol monoethyl ether	PVP	Triangular NPs with edge length 50–200	Darmanin et al. (2012)
Flower shaped NPs	AgNO$_3$	Ascorbic acid	PVP		Cai and Zhai (2010)
		Ascorbic acid	CTAB		Zaheer (2011)

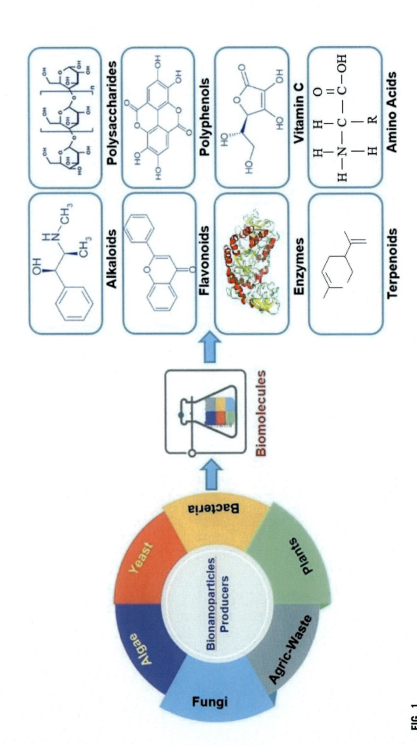

FIG. 1

Various phytochemicals and microbial biomolecules that serve as capping, reducing, and stabilizing agents in the fabrication of silver nanoparticles. For example, polysaccharides, phenolic compounds, proteins, enzymes, vitamins, nucleic acid, and other chelating agents are used.

synthesis of nanoparticles within the cells of microorganisms and the reduction and stabilization of metal NPs is brought about by the intracellular proteins or enzymes responsible for different metabolic pathways required for the survival of the microorganisms. This method requires no pretreatment of the microorganisms. The enzymes participating in different electron transport systems (ETS) or existing in different cell organelles and cell walls participate in intracellular synthesis of NPs. The limitation of intracellular synthesis of AgNPs is the recovery and purification of NPs after the biosynthesis, which is a tedious process. In extracellular synthesis, different by-products of cell metabolism, like metabolites, ions, pigments, hormones, numerous proteins (enzymes), and nonproteins like DNA and RNA, participate in the biosynthesis of AgNPs (Khanna et al., 2019). The extracellular method of biosynthesis requires pretreatment of the microorganisms, such as washing, drying, and blending. Table 3 represents the use of different microorganisms for synthesis of AgNPs.

Green synthesis is the basis of biosynthesis of AgNPs using microorganisms or plant extracts, which is both economic and environment friendly. But a few key factors regulate the formation of NPs, like choice of a suitable reaction medium, selection of biological reducing agent, and choice of nontoxic capping or stabilizing agents (El-Shishtawy et al., 2011). For green synthesis of NPs, mainly aqueous medium is required and other conditions like temperature and pH of the medium affect the morphology of NPs. So, the reaction conditions must be optimized to synthesize NPs of desired size and shape.

Biosynthesis of metal nanoparticles takes place via three steps (Makarov et al., 2014):

(i) *Activation phase* involves the reduction of metal ion (M^{n+}) by bioreductant and nuclei formation of reduced metal is initiated (M^0);
(ii) *Growth phase* involves fast coagulation of smaller NPs to grow as larger NPs with increase in thermodynamic stability by Ostwald ripening;
(iii) *Termination phase* includes the final formation of the shape of the NPs.

Although plant extracts have been used extensively for synthesis of AgNPs, microorganism-based synthesis is also getting encouraging outcomes and is an interesting challenge for biochemists. Microorganisms used for biosynthesis of AgNPs are algae, fungi, bacteria, and yeasts, as represented in Fig. 2. The mechanism of reduction of metal ions M^{n+} to M^0 requires some enzymes, surface proteins, and metabolites, which can cause the reduction and stabilization of the synthesized NPs (Almatroudi, 2020). Enzymes play a major role in bioreduction and stabilization of metal NPs. The synthesis of AgNPs may occur by extracellular process or intracellularly. The extracellular synthesis of AgNPs depends on the enzymes or proteins secreted or extracted from the microorganisms, whereas the intracellular mechanism of biosynthesis of AgNPs involves the reduction of metal ions within the cellular structure of the microorganisms (Mikhailova, 2020).

(i) *Biosynthesis of AgNPs by algae*: Algae perform photosynthesis to obtain their nutritional requirements (Choudhary et al., 2016). The cellular reductase in

Table 3 Synthesis of silver NPs with microbial biosynthesis.

Name of the microorganism	Size of nanoparticles (nm)	Shape of the nanoparticles	Extracellular/intracellular	References
Acinetobacter calcoaceticus	8–12	Sphere shaped	Extracellular	Singh et al. (2013)
Acinetobacter haemolyticus	4–40		Extracellular	Gaidhani et al. (2014)
Aeromonas sp.	6.4		Extracellular	Wang et al. (2012)
Bordetella sp.	63–90		Extracellular	Thomas et al. (2012)
Enterobacter erogenes	25–35	Sphere shaped	Extracellular	Karthik and Radha (2012)
Gluconobacter roseus	10		Intracellular	Krishnaraj and Berchmans (2013)
Idiomarina sp.	25		Intracellular	Seshadri et al. (2012)
Klebsiella pneumonia	15–37	Sphere shaped	Extracellular	Kalpana and Lee (2013)
Morganella sp.	10–40	Quasispherical	Extracellular	Parikh et al. (2008)
Proteus mirabilis	10–20	Sphere shaped	Extracellular and intracellular	Samadi et al. (2009)
Pseudomonas aeruginosa	6.3 ± 4.9	Sphere shaped, disk shaped	Extracellular and intracellular	Srivastava and Constanti (2012)
	8–24	Sphere shaped	Extracellular	Kumar and Mamidyala (2011)
Pseudomonas deceptionensis DC5	10–30	Sphere shaped	Extracellular	Jo et al. (2016)
Pseudomonas sp. THG-LS1.4 strain	10–40	Irregular shaped	Extracellular	Singh et al. (2018)
Rhodobacter sphaeroides	3–15	Sphere shaped	Extracellular	Bai et al. (2011)
Stenotrophomonas maltophilia	93	Cuboidal	Extracellular	Oves et al. (2013)

Vibrio alginolyticus	50–100	Sphere shaped	Extracellular and intracellular	Rajeshkumar et al. (2013)
Xanthomonas oryzae	14.86	Sphere, triangle, and rod-shaped	Extracellular	Narayanan and Sakthivel (2013)
Yersinia enterocolitica	10–80	Sphere shaped	Extracellular	Pourali et al. (2012)
Bacillus flexus	12 and 65	Sphere shaped and triangular	Extracellular	Priyadarshini et al. (2013)
Bacillus licheniformis Dahb1	18.69–63.42	Sphere shaped	Extracellular	Shanthi et al. (2016)
Bacillus safensis	5–30	Sphere shaped	Extracellular	Lateef et al. (2015)
Bacillus sp. MB353	49–53	Crystalline	Extracellular	Khan et al. (2020)
Bacillus subtilis spizizenni (NCIM 2063)	20–70	Quasispherical	Extracellular and intracellular	Sable et al. (2020)
Bacillus brevis (NCIM 2533)	41–68	Sphere shaped	Extracellular	Saravanan et al. (2018)
Exiguobacterium sp.	5–50	Sphere shaped	Extracellular	Tamboli and Lee (2013)
Geobacillus stearothermophilus	5–35	Sphere shaped	Extracellular	Fayaz et al. (2011)
Lactobacillus sp.	3–35	Sphere shaped	Extracellular	Matei et al. (2020)
Lactobacillus reuteri	40–90	Quasispherical	Extracellular	Tharani et al. (2020)
Sphingobium sp.	7–22	Sphere shaped	Extracellular	Akter and Huq (2020)
Sporosarcina koreensis	30–50	Sphere shaped	Extracellular	Singh et al. (2016)
Escherichia coli, Exiguobacterium aurantiacumm, and Brevundimonas diminuta	5–50	Sphere shaped	Extracellular	Saeed et al. (2020)
Cupriavidus sp.	10–50	Sphere shaped	Extracellular	Ameen et al. (2020)
Enterococcus faecalis S7	10–16	Sphere shaped	Extracellular	El Shanshoury et al. (2020)
Pseudoduganella eburnea MAHUQ-39	8–24	Sphere shaped	Extracellular	Huq (2020)
Rhodococcus sp.	5–50	Sphere shaped	Intracellular	Otari et al. (2015)
Proteus mirabilis	10–20	Sphere shaped	Extracellular and intracellular	Samadi et al. (2009)
Idiomarina sp. PR58-8	26	Sphere shaped	Intracellular	Seshadri et al. (2012)

Continued

Table 3 Synthesis of silver NPs with microbial biosynthesis—cont'd

Name of the microorganism	Size of nanoparticles (nm)	Shape of the nanoparticles	Extracellular/intracellular	References
Fungus-mediated biosynthesis				
Aspergillus flavus	8.92	Sphere shaped	Intracellular	Vigneshwaran et al. (2007)
Aspergillus terreus	1–20; ~25	Sphere shaped	Extracellular	Li et al. (2012a) and Kumari et al. (2020)
Aspergillus niger and Fusarium semitectum		Sphere and ellipsoid shaped	Extracellular	Madakka et al. (2018)
Coriolus versicolor	25–75, 444–491	Sphere shaped	Intracellular	Sanghi and Verma (2009)
Humicola sp.	5–25	Sphere shaped	Extracellular	Syed et al. (2013)
Macrophomina phaseolina	5–40	Sphere shaped	Extracellular	Chowdhury et al. (2014)
Penicillium fellutanum	5–25	Sphere shaped	Extracellular	Kathiresan et al. (2009)
Penicillium nalgiovense	25 ± 2.8	Sphere shaped	Extracellular	Maliszewska et al. (2014)
Phaenerochaete chrysosporium	5–200	Pyramidal	Extracellular	Vigneshwaran et al. (2006)
Phoma glomerata	60–80	Sphere shaped	Extracellular	Birla et al. (2009)
Pleurotus ostreatus	<40	Sphere shaped	Extracellular	Al-Bahrani et al. (2017)
Trichoderma asperellum	13–18	Crystalline	Extracellular	Mukherjee et al. (2008)
Trichoderma harzianum	57–82	Sphere shaped	Extracellular	Guilger-Casagrande et al. (2019)

Trichoderma sp.	150–200	Sphere shaped	Extracellular	Ramos et al. (2020)
Rhizopus stolonifer	3–10	Sphere shaped	Extracellular	AbdelRahim et al. (2017)
Sporotrichum thermophile	40–70	Sphere shaped	Extracellular	Shankar et al. (2020)
Tritirachium oryzae W5H	5–22	Sphere shaped	Extracellular	Qaralleh et al. (2019)
Beauveria bassiana	10–50	Triangular, circular, hexagonal	Extracellular	Tyagi et al. (2019)
Schizophyllum commune	54–99	Sphere shaped	Intracellular	Arun et al. (2014)
Meyerozyma guilliermondii KX008616	2.5–30	Sphere shaped	Intracellular	Alamri et al. (2018)
Fusarium oxysporum	25–50	Sphere shaped	Intracellular	Korbekandi et al. (2013)
Cyanobacteria (blue-green algae)-mediated biosynthesis				
Cylindrospermum stagnale	38–88	Pentagonal	Extracellular	Husain et al. (2015)
Microchaete	60–80	Sphere shaped	Extracellular	Husain et al. (2019)
Oscillatoria limnetica	3.3–18	Quasispherical	Extracellular	Hamouda et al. (2019)
Green algae mediated biosynthesis				
Chlorella vulgaris	8–20		Intracellular	Satapathy et al. (2015)
Euglena intermedia	6–24	Sphere shaped	Intracellular	Li et al. (2015)
Ulva fasciata	28–41	Sphere shaped	Extracellular	Rajesh et al. (2012)
	7–20	Sphere shaped	Extracellular	El-Rafie et al. (2013)
Chaetomorpha linum	3–44		Extracellular	Kannan et al. (2013)
Caulerpa serrulata	10 ± 2	Sphere shaped	Extracellular	Aboelfetoh et al. (2017)
Ulva armoricana	Variable	Sphere shaped	Extracellular	Massironi et al. (2019)
Chlamydomonas reinhardtii	3–25	Sphere shaped	Extracellular	Rahman et al. (2019)
Desmodesmus sp.	15–30	Sphere shaped	Intracellular	Dağlıoğlu and Öztürk (2019)
	3–30	Sphere shaped	Extracellular and intracellular	Öztürk (2019)

Continued

Table 3 Synthesis of silver NPs with microbial biosynthesis—cont'd

Name of the microorganism	Size of nanoparticles (nm)	Shape of the nanoparticles	Extracellular/intracellular	References
Red algae-mediated biosynthesis				
Gracilaria edulis	12.5–100	Sphere shaped	Extracellular	Murugesan et al. (2011)
Acanthophora spicifera	48	Sphere shaped	Extracellular	Kumar et al. (2012a)
Pterocladia capillacae	7	Sphere shaped	Extracellular	El-Rafie et al. (2013)
Jania rubins	12	Sphere shaped	Extracellular	El-Rafie et al. (2013)
Gracilaria birdiae	20.3	Sphere shaped	Extracellular	de Aragao et al. (2019)
Brown algae-mediated biosynthesis				
Sargassum wightii grevilli	8–27	Sphere shaped	Extracellular	Govindaraju et al. (2009)
Sargassum ilicifolium	33–40	Sphere shaped	Extracellular	Kumar et al. (2012b)
Sargassum plagiophyllum	20–50	Sphere shaped	Extracellular	Dhanalakshmi et al. (2012)
Sargassum vulgare	10	Sphere shaped	Extracellular	Govindaraju et al. (2015)
Turbinaria conoides	2–17	Sphere shaped	Extracellular	Krishnan et al. (2015)
Sargassum muticum	43–79	Sphere shaped	Extracellular	Madhiyazhagan et al. (2015)
Hydroclathrus clathratus	7–83	Irregular shaped	Extracellular	Alzahrani et al. (2020)
Padina sp.	25–60	Sphere shaped	Extracellular	Bhuyar et al. (2020)

2 Methods of preparation of silver nanoparticles

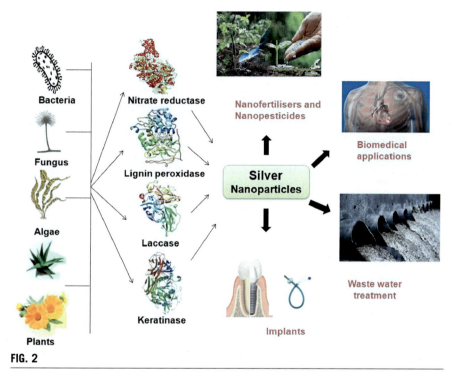

FIG. 2

Schematic representation of green synthesis of silver nanoparticles using various enzymes extracted from different biological sources and their applications.

different types of algae causes the bioreduction of Ag^+ to Ag^0 and subsequently formation of AgNPs. Table 3 illustrates the reports of extracellular and intracellular methods used for biosynthesis of AgNPs from cyanobacteria or blue-green algae, green, red, and brown algae, etc. (Govindaraju et al., 2009; Murugesan et al., 2011; Kumar et al., 2012a,b; Dhanalakshmi et al., 2012; Rajesh et al., 2012; El-Rafie et al., 2013; Kannan et al., 2013; Husain et al., 2015; Satapathy et al., 2015; Li et al., 2015; Govindaraju et al., 2015; Krishnan et al., 2015; Madhiyazhagan et al., 2015; Aboelfetoh et al., 2017; Husain et al., 2019; Hamouda et al., 2019; Massironi et al., 2019; de Aragao et al., 2019).

(ii) *Biosynthesis of AgNPs by fungi*: Fungi have a very strong potential for biosynthesis of AgNPs even in the presence of different abiotic and biotic stress conditions. Fungi can synthesize AgNPs by both intracellular and extracellular enzymes (Khandel and Shahi, 2018). The intracellular method of biosynthesis of AgNPs involves binding of Ag^+ to surface enzymes or proteins present on the cell wall due to electrostatic attraction and the enzymes bring about the reduction of Ag^+ to Ag^0 for synthesis of AgNPs. Whereas in extracellular synthesis, the enzymes secreted by fungi cause the reduction and stabilization of AgNPs. Both intra- and extracellular methods of biosynthesis of AgNPs by fungi are ecofriendly and can be easily used for large scale production of AgNPs (Siddiqi et al., 2018).

(iii) *Biosynthesis of AgNPs by bacteria*: Prokaryotic bacteria can also be used for intracellular and extracellular biosynthesis of AgNPs. The mechanism of intracellular synthesis is similar to fungi, i.e., the metal ions bind to cell wall surface enzymes and undergo reduction for formation of AgNPs, whereas extracellular biosynthesis involves the interaction of metal ions with enzymes secreted by bacterial cells outside the cells (Ovais et al., 2018). AgNPs synthesized by the intracellular mechanism are accumulated in the periplasm, cytoplasm, and cell wall components of bacteria, so are difficult for recovery. Extracellular synthesis of AgNPs by bacteria can be easily recovered (Siddiqi et al., 2018).

The key ingredients participating in the biosynthesis of AgNPs using microorganisms are the intracellular and extracellular enzymes like nitrate reductase, peroxidise, laccase, cellulose, keratinase, and xylanase. The following section discusses the mechanisms involved and reports on enzyme-assisted biosynthesis of AgNPs.

3 Enzyme-assisted biosynthesis of silver nanoparticles

One of the most recent advances in nanotechnology is the use of enzymes to synthesize various types of nanoparticles. In most cases, active enzymes catalyze nanoparticle formation, but in some cases, enzymes are denatured to release amino acids, which act as reducing and stabilizing agents during nanoparticle synthesis. Biological methods of synthesis of NPs can be achieved by use of plant extracts that contain phytomolecules, which act as active reducing and stabilizing agents, whereas microorganisms like algae, fungi, yeasts, and bacteria contain different enzymes, which may be released extracellularly, or some electron shuttles present on the cell wall and the intracellular enzymes cause reduction and stabilization of synthesized NPs (Oza et al., 2020).

Microbe-assisted biosynthesis of silver nanoparticles takes place by entrapment of toxic ions into the cellular compartments and cell organelles. The metal ions interact within the cell via any one of the molecules like glutathione, phytochelatins, or cysteine enriched metallothioneins (Oza et al., 2020). Apart from these, the enzymes present inside the cells of the microorganism and excreted to the outer environment bring about reduction of silver ion and synthesize the metal nanoparticles. Mostly NADH-dependent nitrate reductase enzyme, i.e., an electron shuttle, transfers the electrons from the nitrate molecule to the Ag^+ and forms the AgNPs (Ag^0). Other enzymes involved in extracellular and intracellular synthesis of AgNPs are keratinase, lignin peroxidase, laccase, cellulase, and xylanase (Table 4).

(i) *NAD(P)H-dependent nitrate reductase*: NAD(P)H-dependent nitrate reductase is a type of oxidoreductase enzyme. In synthesis of AgNPs, the nitrate reductase causes reduction of Ag^+ to Ag^0, which is simultaneously

Table 4 Molecular formula of some extracellular and intracellular enzymes involved in the synthesis of AgNPs.

Name of the enzyme	Chemical structure of the enzyme
Nitrate reductase EC 1.5.1.20—methylenetetrahydrofolate reductase [NAD(P)H]	

Continued

Table 4 Molecular formula of some extracellular and intracellular enzymes involved in the synthesis of AgNPs—cont'd

Name of the enzyme	Chemical structure of the enzyme
Laccase, $C_{44}H_{69}N_{11}O_{20}$	

EC 1.11.1.14—lignin peroxidase

β-cellotriose, C₁₈H₃₂

Xylanase, C₃₄H₅₀

Continued

Table 4 Molecular formula of some extracellular and intracellular enzymes involved in the synthesis of AgNPs—cont'd

Name of the enzyme	Chemical structure of the enzyme
EC 3.2.1.1 alpha amylase $C_9H_{14}N_4O_3$	

3 Enzyme-assisted biosynthesis of silver nanoparticles

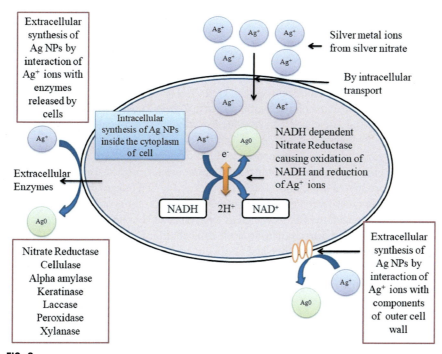

FIG. 3

Mechanism of extracellular and intracellular enzymatic synthesis of silver nanoparticles.

accompanied with oxidation of nitrate to nitrite catalyzed by NADH to NAD^+, as illustrated in Fig. 3 (Roy et al., 2019).

Mukherjee et al. reported intracellular biosynthesis of AgNPs using *Bacillus clausii*, in which the nitrate reductase enzyme present in the periplasm or inner cell wall helps in reduction of silver ions and formation of AgNPs. The AgNPs formed were spherical shaped with an average size of 150 nm and computational profiling of the enzyme was performed (Mukherjee et al., 2018).

El-Bendary et al. reported extracellular synthesis of variable shaped AgNPs, including crystalline, spherical, cuboid, hexagonal, rod, and anisotropic shaped, using the cell-free extract of mosquitocidal strains of *Lysinibacillus sphaericus* and *Bacillus thuringiensis israelensis*. The variable shaped AgNPs have an average size of 15–21 nm and showed significant antimicrobial activities against mosquitoes, bacteria, fungi, and viruses (El-Bendary et al., 2020). Sable et al. reported both extracellular and intracellular synthesis of AgNPs using *Bacillus subtilis spizizenni*. The nitrate reductase assay proved the involvement of the enzyme nitrate reductase in the intracellular and extracellular synthesis of AgNPs (Sable et al., 2020). Popli et al. reported an extracellular synthesis of AgNPs using the aqueous extract of an endophytic fungi *Cladosporium* species. The antioxidant activity study by

DPPH radical scavenging assay showed the involvement of NAD(P)H-dependent nitrate reductase in the biosynthesis of AgNPs (Popli et al., 2018). Hamedi et al. reported the nitrate reductase involvement in the extracellular synthesis of AgNPs using the cell-free extract of a fungus *Fusarium oxysporum*. The activity of nitrate reductase in biosynthesis of AgNPs was assayed by Harley method (Hamedi et al., 2017). Zomorodian et al. performed a comparative extracellular synthesis study of AgNPs using different species of *Aspergillus*. Efficacy of nitrate reductase enzyme in extracellular synthesis of AgNPs by *Aspergillus fumigatus*, *Aspergillus flavus*, *Aspergillus clavatus*, and *Aspergillus niger* were compared and the maximum efficacy was found with *Aspergillus fumigatus* as compared to minimum efficacy of *Aspergillus flavus* (Zomorodian et al., 2016). Talekar et al. used NADH-dependent nitrate reductase cross linked enzyme aggregates (CLEAs) for synthesis of AgNPs at pH 7.2, in which the CLEAs acted as catalysts and NADH acted as an electron source with 8-hydroxyquinoline as electron shuttle. The AgNPs synthesized by this method were crystalline and of dimension 5–7 nm and the CLAEs could be reused even after five cycles of synthesis of AgNPs (Talekar et al., 2016). The same group of scientists reported the preparation of CLEAs by immobilizing the NADH-dependent nitrate reductase from cell free extract of *Fusarium oxysporum* and use of these immobilized CLEAs for biosynthesis of AgNPs (Talekar et al., 2014).

(ii) *Cellulase*: Cellulases are the enzymes that hydrolyze β-1,4 linkages in cellulose chains. The hydrolysis of cellulose to glucose is a multistep process in which the enzyme plays an important role. Cellulase is produced by bacteria, fungi, protozoans, animals, and plants (Zhang and Zhang, 2013). This enzyme has also been used for biosynthesis of AgNPs. Mishra and Sardar isolated the enzyme cellulase from *Aspergillus niger* and used the same for biosynthesis of AgNPs and also produced immobilized cellulase on AgNPs. Cellulase activity in free enzyme and immobilized cellulase on AgNPs were measure by enzyme-substrate reaction, where caboxy methyl cellulose (CMC) was used as a substrate (Mishra and Sardar, 2015). Rai and Panda reported the cellulase-assisted biosynthesis of AgNPs isolated from *Trichoderma ressei* ATCC. The biosynthesis was carried out in both aqueous and methanolic medium. The study concluded that the enzyme contains 344 amino acid residues in which 8 are cysteine and 1 is methionine. So, addition of silver nitrate to alkaline medium causes coordination of Ag^+ to -NH, -SH and -OH groups of amino acid residues and reduction and formation of AgNPs (Rai and Panda, 2015).

(iii) *Laccase*: Laccase is an oxidoreductase type copper containing enzyme (1, 4—benzenediol oxygen oxidoreductases [EC 1.10.3.2]) found in higher plants and microorganisms (Jini and Rohiniraj, 2017). Laccase oxidizes the substrates, liberating one electron, which helps in reduction of Ag^+ to Ag^0 (Debnath and Saha, 2020). Jini and Rohiniraj isolated laccase enzyme from mushrooms (*Pleurotus ostreatus*, *Pleurotus florida*). The extracellular enzyme was purified and used for biosynthesis of AgNPs (Jini and Rohiniraj, 2017). Lateef and Adeeyo reported the extracellular synthesis of walnut-shaped,

50–100 nm AgNPs using laccase isolated from *Lentinus edodes*. The biosynthesized AgNPs showed significant antibacterial activity against *Escherichia coli*, *Pseudomonas aeruginosa*, and *Klebsiella pneumonia* (Lateef et al., 2018). Duran et al. reported extracellular biosynthesis of AgNPs by laccase isolated from *Trametes versicolor*. A complete characterization showed a complex of AgNPs with Ag^+ chemical structure. The research group proposed two possible mechanisms for synthesis of AgNPs; one involves the interaction of cysteine in laccase with Ag^+ ions, which oxidizes cysteine and reduced Ag^+ to Ag^0, whereas the second possible mechanism explained the direct interaction of Ag^+ with the Cys-Cys moiety of laccase, generating silver nanoparticles, and the sulfhydryl moiety bound to silver nanoparticles as capped proteins (Durán et al., 2014).

(iv) *Alpha amylase*: Alpha amylase enzyme catalyzes the hydrolysis of internal α-1,4-glycosidic linkages in polysaccharides. The enzymes involved in the biosynthesis of metal nanoparticles involve the interaction of thiol group of cysteine with the metal ion. The interaction results in reduction of Ag^+ to Ag^0 and formation of stable AgNPs (Mishra and Sardar, 2012). Mishra et al. reported the synthesis of AgNPs using alpha amylase enzyme. The AgNPs are monodispersing in nature with triangular or hexagonal shape of 22–44 nm (Mishra and Sardar, 2012). Mishra et al. reported the synthesis of AgNPs using neem leaves extract. Neem leaves contain both alpha and beta amylase, so after formation of AgNPs, 85% amylase activity was retained (Mishra et al., 2013). In considering five metal ions, including Cu^{2+}, Se^{4+}, Bi^{4+}, Au^{3+}, and Ag^+, gold nanoparticles (AuNPs) and silver nanoparticles (AgNP) were synthesized successfully by Moshfegh et al., who investigated the use of α-amylase as a reducing substance for the formation of metal nanoparticles. The synthesized AuNPs, AgNPs and Au/AgNPs showed maximum absorption at 530, 440, and 458 nm wavelengths. Their respective sizes were 89, 37, and 63 nm (Moshfegh et al., 2011).

(v) *Lignin peroxidase*: Peroxidase enzymes belong to oxidoreductase and act on peroxide as an electron acceptor or electron withdrawing groups and broadly categorized as ligninases. Ligninases or lignin-modifying enzymes are mostly produced by fungi and bacteria that help in the breakdown of lignin, found in cell walls of plants (Falade et al., 2017). Even so, there have been relatively few experiments on the synthesis of nanoparticles using the peroxidase enzyme. Some researcher like Singh et al. isolated purified lignin peroxidase from secreted enzyme extract of *Acinetobacter* sp. and biosynthesized spherical polydisperse AgNPs of ~50 nm size (Singh et al., 2017). Cicek et al. isolated peroxidase enzyme in partially purified form *Euphorbia* (*Euphorbia amygdaloides*) plant. The AgNPs synthesized using the peroxidase enzyme were spherical, well dispersed, with sizes of 7–20 nm, and significant antibacterial activity (Cicek et al., 2015).

(vi) *Xylanase*: Xylanase degrades the β-1,4 glycosidic linkages in polysaccharides and produce xylose and xylan (Kamble and Jadhav, 2012). Dhivahar et al. reported extracellular biosynthesis of AgNPs using the thermoalkali-stable xylanase from *Pseudomonas nitroreducens*. The dimension of xylanase-reduced AgNPs were spherical with an average particle size of 100 nm (Dhivahar et al., 2020). Elegbede et al. reported the biosynthesis of AgNPs using fungal xylanase isolated from *Aspergillus niger* and *Trichoderma longibrachiatum*. The enzyme-assisted synthesis produced spherical, cylindrical, and oval shaped AgNPs of size 15–78 nm (Elegbede et al., 2018).

(vii) *Keratinase*: Keratinase is a universal protease that is present in hair, nails, feathers, and skin of animals, birds, and human beings. The enzyme is gaining importance due to its application in various fields of biotechnology, pharmaceuticals, and nanotechnology (Nnolim et al., 2020). Jang and his team reported the extracellular synthesis of AgNPs using the enzyme keratinase from *Stenotrophomonas maltophilia* R13 at pH 9 and produced spherical AgNPs of size ~8.4 nm. The keratinase-assisted AgNPs showed significant antimicrobial, antioxidant, and anticollagenase activity (Jang et al., 2018). Tao et al. reported the expression of keratinase enzyme mined from soil metagenomes in *Bacillus subtilis* and used the same for biosynthesis of AgNPs (Tao et al., 2018). Lateef et al. reported the synthesis of AgNPs using the keratinase isolated from cell-free extract of *Bacillus safensis* LAU 13. The spherical AgNPs formed with size range 5–95 nm showed anticandida, anticoagulant, and thrombolytic activity (Lateef et al., 2016).

The presence of thiol groups and disulfide bridge moieties of enzymes as reaction sites for nanoparticle formation was demonstrated by Durán et al. (2014). Denatured enzymes' S-S and S-H moieties converted metallic ions to nanoparticles. As a result, both enzymatic and nonenzymatic processes were employed to produce different types of nanoparticles (Table 5).

Table 5 Green synthesis of silver nanoparticles by some enzymes.

Enzyme	Reaction condition	Type of NPs	Size (nm)	NPs shape	References
α-Amylase	40 mL α-amylase solution + 60 mL of 1 mM AgNO$_3$ at 25°C	AgNPs	22–44	Triangular and hexagonal	Mishra and Sardar (2012)
	2 mL of enzyme solution + HAuCl$_4$/AgNO$_3$ of mole ratio 1:1 at 70°C/40°C for 48 h	Au/AgNPs	63	Spherical	Moshfegh et al. (2011)
	40 mL of α-amylase solution (in phosphate buffer of	AgNPs	63–142	Cubic	Pandey et al. (2018)

Table 5 Green synthesis of silver nanoparticles by some enzymes—cont'd

Enzyme	Reaction condition	Type of NPs	Size (nm)	NPs shape	References
Cellulase	pH 6.8) + 60 mL of AgNO$_3$ at 30°C				
	10 mL cellulase + 90 mL 1 mM AgNO$_3$/1 Mm HAuCl$_4$ at 25°C	Ag/AuNPs	5–25, 5–20	Spherical, cubic	Mishra and Sardar (2015)
	100 μL of 5 mg/mL cellulase + 10% glycerol + 10 mL of 0.5 mm AgNO$_3$ for 24 h	AgNPs	5.04 ± 3.50	Spherical	Rai and Panda (2015)
Nitrate reductase	Nitrate reductase + 10 mL of 1 mM AgNO$_3$ + 1 mM 8-hydroxyquinoline + 1 mm NADH at pH 7.2, 35°C and 150 rpm	AgNPs	10–20	Spherical	Talekar et al. (2014)
	Microbial filtrate + 1 mM AgNO$_3$ incubated at 28 ± 2°C for 48 h under light	AgNPs	10–33	Spherical, oval, cuboid	El-Bendary et al. (2020)
Keratinase	1 mL of keratinase (50 U/mL + 50 mL of 1 mM AgNO$_3$ at 30 ± 2°C for 2 h	AgNPs	5–30	Spherical	Lateef et al. (2015)
	50 mL 1 mM AgNO$_3$ + 50 mL crude enzyme solution (79.3 U/mL of keratinase activity) incubated at 30°C for 24 h in the dark	AgNPs	5–25	Spherical	Jang et al. (2018)
Laccase	12 mL of laccase + 30 mL of 1 mm AgNO$_3$ at 28°C ± 2°C for 2 h	AgNPs	50–100	Spherical	Lateef et al. (2018)
	200 mL of laccase + 1 mM AgNO$_3$, at 30–50°C	AgNPs	<100	Spherical	Durán et al. (2014)

Continued

Table 5 Green synthesis of silver nanoparticles by some enzymes—cont'd

Enzyme	Reaction condition	Type of NPs	Size (nm)	NPs shape	References
Xylanase	1 mL of xylanase +19 mL of 0.1 M AgNO$_3$ solutions under sunlight	AgNPs	100	Spherical	Dhivahar et al. (2020)
	1 mL of crude enzyme+50 mL of 1 mM AgNO$_3$ solution at ambient temperature (30±2°C)	AgNPs	15–78	Spherical, cylindrical, oval	Elegbede et al. (2018)
Peroxidase	0.07 mg enzyme (0.1 U of lignin peroxidase activity) +1 mM AgNO$_3$ solution in total reaction volume of 1 mL	AgNPs	~50	Spherical	Singh et al. (2017)
	10 μL of purified peroxidase+2.9 mL of 10 Mm AgNO$_3$ incubated at 4 h	AgNPs	7–20	Spherical	Falade et al. (2017)

4 Characterization of silver nanoparticles

Various spectroscopic and analytical methods are utilized to characterize the shape, size, distribution, aggregation, rate of stability, surface properties, structural confirmation, configuration, elemental composition, and variety of dispersity of nanoparticles (Lin et al., 2014b; Mourdikoudis et al., 2018). The production of silver nanoparticles is important for different biological activities, as it acts by combating different disorders, and also essential in agricultural fields and nanofabrication development, so the characterization of silver nanoparticle is the foremost task to improve its usage in several sectors of society. It can be characterized using different techniques such as UV-visible spectroscopy, Fourier transform infrared spectroscopy (FTIR), X-ray diffractometry (XRD), scanning electron microscopy (SEM), transmission electron microscopy (TEM), X-ray photoelectron spectroscopy (XPS), atomic force microscopy (AFM), etc. Some of the essential techniques used for characterization of silver nanoparticles are discussed in the following sections with their advantages and limitations.

(i) *UV-visible spectroscopy*: This is the most useful technique for evaluation of stability of silver nanoparticles during synthesis of formed nanoparticles at the initial stage. The absorption wavelength for silver nanoparticles is generally found between 420 and 430 nm. The width of the peak defines the particle size. The formation of larger particle size of silver nanoparticles generally has red shift (longer wavelength) and broader peak whereas smaller nanoparticles absorb near range 400 nm with narrow peaks. This strongly indicates the stability of silver nanoparticles as it gradually decreases with the intensity and broadens the peak for the appearance of secondary peak at higher wavelength due to the aggregation of the particle. Likewise, widening of peak also describes the distribution of particle size of AgNPs, as the broader peak indicates formation of a broader size range of nanoparticle with wide dispersity and vice versa (Paramelle et al., 2014). The optical properties of AgNPs, i.e., the change in color during synthesis, also characterize the size and shape of nanoparticles (Huang and Xu, 2010). This method is relatively considered as a fast, easy, efficacious, sensitive, and selective method for the interpretation of synthesis of silver nanoparticles. In the formation of silver nanoparticles, the conduction band is also found near to valence bond and the electrons move freely resulting in production of surface plasmon resonance (SPR) band. The characteristics of SPR band highly depends on size of particle, dielectric constant of the medium, and chemical environment (Das et al., 2009; He et al., 2002).

(ii) *Fourier transform infrared spectroscopy* (*FTIR*): This characterizes the functional groups of reducing agents and stabilizers that are involved in the process of reduction and capping of silver nanoparticles. It also provides information regarding the presence of chemical residue on the surface of AgNPs by showing the interaction of infrared electromagnetic radiations with bonds in molecules in the form of stretching and bending vibrations in the 4000–400 cm^{-1} region (Zhang et al., 2016). It is a very simple, easy, highly reproducible, noninvasive, nondestructive, and cost-effective process. This detects the functional groups of biomolecules involved in the reduction of Ag^+ to Ag^0 (Lin et al., 2014a; Zhang et al., 2016). This technique is highly sensitive, as slight variation in the range of absorbance, i.e., 10^{-3}, interprets the synthesis of AgNPs. It also distinguishes the involvement of different functional groups such as hydroxyl (-OH), carbonyl (-CO), and amide (-CO-NH_2), which are accountable for stability, reduction, and capping of AgNPs (Das et al., 2019; Haggag et al., 2019; Hamouda et al., 2019; Zahed et al., 2018). Moreover, the recent technique attenuated total reflection (ATR)-FTIR directly interprets the nanostructures irrespective of the state and also avoids sample preparation (Hind et al., 2001).

(iii) *X-ray diffraction* (*XRD*): This is a widely used technique, which determines the size, shape, elemental composition, spatial arrangement in the molecule, and three-dimensional electron density distribution. It also identifies the crystallinity state of the biomaterial. Generally, nanoparticles reside between crystalline and noncrystalline and their surface geometry acts like a grating, which develops XRD patterns and results in Bragg's peak. Since all NPs are noncrystalline in nature, so sharp Bragg's peaks are difficult to produce. X-ray diffractometry also reveals the three-dimensional arrangement and impurity in phase of the NPs. Traces of silver oxide found in synthesized AgNPs may be found due to oxidation of AgNPs in a hot air oven (Manikandan et al., 2017). The obtained diffraction peaks calculate the size of crystallite of AgNPs by measuring broadening as full width at half maximum (FWHM) via Scherrer equation (Holzwarth and Gibson, 2011). As the line broadening is inversely proportional to the size of crystallite, so the obtained size of nanoparticles will be smaller and vice versa (Giannini et al., 2016).

(iv) *X-ray photon electron spectroscopy* (*XPS*): This is a surface-sensitive technique used for quantitative analysis of surface chemical properties. It is also known as electron spectroscopy for chemical analysis (ESCA). The main principle of XPS includes measurement of kinetic energy and number of escaping electrons when sample is irradiated from high vacuum. Thus, XPS spectra reproduce information about the elemental composition, chemical state, and electronic state of the elements presents in the sample and empirical formula of the NPs (Zhang et al., 2016).

(v) *Scanning electron microscopy* (*SEM*): This mainly characterizes surface morphology, particle size, state of aggregation, and distribution of nanoparticles, when an electron beam is used as imaging probe. It produces high resolution images at nanometer scale and its encapsulation property measures the particle size of AgNPs (Gupta et al., 2019; Fissan et al., 2014). However, for this analysis, dry and conductive samples are required and sometimes it lacks in accuracy compared to TEM due to van der Waals clusters of small entities (Theivasanthi and Alagar, 2011). It is unable to represent the internal structure of the sample but estimates the purity and extent of particle aggregation in the NPs. Recently, X-ray spectroscopy (EDX)-SEM, hyphenated with high dispersive energy has been designed, which provides information regarding the presence of different atoms present in the nanomaterial (Zhang et al., 2016).

(vi) *Transmission electron microscopy* (*TEM*): This is another useful electronic spectroscopic imaging technique used for characterization of nanoparticles, which provides more highly reproducible and accurate spatial resolution and analytical measurements than SEM. It also determines particle size, shape, morphology, distribution, and aggregation (Mourdikoudis et al., 2018; Lin et al., 2014a). The limitations of TEM are that it requires high vacuum facilities, thin sample section, and sample preparation, which is a time consuming process. Moreover, HRTEM (high resolution transmission electron microscopy), which is a phase contrast microscopy technique of TEM, produces the highest resolution images of nanoparticles and visualizes the atomic lattice in the image of crystal defects (Selvaraj et al., 2018). Similarly, selected area electron diffraction (SAED), also called electron diffraction, is a method performed in TEM, which measures the lattice parameters, crystal structure, and extent of crystallinity of nanoparticles from the diffraction technique, in which the sample is targeted with a parallel beam of high energy electrons. It also authenticates XRD results by calculating d-spacing from circular rings from patterns of diffraction (Hall et al., 2007).

(vii) *Dynamic light scattering* (*DLS*): This is a technique used for determination of qualitative properties. It provides information about the size distribution of the atoms present in nanomaterials ranging from submicron to nanometer. In this method, the radiation of monochromatic laser beam light focuses on the suspensions and colloidal solution of NPs, then the source of light interacts with the nanoparticles and the size distribution of the sample produces Rayleigh scattering at different intensities and the particle size is measured using Stokes-Einstein relationship. The obtained scattered light modulates the particle size distribution. The main disadvantage of this technique is that the particle size determined by DLS is usually larger than TEM due to Brownian motion within the sample (Tomaszewska et al., 2013; Lim et al., 2013). DLS is a nondestructive method and is usually used for determination of nanoparticle diameter in liquids. It also determines the zeta potential value of AgNPs and estimates the electrostatic stabilization of nanoparticles in larger magnitude (Gakiya-Teruya et al., 2019; Helmlinger et al., 2016). It also measures the hydrodynamic diameter of NPs affected by surfactants, stabilizing, and capping agents, etc., and the electrical double layer present on surface of NPs and also works best at monodispersed nanoparticles (Zanetti-Ramos et al., 2009).

(viii) *Atomic force microscopy*: This is an advanced three-dimensional technology of NPs that characterizes the topological properties and surface chemistry of nanoparticles at a nanometer scale by using a cantilever with a mechanical probe for its measurement. This technique possesses several advantages over SEM, TEM, XRD, and DLS as it can provide information about height and volume of the sample, topology of NPs, dispersity, variability in size and shape, and envisage hydrated nanoparticles (Zhang et al., 2016; He et al., 2002). However, it is a very time-consuming process and determines data on only exterior structure of the NPs (Lin et al., 2014a; Bhushan and Marti, 2004).

5 Applications of biogenic silver nanoparticles

(i) *Applications in agriculture*

Different phytopathogens such as fungi, bacteria, nematodes, and viruses become the key limiting factor for the growth and manufacturing of food material. Several methods like usage of pesticides and insecticides are utilized to control these pathogens but do not kill these pathogens effectively. However, the existence of nanotechnology formulations plays an important role in management and prevention of plant pathogens (Nangmenyi and Economy, 2009). AgNPs exhibit the highest agricultural benefits globally, showing antimicrobial and insecticidal properties that strongly affects and kills several phytopathogens (Clement and Jarrett, 1994). The AgNPs/PVP was tested against different yeasts and molds and possesses higher antimicrobial effect against in *Candida albicans*, *Candida glabrata*, *Candida krusei*, *Candida tropicalis*, and *Aspergillus brasiliensis* (Bryaskova et al., 2011). The antifungal activities of AgNPs can be characterized using different techniques such as Raman spectroscopy, SEM, and microbiological plating to observe the changes in morphology of the fungus, cellular composition, and rate of cell death. Aziz et al. investigated biogenic AgNPs' effect against different pathogenic fungal organisms such as *Fusarium oxysporum*, *Candida albicans*, and *Aspergillus flavus* and showed synergistic effect when combined with established fungicides, i.e., amphotericin B, ketoconazole, and fluconazole by offering greater resistance to microbial growth (Aziz et al., 2016). AgNPs synthesized from *Solanum lycopersicum* extract produced antimicrobial activity against rose aphid at 50 ppm concentration showing up to 100% death rate (Bhattacharyya et al., 2016; Balakumaran et al., 2016). In another study, AgNPs based on silica (nano silica-silver) produced the greatest antibacterial activity against various plant pathogens, viz., *Colletotrichum* sp., *Pythium* sp., *Xanthomonas compestris*, and *Pseudomonas syringae*; it also destroyed the powdery mildew of pumpkin at a very low concentration (0.3 mg/mL) with 100% mortality.

AgNPs also produce antibacterial effect in agricultural soil by affecting multiple bacterial communities (Park et al., 2006). Chen et al. studied the surfactant stabilized AgNPs (SBS-Tween 80-CTAB-PVP) possessed both bacteriostatic and bactericidal properties against fungi *Ralstonia solanacearum* causing severe bacterial wilt in tobacco plant (Chen et al., 2016). Ocsoy et al. revealed that nanocomposite DNA formulation of AgNPs (Ag@dsDNA@GO) grown on graphene oxides effectively reduced the cell viability of *X. perforans* in plants and culture at a very low concentration (16 ppm) and showed maximal antibacterial ability with improved stability and greater antibacterial activity (Ocsoy et al., 2013). This is due to the high-density solution of AgNPs, which saturate and cohere to the fungal hyphae, resulting in degradation of plant pathogenic fungi (Kim et al., 2012). AgNPs also exhibited nanopesticidal and insecticidal properties, which possessed the capability of destroying undesired microorganisms in hydroponic systems and planter soils. They were also applied in foliar spray to control spreading of different molds, rot, fungi, and other plant diseases (Mishra and Singh, 2015).

The mechanism of AgNPs for antibacterial activity includes interference in cell permeability, release of toxins, which leads to disruption of DNA synthesis, interruption with uptake of nutrients, electron transport chain, and protein oxidation system, and generation of oxidative stress-mediated cellular and membrane damage (Banerjee et al., 2010; Pal et al., 2007). Silver nanoparticles have extensive application for insecticidal and larvicidal activity by killing mosquitoes and fleas. Many studies reported that biogenic synthesized AgNPs possesses efficient larvicidal and pupicidal properties as they kill different mosquitoes and larvae of different organisms such as *Culex quinquefasciatus*, *Anopheles subpictus*, *Heteroscodra maculate*, *Aedes aegypti*, *Anopheles stephensi*, *Rhipicephalus microplus*, and *Aphis nerii*

(Jayaseelan et al., 2011; Rouhani et al., 2013; Suman et al., 2013; Mondal et al., 2014). Silver nanoparticles also deal in controlled, slow, and sustained release of agrochemicals, fertilizers, and pesticides in agricultural practices. They possess efficient therapeutic activity on target site with optimum concentration, time-controlled release, and reduced toxicity. This also reduced the risk of nonspecific chemical contamination in the agricultural field (Tsuji, 2001; Chowdappa and Gowda, 2013).

A combination of three nutrients mainly considered as inorganic fertilizers (i.e., nitrogen, potassium, and phosphorus) with AgNPs boosts nutrients management profile of different plant products. It also acts as plant regulator and stimulator, which is a positive effect, but these nanoparticles also have negative effect on some plants. The biological/biogenic synthesized AgNPs significantly produce positive effect on *Bacopa monnieri* and *Boswellia ovalifoliolata* by enhancing its seed germination, seedling growth, synthesis of carbohydrates and proteins, and decreasing total phenolic components, catalase, and peroxidase activities (Krishnaraj et al., 2012). It also increases plants' morphological features like size, leaf, root and shoot length, and biochemical attributes like carbohydrate content, chlorophyll, antioxidant enzymes, and protein content of common bean, corn, and *Brassica juncea* (Salama, 2012; Sharma et al., 2012). However, some reports also show that AgNPs significantly increases the root length of barley but produce negative effect on lettuce (Gruyer et al., 2013). Yin et al. also studied the effect of AgNPs on 11 wetland plant species and found that it enhances the germination rate of only one species, i.e., *Eutrochium fistulosum* (Yin et al., 2012). It was also observed that treatment of AgNPs with a magnetic field improved the quantitative yield of plant *Zea mays* (Berahmand et al., 2012).

It is reported that the variation in particle size confers the different effects of AgNPs. The size ranging from 200 to 800 nm was observed to have significant plant growth profile, whereas 35–40 nm increases the length of shoot and root of diverse plant species (Jasim et al., 2017; Mehta et al., 2016). AgNPs of particle size <30 nm, when used in high concentration, possess negative effects, i.e., hampering the development of shoot and root of different plant products (Dimkpa et al., 2013; Vinković et al., 2017). Treatment with 100 ppm of biogenic AgNPs in *Lupinus termis* improved the growth profile, whereas 300 and 500 ppm results in considerable decrease in both growth indices and growth parameters (Al-Huqail et al., 2018). Silver nanoparticles also possesses the capacity of detecting plant pathogens by an electrochemical techniques i.e., stripping voltametrys which uses different inorganic crystals like PbS, ZnS, and CdS for analytical detection (Khiyami et al., 2014; Upadhyayula, 2012). Schwenkbier et al. designed a helicase-dependent isothermal amplification with on-chip hybridization, which significantly detected *Phytophthora* species by targeting the amplification power on GTP-binding protein (Y pt 1) under same temperature condition in a miniature heating device. The assay was determined by deposition of AgNPs and hybridization of on-chip DNA, which visualized the end-point signals. This technology was implemented for future use with additional combined techniques for a reliable detection of plant pathogens (Schwenkbier et al., 2015).

AgNPs are also employed for food packaging purpose due to their active antimicrobial, antibacterial, and antiviral effects (Carbone et al., 2016). AgNPs were developed with hybrid cellulose nanofiber and observed to have potent antimicrobial properties, and AgNPs significantly increased the efficacy of nanofiber by reducing 100% growth of *Escherichia coli* and *Staphylococcus aureus* at 20 ppm concentration (Amini et al., 2016). It is also reported that the AgNPs-conjugated biofilms improved mechanical properties efficiently and prolonged the shelf-life of food products. It is also widely used in fresh food bags for its food sanitization property (Kraśniewska et al., 2020).

(ii) *Biomedical applications*

Nanoparticles exhibit a promising effect as alternative drugs for combating different disorders involved in health conditions of human beings. AgNPs produced by biosynthesis were found to be a convincing anti-HIV drug by inhibiting the early stage of reverse transcriptase in viral cells (Suriyakalaa et al., 2013). AgNPs have multiple binding sites for gp120 viral membrane, which significantly disrupts the functions of cells, causing cell death, and acts as a potent virucidal agent against HIV or cell-associated virus (Sun et al., 2005). Many in vitro studies also revealed potent antiviral activity of AgNPs by acting against HIV-1 and monkey pox virus (Lara et al., 2010; Rogers et al., 2008). In another study, when AgNPs were exposed to Tacaribe virus, they significantly inhibited viral RNA production and fascination of uptake of virus into the host cells, thereby resulting in reduction of growth and virus infection (Speshock et al., 2010). AgNPs significantly possess antitumor property by inhibiting lymphoma cells in in vitro and in vivo models by producing dose-dependent cytotoxicity and increasing the rate of apoptosis in the tumor cells. In addition, they also increased the survival rate of mouse and decreased the volume of ascitic fluid in tumor-bearing mice (Sriram et al., 2010). Silver nanoparticles also produced potent activity against human lung epithelial cell line by arresting the cell cycle G2/M phase (Foldbjerg et al., 2012). They also induced cell autophagy in cancer cell line by activating PtdIns3k signaling pathway (Lin et al., 2014a).

Green synthesis of silver NPs also produced significant dose-dependent effects in a human lung cancer study (Sankar et al., 2013). Moreover, combination of AgNPs with wortmannin (inhibitor of autophagy) exhibited enhanced antitumor activity against melanoma cell model (Lin et al., 2014a). Leaves of *Panax ginseng* loaded with AgNPs exhibited cytotoxic and oxidative effect on human cancer cell lines (Castro-Aceituno et al., 2016). The cytotoxicity to cancer cell increases with the increased concentration of AgNPs (Khateef et al., 2019). AgNPs also significantly improve the drug delivery systems of various anticancer drugs such as doxorubicin and alendronate, thereby resulting in higher therapeutic indices of both drugs (Benyettou et al., 2015). It is also reported that hybrid molecular units with AgNPs increases the target system of many inflammatory and other drugs. AgNPs also possess powerful antiparasitic action against the dengue vectors, i.e., *Culex quinquefasciatus* and *Aedes aegypti* (Suresh et al., 2014; Tarannum and Divya Gautam, 2019). Allahverdiyev et al. also conducted a study and found that AgNPs act as a potent antileishmanial drug as it inhibits the proliferation activity of parasite *Leishmania tropica* (Allahverdiyev et al., 2011). It also inhibits the survival of parasite amastigotes in host cells and oocyte viability of *Cryptosporidium parvum* and *Entamoeba histolytica* (Saad et al., 2015). AgNPs commonly promote wound healing, either alone or in combination with other antibacterial drugs. AgNPs-based dressings have been applied to fibroblast cell culture but did not show significant effect; however, when applied in combination with antibacterial agent tetracycline, it efficiently increased the fibroblasts and keratinocytes tissues with relevant establishment of normal skin (Rigo et al., 2013; Ahmadi and Adibhesami, 2017). Furthermore, Rujitanaroj et al. designed a randomized controlled clinical trial study in nursing clients with ulcerated legs and compared the efficacy of two antimicrobial agents, i.e., cadexomer iodine and noncrystalline silver. The study revealed that the use of AgNPs caused faster healing rate than cadexomer iodine (Rujitanaroj et al., 2008; Miller et al., 2010).

(iii) *Nanofabrication*

AgNPs have been demonstrated to possess beneficial applications in dentistry and endodontics due to their effective property of killing and inhibiting different microbes and bacterial strains (Palanisamy et al., 2014; Nam, 2011; Fabrega et al., 2009). In a study, silver oxide (Ag_2O) nanoparticles synthesized from root extract of *Ficus benghalensis* were reported and evaluated for powerful antibacterial activity against dental bacterial strains (Manikandan et al., 2017). AgNPs exhibited an effective role in dental caries by fabricating bactericidal activity against *Streptococcus mutans* (Pérez-Díaz et al., 2015). They also effectively killed *Streptococcus mutans* biofilms and inhibited clinically isolated planktonic *Streptococcus mutans* (Santos et al., 2013). In bone healing/orthopedic implants, AgNPs-based nanodevices were employed because of lower risk of incidence. AgNPs-based stainless steel was used to decrease the infections associated with orthopedic implants (Bharti et al., 2016). Unique hydroxyapatite (HAp), when combined with AgNPs, provided effective results in orthopedic implants and was reported to be well-suited for implantation.

In another study, Ag NPs doped HAp shells showed unique antibacterial activity in preventing and treating bacterial infections associated with bone implants (Zhou et al., 2015). It was also observed that it improved biocompatibility and increased the cell viability by activating murine macrophages (Ciobanu et al., 2013). Different catheters used in hospitals have a high chance of contamination, leading to different complications. However, AgNPs-coated catheters have potential ability to decrease biofilm development by activating powerful antibacterial effect, and productively, this reduces the bacteria count within 72 h (Roe et al., 2008; Chou et al., 2008). In a clinical pilot study, it was found that 19 patients who received AgNPs-coated catheters had reduced frequency of catheter-associated ventriculitis (CAV) and no effect on cerebrospinal fluid (Lackner et al., 2008). AgNPs-based devices—mainly AgNPs multilayer films—were used in cardiovascular implants due to their efficient anticoagulant and antibacterial activity. They were also used in surface modification of medical devices, particularly those used for cardiovascular implants. In another study, AgNPs significantly reduced left ventricle pressure and perfusion pressure of hypertension leading to cardiotoxicity (Ramirez-Lee et al., 2018). AgNPs are significantly incorporated into different medical devices for their safe, nontoxic, and high antibacterial coating properties (Ge et al., 2014). Nanocomposites of AgNPs and diamond-like carbon coatings are widely used for heart stents and valves due to their potent antibacterial and antithrombogenic properties.

(iv) *Environmental applications*
   (a) *AgNPs as sensors*: AgNPs possess relative sensing/detecting properties, which are utilized in finding various types of metallic substances present in the environment. AgNPs significantly detect the metals Hg (II), Ni, Co, and S ions due to their colorimetric sensing property (Teodoro et al., 2019; Oluwafemi et al., 2019; Mochi et al., 2018; Zhao et al., 2017). Triangular silver nanoplates possess improved lightning rod and anisotropy effect. Due to this effect, they are used as plasmon sensors to identify Hg^{2+} ion in the solution and blue shift of the wavelength symbolizes increase in the concentration of Hg^{2+} ions (Zannotti et al., 2020). AgNPs were doped in tris (4,7-diphenyl-1,10-phenanthroline) ruthenium (II) dichloride complex solution and encapsulated with plasticized polymethyl methacrylate, acting as a biofilm and radiometric sensor, which efficiently measured the amount of dissolved oxygen in the aqueous solution (Jiang et al., 2017). Electrochemical sensors of AgNPs can detect common herbicide atrazine activity (Zahran et al., 2020). SERS sensors are industrialized by in situ development of AgNPs on polydopamine templated filter papers, which were extremely flexible and rapidly collected and detected malachite green residues in production (Zhang et al., 2020). The silver nanoparticles also have tremendous use in the field of pharmaceuticals, food, environment, healthcare, etc. for detection and identification of different chemicals and contaminants in numerous samples. Ngeontae et al. investigated the use of AgNPs in detection of glucose in various food and beverage items by potentiometric redox marker. It was also responsible in detecting mismatch of single base pair in DNA hybridization with rapid sensitivity (Ngeontae et al., 2009). AgNPs assisted in biosensing of the DNA concentration from morbific bacteria even at 1 ZM (Zepto-molar). AgNPs-silk-fibroin-based bio sensing technique is widely applicable for on-site detection of metal mercury ion at ppb level (Mane et al., 2019). It was widely used in detection of pathogenic microbes, for example, AgNPs graphene quantum dots which were utilized for detection of *Legionella*

pneumophila causing legionnaires disease and pontiac fever (Mobed et al., 2020).

(b) *Waste water treatment:* Sometimes, water develops bacterial membranes that need to be purified before its use, and AgNPs have promising applications in wastewater treatment. AgNPs also have potential effects on eradicating certain fatal pesticides and different metal contents such as lead, arsenic, iron, etc. from water (Candido et al., 2019). AgNPs, when used as catalyst, possess potential remediation against different pollutants like heavy metals, antibiotics, and pesticides (Malakootian et al., 2019). In another study, colloidal formulation of AgNPs degraded toxic dye, i.e., methylene blue, by reducing it by 100% within the short span of 10 min as AgNPs acted as a mediator of electron transfer of methylene blue dye during catalytic degradation (Saha et al., 2017). Allam et al. biosynthesized AgNPs of oval shape and particle size ranging from 4 to 20 nm from bacterial strains like *Bacillus pumilus*, *Sphingomonas paucimobilis*, and *Bacillus paralicheniformis*, which produced dose-dependent dye degradation from the water and maximal degradation was 97% at 1.5×10^{-5} mol/L (Allam et al., 2019). Romeh et al. examined AgNPs obtained from different medicinal plants, *Brassica*, *Camellia*, *Plantago*, and *Ipomoea*, which significantly degraded the toxic pesticide Fipronil, associated with neurotoxicity, endocrine disruption, and carcinogenicity. The results reported that *Brassica*-AgNPs efficiently degrade up to 95.45%, whereas *Ipomoea*-AgNPs showed 90.15% of degradation from water within 48 h of treatment. Furthermore, when AgNPs were applied to the contaminated soil with Fipronil, the results showed that *Brassica*-AgNPs were more efficient to degrade the pesticide from soil, up to 68.8% within 6 days (Romeh, 2018).

AgNPs biosynthesized with 1 mg/L on nitrocellulose membrane cellulose showed potential inhibition of different bacterial communities like *Escherichia coli*, *Pseudomonas aeruginosa*, *Staphylococcus aureus*, and *Enterococcus faecalis* suspensions, and from this the power of inactivation of *Staphylococcus aureus* and *Escherichia coli* reached up to 5.2 and 6 order of magnitude (Fernández et al., 2016). AgNPs, when combined with zwitterionic sulfobetaine methacrylate (SBMA) grafted on polyamide membrane, showed significant antimicrobial and antifouling properties of the membrane (Zhang et al., 2018). AgNPs could simply detach the beads and inhibited the microbial growth in real water samples (Abu-Saied et al., 2018; Beisl et al., 2019). The polyacrylonitrile used as water sorbent mainly stores pathogenic microorganisms on the water surface, but when it was treated with AgNPs solution, the pathogenic biofilm was not developed in the water surface and effectively it treated and purified the wastewater (Ali et al., 2017). Sylvia Devi et al. also examined AgNPs (size 7–15 nm, spherical shaped) with reducing and stabilizing agent tannic acid, which degraded significantly inorganic pollutant, i.e., pesticide Chlorpyrifos and methylene blue, up to 75-fold by following first order kinetic mechanism (Sylvia Devi et al., 2015). In another study, Ali et al. synthesized AgNPs impregnated with cotton fibers to study elimination of different metal ions, i.e., Hg, Ni, Co, Pb, and Cr, discretely from synthetic wastewater. It was observed that silver nanoparticles showed significant rate of removal following pseudo second order kinetics and was faster in Hg and Ni metal ion than other heavy metal ions. The adsorption phenomenon of AgNPs impregnated cotton fiber followed the sequence $Hg > Ni > Cr > Co > Pb$ and, according to its adsorption property, it follows the rate of degradation (Ali et al., 2018).

(c) *Pollutant degradation*: Due to the prominent catalytic activity of AgNPs, they possess the capacity to degrade various pollutants affecting the environment. For example: AgNPs act as a heterogeneous catalyst in reducing different halogenated organic pollutants like methyl orange, 4-nitrophenol, 4-aminophenol, and methylene blue by sodium borohydride ($NaBH_4$) (Zhao and Stevens, 1998). The photocatalytic degradation of pollutants like methylene blue, crystal violet, and malachite green with AgNPs occurs by shifting metal oxides to visible region absorption (Jaffri and Ahmad, 2020). It was reported that ZnO/Ag, Ni/Ag, and TiO_2/Ag have better photocatalytic activity than other metal oxides, along with rate of degradation to various environmental pollutants such as toxic colorless 4-chlorophenol and colorful methyl orange (Ziashahabi et al., 2019; Karimi-Maleh et al., 2020).

6 Challenges and limitations of enzyme assisted silver nanoparticles

Silver nanoparticles exhibit significant antibacterial and antimicrobial properties, which recommends their usage in the coming years. The physicochemical properties of AgNPs enhance the efficacy of silver nanoparticles in numerous fields. But there are some limitations of silver nanoparticles, in that they can induce toxicity when used at higher concentrations and affects health hazards causing different health problems. It was also revealed that they can induce several ecological problems and deteriorate the ecosystem if released into the environment. To evaluate the toxicity, various animal models were conducted to check their effect on physiology and tissue architecture. Ag^+ ions when absorbed in the animals produce a nonclassical permeability, which increases the permeability of mitochondrial inner membrane and results in increased permeability leading to mitochondrial swelling, abnormal metabolism, and lastly, cell apoptosis (Almofti et al., 2003). In another study, AgNPs also decreased the level of glutathione and mitochondrial membrane potential, and further increased the reactive oxygen species (Hussain et al., 2005). This produces a cytotoxicity effect in the liver cells in a size range of 15–100 nm, which is probably facilitated via oxidative stress mechanisms (Braydich-Stolle et al., 2005). It is also responsible for causing toxicity to fish embryos (Laban et al., 2010). Prolonged exposure to AgNPs can induce changes in liver functions as well as decreases the tidal volume and causes other inflammatory responses (Sung et al., 2009).

The challenge associated with green synthesis of AgNPs using enzymes from microorganisms is the selection and availability of suitable strains. In the process of biosynthesis, growth or culture of the microorganism in a suitable media affects the amount of growth and consequently the amount of extracellular and intracellular enzyme availability for the biosynthesis. The optimization of parameters for the enzyme-assisted synthesis of AgNPs plays an important role in obtaining NPs of suitable shape and size. As the dimension and morphology determines the physicochemical properties of NPs, so the reaction conditions like concentration of silver nitrate, pH and temperature of the reaction medium, concentration of bioreductant, and the capping agents, etc., decides the quality and yield of AgNPs. Again, in intracellular synthesis of AgNPs, it becomes very difficult to recover the synthesized AgNP inside the cell of microorganisms. So, the extracellular biosynthesis has been preferred over intracellular synthesis of AgNPs. As the AgNPs show significant biological activities, its biosynthesis has become a promising challenge for scientists.

7 Conclusion and future perspective

Recent studies develop different biogenic synthesis of AgNPs using various modernized techniques like chemical, physical, and biological methods, which offer several advantages and make a promising potential for the different applications in the areas of agriculture, health, and the environment. The efficient antimicrobial properties of silver nanoparticles offer control of different infectious diseases by combating microorganisms, use in wastewater treatment by remediating heavy metal ions, and also eliminating plant pathogens. AgNPs are also widely used in the production of catheters, cardiovascular implants, orthopedic implants, and play a major role in elimination of different bacteria and microbes from dental carries. Silver nanoparticles possess various applications in many sectors, however, drug delivery systems of AgNPs are lacking and can be implemented in the near future due to their nano-size and can be significantly utilized to treat different incurable diseases like HIV, cancers, etc. without harming healthy cells. AgNPs can also be utilized for washing and staining clothes due to their antimicrobial properties. As it also causes some toxicity to living organisms and the environment, researchers have highlighted that the usage of low concentrations of nano-silver is safe and also has potential efficacy with wider applications. Furthermore, using advanced tools and technologies, the implementation of AgNPs can be enhanced in the near future.

References

AbdelRahim, K., Mahmoud, S.Y., Ali, A.M., Almaary, K.S., Mustafa, A.E.Z.M., Husseiny, S. M., 2017. Extracellular biosynthesis of silver nanoparticles using *Rhizopus stolonifer*. Saudi. J. Biol. Sci. 24 (1), 208–216.

Aboelfetoh, E.F., El-Shenody, R.A., Ghobara, M.M., 2017. Eco–friendly synthesis of silver nanoparticles using green algae (*Caulerpa serrulata*): reaction optimization, catalytic and antibacterial activities. Environ. Monit. Assess. 189 (7), 1–15.

Abu-Saied, M.A., Taha, T.H., El-Deeb, N.M., Hafez, E.E., 2018. Polyvinyl alcohol/sodium alginate integrated silver nanoparticles as probable solution for decontamination of microbes contaminated water. Int. J. Biol. Macromol. 107, 1773–1781.

Aguilar-Méndez, M., San Martín-Martínez, E., Ortega-Arroyo, L., Cobián-Portillo, G., Sánchez-Espíndola, E., 2010. Synthesis and characterization of silver nanoparticles: effect on phytopathogen *Colletotrichum gloesporioides*. J. Nanopart. Res. 13 (6), 2525–2532.

Ahmadi, M., Adibhesami, M., 2017. The effect of silver nanoparticles on wounds contaminated with Pseudomonas aeruginosa in mice: an experimental study. Iran J. Pharma Res. 16, 661.

Akter, S., Huq, M.A., 2020. Biologically rapid synthesis of silver nanoparticles by *Sphingobium* sp. MAH–11T and their antibacterial activity and mechanisms investigation against drug–resistant pathogenic microbes. Artif. Cells Nanomed. Biotechnol. 48 (1), 672–682.

Alamri, S.A., Hashem, M., Nafady, N.A., Sayed, M.A., Alshehri, A.M., El–Shaboury, G.A., 2018. Controllable biogenic synthesis of intracellular silver/silver chloride nanoparticles by *Meyerozyma guilliermondii* KX008616. J. Microbiol. Biotechnol. 28 (6), 917–930.

Al-Bahrani, R., Raman, J., Lakshmanan, H., Hassan, A.A., Sabaratnam, V., 2017. Green synthesis of silver nanoparticles using tree oyster mushroom *Pleurotus ostreatus* and its inhibitory activity against pathogenic bacteria. Mater. Lett. 186, 21–25.

Al-Huqail, A.A., Hatata, M.M., Ibrahim, M.M., 2018. Preparation, characterization of silver phyto nanoparticles and their impact on growth potential of *Lupinus termis* L. seedlings. Saudi J. Biol. Sci. 25, 313–319.

Ali, Q., Ahmed, W., Lal, S., Sen, T., 2017. Novel multifunctional carbon nanotube containing silver and iron oxide nanoparticles for antimicrobial applications in water treatment. Mater. Today. Proc. 4, 57–64. https://doi.org/10.1016/j.matpr.2017.01.193.

Ali, A., Mannan, A., Hussain, I., Hussain, I., Zia, M., 2018. Effective removal of metal ions from aquous solution by silver and zinc nanoparticles functionalized cellulose: isotherm, kinetics and statistical supposition of process. Environ. Nanotechnol. Monit. Manag. 9, 1–11.

Allahverdiyev, A.M., Abamor, E.S., Bagirova, M., Ustundag, C.B., Kaya, C., Kaya, F., Rafailovich, M., 2011. Antileishmanial effect of silver nanoparticles and their enhanced antiparasitic activity under ultraviolet light. Int. J. Nanomedicine 6, 2705.

Allam, N.G., Ismail, G.A., El-Gemizy, W.M., Salem, M.A., 2019. Biosynthesis of silver nanoparticles by cell–free extracts from some bacteria species for dye removal from wastewater. Biotechnol. Lett. 41, 379–389.

Almatroudi, A., 2020. Silver nanoparticles: synthesis, characterisation and biomedical applications. Open. Life Sci. 15 (1), 819–839.

Almofti, M.R., Ichikawa, T., Yamashita, K., Terada, H., Shinohara, Y., 2003. Silver ion induces a cyclosporine a–insensitive permeability transition in rat liver mitochondria and release of apoptogenic cytochrome C. J. Biochem. 134, 43–49.

ALThabaiti, S.A., Malik, M.A., Al-Youbi, A.A.O., Khan, Z., Hussain, J.I., 2013. Effects of surfactant and polymer on the morphology of advanced nanomaterials in aqueous solution. Int. J. Electrochem. Sci. 8, 204–218.

Alzahrani, R.R., Alkhulaifi, M.M., Al-Enazi, N.M., 2020. In vitro biological activity of *Hydroclathrus clathratus* and its use as an extracellular bioreductant for silver nanoparticle formation. Green. Process. Synth. 9 (1), 416–428.

Ameen, F., AlYahya, S., Govarthanan, M., ALjahdali, N., Al-Enazi, N., Alsamhary, K., Alshehri, W.A., Alwakeel, S.S., Alharbi, S.A., 2020. Soil bacteria *Cupriavidus* sp. mediates the extracellular synthesis of antibacterial silver nanoparticles. J. Mol. Struct. 1202, 127233.

Amini, E., Azadfallah, M., Layeghi, M., Talaei-Hassanloui, R., 2016. Silver-nanoparticle-impregnated cellulose nanofiber coating for packaging paper. Cellulose 23, 557–570.

Arun, G., Eyini, M., Gunasekaran, P., 2014. Green synthesis of silver nanoparticles using the mushroom fungus *Schizophyllum commune* and its biomedical applications. Biotechnol. Bioprocess Eng. 19 (6), 1083–1090.

Aziz, N., Pandey, R., Barman, I., Prasad, R., 2016. Leveraging the attributes of Mucor hiemalis–derived silver nanoparticles for a synergistic broad–spectrum antimicrobial platform. Front. Microbiol. 7, 1984.

Bai, J., Li, Y., Du, J., Wang, S., Zheng, J., Yang, Q., Chen, X., 2007. One–pot synthesis of polyacrylamide–gold nanocomposite. Mater. Chem. Phys. 106 (2–3), 412–415.

Bai, H., Yang, B., Chai, C., Yang, G., Jia, W., Yi, Z., 2011. Green synthesis of silver nanoparticles using *Rhodobacter Sphaeroides*. World J. Microbiol. Biotechnol. 27, 2723–2728.

Balakumaran, M.D., Ramachandran, R., Balashanmugam, P., Mukeshkumar, D.J., Kalaichelvan, P.T., 2016. Mycosynthesis of silver and gold nanoparticles: optimization, characterization and antimicrobial activity against human pathogens. Microbiol. Res. 182, 8–20.

Banerjee, M., Mallick, S., Paul, A., Chattopadhyay, A., Ghosh, S.S., 2010. Heightened reactive oxygen species generation in the antimicrobial activity of a three–component iodinated chitosan–silver nanoparticle composite. Langmuir 26, 5901–5908.

Behera, A., Mittu, B., Padhi, S., Patra, N., Singh, J., 2020. Bimetallic nanoparticle: green synthesis, applications and future perspectives. In: Abd-Elsalam, K.A. (Ed.), Multifunctional Hybrid Nanomaterials for Sustainable Agri-Food and Ecosystems. Elsevier Publications, pp. 639–681.

Beisl, S., Monteiro, S., Santos, R., Figueiredo, A.S., Sánchez-Loredo, M.G., Lemos, M.A., Lemos, F., Minhalma, M., De Pinho, M.N., 2019. Synthesis and bactericide activity of nanofiltration composite membranes-cellulose acetate/silver nanoparticles and cellulose acetate/silver ion exchanged zeolites. Water Res. 149, 225–231.

Benyettou, F., Rezgui, R., Ravaux, F., Jaber, T., Blumer, K., Jouiad, M., Motte, L., Olsen, J.C., Platas-Iglesias, C., Magzoub, M., Trabolsi, A., 2015. Synthesis of silver nanoparticles for the dual delivery of doxorubicin and alendronate to cancer cells. J. Mater. Chem. B 3, 7237–7245.

Berahmand, A.A., Panahi, A.G., Sahabi, H., Feizi, H., Moghaddam, P.R., Shahtahmassebi, N., Fotovat, A., Karimpour, H., Gallehgir, O., 2012. Effects silver nanoparticles and magnetic field on growth of fodder maize (*Zea mays* L.). Biol. Trace Elem. Res. 149, 419–424.

Bharti, A., Singh, S., Meena, V.K., Goyal, N., 2016. Structural characterization of silver–hydroxyapatite nanocomposite: a bone repair biomaterial. Mater. Today. Proc. 3, 2113–2120.

Bhattacharyya, A., Prasad, R., Buhroo, A.A., Duraisamy, P., Yousuf, I., Umadevi, M., Bindhu, M.R., Govindarajan, M., Khanday, A.L., 2016. One–pot fabrication and characterization of silver nanoparticles using *Solanum lycopersicum*: an eco-friendly and potent control tool against rose aphid, *Macrosiphum rosae*. J. Nanosci. 2016, 1–7.

Bhushan, B., Marti, O., 2004. Scanning probe microscopy—principle of operation, instrumentation, and probes. In: Bhushan, B. (Ed.), Springer Handbook of Nanotechnology. Springer Verlag, Berlin, pp. 325–369.

Bhuyar, P., Rahim, M.H.A., Sundararaju, S., Ramaraj, R., Maniam, G.P., Govindan, N., 2020. Synthesis of silver nanoparticles using marine macroalgae *Padina* sp. and its antibacterial activity towards pathogenic bacteria. Beni-Suef Univ. J. Basic Appl. Sci. 9 (1), 1–15.

Birla, S.S., Tiwari, V.V., Gade, A.K., Ingle, A.P., Yadav, A.P., Rai, M.K., 2009. Fabrication of silver nanoparticles by Phoma glomerata and its combined effect against *Escherichia coli, Pseudomonas aeruginosa* and *Staphylococcus aureus*. Lett. Appl. Microbiol. 43 (2), 173–179.

Braydich-Stolle, L., Hussain, S., Schlager, J.J., Hofmann, M.C., 2005. In vitro cytotoxicity of nanoparticles in mammalian germline stem cells. Toxicol. Sci. 88, 412–419.

Bryaskova, R., Pencheva, D., Nikolov, S., Kantardjiev, T., 2011. Synthesis and comparative study on the antimicrobial activity of hybrid materials based on silver nanoparticles (AgNPs) stabilized by polyvinylpyrrolidone (PVP). J. Chem. Biol. 4, 185–191.

Cai, X., Zhai, A., 2010. Preparation of microsized silver crystals with different morphologies by a wet–chemical method. Rare Metals 29 (4), 407–412.

Candido, I.C.M., Soares, J.M.D., Barbosa, J.D.A.B., de Oliveira, H.P., 2019. Adsorption and identification of traces of dyes in aqueous solutions using chemically modified eggshell membranes. Bioresour. Technol. Rep. 7, 100267.

Carbone, M., Donia, D.T., Sabbatella, G., Antiochia, R., 2016. Silver nanoparticles in polymeric matrices for fresh food packaging. J. King Saud Univ. Sci. 28, 273–279.

Castro-Aceituno, V., Ahn, S., Simu, S.Y., Singh, P., Mathiyalagan, R., Lee, H.A., Yang, D., 2016. Anticancer activity of silver nanoparticles from *Panax ginseng* fresh leaves in human cancer cells. Biomed. Pharmacother. 84, 158–165.

Chen, D., Qiao, X., Qiu, X., Chen, J., Jiang, R., 2011. Large–scale synthesis of silver nanowires via a solvothermal method. J. Mater. Sci. Mater. Electron. 22 (1), 6–13.

Chen, J., Li, S., Luo, J., Wang, R., Ding, W., 2016. Enhancement of the antibacterial activity of silver nanoparticles against phytopathogenic bacterium *Ralstonia solanacearum* by stabilization. J. Nanomater. 2016, 7135852.

Chou, C.W., Hsu, S.H., Wang, P.H., 2008. Biostability and biocompatibility of poly (ether) urethane containing gold or silver nanoparticles in a porcine model. J. Biomed. Mater. Res. A 84, 785–794.

Choudhary, P., Prajapati, S.K., Malik, A., 2016. Screening native microalgal consortia for biomass production and nutrient removal from rural wastewaters for bioenergy applications. Ecol. Eng. 91, 221–230.

Chouhan, N., Meena, R.K., 2015. Biosynthesis of silver nanoparticles using *Trachyspermum ammi* and evaluation of their antibacterial activities. Int. J. Pharm. Biol. Sci. 62, 1077–1086.

Chowdappa, P., Gowda, S., 2013. Nanotechnology in crop protection: status and scope. Pest. Manag. Horticult. Ecosyst. 19, 131–151.

Chowdhury, S., Basu, A., Kundu, S., 2014. Green synthesis of protein capped silver nanoparticles from phytopathogenic fungus *Macrophomina phaseolina* (Tassi) Goid with antimicrobial properties against multidrug–resistant bacteria. Nanoscale Res. Lett. 9 (1), 1–11.

Cicek, S., Gungor, A.A., Adiguzel, A., Nadaroglu, H., 2015. Biochemical evaluation and green synthesis of nano silver using peroxidase from Euphorbia (*Euphorbia amygdaloides*) and its antibacterial activity. J. Chem. 2015, 486948.

Ciobanu, C.S., Iconaru, S.L., Pasuk, I., Vasile, B.S., Lupu, A.R., Hermenean, A., Dinischiotu, A., Predoi, D., 2013. Structural properties of silver doped hydroxyapatite and their biocompatibility. Mater. Sci. Eng. C 33, 1395–1402.

Clement, J.L., Jarrett, P.S., 1994. Antibacterial silver. Met. Based Drug. 1, 467–482.

Cong, F.Z., Wei, H., Tian, X.R., Xu, H.X., 2012. A facile synthesis of branched silver nanowire structures and its applications in surface–enhanced Raman scattering. Front. Physiol. 7 (5), 521–526.

Dağlıoğlu, Y., Öztürk, B.Y., 2019. A novel intracellular synthesis of silver nanoparticles using *Desmodesmus* sp. (Scenedesmaceae): different methods of pigment change. Rendiconti Lincei. Scienze Fisiche e Naturali 30 (3), 611–621.

Dakal, T., Kumar, A., Majumdar, R., Yadav, V., 2016. Mechanistic basis of antimicrobial actions of silver nanoparticles. Front. Microbiol. 7 (7), 1831.

Darmanin, T., Nativo, P., Gilliland, D., Ceccone, G., Pascual, C., De Berardis, B., Guittard, F., Rossi, F., 2012. Microwave–assisted synthesis of silver nanoprisms/nanoplates using a "modified polyol process". Colloids Surf. A Physicochem. Eng. Asp. 395, 145–151.

Das, R., Nath, S.S., Chakdar, D., Gope, G., Bhattacharjee, R., 2009. Preparation of silver nanoparticles and their characterization. J. Nanotechnol. 5, 1–6.

Das, G., Patra, J.K., Nagaraj Basavegowda, C.N.V., Shin, H.S., 2019. Comparative study on antidiabetic, cytotoxicity, antioxidant and antibacterial properties of biosynthesized silver nanoparticles using outer peels of two varieties of *Ipomoea batatas* (L.) Lam. Int. J. Nanomedicine 14, 4741.

de Aragao, A.P., de Oliveira, T.M., Quelemes, P.V., Perfeito, M.L.G., Araujo, M.C., Santiago, J.D.A.S., Cardoso, V.S., Quaresma, P., de Almeida, J.R.D.S., da Silva, D.A., 2019. Green synthesis of silver nanoparticles using the seaweed *Gracilaria birdiae* and their antibacterial activity. Arab. J. Chem. 12 (8), 4182–4188.

Debnath, R., Saha, T., 2020. An insight into the production strategies and applications of the ligninolytic enzyme laccase from bacteria and fungi. Biocatal. Agric. Biotechnol. 26, 101645.

Dhanalakshmi, P.K., Azeez, R., Rekha, R., Poonkodi, S., Nallamuthu, T., 2012. Synthesis of silver nanoparticles using green and brown seaweeds. Phykos 42, 39–45.

Dhivahar, J., Khusro, A., Elancheran, L., Agastian, P., Al-Dhabi, N.A., Esmail, G.A., Arasu, M.V., Kim, Y.O., Kim, H., Kim, H.J., 2020. Photo–mediated biosynthesis and characterization of silver nanoparticles using bacterial xylanases as reductant: role of synthesized product (Xyl–AgNPs) in fruits juice clarification. Surf. Interface 21, 100747.

Dimkpa, C.O., McLean, J.E., Martineau, N., Britt, D.W., Haverkamp, R., Anderson, A.J., 2013. Silver nanoparticles disrupt wheat (*Triticum aestivum* L.) growth in a sand matrix. Environ. Sci. Technol. 47, 1082–1090.

Dong, X., Ji, X., Jing, J., Li, M., Li, J., Yang, W., 2010. Synthesis of triangular silver nanoprisms by stepwise reduction of sodium borohydride and trisodium citrate. J. Phys. Chem. C 114 (5), 2070–2074.

Durán, N., Cuevas, R., Cordi, L., Rubilar, O., Diez, M.C., 2014. Biogenic silver nanoparticles associated with silver chloride nanoparticles (Ag@ AgCl) produced by laccase from *Trametes versicolor*. SpringerPlus 3 (1), 1–7.

El Shanshoury, A.E.R., Sabae, S.Z., El Shouny, W.A., Abu Shady, A.M., Badr, H.M., 2020. Extracellular Biosynthesis of silver nanoparticles using aquatic bacterial isolate and its antibacterial and antioxidant potentials. Egypt. J. Aquat. Biol. Fisher 24 (7-Special issue), 183–201.

El-Bendary, M.A., Moharam, M.E., Abdelraof, M., Allam, M.A., Roshdy, A.M., Shaheen, M.N., Elmahdy, E.M., Elkomy, G.M., 2020. Multi–bioactive silver nanoparticles synthesized using mosquitocidal bacilli and their characterization. Arch. Microbiol. 202 (1), 63–75.

Elegbede, J.A., Lateef, A., Azeez, M.A., Asafa, T.B., Yekeen, T.A., Oladipo, I.C., Adebayo, E.A., Beukes, L.S., Gueguim–Kana, E.B., 2018. Fungal xylanases–mediated synthesis of silver nanoparticles for catalytic and biomedical applications. IET Nanobiotechnol. 12 (6), 857–863.

El-Khatib, A., Doma, A., Abo-Zaid, G., Badawi, M., Mohamed, M., Mohamed, A., 2020. Antibacterial activity of some nanoparticles prepared by double arc discharge method. Nanostruct. Nano-objects. 23, 100473.

El-Rafie, H.M., El-Rafie, M., Zahran, M.K., 2013. Green synthesis of silver nanoparticles using polysaccharides extracted from marine macro algae. Carbohydr. Polym. 96 (2), 403–410.

El-Shishtawy, R.M., Asiri, A.M., Al-Otaibi, M.M., 2011. Synthesis and spectroscopic studies of stable aqueous dispersion of silver nanoparticles. Spectrochim. Acta A 79 (5), 1505–1510.

Fabrega, J., Renshaw, J.C., Lead, J.R., 2009. Interactions of silver nanoparticles with *Pseudomonas putida* biofilms. Environ. Sci. Technol. 43, 9004–9009.

Falade, A.O., Nwodo, U.U., Iweriebor, B.C., Green, E., Mabinya, L.V., Okoh, A.I., 2017. Lignin peroxidase functionalities and prospective applications. Microbiol. Open 6, e00394.

Fayaz, A.M., Girilal, M., Rahman, M., Venkatesan, R., Kalaichelvan, P.T., 2011. Biosynthesis of silver and gold nanoparticles using thermophilic bacterium *Geobacillus stearothermophilus*. Process Biochem. 46 (10), 1958–1962.

Fernández, J.G., Almeida, C.A., Fernández-Baldo, M.A., Felici, E., Raba, J., Sanz, M.I., 2016. Development of nitrocellulose membrane filters impregnated with different biosynthesized silver nanoparticles applied to water purification. Talanta 146, 237–243.

Fissan, H., Ristig, S., Kaminski, H., Asbach, C., Epple, M., 2014. Comparison of different characterization methods for nanoparticle dispersions before and after aerosolization. Anal. Methods 6, 7324.

Foldbjerg, R., Irving, E.S., Hayashi, Y., Sutherland, D.S., Thorsen, K., Autrup, H., Beer, C., 2012. Global gene expression profiling of human lung epithelial cells after exposure to nanosilver. Toxicol. Sci. 130, 145–157.

Gaidhani, S.V., Raskar, A.V., Poddar, S., Gosavi, S., Sahu, P.K., Pardesi, K.R., Bhide, S.V., Chopade, B.A., 2014. Time dependent enhanced resistance against antibiotics & metal salts by planktonic & biofilm form of Acinetobacter haemolyticus MMC 8 clinical isolate. Indian J. Med. Res. 140 (5), 665.

Gakiya-Teruya, M., Palomino-Marcelo, L., Rodriguez-Reyes, J.C.F., 2019. Synthesis of highly concentrated suspensions of silver nanoparticles by two versions of the chemical reduction method. Methods Protoc. 2 (1), 3.

Garg, D., Sarkar, A., Chand, P., Bansal, P., Gola, D., Sharma, S., Khantwal, S., Surabhi, Mehrotra, R., Chauhan, N., Bharti, R., 2020. Synthesis of silver nanoparticles utilizing various biological systems: mechanisms and applications—a review. Prog. Biomater. 9 (3), 81–95.

Ge, L., Li, Q., Wang, M., Ouyang, J., Li, X., Xing, M.M., 2014. Nanosilver particles in medical applications: synthesis, performance, and toxicity. Int. Nanomedicine 9, 2399.

Giannini, C., Ladisa, M., Altamura, D., Siliqi, D., Sibillano, T., De Caro, L., 2016. X–ray diffraction: a powerful technique for the multiple–length–scale structural analysis of nanomaterials. Crystals 6, 87.

Gong, H., Zhou, L., Su, X., Xiao, S., Liu, S., Wang, Q., 2009. Illuminating dark plasmons of silver nanoantenna rings to enhance exciton–plasmon interactions. Adv. Funct. Mater. 19 (2), 298–303.

Goudarzi, M., Mir, N., Mousavi-Kamazani, M., Bagheri, S., Salavati-Niasari, M., 2016. Biosynthesis and characterization of silver nanoparticles prepared from two novel natural precursors by facile thermal decomposition methods. Sci. Rep. 6 (1), 32539.

Goulet, P., Lennox, R., 2010. New insights into Brust–Schiffrin metal nanoparticle synthesis. J. Am. Chem. Soc. 132 (28), 9582–9584.

Govindaraju, K., Kiruthiga, V., Kumar, V.G., Singaravelu, G., 2009. Extracellular synthesis of silver nanoparticles by a marine alga, *Sargassum wightii* Grevilli and their antibacterial effects. J. Nanosci. Nanotechnol. 9 (9), 5497–5501.

Govindaraju, K., Krishnamoorthy, K., Alsagaby, S.A., Singaravelu, G., Premanathan, M., 2015. Green synthesis of silver nanoparticles for selective toxicity towards cancer cells. IET Nanobiotechnol. 9 (6), 325–330.

Gruyer, N., Dorais, M., Bastien, C., Dassylva, N., Triffault–Bouchet, G., 2013. Interaction between silver nanoparticles and plant growth. In: International Symposium on New Technologies for Environment Control, Energy-Saving and Crop Production in Greenhouse and Plant Factory. vol. 1037, pp. 795–800.

Gu, X., Nie, C., Lai, Y., Lin, C., 2006. Synthesis of silver nanorods and nanowires by tartrate–reduced route in aqueous solutions. Mater. Chem. Phys. 96 (2–3), 217–222.

Guilger-Casagrande, M., Germano-Costa, T., Pasquoto-Stigliani, T., Fraceto, L.F., de Lima, R., 2019. Biosynthesis of silver nanoparticles employing *Trichoderma harzianum* with enzymatic stimulation for the control of Sclerotinia sclerotiorum. Sci. Rep. 9 (1), 1–9.

Gupta, A., Koirala, A.R., Gupta, B., Parajuli, N., 2019. Improved method for separation of silver nanoparticles synthesized using the *Nyctanthes arbortristis* shrub. Acta Chem. Malays. 3, 35–42.

Haggag, E.G., Elshamy, A.M., Rabeh, M.A., Gabr, N.M., Salem, M., Youssif, K.A., Samir, A., Muhsinah, A.B., Alsayari, A., Abdelmohsen, U.R., 2019. Antiviral potential of green synthesized silver nanoparticles of *Lampranthus coccineus* and *Malephora lutea*. Int. J. Nanomedicine 14, 6217.

Hall, J.B., Dobrovolskaia, M.A., Patri, A.K., McNeil, S.E., 2007. Characterization of nanoparticles for therapeutics. Nanomed. Nanotechnol. Biol. Med. 2, 789–803.

Hamedi, S., Ghaseminezhad, M., Shokrollahzadeh, S., Shojaosadati, S.A., 2017. Controlled biosynthesis of silver nanoparticles using nitrate reductase enzyme induction of filamentous fungus and their antibacterial evaluation. Artif. Cell Nanomed. Biotechnol. 45 (8), 1588–1596.

Hamouda, R.A., Hussein, M.H., Abo-Elmagd, R.A., Bawazir, S.S., 2019. Synthesis and biological characterization of silver nanoparticles derived from the cyanobacterium *Oscillatoria limnetica*. Sci. Rep. 9 (1), 1–17.

He, R., Qian, X., Yin, J., Zhu, Z., 2002. Preparation of polychrome silver nanoparticles in different solvents. J. Mater. Chem. 12, 3783–3786.

Helmlinger, J., Sengstock, C., Groß-Heitfeld, C., Mayer, C., Schildhauer, T.A., Köller, M., Epple, M., 2016. Silver nanoparticles with different size and shape: equal cytotoxicity, but different antibacterial effects. RSC Adv. 6 (22), 18490–18501.

Hind, A.R., Bhargava, S.K., McKinnon, A., 2001. At the solid/liquid interface: FTIR/ATR-the tool of choice. Adv. Colloid Interface Sci. 93, 91–114.

Hochella, M., 2002. There's plenty of room at the bottom: nanoscience in geochemistry. Geochim. Cosmochim. Acta 66 (5), 735–743.

Holzwarth, U., Gibson, N., 2011. The Scherrer equation versus the'Debye–Scherrer equation'. Nat. Nanotechnol. 6, 534.

Hsieh, C.T., Tzou, D.Y., Pan, C., Chen, W.Y., 2012. Microwave–assisted deposition, scalable coating, and wetting behavior of silver nanowire layers. Surf. Coat. Technol. 207, 11–18.

Hu, J., Chen, Q., Xie, Z., Han, G., Wang, R., Ren, B., Zhang, Y., Yang, Z., Tian, Z., 2004. A simple and effective route for the synthesis of crystalline silver nanorods and nanowires. Adv. Funct. Mater. 14 (2), 183–189.

Hu, Z., Hung, F., Chang, S., Hsieh, W., Chen, K., 2012. Align Ag nanorods via oxidation reduction growth using RF–sputtering. J. Nanomater. 2012, 1–6.

Huang, T., Xu, X.H.N., 2010. Synthesis and characterization of tunable rainbow colored colloidal silver nanoparticles using single–nanoparticle plasmonic microscopy and spectroscopy. J. Mater. Chem. 20, 9867–9876.

Huq, M., 2020. Green synthesis of silver nanoparticles using *Pseudoduganella eburnea* MAHUQ–39 and their antimicrobial mechanisms investigation against drug resistant human pathogens. Int. J. Mol. Sci. 21 (4), 1510.

Husain, S., Sardar, M., Fatma, T., 2015. Screening of cyanobacterial extracts for synthesis of silver nanoparticles. World J. Microbiol. Biotechnol. 31 (8), 1279–1283.

Husain, S., Afreen, S., Yasin, D., Afzal, B., Fatma, T., 2019. Cyanobacteria as a bioreactor for synthesis of silver nanoparticles–an effect of different reaction conditions on the size of nanoparticles and their dye decolorization ability. J. Microbiol. Methods 162, 77–82.

Hussain, S.M., Hess, K.L., Gearhart, J.M., Geiss, K.T., Schlager, J.J., 2005. *In vitro* toxicity of nanoparticles in BRL 3A rat liver cells. Toxicol. In Vitro 19 (7), 975–983.

Ishida, Y., Nakabayashi, R., Corpuz, R., Yonezawa, T., 2017. Water–dispersible fluorescent silver nanoparticles via sputtering deposition over liquid polymer using a very short thiol ligand. Colloids Surf. A Physicochem. Eng. Asp. 518, 25–29.

Jabłońska, J., Jankowski, K., Tomasik, M., Cykalewicz, D., Uznański, P., Całuch, S., Szybowicz, M., Zakrzewska, J., Mazurek, P., 2021. Preparation of silver nanoparticles in a high voltage AC arc in water. SN Appl. Sci. 3 (2), 244.

Jaffri, S., Ahmad, K., 2020. Biomimetic detoxifier *Prunus cerasifera* Ehrh. Silver nanoparticles: innate green bullets for morbific pathogens and persistent pollutants. Environ. Sci. Pollut. Res. 27, 9669–9685.

Jain, A., Ahmad, F., Gola, D., Malik, A., Chauhan, N., Dey, P., Tyagi, P., 2020. Multi dye degradation and antibacterial potential of papaya leaf derived silver nanoparticles. Environ. Nanotechnol. Monit. Manag. 14, 100337.

Jana, N., Gearheart, L., Murphy, C., 2001. Wet chemical synthesis of silver nanorods and nanowires of controllable aspect ratio. Chem. Commun. (7), 617–618.

Jang, E.Y., Son, Y.J., Park, S.Y., Yoo, J.Y., Cho, Y.N., Jeong, S.Y., Liu, S., Son, H.J., 2018. Improved biosynthesis of silver nanoparticles using keratinase from *Stenotrophomonas maltophilia* R13: reaction optimization, structural characterization, and biomedical activity. Bioprocess Biosyst. Eng. 41 (3), 381–393.

Jasim, B., Thomas, R., Mathew, J., Radhakrishnan, E.K., 2017. Plant growth and diosgenin enhancement effect of silver nanoparticles in fenugreek (*Trigonella foenumgraecum* L.). Saudi Pharma. J. 25, 443–447.

Jayaseelan, C., Rahuman, A.A., Rajakumar, G., Kirthi, A.V., Santhoshkumar, T., Marimuthu, S., Bagavan, A., Kamaraj, C., Zahir, A.A., Elango, G., 2011. Synthesis of pediculocidal and larvicidal silver nanoparticles by leaf extract from heartleaf moonseed plant, *Tinospora cordifolia* Miers. Parasitol. Res. 109, 185–194.

Jeevanandam, P., Srikanth, C., Dixit, S., 2010. Synthesis of monodisperse silver nanoparticles and their self–assembly through simple thermal decomposition approach. Mater. Chem. Phys. 122 (2–3), 402–407.

Jiang, P., Li, S., Xie, S., Gao, Y., Song, L., 2004. Machinable long PVP–stabilized silver nanowires. Chem. A Eur. J. 10 (19), 4817–4821.

Jiang, Z., Yu, X., Zhai, S., Hao, Y., 2017. Ratiometric dissolved oxygen sensors based on ruthenium complex doped with silver nanoparticles. Sensors 17, 548.

Jini, J., Rohiniraj, N.R., 2017. Production and partial purification of extracellular laccase from mushroom *Pleurotus* sp. and its application in green synthesis of silver nanoparticle. Int. J. Res. Appl. Sci. Eng. Technol. 5 (12), 2611–2619.

Jo, J.H., Singh, P., Kim, Y.J., Wang, C., Mathiyalagan, R., Jin, C.G., Yang, D.C., 2016. *Pseudomonas deceptionensis* DC5–mediated synthesis of extracellular silver nanoparticles. Artif. Cells Nanomed. Biotechnol. 44 (6), 1576–1581.

Jung, J., Oh, H., Ji, J., Kim, S., 2007. In-situ gold nanoparticle generation using a small–sized ceramic heater with a local heating area. Mater. Sci. Forum 544–545, 1001–1004.

Kalpana, D., Lee, Y.S., 2013. Synthesis and characterization of bactericidal silver nanoparticles using cultural filtrate of simulated microgravity grown *Klebsiella pneumoniae*. Enzyme Microb. Technol. 52 (3), 151–156.

Kamble, R.D., Jadhav, A.R., 2012. Isolation, purification, and characterization of xylanase produced by a new species of Bacillus in solid state fermentation. Int. J. Microbiol 2012, 1–8.

Kannan, R.R.R., Arumugam, R., Ramya, D., Manivannan, K., Anantharaman, P., 2013. Green synthesis of silver nanoparticles using marine macroalga *Chaetomorpha linum*. Appl. Nanosci. 3 (3), 229–233.

Karimi-Maleh, H., Kumar, B.G., Rajendran, S., Qin, J., Vadivel, S., Durgalakshmi, D., Gracia, F., Soto-Moscoso, M., Orooji, Y., Karimi, F., 2020. Tuning of metal oxides photocatalytic performance using Ag nanoparticles integration. J. Mol. Liq. 314, 113588.

Karthik, C., Radha, K.V., 2012. Biosynthesis and characterization of silver nanoparticles using Enterobacter aerogenes: a kinetic approach. Dig. J. Nanomater. Biostruct. 7 (3), 1007–1014.

Kathiresan, K., Manivannan, S., Nabeel, M.A., Dhivya, B., 2009. Studies on silver nanoparticles synthesized by a marine fungus, *Penicillium fellutanum* isolated from coastal mangrove sediment. Colloids Surf. B. Biointerfaces 71 (1), 133–137.

Kelly, J., Keegan, G., Brennan-Fournet, M., 2012. Triangular silver nanoparticles: their preparation, functionalisation and properties. Acta Phys. Pol. A 122 (2), 337–345.

Khan, T., Yasmin, A., Townley, H.E., 2020. An evaluation of the activity of biologically synthesized silver nanoparticles against bacteria, fungi and mammalian cell lines. Colloids Surf. B Biointerfaces 194, 111156.

Khandel, P., Shahi, S.K., 2018. Mycogenic nanoparticles and their bio–prospective applications: current status and future challenges. J. Nanostructure. Chem. 8 (4), 369–391.

Khanna, P., Kaur, A., Goyal, D., 2019. Algae–based metallic nanoparticles: synthesis, characterization and applications. J. Microbiol. Methods 163, 105656.

Khateef, R., Khadri, H., Almatroudi, A., Alsuhaibani, S.A., Mobeen, S.A., Khan, R.A., 2019. Potential in vitro anti–breast cancer activity of green–synthesized silver nanoparticles preparation against human MCF-7 cell-lines. Adv. Nat. Sci. Nanosci. Nanotechnol. 10 (4), 045012.

Khiyami, M.A., Almoammar, H., Awad, Y.M., Alghuthaymi, M.A., Abd-Elsalam, K.A., 2014. Plant pathogen nanodiagnostic techniques: forthcoming changes? Biotechnol. Biotechnol. Equip. 28, 775–785.

Khodashenas, B., Ghorbani, H., 2019. Synthesis of silver nanoparticles with different shapes. Arab. J. Chem. 12 (8), 1823–1838.

Kim, S.W., Jung, J.H., Lamsal, K., Kim, Y.S., Min, J.S., Lee, Y.S., 2012. Antifungal effects of silver nanoparticles (AgNPs) against various plant pathogenic fungi. Mycobiology 40, 53–58.

Korbekandi, H., Ashari, Z., Iravani, S., Abbasi, S., 2013. Optimization of biological synthesis of silver nanoparticles using *Fusarium oxysporum*. Iran J. Pharma. Res. 12 (3), 289.

Kraśniewska, K., Galus, S., Gniewosz, M., 2020. Biopolymers–based materials containing silver nanoparticles as active packaging for food applications-a review. Int. J. Mol Sci. 21, 698.

Krishnan, M., Sivanandham, V., Hans-Uwe, D., Murugaiah, S.G., Seeni, P., Gopalan, S., Rathinam, A.J., 2015. Antifouling assessments on biogenic nanoparticles: a field study from polluted offshore platform. Mar. Pollut. Bull. 101 (2), 816–825.

Krishnaraj, R.N., Berchmans, S., 2013. In vitro antiplatelet activity of silver nanoparticles synthesized using the microorganism *Gluconobacter roseus*: an AFM-based study. RSC Adv. 3 (23), 8953–8959.

Krishnaraj, C., Jagan, E.G., Ramachandran, R., Abirami, S.M., Mohan, N., Kalaichelvan, P.T., 2012. Effect of biologically synthesized silver nanoparticles on *Bacopa monnieri* (Linn.) Wettst. Plant growth metabolism. Process Biochem. 47, 651–658.

Krutyakov, Y., Kudrinskiy, A., Olenin, A., Lisichkin, G., 2008. Synthesis and properties of silver nanoparticles: advances and prospects. Russ. Chem. Rev. 77 (3), 233–257.

Kshirsagar, P., Sangaru, S.S., Malvindi, M.A., Martiradonna, L., Cingolani, R., Pompa, P.P., 2011. Synthesis of highly stable silver nanoparticles by photoreduction and their size fractionation by phase transfer method. Colloids Surf. A Physicochem. Eng. Asp. 392 (1), 264–270.

Kumar, C.G., Mamidyala, S.K., 2011. Extracellular synthesis of silver nanoparticles using culture supernatant of *Pseudomonas aeruginosa*. Colloids Surf. B. Biointerfaces 84 (2), 462–466.

Kumar, P., Senthamilselvi, S., Lakshmipraba, A., Premkumar, K., Muthukumaran, R., Visvanathan, P., Ganeshkumar, R.S., Govindaraju, M., 2012a. Efficacy of bio–synthesized silver nanoparticles using *Acanthophora spicifera* to encumber biofilm formation. Dig. J. Nanomater. Biostruct. 7, 511–522.

Kumar, P., Selvi, S.S., Praba, A.L., Selvaraj, M., 2012b. Antibacterial activity and in–vitro cytotoxicity assay against brine shrimp using silver nanoparticles synthesized from *Sargassum ilicifolium*. Dig. J. Nanomater. Biostruct. 7, 1447–1455.

Kumari, R.M., Kumar, V., Kumar, M., Pareek, N., Nimesh, S., 2020. Assessment of antibacterial and anticancer capability of silver nanoparticles extracellularly biosynthesized using *Aspergillus terreus*. Nano Express 1 (3), 030011.

Laban, G., Nies, L.F., Turco, R.F., Bickham, J.W., Sepúlveda, M.S., 2010. The effects of silver nanoparticles on fathead minnow (*Pimephales promelas*) embryos. Ecotoxicology 19, 185–195.

Lackner, P., Beer, R., Broessner, G., Helbok, R., Galiano, K., Pleifer, C., Pfausler, B., Brenneis, C., Huck, C., Engelhardt, K., Obwegeser, A.A., 2008. Efficacy of silver nanoparticles–impregnated external ventricular drain catheters in patients with acute occlusive hydrocephalus. Neurocrit. Care 8, 360–365.

Lara, H.H., Ayala-Nuñez, N.V., Ixtepan-Turrent, L., Rodriguez-Padilla, C., 2010. Mode of antiviral action of silver nanoparticles against HIV–1. J. Nanobiotechnol. 8, 1–10.

Lateef, A., Adelere, I.A., Gueguim-Kana, E.B., Asafa, T.B., Beukes, L.S., 2015. Green synthesis of silver nanoparticles using keratinase obtained from a strain of *Bacillus safensis* LAU 13. Int. Nano Lett. 5 (1), 29–35.

Lateef, A., Ojo, S.A., Oladejo, S.M., 2016. Anti–candida, anti–coagulant and thrombolytic activities of biosynthesized silver nanoparticles using cell–free extract of *Bacillus safensis* LAU 13. Process Biochem. 51, 1406–1412.

Lateef, A., Ojo, S.A., Elegbede, J.A., Akinola, P.O., Akanni, E.O., 2018. Nanomedical applications of nanoparticles for blood coagulation disorders. In: Dasgupta, N., Ranjan, S., Lichtfouse, E. (Eds.), Environmental Nanotechnology. Environmental Chemistry for a Sustainable World, vol. 14. Springer, Cham, pp. 243–277.

Lee, S., Jun, B., 2019. Silver nanoparticles: synthesis and application for nanomedicine. Int. J. Mol. Sci. 20 (4), 865.

Lee, D., Kang, Y., 2004. Synthesis of silver nanocrystallites by a new thermal decomposition method and their characterization. ETRI J. 26 (3), 252–256.

Li, G., He, D., Qian, Y., Guan, B., Gao, S., Cui, Y., Yokoyama, K., Wang, L., 2012a. Fungus–mediated green synthesis of silver nanoparticles using *Aspergillus terreus*. Int. J. Mol. Sci. 13 (1), 466–476.

Li, K., Jia, X., Tang, A., Zhu, X., Meng, H., Wang, Y., 2012b. Preparation of spherical and triangular silver nanoparticles by a convenient method. Integr. Ferroelectr. 136 (1), 9–14.

Li, X., Schirmer, K., Bernard, L., Sigg, L., Pillai, S., Behra, R., 2015. Silver nanoparticle toxicity and association with the alga *Euglena gracilis*. Environ. Sci. 2 (6), 594–602.

Liang, H., Wang, W., Huang, Y., Zhang, S., Wei, H., Xu, H., 2010. Controlled synthesis of uniform silver nanospheres. J. Phys. Chem. C 114 (16), 7427–7431.

Lim, J., Yeap, S., Che, H., Low, S., 2013. Characterization of magnetic nanoparticle by dynamic light scattering. Nanoscale Res. Lett. 8 (1), 381.

Lin, J., Huang, Z., Wu, H., Zhou, W., Jin, P., Wei, P., Zhang, Y., Zheng, F., Zhang, J., Xu, J., Hu, Y., 2014a. Inhibition of autophagy enhances the anticancer activity of silver nanoparticles. Autophagy 10, 2006–2020.

Lin, P.C., Lin, S., Wang, P.C., Sridhar, R., 2014b. Techniques for physicochemical characterization of nanomaterials. Biotechnol. Adv. 32, 711–726.

Madakka, M., Jayaraju, N., Rajesh, N., 2018. Mycosynthesis of silver nanoparticles and their characterization. MethodsX 5, 20–29.

Madhiyazhagan, P., Murugan, K., Kumar, A.N., Nataraj, T., Dinesh, D., Panneerselvam, C., Subramaniam, J., Kumar, P.M., Suresh, U., Roni, M., Nicoletti, M., 2015. *Sargassum muticum*–synthesized silver nanoparticles: an effective control tool against mosquito vectors and bacterial pathogens. Parasitol. Res. 114 (11), 4305–4317.

Makarov, V.V., Love, A.J., Sinitsyna, O.V., Makarova, S.S., Yaminsky, I.V., Taliansky, M.E., Kalinina, N.O., 2014. "Green" nanotechnologies: synthesis of metal nanoparticles using plants. Acta Nat. 6 (1), 35–44.

Malakootian, M., Yaseri, M., Faraji, M., 2019. Removal of antibiotics from aqueous solutions by nanoparticles: a systematic review and meta-analysis. Environ. Sci. Pollut. Res. 26, 8444–8458.

Maliszewska, I., Juraszek, A., Bielska, K., 2014. Green synthesis and characterization of silver nanoparticles using ascomycota fungi *Penicillium nalgiovense* AJ12. J. Clust. Sci. 25 (4), 989–1004.

Mane, P., Chaudhari, R., Qureshi, N., 2019. Silver nanoparticlessilk fibroin nanocomposite based colorimetric bio–interfacial sensor for on–site ultra–trace impurity detection of mercury ions. J. Nanosci. Nanotechnol. 20, 2122–2129.

Manikandan, V., Velmurugan, P., Park, J.H., Chang, W.S., Park, Y.J., Jayanthi, P., Cho, M., Oh, B.T., 2017. Green synthesis of silver oxide nanoparticles and its antibacterial activity against dental pathogens. 3 Biotech 7, 72.

Massironi, A., Morelli, A., Grassi, L., Puppi, D., Braccini, S., Maisetta, G., Esin, S., Batoni, G., Della Pina, C., Chiellini, F., 2019. Ulvan as novel reducing and stabilizing agent from renewable algal biomass: application to green synthesis of silver nanoparticles. Carbohydr. Polym. 203, 310–321.

Matei, A., Matei, S., Matei, G.M., Cogălniceanu, G., Cornea, C.P., 2020. Biosynthesis of silver nanoparticles mediated by culture filtrate of lactic acid bacteria, characterization and antifungal activity. EuroBiotech J. 4 (2), 97–103.

Mehta, C.M., Srivastava, R., Arora, S., Sharma, A.K., 2016. Impact assessment of silver nanoparticles on plant growth and soil bacterial diversity. 3 Biotech 6, 1–10.

Mikhailova, E.O., 2020. Silver nanoparticles: mechanism of action and probable bio–application. J. Funct. Biomater. 11 (4), 84.

Miller, C.N., Newall, N., Kapp, S.E., Lewin, G., Karimi, L., Carville, K., Gliddon, T., Santamaria, N.M., 2010. A randomized-controlled trial comparing cadexomer iodine and nanocrystalline silver on the healing of leg ulcers. Wound Repair Regen. 18, 359–367.

Mishra, A., Sardar, M., 2012. Alpha-amylase mediated synthesis of silver nanoparticles. Sci. Adv. Mater. 4 (1), 143–146.

Mishra, A., Sardar, M., 2015. Cellulase assisted synthesis of nano–silver and gold: application as immobilization matrix for biocatalysis. Int. J. Biol. Macromol. 77, 105–113.

Mishra, S., Singh, H.B., 2015. Biosynthesized silver nanoparticles as a nanoweapon against phytopathogens: exploring their scope and potential in agriculture. Appl. Microbiol. Biotechnol. 99, 1097–1107.

Mishra, A., Ahmad, R., Singh, V., Gupta, M.N., Sardar, M., 2013. Preparation, characterization and biocatalytic activity of a nanoconjugate of alpha amylase and silver nanoparticles. J. Nanosci. Nanotechnol. 13 (7), 5028–5033.

Mobed, A., Hasanzadeh, M., Shadjou, N., Hassanpour, S., Saadati, A., Agazadeh, M., 2020. Immobilization of ssDNA on the surface of silver nanoparticles–graphene quantum dots modified by gold nanoparticles towards biosensing of microorganism. Microchem. J. 152, 104286.

Mochi, F., Burratti, L., Fratoddi, I., Venditti, I., Battocchio, C., Carlini, L., Iucci, G., Casalboni, M., De Matteis, F., Casciardi, S., Nappini, S., 2018. Plasmonic sensor based on interaction between silver nanoparticles and Ni^{2+} or Co^{2+} in water. Nanomaterials 8, 488.

Moghimi-Rad, J., Isfahani, T., Hadi, I., Ghalamdaran, S., Sabbaghzadeh, J., Sharif, M., 2011. Shape-controlled synthesis of silver particles by surfactant self-assembly under ultrasound radiation. Appl. Nanosci. 1 (1), 27–35.

Mondal, N.K., Chowdhury, A., Dey, U., Mukhopadhya, P., Chatterjee, S., Das, K., Datta, J.K., 2014. Green synthesis of silver nanoparticles and its application for mosquito control. Asian Pacif J. Trop. Dis. 4, S204–S210.

Moshfegh, M., Forootanfar, H., Zare, B., Shahverdi, A.R., Zarrini, G., Faramarzi, M.A., 2011. Biological synthesis of Au, Ag and Au–Ag bimetallic nanoparticles by α–amylase. Dig. J. Nanomater. Biostruct. 6, 1419–1426.

Mostafa, A., Menazea, A., 2020. Polyvinyl alcohol/silver nanoparticles film prepared via pulsed laser ablation: an eco–friendly nano–catalyst for 4–nitrophenol degradation. J. Mol. Struct. 1212, 128125.

Mourdikoudis, S., Pallares, R.M., Thanh, N.T., 2018. Characterization techniques for nanoparticles: comparison and complementarity upon studying nanoparticle properties. Nanoscale 10, 12871–12934.

Mukherjee, P., Roy, M., Mandal, B.P., Dey, G.K., Mukherjee, P.K., Ghatak, J., Tyagi, A.K., Kale, S.P., 2008. Green synthesis of highly stabilized nanocrystalline silver particles by a non–pathogenic and agriculturally important fungus *T. asperellum*. Nanotechnology 19 (7), 075103.

Mukherjee, K., Gupta, R., Kumar, G., Kumari, S., Biswas, S., Padmanabhan, P., 2018. Synthesis of silver nanoparticles by *Bacillus clausii* and computational profiling of nitrate reductase enzyme involved in production. J. Genet. Eng. Biotechnol. 16 (2), 527–536.

Murugesan, S., Elumalai, M., Dhamotharan, R., 2011. Green synthesis of silver nano particles from marine alga Gracilaria edulis. Bosci. Biotech. Res. Comm 4 (1), 105–110.

Nam, K.Y., 2011. In vitro antimicrobial effect of the tissue conditioner containing silver nanoparticles. J. Adv. Prosthodont. 3, 20.

Nangmenyi, G., Economy, J., 2009. Nonmetallic particles for oligodynamic microbial disinfection. In: Street, A., Sustich, R., Duncan, J., Savage, N. (Eds.), Nanotechnology Application for Clean Water. William Andrew, Norwich, NY, pp. 3–15.

Narayanan, K.B., Sakthivel, N., 2013. Biosynthesis of silver nanoparticles by phytopathogen *Xanthomonas oryzae* pv. *oryzae* strain BXO8. J. Microbiol. Biotechnol. 23 (9), 1287–1292.

Navaladian, S., Viswanathan, B., Viswanath, R., Varadarajan, T., 2006. Thermal decomposition as route for silver nanoparticles. Nanoscale Res. Lett. 2 (1), 44–48.

Ngeontae, W., Janrungroatsakul, W., Maneewattanapinyo, P., Ekgasit, S., Aeungmaitrepirom, W., Tuntulani, T., 2009. Novel potentiometric approach in glucose biosensor using silver nanoparticles as redox marker. Sens. Actuators B 137, 320–326.

Nghia, N.V., Truong, N.N., Thong, N.M., Hung, N.P., 2012. Synthesis of nanowire–shaped silver by polyol process of sodium chloride. Int. J. Mater. Chem. 2 (2), 75–78.

Nnolim, N.E., Udenigwe, C.C., Okoh, A.I., Nwodo, U.U., 2020. Microbial keratinase: next generation green catalyst and prospective applications. Front. Microbiol. 11, 3280.

Ocsoy, I., Paret, M.L., Ocsoy, M.A., Kunwar, S., Chen, T., You, M., Tan, W., 2013. Nanotechnology in plant disease management: DNA–directed silver nanoparticles on graphene oxide as an antibacterial against *Xanthomonas perforans*. ACS Nano 7, 8972–8980.

Oluwafemi, O.S., Anyik, J.L., Zikalala, N.E., 2019. Biosynthesis of silver nanoparticles from water hyacinth plant leaves extract for colourimetric sensing of heavy metals. Nano-Str. Nano-Obj. 20, 100387.

Otari, S.V., Patil, R.M., Ghosh, S.J., Thorat, N.D., Pawar, S.H., 2015. Intracellular synthesis of silver nanoparticle by actinobacteria and its antimicrobial activity. Spectrochim. Acta A Mol. Biomol. Spectrosc. 136, 1175–1180.

Ovais, M., Khalil, A.T., Ayaz, M., Ahmad, I., Nethi, S.K., Mukherjee, S., 2018. Biosynthesis of metal nanoparticles via microbial enzymes: a mechanistic approach. Int. J. Mol. Sci. 19 (12), 4100.

Oves, M., Khan, M.S., Zaidi, A., Ahmed, A.S., Ahmed, F., Ahmad, E., Sherwani, A., Owais, M., Azam, A., 2013. Antibacterial and cytotoxic efficacy of extracellular silver nanoparticles biofabricated from chromium reducing novel OS4 strain of Stenotrophomonas maltophilia. PLoS One 8 (3), e59140.

Oza, G., Reyes-Calderón, A., Mewada, A., Arriaga, L.G., Cabrera, G.B., Luna, D.E., Iqbal, H. M., Sharon, M., Sharma, A., 2020. Plant–based metal and metal alloy nanoparticle synthesis: a comprehensive mechanistic approach. J. Mater. Sci. 55 (4), 1309–1330.

Öztürk, B.Y., 2019. Intracellular and extracellular green synthesis of silver nanoparticles using *Desmodesmus* sp.: their antibacterial and antifungal effects. Caryologia Int. J. Cytol. Cytosystematic. Cytogenet. 72 (1), 29–43.

Padhi, S., Mirza, M., Verma, D., Khuroo, T., Panda, A., Talegaonkar, S., Khar, R., Iqbal, Z., 2015. Revisiting the nanoformulation design approach for effective delivery of topotecan in its stable form: an appraisal of its in vitro behavior and tumor amelioration potential. Drug Deliv. 23 (8), 2827–2837.

Padhi, S., Kapoor, R., Verma, D., Panda, A., Iqbal, Z., 2018. Formulation and optimization of topotecan nanoparticles: in vitro characterization, cytotoxicity, cellular uptake and pharmacokinetic outcomes. J. Photochem. Photobiol. B 183, 222–232.

Pal, S., Tak, Y.K., Song, J.M., 2007. Does the antibacterial activity of silver nanoparticles depend on the shape of the nanoparticle? A study of the gram–negative bacterium Escherichia coli. Appl. Environ. Microbiol. 73, 1712.

Palanisamy, N.K., Ferina, N., Amirulhusni, A.N., Mohd-Zain, Z., Hussaini, J., Ping, L.J., Durairaj, R., 2014. Antibiofilm properties of chemically synthesized silver nanoparticles found against *Pseudomonas aeruginosa*. J. Nanobiotechnol. 12, 1–7.

Pandey, A., Shankar, S., Arora, N.K., 2018. Amylase–assisted green synthesis of silver nanocubes for antibacterial applications. Bioinsp. Biomimetic Nanobiomater 8, 161–170.

Paramelle, D., Sadovoy, A., Gorelik, S., Free, P., Hobley, J., Fernig, D.G., 2014. A rapid method to estimate the concentration of citrate capped silver nanoparticles from UV–visible light spectra. Analyst 139, 4855–4861.

Parikh, R.Y., Singh, S., Prasad, B.L.V., Patole, M.S., Sastry, M., Shouche, Y.S., 2008. Extracellular synthesis of crystalline silver nanoparticles and molecular evidence of silver resistance from Morganella sp.: towards understanding biochemical synthesis mechanism. ChemBio. Chem. 9 (9), 1415–1422.

Park, H.J., Kim, S.H., Kim, H.J., Choi, S.H., 2006. A new composition of nanosized silica–silver for control of various plant diseases. Plant Pathol. J. 22, 295–302.

Pérez-Díaz, M.A., Boegli, L., James, G., Velasquillo, C., Sanchez-Sanchez, R., Martinez-Martinez, R.E., Martínez-Castañón, G.A., Martinez-Gutierrez, F., 2015. Silver nanoparticles with antimicrobial activities against *Streptococcus mutans* and their cytotoxic effect. Mater. Sci. Eng. C 55, 360–366.

Pillai, Z., Kamat, P., 2003. What factors control the size and shape of silver nanoparticles in the citrate ion reduction method? J. Phys. Chem. B 108 (3), 945–951.

Popli, D., Anil, V., Subramanyam, A.B., Namratha, M.N., Ranjitha, V.R., Rao, S.N., Rai, R. V., Govindappa, M., 2018. Endophyte fungi, Cladosporium species–mediated synthesis of silver nanoparticles possessing in vitro antioxidant, anti–diabetic and anti–Alzheimer activity. Artif. Cell Nanomed. Biotechnol. 46 (Suppl.1), 676–683.

Pourali, P., Baserisalehi, M., Afsharnezhad, S., Behravan, J., Alavi, H., Hosseini, A., 2012. Biological synthesis of silver and gold nanoparticles by bacteria in different temperatures (37 °C and 50 °C). J. Pure Appl. Microbiol. 6, 757–763.

Priyadarshini, S., Gopinath, V., Priyadharsshini, N.M., MubarakAli, D., Velusamy, P., 2013. Synthesis of anisotropic silver nanoparticles using novel strain, *Bacillus flexus* and its biomedical application. Colloids Surf. B 102, 232–237.

Pyatenko, A., Shimokawa, K., Yamaguchi, M., Nishimura, O., Suzuki, M., 2004. Synthesis of silver nanoparticles by laser ablation in pure water. Appl. Phys. A Mater. Sci. Process. 79 (4–6), 803–806.

Qaralleh, H., Khleifat, K.M., Al-Limoun, M.O., Alzedaneen, F.Y., Al-Tawarah, N., 2019. Antibacterial and synergistic effect of biosynthesized silver nanoparticles using the fungi *Tritirachium oryzae* W5H with essential oil of *Centaurea damascena* to enhance conventional antibiotics activity. Adv. Nat. Sci. Nanosci. Nanotechnol. 10 (2), 025016.

Qin, Y., Ji, X., Jing, J., Liu, H., Wu, H., Yang, W., 2010. Size control over spherical silver nanoparticles by ascorbic acid reduction. Colloids Surf. A Physicochem. Eng. Asp. 372 (1–3), 172–176.

Quintero-Quiroz, C., Acevedo, N., Zapata-Giraldo, J., Botero, L., Quintero, J., Zárate-Triviño, D., Saldarriaga, J., Pérez, V., 2019. Optimization of silver nanoparticle synthesis by chemical reduction and evaluation of its antimicrobial and toxic activity. Biomater. Res. 23 (1), 27.

Rafique, M., Sadaf, I., Rafique, M., Tahir, M., 2016. A review on green synthesis of silver nanoparticles and their applications. Artif. Cell Nanomed. B. 45 (7), 1272–1291.

Rahman, A., Kumar, S., Bafana, A., Dahoumane, S.A., Jeffryes, C., 2019. Individual and combined effects of extracellular polymeric substances and whole cell components of *Chlamydomonas reinhardtii* on silver nanoparticle synthesis and stability. Molecules 24 (5), 956.

Rai, T., Panda, D., 2015. An extracellular enzyme synthesizes narrow–sized silver nanoparticles in both water and methanol. Chem. Phys. Lett. 623, 108–112.

Rajesh, S., Raja, D.P., Rathi, J.M., Sahayaraj, K., 2012. Biosynthesis of silver nanoparticles using *Ulva fasciata* (Delile) ethyl acetate extract and its activity against *Xanthomonas campestris* pv. *malvacearum*. J. Biopest. 5, 119–128.

Rajeshkumar, S., Malarkodi, C., Paulkumar, K., Vanaja, M., Gnanajobitha, G., Annadurai, G., 2013. Intracellular and extracellular biosynthesis of silver nanoparticles by using marine bacteria *Vibrio alginolyticus*. Nanosci. Nanotechnol. 3 (1), 21–25.

Ramirez-Lee, M.A., Aguirre-Bañuelos, P., Martinez-Cuevas, P.P., Espinosa-Tanguma, R., Chi-Ahumada, E., Martinez-Castañon, G.A., Gonzalez, C., 2018. Evaluation of cardiovascular responses to silver nanoparticles (AgNPs) in spontaneously hypertensive rats. Nanomed. Nanotechnol. Biol. Med. 14, 385–395.

Ramos, M.M., Morais, E.D.S., Sena, I.D.S., Lima, A.L., de Oliveira, F.R., de Freitas, C.M., Fernandes, C.P., de Carvalho, J.C.T., Ferreira, I.M., 2020. Silver nanoparticle from whole cells of the fungi Trichoderma species isolated from Brazilian Amazon. Biotechnol. Lett. 42, 833–843.

Ranoszek-Soliwoda, K., Tomaszewska, E., Socha, E., Krzyczmonik, P., Ignaczak, A., Orlowski, P., Krzyzowska, M., Celichowski, G., Grobelny, J., 2017. The role of tannic acid and sodium citrate in the synthesis of silver nanoparticles. J. Nanopart. Res. 19 (8), 273.

Rashid, M., Bhuiyan, M., Quayum, M., 2013. Synthesis of silver Nano particles (Ag–NPs) and their uses for quantitative analysis of vitamin C tablets. Dhaka Univ. J. Pharm. Sci. 12 (1), 29–33.

Rigo, C., Ferroni, L., Tocco, I., Roman, M., Munivrana, I., Gardin, C., Cairns, W.R., Vindigni, V., Azzena, B., Barbante, C., Zavan, B., 2013. Active silver nanoparticles for wound healing. Int. J. Mol. Sci. 14, 4817–4840.

Roe, D., Karandikar, B., Bonn-Savage, N., Gibbins, B., Roullet, J.B., 2008. Antimicrobial surface functionalization of plastic catheters by silver nanoparticles. J. Antimicrob. Chemother. 61, 869–876.

Rogers, J.V., Parkinson, C.V., Choi, Y.W., Speshock, J.L., Hussain, S.M., 2008. A preliminary assessment of silver nanoparticle inhibition of monkeypox virus plaque formation. Nanoscale Res. Lett. 3, 129–133.

Romeh, A., 2018. Green silver nanoparticles for enhancing the phytoremediation of soil and water contaminated by fipronil and degradation products. Water Air Soil Pollut. 229, 147.

Rouhani, M., Samih, M.A., Kalantari, S., 2013. Insecticidal effect of silica and silver nanoparticles on the cowpea seed beetle, *Callosobruchus maculatus* F (col: Bruchidae). J. Entomol. Res. Soc. 4, 297–305.

Roy, A., Bulut, O., Some, S., Mandal, A., Yilmaz, M., 2019. Green synthesis of silver nanoparticles: biomolecule–nanoparticle organizations targeting antimicrobial activity. RSC Adv. 9 (5), 2673–2702.

Rujitanaroj, P., Pimpha, N., Supaphol, P., 2008. Wound–dressing materials with antibacterial activity from electrospun gelatin fiber mats containing silver nanoparticles. Polymer 49, 4723–4732.

Saad, A.H.A., Soliman, M.I., Azzam, A.M., Mostafa, A.B., 2015. Antiparasitic activity of silver and copper oxide nanoparticles against *Entamoeba histolytica* and *Cryptosporidium parvum* cysts. J. Egypt. Soc. Parasitol. 45, 593–602.

Sable, S.V., Kawade, S., Ranade, S., Joshi, S., 2020. Bioreduction mechanism of silver nanoparticles. Mater. Sci. Eng. C 107, 110299.

Sadrolhesseini, A., Noor, A., Mahdi, M., Kharazmi, A., Zakaria, A., Yunus, W., Huang, N., 2013. Laser ablation synthesis of silver nanoparticle in graphene oxide and thermal effusivity of nanocomposite. In: 2013 IEEE 4th International Conference on Photonics (ICP), pp. 62–65.

Saeed, S., Iqbal, A., Ashraf, M.A., 2020. Bacterial-mediated synthesis of silver nanoparticles and their significant effect against pathogens. Environ. Sci. Pollut. Res. 27 (30), 37347–37356.

Saha, J., Begum, A., Mukherjee, A., Kumar, S., 2017. A novel green synthesis of silver nanoparticles and their catalytic action in reduction of methylene blue dye. Sustain. Environ. Res. 27, 245–250.

Sakamoto, M., Fujistuka, M., Majima, T., 2009. Light as a construction tool of metal nanoparticles: synthesis and mechanism. J. Photochem. Photobiol. C 10 (1), 33–56.

Salama, H.M., 2012. Effects of silver nanoparticles in some crop plants, common bean (*Phaseolus vulgaris* L.) and corn (*Zea mays* L.). Int. Res. J. Biotechnol. 3, 190–197.

Samadi, N., Golkaran, D., Eslamifar, A., Jamalifar, H., Fazeli, M.R., Mohseni, F.A., 2009. Intra/extracellular biosynthesis of silver nanoparticles by an autochthonous strain of *Proteus mirabilis* isolated from photographic waste. J. Biomed. Nanotechnol. 5 (3), 247–253.

Sanghi, R., Verma, P., 2009. Biomimetic synthesis and characterisation of protein capped silver nanoparticles. Bioresour. Technol. 100 (1), 501–504.

Sankar, R., Karthik, A., Prabu, A., Karthik, S., Shivashangari, K.S., Ravikumar, V., 2013. Origanum vulgare mediated biosynthesis of silver nanoparticles for its antibacterial and anticancer activity. Colloids Surf. B Biointerfaces 108, 80–84.

Santos, V.E., Targino, A.G.R., Flores, M.A.P., Pessoa, H.D.L.F., Galembeck, A., Rosenblatt, A., 2013. Antimicrobial activity of silver nanoparticles in treating dental caries. Revista da Faculdade de Odontologia–UPF 18, 312–315.

Saravanan, M., Barik, S.K., MubarakAli, D., Prakash, P., Pugazhendhi, A., 2018. Synthesis of silver nanoparticles from *Bacillus brevis* (NCIM 2533) and their antibacterial activity against pathogenic bacteria. Microb. Pathog. 116, 221–226.

Satapathy, S., Shukla, S.P., Sandeep, K.P., Singh, A.R., Sharma, N., 2015. Evaluation of the performance of an algal bioreactor for silver nanoparticle production. J. Appl. Phycol. 27 (1), 285–291.

Schwenkbier, L., Pollok, S., König, S., Urban, M., Werres, S., Cialla-May, D., Weber, K., Popp, J., 2015. Towards on–site testing of *Phytophthora* species. Anal. Methods 7, 211–217.

Selvaraj, S., Thangam, R., Fathima, N.N., 2018. Electrospinning of casein nanofibers with silver nanoparticles for potential biomedical applications. Int. J. Biol. Macromol. 120, 1674–1681.

Seshadri, S., Prakash, A., Kowshik, M., 2012. Biosynthesis of silver nanoparticles by marine bacterium, *Idiomarina* sp. PR58–8. Bull. Mater. Sci. 35 (7), 1201–1205.

Shankar, A., Kumar, V., Kaushik, N.K., Kumar, A., Malik, V., Singh, D., Singh, B., 2020. *Sporotrichum thermophile* culture extract–mediated greener synthesis of silver nanoparticles: eco–friendly functional group transformation and anti–bacterial study. Curr. Res. Green Sustain. Chem. 3, 100029.

Shanthi, S., Jayaseelan, B.D., Velusamy, P., Vijayakumar, S., Chih, C.T., Vaseeharan, B., 2016. Biosynthesis of silver nanoparticles using a probiotic *Bacillus licheniformis* Dahb1 and their antibiofilm activity and toxicity effects in *Ceriodaphnia cornuta*. Microb. Pathog. 93, 70–77.

Sharma, P., Bhatt, D., Zaidi, M.G.H., Saradhi, P.P., Khanna, P.K., Arora, S., 2012. Silver nanoparticle–mediated enhancement in growth and antioxidant status of *Brassica juncea*. Appl. Biochem. Biotechnol. 167, 2225–2233.

Sharma, G., Kumar, A., Sharma, S., Naushad, M., Prakash Dwivedi, R., ALOthman, Z., Mola, G., 2019. Novel development of nanoparticles to bimetallic nanoparticles and their composites: a review. J. King Saud Univ. Sci. 31 (2), 257–269.

Siddiqi, K.S., Husen, A., Rao, R.A., 2018. A review on biosynthesis of silver nanoparticles and their biocidal properties. J. Nanobiotechnol. 16 (1), 1–28.

Siegel, J., Kvítek, O., Ulbrich, P., Kolská, Z., Slepička, P., Švorčík, V., 2012. Progressive approach for metal nanoparticle synthesis. Mater. Lett. 89, 47–50.

Singh, R., Wagh, P., Wadhwani, S., Gaidhani, S., Kumbhar, A., Bellare, J., Chopade, B.A., 2013. Synthesis, optimization, and characterization of silver nanoparticles from *Acinetobacter calcoaceticus* and their enhanced antibacterial activity when combined with antibiotics. Int. J. Nanomedicine 8, 4277.

Singh, P., Singh, H., Kim, Y.J., Mathiyalagan, R., Wang, C., Yang, D.C., 2016. Extracellular synthesis of silver and gold nanoparticles by *Sporosarcina koreensis* DC4 and their biological applications. Enzyme Microb. Technol. 86, 75–83.

Singh, R., Shedbalkar, U.U., Nadhe, S.B., Wadhwani, S.A., Chopade, B.A., 2017. Lignin peroxidase mediated silver nanoparticle synthesis in *Acinetobacter* sp. AMB Express 7 (1), 1–10.

Singh, H., Du, J., Singh, P., Yi, T.H., 2018. Extracellular synthesis of silver nanoparticles by *Pseudomonas* sp. THG–LS1. 4 and their antimicrobial application. J. Pharm. Anal. 8 (4), 258–264.

Skrabalak, S., Au, L., Li, X., Xia, Y., 2007. Facile synthesis of Ag nanocubes and Au nanocages. Nat. Protoc. 2 (9), 2182–2190.

Speshock, J.L., Murdock, R.C., Braydich-Stolle, L.K., Schrand, A.M., Hussain, S.M., 2010. Interaction of silver nanoparticles with Tacaribe virus. J. Nanobiotechnol. 8, 1–9.

Sportelli, M., Clemente, M., Izzi, M., Volpe, A., Ancona, A., Picca, R., Palazzo, G., Cioffi, N., 2018. Exceptionally stable silver nanoparticles synthesized by laser ablation in alcoholic organic solvent. Colloids Surf. A Physicochem. Eng. Asp. 559, 148–158.

Sriram, M.I., Kanth, S.B.M., Kalishwaralal, K., Gurunathan, S., 2010. Antitumor activity of silver nanoparticles in Dalton's lymphoma ascites tumor model. Int. J. Nanomedicine 5, 753.

Srivastava, S.K., Constanti, M., 2012. Room temperature biogenic synthesis of multiple nanoparticles (Ag, Pd, Fe, Rh, Ni, Ru, Pt, Co, and Li) by *Pseudomonas aeruginosa* SM1. J. Nanopart. Res. 14 (4), 1–10.

Staszek, M., Siegel, J., Polívková, M., Švorčík, V., 2017. Influence of temperature on silver nanoparticle size prepared by sputtering into PVP–glycerol system. Mater. Lett. 186, 341–344.

Suman, T.Y., Elumalai, D., Kaleena, P.K., Rajasree, S.R., 2013. GC–MS analysis of bioactive components and synthesis of silver nanoparticle using *Ammannia baccifera* aerial extract and its larvicidal activity against malaria and filariasis vectors. Ind. Crop Prod. 47, 239–245.

Sun, R.W.Y., Chen, R., Chung, N.P.Y., Ho, C.M., Lin, C.L.S., Che, C.M., 2005. Silver nanoparticles fabricated in Hepes buffer exhibit cytoprotective activities toward HIV–1 infected cells. Chem. Commun., 5059–5061.

Sun, X., Chen, S., Liu, J., Zhao, S., Yoon, J., 2020. Hydrodynamic cavitation: a promising technology for industrial–scale synthesis of nanomaterials. Front. Chem. 8, 259.

Sung, J.H., Ji, J.H., Park, J.D., Yoon, J.U., Kim, D.S., Jeon, K.S., Song, M.Y., Jeong, J., Han, B.S., Han, J.H., Chung, Y.H., 2009. Subchronic inhalation toxicity of silver nanoparticles. Toxicol. Sci. 108, 452–461.

Suresh, G., Gunasekar, P.H., Kokila, D., Prabhu, D., Dinesh, D., Ravichandran, N., Ramesh, B., Koodalingam, A., Siva, G.V., 2014. Green synthesis of silver nanoparticles using *Delphinium denudatum* root extract exhibits antibacterial and mosquito larvicidal activities. Spectrochim. Acta A Mol. Biomol. Spectrosc. 127, 61–66.

Suriyakalaa, U., Antony, J.J., Suganya, S., Siva, D., Sukirtha, R., Kamalakkannan, S., Pichiah, P.T., Achiraman, S., 2013. Hepatocurative activity of biosynthesized silver nanoparticles fabricated using *Andrographis paniculata*. Colloids Surf. B Biointerfaces 102, 189–194.

Syed, A., Saraswati, S., Kundu, G.C., Ahmad, A., 2013. Biological synthesis of silver nanoparticles using the fungus Humicola sp. and evaluation of their cytoxicity using normal and cancer cell lines. Spectrochim. Acta A Mol. Biomol. Spectrosc. 114, 144–147.

Sylvia Devi, H., Rajmuhon Singh, N., David Singh, T., 2015. A benign approach for synthesis of silver nanoparticles and their application in treatment of organic pollutant. Arab. J. Sci. Eng. 41, 2249–2256.

Talekar, S., Joshi, G., Chougle, R., Nainegali, B., Desai, S., Joshi, A., Kambale, S., Kamat, P., Haripurkar, R., Jadhav, S., Nadar, S., 2014. Preparation of stable cross–linked enzyme aggregates (CLEAs) of NADH–dependent nitrate reductase and its use for silver nanoparticle synthesis from silver nitrate. Cat. Com. 53, 62–66.

Talekar, S., Joshi, A., Chougle, R., Nakhe, A., Bhojwani, R., 2016. Immobilized enzyme mediated synthesis of silver nanoparticles using cross–linked enzyme aggregates (CLEAs) of NADH–dependent nitrate reductase. Nano-Str. Nano-Obj. 6, 23–33.

Tamboli, D.P., Lee, D.S., 2013. Mechanistic antimicrobial approach of extracellularly synthesized silver nanoparticles against gram positive and gram-negative bacteria. J. Hazard. Mater. 260, 878–884.

Tao, A., Sinsermsuksakul, P., Yang, P., 2006. Polyhedral silver nanocrystals with distinct scattering signatures. Angew. Chem. 118 (28), 4713–4717.

Tao, L.Y., Gong, J.S., Su, C., Jiang, M., Li, H., Li, H., Lu, Z.M., Xu, Z.H., Shi, J.S., 2018. Mining and expression of a metagenome–derived keratinase responsible for biosynthesis of silver nanoparticles. ACS Biomater Sci. Eng. 4, 1307–1315.

Tarannum, N., Divya Gautam, Y.K., 2019. Facile green synthesis and applications of silver nanoparticles: a state–of–the–art review. RSC Adv. 9, 34926–34948.

Teodoro, K.B., Shimizu, F.M., Scagion, V.P., Correa, D.S., 2019. Ternary nanocomposites based on cellulose nanowhiskers, silver nanoparticles and electrospun nanofibers: use in an electronic tongue for heavy metal detection. Sens. Actuators B 290, 387–395.

Tharani, S., Bharathi, D., Ranjithkumar, R., 2020. Extracellular green synthesis of chitosan-silver nanoparticles using Lactobacillus reuteri for antibacterial applications. Biocatal. Agric. Biotechnol. 30, 101838.

Theivasanthi, T., Alagar, M., 2011. Nano sized copper particles by electrolytic synthesis and characterizations. Int. J. Phys. Sci. 6, 3662–3671.

Thomas, R., Jasim, B., Mathew, J., Radhakrishnan, E.K., 2012. Extracellular synthesis of silver nanoparticles by endophytic *Bordetella* sp. isolated from Piper nigrum and its antibacterial activity analysis. Nano. Biomed. Eng. 4 (4), 183–187.

Tien, D., Tseng, K., Liao, C., Huang, J., Tsung, T., 2008. Discovery of ionic silver in silver nanoparticle suspension fabricated by arc discharge method. J. Alloys Compd. 463 (1–2), 408–411.

Togashi, T., Tsuchida, K., Soma, S., Nozawa, R., Matsui, J., Kanaizuka, K., Kurihara, M., 2020. Size–tunable continuous–seed–mediated growth of silver nanoparticles in alkylamine mixture via the stepwise thermal decomposition of silver oxalate. Chem. Mater. 32 (21), 9363–9370.

Tomaszewska, E., Soliwoda, K., Kadziola, K., Tkacz-Szczesna, B., Celichowski, G., Cichomski, M., Szmaja, W., Grobelny, J., 2013. Detection limits of DLS and UV–vis spectroscopy in characterization of polydisperse nanoparticles colloids. J. Nanomater. 2013.

Tran, Q., Nguyen, V., Le, A., 2013. Silver nanoparticles: synthesis, properties, toxicology, applications and perspectives. Adv. Nat. Sci. Nanosci. Nanotechnol. 4 (3), 033001.

Tsuji, K., 2001. Microencapsulation of pesticides and their improved handling safety. J. Microencapsul. 18, 137–147.

Turkevich, J., 1985. Colloidal gold. Part I. Gold Bull. 18 (3), 86–91.

Turkevich, J., Kim, G., 1970. Palladium: preparation and catalytic properties of particles of uniform size. Science 169 (3948), 873–879.

Tyagi, S., Tyagi, P.K., Gola, D., Chauhan, N., Bharti, R.K., 2019. Extracellular synthesis of silver nanoparticles using entomopathogenic fungus: characterization and antibacterial potential. SN Appl. Sci. 1 (12), 1–9.

Tyagi, P., Mishra, R., Khan, F., Gupta, D., Gola, D., 2020. Antifungal effects of silver nanoparticles against various plant pathogenic fungi and its safety evaluation on *Drosophila melanogaster*. Biointerface. Res. Appl. Chem. 10 (6), 6587–6596.

Tylkowski, B., Trojanowska, A., Nowak, M., Marciniak, L., Jastrzab, R., 2017. Applications of silver nanoparticles stabilized and/or immobilized by polymer matrixes. Phys. Sci. Rev. 2 (7), 1–16.

Upadhyayula, V.K., 2012. Functionalized gold nanoparticle supported sensory mechanisms applied in detection of chemical and biological threat agents: a review. Anal. Chim. Acta 715, 1–18.

Valverde-Alva, M., García-Fernández, T., Villagrán-Muniz, M., Sánchez-Aké, C., Castañeda-Guzmán, R., Esparza-Alegría, E., Sánchez-Valdés, C., Llamazares, J., Herrera, C., 2015. Synthesis of silver nanoparticles by laser ablation in ethanol: a pulsed photoacoustic study. Appl. Surf. Sci. 355, 341–349.

Vigneshwaran, N., Kathe, A.A., Varadarajan, P.V., Nachane, R.P., Balasubramanya, R.H., 2006. Biomimetics of silver nanoparticles by white rot fungus, *Phaenerochaete chrysosporium*. Colloids Surf. B. Biointerfaces 53 (1), 55–59.

Vigneshwaran, N., Ashtaputre, N.M., Varadarajan, P.V., Nachane, R.P., Paralikar, K.M., Balasubramanya, R.H., 2007. Biological synthesis of silver nanoparticles using the fungus *Aspergillus flavus*. Mater. Lett. 61 (6), 1413–1418.

Vinković, T., Novák, O., Strnad, M., Goessler, W., Jurašin, D.D., Parađiković, N., Vrček, I.V., 2017. Cytokinin response in pepper plants (*Capsicum annuum* L.) exposed to silver nanoparticles. Environ. Res. 156, 10–18.

Wang, H., Chen, H., Wang, Y., Huang, J., Kong, T., Lin, W., Zhou, Y., Lin, L., Sun, D., Li, Q., 2012. Stable silver nanoparticles with narrow size distribution non–enzymatically synthesized by Aeromonas sp. SH10 cells in the presence of hydroxyl ions. Curr. Nanosci. 8 (6), 838–846.

Wiley, B., Im, S., Li, Z., McLellan, J., Siekkinen, A., Xia, Y., 2006. Maneuvering the surface plasmon resonance of silver nanostructures through shape–controlled synthesis. J. Phys. Chem. B 110 (32), 15666–15675.

Xu, J., Cheng, G., Zheng, R., 2010. Controllable synthesis of highly ordered ag nanorod arrays by chemical deposition method. Appl. Surf. Sci. 256 (16), 5006–5010.

Yin, L., Colman, B.P., McGill, B.M., Wright, J.P., Bernhardt, E.S., 2012. Effects of silver nanoparticle exposure on germination and early growth of eleven wetland plants. PLoS One 7, e47674.

Young, K., Xia, Y., Lu, X., 2006. Synthesis and galvanic replacement reaction of silver nanocubes in organic medium. Materials 16, 90–91.

Zaarour, M., El Roz, M., Dong, B., Retoux, R., Aad, R., Cardin, J., Dufour, C., Gourbilleau, F., Gilson, J.P., Mintova, S., 2014. Photochemical preparation of silver nanoparticles supported on zeolite crystals. Langmuir 30 (21), 6250–6256.

Zahed, F., Hatamluyi, B., Lorestani, F., Es'haghi, Z., 2018. Silver nanoparticles decorated polyaniline nanocomposite based electrochemical sensor for the determination of anticancer drug 5–fluorouracil. J. Pharm. Biomed. Anal. 161, 12–19.

Zaheer, Z., 2011. Multi–branched flower–like silver nanoparticles: preparation and characterization. Colloids Surf. A Physicochem. Eng. Asp. 384 (1–3), 427–431.

Zahran, M., Khalifa, Z., Zahran, M.A.H., Abdel Azzem, M., 2020. Dissolved organic matter–capped silver nanoparticles for electrochemical aggregation sensing of atrazine in aqueous systems. ACS Appl. Nano Mater. 3, 3868–3875.

Zanetti-Ramos, B.G., Fritzen-Garcia, M.B., de Oliveira, C.S., Pasa, A.A., Soldi, V., Borsali, R., Creczynski-Pasa, T.B., 2009. Dynamic light scattering and atomic force microscopy techniques for size determination of polyurethane nanoparticles. Mater. Sci. Eng. C 29 (2), 638–640.

Zannotti, M., Vicomandi, V., Rossi, A., Minicucci, M., Ferraro, S., Petetta, L., Giovannetti, R., 2020. Tuning of hydrogen peroxide etching during the synthesis of silver nanoparticles. An application of triangular nanoplates as plasmon sensors for Hg^{2+} in aqueous solution. J. Mol Liq. 309, 113238.

Zhang, X.Z., Zhang, Y.H.P., 2013. Cellulases: characteristics, sources, production, and applications. In: Yang, S., El-Enshasy, H., Thongchul, N. (Eds.), Bioprocessing Technologies in Biorefinery for Sustainable Production of Fuels, Chemicals, and Polymers. 1. Wiley, pp. 131–146.

Zhang, W., Chen, P., Gao, Q., Zhang, Y., Tang, Y., 2008. High–concentration preparation of silver nanowires: restraining in situ nitric acidic etching by steel–assisted polyol method. Chem. Mater. 20 (5), 1699–1704.

Zhang, Q., Li, W., Wen, L., Chen, J., Xia, Y., 2010. Facile synthesis of Ag Nanocubes of 30 to 70 nm in edge length with CF3COOAg as a precursor. Chem. A Eur. J. 16 (33), 10234–10239.

Zhang, X., Liu, Z., Shen, W., Gurunathan, S., 2016. Silver nanoparticles: synthesis, characterization, properties, applications, and therapeutic approaches. Int. J. Mol. Sci. 17 (9), 1534.

Zhang, H., Zou, G., Liu, L., Tong, H., Li, Y., Bai, H., Wu, A., 2017. Synthesis of silver nanoparticles using large–area arc discharge and its application in electronic packaging. J. Mater. Sci. 52 (6), 3375–3387.

Zhang, D.Y., Hao, Q., Liu, J., Shi, Y.S., Zhu, J., Su, L., Wang, Y., 2018. Antifouling polyimide membrane with grafted silver nanoparticles and zwitterion. Sep. Purif. Technol. 192, 230–239.

Zhang, L., Liu, J., Zhou, G., Zhang, Z., 2020. Controllable in-situ growth of silver nanoparticles on filter paper for flexible and highly sensitive SERS sensors for malachite green residue detection. Nanomaterials 10, 826.

Zhao, G., Stevens, S.E., 1998. Multiple parameters for the comprehensive evaluation of the susceptibility of *Escherichia coli* to the silver ion. Biometals 11, 27–32.

Zhao, L., Miao, Y., Liu, C., Zhang, C., 2017. A colorimetric sensor for the highly selective detection of sulfide and 1, 4–dithiothreitol based on the in–situ formation of silver nanoparticles using dopamine. Sensors 17, 626.

Zhou, K., Dong, C., Zhang, X., Shi, L., Chen, Z., Xu, Y., Cai, H., 2015. Preparation and characterization of nanosilver–doped porous hydroxyapatite scaffolds. Ceram. Int. 41, 1671–1676.

Zhu, J.J., Kan, C.X., Wan, J.G., Han, M., Wang, G.H., 2011. High–yield synthesis of uniform ag nanowires with high aspect ratios by introducing the long–chain PVP in an improved polyol process. J. Nanomater. 2011, 982547.

Ziashahabi, A., Prato, M., Dang, Z., Poursalehi, R., Naseri, N., 2019. The effect of silver oxidation on the photocatalytic activity of Ag/ZnO hybrid plasmonic/metal-oxide nanostructures under visible light and in the dark. Sci. Rep. 9, 1–12.

Zomorodian, K., Pourshahid, S., Sadatsharifi, A., Mehryar, P., Pakshir, K., Rahimi, M.J., Arabi Monfared, A., 2016. Biosynthesis and characterization of silver nanoparticles by *Aspergillus* species. Biomed. Res. Int. 2016, 1–6.

Further reading

Borah, D., Das, N., Das, N., Bhattacharjee, A., Sarmah, P., Ghosh, K., Chandel, M., Rout, J., Pandey, P., Ghosh, N.N., Bhattacharjee, C.R., 2020. Alga-mediated facile green synthesis of silver nanoparticles: photophysical, catalytic and antibacterial activity. Appl. Organomet. Chem. 34 (5), e5597.

Manikandan, R., Anjali, R., Beulaja, M., Prabhu, N.M., Koodalingam, A., Saiprasad, G., Chitra, P., Arumugam, M., 2019. Synthesis, characterization, anti-proliferative and wound healing activities of silver nanoparticles synthesized from *Caulerpa Scalpelliformis*. Process Biochem. 79, 135–141.

Index

Note: Page numbers followed by *f* indicate figures and *t* indicate tables.

A

Actinobacteria, synthesis of silver nanoparticles (AgNPs)
    antimicrobial properties, 501–505
    application, 485–487
    bioactive molecules, 479
    cell filtrate, 483
    cell-free extract, 481
    cell-free supernatant, 499–500
    factors, 484–485
    Fourier-transform infrared spectroscopy, 500–501
    gram-positive aerobic bacterium, 498
    intracellular synthesis, 480–481
    list, actinobacterial strains, 481–483, 482*t*
    nanotechnology, 479
    optimal growth conditions, 480
    photo-irradiation method, 501
    pigments and bio-flocculant, 483–484
    secondary metabolites, 479
    silver nitrate reduction, 498–499
    steps, 480, 480*f*
    Streptomyces genus, 483, 498–499
    transporting the synthesized nanoparticles, 481
    wound healing properties, 479
Actinomycetes
    antibacterial activity
        agar well diffusion method, 559–560
        disk diffusion method, 559–560
        synergistic effect, 560
    biological activity, 558–559*t*
        antibacterial, 559–560
        antibiofilm, 561
        anticancer, 557, 560–561
        antifungal, 557, 561
        antimicrobial, 557
        antioxidant, 557, 561
    concentration effect, 686
    extracellular biosynthesis, 552–557
    gram-positive bacteria, 551
    intracellular biosynthesis, 557
    preparation, 553–556*t*
    producers of active substances, 551
    properties, 551–552

AFM. *See* Atomic force microscopy (AFM)
Agar well diffusion method, 559–560
Agro-ecosystems, 230–233
Agro-waste, 15
Algae-mediated silver nanoparticles
    advantages, 526–527
    antimicrobial agents, 538
    antioxidant properties, 537–538
    applications, 526–527, 538–539, 539*f*
    biosynthesis methods, 527–530
    culture of microalgae, 528–529
    cytotoxicity and antimicrobial activity, 537
    green nanotechnology, advantages, 540*f*
    properties of, 533–534
    reducing agents, 527–528*f*
Algal extract composition, 532
Algal extract concentration, 531
Algal-mediated synthesis, 450–452
Alkaloids, 285–286
Alpha amylase, 721
Annelids, 636
Anogeissus latifolia, 201
Anti-arthropods activity, 171
Antibacterial activity, 159–161, 167–168, 289–295
    Ag^+ ions, 161
    cell membranes, 161–162
    nanoparticles, 158–159
Anticancer activity, 296–299
Anticancer bio-efficacy, 109–111
Antifungal activity, 169–170, 301–302
Antiinflammatory bio-efficacy, 106
Antimicrobial activity, 135–136, 341–343
Antimicrobial bio-efficacy, 107
Antineoplastic properties, 343–344
Antineoplastic strategy, 344
Antioxidant activity, 299–300
Antioxidant bio-efficacy, 108
Antioxidant/medicinal properties, 339–340
Anti-parasitic action, 171
Antiprotozoal activity, 171
Antiviral activity, 170, 302–303

758 Index

Applications, mushroom-mediated synthesis
  agriculture, 428–429
  biological activity
    antibacterial, 421–423
    anticancer, 423–424
    antidiabetic, 425–426
    antifungal, 423
  environment
    nanophotocatalyst, 426–427
    nanosensor, 427–428
  food, 429–430
Arc discharge method, 607–608
Arthroderma fulvum, 402–403
Astragalus gummifer, 202–203
Atomic force microscopy (AFM), 329, 369, 608–609, 620, 727

B

Bacteria, synthesis of silver nanoparticles
  advantages, 496
  Cytobacillus firmus, 498
  extracellular synthesis, 496
  intracellular synthesis, 496
  pseudomonas, 498
  use of *Bacillus*, 496–498
Bacterial infections, 547
Bacterial resistance, 341
Bacteria-mediated synthesis, 573–576
Bactericidal mechanisms, 403
Basidiomycota fungus/mushrooms, 400–401
Bilberry waste (BW), 288–289
Bimetallic nanoparticles, 103–104
Bioactive compounds/active phytochemicals, 362–363
Biochemistry, 680–684
Biodegradable gum
  applications, 192–193
  bio-fabrication, 193–210
Bio-efficacies, 106–111
Biogenic silver nanoparticles, 137–138
Biological applications
  anti-arthropods activity, 171
  antibacterial activity, 167–168
  antifungal activity, 169–170
  anti-parasitic action, 171
  antiprotozoal activity, 171
  antiviral activity, 170
  silver nanoparticles
    antibacterial coatings, 175
    antioxidant activity, 176
    catalytic activity, 172–174
    diagnosis and imaging, 174–175
    food packaging, 177
    wound dressing, 175–176
Biological organisms, 571
Biomedical application, core-shell silver nanoparticles (CS-AgNPs)
  antibacterial, 85–87
  antifungal, 88
  antiviral, 87
  drug delivery system, 89
  wound healing, 88–89
Biomolecules, 8–9
Bioorganic reductants, 253–255
Bio-reduction, 4
Biosynthesis
  algal extract composition, 532
  effect of
    contact time, 532
    extract concentration, 531
    pH, 532–533
    pressure, 532
    temperature, 531–532
  methods/techniques, 531
  metal nanoparticles, 707
  silver nanoparticles, 194–199
Bivalves, 635–636
Bottom-up approach, 287, 363, 525

C

Capping agents, 40–41, 548–550
Cashew gum, 205–206
Catla catla, 632
Cell-free extract, actinobacteria, 481
Cellulase, 720
Cellulose-silver nanocomposites, 57
Chemical method
  chitosan/cellulose-silver nanocomposites, 57
  clay-silver nanocomposites, 57
  silver graphene nanocomposite, 56–57
  silver nanocomposites, 58
  silver nanoparticles, 56
Chemical redox reaction, 322, 322*f*
Chemical reduction methods, 30–43, 495
Chikungunya virus, 302–303
Chitosan nanocomposites, 57
Chlorination, 657
Clay-silver nanocomposites, 57
Contact time, 532
Core-shell silver nanoparticles (CS-AgNPs)
  biomedical application, 85–89
  characterization, 79–84
  synthesis, 76–79
Cryptosporidium parvum (CP), 171

Cyanobacteria, 441
Cyprinus carpio, 631–632
Cytotoxic effects, 344, 571
Cytotoxicity, graphene-based materials, 130–131

D

Dengue virus, 302–303
Dental root-canal disinfection, 403
Direct chlorination, 657
Disk diffusion method, 559–560
DLS. *See* Dynamic light scattering (DLS)
Downstream processing, 401
Dwarf, 589
Dynamic light scattering (DLS), 167, 369, 608–609, 619, 727

E

Electrochemical methods, 495
Electrochemical reduction method, 43–45
Electron energy loss spectroscopy (EELS), 620
Electrostatic coating, 653
Electrostatic repulsion, 651–652
Endophytic actinobacteria, 481
Endophytic bacteria, 441–442
Energy dispersive spectroscopy (EDS), 619
Energy dispersive X-ray (EDX) spectroscopy, 368–369, 455–456, 572–573, 620
Environment-related chlorination, 657
Enzymatic synthesis, silver nanoparticles (AgNPs)
  applications
    in agriculture, 727
    in biomedical applications, 730
    in environmental applications, 732–734
    in nanofabrication, 731
  biological methods, 702
  biosynthesis
    by algae, 707–713
    by bacteria, 714
    by fungi, 713–714
  challenges, 734–735
  characterization of
    atomic force microscopy, 727
    dynamic light scattering, 727
    Fourier transform infrared spectroscopy, 725
    scanning electron microscopy, 726
    transmission electron microscopy, 726–727
    ultraviolet-visible (UV-vis) spectrophotometry, 724–725
    X-ray diffraction, 725–726
    X-ray photon electron spectroscopy, 726
  chemical methods, 701–702
  chemical reduction method, 703–705t
  enzyme-assisted biosynthesis (*see* Enzyme-assisted biosynthesis)
  geen synthesis, 707
  limitations, 734–735
  microbial biosynthesis, 708–712t
  microorganisms, 702–707
  photochemical method, 702
  physical methods, 700–701, 701t
  stabilizing/capping agents, 702
  top-down methods, 700
Enzyme-assisted biosynthesis
  alpha amylase, 721
  cellulase, 720
  extracellular and intracellular enzymes
    formula, 715–718t
    mechanism of, 719f
  green synthesis, 722–724t
  keratinase, 722
  laccase, 720–721
  lignin peroxidase, 721
  NAD(P)H-dependent nitrate reductase, 714–719
  xylanase, 721–722
EPS. *See* Exopolysaccharides (EPS)
Euclidean dendrogram, 329–337
Eukaryotic organisms, 447–449
Evaporation-condensation methods, 571
Evernia mesomorpha, 514–515
Exopolysaccharides (EPS), 528–529
Extracellular biosynthesis, 552

F

Ferric reducing antioxidant power (FRAP), 299–300
Field emission scanning electron microscopy (FE-SEM), 498–499
Flavonoids, 106–107, 285–286, 324–326
Food contamination, 343
Fourier-transform infrared spectroscopy (FTIR), 164–165, 329, 454–455, 500–501, 533, 572–573, 619, 725
FRAP. *See* Ferric reducing antioxidant power (FRAP)
Free aldehyde groups, 326
Fungal-mediated synthesis, 447–449, 573
Fungi biomasses, 398

G

Gellan gum, 206
Generally regarded as safe (GRAS), 194–199, 202
Genetic engineering, 680–684
Gibbs free energy, 326–327

Gram-negative bacteria, 285–286
Gram-positive bacteria, 285–286
Graphene oxide-silver (Ag) nanocomposites
   green synthesis, 133–141
   silver and carbon composites, 139–141
Graphite, 130
Green biosynthesis
   application, 671f
   biological cells, 673–676t
   challenges, 687–688
   characteristics of, 685t
   commercial quantities, 689
   development of information
      in biochemistry, 680–684
      genetic engineering, 680–684
      molecular biology, 680–684
      strategies for scaling up, 690f
   food waste used, 681–683t
   nanoscale materials, 671–672
   physicochemical factors
      $AgNO_3$ concentration effect, 686
      pH effect, 685–686
      reaction time effect, 686
      temperature effect, 684–685
   selection of
      biological cell, 672–678
      culture media, 679–680
Green chemical and bio-mediated methods, 222–225
Green chemistry principles, 320–321
Green nanotechnology, 525, 571
Green physical vapor deposition methods, 221
Green production
   polymer-silver nanocomposites, 60–62
   silver-graphene nanocomposites, 62
   silver nanocomposites, 62–63
   silver nanoparticles, 59–60
Green synthesis, 3–4, 414–415, 607
   advantages, 9–10
   biomolecules, 8–9
   bottom-up methods, 320
   catalysis, 261–265
   challenges, 10–13
   chemical and biochemical analysis, 258–260
   chemical redox reaction, 322, 322f
   demolition worker and brickmaker approaches, 320f
   flavonoids, 324–326
   free aldehyde groups, 326
   graphene oxide-ag nanocomposites, 133–141
   green chemistry principles, 320–321
   hydrophytes/aquatic plants, 324
   keto-enol tautomeric equilibrium, 324
   mechanism of (phyto)synthesis, 324
   microorganisms, 6–8
   phytomolecules, 322–323, 323f
   top-down methods, 320
   use of plant extracts, 322
Guar gum, 208–209
Gum acacia, 201–202
Gum arabic. See Gum acacia
Gum ghatti, 201
Gum karaya, 194–200
Gum kondagogu, 204–205
Gum Tragacanth, 202–203
Gymnema sylvestre, 366

H

Herpes simplex virus, 170
Hesperidin, 288–289
Hydrazine (N_2H_4), 32
Hypophthalmichthys molitrix (silver carp), 631

I

Inert gas condensation methods, 495
Intracellular biosynthesis, 480–481, 557
Irradiation-assisted methods, 225, 495
Isopropanol, 255

K

Keratinase, 722
Ketoconazole (KTZ), 169–170
Keto-enol tautomeric equilibrium, 324
Killing effect, 341
KTZ. See Ketoconazole (KTZ)

L

Labeo rohita, 632
Laccase, 720–721
Large-scale biosynthesis, 6–7
Laser ablation methods, 495, 571
Leaf extracts
   aerial parts, 317–318
   antimicrobial activity, 341–343
   antineoplastic properties, 343–344
   antioxidant/medicinal properties, 339–345
   applications, 331–336t
   effect of the concentration, 338
   mechanisms of reduction, 338
   Mentha piperita, 337
   molecular sensing, 344–345
   photocatalytic properties, 344–345
   phytosynthetic approach, 327
      advantages, 329
      effect of, 329
      size distribution, 329–337

properties, 331–336t
reducing and stabilizing agents, 331–336t
scopus database, 327
stabilization of, 338–339
steps, 327–329
synthesis conditions, 337
Leukemic cells, 296
Lichen-assisted biomechanochemical solid-state synthesis, 514–515
Lichens
  algal/cyanobacterial species, 514
  antibacterial activity, 514–515
  applications, 514, 514t
  biomechanochemical solid-state synthesis, 514–515, 516f
  biomedical applications, 518–519
  different species of, 514–515
  disk diffusion technique, 514–515
  growth forms, 514
  transmission electron microscopy image, 515, 518f
  X-ray diffraction pattern, 515, 517f
Lignin peroxidase, 721
Lithography methods, 571
Locust bean gum, 207–208

M

MDR. *See* Multidrug-resistant (MDR)
Mechanisms, silver nanoparticles (AgNPs)
  bioreduction methods, 608
  biosynthesis process
    bacterial-mediated synthesis, 614
    fungal-mediated synthesis, 614–617
    fungal species, 613t
    other synthesis, 617–618
  characterization methods
    dynamic light scattering, 619
    energy dispersive spectroscopy, 619
    Fourier transform infrared analysis, 619
    scanning or transmission electron microscopy, 619
    spectrophotometry, 619
    X-ray diffraction method, 619
  chemical synthesis
    chemical reduction, 611
    electrochemical methods, 611
    photochemical method, 611
  downstreaming, 618
  electron transfer, 608
  formation of nanoparticles, 615f
  light scattering systems, 608–609
  microscopic techniques, 608–609
  nanocage-assisted synthesis, 611–612
  nanoparticle characterization, 608–609
  physical approach
    arc discharge method, 609
    laser ablation method, 609
    physical vapor condensation method, 609
  purification, 618
  secrete enzymes, 612
  solid-assisted synthesis, 611–612
Medicinal plant extract
  antibiotics/antimicrobial substances, 370
  bioactive compounds/active phytochemicals, 362–363
  characterization
    atomic force microscopy, 369
    dynamic light scattering, 369
    energy dispersive X-ray spectroscopy, 368–369
    photon correlation spectroscopy, 369
    scanning electron microscopy, 369
    surface plasmon resonance, 368–369
    transmission electron microscopy, 369
    X-ray diffraction, 369
  classes of antibiotics, 371t
  mechanisms of bacterial resistance, 371–373
  multidrug-resistance pathogen, 373–377
  overview, 361–363
  plausible mechanism, 368
  protocol of synthesis, 366–367
  types of synthesis methods, 363–365
Metallic nanoparticles (MNPs), 151, 359, 699
Metal vapor synthesis (MVS), 250–253, 255–258
Microalgae, 629–630
Microbes, 3–4
Microbes-mediated synthesis, 484
Microbial-mediated, silver nanoparticles (AgNPs)
  algae, 444
  applications
    in agriculture, 461–462
    biosensing applications, 458–459
    in food sector, 460–461
    in textiles, 460
    wound healing applications, 458
  bacteria-mediated, 440–442
    cyanobacteria, 441
    endophytic bacteria, 441–442
  bioactivity studies, 452–453
  characterization of
    energy dispersive X-ray spectroscopy, 455–456
    Fourier transform infrared spectroscopy, 454–455

Microbial-mediated, silver nanoparticles (AgNPs) *(Continued)*
  scanning electron microscopy, 455–456
  ultraviolet-visible spectroscopy, 454
  X-ray diffraction, 456
 fungi, 442–443
 mechanism
  algal-mediated synthesis, 450–452
  bacterial-mediated synthesis
   nitrate reductase enzyme, 446–447
  fungal-mediated synthesis, 447–449
 types of bioactivity, 439–440
Microbiological object, 251–252*t*
Microorganisms, 551
 advantages, antibacterial activity, 580–581
 antibacterial activity, 579
 antibacterial mechanism, 579–580
 bacteria-mediated synthesis, 573–576
 biomass and supernatant, 572–573
 color changes, 572–573
 developments in biosynthetic methods, 581
 extracellular synthesis, 572–573
 fungi-mediated synthesis, 573
 mechanism of, 577–578
 other microorganisms, 576–577
 synthetic route, 572*f*
Microwave-assisted (MWA) method, 45, 153–154, 226, 495
Minimum inhibitory concentrations (MICs), 402–403
Molecular biology, 680–684
Molecular sensing, 344–345
Multicellular eukaryotic organisms, 392
Multidrug-resistant (MDR) bacteria, 361
Multimetallic nanoparticles, 103–104
Mushroom-mediated synthesis
 application (*see* Application, mushroom-mediated synthesis)
 capping agent, 417
 carbonyl and hydroxyl groups, 419
 Ganoderma lucidum, 417
 in nanoparticle production, 416
 Ostwald ripening/coalescence, 417
 Pleurotus florida, 417
 polysaccharide glucan, 418–419
 reducing and stabilizing agent, 417
 safety protocol, 419–421
MVS. *See* Metal vapor synthesis (MVS)
MWA. *See* Microwave-assisted (MWA)
Mycosynthesis of silver nanoparticles
 advantages, 402
 application, 402–405
 cost-effective, 391
 defense mechanism, 398–399
 development of, 405–406
 elimination of toxic chemicals, 399
 extracellular synthesis, 400–401
 fungal species, 392–397
 fungi biomasses, 398
 intracellular synthesis, 399
 physicochemical properties, 392
 role of, 405
 size-controlled synthesis, 391
 types of fungi, 394–398*t*

N

N, N-Dimethylformamide (DMF), 41–42
NAD(P)H-dependent nitrate reductase, 714–719
Nanocomposites
 chemical synthesis
  chitosan/cellulose-silver, 57
  clay-silver, 57
  polymer-silver, 60–62
  silver-graphene, 56–57
 green synthesis
  polymer-silver, 60–62
  silver-graphene, 62
Nano-effect, 318, 319*f*
Nanofabrication, 699–700
Nanomanufacturing, 669
Nanomaterials (NMs), 669
Nanoparticles (NPs), 318, 359
 applications, 360*f*, 414*f*
Nanophotocatalyst, 426–427
Nanosensor, 427–428
Nanotechnology, 318
Natural ecosystems
 aggregation, 658–660
 bioaccumulation, 649–650
 chlorination, 657
 dissolution, 658–660
 photoinduced transformation, 660
 sink, 649–650
 sources of, 650–651
 sulfidation, 655–656
 transformations of, 653–654
 transport of, 651–653
Neem gum, 209–210
Nitrate reductase enzyme, 446–447
Nuclear magnetic resonance (NMR), 620

O

Oilbanum gum, 206–207
Optical nano-effect, 318–319
Oxysulfidation, 655

P

Parmeliopsis ambigua, 514–515
PEO. *See* Poly (ethylene oxide) (PEO)
Phenols, 285–286
Photocatalytic properties, 344–345
Photochemical synthesis, 42–43, 152
Photoinduced reduction methods, 495
Photoinduced transformation, 660
Photo-irradiation method, 501
Photoluminescence, 329
Photon correlation spectroscopy (PCS), 369
Physical method
   condensation, 591
   electrical irradiation, 591
   evaporation, 591
   gamma irradiation, 591
   laser ablation, 591
   microwave irradiation, 591
Phyto-fabricated silver nanoparticles, 103
   bio-efficacies, 106–111
   characterization, 104–106
Phytomolecules, 322–323, 323*f*
Phytoplankton, 628–629
Phytosynthetic method, 327
Pistia stratiotes, 366
Plant-based synthesis, 190–191
   sources, 192
Plant extract, 59
   flowers, 250*t*
   fruits or seeds, 247–249*t*
   leaves, 242–246*t*
   microbiological objects, 251–252*t*
Plant-mediated bimetallic nanoparticles, 105*t*
Plant-mediated synthesis, 223*f*
Plants and agricultural waste
   applications
      antibacterial activity, 289–295
      anticancer activity, 296–299
      antifungal activity, 301–302
      antioxidant activity, 299–300
      antiviral activity, 302–303
   biological resources, 287
   bottom-up approach, 287
   mechanism of, 289
      antibacterial activity, 294–295
      anticancer activity, 296–299
      antifungal activity, 301–302
   reaction temperature, 288
   reducing and stabilizing agents, 287–289
   stabilization of silver nanoparticles, 287–288
   top-down approach, 287
Plausible mechanism, 368
Poly (ethylene oxide) (PEO), 60–62
Poly-D,L-lactide-co-glycolide (PLGA), 60–62
Polymer matrix nanocomposites, 129–130
Polymer-silver nanocomposites
   characterization, 63–64
   green production, 60–62
Polyol synthesis, 37–39
Polysaccharide method, 40–41
Porous media, 653
Pressure, 532
Punctelia subrudecta, 514–515

R

Reaction time effect, 686
Reactive oxygen species (ROS), 294, 518–519, 579–580, 627–628
Red currant waste (RCW), 288–289
Reducing agents, 591
Replication process, 403
Response surface methodology (RSM), 514–515
Rhamnogalacturonan gum. *See* Gum kondagogu
ROS. *See* Reactive oxygen species (ROS)

S

SAED. *See* Selected area electron diffraction (SAED) patterns
Salmalia malabarica gum (SMG), 168, 203–204
Scanning electron microscopy (SEM), 369, 455–456, 572–573, 608–609, 726
Scanning/transmission electron microscopy (SEM/TEM), 329, 619
Scanning tunneling microscopy (STM), 620
Selected area electron diffraction (SAED), 515, 620
Shrimps, 636
Silver graphene nanocomposites, 56–57
Silver nanocomposites, 58
   chemical production, 56
   green production, 59–63
Silver nanoparticles (AgNPs), 547
   advantages of biological *vs.* chemical methods, 495
   antibacterial activity, 158–162
   antimicrobial activity, 415, 493–495
   antimicrobial properties, 415–416, 607
   applications, 493–495, 494*f*, 547, 589, 597–598, 607
   bacterial strains, 547
   biological methods, 548–550, 549*t*
   biological synthesis, 153
   catalytic activity, 172–174
   chemical methods, 24–29, 152

Silver nanoparticles (AgNPs) *(Continued)*
  chemical reduction, 30–43
  disadvantages, 607
  exponential growth, 415
  fabrication, 100–103
  formation mechanisms, 29
  green synthesis, 548
  mechanisms, 64–66
  medicinal property, 415
  microorganisms, 547–548, 551
  mycosynthesis, 415–416
  nanoparticles, 493
  photochemical methods, 152
  physical methods, 151–152
  plants, 5–6
  sensing property, 415
  synthesis, 22–23, 495, 547–548
  synthesis by
    actinomycetes (*see* Actinomycetes, silver nanoparticles)
    plant extracts, 550
Small-angle X-ray scattering (SAXS), 620
SMG. *See* Salmalia malabarica gum (SMG)
Sodium borohydride (NaBH$_4$), 31–32
Sol gel methods, 495
Solvents, 37
Spectrophotometry, 619
SPR. *See* Surface plasmon resonance (SPR)
Stabilization process, 419–420
Stabilizing agents, 591
Sterculia gum. *See* Gum karaya
Streptomyces albidoflavus strain, 481
Sulfidation with metal sulfides, 655
Sulfidation without oxygen, 655
Surface-enhanced Raman scattering (SERS), 620
Surface plasmon resonance (SPR), 318–319, 368–369
Surface plasmons, 21–22
Synthesis, silver nanoparticles (AgNPs)
  by actinobacteria (*see* Actinobacteria, synthesis of silver nanoparticles)
  algae (*see* Algae-mediated silver nanoparticles)
  applications, 597–598
  by bacteria (*see* Bacteria, synthesis of silver nanoparticles)
  bacteria-mediated synthesis, 595, 596t
  biological methods, 591
  bottom-up approach, 590
  chemical methods, 591
  fungi, 595, 597t
  leaf extracts (*see* Leaf extracts)
  by lichens (*see* Lichens)
  medicinal plant extract (*see* Medicinal plant extract)
  by microorganisms (*see* Microorganisms)
  mushroom (*see* Mushroom)
  nitrate reductase enzyme, 595
  photolithography approach, 589–590
  physical methods, 591
  by plants and agricultural waste (*see* Plants and agricultural waste)
  top-down approaches, 590
  ultrasonic spray pyrolysis, 590
  use of leaf extract, 594
  use of plant parts, 591–594, 592–594t

T

TEM. *See* Transmission electron microscopy (TEM)
Temperature, 531–532
Temperature effect, 684–685
Thermal decomposition method, 151–152
Tollens method, 39–40
Top-down approach, 287, 363, 525
Toxicity, aquatic system
  annelids, 636
  bivalves, 635–636
  causes of, 628
  fate and accumulation, 637–638
  to fish (*see* Toxicity to fish)
  LC50 values, 636–637
  mechanism, 638–640
  microalgae, 629–630
  phytoplankton, 628–629
  routes of entry, 628
  shrimps, 636
Toxicity to fish
  blood-induced silver nanoparticles, 634
  Catla catla, 632
  chronic toxicity test, 634
  Cyprinus carpio, 631–632
  glucose concentration, 633
  heamotoxicity, 632
  hemorrhages to parts, 634
  histotoxicity, 632
  Hypophthalmichthys molitrix, 631
  Japanese rice fish, 633
  Labeo rohita, 632
  Nile tilapia, 634
  optic tectum width, 633–634
  oxidative stress, 632
  oxygen diffusion, 632
  Piaractus mesopotamicus, 635
  pigmentation, 633–634
  Rhamdia quelen, 635
  serum level, 633

Trametes trogii, 393–394
Transmission electron microscopy (TEM), 166–167, 369, 498–499, 515, 572–573, 608–609, 726–727
Trichoderma longibrachiatum, 393
Trisodium citrate, 34–37

U

Ultra-sonication methods, 495
Ultraviolet-visible (UV-vis) spectrophotometry, 162, 454, 533, 572–573, 608–609, 724–725
United States Environmental Protection Agency (USEPA), 421

W

Wound healing, 340
  bio-efficacy, 106–107

X

Xanthoparmelia plitti, 514–515
Xerophytes, 324
X-ray diffraction (XRD), 165–166, 369, 456, 515, 572–573, 608–609, 619, 725–726
X-ray photon electron spectroscopy (XPS), 620, 726
Xylanase, 721–722

Printed in the United States
by Baker & Taylor Publisher Services